Second Edition

Quantum Hall Effects

Field Theoretical Approach and Related Topics

Zyun F. Ezawa

NEW JERSEY • LONDON • SINGAPORE • BEIJING • SHANGHAI • HONG KONG • TAIPEI • CHENNAI

Published by

World Scientific Publishing Co. Pte. Ltd.
5 Toh Tuck Link, Singapore 596224
USA office: 27 Warren Street, Suite 401-402, Hackensack, NJ 07601
UK office: 57 Shelton Street, Covent Garden, London WC2H 9HE

British Library Cataloguing-in-Publication Data
A catalogue record for this book is available from the British Library.

QUANTUM HALL EFFECTS (2nd Edition)
Field Theoretical Approach and Related Topics

ISBN-13 978-981-270-032-2
ISBN-10 981-270-032-3

Printed by FuIsland Offset Printing (S) Pte Ltd, Singapore

To Shizu & Motohiko

PREFACE TO THE SECOND EDITION

The quantum Hall (QH) effect is one of the most fascinating and beautiful phenomena in all branches of physics. Composite bosons, composite fermions and fractional charged excitations (anyons) were among the distinguishing ideas in the original edition. Seven years have passed since the original edition. Tremendous theoretical and experimental developments are still being made in this sphere. Many novel ideas have been proposed to understand various novel experimental results, which we have included in this new edition.

First, physics in higher Landau levels is quite different from that in the lowest Landau level because the effective Coulomb interactions are different. Charge density waves such as stripe states emerge, and indeed have been observed experimentally. Second, unconventional QH effects were discovered in graphene (a single atomic layer graphite), which immediately triggered enormous theoretical and experimental studies. It is remarkable that the electron dynamics is governed by the relativistic Dirac theory and that even supersymmetric quantum mechanics plays a key role. Third, intriguing phenomena associated with the interlayer phase coherence and SU(4) QH ferromagnets in the bilayer system have been fully revealed. They include the anomalous Hall resistivity in counter flow experiments and the anomalous diagonal resistivity near the commensurate-incommensurate phase transition point. The latter would signal the formation of a soliton lattice made of sine-Gordon solitons between the two layers. They also include an SU(4) skyrmion as a quasiparticle, which changes its shape from a pseudospin SU(2) texture to a spin SU(2) texture as the density imbalance is controlled between the two layers. Fourth, the microscopic theory of the QH effect is formulated entirely within the theory of noncommutative geometry. Thus, quasiparticles are noncommutative solitons in QH ferromagnets.

This new edition provides an instructive, comprehensive and self-contained overview of the QH effect including recent developments. It is also suitable for an introduction to quantum field theory with vivid applications. For instance, the Dirac theory of electrons and holes has a remarkable realization in QH effects in graphene. A fantastic world of noncommutative geometry together with noncommutative solitons has a concrete realization in QH systems, where various imaginative ideas can be explored theoretically and tested experimentally. QH effects have proved to be so special in condensed matter physics that they are deeply connected with fundamental principles of physics and mathematics. This book is ideal for students and researchers in condensed matter physics, particle physics, theoretical physics and mathematical physics.

In Part I, a new chapter is added for Dirac electrons, holes and supersymmetry. In Part II, two chapters are added for charge density wave states in higher Landau levels and unconventional QH effects in graphene. Part III is revised fully to meet up-to-date achievements in bilayer QH systems. Part IV is rewritten anew in view of noncommutative geometry. Furthermore, almost all parts are retouched for improvement, and a number of misprints have been corrected.

To complete the second edition I have benefited from fruitful discussions on the subject with A. Sawada, Y. Hirayama, N. Kumada, A. Fukuda, D. Terasawa, M. Morino, K. Iwata, S. Kozumi, K. Hasebe, S. Suzuki, K. Ishii, G. Tsitsishvili and others. In particular, the collaboration with A. Sawada and G. Tsitsishvili was indispensable to complete this revision. Special thanks are due to N. Shibata and M. Ezawa for providing me with the contents of Chapter 19 and 20, respectively. Further, I am grateful to N. Shibata for careful reading of the manuscript, and to M. Ezawa for allowing me to use an illustration of the Dirac cone in the cover of this book.

Zyun Francis Ezawa
Sendai
January 2007

PREFACE TO THE FIRST EDITION

Quantum Hall (QH) effects are remarkable macroscopic quantum phenomena observed in the 2-dimensional electron system. The integer QH effect was discovered in 1980 by K. von Klitzing, a century after the discovery of the classical Hall effect, for which he received the Nobel prize in 1985. The fractional QH effect was discovered in 1982 by D. Tsui, H. Störmer and A.C. Gossard. It was predicted by B. Laughlin that a quasiparticle is an anyon carrying electric charge e/m at the filling factor $\nu = 1/m$. In 1997 a direct observation of fractional charges was successfully carried out at $\nu = 1/3$ by measuring a back scattering current noise in Hall-bar experiments. B. Laughlin, D. Tsui and H. Störmer received the Nobel prize in 1998.

QH effects are so special in condensed matter physics that they are deeply connected with the fundamental principles of physics. Moreover, they present concrete realizations of various modern concepts related with topological investigations not only in physics but also in mathematics. The QH system provides us with a rare opportunity to enjoy the interplay between condensed matter physics and particle physics. It is worthwhile to make the subject as a part of the training for all graduate students in physics.

Many fancy ideas appear in QH effects. Composite particle (boson or fermion) is one of them. It is an electron bound to flux quanta. Laughlin started his seminal paper by saying that "The $\frac{1}{3}$ effect, recently discovered by Tsui, Störmer and Gossard, results from the condensation of the two-dimensional electron gas in a GaAs-Ga_xAl_{1-x} heterostructure into a new type of collective ground state". It is our present understanding that the QH state is a condensate of composite bosons. Intriguingly a single electron is converted into a boson by acquiring flux quanta in QH states. Such a statistical transmutation is allowed in the planar geometry due to its intrinsic topological structure. Despite their bosonic low-energy properties, composite bosons obey the fermionlike exclusion principle. The hierarchy of fractional QH states is understood by the use of composite fermions.

Topological solitons play the leading role in QH effects. Indeed, charged excitations (quasiparticles) are topological solitons in the QH condensate. When quantum coherence develops with the spin wave as a Goldstone mode, quasiparticles are skyrmion O(3)-spin textures. Skyrmions were originally proposed in nuclear physics, where they are O(4)-isospin textures to be identified with nucleons. Though their relevance is still unclear in nuclear physics, their existence is firmly established in the QH system.

Edge excitations are described by the chiral Tomonaga-Luttinger model. Electrons are topological excitations in this model, and obey a relativistic field equation. When tunneling interactions are allowed between the opposite edges, topological excitations turn into sine-Gordon solitons. Physics on the edge is "exactly solvable", due to the existence of infinitely many conservation rules. It can be a laboratory to test various results of conformal field theories.

The bilayer QH system consists of two quantum wells separated by a barrier with $\sim$30 nm width. A phase transition has been observed between two distinguishable phases at a fixed filling factor by changing the electron density. When the tunneling gap is not too large, interlayer coherence develops, as is a reminiscence of the superconductor Josephson junction. Pseudoparticles are CP^3 skyrmions.

Physics confined to the lowest Landau level is curious, because the x and y coordinates of the electron position become noncommutative. It is a simplest physical system subject to noncommutative geometry. The system is characterized by the Moyal algebra or the W_∞ algebra. When restricted to the edge of the QH droplet, the Kac-Moody algebra emerges. The excitation spectrum of the QH system forms a representation of the algebra.

This book is intended to give a pedagogical and self-contained introduction to these new concepts in QH effects. It is accessible, and will be of interest, to students and researchers in condensed matter physics, particle and mathematical physics. It comprises four parts. Part I is a quick summary of quantum field theory, where I explain various concepts necessary to understand QH effects, namely, canonical quantization, quantum coherence, topological solitons, anyons and so on. I hope that this part is useful to those who are not familiar to these concepts. Readers may skip this part and come back to relevant places when necessary. Part II is devoted to monolayer QH systems, while Part III to bilayer QH systems. Some algebraic aspects of the QH system are reviewed in Part IV, where I derive some basic formulas used in Part II and Part III.

I have benefited from fruitful collaborations on the subject with A. Sawada, A. Iwazaki, H. Ohno, Y. Hirayama, K. Muraki, Y. Horikoshi, Y. Ohno, T. Saku, N. Kumada, M. Hotta, K. Sasaki, K. Hasebe, Y-S. Wu and others. I am grateful to H. Aoki, D. Yoshioka, A. MacDonald, S. Murphy and J. Eisenstein for valuable discussions. Special thanks are due to K. Shizuya, S. Katsumoto and K. Takahashi for careful reading of the manuscript. Their comments were very helpful for me to improve the manuscript.

Zyun Francis Ezawa
Sendai
January 2000

Quantum Hall Effect

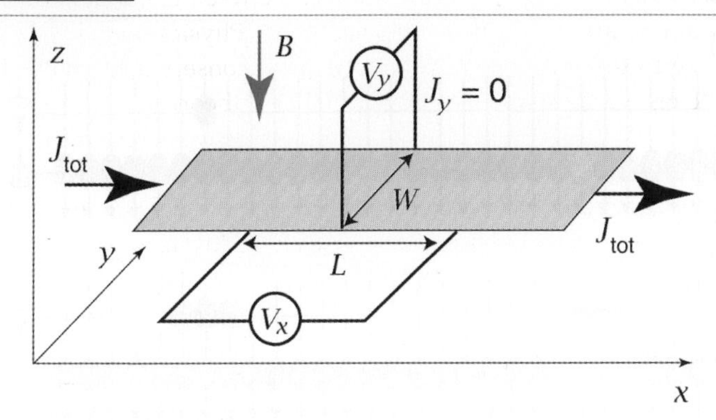

Fig. 1 Hall and diagonal resistivities R_{xy} and R_{xx} are independent of sample properties and given by $R_{xy} = V_y/J_{\rm tot} \longrightarrow \nu^{-1}(2\pi\hbar/e^2)$, $R_{xx} = -WV_x/LJ_{\rm tot} \longrightarrow 0$, in quantum Hall (QH) states. Here, ν is the filling factor.

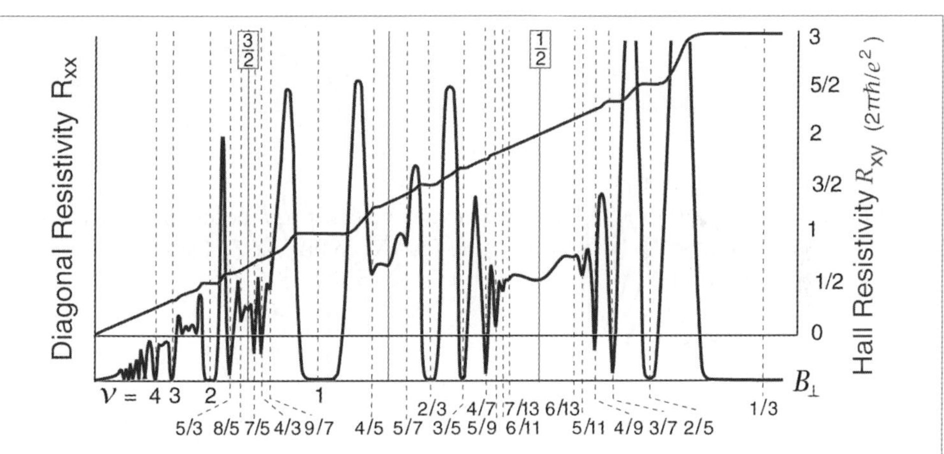

Fig. 2 QH states are detected by plateaux developed in the Hall resistivity R_{xy} or dips in the diagonal resistivity R_{xx}.

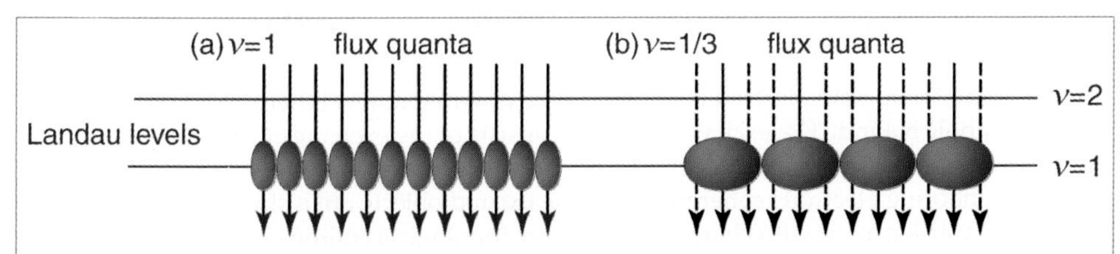

Fig. 3 There are m flux quanta (vertical lines) per electron at $\nu = 1/m$. A composite particle (fermion or boson) is an electron bound to ($m-1$ or m) flux quanta. It behaves as a "fat" electron in the fractional QH state. Here, (a) $m = 1$ and (b) $m = 3$.

Skyrmions

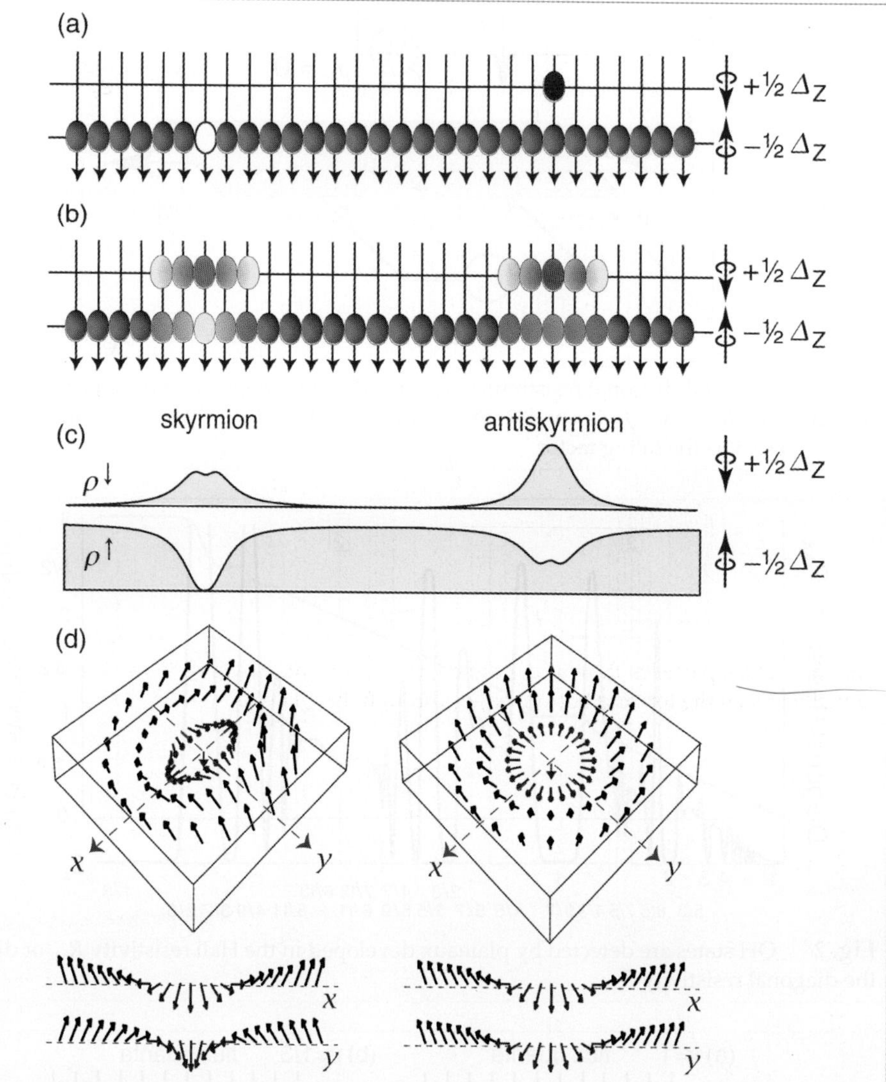

Fig. 4 (a) An up-spin electron is excited to the down-spin level thermally at $\nu = 1$, leaving a quasihole behind. The excitation energy consists of the exchange Coulomb energy, the direct Coulomb energy and the Zeeman energy. (b) The direct Coulomb is reduced by exciting neighboring electrons coherently. (c) The excited objects are a skyrmion with charge $+e$ and an antiskyrmion with charge $-e$. They are topological solitons in QH ferromagnets. The size is determined so as to optimize the Coulomb and Zeeman energies. (c) The electron densities $\rho^{\uparrow}(\boldsymbol{x})$ and $\rho^{\downarrow}(\boldsymbol{x})$ are illustrated around them. (d) The electron spin is reversed at the center of the skyrmion and the antiskyrmion. The normalized spin density is depicted.

Edge States

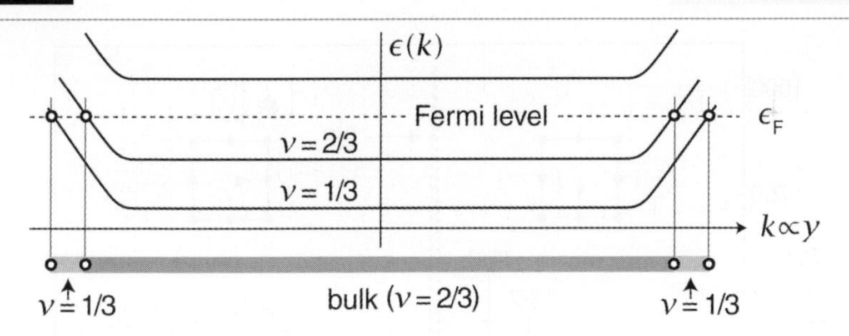

Fig. 5 Dispersion relation is illustrated in a QH bar as a function of the one-dimensional momentum $\hbar k$, related to the electron position as $y = \ell_B^2 k$ in the Landau gauge. Gapless edge channels appear when the Fermi level crosses energy levels.

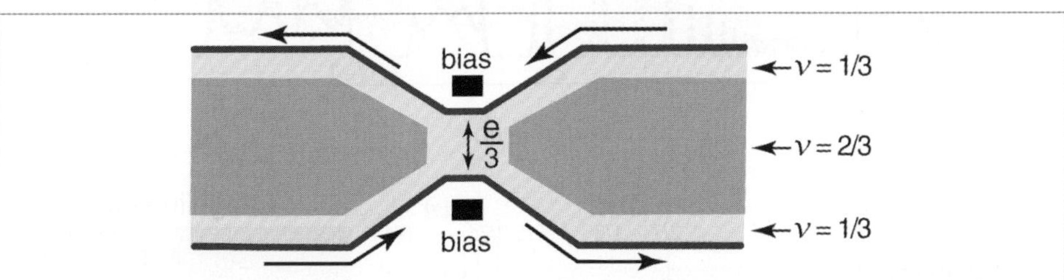

Fig. 6 Quasiparticles flow along the edge. With weak pinch-off by bias voltages, they tunnel between the top and bottom edges through the QH liquid.

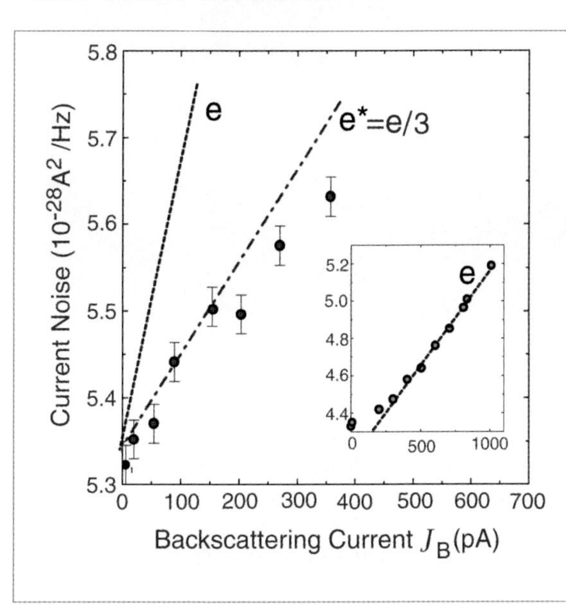

Fig. 7 The charge of quasiparticles can be determined by the Shchottky noise formula, when they tunnel one by one as in Fig.6. Tunneling noise was measured in the QH state at the bulk filling factor $\nu = 2/3$, and also at $\nu = 4$ (given in the inset). The charge $e/3$ of quasiparticles was clearly observed at $\nu = 2/3$. Data are taken from L. Saminadayar et al.[419]

Stripes and Bubbles

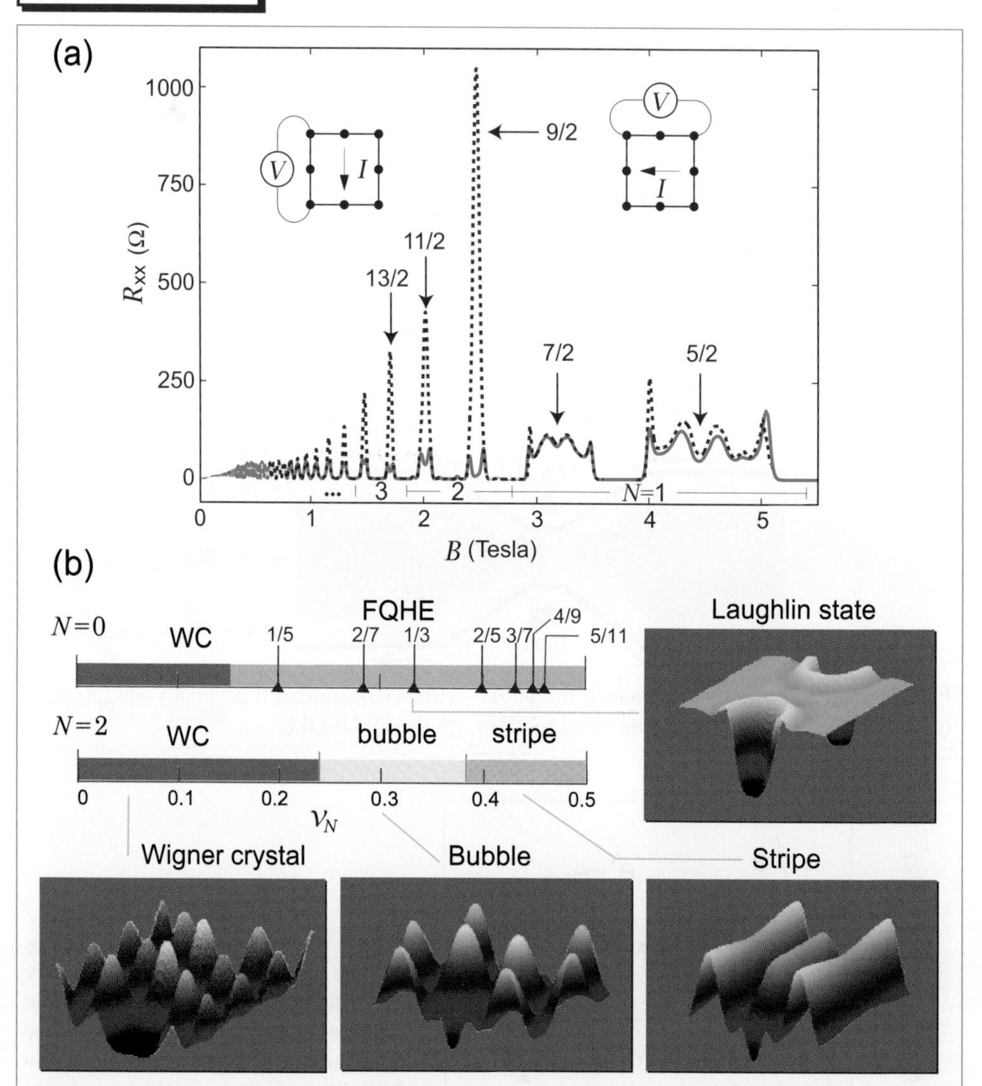

Fig. 8 (a) A strong anisotropy has been observed in the diagonal resistivity R_{xx} at $\nu = n/2$ with $n = 5, 7, 9, 11, \cdots$ by changing the direction of current through the sample. Data are taken from M. Lilly et al.[313] (b) We expect a variety of charge density waves (CDWs) in QH systems. Electrons form a Wigner crystal at low density. The Laughlin state emerges in the lowest Landau level ($N = 0$), while stripes and bubbles will appear in higher Landau levels ($N > 0$), where the Coulomb potential is not a monotonically decreasing function. In these CDW states electrons form clusters with other electrons. The stripe state leads to a strong anisotropy in the diagonal resistivity R_{xx} to be identified with the experimentally observed one. These figures are taken from N. Shibata et al.[436]

QH Effects in Graphene

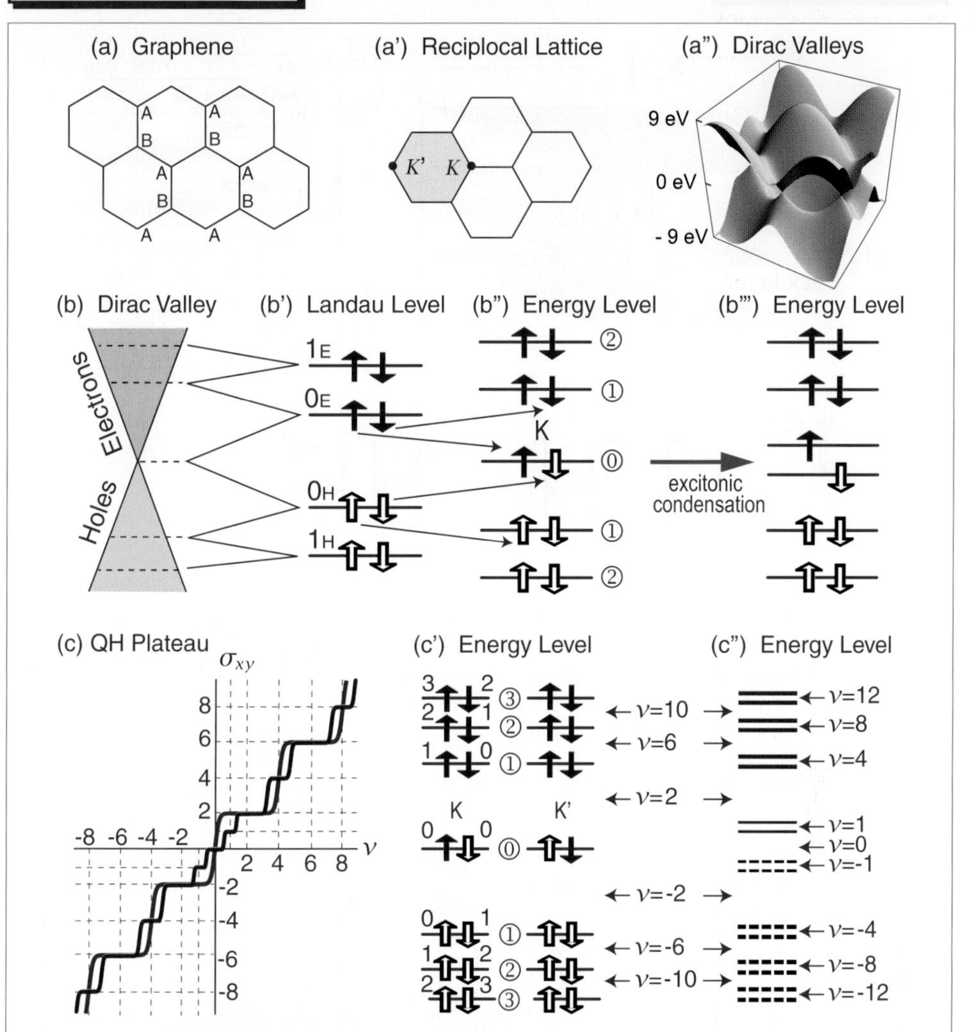

Fig. 9 Unconventional QH effects have been found in graphene. (a) The honeycomb lattice structure is illustrated for graphene. (a′) The reciplocal lattice is also a honeycomb lattice with inequivalent K and K′ points. (a″) The dispersion relation, showing a valley-like structure, is that of Dirac electrons near these points. (b) Electrons and holes are charged objects in the graphene system with all negative-energy states filled up. (b′) Electrons (holes) are quantized into Landau levels under magnetic field. (b″) Two Landau levels are mixed to create a supersymmetric spectrum due to the intrinsic Zeeman effect. (b‴) An electron and a hole in the zero-energy level make excitonic condensation, which opens a gap between the electron and hole states. (c) The QH plateau is illustrated without (black curve) and with (red curve) Coulomb interactions. The energy spectrum is given without (c′) and with (c″) Coulomb interactions. These illustrations are taken from M. Ezawa.[132]

Bilayer QH System

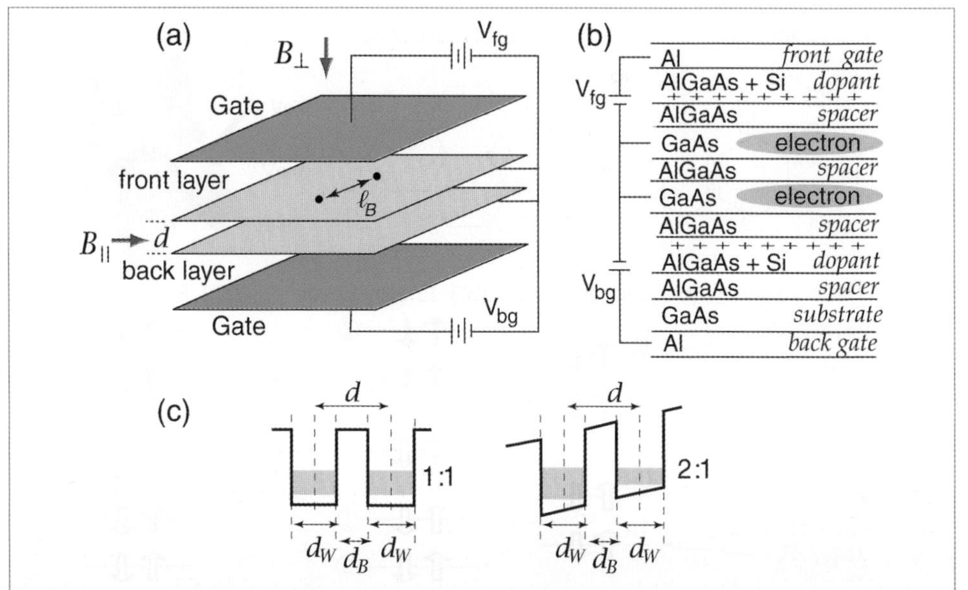

Fig. 10 (a) The bilayer system has four scale parameters, the magnetic length ℓ_B, the Zeeman gap Δ_Z, the interlayer distance d and the tunneling gap Δ_{SAS}. A parallel magnetic field $B_{\|}$ may be additionally applied to the system by tilting the sample. (b) A typical sample structure is illustrated. Electrons are provided by two remote δ-doped donor layers separated by AlGaAs spacers. (c) The number density ρ^α in each quantum well is controlled by applying gate bias voltages.

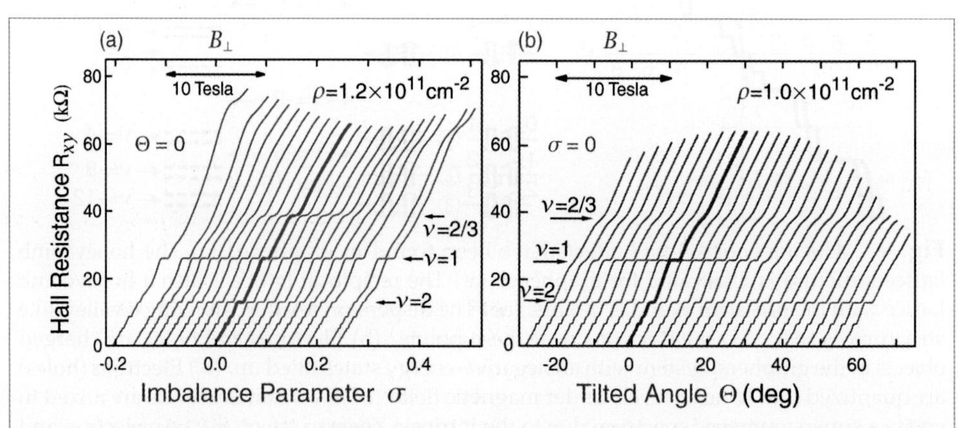

Fig. 11 Hall resistivity is plotted at a fixed total density ρ: Each curve is obtained (a) for a fixed value of the imbalance parameter σ, and (b) for a fixed value of the tilted angle Θ. The imbalance parameter is defined by $\sigma = (\rho^f - \rho^b)/\rho$ with ρ^f (ρ^b) the density in the front (back) layer. The data are taken from A. Sawada et al.[422,424]

Pseudospins

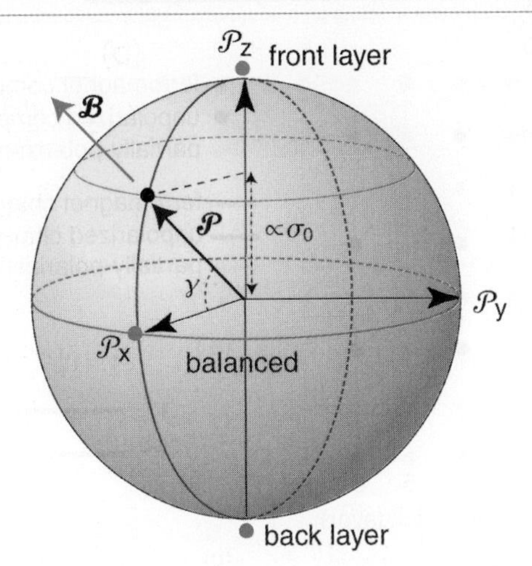

Fig. 12 The pseudospin $\mathscr{P} = (\mathscr{P}_x, \mathscr{P}_y, \mathscr{P}_z)$ is defined by assigning $\mathscr{P}_z = \pm 1/2$ to electrons in the front or back layer. The pseudomagnetic field $\mathscr{B} = (\Delta_{\text{SAS}}, 0, eV_{\text{bias}})$ is "tilted" by applying the bias voltage V_{bias}, as rotates the pseudospin $\mathscr{P}$ so that the pseudo-Zeeman energy $H_{\text{PZ}} \propto -\mathscr{B}\cdot\mathscr{P}$ is minimized. This is the process of a charge transfer between the two layers. When a bilayer QH state is stable (unstable) against this charge transfer, it belongs to a continuous (discrete) spectrum of $\mathscr{P}_z$.

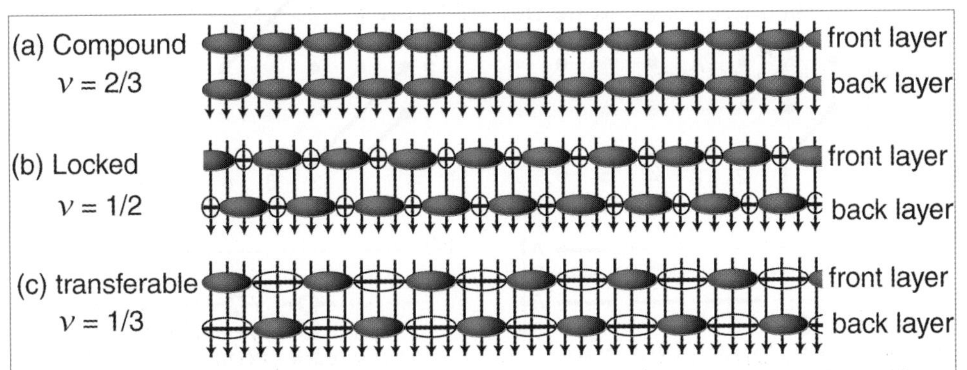

Fig. 13 Various bilayer QH states are illustrated in terms of composite particles. In these three figures the magnetic field is the same but the electron densities are different. (a) Compound state at $\nu = 2/3$: The QH state develops independently in each layer. (b) Bilayer-locked state at $\nu = 1/2$: The Coulomb repulsive force repels electrons on the opposite layer and creates a "hole" as indicated by an open oval. (c) Charge-transferable state at $\nu = 1/3$: Electrons do not have home layers since they tunnel freely between the two layers. Due to the interlayer exchange interaction the interlayer coherence may develop spontaneously on this state.

Compound and Charge-Transferable States (Theory)

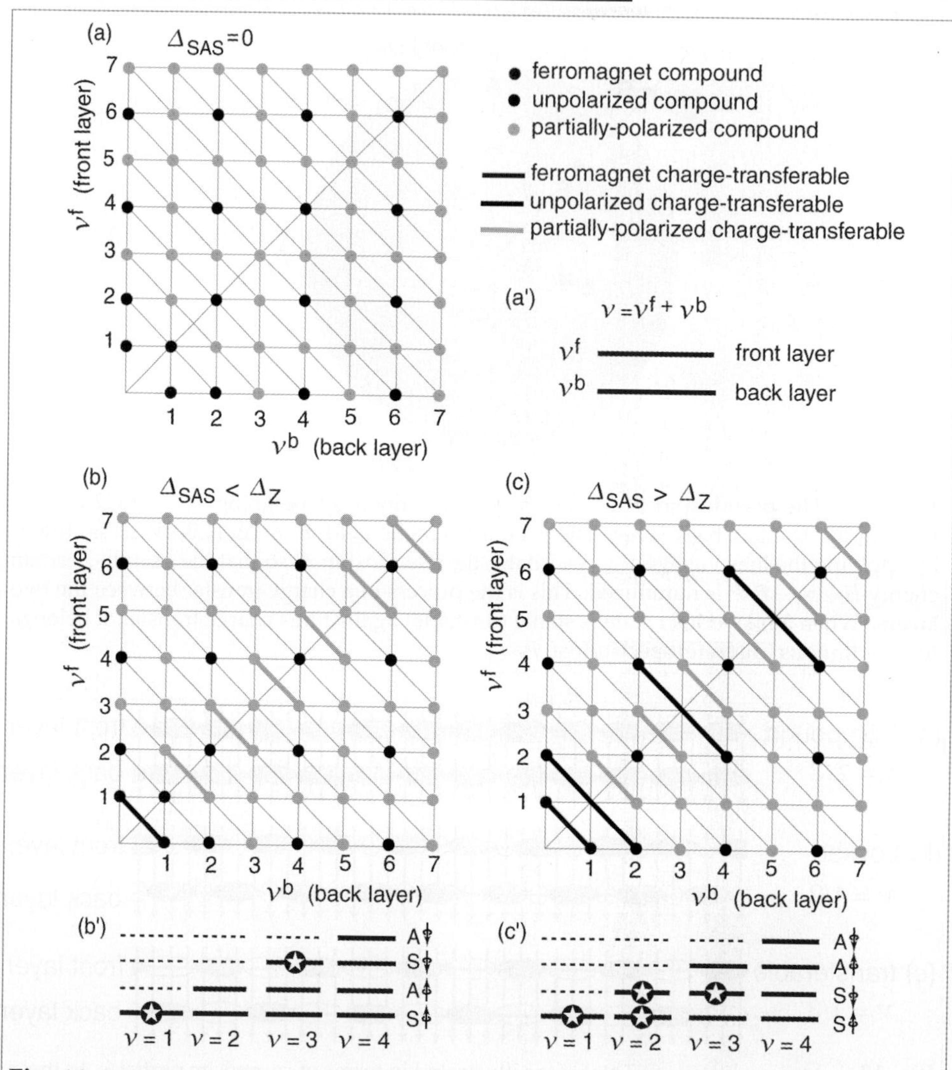

Fig. 14 A bilayer QH state belongs either to a discrete spectrum or a continuous spectrum of the pseudospin component $\mathcal{P}_z$. A solid circle on lattice point (ν^b, ν^f) denotes a *compound state* at $\nu = \nu^b + \nu^f$, which is made of the $\nu = \nu^b$ and $\nu = \nu^f$ monolayer QH states. A heavy line indicates a charge-transferable (*coherent* or *resonant*) state. Charge transfer is forbidden in the pseudospin singlet state. Let Δ_Z and Δ_{SAS} the Zeeman and tunneling energies. Various spectra appear (a) for $\Delta_{SAS} = 0$, (b) for $0 \neq \Delta_{SAS} < \Delta_Z$ and (c) for $\Delta_{SAS} > \Delta_Z$. (d) One Landau level is split into four energy levels, the up-spin symmetric (S↑) and antisymmetric (A↑) states, the down-spin symmetric (S↓) and antisymmetric (A↓) states. A heavy line indicates a filled level. The symbol ✪ indicates a charge-transferable level.

Compound and Charge-Transferable States (Experiments)

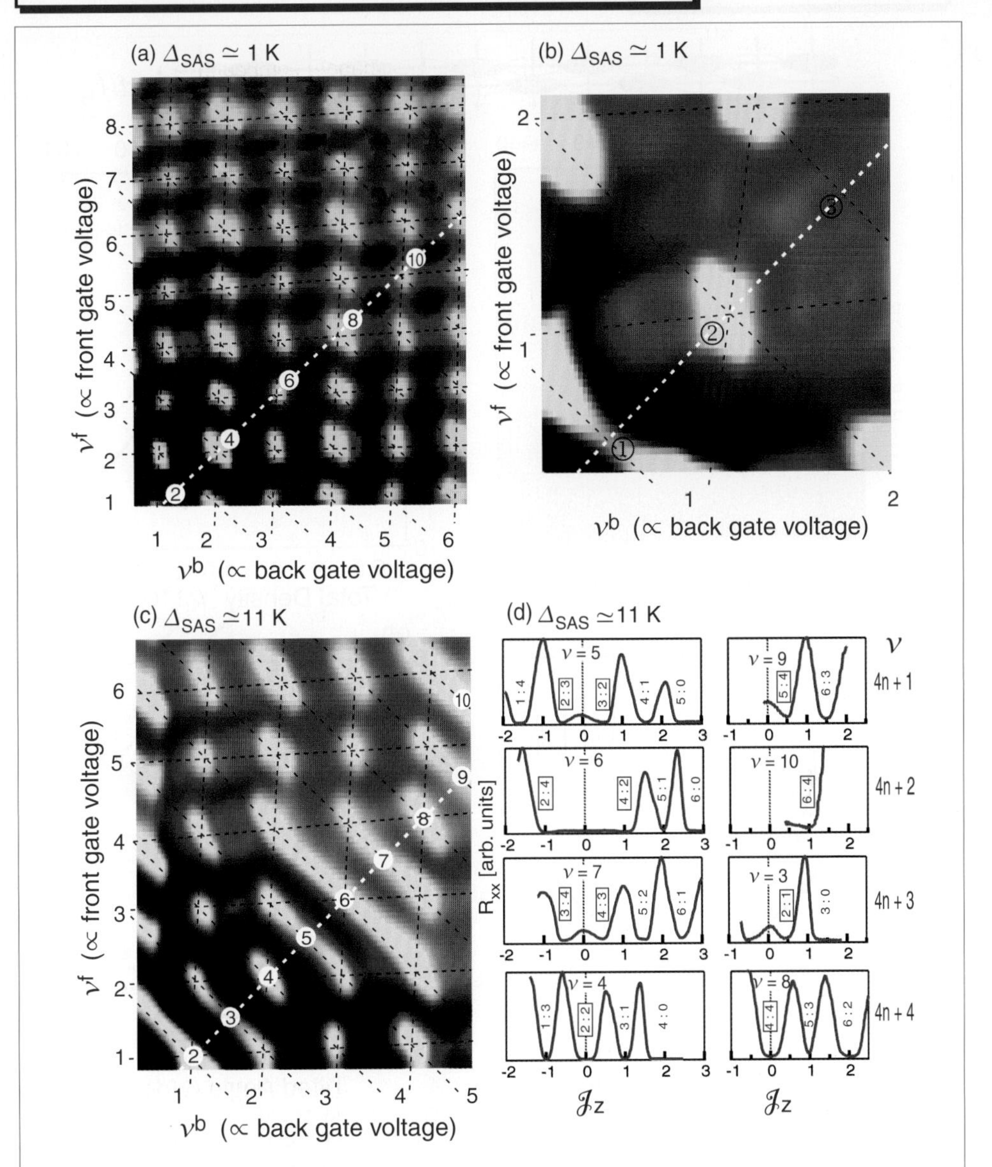

Fig. 15 Diagonal resistance was measured by scanning both the front and back gate voltages in a bilayer system. The bright regions correspond to small resistance, indicating QH states [See Fig.14]. (a) In a sample with $\Delta_{SAS} \simeq 1$ K, only compound states are seen for $\nu \geq 2$. (b) In the same sample, the $\nu = 1$ state is elongated, indicating a coherent state. (c) & (d) In a sample with $\Delta_{SAS} \simeq 11$ K, a characteristic pattern {1,2,1,0} of charge transferable states is prominent besides compound states. The data are taken from K. Muraki et al.[357]

Bilayer QH States at $\nu = 1$

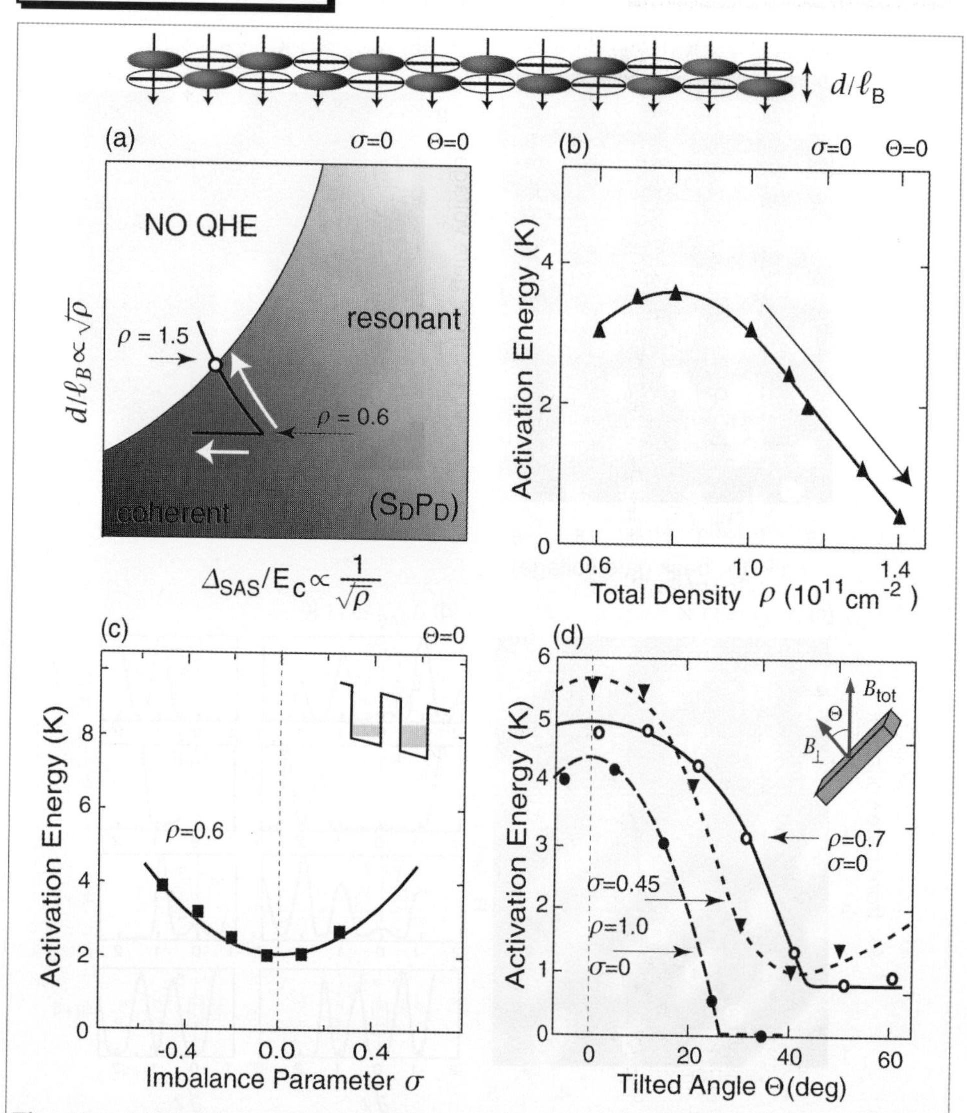

Fig. 16 (a) A phase diagram is illustrated at $\nu = 1$, where QH state exists only when the total density ρ is smaller than a certain value ($\rho > \rho_{\text{QH}}$) in a given sample. Interlayer coherence will develop provided $\Delta_{\text{SAS}}/E_{\text{C}}^0$ is small. The horizontal segment at $\rho = 0.6$ indicates that the sample was tilted at this density. (b) Activation energy decreases as ρ increases, showing that $\rho_{\text{QH}} = 1.5 \times 10^{11}\text{cm}^{-2}$. Here, sample parameters are $d = 23$ nm and $\Delta_{\text{SAS}} = 6.7$ K. (c) Activation energy increases as σ increases, as will be due to the increase of the capacitance energy [See Fig.18]. (d) Activation energy decreases anomalously in the commensurate phase, as the sample is tilted. The data are taken from A. Sawada et al.[422,424]

Bilayer QH States at $\nu = 2$

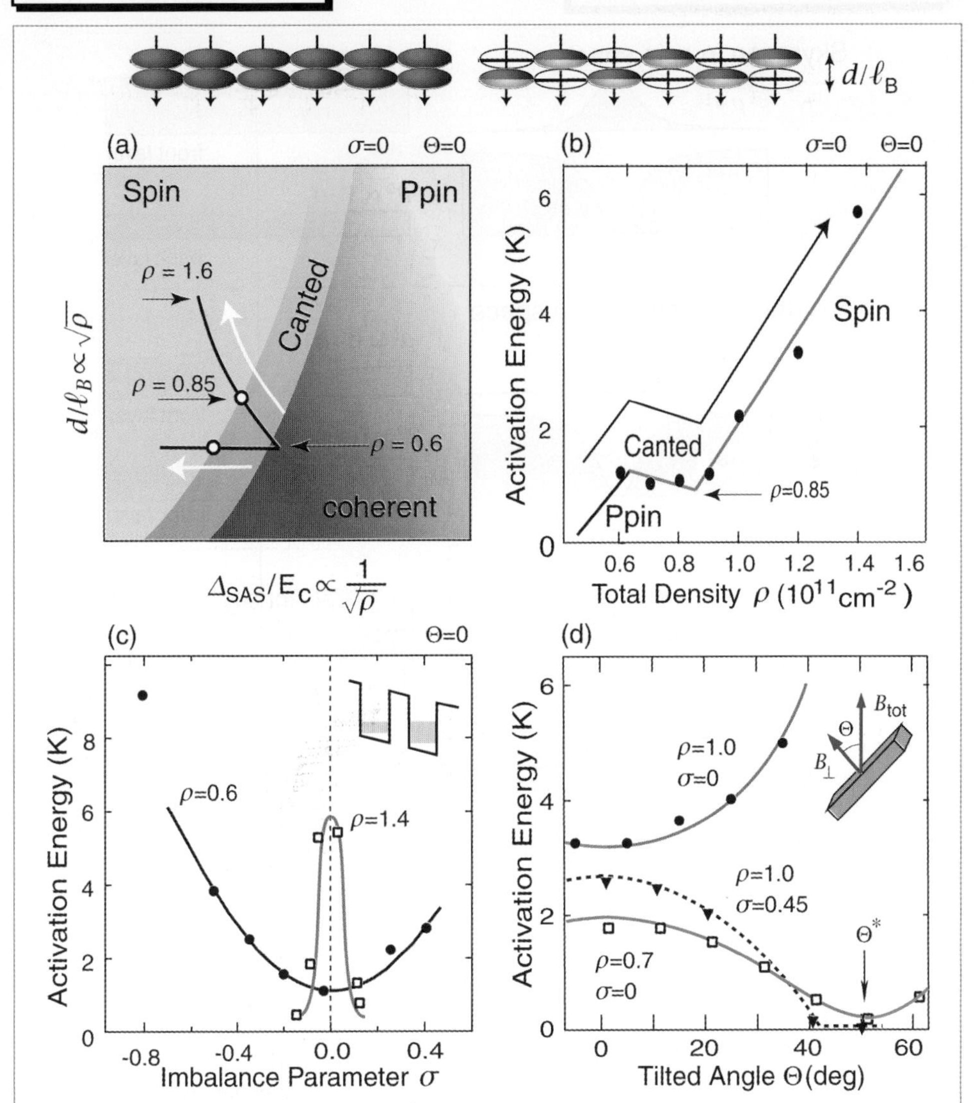

Fig. 17 (a) A phase diagram is illustrated at $\nu = 2$, where there are three phases: the spin ferromagnet, the canted antiferromagnet, and the pseudospin (ppin) ferromagnet phases. Interlayer coherence develops provided Δ_{SAS}/E_C^0 is small. (b) The observed activation energy is shown as a function of ρ together with a theoretical expectation. (c) As σ increases, activation energy increases smoothly in the ppin phase but decreases rapidly in the spin phase. (d) As the sample is tilted, activation energy decreases anomalously in the charge-transferable state (the canted and ppin phases). On the other hand, it increases in the spin phase due to the Zeeman effect. The data are taken from A. Sawada et al.[422,424]

Skyrmions in Bilayer QH States

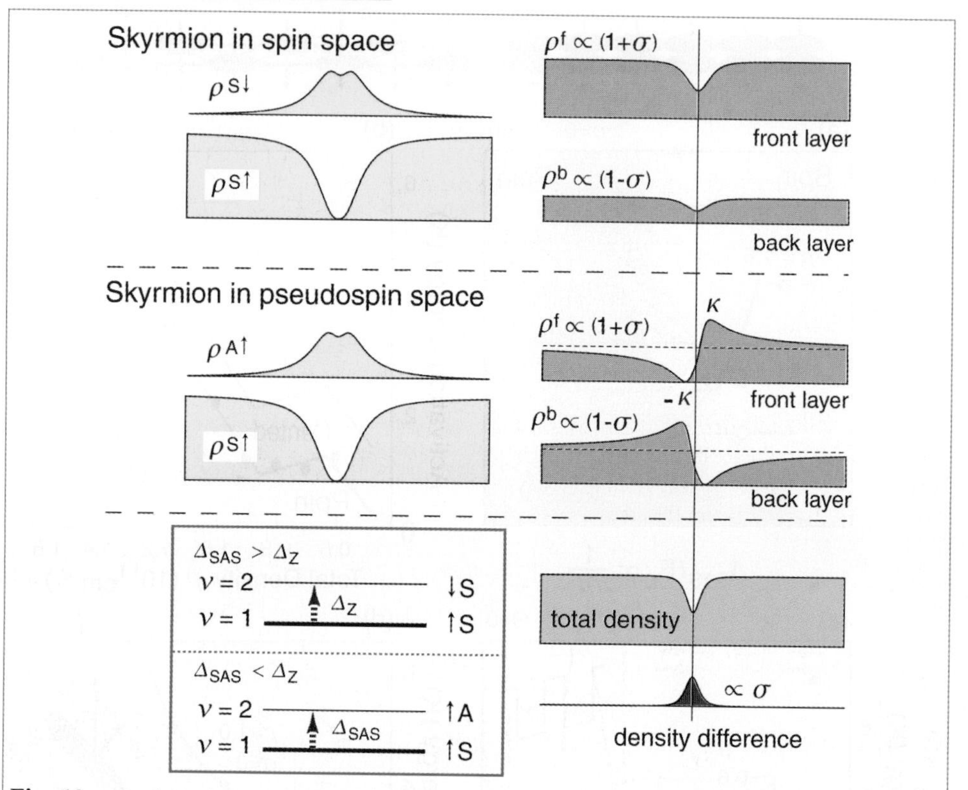

Fig. 18 A skyrmion excitation in the bilayer system modulates electron densities in both layers. The capacitance energy is proportional to σ^2, where σ is the imbalance parameter.

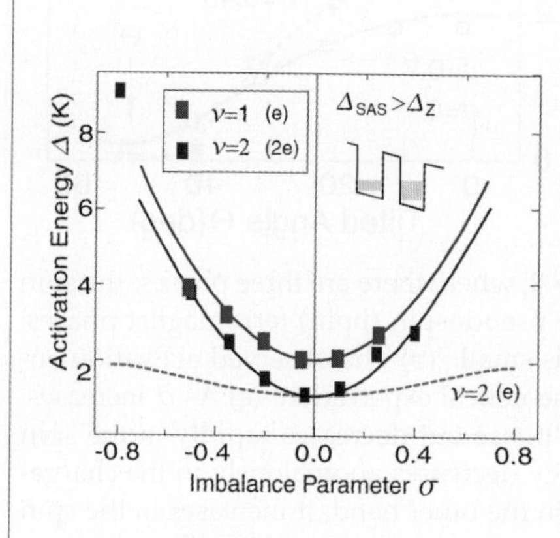

Fig. 19 Activation energy is shown as a function of σ. At $\nu = 1$ it is explained by the capacitance energy of a skyrmion with charge e, while at $\nu = 2$ by the capacitance and tunneling energies of a skyrmion with charge $2e$. The data are taken from A. Sawada et al.[423]

Counterflow and Drag Experiments

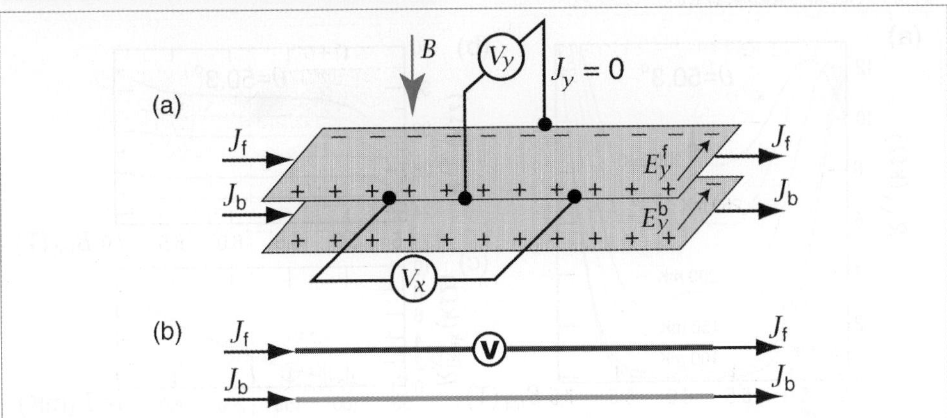

Fig. 20 (a) Hall currents are injected to the front and back layers independently. The Hall and diagonal resistances are measured only in one of the layers. (b) A simplified picture is given to represent the same measurement, where the symbol V in a circle indicates that the measurement is done on this layer.

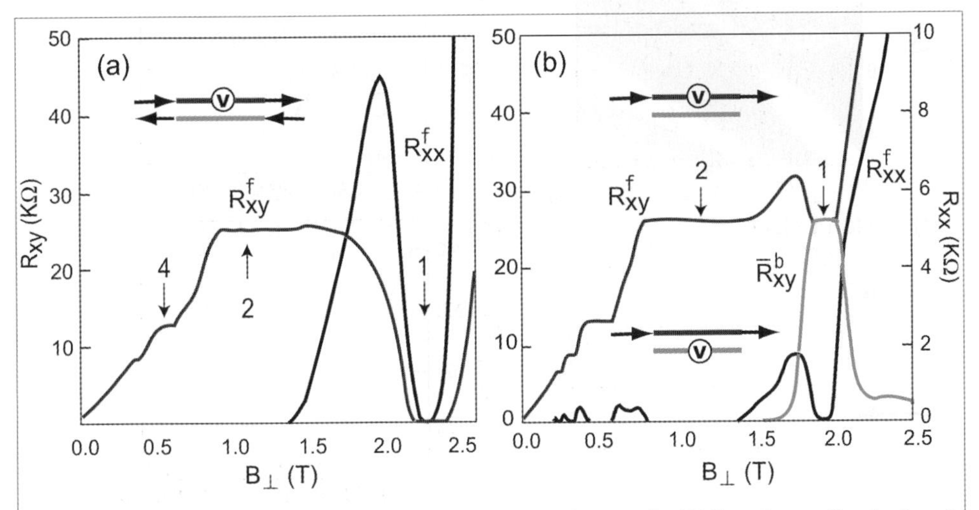

Fig. 21 By feeding different currents into the two layers, the Hall resitance R_{xy} (red and blue curves) and the diagonal resistance R_{xx} (black curve) are measured on one of the layers (indicated by V in a circle). (a) In the counterflow experiment the opposite amounts of currents are fed to the two layers, where $R^f_{xx} = 0$ as expected, but $R^f_{xy} = 0$ anomalously instead of developing a Hall plateau at $\nu = 1$. Data are taken from E. Tutuc et al.[481] (b) In the drag experiment the current is fed only to the front layer, where the value R^f_{xy} of the Hall plateau takes anomalously the same value at $\nu = 1$ and $\nu = 2$. It is also remarkable that a Hall plateau develops in $\bar{R}^b_{xy}$ and $\bar{R}^b_{xy} = R^f_{xy}$ at $\nu = 1$, where $\bar{R}^b_{xy} \equiv E^b_y / \mathcal{J}^f_x$. Data are taken from M. Kellogg et al.[267]

Soliton Lattice

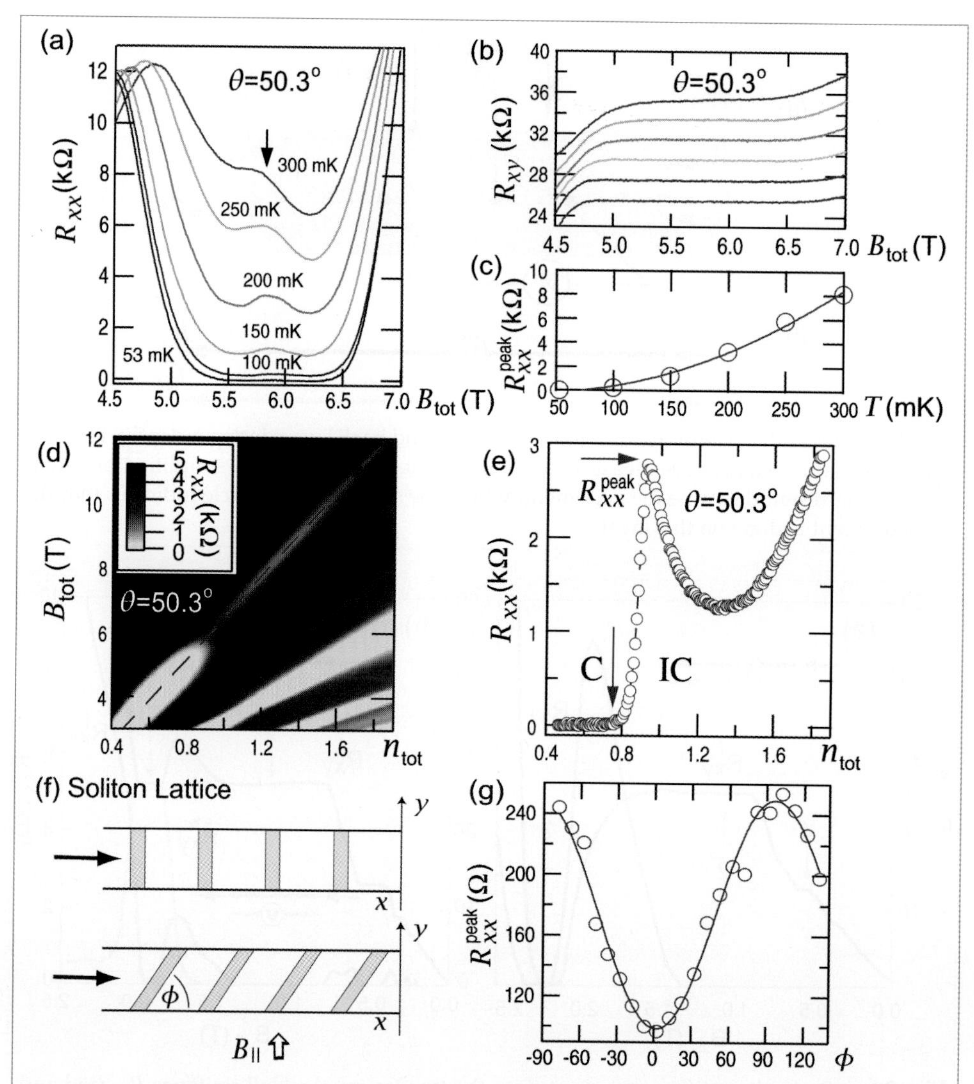

Fig. 22 (a,b) Diagonal and Hall resistances are measured by changing the total magnetic field B_{tot} in a tilted sample at 50.3°. There is a local maximum in R_{xx} in the range where the Hall plateau develops well. (c) The peak value of R_{xx}^{peak} decreases as the temperature T decreases. (d) The image plot of R_{xx} as a function of B_{tot} and the electron density $\rho = n_{\text{tot}} \times 10^{15}\text{cm}^{-2}$ shows a commensurate-incommensurate (C-IC) phase transition. (e) There is a sharp peak (R_{xx}^{peak}) in R_{xx} just beyond the phase transition point inside the IC phase. (f) The peak is understood to arise due to a backscattering of the Hall current against the soliton lattice made of the parallel magnetic field $B_{\parallel}$. (g) This picture is confirmed by the change of R_{xx}^{peak} as the sample is rotated in $B_{\parallel}$. Data are taken from A. Fukuda et al.[181]

CONTENTS

Preface vii

Part. I Quantum Field Theory 1

1. Quantum Mechanics 3

1.1 Hilbert Spaces ... 3
1.2 Canonical Formalism ... 4
1.3 Creation and Annihilation Operators ... 6
1.4 Uncertainty Principle ... 10
1.5 Coherent States and von Neumann Lattice ... 11
1.6 Squeezed Coherent State ... 14
1.7 Particle Number and Phase ... 16
1.8 Macroscopic Coherence ... 18

2. Quantum Field Theory 20

2.1 One-Body Hamiltonian ... 20
2.2 Many-Body Hamiltonian ... 21
2.3 Boson Field Operators ... 23
2.4 Quantum Field Theory ... 25
2.5 Fermion Field Operators ... 27
2.6 Electrons Interacting with Electromagnetic Field ... 29

3. Canonical Quantization 31

3.1 Relativistic Particles and Waves ... 31
3.2 Schrödinger Field ... 32
3.3 Real Klein-Gordon Field ... 37
3.4 Complex Klein-Gordon Field ... 42

3.5 Nöther Currents 44

4. **Spontaneous Symmetry Breaking** 47

4.1 Ferromagnets 47
4.2 Real Klein-Gordon Field 51
4.3 Complex Klein-Gordon Field 55
4.4 Sigma Model 56
4.5 Schrödinger Field 58
4.6 Superfluidity 62
4.7 Goldstone Theorem 64

5. **Electromagnetic Field** 67

5.1 Maxwell Equations 67
5.2 Canonical Quantization 70
5.3 Interaction with Matter Field 77
5.4 Anderson-Higgs Mechanism 79
5.5 Massive Vector Field 81
5.6 Superconductivity 83
5.7 Aharonov-Bohm Effect 89

6. **Dirac Field** 94

6.1 Dirac Equation 94
6.2 Plane Wave Solutions 97
6.3 Canonical Quantization 99
6.4 Interaction with Electromagnetic Field 103
6.5 Weyl Field (Massless Dirac Field) 105
6.6 Dirac Electrons in Magnetic Field 108

7. **Topological Solitons** 113

7.1 Topological Sectors 113
7.2 Classical Fields 115
7.3 Solitary Waves, Kinks and Solitons 117
7.4 Sine-Gordon Solitons 119
7.5 Vortex Solitons 123
7.6 Homotopy Classes 127
7.7 O(3) Skyrmions 132
7.8 CP^{N-1} Skyrmions 138

8. **Anyons** 144

8.1 Spin and Statistics 144
8.2 Fractional Statistics 148
8.3 Quantum Mechanics 150

8.4 Chern-Simons Gauge Theory 153
8.5 Anyon Field Operators 157

Part. II Monolayer Quantum Hall Systems **161**

9. Overview of Monolayer QH Systems 163

10. Landau Quantization 177

10.1 Planar Electrons 177
10.2 Cyclotron Motion 181
10.3 Symmetric Gauge 187
10.4 Landau Gauge 190
10.5 von Neumann Lattice 193
10.6 Electrons in *N*th Landau Level 195
10.7 Hall Current 199

11. Quantum Hall Effects 204

11.1 Incompressibility 204
11.2 Integer Quantum Hall Effects 205
11.3 Fractional Quantum Hall Effects 207
11.4 Quasiparticles 210
11.5 Hall Plateaux 213

12. Quasiparticles and Activation Energy 215

12.1 Impurity Potentials 215
12.2 Gap Energies 216
12.3 Dispersion Relation 219
12.4 Thermal Activation 222

13. Field Theory of Composite Particles 227

13.1 Composite Particles 227
13.2 Statistical Transmutation 230
13.3 Effective Magnetic Field 233
13.4 Dressed Composite Particles 235
13.5 Composite Particles in Lowest Landau Level 239
13.6 Composite Fermions in Lowest Landau Level 242

14. Composite Bosons and Semiclassical Analysis 246

14.1 Ground State and Laughlin Wave Function 246

14.2 Perturbative Excitations 248
14.3 Vortex Excitations 250
14.4 Field Theory of Vortex Solitons 255
14.5 Haldane-Halperin Hierarchy 258

15. Quantum Hall Ferromagnets 261

15.1 Spin Coherence 261
15.2 Spin Degree of Freedom 263
15.3 Composite Bosons and Spin-Charge Separation 265
15.4 Spin Field, Sigma Field and CP^1 Field 268
15.5 Effective Hamiltonian 271

16. Spin Textures 274

16.1 Spin Excitations 274
16.2 Factorizable Skyrmions 277
16.3 Skyrmion Excitation Energy 284
16.4 Experimental Evidence 287

17. Hierarchy of Fractional QH States 291

17.1 Jain Hierarchy 291
17.2 Landau Levels of Composite Fermions 294
17.3 Beyond Principal Sequences 296
17.4 Spin Polarization 298
17.5 Gap Energies 301

18. Edge Effects 307

18.1 Edge Currents and Bulk Currents 307
18.2 Shot Noises of Fractional Charges 308
18.3 Chiral Edge Excitations 311
18.4 Chiral Tomonaga-Luttinger Liquid 312
18.5 Electrodynamics on Edge 316
18.6 Edge Tunneling and Sine-Gordon Solitons 320

19. Stripes and Bubbles in Higher Landau Levels 325

19.1 Higher Landau Levels 325
19.2 Haldane's Pseudopotentials 327
19.3 Effective Coulomb Interactions 328
19.4 Density Matrix Renormalization Group Method 331
19.5 Stripes, Bubbles and Wigner Crystal 332

20. Quantum Hall Effects in Graphene 335

20.1 Unconventional QH Effects 335

20.2 Graphene and Dirac Electrons 338
20.3 Dirac Hamiltonian and Supersymmetry 343
20.4 Effective Coulomb Interactions 346
20.5 Excitonic Condensation 352
20.6 Valley Polarization 355
20.7 Multilayer Graphene Systems 360
20.8 Berry Phase and Index Theorem 365

Part. III Bilayer Quantum Hall Systems **367**

21. Overview of Bilayer QH Systems 369

22. SU(2) Pseudospin Structure 383

22.1 Bilayer Planar Electrons 383
22.2 Pseudospins 385
22.3 Tunneling Interaction 387
22.4 Imbalanced Configuration 388
22.5 Capacitance Energy 392
22.6 Compound States 393
22.7 Charge-Transferable States 394

23. Bilayer-Locked States 400

23.1 Composite-Boson Field 400
23.2 Wave Functions 404
23.3 Ground State 405
23.4 Vortex Excitations 406

24. Interlayer Coherence 409

24.1 Pseudospin Ferromagnet 409
24.2 Effective Hamiltonian 413
24.3 Pseudospin Waves 415
24.4 Anomalous Bilayer QH Currents 417
24.5 Pseudospin Texture 424

25. SU(4) Quantum Hall Ferromagnets 428

25.1 SU(4) Isospin Structure 428
25.2 SU(4) Isospin Fields 431
25.3 SU(4) Isospin Waves 433
25.4 SU(4) Isospin Textures 436

25.5 Excitation Energy of SU(4) Skyrmions 440
25.6 Activation Energy Anomaly 446

26. Bilayer Quantum Hall Systems at $\nu = 2$ 454

26.1 Spin Phase, Ppin Phase and Canted Phase 454
26.2 Ground-State Energy 457
26.3 Ground-State Structure 460
26.4 Phase Diagrams 463
26.5 Experimental Data 466
26.6 SU(4) Breaking and Grassmannian Fields 471
26.7 Grassmannian $G_{4,2}$ Solitons 475
26.8 Genuine Bilayer versus Two-Monolayer Systems 479
26.9 Experimental Indication of Biskyrmions 482

27. Bilayer Quantum-Hall Junction 485

27.1 Josephson-Like Phenomena 485
27.2 Parallel Magnetic Field 487
27.3 Effective Hamiltonian 493
27.4 Commensurate-Incommensurate Phase Transition 495
27.5 Soliton Lattice 499
27.6 Anomalous Diagonal Resistivity 508
27.7 Plasmon Excitations 514
27.8 Josephson-Like Effects 516

Part. IV Microscopic Theory 521

28. Overview of Microscopic Theory 523

29. Noncommutative Geometry 534

29.1 Noncommutative Coordinate 534
29.2 Weyl Operator and Symbol 535
29.3 Magnetic Translation 541
29.4 Density Operators 543
29.5 SU(N)-Extended W_∞ Algebra 549
29.6 Classical Fields 552
29.7 Topological Charge Density 557
29.8 Kac-Moody Algebra on Edges 559

30. Landau Level Projection 562

30.1 Projected Coulomb Interactions 562
30.2 Monolayer QH System 565
30.3 Electron-Hole Pair Excitations 569
30.4 Electron Excitation and Hole Excitation 570
30.5 Bilayer System without Spin ($\nu = 1$) 573
30.6 Bilayer System with Spin ($\nu = 1$) 575
30.7 Bilayer System with Spins ($\nu = 2$) 578

31. Noncommutative Solitons 584

31.1 Topological Charge and Electric Charge 584
31.2 Microscopic Skyrmion States 585
31.3 Noncommutative CP^1 Skyrmion 589
31.4 Hole and Skyrmion 592
31.5 Skyrmion Wave Functions 593
31.6 Hardcore Interaction 595
31.7 Coulomb Interaction 600
31.8 SU(4) Skyrmions 605

32. Exchange Interactions and Effective Theory 607

32.1 Exchange Hamiltonian 607
32.2 Decomposition Formula 612
32.3 Spontaneous Symmetry Breaking 615
32.4 Classical Equations of Motion 616
32.5 Four-Layer Condenser Model 618
32.6 Derivative Expansion and Effective Theory 622
32.7 Noncommutative CP^{N-1} Model 627
32.8 Equations of Motion and Hall Currents 631
32.9 Hall Currents in Pseudospin QH Ferromagnet 634

Appendices 643

A Energy Scales 643
B Hausdorff Formulas 645
C Group SU(2) and Pauli Matrices 647
D Groups SU(N) and SU(2N) 647
E Cauchy-Riemann Equations 650
F Green Function 651
G Bogoliubov Transformation 652
H Energy-Momentum Tensor 654
I Exchange Interaction 655
J Mermin-Wagner Theorem 659
K Lorentz Transformation 663
L One-Dimensional Soliton Solutions 666

M Field-Theoretical Vortex Operators 669
N Bosonization in One-Dimensional Space 672
O Coulomb Energy Formulas 677
P W_∞ Algebra 679

References 681

Index 699

Part 1

Quantum Field Theory

CHAPTER 1

QUANTUM MECHANICS

Macroscopic quantum phenomena are exceedingly attractive and important in modern physics. A minimum review of quantum mechanics is given from this point of view in this Section. Quantum field theory is constructed on the basis of creation and annihilation operators. An emphasis is placed on the minimum-uncertainty state, that is a coherent state, where the particle number and the conjugate phase are simultaneously measurable most accurately as a quantum system.

1.1 HILBERT SPACES

A state is represented by an element $|\mathfrak{S}\rangle$ in a Hilbert space $\mathbb{H}$. A Hilbert space $\mathbb{H}$ is a vector space characterized by the following properties:

(1) *Principle of superposition*
If $|\mathfrak{S}_1\rangle$ and $|\mathfrak{S}_2\rangle$ are elements of the space $\mathbb{H}$ then so is $\lambda_1|\mathfrak{S}_1\rangle + \lambda_2|\mathfrak{S}_2\rangle$ for arbitrary complex numbers λ_1 and λ_2.

(2) *Inner product*
A complex number $\langle\mathfrak{S}_2|\mathfrak{S}_1\rangle$, called *inner product*, is assigned to any pair of states. It obeys

$$\langle\mathfrak{S}_2|\mathfrak{S}_1\rangle = \langle\mathfrak{S}_1|\mathfrak{S}_2\rangle^*, \tag{1.1.1a}$$

$$\langle\mathfrak{S}_3|\,(\lambda_1|\mathfrak{S}_1\rangle + \lambda_2|\mathfrak{S}_2\rangle) = \lambda_1\langle\mathfrak{S}_3|\mathfrak{S}_1\rangle + \lambda_2\langle\mathfrak{S}_3|\mathfrak{S}_2\rangle, \tag{1.1.1b}$$

$$\langle\mathfrak{S}|\mathfrak{S}\rangle \geq 0. \tag{1.1.1c}$$

The inner product $\langle\mathfrak{S}|\mathfrak{S}\rangle$ is called the norm of the state $|\mathfrak{S}\rangle$. It vanishes if and only if $|\mathfrak{S}\rangle = 0$.

When an operator $\mathcal{O}$ is given, a complex number $\langle\mathfrak{S}_2|\mathcal{O}|\mathfrak{S}_1\rangle$ is determined for any pair of states $|\mathfrak{S}_1\rangle$ and $|\mathfrak{S}_2\rangle$ as the inner product of two states $|\mathfrak{S}_2\rangle$ and $\mathcal{O}|\mathfrak{S}_1\rangle$. Conversely, the operator $\mathcal{O}$ is uniquely defined when a complex number $\langle\mathfrak{S}_2|\mathcal{O}|\mathfrak{S}_1\rangle$ is given to any pair of states. Thus, a new operator $\mathcal{O}^\dagger$ is defined from

$\mathcal{O}$ by requiring

$$\langle \mathfrak{S}_2 | \mathcal{O}^\dagger | \mathfrak{S}_1 \rangle = \langle \mathfrak{S}_1 | \mathcal{O} | \mathfrak{S}_2 \rangle^* \tag{1.1.2}$$

for any pair of states, where $\langle \mathfrak{S}_1 | \mathcal{O} | \mathfrak{S}_2 \rangle^*$ is the complex conjugate of $\langle \mathfrak{S}_1 | \mathcal{O} | \mathfrak{S}_2 \rangle$. The operator $\mathcal{O}^\dagger$ is called the *Hermitian conjugate* of $\mathcal{O}$.

A measurement of the operator $\mathcal{O}$ made on the state $|\mathfrak{S}\rangle$ yields an expectation value $\langle \mathfrak{S} | \mathcal{O} | \mathfrak{S} \rangle$. Since we get a real number from the measurement of a physical quantity, the physical operator should satisfy $\mathcal{O}^\dagger = \mathcal{O}$: such an operator is called a *Hermitian operator*. On the other hand, when an operator $\mathcal{O}$ satisfies $\mathcal{O}^\dagger = \mathcal{O}^{-1}$, where $\mathcal{O}^{-1}$ is the inverse of $\mathcal{O}$, it is called a *unitary operator*.

For any Hermitian operator H, solutions of the eigenvalue problem,

$$H | \mathfrak{S}_i \rangle = \varepsilon_i | \mathfrak{S}_i \rangle, \tag{1.1.3}$$

can be chosen to satisfy the orthonormality condition

$$\langle \mathfrak{S}_j | \mathfrak{S}_k \rangle = \delta_{jk}, \tag{1.1.4}$$

and the completeness condition

$$\sum_j | \mathfrak{S}_j \rangle \langle \mathfrak{S}_j | = 1. \tag{1.1.5}$$

Such a set of states $|\mathfrak{S}_j\rangle$ form a basis of a Hilbert space $\mathbb{H}$. Namely, based on (1.1.4) and (1.1.5), any state is expanded in terms of them as

$$|\mathfrak{S}\rangle = \sum_j |\mathfrak{S}_j\rangle \langle \mathfrak{S}_j | \mathfrak{S} \rangle = \sum_j \lambda_j |\mathfrak{S}_j\rangle, \tag{1.1.6}$$

where $\lambda_j = \langle \mathfrak{S}_j | \mathfrak{S} \rangle$ is the probability amplitude to find the state $|\mathfrak{S}_j\rangle$ in $|\mathfrak{S}\rangle$.

1.2 Canonical Formalism

The dynamics of a system is determined when the Lagrangian L is given. It is a function of a collection of "coordinates", $q \equiv \{q_1, q_2, \cdots, q_N\}$, with their "velocities" $\dot{q}$ at time t. The action is defined by

$$S = \int_{t_1}^{t_2} dt\, L[q(t), \dot{q}(t)]. \tag{1.2.1}$$

The principle of the least action implies

$$\delta S = \int_{t_1}^{t_2} dt\, \{L[q(t) + \delta q(t), \dot{q}(t) + \delta \dot{q}(t)] - L[q(t), \dot{q}(t)]\} = 0, \tag{1.2.2}$$

as yields the Euler-Lagrange equation,

$$\frac{\delta S}{\delta q(t)} \equiv \frac{\partial L}{\partial q(t)} - \frac{d}{dt}\frac{\partial L}{\partial \dot{q}(t)} = 0. \tag{1.2.3}$$

The conjugate "momentum" (canonical momentum) is defined by

$$p_i = \frac{\partial L(q,\dot{q})}{\partial \dot{q}_i}. \tag{1.2.4}$$

The Hamiltonian is then obtained by a Legendre transformation,

$$H(p,q) = \sum_i p_i \dot{q}_i - L[q, \dot{q}(p,q)], \tag{1.2.5}$$

which is a function of the coordinate and the momentum.[1] The physical meaning of the Hamiltonian is the energy of the system. It is a Hermitian operator.

The system is quantized when the canonical commutation relation is postulated between the coordinate and the canonical momentum,

$$[q_i, p_j] = i\hbar\delta_{ij}. \tag{1.2.7}$$

The Hamiltonian is promoted to an operator acting on states $|\mathfrak{S}\rangle$. The dynamics of a quantum system is determined when the Hamiltonian and the canonical commutation relation are given. This procedure is customarily called the *first quantization*.

The Schrödinger equation is

$$i\hbar\frac{d}{dt}|\mathfrak{S}(t)\rangle = H|\mathfrak{S}(t)\rangle. \tag{1.2.8}$$

The time-evolution operator $U(t)$ is defined by

$$|\mathfrak{S}(t)\rangle = U(t)|\mathfrak{S}(0)\rangle, \tag{1.2.9}$$

and obeys the differential equation

$$i\hbar\frac{d}{dt}U(t) = HU(t) \tag{1.2.10}$$

together with the boundary condition $U(0) = 1$. It is a unitary operator. When the Hamiltonian is independent of time, it is formally solved as

$$U(t) = e^{-iHt/\hbar}. \tag{1.2.11}$$

The state evolves according to the formula (1.2.9), while the operator remains unchanged. This is the *Schrödinger picture.*

Although the Schrödinger picture is familiar in quantum mechanics, the *Heisenberg picture* is more convenient in quantum field theory. These two pictures are

---

[1] To see this, we take the external derivative. Since the Lagrangian is a function of q and $\dot{q}$, we have $dL = (\partial_q L)dq + (\partial_{\dot{q}} L)d\dot{q}$. Now,

$$dH = d(p\dot{q}) - dL = \dot{q}dp + pd\dot{q} - (\partial_q L)dq - (\partial_{\dot{q}} L)d\dot{q} = \dot{q}dp - \dot{p}dq, \tag{1.2.6}$$

where use was made of the Euler-Lagrange equation (1.2.3) and the definition of the momentum (1.2.4). It implies that H is a function of q and p with the canonical equations of motion, $\dot{q} = \partial_p H$ and $\dot{p} = -\partial_q H$.

related by way of the time-evolution operator $U(t)$. The operator $A_\mathrm{H}(t)$ and the state $|\mathfrak{S}_\mathrm{H}\rangle$ in the Heisenberg picture are defined by

$$A_\mathrm{H}(t) \equiv U^\dagger(t)AU(t), \qquad |\mathfrak{S}_\mathrm{H}\rangle \equiv U^\dagger(t)|\mathfrak{S}(t)\rangle = |\mathfrak{S}(0)\rangle, \tag{1.2.12}$$

from the operator A and the state $|\mathfrak{S}(t)\rangle$ in the Schrödinger picture. The Heisenberg state $|\mathfrak{S}_\mathrm{H}\rangle$ is independent of time,

$$\frac{d}{dt}|\mathfrak{S}_\mathrm{H}\rangle = 0, \tag{1.2.13}$$

and the Heisenberg operator $A_\mathrm{H}(t)$ obeys the differential equation

$$i\hbar\frac{d}{dt}A_\mathrm{H}(t) = U^\dagger(t)[A, H(\mathcal{O})]U(t) = [A_\mathrm{H}(t), H(\mathcal{O}_\mathrm{H}(t))]. \tag{1.2.14}$$

Here, $\mathcal{O}$ stands for a generic operator (such as a momentum) involved in the Hamiltonian. The Hamiltonian $H(\mathcal{O}_\mathrm{H}(t))$ is given simply by replacing $\mathcal{O}$ with the Heisenberg operator $\mathcal{O}_\mathrm{H}(t)$ in $H(\mathcal{O})$, since

$$\begin{aligned} U^\dagger H(\mathcal{O})U &= U^\dagger(b_0 + b_1\mathcal{O} + b_2\mathcal{O}^2 + \cdots)U \\ &= b_0 + b_1\mathcal{O}_\mathrm{H} + b_2\mathcal{O}_\mathrm{H}^2 + \cdots = H(\mathcal{O}_\mathrm{H}), \end{aligned} \tag{1.2.15}$$

where $H(\mathcal{O})$ is expanded in polynomials of $\mathcal{O}$ with coefficients b_i. Equation (1.2.14) is called the Heisenberg equation of motion. In the Heisenberg picture the state remains unchanged while the operator evolves as time passes.

In the Schrödinger picture, the observed value of the operator A in the state $|\mathfrak{S}(t)\rangle$ is

$$\langle A\rangle = \langle\mathfrak{S}(t)|A|\mathfrak{S}(t)\rangle, \tag{1.2.16}$$

which reads

$$\langle A\rangle = \langle\mathfrak{S}_\mathrm{H}|U^\dagger U A_\mathrm{H}(t)U^\dagger U|\mathfrak{S}_\mathrm{H}\rangle = \langle\mathfrak{S}_\mathrm{H}|A_\mathrm{H}(t)|\mathfrak{S}_\mathrm{H}\rangle \tag{1.2.17}$$

in the Heisenberg picture. Namely, the observed value $\langle A\rangle$ of the operator A is identical both in the Schrödinger and Heisenberg pictures, as should be the case. We omit the index H of the Heisenberg operator A_H in what follows.

1.3 Creation and Annihilation Operators

We introduce the creation and annihilation operators of quanta, that is, particles or excitation modes. Quanta obey either the Bose statistics or the Fermi statistics in the 3-dimensional space.[2] *Bose* quanta are characterized by the property that one quantum state can contain an arbitrary number of identical quanta. *Fermi* quanta obey the Pauli exclusion principle stating that no two quanta of the same type can

[2] A generalized statistics is possible in the 2-dimensional space, as will be reviewed in Chapter 8.

occupy a single quantum state. They are essential ingredients of quantum field theory.

Bose Quanta

Bose quanta are described by a pair of the operator a and its Hermitian conjugate $a^\dagger$ satisfying the commutation relation

$$[a, a^\dagger] = 1. \tag{1.3.1}$$

We construct a Hilbert space $\mathbb{H}$ realizing this commutation relation. We define the number operator N by

$$N = a^\dagger a, \tag{1.3.2}$$

and the vacuum state $|0\rangle$ by

$$a|0\rangle = 0. \tag{1.3.3}$$

It contains no quanta, $N|0\rangle = 0$, and is normalized as $\langle 0|0\rangle = 1$.

It follows from (1.3.1) and (1.3.2) that

$$a^\dagger|n\rangle = \sqrt{n+1}|n+1\rangle, \qquad a|n\rangle = \sqrt{n}|n-1\rangle. \tag{1.3.4}$$

Thus, $a^\dagger$ and a are the creation and annihilation operators. One Bose quanta is created by operating $a^\dagger$ onto the vacuum state $|0\rangle$, $a^\dagger|0\rangle = |1\rangle$. In general, the normalized state containing n Bose quanta is

$$|n\rangle = \frac{1}{\sqrt{n!}}(a^\dagger)^n|0\rangle, \tag{1.3.5}$$

where

$$N|n\rangle = n|n\rangle, \qquad \langle n|m\rangle = \delta_{nm}. \tag{1.3.6}$$

All the states, $\{|n\rangle; n = 0, 1, 2, 3, \cdots\}$, constitute a Hilbert space $\mathbb{H}$ together with the orthonormality condition (1.3.6) and the completeness condition

$$\sum_n |n\rangle\langle n| = 1. \tag{1.3.7}$$

It is called the occupation-number space or the *Fock space.* The vacuum $|0\rangle$ is called the *Fock vacuum.*

If one quantum carries the energy ε, the Hamiltonian is postulated as

$$H = \varepsilon N = \varepsilon a^\dagger a, \tag{1.3.8}$$

so that the energy of the state $|n\rangle$ is $n\varepsilon$,

$$H|n\rangle = n\varepsilon|n\rangle. \tag{1.3.9}$$

The Heisenberg equation reads

$$i\hbar\frac{\partial}{\partial t}a = [a, H] = \varepsilon a. \tag{1.3.10}$$

Its solution is given by

$$a(t) = e^{-i\omega t}a(0), \tag{1.3.11}$$

where

$$\varepsilon = \hbar\omega. \tag{1.3.12}$$

It describes a noninteracting system of one kind of Bose quanta.

When the system contains several kinds of quanta, the annihilation and creation operators are introduced with respect to each of them,

$$[a_j, a_k^\dagger] = \delta_{jk}, \qquad [a_j, a_k] = 0. \tag{1.3.13}$$

They satisfies

$$N_j a_k^\dagger = a_k^\dagger(N_j + \delta_{jk}), \qquad N_j a_k = a_k(N_j - \delta_{jk}), \tag{1.3.14}$$

with the number operator of the jth quantum being

$$N_j = a_j^\dagger a_j. \tag{1.3.15}$$

The Fock vacuum $|0\rangle$ satisfies $a_j|0\rangle = 0$ with $\langle 0|0\rangle = 1$. The state $|n_1, n_2, \cdots\rangle$ containing these quanta is given by a product of the states (1.3.5),

$$|n_1, n_2, \cdots\rangle = \prod_j \frac{1}{\sqrt{n_j!}}(a_j^\dagger)^{n_j}|0\rangle. \tag{1.3.16}$$

When the energy of the jth quantum is ε_j, the Hamiltonian is given by

$$H = \sum_j \varepsilon_j a_j^\dagger a_j. \tag{1.3.17}$$

It satisfies

$$H|n_1, n_2, \cdots\rangle = E|n_1, n_2, \cdots\rangle, \tag{1.3.18}$$

with

$$E = \sum_j \varepsilon_j n_j. \tag{1.3.19}$$

This is the energy of the state $|n_1, n_2, \cdots\rangle$.

For later convenience we define *normal ordering* of operators. Normal ordering, denoted by a double-dot symbol, dictates that all creation operators are to be placed to the left of all annihilation operators. An example reads

$$:a_1 a_2 a_3^\dagger a_1^\dagger: = :a_1^\dagger a_1 a_2 a_3^\dagger: = a_1^\dagger a_3^\dagger a_1 a_2. \tag{1.3.20}$$

Note that the conjugate operators a_1 and $a_1^\dagger$ commute freely within the normal ordering symbol.

Fermi Quanta

Fermi quanta are described by a set of operators c and its Hermitian conjugate $c^\dagger$ satisfying the anticommutation relations

$$\{c, c^\dagger\} = 1, \qquad \{c, c\} = \{c^\dagger, c^\dagger\} = 0. \tag{1.3.21}$$

Here, $\{A, B\} \equiv AB + BA$. We define the number operator by

$$N = c^\dagger c, \tag{1.3.22}$$

and the Fock vacuum $|0\rangle$ by

$$c|0\rangle = 0. \tag{1.3.23}$$

Because

$$N^2 = c^\dagger c c^\dagger c = c^\dagger (1 - c^\dagger c) c = c^\dagger c = N, \tag{1.3.24}$$

the eigenvalues of N are 0 and 1. Namely, there are only two states $|0\rangle$ and $|1\rangle = c^\dagger|0\rangle$ in the Hilbert space $\mathbb{H}$. No two quanta can occupy the same state. Indeed, $c^\dagger|1\rangle = c^\dagger c^\dagger|0\rangle = 0$. It is referred to as the Pauli exclusion principle.

When there are several kinds of Fermi quanta we introduce the creation operator $c_j^\dagger$ and the annihilation operator c_j to each of them,

$$\{c_j, c_k^\dagger\} = \delta_{jk}, \qquad \{c_j, c_k\} = 0. \tag{1.3.25}$$

Note that $(c_j c_k)^\dagger = c_k^\dagger c_j^\dagger$. Many-quantum states are given by

$$|n_1, n_2, \cdots\rangle = (c_1^\dagger)^{n_1}(c_2^\dagger)^{n_2}\cdots|0\rangle, \tag{1.3.26}$$

where $n_j = 0$ or $n_j = 1$.

Normal ordering of fermion operators is defined in the same way as boson operators. However, to be consistent with the anticommutation relations, it is understood that any two operators anticommute freely within the normal ordering symbol. An example reads

$$:c_1 c_2 c_3^\dagger c_1^\dagger: = -:c_1^\dagger c_1 c_2 c_3^\dagger: = -c_1^\dagger c_3^\dagger c_1 c_2. \tag{1.3.27}$$

Here, we first moved $c_1^\dagger$. The minus sign arises since it was anticommuted with three fermion operators including c_1. We then moved $c_3^\dagger$ by making it anticommute with two fermion operators.

1.4 Uncertainty Principle

We define the Hermitian operators p and q by

$$a = \frac{1}{\sqrt{2M\hbar\omega}}(M\omega q + ip), \qquad a^\dagger = \frac{1}{\sqrt{2M\hbar\omega}}(M\omega q - ip). \tag{1.4.1}$$

The commutation relation (1.3.1) is equivalent to

$$[q, p] = i\hbar. \tag{1.4.2}$$

Comparing this with (1.2.7) we identify q and p as the coordinate and the momentum. The Hamiltonian (1.3.8) is rewritten as

$$H = \frac{1}{2M}p^2 + \frac{M\omega^2}{2}q^2, \tag{1.4.3}$$

with $\varepsilon = \hbar\omega$. This is the Hamiltonian of the harmonic oscillator with frequency ω.

In classical theory a simultaneous measurement is possible of coordinate q and momentum p with arbitrary accuracy. This is impossible in quantum theory. The commutation relation (1.4.2) leads to the Heisenberg uncertainty principle, which is expressed as

$$\Delta q \Delta p \geq \frac{\hbar}{2}. \tag{1.4.4}$$

Its precise meaning is

$$\langle(\Delta q)^2\rangle\langle(\Delta p)^2\rangle \geq \frac{\hbar^2}{4}, \tag{1.4.5}$$

where $\langle\cdots\rangle$ stands for the expectation value with respect to a generic state, with

$$\Delta q = q - \langle q\rangle, \qquad \Delta p = p - \langle p\rangle. \tag{1.4.6}$$

Here, $\langle q\rangle$ and $\langle p\rangle$ are the averages, while $\sqrt{\langle(\Delta q)^2\rangle}$ and $\sqrt{\langle(\Delta p)^2\rangle}$ are the variances. When the coordinate is fixed, $\Delta q = 0$, the momentum is completely ambiguous, $\Delta p = \infty$, from (1.4.4). The momentum p is not a good quantum number in an eigenstate of the coordinate q.

The uncertainly relation (1.4.4) makes a simultaneous measurement of the coordinate and the momentum impossible. Nevertheless, there exist quantum systems where it is practically possible. They are *coherent states* and *squeezed coherent states.* These states are very important in the study of quantum Hall (QH) effects. QH states are coherent states with the spin coherence, when the Zeeman gap is not too large. Bilayer QH states are squeezed coherent states with the interlayer coherence, when the tunneling gap is not too large.

1.5 Coherent States and von Neumann Lattice

There exists a class of states where the minimum uncertainty holds,

$$\langle(\Delta q)^2\rangle\langle(\Delta p)^2\rangle = \frac{\hbar^2}{4}. \tag{1.5.1}$$

They are the ones closest to the classical states.

We introduce the *coherent state.* By definition it is an eigenstate of the bosonic annihilation operator. We consider an eigenstate $|v\rangle$ of the annihilation operator a,

$$a|v\rangle = v|v\rangle, \qquad \langle v|a^\dagger = \langle v|v^* \tag{1.5.2}$$

The eigenvalue v is a complex number. An intriguing feature is that a has a classical counterpart $v = \langle v|a|v\rangle$ on the coherent state $|v\rangle$ though it is not a Hermitian operator.[3]

We then introduce the operator η by

$$a = v + \eta, \tag{1.5.3}$$

obeying $[\eta, \eta^\dagger] = 1$. The coherent state $|v\rangle$ is the Fock vacuum of the new operator η,

$$\eta|v\rangle = 0. \tag{1.5.4}$$

We evaluate the variances of the operators q and p defined by (1.4.1). It is easy to see

$$\Delta q = \sqrt{\frac{\hbar}{2\omega}}(\eta^\dagger + \eta), \qquad \Delta p = i\sqrt{\frac{\hbar\omega}{2}}(\eta^\dagger - \eta), \tag{1.5.5}$$

and

$$\begin{aligned}\langle(\Delta q)^2\rangle &= \frac{\hbar}{2\omega}\langle(\eta^\dagger + \eta)^2\rangle = \frac{\hbar}{2\omega},\\ \langle(\Delta p)^2\rangle &= -\frac{\hbar\omega}{2}\langle(\eta^\dagger - \eta)^2\rangle = \frac{\hbar\omega}{2},\end{aligned} \tag{1.5.6}$$

where $\langle\cdots\rangle$ stands for the expectation value with respect to the state $|v\rangle$. Hence, the minimum uncertainty (1.5.1) holds on the coherent state.

To construct explicitly the coherent state, we make use of the displacement operator,[4]

$$D(v) \equiv e^{va^\dagger - v^*a} = e^{-|v|^2/2}e^{va^\dagger}e^{-v^*a}. \tag{1.5.7}$$

Since it has the property $D^\dagger(v) = D(-v) = D^{-1}(v)$, it is a unitary operator,

$$D(v)D^\dagger(v) = D^\dagger(v)D(v) = 1. \tag{1.5.8}$$

---

[3] In physical applications we use the quantum field $\phi(t, x)$ for the operator a. It is important that the classical field $\phi^{\rm cl}(t, x)$ is defined on the coherent state by $\phi^{\rm cl}(t, x) = \langle v|\phi(t, x)|v\rangle$. See Chapter 4 and Chapter 7 for details.

[4] We use the relation $e^{A+B} = e^A e^B e^{-[A,B]/2}$ with $A = va^\dagger$ and $B = v^*a$ to derive the last term.

We take the derivative of $D^\dagger(v)aD(v)$ with respect to v,

$$\frac{\partial}{\partial v}\left[D^\dagger(v)\hat{a}D(v)\right] = D^\dagger(v)\left[\hat{a},\hat{a}^\dagger\right]D(v) = D^\dagger(v)D(v) = 1,$$

noticing that v and v^* are independent variables. We then integrate it over v from $v = 0$ to $v = 1$,

$$D^\dagger(v)\hat{a}D(v) = \hat{a} + v.$$

Similarly we can derive

$$D^\dagger(v)a^\dagger D(v) = a^\dagger + v^*. \tag{1.5.9}$$

Thus it shifts the annihilation operator a by v, and the creation operator $a^\dagger$ by v^*. We now take the state

$$|v\rangle \equiv D(v)|0\rangle = e^{-|v|^2/2}e^{va^\dagger}|0\rangle, \tag{1.5.10}$$

and find

$$a|v\rangle = D(v)D^\dagger(v)aD(v)|0\rangle = D(v)\left(a+v\right)|0\rangle = v|v\rangle. \tag{1.5.11}$$

Hence the state $|v\rangle \equiv D(v)|0\rangle$ is the coherent state by the definition (1.5.2).

The coherent state (1.5.10) is expanded in terms of the eigenstates $|n\rangle$ of the number operator N,

$$|v\rangle = e^{-|v|^2/2}\sum_{n=0}^{\infty}\frac{1}{n!}(va^\dagger)^n|0\rangle = e^{-|v|^2/2}\sum_{n=0}^{\infty}\frac{v^n}{\sqrt{n!}}|n\rangle. \tag{1.5.12}$$

Using the orthonormality condition $\langle n|m\rangle = \delta_{nm}$, we find

$$\langle n|v\rangle = e^{-|v|^2/2}\frac{v^n}{\sqrt{n!}}, \tag{1.5.13}$$

or

$$P_n \equiv |\langle n|v\rangle|^2 = \frac{\bar{n}^n}{n!}e^{-\bar{n}^2}, \tag{1.5.14}$$

where $\bar{n} \equiv \langle v|N|v\rangle = v^2$ is the average number of bosons in $|v\rangle$. Namely, the coherent state is such a state where the probability of finding n bosons obeys the Poisson distribution.

Since a is not a Hermitian operator, there is no reason that its eigenstates span an orthonormal complete set. Indeed, the scalar product of two coherent states $|u\rangle$ and $|v\rangle$ are

$$\langle u|v\rangle = \exp\left(-\frac{1}{2}|u|^2 - \frac{1}{2}|v|^2 + u^*v\right). \tag{1.5.15}$$

They are not orthogonal one to another. On the other hand, the completeness condition holds,

$$\frac{1}{\pi}\int d^2v\ |v\rangle\langle v| = 1, \tag{1.5.16}$$

because, setting $v = re^{i\theta}$, we find

$$\frac{1}{\pi}\int d^2v\, \langle n|v\rangle\langle v|m\rangle = \frac{1}{\sqrt{n!}\sqrt{m!}}\int_0^\infty dr^2\, r^{n+m}e^{-r^2}\int_0^{2\pi}\frac{d\theta}{2\pi}\, e^{i(m-n)\theta} = \delta_{nm} \quad (1.5.17)$$

for any two states $|m\rangle$ and $|n\rangle$, where we have used (1.5.13) and $\int_0^\infty s^n e^{-s} ds = n!$.

It is notable that all coherent states $|v\rangle$ participate to form a complete set in (1.5.16). However, the number of coherent states $|v\rangle$ is "much larger" than that of the occupation-number states $|n\rangle$ because the coherent state $|v\rangle$ is defined for an arbitrary complex number v. Thus, the set of all coherent states must be "overcomplete". It is expected that a countable number of coherent states exist to form a complete set, as was first addressed by von Neumann.[369]

Let us construct a subset by choosing only those states $|v_{nm}\rangle$ with discrete values of v_{mn} of the parameter v in (1.5.2) such that

$$v_{mn} = \sqrt{\pi}\ell(m + in), \quad (1.5.18)$$

where m and n are integers, and ℓ is a real number. They form a lattice with the lattice spacing $\sqrt{\pi}\ell$ in the v-space. From (1.4.1) and (1.5.2) we find

$$\langle q\rangle = \sqrt{\frac{2\pi\hbar}{M\omega}}\ell m, \qquad \langle p\rangle = \sqrt{2\pi M\hbar\omega}\ell n. \quad (1.5.19)$$

The unit cell has the area $2\pi\hbar\ell^2$ in the (q, p) phase space. It has been proved[389,62] that the condition for the states $|v_{mn}\rangle$ to form a complete set is $\ell \le 1$. Furthermore, a minimum complete set of coherent states is obtained by setting $\ell = 1$, though it is still overcomplete by having just one linear dependent state, $\sum_{m,n}|v_{mn}\rangle = 0$. However, the removal of this dependent state turns out to change the classical character and spoil the coherence.[74]

We have taken a square lattice as a simplest example. We may consider a wide class of lattices spanned by two complex numbers ℓ_q and ℓ_p,

$$v_{nm} = \sqrt{\pi}\,(\ell_q m + \ell_p n)\,. \quad (1.5.20)$$

The area spanned by a unit cell is

$$\left|\vec{\ell}_q \times \vec{\ell}_p\right| = \left|\ell_q^x\ell_p^y - \ell_q^y\ell_p^x\right| = \left|\mathrm{Im}[\ell_q^*\ell_p]\right|, \quad (1.5.21)$$

where we have set $\ell_j = \ell_j^x + i\ell_j^y$ and $\vec{\ell}_j = (\ell_j^x, \ell_j^y)$ for $j = q$ and p. We require the unit cell to have the area $2\pi\hbar$, as corresponds to the uncertainty relation $[q, p] = i\hbar$. Such a lattice is called a von Neumann lattice.[74] The states $|v_{nm}\rangle$ on a von Neumann lattice form a minimum complete set of coherent states. See also Section 10.5.

1.6 Squeezed Coherent State

We construct a general class of coherent states by redistributing quantum fluctuations between the conjugate variables q and p [Fig.1.1]. For this purpose we make a *Bogoliubov transformation*, which is a linear transformation from a set of the annihilation and creation operators into another set, $(\eta, \eta^\dagger) \to (\zeta, \zeta^\dagger)$,

$$\begin{aligned} \zeta &= \eta \cosh\tau + \eta^\dagger e^{i\beta} \sinh\tau, \\ \zeta^\dagger &= \eta^\dagger \cosh\tau + \eta e^{-i\beta} \sinh\tau, \end{aligned} \tag{1.6.1}$$

where β and τ are real constants. It is a canonical transformation keeping the commutation relation $[\zeta, \zeta^\dagger] = 1$. It is essentially a "rotation" with an angle τ. The inverse transformation is a rotation with the angle $-\tau$,

$$\begin{aligned} \eta &= \zeta \cosh\tau - \zeta^\dagger e^{i\beta} \sinh\tau, \\ \eta^\dagger &= \zeta^\dagger \cosh\tau - \zeta e^{-i\beta} \sinh\tau. \end{aligned} \tag{1.6.2}$$

The Bogoliubov transformation (1.6.1) is expressed as (see Appendix G),

$$\zeta = e^{-iG} \eta e^{iG}, \qquad \zeta^\dagger = e^{-iG} \eta^\dagger e^{iG} \tag{1.6.3}$$

with the generator

$$G = \frac{i}{2}\tau(e^{-i\beta}\eta\eta - e^{i\beta}\eta^\dagger\eta^\dagger). \tag{1.6.4}$$

We consider the state defined by

$$|v\rangle\!\rangle = e^{-iG}|v\rangle, \tag{1.6.5}$$

and the operator defined by

$$b = e^{-iG} a e^{iG} = v + \zeta. \tag{1.6.6}$$

It is easy to see that

$$b|v\rangle\!\rangle = e^{-iG} a e^{iG} e^{-iG}|v\rangle = v|v\rangle\!\rangle, \tag{1.6.7}$$

and

$$\zeta|v\rangle\!\rangle = 0. \tag{1.6.8}$$

The state $|v\rangle\!\rangle$ is a coherent state of the b quantum and the Fock vacuum of the ζ quantum. It will turn out to be a squeezed state.

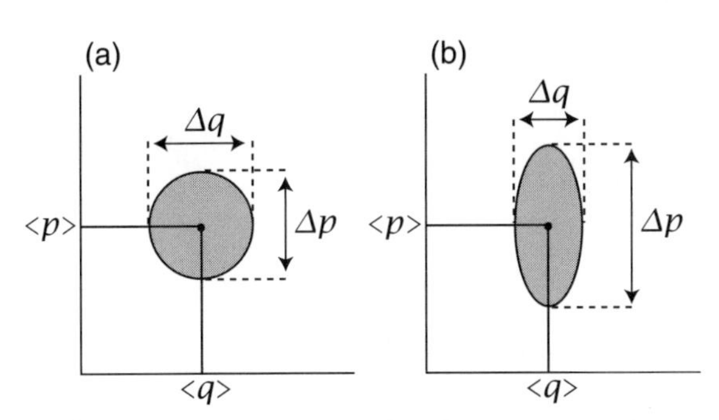

Fig. 1.1 (a) Due to the commutation relation $[q, p] = i\hbar$, we can determine position q and momentum p only with certain uncertainties Δp and Δq. The minimum uncertainty holds in the coherent state, where $(\Delta p)(\Delta q) = \hbar/2$. We have illustrated an uncertainty domain with $\Delta p \simeq \Delta q$. (b) In the squeezed state, an uncertainty domain may be squeezed without spoiling the minimum uncertainty, $(\Delta p)(\Delta q) = \hbar/2$. The position can be determined more accurately than the momentum in this instance.

We evaluate the variance (1.5.5) of the operators q and p. Since

$$\Delta q = \sqrt{\frac{\hbar}{2\omega}}(\eta^\dagger + \eta) = \sqrt{\frac{\hbar}{2\omega}}\left[(\cosh\tau - e^{i\beta}\sinh\tau)\zeta^\dagger + (\cosh\tau - e^{-i\beta}\sinh\tau)\zeta\right],$$
$$\Delta p = i\sqrt{\frac{\hbar\omega}{2}}(\eta^\dagger - \eta) = i\sqrt{\frac{\hbar\omega}{2}}\left[(\cosh\tau + e^{i\beta}\sinh\tau)\zeta^\dagger - (\cosh\tau + e^{-i\beta}\sinh\tau)\zeta\right], \tag{1.6.9}$$

we find

$$\langle\!\langle(\Delta q)^2\rangle\!\rangle = \frac{\hbar}{2\omega}\langle\!\langle(\eta^\dagger + \eta)^2\rangle\!\rangle = \frac{\hbar}{2\omega}(\cosh\tau - e^{i\beta}\sinh\tau)(\cosh\tau - e^{-i\beta}\sinh\tau),$$
$$\langle\!\langle(\Delta p)^2\rangle\!\rangle = -\frac{\hbar\omega}{2}\langle\!\langle(\eta^\dagger - \eta)^2\rangle\!\rangle = \frac{\hbar\omega}{2}(\cosh\tau + e^{i\beta}\sinh\tau)(\cosh\tau + e^{-i\beta}\sinh\tau), \tag{1.6.10}$$

and

$$\langle\!\langle(\Delta q)^2\rangle\!\rangle\langle\!\langle(\Delta p)^2\rangle\!\rangle = \frac{\hbar^2}{4}(\cosh^2\tau - e^{2i\beta}\sinh^2\tau)(\cosh^2\tau - e^{-2i\beta}\sinh^2\tau). \tag{1.6.11}$$

Here, $\langle\!\langle\cdots\rangle\!\rangle$ stands for the expectation value with respect to the state $|v\rangle\!\rangle$. The uncertainties Δq and Δp in the conjugate variables q and p are controlled by angles τ and β. The minimum uncertainty holds only when $\beta = 0, \pm\pi$.

We need to use the Bogoliubov transformation with $\beta = 0$ in later chapters. In this case, (1.6.9) yields

$$\zeta = \sqrt{\frac{\omega}{2\hbar}} e^{\tau} \Delta q + i \sqrt{\frac{1}{2\hbar\omega}} e^{-\tau} \Delta p. \tag{1.6.12}$$

Compare this with (1.5.5), or

$$\eta = \sqrt{\frac{\omega}{2\hbar}} \Delta q + i \sqrt{\frac{1}{2\hbar\omega}} \Delta p. \tag{1.6.13}$$

The Bogoliubov transformation from the field η to the field ζ is to insert weights $e^{\pm\tau}$ to the canonical conjugate variables Δq and Δp. The uncertainties are

$$\langle\!\langle (\Delta q)^2 \rangle\!\rangle = \frac{\hbar}{2\omega} e^{-2\tau}, \qquad \langle\!\langle (\Delta p)^2 \rangle\!\rangle = \frac{\hbar\omega}{2} e^{2\tau}, \tag{1.6.14}$$

and

$$\langle\!\langle (\Delta q)^2 \rangle\!\rangle \langle\!\langle (\Delta p)^2 \rangle\!\rangle = \frac{\hbar^2}{4}. \tag{1.6.15}$$

The state $|v\rangle\!\rangle$ is called a *squeezed state,*[35,34] since one of the uncertainties is squeezed. By definition (1.5.2), it is not a coherent state of the a quantum. Nevertheless, it has a classical counterpart,

$$\langle\!\langle v | a | v \rangle\!\rangle = v, \tag{1.6.16}$$

which is the key property of the coherent state. Hence, it is also called a *squeezed coherent state.*

1.7 Particle Number and Phase

The phase operator Θ is defined by[117]

$$a \equiv e^{i\Theta} \sqrt{N}, \qquad a^{\dagger} \equiv \sqrt{N} e^{-i\Theta}, \tag{1.7.1}$$

where $N = a^{\dagger} a$ is the number operator. From the commutation relation between a and $a^{\dagger}$, we find that

$$[e^{i\Theta}, N] = e^{i\Theta}, \tag{1.7.2}$$

from which we obtain[5]

$$[N, \Theta] = i. \tag{1.7.3}$$

The Heisenberg uncertainty principle follows from (1.7.3),

$$\Delta n \Delta \theta \geq \frac{1}{2}, \tag{1.7.4}$$

[5] Equation (1.7.3) is the first order term in (1.7.2) when the power expansion, $e^{i\Theta} = 1 + i\Theta + \cdots$, is made. Conversely, (1.7.2) is derived from (1.7.3) by setting $A = i\Theta$ and $B = N$ in (B.1) in Appendix A.

where $\Delta n = \sqrt{\langle(\Delta N)^2\rangle}$ and $\Delta\theta = \sqrt{\langle(\Delta\Theta)^2\rangle}$. They act on the eigenstates of the number operator as

$$N|n\rangle = n|n\rangle, \qquad e^{i\Theta}|n\rangle = |n-1\rangle, \tag{1.7.5}$$

and hence

$$\langle n|e^{i\Theta}|n\rangle = 0. \tag{1.7.6}$$

It tells physically that no measurement of the phase Θ is possible on the eigenstate $|n\rangle$ of the number operator N: The phase oscillates so rapidly that its exponentiation $e^{i\Theta}$ vanishes.

The eigenstate $|\theta\rangle$ of the phase operator Θ may be formally constructed,

$$|\theta\rangle = \sum_n e^{i\theta n}|n\rangle, \tag{1.7.7}$$

as a superposition of the eigenstates $|n\rangle$ of the number operator. To see this, we realize the commutation relation (1.7.3) by

$$\Theta = -i\frac{\partial}{\partial n} \tag{1.7.8}$$

in the representation where N is diagonalized. It acts on $|\theta\rangle$ as

$$\Theta|\theta\rangle = -i\sum_n \frac{\partial}{\partial n}e^{i\theta n}|n\rangle = \theta|\theta\rangle. \tag{1.7.9}$$

It is normalized as

$$\langle\theta'|\theta\rangle = \sum_{n,m} e^{i\theta n - i\theta' m}\langle m|n\rangle = \sum_n e^{in(\theta-\theta')} = 2\pi\delta(\theta'-\theta). \tag{1.7.10}$$

The inverse relation is

$$|n\rangle = \int_0^{2\pi}\frac{d\theta}{2\pi}e^{-i\theta n}|\theta\rangle. \tag{1.7.11}$$

The state $|\theta\rangle$ has a periodic property, $|\theta + 2\pi\rangle = |\theta\rangle$, as is clear in the definition (1.7.7).

A careful treatment is needed for the phase operator. It follows from (1.7.1) that $e^{i\Theta} = aN^{-1/2}$ and $e^{-i\Theta} = N^{-1/2}a^\dagger$. On one hand, we have

$$e^{i\Theta}e^{-i\Theta}|n\rangle = aN^{-1}a^\dagger|n\rangle = \sqrt{n+1}aN^{-1}|n+1\rangle = |n\rangle \tag{1.7.12}$$

for any state $|n\rangle$. On the other hand, we have

$$e^{-i\Theta}e^{i\Theta}|n\rangle = N^{-1/2}a^\dagger aN^{-1/2}|n\rangle = \frac{1}{\sqrt{n}}N^{-1/2}a^\dagger a|n\rangle = |n\rangle \tag{1.7.13}$$

for any state $|n\rangle \neq |0\rangle$, but

$$e^{-i\Theta}e^{i\Theta}|0\rangle = 0 \tag{1.7.14}$$

by (1.7.5). Hence, we conclude

$$e^{i\Theta}e^{-i\Theta} = 1, \qquad e^{-i\Theta}e^{i\Theta} = 1 - |0\rangle\langle 0|. \tag{1.7.15}$$

Strictly speaking, the phase Θ is not a Hermitian operator, though it is practically so in a macroscopic system. See literature[466,238,387,80,179] for more details on this point.

1.8 Macroscopic Coherence

We evaluate the number and phase operators on the coherent state. See Section 4.5 with respect to the squeezed state. The coherent state is characterized by

$$\langle v|a|v\rangle = v, \qquad \langle v|N|v\rangle = |v|^2 \equiv n. \tag{1.8.1}$$

Thus, the particle number is measurable. However, it is not an eigenstate of the number operator,

$$N|v\rangle = \left(n + v\eta^\dagger\right)|v\rangle. \tag{1.8.2}$$

It is an eigenstate of the annihilation operator,

$$a|v\rangle = e^{i\theta}\sqrt{n}|v\rangle, \tag{1.8.3}$$

where we have set $v = e^{i\theta}\sqrt{n}$ with a certain phase θ.

The standard deviation Δn is

$$(\Delta n)^2 \equiv \langle(\Delta N)^2\rangle = \langle v|(a^\dagger a - n)^2|v\rangle = n, \tag{1.8.4}$$

as is easily verified by using $a = v + \eta$. Because of this large uncertainty in a macroscopic system ($n \to \infty$), the conjugate phase is measurable simultaneously,

$$\langle v|e^{i\Theta}|v\rangle \simeq e^{i\theta}. \tag{1.8.5}$$

We can derive this as follows. It follows from (1.7.1) that

$$e^{i\Theta} = aN^{-1/2} = \frac{v}{\sqrt{n}}\left(1 + \frac{\eta}{v}\right)\left(1 - \frac{\eta^\dagger}{v^*}\right)^{-1/2} = e^{i\theta}\left(1 + e^{-i\theta}\frac{\eta}{\sqrt{n}}\right)\left(1 - e^{i\theta}\frac{\eta^\dagger}{\sqrt{n}}\right)^{-1/2}. \tag{1.8.6}$$

We evaluate it by making a power expansion. Since (1.5.4) implies $\langle v|\eta|v\rangle = \langle v|(\eta^\dagger)^k|v\rangle = 0$ for an arbitrary integer k, we obtain

$$\langle v|e^{i\Theta}|v\rangle = e^{i\theta}\langle v|\left(1 + \frac{\eta\eta^\dagger}{2n}\right)|v\rangle = e^{i\theta}\left(1 + \frac{1}{2n}\right) \simeq e^{i\theta} \tag{1.8.7}$$

in a macroscopic system ($n \to \infty$).

We make a heuristic argument to show that the minimum uncertainty (1.7.4) holds on a macroscopic coherent state. In (1.5.3) we regard η as a small deviation of a from the average value, $v = \langle a \rangle \equiv \langle v|a|v \rangle$. Using the parametrization (1.7.1), we find

$$a = v + \eta = e^{i(\langle\Theta\rangle + \Delta\Theta)}\sqrt{\langle N \rangle + \Delta N} = \left(1 + i\Delta\Theta + \frac{\Delta N}{2\langle N \rangle}\right) v + \cdots, \tag{1.8.8}$$

with

$$v = e^{i\theta}\sqrt{n}, \qquad \eta \simeq \left(i\Delta\Theta + \frac{\Delta N}{2n}\right) v. \tag{1.8.9}$$

We solve this equation for ΔN and $\Delta\Theta$,

$$\Delta N \simeq v\eta^\dagger + v^*\eta, \qquad \Delta\Theta \simeq \frac{i}{2n}\left(v\eta^\dagger - v^*\eta\right), \tag{1.8.10}$$

and obtain

$$(\Delta n)^2 \equiv \langle (\Delta N)^2 \rangle \simeq n, \tag{1.8.11a}$$

$$(\Delta\theta)^2 \equiv \langle (\Delta\Theta)^2 \rangle \simeq \frac{1}{4n} \tag{1.8.11b}$$

by the use of (1.5.3). Actually, (1.8.11a) is a rigorous formula given by (1.8.4). The minimum uncertainty follows,

$$\Delta n \Delta\theta \simeq \frac{1}{2}. \tag{1.8.12}$$

It is concluded that

$$\frac{\Delta n}{n} \simeq \frac{1}{\sqrt{n}} \to 0, \qquad \Delta\theta \simeq \frac{1}{\sqrt{n}} \to 0 \tag{1.8.13}$$

as $n \to \infty$. Both the particle number and the phase are measurable very accurately on macroscopic coherent states. Indeed, if $n \simeq 10^{22}$, the uncertainties are only $\Delta n/n \simeq 10^{-11}$ and $\Delta\theta \simeq 10^{-11}$, which are negligible experimentally. Both the particle number and the phase are practically good quantum numbers for macroscopic coherent states.

CHAPTER 2

QUANTUM FIELD THEORY

Quantum field theory is formulated by rewriting a many-body quantum mechanics. The basic objects are the field operator $\phi(t,\boldsymbol{x})$ and its Hermitian conjugate $\phi^\dagger(t,\boldsymbol{x})$ that annihilates and creates a particle at a space-time point $(t,\boldsymbol{x})$, respectively.

2.1 ONE-BODY HAMILTONIAN

We consider a system of one particle with mass m in a potential $V(\boldsymbol{x})$. For definiteness, we consider the 3-dimensional space with vector $\boldsymbol{x} = (x_1, x_2, x_3)$, though all formulas are valid in the 2-dimensional space as well. The Lagrangian consists of the kinetic and potential terms,

$$L = \frac{m}{2}\dot{\boldsymbol{x}}^2 - V(\boldsymbol{x}). \tag{2.1.1}$$

According to the standard prescription we define the momentum by

$$p_j = \frac{\partial L}{\partial \dot{x}_j} = m\dot{x}_j, \tag{2.1.2}$$

and the Hamiltonian by

$$H = \boldsymbol{p}\cdot\dot{\boldsymbol{x}} - L = \frac{1}{2m}\boldsymbol{p}^2 + V(\boldsymbol{x}). \tag{2.1.3}$$

We postulate the canonical commutation relation between the coordinate and the momentum,

$$[x_i, p_j] = i\hbar\delta_{ij}. \tag{2.1.4}$$

The Hamiltonian is now an operator acting on one-body states $|\mathfrak{S}\rangle$.

The commutation relation (2.1.4) is realized in the coordinate representation by

$$\boldsymbol{p} = -i\hbar\boldsymbol{\nabla}. \tag{2.1.5}$$

The Hamiltonian (2.1.3) reads

$$H = -\frac{\hbar^2}{2m}\nabla^2 + V(\boldsymbol{x}). \tag{2.1.6}$$

The time-dependent Schrödinger equation is

$$i\hbar\frac{\partial}{\partial t}\mathfrak{S}(t,\boldsymbol{x}) = \left(-\frac{\hbar^2}{2m}\nabla^2 + V(\boldsymbol{x})\right)\mathfrak{S}(t,\boldsymbol{x}), \tag{2.1.7}$$

where $\mathfrak{S}(t,\boldsymbol{x})$ is the wave function in the Schrödinger picture.

We work in the Heisenberg picture, where the state is independent of time evolution. The state $|\mathfrak{S}\rangle$ is described by a wave function $\mathfrak{S}(\boldsymbol{x}) \equiv \mathfrak{S}(t = 0,\boldsymbol{x})$ in the coordinate representation,

$$\mathfrak{S}(\boldsymbol{x}) = \langle \boldsymbol{x}|\mathfrak{S}\rangle. \tag{2.1.8}$$

The completeness and orthonormality conditions for the set of states $\{|\boldsymbol{x}\rangle\}$are

$$\int d^3x|\boldsymbol{x}\rangle\langle\boldsymbol{x}| = 1, \qquad \langle\boldsymbol{x}|\boldsymbol{x}'\rangle = \delta(\boldsymbol{x}-\boldsymbol{x}'). \tag{2.1.9}$$

The eigenvalue equation (1.1.3) is reduced to the static Schrödinger equation,

$$\left(-\frac{\hbar^2}{2m}\nabla^2 + V(\boldsymbol{x})\right)\mathfrak{S}_i(\boldsymbol{x}) = \varepsilon_i\mathfrak{S}_i(\boldsymbol{x}). \tag{2.1.10}$$

The eigenfunctions can be chosen to form a basis obeying the orthonormality condition

$$\langle\mathfrak{S}_j|\mathfrak{S}_k\rangle = \int d^3x\langle\mathfrak{S}_j|\boldsymbol{x}\rangle\langle\boldsymbol{x}|\mathfrak{S}_k\rangle = \int d^3x\mathfrak{S}_j^*(\boldsymbol{x})\mathfrak{S}_k(\boldsymbol{x}) = \delta_{jk}, \tag{2.1.11}$$

and the completeness condition

$$\sum_j\langle\boldsymbol{x}|\mathfrak{S}_j\rangle\langle\mathfrak{S}_j|\boldsymbol{x}'\rangle = \sum_j\mathfrak{S}_j^*(\boldsymbol{x}')\mathfrak{S}_j(\boldsymbol{x}) = \delta(\boldsymbol{x}-\boldsymbol{x}'). \tag{2.1.12}$$

They correspond to the conditions (1.1.4) and (1.1.5).

2.2 Many-Body Hamiltonian

We consider a system of N noninteracting identical particles,

$$H = \sum_{i=1}^{N}\left(-\frac{\hbar^2}{2m}\nabla_i^2 + V(\boldsymbol{x}_i)\right). \tag{2.2.1}$$

The eigenvalue equation is

$$\sum_{i=1}^{N}\left(-\frac{\hbar^2}{2m}\nabla_i^2 + V(\boldsymbol{x}_i)\right)\mathfrak{S}(\boldsymbol{x}_1,\cdots,\boldsymbol{x}_N) = E\mathfrak{S}(\boldsymbol{x}_1,\cdots,\boldsymbol{x}_N). \tag{2.2.2}$$

We can solve this by making a decomposition of variables,

$$\mathfrak{S}(x_1, \cdots, x_N) = \mathfrak{S}^1(x_1) \cdots \mathfrak{S}^N(x_N). \tag{2.2.3}$$

This is a solution of (2.2.2) provided each factor $\mathfrak{S}^j(x)$ is a solution of the one-body Schrödinger equation (2.1.10). The eigenvalue is

$$E = \sum_{j=1}^{\infty} n_j \varepsilon_j \tag{2.2.4}$$

with $\sum_{j=1}^{\infty} n_j = N$, when there are n_j particles in the one-body state $|\mathfrak{S}_j\rangle$ [Fig.2.1], by assuming $|\mathfrak{S}_j\rangle \neq |\mathfrak{S}_k\rangle$ for $j \neq k$.

The solution of the form (2.2.3) is valid only if particles are distinguishable. It is modified appropriately due to the statistics of particles. We consider the case of bosons in this section. Since all particles are indistinguishable, the wave function $\mathfrak{S}(x_1, \cdots, x_N)$ is invariant under any permutation of coordinates x_i and x_j, as is the statement of the Bose statistics. For instance, if the system contains two particles in two different states $|\mathfrak{S}_1\rangle$ and $|\mathfrak{S}_2\rangle$, the wave function is

$$\mathfrak{S}(x_1, x_2) = \frac{1}{\sqrt{2!}} \left(\mathfrak{S}_1(x_1)\mathfrak{S}_2(x_2) + \mathfrak{S}_1(x_2)\mathfrak{S}_2(x_1)\right). \tag{2.2.5}$$

If they are in the same state, $|\mathfrak{S}_1\rangle = |\mathfrak{S}_2\rangle \equiv |\mathfrak{S}\rangle$, we have

$$\mathfrak{S}(x_1, x_2) = \mathfrak{S}(x_1)\mathfrak{S}(x_2). \tag{2.2.6}$$

For the N-body system containing all particles in different states, it is necessary to make all possible permutations σ of the coordinates in the wave function,

$$\sigma = \begin{pmatrix} 1, & 2, & \cdots, & N \\ \sigma_1, & \sigma_2, & \cdots, & \sigma_N \end{pmatrix}, \tag{2.2.7}$$

where there are $N!$ permutations,

$$\mathfrak{S}(x_1, \cdots, x_N) = \frac{1}{\sqrt{N!}} \sum_{\sigma}^{N!} \mathfrak{S}_1(x_{\sigma_1}) \cdots \mathfrak{S}_N(x_{\sigma_N}). \tag{2.2.8}$$

When n_i states are identical ($n_1 + n_2 + \cdots = N$), the normalization factor should read $\sqrt{n_1!n_2!\cdots}/\sqrt{N!}$ in this formula. The energy is still given by (2.2.4).

Many-body states are given by (1.3.16), or

$$|n_1, n_2, \cdots\rangle = \frac{1}{\sqrt{n_1!n_2!\cdots}} (a_1^\dagger)^{n_1} (a_2^\dagger)^{n_2} \cdots |0\rangle. \tag{2.2.9}$$

Here, we have introduced the creation operator $a_j^\dagger$ and the annihilation operator a_j of the particle in the jth eigenstate $|\mathfrak{S}_j\rangle$ of the one-body Hamiltonian (2.1.10). The state $|0\rangle$ is the Fock vacuum and defined by

$$a_j|0\rangle = 0 \qquad \text{for all } j. \tag{2.2.10}$$

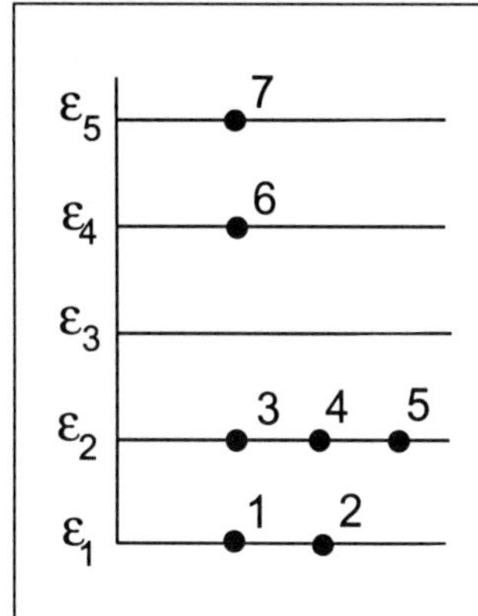

Fig. 2.1 In this example the system contains 7 particles. The kth particle is in the energy level ε_{s_k}, where $\varepsilon_{s_1} = \varepsilon_1$, $\varepsilon_{s_2} = \varepsilon_1$, $\varepsilon_{s_3} = \varepsilon_2$, $\varepsilon_{s_4} = \varepsilon_2$, $\varepsilon_{s_5} = \varepsilon_2$, $\varepsilon_{s_6} = \varepsilon_4$, and $\varepsilon_{s_7} = \varepsilon_5$. We cannot actually say in this way, when the particle is indistinguishable. We can only say that 2 particles are in the energy level ε_1, 3 particles are in the level ε_2, and so on. The state is given by $(a_1^\dagger)^2(a_2^\dagger)^3 a_4^\dagger a_5^\dagger|0\rangle$ up to the normalization, with the total energy $E = 2\varepsilon_1 + 3\varepsilon_2 + \varepsilon_4 + \varepsilon_5$.

The many-body states (2.2.9) form a basis of the Fock space. When we define the Hamiltonian

$$H = \sum_{j=1}^{\infty} \varepsilon_j a_j^\dagger a_j, \tag{2.2.11}$$

it satisfies

$$H|n_1, n_2, \cdots\rangle = \sum n_j \varepsilon_j |n_1, n_2, \cdots\rangle, \tag{2.2.12}$$

as yields the eigenvalue (2.2.4). In quantum mechanics we can only say that there are n_j particles in the state $|\mathfrak{S}_j\rangle$ since they are indistinguishable. Consequently the Hamiltonians (2.2.1) and (2.2.11) are physically equivalent.

2.3 Boson Field Operators

We can rewrite the Hamiltonian (2.2.11) so that it does not contain quantities depending on the one-body wave function $\mathfrak{S}_j(\boldsymbol{x})$ and its eigenvalue ε_j explicitly. For this purpose we define a field operator,

$$\phi(t, \boldsymbol{x}) = \sum_j a_j(t)\mathfrak{S}_j(\boldsymbol{x}) = \sum_j a_j(0)\mathfrak{S}_j(\boldsymbol{x})e^{-i\varepsilon_j t/\hbar}, \tag{2.3.1}$$

where (1.3.11) was used. The wave functions satisfy the orthonormality condition (1.1.4) and the completeness condition (1.1.5),

$$\int d^3x\, \langle\mathfrak{S}_j|\boldsymbol{x}\rangle\langle\boldsymbol{x}|\mathfrak{S}_k\rangle = \int d^3x\, \mathfrak{S}_j^*(\boldsymbol{x})\mathfrak{S}_k(\boldsymbol{x}) = \delta_{jk}, \tag{2.3.2}$$

$$\sum_j \langle\boldsymbol{x}|\mathfrak{S}_j\rangle\langle\mathfrak{S}_j|\boldsymbol{y}\rangle = \sum_j \mathfrak{S}_j(\boldsymbol{x})\mathfrak{S}_j^*(\boldsymbol{y}) = \delta(\boldsymbol{x} - \boldsymbol{y}). \tag{2.3.3}$$

The operator (2.3.1) satisfies the commutation relations,

$$[\phi(t,\mathbf{x}),\phi(t,\mathbf{y})]=[\phi^\dagger(t,\mathbf{x}),\phi^\dagger(t,\mathbf{y})]=0,$$
$$[\phi(t,\mathbf{x}),\phi^\dagger(t,\mathbf{y})]=\delta(\mathbf{x}-\mathbf{y}), \tag{2.3.4}$$

as can be checked explicitly by using the completeness condition (2.3.3) and the commutation relations (1.3.13). Field (2.3.1) is customarily called the second-quantized operator.

We define the density operator by

$$\rho(t,\mathbf{x})=\phi^\dagger(t,\mathbf{x})\phi(t,\mathbf{x}), \tag{2.3.5}$$

so that the integration over space is the number operator,

$$\int d^3x\,\rho(t,\mathbf{x})=\sum_{ij}a_i^\dagger(t)a_j(t)\int d^3x\,\mathfrak{S}_i^*(\mathbf{x})\mathfrak{S}_j(\mathbf{x})=\sum_i a_i^\dagger a_i=N. \tag{2.3.6}$$

To understand the physical meaning of these operators, we notice the commutation relations,

$$\rho(t,\mathbf{x})\phi^\dagger(t,\mathbf{y})=\phi^\dagger(t,\mathbf{y})[\rho(t,\mathbf{x})+\delta(\mathbf{x}-\mathbf{y})],$$
$$\rho(t,\mathbf{x})\phi(t,\mathbf{y})=\phi(t,\mathbf{y})[\rho(t,\mathbf{x})-\delta(\mathbf{x}-\mathbf{y})]. \tag{2.3.7}$$

We compare (2.3.7) with (1.3.14). The physical meaning of (1.3.14) is that $a_k^\dagger$ creates one kth-type Bose quanta by increasing its number N_k. Similarly, $\phi^\dagger(t,\mathbf{y})$ creates one Bose quanta at position $\mathbf{y}$ by increasing the density $\rho(t,\mathbf{x})$ at $\mathbf{x}=\mathbf{y}$. We may regard $\phi^\dagger(t,\mathbf{x})$ and $\phi(t,\mathbf{x})$ as the creation operator and the annihilation operator of the particle at space point $\mathbf{x}$ at time t.

We are ready to show that the Hamiltonian (2.2.11) is equivalent to

$$H=\int d^3x\left\{\frac{\hbar^2}{2m}\nabla\phi^\dagger(t,\mathbf{x})\nabla\phi(t,\mathbf{x})+V(\mathbf{x})\phi^\dagger(t,\mathbf{x})\phi(t,\mathbf{x})\right\}. \tag{2.3.8}$$

Indeed, substituting (2.3.1) into this, we find (2.2.11), or

$$H=\sum_{jk}\int d^3x\,\mathfrak{S}_j^\dagger(\mathbf{x})\left(-\frac{\hbar^2}{2m}\nabla^2\mathfrak{S}_k(\mathbf{x})+V(\mathbf{x})\mathfrak{S}_k(\mathbf{x})\right)a_j^\dagger a_k$$
$$=\sum_{jk}\varepsilon_k a_j^\dagger a_k\int d^3x\mathfrak{S}_j^\dagger(\mathbf{x})\mathfrak{S}_k(\mathbf{x})=\sum_j\varepsilon_j a_j^\dagger a_j, \tag{2.3.9}$$

where we have used the one-body Schrödinger equation (2.1.10) and the orthonormality condition (2.3.2).

2.4 QUANTUM FIELD THEORY

We have shown that the many-body quantum-mechanical Hamiltonian (2.2.1) is equivalent to the field-theoretical Hamiltonian (2.3.8), or

$$H = \frac{1}{i\hbar}\int d^3x\, \pi(t,\boldsymbol{x})\left[\frac{1}{2m}\boldsymbol{p}^2 + V(\boldsymbol{x})\right]\phi(t,\boldsymbol{x}), \tag{2.4.1}$$

where $\boldsymbol{p} = -i\hbar\boldsymbol{\nabla}$, and we have defined

$$\pi(t,\boldsymbol{x}) = i\hbar\phi^\dagger(t,\boldsymbol{x}). \tag{2.4.2}$$

Field operators $\phi(t,\boldsymbol{x})$ and $\pi(t,\boldsymbol{x})$ are characterized by commutation relations (2.3.4), or

$$[\phi(t,\boldsymbol{x}),\phi(t,\boldsymbol{y})] = [\pi(t,\boldsymbol{x}),\pi(t,\boldsymbol{y})] = 0,$$
$$[\phi(t,\boldsymbol{x}),\pi(t,\boldsymbol{y})] = i\hbar\delta(\boldsymbol{x}-\boldsymbol{y}). \tag{2.4.3}$$

The last relation is an analogue of (2.1.4). The original definition (2.3.1) should be regarded merely as a representation of the field operator ϕ in terms of eigenfunctions $\mathfrak{S}_j$ of the one-body Hamiltonian. The field-theoretical Hamiltonian is customarily called the second-quantized Hamiltonian.

According to the formula (2.4.1) the field-theoretical Hamiltonian H^{FT} is constructed by sandwiching the quantum-mechanical Hamiltonian H^{QM} with field operators $\phi(x)$ and $\pi(x)$,

$$H^{\text{FT}} = \frac{1}{i\hbar}\int d^3x\, \pi(t,\boldsymbol{x})H^{\text{QM}}(\boldsymbol{x})\phi(t,\boldsymbol{x}), \tag{2.4.4}$$

where $H^{\text{QM}}(\boldsymbol{x})$ is given by (2.1.6). It implies also that the field-theoretical momentum $\boldsymbol{p}^{\text{FT}}$ is constructed by sandwiching the quantum-mechanical momentum $\boldsymbol{p}^{\text{QM}} = -i\hbar\boldsymbol{\nabla}$ with field operators $\phi(x)$ and $\pi(x)$,

$$\boldsymbol{p}^{\text{FT}} = \frac{1}{i\hbar}\int d^3x\, \pi(t,\boldsymbol{x})\boldsymbol{p}^{\text{QM}}(\boldsymbol{x})\phi(t,\boldsymbol{x}). \tag{2.4.5}$$

We shall see later that this is indeed the case: See (3.2.19) and (3.3.25).

The Heisenberg equation is calculated by the use of the commutation relation (2.4.3),

$$\begin{aligned} i\hbar\partial_t\phi(t,\boldsymbol{x}) &= [\phi(t,\boldsymbol{x}),H^{\text{FT}}] = H^{\text{QM}}(\boldsymbol{x})\phi(t,\boldsymbol{x}) \\ &= \left(-\frac{\hbar^2}{2m}\boldsymbol{\nabla}^2 + V(\boldsymbol{x})\right)\phi(t,\boldsymbol{x}). \end{aligned} \tag{2.4.6}$$

This has formally the same expression as the one-body Schrödinger equation (2.1.7); the difference is that ϕ is an operator and not a wave function. Because of this fact it is customarily said that quantum field theory is obtained by *second quantization* from quantum mechanics, that is, by promoting the wave function $\mathfrak{S}(t,\boldsymbol{x})$ to the field operator $\phi(t,\boldsymbol{x})$ obeying the commutation relations (2.4.3). Actually in the nonrelativistic field theory, as we have seen, the field-theoretical Hamiltonian is equivalent to the many-body quantum-mechanical Hamiltonian. This is in sharp contrast to the case of quantum mechanics which is constructed from classical mechanics by making a new postulation (quantization), as was reviewed in Section 1.2. It may be appropriate to call quantum field theory the second-quantized scheme; the first quantized scheme is quantum mechanics.

A comment is in order. In the case of the Schrödinger field, since $\pi(t, \boldsymbol{x}) = i\hbar\phi^\dagger(t, \boldsymbol{x})$, we may rewrite (2.4.4) as

$$H^{\text{FT}} = \langle\phi|H^{\text{QM}}|\phi\rangle, \tag{2.4.7}$$

where we have set

$$\phi(\boldsymbol{x}) = \langle\boldsymbol{x}|\phi\rangle \tag{2.4.8}$$

for the field operator $\phi(\boldsymbol{x})$, or

$$|\phi\rangle = \sum_j |\mathfrak{S}_j\rangle a_j(t). \tag{2.4.9}$$

Indeed,

$$H^{\text{FT}} = \int d^3x d^3y\, \langle\phi|\boldsymbol{x}\rangle\langle\boldsymbol{x}|H^{\text{QM}}|\boldsymbol{y}\rangle\langle\boldsymbol{y}|\phi\rangle = \int d^3x\, \phi^\dagger(\boldsymbol{x})H^{\text{QM}}(\boldsymbol{x})\phi(\boldsymbol{x}), \tag{2.4.10}$$

where the completeness condition (2.1.9) is used. It agrees with (2.4.1).

Interaction between Particles

We introduce interactions between particles. We only consider the case where the interaction depends on the distance between two particles. The total Hamiltonian is given by

$$H = \sum_{i=1}^{N}\left(-\frac{\hbar^2}{2m}\nabla_i^2 + V(\boldsymbol{x}_i)\right) + \frac{1}{2}\sum_{ij} U(|\boldsymbol{x}_i - \boldsymbol{x}_j|). \tag{2.4.11}$$

An example is given by the Coulomb interaction when these particles carry electric charges. Another example is given by a repulsive contact interaction between two particles,

$$U(|\boldsymbol{x}_i - \boldsymbol{x}_j|) = g\delta(\boldsymbol{x}_i - \boldsymbol{x}_j), \qquad g > 0. \tag{2.4.12}$$

We use both of these interactions extensively throughout this book.

We prove that this quantum-mechanical system is equivalent to the field-theoretical system given by

$$\begin{aligned} H^{\text{FT}} = &\int d^3x \left\{\frac{\hbar^2}{2m}\nabla\phi^\dagger(t,\boldsymbol{x})\nabla\phi(t,\boldsymbol{x}) + V(\boldsymbol{x})\phi^\dagger(t,\boldsymbol{x})\phi(t,\boldsymbol{x})\right\} \\ &+ \frac{1}{2}\int d^3x \int d^3y\, \phi^\dagger(t,\boldsymbol{x})\phi^\dagger(t,\boldsymbol{y})U(|\boldsymbol{x}-\boldsymbol{y}|)\phi(t,\boldsymbol{y})\phi(t,\boldsymbol{x}) \end{aligned} \tag{2.4.13}$$

together with the commutation relation (2.3.4) for $\phi(t, \mathbf{x})$.

The wave function of the N-body state $|\mathfrak{S}\rangle$ is defined by

$$\mathfrak{S}(t, \mathbf{x}_1, \cdots, \mathbf{x}_N) = \frac{1}{\sqrt{N!}} \langle 0|\phi(x_1)\phi(x_2)\cdots\phi(x_N)|\mathfrak{S}\rangle, \tag{2.4.14}$$

where we have set $x_j = (t, \mathbf{x}_j)$ for notational simplicity. The factor $1/\sqrt{N!}$ is inserted for normalization, $\prod_i \int d^3x_i\, \mathfrak{S}^*\mathfrak{S} = 1$. The Heisenberg equation is

$$i\hbar\frac{\partial}{\partial t}\phi(x) = [\phi(x), H^{\rm FT}], \tag{2.4.15}$$

which gives

$$\begin{aligned} i\hbar\frac{\partial}{\partial t}\mathfrak{S}(t, \mathbf{x}_1, \cdots, \mathbf{x}_N) &= i\hbar\sum_j \langle 0|\phi(x_1)\cdots\dot{\phi}(x_j)\cdots\phi(x_N)|\mathfrak{S}\rangle \\ &= \sum_j \langle 0|\phi(x_1)\cdots[\phi(x_j), H^{\rm FT}]\cdots\phi(x_N)|\mathfrak{S}\rangle. \end{aligned} \tag{2.4.16}$$

Using the explicit form of the Heisenberg equation, we find

$$[\phi(t, \mathbf{x}), H^{\rm FT}] = \left(-\frac{\hbar^2}{2m}\nabla^2 + V(\mathbf{x})\right)\phi(x) + \int d^3y\, \phi^\dagger(y)U(|\mathbf{x}-\mathbf{y}|)\phi(y)\phi(x), \tag{2.4.17}$$

and

$$\begin{aligned} \langle 0|\phi(x_1)\cdots[\phi(x_j), H^{\rm FT}]\cdots\phi(x_N)|\mathfrak{S}\rangle &= \left(-\frac{\hbar^2}{2m}\nabla_j^2 + V(\mathbf{x}_j)\right)\mathfrak{S}(t, \mathbf{x}_1, \cdots, \mathbf{x}_N) \\ &+ \int d^3y U(|\mathbf{x}_j - \mathbf{y}|)\langle 0|\phi(x_1)\cdots\phi^\dagger(y)\phi(y)\phi(x_j)\cdots\phi(x_N)|\mathfrak{S}\rangle. \end{aligned} \tag{2.4.18}$$

Now, with the aid of $[\phi(x), \phi^\dagger(y)\phi(y)] = \delta(\mathbf{x}-\mathbf{y})\phi(y)$ and $\langle 0|\phi^\dagger(y) = 0$, the Schrödinger equation follows,

$$i\hbar\frac{\partial}{\partial t}\Psi(t, \mathbf{x}_1, \cdots, \mathbf{x}_N) = H^{\rm QM}\Psi(t, \mathbf{x}_1, \cdots, \mathbf{x}_N), \tag{2.4.19}$$

where $H^{\rm QM}$ is given by (2.4.11). The equivalence is established.

2.5 Fermion Field Operators

We discuss the case of fermions. The N-body Schrödinger equation (2.2.2) is the same for bosons and for fermions; the difference appears in the condition on the wave function corresponding to the Bose statistics or the Fermi statistics. In quantum field theory, to describe bosons we have introduced the field operator ϕ with the use of the boson creation and annihilation operators $a_j^\dagger$ and a_j. In the fermion case we use the fermion creation and annihilation operators $c_j^\dagger$ and c_j to define the field operator,

$$\psi(t,\boldsymbol{x}) = \sum_j c_j(t)\mathfrak{S}_j(\boldsymbol{x}) = \sum_j c_j(0)\mathfrak{S}_j(\boldsymbol{x})e^{-i\varepsilon_j t/\hbar}. \tag{2.5.1}$$

The anticommutation relations follow,

$$\begin{aligned}\{\psi(t,\boldsymbol{x}),\psi(t,\boldsymbol{y})\} &= \{\psi^\dagger(t,\boldsymbol{x}),\psi^\dagger(t,\boldsymbol{y})\} = 0,\\ \{\psi(t,\boldsymbol{x}),\psi^\dagger(t,\boldsymbol{y})\} &= \delta(\boldsymbol{x}-\boldsymbol{y}),\end{aligned} \tag{2.5.2}$$

instead of the commutation relations (2.3.4). The operator $\psi(t,\boldsymbol{x})$ is characterized by these anticommutation relations, and (2.5.1) is to be considered as an expansion in terms of the eigenfunctions of the one-body Schrödinger equation.

The field-theoretical Hamiltonian is given by

$$\begin{aligned}H_0 =& \int d^3x \left\{\frac{\hbar^2}{2m}\nabla\psi^\dagger(t,\boldsymbol{x})\nabla\psi(t,\boldsymbol{x}) + V(\boldsymbol{x})\psi^\dagger(t,\boldsymbol{x})\psi(t,\boldsymbol{x})\right\}\\ &+\frac{1}{2}\int d^3x\int d^3y\ \psi^\dagger(t,\boldsymbol{x})\psi^\dagger(t,\boldsymbol{y})U(|\boldsymbol{x}-\boldsymbol{y}|)\psi(t,\boldsymbol{y})\psi(t,\boldsymbol{x}),\end{aligned} \tag{2.5.3}$$

as in the boson case; see from (2.4.11) to (2.4.19).

The Fock space consists of the many-body state (1.3.26), or

$$|n_1,n_2,\cdots\rangle = (c_1^\dagger)^{n_1}(c_2^\dagger)^{n_2}\cdots|0\rangle, \tag{2.5.4}$$

where the number n_j is either 0 or 1 due to the Fermi statistics. The many-body wave function is constructed with the Fermi statistics taken into account. The two-body wave function (2.2.5) becomes

$$\mathfrak{S}(\boldsymbol{x}_1,\boldsymbol{x}_2) = \frac{1}{\sqrt{2!}}\left(\mathfrak{S}_1(\boldsymbol{x}_1)\mathfrak{S}_2(\boldsymbol{x}_2) - \mathfrak{S}_1(\boldsymbol{x}_2)\mathfrak{S}_2(\boldsymbol{x}_1)\right). \tag{2.5.5}$$

For the N-body system, there are $N!$ permutations, and we obtain

$$\mathfrak{S}(\boldsymbol{x}_1,\cdots,\boldsymbol{x}_N) = \frac{1}{\sqrt{N!}}\sum_\sigma \mathrm{sgn}(\sigma)\mathfrak{S}_1(\boldsymbol{x}_{\sigma_1})\cdots\mathfrak{S}_N(\boldsymbol{x}_{\sigma_N}), \tag{2.5.6}$$

where $\mathrm{sgn}(\sigma) = \pm$ for even (odd) permutation σ. This is explicitly written as

$$\mathfrak{S}(\boldsymbol{x}_1,\cdots,\boldsymbol{x}_N) = \frac{1}{\sqrt{N!}}\begin{vmatrix}\mathfrak{S}_1(\boldsymbol{x}_1) & \mathfrak{S}_1(\boldsymbol{x}_2) & \cdots & \mathfrak{S}_1(\boldsymbol{x}_N)\\ \mathfrak{S}_2(\boldsymbol{x}_1) & \mathfrak{S}_2(\boldsymbol{x}_2) & \cdots & \mathfrak{S}_2(\boldsymbol{x}_N)\\ \vdots & \vdots & \ddots & \vdots\\ \mathfrak{S}_N(\boldsymbol{x}_1) & \mathfrak{S}_N(\boldsymbol{x}_2) & \cdots & \mathfrak{S}_N(\boldsymbol{x}_N)\end{vmatrix}, \tag{2.5.7}$$

which is known as the Slater determinant.

2.6 Electrons Interacting with Electromagnetic Field

We study an electron interacting with the electromagnetic field. Let the charge of an electron be $-e$ ($e > 0$). The classical equation of motion is

$$m\ddot{\boldsymbol{x}} = -e\left(\boldsymbol{E} + \dot{\boldsymbol{x}} \times \boldsymbol{B}\right), \tag{2.6.1}$$

where the last term describes the Lorentz force.

We construct the Lagrangian leading to this equation of motion. It is given by

$$L = \frac{m}{2}\dot{\boldsymbol{x}}^2 - e\boldsymbol{A}(t,\boldsymbol{x}) \cdot \dot{\boldsymbol{x}} + e\varphi(t,\boldsymbol{x}), \tag{2.6.2}$$

where $\boldsymbol{A}(t,\boldsymbol{x})$ and $\varphi(t,\boldsymbol{x})$ are the vector and scalar potentials. The electric and magnetic fields are

$$E_k = -\partial_t A_k - \partial_k \varphi, \qquad B_k = \epsilon_{kij}\partial_i A_j, \tag{2.6.3}$$

where ε_{ijk} is the antisymmetric tensor with $\varepsilon_{123} = 1$. Indeed, the Euler-Lagrange equation reads

$$\frac{d}{dt}\frac{\partial L}{\partial \dot{x}_j} - \frac{\partial L}{\partial x_j} = 0, \tag{2.6.4}$$

which yields

$$m\ddot{x}_j = -eE_j - e\epsilon_{jki}\dot{x}_j B_i, \tag{2.6.5}$$

or the equation of motion (2.6.1).

Let us quantize electrons, while keeping the electromagnetic field as a classical field.[1] The canonical momentum is

$$p_j = \frac{\partial L}{\partial \dot{x}_j} = m\dot{x}_j - eA_j. \tag{2.6.6}$$

We impose the canonical commutation relation,

$$[x_i, p_j] = i\hbar\delta_{ij}. \tag{2.6.7}$$

The momentum operator is represented as

$$\boldsymbol{p} = -i\hbar\nabla. \tag{2.6.8}$$

The quantum-mechanical Hamiltonian (2.1.3) reads

$$H^{\text{QM}} = p_j\dot{x}_j - L = \frac{m}{2}\dot{\boldsymbol{x}}^2 - e\varphi \equiv \frac{1}{2m}\boldsymbol{P}^2 - e\varphi, \tag{2.6.9}$$

where

$$\boldsymbol{P} = m\dot{x}_j = \boldsymbol{p} + e\boldsymbol{A} = -i\hbar\nabla + e\boldsymbol{A}. \tag{2.6.10}$$

---

[1]Quantization of the electromagnetic field is carried out in Chapter 5.

It is the kinetic momentum, also called the covariant momentum, as is explained in (5.3.10).

The field-theoretical Hamiltonian is given by (2.4.4), or

$$H^{\mathrm{FT}} = \int d^3x\, \psi^\dagger(t,\boldsymbol{x}) \left[\frac{1}{2m} (-i\hbar\nabla + e\boldsymbol{A})^2 - e\varphi(t,\boldsymbol{x}) \right] \psi(t,\boldsymbol{x}), \tag{2.6.11}$$

where ψ is the field operator describing electrons: Note that $\pi(t,\boldsymbol{x}) = i\hbar\psi^\dagger(t,\boldsymbol{x})$ as in (2.4.2).

CHAPTER 3

CANONICAL QUANTIZATION

Canonical quantization is a generic scheme applicable to any system once the Lagrangian is given. In this chapter we analyze the Schrödinger field and the Klein-Gordon field. In separate chapters we analyze the electromagnetic field and the Dirac field. The Nöther theorem assures the existence of a conserved current in the presence of a continuous symmetry. The conserved charge distinguishes the antiparticle from the particle.

3.1 RELATIVISTIC PARTICLES AND WAVES

Quantum field theory of nonrelativistic particles has been formulated by rewriting many-body quantum mechanics. However, classical waves are not quantized in this way. Furthermore, this method is not applicable to relativistic particles because of the negative-energy problem. A relativistic particle is subject to the Einstein relation, $E^2 = c^2\boldsymbol{p}^2 + m^2c^4$, which allows not only the positive-energy state but also the negative-energy state, $E = \pm\sqrt{c^2\boldsymbol{p}^2 + m^2c^4}$. Quantum field theory is necessary to overcome this problem.

Relativistic particles are relevant also to condensed matter physics. When the mass of the particle is zero, they should be treated as relativistic particles. Thus, photons which are quanta associated with the electromagnetic wave are relativistic particles. Edge excitations in QH systems behave as if they were relativistic particles. Recently, much attention has been paid to electrons in graphene, since they show unconventional QH effects.[374,538,376,340] They are known[434] to obey the Dirac equation. We also note that the relativistic field theory gives a good exercise to learn about quantum field theory, because it is simpler than the nonrelativistic field theory in various aspects.

There is a resemblance between classical waves and relativistic fields. The classical wave equation is given by

$$\frac{\partial^2}{\partial t^2}\phi(t,\boldsymbol{x}) - C^2\nabla^2\phi(t,\boldsymbol{x}) = 0, \tag{3.1.1}$$

where C stands for the velocity of the wave. The field ϕ may describe a sound wave. It may be a component of the electromagnetic field, where C is the speed of light ($C = c$) in the vacuum. We may add a mass term to the wave equation,

$$\frac{\partial^2}{\partial t^2}\phi(t,\boldsymbol{x}) - C^2\nabla^2\phi(t,\boldsymbol{x}) + \frac{m^2C^4}{\hbar^2}\phi(t,\boldsymbol{x}) = 0. \tag{3.1.2}$$

Indeed, photons acquire a mass when they penetrate into superconductor. When $C = c$ this equation is known as the Klein-Gordon equation, which is the fundamental equation for the relativistic field.

Quantization is carried out once the Lagrangian density $\mathcal{L}$ is given. The field equation is derived from the action as an Euler-Lagrange equation as in (1.2.2), or $\delta S/\delta\phi(x) = 0$, where $S = \int dtd^3x\, \mathcal{L}$. Conversely, when a field equation is given, we may guess the action so that it is the Euler-Lagrange equation. For instance, we find

$$S = \int dtd^3x\, \phi(t,\boldsymbol{x})\left(\frac{\partial^2}{\partial t^2}\phi(t,\boldsymbol{x}) - C^2\nabla^2\phi(t,\boldsymbol{x}) + \frac{m^2C^4}{\hbar^2}\phi(t,\boldsymbol{x})\right) \tag{3.1.3}$$

from (3.1.2), where we have assumed that $\phi(t,\boldsymbol{x})$ is a real field.

There exist two main quantization schemes of any classical system, the canonical quantization and the path-integral quantization. We use the *canonical quantization*. According to the standard prescription, a classical field is promoted to a quantum field by imposing the *canonical commutation relation* between a field ("coordinate") and its canonical conjugate field ("momentum"). We confirm explicitly that the method reproduces the previous results for the quantization of the Schrödinger field. In this chapter we analyze the Schrödinger field and the Klein-Gordon field. In a separate chapter we analyze the electromagnetic field.

The symmetry of the Lagrangian is very important since it determines the basic structure of the system. Related topics are the Nöther theorem (this chapter and Appendix H), Goldstone modes (Chapter 4), and topological solitons (Chapter 7). The Nöther theorem assures the existence of a conserved current in any system possessing a continuous symmetry. We use the theorem to define the current and the energy-momentum tensor.

3.2 SCHRÖDINGER FIELD

First we discuss the case of nonrelativistic particles to ensure the method. Regarding the Heisenberg equation (2.4.6) as a classical field equation, we construct an action,

$$S \equiv \int dtd^3x\, \mathcal{L} = \int dtd^3x\, \phi^\dagger(t,\boldsymbol{x})\left(E - \frac{\boldsymbol{p}^2}{2m} - V(\boldsymbol{x})\right)\phi(t,\boldsymbol{x}), \tag{3.2.1}$$

where $\mathcal{L}$ is the Lagrangian density, and

$$E = i\hbar\partial_t, \qquad \boldsymbol{p} = -i\hbar\nabla. \tag{3.2.2}$$

Indeed, the Heisenberg equation (2.4.6) follows from (3.2.1) as the Euler-Lagrange equation, $\delta S/\delta\phi^\dagger = 0$. After a partial integration we have

$$S = \int dt d^3x \left(i\hbar\phi^\dagger(t,\boldsymbol{x})\frac{\partial}{\partial t}\phi(t,\boldsymbol{x}) - \frac{\hbar^2}{2m}\nabla\phi^\dagger(t,\boldsymbol{x})\nabla\phi(t,\boldsymbol{x}) \right.$$
$$\left. - V(\boldsymbol{x})\phi^\dagger(t,\boldsymbol{x})\phi(t,\boldsymbol{x}) \right). \tag{3.2.3}$$

The Lagrangian density is

$$\mathcal{L} = i\hbar\phi^\dagger(t,\boldsymbol{x})\frac{\partial}{\partial t}\phi(t,\boldsymbol{x}) - \frac{\hbar^2}{2m}\nabla\phi^\dagger(t,\boldsymbol{x})\nabla\phi(t,\boldsymbol{x}) - V(\boldsymbol{x})\phi^\dagger(t,\boldsymbol{x})\phi(t,\boldsymbol{x}). \tag{3.2.4}$$

The standard procedure of quantization follows from the Lagrangian density. We regard $\phi(t,\boldsymbol{x})$ as the "coordinate" and $\dot{\phi}(t,\boldsymbol{x}) \equiv \partial\phi/\partial t$ as the "velocity". The "canonical momentum" is

$$\pi(x) = \frac{\delta\mathcal{L}(x)}{\delta\dot{\phi}(x)} = i\hbar\phi^\dagger(x), \tag{3.2.5}$$

which is nothing but (2.4.2). We also call $\pi(t,\boldsymbol{x})$ the canonical conjugate of $\phi(t,\boldsymbol{x})$. The Hamiltonian density is a Legendre transform of the Lagrangian density,

$$\mathcal{H} = \pi(t,\boldsymbol{x})\dot{\phi}(t,\boldsymbol{x}) - \mathcal{L}. \tag{3.2.6}$$

The Hamiltonian is

$$H = \int d^3x \left(\frac{\hbar^2}{2m}\nabla\phi^\dagger(t,\boldsymbol{x})\nabla\phi(t,\boldsymbol{x}) + V(\boldsymbol{x})\phi^\dagger(t,\boldsymbol{x})\phi(t,\boldsymbol{x}) \right)$$
$$= \frac{1}{i\hbar}\int d^3x\, \pi(t,\boldsymbol{x})H^{\text{QM}}(\boldsymbol{x})\phi(t,\boldsymbol{x}) \tag{3.2.7}$$

with (2.1.6) for $H^{\text{QM}}(\boldsymbol{x})$. This agrees with the Hamiltonian (2.4.4).

The field $\phi(t,\boldsymbol{x})$ is so far a classical one. We now promote it to a field operator. The field $\phi(t,\boldsymbol{x})$ may describe bosons or fermions. When they are bosons (fermions), *canonical quantization* is a procedure to postulate the following equal-time commutation (anticommutation) relations between the field $\phi(t,\boldsymbol{x})$ and the canonical conjugate field $\pi(t,\boldsymbol{x})$,

$$[\phi(t,\boldsymbol{x}),\phi(t,\boldsymbol{y})]_\pm = [\pi(t,\boldsymbol{x}),\pi(t,\boldsymbol{y})]_\pm = 0,$$
$$[\phi(t,\boldsymbol{x}),\pi(t,\boldsymbol{y})]_\pm = i\hbar\delta(\boldsymbol{x}-\boldsymbol{y}), \tag{3.2.8}$$

where $[A,B]_-$ stands for the commutator and $[A,B]_+$ for the anticommutator, $[A,B]_\pm = AB \pm BA$, for notational convenience. We call them the canonical commutation (anticommutation) relations. They amount to

$$[\phi(t,\boldsymbol{x}),\phi(t,\boldsymbol{y})]_\pm = [\phi^\dagger(t,\boldsymbol{x}),\phi^\dagger(t,\boldsymbol{y})]_\pm = 0,$$
$$[\phi(t,\boldsymbol{x}),\phi^\dagger(t,\boldsymbol{y})]_\pm = \delta(\boldsymbol{x}-\boldsymbol{y}), \tag{3.2.9}$$

and are identical to the commutation relations (2.3.4) or the anticommutation relations (2.5.2).

The next procedure is to expand the field operator in terms of an orthonormal complete set. It is convenient to choose the eigenstates of the quantum mechanical Hamiltonian H^{QM}. It is given by (2.3.1), or

$$\phi(t, \boldsymbol{x}) = \sum_j a_j \mathfrak{S}_j(\boldsymbol{x}) e^{-i\varepsilon_j t/\hbar}. \tag{3.2.10}$$

Substituting this into (3.2.7) we readily reproduce (2.2.11), or

$$H = \sum_{j=1}^{\infty} \varepsilon_j a_j^\dagger a_j. \tag{3.2.11}$$

Hence, canonical quantization is a consistent procedure.

We make two comments. When we choose ϕ as the independent variable, $\phi^\dagger$ is its canonical momentum as in (3.2.5). We cannot choose both ϕ and $\phi^\dagger$ as two independent canonical variables. The second comment is on the hermiticity of the Lagrangian density (3.2.4). It is not Hermitian,

$$\mathcal{L}^\dagger = \mathcal{L} - i\hbar \frac{\partial}{\partial t}(\phi^\dagger \phi). \tag{3.2.12}$$

We can make it Hermitian by adding a total derivative term to the Lagrangian density; see (5.3.4). This does not modify the physics since the action is invariant.

Momentum Expansion

When an interaction term is included, we obtain the Lagrangian density,

$$\begin{aligned}\mathcal{L} =& i\hbar\phi^\dagger(t, \boldsymbol{x})\frac{\partial}{\partial t}\phi(t, \boldsymbol{x}) - \frac{\hbar^2}{2m}\nabla\phi^\dagger(t, \boldsymbol{x})\nabla\phi(t, \boldsymbol{x}) - V(\boldsymbol{x})\phi^\dagger(t, \boldsymbol{x})\phi(t, \boldsymbol{x}) \\ & - \frac{1}{2}\int d^3y\, \phi^\dagger(t, \boldsymbol{x})\phi^\dagger(t, \boldsymbol{y})U(|\boldsymbol{y} - \boldsymbol{x}|)\phi(t, \boldsymbol{y})\phi(t, \boldsymbol{x}).\end{aligned} \tag{3.2.13}$$

Canonical quantization proceeds precisely as before. The Hamiltonian is

$$\begin{aligned}H =& \int d^3x \left(\frac{\hbar^2}{2m}\nabla\phi^\dagger(t, \boldsymbol{x})\nabla\phi(t, \boldsymbol{x}) + V(\boldsymbol{x})\phi^\dagger(t, \boldsymbol{x})\phi(t, \boldsymbol{x})\right) \\ & + \frac{1}{2}\int d^3x \int d^3y\, \phi^\dagger(t, \boldsymbol{x})\phi^\dagger(t, \boldsymbol{y})U(|\boldsymbol{x} - \boldsymbol{y}|)\phi(t, \boldsymbol{y})\phi(t, \boldsymbol{x}).\end{aligned} \tag{3.2.14}$$

However, it is impossible to extract the quantum-mechanical Hamiltonian and to construct its eigenstates as in the noninteracting theory.

Nevertheless, we are able to expand the field operator ϕ in terms of any complete set of functions. It is convenient to use the plane wave, $e^{i\boldsymbol{kx}}$, for the expansion,

$$\phi(t, \boldsymbol{x}) = \int \frac{d^3k}{\sqrt{(2\pi)^3}} e^{i\boldsymbol{kx}} a_{\boldsymbol{k}}(t). \tag{3.2.15}$$

This is a Fourier transformation, and its inverse transformation is

$$a_{\boldsymbol{k}}(t) = \int \frac{d^3x}{\sqrt{(2\pi)^3}} e^{-i\boldsymbol{k}\boldsymbol{x}} \phi(t, \boldsymbol{x}). \tag{3.2.16}$$

The commutation relations between $a_{\boldsymbol{k}}(t)$ and $a_{\boldsymbol{l}}^{\dagger}(t)$ follow,

$$[a_{\boldsymbol{k}}(t), a_{\boldsymbol{l}}(t)]_{\pm} = [a_{\boldsymbol{k}}^{\dagger}(t), a_{\boldsymbol{l}}^{\dagger}(t)]_{\pm} = 0, \tag{3.2.17}$$

and

$$\begin{aligned} [a_{\boldsymbol{k}}(t), a_{\boldsymbol{l}}^{\dagger}(t)]_{\pm} &= \int \frac{d^3x d^3y}{(2\pi)^3} e^{-i\boldsymbol{k}\boldsymbol{x}+i\boldsymbol{l}\boldsymbol{y}} [\phi(t, \boldsymbol{x}), \phi^{\dagger}(t, \boldsymbol{y})]_{\pm} \\ &= \int \frac{d^3x}{(2\pi)^3} e^{-i(\boldsymbol{k}-\boldsymbol{l})\boldsymbol{x}} = \delta(\boldsymbol{k} - \boldsymbol{l}), \end{aligned} \tag{3.2.18}$$

from the (anti)commutation relations (3.2.9). The operators $a_{\boldsymbol{k}}(t)$ and $a_{\boldsymbol{k}}^{\dagger}(t)$ are interpreted as the annihilation and creation operators of a particle with wave vector $\boldsymbol{k}$ at time t. The total momentum is given by (2.4.5), or

$$\boldsymbol{p} = -i\hbar \int d^3x\, \phi^{\dagger}(t, \boldsymbol{x}) \nabla \phi(t, \boldsymbol{x}) = \int d^3k\, \hbar \boldsymbol{k} a_{\boldsymbol{k}}^{\dagger} a_{\boldsymbol{k}}, \tag{3.2.19}$$

which shows that each mode with wave vector $\boldsymbol{k}$ carries momentum $\hbar\boldsymbol{k}$, as expected.

We encounter an infinity when we use the plane wave in an infinitely large volume. Setting $\boldsymbol{k} = \boldsymbol{l}$ in the δ-function normalization (3.2.18), we get

$$\delta(0) = \frac{V}{(2\pi)^3}, \tag{3.2.20}$$

with V the volume of the system, $V = \int d^3x \to \infty$. This divergence is avoided by employing the plane wave confined in a finite volume $V = L^3$ with the periodic boundary condition, $\phi(t, \boldsymbol{x} + \boldsymbol{L}) = \phi(t, \boldsymbol{x})$. In this *box normalization* the field is expanded in a Fourier series,

$$\phi(t, \boldsymbol{x}) = \frac{1}{\sqrt{V}} \sum_{\boldsymbol{k}} e^{i\boldsymbol{k}\boldsymbol{x}} \tilde{a}_{\boldsymbol{k}}(t), \tag{3.2.21}$$

which replaces (3.2.15), where each momentum component takes discrete values,

$$k_i = \frac{2\pi}{L} n_i, \qquad n_i = 0, \pm 1, \pm 2, \cdots. \tag{3.2.22}$$

The inverse transformation is

$$\tilde{a}_{\boldsymbol{k}}(t) = \frac{1}{\sqrt{V}} \int_V d^3x\, e^{-i\boldsymbol{k}\boldsymbol{x}} \phi(t, \boldsymbol{x}), \tag{3.2.23}$$

in place of (3.2.16). The commutation relation reads

$$[\tilde{a}_{\boldsymbol{k}}(t), \tilde{a}_{\boldsymbol{l}}^{\dagger}(t)]_{\pm} = \frac{1}{V} \int_V d^3x\, e^{-i(\boldsymbol{k}-\boldsymbol{l})\boldsymbol{x}} = \delta_{\boldsymbol{k},\boldsymbol{l}}. \tag{3.2.24}$$

The Fourier expansion of the c-number potential $U(\boldsymbol{x})$ is

$$\tilde{U}_{\boldsymbol{k}} = \frac{1}{V}\int_V d^3x\, e^{-i\boldsymbol{k}\boldsymbol{x}} U(\boldsymbol{x}). \tag{3.2.25}$$

The essential difference is that the quantity in the momentum space (such as $\tilde{a}_{\boldsymbol{k}}$) is dimensionless in the box normalization but the corresponding one (such as $a_{\boldsymbol{k}}$) is dimensional in the δ-function normalization. The dimension is taken care of by the volume V of the system. It is trivial to change the normalization convention from one to another by the following rule,

$$\int d^3k \Longleftrightarrow \frac{(2\pi)^3}{V}\sum_{\boldsymbol{k}}, \qquad \delta(\boldsymbol{k}-\boldsymbol{l}) \Longleftrightarrow \frac{V}{(2\pi)^3}\delta_{\boldsymbol{k},\boldsymbol{l}}, \tag{3.2.26a}$$

$$a_{\boldsymbol{k}} \Longleftrightarrow \sqrt{\frac{V}{(2\pi)^3}}\tilde{a}_{\boldsymbol{k}}, \qquad U_{\boldsymbol{k}} \Longleftrightarrow \frac{V}{(2\pi)^3}\tilde{U}_{\boldsymbol{k}}. \tag{3.2.26b}$$

We mainly use the δ-function normalization for calculation since it is easier. We convert the result into the corresponding one in the box normalization when necessary, but we omit the tilde from all symbols for simplicity.

We may evaluate the Hamiltonian (2.4.13) by substituting the momentum expansion (3.2.15) at $t = 0$, since it is independent of time,

$$\begin{aligned} H = &\int d^3k\, \frac{(\hbar\boldsymbol{k})^2}{2m} a_{\boldsymbol{k}}^\dagger a_{\boldsymbol{k}} + \int d^3k d^3l\, V_{\boldsymbol{l}-\boldsymbol{k}} a_{\boldsymbol{l}}^\dagger a_{\boldsymbol{k}} \\ &+ \frac{1}{2}\int d^3k d^3l d^3p d^3q\, \delta(\boldsymbol{k}'+\boldsymbol{l}'-\boldsymbol{k}-\boldsymbol{l}) U_{\boldsymbol{k}-\boldsymbol{k}'} a_{\boldsymbol{k}'}^\dagger a_{\boldsymbol{l}'}^\dagger a_{\boldsymbol{k}} a_{\boldsymbol{l}}, \end{aligned} \tag{3.2.27}$$

with

$$V_{\boldsymbol{k}} = \int \frac{d^3x}{(2\pi)^3} e^{-i\boldsymbol{k}\boldsymbol{x}} V(\boldsymbol{x}), \qquad U_{\boldsymbol{k}} = \int \frac{d^3x}{(2\pi)^3} e^{-i\boldsymbol{k}\boldsymbol{x}} U(|\boldsymbol{x}|). \tag{3.2.28}$$

In particular, $U_{\boldsymbol{k}} = g/(2\pi)^3$ for the contact interaction (2.4.12). In the box normalization it reads,

$$H = \sum_{\boldsymbol{k}} \frac{(\hbar\boldsymbol{k})^2}{2m} a_{\boldsymbol{k}}^\dagger a_{\boldsymbol{k}} + \sum_{\boldsymbol{k},\boldsymbol{l}} V_{\boldsymbol{l}-\boldsymbol{k}} a_{\boldsymbol{l}}^\dagger a_{\boldsymbol{k}} + \frac{1}{2}\sum_{\boldsymbol{k}',\boldsymbol{l}'}\sum_{\boldsymbol{k},\boldsymbol{l}} \delta_{\boldsymbol{k}'+\boldsymbol{l}',\boldsymbol{k}+\boldsymbol{l}} U_{\boldsymbol{k}-\boldsymbol{k}'} a_{\boldsymbol{k}'}^\dagger a_{\boldsymbol{l}'}^\dagger a_{\boldsymbol{k}} a_{\boldsymbol{l}}, \tag{3.2.29}$$

as is easily checked based on the above conversion rule: Note that we have omitted the tilde from all symbols.

Each term in the Hamiltonian (3.2.27) has a graphical interpretation [Fig.3.1]. In the Schrödinger picture*Schrödinger picture* the time-evolution of the state is governed by the total Hamiltonian according to the formula (1.2.9). The first term of the right hand side of (3.2.27) is the kinetic energy of a particle with momentum $\hbar\boldsymbol{k}$. Since one particle with momentum $\hbar\boldsymbol{k}$ is annihilated and succeedingly created, the state remain unchanged. The second term represents an interaction with the external potential V. It scatters a particle with momentum $\hbar\boldsymbol{k}$ and change it to a particle with momentum $\hbar\boldsymbol{l}$: The momentum $\hbar(\boldsymbol{k}-\boldsymbol{l})$ is absorbed by the potential.

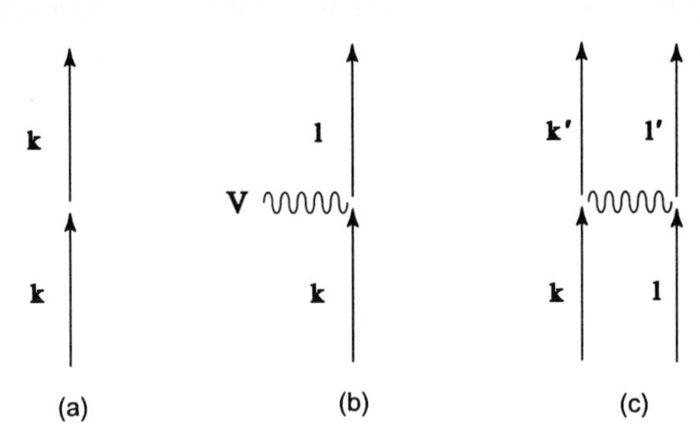

Fig. 3.1 Each term in the Hamiltonian (3.2.27) has a graphical representation: (a) for the kinetic energy with momentum $\hbar\boldsymbol{k}$; (b) for an interaction with the external potential V, which scatters a particle with momentum $\hbar\boldsymbol{k}$ and changes it to a particle with momentum $\hbar\boldsymbol{l}$; (c) for a scattering of two particles with momenta $\hbar\boldsymbol{k}$ and $\hbar\boldsymbol{l}$ into two particles with momenta $\hbar\boldsymbol{k}'$ and $\hbar\boldsymbol{l}'$ by exchanging a momentum.

The last term represents that two particles with momenta $\hbar\boldsymbol{k}$ and $\hbar\boldsymbol{l}$ scatter to two particles with momenta $\hbar\boldsymbol{k}'$ and $\hbar\boldsymbol{l}'$ by the potential U: The δ function implies the momentum conservation.

3.3 Real Klein-Gordon Field

We analyze the simplest model of relativistic particles. A relativistic particle has the energy-momentum relation called the Einstein relation,

$$E^2 = c^2\boldsymbol{p}^2 + m^2c^4. \tag{3.3.1}$$

Regarding this as an operator identity with

$$E = i\hbar\partial_t, \qquad \boldsymbol{p} = -i\hbar\nabla, \tag{3.3.2}$$

we obtain a relativistic wave equation called the Klein-Gordon equation,

$$(\hbar^2\partial_t^2 - \hbar^2c^2\nabla^2 + m^2c^4)\phi(t,\boldsymbol{x}) = 0. \tag{3.3.3}$$

We use a relativistic convention with the metric

$$g_{\mu\nu} = \begin{pmatrix} -1 & 0 & 0 & 0 \\ 0 & 1 & 0 & 0 \\ 0 & 0 & 1 & 0 \\ 0 & 0 & 0 & 1 \end{pmatrix}. \tag{3.3.4}$$

The time and spatial coordinates are set into four-vectors,

$$x^\mu = (ct, \boldsymbol{x}), \qquad x_\mu = g_{\mu\nu}x^\nu = (-ct, \boldsymbol{x}), \tag{3.3.5}$$

which are called the contravariant and covariant vectors, respectively. We understand a summation convention over repeated indices; thus,

$$x^\mu x_\mu = \sum_\mu x^\mu x_\mu = \boldsymbol{x}^2 - (ct)^2. \tag{3.3.6}$$

We also set

$$p^\mu = (E/c, \boldsymbol{p}), \qquad p_\mu = g_{\mu\nu}p^\nu = (-E/c, \boldsymbol{p}). \tag{3.3.7}$$

The quantum-mechanical relations (3.3.2) are combined into

$$p_\mu = -i\hbar\partial_\mu. \tag{3.3.8}$$

The derivative with respect to the space coordinate is

$$\boldsymbol{\nabla} = \frac{\partial}{\partial \boldsymbol{x}} = \left(\frac{\partial}{\partial x_1}, \frac{\partial}{\partial x_2}, \frac{\partial}{\partial x_3}\right) = (\partial^1, \partial^2, \partial^3) = (\partial_1, \partial_2, \partial_3), \tag{3.3.9}$$

while that with respect to the time is

$$\partial_t \equiv \frac{\partial}{\partial t} = c\frac{\partial}{\partial x^0} = c\partial_0 = -c\partial^0. \tag{3.3.10}$$

Because

$$p^\mu p_\mu = g^{\mu\nu}p_\mu p_\nu = \boldsymbol{p}^2 - \frac{E^2}{c^2}, \tag{3.3.11}$$

the energy-momentum relation (3.3.1) is rewritten as

$$p^\mu p_\mu + m^2c^2 = 0 \qquad \text{with} \quad p_\mu = -i\hbar\partial_\mu, \tag{3.3.12}$$

and the Klein-Gordon equation (3.3.3) as

$$\left(\hbar^2\partial^\mu\partial_\mu - m^2c^2\right)\phi(x) = 0. \tag{3.3.13}$$

They have manifestly Lorentz-covariant forms.

Comments are in order. The Klein-Gordon equation (3.3.3) allows the energy of a particle to take a negative value as well as a positive value ($E = \pm c\sqrt{\boldsymbol{p}^2 + m^2c^2}$). When the positive-energy part is taken, the Klein-Gordon equation is reduced to the Schrödinger equation in a small-momentum limit ($|\boldsymbol{p}| \ll mc$) via the expansion of the dispersion relation

$$E = mc^2\sqrt{1 + \frac{\boldsymbol{p}^2}{m^2c^2}} = mc^2 + \frac{\boldsymbol{p}^2}{2m} + \cdots. \tag{3.3.14}$$

One is tempted to postulate the second-quantized Hamiltonian (2.4.4),

$$H = \frac{c}{i\hbar}\int d^3x\, \pi(t, \boldsymbol{x})\sqrt{-\hbar^2\boldsymbol{\nabla}^2 + m^2c^2}\phi(t, \boldsymbol{x}). \tag{3.3.15}$$

However, this leads to a nonlocal field theory violating the causality[1] since the expansion (3.3.14) involves an infinite series of derivative ∇^2. Nevertheless, when both the positive and negative parts are taken simultaneously, as we see soon, canonical quantization yields a self-consistent quantum field theory of relativistic particle. Finally, the Klein-Gordon field can be massless ($m = 0$), where it describes a wave and not the Schrödinger field in any limit.

We postulate an action,

$$S \equiv \int d^4x\, \mathcal{L} = \frac{1}{2}\int d^4x\, \phi(x)\left(\hbar^2\partial^\mu\partial_\mu - m^2c^2\right)\phi(x), \tag{3.3.16}$$

so that the field equation (3.3.13) follows as the Euler-Lagrange equation, $\delta S/\delta\phi(x) = 0$. The Lagrangian density is

$$\mathcal{L} = -\frac{\hbar^2}{2}(\partial^\mu\phi)(\partial_\mu\phi) - \frac{m^2c^2}{2}\phi^2 = \frac{\hbar^2}{2c^2}\dot{\phi}^2 - \frac{\hbar^2}{2}(\nabla\phi)^2 - \frac{m^2c^2}{2}\phi^2, \tag{3.3.17}$$

by making a partial integration in (3.3.16). Mathematically, it is identical to the system (3.1.3) by setting $C = c$ (light velocity) and by changing the normalization of the field ϕ. Moreover, we may simply set $m = 0$ for the quantization of classical waves.

The canonical momentum $\pi(x)$ of the field $\phi(x)$ is

$$\pi(x) = \frac{\partial\mathcal{L}(x)}{\partial\dot{\phi}(x)} = \frac{\hbar^2}{c^2}\dot{\phi}(x). \tag{3.3.18}$$

The Hamiltonian density is given by a Legendre transformation,

$$\mathcal{H} = \pi\dot{\phi} - \mathcal{L} = \frac{\hbar^2}{2c^2}\dot{\phi}^2 + \frac{\hbar^2}{2}(\nabla\phi)^2 + \frac{m^2c^2}{2}\phi^2. \tag{3.3.19}$$

The canonical commutation relations yield

$$[\phi(t,\boldsymbol{x}),\phi(t,\boldsymbol{y})] = [\dot{\phi}(t,\boldsymbol{x}),\dot{\phi}(t,\boldsymbol{y})] = 0, \tag{3.3.20}$$

and

$$[\phi(t,\boldsymbol{x}),\dot{\phi}(t,\boldsymbol{y})] = i\frac{c^2}{\hbar}\delta(\boldsymbol{x}-\boldsymbol{y}). \tag{3.3.21}$$

The Heisenberg equation for the operator $\dot{\phi}$ is calculated as

$$i\hbar\partial_t\dot{\phi} = [\dot{\phi}, H] = i\hbar c^2\nabla^2\phi - i\frac{m^2c^4}{\hbar}\phi, \tag{3.3.22}$$

which is the Klein-Gordon equation (3.3.3).

---

[1] The equal-time commutation relation between $\pi(t,\boldsymbol{x})$ and the Hamiltonian density $\mathcal{H}(t,\boldsymbol{y})$ reads

$$[\pi(t,\boldsymbol{x}),\mathcal{H}(t,\boldsymbol{y})] = -c\,\pi(t,\boldsymbol{y})\sqrt{-\hbar^2\nabla^2 + m^2c^2}\,\delta(\boldsymbol{x}-\boldsymbol{y}),$$

which does not vanish for the space-like separation, $(x-y)^\mu(x-y)_\mu = (\boldsymbol{x}-\boldsymbol{y})^2 > 0$, as implies the violation of the causality.

To obtain the momentum we recall the second-quantization formula (2.4.5), or

$$\boldsymbol{p} = -\int d^3x\, \pi(t,\boldsymbol{x})\nabla\phi(t,\boldsymbol{x}) = -\frac{\hbar^2}{c^2}\int d^3x\, \dot{\phi}(x)\nabla\phi(x). \tag{3.3.23}$$

We can confirm this formula by invoking the energy-momentum tensor $\Theta^{\mu\nu}$. It is defined in association with the space-time translational invariance based on the Nöther theorem (see Appendix H),

$$\Theta^{\mu\nu} \equiv -\frac{\partial\mathcal{L}}{\partial(\partial_\mu\phi)}\partial^\nu\phi + g^{\mu\nu}\mathcal{L} = \hbar^2\partial^\mu\phi\partial^\nu\phi + g^{\mu\nu}\mathcal{L}. \tag{3.3.24}$$

It is easy to check that Θ^{00} is the Hamiltonian density (3.3.19). The momentum density is defined by

$$p_j = \frac{1}{c}\Theta^{0j} = \frac{\hbar^2}{c}\partial^0\phi(x)\partial^j\phi(x) = -\frac{\hbar^2}{c^2}\dot{\phi}(x)\partial^j\phi(x), \tag{3.3.25}$$

which agrees with (3.3.23). The factor c^{-1} is necessary to define the momentum in accord with the energy-momentum four vector (3.3.7).

Momentum Expansion

The eigenfunctions of the Klein-Gordon equation (3.3.3) are given by the plane waves. We choose them to expand the field operator $\phi(x)$,

$$\phi(x) = c\int \frac{d^3k}{\sqrt{(2\pi)^3}}\frac{1}{\sqrt{2\hbar\omega_k}}\left(a_{\boldsymbol{k}}e^{ikx} + a_{\boldsymbol{k}}^\dagger e^{-ikx}\right), \tag{3.3.26}$$

where $kx \equiv \boldsymbol{kx} - \omega_k t$ with $k^0 \equiv \omega_k/c > 0$ and

$$\hbar\omega_k = c\sqrt{\hbar^2\boldsymbol{k}^2 + m^2c^2}. \tag{3.3.27}$$

The term proportional to $a_{\boldsymbol{k}}^\dagger$ in (3.3.26) is necessary to ensure the reality of the field, $\phi^\dagger(x) = \phi(x)$.

The field $\phi(x)$ is decomposed into the positive-frequency component $\phi_+(x)$ and the negative-frequency component $\phi_-(x)$,

$$\phi_+(x) = c\int \frac{d^3k}{\sqrt{(2\pi)^3}}\frac{1}{\sqrt{2\hbar\omega_k}}a_{\boldsymbol{k}}e^{ikx}, \tag{3.3.28a}$$

$$\phi_-(x) = c\int \frac{d^3k}{\sqrt{(2\pi)^3}}\frac{1}{\sqrt{2\hbar\omega_k}}a_{\boldsymbol{k}}^\dagger e^{-ikx}, \tag{3.3.28b}$$

with $\phi_+^\dagger = \phi_-$. If $\phi(x)$ were a wave function in quantum mechanics, since

$$Ee^{-ikx} = -\hbar\omega_k e^{-ikx}, \tag{3.3.29}$$

with $E = i\hbar\partial_t$ the energy operator (3.3.2), the component $\phi_-(x)$ would describe a negative-energy state.

We now require the canonical commutation relation. We note that

$$\dot{\phi}(x) = -ic \int \frac{d^3k}{\sqrt{(2\pi)^3}} \sqrt{\frac{\omega_k}{2\hbar}} \left(a_{\boldsymbol{k}} e^{ikx} - a_{\boldsymbol{k}}^\dagger e^{-ikx} \right). \tag{3.3.30}$$

Combining (3.3.26) and (3.3.30), we are able to solve for the Fourier components $a_{\boldsymbol{k}}$ and $a_{\boldsymbol{k}}^\dagger$ in terms of the fields $\phi(\boldsymbol{x})$ and $\dot{\phi}(\boldsymbol{x})$ at $t = 0$,

$$a_{\boldsymbol{k}} = \sqrt{\frac{\hbar}{2\omega_{\boldsymbol{k}} c^2}} \int \frac{d^3x}{\sqrt{(2\pi)^3}} e^{-ikx} [\omega_{\boldsymbol{k}} \phi(\boldsymbol{x}) + i\dot{\phi}(\boldsymbol{x})], \tag{3.3.31a}$$

$$a_{\boldsymbol{k}}^\dagger = \sqrt{\frac{\hbar}{2\omega_{\boldsymbol{k}} c^2}} \int \frac{d^3x}{\sqrt{(2\pi)^3}} e^{ikx} [\omega_{\boldsymbol{k}} \phi(\boldsymbol{x}) - i\dot{\phi}(\boldsymbol{x})]. \tag{3.3.31b}$$

It is easy to derive

$$[a_{\boldsymbol{k}}, a_{\boldsymbol{l}}] = [a_{\boldsymbol{k}}^\dagger, a_{\boldsymbol{l}}^\dagger] = 0, \qquad [a_{\boldsymbol{k}}, a_{\boldsymbol{l}}^\dagger] = \delta(\boldsymbol{k} - \boldsymbol{l}) \tag{3.3.32}$$

from (3.3.20) and (3.3.21). These commutation relations imply that $a_{\boldsymbol{k}}$ and $a_{\boldsymbol{k}}^\dagger$ are the annihilation and creation operators of a particle with momentum $\boldsymbol{k}$.

The one-body state is defined by $|\boldsymbol{k}\rangle = a_{\boldsymbol{k}}^\dagger|0\rangle$. Its wave function reads

$$\mathfrak{S}_{\boldsymbol{k}}(x) = \langle 0|\phi(x)|\boldsymbol{k}\rangle = \langle 0|\phi(x)a_{\boldsymbol{k}}^\dagger|0\rangle = \frac{1}{\sqrt{(2\pi)^3}} \frac{c}{\sqrt{2\hbar\omega_k}} e^{ikx}. \tag{3.3.33}$$

The state has a positive energy, $E\mathfrak{S}_{\boldsymbol{k}}(x) = \hbar\omega_k \mathfrak{S}_{\boldsymbol{k}}(x)$, with $E = i\hbar\partial_t$. Many-body states, being given as in (1.3.16), have positive energies as well.

Consequently, the positive-frequency and negative-frequency components ϕ_+ and ϕ_- are the annihilation and creation operators of a positive-energy particle, respectively. *We have overcome the negative-energy problem of the relativistic quantum mechanics by interpreting the negative energy component as the creation operator.*

We represent the Hamiltonian (3.3.19) in terms of the number operators $a_{\boldsymbol{k}}^\dagger a_{\boldsymbol{k}}$ with the use of the expansions (3.3.26) and (3.3.30),

$$H = \int d^3x\, \mathcal{H}(x) = \frac{1}{2} \int d^3k\, \hbar\omega_k \left(a_{\boldsymbol{k}}^\dagger a_{\boldsymbol{k}} + a_{\boldsymbol{k}} a_{\boldsymbol{k}}^\dagger \right) \tag{3.3.34a}$$

$$= \int d^3k\, \hbar\omega_k \left(a_{\boldsymbol{k}}^\dagger a_{\boldsymbol{k}} + \frac{1}{2}\delta(0) \right). \tag{3.3.34b}$$

The terms $a_{\boldsymbol{k}} a_{\boldsymbol{l}}$ and $a_{\boldsymbol{k}}^\dagger a_{\boldsymbol{l}}^\dagger$ do not appear since their coefficients are proportional to $(\omega_k^2 - c^2\boldsymbol{k}^2 - m^2c^4/\hbar^2)$ and vanish due to the relation (3.3.27). Note that we have used the commutation relation (3.3.32) only to derive (3.3.34b) from (3.3.34a). Similarly, for the momentum operator (3.3.23) we find

$$\boldsymbol{p} = \int d^3k\, \hbar\boldsymbol{k}\, a_{\boldsymbol{k}}^\dagger a_{\boldsymbol{k}}, \tag{3.3.35}$$

where we have used $\int d^3k\, \boldsymbol{k} a_{\boldsymbol{k}} a_{-\boldsymbol{k}} = 0$ and $\int d^3k\, \boldsymbol{k} = 0$.

The Fock vacuum is given by $a_{\boldsymbol{k}}|0\rangle = 0$, for which the energy (3.3.34b) diverges,

$$\langle 0|H|0\rangle = \delta(0)\int d^3k\,\frac{\hbar\omega_k}{2} = V\int\frac{d^3k}{(2\pi)^3}\frac{\hbar\omega_k}{2}, \tag{3.3.36}$$

with V being the volume of the system. See (3.2.20) for the meaning of $\delta(0)$. This is called the zero-point energy, whose divergence is a general feature of a quantum field theory. It is not an observable quantity. The physical energy is given by

$$\hat{H} = H - \langle 0|H|0\rangle = \int d^3k\,\hbar\omega_k a_{\boldsymbol{k}}^\dagger a_{\boldsymbol{k}}. \tag{3.3.37}$$

This subtraction is done systematically by normal ordering defined in (1.3.20). The zero-point divergence is absent if normal ordering is taken in the Hamiltonian, since it arises when we interchange $a_{\boldsymbol{k}}$ with $a_{\boldsymbol{k}}^\dagger$ to get (3.3.34b) from (3.3.34a). We obtain

$$H = \int d^3x\ :\mathcal{H}(x): = \frac{1}{2}\int d^3k\,\hbar\omega_k : \left(a_{\boldsymbol{k}}^\dagger a_{\boldsymbol{k}} + a_{\boldsymbol{k}} a_{\boldsymbol{k}}^\dagger\right): = \int d^3k\,\hbar\omega_k a_{\boldsymbol{k}}^\dagger a_{\boldsymbol{k}}, \tag{3.3.38}$$

in place of (3.3.34b), which agrees with (3.3.37).

The dispersion relation is linear for the massless relativistic particle,

$$E_{\boldsymbol{k}} = \hbar\omega_k = c\hbar|\boldsymbol{k}|. \tag{3.3.39}$$

As we have stated, the result for the wave system (3.1.1) is obtained by replacing the light velocity c with the wave velocity C,

$$E_{\boldsymbol{k}} = \hbar\omega_k = C\hbar|\boldsymbol{k}|. \tag{3.3.40}$$

The wave has the relativistic dispersion relation.

3.4 Complex Klein-Gordon Field

We consider a complex field ϕ, which is equal to considering a pair of two real fields ϕ_1 and ϕ_2,

$$\phi = \frac{1}{\sqrt{2}}(\phi_1 + i\phi_2), \qquad \phi^\dagger = \frac{1}{\sqrt{2}}(\phi_1 - i\phi_2). \tag{3.4.1}$$

The Lagrangian density is

$$\mathcal{L} = -\sum_{j=1}^{2}\left(\frac{\hbar^2}{2}(\partial^\mu\phi_j)(\partial_\mu\phi_j) + \frac{m^2c^2}{2}\phi_j^2\right) = -\hbar^2(\partial^\mu\phi^\dagger)(\partial_\mu\phi) - m^2c^2\phi^\dagger\phi. \tag{3.4.2}$$

The field equation is

$$\left(\hbar^2\partial^\mu\partial_\mu - m^2c^2\right)\phi(x) = 0. \tag{3.4.3}$$

The canonical momenta are

$$\pi(x) = \frac{\partial\mathcal{L}(x)}{\partial\dot{\phi}(x)} = \frac{\hbar^2}{c^2}\dot{\phi}^\dagger(x), \qquad \pi^\dagger(x) = \frac{\partial\mathcal{L}(x)}{\partial\dot{\phi}^\dagger(x)} = \frac{\hbar^2}{c^2}\dot{\phi}(x). \tag{3.4.4}$$

Since ϕ and $\phi^\dagger$ are independent fields, the Hamiltonian density is

$$\mathcal{H} = \pi\dot{\phi} + \dot{\phi}^\dagger\pi^\dagger - \mathcal{L} = \frac{\hbar^2}{c^2}\dot{\phi}^\dagger\dot{\phi} + \hbar^2\nabla\phi^\dagger\nabla\phi + m^2c^2\phi^\dagger\phi. \tag{3.4.5}$$

The energy-momentum tensor density (Appendix H) is

$$\Theta_{\mu\nu} = \hbar^2\left(\partial_\mu\phi^\dagger\partial_\nu\phi + \partial_\nu\phi^\dagger\partial_\mu\phi\right) + g_{\mu\nu}\mathcal{L}. \tag{3.4.6}$$

The canonical commutation relations are

$$[\phi(t,\boldsymbol{x}), \dot{\phi}^\dagger(t,\boldsymbol{y})] = [\phi^\dagger(t,\boldsymbol{x}), \dot{\phi}(t,\boldsymbol{y})] = i\frac{c^2}{\hbar}\delta(\boldsymbol{x} - \boldsymbol{y}) \tag{3.4.7}$$

together with the other trivial commutation relations: They are identical to two sets of the commutation relations (3.3.20) and (3.3.21) for two real fields ϕ_1 and ϕ_2.

Momentum Expansion

We expand the field operator in terms of plane waves,

$$\phi(x) = c\int\frac{d^3k}{\sqrt{(2\pi)^3}}\frac{1}{\sqrt{2\hbar\omega_k}}\left(b_{\boldsymbol{k}}e^{ikx} + c_{\boldsymbol{k}}^\dagger e^{-ikx}\right), \tag{3.4.8}$$

where ω_k is given by (3.3.27) and

$$b_{\boldsymbol{k}} = \frac{1}{\sqrt{2}}\left(a_{1\boldsymbol{k}} + ia_{2\boldsymbol{k}}\right), \qquad c_{\boldsymbol{k}} = \frac{1}{\sqrt{2}}\left(a_{1\boldsymbol{k}} - ia_{2\boldsymbol{k}}\right). \tag{3.4.9}$$

Here, $a_{1\boldsymbol{k}}$ and $a_{2\boldsymbol{k}}$ are the annihilation operators of two real fields $\phi_1(x)$ and $\phi_2(x)$, as defined by (3.3.26). It is clear that $b_{\boldsymbol{k}}$ and $c_{\boldsymbol{k}}$ are independent operators. They satisfy

$$[b_{\boldsymbol{k}}, b_{\boldsymbol{l}}^\dagger] = [c_{\boldsymbol{k}}, c_{\boldsymbol{l}}^\dagger] = \delta(\boldsymbol{k} - \boldsymbol{l}) \tag{3.4.10}$$

and other trivial commutation relations.

We write the Hamiltonian and the momentum in terms of the operators $c_{\boldsymbol{k}}$ and $b_{\boldsymbol{k}}$ by substituting (3.4.8) into them. After normal ordering we see

$$H = \int d^3x :\mathcal{H}:= \int d^3k\,\hbar\omega_k(b_{\boldsymbol{k}}^\dagger b_{\boldsymbol{k}} + c_{\boldsymbol{k}}^\dagger c_{\boldsymbol{k}}), \tag{3.4.11}$$

$$\boldsymbol{p} = \frac{1}{c}\int d^3x :\Theta^{j0}:= \int d^3k\,\hbar\boldsymbol{k}(b_{\boldsymbol{k}}^\dagger b_{\boldsymbol{k}} + c_{\boldsymbol{k}}^\dagger c_{\boldsymbol{k}}). \tag{3.4.12}$$

From (3.4.11) and (3.4.12) it follows that the b-quantum and c-quantum have the same energy and momentum, and hence they are degenerate.

We can distinguish the b-quantum and c-quantum by way of the charge operator defined by

$$Q = \int d^3k\,(b_{\boldsymbol{k}}^\dagger b_{\boldsymbol{k}} - c_{\boldsymbol{k}}^\dagger c_{\boldsymbol{k}}). \tag{3.4.13}$$

The b-quantum has the positive charge while the c-quantum has the negative charge. We interpret that the c-quantum is the *antiparticle* of the b-quantum: The particle and the antiparticle are distinguished by their opposite charges. The complex field operator $\phi(x)$ annihilates a particle and creates an antiparticle at the space-time point x. Thus, *we have overcome the negative-energy problem of the relativistic quantum mechanics by interpreting the negative energy component as the creation operator of the antiparticle.*

It is trivial to see that Q is a constant of motion since

$$i\hbar\frac{dQ}{dt} = [Q, H] = 0 \tag{3.4.14}$$

with the Hamiltonian (3.4.11). This proof is valid in the free field theory. The Hamiltonian is in general very complicated in the presence of interactions. Nevertheless, the charge conservation follows irrespective of interactions, as we shall see in the next section.

3.5 NÖTHER CURRENTS

The particle and the antiparticle are distinguished by the conserved charge (3.4.13). The charge conservation is a general property due to the Nöther theorem. The Nöther theorem says that *when there is a continuous symmetry there is a conservation law.* Well-known examples are the conservations of the momentum and the angular momentum, which are results of the invariance of the Lagrangian under a translation and a rotation, respectively.

We consider a system consisting of a complex field ϕ. Since the Lagrangian is real, it must be invariant under a phase transformation,[2]

$$\phi(x) \to e^{if}\phi(x), \tag{3.5.1}$$

where f is an arbitrary real constant. This is indeed the case in the complex Klein-Gordon system as well as in the Schrödinger system. We now present a formal proof of the charge conservation.[3] See also Appendix H for the conservation of the energy-momentum tensor.

The transformation (3.5.1) generates a continuous symmetry, since f is arbitrary. In particular, we may take an infinitesimal value ($|f| \ll 1$). Then, we can expand $e^{i\delta f} = 1 + i\delta f + \cdots$. The Lagrangian density changes as

$$\mathcal{L} \to \mathcal{L}_f = \mathcal{L}(\phi + \delta\phi, \partial_\mu\phi + \delta\partial_\mu\phi, \cdots) = \mathcal{L} + \delta\mathcal{L}_f, \tag{3.5.2}$$

---

[2] The phase transformation (3.5.1) is also called, after Pauli, the gauge transformation of the first kind.

[3] The proof is made with the use of the equation of motion, where we pay no attention to renormalization. To regularize infinities encountered in quantum theory it is necessary to introduce some sort of regulators and renormalize them. But, regulators may violate a symmetry of the theory at the classical level. At the end of the calculation, though regulators are removed, they may leave traces of this symmetry violation on rare occasions. Then, there is no conserved current associated with the symmetry: The symmetry is said anomalous. Anomalies can be physical and experimentally observable,[432,38,68] involving a deeper aspect of quantum field theory. We shall encounter such an anomaly in the electron-number conservation on the edge of the QH droplet in Section 18.5.

where

$$\delta\mathcal{L}_f = \frac{\partial\mathcal{L}}{\partial\phi}\phi + \frac{\partial\mathcal{L}}{\partial(\partial_\mu\phi)}\partial_\mu\phi + \phi^\dagger\frac{\partial\mathcal{L}}{\partial\phi^\dagger} + \partial_\mu\phi^\dagger\frac{\partial\mathcal{L}}{\partial(\partial_\mu\phi^\dagger)} = 0, \tag{3.5.3}$$

with

$$\delta\phi(x) = i\phi(x)\delta f, \qquad \delta\partial_\mu\phi(x) = i\partial_\mu\phi(x)\delta f \tag{3.5.4}$$

up to the first order in δf. By this transformation, however, the Lagrangian density is invariant. It implies $\delta\mathcal{L}_f/\delta f = 0$, or

$$\frac{\partial\mathcal{L}}{\partial\phi}\phi + \frac{\partial\mathcal{L}}{\partial(\partial_\mu\phi)}\partial_\mu\phi - \phi^\dagger\frac{\partial\mathcal{L}}{\partial\phi^\dagger} - \partial_\mu\phi^\dagger\frac{\partial\mathcal{L}}{\partial(\partial_\mu\phi^\dagger)} = 0. \tag{3.5.5}$$

Here we use the Euler-Lagrange equation,

$$\frac{\partial\mathcal{L}}{\partial\phi} - \partial_\mu\frac{\partial\mathcal{L}}{\partial(\partial_\mu\phi)} = 0, \tag{3.5.6}$$

and its Hermitian conjugate equation. We eliminate $\partial\mathcal{L}/\partial\phi$ and $\partial\mathcal{L}/\partial\phi^\dagger$ in (3.5.5) by using the Euler-Lagrange equation. The result is the continuity equation,

$$\partial^\mu j_\mu(x) = 0, \tag{3.5.7}$$

where[4]

$$j_\mu \equiv \frac{i}{\hbar}\left(\phi^\dagger\frac{\partial\mathcal{L}}{\partial(\partial^\mu\phi^\dagger)} - \frac{\partial\mathcal{L}}{\partial(\partial^\mu\phi)}\phi\right). \tag{3.5.8}$$

This is referred to as the Nöther current. We stress that the continuity equation (3.5.7) has followed from the invariance of the Lagrangian under an infinitesimal transformation, which is an intrinsic feature of the continuous symmetry.

The charge is defined by

$$Q = \frac{1}{c}\int d^3x j^0(t,\mathbf{x}). \tag{3.5.9}$$

It is independent of time,

$$\frac{d}{dt}Q = \int d^3x\partial_0 j^0(t,\mathbf{x}) = -\int d^3x\partial_k j_k(t,\mathbf{x}) = 0, \tag{3.5.10}$$

since there should be no current at spatial infinity, $j_k(t,\mathbf{x}) \to 0$ as $|\mathbf{x}| \to \infty$. Hence, the charge conservation is a result of phase invariance.

For the complex Klein-Gordon field with the Lagrangian density (3.4.2), the Nöther current (3.5.8) is

$$j_\mu = -i\hbar\left\{\phi^\dagger(\partial_\mu\phi) - (\partial_\mu\phi^\dagger)\phi\right\}. \tag{3.5.11}$$

The charge Q is explicitly calculated by substituting (3.4.8) into (3.5.9), which agrees with (3.4.13).

---

[4] We have introduced a normalization factor to reproduce (3.4.13) for the charge.

For the Schrödinger field with the Lagrangian density (3.2.4), the Nöther current (3.5.8) is $j^\mu = (c\rho, j^k)$ with

$$\rho = \phi^\dagger \phi, \qquad j_k = \frac{\hbar}{2mi}\left\{\phi^\dagger(\partial_k \phi) - (\partial_k \phi^\dagger)\phi\right\}. \tag{3.5.12}$$

Multiplying the Nöther current by the electric charge of the particle, we obtain the electric current; see (5.3.15) and (5.3.14).

It is instructive to rewrite the Nöther current in association with a local phase transformation,

$$\phi(x) \to e^{if(x)}\phi(x), \tag{3.5.13}$$

where $f(x)$ is a real function. By an infinitesimal transformation with $\delta f(x)$ the Lagrangian density changes as in (3.5.2), but now with

$$\delta\phi(x) = i\phi(x)\delta f, \qquad \delta\partial_\mu\phi(x) = i\partial_\mu\phi(x)\delta f + i\phi(x)(\partial_\mu \delta f). \tag{3.5.14}$$

It is clear from (3.5.3) that the Nöther current (3.5.8) is expressed as

$$j_\mu \equiv \frac{1}{\hbar}\frac{\partial \mathcal{L}_f}{\partial(\partial^\mu \delta f)}\bigg|_{\delta f \to 0} \tag{3.5.15}$$

where the limit $\delta f \to 0$ is taken after the derivation is taken.

CHAPTER 4

SPONTANEOUS SYMMETRY BREAKING

Ground states are degenerate in the presence of a symmetry. Bose condensation occurs spontaneously into one of the degenerate ground states at temperature $T = 0$, as leads to a spontaneous breaking of the symmetry. A gapless mode arises necessarily from the spontaneous breaking of a continuous global symmetry: It is called the Goldstone mode.

4.1 FERROMAGNETS

We consider electrons localized on lattice points [Fig.4.1]. The Coulomb interaction gives rise to the exchange interaction, which is summarized in a spin-spin interaction (see Appendix I),

$$H_X = -4\sum_{\langle ij\rangle} J_{ij}[\boldsymbol{S}(i)\cdot\boldsymbol{S}(j) + \frac{1}{4}\rho(i)\rho(j)], \tag{4.1.1}$$

where $\rho(i)$ and $\boldsymbol{S}(i)$ are the electron number and the spin at the lattice point indexed by i, and the sum runs over spin pairs ($i \neq j$). Here, J_{ij} is an exchange integral of wave functions localized at lattice points i and j, and the spin-spin interaction is short ranged.

This Hamiltonian has the global O(3) symmetry with respect to the spin degree of freedom: It is invariant when all spins are rotated simultaneously. The product $\boldsymbol{S}(i)\cdot\boldsymbol{S}(j)$ takes the maximum value, $\boldsymbol{S}(i)\cdot\boldsymbol{S}(j) = 1/4$, when $\boldsymbol{S}(i) = \boldsymbol{S}(j)$. Hence, provided $J_{ij} > 0$, all spin are spontaneously polarized in order to minimize the exchange energy (4.1.1), where the direction of polarization is arbitrary ($\boldsymbol{S}(i) = \boldsymbol{S}$ for all points i but $\boldsymbol{S}$ is arbitrary), as illustrated in Fig.4.2(a) and (b). The ground-state energy is

$$\langle H_X\rangle_g = -2\sum_{\langle ij\rangle} J_{ij} = -N\sum_{j\neq i} J_{ij} \equiv -\Delta_X N, \tag{4.1.2}$$

where N is the number of electrons, and $\Delta_X \equiv \sum_{j\neq i} J_{ij}$ with i fixed arbitrary; the sum runs over all lattice points j for $J_{ij} \neq 0$ and $j \neq i$. It is clear in (4.1.1) that the loss of the exchange energy is $2\Delta_X$ when one electron is removed from the filled

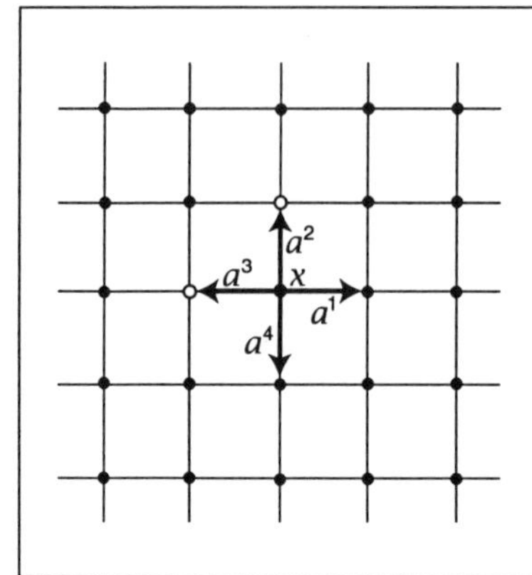

Fig. 4.1 Electrons localized on lattice points make exchange interactions with other electrons. The interaction strength J_{ij} is determined by the exchange integral of the two wave functions localized on lattice points i and j. In this 2-dimensional example, the lattice points in the nearest neighborhood of the point $\boldsymbol{x}$ are spanned by four lattice vectors $\boldsymbol{a}^\alpha$, and designated by open circles.

lattice [Fig.4.1]. This is equal to the energy necessary to flip one spin in the system (4.1.1). It is also clear that the exchange energy is unchanged when one electron is added to the filled lattice:[1] See also (30.4.2) and (30.3.9) in Chapter 30.

Ferromagnetism is a good example of the Bose condensation and the spontaneous symmetry breaking. Let us explain the terminology. First, it is referred to as the *Bose condensation* when one quantum state contains a macroscopic number of quanta proportional to the volume of the system. In a ferromagnet all spins are polarized into a single direction, namely, all electrons are in a single spin state. Second, it is referred to as the *spontaneous symmetry breaking* when the ground state transforms into another ground state under a symmetry operation of the Hamiltonian. A ground state of the ferromagnet is a definite spin state [Fig.4.2(a)]. Under a rotation it is transformed into another ground state, which is another spin state [Fig.4.2(b)]. Thus, the ground state is not invariant under rotation though the Hamiltonian is invariant. Since all these states have the same energy, the direction of the magnetization is not determined by the Hamiltonian but chosen spontaneously.

We take the field-theoretical limit of the exchange energy (4.1.1). Since the exchange interaction is short-ranged, we assume the nearest-neighbor interaction, for which we set $J_s \equiv J_{ij}$. Let the lattice point be specified by lattice vector $\boldsymbol{a}^\alpha$ with $\sum_\alpha \boldsymbol{a}^\alpha = 0$. Since the exchange interaction is short ranged, we make the Taylor

[1] Let us assume that all lattice points are filled with up-spin electrons. We are able to add an electron with down spin. When it is added to a filled lattice point i, the spin of this site becomes singlet $[\boldsymbol{S}(i) = 0]$ and hence the spin part $\boldsymbol{S}(i) \cdot \boldsymbol{S}(j)$ of the Hamiltonian decreases. However, the decrease is precisely cancelled by the increase of the number part $\rho(i)\rho(j)$. Physically, it implies that no exchange energy is needed to put an electron to the empty down-spin level.

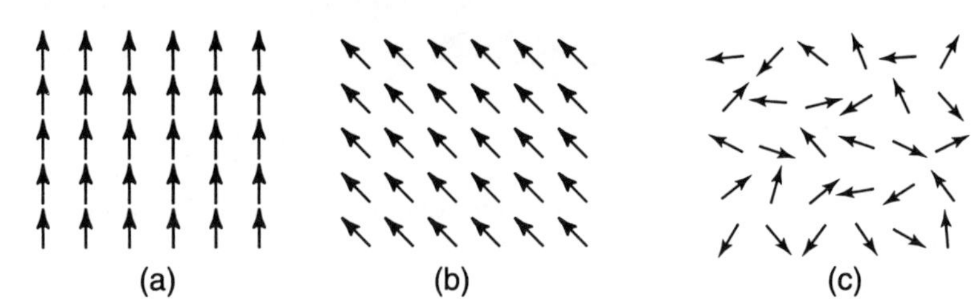

Fig. 4.2 All the spins are aligned in a ferromagnet. The direction of magnetization is chosen spontaneously. Two states (a) and (b) have the same energy and are related by a global spin rotation. Above the critical temperature each spin takes an arbitrary direction due to thermal fluctuations, as in the state (c).

expansion of the spin field and keep the nontrivial lowest-order term,

$$\sum_{\langle ij\rangle} \boldsymbol{S}(i)\cdot\boldsymbol{S}(j) = \frac{1}{2}\sum_{\boldsymbol{x}}\sum_{\alpha} \boldsymbol{\mathcal{S}}(\boldsymbol{x})\cdot\boldsymbol{\mathcal{S}}(\boldsymbol{x}+\boldsymbol{a}^{\alpha})$$

$$\simeq \frac{1}{2}\sum_{\boldsymbol{x}}\sum_{\alpha}\left[\boldsymbol{\mathcal{S}}(\boldsymbol{x})^2 - \frac{1}{2}a_i^{\alpha}a_j^{\alpha}\partial_i\boldsymbol{\mathcal{S}}(\boldsymbol{x})\cdot\partial_j\boldsymbol{\mathcal{S}}(\boldsymbol{x})\right], \tag{4.1.3}$$

where a partial integration was made. This approximation is valid for sufficiently smooth spin field $\boldsymbol{\mathcal{S}}(\boldsymbol{x})$. Thus, the resulting Hamiltonian describes phenomena correctly whose typical size is sufficiently larger than the spacing a. The exchange Hamiltonian (4.1.1) yields

$$H_{\rm X} \simeq J_s\sum_{\alpha}\left(a_i^{\alpha}a_j^{\alpha}\sum_{\boldsymbol{x}}\partial_i\boldsymbol{\mathcal{S}}(\boldsymbol{x})\cdot\partial_i\boldsymbol{\mathcal{S}}(\boldsymbol{x}) - N\right), \tag{4.1.4}$$

where $\boldsymbol{\mathcal{S}}(\boldsymbol{x})$ is the spin field normalized as $\boldsymbol{\mathcal{S}}^2 = 1/4$:[2] See also (30.2.30). The parameter J_s is called the spin stiffness.

We consider a simple cubic lattice as a simplest example [Fig.4.1], where $\sum_{\alpha} a_i^{\alpha}a_j^{\alpha} = 2a^2\delta_{ij}$ and $\sum_{\boldsymbol{x}} = a^{-m}\int d^m x$ in the m-dimensional space. The Hamiltonian (4.1.1) yields the O(3) nonlinear sigma model[3]

$$H_{\rm X} = \frac{2J_s}{a}\int d^3x\, \partial_i\boldsymbol{\mathcal{S}}(x)\cdot\partial_i\boldsymbol{\mathcal{S}}(x) - \Delta_{\rm X}N \tag{4.1.5a}$$

---

[2] It is suggested that $\boldsymbol{\mathcal{S}}^2 = 3/4$ as a Casimir invariant. However, here we should set $\boldsymbol{\mathcal{S}}^2 = 1/4$ since it arises from a continuum limit of $\boldsymbol{S}(i)\cdot\boldsymbol{S}(j) = 1/4$.

[3] It is also known as the SU(2) sigma model[124] or the CP^1 sigma model.[110] Two groups O(3) and SU(2) are equivalent locally. More precisely, O(3)=SU(2)/Z_2, where Z_2 is the center of SU(2). The symmetry group of the spin system is O(3), but the symmetry group of the two-component (up-spin and down-spin) electron system is SU(2). See Section 7.7 and Section 7.8 for related topics.

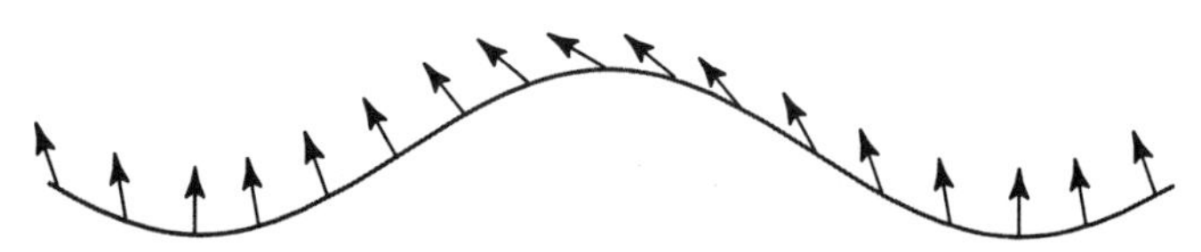

Fig. 4.3 A spin wave induces a smooth modulation of the spin direction in a ferromagnet. As the smooth modulation becomes infinitesimal, the energy becomes infinitesimal, as implies that the spin waveis a gapless mode.

with $\Delta_X = 6J_s$ in the 3-dimensional space, and

$$H_X = 2J_s \int d^2x\, \partial_i \boldsymbol{\mathcal{S}}(x) \cdot \partial_i \boldsymbol{\mathcal{S}}(x) - \Delta_X N \tag{4.1.5b}$$

with $\Delta_X = 4J_s$ in the 2-dimensional space. The normalized spin field $\boldsymbol{\mathcal{S}}(\boldsymbol{x})$ is called the nonlinear sigma field or simply the sigma field. In general, the Hamiltonian (4.1.5a) or (4.1.5b) follows in any lattice model, provided the system has translational and rotational invariance at a long-distance scale.

We proceed to consider a spin wave of long wavelength [Fig.4.3]. It induces a smooth modulation of the spin direction over the sample. When a modulation is sufficiently smooth and infinitesimal, it is clear that the associated energy (4.1.5) is also infinitesimal. Consequently the spin wave is a gapless mode. This is a general phenomenon associated with spontaneous symmetry breaking of a global continuous symmetry, known as the Goldstone theorem.[199,368,200] The Goldstone theorem assures the existence of a gapless mode called the *Goldstone mode*, when a global continuous symmetry is spontaneously broken.

Spontaneous symmetry breaking is a central issue of modern physics, since it is accompanied by various physical phenomena such as Bose condensation, quantum coherence and additionally topological excitations in some systems.

There exists a subtle theorem on spontaneous symmetry breaking in lower dimensional spaces, known as the Mermin-Wagner theorem.[345,346] Quantum fluctuations of a gapless mode, when compounded by thermal fluctuations, are so violent that the condensation is destroyed for $T \geq 0$ in the 1-dimensional space and for $T > 0$ in the 2-dimensional space. Hence, there exists no spontaneous symmetry breaking in these cases. See Appendix J for details on the theorem.

The nonlinear sigma model is very important since it describes quantum coherence in QH systems. We use it to analyze QH ferromagnets together with topological excitations in monolayer systems (Part II), and to describe an interlayer coherence together with topological excitations in bilayer systems (Part III). See Appendix J for discussions how we are allowed to use these concepts in the 2-dimensional space in spite of the Mermin-Wagner theorem. A main theme of Part IV is to derive the nonlinear sigma model in multicomponent QH systems. In the rest of this chapter we study simple field-theoretical models for the Bose condensation and the spontaneous symmetry breaking.

4.2 Real Klein-Gordon Field

The simplest model showing Bose condensation and spontaneous symmetry breaking is given by

$$\mathcal{L} = -\frac{\hbar^2}{2}(\partial^\mu\phi)(\partial_\mu\phi) + \frac{\gamma}{2}\phi^2 - \frac{g}{4}\phi^4, \tag{4.2.1}$$

where $\phi(x)$ is the real Klein-Gordon field, and γ is just a real parameter. When $\gamma < 0$ this is the real Klein-Gordon model (3.3.17) with a contact interaction term $\propto \phi^4$ included additionally [see (2.4.12)]. The Lagrangian is invariant under a discrete transformation,

$$\phi(x) \longrightarrow -\phi(x). \tag{4.2.2}$$

When $\gamma > 0$, as we shall see, this is a model where Bose condensation occurs without a gapless mode.

The Hamiltonian density is given by

$$\mathcal{H} = \frac{\hbar^2}{2c^2}(\dot{\phi})^2 + \frac{\hbar^2}{2}(\nabla\phi)^2 - \frac{\gamma}{2}\phi^2 + \frac{g}{4}\phi^4. \tag{4.2.3}$$

The interaction should be repulsive ($g > 0$). Otherwise, the system is ill defined because the energy is unbounded from below: The energy is decreased without limit by increasing the number of ϕ quanta.

In the absence of interaction ($g = 0$) and if $\gamma < 0$, it describes free Bose particles with mass $m_\phi \equiv \sqrt{|\gamma|}/c$. The Fock vacuum $|0\rangle$ is defined by $\langle 0|\phi|0\rangle = 0$, or

$$a_{\mathbf{k}}|0\rangle = 0, \tag{4.2.4}$$

where $a_{\mathbf{k}}$ is the annihilation operator defined in the momentum expansion (3.3.26). It is the nondegenerate state with the minimum energy. The ground state $|0\rangle$ is invariant under the transformation (4.2.2). This is true even if we switch on a small interaction ($g > 0$) provided $\gamma < 0$, as illustrated in Fig.4.4(a).

We proceed to the case where $g > 0$ and $\gamma > 0$. Note that $\gamma\phi^2$ is no longer the mass term of the ϕ field in the Lagrangian (4.2.1). In this case the energy of the system is minimized with a certain density of ϕ quanta present in the ground state, $\langle 0|\phi|0\rangle \neq 0$. The ground state $|0\rangle$ is no longer given by the state $|\underline{0}\rangle$ satisfying

$$a_{\mathbf{k}}|\underline{0}\rangle = 0, \tag{4.2.5}$$

where $\langle\underline{0}|\phi|\underline{0}\rangle = 0$. It is a "false vacuum". Fluctuations around $\phi = 0$ are unstable [Fig.4.4(b)].

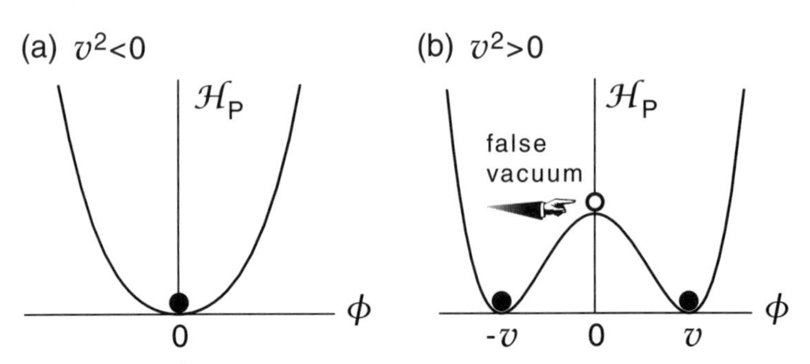

Fig. 4.4 The potential of the real scalar model (4.2.1) is illustrated. (a) When $\gamma < 0$ the ground state is unique with the vanishing number density, $\langle\phi^\dagger\phi\rangle = 0$. (b) When $\gamma > 0$ there are two classical ground states ($\phi = \pm v$) with nonzero number density, $\langle\phi^\dagger\phi\rangle \equiv v^2 \neq 0$, where Bose condensation occurs. Since one of them is chosen as a ground state, the discrete symmetry (4.2.2) is spontaneously broken. The point $\phi = 0$ corresponds to the "false vacuum" (4.2.5). A fluctuation is unstable around $\phi = 0$, but stable around $\phi = v$.

To find the ground state we rewrite the potential term as

$$\mathcal{H}_\text{P} = -\frac{\gamma}{2}\phi^2 + \frac{g}{4}\phi^4 = \frac{g}{4}(\phi^2 - v^2)^2 - \frac{\gamma^2}{4g}, \tag{4.2.6}$$

where

$$v = \sqrt{\frac{\gamma}{g}}. \tag{4.2.7}$$

This potential is called the *Higgs potential*, and is depicted in Fig.4.4(b). The Hamiltonian is minimized by the choice of $\phi = \pm v$. We call $\phi = \pm v$ the classical vacuum. It is necessary to choose one of them to carry out canonical quantization. We choose $\phi = v$ without loss of generality, and analyze perturbative fluctuations around it by setting

$$\phi(x) = v + \eta(x). \tag{4.2.8}$$

We substitute it into the Hamiltonian density (4.2.3),

$$\mathcal{H} = \frac{\hbar^2}{2c^2}(\dot{\eta})^2 + \frac{\hbar^2}{2}(\nabla\eta)^2 + \frac{g}{4}(\eta^2 + 2v\eta)^2 - \frac{\gamma^2}{4g}. \tag{4.2.9}$$

It is free from unphysical modes, because the Hamiltonian density is positive semidefinite apart from a c-number term $-\gamma^2/4g$.

The absence of unphysical modes is intuitively clear. Since the classical vacuum minimizes the energy, any small deviation $\eta(\boldsymbol{x})$ from it increases the energy [Fig.4.4(b)], as implies that excitations are gapful. The present model does not contain a Goldstone mode.

Rewriting the Lagrangian density (4.2.1) in terms of the field η we obtain

$$\begin{aligned}\mathcal{L} &= -\frac{\hbar^2}{2}(\partial^\mu\eta)(\partial_\mu\eta) - \frac{g}{4}(\eta^2 + 2v\eta)^2 \\ &= -\frac{\hbar^2}{2}(\partial^\mu\eta)(\partial_\mu\eta) - \gamma\eta^2 - \frac{\gamma}{v}\eta^3 - \frac{\gamma}{4v^2}\eta^4,\end{aligned} \tag{4.2.10}$$

where the irrelevant c-number term has been discarded. The field $\eta(x)$ is a real scalar field obeying the canonical commutation relation. The Lagrangian describes an interacting system of one kind of particles with mass $m_\eta \equiv \sqrt{2\gamma}/c$. The terms involving η^3 and η^4 describe interactions.

We have quantized a small perturbation $\eta(x)$ about the classical vacuum $\phi(x) = v$. Quantization around the other classical vacuum $\phi(x) = -v$ gives the identical result. However, since one of the vacua is chosen, the system is no longer invariant under the discrete symmetry (4.2.2). The symmetry is spontaneously broken. We comment that it is possible to consider a system containing these two different vacua simultaneously. It is no longer a ground state but describes a topological object called kink [see Fig.7.4 (page 119)].

The free part of the Hamiltonian density is

$$\mathcal{H}_{\text{free}} = \frac{\hbar^2}{2c^2}(\dot{\eta})^2 + \frac{\hbar^2}{2}(\nabla\eta)^2 + \gamma\eta^2, \tag{4.2.11}$$

which agrees with the Hamiltonian density (3.3.19) with mass $m_\eta = \sqrt{2\gamma}/c$. The canonical commutation relations are

$$\begin{aligned}&[\eta(t,\boldsymbol{x}), \eta(t,\boldsymbol{y})] = [\dot{\eta}(t,\boldsymbol{x}), \dot{\eta}(t,\boldsymbol{y})] = 0, \\ &[\eta(t,\boldsymbol{x}), \dot{\eta}(t,\boldsymbol{y})] = i\frac{c^2}{\hbar}\delta(\boldsymbol{x} - \boldsymbol{y}).\end{aligned} \tag{4.2.12}$$

We make the momentum expansion of field $\eta(x)$ as in (3.3.26),

$$\eta(x) = c\int \frac{d^3k}{\sqrt{(2\pi)^3}}\frac{1}{\sqrt{2\hbar\omega_k}}\left(\eta_{\boldsymbol{k}}e^{ikx} + \eta_{\boldsymbol{k}}^\dagger e^{-ikx}\right) \tag{4.2.13}$$

with $\hbar\omega_k = c\sqrt{\hbar^2\boldsymbol{k}^2 + m_\eta^2c^2}$. The Hamiltonian is diagonalized,

$$H_{\text{free}} = \int d^3k\,\hbar\omega_k\left(\eta_{\boldsymbol{k}}^\dagger\eta_{\boldsymbol{k}} + \frac{1}{2}\delta(0)\right), \tag{4.2.14}$$

where

$$\langle 0|H_{\text{free}}|0\rangle = \delta(0)\int d^3k\frac{\hbar\omega_k}{2} = V\int\frac{d^3k}{(2\pi)^3}\frac{\hbar\omega_k}{2} \tag{4.2.15}$$

is the zero-point energy of the vacuum: See (3.3.34b) on this point. The ground state is the Fock vacuum $|0\rangle$ of the η quantum,

$$\eta_{\boldsymbol{k}}|0\rangle = 0, \tag{4.2.16}$$

which is to be compared with the false vacuum (4.2.5).

We decompose the fields ϕ and η into the positive and negative frequency components as in (3.3.28),

$$\phi^{(\pm)}(x) = \frac{1}{2}v + \eta^{(\pm)}(x), \tag{4.2.17}$$

where $\phi^{(+)\dagger} = \phi^{(-)}$. It follows from (4.2.16) that

$$\phi^{(+)}(x)|0\rangle = \frac{1}{2}v|0\rangle. \tag{4.2.18}$$

The ground state $|0\rangle$ is a coherent state since it is an eigenstate of the annihilation operator $\phi^{(+)}$. Integrating (4.2.18) over the volume of the system, $\int d^3x\, \phi^{(+)}(x)|0\rangle = \frac{1}{2}vV|0\rangle$ with $V \equiv \int d^3x$, we find

$$a_0|0\rangle = vV\sqrt{\frac{m_\eta}{2(2\pi)^3}}|0\rangle, \qquad a_{\boldsymbol{k}}|0\rangle = 0 \quad \text{for} \quad \boldsymbol{k} \neq 0. \tag{4.2.19}$$

It is actually a coherent state of the a_0 mode, i.e., the infrared component of the ϕ quantum. The coherent state is a Bose condensate, where the number of the quanta is proportional to the volume of the system.

It is convenient to characterize the ground state by the condition[4]

$$\langle 0|\phi(x)|0\rangle = v \neq 0 \tag{4.2.20}$$

rather than (4.2.18). This is to be contrasted with

$$\langle \underline{0}|\phi(x)|\underline{0}\rangle = 0, \tag{4.2.21}$$

which characterizes the "false vacuum" $|\underline{0}\rangle$.

The static Euler-Lagrange equation follows from (4.2.10),

$$\hbar^2\nabla^2\eta = 2\gamma\eta + 3\frac{\gamma}{v}\eta^2 + \frac{\gamma}{v^2}\eta^3. \tag{4.2.22}$$

As far as small fluctuations are concerned we may neglect the η^2 and η^3 terms,

$$\xi^2\nabla^2\eta - \eta \simeq 0, \tag{4.2.23}$$

where the parameter ξ is known as the *coherence length*,

$$\xi = \frac{\hbar}{\sqrt{2\gamma}} = \frac{\hbar}{c}\frac{1}{m_\eta}. \tag{4.2.24}$$

The physical meaning reads as follows. Due to the field equation (4.2.23), the two-point function decreases asymptotically as

$$\langle 0|\eta(\boldsymbol{x})\eta(\boldsymbol{y})|0\rangle = \langle 0|\eta(\boldsymbol{x}-\boldsymbol{y})\eta(0)|0\rangle \propto e^{-|\boldsymbol{x}-\boldsymbol{y}|/\xi}. \tag{4.2.25}$$

The correlation mediated by the η quantum decreases rapidly when the two points are spatially separated far more than the coherence length ξ.

---

[4] This condition is used to verify the Goldstone theorem, as is an essential result of spontaneous symmetry breaking: See (4.7.1).

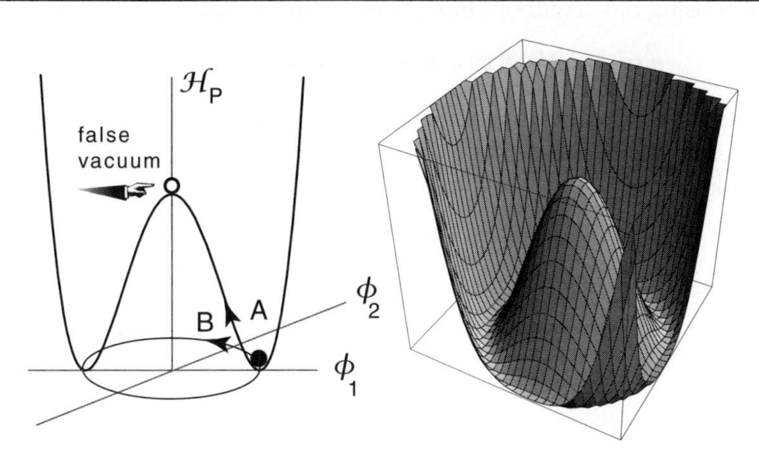

Fig. 4.5 The Higgs potential of the complex scalar model looks like the bottom of a wine bottle. The bottom is a circle, and each point on it represents a classical vacuum. There exist two oscillation modes indicated by arrows A and B. One mode A climbs a hill, representing a gapful mode since it costs an energy. The other mode B moves along the circle, representing a gapless mode since it costs no energy. The false vacuum is at the local maximum point of the potential at $\phi_1 = \phi_2 = 0$.

4.3 COMPLEX KLEIN-GORDON FIELD

We consider a relativistic model of the complex scalar field,

$$\mathcal{L} = -\hbar^2(\partial^\mu \phi^\dagger)(\partial_\mu \phi) + \gamma \phi^\dagger(x)\phi(x) - \frac{g}{2}\left(\phi^\dagger(x)\phi(x)\right)^2 \tag{4.3.1}$$

with a repulsive interaction $g > 0$. The Lagrangian is invariant under a phase transformation,

$$\phi(x) \longrightarrow e^{i\alpha}\phi(x) \tag{4.3.2}$$

with α a real constant. It is a global continuous transformation since α is an arbitrary constant. The Hamiltonian density is given by

$$\mathcal{H} = \frac{\hbar^2}{c^2}\dot{\phi}^\dagger\dot{\phi} + \hbar^2\nabla\phi^\dagger\nabla\phi - \gamma\phi^\dagger(x)\phi(x) + \frac{g}{2}\left(\phi^\dagger(x)\phi(x)\right)^2. \tag{4.3.3}$$

When $\gamma < 0$ it describes an interacting system of one kind of particles and antiparticles with mass $m_\phi \equiv \sqrt{|\gamma|}/c$.

We study the case with $\gamma > 0$. The Hamiltonian has a Higgs potential $\mathcal{H}_P$, as in Fig.4.5. It is minimized by the classical vacuum $\phi = e^{i\alpha}v$, where α is an arbitrary real constant, and $v = \sqrt{\gamma/g}$. There are infinitely many classical vacua labeled by α. To make canonical quantization it is necessary to choose one of the degenerate vacua. Choosing $\alpha = 0$ without loss of generality, we set

$$\phi(x) = e^{i\chi(x)/v}\,(v + \eta(x)) = v + \eta(x) + i\chi(x) + \cdots, \tag{4.3.4}$$

and substitute it into the Hamiltonian density. The Hamiltonian density becomes positive semidefinite apart from a trivial c-number $-\gamma^2/2g$. The Fock vacuum $|0\rangle$ of the η quantum satisfies,

$$\langle 0|\phi(x)|0\rangle = v \neq 0. \tag{4.3.5}$$

It is a coherent state. The global phase symmetry (4.3.2) is spontaneously broken by the ground state $|0\rangle$.

The Lagrangian density reads

$$\mathcal{L} = -\hbar^2(\partial^\mu\eta)(\partial_\mu\eta) - 2\gamma\eta^2 - \hbar^2(\partial^\mu\chi)(\partial_\mu\chi) + \frac{\gamma^2}{2g} + \mathcal{L}_{\text{int}}, \tag{4.3.6}$$

where $\mathcal{L}_{\text{int}}$ represents the interaction terms involving higher order terms in η. Fields $\eta(x)$ and $\chi(x)$ are two independent real scalar fields, each of which can be quantized canonically. Therefore, the Lagrangian describes an interacting system of one kind of particles with mass $m_\eta \equiv \sqrt{2\gamma}/c$ and one kind of massless particles.

The free part of the Hamiltonian is diagonalized,

$$H_{\text{free}} = \int d^3k E_\eta(\boldsymbol{k})\eta^\dagger_{\boldsymbol{k}}\eta_{\boldsymbol{k}} + \int d^3k E_\chi(\boldsymbol{k})\chi^\dagger_{\boldsymbol{k}}\chi_{\boldsymbol{k}}. \tag{4.3.7}$$

Their dispersion relations are

$$E_\eta(\boldsymbol{k}) = \hbar c\sqrt{2\gamma + \boldsymbol{k}^2}, \qquad E_\chi(\boldsymbol{k}) = \hbar c|\boldsymbol{k}|. \tag{4.3.8}$$

The χ field is a gapless mode with a linear dispersion relation. Such a mode is a superfluid mode,[21] as we shall see in Section 4.6.

The complex Klein-Gordon model gives the simplest example with a spontaneous symmetry breaking accompanied by a gapless mode.

4.4 Sigma Model

The complex Klein-Gordon model consists of a set of two real fields. We may consider a set of N real scalar fields, $\boldsymbol{\phi} = (\phi_1, \phi_2, \cdots, \phi_N)$, in the Higgs potential,

$$\mathcal{L} = -\frac{\hbar^2}{2}(\partial^\mu\boldsymbol{\phi})(\partial_\mu\boldsymbol{\phi}) - \frac{g}{4}(\boldsymbol{\phi}^2 - v^2)^2, \tag{4.4.1}$$

with positive constants g and v. This is invariant under a linear O(N) transformation of the $\boldsymbol{\phi}$ field,

$$\boldsymbol{\phi} \to U\boldsymbol{\phi}, \tag{4.4.2}$$

where U is a real $N \times N$ matrix obeying $U^T U = 1$. It is a global continuous transformation.

We substitute $\phi_a = v_a + \eta_a$ into the Hamiltonian density. When we choose

$$\sum_{a=1}^{N} v_a^2 = v^2, \tag{4.4.3}$$

the c-number part of the Hamiltonian density is minimized and the Hamiltonian density becomes positive semidefinite. Each choice of vector $\boldsymbol{v} = (v_1, v_2, \cdots, v_N)$ gives a classical vacuum. Ground states are infinitely degenerate. Without loss of generality we may choose

$$\langle 0|\phi_a(\mathbf{x})|0\rangle = v\delta_{aN}. \tag{4.4.4}$$

It is customary to write

$$\phi_N = v + \sigma, \qquad \phi_a = \pi_a \quad \text{for} \quad a = 1, \cdots, N-1. \tag{4.4.5}$$

The ground state $|0\rangle$ is the Fock vacuum with respect to the σ and π_a quanta, $\langle 0|\sigma|0\rangle = \langle 0|\pi_a|0\rangle = 0$.

Substituting (4.4.5) into the Lagrangian (4.4.1), we obtain,

$$\mathcal{L} = -\frac{\hbar^2}{2}(\partial^\mu \boldsymbol{\pi})(\partial_\mu \boldsymbol{\pi}) - \frac{\hbar^2}{2}\partial^\mu\sigma\partial_\mu\sigma - \frac{m_\sigma^2 c^2}{2}\sigma^2 + \cdots, \tag{4.4.6}$$

where the σ quantum has a mass $m_\sigma = \sqrt{2g}v/c$ but the π_a quantum is massless. Although the Hamiltonian has the O(N) symmetry, it is spontaneously broken by the ground state. Fields π_a describe Goldstone bosons associated with the spontaneous symmetry breaking of the O(N) symmetry to the O($N-1$) symmetry. By choosing $N = 4$ we may identify three massless fields π_a with the *pi mesons* and the massive field σ with the *sigma meson* in nuclear theory.[184,25] With this choice the Lagrangian (4.4.1) gives an effective theory of the meson part of the low-energy nuclear theory. Because of the notation σ used for the sigma field and because of the linear O(N) invariance, the Lagrangian (4.4.1) is known as the O(N) linear sigma model.

It is interesting to take the limit $g \to \infty$ with v kept fixed in the Lagrangian (4.4.1). To keep the Lagrangian finite except for an irrelevant c-number term, it is necessary to impose

$$\boldsymbol{\phi}^2 = v^2, \tag{4.4.7}$$

and the Lagrangian is simplified as

$$\mathcal{L} = -\frac{\hbar^2}{2}(\partial^\mu \boldsymbol{\phi}) \cdot (\partial_\mu \boldsymbol{\phi}). \tag{4.4.8}$$

Ground states are infinitely degenerate as before. Without loss of generality we may choose $v_a = v\delta_{aN}$ to define the ground state $|0\rangle$ as in (4.4.4).

Due to the constraint (4.4.7) only $N-1$ fields are independent. We may choose $\pi_a \equiv \phi_a$, $a = 1, 2, \cdots, N-1$, as the independent fields, and solve (4.4.7) for

$$\phi_N = \sqrt{v^2 - \sum_{a=1}^{N-1} \phi_a^2}, \tag{4.4.9}$$

and rewrite the Lagrangian in terms of $N-1$ fields π_a. Field ϕ_N (or field σ) has been eliminated. It is physically reasonable since the σ meson has not been observed experimentally. The perturbative quantization is carried out by treating π_i as fluctuation fields. This model involves merely the pi mesons which are Goldstone bosons. Although there are only $N-1$ independent fields, the Lagrangian (4.4.8) has the symmetry group O(N): The $N-1$ fields transform nonlinearly under the O(N) transformation. This model is called the O(N) nonlinear sigma model.

By choosing $N = 3$ we may identify the field $\phi_i(\boldsymbol{x})$ as the normalized spin field $\mathcal{S}_i(\boldsymbol{x})$ in (4.1.5a). The choice (4.4.4) of the ground state reads,

$$\langle 0|\phi_1(\boldsymbol{x})|0\rangle = 0, \quad \langle 0|\phi_2(\boldsymbol{x})|0\rangle = 0, \quad \langle 0|\phi_3(\boldsymbol{x})|0\rangle = v, \tag{4.4.10}$$

as corresponds to the spontaneous magnetization into the 3rd axis.

4.5 Schrödinger Field

We finally consider the Schrödinger field in the Higgs potential. The Hamiltonian density is

$$\mathcal{H} = \frac{\hbar^2}{2m}\nabla\phi^\dagger(x)\nabla\phi(x) + \frac{g}{2}\left(\phi^\dagger\phi - v^2\right)^2, \tag{4.5.1}$$

where $g > 0$ and $v > 0$. The Hamiltonian has the global U(1) phase symmetry,

$$\phi(x) \longrightarrow e^{i\alpha}\phi(x), \tag{4.5.2}$$

with α an arbitrary real constant. The classical Hamiltonian is minimized by

$$\rho \equiv \phi^\dagger\phi = v^2, \tag{4.5.3}$$

where ρ is the number density of the ϕ quantum. This is the condition that the chemical potential of the system vanishes,

$$\mu \equiv \frac{\partial\mathcal{H}}{\partial\rho} = g\left(\phi^\dagger\phi - v^2\right) = 0. \tag{4.5.4}$$

We may interpret $\rho_0 \equiv v^2$ as the average density. There are many classical vacua, $\phi = e^{i\alpha}v$, among which we may choose $\phi = v$ without loss of generality. Vacuum $|0\rangle$ is a Bose condensate. The choice violates the U(1) symmetry (4.5.2) of the Hamiltonian spontaneously.

We consider a small fluctuation around the classical vacuum as before,

$$\phi(x) = v + \eta(x). \tag{4.5.5}$$

The canonical commutation relations are

$$[\eta(t,\boldsymbol{x}),\eta(t,\boldsymbol{y})]=[\eta(t,\boldsymbol{x}),\eta(t,\boldsymbol{y})]=0,$$
$$[\eta(t,\boldsymbol{x}),\eta^{\dagger}(t,\boldsymbol{y})]=\delta(\boldsymbol{x}-\boldsymbol{y}). \tag{4.5.6}$$

The Hamiltonian density reads

$$\mathcal{H}=\frac{\hbar^2}{2m}\nabla\eta^{\dagger}\nabla\eta+\frac{gv^2}{2}(2\eta^{\dagger}\eta+\eta^{\dagger 2}+\eta^2)+\mathcal{H}_{\text{int}}, \tag{4.5.7}$$

$$\mathcal{H}_{\text{int}}=gv\left(\eta^{\dagger}\eta(\eta^{\dagger}+\eta)+\frac{1}{2v}(\eta^{\dagger}\eta)^2\right). \tag{4.5.8}$$

We expand $\eta(x)$ in terms of plane waves,

$$\eta(x)=\int\frac{d^3k}{\sqrt{(2\pi)^3}}e^{-ikx}\eta_{\boldsymbol{k}}, \tag{4.5.9}$$

where $\eta_{\boldsymbol{k}}$ is the annihilation operator, obeying

$$[\eta_{\boldsymbol{k}},\eta_{\boldsymbol{l}}^{\dagger}]=\delta(\boldsymbol{k}-\boldsymbol{l}),\qquad [\eta_{\boldsymbol{k}},\eta_{\boldsymbol{l}}]=[\eta_{\boldsymbol{k}}^{\dagger},\eta_{\boldsymbol{l}}^{\dagger}]=0. \tag{4.5.10}$$

The Fock vacuum $|0\rangle$ satisfies

$$\eta_{\boldsymbol{k}}|0\rangle=0. \tag{4.5.11}$$

It is characterized by

$$\langle 0|\phi(x)|0\rangle=v,\qquad \langle 0|\rho(x)|0\rangle=v^2. \tag{4.5.12}$$

The "free" part of the Hamiltonian density reads

$$\mathcal{H}_{\text{free}}(\boldsymbol{k})=\varepsilon_{\boldsymbol{k}}\eta_{\boldsymbol{k}}^{\dagger}\eta_{\boldsymbol{k}}+U_{\boldsymbol{k}}\left(2\eta_{\boldsymbol{k}}^{\dagger}\eta_{\boldsymbol{k}}+\eta_{\boldsymbol{k}}^{\dagger}\eta_{-\boldsymbol{k}}^{\dagger}+\eta_{\boldsymbol{k}}\eta_{-\boldsymbol{k}}\right), \tag{4.5.13}$$

where

$$\varepsilon_{\boldsymbol{k}}=\frac{\hbar^2\boldsymbol{k}^2}{2m},\qquad U_{\boldsymbol{k}}=\frac{gv^2}{2}. \tag{4.5.14}$$

The Fock vacuum $|0\rangle$ is not the ground state when $v\neq 0$.

It is necessary to diagonalize the Hamiltonian (4.5.13). We use the Bogoliubov transformation,

$$\zeta_{\boldsymbol{k}}=c_{\boldsymbol{k}}\eta_{\boldsymbol{k}}+s_{\boldsymbol{k}}\eta_{-\boldsymbol{k}}^{\dagger},\qquad \zeta_{\boldsymbol{k}}^{\dagger}=c_{\boldsymbol{k}}\eta_{\boldsymbol{k}}^{\dagger}+s_{\boldsymbol{k}}\eta_{-\boldsymbol{k}}, \tag{4.5.15}$$

where

$$c_{\boldsymbol{k}}=\cosh\tau_{\boldsymbol{k}},\qquad s_{\boldsymbol{k}}=\sinh\tau_{\boldsymbol{k}}. \tag{4.5.16}$$

It is a canonical transformation since $\zeta_{\boldsymbol{k}}$ and $\zeta_{\boldsymbol{k}}^{\dagger}$ satisfy

$$[\zeta_{\boldsymbol{k}},\zeta_{\boldsymbol{l}}^{\dagger}]=\delta(\boldsymbol{k}-\boldsymbol{l}),\qquad [\zeta_{\boldsymbol{k}},\zeta_{\boldsymbol{l}}]=[\zeta_{\boldsymbol{k}}^{\dagger},\zeta_{\boldsymbol{l}}^{\dagger}]=0. \tag{4.5.17}$$

The inverse transformation is

$$\eta_k = c_k\zeta_k - s_k\zeta^\dagger_{-k}, \qquad \eta^\dagger_k = c_k\zeta^\dagger_k - s_k\zeta_{-k}. \tag{4.5.18}$$

We substitute this into the Hamiltonian (4.5.13),

$$\begin{aligned}\mathcal{H}_{\text{free}}(\boldsymbol{k}) =&[(\varepsilon_k + 2U_k)(c_k^2 + s_k^2) - 4U_kc_ks_k]\zeta^\dagger_k\zeta_k \\ &+ [U_k(c_k^2 + s_k^2) - (\varepsilon_k + 2U_k)c_ks_k](\zeta^\dagger_k\zeta^\dagger_{-k} + \zeta_k\zeta_{-k}),\end{aligned} \tag{4.5.19}$$

apart from an additive c-number constant. In deriving this formula we have used the fact that ε_k, U_k, c_k and s_k are even functions of $\boldsymbol{k}$.

The Hamiltonian density (4.5.19) has a diagonalized form provided that

$$U_k(c_k^2 + s_k^2) - (\varepsilon_k + 2U_k)c_ks_k = 0. \tag{4.5.20}$$

Solving this equation we determine the coefficients in (4.5.15) as

$$c_k^2 = \frac{1}{2}\left(\frac{\varepsilon_k + 2U_k}{E_k} + 1\right), \qquad s_k^2 = \frac{1}{2}\left(\frac{\varepsilon_k + 2U_k}{E_k} - 1\right), \qquad c_ks_k = \frac{U_k}{E_k}, \tag{4.5.21}$$

where

$$E_k = \sqrt{\varepsilon_k^2 + 4\varepsilon_kU_k}. \tag{4.5.22}$$

By substituting these values into (4.5.19) the Hamiltonian reads,

$$H_{\text{free}} = \int d^3kE_k\left(\zeta^\dagger_k\zeta_k + \frac{1}{2}\delta(0)\right). \tag{4.5.23}$$

The ground state $|0\rangle\!\rangle$ is the Fock vacuum of the ζ quantum,

$$\zeta_k|0\rangle\!\rangle = 0. \tag{4.5.24}$$

It is a squeezed state with the squeezing factor e^{τ_k}, as we see in (4.5.35).

The dispersion relation exhibits a linear dependence,

$$E_k = \hbar|\boldsymbol{k}|\sqrt{\frac{gv^2}{m} + \frac{\hbar^2\boldsymbol{k}^2}{4m^2}} \simeq \sqrt{\frac{g}{m}}v\hbar|\boldsymbol{k}|, \tag{4.5.25}$$

for $|\boldsymbol{k}| \simeq 0$. It is to be contrasted with the standard dispersion relation $E_k \propto \boldsymbol{k}^2$ of the nonrelativistic particles. Hence, as far as the low-energy component is concerned, the excitation has the relativistic dispersion relation (3.3.40) with the velocity $C = v\sqrt{g/m}$. The mode with a linear dispersion relation is a superfluid mode,[21] as will be discussed in Section 4.6.

The Fock vacuum $|0\rangle$ of the η quantum and the Fock vacuum $|0\rangle\!\rangle$ of the ζ quantum, defined by (4.5.11) and (4.5.24), are related one another by a unitary operator e^{iG},

$$\zeta_k = e^{-iG}\eta_ke^{iG}, \qquad \zeta^\dagger_k = e^{-iG}\eta^\dagger_ke^{iG}. \tag{4.5.26}$$

It follows from (4.5.24) that $\eta_k e^{iG}|0\rangle\!\rangle = 0$. Comparing this with (4.5.11), we obtain

$$|0\rangle\!\rangle = e^{-iG}|0\rangle, \tag{4.5.27}$$

since the state $|0\rangle$ is unique. The Hermitian operator G is given by

$$G = \frac{i}{2}\int d^3k\, \vartheta_k(\eta_k\eta_{-k} - \eta_k^\dagger\eta_{-k}^\dagger), \tag{4.5.28}$$

as is shown in Appendix G. Compare (4.5.26) ~ (4.5.28) with (1.6.3) ~ (1.6.5).

We can derive the diagonalized Hamiltonian (4.5.23) also in the following way.[5] We parametrize the Schrödinger field $\phi(x)$ as

$$\phi(x) = v\sqrt{1+\sigma(x)}e^{i\vartheta(x)/2} \simeq v\left(1 + \frac{\sigma(x) + i\vartheta(x)}{2}\right), \tag{4.5.29}$$

by introducing the phase field $\vartheta(x)$ and the normalized density $\sigma(x)$, or

$$\eta(x) \simeq \frac{v}{2}\left(\sigma(x) + i\vartheta(x)\right). \tag{4.5.30}$$

The canonical commutation relation (4.5.10) reads

$$\frac{v^2}{2}[\sigma_k, \vartheta_l] = i\delta(\boldsymbol{k}+\boldsymbol{l}), \tag{4.5.31}$$

while the Hamiltonian density (4.5.13) reads

$$\begin{aligned}\mathcal{H}_{\text{free}}(\boldsymbol{k}) &= \frac{v^2\hbar^2\boldsymbol{k}^2}{8m}\vartheta_k^\dagger\vartheta_k + \frac{v^2}{2}\left(\frac{\hbar^2\boldsymbol{k}^2}{4m} + gv^2\right)\sigma_k^\dagger\sigma_k \\ &= \frac{1}{2M_k}\vartheta_k^\dagger\vartheta_k + \frac{M_k}{8}E_k^2v^4\sigma_k^\dagger\sigma_k,\end{aligned} \tag{4.5.32}$$

where $1/M_k = v^2\hbar^2\boldsymbol{k}^2/4m$, and E_k is given by (4.5.25). This is diagonalized as

$$\mathcal{H}_{\text{free}}(\boldsymbol{k}) = E_k\left(\zeta_k^\dagger\zeta_k + \frac{1}{2}\delta(0)\right), \tag{4.5.33}$$

by setting

$$\zeta_k = \frac{v}{2}\left(\sqrt{G_k}\sigma_k + i\frac{1}{\sqrt{G_k}}\vartheta_k\right), \quad \zeta_k^\dagger = \frac{v}{2}\left(\sqrt{G_k}\sigma_k^\dagger - i\frac{1}{\sqrt{G_k}}\vartheta_k^\dagger\right) \tag{4.5.34}$$

with $G_k = v^2M_kE_k/2$. The result agrees with the diagonalized Hamiltonian (4.5.23), and (4.5.34) is nothing but the Bogoliubov transformation (4.5.15).

We now show that the ground state $|0\rangle\!\rangle$ is a squeezed coherent state. By comparing (4.5.34) with (4.5.30), the squeezing factor is found to be [recall (1.6.13) and (1.6.12)]

$$e^{\tau_k} = \sqrt{G_k} = \left(1 + 4\frac{U_k}{\varepsilon_k}\right)^{1/4} = \left(1 + \frac{4m\gamma}{\hbar^2\boldsymbol{k}^2}\right)^{1/4}. \tag{4.5.35}$$

[5] These formulas will be used to analyze a Goldstone mode in bilayer QH systems (see Section 24.3).

The phase and density uncertainties are

$$\langle\Delta\vartheta_k\rangle = \sqrt{\frac{(2\pi)^3}{V}}\sqrt{M_k}\langle\Delta p\rangle = \frac{1}{v}\sqrt{\frac{(2\pi)^3}{V}}e^{2T_k},$$
$$\langle\Delta\sigma_k\rangle = \sqrt{\frac{(2\pi)^3}{V}}\frac{2\langle\Delta q\rangle}{\hbar v^2\sqrt{M_k}} = \frac{1}{v}\sqrt{\frac{(2\pi)^3}{V}}e^{-2T_k}, \tag{4.5.36}$$

where use was made of (1.6.14). In a thermodynamical limit ($V \to \infty$), the uncertainties vanishes for $\boldsymbol{k} \neq 0$. However, in the ground state ($\boldsymbol{k} \to 0$), the density has a fixed value ($\langle\Delta\sigma_k\rangle = 0$) but the phase is totally ambiguous ($\langle\Delta\vartheta_k\rangle = \infty$). A squeezing occurs due to the term $(\gamma v^2/8)\sigma_k^\dagger\sigma_k$ in the Hamiltonian (4.5.32), since it introduces an imbalance between the conjugate fields ϑ and σ.

4.6 SUPERFLUIDITY

The ground state is given by $\langle\phi(\boldsymbol{x})\rangle = v$ in the Schrödinger model (4.5.1). Let us suppose that the Bose field has a $\boldsymbol{x}$-dependent phase,

$$\langle\phi(\boldsymbol{x})\rangle = ve^{i\chi(\boldsymbol{x})}. \tag{4.6.1}$$

This is a higher energy state with the energy density $\langle\mathcal{H}\rangle = (\hbar^2 v^2/2m)(\nabla\chi)^2$. The current (3.5.12) flows to decrease the energy by reducing the phase inhomogeneity,

$$\langle\boldsymbol{j}(\boldsymbol{x})\rangle = -\frac{\hbar v^2}{m}\nabla\chi(\boldsymbol{x}). \tag{4.6.2}$$

Let us call it the phase current.[6] We now show that it is a superfluid.

We consider a particle in the Bose condensate. Viscosity never emerges if Goldstone bosons never interact with any particle.[21] In such a case it is a superfluid.

A model Hamiltonian consists of the nonrelativistic particle and the Goldstone boson (called phonon) together with their interaction,

$$H = \int d^3K\frac{(\hbar\boldsymbol{K})^2}{2M}a_{\boldsymbol{K}}^\dagger a_{\boldsymbol{K}} + \int d^3kE_{\boldsymbol{k}}\zeta_{\boldsymbol{k}}^\dagger\zeta_{\boldsymbol{k}} + H_{\text{int}} \tag{4.6.3}$$

with M the mass of the particle and $E_{\boldsymbol{k}} = C\hbar|\boldsymbol{k}|$ for the phonon energy. The simplest interaction is given by

$$H_{\text{int}} = g\int d^3x\,\phi^\dagger(x)\phi(x)\zeta(x) \tag{4.6.4}$$

in terms of the Schrödinger field $\phi(x)$ and the phonon field $\zeta(x)$. Substituting the momentum expansions (3.2.15) for the Schrödinger field and (3.3.26) for the phonon field we find that

$$H_{\text{int}} = \frac{gC}{\sqrt{(2\pi)^3}}\int\frac{d^3Kd^3k}{\sqrt{2E_{\boldsymbol{k}}}}(a_{\boldsymbol{K}-\boldsymbol{k}}^\dagger a_{\boldsymbol{K}}\zeta_{\boldsymbol{k}}^\dagger + a_{\boldsymbol{K}+\boldsymbol{k}}^\dagger a_{\boldsymbol{K}}\zeta_{\boldsymbol{k}}), \tag{4.6.5}$$

---

[6]The phase current plays an important role in the interlayer-coherent state in the bilayer QH system.

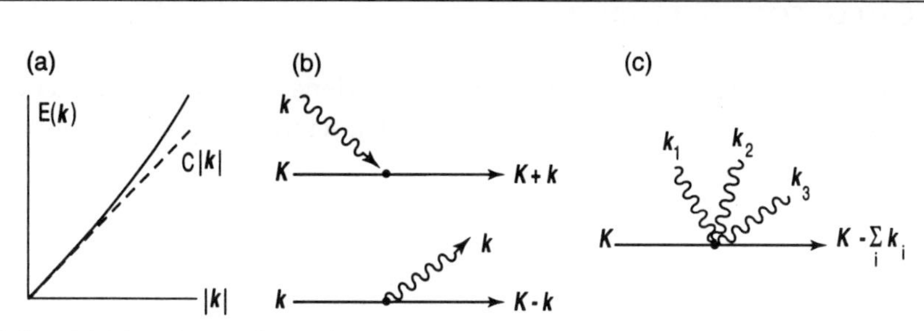

Fig. 4.6 (a) The phonon has a dispersion relation E_k in a medium, which is well approximated by a linear behavior $E_k \simeq C\hbar|\boldsymbol{k}|$. (b) A particle loses its velocity when it emits a phonon. This process is forbidden when the particle velocity is less than the critical one. (c) A multiple scattering is also forbidden when the particle velocity is less than the critical one.

which describes one-phonon emission and absorption [Fig.4.6(b)].

Although we have written down the Hamiltonian, we do not need its details for the following discussion. We only use the energy-momentum conservation,

$$\frac{\hbar^2 \boldsymbol{K}^2}{2M} = \frac{\hbar^2(\boldsymbol{K} \pm \boldsymbol{k})^2}{2M} + C\hbar|\boldsymbol{k}|, \tag{4.6.6}$$

for the phonon emission ($-\boldsymbol{k}$) and the phonon absorption ($+\boldsymbol{k}$). This amounts to

$$\frac{C|\boldsymbol{k}|}{\hbar} = \frac{\pm 2\boldsymbol{K}\boldsymbol{k} - \boldsymbol{k}^2}{2M} \leq \frac{V|\boldsymbol{k}|\cos\theta}{\hbar} \leq \frac{V|\boldsymbol{k}|}{\hbar}, \tag{4.6.7}$$

or $C \leq V$, where $V = \hbar|\boldsymbol{K}|/M$ is the initial velocity of the particle. Namely, the phonon emission is forbidden if the velocity V of the particle is less than the critical velocity C.

We have considered the simplest interaction [Fig.4.6(b)]. For a multiple scattering of phonons [Fig.4.6(c)], we can prove that

$$C\sum|\boldsymbol{k}_j| \leq V\sum|\boldsymbol{k}_j|, \tag{4.6.8}$$

from which we get the same condition for the critical velocity. Since the phonon emission is impossible, the particle makes no interaction in the medium (Bose condensate). Namely, the medium exhibits superfluidity for a particle whose velocity is less than the critical one.

Comments are in order. First, if the dispersion relation is $E_k \propto |\boldsymbol{k}|^2$ instead of the linear one, the phonon emission is possible for a particle with arbitrary small velocity. The medium exhibits no superfluidity. Second, it exhibits superfluidity if all excitations of the medium are gapful. This is the case for the supercurrent where the Goldstone boson disappears by the Anderson-Higgs mechanism (see Section 5.6).

4.7 Goldstone Theorem

We have studied four models where Bose condensation and spontaneous symmetry breaking occur. Bose condensation is characterized by the property of the ground state,

$$\langle 0|\phi(x)|0\rangle = v \neq 0, \tag{4.7.1}$$

which contrasts with the false vacuum state $|\underline{0}\rangle$,

$$\langle \underline{0}|\phi(x)|\underline{0}\rangle = 0. \tag{4.7.2}$$

There are gapless modes in three of the models: The gapless mode is the excitation mode whose energy vanishes as the momentum goes to zero. It is a characteristic feature of these three models that the Hamiltonian is invariant under a *continuous* global transformation. In the Schrödinger model and the complex Klein-Gordon model it is the U(1) phase transformation,

$$\phi(x) \to e^{i\alpha}\phi(x), \tag{4.7.3}$$

where α is a real constant. In the sigma model it is the O(N) rotation.

It is a generic phenomenon, known as the Goldstone theorem, that spontaneous symmetry breaking of a continuous symmetry is accompanied by a gapless mode. The gapless mode is called the Goldstone boson.

To make a formal study of spontaneous symmetry breaking we take an example of the phase transformation. Let Q be its generator,

$$e^{-i\alpha Q}\phi(x)e^{i\alpha Q} = e^{i\alpha}\phi(x). \tag{4.7.4}$$

For an infinitesimal global transformation ($|\alpha| \ll 1$), we get

$$(1 - i\alpha Q)\phi(x)(1 + i\alpha Q) = (1 + i\alpha)\phi(x), \tag{4.7.5}$$

or

$$[\phi(x), Q] = \phi(x). \tag{4.7.6}$$

Operator Q is the conserved charge of the Nöther current (3.5.8) associated with the phase transformation (4.7.3),

$$Q = \frac{1}{c}\int d^3x j^0(x). \tag{4.7.7}$$

Explicit expressions are given by (3.5.11) and (3.5.12) for the complex Klein-Gordon model and the Schrödinger model, respectively.

The key observation is that the ground state $|0\rangle$ is not invariant under the symmetry operation. As illustrated in Fig.4.7, the state is transformed into another state,

$$|\alpha\rangle \equiv e^{i\alpha Q}|0\rangle \neq |0\rangle, \tag{4.7.8}$$

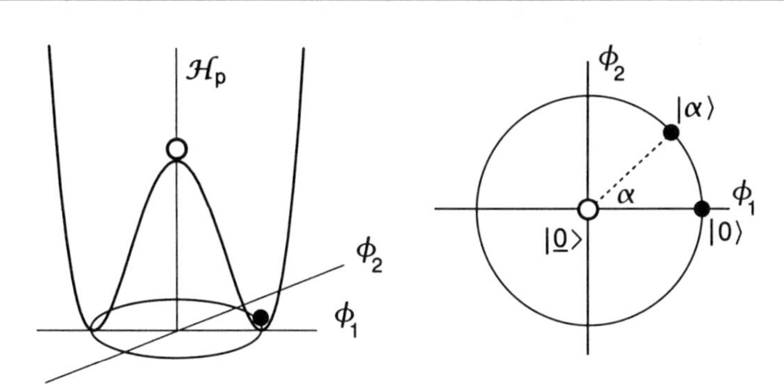

Fig. 4.7 After Bose condensation there are degenerate ground states $|\alpha\rangle$ labeled by a continuous parameter α. On the contrary the false vacuum $|\underline{0}\rangle$ is unique.

because

$$\langle\alpha|\phi(x)|\alpha\rangle = \langle 0|e^{-i\alpha Q}\phi(x)e^{i\alpha Q}|0\rangle = e^{i\alpha}\langle 0|\phi(x)|0\rangle = e^{i\alpha}v. \tag{4.7.9}$$

It follows from (4.7.8) that

$$Q|0\rangle \neq 0. \tag{4.7.10}$$

This contrasts with the condition of the false vacuum,

$$Q|\underline{0}\rangle = 0. \tag{4.7.11}$$

Equation (4.7.10) is the formal statement of spontaneous symmetry breaking. Namely, the symmetry is spontaneously broken when its generator Q does not annihilate the vacuum.

Let us prove the Goldstone theorem by using the above example though the proof is general. From (4.7.6) we obtain

$$\int d^3y\langle 0|[\phi(x), j^0(y)]|0\rangle = \langle 0|\phi(x)|0\rangle = v \neq 0. \tag{4.7.12}$$

This equation reads

$$I_+ - I_- = v, \tag{4.7.13}$$

where

$$I_+ = \int d^3y\langle 0|\phi(x)j^0(y)|0\rangle, \qquad I_- = \int d^3y\langle 0|j^0(y)\phi(x)|0\rangle. \tag{4.7.14}$$

Note that they vanish if the symmetry is not spontaneously broken, $I_+ = I_- = 0$, due to the property (4.7.11). We substitute the completeness condition

$$\int d\xi \int dE \int d^3k|E, \boldsymbol{k}, \xi\rangle\langle E, \boldsymbol{k}, \xi| = 1, \tag{4.7.15}$$

into I_+ and I_-. Here, E and $\boldsymbol{k}$ are the total energy and momentum of the state, and ξ denotes all the other quantum numbers (discrete or continuous) labeling the state. Now, the translational invariance implies

$$\phi(x) = e^{-ipx}\phi(0)e^{ipx}, \qquad j^0(y) = e^{-ipy}j^0(0)e^{ipy}. \tag{4.7.16}$$

Hence, we find

$$\begin{aligned} I_+ &= \int d\xi \int dE \int d^3k \int d^3y \langle 0|\phi(x)|E,\boldsymbol{k},\xi\rangle\langle E,\boldsymbol{k},\xi|j^0(y)|0\rangle \\ &= \int dE \int d^3k \int d^3y e^{ik(x-y)-iE\Delta t}\sigma_+(\boldsymbol{k},E) \end{aligned} \tag{4.7.17}$$

with $\Delta t = t_x - t_y$. Here,

$$\sigma_+(\boldsymbol{k},E) \equiv \int d\xi \langle 0|\phi(0)|E,\boldsymbol{k},\xi\rangle\langle E,\boldsymbol{k},\xi|j^0(0)|0\rangle \tag{4.7.18}$$

is called the spectral function. We have used that the energy and momentum are zero on $|0\rangle$. Integrating over $\boldsymbol{y}$ and $\boldsymbol{k}$ in (4.7.17) we find

$$I_\pm = (2\pi)^3 \int dE e^{\mp iE\Delta t}\sigma_\pm(\boldsymbol{k}=0,E), \tag{4.7.19}$$

where we have made a similar analysis also for I_-. Making a Fourier inversion of (4.7.13) with (4.7.19), we get

$$\sigma_+(\boldsymbol{k}=0,E) - \sigma_-(\boldsymbol{k}=0,E) = \frac{v\delta(E)}{(2\pi)^3}. \tag{4.7.20}$$

Therefore, at least one of the spectral functions, $\sigma_+(\boldsymbol{k} = 0,E)$ or $\sigma_-(\boldsymbol{k} = 0,E)$, contains a term proportional $\delta(E)$. This implies that there exists a gapless mode ($E = 0$ when $\boldsymbol{k} = 0$) in the system.

CHAPTER 5

ELECTROMAGNETIC FIELD

The electromagnetic interaction is an essential part of condensed matter physics. It is invariant under a gauge transformation. To quantize the electromagnetic field it is necessary to fix a gauge. It interacts with the matter field via the minimal coupling. In the broken phase of the phase symmetry the electromagnetic field absorbs the Goldstone mode and gets massive, which is known as the Anderson-Higgs mechanism.

5.1 MAXWELL EQUATIONS

We review the classical field theory of electromagnetism: See also Section 2.6. The dynamics is governed by the Maxwell equations. The first set reads

$$\nabla \times \boldsymbol{E} = -\partial_t \boldsymbol{B}, \qquad \nabla \cdot \boldsymbol{B} = 0, \tag{5.1.1}$$

while the second set reads

$$\nabla \times \boldsymbol{H} = \partial_t \boldsymbol{D} + \boldsymbol{J}, \qquad \nabla \cdot \boldsymbol{D} = \rho_e. \tag{5.1.2}$$

Here, $\boldsymbol{E}$ and $\boldsymbol{H}$ are the electric and magnetic fields; $\boldsymbol{D}$ and $\boldsymbol{B}$ are the electric and magnetic flux densities,[1]

$$\boldsymbol{D} = \varepsilon \boldsymbol{E}, \qquad \boldsymbol{B} = \mu_{\mathrm{m}} \boldsymbol{H}, \tag{5.1.3}$$

with the dielectric constant ε and the magnetic permeability μ_{m}; $\rho_e(t, \boldsymbol{x})$ and $\boldsymbol{J}(t, \boldsymbol{x})$ are the electric charge density and the electric current density.

The first set of the Maxwell equations allow us to introduce the scalar potential φ and the vector potential $\boldsymbol{A}$ by

$$\boldsymbol{B} = \nabla \times \boldsymbol{A}, \qquad \boldsymbol{E} = -\partial_t \boldsymbol{A} - \nabla \varphi. \tag{5.1.4}$$

We rewrite these in a relativistic notation. The 4-dimensional potential and current are

$$A^\mu = (c^{-1}\varphi, \boldsymbol{A}), \qquad J^\mu = (c\rho_e, \boldsymbol{J}). \tag{5.1.5}$$

[1] For brevity we call $\boldsymbol{B}$ also the magnetic field.

Thus $A_0 = -\varphi/c$. The electromagnetic field tensor is

$$F_{\mu\nu} = \partial_\mu A_\nu - \partial_\nu A_\mu, \tag{5.1.6}$$

which is antisymmetric, $F_{\mu\nu} = -F_{\nu\mu}$. The formula (5.1.4) corresponds to

$$B_x = F_{yz}, \quad B_y = F_{zx}, \quad B_z = F_{xy}, \tag{5.1.7}$$

and

$$E_k = -\partial_t A_k - \partial_k \varphi = c(\partial_k A_0 - \partial_0 A_k) = cF_{k0}. \tag{5.1.8}$$

Equation (5.1.7) is equivalent to $F_{ij} = \varepsilon_{ijk} B_k$, where ε_{ijk} is the antisymmetric tensor with $\varepsilon_{123} = 1$. From the definition (5.1.6) an identity follows,

$$\partial_\mu F_{\nu\sigma} + \partial_\nu F_{\sigma\mu} + \partial_\sigma F_{\mu\nu} = 0, \tag{5.1.9}$$

which is nothing but the first set of the Maxwell equations (5.1.1). The second set of the Maxwell equations (5.1.2) is equivalent to

$$\partial_i(\mu_{\text{m}}^{-1} F_{ki}) = c\partial_t(\varepsilon F_{k0}) + J_k, \qquad c\partial_i(\varepsilon F_{i0}) = \rho_e, \tag{5.1.10}$$

where ε and μ_{m} may depend on t and $\boldsymbol{x}$.

The physical field is the electromagnetic field $F_{\mu\nu}$, and not the potential A_μ. However, it is necessary to use the gauge potential to carry out canonical quantization.[2] It is also necessary to use the gauge potential to describe the interaction between the charged particle and the electromagnetic field, as will be seen in Section 5.3. Furthermore, the field $F_{\mu\nu}$ does not possess the global property of the gauge potential, as will be discussed for the Aharonov-Bohm effect in Section 5.7.

Choosing A_μ as independent variables, we may construct the Lagrangian density so that the Maxwell equations follows as the Euler-Lagrange equations from it,

$$\mathcal{L} = \mathcal{L}_{\text{EM}} - J^\mu A_\mu, \tag{5.1.11}$$

with

$$\mathcal{L}_{\text{EM}} = -\frac{c^2\varepsilon}{2} F_{0i} F^{0i} - \frac{1}{4\mu_{\text{m}}} F_{ij} F^{ij} = \frac{1}{2}\left(\varepsilon \boldsymbol{E}^2 - \frac{\boldsymbol{B}^2}{\mu_{\text{m}}}\right). \tag{5.1.12}$$

In particular, the Euler-Lagrange equation with respect to the scalar potential A_0 is the second equation of (5.1.2),

$$\partial_k (\varepsilon E_k) = \rho_e(x). \tag{5.1.13}$$

This is a constraint equation, since there is no time derivative of A_0 in the Lagrangian density.

---

[2] There exists an attempt to formulate the electromagnetism without using the gauge potential.[330] However, the attempt does not seem to be so successful.

We define the canonical momentum according to the standard procedure,

$$\pi_0(x) = \frac{\partial \mathcal{L}(x)}{\partial \dot{A}^0(x)} = 0, \quad \pi_j(x) = \frac{\partial \mathcal{L}(x)}{\partial \dot{A}^j(x)} = c\varepsilon F_{0j}. \tag{5.1.14}$$

The Hamiltonian density reads

$$\mathcal{H} = -\varepsilon E_k \dot{A}^k(x) - \mathcal{L} = \frac{1}{2}\varepsilon \boldsymbol{E}^2 + \frac{1}{2}\frac{\boldsymbol{B}^2}{\mu_{\rm m}} + cA_0\left[\partial_k(\varepsilon E_k) - \rho_e(x)\right] + A_k J_k(x). \tag{5.1.15}$$

The electric field E_k is not a dynamical field since it is given by the constraint equation (5.1.13). We solve it as

$$\varepsilon E_k = \partial_k \frac{1}{\Delta}\rho_e(x) = -\frac{1}{4\pi}\partial_k \int d^3y\, \frac{\rho_e(t,\boldsymbol{y})}{|\boldsymbol{x}-\boldsymbol{y}|}, \tag{5.1.16}$$

by means of the Green function (see Appendix F),

$$\Delta \frac{1}{|\boldsymbol{x}-\boldsymbol{y}|} = -4\pi\delta(\boldsymbol{x}-\boldsymbol{y}). \tag{5.1.17}$$

Substituting (5.1.16) into the Hamiltonian density (5.1.15) we obtain

$$\mathcal{H} = \frac{1}{2}\int d^3x d^3y\, \rho_e(t,\boldsymbol{x})\frac{1}{4\pi\varepsilon|\boldsymbol{x}-\boldsymbol{y}|}\rho_e(t,\boldsymbol{y}) + \frac{1}{2}\frac{\boldsymbol{B}^2}{\mu_{\rm m}} + A_k J_k(x). \tag{5.1.18}$$

The long-range Coulomb interaction arises instead of the energy $\frac{1}{2}\varepsilon \boldsymbol{E}^2$ of the electric field.

There arises a problem to carry out the canonical quantization by choosing A_μ as independent variables, since the scalar potential A_0 has no canonical momentum as in (5.1.14). This problem is related with the fact that the gauge potential is not determined uniquely by the electromagnetic field. Indeed, the field $F_{\mu\nu}$ remains invariant when we make a transformation,

$$F_{\mu\nu} \to F_{\mu\nu} \quad \text{under} \quad A_\mu \longrightarrow A_\mu + \partial_\mu f(x), \tag{5.1.19}$$

since $\partial_\mu\partial_\nu f(x) = \partial_\nu\partial_\mu f(x)$ for an arbitrary single-valued smooth function $f(x)$. It is called the gauge transformation.[3] We need to impose a gauge condition to fix the redundant gauge degree of freedom in quantizing the gauge potential. We mention two gauge choices.

The first choice is the *Coulomb gauge*. When A'_μ is a gauge potential, the field $A_\mu = A'_\mu + \partial_\mu f(x)$ is also a gauge potential yielding the same field $F_{\mu\nu}$. We may choose f to satisfy $\partial_k A'_k + \Delta f(x) = 0$. We then have

$$\nabla \cdot \boldsymbol{A} = 0, \tag{5.1.20}$$

which is referred to as the Coulomb-gauge condition. The second equation in (5.1.2) is reduced to the Poisson equation,

$$\Delta(c\varepsilon A_0) = \rho_e. \tag{5.1.21}$$

[3] It is also called, after Pauli, the gauge transformation of the second kind.

The component A_0 is determined in terms of ρ_e,

$$A_0(x) = \frac{1}{c\varepsilon}\frac{1}{\Delta}\rho_e(x) = -\frac{1}{4\pi c\varepsilon}\int d^3y\, \frac{\rho_e(t,y)}{|x-y|}, \tag{5.1.22}$$

by means of the Green function (5.1.17). When the external source is absent, $\rho_e = 0$, we may set $A_0 = 0$ from (5.1.22). The special property of the Coulomb gauge is that the gauge degree of freedom is completely fixed. The electromagnetic field has two independent degrees of freedom, because there are three components A_1, A_2 and A_3 with one constraint condition (5.1.20).

The other choice is the *Lorentz gauge*. In the vacuum ($\mu_m = \mu_0$), where $c^2\varepsilon\mu_0 = 1$, the Lagrangian density (5.1.12) reads

$$\mathcal{L}_{EM} = -\frac{1}{4\mu_0}F_{\mu\nu}F^{\mu\nu}. \tag{5.1.23}$$

This is invariant under the Lorentz transformation. The field equations (5.1.10) are combined into

$$\partial^\nu F_{\mu\nu} = \mu_0 J_\mu. \tag{5.1.24}$$

They are invariant under the Lorentz transformation. It is preferable that the gauge condition has also the Lorentz invariance. When A'_μ is a gauge potential, the field $A_\mu = A'_\mu + \partial_\mu f(x)$ is also a gauge potential yielding the same field $F_{\mu\nu}$. We may choose f to satisfy $\partial^\mu A'_\mu + \partial^\mu\partial_\mu f(x) = 0$. We then have

$$\partial^\mu A_\mu = 0. \tag{5.1.25}$$

This is referred to as the Lorentz gauge condition.

5.2 Canonical Quantization

We quantize the electromagnetic field in the absence of the external current ($J_\mu = 0$). Though the electromagnetic field has only two independent degrees of freedom, we have three fields A_i in the Coulomb gauge and four fields A_μ in the Lorentz gauge. Canonical quantization is not trivial due to this problem.

Coulomb Gauge

We choose the Coulomb gauge, where

$$A_0 = 0, \qquad \nabla\cdot\boldsymbol{A} = 0. \tag{5.2.1}$$

We treat them as operator identities. The second Maxwell equation (5.1.10) is reduced to

$$c^2\varepsilon\mu_m\partial^0\partial_0 A_k + \partial^j\partial_j A_k = 0. \tag{5.2.2}$$

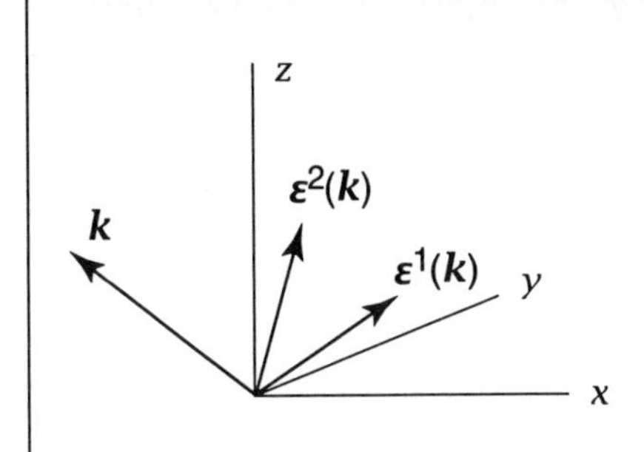

Fig. 5.1 Each photon has two independent degrees of freedom characterized by two polarization vectors $\epsilon^1(\boldsymbol{k})$ and $\epsilon^2(\boldsymbol{k})$ orthogonal to its wave vector $\boldsymbol{k}$.

According to the standard scheme we define the canonical momentum by

$$\pi_j(x) = \frac{\partial \mathcal{L}(x)}{\partial \dot{A}^j(x)} = c\varepsilon F_{0j} = \varepsilon \dot{A}_j = -\varepsilon E_j, \tag{5.2.3}$$

and write down the canonical commutation relations. The trivial commutation relations are

$$[A_j(t,\boldsymbol{x}), A_k(t,\boldsymbol{y})] = 0, \qquad [\dot{A}_j(t,\boldsymbol{x}), \dot{A}_k(t,\boldsymbol{y})] = 0. \tag{5.2.4}$$

It is tempting to set the nontrivial commutation relation to be

$$[A_j(t,\boldsymbol{x}), \pi_k(t,\boldsymbol{y})] = i\hbar\delta_{jk}\delta(\boldsymbol{x}-\boldsymbol{y}), \tag{5.2.5}$$

or

$$[A_j(t,\boldsymbol{x}), \dot{A}_k(t,\boldsymbol{y})] = \frac{i\hbar}{\varepsilon}\delta_{jk}\delta(\boldsymbol{x}-\boldsymbol{y}). \tag{5.2.6}$$

However, it contradicts the gauge condition $\boldsymbol{\nabla}\cdot\boldsymbol{A} = 0$ in (5.2.1). We should have expected it: Three independent commutation relations cannot be imposed when only two components of the potential A_j are independent variables. We modify (5.2.6) as

$$[A_j(t,\boldsymbol{x}), \dot{A}_k(t,\boldsymbol{y})] = \frac{i\hbar}{\varepsilon}\left(\delta_{jk} - \frac{\partial_j\partial_k}{\Delta}\right)\delta(\boldsymbol{x}-\boldsymbol{y}), \tag{5.2.7}$$

so that the Coulomb gauge condition $\boldsymbol{\nabla}\cdot\boldsymbol{A} = 0$ is obeyed. Here, the symbol $1/\Delta$ represents the integration operator as in (5.1.22), or equivalently

$$\delta_{ij} - \frac{\partial_i\partial_j}{\Delta} \Rightarrow \delta_{ij} - \frac{k_i k_j}{\boldsymbol{k}^2}, \tag{5.2.8}$$

in the momentum space.

We expand the potential A_j in terms of plane waves as

$$A_j(x) = \sqrt{\frac{\hbar}{\varepsilon}} \int \frac{d^3k}{\sqrt{(2\pi)^3}} \frac{1}{\sqrt{2\omega_k}} \sum_{\lambda=1}^{2} \epsilon_j^\lambda(\boldsymbol{k}) \left(a_{\boldsymbol{k}}^\lambda e^{ikx} + a_{\boldsymbol{k}}^{\lambda\dagger} e^{-ikx}\right), \tag{5.2.9}$$

with $kx = -\omega_k t + \boldsymbol{kx}$, as in the case (3.3.26) of the real scalar field. Substituting this into the Maxwell equation (5.2.2), we obtain

$$\omega_k = \frac{1}{\sqrt{\varepsilon\mu_{\text{m}}}}|\boldsymbol{k}|. \tag{5.2.10}$$

Because there are only two independent fields, there are only two independent "amplitudes" $a_{\boldsymbol{k}}^{\lambda}$ with the polarization vectors $\boldsymbol{\epsilon}^{\lambda}(\boldsymbol{k})$, where $\lambda = 1, 2$. We shall interpret $a_{\boldsymbol{k}}^{\lambda}$ and $a_{\boldsymbol{k}}^{\lambda\dagger}$ as the annihilation and creation operators of one photon with polarization $\boldsymbol{\epsilon}^{\lambda}(\boldsymbol{k})$ and wave vector $\boldsymbol{k}$ in quantum field theory [see (5.2.15)].

The polarization vectors are determined as follows [see Fig.5.1]. The Coulomb gauge condition (5.1.20) implies

$$\boldsymbol{k}\cdot\boldsymbol{\epsilon}^{\lambda}(\boldsymbol{k}) \equiv k_j\epsilon_j^{\lambda}(\boldsymbol{k}) = 0. \tag{5.2.11}$$

The polarization vectors are orthogonal to the wave vector $\boldsymbol{k}$ of the photon. The orthonormality condition is

$$\sum_{j=1}^{3}\epsilon_j^{\lambda}(\boldsymbol{k})\epsilon_j^{\lambda'}(\boldsymbol{k}) = \delta_{\lambda\lambda'}, \tag{5.2.12}$$

while the completeness condition is

$$\sum_{\lambda=1}^{2}\epsilon_j^{\lambda}(\boldsymbol{k})\epsilon_k^{\lambda}(\boldsymbol{k}) = \delta_{jk} - \frac{k_j k_k}{\boldsymbol{k}^2} \tag{5.2.13}$$

in consistent with the gauge condition (5.2.11). These three conditions determine the polarization vectors. We give an explicit example. We take a frame in which a photon is moving along the z direction by choosing $\boldsymbol{k} = (0, 0, k)$. We then have

$$\boldsymbol{\epsilon}^1(\boldsymbol{k}) = (1,0,0), \qquad \boldsymbol{\epsilon}^2(\boldsymbol{k}) = (0,1,0). \tag{5.2.14}$$

They are the two basis vectors spanning the xy-plane.

The potential $A_j(x)$ is a real field and its momentum expansion (5.2.9) is quite the same as (3.3.26) in the theory of the real Klein-Gordon field. In the same manner as used there, the commutation relations are derived,

$$[a_{\boldsymbol{k}}^{\lambda}, a_{\boldsymbol{l}}^{\lambda'\dagger}] = \delta_{\lambda\lambda'}\delta(\boldsymbol{k}-\boldsymbol{l}), \qquad [a_{\boldsymbol{k}}^{\lambda}, a_{\boldsymbol{l}}^{\lambda'}] = [a_{\boldsymbol{k}}^{\lambda\dagger}, a_{\boldsymbol{l}}^{\lambda'\dagger}] = 0. \tag{5.2.15}$$

We can interpret $a_{\boldsymbol{k}}^{\lambda}$ and $a_{\boldsymbol{k}}^{\lambda\dagger}$ as the annihilation and creation operators of the photon with polarization vector ϵ^{λ} and wave vector $\boldsymbol{k}$. The Fock vacuum is defined by

$$a_{\boldsymbol{k}}^{\lambda}|0\rangle = 0, \tag{5.2.16}$$

which is the state containing no photons.

We represent the Hamiltonian and the momentum in terms of the creation and annihilation operators $a_{\boldsymbol{k}}^{\lambda\dagger}$ and $a_{\boldsymbol{k}}^{\lambda}$. From the Lagrangian density (5.1.12) the Hamiltonian density is derived,

$$\mathcal{H}_{\text{EM}} = \varepsilon \dot{A}_k \dot{A}_k - \mathcal{L}_{\text{EM}} = \frac{1}{2}\left(\varepsilon \boldsymbol{E}^2 + \frac{\boldsymbol{B}^2}{\mu_{\text{m}}}\right). \tag{5.2.17}$$

The Hamiltonian H_{EM}^{C} in the Coulomb gauge reads

$$\begin{aligned} H_{\text{EM}}^{\text{C}} &= \int d^3x \; : \mathcal{H} : \\ &= \frac{1}{2}\int d^3x \; :\left(\varepsilon \dot{\boldsymbol{A}}^2 + \frac{1}{\mu_{\text{m}}}(\boldsymbol{\nabla}\times\boldsymbol{A})^2\right): = \frac{1}{2}\int d^3x \; :\left(\varepsilon \dot{\boldsymbol{A}}^2 - \frac{1}{\mu_{\text{m}}}\boldsymbol{A}\boldsymbol{\nabla}^2\boldsymbol{A}\right): \\ &= \sum_{\lambda=1}^{2}\int d^3k \; E_{\boldsymbol{k}} a_{\boldsymbol{k}}^{\lambda\dagger} a_{\boldsymbol{k}}^{\lambda}, \end{aligned} \tag{5.2.18}$$

with the dispersion relation

$$E_{\boldsymbol{k}} = \hbar\omega_{\boldsymbol{k}} = \frac{\hbar}{\sqrt{\varepsilon\mu_{\text{m}}}}|\boldsymbol{k}|. \tag{5.2.19}$$

We have $E_{\boldsymbol{k}} = \hbar c|\boldsymbol{k}|$ in the vacuum where $c^2\varepsilon\mu_{\text{m}} = 1$. The dispersion relation implies that the velocity of photons is

$$v = \frac{c}{\sqrt{c^2\varepsilon\mu_{\text{m}}}}, \tag{5.2.20}$$

and is different from c in medium with $c^2\varepsilon\mu_{\text{m}} \neq 1$: $n = \sqrt{c^2\varepsilon\mu_{\text{m}}}$ is the refractive index.

The energy-momentum tensor density is

$$\Theta^{\mu\nu} = -\frac{\delta\mathcal{L}}{\delta(\partial_\mu A^\lambda)}\partial^\nu A^\lambda + g^{\mu\nu}\mathcal{L}_{\text{EM}}, \tag{5.2.21}$$

which gives $\Theta^{00} = \mathcal{H}_{\text{EM}}$ and

$$p_j = \frac{1}{c}\int d^3x \; : \theta^{0j} : = \varepsilon\int d^3x \; : \boldsymbol{E}\cdot\partial_j\boldsymbol{A} : = \varepsilon\int d^3x \; : (\boldsymbol{E}\times\boldsymbol{B})_j :, \tag{5.2.22}$$

where use was made of the Coulomb gauge condition $\partial_k A_k = 0$. Substituting the momentum expansion (5.2.9), we obtain

$$\boldsymbol{p} = \sum_{\lambda=1}^{2}\int d^3k \; \hbar\boldsymbol{k} a_{\boldsymbol{k}}^{\lambda\dagger} a_{\boldsymbol{k}}^{\lambda}. \tag{5.2.23}$$

Both the Hamiltonian (5.2.18) and the momentum (5.2.23) are precisely what we have expected. The electromagnetic field is quantized as photons with two polarizations each of which carries the energy quantum $\hbar\omega_{\boldsymbol{k}}$ and the momentum $\hbar\boldsymbol{k}$. Thus the present quantization scheme must be the right one.

Lorentz Gauge

We proceed to choose the Lorentz gauge, where

$$\partial^\mu A_\mu = 0. \tag{5.2.24}$$

We quantize the electromagnetic field in the vacuum since the merit of this gauge is the Lorentz invariance. We start with the Lagrangian density,

$$\mathcal{L}_{\rm EM} = -\frac{1}{4\mu_0} F_{\mu\nu}F^{\mu\nu} - \frac{1}{2}\left(\partial^\mu A_\mu\right)^2, \tag{5.2.25}$$

which is reduced to the original Lagrangian (5.1.23) together with the Lorentz gauge condition.

The canonical momentum is given by

$$\pi_0(x) = \frac{\partial \mathcal{L}(x)}{\partial \dot{A}^0(x)} = \partial^\mu A_\mu, \tag{5.2.26}$$

$$\pi_i(x) = \frac{\partial \mathcal{L}(x)}{\partial \dot{A}^i(x)} = \partial_i A^0 - \dot{A}_i. \tag{5.2.27}$$

However, $\pi_0(x)$ is zero if the Lorentz gauge condition is imposed, as should be the case. To carry out quantization we give up the gauge condition (5.2.24) as an operator identity. We shall later impose it on the state.

The canonical momentum is set up without any problem,

$$[A_\mu(t,\boldsymbol{x}), \pi_\nu(t,\boldsymbol{y})] = i\hbar g_{\mu\nu}\delta(\boldsymbol{x}-\boldsymbol{y}), \tag{5.2.28}$$

$$[A_\mu(t,\boldsymbol{x}), A_\nu(t,\boldsymbol{y})] = [\pi_\mu(t,\boldsymbol{x}), \pi_\nu(t,\boldsymbol{y})] = 0. \tag{5.2.29}$$

We make the plane wave expansion akin to (5.2.9),

$$A_\mu(x) = \sqrt{\frac{\hbar}{\varepsilon}} \int \frac{d^3k}{\sqrt{(2\pi)^3}} \frac{1}{\sqrt{2\omega_k}} \sum_{\lambda=0}^{3} \epsilon_\mu^\lambda(\boldsymbol{k}) \left(a_{\boldsymbol{k}}^\lambda e^{ikx} + a_{\boldsymbol{k}}^{\lambda\dagger} e^{-ikx} \right). \tag{5.2.30}$$

The difference is that there are four polarization vectors $\epsilon_\mu^\lambda(\boldsymbol{k})$ at each momentum $\boldsymbol{k}$. We may choose them so that $\boldsymbol{\epsilon}^1(\boldsymbol{k})$ and $\boldsymbol{\epsilon}^2(\boldsymbol{k})$ are orthogonal to the four momentum $k^\mu = (k^0, \boldsymbol{k})$,

$$k^\mu \epsilon_\mu^1(\boldsymbol{k}) = k^\mu \epsilon_\mu^2(\boldsymbol{k}) = 0, \tag{5.2.31}$$

and $\boldsymbol{\epsilon}^0(\boldsymbol{k})$ and $\boldsymbol{\epsilon}^3(\boldsymbol{k})$ to satisfy

$$k^\mu \epsilon_\mu^0(\boldsymbol{k}) = k^\mu \epsilon_\mu^3(\boldsymbol{k}). \tag{5.2.32}$$

The orthonormalization condition is

$$g^{\mu\nu}\epsilon_\mu^\lambda(\boldsymbol{k})\epsilon_\nu^\lambda(\boldsymbol{k}) = -\epsilon_0^\lambda(\boldsymbol{k})\epsilon_0^\lambda(\boldsymbol{k}) + \boldsymbol{\epsilon}^\lambda(\boldsymbol{k})\boldsymbol{\epsilon}^\lambda(\boldsymbol{k}) = 0. \tag{5.2.33}$$

To construct such four vectors, taking a frame in which $\boldsymbol{k}$ is in the z-axis, where $k^\mu = (k, 0, 0, k)$, we choose

$$\epsilon^0_\mu = \begin{pmatrix}1\\0\\0\\0\end{pmatrix}, \quad \epsilon^1_\mu = \begin{pmatrix}0\\1\\0\\0\end{pmatrix}, \quad \epsilon^2_\mu = \begin{pmatrix}0\\0\\1\\0\end{pmatrix}, \quad \epsilon^3_\mu = \begin{pmatrix}0\\0\\0\\1\end{pmatrix}. \tag{5.2.34}$$

All conditions (5.2.31) ~ (5.2.32) hold explicitly in this frame. Then they holds in any Lorentz frame.

Substituting the field expansion into the canonical commutation relation and using the orthonormalization condition we obtain

$$[a^\lambda_{\boldsymbol{k}}, a^{\lambda'\dagger}_{\boldsymbol{k}'}] = g^{\lambda\lambda'}\delta(\boldsymbol{k}-\boldsymbol{k}'). \tag{5.2.35}$$

The Hamiltonian is calculated as

$$H^{\mathrm{L}}_{\mathrm{EM}} = \int d^3k\, E_{\boldsymbol{k}} \left(\sum_{\lambda=0}^{3} a^{\lambda\dagger}_{\boldsymbol{k}} a^\lambda_{\boldsymbol{k}} - a^{0\dagger}_{\boldsymbol{k}} a^0_{\boldsymbol{k}}\right). \tag{5.2.36}$$

There are some problems. First, the time component is anomalous,

$$[a^0_{\boldsymbol{k}}, a^{0\dagger}_{\boldsymbol{k}'}] = -\delta(\boldsymbol{k}-\boldsymbol{k}'). \tag{5.2.37}$$

The norm of one-photon state $|\boldsymbol{k}\rangle = a^{0\dagger}_{\boldsymbol{k}}|0\rangle$ is negative,

$$\langle\boldsymbol{k}|\boldsymbol{k}\rangle = \langle 0|a^0_{\boldsymbol{k}} a^{0\dagger}_{\boldsymbol{k}}|0\rangle = -\langle 0|0\rangle + \langle 0|a^{0\dagger}_{\boldsymbol{k}} a^0_{\boldsymbol{k}}|0\rangle = -1. \tag{5.2.38}$$

Hence, strictly speaking, $|\boldsymbol{k}\rangle$ cannot be a member of a Hilbert space. Furthermore, the energy of this state is negative.

We have so far ignored the Lorentz gauge condition (5.2.24). We now employ it to define the physical state,

$$\partial^\mu A^{(+)}_\mu(x)|\mathrm{phys}\rangle \equiv i\sqrt{\frac{\hbar}{\varepsilon}}\int \frac{d^3k}{\sqrt{(2\pi)^3}}\frac{1}{\sqrt{2\omega_{\boldsymbol{k}}}}\sum_{\lambda=0}^{3} k^\mu \epsilon^\lambda_\mu(\boldsymbol{k}) a^\lambda_{\boldsymbol{k}} e^{ikx}|\mathrm{phys}\rangle = 0, \tag{5.2.39}$$

where $A^{(+)}_\mu$ is the annihilation part of A_μ. It thus follows that $\langle\mathrm{phys}'|A_\mu(x)|\mathrm{phys}\rangle = 0$. For this condition to hold for any state $|\mathrm{phys}\rangle$ at any momentum $\boldsymbol{k}$ it is necessary that

$$\sum_{\lambda=0}^{3} k^\mu \epsilon^\lambda_\mu(\boldsymbol{k}) a^\lambda_{\boldsymbol{k}}|\mathrm{phys}\rangle = 0. \tag{5.2.40}$$

This becomes

$$k^\mu\left[\epsilon^0_\mu(\boldsymbol{k}) a^0_{\boldsymbol{k}} - \epsilon^3_\mu(\boldsymbol{k}) a^3_{\boldsymbol{k}}\right]|\mathrm{phys}\rangle = 0 \tag{5.2.41}$$

due to the transverse-wave condition (5.2.31). We then use (5.2.32) to find

$$\left(a^0_{\boldsymbol{k}} - a^3_{\boldsymbol{k}}\right)|\mathrm{phys}\rangle = 0. \tag{5.2.42}$$

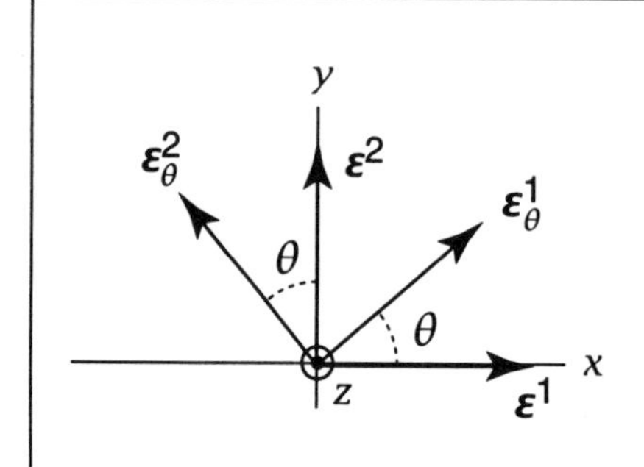

Fig. 5.2 Rotating the polarization vector ϵ^i around the z axis by angle θ, we obtain the new polarization vector ϵ_θ^i.

Then

$$\langle \text{phys}'| \left(a_{\boldsymbol{k}}^{3\dagger} a_{\boldsymbol{k}}^3 - a_{\boldsymbol{k}}^{0\dagger} a_{\boldsymbol{k}}^0\right) |\text{phys}\rangle = \langle \text{phys}'| \left(a_{\boldsymbol{k}}^{0\dagger} a_{\boldsymbol{k}}^0 - a_{\boldsymbol{k}}^{0\dagger} a_{\boldsymbol{k}}^0\right) |\text{phys}\rangle = 0, \quad (5.2.43)$$

and hence

$$\langle \text{phys}'|H_{\text{EM}}^{\text{L}}|\text{phys}\rangle = \langle \text{phys}'|H_{\text{EM}}^{\text{C}}|\text{phys}\rangle. \quad (5.2.44)$$

Namely, the Hamiltonian is identical in the Coulomb gauge and in the Lorentz gauge within the physical subspace, which is constructed by collecting all physical states satisfying (5.2.39). We can similarly prove all physical variables are identical in this sense.

Spin of Photons

The photon has two independent degrees of freedom corresponding two polarization vectors (5.2.14). We investigate the physical meaning of $\boldsymbol{\epsilon}^1$ and $\boldsymbol{\epsilon}^2$ a little more closely. We rotate the gauge potential (5.2.9) around the z axis by angle θ. By the rotation, the plane wave part $e^{\pm ikx}$ is invariant: Only the polarization vectors are affected. We rotate them around the z axis by angle θ. It is clear as in Fig.5.2 that the new polarization vectors read

$$\begin{aligned}\boldsymbol{\epsilon}_\theta^1 &\equiv e^{i\theta J_z/\hbar}\boldsymbol{\epsilon}^1 = \boldsymbol{\epsilon}^1\cos\theta + \boldsymbol{\epsilon}^2\sin\theta,\\ \boldsymbol{\epsilon}_\theta^2 &\equiv e^{i\theta J_z/\hbar}\boldsymbol{\epsilon}^2 = -\boldsymbol{\epsilon}^1\sin\theta + \boldsymbol{\epsilon}^2\cos\theta,\end{aligned} \quad (5.2.45)$$

where J_z is the z component of the angular-momentum operator. Recall that J_z is the generator of rotation around the z axis.

We define

$$\boldsymbol{\epsilon}^\pm = \frac{1}{2}(\boldsymbol{\epsilon}^1 \mp i\boldsymbol{\epsilon}^2). \quad (5.2.46)$$

It follows from (5.2.45) that these two vectors transform under the rotation as

$$\boldsymbol{\epsilon}_\theta^+ \equiv e^{i\theta J_z/\hbar}\boldsymbol{\epsilon}^+ = e^{i\theta}\boldsymbol{\epsilon}^+, \qquad \boldsymbol{\epsilon}_\theta^- \equiv e^{i\theta J_z/\hbar}\boldsymbol{\epsilon}^- = e^{-i\theta}\boldsymbol{\epsilon}^-. \quad (5.2.47)$$

We find

$$J_z \epsilon^+ = \hbar \epsilon^+, \qquad J_z^- \epsilon^- = -\hbar \epsilon^-. \tag{5.2.48}$$

It implies that the photon with polarization $\epsilon^{\pm}$ is an eigenstate of the angular-momentum component with $J_z = \pm\hbar$. Consequently, the photon has spin 1 and has two components corresponding to $J_z = \pm\hbar$. It has no spin component $J_z = 0$.

5.3 Interaction with Matter Field

We analyze the interaction between the electromagnetic field and the matter field. We first study the Schrödinger field. The Hamiltonian is the sum of those of the electromagnetic field and of the matter field in the electromagnetic field,

$$\mathcal{H} = \mathcal{H}_{\text{EM}} + \mathcal{H}_{\text{matter}} + \mathcal{H}_{\text{I}}. \tag{5.3.1}$$

Here $\mathcal{H}_{\text{EM}}$ is given by (5.2.17). The matter Hamiltonian in the presence of the electromagnetic field is given by (2.6.11), or

$$\mathcal{H}_{\text{matter}} + \mathcal{H}_{\text{I}} = \frac{1}{2m}\phi^\dagger(x)\boldsymbol{P}^2\phi(x) + V(\phi^\dagger\phi) \tag{5.3.2}$$

with

$$P_k = -i\hbar\partial_k + eA_k. \tag{5.3.3}$$

This Hamiltonian follows from the Lagrangian $\mathcal{L} = \mathcal{L}_{\text{EM}} + \mathcal{L}_{\text{I}}$ with

$$\begin{aligned}\mathcal{L}_{\text{matter}} + \mathcal{L}_{\text{I}} =& \frac{i}{2}\left(\phi^\dagger\left(\hbar\partial_t + iceA_0\right)\phi - \left(\hbar\partial_t - iceA_0\right)\phi^\dagger\cdot\phi\right) \\ &- \frac{1}{2M}\left(\hbar\partial_k - ieA_k\right)\phi^\dagger\left(\hbar\partial_k + ieA_k\right)\phi - V(\phi^\dagger\phi),\end{aligned} \tag{5.3.4}$$

where we have symmetrized the kinetic term and $A_0 = -\varphi/c$.

It is a prominent feature of this Lagrangian that it is invariant under a simultaneous transformation of the matter field ϕ and the gauge potential A_μ,

$$\phi(x) \to e^{if(x)}\phi(x), \tag{5.3.5a}$$

$$A_\mu(x) \to A_\mu(x) - \frac{\hbar}{e}\partial_\mu f(x). \tag{5.3.5b}$$

This set of transformations is also called the gauge transformation.

It is convenient to define

$$D_\mu \equiv \partial_\mu + i\frac{e}{\hbar}A_\mu. \tag{5.3.6}$$

Under the phase transformation (5.3.5a), $D_\mu\phi$ transforms *covariantly*,

$$D_\mu\phi \longrightarrow D_\mu(e^{if}\phi) = e^{if}D_\mu\phi, \tag{5.3.7}$$

though $\partial_\mu \phi$ changes in a complicated way,

$$\partial_\mu \phi \longrightarrow \partial_\mu (e^{if} \phi) = e^{if} (\partial_\mu \phi + i\phi \partial_\mu f). \tag{5.3.8}$$

In particular, $(D_\mu \phi)^\dagger (D_\mu \phi)$ is invariant,

$$(D_\mu \phi)^\dagger (D_\mu \phi) \longrightarrow (D_\mu \phi)^\dagger (D_\mu \phi). \tag{5.3.9}$$

Thus, $D_\mu \phi$ is called the *covariant derivative*. Accordingly, $P_\mu = -i\hbar D_\mu$ is called the *covariant momentum*.

Consequently, when we replace the canonical momentum p_μ with the covariant momentum P_μ in the pure matter Lagrangian,

$$p_\mu = -i\hbar \partial_\mu \quad \longrightarrow \quad P_\mu = -i\hbar D_\mu = -i\hbar \partial_\mu + eA_\mu, \tag{5.3.10}$$

we obtain the gauge-invariant Lagrangian density. It is called the minimal substitution. This is a general feature provided the matter field is a complex field. By this substitution an interaction between the the matter field and the electromagnetic field is determined. The interaction of this type is called the *minimal interaction*. The electromagnetic interaction is a minimal interaction.

We consider next the complex Klein-Gordon field, whose Lagrangian density is given by

$$\mathcal{L}_{\text{matter}} = -\hbar^2 (\partial^\mu \phi^\dagger)(\partial_\mu \phi) - V(\phi^\dagger \phi). \tag{5.3.11}$$

We make the minimal substitution,

$$\mathcal{L}_{\text{matter}} + \mathcal{L}_{\text{I}} = -\,(\hbar \partial^\mu - ieA^\mu)\, \phi^\dagger \,(\hbar \partial_\mu + ieA_\mu)\, \phi - V(\phi^\dagger \phi). \tag{5.3.12}$$

It determines the interaction between the Klein-Gordon field and the electromagnetic field. The total Lagrangian density $\mathcal{L} = \mathcal{L}_{\text{EM}} + \mathcal{L}_{\text{I}}$ is gauge invariant.

The Maxwell equation is the Euler-Lagrange equation, $\partial \mathcal{L} / \partial A^\mu = 0$, or

$$\frac{\partial \mathcal{L}_{\text{EM}}}{\partial A^\mu} = -\frac{\partial \mathcal{L}_{\text{I}}}{\partial A^\mu} \equiv -J_\mu. \tag{5.3.13}$$

It agrees with (5.1.10) in medium. For the Schrödinger field the current reads

$$J^0 = -ce\phi^\dagger \phi, \qquad J_k = -\frac{e\hbar}{2Mi} \left\{ \phi^\dagger (D_k \phi) - (D_k \phi)^\dagger \phi \right\}, \tag{5.3.14}$$

with the use of the Lagrangian density (5.3.4). Recall that $J^\mu = (c\rho_e, J_k)$. The conserved current (5.3.13) is the Nöther current, which is equal to the Nöther current (3.5.12) with the replacement of ∂_μ by D_μ. See related topics associated with superconductivity in Section 5.6. On the other hand, the current (5.3.13) for the complex Klein-Gordon field reads

$$J_\mu = -\frac{e\hbar}{i} \phi^\dagger \overleftrightarrow{\partial_\mu} \phi - 2e^2 A_\mu \phi^\dagger \phi = -\frac{e\hbar}{i} \left\{ \phi^\dagger D_\mu \phi - (D_\mu \phi)^\dagger \phi \right\}. \tag{5.3.15}$$

It is identical to the Nöther current (3.5.11) with the replacement of ∂_μ by D_μ.

Indeed, the electric current (5.3.13) is the Nöther current, leading to the current conservation,

$$\partial^\mu J_\mu = \partial_t \rho_e + \nabla \cdot J = 0. \tag{5.3.16}$$

The Nöther current is defined on the basis of the invariance of the Lagrangian under the global phase transformation, and this definition is not modified by the presence of the electromagnetic potential. Since the electromagnetic potential has been introduced to cancel the local phase transformation, when we make the local phase transformation with A_μ fixed, the term $\partial_\mu f$ appears in the combination $\partial_\mu f + (e/\hbar)A_\mu$ in (3.5.2). Therefore, the formula (3.5.15) is rewritten as

$$j_\mu = \frac{1}{\hbar}\frac{\partial \mathcal{L}_f}{\partial(\partial^\mu f)}\bigg|_{f\to 0} = \frac{1}{e}\frac{\partial \mathcal{L}_\mathrm{I}}{\partial A^\mu}. \tag{5.3.17}$$

Combining this with (5.3.13) we obtain $J_\mu = -e j_\mu$. It turns out that the field ϕ carries the electric charge $-e$ by adopting the covariant derivative (5.3.6).

5.4 ANDERSON-HIGGS MECHANISM

We choose the potential term $V(\phi^\dagger\phi)$ appropriately in the Lagrangian density so that Bose condensation occurs. For brevity we take the Klein-Gordon field coupled with the electromagnetic field in the vacuum,

$$\mathcal{L} = -\frac{1}{4\mu_0}F_{\mu\nu}F^{\mu\nu} - (\hbar\partial^\mu - ieA^\mu)\,\phi^\dagger\,(\hbar\partial_\mu + ieA_\mu)\,\phi + \gamma\phi^\dagger\phi - \frac{g}{2}(\phi^\dagger\phi)^2, \tag{5.4.1}$$

where $\gamma > 0$ and $g > 0$. The classical vacuum is the configuration minimizing the classical Hamiltonian. An obvious one is given by

$$\phi(x) = v, \qquad A_\mu(x) = 0 \tag{5.4.2}$$

with $v = \sqrt{\gamma/g}$. Actually there are infinitely many vacua,

$$\phi(x) = v e^{if(x)}, \qquad A_\mu(x) = -\frac{\hbar}{e}\partial_\mu f(x), \tag{5.4.3}$$

where $f(x)$ is an arbitrary real function. See Section 5.6 in the case of the Schrödinger field.

In order to perform canonical quantization we consider small fluctuations around the classical vacuum (5.4.2). We make a change of variables from a complex field $\phi(x)$ to two real fields $\eta(x)$ and $\chi(x)$, and from A_μ to U_μ,

$$\phi(x) = e^{i\chi(x)}\,(v + \eta(x)), \tag{5.4.4a}$$

$$A_\mu(x) = U_\mu(x) - \frac{\hbar}{e}\partial_\mu\chi(x), \tag{5.4.4b}$$

and treat $\eta(x)$ and U_μ as small variables. This is an operator gauge transformation. The kinetic term of the matter part yields

$$(\hbar\partial^\mu - ieA^\mu)\,\phi^\dagger\,(\hbar\partial_\mu + ieA_\mu)\,\phi$$
$$= (\hbar\partial^\mu\eta - ie(v+\eta)U^\mu)\,(\hbar\partial_\mu\eta + ie(v+\eta)U_\mu)\,. \tag{5.4.5}$$

The Lagrangian density reads

$$\mathcal{L} = -\frac{1}{4\mu_0}G^{\mu\nu}G_{\mu\nu} - e^2v^2U^\mu U_\mu - \hbar^2(\partial^\mu\eta)(\partial_\mu\eta) - 2\gamma\eta^2 + \cdots, \tag{5.4.6}$$

where equations are explicitly written up to the second order of $U_\mu(x)$ and $\eta(x)$, with $G^{\mu\nu} = \partial^\mu U^\nu - \partial^\nu U^\mu$.

We work in the Lorentz gauge (5.1.25),

$$\partial^\mu A_\mu = \partial^\mu U_\mu - \frac{\hbar}{e}\partial^\mu\partial_\mu\chi = 0. \tag{5.4.7}$$

Because the Goldstone mode $\chi(x)$ is decoupled from the Lagrangian (5.4.6), it may be assumed to be a free field obeying $\partial^\mu\partial_\mu\chi = 0$. The Euler-Lagrange equations are

$$\partial^\nu\partial_\nu U_\mu - \frac{m_A^2c^2}{\hbar^2}U_\mu + \cdots = 0, \tag{5.4.8a}$$

$$\partial^\nu\partial_\nu\eta - \frac{m_\phi^2c^2}{\hbar^2}\eta + \cdots = 0, \tag{5.4.8b}$$

where equations are explicitly written up to the first order of $U_\mu(x)$ and $\eta(x)$, with $m_A \equiv ev\hbar\sqrt{2\mu_0}/c$ and $m_\eta \equiv \sqrt{2\gamma}/c$. The system contains one massive scalar field η with mass m_η and one massive vector field U_ν with mass m_A. The parameters ξ and λ defined by

$$\xi = \frac{\hbar}{c}\frac{1}{m_\eta} = \frac{\hbar}{\sqrt{2\gamma}}, \qquad \lambda = \frac{\hbar}{c}\frac{1}{m_A} = \frac{1}{ev\sqrt{2\mu_0}} \tag{5.4.9}$$

are referred to as the coherence length and the penetration depth. See (4.2.24) for the coherence length and (5.6.13) for the penetration depth.

We count the number of the independent degree of freedom the massive vector field has. On one hand, when Bose condensation does not occur ($\gamma < 0$), the system contains two massive real scalar fields and one massless vector field. Since one massless vector field contains two physical modes, there are four physical modes in total. On the other hand, when Bose condensation occurs ($\gamma > 0$), the system contains one massive real scalar field and one massive vector field. Hence, we expect that one massive vector field has three physical modes, which we verify in the next section: The total physical degrees of freedom are not changed by Bose condensation.

The gauge field has been made massive without breaking the gauge symmetry explicitly: It has become massive by Bose condensation, and furthermore acquired one additional physical degree of freedom by absorbing the Goldstone mode into the longitudinal component. This mechanism is called the *Anderson-Higgs mechanism.*[231] Often it is said as follows: *Bose condensation yields a Goldstone mode, which gauge bosons eat and get massive.* See also an instance of superconductivity in Section 5.6.

5.5 MASSIVE VECTOR FIELD

We study a field theory of the massive vector boson, extracting the free-field part of U_μ from (5.4.6),

$$\mathcal{L} = -\frac{1}{4\mu_0} G^{\mu\nu} G_{\mu\nu} - \frac{m_A^2 c^2}{2\hbar^2 \mu_0} U^\mu U_\mu. \tag{5.5.1}$$

The Euler-Lagrange equation for the field U_μ is

$$\partial^\nu G_{\mu\nu} + \frac{m_A^2 c^2}{\hbar^2} U_\mu = 0. \tag{5.5.2}$$

This is called the Procaequation. The massive vector field is called the Procafield. Taking the derivative (∂^μ) of this equation we obtain the Lorentz condition,

$$\partial^\mu U_\mu = 0. \tag{5.5.3}$$

The Lorentz condition is a consistency condition. Using this in (5.5.2) we have,

$$\partial^\nu \partial_\nu U_\mu - \frac{m_A^2 c^2}{\hbar^2} U_\mu = 0, \tag{5.5.4}$$

which describes real Klein-Gordon bosons with mass m_A.

The Lagrangian is not invariant under the gauge transformation,

$$U_\mu(x) \to U_\mu(x) - \frac{\hbar}{e} \partial_\mu \chi(x), \tag{5.5.5}$$

due to the mass term. Hence, the Lagrangian system (5.5.1) is not a gauge theory.

Canonical quantization is straightforward because there is no problem associated with the gauge transformation. Since there are three independent fields, we choose U_k the canonical variables. The conjugate momentum is

$$\pi_k(x) = \frac{\delta \mathcal{L}(x)}{\delta \dot{U}^k(x)} = \frac{1}{c\mu_0} G_{0k}. \tag{5.5.6}$$

The field U_0 is not a canonical field, because U_0 is determined as

$$U_0 = \frac{\hbar^2}{m_A^2 c^2} \partial^\mu \partial_\mu U_0 = -\frac{\hbar^2}{m_A^2 c^2} \partial_k U_{0k} = -\frac{\hbar^2}{m_A^2 c} \partial_k \pi_k, \tag{5.5.7}$$

where use was made of the Procaequation (5.5.4) and the Lorentz condition, and $\dot{U}_0$ is determined as

$$\dot{U}_0 = c\partial_k U_k \tag{5.5.8}$$

from the Lorentz condition.

The canonical commutation relations consist of

$$[U_j(t,\boldsymbol{x}),\pi_k(t,\boldsymbol{y})] = i\hbar\delta_{jk}\delta(\boldsymbol{x}-\boldsymbol{y}), \tag{5.5.9}$$

and all other trivial ones. Equation (5.5.9) is equivalent to

$$[U_\mu(t,\boldsymbol{x}),\dot{U}_\nu(t,\boldsymbol{y})] = i\hbar c^2\mu_0\left(g_{\mu\nu} - \frac{\hbar^2}{m_A^2c^2}\partial_\mu\partial_\nu\right)\delta(\boldsymbol{x}-\boldsymbol{y}), \tag{5.5.10}$$

which is compatible with the Lorentz condition (5.5.3).

The momentum expansion is done precisely as in the case of the gauge field in the Coulomb gauge but with three independent polarization vectors $\varepsilon_\mu^\lambda(\boldsymbol{k})$, $\lambda = 1,2,3$,

$$U_\mu(x) = c\sqrt{\hbar\mu_0}\int\frac{d^3k}{\sqrt{(2\pi)^3}}\frac{1}{\sqrt{2\omega_k}}\sum_{\lambda=1}^{3}\varepsilon_\mu^\lambda(\boldsymbol{k})\left(a_{\boldsymbol{k}}^\lambda e^{ikx} + a_{\boldsymbol{k}}^{\lambda\dagger}e^{-ikx}\right), \tag{5.5.11}$$

where $\omega_k = c\sqrt{\boldsymbol{k}^2 + m_A^2c^2/\hbar^2}$ and

$$[a_{\boldsymbol{k}}^\lambda, a_{\boldsymbol{l}}^{\lambda'\dagger}] = \delta^{\lambda\lambda'}\delta(\boldsymbol{k}-\boldsymbol{l}), \qquad [a_{\boldsymbol{k}}^\lambda, a_{\boldsymbol{l}}^{\lambda'}] = [a_{\boldsymbol{k}}^{\lambda\dagger}, a_{\boldsymbol{l}}^{\lambda'\dagger}] = 0. \tag{5.5.12}$$

The polarization vectors satisfy the Lorentz condition,

$$k^\mu\varepsilon_\mu^\lambda(\boldsymbol{k}) = 0, \tag{5.5.13}$$

the orthonormality condition,

$$g^{\mu\nu}\varepsilon_\mu^\lambda(\boldsymbol{k})\varepsilon_\nu^{\lambda'}(\boldsymbol{k}) = \delta^{\lambda\lambda'}, \tag{5.5.14}$$

and the completeness condition,

$$\sum_{\lambda=1}^{3}\varepsilon_\mu^\lambda(\boldsymbol{k})\varepsilon_\nu^\lambda(\boldsymbol{k}) = g_{\mu\nu} + \frac{\hbar^2}{m_A^2c^2}k_\mu k_\nu. \tag{5.5.15}$$

We give an explicit example of the polarization vectors. We consider a Lorentz frame in which a massive vector boson is moving along the z axis, where $k^\mu = (\omega_k/c, 0, 0, k)$ and $k > 0$. In this frame we may choose the polarization vectors to be

$$\varepsilon_\mu^1 = (0,1,0,0), \quad \varepsilon_\mu^2 = (0,0,1,0), \quad \varepsilon_\mu^3 = \frac{\hbar}{m_A c}\left(-k,0,0,\frac{\omega_k}{c}\right), \tag{5.5.16}$$

as satisfy (5.5.13) $\sim$ (5.5.15). The three independent polarization corresponds to spin 1. Thus, the massive vector particle has spin 1 with three components -1, 0, +1. Clearly, the particle with polarization (5.2.46) has the component $J_z = \pm\hbar$ as in the case of photons, and the one with ε_μ^3 has the component $J_z = 0$ since it is invariant under the rotation around the z axis.

Because the Lagrangian (5.5.1) is identical to the vector-field part of the Lagrangian (5.4.6), the photon field becomes the Procafield in a Bose condensate. This is correct only partially since the photon field is a gauge field but the Procafield is not. The gauge symmetry is broken explicitly in the massive vector-field theory (5.5.1), while it is broken spontaneously in the electromagnetic theory (5.4.1).

5.6 SUPERCONDUCTIVITY

Superconductivity is a good example of Bose condensation and spontaneous symmetry breaking.[4] In the BCS superconductor electrons feel an attractive force mediated by phonon exchange corresponding to lattice distortion induced by electrostatic force. This attractive force can overcome the repulsive Coulomb potential, since the charges of electrons are screened by the background charges on lattice. The resulting net attractive potential can be approximated by a contact interaction.

The Hamiltonian density is modeled by

$$\mathcal{H}(\boldsymbol{x}) = \frac{\hbar^2}{2m}\nabla\psi_\uparrow^\dagger(\boldsymbol{x})\nabla\psi_\uparrow(\boldsymbol{x}) + \frac{\hbar^2}{2m}\nabla\psi_\downarrow^\dagger(\boldsymbol{x})\nabla\psi_\downarrow(\boldsymbol{x}) - g\psi_\downarrow^\dagger(\boldsymbol{x})\psi_\uparrow^\dagger(\boldsymbol{x})\psi_\uparrow(\boldsymbol{x})\psi_\downarrow(\boldsymbol{x}), \tag{5.6.1}$$

where $\psi_\uparrow(\boldsymbol{x})$ and $\psi_\downarrow(\boldsymbol{x})$ represent the electron fields with up-spin and down-spin. Due to the attractive force between electrons ($g > 0$) we expect the emergence of a particle described by the composite field

$$\phi(\boldsymbol{x}) = g\psi_\uparrow(\boldsymbol{x})\psi_\downarrow(\boldsymbol{x}). \tag{5.6.2}$$

The potential energy, which now reads $-g^{-1}|\phi(\boldsymbol{x})|^2$, becomes lower as $|\phi(\boldsymbol{x})|$ becomes larger, and hence we expect that the ground state is characterized by the formation of a condensate of paired electrons (Cooper pairs), $\langle 0|\phi(\boldsymbol{x})|0\rangle \neq 0$.

The Lagrangian density for the composite field $\phi(\boldsymbol{x})$ may be derived. After introducing the coupling with the electromagnetic field by the minimal interaction, we find

$$\begin{aligned}\mathcal{L} = \mathcal{L}_{\text{EM}} &+ \frac{i}{2}\left\{\phi^\dagger\left(\hbar\partial_t + ice^*A_0\right)\phi - \left(\hbar\partial_t - ice^*A_0\right)\phi^\dagger\cdot\phi\right\} \\ &- \frac{1}{2M}\left(\hbar\partial_k - ie^*A_k\right)\phi^\dagger\left(\hbar\partial_k + ie^*A_k\right)\phi - \frac{g}{2}\left(\phi^\dagger\phi - v^2\right)^2,\end{aligned} \tag{5.6.3}$$

where $M = 2m$ and $e^* = 2e$ are the mass and the charge of a Cooper pair; $\mathcal{L}_{\text{EM}}$ is the Maxwell term (5.1.12). The Hamiltonian density is derived as

$$\begin{aligned}\mathcal{H} =& \frac{\varepsilon}{2}\boldsymbol{E}^2 + \frac{1}{2\mu_{\text{m}}}\boldsymbol{B}^2 + \frac{1}{2M}\left(\hbar\partial_k - ie^*A_k\right)\phi^\dagger\left(\hbar\partial_k + ie^*A_k\right)\phi \\ &+ \frac{g}{2}\left(\phi^\dagger\phi - v^2\right)^2 + ceA_0\phi^\dagger\phi.\end{aligned} \tag{5.6.4}$$

[4] There are many common features between superconductors and QH systems. It is intriguing how QH systems are similar to and different from superconductors. See textbooks[32,13] for references and related topics on superconductors.

The Maxwell equations are

$$\nabla \times \boldsymbol{H} - \partial_t \boldsymbol{D} = \boldsymbol{J}, \tag{5.6.5a}$$

$$\nabla \boldsymbol{D} = \rho_e, \tag{5.6.5b}$$

where

$$\boldsymbol{B} = \nabla \times \boldsymbol{A}, \qquad \boldsymbol{E} = -\partial_t \boldsymbol{A} + c\nabla A_0 \tag{5.6.6}$$

with $cA_0 = -\varphi$ (the scalar potential). Charge ρ_e and current $\boldsymbol{J}$ are given by

$$\rho_e \equiv \frac{\delta \mathcal{L}_I}{c\delta A_0} = -e^* \phi^\dagger \phi, \tag{5.6.7}$$

$$J_k \equiv \frac{\delta \mathcal{L}_I}{\delta A^k} = -\frac{e^*\hbar}{2Mi}\left\{\phi^\dagger (D_k\phi) - (D_k\phi)^\dagger \phi\right\}$$

$$= -\frac{e^*\hbar}{2Mi}\left\{\phi^\dagger \partial_k \phi - (\partial_k\phi)^\dagger \phi\right\} - \frac{e^{*2}}{M}\phi^\dagger \phi A_k. \tag{5.6.8}$$

Current J_k is the Nöther current (5.3.14) with the replacement of e by e^*.

Supercurrent

In the rest of this section we use the classical field. In the Coulomb gauge the classical vacuum is given by

$$\phi = v, \qquad \boldsymbol{A} = 0, \qquad A_0 = 0. \tag{5.6.9}$$

It implies that the ground state is a Bose condensate of Cooper pairs. We are interested in properties of the electromagnetic field in the Bose condensate.

Let us suppose that the Bose field has a $\boldsymbol{x}$-dependent phase,

$$\phi(\boldsymbol{x}) = ve^{i\chi(\boldsymbol{x})}, \qquad \boldsymbol{A} = 0, \qquad A_0 = 0, \tag{5.6.10}$$

as corresponds to (4.6.1) in the case of superfluid. This is a higher energy state with the energy density

$$\mathcal{H} = \frac{\hbar^2}{2M}(\nabla\chi)^2. \tag{5.6.11}$$

The phase current (5.6.8) is induced to decrease the energy by reducing the phase inhomogeneity,

$$\boldsymbol{J}(\boldsymbol{x}) = -\frac{1}{\lambda^2 e^* \mu_{\mathrm{m}}}\nabla\chi(\boldsymbol{x}), \tag{5.6.12}$$

where we have set

$$\lambda = \frac{1}{e^* v}\sqrt{\frac{M}{\mu_{\mathrm{m}}}}. \tag{5.6.13}$$

It corresponds to (4.6.2) in the case of superfluid. It flows even in the absence of the electric field $\boldsymbol{E} = 0$. By the Ohm law, $\boldsymbol{E} = R\boldsymbol{J}$, we find that $R = 0$. Moreover,

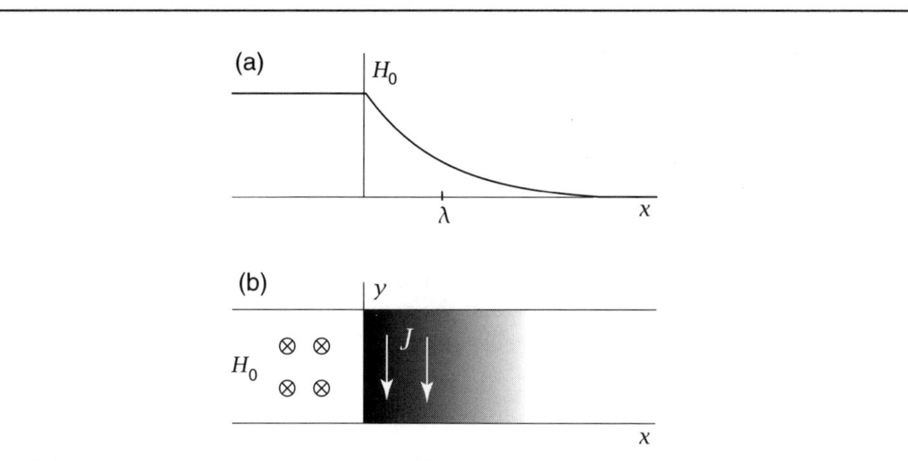

Fig. 5.3 (a)The magnetic field outside the superconductor can penetrate into it within an order of the penetration depth λ. (b) This screening is generated by the supercurrent flowing within this region.

there are no gapless modes in the system. It is a supercurrent for these reasons. Recall arguments for superfluidity in Section 4.6.

We next apply a static magnetic field but not electric field to the ground state (5.6.9),

$$\phi = v, \qquad \partial_t \boldsymbol{A}(\boldsymbol{x}) = 0, \qquad A_0 = 0. \tag{5.6.14}$$

It increases the energy density to

$$\mathcal{H} = \frac{1}{2\mu_{\mathrm{m}}}\left(\boldsymbol{B}^2 + \lambda^2 \boldsymbol{A}^2\right). \tag{5.6.15}$$

The electric current (5.6.8) flows,

$$\boldsymbol{J}(\boldsymbol{x}) = -\frac{1}{\lambda^2 \mu_{\mathrm{m}}} \boldsymbol{A}(\boldsymbol{x}). \tag{5.6.16}$$

This is known as the London equation.[314] It is a supercurrent for the same reasons as given just above.

Meissner Effect

We place a superconductor in a uniform magnetic field parallel to the z axis. The magnetic field is taken as

$$B_z = -B_0, \qquad B_x = B_y = 0. \tag{5.6.17}$$

We assume that the superconductor occupies the half space ($x > 0$). We examine how the magnetic flux penetrates into the superconductor [Fig.5.3]. We minimize

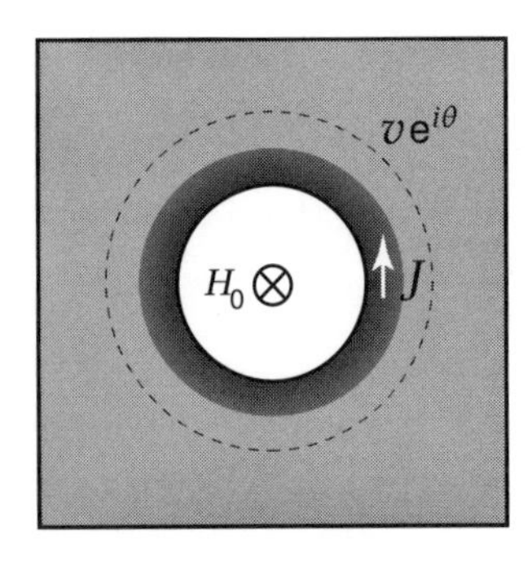

Fig. 5.4 When a superconductor has a hole, the magnetic flux applied perpendicularly is confined within the hole: It can penetrate into the superconductor only within the penetration depth. A supercurrent J flows within this depth. The confined flux is quantized.

the Hamiltonian (5.6.4) by requiring the boundary condition (5.6.17) at $x = 0$. We can set $\phi = v$ and $\boldsymbol{E} = 0$. Then, together with the London equation(5.6.16), the Maxwell equation (5.6.5a) yields

$$\nabla^2 \boldsymbol{A} = \lambda^{-2}\boldsymbol{A}, \tag{5.6.18}$$

or

$$\nabla^2 \boldsymbol{B} = \lambda^{-2}\boldsymbol{B}. \tag{5.6.19}$$

Solving this we find

$$B_z(x) = -B_0 e^{-x/\lambda}, \tag{5.6.20}$$

for $x \geq 0$. The magnetic field penetrates into superconductor only within the length scale λ. Because of this fact, λ is called the penetration depth. This phenomenon is known as the *Meissner effect*. Since $\boldsymbol{B} = \mu_{\rm m}\boldsymbol{H}$, current $\boldsymbol{J}$ is calculated from (5.6.5a),

$$J_x = J_z = 0, \qquad J_y = -\frac{H_0}{\lambda} e^{-x/\lambda}. \tag{5.6.21}$$

This is a supercurrent flowing in the y direction within the penetration depth from the boundary of the superconductor.

Flux Quantization

We next assume that the superconductor has a cylindrical hole (with radius R) parallel to the magnetic field [Fig.5.4]. We take the center of the hole at the origin of the xy plane.

We consider the field configuration of the form

$$\phi(\boldsymbol{x}) = ve^{-i\chi(\boldsymbol{x})}, \qquad A_k(\boldsymbol{x}) = U_k(\boldsymbol{x}) + \frac{\hbar}{e*}\partial_k \chi(\boldsymbol{x}) \tag{5.6.22}$$

for $r > R$. The Hamiltonian is given by

$$\mathcal{H} = \frac{1}{2\mu_{\rm m}}\left(\boldsymbol{B}^2 + \lambda^2 \boldsymbol{U}^2\right) = \frac{1}{2\mu_{\rm m}}\left[(\nabla \times \boldsymbol{U})^2 + \lambda^2 \boldsymbol{U}^2\right]. \tag{5.6.23}$$

The boundary condition is $\boldsymbol{U}(\boldsymbol{x}) = 0$ as $r = |\boldsymbol{x}| \to \infty$. We assume the phase $\chi(\boldsymbol{x})$ to be a multivalued function,

$$\chi(\boldsymbol{x}) = q\theta \tag{5.6.24}$$

with θ the azimuthal angle. It is necessary that q is an integer for the single-valued of $\phi(\boldsymbol{x})$, i.e., $\exp(i2\pi q) = 1$.

The current (5.6.8) becomes

$$\boldsymbol{J}(\boldsymbol{x}) = -\frac{1}{\lambda^2 \mu_{\text{m}}} \boldsymbol{U}(\boldsymbol{x}), \tag{5.6.25}$$

corresponding to the London equation (5.6.16). The Maxwell equation (5.6.5a) yields

$$\nabla^2 \boldsymbol{U} = \lambda^{-2} \boldsymbol{U}, \tag{5.6.26}$$

or

$$\nabla^2 \boldsymbol{B} = \lambda^{-2} \boldsymbol{B}. \tag{5.6.27}$$

Since

$$\nabla \times \boldsymbol{B} = \nabla \times (\nabla \times \boldsymbol{U}) = -\nabla^2 \boldsymbol{U} = -\lambda^{-2} \boldsymbol{U}, \tag{5.6.28}$$

the current (5.6.25) is

$$\mu_{\text{m}} \boldsymbol{J}(\boldsymbol{x}) = -\frac{1}{\lambda^2} \boldsymbol{U}(\boldsymbol{x}) = \nabla \times \boldsymbol{B}, \tag{5.6.29}$$

or

$$J_i = \mu_{\text{m}}\, \varepsilon_{ij} \partial_j B_z(r). \tag{5.6.30}$$

This is a supercurrent flowing around the hole within the penetration depth λ from the boundary of the superconductor ($r = R$).

We evaluate the total magnetic flux. We take a large circle deep inside the superconductor [Fig.5.4]. Since $U_k = 0$ or $A_k = (\hbar q/e^*)\partial_k \theta$ from (5.6.22), it follows that

$$\Phi = \int d^2x\, B(\boldsymbol{x}) = \oint dx_k A_k = \frac{q\Phi_{\text{L}}}{2\pi} \int_0^{2\pi} d\theta = q\Phi_{\text{L}}. \tag{5.6.31}$$

The total flux is quantized in units of the London flux unit $\Phi_{\text{L}} \equiv 2\pi\hbar/e^* = \pi\hbar/e$. The flux quantization was first pointed out by London,[26] though his proposed flux unit was the Dirac flux unit $\Phi_{\text{D}} \equiv 2\Phi_{\text{L}}$ originally, corresponding to the condensation of a single charge e.

<u>Vortices</u>

We consider the limit $R \to 0$ in the above problem. We solve the equation of motion (5.6.27) in the cylindrical coordinate,

$$\nabla^2 = \frac{\partial^2}{\partial r^2} + \frac{1}{r}\frac{\partial}{\partial r} + \frac{1}{r^2}\frac{\partial^2}{\partial \theta^2}. \tag{5.6.32}$$

For a cylindrical symmetric configuration (5.6.27) reads

$$r^2 \frac{\partial^2}{\partial r^2} B_z(r) + r \frac{\partial}{\partial r} B_z(r) - \frac{r^2}{\lambda^2} B_z(r) = 0. \tag{5.6.33}$$

The solution is a modified Bessel function,

$$B_z(r) = \alpha K_0 \left(\frac{r}{\lambda}\right) \tag{5.6.34}$$

with a numerical constant α. The asymptotic behaviors are

$$K_0(x) \to \begin{cases} -(\ln x + \gamma_{\rm E} - \ln 2) & \text{as} \quad x \to 0 \\ \sqrt{\pi/2x} e^{-x} & \text{as} \quad x \to \infty \end{cases}, \tag{5.6.35}$$

with the Euler constant $\gamma_{\rm E}$. The function $U_j(\boldsymbol{x})$ is singular as $r \to 0$, where

$$U_j = -\lambda^2 \varepsilon_{ij} \partial_j B_z \to -\alpha \lambda^2 \varepsilon_{ij} \partial_j \ln \frac{r}{\lambda} = \alpha \lambda^2 \varepsilon_{ij} \frac{x_j}{r^2}. \tag{5.6.36}$$

Let us examine the limit $r \to 0$ of the potential $A_j(\boldsymbol{x})$ given by (5.6.22),

$$\lim_{r\to 0} A_j(\boldsymbol{x}) = \lim_{r\to 0} \left[U_j(\boldsymbol{x}) + \frac{q\Phi_{\rm L}}{2\pi} \partial_j \theta \right] = \lim_{r\to 0} \left(\alpha\lambda^2 - \frac{q\Phi_{\rm L}}{2\pi} \right) \varepsilon_{ij} \frac{x_j}{r^2}. \tag{5.6.37}$$

Since the potential $A_j(\boldsymbol{x})$ is not singular it is necessary that

$$\alpha = \frac{q\Phi_{\rm L}}{2\pi\lambda^2}. \tag{5.6.38}$$

Hence

$$B_z(r) = \frac{q\Phi_{\rm L}}{2\pi\lambda^2} K_0 \left(\frac{r}{\lambda}\right). \tag{5.6.39}$$

This is the magnetic flux density around the origin.

We can rewrite the Hamiltonian density as

$$\mathcal{H} = \frac{1}{2\mu_{\rm m}} \left(\boldsymbol{B}^2 + \lambda^2 \boldsymbol{U}^2 \right) = \frac{1}{2\mu_{\rm m}} \left[\boldsymbol{B}^2 + \lambda^2 (\nabla \times \boldsymbol{B})^2 \right] \tag{5.6.40}$$

with the aid of (5.6.29). We evaluate the energy of a vortex excluding the origin,

$$\begin{aligned} E =& \frac{1}{2\mu_{\rm m}} \int_{r>0} d^2x \left[B_z^2 + \lambda^2 (\partial_x B_z)^2 + \lambda^2 (\partial_y B_z)^2 \right] \\ =& \frac{1}{2\mu_{\rm m}} \int_{r>0} d^2x \, B_z \left[B_z + \lambda^2 \left(\partial_x^2 B_z + \partial_y^2 B_z \right) \right] + \frac{\lambda^2}{2\mu_{\rm m}} \oint dx_i \left(B_z \varepsilon_{ij} \partial_j B_z \right), \end{aligned} \tag{5.6.41}$$

where we have made a partial integration. The bulk term vanishes because of the equation of motion (5.6.27), and only the surface term remains. The magnetic flux density $B_z(r)$ decreases rapidly as $r \to \infty$, and hence the surface contribution comes only from an infinitesimal circle around $r = 0$. It is estimated as

$$E = \frac{\lambda^2}{2\mu_{\rm m}} \left(\frac{q\Phi_{\rm L}}{2\pi\lambda^2} \right)^2 \lim_{r\to 0} \oint r d\theta \left(\frac{1}{r} \ln \frac{\lambda}{r} \right) = \frac{1}{\mu_{\rm m}} \frac{q^2 \Phi_{\rm L}^2}{4\pi\lambda^2} \ln \frac{\lambda}{\xi}, \tag{5.6.42}$$

where we have introduced the cut-off parameter ξ since it is divergent.

The origin of the divergence is the oversimplification of the field configuration (5.6.22). When the magnetic flux penetrates into the superconductor, it is not a good assumption that the matter field ϕ changes only by a phase from the ground-state value. We should set

$$\phi(\boldsymbol{x}) = v(r)e^{-iq\theta(\boldsymbol{x})}, \qquad A_j(\boldsymbol{x}) = U_j(\boldsymbol{x}) + \frac{q\Phi_{\rm L}}{2\pi}\partial_j\theta(\boldsymbol{x}) \tag{5.6.43}$$

instead of (5.6.22). Since $e^{-iq\theta(\boldsymbol{x})}$ is ill-defined at $r = 0$, it is necessary that $v(0) = 0$ so that $\phi(\boldsymbol{x})$ is well defined there. Hence the condensation is resolved at the origin. Actually the condensation is resolved for $r < \xi$, where ξ is the coherence length. We analyze such a solution in Section 7.5.

The vortex energy is proportional to q^2. It is favorable to create q vortices each of which carries $\Phi_{\rm L}$, since the total vortex energy becomes proportional to q. Hence we take $q = \pm 1$ in what follows.

When there are two vortices with the flux q_i at $\boldsymbol{r}_i$, the magnetic flux density is the sum of (5.6.39),

$$B_z(r) = \frac{q_1\Phi_{\rm L}}{2\pi\lambda^2}K_0\left(\frac{|\boldsymbol{r}-\boldsymbol{r}_1|}{\lambda}\right) + \frac{q_2\Phi_{\rm L}}{2\pi\lambda^2}K_0\left(\frac{|\boldsymbol{r}-\boldsymbol{r}_2|}{\lambda}\right). \tag{5.6.44}$$

The energy is calculated by substituting this into (5.6.41). It consists of the self-energies of two vortices (5.6.42) and the extra term. The extra term describes the interaction energy, and given by

$$E_{\rm extra}(r_{12}) = \frac{q_1q_2\lambda^2}{2\mu_{\rm m}}B_z(r_{12})\oint dx_i\left(\varepsilon_{ij}\partial_jB_z\right) = \frac{q_1q_2}{\mu_{\rm m}}\frac{\Phi_{\rm L}^2}{2\pi\lambda^2}K_0\left(\frac{r_{12}}{\lambda}\right), \tag{5.6.45}$$

where the loop integration $\oint$ is made around two vortex centers. It is interpreted as the energy of the second flux in the magnetic field made by the first flux and vice versa. Note that $K_0(x)$ is a monotonically decreasing function of x. Since $K_0(x) > 0$, the interaction between two vortices is repulsive for $q_1q_2 = 1$ and attractive for $q_1q_2 = -1$. The force on the second vortex at is given by

$$\boldsymbol{f}(\boldsymbol{r}_2) = -\nabla E_{\rm extra}(r_{12}), \tag{5.6.46}$$

where the derivative acts on $\boldsymbol{r}_2$. We may rewrite it as

$$f_i(\boldsymbol{r}_2) = -\frac{q_1q_2\Phi_{\rm L}}{\mu_{\rm m}}\partial_iB_z(|\boldsymbol{r}_2-\boldsymbol{r}_1|) = q_1q_2\Phi_{\rm L}\varepsilon_{ij}J_j(\boldsymbol{r}_2) \tag{5.6.47}$$

with the use of (5.6.30). The force at the second vortex is given by the current made by the first vortex with its direction perpendicular to the current.

5.7 Aharonov-Bohm Effect

The physical field is the electromagnetic field $F_{\mu\nu}$ and not the potential A_μ as a local quantity. However, the potential possesses a certain global physical in-

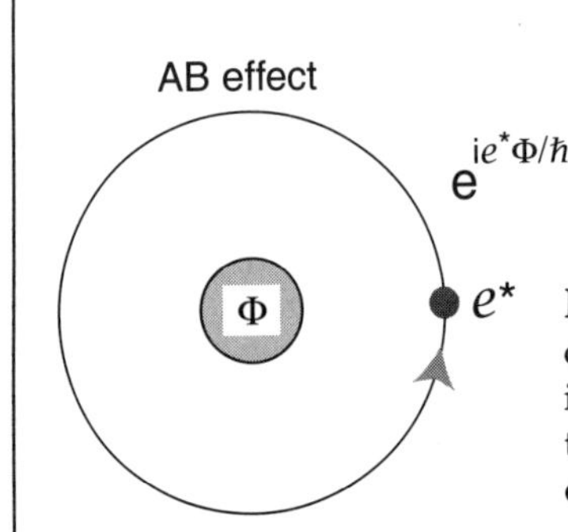

Fig. 5.5 A magnetic flux is confined within a cylindrical domain. Nevertheless, a charged particle feels its existence since there exists a nontrivial electromagnetic potential outside the domain. When it encircles the domain once it acquires an Aharonov-Bohm phase.

formation which the electromagnetic field does not. To show this, we study the *Aharonov-Bohm effect.*[39] We take a particle with electric charge e^*, and magnetic flux Φ parallel to the z axis confined within a cylindrical domain [Fig.5.5]. Although no flux exists outside the cylinder, the electromagnetic potential does not vanish identically. Indeed, integrating the electromagnetic potential along a path encircling the cylinder in the xy plane and using the Stokes theorem, we obtain

$$\oint dx_k A_k = \int d^2x\, \varepsilon_{ij}\partial_i A_j = \int d^2x\, B_z = \Phi, \tag{5.7.1}$$

where $\varepsilon_{xy} = -\varepsilon_{yx} = 1$, $\varepsilon_{xx} = \varepsilon_{yy} = 0$, and Φ is the total flux passing through the cylinder. We cannot set $A_k(\boldsymbol{x}) = 0$ identically when $\Phi \neq 0$.

A simplest choice of the electromagnetic potential is

$$A_k(\boldsymbol{x}) = \frac{\Phi}{2\pi}\partial_k\theta(\boldsymbol{x}) = -\frac{\Phi}{2\pi}\varepsilon_{kj}\frac{x_j}{r^2} \tag{5.7.2}$$

outside the cylinder, where θ is the angle around the cylinder. If the magnetic flux is uniform within the cylinder of radius R_0,

$$B_r = B_\theta = 0, \qquad B_z = \begin{cases} B & \text{for} \quad r < R_0 \\ 0 & \text{for} \quad r > R_0 \end{cases}, \tag{5.7.3}$$

the electromagnetic potential may be taken as

$$A_r = A_z = 0, \qquad A_\theta = \begin{cases} \dfrac{1}{2}Br & \text{for} \quad r < R_0 \\ \dfrac{1}{2\pi r}\Phi & \text{for} \quad r > R_0 \end{cases}, \tag{5.7.4}$$

where $\Phi = \pi R_0^2 B$ is the total flux; note that $A_x = -\sin\theta\, A_\theta$ and $A_y = \cos\theta\, A_\theta$.

Because the field operator $\phi(\boldsymbol{x})$ of a charged particle is not invariant under a gauge transformation, strictly speaking, it is not a physical quantity, as stressed by

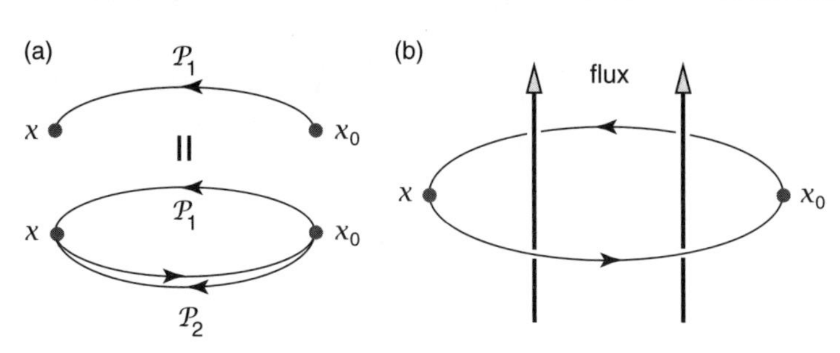

Fig. 5.6 A gauge invariant wave function is defined by the formula (5.7.5) together with a phase involving a path integration over the electromagnetic potential A_k. It depends on the path. (a) The path $\mathcal{P}_1$ is equivalent to the path $\mathcal{P}_2$ and a loop. (b) The loop encircles a flux and produces the Aharonov-Bohm phase to the wave function as in (5.7.7).

Dirac. A gauge invariant field is constructed[5] as [Fig.5.6(a)]

$$\phi_{\mathcal{P}}(\boldsymbol{x}) = \exp\left(i\frac{e^*}{\hbar}\int_{\boldsymbol{x}_0}^{\boldsymbol{x}} dy_k A_k\right)\phi(\boldsymbol{x}), \tag{5.7.5}$$

where the integration is made along an arbitrary path $\mathcal{P}$ starting from an arbitrary fixed point $\boldsymbol{x}_0$ and ending at the point $\boldsymbol{x}$. It is notable that

$$\partial_k\phi_{\mathcal{P}}(\boldsymbol{x}) = \exp\left(i\frac{e^*}{\hbar}\int_{\boldsymbol{x}_0}^{\boldsymbol{x}} dy_k A_k\right)D_k\phi(\boldsymbol{x}), \tag{5.7.6}$$

where $D_k \equiv \partial_k + i(e^*/\hbar)A_k$ is the covariant derivative (5.3.6). However, the gauge-invariant field (5.7.5) depends on the path in general. We take two paths $\mathcal{P}_i$ connecting the points $\boldsymbol{x}_0$ and $\boldsymbol{x}$. Two fields $\phi_{\mathcal{P}_i}(\boldsymbol{x})$ are related as

$$\phi_{\mathcal{P}_1}(\boldsymbol{x}) = \exp\left(i\frac{e^*}{\hbar}\oint dy_k A_k\right)\phi_{\mathcal{P}_2}(\boldsymbol{x}) = \exp\left(i\frac{e^*\Phi}{\hbar}\right)\phi_{\mathcal{P}_2}(\boldsymbol{x}), \tag{5.7.7}$$

where the loop integration is made along a closed path made of $\mathcal{P}_1$ and $\mathcal{P}_2$, and Φ is the magnetic flux through the loop [Fig.5.6(b)]. The relation (5.7.7) is reminiscent of the characteristic property (8.2.3) of the exchange phase of anyons, as we shall discuss later. The phase $e^*\Phi/\hbar$ generated is called the Aharonov-Bohm phase.

We make an important notice on the Hamiltonian. Based on the minimal substitution (5.3.10), it is given by

$$H = \int d^3x\left\{-\frac{1}{2m}\left[D_k\phi(\boldsymbol{x})\right]^\dagger D_k\phi(\boldsymbol{x}) + V(\boldsymbol{x})\phi^\dagger(\boldsymbol{x})\phi(\boldsymbol{x})\right\}. \tag{5.7.8}$$

We may rewrite this in terms of the path-dependent field $\phi_{\mathcal{P}}(\boldsymbol{x})$ with the use of the

---

[5] The field (5.7.5) is invariant under the gauge transformation (5.3.5) such that $f(\boldsymbol{x}_0) = 0$.

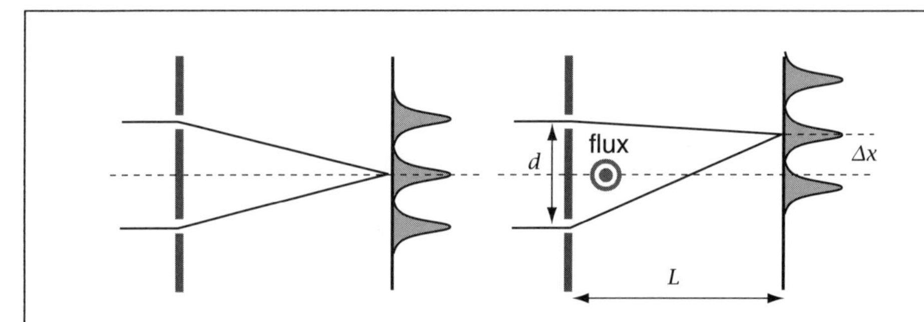

Fig. 5.7 The Aharonov-Bohm effect is observed in the 2-slit interference experiment, where electrons produce a characteristic interference pattern. Magnetic flux Φ is confined inside the solenoid. Let δ be the Aharonov-Bohm phase in (5.7.7). The interference pattern is shown to move by an amount $\Delta x = (L\lambda/2\pi d)\delta = (L\lambda e^*/2\pi d\hbar)\Phi$, which has been experimentally observed.

relation (5.7.6),

$$H = \int d^3x \left\{ -\frac{1}{2m}\partial_k \phi^\dagger_{\mathcal{P}}(\boldsymbol{x})\partial_k \phi_{\mathcal{P}}(\boldsymbol{x}) + V(\boldsymbol{x})\phi^\dagger_{\mathcal{P}}(\boldsymbol{x})\phi_{\mathcal{P}}(\boldsymbol{x}) \right\}. \tag{5.7.9}$$

Thus the gauge field is disappeared from the Hamiltonian. This is natural since $\phi_{\mathcal{P}}(\boldsymbol{x})$ is a gauge invariant field.

The Aharonov-Bohm phase has a physical effect. Electrons produce an interference pattern in the 2-slit interference experiment [Fig.5.7]. The idea of Aharonov and Bohm was to introduce a small solenoid behind the wall between the two slits. Magnetic flux Φ is confined inside the solenoid so that electrons never see the magnetic flux. Nevertheless, the interference pattern is found to move as the magnetic flux Φ changes. More sophisticated experiments[33] were proposed and successfully carried out to prove the Aharonov-Bohm effect. Consequently, although the electromagnetic potential A_k is not a local observable, its global property has a physical effect.

We give a simple dynamical model demonstrating the Aharonov-Bohm effect. We consider a particle moving in a potential $V(r)$ with r the distance from the center of a cylinder. The quantum-mechanical Hamiltonian is given by

$$H = \frac{1}{2M}\left(\boldsymbol{p} + e^*\boldsymbol{A}(\boldsymbol{x})\right)^2 + V(r) \tag{5.7.10}$$

with

$$\boldsymbol{A} = \frac{\Phi}{2\pi}\nabla\theta(\boldsymbol{x}). \tag{5.7.11}$$

We solve the Schrödinger equation by assuming that the particle is restricted to

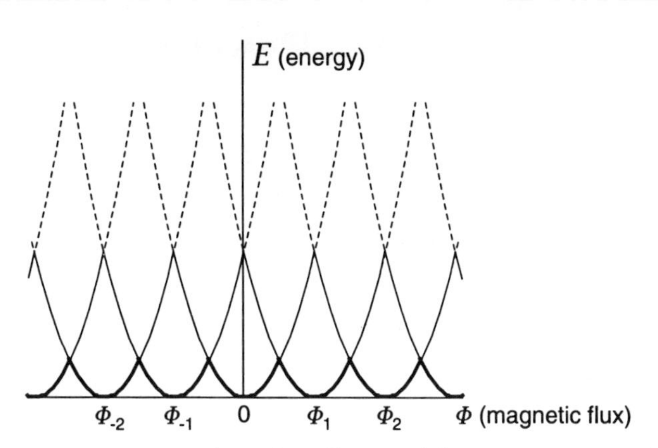

Fig. 5.8 The energy spectrum (5.7.13) is illustrated for a particle encircling a magnetic flux. The horizontal axis represents the flux Φ. The energy $E(\Phi)$ vanishes at $\Phi_n = 2\pi\hbar n/e^*$. The thick curve is for the ground state, the thin curve for the first excited state, and the dotted curve for the second excited state. The energy $E(\Phi)$ is a periodic function of Φ.

move along a circle with a fixed radius R. The solution is given by

$$\mathfrak{S}(\theta) = \frac{1}{2\pi} e^{in\theta}, \tag{5.7.12}$$

where n is an integer due to the single-valued condition. We find the energy of a particle as a function of flux Φ,

$$E = \frac{\hbar^2}{2MR^2} \left(n + \frac{e^*\Phi}{2\pi\hbar} \right)^2, \tag{5.7.13}$$

by evaluating $H\mathfrak{S}(\theta) = E\mathfrak{S}(\theta)$: See Fig.5.8. The ground-state energy is given by choosing n so that E takes the minimum value.

Although the magnetic field is absent in the path of the particle, the energy depends on the magnetic flux Φ which the path encircles. The Aharonov-Bohm phase produces effectively an angular momentum. It is interesting that we can use the energy formula (5.7.13) to determine the effective mass of composite fermions in antidot lattices.[370]

CHAPTER 6

DIRAC FIELD

An electron, moving at high speed, becomes relativistic and is described by the Dirac equation. The Dirac field has 4 components, the up-spin and down-spin states for a particle and an antiparticle. An antiparticle is identified with a hole in condensed matter physics. It is intriguing that electrons in graphene are described by the massless Dirac equation, though their speed is given by the Fermi velocity. It is also interesting that edge excitations in a QH droplet are described by Weyl fermions.

6.1 DIRAC EQUATION

Relativistic equations contain negative-energy solutions due to the Einstein relation, $E^2 = c^2\boldsymbol{p}^2 + m^2c^4$. Historically, to avoid the negative-energy problem, Dirac invented the first-order differential equation called the Dirac equation. As we shall see, the Dirac equation yields also negative energy solutions. This problem is resolved by making the second quantization.

The Dirac equation is a first-order differential equation given by

$$(-i\hbar\gamma^\mu\partial_\mu + mc)\psi(x) = 0, \tag{6.1.1}$$

or

$$(\gamma^\mu p_\mu + mc)\psi(p) = 0. \tag{6.1.2}$$

Here, γ^μ is a constant factor but it cannot be a c-number. Multiplying $(-\gamma^\nu p_\nu + mc)$ to this equation, and keeping the ordering between γ^ν and γ^μ, we obtain a second order differential equation,

$$(-\gamma^\nu p_\nu + mc)(\gamma^\mu p_\mu + mc)\psi(p) = \left(-\gamma^\nu\gamma^\mu p_\nu p_\mu + m^2c^2\right)\psi(p) = 0. \tag{6.1.3}$$

A relativistic particle should satisfy the Einstein relation. Namely the field ψ

should satisfy the Klein-Gordon equation, $(p^\mu p_\mu + m^2c^2)\psi(p) = 0$, as implies

$$\gamma^\nu\gamma^\mu + \gamma^\mu\gamma^\nu = -2g^{\mu\nu} \equiv -2\begin{pmatrix} -1 & 0 & 0 & 0 \\ 0 & 1 & 0 & 0 \\ 0 & 0 & 1 & 0 \\ 0 & 0 & 0 & 1 \end{pmatrix}. \tag{6.1.4}$$

The factor γ^μ is said to form the Clifford algebra.

Since $\gamma^0\gamma^0 = 1$, we rewrite the Dirac equation (6.1.1) as

$$i\hbar\partial_t\psi = H\psi, \tag{6.1.5}$$

with

$$H = -i\hbar c\gamma^0\gamma_j\partial_j + mc^2\gamma^0 = c\left(\alpha_j\frac{\hbar}{i}\partial_j + mc\beta\right), \tag{6.1.6}$$

where

$$\alpha_j = \gamma^0\gamma_j, \qquad \beta = \gamma^0. \tag{6.1.7}$$

We may regard this as the Schrödinger equation with the Hamiltonian (6.1.6), or[1]

$$H = c\boldsymbol{\alpha}\cdot\boldsymbol{p} + mc^2\beta. \tag{6.1.8}$$

It is necessary that $\beta^\dagger = \beta$ and $\alpha_j^\dagger = \alpha_j$, or

$$\gamma^{\mu\dagger} = \gamma^0\gamma^\mu\gamma^0, \tag{6.1.9}$$

for the Hermiticity of H.

There exists a set of 4×4 matrices satisfying the algebra (6.1.4) and the hermiticity property (6.1.9),

$$\gamma^0 = \begin{pmatrix} 1 & 0 \\ 0 & -1 \end{pmatrix}, \qquad \gamma^i = \begin{pmatrix} 0 & \sigma_i \\ -\sigma_i & 0 \end{pmatrix}, \qquad \gamma_5 = \begin{pmatrix} 0 & 1 \\ 1 & 0 \end{pmatrix}, \tag{6.1.10}$$

or

$$\alpha_j = \gamma^0\gamma_j = \begin{pmatrix} 0 & \sigma_j \\ \sigma_j & 0 \end{pmatrix}, \qquad \beta = \gamma^0 = \begin{pmatrix} 1 & 0 \\ 0 & -1 \end{pmatrix}, \tag{6.1.11}$$

where σ^i is the Pauli matrix. (We may equivalently use $i = 1,2,3$ or $i = x,y,z$.) They are called the Dirac matrices in the standard (or Dirac) representation.

There are 16 linear independent 4×4 matrices. They are explicitly given by

$$1, \qquad \gamma_\mu, \qquad \sigma_{\mu\nu} \equiv \frac{i}{2}[\gamma_\mu, \gamma_\nu], \qquad \gamma_\mu\gamma_5, \qquad \gamma_5, \tag{6.1.12}$$

---

[1]The velocity of light c in (6.1.8) is characteristic to the relativistic particle. Replacing it with the Fermi velocity (the velocity of electron in materials) we obtain the model Hamiltonian for electrons near the Fermi point in graphene: See Section 20.2 for more details on this point.

where we have defined

$$\gamma_5 = i\gamma^0\gamma^1\gamma^2\gamma^3. \tag{6.1.13}$$

An important property is that

$$M_{\mu\nu} \equiv \frac{1}{2}\sigma_{\mu\nu} \tag{6.1.14}$$

is the generator of the Lorentz transformation (4-dimentional rotation), about which we summarize in Appendix K.

The angular momentum is given by

$$S_i \equiv \frac{\hbar}{2}\varepsilon_{ijk}M_{jk} = \frac{i\hbar}{4}\varepsilon_{ijk}[\gamma_j, \gamma_k]. \tag{6.1.15}$$

Indeed, in the standard representation (6.1.10), it is easy to make a check of the relation for the angular momentum,

$$[S_x, S_y] = \hbar^2[M_{yz}, M_{zx}] = -i\hbar M_{yx} = i\hbar S_z, \tag{6.1.16}$$

and

$$S_x^2 + S_x^2 + S_x^2 = \frac{1}{2}\left(1 + \frac{1}{2}\right)\hbar^2. \tag{6.1.17}$$

Furthermore we have

$$S_z = \frac{i\hbar}{4}[\gamma_x, \gamma_y] = \frac{\hbar}{2}\begin{pmatrix} \sigma_z & 0 \\ 0 & \sigma_z \end{pmatrix}. \tag{6.1.18}$$

Hence, S_i describes the spin with the eigenvalues $\pm\frac{1}{2}\hbar$.

We construct the Lagrangian density to reproduce the Dirac equation (6.1.1). We naively expect it to be $\psi^\dagger(i\hbar\gamma^\mu\partial_\mu - mc)\psi$, but this is not a Hermitian quantity due to the relation (6.1.9). We introduce the Dirac conjugate of the field ψ by

$$\bar{\psi} = \psi^\dagger\gamma^0, \tag{6.1.19}$$

and define the Lagrangian density as[2]

$$\mathcal{L} = c\bar{\psi}(x)(i\hbar\gamma^\mu\partial_\mu - mc)\psi(x). \tag{6.1.20}$$

Since

$$(\bar{\psi}\gamma^\mu p_\mu\psi)^\dagger = (\psi^\dagger\gamma^0\gamma^\mu p_\mu\psi)^\dagger = \psi^\dagger p_\mu\gamma^{\mu\dagger}\gamma^0\psi = \bar{\psi}\gamma^\mu p_\mu\psi, \tag{6.1.21a}$$

$$(\bar{\psi}\psi)^\dagger = (\psi^\dagger\gamma^0\psi)^\dagger = \psi^\dagger\gamma^0\psi = \bar{\psi}\psi, \tag{6.1.21b}$$

the Lagrangian thus defined is Hermitian.

The Lagrangian is invariant under a global phase transformation with a real constant f,

$$\psi(x) \to e^{if}\psi(x). \tag{6.1.22}$$

[2]We have inserted the factor c in the Lagrangian density so that the anticommutation relation for ψ becomes (6.3.6).

The associated Nöther current is given by

$$j^\mu = \bar{\psi}\gamma^\mu\psi. \tag{6.1.23}$$

It conserves by way of the Dirac equation.

It is interesting to study the chiral transformation defined by the following transformation with a real constant f,

$$\psi(x) \to e^{if\gamma_5}\psi(x). \tag{6.1.24}$$

The Lagrangian (6.1.20) transforms as

$$\mathcal{L} \to \mathcal{L}' = c\bar{\psi}(x)(i\hbar\gamma^\mu\partial_\mu - mce^{2if\gamma_5})\psi(x). \tag{6.1.25}$$

Hence it is invariant if the mass term is absent in the Lagrangian ($m = 0$). The associated Nöther current, called the peudo-current, is given by

$$j_5^\mu = \bar{\psi}\gamma^\mu\gamma_5\psi. \tag{6.1.26}$$

It conserves only in the massless Dirac theory.

6.2 Plane Wave Solutions

We search for a complete set of solutions of the Dirac equation

$$i\hbar\partial_t\psi = H\psi, \tag{6.2.1}$$

where the Hamiltonian (6.1.8) is explicitly given by

$$H = c\begin{pmatrix} mc & \boldsymbol{\sigma}\cdot\boldsymbol{p} \\ \boldsymbol{\sigma}\cdot\boldsymbol{p} & -mc \end{pmatrix} \tag{6.2.2}$$

in the standard representation. To diagonalize this we note that

$$H^2 = c^2\begin{pmatrix} m^2c^2 + (\boldsymbol{\sigma}\cdot\boldsymbol{p})^2 & 0 \\ 0 & m^2c^2 + (\boldsymbol{\sigma}\cdot\boldsymbol{p})^2 \end{pmatrix}. \tag{6.2.3}$$

The eigenvalues are $\pm c\sqrt{m^2c^2 + \boldsymbol{p}^2}$ since $(\boldsymbol{\sigma}\cdot\boldsymbol{p})^2 = \boldsymbol{p}^2$. The eigen equations are

$$H\psi = \pm c\sqrt{m^2c^2 + \boldsymbol{p}^2}\psi. \tag{6.2.4}$$

It is necessary to discuss the massive and massless Dirac theories separately.

We first analyze the massive Dirac theory ($m \neq 0$): See Sections 6.5 and 6.6 for the massless Dirac theory. Taking advantage of the Lorentz invariance we choose the rest frame, where $\boldsymbol{p} = 0$. We may arrange the diagonalized Hamiltonian as

$$H = mc^2\begin{pmatrix} 1 & 0 & 0 & 0 \\ 0 & 1 & 0 & 0 \\ 0 & 0 & -1 & 0 \\ 0 & 0 & 0 & -1 \end{pmatrix}. \tag{6.2.5}$$

Obviously there are four independent solutions for the eigen equation (6.2.4) with (6.2.5), which we set

$$u_\uparrow(0)=\begin{pmatrix}1\\0\\0\\0\end{pmatrix},\quad u_\downarrow(0)=\begin{pmatrix}0\\1\\0\\0\end{pmatrix},\quad v_\downarrow(0)=\begin{pmatrix}0\\0\\1\\0\end{pmatrix},\quad v_\uparrow(0)=\begin{pmatrix}0\\0\\0\\1\end{pmatrix}. \tag{6.2.6}$$

These are eigenstates of the quantum-mechanical spin S_z of electrons (6.1.18) with eigenvalues $+\frac{1}{2}\hbar, -\frac{1}{2}\hbar, +\frac{1}{2}\hbar, -\frac{1}{2}\hbar$: See also (6.3.15). As we shall see later in (6.3.22), the spin index $\sigma=\uparrow$ and $\downarrow$ of v_σ represent the eigenvalues $+\frac{1}{2}\hbar$ and $-\frac{1}{2}\hbar$, respectively, for holes. The Dirac equation (6.2.1) is now solved as

$$\psi_\sigma^+(t)=\frac{1}{\sqrt{c}}e^{-i\frac{mc^2t}{\hbar}}u_\sigma(0),\quad \psi_\sigma^-(t)=\frac{1}{\sqrt{c}}e^{+i\frac{mc^2t}{\hbar}}v_\sigma(0), \tag{6.2.7}$$

where ψ_σ^+ and ψ_σ^- are positive-energy and negative-energy solutions, respectively.

We move to a general Lorentz frame. Due to the Lorentz invariance the rest-frame solutions (6.2.7) become the plane wave solutions,

$$\psi_\sigma^+(x)=c^{-1/2}e^{ipx/\hbar}u_\sigma(\boldsymbol{p}),\qquad \psi_\sigma^-(x)=c^{-1/2}e^{-ipx/\hbar}v_\sigma(\boldsymbol{p}). \tag{6.2.8}$$

Here

$$px=\boldsymbol{px}-tE_{\boldsymbol{p}},\qquad E_{\boldsymbol{p}}=c\sqrt{m^2c^2+\boldsymbol{p}^2}. \tag{6.2.9}$$

From the Dirac equation we find the wave functions to satisfy

$$\begin{aligned}(-i\hbar\gamma^\mu\partial_\mu+mc)\psi_\sigma^+(x)&=e^{ipx/\hbar}(\gamma^\mu p_\mu+mc)u_\sigma(\boldsymbol{p})=0,\\(-i\hbar\gamma^\mu\partial_\mu+mc)\psi_\sigma^-(x)&=-e^{-ipx/\hbar}(\gamma^\mu p_\mu-mc)v_\sigma(\boldsymbol{p})=0.\end{aligned} \tag{6.2.10}$$

Hence

$$\begin{aligned}(\gamma^\mu p_\mu+mc)u_\sigma(\boldsymbol{p})&=0,\qquad \bar{u}_\sigma(\boldsymbol{p})(\gamma^\mu p_\mu+mc)=0,\\(\gamma^\mu p_\mu-mc)v_\sigma(\boldsymbol{p})&=0,\qquad \bar{v}_\sigma(\boldsymbol{p})(\gamma^\mu p_\mu-mc)=0.\end{aligned} \tag{6.2.11}$$

We adopt the Lorentz invariant normalization condition[3],

$$\begin{aligned}\bar{u}_\sigma(\boldsymbol{p})u_{\sigma'}(\boldsymbol{p})&=\delta_{\sigma\sigma'},\qquad \bar{v}_\sigma(\boldsymbol{p})v_{\sigma'}(\boldsymbol{p})=-\delta_{\sigma\sigma'},\\\bar{u}_\sigma(\boldsymbol{p})v_{\sigma'}(\boldsymbol{p})&=0,\qquad \bar{v}_\sigma(\boldsymbol{p})u_{\sigma'}(\boldsymbol{p})=0.\end{aligned} \tag{6.2.12}$$

The completeness condition is

$$\sum_\sigma u_\sigma(\boldsymbol{p})\bar{u}_\sigma(\boldsymbol{p})-\sum_\sigma v_\sigma(\boldsymbol{p})\bar{v}_\sigma(\boldsymbol{p})=1. \tag{6.2.13}$$

[3]There exist two types of normalization. One is the Lorentz invariant one such that $\bar{u}_\sigma(\boldsymbol{p})u_{\sigma'}(\boldsymbol{p})=\delta_{\sigma\sigma'}$. The other is such that $u_\sigma^\dagger(\boldsymbol{p})u_{\sigma'}(\boldsymbol{p})=\delta_{\sigma\sigma'}$. The Lorentz invariant one is easy to handle with, since it is enough to check it in the rest frame. However, we cannot take the massless limit, as is clear in (6.2.15). Hence, we adopt the latter one in the case of the massless Dirac field.

These conditions hold in the rest frame (6.2.7) with the use of (6.2.6), and hence they hold in any frame.

The wave functions $u_\sigma(\boldsymbol{p})$ and $v_\sigma(\boldsymbol{p})$ are constructed by making a Lorentz transformation from $u_\sigma(0)$ and $v_\sigma(0)$. Actually, it is easier to construct them by requiring them to satisfy (6.2.11). Namely, using

$$(\gamma^\mu p_\mu - mc)(\gamma^\mu p_\mu + mc) = (\gamma^\mu p_\mu + mc)(\gamma^\mu p_\mu - mc) = 0, \tag{6.2.14}$$

we obtain

$$\begin{aligned} u_\sigma(\boldsymbol{p}) &= \frac{mc - \gamma^\mu p_\mu}{\sqrt{2m(mc^2 + \hbar\omega_p)}} u_\sigma(0), \\ v_\sigma(\boldsymbol{p}) &= \frac{mc + \gamma^\mu p_\mu}{\sqrt{2m(mc^2 + \hbar\omega_p)}} v_\sigma(0). \end{aligned} \tag{6.2.15}$$

The normalization factor has been chosen to satisfy (6.2.12).

6.3 Canonical Quantization

To carry out canonical quantization, we introduce the canonical momentum

$$\pi = \frac{\partial \mathcal{L}}{\partial \dot{\psi}} = i\hbar\psi^\dagger \tag{6.3.1}$$

with the use of the Lagrangian density (6.1.20). The Hamiltonian reads

$$H = \int d^3x \left[\frac{\partial \mathcal{L}}{\partial \dot{\psi}} \dot{\psi} - \mathcal{L} \right] = c \int d^3x\, \bar{\psi}(-i\hbar\gamma^j \partial_j + mc)\psi. \tag{6.3.2}$$

As we shall see in (6.3.6), we impose the canonical anticommutation relation rather than the canonical commutation relation between the Dirac field and its canonical momentum.

We expand the Dirac field in terms of plane waves,

$$\psi(x) = \sum_\sigma \int \frac{d^3k}{\sqrt{(2\pi)^3}} \sqrt{\frac{mc^2}{\hbar\omega_k}} \left(c_\sigma(\boldsymbol{k}) u_\sigma(\boldsymbol{k}) e^{ikx} + d_\sigma^\dagger(\boldsymbol{k}) v_\sigma(\boldsymbol{k}) e^{-ikx} \right), \tag{6.3.3a}$$

$$\bar{\psi}(x) = \sum_\sigma \int \frac{d^3k}{\sqrt{(2\pi)^3}} \sqrt{\frac{mc^2}{\hbar\omega_k}} \left(c_\sigma^\dagger(\boldsymbol{k}) \bar{u}_\sigma(\boldsymbol{k}) e^{-ikx} + d_\sigma(\boldsymbol{k}) \bar{v}_\sigma(\boldsymbol{k}) e^{ikx} \right), \tag{6.3.3b}$$

where

$$\hbar\omega_k = c\sqrt{\hbar^2 \boldsymbol{k}^2 + m^2c^2}. \tag{6.3.4}$$

Substituting the expansion (6.3.3) into the Hamiltonian (6.3.2), and using the orthonormal condition (6.2.12) of the plane waves, we obtain

$$H = \sum_s \int d^3k\, \hbar\omega_k \left[c_\sigma^\dagger(\boldsymbol{k}) c_\sigma(\boldsymbol{k}) - d_\sigma(\boldsymbol{k}) d_\sigma^\dagger(\boldsymbol{k}) \right]. \tag{6.3.5}$$

In this derivation the ordering of operators has not been interchanged. In classical theory $c_\sigma(\boldsymbol{k})$ and $d_\sigma(\boldsymbol{k})$ are complex numbers interpreted as the probability amplitude, and this energy formula would lead to the negative-energy problem.

To overcome the negative-energy problem, we impose the canonical anticommutation relation $\{\psi_\alpha, \pi_\beta\} = i\hbar\delta_{\alpha\beta}\delta(\boldsymbol{x}-\boldsymbol{y})$, or

$$\{\psi_\alpha(t,\boldsymbol{x}), \psi_\beta(t,\boldsymbol{y})\} = \{\psi^\dagger_\alpha(t,\boldsymbol{x}), \psi^\dagger_\beta(t,\boldsymbol{y})\} = 0, \tag{6.3.6a}$$

$$\{\psi_\alpha(t,\boldsymbol{x}), \psi^\dagger_\beta(t,\boldsymbol{y})\} = \delta_{\alpha\beta}\delta(\boldsymbol{x}-\boldsymbol{y}). \tag{6.3.6b}$$

Substituting the expansion (6.3.3) into (6.3.6), they are checked to hold provided that

$$\{c_\sigma(\boldsymbol{k}), c_{\sigma'}(\boldsymbol{k}')\} = \{d_\sigma(\boldsymbol{k}), d_{\sigma'}(\boldsymbol{k}')\} = 0, \tag{6.3.7a}$$

$$\{c_\sigma(\boldsymbol{k}), c^\dagger_{\sigma'}(\boldsymbol{k}')\} = \{d_\sigma(\boldsymbol{k}), d^\dagger_{\sigma'}(\boldsymbol{k}')\} = \delta_{\sigma\sigma'}\delta(\boldsymbol{k}-\boldsymbol{k}'). \tag{6.3.7b}$$

We define the Fock vacuum by

$$c(\boldsymbol{k},s)|0\rangle = d(\boldsymbol{k},s)|0\rangle = 0, \tag{6.3.8}$$

upon which we construct the Fock space.

With the use of the anticommutation relation (6.3.7) the Hamiltonian (6.3.5) is rewritten as

$$H = \sum_\sigma \int d^3k\, \hbar\omega_k \left[c^\dagger_\sigma(\boldsymbol{k})c_\sigma(\boldsymbol{k}) + d^\dagger_\sigma(\boldsymbol{k})d_\sigma(\boldsymbol{k})\right] - \delta(0)\sum_s \int d^3k\, \hbar\omega_k. \tag{6.3.9}$$

The infinitely large c-number quantity $-\delta(0)\sum_s \int d^3k\, \hbar\omega_k$ is a zero-point energy[4] encountered also in the Klein-Gordon theory. We can drop it out by defining the Hamiltonian together with normal ordering. The momentum operator is given by

$$\boldsymbol{P} = \int d^3x\, \bar{\psi}(-i\hbar\nabla)\psi = \sum_\sigma \int d^3k\, \hbar\boldsymbol{k} \left[c^\dagger_\sigma(\boldsymbol{k})c_\sigma(\boldsymbol{k}) + d^\dagger_\sigma(\boldsymbol{k})d_\sigma(\boldsymbol{k})\right]. \tag{6.3.10}$$

It follows that $c_\sigma(\boldsymbol{k})$ and $d_\sigma(\boldsymbol{k})$ are the annihilation operators of fermions with the energy $\hbar\omega_k$ and the momentum $\hbar\boldsymbol{k}$.

We can distinguish the c-quantum and d-quantum by way of the charge operator. The charge density is given by $J^0 = \bar{\psi}\gamma^0\psi = \psi^\dagger\psi$ with (6.1.23). The total charge is

$$Q = \int d^3x\, \psi^\dagger\psi = \sum_\sigma \int d^3k \left[c^\dagger_\sigma(\boldsymbol{k})c_\sigma(\boldsymbol{k}) - d^\dagger_\sigma(\boldsymbol{k})d_\sigma(\boldsymbol{k})\right]. \tag{6.3.11}$$

The c-quantum has the positive charge while the d-quantum has the negative charge. We interpret that the d-quantum is the *antiparticle* of the c-quantum: The

---

[4]It is intriguing that the zero-point energies are positive in the Hamiltonian H_B of the complex Klein-Gordon field (3.4.11) and negative in the Hamiltonian H_F of the Dirac field (6.3.9). They are canceled out in $H_{BF} = H_B + H_F$, provided the masses of bosons and fermions are equal. This is reminiscent of the supersymmetry in the boson-fermion system H_{BF}.

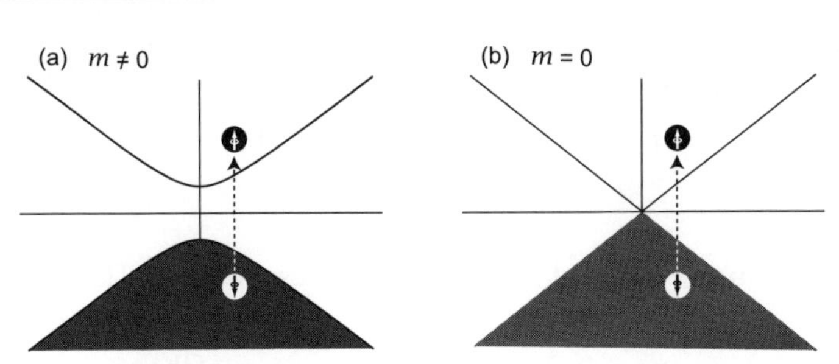

Fig. 6.1 The Dirac sea together with an electron-hole pair excitation is illustrated for the massive ($m \neq 0$) and massless ($m = 0$) cases. All the negative-energy states are filled up in the Dirac sea. A hole is left behind, when a negative energy electron is excited into a postive energy state. We renormalize the quantum number of the Dirac sea to be zero, by taking it as the ground state. Then, the quantum numbers of a hole become opposite to those of an electron.

particle and the antiparticle are distinguished by their opposite charges. When the c-quantum represents the electron, the d-quantum represents the *hole* in condensed matter physics while the *positron* in high energy physics.

Dirac Sea

We explain the original idea of Dirac to overcome the negative energy problem by introducing the *Dirac sea* and the hole (positron) state [Fig.6.1]. Let us quantize the Dirac field as if it were a nonrelativistic field as in (3.2.15). Since there exist both the positive and negative energy solutions we expand the Dirac field as

$$\psi(x) = \sum_{\sigma} \int \frac{d^3k}{\sqrt{(2\pi)^3}} \sqrt{\frac{mc^2}{\hbar\omega_k}} \left(c_\sigma^+(\boldsymbol{k}) u_\sigma^+(\boldsymbol{k}) e^{i\boldsymbol{kx}-i\omega_k t} + c_\sigma^-(\boldsymbol{k}) u_\sigma^-(\boldsymbol{k}) e^{i\boldsymbol{kx}+i\omega_k t} \right). \tag{6.3.12}$$

Then,

$$H = \sum_{\sigma} \int d^3k \, E_{\boldsymbol{k}} \left[c_\sigma^{+\dagger}(\boldsymbol{k}) c_\sigma^+(\boldsymbol{k}) - c_\sigma^{-\dagger}(\boldsymbol{k}) c_\sigma^-(\boldsymbol{k}) \right], \tag{6.3.13}$$

$$\boldsymbol{P} = \sum_{\sigma} \int d^3k \, \hbar\boldsymbol{k} \left[c_\sigma^{+\dagger}(\boldsymbol{k}) c_\sigma^+(\boldsymbol{k}) + c_\sigma^{-\dagger}(\boldsymbol{k}) c_\sigma^-(\boldsymbol{k}) \right]. \tag{6.3.14}$$

Hence $c_\sigma^\pm(\boldsymbol{k})$ annihilates the electron having the energy $\pm E_{\boldsymbol{k}}$ and the momentum $\hbar\boldsymbol{k}$.

We study the spin. The quantum-mechanical spin S_z is given by (6.1.18), whose eigenstates are

$$u_\uparrow^+(0) = \begin{pmatrix} 1 \\ 0 \\ 0 \\ 0 \end{pmatrix}, \quad u_\downarrow^+(0) = \begin{pmatrix} 0 \\ 1 \\ 0 \\ 0 \end{pmatrix}, \quad u_\uparrow^-(0) = \begin{pmatrix} 0 \\ 0 \\ 1 \\ 0 \end{pmatrix}, \quad u_\downarrow^-(0) = \begin{pmatrix} 0 \\ 0 \\ 0 \\ 1 \end{pmatrix} \tag{6.3.15}$$

with the eigenvalues $+\frac{1}{2}\hbar$, $-\frac{1}{2}\hbar$, $+\frac{1}{2}\hbar$ and $-\frac{1}{2}\hbar$, respectively. The spin density operator is defined by

$$\mathcal{S}_i(x) = \psi^\dagger(x) S_i \psi(x), \tag{6.3.16}$$

with (6.1.15). The total spin is

$$\mathcal{S}_i = \int d^3x\, \mathcal{S}_i(x). \tag{6.3.17}$$

We are interested in the spin of the state $c_\sigma^{\pm\dagger}(\boldsymbol{k})|0\rangle$ in the rest frame,

$$\langle 0|c_\sigma^\pm(0)\mathcal{S}_z c_\sigma^{\pm\dagger}(0)|0\rangle = \frac{\hbar}{2} u_\sigma^{\pm\dagger}(0) \begin{pmatrix} \sigma_z & 0 \\ 0 & \sigma_z \end{pmatrix} u_\sigma^\pm(0). \tag{6.3.18}$$

Hence, $c_\sigma^\pm(\boldsymbol{k})$ annihilates the electron having the spin $\sigma = \uparrow, \downarrow$ in the rest frame.

We assume that the ground state is such that all negative-energy states are occupied and all positive-energy states are empty. It is called the Dirac sea. Since electrons are fermions obeying the exclusion principle, no additional negative-energy electrons are created. The ground state is defined by

$$c_\sigma^+(\boldsymbol{k})|0\rangle = c_\sigma^{-\dagger}(\boldsymbol{k})|0\rangle = 0. \tag{6.3.19}$$

The ground-state energy is $\langle 0|H|0\rangle = -\delta(0)\sum_s \int d^3k\, \hbar\omega_k$, which is the constant term in (6.3.9).

When we remove one negative-energy electron from the Dirac sea, it is a hole state. If the negative electron has the momentum $\hbar\boldsymbol{k}$ and the spin σ, the remaining hole must have the momentum $-\hbar\boldsymbol{k}$ and the spin $\bar{\sigma}$, where $\bar{\sigma} = (\downarrow, \uparrow)$ corresponding to $\sigma = (\uparrow, \downarrow)$. Namely, as illustrated in Fig.6.1, we should identify

$$c_\uparrow(\boldsymbol{k}) = c_\uparrow^+(\boldsymbol{k}), \quad c_\downarrow(\boldsymbol{k}) = c_\downarrow^-(\boldsymbol{k}), \quad d_\uparrow^\dagger(\boldsymbol{k}) = c_\downarrow^-(-\boldsymbol{k}), \quad d_\downarrow^\dagger(\boldsymbol{k}) = c_\uparrow^-(-\boldsymbol{k}). \tag{6.3.20}$$

We change the integration variable as $\boldsymbol{k} \to -\boldsymbol{k}$ in the negative-energy part of the field operator (6.3.12), and substitute this identification,

$$\psi(x) = \sum_\sigma \int \frac{d^3k}{\sqrt{(2\pi)^3}} \sqrt{\frac{mc^2}{\hbar\omega_k}} \left(c_\sigma(\boldsymbol{k}) u_\sigma^+(\boldsymbol{k}) e^{i\boldsymbol{kx} - i\omega_k t} + d_\sigma^\dagger(\boldsymbol{k}) u_{\bar{\sigma}}^-(-\boldsymbol{k}) e^{-i\boldsymbol{kx} + i\omega_k t} \right). \tag{6.3.21}$$

Comparing this with (6.3.3) we have

$$u_\uparrow(\boldsymbol{k}) = u_\uparrow^+(\boldsymbol{k}), \quad u_\downarrow(\boldsymbol{k}) = u_\downarrow^-(\boldsymbol{k}), \quad v_\uparrow(\boldsymbol{k}) = u_\downarrow^-(-\boldsymbol{k}), \quad v_\downarrow(\boldsymbol{k}) = u_\uparrow^-(-\boldsymbol{k}). \tag{6.3.22}$$

Indeed, with this identification we reproduce (6.2.6) from (6.3.15) in the rest frame. Consequently, $d_\sigma^\dagger(\boldsymbol{k})$ creates a hole with the spin σ in the rest frame. This is the reason why we have chosen the spin index as in (6.2.6).

6.4 Interaction with Electromagnetic Field

The coupling between the Dirac fermion and the electromagnetic potential is determined by employing the minimal substitution (5.3.10), $\partial_\mu \to \partial_\mu + i\frac{e}{\hbar}A_\mu$. The total Lagrangian becomes

$$\mathcal{L} = \mathcal{L}_{\text{EM}} + \mathcal{L}_{\text{matter}} + \mathcal{L}_{\text{I}} = -\frac{1}{4\mu_{\text{m}}}F_{\mu\nu}F^{\mu\nu} + c\bar{\psi}(x)\left[i\hbar\gamma^\mu\left(\partial_\mu + i\frac{e}{\hbar}A_\mu\right) - mc\right]\psi(x). \tag{6.4.1}$$

The interaction between the electron and the photon is described by the interaction Hamiltonian

$$\mathcal{H}_{\text{I}} = -\mathcal{L}_{\text{I}} = ceA_\mu(x)\bar{\psi}(x)\gamma^\mu\psi(x). \tag{6.4.2}$$

The 4-current is

$$J_\mu = \frac{\partial \mathcal{L}_{\text{I}}}{\partial A^\mu} = -ce\,\bar{\psi}(x)\gamma_\mu\psi(x) \tag{6.4.3}$$

as in (5.3.13). The Maxwell equation reads

$$\frac{\partial \mathcal{L}_{\text{EM}}}{\partial A^\mu} = -J_\mu = ce\,\bar{\psi}(x)\gamma_\mu\psi(x), \tag{6.4.4}$$

or

$$\frac{1}{\mu_{\text{m}}}\partial^\nu F_{\mu\nu} = J_\mu = -ce\,\bar{\psi}(x)\gamma_\mu\psi(x). \tag{6.4.5}$$

Thus, $J_\mu = -ej_\mu$ with (6.1.23). Recall that we have used the convention that the electron charge is $-e$.

We evaluate the interaction Hamiltonian (6.4.2) for a static external field $A_\mu^{\text{ext}}(\boldsymbol{x})$ and two electron states $|\boldsymbol{p},\sigma\rangle = c_\sigma^\dagger(\boldsymbol{p})|0\rangle$ and $|\boldsymbol{p}',\sigma'\rangle = c_{\sigma'}^\dagger(\boldsymbol{p}')|0\rangle$,

$$\langle \boldsymbol{p}',\sigma'|H_{\text{I}}|\boldsymbol{p},\sigma\rangle_{\text{e}} = \frac{mec^3}{\sqrt{\hbar^2\omega_p\omega_{p'}}}A_\mu^{\text{ext}}(\boldsymbol{k})\bar{u}_{\sigma'}(\boldsymbol{p}')\gamma^\mu u_\sigma(\boldsymbol{p}) \tag{6.4.6}$$

with $\hbar\boldsymbol{k} = \boldsymbol{p}' - \boldsymbol{p}$. From the Clifford algebra (6.1.4) and the definition (6.1.12) of $\sigma_{\mu\nu}$, we get

$$\gamma^\mu\gamma^\nu = -g^{\mu\nu} - i\sigma^{\mu\nu}. \tag{6.4.7}$$

Multiplying p_ν and using the Dirac equation (6.2.11) for $u(\boldsymbol{p},s)$ we obtain

$$\gamma^\mu u_\sigma(\boldsymbol{p}) = \frac{1}{mc}(p^\mu + i\sigma^{\mu\nu}p_\nu)u_\sigma(\boldsymbol{p}). \tag{6.4.8}$$

Similarly we have

$$\bar{u}_{\sigma'}(\boldsymbol{p}')\gamma^\mu = \frac{1}{mc}\bar{u}_{\sigma'}(\boldsymbol{p}')(p'^\mu - i\sigma^{\mu\nu}p'_\nu). \tag{6.4.9}$$

Combining these we get

$$\begin{aligned}&\langle \boldsymbol{p}',\sigma'|H_{\rm I}|\boldsymbol{p},\sigma\rangle_{\rm e}\\ &=\frac{mc^2}{\sqrt{\hbar^2\omega_p\omega_{p'}}}\bar{u}_{\sigma'}(\boldsymbol{p}')\left[\frac{e}{2m}\left(p'^\mu + p^\mu\right)A^{\rm ext}_\mu(\boldsymbol{k}) + i\frac{e}{2m}\sigma^{\mu\nu}k_\nu A^{\rm ext}_\mu(\boldsymbol{k})\right]u_\sigma(\boldsymbol{p})\end{aligned} \tag{6.4.10}$$

for electrons.

In particular, for constant external magnetic field, where $A_0^{\rm ext} = 0$ and $\nabla\times\boldsymbol{A}^{\rm ext} = \boldsymbol{B}$, only the space component is relevant. Making use of

$$\sigma^{ij} = -2i\varepsilon^{ijk}\sigma_k\begin{pmatrix}1 & 0\\ 0 & -1\end{pmatrix}, \tag{6.4.11}$$

we derive the Zeeman interaction,

$$\langle \boldsymbol{p}',\sigma'|H_{\rm I}|\boldsymbol{p},\sigma\rangle_{\rm e} = \frac{mc^2}{\sqrt{\hbar^2\omega_p\omega_{p'}}}\bar{u}_{\sigma'}(\boldsymbol{p}')\left[\frac{e}{2m}\left(\boldsymbol{p}' + \boldsymbol{p}\right)\boldsymbol{A}^{\rm ext} + g\mu_B\boldsymbol{S}\cdot\boldsymbol{B}\right]u_\sigma(\boldsymbol{p}), \tag{6.4.12}$$

where $S_j = \frac{1}{2}\hbar\sigma_j$ is the spin, and we have set

$$\mu_B \equiv \frac{e\hbar}{2m} = 5.79\times 10^{-9}\text{eV/G}, \tag{6.4.13}$$

and

$$g = 2. \tag{6.4.14}$$

The parameters μ_B and g are called the Bohr magneton and the gyromagnetic factor, respectively.

Similarly, we treat the interaction Hamiltonian (6.4.2) for a static external field $A^{\rm ext}_\mu(\boldsymbol{x})$ and two positron (hole) states $|\boldsymbol{p},\sigma\rangle = d^\dagger_\sigma(\boldsymbol{p})|0\rangle$ and $|\boldsymbol{p}',\sigma'\rangle = d^\dagger_{\sigma'}(\boldsymbol{p}')|0\rangle$. In so doing we require the normal ordering of the interaction Hamiltonian,

$$\begin{aligned}:\bar{\psi}(x)\gamma^\mu\psi(x): \simeq\, &:\left(c^\dagger_{\sigma'}\bar{u}_{\sigma'}e^{-ikx} + d_{\sigma'}\bar{v}_{\sigma'}e^{ikx}\right)\gamma^\mu\left(c_\sigma u_\sigma e^{ikx} + d^\dagger_\sigma v_\sigma e^{-ikx}\right):\\ =&\bar{u}_{\sigma'}\gamma^\mu u_\sigma c^\dagger_{\sigma'}c_\sigma - \bar{v}_{\sigma'}\gamma^\mu v_\sigma d^\dagger_\sigma d_{\sigma'} + \cdots,\end{aligned} \tag{6.4.15}$$

and we find

$$\langle \boldsymbol{p},\sigma|:H_{\rm I}:|\boldsymbol{p}',\sigma'\rangle_{\rm hole} = \frac{mc^2}{\sqrt{\hbar^2\omega_p\omega_{p'}}}\bar{v}_{\sigma'}(\boldsymbol{p}')\left[\frac{e}{2m}\left(\boldsymbol{p}' + \boldsymbol{p}\right)\boldsymbol{A}^{\rm ext} + 2\mu_B\boldsymbol{S}\cdot\boldsymbol{B}\right]v_\sigma(\boldsymbol{p}). \tag{6.4.16}$$

By comparing this with (6.4.12), the hole (positron) has the positive charge, while the Zeeman energy is opposite to that of the electron with the same spin.

Electrons and Holes in Nonrelativistic Limit

We study the nonrelativistic limit of the Dirac Hamiltonian in the presence of the magnetic field. We make the minimal substitution (5.3.10) in the Hamiltonian (6.2.2) as

$$H = c\begin{pmatrix} mc & \boldsymbol{\sigma}\cdot\boldsymbol{P} \\ \boldsymbol{\sigma}\cdot\boldsymbol{P} & -mc \end{pmatrix}, \tag{6.4.17}$$

from which we obtain

$$H^2 = c^2\begin{pmatrix} (\boldsymbol{\sigma}\cdot\boldsymbol{P})^2 + m^2c^2 & 0 \\ 0 & -(\boldsymbol{\sigma}\cdot\boldsymbol{P})^2 + m^2c^2 \end{pmatrix} \tag{6.4.18}$$

in place of (6.2.3). The eigenvalues are $\pm c\sqrt{(\boldsymbol{\sigma}\cdot\boldsymbol{P})^2 + m^2c^2}$.

In the nonrelativistic limit, where $mc \gg |\boldsymbol{\sigma}\cdot\boldsymbol{P}|$, we find

$$c\sqrt{(\boldsymbol{\sigma}\cdot\boldsymbol{P})^2 + m^2c^2} = mc^2\left(1 + \frac{(\boldsymbol{\sigma}\cdot\boldsymbol{P})^2}{m^2c^2}\right)^{1/2} \simeq mc^2 + \frac{1}{2m}(\boldsymbol{\sigma}\cdot\boldsymbol{P})^2. \tag{6.4.19}$$

This is the Pauli Hamiltonian,

$$H_{\mathrm{P}} = \frac{1}{2m}(\boldsymbol{\sigma}\cdot\boldsymbol{P})^2, \tag{6.4.20}$$

apart from a trivial constant. Namely, the Dirac Hamiltonian consists of two Pauli Hamiltonians,

$$H \simeq \begin{pmatrix} H_{\mathrm{P}} & 0 \\ 0 & -H_{\mathrm{P}} \end{pmatrix}. \tag{6.4.21}$$

It is straightforward to find

$$H_{\mathrm{P}} = \frac{1}{2m}\left[\sigma_j\left(p_j + eA_j\right)\right]^2 = \frac{1}{2m}\left(p_j + eA_j\right)^2 + \frac{e\hbar}{m}S_jB_j, \tag{6.4.22}$$

with the use of the relation $\sigma_i\sigma_j = \delta_{ij} + i\varepsilon_{ijk}\sigma_k$. Extracting the electron-photon interaction part from (6.4.22), we express it as

$$H_{\mathrm{I}} = \frac{e}{2m}(\boldsymbol{pA} + \boldsymbol{Ap}) + g\mu_B S_j B_j, \tag{6.4.23}$$

where μ_B is the Bohr magneton (6.4.13), and $g = 2$ is the magnetic g-factor. This interaction Hamiltonian agrees with the nonrelativistic limit of (6.4.12).

6.5 WEYL FIELD (MASSLESS DIRAC FIELD)

It is necessary to analyze the massless Dirac theory separately,

$$\mathcal{L} = i\hbar\bar{\psi}(x)\gamma^\mu\partial_\mu\psi(x), \tag{6.5.1}$$

since we cannot take the rest frame. As we have noticed in (6.1.25), it is invariant under the chiral transformation,

$$\psi(x) \to e^{if\gamma^5}\psi(x), \tag{6.5.2}$$

and the chiral current

$$j_5^\mu = \bar{\psi}\gamma^\mu\gamma_5\psi \tag{6.5.3}$$

conserves. Consequently the Hamiltonian commutes with γ_5.

More explicitly, in the standard representation the Hamiltonian (6.2.2) reads

$$H = c\begin{pmatrix} 0 & \boldsymbol{\sigma}\cdot\boldsymbol{p} \\ \boldsymbol{\sigma}\cdot\boldsymbol{p} & 0 \end{pmatrix} = c\boldsymbol{\sigma}\cdot\boldsymbol{p}\gamma_5. \tag{6.5.4}$$

It is obvious that $[\gamma_5, H] = 0$.

Accordingly we can label the eigenstate of the Hamiltonian by the eigenvalue of γ_5. The eigenvalues of γ_5 are ± 1, since $(\gamma_5)^2 = 1$. We call the eigenstate with $\gamma_5 = +1$ (-1) the right-handed (left-handed) chirality state. The Dirac equation (6.2.1) yields

$$i\hbar\partial_t\psi_{\mathrm{R}} = c\boldsymbol{\sigma}\cdot\boldsymbol{p}\psi_{\mathrm{R}} = -i\hbar c\boldsymbol{\sigma}\cdot\nabla\psi_{\mathrm{R}}, \tag{6.5.5a}$$

$$i\hbar\partial_t\psi_{\mathrm{L}} = -c\boldsymbol{\sigma}\cdot\boldsymbol{p}\psi_{\mathrm{L}} = i\hbar c\boldsymbol{\sigma}\cdot\nabla\psi_{\mathrm{L}} \tag{6.5.5b}$$

for the right-handed (ψ_{R}) and left-handed (ψ_{L}) wave functions, each of which is a two-component equation. They are called the Weyl equations. It is notable that the right-handed and left-handed components do not mix one to another. Thus it is possible to consider the world consisting solely of the right-handed or left-handed fermions. They are called Weyl fermions. Neutrinos are left-handed Weyl fermions provided they are massless.

The eigen equation (6.2.4) reads

$$H\psi = \pm c|\boldsymbol{p}|\psi \tag{6.5.6}$$

with (6.5.4). There are four independent eigenfunctions, obeying

$$\boldsymbol{\sigma}\cdot\boldsymbol{p}u_{\mathrm{R}}(\boldsymbol{p}) = |\boldsymbol{p}|u_{\mathrm{R}}(\boldsymbol{p}), \qquad \boldsymbol{\sigma}\cdot\boldsymbol{p}v_{\mathrm{R}}(\boldsymbol{p}) = -|\boldsymbol{p}|v_{\mathrm{R}}(\boldsymbol{p}), \tag{6.5.7a}$$

$$\boldsymbol{\sigma}\cdot\boldsymbol{p}u_{\mathrm{L}}(\boldsymbol{p}) = -|\boldsymbol{p}|u_{\mathrm{L}}(\boldsymbol{p}), \qquad \boldsymbol{\sigma}\cdot\boldsymbol{p}v_{\mathrm{L}}(\boldsymbol{p}) = |\boldsymbol{p}|v_{\mathrm{L}}(\boldsymbol{p}). \tag{6.5.7b}$$

They are explicitly constructed as

$$u_{\mathrm{R}}(\boldsymbol{p}) = \alpha_{\mathrm{R}}\left(1 + \frac{\boldsymbol{\sigma}\cdot\boldsymbol{p}}{|\boldsymbol{p}|}\right)\begin{pmatrix}1\\0\end{pmatrix}_{\mathrm{R}}, \qquad v_{\mathrm{R}}(\boldsymbol{p}) = \beta_{\mathrm{R}}\left(1 - \frac{\boldsymbol{\sigma}\cdot\boldsymbol{p}}{|\boldsymbol{p}|}\right)\begin{pmatrix}0\\1\end{pmatrix}_{\mathrm{R}}, \tag{6.5.8a}$$

$$u_{\mathrm{L}}(\boldsymbol{p}) = \alpha_{\mathrm{L}}\left(1 - \frac{\boldsymbol{\sigma}\cdot\boldsymbol{p}}{|\boldsymbol{p}|}\right)\begin{pmatrix}1\\0\end{pmatrix}_{\mathrm{L}}, \qquad v_{\mathrm{L}}(\boldsymbol{p}) = \beta_{\mathrm{L}}\left(1 + \frac{\boldsymbol{\sigma}\cdot\boldsymbol{p}}{|\boldsymbol{p}|}\right)\begin{pmatrix}0\\1\end{pmatrix}_{\mathrm{L}}, \tag{6.5.8b}$$

where the normalization constants α_i and β_i are chosen to satisfy

$$u_R^\dagger(\boldsymbol{p})u_R(\boldsymbol{p}) = v_R^\dagger(\boldsymbol{p})v_R(\boldsymbol{p}) = 1, \tag{6.5.9a}$$

$$u_L^\dagger(\boldsymbol{p})u_L(\boldsymbol{p}) = v_L^\dagger(\boldsymbol{p})v_L(\boldsymbol{p}) = 1. \tag{6.5.9b}$$

The plane wave solutions of the Weyl equations are

$$u_R(\boldsymbol{p})e^{ipx/\hbar}, \qquad v_R(\boldsymbol{p})e^{-ipx/\hbar}, \tag{6.5.10a}$$

$$u_L(\boldsymbol{p})e^{ipx/\hbar}, \qquad v_L(\boldsymbol{p})e^{-ipx/\hbar}. \tag{6.5.10b}$$

It follows from (6.5.7a) that the direction of the spin $\frac{1}{2}\hbar\boldsymbol{\sigma}$ and the momentum $\boldsymbol{p}$ are identical for $u_R(\boldsymbol{p})$ and opposite for $v_R(\boldsymbol{p})$. It is said that $u_R(\boldsymbol{p})$ and $v_R(\boldsymbol{p})$ describe the positive-helicity state and the negative-helicity state, respectively. It is natural that the quantization axis of spin is the direction of the momentum.

The second-quantized field operator for the right-chirarity particle is

$$\psi_R(x) = \sqrt{c}\int \frac{d^3k}{\sqrt{(2\pi)^3}}\left[c_R(\boldsymbol{k})u_R(\boldsymbol{k})e^{ikx} + d_R^\dagger(\boldsymbol{k})v_R(\boldsymbol{k})e^{-ikx}\right], \tag{6.5.11}$$

together with

$$\{c_R(\boldsymbol{k}), c_R^\dagger(\boldsymbol{k}')\} = \{d_R(\boldsymbol{k}), d_R^\dagger(\boldsymbol{k}')\} = \delta(\boldsymbol{k}-\boldsymbol{k}'). \tag{6.5.12}$$

The operator $c_R(\boldsymbol{k})$ annihilates a particle of the energy $\hbar c|\boldsymbol{k}|$ with the positive helicity, while $d_R(\boldsymbol{k})$ annihilates an antiparticle of the energy $\hbar c|\boldsymbol{k}|$ with the negative helicity. We can verify that

$$\{\psi_{R\alpha}(t,\boldsymbol{x}), \psi_{R\beta}(t,\boldsymbol{y})\} = \{\psi_{R\alpha}^\dagger(t,\boldsymbol{x}), \psi_{R\beta}^\dagger(t,\boldsymbol{y})\} = 0, \tag{6.5.13a}$$

$$\{\psi_{R\alpha}(t,\boldsymbol{x}), \psi_{R\beta}^\dagger(t,\boldsymbol{y})\} = c\delta_{\alpha\beta}\delta(\boldsymbol{x}-\boldsymbol{y}), \tag{6.5.13b}$$

with the aid of (6.5.7a) and (6.5.9a).

Similarly, the second-quantized field operator for the left-chirarity particle is

$$\psi_L(x) = \sqrt{c}\int \frac{d^3k}{\sqrt{(2\pi)^3}}\left[c_L(\boldsymbol{k})u_L(\boldsymbol{k})e^{ikx} + d_L^\dagger(\boldsymbol{k})v_L(\boldsymbol{k})e^{-ikx}\right], \tag{6.5.14}$$

together with

$$\{c_L(\boldsymbol{k}), c_L^\dagger(\boldsymbol{k}')\} = \{d_L(\boldsymbol{k}), d_L^\dagger(\boldsymbol{k}')\} = \delta(\boldsymbol{k}-\boldsymbol{k}'), \tag{6.5.15}$$

and

$$\{\psi_{L\alpha}(t,\boldsymbol{x}), \psi_{L\beta}(t,\boldsymbol{y})\} = \{\psi_{L\alpha}^\dagger(t,\boldsymbol{x}), \psi_{L\beta}^\dagger(t,\boldsymbol{y})\} = 0, \tag{6.5.16a}$$

$$\{\psi_{L\alpha}(t,\boldsymbol{x}), \psi_{L\beta}^\dagger(t,\boldsymbol{y})\} = c\delta_{\alpha\beta}\delta(\boldsymbol{x}-\boldsymbol{y}). \tag{6.5.16b}$$

The operator $c_L(\boldsymbol{k})$ annihilates a particle of the energy $\hbar c|\boldsymbol{k}|$ with the negative helicity, while $d_L(\boldsymbol{k})$ annihilates an antiparticle of the energy $\hbar c|\boldsymbol{k}|$ with the positive helicity.

Finally we mention that, to study the Weyl field, it is customary to choose a representation where the matrix γ_5 is diagonalized. Such a set of Dirac matrices are given by

$$\gamma^0 = -\gamma_0 = \begin{pmatrix} 0 & 1 \\ 1 & 0 \end{pmatrix}, \qquad \gamma^i = \gamma_i = \begin{pmatrix} 0 & -\sigma^i \\ \sigma^i & 0 \end{pmatrix}, \tag{6.5.17}$$

with

$$\gamma_5 = i\gamma^0\gamma^1\gamma^2\gamma^3 = \begin{pmatrix} 1 & 0 \\ 0 & -1 \end{pmatrix}. \tag{6.5.18}$$

They are called the Dirac matrices in the chiral (or Weyl) representation. In this representation the Hamiltonian (6.1.8) reads

$$H = c\boldsymbol{\alpha} \cdot \boldsymbol{p} = c \begin{pmatrix} \boldsymbol{\sigma}\cdot\boldsymbol{p} & 0 \\ 0 & -\boldsymbol{\sigma}\cdot\boldsymbol{p} \end{pmatrix}, \tag{6.5.19}$$

from which the eigen equations (6.5.7) follows trivially.

6.6 Dirac Electrons in Magnetic Field

Massless electrons obeying the Dirac equation are known[434] to emerge in graphene. Graphene is a two-dimensional system made of carbon atoms: See Chapter 20 for more details. As we have shown, massless Dirac fermions are Weyl fermions, where the quantization axis of spin is chosen to be the direction of the moving fermion. However, when there exists external magnetic field $\boldsymbol{B}$, the quantization axis should be the direction of the field $\boldsymbol{B}$. It is proper to treat them as massless 4-component Dirac electrons.

We consider electrons confined within the xy plane in the presence of the uniform magnetic field along the z axis, $\boldsymbol{B} = (0, 0, -B)$ with $B > 0$. The following arguments are valid whether electrons are massless ($m = 0$) or massive ($m \neq 0$).

The Hamiltonian (6.4.17) is rewritten as

$$H = \begin{pmatrix} mc^2 & Q \\ Q & -mc^2 \end{pmatrix}, \tag{6.6.1}$$

where

$$Q = c\left(\sigma_x P_x + \sigma_y P_y\right) = c \begin{pmatrix} 0 & P_x - iP_y \\ P_x + iP_z & 0 \end{pmatrix} = \begin{pmatrix} 0 & A^\dagger \\ A & 0 \end{pmatrix}, \tag{6.6.2}$$

with $A = \hbar\omega_c a$, $\omega_c = c\sqrt{2eB/\hbar}$ and

$$a = \frac{\ell_B}{\sqrt{2}\hbar}(P_x + iP_y), \qquad a^\dagger = \frac{\ell_B}{\sqrt{2}\hbar}(P_x - iP_y). \tag{6.6.3}$$

Here, $\ell_B = \sqrt{\hbar/eB}$ is the magnetic length. The commutation relation $[a, a^\dagger] = 1$ follows from $[P_x, P_y] = i\hbar^2/\ell_B^2$: See (10.2.5) in Section 10.2 for more details. As we

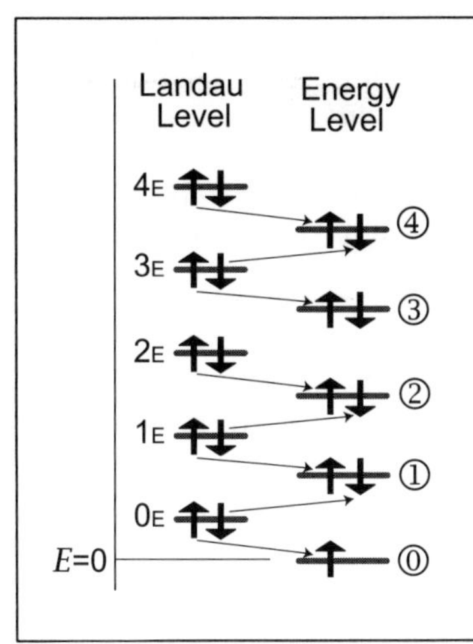

Fig. 6.2 The energy level (number in circle) and the Landau level (number with index E) are illustrated for electrons in the Pauli Hamiltonian. The spin is indicated by an arrow. This energy spectrum is a manifestation of SUSY. The Nth energy level contains up-spin electrons from the Nth Landau level and down-spin electrons from the (N-1)th Landau level for $N \geq 1$. The zero-energy level ($N = 0$) contains only up-spin electrons.

explain there, a and $a^\dagger$ are the Landau-level ladder operators. The Fock state

$$|N\rangle = \frac{1}{\sqrt{N!}}(a^\dagger)^N|0\rangle \tag{6.6.4}$$

describes the Nth Landau level.

The diagonalized Hamiltonian (6.4.18) reads

$$H = \begin{pmatrix} \sqrt{QQ + m^2c^4} & 0 \\ 0 & -\sqrt{QQ + m^2c^4} \end{pmatrix} \tag{6.6.5}$$

with

$$QQ = \begin{pmatrix} A^\dagger A & 0 \\ 0 & AA^\dagger \end{pmatrix} = (\hbar\omega_c)^2 \begin{pmatrix} a^\dagger a & 0 \\ 0 & aa^\dagger \end{pmatrix}. \tag{6.6.6}$$

Since $a^\dagger a$ is the number operator, the energy spectrum of the Dirac Hamiltonian follows immediately,

$$H = \begin{pmatrix} \mathcal{E}(N) & 0 & 0 & 0 \\ 0 & \mathcal{E}(N+1) & 0 & 0 \\ 0 & 0 & -\mathcal{E}(N) & 0 \\ 0 & 0 & 0 & -\mathcal{E}(N+1) \end{pmatrix}, \tag{6.6.7}$$

where $\mathcal{E}(N) = \sqrt{(\hbar\omega_c)^2 N + m^2c^4}$ and $N = 0, 1, 2, 3, \cdots$: The eigenstate is the Fock state (6.6.4). Here, the diagonal elements correspond to the up-spin electron, down-spin electron, up-spin electron with negative energy (down-spin hole), down-spin electron with negative energy (up-spin hole) in this order. It is remarkable that there exists a degeneracy in the energy spectrum and that there exists a zero-energy state for electrons and holes when $m = 0$.

We investigate the Hamiltonian

$$H_P = QQ = c^2(\boldsymbol{\sigma}\cdot\boldsymbol{P})^2 = c^2\left(-i\hbar\nabla + e\boldsymbol{A}\right)^2 - e\hbar c^2\sigma_z B, \tag{6.6.8}$$

which is the building blocks of the Dirac Hamiltonian (6.6.5). Since this has the same form as the Pauli Hamiltonian (6.4.20) with the mass $m^* = 1/2c^2$ except for the dimension, we call it also the Pauli Hamiltonian for brevity. The salient feature of the *relativistic* Dirac Hamiltonian is that its spectrum is mapped from that of the *nonrelativistic* Pauli Hamiltonian. Thus, the energy spectrum $\mathcal{E}(N)$ of the Dirac Hamiltonian H is constructed as $\mathcal{E}(N) = \pm\sqrt{E(N) + m^2c^4}$ from the energy spectrum $E(N)$ of the Pauli Hamiltonian H_P.

In the Pauli Hamiltonian (6.6.8), the first term is the kinetic term while the second term is the Zeeman term. The Zeeman effect is fixed uniquely as an intrinsic property of the Dirac theory. Hence we call it the intrinsic Zeeman effect. The Landau level is created by electrons making cyclotron motion, as we discuss in Section 10.2 in details. According to the Pauli Hamiltonian (6.6.8), the intrinsic Zeeman energy is precisely one half of the cyclotron energy for Dirac electrons, and two Landau levels mix to create one energy level. Namely, one energy level contains two kinds of electrons coming from different Landau levels, as illustrated in Fig.6.2. It is interesting that the energy level and the Landau level are entirely different objects for Dirac electrons.

We may carry out the second quantization of the Dirac Hamiltonian (6.6.7) by inspecting the energy spectrum. We shall explicitly do this when we need the second-quantized field operator to analyze the QH effect in graphene: See (20.3.15) in Section 20.3.

SUSY Quantum Mechanics

The intrinsic structure of the energy spectrum (6.6.7) can be understood on the basis of the SUSY quantum mechanics[508,31]. We investigate the generalized Pauli Hamiltonian

$$H_P = QQ = \begin{pmatrix} A^\dagger A & 0 \\ 0 & AA^\dagger \end{pmatrix} \equiv \begin{pmatrix} H^\uparrow & 0 \\ 0 & H^\downarrow \end{pmatrix}, \tag{6.6.9}$$

with

$$Q = \begin{pmatrix} 0 & A^\dagger \\ A & 0 \end{pmatrix}. \tag{6.6.10}$$

Here, A is an arbitrary operator such that $A \neq A^\dagger$. (In the case of $A = \hbar\omega_c a^m$ it describes the m-layer graphene in magnetic field, as we discuss in Section 20.7.)

We consider the operators

$$Q_S = \begin{pmatrix} 0 & 0 \\ A & 0 \end{pmatrix}, \qquad Q_S^\dagger = \begin{pmatrix} 0 & A^\dagger \\ 0 & 0 \end{pmatrix}. \tag{6.6.11}$$

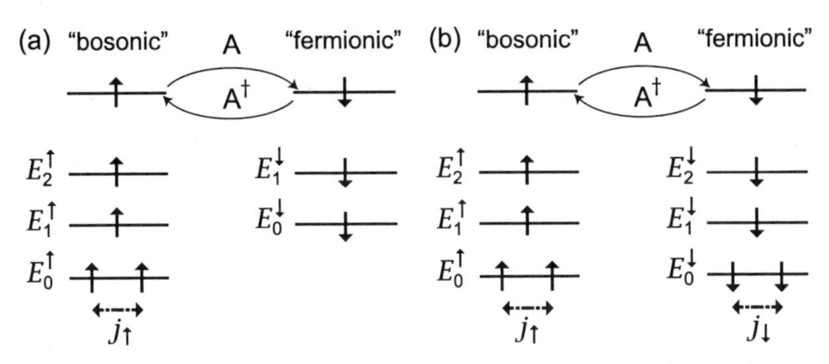

Fig. 6.3 The energy spectrum of the superpartner Hamiltonians $H^\uparrow$ and $H^\downarrow$ for electrons. Two states on the same horizontal line have the same energy, making a supermultiplet, except for the zero-energy state ($E_0^\uparrow = 0$). The zero-energy state is $j_\uparrow$-fold ($j_\downarrow$-fold) degenerate for up-spin (down-spin) electrons. (a) The case with $j_\downarrow = 0$. We have $j_\uparrow = m$ in the m-layer graphene, as we shall see in Section 20.7. (b) The case with $j_\downarrow \neq 0$.

The Hamiltonian is expressed as

$$H_P = \{Q_S, Q_S^\dagger\} = Q_S Q_S^\dagger + Q_S^\dagger Q_S = \begin{pmatrix} A^\dagger A & 0 \\ 0 & AA^\dagger \end{pmatrix}. \tag{6.6.12}$$

It is easy to check the following commutation and anticommutation relations,

$$[H_P, Q_S] = [H_P, Q_S^\dagger] = 0, \quad \{Q_S, Q_S^\dagger\} = H_P, \quad \{Q_S, Q_S\} = \{Q_S^\dagger, Q_S^\dagger\} = 0. \tag{6.6.13}$$

The operators Q_S, $Q_S^\dagger$ and H_P form the closed superalgebra $sl(1/1)$.

The fact that Q_S commutes with H_P implies the existence of a symmetry and a degeneracy in the spectra of the two Hamiltonians $H^\uparrow$ and $H^\downarrow$. The symmetry is referred to as a supersymmetry with Q_S being the supercharge. Since $Q_S Q_S = Q_S^\dagger Q_S^\dagger = 0$, they are fermionic operators. Hence, they interchange the bosonic and fermionic sectors. It is customary to identify[508] $H^\uparrow$ and $H^\downarrow$ as the bosonic and fermionic sectors, respectively.

Let us examine the degeneracy in the spectra more closely. We consider separately the eigenvalue problems for the up-spin and down-spin components,

$$H^\uparrow |\psi_n^\uparrow\rangle = E_n^\uparrow |\psi_n^\uparrow\rangle, \tag{6.6.14a}$$

$$H^\downarrow |\psi_n^\downarrow\rangle = E_n^\downarrow |\psi_n^\downarrow\rangle, \tag{6.6.14b}$$

where we have arranged such that $E_{n+1}^\uparrow > E_n^\uparrow \geq E_0^\uparrow$ and $E_{n+1}^\downarrow > E_n^\downarrow \geq E_0^\downarrow$, as illustrated in Fig.6.3. Using the relations

$$AH^\uparrow = AA^\dagger A = H^\downarrow A, \tag{6.6.15}$$

we obtain

$$H^{\downarrow}A|\psi_n^{\uparrow}\rangle = AH^{\uparrow}|\psi_n^{\uparrow}\rangle = E_n^{\uparrow}A|\psi_n^{\uparrow}\rangle, \tag{6.6.16a}$$

$$H^{\uparrow}A^{\dagger}|\psi_n^{\downarrow}\rangle = A^{\dagger}H^{\downarrow}|\psi_n^{\downarrow}\rangle = E_n^{\downarrow}A^{\dagger}|\psi_n^{\downarrow}\rangle. \tag{6.6.16b}$$

We compare (6.6.14) and (6.6.16) for the states having the same eigenvalue.

If $E_n^{\uparrow} \neq 0$, we must have $A|\psi_n^{\uparrow}\rangle \neq 0$ since $H^{\uparrow} = A^{\dagger}A$. Then, $A|\psi_n^{\uparrow}\rangle$ is an eigenstate of $H^{\downarrow}$ with the same eigenvalue $E_n^{\uparrow}$. Hence, $|\psi_n^{\uparrow}\rangle$ and $A|\psi_n^{\uparrow}\rangle$ are degenerate. Similarly, if $E_n^{\downarrow} \neq 0$, $|\psi_n^{\downarrow}\rangle$ and $A^{\dagger}|\psi_n^{\downarrow}\rangle$ are degenerate. Thus, there is one-to-one correspondence between the up-spin eigenstate and the down-spin eigenstate for nonzero-energy states [Fig.6.3]. The corresponding states are said to be superpartners. On the other hand, if $E_0^{\uparrow} = 0$, we must have $A|\psi_0^{\uparrow}\rangle = 0$. There is no superpartner for the zero-energy state.

To establish the correspondence explicitly, it is necessary to examine the cases $E_0^{\downarrow} \neq 0$ and $E_0^{\downarrow} = 0$, separately. If $E_0^{\downarrow} \neq 0$, the lowest energy eigenstate of $H^{\downarrow}$ is $|\psi_0^{\downarrow}\rangle \propto A|\psi_1^{\uparrow}\rangle$. Then, the states $|\psi_{n+1}^{\uparrow}\rangle$ and $|\psi_n^{\downarrow}\rangle$ are superpartners having the same energy $E_{n+1}^{\uparrow} = E_n^{\downarrow}$. Namely, the superpartners are [Fig.6.3(a)]

$$|\psi_n^{\downarrow}\rangle = \frac{1}{\sqrt{E_{n+1}^{\uparrow}}}A|\psi_{n+1}^{\uparrow}\rangle \quad \text{and} \quad |\psi_{n+1}^{\uparrow}\rangle = \frac{1}{\sqrt{E_n^{\downarrow}}}A^{\dagger}|\psi_n^{\downarrow}\rangle \qquad \text{for} \quad n \geq 0. \tag{6.6.17a}$$

On the other hand, if $E_0^{\downarrow} = 0$, for which $A^{\dagger}|\psi_0^{\uparrow}\rangle = 0$, the superpartners are [Fig.6.3(b)]

$$|\psi_n^{\downarrow}\rangle = \frac{1}{\sqrt{E_n^{\uparrow}}}A|\psi_n^{\uparrow}\rangle \quad \text{and} \quad |\psi_n^{\uparrow}\rangle = \frac{1}{\sqrt{E_n^{\downarrow}}}A^{\dagger}|\psi_n^{\downarrow}\rangle \qquad \text{for} \quad n \geq 1. \tag{6.6.18a}$$

We cannot say anything about the degeneracy of the zero-energy state by this general argument. See Section 20.7 for an explicit application to the QH effect in graphene.

CHAPTER 7

TOPOLOGICAL SOLITONS

Certain nonlinear classical equations possess classical solutions describing extended object known as solitons. In quantum field theory they are realized as coherent states describing collective excitations of the basic field. Solitons are stable particle-like objects with finite energy. When their stability is guaranteed topologically, they are said to be topological solitons. As typical examples we study sine-Gordon solitons, vortices and skyrmions.

7.1 TOPOLOGICAL SECTORS

We denote by $\phi(\boldsymbol{x})$ a generic field in a Lagrangian. We have analyzed small fluctuations around the classical vacuum $\phi(\boldsymbol{x}) = v$, and quantized them as point particles. There exist a new type of excitations. Certain nonlinear field equations possess classical solutions describing extended particle-like objects. They represent stable objects with finite energy, and scatter among themselves just as ordinary particles do. They are called *solitons*. An important class of solitons are *topological solitons* whose stability is guaranteed topologically. Well-known examples are magnetic vortices observed in superconductors. Topological solitons (vortices, skyrmions, sine-Gordon solitons) emerge as quasiparticles in QH systems. See a textbook[28] for references on solitons and related topics before 1980. (We do not review *nontopological solitons* in spite of their attractive features.[239,136])

In this chapter all fields are treated as classical fields . We consider a collection of all smooth functions (classical field configurations) $\phi(\boldsymbol{x})$, which we call a field manifold: A field manifold consists of mappings from the real space to the group space in which the field $\phi(\boldsymbol{x})$ takes values, as we shall explain in Section 7.6. We associate the classical energy $E(\phi)$ with each function by evaluating the Hamiltonian classically. We then consider a submanifold consisting of all functions that have finite energies. The energy-finiteness condition, $E(\phi) < \infty$, imposes a certain boundary condition ($\lim_{|\boldsymbol{x}|\to\infty} \phi(\boldsymbol{x})$) on the function. We ask whether one function $\phi_1(\boldsymbol{x})$ is continuously deformable into the classical vacuum $\phi(\boldsymbol{x}) = v$ without violating the energy-finiteness condition. Mathematicians have a special word for "continuous deformable"; they call it homotopic as we shall explain in Section 7.6.

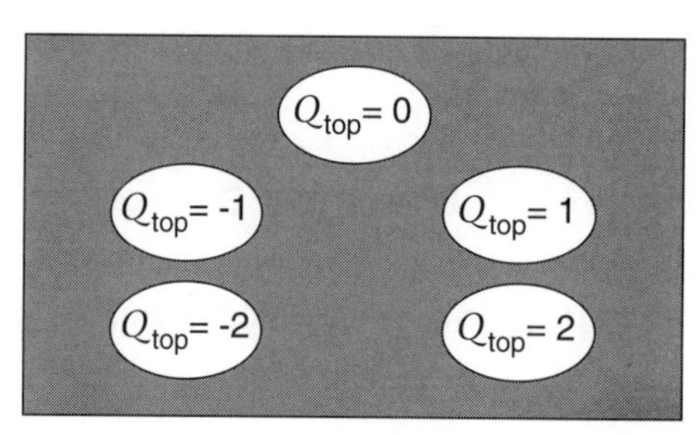

Fig. 7.1 The entire space represents a manifold made of all field configurations. We then consider a submanifold (white domains) where the energy is finite. The submanifold consists of several disjoint sectors labeled by the topological charge Q_{top}. Two field configurations belong to the same sector if and only if they are continuously deformable without violating the energy-finiteness condition. The sector containing the classical vacuum is called the vacuum sector and labeled by $Q_{\text{top}} = 0$. No physical transition is allowed between two different sectors because there is an infinitely high energy barrier between them.

See textbooks for details.[24,29]

If deformable to the classical vacuum $\phi(\boldsymbol{x}) = v$, since it can decay into the vacuum, the object described by the function $\phi_1(\boldsymbol{x})$ is unstable. For instance, let us consider a solution $\phi_1(\boldsymbol{x})$ of the Euler-Lagrange equation. Since it is merely a local minimum of the Hamiltonian, it may decay into the vacuum by way of a tunneling process.[1] We say that $\phi_1(\boldsymbol{x})$ belongs to the *vacuum sector*. In short, the vacuum sector is the submanifold whose element is continuously deformable to the classical vacuum.

If not deformable to the classical vacuum, we say that $\phi_1(\boldsymbol{x})$ belongs to a *topological sector* different from the vacuum sector. The sector consists of all the functions continuously deformable from $\phi_1(\boldsymbol{x})$ without violating the energy-finiteness condition.

Clearly, the field manifold is decomposed into topological sectors by the above criterion [Fig.7.1]. Namely, any two functions belong to the same sector if and only if they are continuously deformable into one another without violating the energy-finiteness condition. A topological sector is labeled by an integer called a *topological charge* Q_{top}. The vacuum sector has $Q_{\text{top}} = 0$. The sector $Q_{\text{top}} = 1$ contains one topological soliton, and the sector $Q_{\text{top}} = -1$ contains one topological antisoliton.[2] Both solitons and antisolitons may be referred to simply as solitons. Solitons with $Q_{\text{top}} = \pm 1$ are stable because, in order to bring them to the classical

[1] The tunneling process may be prohibited physically by the energy-momentum conservation. In this case the object described by the function $\phi_1(\boldsymbol{x})$ is stable though it belongs to the vacuum sector. It is called a *nontopological soliton*.

[2] The sector is characterized by the topological charge. The sector $Q_{\text{top}} = 1$ may contain N+1 solitons and N antisolitons.

vacuum, they must jump over an infinitely high energy barrier: The time evolution never allows such a transition.

An explicit form of the Hamiltonian is not important for the existence of topological solitons. It is used only to determine the explicit form of a topological soliton and to evaluate its energy. The same topological solitons such as vortices manifest themselves in various branches of physics,[3] although Hamiltonians are very much different.

7.2 CLASSICAL FIELDS

Solitons are solutions of classical field equations. Before analyzing various solitons explicitly we question what the classical field means in quantum field theory. The classical field is the expectation value of the quantum field by a certain state,

$$\langle f|\phi(x)|f\rangle = \phi^{\rm cl}(\boldsymbol{x}). \tag{7.2.1}$$

Accordingly, corresponding to a classical field $\phi^{\rm cl}(\boldsymbol{x})$ there must be a quantum state $|f\rangle$.

Such a quantum state is easily constructed in a system where Bose condensation occurs. For instance, we consider the Schrödinger field ϕ, which is expanded as in (3.2.15),

$$\phi(t,\boldsymbol{x}) = \int \frac{d^3k}{\sqrt{(2\pi)^3}} e^{ikx} a_{\boldsymbol{k}}. \tag{7.2.2}$$

A coherent state $|f\rangle$ is an eigenstate of an annihilation operator as in (1.8.3),

$$a_{\boldsymbol{k}}|f\rangle = \alpha_{\boldsymbol{k}}|f\rangle, \qquad \alpha_{\boldsymbol{k}} = e^{i\theta_{\boldsymbol{k}}}\sqrt{n_{\boldsymbol{k}}}, \tag{7.2.3}$$

where $n_{\boldsymbol{k}} = \alpha_{\boldsymbol{k}}^* \alpha_{\boldsymbol{k}}$ is the number of the $\boldsymbol{k}$-component. Hence,

$$\phi(t,\boldsymbol{x})|f\rangle = \phi^{\rm cl}(\boldsymbol{x})|f\rangle, \tag{7.2.4}$$

with

$$\phi^{\rm cl}(\boldsymbol{x}) = \int \frac{d^3k}{\sqrt{(2\pi)^3}} e^{ikx+i\theta_{\boldsymbol{k}}}\sqrt{n_{\boldsymbol{k}}}. \tag{7.2.5}$$

The classical field is a linear superposition of plane waves, each of which has definite phase $\theta_{\boldsymbol{k}}$ and intensity $n_{\boldsymbol{k}}$ simultaneously.

We construct explicitly a coherent state having a given classical field.[133] We start with the false vacuum $|\underline{0}\rangle$,

$$\langle \underline{0}|\phi(x)|\underline{0}\rangle = 0, \qquad \langle \underline{0}|\underline{0}\rangle = 1. \tag{7.2.6}$$

---

[3] Vortices are quasiparticle possessing fractional charges in fractional QH states,[294] as will be extensively studied in Part II. Vortices have been observed in superconductors. New vortex physics has been discussed in a multi-layer structure of high-temperature superconductors.[109] Vortices are called strings in particle physics and cosmology. Three quarks are combined by strings to generate a proton.[482] Cosmic strings might have played an important role in forming galaxies by attracting matters around them.[484]

It is to be contrasted with the ground state $|0\rangle$,

$$\langle 0|\phi(x)|0\rangle = v \neq 0, \qquad \langle 0|0\rangle = 1. \tag{7.2.7}$$

The classical vacuum $\phi^{\text{cl}}(\boldsymbol{x}) = v$ is a trivial example of the classical field. We define an operator

$$Q(t) = \frac{1}{\hbar}\int d^3x\, \phi^{\text{cl}}(\boldsymbol{x})\pi(t,\boldsymbol{x}) \tag{7.2.8}$$

with $\pi(x)$ the conjugate momentum of $\phi(x)$,

$$[\phi(t,\boldsymbol{x}), \pi(t,\boldsymbol{y})] = i\hbar\delta(\boldsymbol{x}-\boldsymbol{y}). \tag{7.2.9}$$

It satisfies

$$[Q(t), \phi(t,\boldsymbol{x})] = -i\phi^{\text{cl}}(\boldsymbol{x}), \tag{7.2.10}$$

and hence

$$e^{iQ}\phi(x)e^{-iQ} = \phi(x) + \phi^{\text{cl}}(\boldsymbol{x}), \tag{7.2.11}$$

as is proved based on the formula (B.1) with $A = Q$ and $B = \phi$ in Appendix B. From (7.2.11) and (7.2.6), we find that

$$\langle \underline{0}|e^{iQ}\phi(x)e^{-iQ}|\underline{0}\rangle = \phi^{\text{cl}}(\boldsymbol{x}). \tag{7.2.12}$$

By comparing this with (7.2.1), the state

$$|f\rangle = e^{-iQ}|\underline{0}\rangle \tag{7.2.13}$$

is the one that has the classical field $\phi^{\text{cl}}(\boldsymbol{x})$.

A special case of the classical field is the classical vacuum. Indeed, the ground state $|0\rangle$ is given by (7.2.13) with $\phi^{\text{cl}}(\boldsymbol{x}) = v$ in (7.2.8). It was determined by minimizing the classical Hamiltonian. There may exist other classical fields $\phi^{\text{cl}}(\boldsymbol{x})$ by minimizing the classical Hamiltonian. The minimum exists in each topological sector, as we have explained. Even if its energy is larger than the ground-state energy, it is stable because there is an infinitely high energy barrier between two different topological sectors. We substitute

$$\phi(x) = \phi^{\text{cl}}(\boldsymbol{x}) + \eta(x) \tag{7.2.14}$$

into the Hamiltonian and analyze small fluctuations $\eta(x)$ around the classical field $\phi^{\text{cl}}(\boldsymbol{x})$. Since $\phi^{\text{cl}}(\boldsymbol{x})$ minimizes the energy by assumption, any fluctuations around it must increase the energy or keep it unchanged. If the energy is unchanged, there exists a gapless mode. We should mention that there exist always gapless modes in fluctuations around a soliton solution. For instance, because of translational invariance, if $\phi^{\text{cl}}(\boldsymbol{x})$ is a solution, $\phi^{\text{cl}}(\boldsymbol{x}-\boldsymbol{x}_0)$ with a fixed $\boldsymbol{x}_0$ is also a solution. Since $\boldsymbol{x}_0$ is arbitrary, it yields a gapless mode associated with the translational invariance. We may argue as follows: Since a particular choice of the classical field $\phi^{\text{cl}}(\boldsymbol{x})$ breaks the translational invariance of the system spontaneously, there appears a

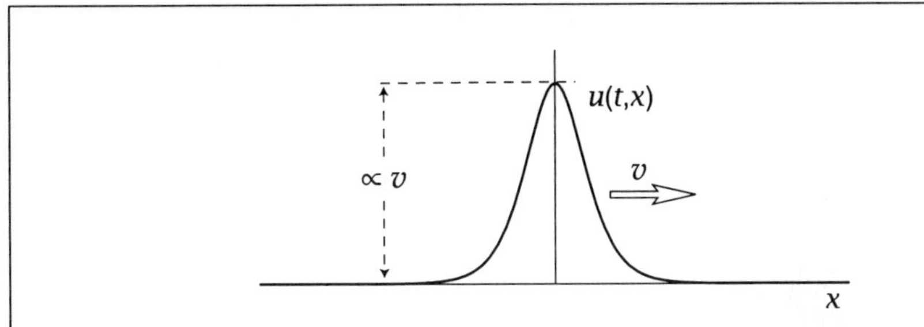

Fig. 7.2 A solution of the KdV equation describes a solitary wave. Its shape and size depend on the velocity v. In particular, the height is proportional to the velocity ($u \propto v$).

Goldstone mode. In the rest of this chapter we omit the index *cl* from the classical field ϕ^{cl} for notational simplicity: All fields stand for classical fields in this chapter.

7.3 Solitary Waves, Kinks and Solitons

Solitary waves were first observed in a canal by J. Scott Russel in 1834. It is amazing that a localized wave packet travels in a canal over miles without changing its shape and size. They are well approximated by solutions of the so-called KdV equation,

$$\frac{\partial u}{\partial t} + \alpha u \frac{\partial u}{\partial x} + \beta \frac{\partial^3 u}{\partial x^3} = 0, \tag{7.3.1}$$

where $u(t,x)$ describes the height of the wave above the water surface; α and β are numerical constants. We can check that the following function satisfies the KdV equation (7.3.1),

$$u(t,x) = \frac{3v}{\alpha} \text{sech}^2 \left[\frac{1}{2} \sqrt{\frac{v}{\beta}} (x - vt) \right], \tag{7.3.2}$$

where v is an integration constant representing the velocity [Fig.7.2]. Solitary waves possess particle-like properties: Namely, several wave packets may collide, merge and reappear after a certain time [Fig.7.3].

A *solitary wave* is defined as a solution of a nonlinear differential equations which describes an object with localized energy and moves without dissipation.[4] A solitary wave is called a *soliton* if it has particle-like properties. The above KdV solitary wave is a soliton but not a topological soliton.

There is an example of a solitary wave, called *kink*, which is stable topologically but not a soliton. Kink solutions exist in the real Klein-Gordon theory (4.2.1) with

[4] It is necessary to consider a nonlinear differential equation. Since the superposition law holds in the linear differential equation, any solution may be decomposed into momentum components which are also solutions, where each momentum component travels with its own velocity. Namely, there is no solution with localized energy that moves without dissipation in the linear differential equation.

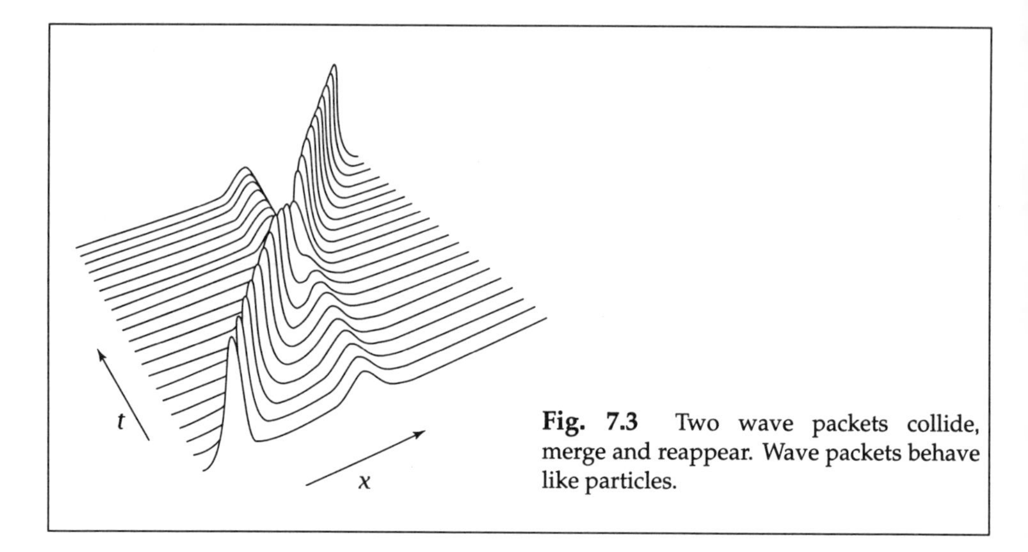

Fig. 7.3 Two wave packets collide, merge and reappear. Wave packets behave like particles.

Bose condensation. The Hamiltonian density is given by

$$\mathcal{H} = \frac{\hbar^2}{2c^2}(\dot{\phi})^2 + \frac{\hbar^2}{2}(\nabla\phi)^2 + \frac{g}{4}(\phi^2 - v^2)^2. \tag{7.3.3}$$

There exist two degenerate vacua $\phi = \pm v$ in the model [Fig.4.4 (page 52)]. In Section 4.2 we have chosen one of the vacua to carry out a perturbative analysis. It is interesting to consider these two vacua simultaneously in a single system, as illustrated in Fig.7.4(a): We have taken one vacuum ($\phi = v$) in the domain $x > 0$ and the other vacuum ($\phi = -v$) in the domain $x < 0$. The junction is at the yz plane located at $x = 0$. This field configuration is unstable since it is not a solution of the Euler-Lagrange equation. We may solve the equation with the boundary condition $\phi \to \pm v$ as $x \to \pm\infty$, as in Fig.7.4(b). The solution is a kink, and given by (see Appendix L)

$$\phi_{\text{kink}}(x) = v \tanh\left(\frac{x}{2\xi}\right), \tag{7.3.4}$$

where ξ is the coherence length (4.2.24). The energy density of the kink is

$$E(x) = \frac{gv^4}{2}\text{sech}^4\left(\frac{x}{2\xi}\right), \tag{7.3.5}$$

which is localized in x as we see in Fig.7.4(c). A time-dependent solution is easily obtained by making a Lorentz boost,

$$\phi_{\text{kink}}(t,x) = v \tanh\left(\frac{1}{2\xi}\frac{x - ut}{\sqrt{1-u^2}}\right), \tag{7.3.6}$$

where $u(\equiv v/c)$ is the normalized velocity, $1 > u > -1$. This describes a solitary wave moving without dissipation.

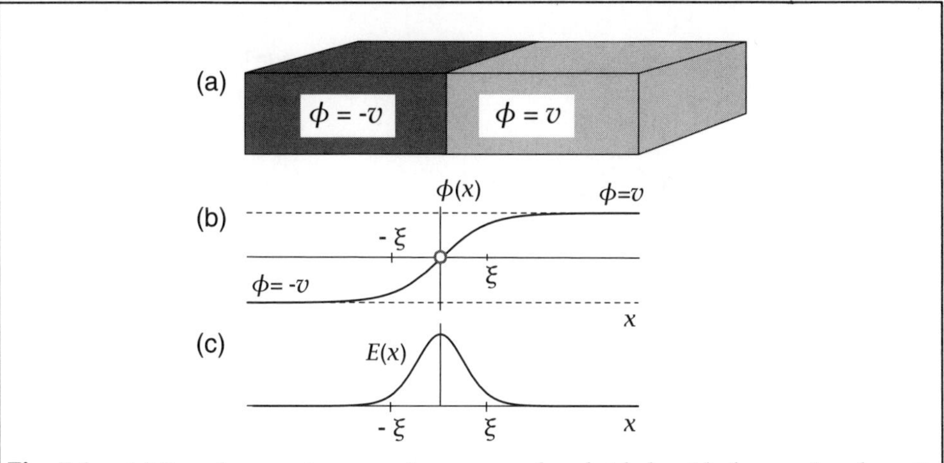

Fig. 7.4 (a) Two degenerate vacua $\phi = \pm v$ are placed side by side for $x > 0$ and $x < 0$. (b) The junction is smoothed to lower the energy, as gives a kink solution. The false vacuum ($\phi = 0$) is realized at the center of the kink. (c) The kink is a solitary wave since the energy density is localized.

The stability of a kink follows from the fact that it interpolates two different vacua. It is impossible to deform a kink solution continuously to one vacuum. This is because the boundary condition is violated and the energy diverges in the process of deforming the kink solution to one vacuum solution. In particular, the time evolution is a continuous deformation respecting the energy-finiteness condition: A kink solution cannot tunnel to the vacuum solution since it is prevented by an infinitely high energy barrier. The kink is topologically stable.

However, a kink is not particle-like for the following reason. A kink appears at the junction $(-v, v)$ between two vacua $\phi = -v$ and $\phi = v$. An antikink appears at the junction $(v, -v)$, and given by $\phi_{\text{antikink}}(x) = \phi_{\text{kink}}(-x)$. It has the same energy density (7.3.5). We may consider a configuration containing one kink and one antikink, which appears at two junctions $(-v, v)$ and $(v, -v)$. However, there is no configuration containing a kink contiguous to another kink. The kink is not a soliton because it is not particle-like in this sense.

7.4 Sine-Gordon Solitons

We analyze the sine-Gordon model in the 1+1 space-time manifold. It is an important model theoretically, since it is a rare model in which topological solitons are quantized rigorously.[104,331] It is also an important model physically, since it describes the dynamics of the parallel magnetic flux penetrating the superconductor Josephson junction and also the bilayer QH system (see Chapter 27). It also appears to describe edge excitations in the QH system (see Chapter 18).

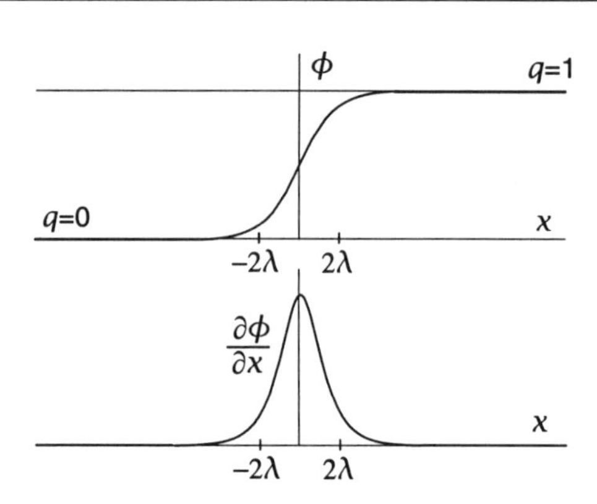

Fig. 7.5 A sine-Gordon soliton ϕ interpolates two classical vacua $q = 0$ and $q = 1$. The energy is localized within a wave packet where $\partial_x \phi \neq 0$. It propagates without dissipation.

The sine-Gordon model describes a self-interacting scalar field $\phi(t, x)$ in a periodic potential. The Lagrangian density is

$$\mathcal{L} = \frac{\hbar}{2c}\left(\frac{\partial\phi}{\partial t}\right)^2 - \frac{c\hbar}{2}\left(\frac{\partial\phi}{\partial x}\right)^2 - \frac{c\hbar}{\lambda^2}(1 - \cos\phi), \tag{7.4.1}$$

where λ is a constant of the length dimension. We have taken the field ϕ to be dimensionless. The Euler-Lagrange equation reads

$$\frac{1}{c^2}\frac{\partial^2\phi}{\partial t^2} - \frac{\partial^2\phi}{\partial x^2} + \frac{1}{\lambda^2}\sin\phi = 0. \tag{7.4.2}$$

It is called the sine-Gordon equation since it is obtained by adding a sine term to the Klein-Gordon equation. The canonical momentum is

$$\pi = \frac{\delta\mathcal{L}}{\delta\dot{\phi}} = \frac{\hbar}{c}\partial_t\phi. \tag{7.4.3}$$

The Hamiltonian density is

$$\mathcal{H} = \frac{\hbar}{2c}\left(\frac{\partial\phi}{\partial t}\right)^2 + \frac{c\hbar}{2}\left(\frac{\partial\phi}{\partial x}\right)^2 + \frac{c\hbar}{\lambda^2}(1 - \cos\phi). \tag{7.4.4}$$

The classical vacuum is easy to obtain,

$$\phi = 2\pi q, \qquad q = 0, \pm 1, \pm 2, \cdots. \tag{7.4.5}$$

The Lagrangian is invariant under a discrete transformation,

$$\phi \longrightarrow \phi + 2\pi, \tag{7.4.6}$$

which is spontaneously broken when we choose one vacuum to carry out perturbative quantization.

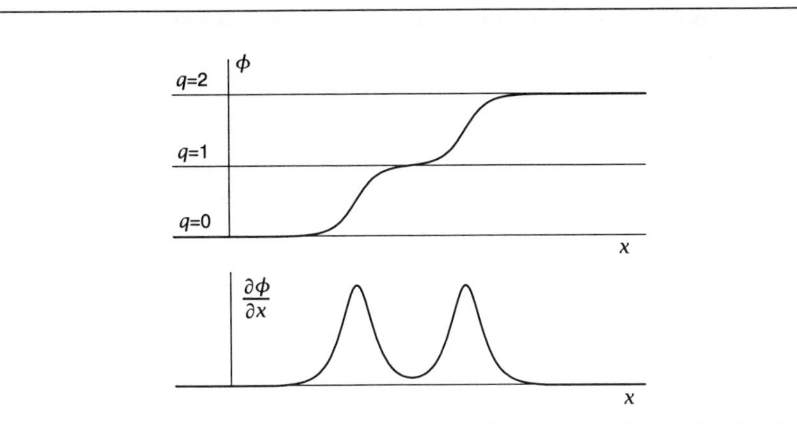

Fig. 7.6 There exists a time-dependent 2-soliton solution interpolating the classical vacua $q = 0$ and $q = 2$. It describes two isolated solitons as $t \to \pm\infty$.

The present model is very different from the previous one since there are infinitely many degenerate ground states as in (7.4.5). We have a solitary wave solution [Fig.7.5], say, at the junction $(0, 2\pi)$ between two different vacua $\phi = 0$ and $\phi = 2\pi$. Furthermore, we have a 2-soliton configuration [Fig.7.6] at two junctions $(0, 2\pi)$ and $(2\pi, 4\pi)$. The solitary wave is a soliton since there is a multi-soliton configuration.

Let us consider a static field configuration $\phi(x)$. The energy-finiteness condition is essential. It requires that the function $\phi(x)$ approaches the classical vacuum asymptotically. The asymptotic values can be different at $x \to -\infty$ and $x \to +\infty$, as illustrated in Fig.7.5. For definiteness we choose

$$\phi(x = -\infty) = 0, \qquad \phi(x = \infty) = 2\pi, \tag{7.4.7}$$

which connects two classical vacua with $q = 0$ and $q = 1$. There exists a configuration $\phi(x)$ that minimizes the Hamiltonian with this boundary condition, as gives a static solution. It is explicitly solved by (Appendix L)

$$\phi(x) = 4 \arctan \exp \left(\frac{x - x_0}{\lambda} \right), \tag{7.4.8}$$

where x_0 is a constant. It describes a soliton with energy localized around the point $x = x_0$. The size of the soliton is of the order λ, as is seen in Fig.7.5. To derive the energy of one soliton we substitute (7.4.8) into (7.4.4),

$$E_{\text{SG}} = c\hbar \int_{-\infty}^{\infty} dx \left[\frac{1}{2} \left(\frac{\partial \phi}{\partial x} \right)^2 + \frac{1}{\lambda^2} (1 - \cos \phi) \right] = \frac{8c\hbar}{\lambda}. \tag{7.4.9}$$

The stability of the sine-Gordon soliton follows from the fact that it interpolates two different vacua.

A moving solution is obtained by making a Lorentz boost,

$$\phi_v(x) = 4\arctan\exp\left(\frac{x - x_0 - vt}{\lambda\sqrt{1-(v/c)^2}}\right), \tag{7.4.10}$$

with v the velocity. This is a solitary wave which travels without dissipation. The energy of a moving soliton is given by substituting (7.4.10) into (7.4.4),

$$E_{\text{SG}}(v) = \frac{mc^2}{\sqrt{1-(v/c)^2}}, \tag{7.4.11}$$

where we have set

$$mc^2 = \frac{8c\hbar}{\lambda} = E_{\text{SG}}. \tag{7.4.12}$$

The Einstein relation $E^2 = c^2p^2 + m^2c^4$ holds, with the relativistic momentum $p = mv/\sqrt{1-(v/c)^2}$. Thus the mass of the soliton is essentially the static energy.

The configuration containing two isolated static solitons gives a very good approximate solution of the sine-Gordon equation: See Fig.16.8. Nevertheless, it is not a solution in a strict sense. It implies the existence of a force between two solitons, however far they are separated from one another. Although no static 2-soliton solution exists, there is a time-dependent 2-soliton solution,

$$\phi(t,x) = 4\arctan\left(\frac{(v/c)\sinh\left(x/\lambda\sqrt{1-(v/c)^2}\right)}{\cosh\left(vt/\lambda\sqrt{1-(v/c)^2}\right)}\right), \tag{7.4.13}$$

which we have depicted in Fig.7.6. It is decomposed into two isolated solitons (7.4.8) as $t \to \pm\infty$.

We should mention that there is a static solution describing a soliton lattice. We shall discuss on it in a bilayer QH system (see Section 27.5 and Appendix L).

It is a generic feature of topological solitons that *there exists a topological current which is conserved trivially*. For instance, in the $1+1$ dimensional space-time we can a priori write down the current

$$J^\mu_{\text{SG}}(t,x) = \frac{1}{2\pi}\varepsilon^{\mu\nu}\partial_\nu\phi(t,x), \tag{7.4.14}$$

where $\varepsilon^{\mu\nu}$ is the antisymmetric tensor with $\varepsilon^{01} = -\varepsilon^{10} = 1$. This current is conserved identically,

$$\partial_\mu J^\mu_{\text{SG}} = \frac{1}{2\pi}\varepsilon^{\mu\nu}\partial_\mu\partial_\nu\phi = 0, \tag{7.4.15}$$

which follows since $\varepsilon^{\mu\nu}$ is antisymmetric while $\partial_\mu\partial_\nu\phi$ is symmetric in the exchange of the indices μ and ν. The topological charge is defined by

$$Q_{\text{SG}} = \int_{-\infty}^{+\infty} dx J^0_{\text{SG}} = \frac{1}{2\pi}\int_{-\infty}^{+\infty} dx\frac{\partial\phi}{\partial x}, \tag{7.4.16}$$

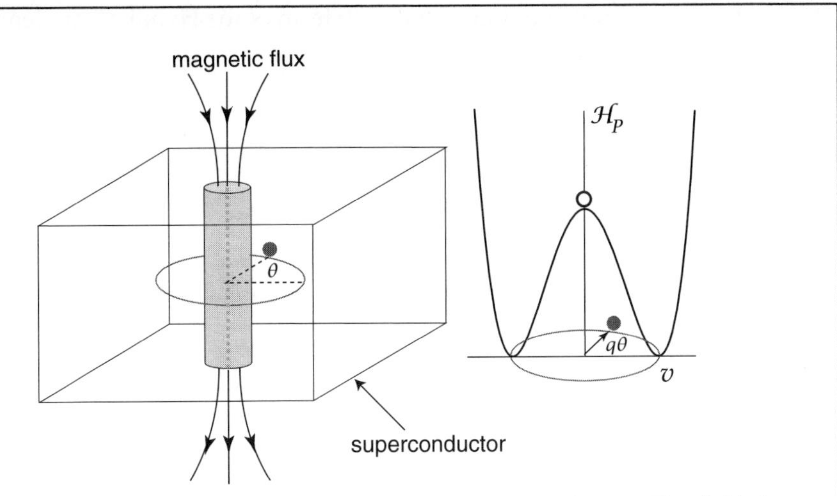

Fig. 7.7 Magnetic flux is squeezed into a vortex in superconductor. The field ϕ interpolates continuously the classical-vacuum value ($|\phi| = v$) outside the vortex and the false-vacuum value ($\phi = 0$) along its center.

which is the difference between the asymptotic values of $\phi(x)$ at $x = \infty$ and $x = -\infty$. We obtain $Q_{SG} = 0$ for the vacuum, and $Q_{SG} = 1$ for the soliton satisfying the boundary condition (7.4.7). The topological sector is labeled by the charge Q_{SG}. The topological current (7.4.14) has nothing to do with the Lagrangian (7.4.1). It is conserved trivially, while the Nöther current is conserved via the Euler-Lagrange equation.

7.5 Vortex Solitons

We consider the complex Klein-Gordon field or the Schrödinger field interacting with the electromagnetic field in the Higgs-type potential. We show that the magnetic field is squeezed into quantized vortices [Fig.7.7], as is a well-known phenomenon in superconductor.[32,13]

Since we are concerned with static classical solutions, the static energy density is relevant,

$$\mathcal{H} = \frac{1}{2\mu_{\rm m}} B^2 + \beta \left(\partial_k - i\frac{e^*}{\hbar}A_k\right)\phi^\dagger \left(\partial_k + i\frac{e^*}{\hbar}A_k\right)\phi + \frac{g}{2}\left(\phi^\dagger\phi - v^2\right)^2, \qquad (7.5.1)$$

where $\beta = \hbar^2/2M$ for the Schrödinger case (5.3.4) and $\beta = \hbar^2$ for the Klein-Gordon case (5.4.1); $\mu_{\rm m}$ is the magnetic permeability; e^* is the charge carried by the field ϕ. We search for a vortex solution sitting along the z axis. For this purpose it is enough to take the planar geometry.

The energy-finiteness condition is essential to characterize the topological properties of a vortex. It requires that each term of the Hamiltonian density (7.5.1) vanishes asymptotically,

$$B_i = \varepsilon_{ijk}\partial_j A_k = 0, \tag{7.5.2a}$$

$$\partial_k \phi + i\frac{e^*}{\hbar}A_k\phi = 0, \tag{7.5.2b}$$

$$|\phi|^2 = v^2, \tag{7.5.2c}$$

as $r \equiv |\boldsymbol{x}| \to \infty$. The gauge potential A_k is a pure gauge from (7.5.2a). We may solve (7.5.2) as

$$\phi(\boldsymbol{x}) = ve^{-if(\boldsymbol{x})}, \qquad A_j(\boldsymbol{x}) = \frac{\hbar}{e^*}\partial_j f(\boldsymbol{x}), \qquad \text{as} \quad r \to \infty. \tag{7.5.3}$$

The energy-finiteness condition (7.5.3) is trivial when $f(\boldsymbol{x})$ is a constant. It characterizes the vacuum sector. It is also trivial when $f(\boldsymbol{x})$ is a regular function, since $f(\boldsymbol{x})$ is eliminated by a gauge transformation.

The boundary condition (7.5.3) yields a nontrivial result when the function $f(\boldsymbol{x})$ is multivalued. A multivalued function is allowed only if the multivalueness does not appear in physical quantities. This is the case provided that $e^{-if(\boldsymbol{x})}$ and $\partial_j f(\boldsymbol{x})$ are single-valued. The boundary condition (7.5.3) is rewritten as

$$\phi^\infty(\theta) \equiv \lim_{r\to\infty} \phi(\boldsymbol{x}) = ve^{-if(\theta)}, \tag{7.5.4}$$

where $f(\theta) = \lim_{r\to\infty} f(\boldsymbol{x})$. The single-valued requires $f(2\pi) = f(0) + 2\pi q$ with integer q. We set

$$\phi(\boldsymbol{x}) = |\phi(\boldsymbol{x})|e^{-iq\theta-ih(\boldsymbol{x})}, \tag{7.5.5}$$

where $h(\boldsymbol{x}) \equiv f(\boldsymbol{x}) - q\theta$ is a single-valued function. It can be continuously gauged away [Fig.7.8], and the boundary condition (7.5.4) is brought to

$$\phi^\infty(\theta) = ve^{-iq\theta}. \tag{7.5.6}$$

Therefore, all classical field configurations are grouped into topological sectors labeled by integer q.

The vacuum sector corresponds to the choice of $q = 0$. When $q \neq 0$, the function $e^{-iq\theta}$ is single-valued except at the origin ($x = y = 0$): It is not well defined at the origin. However, the classical field (7.5.5) is well defined everywhere provided

$$\phi(\boldsymbol{x}) = 0 \qquad \text{at} \quad \boldsymbol{x} = 0. \tag{7.5.7}$$

This is the essential feature of the vortex center. Recall that the U(1) symmetry is spontaneously broken in the present model, where the classical vacuum is $\phi(\boldsymbol{x}) = v$. The U(1) symmetry is restored at the vortex center, where the false vacuum ($\phi(\boldsymbol{x}) = 0$) is realized [Fig.7.7].

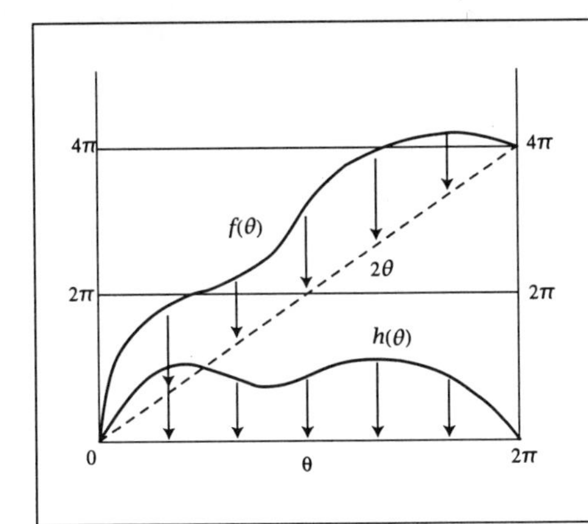

Fig. 7.8 The phase $f(\theta)$ characterizes the vortex sector. In this example $f(\theta)$ is a multivalued function interpolating two classical vacua $q = 0$ and $q = 2$. The singlevalued function $h(\theta) \equiv f(\theta) - 2\theta$ can be continuously gauged away.

We next analyze the gauge potential. The boundary condition (7.5.3) with (7.5.6) implies that $f(\boldsymbol{x}) = q\theta$ and

$$A_k(\boldsymbol{x}) \to \frac{q\hbar}{e^*}\partial_k\theta = -\frac{n\Phi_0}{2\pi}\varepsilon_{kj}\frac{x^j}{r^2} \qquad \text{as} \quad r \to \infty, \tag{7.5.8}$$

where

$$\Phi_0 \equiv \frac{2\pi\hbar}{e^*}. \tag{7.5.9}$$

It is the *Dirac flux unit* when $e^* = e$ and the *London flux unit* when $e^* = 2e$. Using the Stokes theorem and (7.5.8) we find

$$\Phi = \int d^2x\, B = \int d^2x\, \varepsilon_{jk}\partial_j A_k = \oint dx_k A_k = \frac{q\Phi_0}{2\pi}\int_0^{2\pi} d\theta = q\Phi_0, \tag{7.5.10}$$

where the integration $\oint dx_k$ is made around a sufficiently large circle ($r \to \infty$). It is concluded that the magnetic flux is quantized in units of Φ_0.

The topological current is defined by

$$J^\mu_{\text{vor}} = \frac{1}{\Phi_0}\varepsilon^{\mu\nu\lambda}\partial_\nu A_\lambda, \tag{7.5.11}$$

which is conserved trivially, $\partial_\mu J^\mu_{\text{vor}} = \varepsilon^{\mu\nu\lambda}\partial_\mu\partial_\nu A_\lambda = 0$. Here, $\varepsilon^{\mu\nu\lambda}$ is the antisymmetric tensor such that $\varepsilon^{012} = 1$. The topological charge

$$Q_{\text{vor}} = \int d^2x J^0_{\text{vor}} = \frac{1}{\Phi_0}\oint dx_k A_k = q \tag{7.5.12}$$

is the magnetic-flux number that the vortex carries. The vortex carries a quantized magnetic flux.

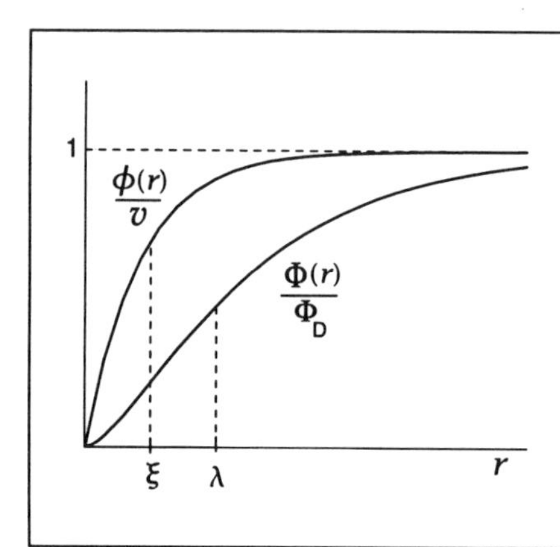

Fig. 7.9 The normalized scalar field $\phi(r)/v$ and the normalized magnetic flux $\Phi(r)/\phi_{\text{D}}$ are depicted as functions of radius r for the vortex soliton solution with $q = 1$. The coherence length ξ and the penetration depth λ give an order of the vortex-core size and the size of the magnetic-flux penetrated region.

We construct an approximate solution [Fig.7.9]. It is convenient to set

$$\phi(x) = (v + \eta(x))\, e^{-iq\theta}, \tag{7.5.13a}$$

$$A_k(x) = U_k(x) + \frac{q\Phi_0}{2\pi}\partial_k\theta, \tag{7.5.13b}$$

as in (5.4.4). This is not a genuine gauge transformation, but a *singular gauge transformation*, since the field strength F_{ij} is not invariant,[5]

$$F_{ij}(\boldsymbol{x}) \longrightarrow \partial_i U_j - \partial_j U_i + q\Phi_0\delta^2(\boldsymbol{x}). \tag{7.5.14}$$

As far as the domain $\boldsymbol{x} \neq 0$ is concerned we may ignore the singularity: The δ-function singularity is equivalent to imposing a boundary condition at $\boldsymbol{x} = 0$. Substituting (7.5.13) into the energy density (7.5.1), we find that small fluctuations of the field η and the field U_k around the classical vacuum obey

$$\xi^2\nabla^2\eta - \eta = 0, \tag{7.5.15a}$$

$$\lambda^2\nabla^2 U_k - U_k = 0, \qquad k = 1, 2, \tag{7.5.15b}$$

where

$$\nabla^2 = \frac{\partial^2}{\partial r^2} + \frac{1}{r}\frac{\partial}{\partial r} + \frac{1}{r^2}\frac{\partial^2}{\partial\theta^2}. \tag{7.5.16}$$

We have defined the coherence length ξ and the penetration depth λ,

$$\xi = \frac{1}{v}\sqrt{\frac{\beta}{2g}}, \qquad \lambda = \frac{\hbar}{ev\sqrt{2\beta\mu_{\text{m}}}}. \tag{7.5.17}$$

They agree with (5.4.9) for the Klein-Gordon field ($\beta = \hbar^2$), and (5.6.13) for the Schrödinger field ($\beta = \hbar^2/2M$).

---

[5] In deriving (7.5.14) we have used $\varepsilon_{ij}\partial_i\partial_j\theta = 2\pi\delta^2(x)$ (see Appendix E).

We can construct an approximate solution in the following way. A vortex configuration should satisfy (7.5.15a) and (7.5.15b) far away from the vortex center. We solve (7.5.15a) for a cylindrical symmetric field $\eta(r)$,

$$\eta(r) \propto e^{-r/\xi}, \qquad \text{for} \quad r \gg \xi. \tag{7.5.18}$$

We extrapolate it into the vortex core so that the boundary condition (7.5.7) is obeyed. In this way we obtain

$$\phi(r) \simeq v(1 - e^{-r/\xi})e^{i\theta}. \tag{7.5.19}$$

We next analyze (7.5.15b). However, this is nothing but the London equation (5.6.26), whose vortex solution has already been studied. The magnetic flux density is given by (5.6.39), or

$$B_z(r) \simeq \frac{q\Phi_L}{2\pi\lambda^2} K_0\left(\frac{r}{\lambda}\right), \tag{7.5.20}$$

from which we find

$$U_k = -\lambda^2 \varepsilon_{kj}\partial_j B_z \simeq -\frac{q\Phi_L}{2\pi}\varepsilon_{kj}\partial_j K_0\left(\frac{r}{\lambda}\right) \tag{7.5.21}$$

by way of (5.6.28). The potential A_k is given by

$$A_k = U_k(\boldsymbol{x}) + \frac{q\Phi_L}{2\pi}\partial_k\theta \simeq -\frac{q\Phi_L}{2\pi}\varepsilon_{kj}\partial_j\left[K_0\left(\frac{r}{\lambda}\right) + \ln\frac{r}{\lambda}\right] = \varepsilon_{kj}\frac{x_j}{2\pi r^2}q\Phi_L\left[\frac{r}{\lambda}K_1\left(\frac{r}{\lambda}\right) - 1\right], \tag{7.5.22}$$

which satisfies the boundary conditions at $r = 0$ and $r = \infty$. We have obtained (7.5.19) and (7.5.22) without knowing any detailed structure of the vortex core. Nevertheless, it gives a reasonable approximate solution for the vortex soliton. See also vortex solutions in QH systems in Section 14.3.

The condensation of the ϕ quanta is partially resolved within the vortex core whose size is of order the coherence length ξ. It is resolved completely at the vortex center. The magnetic flux penetrates around the vortex core whose size is of order the penetration depth λ. We have depicted the field $\phi(r)$ and the magnetic flux $\Phi(r)$ around a vortex soliton in Fig.7.9.

7.6 Homotopy Classes

We have explained the mechanism of topological stability of the vortex soliton. It is worthwhile to recapitulate it because the reasoning is general enough to argue the existence of topological solitons in various field theories. The vortex soliton is characterized by the boundary condition. The boundary value $\phi^\infty(\theta)$ of the field $\phi(\boldsymbol{x})$ is a function defined on a circle S^1, where the variable θ moves from $\theta = 0$ to $\theta = 2\pi$ with these two points identified. As given by (7.5.6), it takes a value in the group space U(1). The group space U(1) is identified with a circle S^1 since

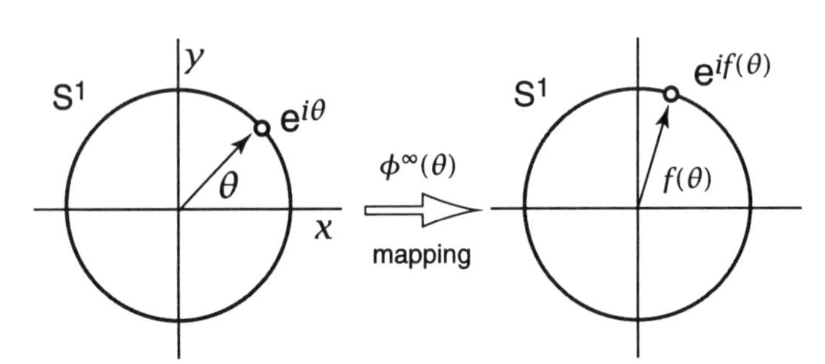

Fig. 7.10 Function $\phi^\infty(\theta)$ defines a continuous mapping from S^1 (the xy plane) to another S^1 (the field space). A point on the xy plane is parametrized as $e^{i\theta}$, which is mapped to $ve^{if(\theta)}$ in the field space. The image of the mapping is a circle when $f(\theta) = \theta$. This implies $\pi_1(S^1) = Z$.

its element is written as e^{if}. Thus, the field $\phi^\infty(\theta)$ defines a continuous mapping from S^1 (the real space) to another S^1 (the field space) as in Fig.7.10,

$$\phi^\infty(\theta): \quad S^1 \longrightarrow S^1. \tag{7.6.1}$$

The mapping is explicitly given by the function $f(\theta)$ in (7.5.4) and illustrated in Fig.7.8. We have shown that all the field configurations are grouped into topological sectors labeled by an integer with (7.5.6) as a representative. "Topological sectors" are properly called *homotopy classes.*[24,29]

This is the homotopy theorem applied to the U(1) gauge theory in the planar geometry. It is rephrased as follows: All continuous mappings of a circle S^1 into another S^1 are grouped into homotopy classes. Mappings within one class are continuously deformable into one another, whereas mappings from two different classes are not continuously deformable into one another. The number of such classes is indexed by the set of integers. This is mathematically expressed as

$$\pi_1(S^1) = \mathbb{Z}, \tag{7.6.2}$$

where $\pi_1(S^1)$ stands for the homotopy group associated with mappings from S^1 to S^1.

To make the idea concrete, it is instructive to consider a mapping of a circle S^1 into a 2-sphere S^2 which is a surface of a ball [Fig.7.11]. Since the image of the mapping is always shrinkable continuously to a point, the mapping is trivial. This is mathematically expressed as $\pi_1(S^2) = 0$.

We can similarly consider all continuous mappings of a 2-sphere S^2 into another S^2, as in Fig.7.12, which are grouped into homotopy classes. The number of such classes are indexed by the set of integers, and mathematically expressed as

$$\pi_2(S^2) = \mathbb{Z}, \tag{7.6.3}$$

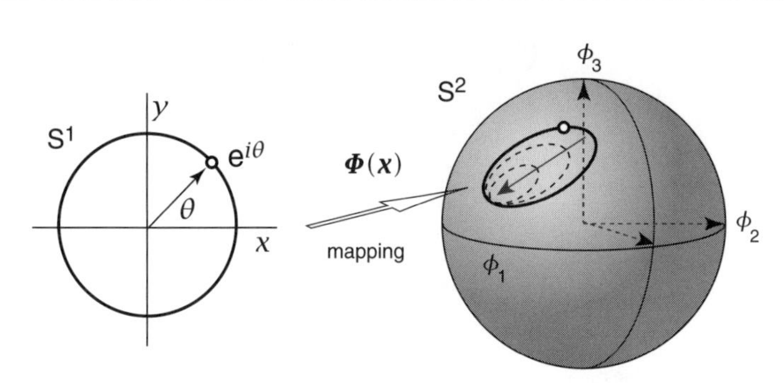

Fig. 7.11 A continuous mapping from S^1 into a 2-sphere S^2 is illustrated. An image of the mapping is contained in a circle on the sphere, which is shrinkable to a point. This implies $\pi_1(S^2) = 0$.

where $\pi_2(S^2)$ stands for the homotopy group associated with mappings from S^2 to S^2. This homotopy group characterizes skyrmion excitations in the O(3) nonlinear sigma model in the planar geometry.

In general, all continuous mappings from an m-sphere to an n-sphere ($S^m \longrightarrow S^n$) are grouped into homotopy classes, which defines the homotopy group $\pi_m(S^n)$. We cite some homotopy groups,[24,29]

$$\pi_n(S^n) = \mathbb{Z}, \tag{7.6.4a}$$

$$\pi_n(S^m) = 0 \quad \text{for} \quad n < m, \tag{7.6.4b}$$

$$\pi_n(S^1) = 0 \quad \text{for} \quad n > 1. \tag{7.6.4c}$$

Homotopy groups $\pi_n(S^m)$ are nontrivial for $n > m$; for instance, $\pi_3(S^2) = \mathbb{Z}$.

Skyrmions

We apply the homotopy theorem to the existence of topological solitons in the O(N) nonlinear sigma model. The O(N) nonlinear sigma model is defined by

$$\mathcal{L} = -\frac{J_s}{2}(\partial^\mu \boldsymbol{\phi}) \cdot (\partial_\mu \boldsymbol{\phi}), \tag{7.6.5}$$

where J_s is the "spin stiffness" whose dimension is taken so that $\boldsymbol{\phi}$ is dimensionless. The field $\boldsymbol{\phi} = (\phi_1, \phi_2, \cdots, \phi_N)$ is subject to the constraint

$$\boldsymbol{\phi}^2 = \sum_{a=1}^{N} \phi_a \phi_a = 1. \tag{7.6.6}$$

In the m-dimensional space the energy of a static classical configuration is

$$E = \frac{J_s}{2} \int d^m x \, (\nabla \boldsymbol{\phi})^2. \tag{7.6.7}$$

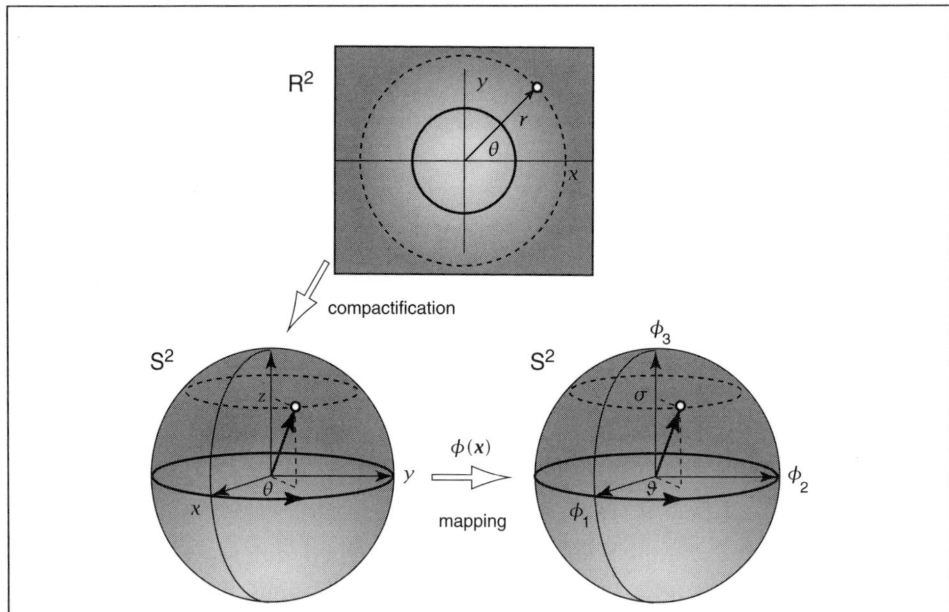

Fig. 7.12 The xy plane (R^2) is compactified into the 2-sphere (S^2) by identifying all points at infinity. All points within (outside) a circle are mapped to the lower (upper) semisphere: The infinity is mapped to the north pole ($z = 1$). The sigma field $\boldsymbol{\phi}(\boldsymbol{x})$ defines a continuous mapping from S^2 (the compactified xy plane) to another S^2 (the field space).

To make the energy finite, it is necessary to impose

$$\lim_{r\to\infty} r^{m-1}|\nabla\boldsymbol{\phi}(\boldsymbol{x})| = 0, \tag{7.6.8}$$

which implies that the boundary value is just a constant,

$$\lim_{r\to\infty} \boldsymbol{\phi}(\boldsymbol{x}) = \boldsymbol{\phi}^\infty. \tag{7.6.9}$$

Since all $\boldsymbol{\phi}(\boldsymbol{x})$ approach one constant configuration $\boldsymbol{\phi}^\infty$ asymptotically, the configuration space R^m is effectively compactified into a spherical surface S^m.

We explain the meaning of compactification by taking an explicit example of the planar geometry ($m = 2$) in Fig.7.12. A point of the xy plane is parametrized as

$$x = r\cos\theta, \qquad y = r\sin\theta. \tag{7.6.10}$$

We consider a 2-sphere S^2 in the xyz space parametrized by

$$x = \sqrt{1-z^2}\cos\theta, \qquad y = \sqrt{1-z^2}\sin\theta, \qquad z = z, \tag{7.6.11}$$

where z is defined by

$$r = \frac{1+z}{1-z}. \tag{7.6.12}$$

Note that $r = 0$ as $z \to -1$, $r = 1$ as $z = 0$ and $r \to \infty$ as $z \to 1$. There is a one-to-one correspondence between a point (7.6.10) on the planar space R^2 and a point (7.6.11) on the 2-sphere S^2 by this mapping, except for the infinity point ($r \to \infty$) on R^2 and the north pole ($z = 1$) on S^2. We now consider any function $\boldsymbol{\phi}(\boldsymbol{x})$ that takes a fixed value $\boldsymbol{\phi}^\infty$ as $r \to \infty$. By making a change of variable, we obtain a function of z and θ, $\boldsymbol{\phi}(\boldsymbol{x}) = \boldsymbol{\phi}(z, \theta)$, which is regular at $z = 1$, $\lim_{z\to 1} \boldsymbol{\phi}(z, \theta) = \boldsymbol{\phi}^\infty$. Namely, the function $\boldsymbol{\phi}(z, \theta)$ is defined and regular on the entire 2-sphere S^2: This is the compactification of R^2 to S^2.

We come back to a general dimension. We consider all field configurations $\boldsymbol{\phi}(\boldsymbol{x})$ that take a fixed value $\boldsymbol{\phi}^\infty$ as $r \to \infty$. Being subject to the constraint (7.6.6), the sigma field $\boldsymbol{\phi}(\boldsymbol{x})$ takes a value on a spherical surface S^{N-1}. Hence, the field $\boldsymbol{\phi}(\boldsymbol{x})$ defines a continuous mapping from the compactified real space S^m to the field space S^{N-1}. According to the homotopy theorem the mappings are classified into the homotopy class $\pi_m(S^{N-1})$. When it is nontrivial, $\pi_m(S^{N-1}) \neq 0$, it defines a topological charge: Topological solitons are called *skyrmions*.

A comment is in order. The Lagrangian (7.6.5) is an isotropic nonlinear sigma model, which is a special limit of an anisotropic nonlinear sigma model

$$\mathcal{L} = -\frac{J_s}{2} \sum_{a=1}^{N} c_a (\partial^\mu \phi_a)(\partial_\mu \phi_a) \tag{7.6.13}$$

with (7.6.6), where $c_a > 0$. It is obvious that the above arguments follow with respect to skyrmions in this model because the boundary condition is given by (7.6.9). We deal with an anisotropic O(3) nonlinear sigma model in bilayer QH systems: See (21.1.7).

Scaling Theorem

The homotopy theorem says that topological solitons can exist in the O(N) nonlinear sigma model in the spatial dimension m if the homotopy class $\pi_m(S^{N-1})$ is nontrivial. However, the theorem does not guarantee its existence as a nontrivial minimum of a particular Hamiltonian. It only guarantees that there is an infinitely high energy barrier between two different topological sectors.

We assume that there is a solution $\boldsymbol{\phi}_0(\boldsymbol{x})$ minimizing the static energy (7.6.7), and we denote the minimized energy by E_0. We consider a configuration, $\boldsymbol{\Phi}(\boldsymbol{x}) = \boldsymbol{\phi}_0(\lambda \boldsymbol{x})$, whose energy is calculated as

$$E_\lambda = \frac{J_s}{2} \int d^m x [\nabla \boldsymbol{\phi}_0(\lambda \boldsymbol{x})]^2 = \frac{J_s}{2\lambda^{m-2}} \int d^m y [\nabla \boldsymbol{\phi}_0(\boldsymbol{y})]^2 = \frac{\lambda^{2-m}}{2} E_0, \tag{7.6.14}$$

where we have made a change of the integration variable, $\boldsymbol{y} = \lambda \boldsymbol{x}$. Unless $m = 2$ the energy E_λ can be made arbitrarily small by changing λ,

$$\lim_{\lambda\to 0} E_\lambda = 0 \quad \text{for} \quad m = 1, \qquad \lim_{\lambda\to\infty} E_\lambda = 0 \quad \text{for} \quad m \geq 3. \tag{7.6.15}$$

The O(N) nonlinear sigma model (7.6.7) possesses topological solitons only in the planar geometry ($m \neq 2$) due to this scaling theorem[115] and only when $N = 3$ due to the homotopy theorem (7.6.4).

Skyrmions were first considered by Skyrme in 1961 in the O(4) nonlinear sigma model in the 3-dimensional space, where the sigma field $\boldsymbol{\phi}$ represents the pi ($\boldsymbol{\pi}$) meson and the sigma (σ) meson. He identified the topological soliton as the proton. However, as we have just seen, there is no soliton in the O(4) nonlinear sigma model (7.6.5) in the 3-dimensional space by the scaling theorem though the homotopy group is nontrivial. It is necessary to modify the Hamiltonian so that the above scaling property is violated. A simplest way is to add a higher derivative term to the nonlinear sigma model (7.6.5). The static energy proposed by Skyrme is[6]

$$
\begin{aligned}
E =& \frac{J_s}{2} \int d^m x [\nabla \boldsymbol{\phi}(\boldsymbol{x})]^2 \\
&+ \kappa \int d^m x \left([\nabla \boldsymbol{\phi}(\boldsymbol{x})]^2 [\nabla \boldsymbol{\phi}(\boldsymbol{x})]^2 - \sum_{ij} [\partial_i \boldsymbol{\phi}(\boldsymbol{x}) \partial_j \boldsymbol{\phi}(\boldsymbol{x})]^2 \right). \qquad (7.6.16)
\end{aligned}
$$

When we make the above scaling, the first term scales as λ^{2-m} while the second term scales as λ^{4-m}. Hence, for $m = 3$ the energy cannot be brought to zero by letting $\lambda \to 0$ or $\lambda \to \infty$. There must be a soliton solution to minimize this static energy. A soliton solution is known only numerically. See a textbook[27] for details on skyrmions in the 3-dimensional space.

7.7 O(3) Skyrmions

The O(3) nonlinear sigma model passes the scaling theorem in the planar geometry, and allows skyrmions.[7] The Lagrangian density is

$$
\mathcal{L} = -\frac{J_s}{2} (\partial^\mu \boldsymbol{\phi}) \cdot (\partial_\mu \boldsymbol{\phi}), \qquad (7.7.1)
$$

where the field $\boldsymbol{\phi} = (\phi_1, \phi_2, \phi_3)$ is subject to the constraint

$$
\boldsymbol{\phi}^2 = \phi_1^2 + \phi_2^2 + \phi_3^2 = 1. \qquad (7.7.2)
$$

To satisfy the boundary condition (7.6.9) we choose

$$
\lim_{r \to \infty} \phi_1(\boldsymbol{x}) = \lim_{r \to \infty} \phi_2(\boldsymbol{x}) = 0, \qquad \lim_{r \to \infty} \phi_3(\boldsymbol{x}) = 1. \qquad (7.7.3)
$$

The O(3) sigma model may describe the spin-spin interaction with the spin stiffness normalized to one. The boundary condition implies that the field approaches the spin-polarized ground state asymptotically.

---

[6] To carry out a consistent field theory it is necessary to give a Lagrangian, from which a Hamiltonian is derived. See the original work[446] of Skyrme for the Lagrangian.

[7] In QH ferromagnets pseudoparticles are O(3) skyrmions: See Chapter 16.

In the planar geometry we can consider a topological current,

$$J^{\mu}_{\text{sky}}(\boldsymbol{x}) = \frac{1}{8\pi}\varepsilon^{\mu\nu\lambda}\varepsilon_{abc}\phi_a\partial_\nu\phi_b\partial_\lambda\phi_c, \tag{7.7.4}$$

where the numerical factor $1/8\pi$ has been introduced so that the topological charge is one for the simplest topological soliton. Let us show that it is conserved trivially. First we obtain

$$\partial_\mu J^{\mu}_{\text{sky}}(\boldsymbol{x}) = \frac{1}{8\pi}\varepsilon^{\mu\nu\lambda}\varepsilon_{abc}\partial_\mu\phi_a\partial_\nu\phi_b\partial_\lambda\phi_c = \frac{3}{4\pi}\varepsilon^{\mu\nu\lambda}\partial_\mu\phi_1\partial_\nu\phi_2\partial_\lambda\phi_3. \tag{7.7.5}$$

From (7.7.2) it follows that

$$\phi_1\partial_\mu\phi_1 + \phi_2\partial_\mu\phi_2 + \phi_3\partial_\mu\phi_3 = 0. \tag{7.7.6}$$

Because of the constraint (7.6.6), one of three $\phi_i(\boldsymbol{x})$ is nonzero at a given point $\boldsymbol{x}$. Let it be $\phi_3(\boldsymbol{x})$. We multiply (7.7.5) by $\phi_3(\boldsymbol{x})$ and use (7.7.6),

$$\phi_3\partial_\mu J^{\mu}_{\text{sky}} = -\frac{3}{4\pi}\varepsilon^{\mu\nu\lambda}\partial_\mu\phi_1\partial_\nu\phi_2(\phi_1\partial_\lambda\phi_1 + \phi_2\partial_\lambda\phi_2) = 0, \tag{7.7.7}$$

from which we find

$$\partial_\mu J^{\mu}_{\text{sky}} = 0. \tag{7.7.8}$$

The topological charge is given by

$$Q_{\text{sky}} = \int d^2x J^0_{\text{sky}} = \frac{1}{8\pi}\int d^2x\, \varepsilon_{abc}\varepsilon_{ij}\phi_a\partial_i\phi_b\partial_j\phi_c. \tag{7.7.9}$$

It is referred to as the Pontryagin number. It counts how many times the group space S^2 is wrapped as the coordinate (x, y) spans the whole planar space R^2, as we shall explain in (7.7.32) with Fig.7.12.

We are able to construct all skyrmion solutions explicitly. In so doing, instead of solving the Euler-Lagrange equation directly, we use the identity

$$\int d^2x\, (\partial_i\phi_a \pm \varepsilon_{ij}\varepsilon_{abc}\phi_b\partial_j\phi_c)^2 \geq 0, \tag{7.7.10}$$

which yields,

$$\int d^2x\, (\partial_i\phi_a)^2 \geq \pm\int d^2x\, \varepsilon_{abc}\varepsilon_{ij}\phi_a\partial_i\phi_b\partial_j\phi_c, \tag{7.7.11}$$

since $\phi_a\phi_a = 1$ and $\phi_a\partial_i\phi_a = 0$. This equation is equivalent to

$$E \geq \pm 4\pi J_s Q_{\text{sky}}, \tag{7.7.12}$$

where E is the classical energy (7.6.7), and Q_{sky} the Pontryagin number (7.7.9). The energy is minimized in each topological sector when the equality holds. Such a soliton is called the Begomol'nyi-Prasad-Sommerfield (BPS) soliton.

Namely, the BPS soliton is obtained by solving $E = \pm 4\pi J_s Q_{\text{sky}}$, or

$$\partial_j\phi_a \pm \varepsilon_{jk}\varepsilon_{abc}\phi_b\partial_k\phi_c = 0. \tag{7.7.13}$$

When we define $\phi = \phi_1 - i\phi_2$, two of these equations for $a = 1$ and $a = 2$ are combined into

$$\partial_j \phi = \mp i \varepsilon_{jk} (\phi \partial_k \phi_3 - \phi_3 \partial_k \phi). \tag{7.7.14}$$

It is convenient to define a new field ω by

$$\omega = \frac{\phi}{1 - \phi_3}. \tag{7.7.15}$$

It satisfies

$$\begin{aligned} \partial_x \omega &= \frac{1}{(1-\phi_3)^2} (\partial_x \phi + \phi \partial_x \phi_3 - \phi_3 \partial_x \phi) \\ &= \frac{1}{(1-\phi_3)^2} (\partial_x \phi \mp i \partial_y \phi) = \mp i \partial_y \omega, \end{aligned} \tag{7.7.16}$$

where use was made of (7.7.14). When we take the upper sign, we find

$$\frac{\partial}{\partial z^*} \omega = 0, \tag{7.7.17}$$

and when we take the lower sign, we find

$$\frac{\partial}{\partial z} \omega = 0. \tag{7.7.18}$$

Here, $z = x + iy$ and

$$\frac{\partial}{\partial z} = \frac{1}{2} \left(\frac{\partial}{\partial x} - i \frac{\partial}{\partial y} \right), \qquad \frac{\partial}{\partial z^*} = \frac{1}{2} \left(\frac{\partial}{\partial x} + i \frac{\partial}{\partial y} \right). \tag{7.7.19}$$

The general solution of (7.7.17) is an arbitrary analytic function $\omega = \omega(z)$ of z describing skyrmions ($Q_{\text{sky}} > 0$), while the general solution of (7.7.18) is an arbitrary analytic function $\omega = \omega(z^*)$ of z^* describing antiskyrmions ($Q_{\text{sky}} < 0$). The field ϕ_a is determined once ω is known. Since $|\phi|^2 = 1 - \phi_3^2$, (7.7.15) yields $|\omega|^2 = (1 + \phi_3)/(1 - \phi_3)$. Therefore, from (7.7.15) the general soliton solution is found to be

$$\phi_3 = \frac{|\omega|^2 - 1}{|\omega|^2 + 1}, \qquad \phi \equiv \phi_1 - i\phi_2 = \frac{2\omega}{|\omega|^2 + 1}. \tag{7.7.20}$$

The energy of the both types of solutions is given by

$$E = 4\pi J_s |Q_{\text{sky}}|, \tag{7.7.21}$$

in terms of the Pontryagin number.

A simple analytic function is given by

$$\omega(z) = \left(\frac{z}{\kappa} \right)^q = \left(\frac{r}{\kappa} \right)^q e^{iq\theta}, \tag{7.7.22}$$

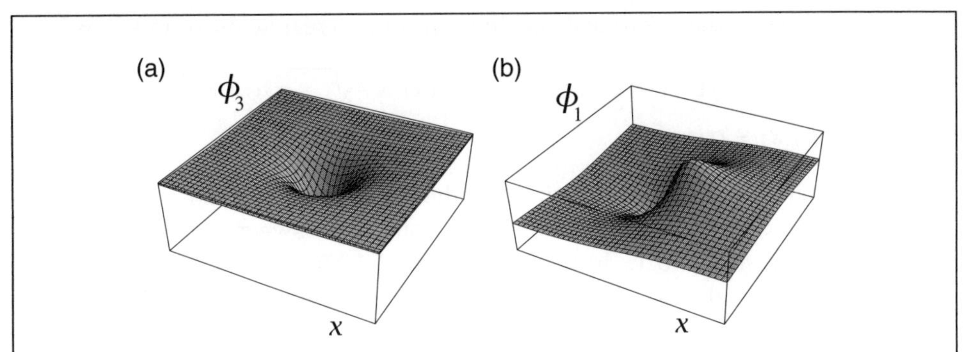

Fig. 7.13 A typical skyrmion configuration is depicted (a) for the component $\phi_3(\boldsymbol{x})$ and (b) for the component $\phi_1(\boldsymbol{x})$. See also Fig.4 in the frontispiece.

which gives a typical skyrmion solution. We can choose κ as a positive constant without loss of generality, while q should be a positive integer to avoid multivalueness in the field $\phi_\alpha(\boldsymbol{x})$. Substituting this into the formula (7.7.20), we may write down the field ϕ_α as

$$\phi_1 = \frac{2\kappa^q r^q}{r^{2q}+\kappa^{2q}}\cos(q\theta), \; \phi_2 = -\frac{2\kappa^q r^q}{r^{2q}+\kappa^{2q}}\sin(q\theta), \; \phi_3 = \frac{r^{2q}-\kappa^{2q}}{r^{2q}+\kappa^{2q}}. \tag{7.7.23}$$

This approaches the classical vacuum (7.7.3) asymptotically ($\phi_a \to \delta_{a3}$), where the magnetization vector points to the positive 3rd axis. The magnetization vector points to the negative 3rd axis ($\phi_a = -\delta_{a3}$) at $r = 0$ [Fig.7.13]. It describes a skyrmion sitting at $z = 0$ with scale κ. Its Pontryagin number is $Q_{\text{sky}} = q > 0$, as we shall see in (7.7.32). The skyrmion sitting at $z = z_0$ is given by $\omega(z - z_0)$. The fact that the scale κ and the position z_0 are arbitrary is a reflection of scale and translational invariance. See also Fig.4 in the frontispiece.

A typical antiskyrmion solution is given by the choice of

$$\omega(z) = \left(\frac{z^*}{\kappa}\right)^q = \left(\frac{r}{\kappa}\right)^q e^{-iq\theta}, \tag{7.7.24}$$

with a positive integer q and a positive constant κ. Substituting this into the formula (7.7.20), we may write down the field ϕ_α as

$$\phi_1 = \frac{2\kappa^q r^q}{r^{2q}+\kappa^{2q}}\cos(q\theta), \quad \phi_2 = \frac{2\kappa^q r^q}{r^{2q}+\kappa^{2q}}\sin(q\theta), \quad \phi_3 = \frac{r^{2q}-\kappa^{2q}}{r^{2q}+\kappa^{2q}}. \tag{7.7.25}$$

The magnetization vector points to the negative 3rd axis ($\phi_a = -\delta_{a3}$) at $r = 0$, and to the positive 3rd axis asymptotically ($\phi_a \to \delta_{a3}$), precisely as for the skyrmion. Nevertheless, its Pontryagin number is $Q_{\text{sky}} = -q < 0$, as we shall see in (7.7.32). See also Fig.4 in the frontispiece.

It is possible to parametrize $\phi_a(\boldsymbol{x})$ in terms of two real fields, σ and ϑ, as

$$\phi_1(\boldsymbol{x}) \equiv \sqrt{1-\sigma^2(\boldsymbol{x})}\cos\vartheta(\boldsymbol{x}), \quad \phi_2(\boldsymbol{x}) \equiv -\sqrt{1-\sigma^2(\boldsymbol{x})}\sin\vartheta(\boldsymbol{x}),$$
$$\phi_3(\boldsymbol{x}) \equiv \sigma(\boldsymbol{x}). \tag{7.7.26}$$

The field manifold S^2 is parametrized by the value of the 3rd axis σ and the azimuthal angle ϑ as in Fig.7.12. It is covered once as ϑ sweeps from 0 to 2π and as σ sweeps from -1 to 1. In the instance of (7.7.23) or (7.7.25), by identifying $\vartheta(\boldsymbol{x}) = \pm q\theta(\boldsymbol{x})$ and $\sigma(\boldsymbol{x}) = \phi_3(\boldsymbol{x})$, when the xy plane is swept once, the field manifold S^2 is swept by $\pm q$ times.

In general this is proved as follows. It is straightforward to show

$$\phi_1[\partial^\nu\phi_2, \partial^\lambda\phi_3] = (1-\sigma^2)[\partial^\lambda\sigma, \partial^\nu\vartheta]\cos^2\vartheta,$$
$$\phi_2[\partial^\nu\phi_3, \partial^\lambda\phi_1] = (1-\sigma^2)[\partial^\lambda\sigma, \partial^\nu\vartheta]\sin^2\vartheta,$$
$$\phi_3[\partial^\nu\phi_1, \partial^\lambda\phi_2] = \sigma^2[\partial^\lambda\sigma, \partial^\nu\vartheta], \tag{7.7.27}$$

where $[\partial^\nu\phi_2, \partial^\lambda\phi_3] \equiv \partial^\nu\phi_2\partial^\lambda\phi_3 - \partial^\nu\phi_3\partial^\lambda\phi_2$ and so on. By using them the topological current (7.7.4) is rewritten as

$$J^\mu_{\text{sky}} = -\frac{1}{4\pi}\varepsilon^{\mu\nu\lambda}\partial_\lambda\sigma\partial_\nu\vartheta = \frac{1}{4\pi}\varepsilon^{\mu\nu\lambda}\partial_\nu\sigma\partial_\lambda\vartheta. \tag{7.7.28}$$

The topological charge (7.7.9) is given by

$$Q_{\text{sky}} = \frac{1}{4\pi}\int d^2x\, \varepsilon_{ij}\partial_i\sigma\partial_j\vartheta \equiv \frac{1}{4\pi}\int d\sigma\wedge d\vartheta. \tag{7.7.29}$$

Here, $\varepsilon_{ij}\partial_i\sigma\partial_j\vartheta$ is the Jacobian from the surface element $dx\wedge dy$ in the real space R^2 to the surface element $d\sigma\wedge d\vartheta$ in the field space S^2. (See a textbook[22] for the differential form $dx\wedge dy$.) The topological charge Q_{sky} counts how many times the field space S^2 is covered ($\int d\sigma d\vartheta$) as the real space is swept ($\int d^2x$).

We calculate the topological charge of the example (7.7.23) or (7.7.25) explicitly. Equating it with (7.7.26), we find that

$$\sigma(\boldsymbol{x}) = \frac{r^{2q}-\kappa^{2q}}{r^{2q}+\kappa^{2q}}, \qquad \vartheta(\boldsymbol{x}) = \pm q\theta, \tag{7.7.30}$$

for skyrmion ($+q$) and antiskyrmion ($-q$) with $q \geq 1$. This defines a mapping from the real space R^2 spanned by (r,θ) to the field space S^2 spanned by (σ,ϑ). At $r=0$ the field σ takes the minimum value $\sigma=-1$, and as r increases it increases monotonically, and at $r=\infty$ it achieves the maximum value $\sigma=1$. The field ϑ is identified with $\pm q$ times the azimuthal angle θ. Thus, as θ rotates once in the real space, ϑ rotates $\pm q$ times in the field space. More explicitly, the charge density (7.7.28) is

$$J^0_{\text{sky}} = \frac{1}{4\pi}\varepsilon_{ij}\partial_i\sigma\partial_j\vartheta = \pm\frac{q}{4\pi r}\frac{d\sigma(r)}{dr}, \tag{7.7.31}$$

and the charge (7.7.29) is

$$Q_{\text{sky}} = \pm\frac{q}{4\pi}\int \frac{d^2x}{r}\frac{d\sigma(r)}{dr} = \pm\frac{q}{2}\int_{-1}^{1} d\sigma = \pm q. \tag{7.7.32}$$

The Pontryagin number of the solution (7.7.30) is $Q_{\text{sky}} = \pm q$.

It is instructive to evaluate the classical energy (7.6.7) explicitly for a skyrmion and an antiskyrmion. First we express the energy in terms of the parametrization (7.7.26),

$$\begin{aligned} E &= \frac{J_s}{2}\sum_{i=1}^{3}\int d^2x\,(\nabla\phi_i)^2 \\ &= \frac{J_s}{2}\int d^2x\left(\frac{1}{1-\sigma^2}(\nabla\sigma)^2 + q^2(1-\sigma^2)(\nabla\theta)^2\right), \end{aligned} \tag{7.7.33}$$

where we have used $\vartheta(\boldsymbol{x}) = \pm q\theta(\boldsymbol{x})$. For the configuration $\sigma(r)$ in (7.7.30) we find

$$\begin{aligned} (\nabla\sigma)^2 &= \frac{q(1-\sigma^2)}{r}\frac{d\sigma}{dr} = 2q(1-\sigma^2)\frac{d\sigma}{dr^2}, \\ (\nabla\theta)^2 &= \frac{1}{r^2} = \frac{2}{q(1-\sigma^2)}\frac{d\sigma}{dr^2}. \end{aligned} \tag{7.7.34}$$

The energy is easily calculated with the aid of these,

$$E = \frac{\pi}{2}(2q+2q)J_s\int_{-1}^{1} d\sigma = 4\pi q J_s. \tag{7.7.35}$$

It is equal to (7.7.21) with $Q_{\text{sky}} = \pm q$.

The O(3)-skyrmion is the quasiparticle in the spin QH ferromagnet as well as in the pseudospin QH ferromagnet, where the O(3) symmetry is explicitly broken, as we discuss in Section 16.2, Section 24.5 and Chapter 31. When the O(3) sigma field represents the spin field in the monolayer QH system, there is a Zeeman coupling Δ_Z, and the O(3) sigma model (7.7.1) is modified as

$$\mathcal{L} = -\frac{J_s}{2}(\partial^\mu\boldsymbol{\phi})\cdot(\partial_\mu\boldsymbol{\phi}) - \frac{\Delta_Z}{2}\phi_3, \tag{7.7.36}$$

where the O(3) symmetry is broken to the O(2) symmetry. When the O(3) sigma field represents the pseudospin field in the spin-frozen bilayer QH system, there is a tunneling coupling Δ_{SAS} and the capacitance coupling ε_{cap}, and the O(3) sigma model (7.7.1) is modified as

$$\mathcal{L} = -\frac{J_s}{2}(\partial^\mu\boldsymbol{\phi})\cdot(\partial_\mu\boldsymbol{\phi}) - \frac{\varepsilon_{\text{cap}}}{4}(\phi_2)^2 - \frac{\Delta_{\text{SAS}}}{2}\phi_3, \tag{7.7.37}$$

where the O(3) symmetry is completely broken. It is important that the field configuration (7.7.3) is a ground state in these broken-O(3) models. In particular, it is the unique ground state of the model (7.7.37). Consequently, topological solitons are skyrmions indexed by the Pontryagin number (7.7.9), since the boundary condition is given by (7.7.3). However, (7.7.23) is no longer a solution. An approximate solution is obtained by substituting (7.7.26) into the Hamiltonians, and by minimizing the energy with respect to the trial function $\sigma(r)$.

7.8 CP^{N-1} SKYRMIONS

The O(N) nonlinear sigma model has topological solitons only when $N = 3$ in the planar geometry. There exists a generalization of the O(3) sigma model so that the new model possess topological solitons for arbitrary N in the planar geometry. It is the CP^{N-1} sigma model,[110] whose group manifold is

$$CP^{N-1} = U(N)/[U(1) \otimes U(N\text{-}1)] = SU(N)/[U(1) \otimes SU(N) \otimes SU(N\text{-}1)]. \quad (7.8.1)$$

The homotopy theorem tells

$$\pi_2(CP^{N-1}) = \mathbb{Z}, \quad (7.8.2)$$

since $\pi_2(G/H) = \pi_1(H)$ (when G is simply connected) and $\pi_n(G \otimes G') = \pi_n(G) \oplus \pi_n(G')$. It is also called the SU(N) sigma model.[124] The O(3) sigma model and the CP^1 sigma model are locally equivalent. CP^3-skyrmions emerge as charged excitations in the bilayer QH system (see Section 25.4).

There exists a further generalization, called the Grassmannian $G_{N,k}$ model,[329] whose group manifold is

$$G_{N,k} = U(N)/[U(k) \otimes U(N\text{-}k)] = SU(N)/[U(1) \otimes SU(N) \otimes SU(N\text{-}k)]. \quad (7.8.3)$$

The Grassmannian field may be regarded as an appropriate combination of k CP^{N-1} fields. The homotopy theorem tells

$$\pi_2(G_{N,k}) = \mathbb{Z} \quad (7.8.4)$$

as in (7.8.2), which implies the existence of topological solitons. We explain the Grassmannian model and its solitons when we discuss the bilayer QH system at the filling factor $\nu = 2$: See Section 26.6 for the Grassmannian model and Section 26.7 for the Grassmannian soliton.

The CP^{N-1} sigma model is defined as follows. We take N complex scalar fields $\boldsymbol{n} = (n_1, n_2, \cdots, n_N)$, which are subject to the constraint equation

$$\boldsymbol{n}^\dagger(x)\boldsymbol{n}(x) = \sum_{a=1}^{N} n_a^*(x) n_a(x) = 1. \quad (7.8.5)$$

The Lagrangian density is given by

$$\mathcal{L} = -2J_s(\partial_\mu + iK_\mu)\boldsymbol{n}^\dagger(\partial^\mu - iK^\mu)\boldsymbol{n}, \quad (7.8.6)$$

where K_μ is an auxiliary field. It is invariant under the global SU(N) group. We consider the model in the planar geometry. The energy of a static classical configuration is

$$E = 2J_s \int d^2x (D_j \boldsymbol{n})^\dagger (D_j \boldsymbol{n}), \tag{7.8.7}$$

where $D_j \boldsymbol{n} = \partial_j \boldsymbol{n} - iK_j \boldsymbol{n}$. The Euler-Lagrange equation with respect to K_μ yields,

$$K_\mu = -\frac{i}{2}\boldsymbol{n}^\dagger \overleftrightarrow{\partial}_\mu \boldsymbol{n} = -i\boldsymbol{n}^\dagger \partial_\mu \boldsymbol{n}, \tag{7.8.8}$$

where use was made of (7.8.5). Since K_μ is determined by this equation, it is not a dynamical field.

Let us count the independent number of real fields. Originally, there are $2N$ real fields, but the constraint equation (7.8.5) removes one of them. The Lagrangian density has the U(1) gauge symmetry such as

$$n_a(x) \longrightarrow n_a(x) \exp[if(x)], \qquad K_\mu(x) \longrightarrow K_\mu(x) + \partial_\mu f(x), \tag{7.8.9}$$

which removes one real field. Namely, we identify two fields $\boldsymbol{n}(x)$ and $\boldsymbol{n}'(x)$ when they are related by

$$\boldsymbol{n}'(x) = \boldsymbol{n}(x) \exp[if(x)]. \tag{7.8.10}$$

Hence there are only $N-1$ independent complex fields.

We can parametrize the CP^{N-1} fieldin terms of N–1 independent complex fields as follows. The generic expression of the CP^{N-1} field is

$$\boldsymbol{n}(x) = \Lambda(x) \left(\omega_1(x), \omega_2(x), \dots, \omega_N(x)\right), \tag{7.8.11}$$

where Λ is a normalization factor to ensure $\boldsymbol{n}^\dagger \boldsymbol{n} = 1$, $\Lambda(x)^{-2} = \sum_{i=1}^N |\omega_i(z)|^2$. In a domain where $n_1(x) \neq 0$, we rewrite this as

$$\boldsymbol{n}(x) = \frac{\omega_1(x)}{|\omega_1(x)|}\Lambda_1(x) \left(1, \frac{\omega_2(x)}{\omega_1(x)}, \dots, \frac{\omega_N(x)}{\omega_1(x)}\right), \tag{7.8.12}$$

with $\Lambda_1(x) = |\omega_1(x)|\Lambda(x)$. Since the overall phase is irrelevant, it is possible to parametrize the CP^{N-1} field as

$$\boldsymbol{n}(x) = \Lambda(x) \left(1, \eta_2(x), \dots, \eta_N(x)\right), \tag{7.8.13}$$

where Λ is a normalization factor to ensure $\boldsymbol{n}^\dagger \boldsymbol{n} = 1$, $\Lambda(\boldsymbol{x})^{-2} = 1 + \sum_{i=2}^N |\eta_i(z)|^2$. Similarly, in the domain where $n_2(x) \neq 0$, we may parametrize it as

$$\boldsymbol{n}(x) = \Lambda(x) \left(\eta_1(x), 1, \eta_3(x), \dots, \eta_N(x)\right). \tag{7.8.14}$$

It is clear that there are $N-1$ independent complex fields. The $\boldsymbol{n}$ field is defined mathematically by the projective representation of (7.8.11). Because of this reason, the $\boldsymbol{n}(x)$ field is called the $N-1$ dimensional *complex projective* field, or the CP^{N-1} field.

We now consider a topological current,

$$J^{\mu}_{\text{sky}} = \frac{1}{2\pi}\varepsilon^{\mu\nu\lambda}\partial_\nu K_\lambda, \tag{7.8.15}$$

as in the vortex case [see (7.5.11)]. The topological charge may be rewritten as

$$Q_{\text{sky}} = -\frac{i}{2\pi}\varepsilon_{jk}\int d^2x\, (D_j\boldsymbol{n})^\dagger(D_k\boldsymbol{n}) = -\frac{i}{2\pi}\varepsilon_{jk}\int d^2x\, \partial_j\boldsymbol{n}^\dagger\partial_k\boldsymbol{n}. \tag{7.8.16}$$

The equivalence is readily checked by using $\boldsymbol{n}^*\boldsymbol{n} = 1$ and the antisymmetry of ε_{jk},

$$\begin{aligned}\varepsilon_{jk}(D_j\boldsymbol{n})^\dagger(D_k\boldsymbol{n}) =&\varepsilon_{jk}[\partial_j\boldsymbol{n}^\dagger\partial_k\boldsymbol{n} - i(\partial_j\boldsymbol{n}^\dagger)\boldsymbol{n}K_k + i\boldsymbol{n}^\dagger(\partial_k\boldsymbol{n})K_j]\\ =&\varepsilon_{jk}\partial_j\boldsymbol{n}^\dagger\partial_k\boldsymbol{n} = i\varepsilon_{jk}\partial_jK_k,\end{aligned} \tag{7.8.17}$$

with the use of (7.8.8). There must be topological solitons characterized by the charge Q_{sky}. We call them also skyrmions since they are reduced to the O(3)-sigma-model skyrmions in the case of the CP^1 sigma model, as we shall see later.

We next search for skyrmion solutions explicitly. We start with the identity

$$\int d^2x\, |D_j\boldsymbol{n} \pm i\varepsilon_{jk}D_k\boldsymbol{n}|^2 \geq 0, \tag{7.8.18}$$

which yields

$$\int d^2x\, (D_j\boldsymbol{n})^\dagger(D_j\boldsymbol{n}) \geq \mp i\int d^2x\, \varepsilon_{jk}(D_j\boldsymbol{n})^\dagger(D_k\boldsymbol{n}). \tag{7.8.19}$$

Since the energy and the topological charge are given by (7.8.7) and (7.8.16), we obtain

$$E \geq \pm 4\pi J_s Q_{\text{sky}}, \tag{7.8.20}$$

as is to be compared with (7.7.12).

The energy is minimized in each topological sector when the equality holds. Hence it is the BPS soliton, given by solving $E = \pm 4\pi J_s Q_{\text{sky}}$, or

$$D_j n_a = \mp i\varepsilon_{jk}D_k n_a. \tag{7.8.21}$$

We substitute $n_a = n_N\omega_a$ into this equation,

$$\omega_a D_j n_N + n_N\partial_j\omega_a = \mp i\varepsilon_{jk}(\omega_a D_k n_N + n_N\partial_k\omega_a). \tag{7.8.22}$$

Using (7.8.21) again we obtain

$$\partial_j\omega_a = \mp i\varepsilon_{jk}\partial_k\omega_a \tag{7.8.23}$$

as in (7.7.16), or

$$\frac{\partial}{\partial z^*}\omega_a = 0 \quad \text{or} \quad \frac{\partial}{\partial z}\omega_a = 0. \tag{7.8.24}$$

The most general solution is an arbitrary analytic function $\omega_a = \omega_a(z)$ of z for which we take the upper sign in (7.8.20), or an arbitrary analytic function $\omega_a = \omega_a(z^*)$ of z^* for which we take the lower sign in (7.8.20). In both cases we find

$$E = 4\pi J_s |Q_{\text{sky}}| \tag{7.8.25}$$

for solutions.

For a later convenience we derive a formula for the topological charge density $J^0_{\text{sky}}(\boldsymbol{x})$. We write down the Cauchy-Riemann equation for the analytic function $\omega^\alpha(z) = \Lambda^{-1}(\boldsymbol{x})n^\alpha(\boldsymbol{x})$ in (7.8.11),

$$0 = \frac{\partial \omega^\alpha(z)}{\partial z^*} = \frac{\partial n^\alpha(\boldsymbol{x})}{\partial z^*}\Lambda^{-1}(\boldsymbol{x}) + n^\alpha(\boldsymbol{x})\frac{\partial \Lambda^{-1}(\boldsymbol{x})}{\partial z^*}. \tag{7.8.26}$$

We multiply it with $\Lambda(\boldsymbol{x})n^{\alpha *}(\boldsymbol{x})$ and take the sum over $i = 1, 2, \cdots, N$,

$$\sum_{\alpha=1}^{N} n^{\alpha *}(\boldsymbol{x})\frac{\partial n^\alpha(\boldsymbol{x})}{\partial z^*} = -\frac{\partial}{\partial z^*}\ln \Lambda^{-1}(\boldsymbol{x}), \tag{7.8.27}$$

or

$$\varepsilon_{kj}K_j(\boldsymbol{x}) = \partial_k \ln \Lambda^{-1}(\boldsymbol{x}) = \partial_k \ln \sqrt{\sum_i |\omega_i(z)|^2}, \tag{7.8.28}$$

where K_k is the auxiliary field (7.8.8). Hence, we conclude

$$J^0_{\text{sky}}(\boldsymbol{x}) = \frac{1}{2\pi}\varepsilon_{kj}\partial_k K_j(\boldsymbol{x}) = \frac{1}{4\pi}\nabla^2 \ln\left[\sum_i |\omega_i(z)|^2\right]. \tag{7.8.29}$$

We shall use this when we discuss skyrmion excitations in both monolayer and bilayer QH systems (see Section 16.2 and Section 25.4).

A nontrivial simplest analytic function is given by

$$\boldsymbol{\omega} = z\boldsymbol{a} + \kappa \boldsymbol{b}, \tag{7.8.30}$$

or

$$\boldsymbol{n} = \frac{\boldsymbol{\omega}}{|\boldsymbol{\omega}|} = \frac{z\boldsymbol{a} + \kappa\boldsymbol{b}}{\sqrt{|z|^2 + \kappa^2}}, \tag{7.8.31}$$

where $\boldsymbol{a}$ and $\boldsymbol{b}$ are constant vectors such that

$$\boldsymbol{a}^\dagger \boldsymbol{a} = \boldsymbol{b}^\dagger \boldsymbol{b} = 1, \qquad \boldsymbol{a}^\dagger \boldsymbol{b} = 0. \tag{7.8.32}$$

For instance,

$$\boldsymbol{a} = (1, 0, 0, \cdots, 0), \quad \boldsymbol{b} = (0, b_2, b_3, \cdots, n_N). \tag{7.8.33}$$

The CP^{N-1} field (7.8.31) behaves asymptotically as

$$\lim_{|\boldsymbol{x}|\to\infty} \boldsymbol{n}(\boldsymbol{x}) \to e^{i\theta}\boldsymbol{a}, \tag{7.8.34}$$

where the U(1) phase factor $z/|z| \equiv e^{i\theta}$ is irrelevant. It describe a skyrmion on the ground state $\boldsymbol{n} = \boldsymbol{a}$, and is indexed by $N-1$ "shape" parameters $\kappa_j \equiv \kappa b_j$. A physical meaning of these parameters is found in the bilayer QH system: See Section 25.4.

The topological charge is calculated easily. Using the Stokes theorem we may rewrite it as

$$Q_{\text{sky}} = \frac{1}{2\pi} \oint dx_j K_j, \tag{7.8.35}$$

as follows from the topological current (7.8.15). Here, the loop integration is made along an infinitely large circle. Using the asymptotic behavior (7.8.34), we find

$$K_j = -i\boldsymbol{n}^\dagger \partial_j \boldsymbol{n} \simeq \partial_j \theta, \qquad \text{as} \quad |\boldsymbol{x}| \to \infty, \tag{7.8.36}$$

and hence $Q_{\text{sky}} = 1$. The above solution has the skyrmion number $Q_{\text{sky}} = 1$.

We prove the equivalence between the O(3) sigma model and the CP[1] sigma model. For this purpose we set

$$\phi_a = \boldsymbol{n}^\dagger \tau_a \boldsymbol{n} \tag{7.8.37}$$

with τ_a the Pauli matrices

$$\tau_1 = \begin{pmatrix} 0 & 1 \\ 1 & 0 \end{pmatrix}, \qquad \tau_2 = \begin{pmatrix} 0 & -i \\ i & 0 \end{pmatrix}, \qquad \tau_3 = \begin{pmatrix} 1 & 0 \\ 0 & -1 \end{pmatrix}. \tag{7.8.38}$$

It follows that

$$\Phi = \sum_{a=1}^{3} \phi_a \tau_a = \begin{pmatrix} n^{1\dagger} n^1 - n^{2\dagger} n^2 & 2n^{2\dagger} n^1 \\ 2n^{1\dagger} n^2 & n^{2\dagger} n^2 - n^{1\dagger} n^1 \end{pmatrix}, \tag{7.8.39}$$

or

$$\Phi^{\alpha\beta} = -\frac{2}{N}\left(\delta^{\alpha\beta} - N n^{\beta\dagger} n^\alpha\right) \tag{7.8.40}$$

with $N = 2$. It is easy to see that

$$\begin{aligned} \sum_{\alpha\beta} \left(\partial_\mu \Phi^{\alpha\beta}\right)\left(\partial^\mu \Phi^{\alpha\beta}\right) &= 8\left(\sum_\alpha \partial_\mu n^{\alpha\dagger} \partial^\mu n^\alpha - (\sum_\alpha n^{\alpha\dagger} \partial_\mu n^\alpha)(\sum_\beta n^{\beta\dagger} \partial_\mu n^\beta)\right) \\ &= 8 \sum_\alpha (\partial_\mu n^{\alpha\dagger} + iK_\mu n^{\alpha\dagger}) \cdot (\partial^\mu n^\alpha + iK^\mu n^\alpha) \end{aligned} \tag{7.8.41}$$

where K_μ is the auxiliary field (7.8.8). We now use the relation

$$\sum_a (\partial_\mu \phi_a)(\partial^\mu \phi_a) = \frac{1}{2} \text{Tr} \partial_\mu \Phi \cdot \partial^\mu \Phi = \frac{1}{2} \sum_{\alpha\beta} (\partial_\mu \Phi^{\alpha\beta})(\partial^\mu \Phi^{\alpha\beta}) \tag{7.8.42}$$

to establish the equivalence between the O(3) sigma model (7.7.1) and the CP[1] model (7.8.6).

It is straightforward to generalize the above analysis to the CP^{N-1} model. Group SU(N) is generated by the Hermitian, traceless, $N \times N$ matrices. There are

(N^2-1) independent matrices. We take a standard basis, λ_a, $a = 1, 2, \cdots, N^2-1$, normalized as $\text{Tr}(\lambda_a\lambda_b) = 2\delta_{ab}$: See Appendix D. They are the generalization of the Pauli matrices. We define the SU(N) sigma field by

$$\phi_a = \boldsymbol{n}^\dagger \lambda_a \boldsymbol{n}, \tag{7.8.43}$$

and set

$$\Phi = \sum_{a=1}^{N^2-1} \phi_a \lambda_a. \tag{7.8.44}$$

It is straightforward to verify the relations (7.8.40)~(7.8.42). Hence, the CP^{N-1} model (7.8.6) is equivalent to

$$\mathcal{L} = -\frac{J_s}{2} \sum_{a=1}^{N^2-1} (\partial^\mu \phi_a)(\partial_\mu \phi_a). \tag{7.8.45}$$

Though it is formally the same as the O(N^2-1) sigma model (7.6.5), it is not so because the number of independent fields is different. It is called the SU(N) sigma model.[124]

Skyrmions of the SU(N) sigma model is given by those of the CP^{N-1} model by way of (7.8.43). In particular, the generic CP^1-skyrmion is expressed as

$$n_1(\boldsymbol{x}) = \frac{\omega(z)}{\sqrt{|\omega(z)|^2 + 1}}, \qquad n_2(\boldsymbol{x}) = \frac{1}{\sqrt{|\omega(z)|^2 + 1}}. \tag{7.8.46}$$

It yields the SU(2)-skyrmion (7.7.20) via the identification (7.8.37). The anti-skyrmion is given by replacing $\omega(z)$ with $\omega(z^*)$.

CHAPTER 8

ANYONS

Particles are either bosons or fermions in the 3-dimensional space. The 2-dimensional space allows exotic particles, called anyons, which have fractional spins and statistics. An anyon may be represented as a flux-carrying boson (or a fermion), where the flux is described by the Chern-Simons gauge theory. They are called composite bosons (or fermions). Field theory of anyons is constructed by using the composite-particle (boson or fermion) field and the Chern-Simons field.

8.1 SPIN AND STATISTICS

In the 3-dimensional space there are only two types of particles, bosons and fermions. They are distinguishable by the following properties:

(1) *Spins*
A particle has an intrinsic spin $s\hbar$. Here, $s = 0, 1, 2, \cdots$ for bosons and $s = \frac{1}{2}, \frac{3}{2}, \cdots$ for fermions.

(2) *Exchange Statistics*
When two identical particles are exchanged, the wave function changes its overall sign for fermions but not for bosons.

(3) *Pauli Exclusion Principle*
Whereas any number of bosons can occupy a single quantum state, a single state cannot be occupied by more than one fermions.

These are the contents of the spin-statistics theorem[1] due to Pauli.[386] We review how they are derived and how they are generalized in the 2-dimensional space.

[1] With respect to the spin-statistics theorem Feynman said as follows:[18] "Why is it that particles with half-integral spin are Fermi particles whose amplitudes add with the minus sign, whereas particles with integral spin are Bose particle whose amplitude add with the positive sign? We apologize for the fact that we cannot give you an elementary explanation. An explanation has been worked out by Pauli from complicated arguments of quantum field theory and relativity. He has shown that the two must necessary go together, but we have not been able to find a way of reproducing his arguments on an elementary level $\cdots$. This probably means that we do not have a complete understanding of the fundamental principle involved $\cdots$." See a collection[17] of original works and also a reference[167] on the spin-statistics theorem.

Why is the spin quantized in the 3-dimensional space? The spin is the internal angular momentum inherent to a particle. There are three spin operators S_j since there are three independent ways of rotations. They satisfy a non-Abelian algebra,

$$[S_i, S_j] = i\hbar\varepsilon_{ijk}S_k, \qquad \text{for} \quad i,j,k = x,y,z. \tag{8.1.1}$$

The algebra not only fixes the normalizations but also leads to the quantization of spin in units of $\frac{1}{2}\hbar$. On the contrary, fractional spin is possible in the 2-dimensional space because there exists only one way of rotation, that is, a rotation around the z axis by embedding the 2-dimensional space into the 3-dimensional space and identifying it with the xy plane. The spin operator is S_z. The eigenvalue of S_z is arbitrary since the rotation group is Abelian.

If two identical particles are exchanged, their wave function $\mathfrak{S}(\boldsymbol{r}_1, \boldsymbol{r}_2)$ may acquire a phase factor $e^{i\alpha\pi}$, since the wave function itself is not an observable quantity,

$$\mathfrak{S}(\boldsymbol{r}_2, \boldsymbol{r}_1) = e^{i\alpha\pi}\mathfrak{S}(\boldsymbol{r}_1, \boldsymbol{r}_2). \tag{8.1.2}$$

If the exchange is made twice, the configuration is reduced to the original one, and hence the phase factor should be cancelled, $e^{2i\alpha\pi} = 1$. The allowed phase factor is $e^{i\alpha\pi} = \pm 1$; $+1$ for bosons and -1 for fermions. Once this is accepted, the anticommutation relation

$$\psi(t, \boldsymbol{y})\psi(t, \boldsymbol{x}) = -\psi(t, \boldsymbol{x})\psi(t, \boldsymbol{y}) \tag{8.1.3}$$

holds for the fermion field operator $\psi(t, \boldsymbol{x})$, from which the exclusion principle $\psi(t, \boldsymbol{x})\psi(t, \boldsymbol{x}) = 0$ follows.

Though the above argument looks general enough, it is not applicable to particles in the 2-dimensional space, as explained in Fig.8.1 and in Section 8.2. The wave function $\mathfrak{S}(\boldsymbol{r}_1, \boldsymbol{r}_2)$ need not be single-valued, and hence it is not required that $e^{i\alpha\pi} = \pm 1$ in (8.1.2). Accordingly the anticommutation relation (8.1.3) is generalized to the anyon commutation relation,[435]

$$\psi(t, \boldsymbol{y})\psi(t, \boldsymbol{x}) = e^{i\alpha\pi}\psi(t, \boldsymbol{x})\psi(t, \boldsymbol{y}), \tag{8.1.4}$$

where the phase factor $e^{i\alpha\pi}$ is arbitrary: $\alpha\pi$ is called the exchange phase and α the statistics parameter.

It is a logical consequence that particles may have fractional spin and statistics in the 2-dimensional space.[310,197,498] These particles were named *anyons* by Wilczek. The relation $s = \frac{1}{2}\alpha$ seems to hold between the spin[2] ($s\hbar$) and the statistics (α) of anyon, as is so in some explicit models.[499,513,135]

How about the Pauli exclusion principle? It is possible to interpolate it from fermions to anyons. The first attempt was made by Haldane,[211] who introduced

[2] Here the spin is the internal angular momentum. It is not necessarily related to the Zeeman effect. Thus, the "spin" of one skyrmion excitation is a function of the magnetic field, though it has a definite exchange statistics in the fractional QH state (see Chapter 16).

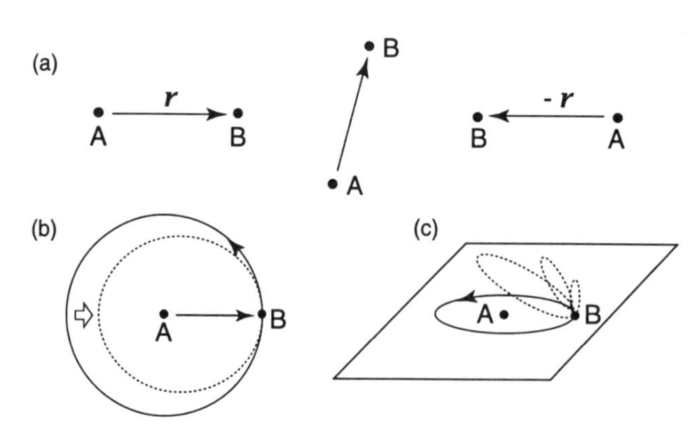

Fig. 8.1 Statistics is a property associated with the exchange of two identical particles. (a) Particles A and B are exchanged by rotating continuously the relative coordinate $\boldsymbol{r}$ by angle π. (b) The net effect of making the interchange twice is to move one particle (at position B) continuously off its original position and bring it back to the same position by drawing a loop. The loop encircles the point A once. (c) The loop is continuously shrinkable to a point in the 3-dimensional space, but not in the 2-dimensional space without crossing particle A.

the concept of *exclusion statistics*. The statistical interaction g (≥ 0) is defined by

$$g = \frac{d_N - d_{N+\Delta N}}{\Delta N}, \tag{8.1.5}$$

where N is the number of particles and d_N is the dimension of the one-particle Hilbert space obtained by holding the coordinate of $N-1$ particles fixed. We have $g = 0$ for bosons since any number of particle can occupy a given state (so that $d_{N+\Delta N} = d_N$), while we have $g = 1$ for fermions since only one particle can occupy a given state (so that $d_{N+\Delta N} = d_N - \Delta N$). The number of quantum states of N identical particles occupying a group of G state is given by[211,515]

$$W = \frac{[G + (N-1)(1-g)]!}{N![G - gN - (1-g)]!}. \tag{8.1.6}$$

For $g = 0$ and $g = 1$, this reproduces the well known formulas

$$W_{\text{boson}} = \frac{(G+N-1)!}{N!(G-1)!}, \qquad W_{\text{fermion}} = \frac{G!}{N!(G-N)!} \tag{8.1.7}$$

for bosons and fermions. There is no periodicity in the statistical strength g, and it makes sense to consider the cases[3] with $g \geq 2$. Attempts have been made to determine the statistical interaction g of quasiparticles in fractional QH systems.[211,518,251,464] The concept of exclusion statistics is valid in any dimensional

[3]Composite particles, to be identified with superelectrons in literature,[515] have the statistics interaction $g = m$ in the $\nu = 1/m$ fractional QH state with m an odd integer. See Fig.13.2 (page 234) and the last subsection 13.6 (page 245) in Section 13.6.

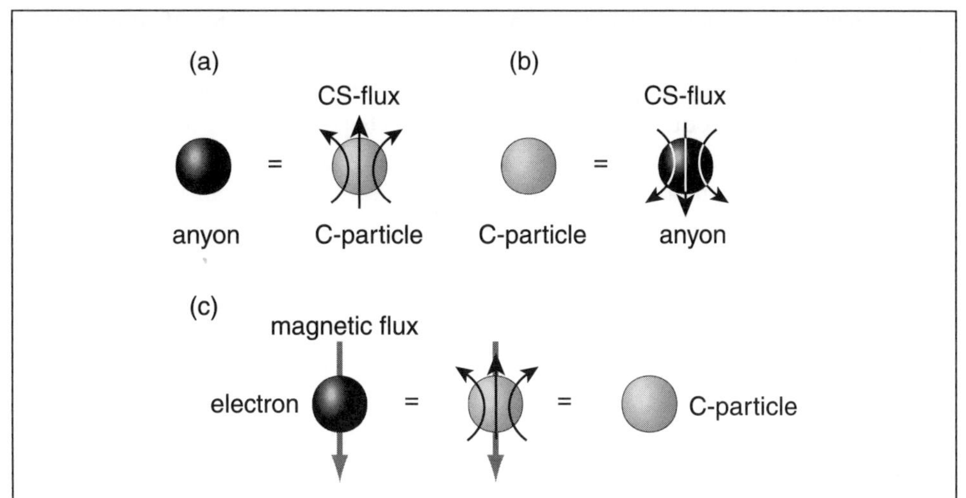

Fig. 8.2 (a) An anyon is regarded as a flux-carrying boson or fermion. The flux is called the Chern-Simons flux (CS-flux), and the boson or fermion is called the composite particle (C-particle). (b) Alternatively, a composite particle is constructed by attaching a Chern-Simons flux to an anyon. (c) A composite particle acquires a physical reality when the Chern-Simons flux is cancelled by the magnetic field, as occurs in the QH system.

space. Excitations in exactly solvable models in one dimension such as the Calgero-Sutherland model[82,467] and the δ-function gas[517,311] were argued[395,361] to obey this statistics. Another form of the exclusion statistics[396] was also suggested.

Dynamics of anyons are extremely complicated[4] except for two-anyon and three-anyon systems. The two-anyon system is reduced to the one-body problem by separating the center-of-mass coordinate, and exactly solvable in a harmonic oscillator potential and in a uniform magnetic field.[250] Exact solutions are known also for three anyons in a harmonic oscillator potential.[512] Though a class of exact solutions for N anyons were obtained in a harmonic oscillator potential[102,64,394] and in a uniform magnetic field,[123,100] they are known to form a very small subset of the full class of solutions. Indeed, the energy of the known solutions varies linearly with the statistical parameter α, but the energy of missing solutions varies nonlinearly with α, as was found numerically.[455,360] See a textbook[8] for details on the subject.

Anyons are characterized uniquely by the exchange property (8.1.2) or by the anyon commutation relation (8.1.4). It is possible to represent[498] an anyon as a flux-carrying boson (or fermion) [Fig.8.2(a)]. Alternatively, a boson (or fermion) emerges by attaching a flux to an anyon [Fig.8.2(b)]. This is known as *statistical transmutation*. We call the boson or the fermion thus obtained the *composite boson* or

[4] This is related to a mathematical observation[511,156,176] that the configuration space of anyons is characterized by the Braid group rather than the permutation group.

the *composite fermion*,generically the *composite particle*, as a matter of convenience. The flux is the Chern-Simons flux to be explained later, and not the magnetic flux.

A variety of composite particles is constructed from one anyon, corresponding to the number of flux quanta attached. A composite particle is a purely mathematical object to represent an anyon, and has nothing to do with physics by itself. Why do we introduce such objects? The first reason is rather technical. It is easy to formulate anyons quantum mechanically as particles (composite particles) making a "statistical interaction", as will be done in Section 8.3. It is also convenient to solve the dynamics of anyons in terms of composite particles since they are ordinary bosons or fermions. The second reason is more fundamental. Electrons are anyons when they are confined to a plane. Hence, an electron is represented as a bound state of a composite particle and the Chern-Simons flux [Fig.8.2(a)]. It is intriguing that a composite particle may acquire a physical reality when the Chern-Simons flux is cancelled by the magnetic flux [Fig.8.2(c)]. As will be shown in Parts II, the fractional QH state at $\nu = 1/(2p+1)$ is regarded as a condensed state of composite bosons carrying $2p+1$ flux quanta, or equivalently as an integer QH state of composite fermions carrying $2p$ flux quanta. The concept of composite particles reveals new physics in QH systems.

Anyons received wide attention after Laughlin'work[294] that quasiparticles carry fractional charges in fractional QH states. They were indeed shown[52] to be anyons. In 1988 Laughlin suggested that anyons could provide a mechanism for the high-T_C superconductor.[297] This gave a great boost[165,98,11] to anyons in physics community. However, whereas the anyon superconductivity violates the discrete symmetry of parity and the time reversal invariance, precision measurements[453,272] showed no indications of such violations in the high-T_C superconductor. Hence, anyons are irrelevant in physics of the high-T_C superconductor, though they play a fundamental role in physics of QH effects.

8.2 Fractional Statistics

Statistics is a property of identical particles associated with the exchange of their positions. Let us take two identical particles. Because the exchanged configuration is physically indistinguishable from the original one, the wave function should be identical except for a phase factor,

$$\mathfrak{S}(\boldsymbol{r}_2, \boldsymbol{r}_1) = e^{i\alpha\pi}\mathfrak{S}(\boldsymbol{r}_1, \boldsymbol{r}_2). \tag{8.2.1}$$

We call α the *statistics parameter*. It is defined mod 2. When their positions are exchanged once again, we have

$$\mathfrak{S}(\boldsymbol{r}_1, \boldsymbol{r}_2) = e^{i2\alpha\pi}\mathfrak{S}(\boldsymbol{r}_1, \boldsymbol{r}_2), \tag{8.2.2}$$

and hence $e^{i\alpha\pi} = \pm 1$. It is concluded naively that either the Bose statistics ($\alpha = 0$) or the Fermi statistics ($\alpha = 1$) is allowed.

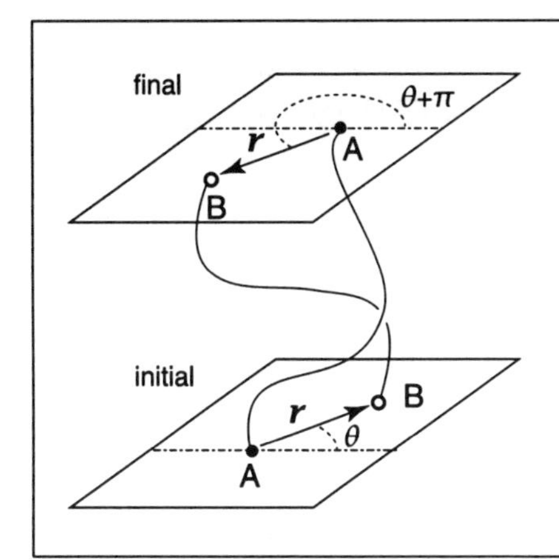

Fig. 8.3 Two particles are placed at points A and B in the xy plane. They make initially an angle θ with respect to the x axis. These two points move continuously so that their points are interchanged finally as in Fig.8.1(a).

In the 2-dimensional space we can have *fractional statistics*. Let us call the 2-dimensional space the xy plane and the direction orthogonal to it the z axis by embedding the space in three dimensions. Fractional statistics means that $e^{i2\alpha\pi} \neq 1$ in (8.2.2). Why is fractional statistics possible geometrically in two dimensions? To study the problem we analyze the process of interchanging the positions of two particles [Fig.8.1]. In so doing we move two particles continuously. The net effect of making the interchange twice is to move one particle continuously off its original position and bring it back to the same position by drawing a loop around the other particle. The phase (8.2.2) generated by this process can depend on the loop in general.

When the loop is moved continuously, the phase remains unchanged if it is an intrinsic property of the particles. In the 3-dimensional space we can shrink the loop continuously to a point by moving it out of the plane [Fig.8.1(c)]. Hence, the phase should be trivial, $e^{i2\alpha\pi} = 1$, and the possible statistics is either boson ($e^{i\alpha\pi} = 1$) or fermion ($e^{i\alpha\pi} = -1$). The situation is entirely different in the 2-dimensional space since the loop is constrained to a plane. It is impossible to shrink the loop without crossing the position of the other particle. Note that fermions cannot cross each others due to the Pauli exclusion principle. We postulate that the particles under consideration possess hardcores to satisfy the exclusion principle. This is a consistent requirement because, as we shall see in (8.3.19), a centrifugal barrier is created between two particles unless they are genuine bosons ($\alpha \neq 0$). Consequently, the parameter α is an arbitrary real number intrinsic to identical particles. When the loop encircles the other particle n times, we have

$$\mathfrak{S}^{(n)}(\boldsymbol{r}_1, \boldsymbol{r}_2) = e^{i2n\alpha\pi}\mathfrak{S}(\boldsymbol{r}_1, \boldsymbol{r}_2). \tag{8.2.3}$$

It is a topological number how many times a point encircles another point without crossing it in the 2-dimensional space. Fractional statistics is a topological property intrinsic to the 2-dimensional space.

8.3 QUANTUM MECHANICS

We formulate the quantum mechanics of anyons. The Lagrangian for ordinary particles is

$$L_0 = \frac{M}{2}\sum_{r=1}^{N}\left(\frac{d\mathbf{x}_r}{dt}\right)^2 - V(\mathbf{x}_1, \cdots, \mathbf{x}_N). \tag{8.3.1}$$

We convert particles into anyons by including a "statistical interaction"[498] between them,

$$L = L_0 - \alpha\hbar \sum_{r<s} \frac{d}{dt}\theta(\mathbf{x}_r - \mathbf{x}_s). \tag{8.3.2}$$

The interaction term involves the azimuthal angle $\theta(\mathbf{x}_r - \mathbf{x}_s)$ of the vector $\mathbf{x}_r - \mathbf{x}_s$ and the x axis. Let us consider two particles A and B [Fig.8.3]. The action between the initial and final states is given by

$$S = \int_{\text{ini}}^{\text{fin}} dtL = \int_{\text{ini}}^{\text{fin}} dtL_0 - \alpha\hbar \sum_{r<s}\left(\theta(\mathbf{x}_r^{\text{fin}} - \mathbf{x}_s^{\text{fin}}) - \theta(\mathbf{x}_r^{\text{ini}} - \mathbf{x}_s^{\text{ini}})\right). \tag{8.3.3}$$

According to the Feynman path-integral prescription of quantum mechanics[19] the amplitude from a particular classical path is given by $e^{iS/\hbar}$,

$$e^{iS/\hbar} = e^{-i\alpha\Delta\theta}\exp\left(\frac{i}{\hbar}\int_{\text{ini}}^{\text{fin}} dtL_0\right), \tag{8.3.4}$$

where

$$\Delta\theta = \theta(\mathbf{x}_r^{\text{fin}} - \mathbf{x}_s^{\text{fin}}) - \theta(\mathbf{x}_r^{\text{ini}} - \mathbf{x}_s^{\text{ini}}). \tag{8.3.5}$$

When the positions of A and B are interchanged by a half rotation ($\Delta\theta = \pi$), it yields

$$e^{iS/\hbar} = e^{-i\alpha\pi}\exp\left(\frac{i}{\hbar}\int_{\text{ini}}^{\text{fin}} dtL_0\right), \tag{8.3.6}$$

since $\theta(\mathbf{x}_A - \mathbf{x}_B) = \theta(\mathbf{x}_B - \mathbf{x}_A) + \pi$. The phase factor $e^{i\alpha\pi}$ in (8.3.6) is precisely the factor (8.2.1) characterizing the anyon statistics. We obtain (8.2.3) when one anyon rotates around the other n times. Consequently, we can regard (8.3.2) as the Lagrangian of anyons.

Once the Lagrangian is obtained, we are ready to apply the standard prescription of quantization. We define the momentum by

$$\mathbf{p}_r = \frac{\delta L}{\delta \dot{\mathbf{x}}_r} = M\dot{\mathbf{x}}_r - \check{e}\mathbf{C}(\mathbf{x}_r), \tag{8.3.7}$$

where we use the relation $(d/dt)\theta(\mathbf{x}) = \dot{\mathbf{x}}\nabla\theta(\mathbf{x})$ to find that

$$C_k(\mathbf{x}_r) = \frac{\alpha\hbar}{\check{e}}\sum_s \partial_k\theta(\mathbf{x}_r - \mathbf{x}_s) = -\frac{\alpha\hbar}{\check{e}}\sum_{r=1}^{N}\varepsilon_{kj}\frac{x_r^j - x_s^j}{|\mathbf{x}_r - \mathbf{x}_s|^2}, \tag{8.3.8}$$

where the point $\boldsymbol{x}_r = \boldsymbol{x}_s$ is not included in the sum. Here, $\check{e}$ is the "charge" introduced for later convenience. (We may set $\check{e} = 1$ without loss of generality. However, we have introduced this symbol since we choose $\check{e} = e$ to discuss QH states.) The Hamiltonian is given by a Legendre transformation,

$$H \equiv \sum_r^N \boldsymbol{p}_r \cdot \dot{\boldsymbol{x}}_r - L$$
$$= \frac{1}{2M} \sum_r^N (\boldsymbol{p}_r + \check{e}\boldsymbol{C}(\boldsymbol{x}_r))^2 + V(\boldsymbol{x}_1, \cdots, \boldsymbol{x}_N). \tag{8.3.9}$$

We then impose the canonical commutation relation between the coordinate $\boldsymbol{x}$ and the momentum $\boldsymbol{p}$, as results in the identification $\boldsymbol{p} = -i\hbar\nabla$. The quantum-mechanical Hamiltonian is

$$H = \frac{1}{2M} \sum_r^N (-i\hbar\nabla_r + \check{e}\boldsymbol{C}(\boldsymbol{x}_r))^2 + V(\boldsymbol{x}_1, \cdots, \boldsymbol{x}_N). \tag{8.3.10}$$

This is formally the one obtained by making the minimal substitution (5.3.10) with the "gauge potential" $\boldsymbol{C}(\boldsymbol{x})$ and the "charge" $\check{e}$: We call $\boldsymbol{C}(\boldsymbol{x})$ the Chern-Simons field. See (8.4.3) for the naming.

The field $\boldsymbol{C}(\boldsymbol{x})$ is actually an auxiliary field given by (8.3.8). We may characterize $C_j(\boldsymbol{x})$ by the constraint equation

$$\varepsilon_{ij}\partial_i C_j(\boldsymbol{x}) = \frac{2\alpha\pi\hbar}{\check{e}}\rho(\boldsymbol{x}), \tag{8.3.11}$$

where

$$\rho(\boldsymbol{x}) = \frac{1}{2\pi} \sum_r^N \varepsilon_{ij}\partial_i\partial_j\theta(\boldsymbol{x} - \boldsymbol{x}_s) = \sum_r^N \delta(\boldsymbol{x} - \boldsymbol{x}_s) \tag{8.3.12}$$

is the number density of anyons.

System of Two Anyons

We solve the system of two anyons with statistics parameters α. The Hamiltonian is

$$H = \frac{1}{2M}\left\{(\boldsymbol{p}_1 + \check{e}\boldsymbol{C}(\boldsymbol{x}_1))^2 + (\boldsymbol{p}_2 + \check{e}\boldsymbol{C}(\boldsymbol{x}_2))^2\right\} + V(r), \tag{8.3.13}$$

where $r = |\boldsymbol{x}_1 - \boldsymbol{x}_2|$, and $\boldsymbol{C}(\boldsymbol{x}_i)$ is the Chern-Simons field explicitly given by

$$\boldsymbol{C}(\boldsymbol{x}_1) = \frac{\alpha\hbar}{\check{e}}\nabla\theta(\boldsymbol{x}_1 - \boldsymbol{x}_2), \qquad \boldsymbol{C}(\boldsymbol{x}_2) = \frac{\alpha\hbar}{\check{e}}\nabla\theta(\boldsymbol{x}_2 - \boldsymbol{x}_1) = -\boldsymbol{C}(\boldsymbol{x}_1). \tag{8.3.14}$$

We separate the relative coordinate $\boldsymbol{r}$ from the center-of-mass coordinate $\boldsymbol{R}$. The Hamiltonian is decomposed into two parts,

$$H_R = \frac{1}{2m_R}\boldsymbol{p}_R^2, \qquad H_r = \frac{1}{2m_r}(\boldsymbol{p}_r - \check{e}\boldsymbol{C})^2 + V(r), \tag{8.3.15}$$

with $m_R = 2M$, $m_r = M/2$ and

$$C(\boldsymbol{r}) = \frac{\alpha\hbar}{\check{e}}\nabla\theta(\boldsymbol{r}). \tag{8.3.16}$$

The center-of-mass Hamiltonian H_R is that of the free one, whose eigenstate is a plain wave. We analyze the relative Hamiltonian H_r, which reads

$$\begin{aligned} H_r &= -\frac{1}{2m_r}[-i\hbar\nabla + \check{e}C(\boldsymbol{r})]^2 + V(r) \\ &= -\frac{\hbar^2}{2m_r}\left[\frac{1}{r}\frac{d}{dr}\left(r\frac{d}{dr}\right) + \frac{1}{r^2}\left(\frac{\partial}{\partial\theta} + i\alpha\right)^2\right] + V(r). \end{aligned} \tag{8.3.17}$$

We denote the eigenfunction by $\check{\mathfrak{S}}$ and set

$$\check{\mathfrak{S}}_l(r,\theta) \equiv e^{il\theta}\check{\mathfrak{S}}_l(r), \tag{8.3.18}$$

with l an integer. The radial part of the Schrödinger equation gives

$$\left[-\frac{1}{r}\frac{d}{dr}\left(r\frac{d}{dr}\right) + \frac{(l+\alpha)^2}{r^2} + \frac{2m_r}{\hbar^2}V(r)\right]\check{\mathfrak{S}}_l(r) = \frac{2m_r}{\hbar^2}E_l\check{\mathfrak{S}}_l(r). \tag{8.3.19}$$

The term proportional to $(l+\alpha)^2$ yields the centrifugal barrier due to the total angular momentum.

Let us explicitly take a harmonic potential well, $V(r) = \frac{1}{4}m_r\omega^2r^2$. The above radial equation is easily solved. The normalizable solutions are found by setting

$$\check{\mathfrak{S}}_l(r) = r^{|l+\alpha|}e^{-\omega m_r r^2/2\hbar}\left(a_0 + a_1 r + a_2 r^2 + \cdots\right), \tag{8.3.20}$$

and by demanding that the series terminates after a finite number of terms. This well-known procedure yields the eigenvalues

$$E_{nl} = \hbar\omega\left(2n + 1 + |l+\alpha|\right), \tag{8.3.21}$$

where $n = 0, 1, 2, \cdots$.

We have derived the wave function $\check{\mathfrak{S}}_l(\boldsymbol{x})$ for the two-anyon system as in (8.3.18). However, strictly speaking, it is not the wave function for anyons, since we have

$$\check{\mathfrak{S}}_l(r,\theta+\pi) = \pm\check{\mathfrak{S}}_l(r,\theta), \tag{8.3.22}$$

depending on whether ℓ is even (+) or odd (−). It is the wave function of *composite bosons* (+) or *fermions* (−). The wave function $\mathfrak{S}_l$ for anyons is given by

$$\mathfrak{S}_l(r,\theta) \equiv e^{i\alpha\theta}\check{\mathfrak{S}}_l(r,\theta) \qquad \text{for} \quad l = 0, \pm 2, \pm 4, \cdots, \tag{8.3.23a}$$

$$\mathfrak{S}_l(r,\theta) \equiv e^{i(\alpha+1)\theta}\check{\mathfrak{S}}_l(r,\theta) \qquad \text{for} \quad l = 1, \pm 3, \pm 5, \cdots, \tag{8.3.23b}$$

so that it has the property

$$\mathfrak{S}(r,\theta+\pi) = e^{i\alpha\pi}\mathfrak{S}(r,\theta) \tag{8.3.24}$$

in accord with (8.2.1). The anyon wave function is not single-valued.

The two-body anyon wave function $\mathfrak{S}(r,\theta)$ is related to the two-body boson wave function $\check{\mathfrak{S}}(r,\theta)$ as in (8.3.23a). The relation is generalized to the N-body anyon wave function. In Section 8.5 we construct the anyon wave function $\mathfrak{S}(\boldsymbol{x}_1,\cdots,\boldsymbol{x}_N)$ by

$$\mathfrak{S}(\boldsymbol{x}_1,\cdots,\boldsymbol{x}_N) \equiv e^{i\alpha\sum_{r<s}\theta(\boldsymbol{x}_r-\boldsymbol{x}_s)}\check{\mathfrak{S}}(\boldsymbol{x}_1,\cdots,\boldsymbol{x}_N) \tag{8.3.25}$$

in terms of the boson eigenfunction $\check{\mathfrak{S}}(\boldsymbol{x}_1,\cdots,\boldsymbol{x}_N)$ of the Hamiltonian (8.3.10).

It follows from (8.3.25) that

$$\check{\mathfrak{S}}(\boldsymbol{x}_1,\cdots,\boldsymbol{x}_N) = e^{-i\alpha\sum_{r<s}\theta(\boldsymbol{x}_r-\boldsymbol{x}_s)}\mathfrak{S}(\boldsymbol{x}_1,\cdots,\boldsymbol{x}_N). \tag{8.3.26}$$

This implies that anyons are converted into bosons or fermions by attaching appropriate Chern-Simons flux to them. They are composite particles, namely, composite bosons or composite fermions [Fig.8.2(b)]. In solving the two-anyon system we have used composite particles to derive the wave function (8.3.20), from which the anyon wave function is constructed as in (8.3.23a) or (8.3.23b). Composite particles are not physical objects; they are introduced merely to represent anyons in a convenient way. However, as will be discussed in Part II, composite particles acquire a physical reality in QH states [Fig.8.2(c)].

8.4 Chern-Simons Gauge Theory

We formulate quantum field theory of anyons. We use the fact that the quantum-mechanical Hamiltonian (8.3.9) of anyons is formally obtained by making a minimal substitution (5.3.10), where the electromagnetic potential eA_k is replaced by the Chern-Simons field $\check{e}C_j$. It is reasonable to make this substitution in the field-theoretical Hamiltonian (3.2.7) of ordinary particles,

$$\begin{aligned}H = &\int d^3x \left(\frac{1}{2M}(-i\hbar\nabla - \check{e}C)\phi^\dagger(x)(-i\hbar\nabla + \check{e}C)\phi(x) + V(\boldsymbol{x})\phi^\dagger(x)\phi(x)\right)\\ &+\frac{1}{2}\int d^3x\int d^3y\phi^\dagger(x)\phi^\dagger(x)U(|\boldsymbol{x}-\boldsymbol{y}|)\phi(x)\phi(x).\end{aligned} \tag{8.4.1}$$

Here $\phi(t,\boldsymbol{x})$ is the boson (or fermion) field to represent anyons. It is the composite-particle field. The Chern-Simons field $C_k(x)$ is defined quantum mechanically by (8.3.11) with the number density $\rho(\boldsymbol{x}) = \phi^\dagger(x)\phi(x)$,

$$\varepsilon_{jk}\partial_j C_k(x) = \frac{2\alpha\pi\hbar}{\check{e}}\phi^\dagger(x)\phi(x). \tag{8.4.2}$$

It is to be emphasized that the "Maxwell term" of the Chern-Simons field is absent in the system: The Chern-Simons field is an auxiliary field subject to the constraint equation (8.4.2) rather than a dynamical field obeying the Maxwell equation.

Lagrangian Formalism

We now justify the above heuristic argument by postulating a Lagrangian from which both the Hamiltonian (8.4.1) and the constraint equation (8.4.2) are derived. Recall that the existence of the Lagrangian guarantees the self-consistency of a field theory. The relevant Lagrangian density is

$$\begin{aligned} L = & \int d^2x \left[\phi^\dagger(x)\,(i\hbar\partial_t + \check{e}C_0(x))\,\phi(x) - \frac{1}{2M}\phi^\dagger(x)\,(i\hbar\partial_k - \check{e}C_k(x))^2\,\phi(x) \right] \\ & - \int d^2x \phi^\dagger(x) V(\boldsymbol{x}) \phi(x) - \frac{1}{2}\int d^2x \int d^2y \phi^\dagger(x)\phi^\dagger(y) U(|\boldsymbol{x}-\boldsymbol{y}|)\phi(x)\phi(y) \\ & - \frac{\check{e}^2}{4\alpha\pi\hbar}\varepsilon^{\mu\nu\lambda}\int d^2x C_\mu(x)\partial_\nu C_\lambda(x), \end{aligned} \tag{8.4.3}$$

where $\varepsilon^{\mu\nu\lambda}$ is the completely antisymmetric tensor with the convention of $\varepsilon^{012} = 1$. This Lagrangian has a well-known expression except for the last term called the *Chern-Simons term.*[15,116] This is the reason why the "gauge potential" C_μ is called the Chern-Simons field. By calculating $\partial L/\partial C_0 = 0$, we obtain

$$\varepsilon_{jk}\partial_j C_k(x) = \frac{2\alpha\pi\hbar}{\check{e}}\phi^\dagger(x)\phi(x). \tag{8.4.4}$$

This is the constraint condition (8.4.2) for C_k. Then, solving $\partial L/\partial C_k = 0$, we find

$$\partial_j C_0(x) = \partial_0 C_j(x) + \frac{2\alpha\pi\hbar}{M\check{e}}\varepsilon_{jk}\phi^\dagger(x)(i\hbar\partial_k - \check{e}C_k)\phi(x). \tag{8.4.5}$$

This is the constraint condition for C_0. The Chern-Simons field C_μ is not a dynamical field but an auxiliary field. The dynamical field is solely given by ϕ, whose canonical momentum is $\partial L/\partial\dot{\phi} = i\hbar\phi^\dagger$. The Hamiltonian is obtained by a Legendre transformation,

$$H = i\int d^2x \phi^\dagger\dot{\phi} - L, \tag{8.4.6}$$

which yields (8.4.1). The canonical commutation relations are the standard ones,

$$\begin{aligned} &[\phi(t,\boldsymbol{x}),\phi(t,\boldsymbol{y})]_\pm = [\phi^\dagger(t,\boldsymbol{x}),\phi^\dagger(t,\boldsymbol{y})]_\pm = 0, \\ &[\phi(t,\boldsymbol{x}),\phi^\dagger(t,\boldsymbol{y})]_\pm = \delta(\boldsymbol{x}-\boldsymbol{y}), \end{aligned} \tag{8.4.7}$$

depending whether $\phi(x)$ is a boson (+) or fermion (−) field. We have constructed a field theory of anyons by using the composite-particle field and the Chern-Simons field.

Equivalence to Quantum-Mechanical System

We next verify the equivalence between the quantum mechanics of anyons described by (8.3.10), or

$$H^{QM} = \frac{1}{2M}\sum_r (\boldsymbol{p}_r + \check{e}\boldsymbol{C}(\boldsymbol{x}_r))^2 + \sum_{r<s} U(|\boldsymbol{x}_r - \boldsymbol{x}_s|)) \tag{8.4.8}$$

with

$$C(\boldsymbol{x}_r) = \frac{\alpha\hbar}{\check{e}} \sum_{s>r} \nabla\theta(\boldsymbol{x}_r - \boldsymbol{x}_s), \tag{8.4.9}$$

and the Chern-Simons gauge theory whose Hamiltonian is given by (8.4.1), or

$$\begin{aligned} H^{FT} =& \frac{1}{2M}\int d^3x : \phi^\dagger(x)(-i\hbar\partial_k + \check{e}C_k)^2\phi(x) : \\ &+ \frac{1}{2}\int d^3x \int d^3y \phi^\dagger(x)\phi^\dagger(y)U(|\boldsymbol{x}-\boldsymbol{y}|)\phi(x)\phi(y) \end{aligned} \tag{8.4.10}$$

with

$$\varepsilon_{jk}\partial_j C_k(x) = \frac{2\alpha\pi\hbar}{\check{e}}\phi^\dagger(x)\phi(x). \tag{8.4.11}$$

Normal ordering, which is to place $\phi^\dagger$ to the left of ϕ, is necessary because the field C_k is a function of ϕ and $\phi^\dagger$. The proof is essentially the same as in the case of the Schrödinger field for the ordinary particle, although calculations are complicated since $C_\mu(x)$ is not an independent variable. Here, we give a sketch of a proof. Readers may skip the rest of this section. In the rest of this section we use the natural unit ($\hbar = c = 1$) and set $\check{e} = 1$.

In evaluating the Heisenberg equation of motion, we need some basic equations. First, solving the constraint equation (8.4.2) we obtain

$$C_k(\boldsymbol{x}) = \alpha\hbar \int d^2y \partial_k\theta(\boldsymbol{x}-\boldsymbol{y})\rho(\boldsymbol{y}), \tag{8.4.12}$$

where $\rho(x) = \phi^\dagger(x)\phi(x)$. Using the canonical commutation relation (8.4.7), we derive

$$[\phi(x), C_k(y)] = \alpha\partial_k^y\theta(\boldsymbol{y}-\boldsymbol{x})\phi(x), \tag{8.4.13a}$$

$$[\phi(x), : C_k^2(y) :] = [\alpha\partial_k\theta(\boldsymbol{y}-\boldsymbol{x})]^2\phi(x) + 2\alpha C_k(y)\phi(x), \tag{8.4.13b}$$

where $\partial_k^x = \partial/\partial x_k$. Hence,

$$\begin{aligned} [\phi(x), \left(i\partial_k^y + : C_k(y) :\right)^2 \phi(y)] =& 2i\alpha\left(C_k(y) + \partial_k^y\theta(\boldsymbol{y}-\boldsymbol{x})\right)\partial_k\phi(y)\phi(x) \\ &+ \left(\alpha\partial_k^y\theta(\boldsymbol{y}-\boldsymbol{x})\right)^2\phi(x)\phi(y). \end{aligned} \tag{8.4.14}$$

We also note that

$$\begin{aligned} [\phi(x), C_0(y)] =& -\frac{\alpha}{M}\partial_k^y\theta(\boldsymbol{y}-\boldsymbol{x})\left(i\partial_k^x + C_k(x)\right)\phi(x) \\ &- \frac{\alpha}{M}\int d^2z \partial_k^y\theta(\boldsymbol{y}-\boldsymbol{z})\phi^\dagger(z)\left(i\partial_k^z + \alpha\partial_k^z\theta(\boldsymbol{z}-\boldsymbol{y})\right)\phi(z)\phi(x). \end{aligned} \tag{8.4.15}$$

Using these we find that

$$\begin{aligned} i\partial_t\phi(x) =&[\phi(x),H] \\ =&\frac{1}{2M}(i\partial_k+:C_k(x):)^2\phi(x)-C_0(x)\phi(x) \\ &+\int d^2y\phi^\dagger(t,\mathbf{y})\phi(t,\mathbf{y})U(|\mathbf{y}-\mathbf{x}|)\phi(t,\mathbf{x}) \\ &+\frac{\alpha^2}{2M}\int d^2y[\partial_k^x\theta(\mathbf{x}-\mathbf{y})]^2\phi^\dagger(y)\phi(y)\phi(x). \end{aligned} \tag{8.4.16}$$

The classical equation of motion, $\partial L/\partial\phi^\dagger=0$, agrees with this equation except for the last term. This is because the Chern-Simons field C_μ is an auxiliary field.

The wave function is defined by

$$\check{\mathfrak{S}}(x_1,\cdots,x_N)=\langle 0|\phi(x_1)\phi(x_2)\cdots\phi(x_N)|N\rangle \tag{8.4.17}$$

for generic state $|N\rangle$ in the N-anyon sector. For the one-anyon state, by sandwiching the Heisenberg equation (8.4.16) with $\langle 0|$ and $|1\rangle$, and by using $\langle 0|\phi^\dagger(x)=0$ and $\langle 0|C_\mu(x)=0$, we obtain that

$$i\partial_t\check{\mathfrak{S}}(x)=\langle 0|[\phi(x),H^{FT}]|1\rangle=-\frac{1}{2M}\nabla^2\check{\mathfrak{S}}(x). \tag{8.4.18}$$

This is the Schrödinger equation for the one-anyon system. One anyon behaves just as an ordinary free particle, as should be the case.

For many-anyon state we start with

$$\begin{aligned} i\partial_t\langle 0|\phi(x_1)\phi(x_2)\cdots\phi(x_N)|N\rangle &= i\sum_j\langle 0|\phi(x_1)\cdots\dot{\phi}(x_j)\cdots\phi(x_N)|N\rangle \\ &=\sum_j\langle 0|\phi(x_1)\cdots[\phi(x_j),H]\cdots\phi(x_N)|N\rangle, \end{aligned} \tag{8.4.19}$$

where use was made of the Heisenberg equation. Let us calculate this explicitly for the $N=2$ case:

$$\begin{aligned} i\partial_t\check{\mathfrak{S}}(x,y) &= \langle 0|[\phi(x),H]\phi(y)|2\rangle+\langle 0|\phi(x)[\phi(y),H]|2\rangle \\ &= -\frac{1}{2M}(\nabla_x^2+\nabla_y^2)\check{\mathfrak{S}}(x,y)+\langle 0|[\phi(x),[\phi(y),H]]|2\rangle. \end{aligned} \tag{8.4.20}$$

Here, we have used the explicit form of the Heisenberg equation (8.4.16). Using the commutation relations (8.4.13a) ~ (8.4.15), we find that

$$\begin{aligned} i\partial_t\check{\mathfrak{S}}(x,y) =&\frac{1}{2M}\left[(i\nabla_x+\alpha\nabla_x\theta(\mathbf{x}-\mathbf{y}))^2+(i\nabla_y+\alpha\nabla_y\theta(\mathbf{y}-\mathbf{x}))^2\right]\check{\mathfrak{S}}(x,y) \\ &+V(\mathbf{x}-\mathbf{y})\mathfrak{S}(x,y). \end{aligned} \tag{8.4.21}$$

In general, we obtain for the N-anyon case that

$$i\partial_t \check{\mathfrak{S}}(x_1, x_2, \cdots, x_N) = H^{QM} \check{\mathfrak{S}}(x_1, x_2, \cdots, x_N), \tag{8.4.22}$$

where H^{QM} is given by (8.4.8). This is the quantum-mechanical Schrödinger equation. We have established the equivalence between the quantum-mechanical anyon system and the quantum-field-theoretical anyon system.

8.5 Anyon Field Operators

The anyon field $\psi(t, \boldsymbol{x})$ is represented by

$$\psi(t, \boldsymbol{x}) = e^{i\alpha\Theta(t,\boldsymbol{x})}\phi(t, \boldsymbol{x}), \tag{8.5.1}$$

in terms of the composite-boson field $\phi(t, \boldsymbol{x})$ with the phase field

$$\Theta(t, \boldsymbol{x}) = \alpha \int d^2y \theta(\boldsymbol{x} - \boldsymbol{y})\rho(t, \boldsymbol{y}), \tag{8.5.2}$$

where $\rho(t, \boldsymbol{x}) = \phi^\dagger(t, \boldsymbol{x})\phi(t, \boldsymbol{x})$. The Chern-Simons field (8.4.12) is expressed as

$$C_k(\boldsymbol{x}) = \frac{\alpha\hbar}{\check{e}}\partial_k \Theta(\boldsymbol{x}) = \frac{\alpha\hbar}{\check{e}} \int d^2y \partial_k \theta(\boldsymbol{x} - \boldsymbol{y})\rho(\boldsymbol{y}). \tag{8.5.3}$$

The relation (8.5.1) is an operator phase transformation, which converts the anyon field $\psi(t, \boldsymbol{x})$ to the boson field operator $\phi(t, \boldsymbol{x})$ by attaching an infinitely thin solenoid.

We examine how the exchange phase (8.2.1) arises for anyons field-theoretically. For brevity we choose a boson field ϕ. We start with the commutation relation between the phase field (8.5.2) and the boson field,

$$[\Theta(\boldsymbol{y}), \phi(\boldsymbol{x})] = \int d^2z \theta(\boldsymbol{y} - \boldsymbol{z})[\rho(\boldsymbol{z}), \phi(\boldsymbol{x})] = -\theta(\boldsymbol{y} - \boldsymbol{x})\phi(\boldsymbol{x}). \tag{8.5.4}$$

We use the Hausdorff formula

$$e^A B e^{-A} = \sum \frac{1}{n!}(\text{ad}_A)^n \cdot B = B + [A, B] + \frac{1}{2}[A, [A, B]] + \cdots, \tag{8.5.5}$$

where $\text{ad}_A \cdot B = [A, B]$. Setting $A = i\alpha\Theta(\boldsymbol{y})$ and $B = \phi(\boldsymbol{x})$ we find that

$$e^{-i\alpha\Theta(\boldsymbol{y})}\phi(\boldsymbol{x})e^{i\alpha\Theta(\boldsymbol{y})} = \phi(\boldsymbol{x}) + i\alpha\theta(\boldsymbol{y} - \boldsymbol{x})\phi(\boldsymbol{x}) + \cdots = e^{i\alpha\theta(\boldsymbol{y}-\boldsymbol{x})}\phi(\boldsymbol{x}). \tag{8.5.6}$$

With this formula we get

$$\begin{aligned} \psi(\boldsymbol{x})\psi(\boldsymbol{y}) &= e^{i\alpha\Theta(\boldsymbol{x})}\phi(\boldsymbol{x})e^{i\alpha\Theta(\boldsymbol{y})}\phi(\boldsymbol{y}) \\ &= e^{i\alpha\theta(\boldsymbol{y}-\boldsymbol{x})}e^{i\alpha\Theta(\boldsymbol{x})}e^{i\alpha\Theta(\boldsymbol{y})}\phi(\boldsymbol{x})\phi(\boldsymbol{y}), \end{aligned} \tag{8.5.7}$$

and

$$\psi(\boldsymbol{y})\psi(\boldsymbol{x}) = e^{i\alpha\theta(\boldsymbol{x}-\boldsymbol{y})}e^{i\alpha\Theta(\boldsymbol{y})}e^{i\alpha\Theta(\boldsymbol{x})}\phi(\boldsymbol{y})\phi(\boldsymbol{x}). \tag{8.5.8}$$

Using the relations $\phi(x)\phi(y) = \phi(y)\phi(x)$ and $\theta(x-y) = \theta(y-x)+\pi$, we compare (8.5.7) and (8.5.8) to find that

$$\psi(y)\psi(x) = e^{i\alpha\pi}\psi(x)\psi(y). \tag{8.5.9}$$

This is the operator version of the interchange of two anyons.

We next examine

$$\psi(x)\psi^\dagger(y) = e^{i\alpha\Theta(x)}\phi(x)\phi^\dagger(y)e^{-i\alpha\Theta(y)}. \tag{8.5.10}$$

We substitute $\phi(x)\phi^\dagger(y) = \phi^\dagger(y)\phi(x) + \delta(x-y)$ into it,

$$\psi(x)\psi^\dagger(y) = e^{i\alpha\Theta(x)}\phi^\dagger(y)\phi(x)e^{-i\alpha\Theta(y)} + \delta(x-y). \tag{8.5.11}$$

We use (8.5.6) for $\phi(x)$ to find

$$\psi(x)\psi^\dagger(y) = e^{-i\alpha\theta(y-x)}e^{i\alpha\Theta(x)}\phi^\dagger(y)e^{-i\alpha\Theta(y)}\phi(x) + \delta(x-y). \tag{8.5.12}$$

On the other hand,

$$\begin{aligned}\psi^\dagger(y)\psi(x) &= \phi^\dagger(y)e^{-i\alpha\Theta(y)}e^{i\alpha\Theta(x)}\phi(x)\\ &= e^{i\alpha\Theta(x)}e^{-i\alpha\Theta(x)}\phi^\dagger(y)e^{i\alpha\Theta(x)}e^{-i\alpha\Theta(y)}\phi(x).\end{aligned} \tag{8.5.13}$$

We use the Hermitian conjugate of (8.5.6) for $\phi^\dagger(y)$,

$$\psi^\dagger(y)\psi(x) = e^{-i\alpha\theta(x-y)}e^{i\alpha\Theta(x)}\phi^\dagger(y)e^{-i\alpha\Theta(y)}\phi(x). \tag{8.5.14}$$

Comparing (8.5.12) and (8.5.14) we obtain

$$\psi^\dagger(y)\psi(x) - e^{i\alpha\pi}\psi(x)\psi^\dagger(y) = \delta(x-y). \tag{8.5.15}$$

The anyon field satisfies the anyon commutation relation (8.5.9) and (8.5.15). In particular, they are reduced to the canonical (anti)commutation relations for bosons (fermions) when α is even (odd).

We proceed to derive the wave function (8.3.25) for anyons. We start with (8.5.7). Since the vacuum $|0\rangle$ contains no electrons, it satisfies $\langle 0|\rho(x) = 0$, or $\langle 0|\Theta(x) = 0$ due to (8.5.2). Hence, we find

$$\langle 0|\psi(x)\psi(y) = e^{i\alpha\theta(x-y)}\langle 0|\phi(x)\phi(y). \tag{8.5.16}$$

In general we obtain

$$\langle 0|\psi(x_1)\psi(x_2)\cdots\psi(x_N) = e^{i\alpha\sum_{r<s}\theta(x_r-x_s)}\langle 0|\phi(x_1)\phi(x_2)\cdots\phi(x_N). \tag{8.5.17}$$

We take the matrix element with an N-body state $|N\rangle$, which is the relation (8.3.25) between the wave functions for anyons and composite bosons.

The exclusion principle follows from (8.5.9),

$$\psi(x)\psi(x) = 0, \tag{8.5.18}$$

for anyons ($\alpha \neq 0 \bmod 2$). We have represented the anyon field $\psi(x)$ in terms of the composite-boson field $\phi(x)$. Composite bosons are not ordinary bosons. This is seen by taking the limit $y \to x$ in (8.5.8). The factor $e^{i\alpha\theta(x-y)}$ gets ill defined, but it is bounded. Therefore, the exclusion principle (8.5.18) implies

$$\phi(x)\phi(x) = \lim_{y\to x} e^{-i\alpha\theta(x-y)} e^{-i2\alpha\Theta(x)} \psi(x)\psi(x) = 0. \tag{8.5.19}$$

Composite bosons obey the exclusion principle.

Part 2

Monolayer Quantum Hall Systems

CHAPTER 9

OVERVIEW OF MONOLAYER QH SYSTEMS

The quantum Hall (QH) effect is one of the most remarkable phenomena discovered in the last century. The integer QH effect was discovered in 1980 by K. von Klitzing, while the fractional QH effect was discovered in 1982 by Tsui, Störmer and Gossard. It was predicted by Laughlin that a quasiparticle is an anyon carrying electric charge e/m at the filling factor $\nu = 1/m$. The QH state forms an incompressible liquid, which is best understood in terms of composite particles. The ground state is a QH ferromagnet at $\nu = 1$, where charged excitations are topological solitons identified with skyrmions. Furthermore, charge density waves such as stripes and bubbles are expected to occur in higher Landau levels. On the other hand, the edge forms a chiral Tomonaga-Luttinger liquid. A most recent topic is the discovery of an unconventional QH effect in graphene, where electrons are described by the Dirac equation rather than the Schrödinger equation. In this section we overview various aspects of the QH effect.

Hall Effects

Electrons, moving with velocity $\boldsymbol{v}$ in an xy plane in magnetic field $\boldsymbol{B}$, obey the equation of motion,

$$M\dot{\boldsymbol{v}} = -e(\boldsymbol{E} + \boldsymbol{v} \times \boldsymbol{B}). \tag{9.1.1}$$

It implies that $\boldsymbol{E} = -\boldsymbol{v} \times \boldsymbol{B}$ for a static current ($\dot{\boldsymbol{v}} = 0$). The current density is $\mathcal{J} = -e\rho_0\boldsymbol{v}$ in a homogeneous electron gas with areal density ρ_0, or

$$\mathcal{J}_x = \frac{e\rho_0}{B_\perp}E_y, \qquad \mathcal{J}_y = -\frac{e\rho_0}{B_\perp}E_x \tag{9.1.2}$$

with $B_z = -B_\perp < 0$. Being driven by the Lorentz force ($e\boldsymbol{v} \times \boldsymbol{B}$), the current $\mathcal{J}$ flows in a direction perpendicular to the electric field $\boldsymbol{E}$.

We take the electric field in the y axis; then $E_x = 0$ and $E_y \neq 0$. It follows from (9.1.2) that the Hall resistivity R_{xy} is

$$R_{xy} \equiv \frac{E_y}{\mathcal{J}_x} = \frac{B_\perp}{e\rho_0} = \frac{1}{\nu}\frac{2\pi\hbar}{e^2} \tag{9.1.3}$$

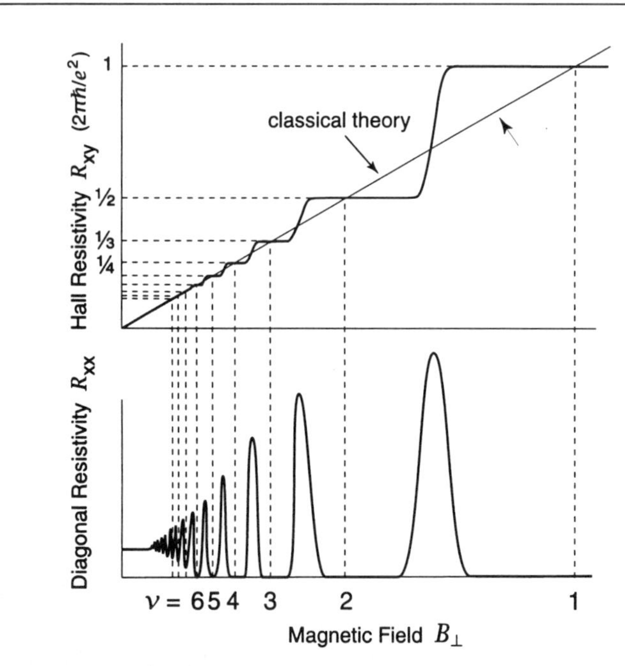

Fig. 9.1 The integer QH effect is illustrated schematically. In a classical electron gas the Hall resistivity R_{xy} is proportional to the magnetic field $B_\perp$, as indicated by a thin line labeled *classical theory*. However, the Hall resistivity R_{xy} shows a stair case in an actual sample, with the plateau crossing the classical line at $\nu = 1, 2, 3, \cdots$. The diagonal resistivity R_{xx} exhibits the Schubnikov-de Haas oscillations, vanishing at these points.

with

$$\nu = \frac{2\pi\hbar\rho_0}{eB_\perp}, \tag{9.1.4}$$

while the diagonal resistivity R_{xx} is

$$R_{xx} \equiv \frac{E_x}{\mathcal{J}_x} = 0. \tag{9.1.5}$$

The Hall resistivity (9.1.3) is classically a linear function of the perpendicular magnetic field $B_\perp$ for fixed density ρ_0, as in Fig.9.1. It is a common knowledge that the resistivity depends sensitively on details of a sample such as its composition, geometry and impurities. Experimental results are strikingly different [Fig.9.1 and Fig.9.2], though they are obtained in dirty solid-state samples. The resistivity is insensitive to details of samples at "magic" values of parameter ν, where the Hall resistivity R_{xy} is quantized and develops a series of plateaux, and the diagonal resistivity R_{xx} shows a series of dips. This is known as the QH effect. The number of the observed Hall plateaux increases as the sample becomes purer.

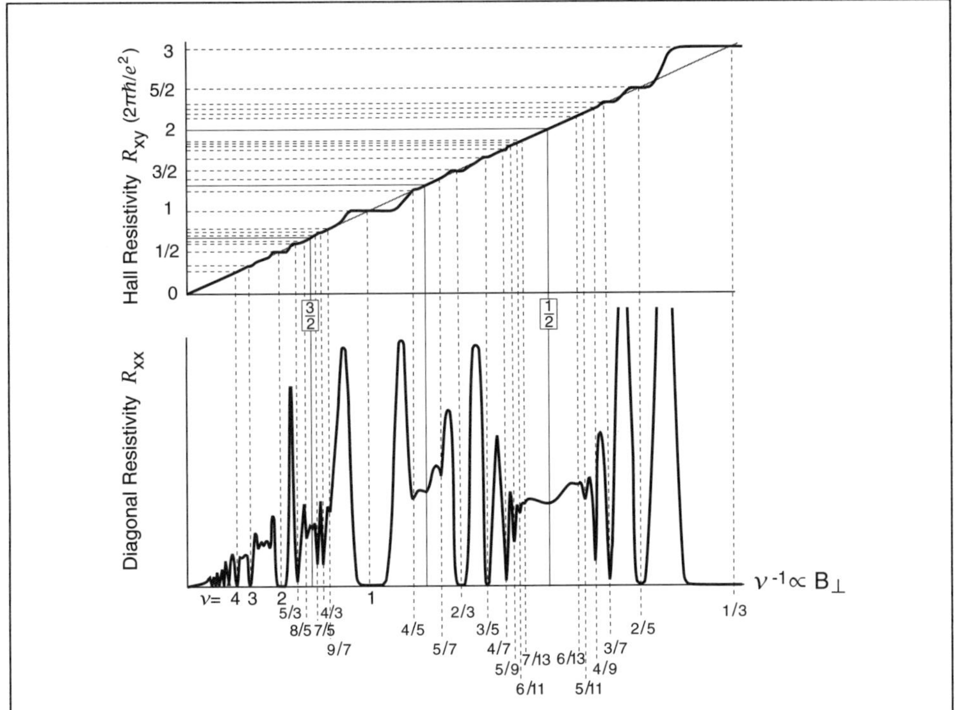

Fig. 9.2 The fractional QH effect is illustrated. Many fractional QH states are observed in pure samples. It is easier to identify them by searching for dips in the diagonal resistance R_{xx} rather than plateaux in the Hall resistance R_{xy}. All fractional QH states in this figure are on the principal sequences $\nu = n/(2n \pm 1)$ or its electron-hole conjugate sequences $\nu = 2 - n/(2n \pm 1)$.

The integer QH effect (at ν = integer) was discovered by von Klitzing[275] in 1980, a century after the discovery of the Hall effect.[213] The discovery was preceded by a theoretical suggestion due to Ando[46] and an experimental indication due to Kawaji,[263] but no one seems to have foreseen the exact quantization of the Hall conductivity. The fractional QH effect (at $\nu = n/m$ with integer n and odd integer m) was discovered by Tsui, Störmer and Gossard[480] in 1982. The exact quantization of the Hall conductivity was explained based on a topological reasoning combined with linear response theory.[292,474,371,277,56,372,54] We list some early works[276,344,94,237] on the QH effect. The reader may profitably consult review articles in the book by Prange and Girvin.[9]

The Hall resistivity is given by (9.1.3), or

$$R_{xy} = \frac{R_K}{\nu} \qquad \text{with} \qquad R_K \equiv \frac{2\pi\hbar}{e^2} \simeq 25812.807\Omega. \tag{9.1.6}$$

Because its precise measurement is possible rather easily in QH systems [Fig.9.3],

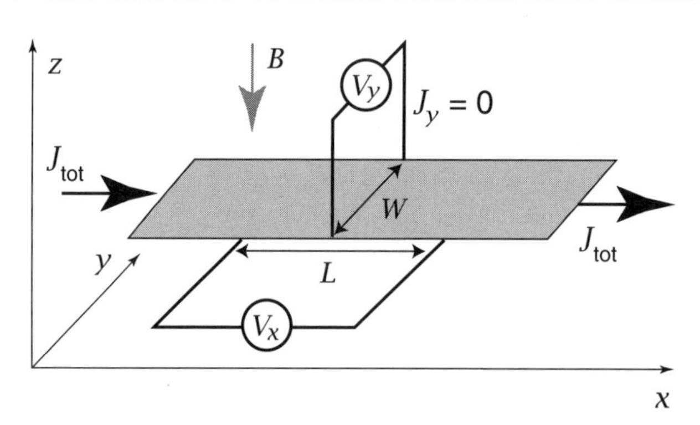

Fig. 9.3 The Hall resistivity R_{xy} and the diagonal resistivity R_{xx} is determined in a rectangular sample by feeding a constant current $\mathcal{J}_{tot}$ along the x axis and measuring the voltage V_y perpendicular to it. It is arranged so that no current flows in the y direction, $\mathcal{J}_y = 0$. The electric field reads

$$E_x = -R_{xx}\mathcal{J}_x = -R_{xx}\frac{\mathcal{J}_{tot}}{W}, \qquad E_y = -R_{yx}\mathcal{J}_x = R_{xy}\frac{\mathcal{J}_{tot}}{W},$$

where $E_y = V_y/W$ and $E_x = V_x/L$. It follows that

$$R_{xy} = \frac{V_y}{\mathcal{J}_{tot}}, \qquad R_{xx} = -\frac{WV_x}{L\mathcal{J}_{tot}}.$$

The Hall resistivity R_{xy} depends only on the current $\mathcal{J}_{tot}$ and the voltage V_y. Hence, it can be determined very accurately. They take peculiar values,

$$R_{xy} = \frac{1}{\nu}\frac{2\pi\hbar}{e^2}, \qquad R_{xx} = 0$$

in the QH state at ν: See Figs.9.1 and 9.2. In weak magnetic field ($B_\perp \lesssim 0.3$ Tesla) the Hall resistivity is described well by the classical theory, $R_{xy} = B_\perp/e\rho_0$, which is used experimentally to determine the electron density ρ_0 of a sample.

since 1990 this has been used as the standard resistance with a definite choice of $R_K = 25812.807\Omega$: It is called the von Klitzing constant.

A measurement of the Hall resistivity is also used as a high-precision determination of the fine structure constant $\alpha = \mu_0 ce^2/4\pi\hbar$. A precise result obtained in this method is[246]

$$\alpha^{-1} = 137.0360037(27) \qquad (0.020 \quad \text{ppm}). \tag{9.1.7}$$

The fine structure constant is one of the fundamental constants of nature characterizing the whole range of physics. The result (9.1.7) is to be compared with those obtained in other experiments such as the ac Josephson effect, the neutron de Broglie wavelength and the electron anomalous magnetic moment. The value

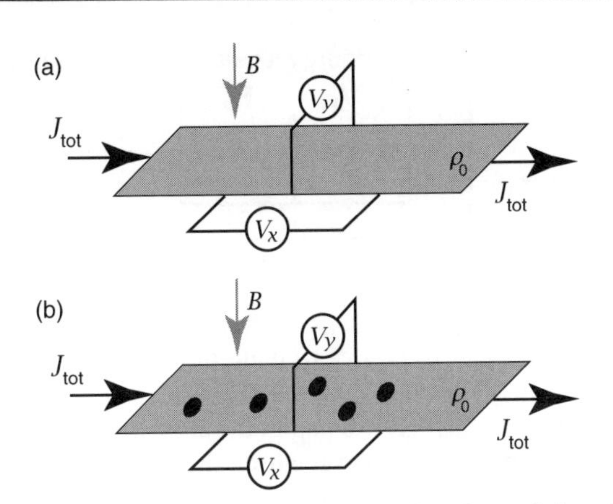

Fig. 9.4 (a) Electrons form an incompressible liquid with gapful excitations at $\nu = 1/m$. The Hall current flows as a superfluid with density ρ_0. (b) quasiparticles are created when the magnetic field changes, but they are localized by impurities and do not contribute to the current at $T = 0$. The Hall current remains unchanged from the one at $\nu = 1/m$, since it flows as a superfluid with the same local density ρ_0. This is the origin of the Hall plateau.

recommended for international use by CODATA (the Commitee of Data for Science and Technology – 1998) is[349]

$$\alpha^{-1} = 137.03599976(50) \qquad (3.7 \quad \text{ppb}). \tag{9.1.8}$$

If the complete consistency of α is not achieved, it possibly indicates the emergence of new physics. A comprehensive review on the fine structure constant is found in literature.[273]

Landau Levels

The physical meaning of the parameter ν in (9.1.4) reads as follows. Due to the Pauli exclusion principle only one electron can occupy one quantum-mechanical state, which we call the *Landau site*. It has the area $\Delta S = 2\pi\ell_B^2$. In each energy level the density of states is

$$\rho_\Phi = \frac{1}{2\pi\ell_B^2} = \frac{B_\perp}{\Phi_D} \tag{9.1.9}$$

with $\Phi_D \equiv 2\pi\hbar/e$ the Dirac flux quantum: It is equal to the density of flux quanta. The filling fraction of the energy level is defined by

$$\nu = \frac{\text{Number of electrons}}{\text{Number of states}} = \frac{\rho_0}{\rho_\Phi} = 2\pi\ell_B^2\rho_0 = \frac{2\pi\hbar\rho_0}{eB_\perp} = \frac{\rho_0\Phi_D}{B_\perp}. \tag{9.1.10}$$

At $\nu = 1/m$ there are m flux quanta per electron, $B_\perp/\rho_0 = m\Phi_D$. The filling fraction ν is customarily called the Landau-level filling factor, but we call it simply the

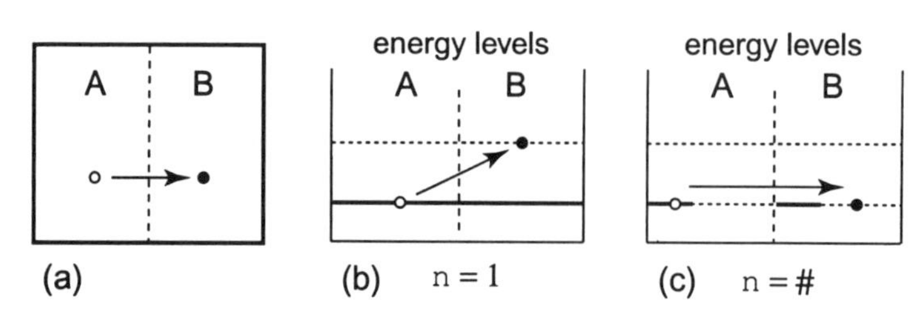

Fig. 9.5 (a) We imagine parts A and B in the Hall system. We question the energy necessary to move one electron from part A to part B. (b) In the integer QH system this is to push one electron from one energy level to a higher level. It costs at least an excitation energy $|g^* \mu_B B|$ or $\hbar\omega_c - |g^* \mu_B B|$, which gives a gap in the chemical potential. (c) In the fractional QH system this is just to move one electron within the same Landau level. At first sight no gap seems to arise in the chemical potential.

filling factor. One Landau level contains two energy levels, the up-spin and down-spin levels. Thus, the lowest Landau level is filled at $\nu = 2$. Because of this fact it is adequate to call ν simply the filling factor.

The electron position is described solely by the guiding center (center of cyclotron motion) $\boldsymbol{X} = (X, Y)$ in each Landau level. Its X and Y coordinates are noncommutative,[1]

$$[X, Y] = -i\ell_B^2. \tag{9.1.11}$$

Due to the Heisenberg uncertainty principle the electron position cannot be determined more accurately than the area $\Delta S = 2\pi\ell_B^2$. It agrees with the area of one Landau site. The noncommutativity (9.1.11) leads to essential physics such as the spin coherence in QH ferromagnets and the Tomonaga-Luttinger liquid governing edge excitations.

QH Liquids

The experimental fact indicates that the 2-dimensional electron gas undergoes a phase transition into a peculiar ordered state at a magic filling factor ν, where the Hall conductivity is accurately quantized. The QH effect comes from the realization of an *incompressible liquid*: It is a macroscopic quantum state where all charged excitations are gapful. Under the Lorentz forcethe electric current flows as an overall flow of an incompressible liquid without dissipation [Fig.9.4(a)]: It is a superfluid called the QH liquid. Moreover, the Hall current is dissipationless because we have $E_y \propto \mathcal{J}_x$ but $E_x = 0$ when the current flows along the x axis.

The incompressibility is easily understood in the integer QH state at $\nu = n$ (in-

[1] The QH system presents a simple example of noncommutative geometry, which we explore extensively in Part IV.

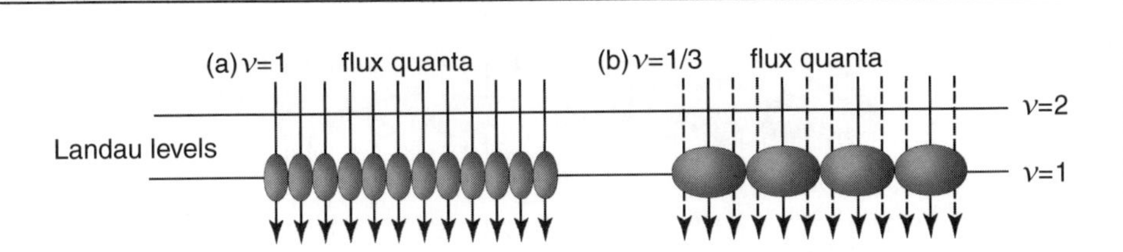

Fig. 9.6 (a) There is one flux quantum per electron at $\nu = 1$. Electrons fill the lowest Landau level, generating the $\nu = 1$ integer QH state. A composite boson is obtained when one flux quantum is attached to an electron. (b) There are m flux quanta per electron at $\nu = 1/m$. Here $m = 3$. A composite fermion is obtained by attaching $2p \equiv m-1$ flux quanta (dotted lines) to each electron. It behaves as a "fat electron" occupying m sites. Composite fermions "fill" the lowest Landau level, generating the $\nu_{\text{eff}} = 1$ integer QH state. It is the $\nu = 1/m$ fractional QH state. A composite boson is obtained when one residual flux quantum (solid line) is attached to a composite fermion, or equivalently all m flux quanta to an electron.

teger), where n energy levels are completely filled with all higher levels empty. We imagine two parts A and B in the system [Fig.9.5]. In order to move one electron from part A to part B, it is necessary to supply an energy (Δ_Z or $\hbar\omega_c - \Delta_Z$) since the electron must be moved to a higher level in part B. (For an actual process it is necessary to supply an additional energy due to a many-body Coulomb effect including the exchange interaction.) It implies a gap in the chemical potential and hence the incompressibility of the integer QH state. The integer QH effect is essentially a one-body effect.

Composite Particles

Why does the incompressibility realize and lead to the fractional QH state at $\nu = 1/(2p+1)$, though there are many vacant states in the lowest Landau level [Fig.9.5(c)]? The concept of *composite particles* is essential for understanding the mechanism of the incompressibility. A composite particle is a flux-carrying electron. The attachment of flux quanta is a natural way to minimize the energy of the system since the associated vortex expels other electrons from its neighborhood and decreases the repulsive interactions between electrons. We may regard a composite particle as a "fat electron" occupying $(2p+1)$ Landau sites at $\nu = 1/(2p+1)$ [Fig.9.6]. There exist two equivalent pictures:

(A) A *composite fermion*[241] is an electron bound to $2p$ flux quanta [Fig.9.6(b)]. Composite fermions make cyclotron motion in reduced effective magnetic field $B_\perp^{\text{eff}} = B_\perp - 2p\Phi_D\rho_0$, fill their own Landau levels, and generate an integer QH state at ν_{eff}. It is the fractional QH state at $\nu = \nu_{\text{eff}}/(2p\nu_{\text{eff}}+1)$. The incompressibility is attributed to the effective cyclotron gap ω_c^{eff}, whose origin is a many-body Coulomb effect. On this understanding, the fractional QH effect may be regarded as if it were a one-body effect of composite fermions.

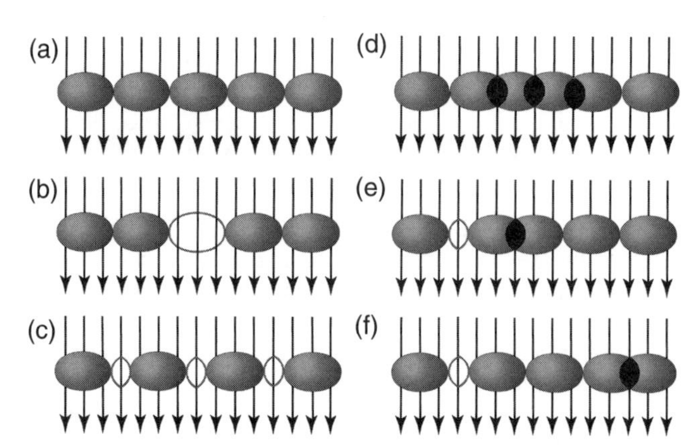

Fig. 9.7 (a) There are m flux quanta per electron at $\nu = 1/m$. Here $m = 3$. Composite bosons, each of which is an electron bound to m flux quanta with odd m, experience no effective magnetic flux and condense into the angular-momentum zero state at $\nu = 1/m$. The fractional QH state is a close packing of these composite bosons. (b) When one electron is removed one big hole is created. (c) To lower the Coulomb energy there appear m "holes", which are quasiholes with positive electric charge e/m. (d) When one electron is added, there appear m "overlaps" which are quasielectrons with negative electric charge $-e/m$. (e) A density fluctuation with the total electron number fixed is generated by a pair creation of quasihole and quasielectron with the creation energy Δ_{pair}. This configuration may describe a magnetoroton. (f) When they are separated far apart, their energy is $\Delta_{\text{pair}} = \Delta_{\text{qe}} + \Delta_{\text{qe}}$, where Δ_{qe} and Δ_{qh} are the creation energy of a quasielectron and a quasihole. quasiparticles are vortices when the Zeeman effect is very large. Otherwise, they are skyrmions exciting up-spin and down-spin electrons coherently.

(B) A *composite boson*[190] is an electron bound to $(2p+1)$ flux quanta. Since composite bosons experience no effective magnetic field, they do not make cyclotron motion, and being bosons, they condense. The condensate is a close packing of composite bosons pierced by flux quanta [Fig.9.7(a)] and incompressible by a many-body Coulomb effect.

When one electron is removed at $\nu = 1/m$, m holes are created and behave as quasiholes [Fig.9.7(c)]; here, m is an odd integer. They are not smeared out because they are topological solitons in the composite-boson picture. A hole itself is associated with a flux quantum, occupying one Landau site. When one electron is added, it turns out that m quasielectrons are created [Fig.9.7(d)].

A quasiparticle (quasihole or quasielectron) is an anyon possessing a fractional charge $\pm e^*$ with[294]

$$e^* = \frac{e}{m}. \tag{9.1.12}$$

It is a topological soliton in the composite-boson theory. Since one quasiparticle has a size of order ℓ_B, the direct Coulomb energy is of order E_C^0/m^2 with (A.2). The exchange Coulomb energy is additionally necessary for excitation. They are the main part of the creation energy of a quasielectron (Δ_{qe}) and a quasihole (Δ_{qh}).

Fractional QH liquids are robust against any perturbative density fluctuation: Density fluctuation is only possible by creation of a quasihole-quasielectron pair [Fig.9.7(e)] at a cost of a creation energy $(\Delta_{qe}+\Delta_{qh})$. The effective cyclotron gap ω_c^{eff} of composite fermions may be identified with this energy. In this way the Coulomb interaction between electrons is the key to turn the 2-dimensional electron gas into an incompressible liquid at $\nu = 1/m$. Hence, the fractional QH effect is essentially a many-body effect.

We have argued that the $\nu = 1/(2p+1)$ state is a fractional QH state. What can we say about the $\nu = 1/2p$ state? According to the composite-fermion theory, since the effective magnetic field vanishes thereat, composite fermions experience no magnetic field and they behave as if they were free electrons possessing a Fermi surface.[218]

Hall Plateaux

The plateau formation around $\nu = 1/m$ is understood also based on the composite-boson theory. In the QH state there is a correspondence between a flux quantum and a quasiparticles as in Fig.9.7. When the external magnetic field $B_\perp$ is increased ($\nu < 1/m$) or decreased ($\nu > 1/m$), quasiholes or quasielectrons are created because the number of the states is increased or decreased in each Landau level. An empty state behaves as a quasihole, and an overlapped state behaves as a quasielectron [Fig.9.7]. If quasiparticles flow together with the QH liquid, the Hall current changes according to the classical formula (9.1.2). Actually, because quasiparticles are charged objects, they are localized by impurities and do not contribute to the current at temperature $T = 0$: They are pinned directly by impurities or subject to the Anderson localization intrinsic to the 2-dimensional space.[44,37] It is feasible that the Anderson localization at zero magnetic field is applicable to quasiparticles because the effective magnetic field vanishes on the QH system. The density of the QH liquid is unchanged [Fig.9.4(b)], and the current flows as a superfluid under the Lorentz force. Consequently, the Hall current remains unchanged from the one at $\nu = 1/m$, and no diagonal current flows at $T = 0$. This is the mechanism of formation of the Hall plateau around $\nu = 1/m$, where the Hall resistivity is quantized and the diagonal resistivity vanishes as in (9.1.3).

Activation Energy

At finite temperature $T > 0$, pairs of quasiparticles are activated thermally in QH states. Their flow generates an Ohmic current with a diagonal resistance

$$R_{xx} \propto \exp\left(-\frac{\Delta_{gap}}{2k_B T}\right), \tag{9.1.13}$$

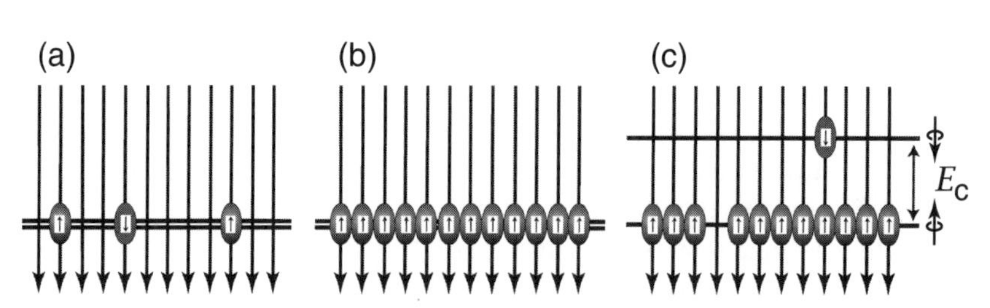

Fig. 9.8 (a) In the vanishing limit of the Zeeman energy ($g^* = 0$) one Landau level contains two degenerate energy levels. Electrons with opposite spins are accommodated within the same energy level. (b) At $\nu = 1$ all spins are polarized spontaneously into one direction due to the exchange interaction: The up-spin level is filled up even if the Zeeman effect is infinitesimal. (c) One electron is excited to the down-spin level, as implies a creation of a quasiparticle pair at a cost of a Coulomb gap energy, $E_C \simeq E_C^0$. quasiparticles may evolve into skyrmions to reduce the direct Coulomb energy.

where k_B is the Boltzmann constant; Δ_{gap} is the excitation energy of a quasiparticle pair and customarily called *activation energy*. It is given by

$$\Delta_{\text{gap}} = \Delta_{\text{pair}} - \Gamma_{\text{offset}} \qquad \text{with} \quad \Delta_{\text{pair}} = \Delta_{\text{qh}} + \Delta_{\text{qe}}, \tag{9.1.14}$$

where Γ_{offset} is a certain sample-dependent offset. The measurement of activation energy provides us with quantitative information of quasiparticles.

QH Ferromagnets

We have so far ignored the spin degree of freedom. This is not justified when the Zeeman gap Δ_Z is small: See (A.5) in Appendix for explicit values of Δ_Z. It is even possible to achieve a vanishingly small Zeeman gap experimentally. Then, one Landau level contains two almost degenerate energy levels [Fig.9.8(a)]. We would naively expect the integer QH state to occur first at $\nu = 2$, where the lowest Landau level is filled. This would be the case if Coulomb interactions were absent between electrons. Actually, there emerges an incompressible state at $\nu = 1$, because a new energy level is created with a Coulomb energy [Fig.9.8(c)]. A similar phenomenon occurs also at $\nu = 1/m$.

A peculiar feature is that the spin coherence develops spontaneously owing to the exchange interaction as in ferromagnets (see Section 4.1). The effective Hamiltonian is a nonlinear sigma model,

$$H_{\text{eff}} = \frac{2}{m} J_s \sum_{a=x,y,z} \int d^2x \, [\partial_k \mathcal{S}_a(\mathbf{x})]^2, \tag{9.1.15}$$

at $\nu = 1/m$, where $\mathcal{S}_a(\mathbf{x})$ is the spin field (sigma field) normalized as $\boldsymbol{\mathcal{S}}^2(\mathbf{x}) = 1/4$, and J_s is the spin stiffness.

The exchange energy (9.1.15) is minimized when all spins are polarized into one arbitrary direction [Fig.9.8(b)]. The effective Hamiltonian (9.1.15) describes a spin wave, which is the Goldstone mode associated with the spontaneous symmetry breaking of the SU(2) symmetry. The system is called the QH ferromagnet.

Strictly speaking, the Mermin-Wagner theorem[345,346] dictates that no spontaneous symmetry breaking is possible in the 2-dimensional space at finite temperature: See Appendix J. This problem is resolved in the QH system since there exists a small Zeeman effect, due to which the spin wave is no longer gapless and the coherence length is finite,

$$\xi_{\text{spin}} = 2\ell_B\sqrt{\frac{\pi m J_s}{\Delta_Z}} \simeq \frac{7.33}{\sqrt{B_\perp}}\ell_B. \tag{9.1.16}$$

It demonstrates explicitly that the exchange interaction ($J_s \neq 0$) is the driving force of the spin coherence.

Skyrmions

Quasiparticles (charged excitations) are topological solitons[2] in the QH liquid. They are vortices[294] in the spin-frozen theory, as advocated by Laughlin, while they are skyrmions[451] in QH ferromagnets described by the effective Hamiltonian (9.1.15). Skyrmions were already reviewed in Section 7.7.

The QH ferromagnet is a macroscopic coherent state, where all spin components $\mathcal{S}_x(\boldsymbol{x})$, $\mathcal{S}_y(\boldsymbol{x})$ and $\mathcal{S}_z(\boldsymbol{x})$ are simultaneously measurable as classical fields. The presence of quantum coherence is confirmed experimentally by observing coherent excitations, namely, skyrmions. A skyrmion is a spin texture[3], rotating several spins coherently to lower the direct Coulomb energy. Skyrmion excitations have been studied experimentally by measuring the spin polarization[63,41], the activation energy[429] and also the heat capacity.[65] Skyrmion excitations are allowed in fractional QH states as well, provided the Zeeman effect is sufficiently small. An experimental evidence[300] exists at $\nu = 1/3$. As the Zeeman effect becomes larger the skyrmion scale becomes smaller, and eventually a skyrmion is reduced to a vortex in its small-size limit.

Hierarchy of Fractional QH States

The fractional QH system at $\nu = n/m$ (m odd, $n \neq 1$) is not understood by a simple condensation of composite bosons. To generate such a QH state, Haldane and Halperin[207,216,302] once suggested a hierarchical condensation of quasiparticle-flux composites. However, this scheme is rather unnatural. For instance, though the $\nu = 6/13$ QH state belongs to the 5th generation in their scheme, its

---

[2] In the present convention the vortex (or skyrmion) corresponds to the quasihole, and the antivortex (or antiskyrmion) corresponds to the quasielectron. Generically we call them simply vortices (or skyrmions).

[3] In the $\nu = 1$ QH ferromagnet, we shall argue in Section 31.2 that a skyrmion state is constructed from a hole state by making a $W_\infty(2)$ rotation, which is an SU(2) rotation in the noncommutative space.

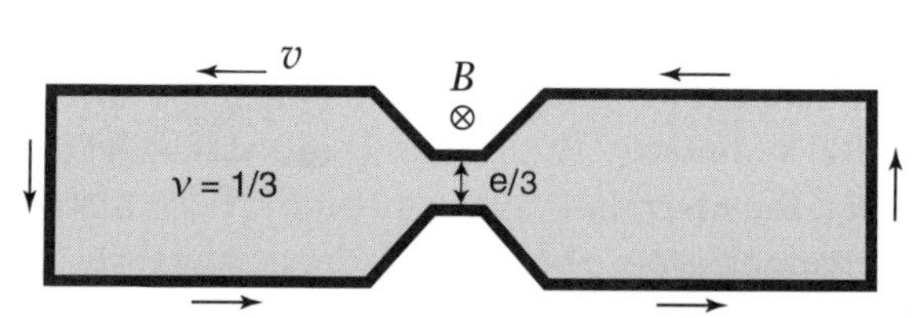

Fig. 9.9 A finite-size Hall bar has edges, where gapless edge excitations propagate. When the two edges are brought into close proximity, quasiparticles tunnel between the two edges. Such an effect is measurable as a backscattering current noise. It reveals the fractional charge $e^* = e/m$ at $\nu = n/m$.

existence is well established experimentally [Fig.9.2]. On the contrary many QH states in earlier generations have not been observed experimentally.

Another hierarchical scheme, due to Jain,[241] is based on the composite-fermion theory. Composite fermions make their own Landau levels called *CF-Landau levels*. They fill a Fermi sea, circle on semiclassical cyclotron orbits and show Shubnikov-de Haas oscillations in a reduced effective magnetic field. Integer QH states are generated when CF-energy levels[4] are filled with composite fermions. When the lowest n CF-energy levels are filled, the filling factor is

$$\nu = \frac{n}{2pn \pm 1}. \tag{9.1.17}$$

They are the principal sequences, along which fractional QH states are observed predominantly [Fig.9.2].

Edge Effects

The QH state is an incompressible liquid confined within a finite domain. Since the bulk is incompressible, the low-energy excitations are deformations of the boundary shape which preserve the area of the liquid. Edge excitations are gapless[214] and move into one direction determined by the sign of the magnetic field [Fig.9.9]. They are not affected by impurities and propagate without dissipation. The edge channel provides us with a unique laboratory[491] to investigate the *chiral Tomonaga-Luttinger liquid* both theoretically and experimentally. The Tomonaga-Luttinger liquid[319] has been known theoretically for a few decades but inaccessible experimentally so far because of difficulty to fabricate clean one-dimensional quantum wires.

In the bulk quasiparticles (vortices) are gapful and pinned in place by impurities. On the contrary, along the edge they are gapless and propagate freely. When the opposite two edges are brought into proximity, quasiparticles backscatter between the right-moving and left-moving edge channels [Fig.9.9]. Such an effect is

[4] One CF-Landau level splits into two energy levels by the spin effect.

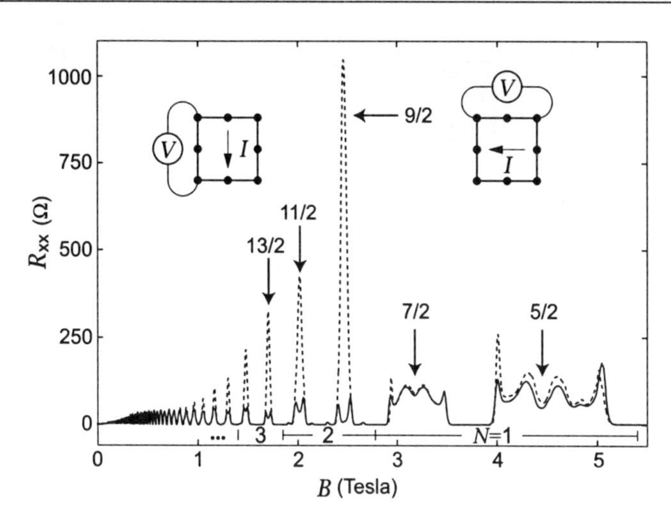

Fig. 9.10 The anisotropy of diagonal resistivity R_{xx} is shown. The two traces result from simply changing the direction of current through the sample; the sample itself is not rotated. This anisotropy would mean that the ground state is in the stripe phase. Data are taken from Lilly et al.[313] See also Fig.8 in the frontispiece.

observable as a backscattering current noise.[259] It has presented a direct measurement of the charge of quasiparticles, $e/3$, in the $\nu = 1/3$ fractional QH state.[391,419]

Stripes and Bubbles in Higher Landau Levels

Electrons occupy only the lowest Landau level in the limit of high magnetic field. With decreasing magnetic field, the number of degeneracy in each Landau level decreases and part of electrons occupy higher Landau levels. The Coulomb interaction shows an intriguing behavior in the Nth Landau level ($N = 0, 1, 2, \cdots$), where the effective Coulomb potential is given by

$$V_N^{\text{eff}}(\boldsymbol{q}) = \frac{e^2}{4\pi\varepsilon|\boldsymbol{q}|}\left[L_N\left(\frac{\ell_B^2\boldsymbol{q}^2}{2}\right)\right]^2 e^{-\ell_B^2\boldsymbol{q}^2/2} \tag{9.1.18}$$

with the Laguerre polynomial $L_N(x)$. It is not a monotonically decreasing function for $N \neq 0$. Fractional QH states are no longer stable for $N \geq 2$. Instead of them we expect charge density waves such as stripes and bubbles[282] emerge. The stripe phase has been observed[313] experimentally by way of a strong anisotropy in the diagonal resistivity R_{xx} at $\nu = \frac{9}{2}, \frac{11}{2}, \frac{13}{2}, \cdots$, as illustrated in Fig.9.10.

Unconventional QH Effects in Graphene

An unconventional QH effect has been revealed in graphene by recent experimental developments.[374,538,376,340] The measured filling factors[374,538] form a quite

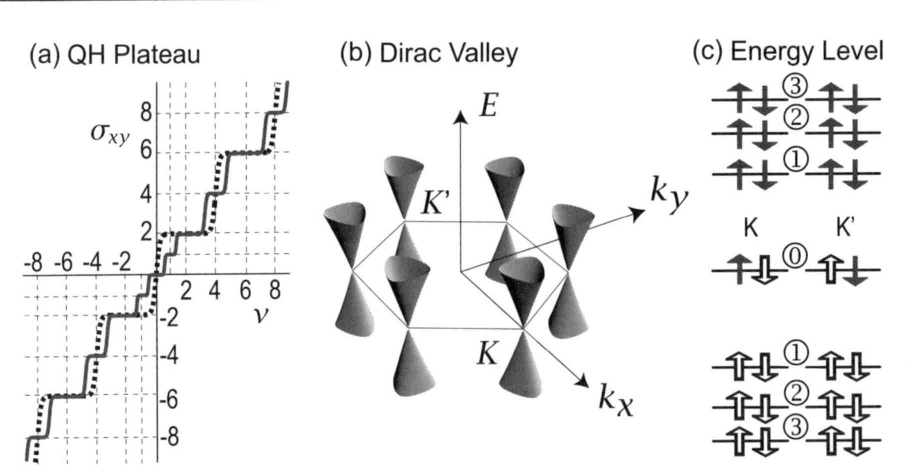

Fig. 9.11 (a) An unconventional QH effect has been observed in graphene. The measured filling factors form a series $\nu = \pm 2, \pm 6, \pm 10, \pm 14, \cdots$, at which Hall plateaux develop in the Hall conductivity σ_{xy}. (b) The low-energy dispersion relation exhibits conically shaped valleys (Dirac valleys) near Brillouin zone corners, where massless Dirac electrons emerge. There exist two inequivalent corners (K and K′ points), producing a twofold valley degeneracy. (c) The energy spectrum of monolayer graphene for electrons at the K and K′ points. The solid (open) arrow indicates the spin of an electron (hole). This spectrum is a manifestation of the supersymmetry. See also Fig.9 in the frontispiece.

mysterious series [Fig.9.11(a)],

$$\nu = \pm 2, \pm 6, \pm 10, \pm 14, \cdots, \tag{9.1.19}$$

where the basic height in the Hall conductance step is $4e^2/h$ except for the first step which is just one half. When larger magnetic field is applied, the series turns out to be [Fig.9.11(a)]

$$\nu = 0, \pm 1, \pm 2, \pm 4, \pm 6, \pm 8, \pm 10, \pm 12, \pm 14, \cdots. \tag{9.1.20}$$

The series (9.1.19) indicates that the lowest Landau level has the 2-fold degeneracy, while all others have the 4-fold degeneracy in monolayer graphene.[47,203,390] The 4-fold degeneracy follows from the spin degeneracy and from the Bloch state degeneracy between two inequivalent points (so-called K and K′ points) in the honeycomb lattice Brillouin zone [Fig.9.11(b)]. Coulomb interactions resolve the degeneracy so as to generate the series (9.1.20).[43,373,193,204,132] It is remarkable[132] that the energy spectrum [Fig.9.11(c)] presents a realization of supersymmetry[508] within noninteracting theory because the Zeeman splitting is as large as the Landau-level separation in this unconventional QH effect. It has been pointed out[132] that the effective Coulomb potential becomes spin dependent in each energy level due to this supersymmetric structure.

CHAPTER 10

LANDAU QUANTIZATION

Electrons are trapped in a thin layer made at the interface between semiconductor and insulator or between semiconductors at sufficiently low temperature. Trapped electrons make a 2-dimensional system. In a perpendicular magnetic field $B_\perp$ the energy of an electron is quantized into Landau levels. The energy difference between two successive Landau levels is $\hbar\omega_c$ with $\omega_c = eB_\perp/M$ the cyclotron frequency. It is remarkable that one-body energy levels are quantized in a macroscopic system.

10.1 PLANAR ELECTRONS

We consider a system where electrons are trapped in a thin layer ($\sim 10nm$) made at the interface between semiconductor and insulator or between semiconductors. A typical device is the GaAs/GaAlAs heterostructure [Fig.10.1]. Let the layer span the xy plane. The thin layer makes a one-dimensional potential well $V(z)$ in the z direction, and the energy levels with respect to this direction are quantized. A transition to higher energy levels is negligible when the potential well is narrow enough and the temperature is low enough. It is a good approximation that electrons are confined to the ground state of the potential well. The wave function of one electron is of the form

$$\mathfrak{S}(x,y,z) = \mathfrak{S}(\mathbf{x})\mathfrak{S}_0(z), \tag{10.1.1}$$

where $\mathbf{x} = (x,y)$ is the 2-dimensional vector and $\mathfrak{S}_0(z)$ is the wave function of the ground state. Since the wave function $\mathfrak{S}_0(z)$ is uniquely fixed, the dynamical problem is only to determine the wave function $\mathfrak{S}(\mathbf{x})$. Consequently, the 3-dimensional problem is reduced to the 2-dimensional one. The system is regarded to be two dimensional, and we refer to these electrons as *planar electrons*. Though the motion of electrons is 2-dimensional, the Coulomb interaction between electrons is a 3-dimensional force given by (5.1.18),

$$H_C = \frac{1}{2}\int d^3x d^3x' \frac{\rho_e(\mathbf{x},z)\rho_e(\mathbf{x}',z')}{4\pi\varepsilon\sqrt{(\mathbf{x}-\mathbf{x}')^2+(z-z')^2}}, \tag{10.1.2}$$

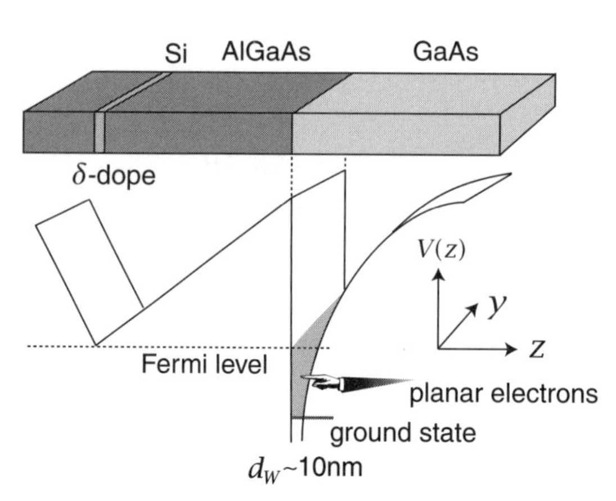

Fig. 10.1 A sample is grown by molecular beam epitaxy and consists of a GaAs quantum well of width $d_W \sim 10nm$. Electrons are provided by a remote δ-doped donor layer separated by a ~ 60 nm AlGaAs spacer. Electron density is adjusted at room temperature, which is fixed by cooling the sample. (Electron density may be increased by way of photo-illumination at low temperature.) The energy levels of the quantum well are quantized with respect to the motion in the z direction. When the potential well is narrow enough, the level mixing is negligible due to a large energy gap at low temperature. We may assume that electrons are confined to the lowest energy level. In this approximation the motion in the z direction is frozen and the system becomes 2-dimensional.

where $\rho_e(\mathbf{x}, z)$ is the electric charge density.

As a simple exercise it is instructive to approximate the narrow potential well $V(z)$ with the δ-function potential,

$$V(z) = -V_0\delta(z). \tag{10.1.3}$$

The Schrödinger equation is easily solved for $\mathfrak{S}_0(z)$ provided the Coulomb potential is neglected. There exists only one energy level, and it is given by

$$\mathfrak{S}_0(z) = \sqrt{\beta}e^{-\beta|z|}, \quad \text{for} \quad z \neq 0, \tag{10.1.4}$$

where $\beta = MV_0/\hbar^2$ with the electron mass M. The energy of the state is $E = -\hbar^2\beta^2/2M$. For a sufficiently deep potential well ($\beta \to \infty$) it may be approximated as

$$\mathfrak{S}_0(z) = \sqrt{\delta(z)}. \tag{10.1.5}$$

In this limit the 3-dimensional system is literally reduced to the 2-dimensional one. The charge density is approximated by

$$\rho_e(\mathbf{x}, z) = -e\rho(\mathbf{x})\delta(z) \tag{10.1.6}$$

with $\rho(\mathbf{x}) = \sum_{r=1}^{N} \delta(\mathbf{x} - \mathbf{x}_r)$, where $\mathbf{x}_1, \mathbf{x}_2, \cdots, \mathbf{x}_N$ denote the positions of N electrons in a plane. The Coulomb energy term is given by

$$H_{\rm C} = \frac{e^2}{2} \int d^2x d^2x' \frac{\rho(\mathbf{x})\rho(\mathbf{x}')}{4\pi\varepsilon|\mathbf{x} - \mathbf{x}'|}. \tag{10.1.7}$$

The Fourier component of the Coulomb potential is

$$V(\mathbf{q}) = \frac{e^2}{4\pi\varepsilon} \int \frac{d^2x}{2\pi} \frac{1}{|\mathbf{x}|} e^{-i\mathbf{q}\mathbf{x}} = \frac{e^2}{4\pi\varepsilon|\mathbf{q}|}. \tag{10.1.8}$$

This is to be contrasted with the behavior $V(\mathbf{q}) \propto |\mathbf{q}|^{-2}$ in the 3-dimensional space.

The charge neutrality holds in any systems, and in the actual QH system it is guaranteed by the positive charge of holes in the remote δ-doped donor layer [Fig.10.1]. The total charge density is

$$\rho_e(\mathbf{x}, z) = -e[\rho(\mathbf{x})\delta(z) - \rho_0\delta(z - d)], \tag{10.1.9}$$

where ρ_0 stands for the homogeneous hole density in the donor layer located at $z = d$. Substituting this into (10.1.2) we obtain

$$H_{\rm C} = \frac{1}{2} \int d^2x d^2y \Delta\rho(\mathbf{x}) V(\mathbf{x} - \mathbf{y}) \Delta\rho(\mathbf{y}), \tag{10.1.10}$$

where an irrelevant constant term has been neglected, and

$$\Delta\rho(\mathbf{x}) = \rho(\mathbf{x}) - \rho_0. \tag{10.1.11}$$

We may interpret this as the modulation of the electron density $\rho(\mathbf{x})$ from the homogeneous one. See Section 32.5 for an explicit derivation of (10.1.10) in an instance of the bilayer QH system.

The total Hamiltonian consists of the kinetic term $H_{\rm K}$, the Coulomb term $H_{\rm C}$ and the Zeeman term $H_{\rm Z}$,

$$H = H_{\rm K} + H_{\rm C} + H_{\rm Z} \tag{10.1.12}$$

with

$$H_{\rm K} = \frac{1}{2M} \sum_{r=1}^{N} \left(-i\hbar\nabla + e\mathbf{A}^{\rm ext}(\mathbf{x}_r)\right)^2, \tag{10.1.13}$$

$$H_{\rm Z} = -\frac{1}{2}\Delta_{\rm Z} \int d^2x \left(\rho^{\uparrow}(\mathbf{x}) - \rho^{\downarrow}(\mathbf{x})\right), \tag{10.1.14}$$

where M is the band mass of electrons, $\mathbf{A}^{\rm ext}(\mathbf{x})$ is the external electromagnetic potential, $\Delta_{\rm Z} \equiv |g^*\mu_B B|$ is the Zeeman gap, $\rho^{\uparrow}(\mathbf{x})$ and $\rho^{\downarrow}(\mathbf{x})$ are the number densities of up-spin and down-spin electrons. We regard the Hamiltonian (10.1.12) as a microscopic Hamiltonian to start with, though it is actually an idealized effective Hamiltonian obtained in solid-state samples. Our problem is to analyze the QH system based on the quantum-mechanical Hamiltonian (10.1.12). However, to reveal the essence of the QH system, we frequently consider the *spinless theory* by

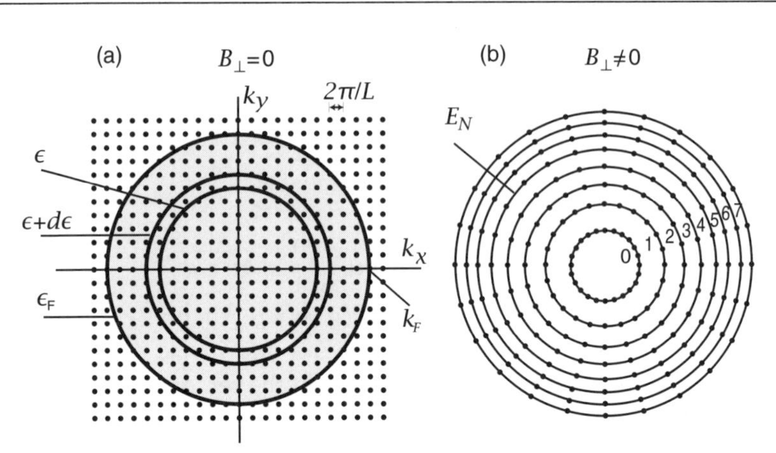

Fig. 10.2 Electron states are given by points on the $\boldsymbol{k}$ plane in the absence and presence of external magnetic field $B_\perp$. The states on each circle are degenerate. (a) When $B_\perp = 0$, there is one state per area $(2\pi/L)^2$. The density of states $D(\epsilon)$ is the number of points in the shell per unit area between the energy ϵ and $\epsilon + d\epsilon$. When the number density is ρ_0, all points within the circle of radius $k_F = \sqrt{4\pi\rho_0}$ are occupied. The Fermi energy is given by $\varepsilon_F = \hbar^2 k_F^2/2M$. (b) When $B_\perp \neq 0$, the electron energy is quantized into Landau levels. Each circle represents one Landau level with energy $E_N = N\hbar\omega_c + (1/2)\hbar\omega_c$. The density of states is the same on every circle, $\rho_{DS} = 1/(2\pi\ell_B^2)$.

ignoring the spin degree of freedom. It is a good approximation to the *spin-frozen* system, where the Zeeman energy is very large and all spins are frozen into their polarized states.

To study the basic nature of planar electrons, we switch off the Coulomb interaction in the rest of this chapter. When there exists no electromagnetic field ($\boldsymbol{A}^{\text{ext}}(\boldsymbol{x}) = 0$), electrons move freely in the xy plane. Their motion is described by the plane wave

$$\mathfrak{S}(\boldsymbol{x}) = \frac{1}{L} e^{i\boldsymbol{k}\cdot\boldsymbol{x}} \tag{10.1.15}$$

in the the box normalization, where $\boldsymbol{x} = (x, y)$ and L is the size of the planar system. The energy of one electron is

$$\varepsilon_{\boldsymbol{k}} = \frac{\hbar^2 \boldsymbol{k}^2}{2M} = \frac{\hbar^2}{2M}(k_x^2 + k_y^2). \tag{10.1.16}$$

By imposing the periodic boundary condition on the wave function, $\mathfrak{S}(x + L, y) = \mathfrak{S}(x, y + L) = \mathfrak{S}(x, y)$, the wave number $\boldsymbol{k}$ is restricted to be

$$k_x = \frac{2\pi}{L} n_x, \qquad k_y = \frac{2\pi}{L} n_y, \tag{10.1.17}$$

with integers n_x and n_y, $(n_i = 0, \pm 1, \pm 2, \cdots)$. The electron state is labeled by the wave vector (k_x, k_y), whose component k_i takes discrete values (10.1.17), as in Fig.10.2. One state occupies the area $(2\pi/L)^2$ in the $\boldsymbol{k}$ plane.

Due to the Pauli exclusion principle, one quantum state can accommodate only one electron. The density of states $D(\epsilon)$ is the number of the electron states per unit area in the energy interval between ϵ and $\epsilon + d\epsilon$,

$$D(\epsilon)d\epsilon = \frac{1}{L^2} \times \frac{L^2}{(2\pi)^2} \times 2\pi|\boldsymbol{k}|d|\boldsymbol{k}|, \tag{10.1.18}$$

or

$$D(\epsilon) = \frac{1}{(2\pi)^2} \times 2\pi|\boldsymbol{k}|\frac{d|\boldsymbol{k}|}{d\epsilon} = \frac{M}{2\pi\hbar^2}, \tag{10.1.19}$$

since $d\epsilon = \hbar^2|\boldsymbol{k}|d|\boldsymbol{k}|/M$ from (10.1.16). Thus, the density of states is a constant in the planar system [Fig.10.3(a)]. Electrons occupy first the lowest-energy state with $\boldsymbol{k} = 0$ and then higher-energy states with $\boldsymbol{k} \neq 0$ up to the *Fermi energy* ϵ_F. It is determined by the electron density ρ_0 by way of

$$\rho_0 = \int_0^{\epsilon_F} D(\epsilon)d\epsilon = \frac{M\epsilon_F}{2\pi\hbar^2}. \tag{10.1.20}$$

Electrons occupy all states such that $|\boldsymbol{k}| \leq k_F$ with

$$k_F = \sqrt{4\pi\rho_0}, \tag{10.1.21}$$

where $\epsilon_F = \hbar^2 k_F^2/2M$.

When the spin degree of freedom is taken into account, these quantities are replaced by

$$D(\epsilon) = \frac{M}{\pi\hbar^2}, \qquad \epsilon_F = \frac{\pi\hbar^2\rho_0}{M}, \qquad k_F = \sqrt{2\pi\rho_0}, \tag{10.1.22}$$

since one quantum state labeled by the wave vector $\boldsymbol{k}$ can accommodate up-spin and down-spin electrons.

10.2 Cyclotron Motion

Electrons make cyclotron motion in an external magnetic field $\boldsymbol{B}$. The Hamiltonian is given by (2.6.9), or

$$H = \frac{1}{2M}\left\{(-i\hbar\partial_x + eA_x^{\text{ext}})^2 + \left(-i\hbar\partial_y + eA_y^{\text{ext}}\right)^2\right\}, \tag{10.2.1}$$

where A_k^{ext} is the external electromagnetic potential describing the external magnetic field $\boldsymbol{B} = (0, 0, -B_\perp)$,

$$B_\perp = -\varepsilon_{jk}\partial_j A_k^{\text{ext}} = \partial_y A_x^{\text{ext}} - \partial_x A_y^{\text{ext}} > 0. \tag{10.2.2}$$

The covariant momentum is

$$P_x \equiv -i\hbar\partial_x + eA_x^{\text{ext}}, \qquad P_y \equiv -i\hbar\partial_y + eA_y^{\text{ext}}. \tag{10.2.3}$$

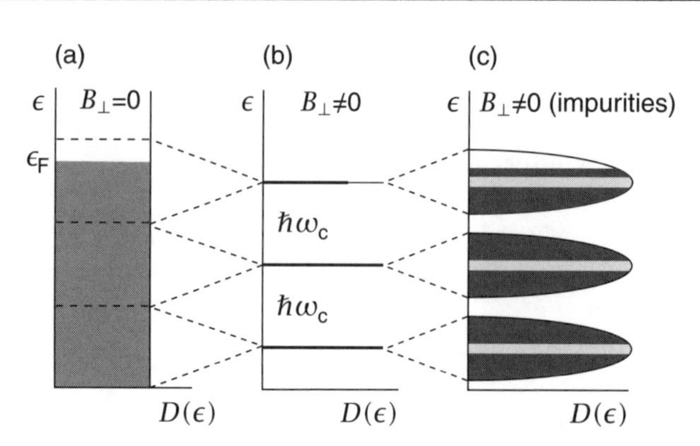

Fig. 10.3 Electrons are put into the planar system. One quantum state accommodates only one electron due to the Fermi statistics. Electrons occupy first the lowest-energy state and then higher-energy states. (a) When $B_\perp = 0$, the density of states is $D(\epsilon) = M/(2\pi\hbar^2)$ for each spin component. (b) When $B_\perp \neq 0$, it is $D(\epsilon) = \rho_{DS}\delta(\epsilon - E_N)$ with $\rho_{DS} = 1/(2\pi\ell_B^2)$ and $E_N = N\hbar\omega_c + (1/2)\hbar\omega_c$. (c) Landau levels are broaden by impurity effects in actual samples. Shaded dark regions represent localized states. Electrons in those regions do not contribute to the current, as results in formation of the Hall plateau. Shaded light regions represent extended states.

We define the guiding-center coordinate by

$$X \equiv x + \frac{1}{eB_\perp}P_y, \qquad Y \equiv y - \frac{1}{eB_\perp}P_x. \tag{10.2.4}$$

Since they satisfy

$$[X,Y] = -i\ell_B^2, \qquad [P_x,P_y] = i\frac{\hbar^2}{\ell_B^2},$$
$$[X,P_x] = [X,P_y] = [Y,P_x] = [Y,P_y] = 0, \tag{10.2.5}$$

the guiding center (X,Y) and the covariant momentum (P_x,P_y) are entirely independent variables. Here we have introduced the magnetic length

$$\ell_B = \sqrt{\frac{\hbar}{eB_\perp}}, \tag{10.2.6}$$

which gives the fundamental scale to the QH system.

The physical meaning of these variables are as follows [Fig.10.4]. The electron coordinate $\boldsymbol{x} = (x,y)$ is decomposed into the guiding center $\boldsymbol{X} = (X,Y)$ and the relative coordinate $\boldsymbol{R} = (R_x,R_y)$ as in (10.2.4) with

$$\boldsymbol{R} = \left(-\frac{1}{eB_\perp}P_y, \frac{1}{eB_\perp}P_x\right). \tag{10.2.7}$$

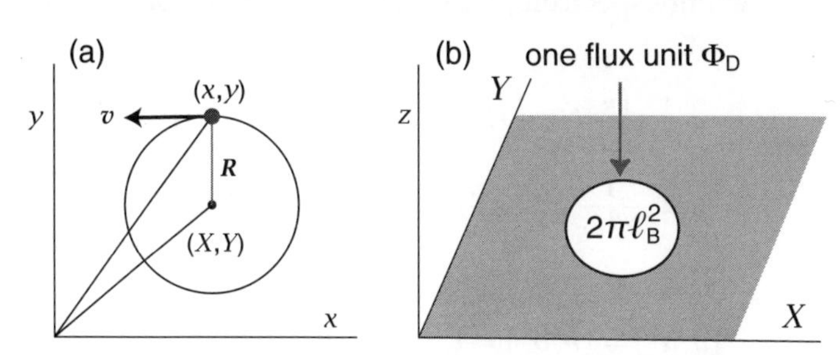

Fig. 10.4 (a) The electron coordinate $\boldsymbol{x}$ is decomposed into the guiding center $\boldsymbol{X}$ and the relative coordinate $\boldsymbol{R}$. An electron jumps to a higher Landau level when a motion in the relative coordinate is excited. (b) The guiding center $\boldsymbol{X} = (X, Y)$ is the only dynamical variable in each Landau level. Quantum mechanically, since the X and Y coordinates do not cummute, $[X, Y] = -i\ell_B^2$, the electron position cannot be determined more accurately than the area $2\pi\ell_B^2$.

Because of the Heisenberg equations of motion,

$$\frac{dX}{dt} = \frac{1}{i\hbar}[X,H] = 0, \qquad \frac{dY}{dt} = \frac{1}{i\hbar}[Y,H] = 0, \tag{10.2.8}$$

there is no motion of an electron if it is at rest initially. Thus the guiding center represents the barycentric coordinate. On the other hand, the relative coordinate $\boldsymbol{R}$ describes the motion around the guiding center. The Heisenberg equations of motion are

$$\frac{dP_x}{dt} = \frac{1}{i\hbar}[P_x, H] = \omega_c P_y, \qquad \frac{dP_y}{dt} = \frac{1}{i\hbar}[P_y, H] = -\omega_c P_x, \tag{10.2.9}$$

with the cyclotron frequency

$$\omega_c = \frac{\hbar}{M\ell_B^2} = \frac{eB_\perp}{M}. \tag{10.2.10}$$

Hence

$$\frac{d}{dt}(R_x + iR_y) = \omega_c R_y - i\omega_c R_x = -i\omega_c (R_x + iR_y). \tag{10.2.11}$$

We integrate this as

$$R_x = R_x^0 \cos\omega_c t + R_y^0 \sin\omega_c t, \qquad R_y = -R_x^0 \sin\omega_c t + R_y^0 \cos\omega_c t, \tag{10.2.12}$$

where R_x^0 and R_y^0 are integration constants. The relative coordinate (10.2.7) makes cyclotron motion [Fig.10.5] in the counterclockwise direction with the cyclotron frequency ω_c.

To derive the energy spectrum, we construct two pairs of operators from these variables,

$$a \equiv \frac{\ell_B}{\sqrt{2}\hbar}(P_x + iP_y), \quad a^\dagger \equiv \frac{\ell_B}{\sqrt{2}\hbar}(P_x - iP_y), \tag{10.2.13a}$$

$$b \equiv \frac{1}{\sqrt{2}\ell_B}(X - iY), \quad b^\dagger \equiv \frac{1}{\sqrt{2}\ell_B}(X + iY), \tag{10.2.13b}$$

obeying

$$[a, a^\dagger] = [b, b^\dagger] = 1, \qquad [a, b] = [a^\dagger, b] = 0. \tag{10.2.14}$$

There are relations

$$a - ib^\dagger = -\frac{i}{\sqrt{2}\ell_B}(x + iy), \qquad a^\dagger + ib = \frac{i}{\sqrt{2}\ell_B}(x - iy). \tag{10.2.15}$$

We have two independent harmonic oscillators: a and b are annihilation (decreasing) operators while $a^\dagger$ and $b^\dagger$ are creation (increasing) operators.

As we have reviewed in Section 1.3, the Fock vacuum is

$$a|0\rangle = 0, \qquad b|0\rangle = 0, \tag{10.2.16}$$

upon which Fock states are constructed,

$$|N, n\rangle = \sqrt{\frac{1}{N!n!}}(a^\dagger)^N (b^\dagger)^n |0\rangle. \tag{10.2.17}$$

The orthonormal completeness condition reads

$$\langle M, m|N, n\rangle = \delta_{MN}\delta_{mn}, \qquad \sum_{N,n} |N, n\rangle\langle N, n| = 1. \tag{10.2.18}$$

The Fock states present the Fock representations of the commutation relations (10.2.14) or equivalently of (10.2.5).

The Hamiltonian (10.2.1) is rewritten as

$$H = (a^\dagger a + aa^\dagger)\frac{\hbar\omega_c}{2} = (a^\dagger a + \frac{1}{2})\hbar\omega_c \tag{10.2.19}$$

with the cyclotron frequency ω_c. The energy levels E_N are those of the harmonic oscillator,

$$E_N = (N + \frac{1}{2})\hbar\omega_c, \tag{10.2.20}$$

with $|N\rangle$ the eigenstates. They are called the Landau levels. A quantum-mechanical state in the Nth Landau level is given by (10.2.17). There exists a degeneracy in each Landau level, corresponding to the quantum mechanical state $|N, n\rangle$, which we call the *Landau site*[1] in the Nth Landau level. The degeneracy is

[1] One Landau site can accommodate one electron in the spinless theory, and two electrons in the theory with spins. It can accommodate four electrons in bilayer systems with spins.

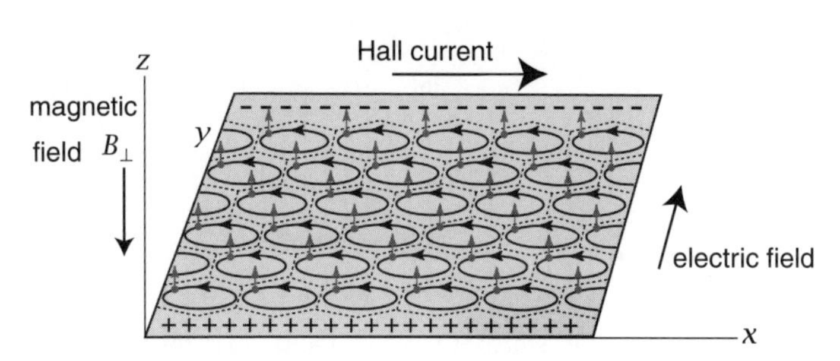

Fig. 10.5 In the magnetic field ($B_z = -B_\perp$) an electron makes a cyclotron motion with radius ℓ_B, occupying an area $2\pi\ell_B^2$ and avoiding all others. At the filling factor $\nu = 1$, spin-polarized electrons fill the lowest Landau level. We may regard them to form a von Neumann lattice with the unit cell area $2\pi\ell_B^2$: See Section 10.5. Each cell provides one Landau site.

proportional to the size of the system. We call a, $a^\dagger$ the Landau-level ladder operators, and b, $b^\dagger$ the angular-momentum ladder operators by the reason given just below (10.3.3) in the succeeding section.

The planar electron system in a magnetic field has been reduced to a simple system of two independent harmonic oscillators, $\{a, a^\dagger\}$ and $\{b, b^\dagger\}$. As will be detailed in Part IV, they lead to "horizontal" W_∞ algebra and "vertical" W_∞ algebra: The horizontal W_∞ algebra characterizes the structure of electron states in each Landau level, while the vertical W_∞ algebra the structure of Landau levels themselves.

The number of Landau sites (which are quantum states in each Landau level) is calculated pictorially [Fig.10.2]. We work in the $\boldsymbol{k}$ space with $\boldsymbol{k} = \boldsymbol{P}/\hbar$. The radius k_n of the nth circle is found from (10.2.7) and (10.2.28) as

$$k_n^2 = \frac{1+2n}{\ell_B^2}. \tag{10.2.21}$$

The area between two successive circles is

$$\Delta S_k = \pi k_{n+1}^2 - \pi k_n^2 = \frac{2\pi}{\ell_B^2}. \tag{10.2.22}$$

Let ρ_Φ be the density of Landau sites in one Landau level. Then, one Landau site occupies area $\Delta S_k/(L^2\rho_\Phi)$ in the $\boldsymbol{k}$ space, since $N_\Phi \equiv L^2\rho_\Phi$ is the number of the Landau sites. This must be equal to the corresponding area $(2\pi/L)^2$ in the absence of magnetic field because the electron number is not changed by the application of magnetic field [Fig.10.3(b)]. From this requirement we obtain

$$\rho_\Phi = \frac{1}{2\pi\ell_B^2}, \tag{10.2.23}$$

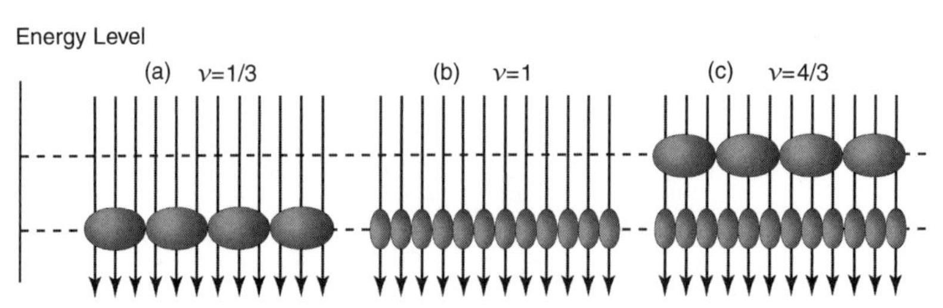

Fig. 10.6 We illustrate the QH states at $\nu = 1/3$, 1 and 4/3 in the same magnetic field. The number of the Landau sites is given by the number N_Φ of the Dirac flux quanta (indicated by arrows) passing through the system. Here, $N_\Phi = 12$. (a) At $\nu = 1/3$ there are 3 flux quanta per electron. (b) At $\nu = 1$ there is one flux quantum per electron. (c) The QH state at $\nu = 4/3$ is regarded as a combination of these two states in the spinless theory.

and hence one Landau site occupies the area

$$\Delta S = 2\pi \ell_B^2 \tag{10.2.24}$$

in the real space. The density of states is rewritten as

$$\rho_\Phi = \frac{1}{2\pi \ell_B^2} = \frac{eB_\perp}{2\pi\hbar} = \frac{B_\perp}{\Phi_D}, \tag{10.2.25}$$

where $\Phi_D \equiv 2\pi\hbar/e$ is the Dirac flux quantum. It is equal to the density of flux quanta.

When there are N_e electrons in the system with area S, the filling factor is

$$\nu = \frac{N_e}{N_\Phi} = \frac{\rho_0}{\rho_\Phi} = 2\pi\ell_B^2\rho_0 = \frac{2\pi\rho_0\hbar}{eB_\perp}. \tag{10.2.26}$$

At $\nu = 1/m \le 1$, there are m flux quanta per electron. When $\nu = 1$ all states in the lowest Landau level are filled up in the spinless theory. When $\nu = 1 + (1/m)$, we may regard the system as a combination of the two states at $\nu = 1$ and $\nu = 1/m$, as illustrated in Fig.10.6.

We investigate semiclassically the cyclotron radius. We rewrite the Hamiltonian (10.2.1) as

$$H = \frac{1}{2M}\boldsymbol{P}^2 = \frac{\hbar\omega_c}{2\ell_B^2}\boldsymbol{R}^2. \tag{10.2.27}$$

We evaluate the expectation value of this operator by the state (10.2.17) and equate it with (10.2.20),

$$\langle \boldsymbol{R}^2 \rangle = (2N+1)\ell_B^2. \tag{10.2.28}$$

The magnetic length ℓ_B gives the radius of the cyclotron motion for the ground state ($N = 0$). It is of order 10nm in a typical magnetic field (~ 10Tesla). We next estimate the cyclotron radius R_c of the electron carrying the Fermi energy ε_F. It is given by (10.2.28) when the electron is in the Nth Landau level. We have $\nu = N+1$ for the highest Landau level. Hence,

$$R_c = \sqrt{2N+1}\ell_B \simeq \sqrt{2\nu}\ell_B = \sqrt{4\pi\rho_0\ell_B^2} = k_F\ell_B^2 = \frac{\hbar k_F}{eB_\perp}, \tag{10.2.29}$$

where $\hbar k_F$ is the Fermi momentum (10.1.21).

10.3 Symmetric Gauge

We proceed to explore quantum-mechanical states of the Hamiltonian (10.2.1) by constructing the wave function explicitly. The motion of an electron within one Landau level is specified by the guiding center $\boldsymbol{X} = (X, Y)$. Since the coordinates X and Y do not commute, we cannot diagonalize both of them simultaneously. It is necessary to choose one of them or an appropriate combination of them to diagonalize.

In this section we diagonalize the symmetric combination $X^2 + Y^2$, which is appropriate to analyze the QH system with disk geometry. It is equivalent to diagonalize the number operator $b^\dagger b$, since it follows from (10.2.13b) that

$$X^2 + Y^2 = (2b^\dagger b + 1)\ell_B^2. \tag{10.3.1}$$

The state $|N, n\rangle$ is an eigenstate of the operator $X^2 + Y^2$,

$$(X^2 + Y^2)|N, n\rangle = (2n + 1)\ell_B^2|N, n\rangle. \tag{10.3.2}$$

The angular-momentum operator is given by

$$L \equiv xp_y - yp_x = \frac{eB_\perp}{2}(X^2 + Y^2) - \frac{1}{2eB_\perp}(P_x^2 + P_y^2) = (b^\dagger b - a^\dagger a)\hbar. \tag{10.3.3}$$

The electron state $|N, n\rangle$ has the angular momentum $(n - N)\hbar$ in a given Landau level N. Thus, b and $b^\dagger$ are the operators increasing or decreasing the angular momentum.

To construct the wave functions for these states we take the *symmetric gauge*, where the vector potential reads

$$A_x^{\text{ext}} = \frac{1}{2}B_\perp y, \qquad A_y^{\text{ext}} = -\frac{1}{2}B_\perp x. \tag{10.3.4}$$

In this gauge the momentum and the guiding center (10.2.4) read

$$P_x = -i\hbar\frac{\partial}{\partial x} + \frac{\hbar}{2\ell_B^2}y, \qquad P_y = -i\hbar\frac{\partial}{\partial y} - \frac{\hbar}{2\ell_B^2}x, \tag{10.3.5a}$$

$$X = \frac{1}{2}x - i\ell_B^2\frac{\partial}{\partial y}, \qquad Y = \frac{1}{2}y + i\ell_B^2\frac{\partial}{\partial x}. \tag{10.3.5b}$$

We introduce the complex number z by

$$z = \frac{1}{2\ell_B}(x + iy), \qquad z^* = \frac{1}{2\ell_B}(x - iy), \tag{10.3.6a}$$

$$\frac{\partial}{\partial z} = \ell_B\left(\frac{\partial}{\partial x} - i\frac{\partial}{\partial y}\right), \qquad \frac{\partial}{\partial z^*} = \ell_B\left(\frac{\partial}{\partial x} + i\frac{\partial}{\partial y}\right). \tag{10.3.6b}$$

The harmonic oscillator operators (10.2.13a) and (10.2.13b) are represented as

$$a = -\frac{i}{\sqrt{2}}\left(z + \frac{\partial}{\partial z^*}\right), \quad a^\dagger = \frac{i}{\sqrt{2}}\left(z^* - \frac{\partial}{\partial z}\right), \tag{10.3.7a}$$

$$b = \frac{1}{\sqrt{2}}\left(z^* + \frac{\partial}{\partial z}\right), \quad b^\dagger = \frac{1}{\sqrt{2}}\left(z - \frac{\partial}{\partial z^*}\right), \tag{10.3.7b}$$

with

$$a - ib^\dagger = -\sqrt{2}iz, \qquad a^\dagger + ib = \sqrt{2}iz^*. \tag{10.3.8}$$

The electron position is represented by the complex number z and its conjugate z^*.

The state in the lowest Landau level ($N = 0$) satisfies $H|\mathfrak{S}\rangle = \frac{1}{2}\hbar\omega_c|\mathfrak{S}\rangle$ with (10.2.19), or

$$a|\mathfrak{S}\rangle = 0, \tag{10.3.9}$$

which reads

$$\langle \boldsymbol{x}|a|\mathfrak{S}\rangle = a(\boldsymbol{x})\langle \boldsymbol{x}|\mathfrak{S}\rangle = -\frac{i}{\sqrt{2}}\left(z + \frac{\partial}{\partial z^*}\right)\mathfrak{S}(\boldsymbol{x}) = 0 \tag{10.3.10}$$

in terms of the wave function $\mathfrak{S}(\boldsymbol{x}) = \langle \boldsymbol{x}|\mathfrak{S}\rangle$. The general solution is

$$\mathfrak{S}(\boldsymbol{x}) = \lambda(z)\exp[-zz^*], \tag{10.3.11}$$

where $\lambda(z)$ is an arbitrary analytic function. All these states are degenerate. We call (10.3.9) the lowest-Landau-level (LLL) condition.

Since we analyze mainly physics taking place in the lowest Landau level, we frequently make an abridgement such that

$$|n\rangle \equiv |0, n\rangle = \sqrt{\frac{1}{n!}}(b^\dagger)^n|0\rangle, \qquad n \geq 0. \tag{10.3.12}$$

The state $|0\rangle$ satisfies

$$\langle \boldsymbol{x}|b|0\rangle = b(\boldsymbol{x})\langle \boldsymbol{x}|0\rangle = \frac{1}{\sqrt{2}}\left(z^* + \frac{\partial}{\partial z}\right)\mathfrak{S}_0(\boldsymbol{x}) = 0, \tag{10.3.13}$$

which is solved as

$$\mathfrak{S}_0(\boldsymbol{x}) = \frac{1}{\sqrt{2\pi\ell_B^2}}e^{-|z|^2} = \frac{1}{\sqrt{2\pi\ell_B^2}}\exp\left(-\frac{r^2}{4\ell_B^2}\right). \tag{10.3.14}$$

The state $|n\rangle$ is described by the wave function,

$$\mathfrak{S}_n(\boldsymbol{x}) = \langle \boldsymbol{x}|n\rangle = \sqrt{\frac{1}{n!}}(b^\dagger)^n \mathfrak{S}_0(\boldsymbol{x}) = \sqrt{\frac{1}{2^n n!}}\left(z - \frac{\partial}{\partial z^*}\right)^n e^{-zz^*} = \sqrt{\frac{2^n}{2\pi\ell_B^2 n!}} z^n e^{-|z|^2}. \tag{10.3.15}$$

The orthonormality condition reads

$$\langle n|m\rangle \equiv \int d^2x\, \langle n|\boldsymbol{x}\rangle \langle \boldsymbol{x}|m\rangle = \int d^2x\, \mathfrak{S}_n^*(\boldsymbol{x})\mathfrak{S}_m(\boldsymbol{x}) = \delta_{nm}. \tag{10.3.16}$$

The wave function of a generic state (10.2.17) is

$$\mathfrak{S}_n^N(\boldsymbol{x}) = \langle \boldsymbol{x}|N,n\rangle = \sqrt{\frac{1}{N!n!}}\langle \boldsymbol{x}|(a^\dagger)^N(b^\dagger)^n|0\rangle \tag{10.3.17a}$$

$$= \sqrt{\frac{1}{N!n!}}(a^\dagger)^N(b^\dagger)^n \mathfrak{S}_0(\boldsymbol{x}) \tag{10.3.17b}$$

with (10.3.7a) and (10.3.7b): In (10.3.17a) $a^\dagger$ and $b^\dagger$ are increasing operators of the Landau level and the angular momentum, while in (10.3.17b) they are the corresponding differential operators as in (10.3.7).

The probability finding the electron at $r = 2\ell_B|z|$ in the lowest Landau level is given by

$$|\mathfrak{S}_n(\boldsymbol{x})|^2 = \frac{2^n}{2\pi\ell_B^2 n!}|z|^{2n}e^{-2|z|^2} \propto r^{2n}\exp\left(-\frac{r^2}{2\ell_B^2}\right), \tag{10.3.18}$$

which has a sharp peak at $r = r_n$ with

$$r_n = \sqrt{2n}\ell_B, \tag{10.3.19}$$

as depicted in Fig.10.7. These states are represented by rings on a disc geometry [Fig.10.7], where a ring is labeled by the angular momentum $n\hbar$. The area of each ring is equal for all n,

$$\Delta S = \pi r_{n+1}^2 - \pi r_n^2 = 2\pi\ell_B^2, \tag{10.3.20}$$

as is the area (10.2.24) of one Landau site. The position of an electron cannot be localized within an area smaller than ΔS. This is because it is confined to the lowest Landau level, where its X and Y coordinates are noncommutative, $[X,Y] = -i\ell_B^2$.

We evaluate the current on the one-body state. The electric current is the Nöther current (5.3.14),

$$J_k(\boldsymbol{x}) = -\frac{e\hbar}{2Mi}\left\{\psi^\dagger(D_k\psi) - (D_k\psi)^\dagger\psi\right\}, \tag{10.3.21}$$

with $D_k = \partial_k + i(e/\hbar)A_k^\perp$. It is easy to find for the state (10.3.15) that

$$\langle \boldsymbol{J}(\boldsymbol{x})\rangle = \mathfrak{S}_n(\boldsymbol{x})^*\boldsymbol{J}(\boldsymbol{x})\mathfrak{S}_n(\boldsymbol{x}) = \frac{2^n}{2\pi\ell_B^2 n!}\frac{e^2B_\perp}{4M}\left(2|z|^2 - n\right)|z|^{2(n-1)}e^{-2|z|^2}(-y,x). \tag{10.3.22}$$

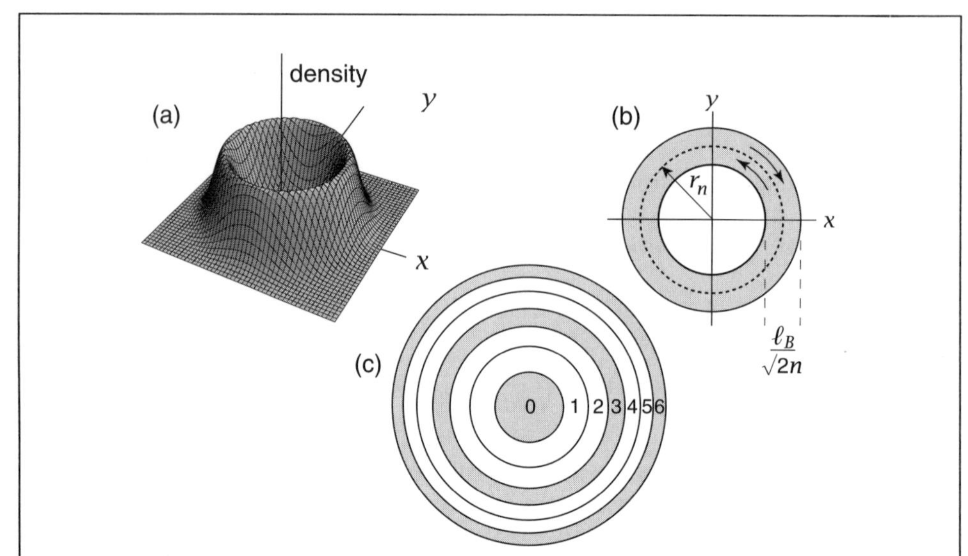

Fig. 10.7 (a) The electron density is illustrated in a state having angular momentum $n\hbar$. It has a sharp peak located at $r_n = \sqrt{2n}\ell_B$ from the origin. (b) A Landau site having angular momentum $n\hbar$ is represented by a ring, which is located at $r_n = \sqrt{2n}\ell_B$ from the origin and has width $\ell_B/\sqrt{2n}$. A circulating current flows in the clockwize direction for $r > r_n$ and into the counterclockwize direction for $r < r_n$. (c) Degenerate Landau sites are represented by rings in a disk geometry, and each ring is labeled by the angular momentum $n\hbar$. In this figure the states with $n = 0, 3, 6 \cdots$ are occupied, and the filling factor is $\nu = 1/3$. We have $\nu = 1$ when all states are filled up. The electron having angular momentum $n\hbar$ encircles the magnetic flux $n\Phi_D$, and acquires the Aharonov-Bohm phase $ne\Phi_D/\hbar = 2\pi n$.

It describes a current density circulating around the origin, which flows into the clockwise direction for $r > \sqrt{2n}\ell_B$ and into the counterclockwise direction for $r < \sqrt{2n}\ell_B$ [see Fig.10.7]. The net current is zero for $n \neq 0$. (A net circular current flows in the $n = 0$ state into the clockwise direction corresponding to cyclotron motion of one electron.)

We comment on a relevance to the Aharonov-Bohm effect reviewed in Section 5.7. The electron with angular momentum $n\hbar$ encircles the area $\pi r_n^2 = 2\pi n\ell_B^2$ and hence the magnetic flux $\Phi = B_\perp \pi r_n^2 = n\Phi_D$. Thus, it acquires the Aharonov-Bohm phase (5.7.7) by amount of $e\Phi/\hbar = 2\pi n$. On the other hand, the wave function yields the phase $2\pi n$ due to the angular momentum in the same process, as is easily seen by setting $z \to e^{2\pi i}z$ in (10.3.15). These two phases are identical.

10.4 LANDAU GAUGE

We continue to explore quantum-mechanical states of the Hamiltonian (10.2.1) by diagonalizing either the coordinate X or Y. Here we diagonalize Y to incorporate the Hall current flowing in the x direction, as we do in Section 10.7. In so doing we take the *Landau gauge*, where the vector potential reads

$$A_x^{\text{ext}} = B_\perp y, \qquad A_y^{\text{ext}} = 0. \tag{10.4.1}$$

The covariant momentum (10.2.3) and the guiding center (10.2.4) are

$$P_x = -i\hbar\frac{\partial}{\partial x} + \frac{\hbar}{\ell_B^2}y, \qquad P_y = -i\hbar\frac{\partial}{\partial y}, \tag{10.4.2a}$$

$$X = x - i\ell_B^2\frac{\partial}{\partial y}, \qquad Y = i\ell_B^2\frac{\partial}{\partial x}. \tag{10.4.2b}$$

This is a good gauge to analyze the QH system with rectangular geometry. The symmetric gauge and the Landau gauge are related by way of a gauge transformation with $\exp(-ixy/2\ell_B^2)$.[2]

We start with the diagonalization of Y. Since $p_x = -i\hbar\partial_x = -(\hbar/\ell_B^2)Y$, the eigenstate of the coordinate Y is that of the canonical momentum p_x,

$$p_x|k\rangle = \hbar k|k\rangle, \qquad Y|k\rangle = -k\ell_B^2|k\rangle. \tag{10.4.3}$$

Here k is a wave number in the x direction.

We next require it to belong to the lowest Landau level. The Landau-level ladder operator (10.2.13a) is

$$a = \frac{1}{\sqrt{2}}\left(\ell_B\frac{\partial}{\partial y} + \frac{y-Y}{\ell_B}\right), \quad a^\dagger = -\frac{1}{\sqrt{2}}\left(\ell_B\frac{\partial}{\partial y} - \frac{y-Y}{\ell_B}\right). \tag{10.4.4}$$

The LLL condition is

$$a|k\rangle = 0. \tag{10.4.5}$$

The conditions (10.4.3) and (10.4.5) read

$$\frac{\partial}{\partial x}\mathfrak{S}_k(\boldsymbol{x}) = ik\mathfrak{S}_k(\boldsymbol{x}), \tag{10.4.6a}$$

$$\left(\ell_B\frac{\partial}{\partial y} + \frac{y+k\ell_B^2}{\ell_B}\right)\mathfrak{S}_k(\boldsymbol{x}) = 0 \tag{10.4.6b}$$

---

[2]The Hamiltonians H^{Lan} and H^{sym} in the two gauges are related as $H^{\text{Lan}} = e^{-ixy/2\ell_B^2}H^{\text{sym}}e^{ixy/2\ell_B^2}$, since the corresponding momenta are related as $P_i^{\text{Lan}} = e^{-ixy/2\ell_B^2}P_i^{\text{sym}}e^{ixy/2\ell_B^2}$. The wave function (10.3.15) in the symmetric gauge satisfies $H^{\text{sym}}\mathfrak{S}_n^{\text{sym}}(\boldsymbol{x}) = \epsilon\mathfrak{S}_n^{\text{sym}}(\boldsymbol{x})$ with $\epsilon = \frac{1}{2}\hbar\omega_c$. Since the wave function $\mathfrak{S}_n^{\text{Lan}}(\boldsymbol{x}) = e^{-ixy/2\ell_B^2}\mathfrak{S}_n^{\text{sym}}(\boldsymbol{x})$ satisfies

$$H^{\text{Lan}}\mathfrak{S}_n^{\text{Lan}}(\boldsymbol{x}) = e^{-ixy/2\ell_B^2}H^{\text{sym}}e^{ixy/2\ell_B^2}\mathfrak{S}_n^{\text{Lan}}(\boldsymbol{x}) = e^{-ixy/2\ell_B^2}H^{\text{sym}}\mathfrak{S}_n^{\text{sym}}(\boldsymbol{x}) = E\mathfrak{S}_n^{\text{Lan}}(\boldsymbol{x}),$$

it is a wave function in the Landau gauge. However, it is highly a nontrivial problem to express $\mathfrak{S}_n^{\text{Lan}}(\boldsymbol{x})$ in terms of the set of wave functions (10.4.7).

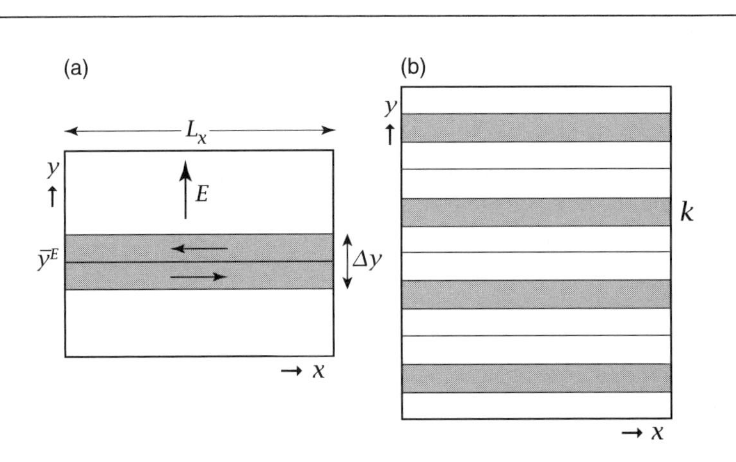

Fig. 10.8 (a) A Landau site having the momentum $\hbar k$ is represented by a strip, which is located at $y = -k\ell_B^2$ and has width $\Delta y = 2\pi\ell_B^2/L_x$. The wave function is a plane wave confined in this region. (b) Degenerate Landau sites are represented by strips in a rectangular geometry, and each strip is labeled by the wave number k. The filling factor is $\nu = 1/3$ in this figure. We have $\nu = 1$ when all states are filled up.

in terms of the wave function, $\mathfrak{S}_k(\boldsymbol{x}) = \langle \boldsymbol{x}|k\rangle$. They are easily solved,

$$\mathfrak{S}_k(\boldsymbol{x}) = \frac{1}{\sqrt{\pi^{1/2}\ell_B}} \exp(ixk) \exp\left[-\frac{1}{2\ell_B^2}(y + k\ell_B^2)^2\right]. \tag{10.4.7}$$

The wave functions obey the orthonormality condition,[3]

$$\langle k|k'\rangle = \int d^2x\, \mathfrak{S}_k^*(\boldsymbol{x})\mathfrak{S}_{k'}(\boldsymbol{x}) = 2\pi\delta(k - k'), \tag{10.4.8}$$

and

$$\int dy\, |\mathfrak{S}_k(\boldsymbol{x})|^2 = 1. \tag{10.4.9}$$

Eigenstates in a higher Landau level are given by

$$\mathfrak{S}_k^N(\boldsymbol{x}) = \langle \boldsymbol{x}|N, k\rangle = \sqrt{\frac{1}{N!}}(a^\dagger)^N \mathfrak{S}_k(\boldsymbol{x}), \tag{10.4.10}$$

where $a^\dagger$ is given by setting $Y = -k\ell_B^2$ in (10.4.4).

The wave function (10.4.7) describes a plane wave propagating in the x direction with momentum $\hbar k$. The probability finding the electron at y has a sharp peak at $y = -k\ell_B^2$. These states are represented by strips on a rectangular geometry [Fig.10.8], where a strip is labeled by the wave number k, located at $y = -k\ell_B^2$

[3]This normalization is made so that the Hall current takes the standard form (10.7.3) for one-particle state.

and has width $\Delta y = \ell_B^2 \Delta k$. Here, $\Delta k = 2\pi/L_x$, because the wave number is quantized as $k = 2\pi n_x/L_x$, as we have seen in (10.1.17), where n_x is an integer and L_x is the x-size of the system. The area of each strip is equal to

$$\Delta S = L_x \times \Delta y = L_x \times \frac{2\pi \ell_B^2}{L_x} = 2\pi \ell_B^2, \tag{10.4.11}$$

as is the area (10.2.24) of one Landau site.

We evaluate the current (10.3.21) on the state (10.4.7),

$$\langle J_x(\mathbf{x})\rangle = -\frac{e\hbar}{\sqrt{\pi} M \ell_B^3}(y + k\ell_B^2)\exp\left[-\frac{1}{\ell_B^2}(y + k\ell_B^2)^2\right], \qquad \langle J_y(\mathbf{x})\rangle = 0, \tag{10.4.12}$$

where we have used (10.4.1). It describes a current density flowing in the negative x axis for $y > \bar{y}$ and in the positive x axis for $y < \bar{y}$, where $\bar{y} = -k\ell_B^2$: The net current is zero.

10.5 von Neumann Lattice

One Landau site accommodates one electron, but the Landau site is not localized around one point in the previous two choices of wave functions [Figs.10.7 and 10.8]. We search for a set of wave functions where an electron is localized around one lattice point as in Fig.10.5. This is achieved by diagonalizing the combination $X - iY$ of the guiding center.

Because of the relations

$$b \equiv \frac{1}{\sqrt{2}\ell_B}(X - iY), \qquad b^\dagger \equiv \frac{1}{\sqrt{2}\ell_B}(X + iY), \tag{10.5.1}$$

it is equivalent to diagonalize the the angular-momentum ladder operator b,

$$b|\beta\rangle = \beta|\beta\rangle. \tag{10.5.2}$$

Because b is an annihilation operator, the state $|\beta\rangle$ is a coherent state by definition: See (1.5.2). It is given by (1.5.10), or

$$|\beta\rangle \equiv e^{\beta b^\dagger - \beta^* b}|0\rangle = e^{-|\beta|^2/2} e^{\beta b^\dagger}|0\rangle, \tag{10.5.3}$$

where $|0\rangle$ is the angular-momentum zero state obeying $b|\beta\rangle = 0$.

The coherent state enjoys the minimum uncertainty subject to the Heisenberg uncertainty relation associated with the noncommutativity (10.2.5) between the coordinates X and Y. The state $|\beta\rangle$ corresponds to the classical state localized around the point

$$x = \sqrt{2}\ell_B \beta_\Re, \qquad y = -\sqrt{2}\ell_B \beta_\Im, \tag{10.5.4}$$

as follows from (10.5.1) and (10.5.2). Because β is an arbitrary complex number, an electron may be localized at any point.

Since each electron occupies an area $2\pi\ell_B^2$, it is reasonable to choose a lattice with the unit cell area $2\pi\ell_B^2$. Then, there is one to one correspondence between the magnetic flux quantum and the lattice site. Such a lattice is nothing but a von Neumann lattice[4] reviewed in Section 1.5. The states on a von Neumann lattice form a minimum complete set in the lowest Landau level.

We consider a square lattice for simplicity. Lattice points are given by $\beta_{mn} = \sqrt{\pi}(m+in)$ as in (1.5.18), or

$$X_m = \sqrt{2\pi}\ell_B m, \qquad Y_n = -\sqrt{2\pi}\ell_B n. \tag{10.5.5}$$

The state (10.5.3) reads

$$|X_m, Y_n\rangle = \exp\left[-\pi(m^2+n^2)\right]\exp\left[\sqrt{\pi}(m+in)b^\dagger\right]|0\rangle \tag{10.5.6}$$

as in (1.5.10). They are not orthogonal,

$$\langle X_{m'}, Y_{n'}|X_m, Y_n\rangle = \exp\left(-\frac{\pi}{2}\left[(m-m')^2+(n-n')^2\right]\right) \tag{10.5.7}$$

as in (1.5.15).

To construct the wave function we use the symmetric gauge (10.3.4),

$$\mathfrak{S}_\beta(\boldsymbol{x}) = \langle\boldsymbol{x}|\beta\rangle = e^{-|\beta|^2/2}e^{\beta b^\dagger}\mathfrak{S}_0(\boldsymbol{x}), \tag{10.5.8}$$

where

$$b^\dagger = \frac{1}{\sqrt{2}}\left(z - \frac{\partial}{\partial z^*}\right) \tag{10.5.9}$$

as in (10.3.7b), and $\mathfrak{S}_0(\boldsymbol{x}) = \langle\boldsymbol{x}|\beta\rangle$ is the wave function for the angular-momentum zero state,

$$\mathfrak{S}_0(\boldsymbol{x}) = \frac{1}{\sqrt{2\pi\ell_B^2}}e^{-|z|^2} = \frac{1}{\sqrt{2\pi\ell_B^2}}\exp\left(-\frac{r^2}{4\ell_B^2}\right) \tag{10.5.10}$$

as in (10.3.14). It describes an electron making cyclotron motion around the center ($\boldsymbol{x}=0$): See Fig.10.7(c). The wave function (10.5.8) is calculated as

$$\begin{aligned}\mathfrak{S}_\beta(\boldsymbol{x}) &= \frac{1}{\sqrt{2\pi\ell_B^2}}e^{-|\beta|^2/2}e^{\sqrt{2}\beta z}e^{-|z|^2} \\ &= \frac{1}{\sqrt{2\pi\ell_B^2}}\exp\left(-\left|z-\frac{1}{\sqrt{2}}\beta\right|^2 + \frac{i}{\sqrt{2}\ell_B}(y\beta_\Re + x\beta_\Im)\right),\end{aligned} \tag{10.5.11}$$

where $\beta = \beta_\Re + i\beta_\Im$. It is obtained by translating $\mathfrak{S}_0(\boldsymbol{x})$ to the point $z = \beta/\sqrt{2}$ in the complex z-plane apart from an irrelevant phase factor. Thus, it describes an

[4] The area of the unit cell is $2\pi\hbar$ corresponding to the commutation relation $[q,p] = i\hbar$ in Section 1.5. Now, it is $2\pi\ell_B^2$ corresponding to the commutation relation $[X,Y] = -i\ell_B^2$.

electron making cyclotron motion around the point $z = \beta/\sqrt{2}$. It yields

$$\mathfrak{S}_{mn}(\mathbf{x}) = \langle \mathbf{x}|X_m, Y_n\rangle = \frac{1}{\sqrt{2\pi\ell_B^2}} \exp\left(-\left|z - \frac{1}{\sqrt{2}}\beta_{mn}\right|^2 + \frac{i\sqrt{\pi}}{\sqrt{2}\ell_B}(ym + xn)\right) \tag{10.5.12}$$

for the electron state (10.5.6).

The von Neumann lattice formulation of the QH effect[235,471] realizes a natural picture of electrons making cyclotron motions in the plane [Fig.10.5]. However it has a demerit that Landau sites $|X_m, Y_n\rangle$ are not orthogonal one to another as in (10.5.7).

10.6 Electrons in *N*th Landau Level

The field-theoretical Hamiltonian is given by (2.6.11), or

$$H = \frac{1}{2M}\int d^2x\, \psi^\dagger(\mathbf{x})(P_x^2 + P_y^2)\psi(\mathbf{x}) \tag{10.6.1a}$$

$$= \frac{1}{2M}\int d^2x\, \psi^\dagger(\mathbf{x})(P_x - iP_y)(P_x + iP_y)\psi(\mathbf{x}) + \frac{N}{2}\hbar\omega_c, \tag{10.6.1b}$$

where

$$N = \int d^2x\, \psi^\dagger\psi \tag{10.6.2}$$

is the total number of electrons. When the cyclotron gap is large enough, excitations across Landau levels are suppressed. It is a good approximation to consider electrons confined to one Landau level, by assuming all lower Landau levels are filled up. For such an electron the kinetic energy is just a constant.

We first study the lowest Landau level. We may rewrite the Hamiltonian as

$$H = \hbar\omega_c \int d^2x\, \psi^\dagger a^\dagger a \psi + \frac{N}{2}\hbar\omega_c, \tag{10.6.3}$$

where

$$a \equiv \frac{\ell_B}{\sqrt{2}\hbar}(P_x + iP_y), \qquad a^\dagger(\mathbf{x}) \equiv \frac{\ell_B}{\sqrt{2}\hbar}(P_x - iP_y). \tag{10.6.4}$$

The LLL condition reads

$$a(\mathbf{x})\psi(\mathbf{x})|\mathfrak{S}\rangle = \frac{\ell_B}{\sqrt{2}\hbar}(P_x + iP_y)\psi(\mathbf{x})|\mathfrak{S}\rangle = 0. \tag{10.6.5}$$

All states satisfying this condition constitute a Hilbert space $\mathbb{H}_{\text{LLL}}$. Any state violating this condition has a component in a higher Landau level.

We have already studied the LLL condition (10.6.5) for the one-body state. We examine it for the N-body state. The wave function is defined by

$$\mathfrak{S}[\mathbf{x}] = \langle \mathbf{x}_1, \mathbf{x}_2, \cdots, \mathbf{x}_N|\mathfrak{S}\rangle \tag{10.6.6}$$

with

$$\langle \mathbf{x}_1, \mathbf{x}_2, \cdots, \mathbf{x}_N| = \langle 0|\psi(\mathbf{x}_1)\psi(\mathbf{x}_2)\cdots\psi(\mathbf{x}_N). \tag{10.6.7}$$

Here, no two coordinates coincide, $\mathbf{x}_r \neq \mathbf{x}_s$ for $r \neq s$, due to the Pauli exclusion principle. Hence, from the LLL condition (10.6.5) we find that

$$a(\mathbf{x}_r)\mathfrak{S}(\mathbf{x}_1, \mathbf{x}_2, \cdots, \mathbf{x}_N) = \langle 0|\psi(\mathbf{x}_1)\cdots a(\mathbf{x}_r)\psi(\mathbf{x}_r)\cdots\psi(\mathbf{x}_N)|\mathfrak{S}\rangle = 0. \tag{10.6.8}$$

The wave function is obtained by solving this equation.

In the symmetric gauge, where

$$a(\mathbf{x}) = -\frac{i}{\sqrt{2}}\left(z + \frac{\partial}{\partial z^*}\right), \qquad a^\dagger(\mathbf{x}) = \frac{i}{\sqrt{2}}\left(z^* - \frac{\partial}{\partial z}\right), \tag{10.6.9}$$

the LLL condition (10.6.8) reads

$$\left(z_r + \frac{\partial}{\partial z_r^*}\right)\mathfrak{S}(\mathbf{x}_1, \mathbf{x}_2, \cdots, \mathbf{x}_N) = 0, \tag{10.6.10}$$

for $r = 1, 2, \cdots, N$. We solve this to obtain the wave function

$$\mathfrak{S}[\mathbf{x}] = \rho_0^{N/2}\lambda[z]e^{-\sum_{r=1}^N |z_r|^2}, \tag{10.6.11}$$

where $\lambda[z]$ is an arbitrary dimensionless polynomial in the N variables z_r. It is required to be totally antisymmetric in the N variables due to the Fermi statistics of electrons. It gives the general solution of the LLL condition.

We proceed to discuss the Nth Landau level. It is convenient to use the notation (2.4.8) for the field operator $\psi(\mathbf{x})$, that is,

$$\psi(\mathbf{x}) = \langle \mathbf{x}|\psi\rangle. \tag{10.6.12}$$

Using the orthonormal completeness condition

$$\sum_{N,n} |N, n\rangle\langle N, n| = 1, \tag{10.6.13}$$

we decompose the field operator as

$$|\psi\rangle = \sum_{N,n} |N, n\rangle\langle N, n|\psi\rangle, \tag{10.6.14}$$

or

$$\psi(\mathbf{x}) = \sum_{N,n} \langle \mathbf{x}|N, n\rangle\langle N, n|\psi\rangle = \sum_{N,n} \mathfrak{S}_n^N(\mathbf{x})c_N(n), \tag{10.6.15}$$

where

$$c_N(n) \equiv \langle N, n|\psi\rangle = \int d^2x\, \langle N, n|\mathbf{x}\rangle\psi(\mathbf{x}) = \int d^2x\, \mathfrak{S}_n^{N*}(\mathbf{x})\psi(\mathbf{x}). \tag{10.6.16}$$

This is the operator annihilating an electron in the Landau site $|N, n\rangle$. The anti-commutation relation

$$\{c_N(n), c_M^\dagger(m)\} = \delta_{NM}\delta_{nm} \tag{10.6.17}$$

is derived from the anticommutation relation

$$\{\psi(\boldsymbol{x}), \psi^\dagger(\boldsymbol{y})\} = \delta(\boldsymbol{x} - \boldsymbol{y}) \tag{10.6.18}$$

and the orthonormality condition (10.2.18).

The N_e-body state in the Nth Landau level is created as in (2.5.4),

$$|\mathfrak{S}^N\rangle = \prod_n c_N^\dagger(n)|0\rangle = \int [d\boldsymbol{x}]\mathfrak{S}^N[\boldsymbol{x}]\psi^\dagger(\boldsymbol{x}_1)\psi^\dagger(\boldsymbol{x}_2)\cdots\psi^\dagger(\boldsymbol{x}_{N_e})|0\rangle, \tag{10.6.19}$$

where $\mathfrak{S}^N[\boldsymbol{x}]$ is given by $\prod_n \mathfrak{S}_n^N(\boldsymbol{x}_n)$ with the positions $\boldsymbol{x}_n$ totally antisymmetrized as in (2.5.7). We have used an abridgement,

$$\int [d\boldsymbol{x}] = \prod_r \int d^2x_r = \int\int\cdots\int d^2x_1 d^2x_2\cdots d^2x_{N_e}. \tag{10.6.20}$$

All these states span the Hilbert space.

In the Nth Landau level the Hilbert space $\mathbb{H}_N$ is spanned by Landau sites $|N, n\rangle$, $n = 1, 2, 3, \cdots$. We may consider the Landau-level projection. The operator

$$\mathbb{P}_N \equiv \sum_{n=0}^{\infty} |N, n\rangle\langle N, n| \tag{10.6.21}$$

is the projection operator to the Nth Landau level, obeying $\mathbb{P}_N\mathbb{P}_N = \mathbb{P}_N$. The field operator for electrons in the Nth Landau level is given by $|\psi_N\rangle = \mathbb{P}_N|\psi\rangle$, or

$$\psi_N(\boldsymbol{x}) \equiv \sum_{n=0}^{\infty} \langle\boldsymbol{x}|N, n\rangle\langle N, n|\psi\rangle = \sum_{n=0}^{\infty} \mathfrak{S}_n^N(\boldsymbol{x})c_N(n) \tag{10.6.22}$$

with (10.6.16). This is the projected field operator to the Nth Landau level.

For definiteness we calculate the anticommutation relation for the field $\psi_0(\boldsymbol{x})$ in the lowest Landau level,

$$\{\psi_0(\boldsymbol{x}), \psi_0^\dagger(\boldsymbol{y})\} = \sum_{n=0}^{\infty} \langle\boldsymbol{x}|n\rangle\langle n|\boldsymbol{y}\rangle = \frac{1}{2\pi\ell_B^2}\exp\left(i\frac{\boldsymbol{x}\wedge\boldsymbol{y}}{2\ell_B^2}\right)\exp\left(-\frac{|\boldsymbol{x}-\boldsymbol{y}|^2}{4\ell_B^2}\right), \tag{10.6.23}$$

with the use of the wave function (10.3.15). Here we see explicitly that

$$\sum_{n=0}^{\infty} |n\rangle\langle n| \neq 1 \tag{10.6.24}$$

in (10.6.23), since $|\boldsymbol{x}\rangle \notin \mathbb{H}_{\text{LLL}}$. (We denote $\mathbb{H}_0$ as $\mathbb{H}_{\text{LLL}}$.) The lack of the completeness condition leads to the nonlocalizability of an electron to a point when it is confined to the lowest Landau level. See Chapters 29 and 30 for more details on this and related topics.

We briefly discuss the field operator $\psi_0(\mathbf{x})$ for the lowest Landau level in the Landau gauge. The Landau site $|k\rangle$ is given by (10.4.7), or

$$\mathfrak{S}_k(\mathbf{x}) = \frac{1}{\sqrt{\pi^{1/2}\ell_B}} \exp(ixk) \exp\left[-\frac{1}{2\ell_B^2}\left(y + k\ell_B^2\right)^2\right], \tag{10.6.25}$$

satisfying the orthonormalization condition (10.4.8), or

$$\langle k|\ell\rangle = 2\pi\delta(k-\ell). \tag{10.6.26}$$

Introducing the operator $c_0(k)$ annihilating an electron in the Landau site $|k\rangle$ with

$$\{c_0(k), c_0^\dagger(\ell)\} = \delta(k-\ell), \tag{10.6.27}$$

we define the field operator as

$$\psi_0(\mathbf{x}) \equiv \int \frac{dk}{\sqrt{2\pi}} c_0(k)\mathfrak{S}_k(\mathbf{x}) = \int \frac{dk}{\sqrt{2\pi}} c_0(k)\langle \mathbf{x}|k\rangle. \tag{10.6.28}$$

The anticommutation relation is given by

$$\{\psi_0(\mathbf{x}), \psi_0^\dagger(\mathbf{x}')\} = \frac{1}{2\pi\ell_B^2} \exp\left(i\frac{xy - x'y'}{2\ell_B^2}\right) \exp\left(i\frac{\mathbf{x}\wedge\mathbf{x}'}{2\ell_B^2}\right) \exp\left(-\frac{|\mathbf{x}-\mathbf{x}'|^2}{4\ell_B^2}\right), \tag{10.6.29}$$

which is slightly different from (10.6.23).

Coherent States

The Hilbert space $\mathbb{H}_N$ is spanned by Landau sites $|N, n\rangle$. The projection operator to the Nth Landau level is given by (10.6.21). The state $|\mathbf{x}\rangle$ is projected as

$$|\mathbf{x}\rangle \to |N, \hat{\mathbf{x}}\rangle \equiv \mathbb{P}_N|\mathbf{x}\rangle = \sum_{n=0}^{\infty} |N, n\rangle\langle N, n|\mathbf{x}\rangle. \tag{10.6.30}$$

For the lowest Landau level, setting $|\hat{\mathbf{x}}\rangle = |0, \hat{\mathbf{x}}\rangle$ we find

$$\langle\hat{\mathbf{x}}|\hat{\mathbf{y}}\rangle = \sum_{n=0}^{\infty} \langle\mathbf{x}|n\rangle\langle n|\mathbf{y}\rangle = \frac{1}{2\pi\ell_B^2} \exp\left(i\frac{\mathbf{x}\wedge\mathbf{y}}{2\ell_B^2}\right) \exp\left(-\frac{|\mathbf{x}-\mathbf{y}|^2}{4\ell_B^2}\right), \tag{10.6.31}$$

and

$$\int d^2x\, |\hat{\mathbf{x}}\rangle\langle\hat{\mathbf{x}}| = \int d^2x \sum_{m,n} |m\rangle\langle m|\mathbf{x}\rangle\langle\mathbf{x}|n\rangle\langle n| = 1 \tag{10.6.32}$$

within the Hilbert space $\mathbb{H}_{\text{LLL}}$. We note that

$$\langle n|\hat{\mathbf{x}}\rangle \equiv \sum_{m=0}^{\infty} \langle n|m\rangle\langle m|\mathbf{x}\rangle = \sum_{m=0}^{\infty} \delta_{nm}\langle m|\mathbf{x}\rangle = \langle n|\mathbf{x}\rangle. \tag{10.6.33}$$

This explains why $\langle\mathbf{x}|n\rangle$ is the wave function of an electron confined to the lowest Landau level though $|\mathbf{x}\rangle$ does not belong to the lowest Landau level.

We study the state $|\hat{x}\rangle$ in more details. Substituting the wave function into (10.6.30) we obtain[5]

$$|\hat{x}\rangle \equiv \sum_{n=0}^{\infty} |n\rangle\langle n|x\rangle = \sqrt{\frac{1}{2\pi\ell_B^2}} \sum_{n=0}^{\infty} \sqrt{\frac{2^n}{n!}} z^n e^{-|z|^2} |n\rangle. \tag{10.6.34}$$

Since it has the same form as (1.5.12) with $v = \sqrt{2}z = (x+iy)/\sqrt{2}\ell_B$, it is expressed as in (1.5.10), or

$$|\hat{x}\rangle = \sqrt{\frac{1}{2\pi\ell_B^2}} e^{-|z|^2} e^{\sqrt{2}zb^\dagger} |0\rangle. \tag{10.6.35}$$

It follows from (1.5.2) that

$$b|\hat{x}\rangle = \frac{(x+iy)}{\sqrt{2}\ell_B}|\hat{x}\rangle. \tag{10.6.36}$$

Thus the projected state $|\hat{x}\rangle$ is a coherent state of b.

10.7 Hall Current

An electron makes cyclotron motion around the guiding center. The guiding center moves and induces a current in the presence of external electric field.

In a constant electric field E_j the quantum-mechanical Hamiltonian reads

$$H = \frac{1}{2M}(P_x^2 + P_y^2) + exE_x + eyE_y. \tag{10.7.1}$$

Within the lowest Landau level, the Heisenberg equation of motions are

$$\begin{aligned} i\hbar\frac{dX}{dt} &= [X,H] = eE_y[X,Y] = -\frac{i\hbar}{B_\perp}E_y, \\ i\hbar\frac{dY}{dt} &= [Y,H] = eE_x[Y,X] = \frac{i\hbar}{B_\perp}E_x. \end{aligned} \tag{10.7.2}$$

The drift velocity is the velocity of the guiding center. When the system consists of a homogeneous distribution of electrons with density ρ_0, the current density is

$$\mathcal{J}_x = -e\rho_0\langle\dot{X}\rangle = \frac{e\rho_0}{B_\perp}E_y, \qquad \mathcal{J}_y = -e\rho_0\langle\dot{Y}\rangle = -\frac{e\rho_0}{B_\perp}E_x, \tag{10.7.3}$$

as agrees with (9.1.2).

---

[5]There is a difference between the coherent states (10.6.34) and (1.5.12) by a normalization factor, $|\hat{x}\rangle = (2\pi\ell_B^2)^{-1/2}|v\rangle$. Thus the completeness condition (1.5.16) reads

$$\frac{1}{\pi}\int d^2v\, |v\rangle\langle v| = \frac{1}{2\pi\ell_B^2}\int d^2x\, |v\rangle\langle v| = \int d^2x\, |\hat{x}\rangle\langle\hat{x}| = 1,$$

with $v = (x+iy)/\sqrt{2}\ell_B$, which is (10.6.32).

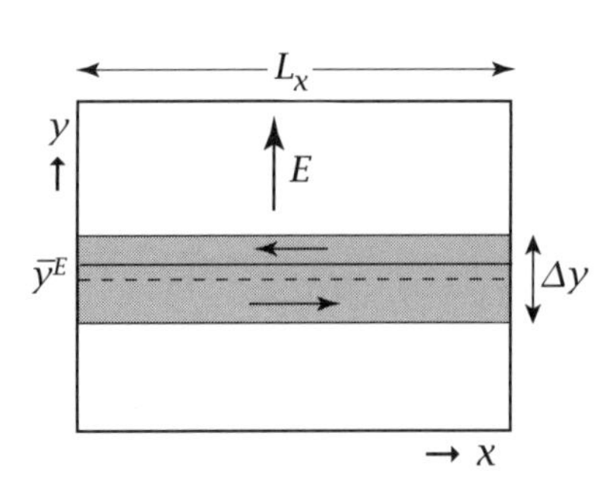

Fig. 10.9 A Landau site having the momentum $\hbar k$ is represented by a strip, which is located at $\bar{y}^E = -k\ell_B^2 + eE\ell_B^2/\hbar\omega_c$ and has width $\Delta y = 2\pi\ell_B^2/L_x$. The wave function is a plane wave confined in this region. A Hall current flows on this strip when the external electric field E is applied along the y axis.

We study the field-theoretical Hamiltonian in the presence of the external electric field $\boldsymbol{E} = c\boldsymbol{\nabla} A_0$,

$$H = H_{\text{K}} + ec\int d^2x\, A_0(\boldsymbol{x})\rho(\boldsymbol{x}). \tag{10.7.4}$$

The electric current is the Nöther current (5.3.14),

$$J_k(\boldsymbol{x}) = -\frac{ie\hbar}{2M}\left\{\psi^\dagger(D_k\psi) - (D_k\psi)^\dagger\psi\right\} \tag{10.7.5}$$

together with the charge density

$$\rho_e(\boldsymbol{x}) = -e\rho(\boldsymbol{x}). \tag{10.7.6}$$

We rewrite the current as

$$J_k = \frac{e}{M}\psi^\dagger P_k\psi + \frac{ie\hbar}{2M}\partial_k\rho. \tag{10.7.7}$$

Let us evaluate the current on the QH state. The last term vanishes since the QH state has a homogeneous density.

If the LLL condition were given by (10.6.5), or

$$\langle J_x + iJ_y\rangle_{\text{QH}} = \frac{e}{M}\langle \text{g}|\psi^\dagger(P_x + iP_y)\psi|\text{g}\rangle = 0, \tag{10.7.8}$$

we would conclude

$$\mathcal{J} \equiv \langle J\rangle_{\text{QH}} = 0, \tag{10.7.9}$$

because the current is real. No current would flow on the state in the lowest Landau level.[189] Is this a paradox because we know that the Hall current flows? It is not: It only says that the Hilbert space $\mathbb{H}_{\text{LLL}}$ is not given by (10.6.5) in the presence of the electric field.

We analyze this problem in the Landau gauge. In the presence of the electric field in the y direction [Figs.9.3 and 10.9], the Hamiltonian is given by

$$H = \frac{1}{2M}\left\{(p_x + eB_\perp y)^2 + p_y^2\right\} + eyE. \tag{10.7.10}$$

We take an eigenstate of the canonical momentum $p_x = -i\hbar\partial_x$, as corresponds to (10.4.3),

$$p_x|k\rangle_E = \hbar k|k\rangle_E, \tag{10.7.11}$$

upon which the Hamiltonian (10.7.10) reads

$$\begin{aligned} H &= \frac{1}{2M}\left\{(\hbar k + eB_\perp y)^2 + p_y^2\right\} + eyE \\ &= \frac{1}{2M}\left\{p_y^2 + \frac{\hbar^2}{\ell_B^4}(y - \bar{y}^E)^2\right\} + e\bar{y}^E E + \frac{(eE\ell_B)^2}{2\hbar\omega_c}, \end{aligned} \tag{10.7.12}$$

where $p_y = -i\hbar\partial_y$, and we have set

$$\bar{y}^E \equiv -k\ell_B^2 - \frac{eE}{\hbar\omega_c}\ell_B^2. \tag{10.7.13}$$

The system is exactly solvable. The wave function is

$$\mathfrak{S}_k^E(\boldsymbol{x}) = \frac{1}{\sqrt{\pi^{1/2}\ell_B}}\exp(ixk)\exp\left[-\frac{1}{2\ell_B^2}(y - \bar{y}^E)^2\right], \tag{10.7.14}$$

with the energy

$$E(k) = -eE\ell_B^2 k - \frac{(eE\ell_B)^2}{2\hbar\omega_c}. \tag{10.7.15}$$

The wave function (10.7.14) approaches to (10.4.7) as $E \to 0$. The state supports a drift current,

$$\begin{aligned} \langle J_x(\boldsymbol{x})\rangle_E &= -\frac{e\hbar}{\sqrt{\pi}M\ell_B^3}\left(y - \bar{y}^E - \frac{eE\ell_B^2}{\hbar\omega_c}\right)\exp\left[-\frac{1}{\ell_B^2}(y - \bar{y}^E)^2\right] \\ &= \frac{eE}{B_\perp}|\mathfrak{S}_k^E(\boldsymbol{x})|^2 - \frac{e\hbar}{\sqrt{\pi}M\ell_B^3}(y - \bar{y}^E)\exp\left[-\frac{1}{\ell_B^2}(y - \bar{y}^E)^2\right], \\ \langle J_y(\boldsymbol{x})\rangle_E &= 0. \end{aligned} \tag{10.7.16}$$

It is clearly seen that there is a net current flowing to the positive x axis on the strip [Fig.10.8]. The total current is

$$\mathcal{J}_x^{\text{total}} = \int dy\, \langle J_x(\boldsymbol{x})\rangle_E = \frac{eE}{B_\perp}, \tag{10.7.17}$$

for one-particle state, as is consistent with (10.7.3).

We comment that the state (10.7.14) violates the LLL condition (10.4.5) defined in the absence of the electric field,

$$a|0,k\rangle_E = -\frac{eE\ell_B}{\sqrt{2}\hbar\omega_c}|0,k\rangle_E \neq 0. \tag{10.7.18}$$

The state $|0,k\rangle_E$ does not belong to the lowest Landau level previously defined unless $E = 0$. It is customary to say[189] that the electric field induces a Landau-level mixing, by way of which the Hall current flows. However, it is better to say that the lowest Landau level needs to be redefined in the presence of the electric field.

In the framework of quantum field theory, the field operator is given by (10.6.28) with the use of the wave function (10.7.14),

$$\psi_0(\boldsymbol{x}) = \frac{1}{\sqrt{\pi^{1/2}\ell_B}}\int\frac{dk}{\sqrt{2\pi}}c_0(k)\exp(ixk)\exp\left[-\frac{1}{2\ell_B^2}\left(y+k\ell_B^2+\frac{eE}{\hbar\omega_c}\ell_B^2\right)^2\right]. \tag{10.7.19}$$

We can calculate the Hall current (10.7.5) by substituting this expression for $\psi(\boldsymbol{x})$, and we obtain

$$\begin{aligned}J_x(\boldsymbol{x}) &= -\frac{e\hbar}{2M}\partial_y\rho(\boldsymbol{x}) + \frac{e^2\ell_B^2}{\hbar}E\rho(\boldsymbol{x}),\\ J_y(\boldsymbol{x}) &= +\frac{e\hbar}{2M}\partial_x\rho(\boldsymbol{x}),\end{aligned} \tag{10.7.20}$$

as operator formulas. For the QH state we have

$$\mathcal{J}_x = \langle J_x\rangle_{\text{QH}} = \nu\frac{e^2}{2\pi\hbar}E, \qquad \mathcal{J}_y = \langle J_y\rangle_{\text{QH}} = 0, \tag{10.7.21}$$

since it has a homogeneous density, $\rho(\boldsymbol{x}) = \rho_0 = \nu/(2\pi\ell_B^2)$ at the filling factor ν. This agrees with the Hall current formula (9.1.3). See Section 32.8 for more details on the Hall current in the field-theoretical framework.

To understand the relation (10.7.13) between the position $\bar{y}^E$of the strip and its wave number k, we study classically how an electron moves in the presence of both the magnetic field $B_\perp$ and the electric field E [Fig.10.10]. An electron has the electric potential energy

$$-e\varphi(y) = eyE + \text{constant}, \tag{10.7.22}$$

with $\varphi(y)$ the scalar potential. The electric force $(-e\partial\varphi/\partial y = eE)$ is balanced with the Lorentz force,

$$eE = evB, \tag{10.7.23}$$

where v is the velocity of the electron. The velocity is related to the dispersion relation $E(k)$,

$$v = \frac{1}{\hbar}\left|\frac{\partial E(k)}{\partial k}\right|. \tag{10.7.24}$$

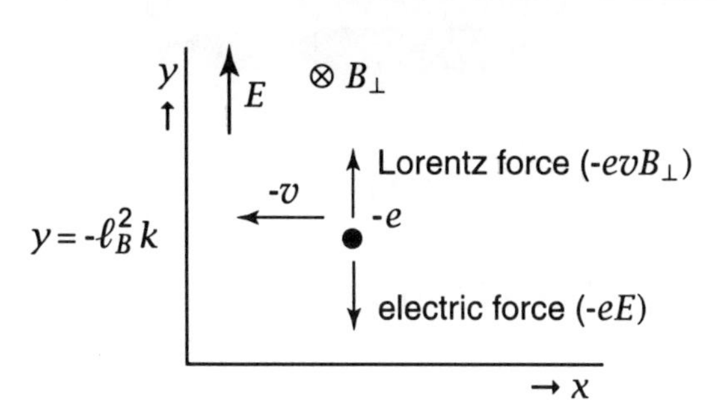

Fig. 10.10 An electron moves classically parallel to the x axis when it is in the perpendicular magnetic field $B_\perp$ and in the electric field E. There exists a relation $y = -\ell_B^2 k$ between the position (x, y) and the wave vector $(k, 0)$ of the electron.

For constant velocity this is solved as

$$E(k) = -\hbar kv + \text{constant} = -e\ell_B^2 Ek + \text{constant}, \tag{10.7.25}$$

where we have eliminated v by using (10.7.23). We may equate the potential energy (10.7.22) with the dispersion relation (10.7.25), and obtain

$$y = -k\ell_B^2 + \text{constant}. \tag{10.7.26}$$

Equations (10.7.25) and (10.7.26) are classical counterparts of the quantum mechanical results (10.7.15) and (10.7.13).

CHAPTER 11

QUANTUM HALL EFFECTS

The QH effect comes from the realization of an incompressible liquid. Charged excitations are quasiparticles (quasiholes and quasielectrons) identified with vortices in spin-frozen theory. They are gapful excitations. quasielectrons do not exist in integer QH states, where electrons are excited. Hall plateaux are generated since quasiparticles are localized by impurities and do not contribute to the Hall current at temperature $T = 0$.

11.1 INCOMPRESSIBILITY

The key feature of the QH state is the incompressibility. Its precise definition reads as follows. The compressibility κ is the relative volume change against the pressure change,

$$\kappa = -\frac{1}{S}\frac{\partial S}{\partial P}\bigg|_N, \tag{11.1.1}$$

where S is the area, P is the pressure and N is the particle number. The system is said to be *incompressible* when $\kappa = 0$. Namely, the area is not changed at all when an infinitesimal pressure P is applied to an incompressible system.

The pressure is the change of energy against the change of area, $P = -\partial E/\partial S$. Hence,

$$\kappa^{-1} \equiv -S\frac{\partial P}{\partial S}\bigg|_N = S\frac{\partial^2 E}{\partial S^2}\bigg|_N. \tag{11.1.2}$$

The energy is an extensive quantity in the thermodynamic limit,

$$E = N\varepsilon(\rho), \tag{11.1.3}$$

where $\varepsilon(\rho)$ is the energy per particle, and ρ is the number density. Using (11.1.3) and $N = S\rho$, we may rewrite (11.1.2) as

$$\kappa^{-1} = \frac{1}{\rho}\frac{d}{d(1/\rho)}\frac{d\varepsilon(\rho)}{d(1/\rho)} = \rho^2\left(2\frac{d\varepsilon(\rho)}{d\rho} + \rho\frac{d^2\varepsilon(\rho)}{d\rho^2}\right) = \rho^2\frac{d^2(\rho\varepsilon)}{d\rho^2}. \tag{11.1.4}$$

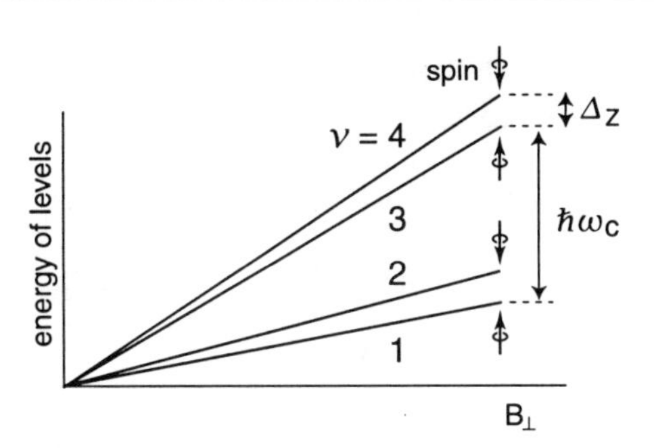

Fig. 11.1 In the magnetic field $B_\perp$ the electron energy is quantized into Landau levels. The one-body gap energy between the two successive Landau levels is $\hbar\omega_c = (e\hbar/M)B_\perp$. The Zeeman term resolves the degeneracy between the electron states with up and down spins by creating a gap energy between them. One Landau level contains two energy levels with the one-body gap energy $\Delta_Z = |g^* \mu_B B|$.

The chemical potential is

$$\mu \equiv \left.\frac{\partial E}{\partial N}\right|_V = \frac{d\mathcal{E}(\rho)}{d\rho} = \frac{d(\rho\varepsilon)}{d\rho}, \tag{11.1.5}$$

where $\mathcal{E}(\rho) = E/S = \rho\varepsilon(\rho)$ is the energy density. From (11.1.4) and (11.1.5) we find that

$$\kappa^{-1} = \rho^2 \frac{d\mu}{d\rho}. \tag{11.1.6}$$

The system is incompressible ($\kappa = 0$) when the chemical potential increases discontinuously as a function of density ($d\mu/d\rho = \infty$).

As we shall see, the area of the QH system can change only in units of $2\pi\ell_B^2$ associated with one Dirac flux quantum. It is accompanied with a creation of a quasiparticle, as costs a nonzero creation energy. This is the mechanism how the QH system is incompressible.

11.2 Integer Quantum Hall Effects

One electron has two spin states, up and down. To avoid unnecessary complications, throughout this chapter, we ignore the dynamical degree of freedom associated with spins by assuming that all spins are polarized and frozen by the Zeeman effect: We call it the spin-frozen system. We illustrate the one-body energy levels in Fig.11.1, by neglecting the Coulomb energy contribution.

It is an easy problem to determine the ground state of the system at the filling factor $\nu = 1$. Since the lowest Landau level is completely filled, the N-body state is uniquely fixed and is given by (10.6.19) where the angular momentum n runs

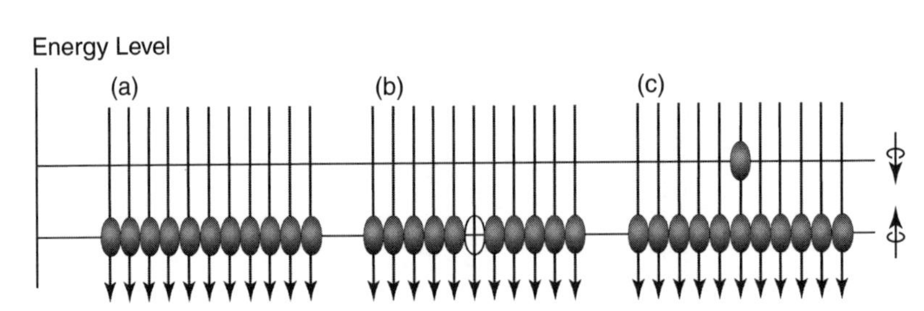

Fig. 11.2 The integer QH state is illustrated at $\nu = 1$. (a) The ground state is a closed packing of electrons pierced by flux quanta. (b) When one electron is removed one quasihole is created. (c) When one electron is added it is placed in a higher energy level.

from 0 to $N-1$,

$$|\mathrm{g}\rangle = \prod_{n=0}^{N-1} c_n^\dagger |0\rangle. \tag{11.2.1}$$

The wave function is the Slater determinant,

$$\begin{aligned}
\mathfrak{S}_{\mathrm{LN}}[\boldsymbol{x}] &= \langle \boldsymbol{x}_1, \cdots, \boldsymbol{x}_N | \mathrm{g}\rangle = \langle 0| \prod_{i=1}^{N} \psi(\boldsymbol{x}_i) \prod_{n=0}^{N-1} c_n^\dagger |0\rangle \\
&= \frac{1}{\sqrt{N!}} \begin{vmatrix} \mathfrak{S}_0(\boldsymbol{x}_1) & \mathfrak{S}_1(\boldsymbol{x}_1) & \cdots & \mathfrak{S}_{N-1}(\boldsymbol{x}_1) \\ \mathfrak{S}_0(\boldsymbol{x}_2) & \mathfrak{S}_1(\boldsymbol{x}_2) & \cdots & \mathfrak{S}_{N-1}(\boldsymbol{x}_2) \\ \vdots & \vdots & \ddots & \vdots \\ \mathfrak{S}_0(\boldsymbol{x}_N) & \mathfrak{S}_1(\boldsymbol{x}_N) & \cdots & \mathfrak{S}_{N-1}(\boldsymbol{x}_N) \end{vmatrix} \\
&= \frac{\rho_0^{N/2}}{\sqrt{N!}} \begin{vmatrix} 1 & z_1 & \cdots & z_1^{N-1} \\ 1 & z_2 & \cdots & z_2^{N-1} \\ \vdots & \vdots & \ddots & \vdots \\ 1 & z_N & \cdots & z_N^{N-1} \end{vmatrix} e^{-\sum_{r=1}^N |z_r|^2} \\
&= \rho_0^{N/2} \prod_{r<s} (z_r - z_s) e^{-\sum_{r=1}^N |z_r|^2},
\end{aligned} \tag{11.2.2}$$

where $\psi(\boldsymbol{x}_i)$ is the electron field operator (10.6.22) and $\mathfrak{S}_n(\boldsymbol{x}_i)$ the one-body wave function (10.3.15). This is the state with the homogeneous electron density ρ_0: See Section 30.2.2.

The Laughlin state is illustrated in Fig.11.2(a). There is one flux quantum per electron at $\nu = 1$, and all the Landau sites are filled up with electrons pierced by flux quanta. When we remove one electron from the system, there appears a hole in it [Fig.11.2(b)]. As will be argued in Section 11.4 and Section 14.3, the hole is not smeared out and behaves as a quasihole which is a vortex soliton carrying the electric charge e. The N-body state containing a hole in the angular-momentum

$n = 0$ site is given by

$$|\mathrm{h}_0\rangle = \prod_{n=1}^{N} c_n^\dagger |0\rangle. \tag{11.2.3}$$

The wave function is

$$\begin{aligned}
\langle \boldsymbol{x}_1, \cdots, \boldsymbol{x}_N | \mathrm{h}_0\rangle &= \langle 0| \prod_{i=1}^{N} \psi(\boldsymbol{x}_i) \prod_{n=1}^{N} c_n^\dagger |0\rangle \\
&= \frac{1}{\sqrt{N!}} \begin{vmatrix} \mathfrak{S}_1(\boldsymbol{x}_1) & \mathfrak{S}_2(\boldsymbol{x}_1) & \cdots & \mathfrak{S}_N(\boldsymbol{x}_1) \\ \mathfrak{S}_1(\boldsymbol{x}_2) & \mathfrak{S}_2(\boldsymbol{x}_2) & \cdots & \mathfrak{S}_N(\boldsymbol{x}_2) \\ \vdots & \vdots & \ddots & \vdots \\ \mathfrak{S}_1(\boldsymbol{x}_N) & \mathfrak{S}_2(\boldsymbol{x}_N) & \cdots & \mathfrak{S}_N(\boldsymbol{x}_N) \end{vmatrix} \\
&= \frac{\rho_0^{N/2}}{\sqrt{N!}} \begin{vmatrix} z_1 & z_1^2 & \cdots & z_1^N \\ z_2 & z_2^2 & \cdots & z_2^N \\ \vdots & \vdots & \ddots & \vdots \\ z_N & z_N^2 & \cdots & z_N^N \end{vmatrix} e^{-\sum_{r=1}^{N-1} |z_r|^2} \\
&= \prod_{i=1}^{N} z_i \mathfrak{S}_{\mathrm{LN}}[\boldsymbol{x}].
\end{aligned} \tag{11.2.4}$$

When we add one electron into the system it is placed in a higher energy level because of the Pauli exclusion principle [Fig.11.2(c)].

The chemical potential is the change of the energy when one electron is added to or removed from the system. There is a jump in the chemical potential when one energy level is filled up. Hence, the system is incompressible ($\kappa = 0$) at the integer value of the filling factor ($\nu = n$).

11.3 Fractional Quantum Hall Effects

The Hall plateau has a similar structure in the integer QH system and the fractional QH system [Fig.9.2]. This suggests the following: At the filling factor $\nu = 1/m$ (m odd) the degeneracy of the ground states is resolved and the system becomes an incompressible liquid with a homogeneous density. The driving force is the Coulomb interaction between electrons because it is the only interaction in the present system. It must resolve the degeneracy by increasing the energy of a state having an inhomogeneous density at $\nu = 1/m$.

The generic N-body state is given by (10.6.11),

$$\mathfrak{S}[\boldsymbol{x}] = \rho_0^{N/2} \lambda[z] e^{-\sum_{r=1}^{N} |z_r|^2}. \tag{11.3.1}$$

Excited states as well as the ground state are specified by the analytic function $\lambda[z]$.

A trial wave function of the ground state was first proposed by Laughlin,[294]

$$\mathfrak{S}_{\rm LN}[\boldsymbol{x}] = \rho_0^{N/2} \prod_{r<s} (z_r - z_s)^m e^{-\sum_{r=1}^N |z_r|^2}. \tag{11.3.2}$$

The factor $\rho_0^{N/2}$ is inserted for a dimensional reason; the complex number z_r is dimensionless.

The Laughlin wave function (11.3.2) has played more roles than just a trial wave function. The theory of the fractional QH effect has developed from this wave function.[9,2,10] It was derived based on the following requirements:

(1) The wave function is antisymmetric under the exchange of any two electron positions.
(2) It is an eigenstate of the total angular momentum, as requires that $\lambda[z]$ is a polynomial in $z_1, \cdots, z_N$ of degree L, where L is the total angular momentum.
(3) The Coulomb repulsion among electrons is included via two-body correlations, as requires that the polynomial part of the wave function is of the Jastrow type, $\lambda[z] = \lambda_0[z] \equiv \prod_{r<s} \lambda_0(z_r - z_s)$.

The only function satisfying the requirements is $\lambda_0(z) = z^m$ with m odd, or $\lambda_0[z] = \prod_{r<s}(z_r - z_s)^m$. The Laughlin wave function (11.3.2) is a particular state in which all electrons, having their own cyclotron motions, stay away from each other as much as possible. See Section 14.1 how it is derived uniquely to describe the nondegenerate homogeneous ground state of electrons semiclassically.

However, the Laughlin state is not an eigenstate of the Coulomb Hamiltonian. It is interesting to construct the interaction Hamiltonian whose eigenstate is the Laughlin state.[207] Let us consider a two-body interaction whose potential is given by $V(\boldsymbol{q})$ in the momentum space. We make the Laguerre expansion of it,

$$V(\boldsymbol{q}) = 2\ell_B^2 \sum_n V_n L_n(q^2\ell_B^2) e^{-q^2\ell_B^2/2}. \tag{11.3.3}$$

Each term is regarded as the projection of the potential $V(\boldsymbol{q})$ to the electron pair state with the relative angular momentum $n\hbar$. Since the relative angular momentum is $m\hbar$ for any pair of electrons in the Laughlin state (11.3.2), only the terms V_n with $n \geq m$ contribute to the state at $\nu = 1/m$. Hence, if we set $V_n = 0$ for $n \geq m$, this interaction gives precisely the zero energy to the state. On the other hand, any other states would acquire energy proportional to V_n with $n < m$. The potential (11.3.3) is called Haldane's pseudopotential, and the parameter V_n the pseudopotential parameter. In particular, considering the potential with $V_n = 0$ for all $n \geq 2$ in (11.3.3), we obtain a good model with short-range potential,

$$V(\boldsymbol{q}) = 2\ell_B^2 V_1 L_1(q^2\ell_B^2) e^{-q^2\ell_B^2/2}. \tag{11.3.4}$$

The Laughlin state is an eigenstate of this interaction for any $\nu = 1/m$.

All pseudopotential parameters V_n are nonzero in the Coulomb interaction. See Fig.19.2 in Section 19.2, where some V_n's are calculated for the Coulomb potential. It is argued that the Laughlin state is the ground state in the lowest Landau level since V_1 is dominant over other terms.

The total angular momentum is the sum of the angular momentum (10.3.3) of individual electrons. It reads $L_{\text{total}} = \hbar \sum_r b_r^\dagger b_r$ for the state in the lowest Landau level. The total angular momentum carried by the Laughlin state (11.3.2) is calculated as follows. A typical term in the Laughlin wave function is

$$z_1^0 z_2^m z_3^{2m} \cdots z_N^{(N-1)m} e^{-\sum_{r=1}^N |z_r|^2}, \tag{11.3.5}$$

where N is the number of electrons. Because the state (10.3.15) has the angular momentum $n\hbar$, it is clear that the above state has the angular moment

$$L_{\text{total}} = m\hbar \sum_{n=0}^{N-1} n = \frac{(N-1)Nm}{2}\hbar. \tag{11.3.6}$$

All other terms in the Laughlin wave function have the same angular momentum. On the other hand, the maximum angular momentum $n_{\max}\hbar$ that one electron can have is given by the maximum power of one of the variables, say, z_N in the wave function (11.3.2),

$$n_{\max} = m(N-1). \tag{11.3.7}$$

It determines the size of the system because all electrons have smaller angular momentum and stay within this area. The area is given by using the formula (10.3.19),

$$S = \pi r_{\max}^2 = 2\pi n_{\max}\ell_B^2 = 2\pi m(N-1)\ell_B^2. \tag{11.3.8}$$

Hence, the Laughlin state is realized at the filling factor

$$\nu = \frac{N\Delta S}{S} = \frac{N}{m(N-1)} = \frac{1}{m} \tag{11.3.9}$$

for $N \gg 1$, where ΔS is the area (10.3.20) occupied by one electron. Thus, there are m flux quanta per electron. The Laughlin state is illustrated in Fig.10.6.

Any excited state in the lowest Landau level is described by the wave function

$$\mathfrak{S}[\boldsymbol{x}] = \omega[z]\mathfrak{S}_{\text{LN}}[\boldsymbol{x}], \tag{11.3.10}$$

where the factor $\omega[z]$ modifies the angular momentum of the Laughlin state: $\omega[z]$ is a dimensionless analytic function symmetric in all N variables, $\omega[z] = \lambda[z]/\lambda_0[z]$. When the Coulomb interactions between electrons are neglected, all these states are degenerate with the Laughlin state (11.3.2). The degeneracy is removed by the Coulomb effect since a nontrivial function $\omega[z]$ means a density inhomogeneity and hence an increase of the Coulomb energy. See Chapter 13 and Chapter 14 for more details on this point.

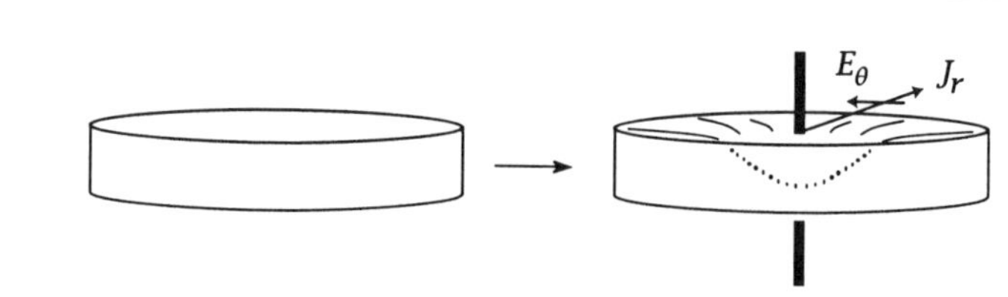

Fig. 11.3 A vortex is created by the insertion of an infinitely thin magnetic solenoid and adiabatically piercing the flux from zero to the flux quanta Φ_D. In so doing the electric field E_θ is generated by the Faraday induction law, as generates the Hall current J_r moving away from the solenoid. Electric charges are accumulated on the edge, causing a chiral edge excitation, as will be described in Section 18.5.

At the filling factor $\nu = n + (1/m)$ the system consists of n filled levels and a partially filled highest level [Fig.10.6]. It is much easier to excite electrons in the highest occupied level. It is a good approximation to regard all electrons in the lower levels inactive. The QH state at $\nu = n + (1/m)$ has the same properties as the QH state at $\nu = 1/m$ except for the Coulomb energy. See Chapter 19 for the Coulomb interaction in higher Landau levels. See also Chapter 17 for a modification due to the spin degree of freedom.

We remark a similarity between the integer QH state and the fractional QH state. Most properties of the spin-frozen integer QH states may be understood by setting $m = 1$ in the analysis of the fractional QH state, provided that we understand the quasielectron is not an antivortex but an electron in a higher energy level. It is worthwhile to emphasize the similarity rather than the difference between them.

11.4 QUASIPARTICLES

Quasiparticles are stable objects carrying electric charges. They are vortex solitons in the spin-frozen theory. We present an intuitive argument due to Laughlin, leaving a detailed analysis to Section 14.3.

We make a thought experiment how to create a vortex [Fig.11.3]. We insert an infinitely thin magnetic solenoid at the origin ($\boldsymbol{x} = 0$) and adiabatically pierce the flux from zero to $\pm\Phi_D$, where Φ_D is the Dirac flux quantum. As the flux is increased, the wave function will evolve adiabatically so that it remains always an eigenstate of the changing Hamiltonian. When the flux becomes $\pm\Phi_D$, it is precisely the one that can be created by a singular gauge transformation,

$$\mathfrak{S}(\boldsymbol{x}) \longrightarrow e^{iq\theta(\boldsymbol{x})}\mathfrak{S}(\boldsymbol{x}), \qquad (q = \pm 1), \tag{11.4.1}$$

where $\theta(\boldsymbol{x})$ is the azimuthal angle around the origin. Hence, the infinitely thin magnetic flux can be gauged away by the inverse singular gauge transformation.

Because the resulting Hamiltonian is the original Hamiltonian, the state left behind is one of its exact eigenstates. Because there is no degeneracy in the ground state, the state is an excited state describing a quasiparticle. quasiparticles are called quasiholes (vortices) for $q = 1$ and quasielectrons (antivortices) for $q = -1$.

The one-body wave function (10.3.15) evolves continuously by the above adiabatic process. At the end of the process the net effect is the singular gauge transformation (11.4.1), which affects the one-body wave function (10.3.15) as $\mathfrak{S}_l(z, z^*) \to e^{iq\theta}\mathfrak{S}_l(z, z^*)$ with its polynomial part transforming

$$\lambda(z) \equiv z^l \longrightarrow \lambda' \equiv e^{iq\theta} z^l, \qquad (q = \pm 1). \tag{11.4.2}$$

However, the state $e^{iq\theta}\mathfrak{S}_l(z, z^*)$ does not belong to the lowest Landau level. According to Laughlin we modify (11.4.2) as

$$\lambda(z) \equiv z^l \longrightarrow \lambda_{\rm qh}(z) = z\lambda(z) \propto e^{i\theta} z^l \quad , \qquad (q = 1), \tag{11.4.3a}$$

$$\lambda(z) \equiv z^l \longrightarrow \lambda_{\rm qe}(z) = \frac{\partial}{\partial z}\lambda(z) \propto e^{-i\theta} z^l, \qquad (q = -1), \tag{11.4.3b}$$

so that the polynomial parts are analytic. This is a LLL projection, about which we discuss extensively later.

The wave function transform as

$$\lambda[z] \longrightarrow \lambda_{\rm qh}[z] = \prod_r z_r \lambda[z] \quad \text{for} \quad q = 1, \tag{11.4.4a}$$

$$\lambda[z] \longrightarrow \lambda_{\rm qe}[z] = \prod_r \frac{\partial}{\partial z_r}\lambda[z] \quad \text{for} \quad q = -1, \tag{11.4.4b}$$

in its polynomial part. Namely, it transforms as

$$\mathfrak{S}[\boldsymbol{x}] \longrightarrow \mathfrak{S}_{\rm qh}[\boldsymbol{x}] \equiv \prod_r b_r^\dagger \mathfrak{S}[\boldsymbol{x}] \quad \text{for} \quad q = 1, \tag{11.4.5a}$$

$$\mathfrak{S}[\boldsymbol{x}] \longrightarrow \mathfrak{S}_{\rm qe}[\boldsymbol{x}] \equiv \prod_r b_r \mathfrak{S}[\boldsymbol{x}] \quad \text{for} \quad q = -1, \tag{11.4.5b}$$

where use was made of the relations

$$b_r \mathfrak{S}[\boldsymbol{x}] = \sqrt{\frac{1}{2}}\frac{\partial}{\partial z_r}\lambda[z] \cdot e^{-\sum_s |z_s|^2}, \qquad b_r^\dagger \mathfrak{S}[\boldsymbol{x}] = \sqrt{2} z_r \lambda[z] \cdot e^{-\sum_s |z_s|^2}. \tag{11.4.6}$$

Here $b_r^\dagger$ and b_r are the angular-momentum ladder operators (10.3.7b) for the rth electron,

$$b_r = \frac{1}{\sqrt{2}}\left(z_r^* + \frac{\partial}{\partial z_r}\right), \qquad b_r^\dagger = \frac{1}{\sqrt{2}}\left(z_r - \frac{\partial}{\partial z_r^*}\right). \tag{11.4.7}$$

They change the power of z_r by one.

We summarize the wave function of one quasiparticle. Any state in the lowest Landau level is described by the wave function (11.3.10),

$$\mathfrak{S}[\boldsymbol{x}] = \omega[z]\mathfrak{S}_{\rm LN}[\boldsymbol{x}], \tag{11.4.8}$$

where the symmetric analytic function $\omega[z]$ describes an excitation on the Laughlin state. The wave function of one vortex (quasihole) is given by the simplest choice,[294]

$$\omega_{\rm qh}[z] = \prod_r z_r, \tag{11.4.9}$$

or

$$\mathfrak{S}_{\rm qh}[\boldsymbol{x}] = \prod_r z_r \mathfrak{S}_{\rm LN}[\boldsymbol{x}]. \tag{11.4.10}$$

We note that this wave function has been derived in the integer QH system as in (11.2.4).

It is suggested[294,296] that the wave function for one antivortex soliton (quasielectron) is given by

$$\mathfrak{S}_{\rm qe}[\boldsymbol{x}] = \prod_r \frac{\partial}{\partial z_r} \mathfrak{S}_{\rm LN}[\boldsymbol{x}] = \omega_{\rm qe}[z]\mathfrak{S}_{\rm LN}[\boldsymbol{x}] \tag{11.4.11}$$

with

$$\omega_{\rm qe}[z] = \frac{\prod_r \frac{\partial}{\partial z_r} \prod_{r<s}(z_r - z_s)^m}{\prod_{r<s}(z_r - z_s)^m}. \tag{11.4.12}$$

In (11.4.11) the derivative $\partial/\partial z_r$ is understood to act only on the polynomial part in the Laughlin wave function. They were argued[209,158] to be good wave functions by calculating numerically the overlap of (11.4.8) or (11.4.11) with the corresponding exact states.

The creation of the quasihole (vortex) with $q = 1$ is to increase the angular momenta of all electrons by one [Fig.10.7 (page 190)]. Its wave function vanishes at the center since the electron at the origin has been pushed outward and is absent in the vortex configuration. On the other hand, the creation of the quasielectron (antivortex) with $q = -1$ is to decrease the angular momentum by one. Its wave function does not vanish at the center since each electron is pushed inward. When an electron exists at the center initially, the configuration should be discarded because otherwise the electron is pushed to the first Landau level at a cost of a large amount of energy $\hbar\omega_c$: This is the LLL projection applied to the antivortex. The integer QH system allows no antivortex, $\omega_{\rm qe}[z] = 0$ in (11.4.12), since all configurations are projected out.

We next study the electric charge of quasiparticles intuitively. We reconsider the process of increasing the flux from zero to $\pm\Phi_{\rm D}$ in the solenoid [Fig.11.3]. The time-dependent magnetic flux gives rise to an electric field according to the Faraday induction law,

$$E_\theta = \frac{d\Phi}{dt}\frac{1}{2\pi r}, \tag{11.4.13}$$

which is directed azimuthally around the solenoid. In the QH state the electric field generates the Hall current (9.1.2) perpendicular to it,

$$J_r = \frac{e\rho_0}{B_\perp} E_\theta, \tag{11.4.14}$$

which holds far away from the solenoid. When the solenoid is removed the total change in the charge is

$$Q = 2\pi r \int dt\, J_r = \frac{e\rho_0}{B_\perp} \int dt\, \frac{d\Phi}{dt} = \pm \frac{e\rho_0}{B_\perp} \Phi_D = \pm \nu e. \tag{11.4.15}$$

Therefore, the quasihole and the quasielectron have fractional electric charges e/m and $-e/m$, respectively, at $\nu = 1/m$. This intuitive argument is justified field theoretically in Appendix M.

A comment is in order. quasiparticles are given by vortices in the spin-frozen theory. However, this is not the case in actual samples with spin, where quasiparticles may be skyrmions. A skyrmion is continuously reduced to a vortex as the Zeeman effect becomes large. However, an antiskyrmion is not reduced to an antivortex, since an antiskyrmion consists mainly of down-spin electrons, but an antivortex consists solely of up-spin electrons. See related topics in Chapter 16 and Chapter 17.

11.5 Hall Plateaux

We have so far studied the QH effect precisely at the filling factor,

$$\nu \equiv \frac{2\pi \rho_0 \hbar}{eB_\perp} = \frac{1}{m}. \tag{11.5.1}$$

It is necessary to examine what happens at $\nu \neq 1/m$, by varying the electron density ρ_0 or the magnetic field $B_\perp$ slightly [Figs.9.7 and 11.4].

When the number density is changed by $\Delta\rho$ with the magnetic field $B_\perp$ fixed, the filling factor is changed as

$$\nu = \frac{2\pi(\rho_0 + \Delta\rho)\hbar}{eB_\perp} = \frac{1}{m}\left(1 + \frac{\Delta\rho}{\rho_0}\right). \tag{11.5.2}$$

Because the electric charge of the quasiparticle is $\pm e/m$, m quasiparticles are created per electron on the Laughlin state at $\nu = 1/m$, as is illustrated in Fig.9.7 (page 170). The number density ρ_{qp} of quasiparticles created is

$$\rho_{qp} = m|\Delta\rho|. \tag{11.5.3}$$

Quasiparticles are quasielectrons (quasiholes) when $\Delta\rho > 0$ ($\Delta\rho < 0$).

Alternatively, we may change the magnetic field by $\Delta B_\perp$ with the number density fixed, as results in the change of the filling factor,

$$\nu = \frac{2\pi \rho_0 \hbar}{e(B_\perp + \Delta B_\perp)} = \frac{1}{m}\left(1 + \frac{\Delta B_\perp}{B_\perp}\right)^{-1}. \tag{11.5.4}$$

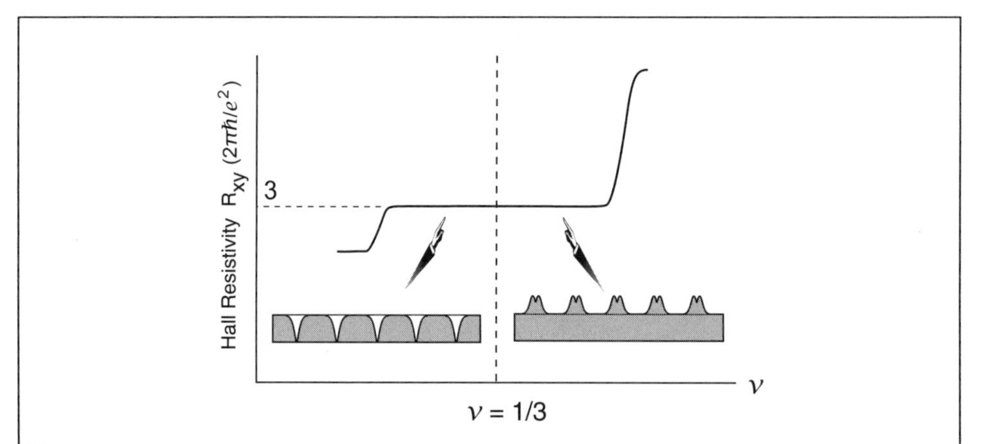

Fig. 11.4 Quasielectrons are excited for $\nu > 1/3$ and quasiholes are excited for $\nu < 1/3$. They are trapped by impurities and do not contribute to the Hall current. The Hall current remains unchanged in the vicinity of $\nu = 1/m$, as is the origin of the Hall plateau.

Since one quasiparticle carries one Dirac flux quanta $\pm\Phi_D$, the number density of quasiparticles created is

$$\rho_{qp} = \frac{|\Delta B_\perp|}{\Phi_D} = \frac{e|\Delta B_\perp|}{2\pi\hbar}. \tag{11.5.5}$$

Quasiparticles are quasielectrons (quasiholes) when $\Delta B_\perp < 0$ ($\Delta B_\perp > 0$).

The filling factor changes according to (11.5.2) or (11.5.4), where quasiparticles are created with density (11.5.3) or (11.5.5). Since the current is carried by both electrons and quasiparticles, the Hall current is given by the classical result (9.1.2) and proportional to the filling factor ν: No Hall plateau is generated. This is true in pure samples without impurities. On the contrary, the observed Hall current is a constant around a quantized magic filling factor [Fig.9.2]. This can be understood as follows. The ground state at the vicinity of $\nu = 1/m$ contains an ensemble of quasiparticles [Fig.11.4]. Since they are particle-like charged objects, they are localized by impurities and do not contribute to the Hall current [Fig.10.3(c)]. As far as the system is described by such a state, the Hall current at $\nu \neq 1/m$ is the same as the one at $\nu = 1/m$ because the local charge density that contributes to the current is still given by $-e\rho_0$, as is illustrated in Fig.9.4 (page 167). This leads to a plateau formation in the Hall resistivity R_{xy} and the vanishing of the diagonal resistivity R_{xx} around $\nu = 1/m$. As the filling factor changes further, a Hall plateau is broken eventually since too many quasiparticles are created to be localized by impurities. .

CHAPTER 12

QUASIPARTICLES AND ACTIVATION ENERGY

Quasiparticles are activated thermally. Their gap energy provides us with an indispensable information on their properties. However, it is not straightforward to compare experimental and theoretical results because the gap energy is reduced considerably due to impurity effects. Furthermore, a tunneling process may become important at low temperature and at strong magnetic field.

12.1 IMPURITY POTENTIALS

Quasiparticles are activated thermally at finite temperature T, and contribute to the diagonal current. It is experimentally known that the diagonal resistance is fitted by the Arrhenius formula,

$$R_{xx} \propto \exp\left(-\frac{\Delta_{\text{gap}}}{2k_B T}\right) \tag{12.1.1}$$

with k_B the Boltzmann constant. The gap energy Δ_{gap} is customarily called the activation energy. Its origin is the creation energy of a pair of quasihole (Δ_{qh}) and quasielectron (Δ_{qe}). Thus, by investigating the gap energy experimentally, we can guess which quasiparticles are excited.

However, the gap energy Δ_{gap} determined by the Arrhenius formula depends on sample qualities and is much smaller than the theoretical value. (There is an attempt to measure the gap energy by the phonon absorption,[343] which is not so sensitive to the sample quality.) A number of theories have been developed to explain the discrepancy by taking account of various effects in real systems. They include (i) finite thickness of the electron layer,[45,534,528,321] (ii) mixing of higher Landau levels,[526] and (iii) disorder in samples.[533,406,323,194,421] We follow the arguments given in reference.[421]

Phenomenologically the gap energy is well fitted by

$$\Delta_{\text{gap}} = \Delta_{\text{pair}} - \Gamma_{\text{offset}} \qquad \text{with} \quad \Delta_{\text{pair}} = \Delta_{\text{qh}} + \Delta_{\text{qe}}, \tag{12.1.2}$$

where the pair-creation energy Δ_{pair} includes a finite-thickness effect of the electron layer, and Γ_{offset} is a sample-dependent offset. The offset consists mainly of the impurity term $\Gamma_{\text{offset}}^{\text{imp}}$,

$$\Gamma_{\text{offset}} = \Gamma_{\text{offset}}^{\text{imp}} + \Gamma_{\text{offset}}^{\text{res}}, \tag{12.1.3}$$

where $\Gamma_{\text{offset}}^{\text{res}}$ is a residual term which may be necessary to fit experimental data in general. The impurity term is attributed to a Landau-level broadening made by disorder due to impurities in the bulk [Fig.10.3(c)], which are provided by the donors situated several hundreds of angstroms away from the electron layer [Fig.10.1]. The impurity term H_{imp} in the Hamiltonian reads

$$H_{\text{imp}} = e\int d^2x\, \rho(\boldsymbol{x})V_{\text{imp}}(\boldsymbol{x}), \tag{12.1.4}$$

where $V_{\text{imp}}(\boldsymbol{x})$ is the Coulomb potential made by impurities. For a single impurity at $\boldsymbol{x} = 0$, it may be approximated by

$$V_{\text{imp}}(\boldsymbol{x}) = \pm\frac{Ze}{4\pi\varepsilon}\frac{1}{\sqrt{|\boldsymbol{x}|^2 + d_{\text{imp}}^2}}, \tag{12.1.5}$$

where $\pm Ze$ is the impurity charge, ε is the dielectric constant ($4\pi\varepsilon \simeq 12.9$), and d_{imp} is the distance from the layer to the impurity in the bulk.

Pair creation of quasihole and quasielectron occurs to minimize the impurity term (12.1.4),

$$\Gamma_{\text{offset}}^{\text{imp}} \simeq e^*|V_{\text{imp}}(0)| = \frac{Zee^*}{4\pi\varepsilon d_{\text{imp}}}, \tag{12.1.6}$$

where a quasiparticle is assumed to be pointlike. Here, e^* is the electric charge of quasiholes, $e^* = e/m$ at the filling factor $\nu = n/m$ with odd m ($m = 1, 3, 5, \cdots$).

12.2 Gap Energies

A dissipationless current flows on QH states at zero temperature because all charged excitations are gapful. They are electrons, vortices, and skyrmions. Being excited at finite temperature, they are detected by magnetotransport experiments according to the Arrhenius formula (12.1.1).

Electron-Hole Excitation

Electrons are excited to a higher Landau level and spins are reversed at $\nu = 2, 4, \cdots$. The excitation energy consists of the cyclotron energy, the Coulomb energy and the Zeeman energy,

$$\Delta_{\text{gap}}^{\nu} = \hbar\omega_c + \alpha_{\text{pair}}^{\nu}E_C^0 - |g^*\mu_B B| - e|V_{\text{imp}}(0)|. \tag{12.2.1}$$

An estimation[254] yields that $\alpha_{\text{pair}}^{\nu=2} = \sqrt{\pi/8} \simeq 0.63$. We have used $\alpha_{\text{pair}}^{\nu=2} = 0.65$ to fit typical data due to Usher et al.[483] in Fig.12.1. The potential $V_{\text{imp}}(0)$ is taken phenomenologically as $e|V_{\text{imp}}(0)| \simeq 37.7$ K. It would imply $d_{\text{imp}}/Z \simeq 330$ Å in (12.1.5).

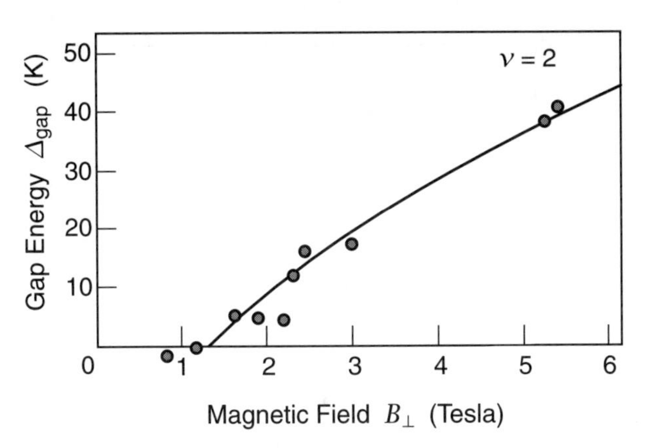

Fig. 12.1 A theoretical result is compared with experimental data at $\nu = 2$. The data are taken from Usher et al.[483] The points plotted are those obtained by subtracting $\hbar\omega_c - \Delta_Z$ from the observed ones. The theoretical curve is just for the Coulomb energy part and the impurity term in the electron-excitation formula (12.2.1). The impurity potential $V_{imp}(0)$ is taken common for all samples.

Vortex Excitation

Vortices are quasiparticles in fractional QH states. They have electric charges $\pm e^*$ at $\nu = n/m$, where $e^* = e/m$. The creation energy of a vortex pair is solely made of the Coulomb energy,

$$\Delta_{pair} = \alpha_{pair}^{1/m} E_C^0, \tag{12.2.2}$$

where $E_C^0 = e^2/4\pi\varepsilon\ell_B$ is the energy unit. There are several independent estimations on the numerical parameter $\alpha_{pair}^{1/3}$: $\alpha_{pair}^{1/3} \simeq 0.056$ according to Laughlin;[296] $\alpha_{pair}^{1/3} \simeq 0.053$ according to Chakraborty;[89] $\alpha_{pair}^{1/3} \simeq 0.094$ according to Morf and Halperin;[353] $\alpha_{pair}^{1/3} \simeq 0.105$ according to Haldane and Rezayi;[209] $\alpha_{pair}^{1/3} \simeq 0.106$ according to Girvin, MacDonald and Platzman.[188] Actual samples have finite layer widths, which decrease the direct Coulomb energy considerably. We treat $\alpha_{pair}^{1/m}$ as a phenomenological parameter to analyze experimental data. The gap energy reads

$$\Delta_{gap}^{1/m} = \alpha_{pair}^{1/m} E_C^0 - \frac{e}{m}|V_{imp}(0)|. \tag{12.2.3}$$

Typical data due to Boebinger et al.[70] are fitted by this formula, who measured the gap energies at $\nu = n/3$ by varying the magnetic field $B_\perp$ as well as the electron density ρ_0. In Fig.12.2 we have used $\alpha_{pair}^{1/3} = 0.056$, and taken the impurity potential common to all samples, $e|V_{imp}(0)| = 20.4$ K. It would imply $d_{imp}/Z \simeq 650$ Å if the impurity potential (12.1.5) is assumed.

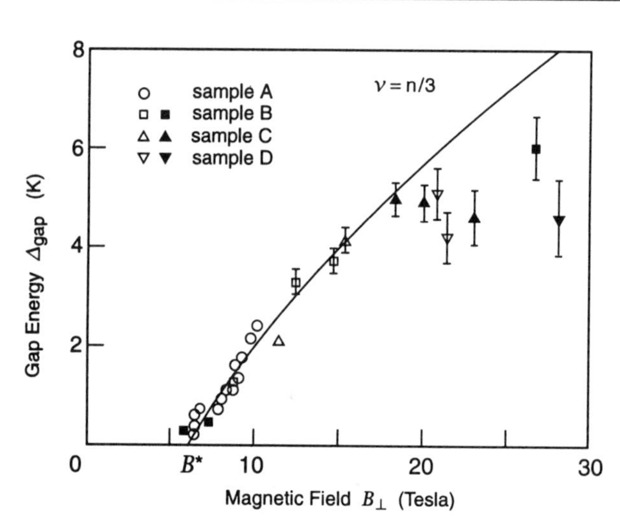

Fig. 12.2 A theoretical result is compared with experimental data for the activation energy at 1/3, 2/3, 4/3 and 5/3. The data are taken from G.S. Boebinger et al.[70] Theoretical curves are based on vortex-excitation formulas (12.2.3). The impurity potential $V_{\text{imp}}(0)$ is taken common for all samples. See also Fig.12.6 for the data at $B_\perp = 8.9$ T and $B_\perp = 20.9$ T.

Skyrmion Excitation

The $\nu = 1$ QH state is a QH ferromagnet, where a skyrmion-antiskyrmion pair is excited since their excitation energy is less than that of an electron-hole pair. The gap energy consists of the Coulomb and Zeeman energies, and summarized as

$$\Delta_{\text{gap}}^{\nu=1} \simeq 2\left\{\sqrt{\frac{\pi}{32}}\alpha + \frac{9\pi^2}{256\kappa}\right\} E_{\text{C}}^0 - e|V_{\text{imp}}(0)|, \tag{12.2.4}$$

where the skyrmion scale κ is determined by a competition between the direct Coulomb energy and the Zeeman energy,

$$\kappa \simeq \frac{1}{2}\left(\frac{3\pi^2}{64}\right)^{1/3}\left\{\tilde{g}\ln\left(\frac{\beta^2\sqrt{2\pi}}{32\tilde{g}} + 1\right)\right\}^{-1/3}, \tag{12.2.5}$$

where $\tilde{g} = \Delta_Z/E_{\text{C}}^0$ with the Zeeman gap $\Delta_Z = |g^*\mu_B B|$. We have introduced two phenomenological parameters α and β. See (16.3.16) and (16.3.15) in Section 16.3. We have used $\alpha \simeq 0.9$ and $\beta \simeq 1.478$ to fit typical data due to Schmeller et al.[429] in Fig.12.3. The potential $V_{\text{imp}}(0)$ is taken phenomenologically as $e|V_{\text{imp}}(0)| \simeq 40 \sim 50$ K. It would imply $d_{\text{imp}}/Z \simeq 260$ Å in (12.1.5).

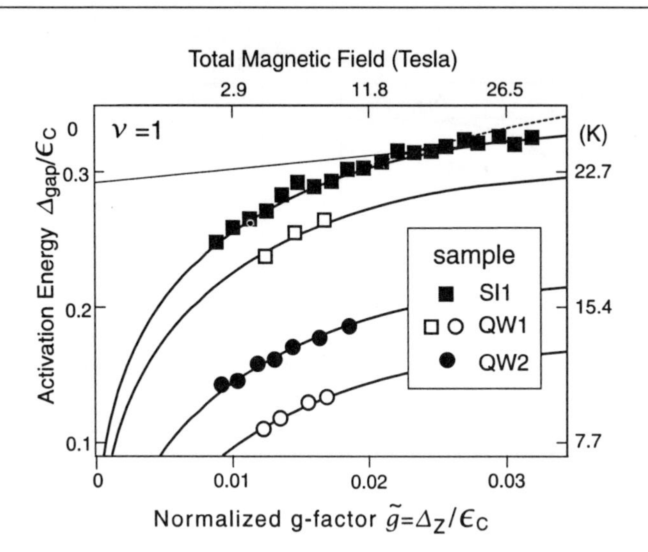

Fig. 12.3 A theoretical result is compared with experimental data taken from Schmeller et al.[429] By the dotted curve we give the activation energy calculated based on the formula (12.2.4). The thin line represents the hole excitation energy. The heavy solid curve is obtained by the numerical analysis made in Section 31.7. The offset Γ_{offset} increases as the mobility decreases. There are two curves for one sample (QW1) but with different mobilities. The mobility changes when electrons are pushed against the wall by a bias voltage, as will result in the increase of the Coulomb energy Γ_{offset} made by impurities.

12.3 Dispersion Relation

Thermal fluctuation activates a quasielectron out of the ground state, leaving behind a quasihole. They are created as electrically neutral objects. Having charges $\pm e^*$ in the magnetic field $B_\perp$, with $e^* = e/m$ at $\nu = n/m$, they feel the Coulomb attractive force as well as the Lorentz force. We examine classically the condition that these two forces are balanced,[254,527] and later check it quantum mechanically. Let $V_{\text{pair}}(r)$ be the potential energy of the quasiparticle pair with a separation r: The attractive force is $\partial V_{\text{pair}}(r)/\partial r$. The Lorentz forceis $e^* vB$ when the pair moves parallel to the x axis with velocity v. They are balanced when

$$\frac{\partial V_{\text{pair}}(r)}{\partial r} = e^* vB. \tag{12.3.1}$$

On the other hand they move with the group velocity

$$v = \frac{\partial E_{\text{pair}}(\boldsymbol{k})}{\hbar \partial k}, \tag{12.3.2}$$

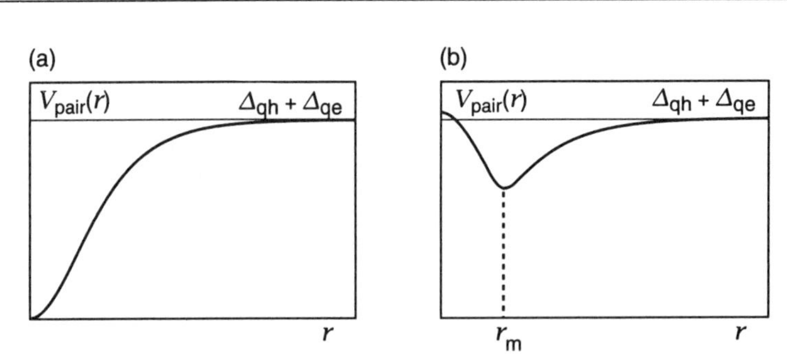

Fig. 12.4 The potential energy $V_{\text{pair}}(r)$ of a quasihole-quasielectron pair is illustrated. It may be regarded as the dispersion relation of a neutral excitation by the use of $r = mk\ell_B^2$, where the wave vector is given by $(k, 0)$. (a) The dispersion relation has a gapless mode, which is the case in QH ferromagnets. (b) The dispersion relation may have a minimum point (at $r = r_m$) describing a magnetoroton, as occurs when a short range repulsive interaction acts between a quasihole and a quasielectron.

where $E_{\text{pair}}(\boldsymbol{k})$ is the dispersion relation with $\boldsymbol{k} = (k, 0)$. Because the kinetic energy is quenched to an unimportant constant, the total energy E_{pair} is equal to the potential energy V_{pair},

$$E_{\text{pair}} = V_{\text{pair}}. \tag{12.3.3}$$

It follows from (12.3.1), (12.3.2) and (12.3.3) that

$$r = mk\ell_B^2. \tag{12.3.4}$$

The dispersion relation $E_{\text{pair}}(\boldsymbol{k})$ of a neutral excitation is obtainable from the potential energy $V_{\text{pair}}(r)$ by the use of this relation.

The potential energy $V_{\text{pair}}(r)$ may be approximated by

$$V_{\text{pair}}(r) \simeq \Delta_{\text{qh}} + \Delta_{\text{qe}} - \frac{e^{*2}}{4\pi\varepsilon r}, \tag{12.3.5}$$

for $r \gg \ell_B$. However, this is a poor approximation for small separation. Indeed, according to this formula, $V_{\text{pair}}(r)$ becomes negative for sufficiently small r. It is necessary to take into account an overlap of quasiparticles. quasiparticles are extended objects, vortices and skyrmions, described by classical fields. We place a quasihole at the origin ($\boldsymbol{x} = 0$) and a quasielectron at the point ($\boldsymbol{x} = \boldsymbol{r}$). The density modulation is $\Delta\rho_{\text{pair}}(\boldsymbol{x}; \boldsymbol{r}) = \Delta\rho_{\text{qh}}(\boldsymbol{x}) + \Delta\rho_{\text{qe}}(\boldsymbol{x} - \boldsymbol{r})$. The Coulomb energy is

$$V_{\text{pair}}(r) = \frac{e^2}{4\pi\varepsilon} \int d^2x d^2x' \frac{\Delta\rho_{\text{pair}}(\boldsymbol{x}; \boldsymbol{r})\Delta\rho_{\text{pair}}(\boldsymbol{x}'; \boldsymbol{r})}{2|\boldsymbol{x} - \boldsymbol{x}'|}. \tag{12.3.6}$$

It depends only on the distance r between two quasiparticles provided they have cylindrically symmetric configurations. This is the direct Coulomb energy of a quasiparticle pair. It is reduced to (12.3.5) when two quasiparticles are sufficiently apart. It is a dynamical problem how $\Delta\rho_{\text{pair}}(\boldsymbol{x};\boldsymbol{r})$ behaves as $\boldsymbol{r} \to 0$. We have $V_{\text{pair}}(r) = 0$ if the quasihole density is precisely cancelled by the quasielectron density, $\Delta\rho_{\text{qh}}(\boldsymbol{x}) = -\Delta\rho_{\text{qe}}(\boldsymbol{x})$, as illustrated in Fig.12.4(a). It implies the existence of a gapless mode in the dispersion relation $E_{\text{pair}}(k)$ via the relations (12.3.3) and (12.3.4).

If the spin degree of freedom is frozen, such a cancellation cannot occur since the QH state is incompressible: Otherwise, a gapless mode which can only exists in the density fluctuation would lead to compressibility. Hence, it must be that $\Delta\rho_{\text{pair}}(\boldsymbol{x};\boldsymbol{r}) \neq 0$ at $\boldsymbol{r} = 0$. When there exists a short-range repulsive interaction between a vortex and an antivortex, the energy $E_{\text{pair}}(r)$ may have a minimum describing a magnetoroton at $r = r_m \simeq \ell_B$ as in Fig.12.4(b).

In QH ferromagnets, on the contrary, the cancellation occurs because the dispersion relation contains a gapless mode [Fig.12.4(a)], as implied by the exchange Coulomb energy (9.1.15): See (15.5.11). A gapless mode develops in the spin fluctuation, and hence QH ferromagnets are incompressible in spite of the existence of a gapless mode. As we shall see later, by neglecting the Zeeman energy, the perturbative dispersion relation is given by (15.5.10),

$$E_{\text{pair}}(\boldsymbol{k}) = \frac{2J_s}{\rho_0}\boldsymbol{k}^2, \tag{12.3.7}$$

which implies

$$V_{\text{pair}}(r) = \frac{2J_s}{m^2\rho_0\ell_B^4}r^2 \qquad \text{at} \quad r \simeq 0, \tag{12.3.8}$$

where J_s is the spin stiffness (15.1.2). The dispersion relation for the gapless mode has also been analyzed numerically[107] at $\nu = 1$.

Quantum-Mechanical Analysis

We analyze the dynamics of a pair of quasihole and quasiparticle in magnetic field $B_\perp$. For simplicity we simulate them by pointlike particles carrying charges $\pm e^* = \pm e/m$. We assume that the pair has a static potential energy $V_{\text{pair}}(r)$ as a function of separation r, as illustrated in Figs.12.4 and 12.5.

The quantum-mechanical Hamiltonian is

$$\begin{aligned} H = &\frac{1}{2M}\left\{(p_{1x} + e^*B_\perp y_1)^2 + p_{1y}^2\right\} \\ &+ \frac{1}{2M}\left\{(p_{2x} - e^*B_\perp y_2)^2 + p_{2y}^2\right\} + V_{\text{pair}}(r) \end{aligned} \tag{12.3.9}$$

in the Landau gauge, where $\boldsymbol{p}_r = -i\hbar\nabla_r$ for each particle ($r = 1,2$). We introduce the center-of-mass coordinate $\boldsymbol{R} = \frac{1}{2}(\boldsymbol{x}_1 + \boldsymbol{x}_2) \equiv (X, Y)$ and the relative coordinate

$\boldsymbol{r} = \boldsymbol{x}_1 - \boldsymbol{x}_2 \equiv (x, y)$. The Hamiltonian is rewritten as

$$\begin{aligned} H = &\frac{1}{2M}\left[-\frac{\hbar^2}{2}\left(\frac{\partial^2}{\partial X^2} + \frac{\partial^2}{\partial Y^2} - \frac{4Y^2}{m^2\ell_B^4}\right)\right] \\ &+ \frac{1}{2M}\left[-2\hbar^2\left(\frac{\partial^2}{\partial x^2} + \frac{\partial^2}{\partial y^2} - \frac{y^2}{4m^2\ell_B^4}\right)\right] \\ &+ \frac{1}{2M}\left[\frac{2i\hbar^2}{m\ell_B^2}\left(\frac{y}{2}\frac{\partial}{\partial X} + 2Y\frac{\partial}{\partial x}\right)\right] + V_{\text{pair}}(\boldsymbol{r}). \end{aligned} \tag{12.3.10}$$

We search for a solution where two particles are bounded by the potential $V_{\text{pair}}(\boldsymbol{r})$ and move parallel to the x axis. Let the pair move with the wave vector $(k, 0)$: Its wave function is proportional to e^{ikX}. There are small fluctuations in the x component of the relative motion and in the Y component of the center-of-mass motion around the average values $\bar{x} = 0$ and $\bar{Y} = 0$. When we neglect them, the Hamiltonian is approximated by

$$H = \frac{\hbar^2}{M}\left[-\frac{\partial^2}{\partial y^2} + \frac{1}{4m^2\ell_B^4}(y + mk\ell_B^2)^2\right] + V_{\text{pair}}(y). \tag{12.3.11}$$

We are interested in physics taking a place in the lowest Landau level. The kinetic term is quenched, as requires the wave function to be

$$\mathfrak{S}(y) = e^{ikX}\exp\left[-\frac{1}{4m\ell_B^2}(y - \bar{y})^2\right], \tag{12.3.12}$$

with

$$\bar{y} = -mk\ell_B^2. \tag{12.3.13}$$

See also (10.4.7). The classical energy of the system is given by

$$E_{\text{pair}} = \langle H \rangle = V_{\text{pair}}(\bar{y}). \tag{12.3.14}$$

They agree with the classical relations (12.3.4) and (12.3.3).

12.4 Thermal Activation

We study thermal creation of quasiparticle pairs in QH ferromagnets with a gapless dispersion relation [Fig.12.4(a)]. We consider two cases. First we analyze a purely thermal process. We then include a tunneling process. The present analysis is applicable also to the system where quasiparticles are vortices without gapless modes [Fig.12.4(b)]. It is applicable also to certain integer QH systems, say at $\nu = 2, 4, \cdots$, where electrons are activated with quasiholes left behind.

Thermal Process

At a finite temperature T, thermal spin fluctuations occur with the rate proportional to the Boltzmann factor $\exp[-E_{\text{pair}}(\boldsymbol{k})/k_B T]$. A well-separated quasiparticle

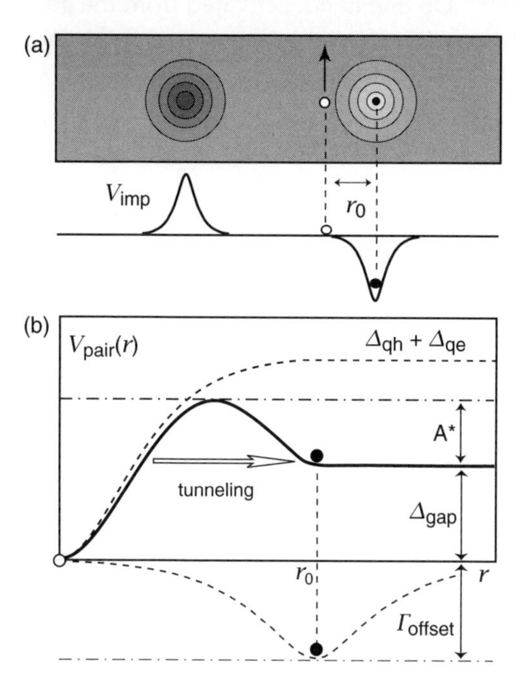

Fig. 12.5 (a) An impurity charge creates a Coulomb potential around it. It enhances a thermal activation of a quasiparticle-quasihole pair. A quasielectron is attracted and trapped by the Coulomb potential due to a positive impurity charge, while a quasihole is expelled by it. A quasihole contributes to an Ohmic current. (b) The creation energy $E_{pair}(r)$ of a quasiparticle pair is considerably reduced by the Coulomb potential due to an impurity charge. The gap energy of one pair is given by $\Delta_{gap} \simeq \Delta_{qh} + \Delta_{qe} - \Gamma_{offset}$, where Γ_{offset} is the energy gain due to the impurity potential. The effective range of an impurity is denoted by r_0.

pair ($r \to \infty$) is created with rate $\exp[-(\Delta_{qh} + \Delta_{qe})/k_BT]$, where use was made of (12.3.5).

Thermal activation of quasiparticles is greatly enhanced in the presence of impurities bearing electric charges [Fig.12.5]. An impurity charge creates a Coulomb potential around it. For definiteness we assume that it has a positive charge. As we have seen in Section 12.3, a thermal spin fluctuation creates a quasihole-quasielectron pair. The pair may be broken near an impurity because a quasielectron is attracted by the Coulomb force due to the impurity charge and a quasihole is expelled by it. The activation energy is given by (12.1.2), where Γ_{offset} is the energy gain (12.1.6) when the quasiparticle is trapped by a impurity potential [Fig.12.5(b)]. When a quasielectron is trapped by an impurity, only a quasihole moves and contributes to an Ohmic current [Fig.12.5(a)].

We estimate the number density of quasiparticles at temperature T, using a standard technique.[20] On one hand, activated from the ground state near an impurity, a quasiparticle is transferred to the center of the impurity [Fig.12.5]. The height of the potential barrier to jump over is $A^* + \Delta_{\text{gap}}$. The transition rate is

$$R_\uparrow = c\rho_0 \exp\left(-\frac{A^* + \Delta_{\text{gap}}}{k_B T}\right), \tag{12.4.1}$$

where c is a constant depending on the density of impurities. On the other hand, recombined with a quasihole, a quasielectron is transferred back to the ground state. The height of the potential barrier to jump over is A^*. The transition rate is

$$R_\downarrow = n_{\text{qh}} n_{\text{qe}} \sigma_{\text{pair}} \exp\left(-\frac{A^*}{k_B T}\right), \tag{12.4.2}$$

where n_{qh} and n_{qe} are the number densities of quasiholes and quasielectrons; σ_{pair} is a certain cross section. When the system is at thermal equilibrium there exists a detailed balance between these two transitions, $R_\uparrow = R_\downarrow$, from which we derive

$$n_{\text{qh}} n_{\text{qe}} = \frac{c\rho_0}{\sigma_{\text{pair}}} \exp\left(-\frac{\Delta_{\text{gap}}}{k_B T}\right). \tag{12.4.3}$$

Since quasiholes and quasiparticles are activated in pairs, we find

$$n_{\text{qh}} = n_{\text{qe}} = n_0 \exp\left(-\frac{\Delta_{\text{gap}}}{2k_B T}\right) \tag{12.4.4}$$

at the center of the Hall plateau, where $n_0 = \sqrt{c\rho_0/\sigma_{\text{pair}}}$. The Ohmic current is described by the formula (12.1.1) with (12.1.2) since it is proportional to the number density of quasiparticles.

The QH system is unstable when the gap energy Δ_{gap} becomes negative. QH states break down when

$$\Delta_{\text{qh}} + \Delta_{\text{qe}} < \Gamma_{\text{offset}}. \tag{12.4.5}$$

The creation energy of the pair is in general given by

$$\Delta_{\text{qh}} + \Delta_{\text{qe}} = \alpha B_\perp + \beta\sqrt{B_\perp}, \tag{12.4.6}$$

where α and β are numerical constants: See (12.2.2), (12.2.4) and (12.2.1). It decreases as the magnetic field decreases. The critical magnetic field $B_\perp^*$ is derived by solving $\alpha B_\perp + \beta\sqrt{B_\perp} = \Gamma_{\text{offset}}$. QH states do not exist for $B < B_\perp^*$, as is consistent with typical data [Figs.12.2, 12.3 and 12.1]. Breakdown of QH states has been studied in literature.[279,265,73,440]

Tunneling Process

We have so far considered a purely thermal process of pair creation. However, tunneling process enhances thermal activation at sufficiently low temperature. When a pair of quasiparticles acquires an energy Δ_{gap} thermally, it can tunnel

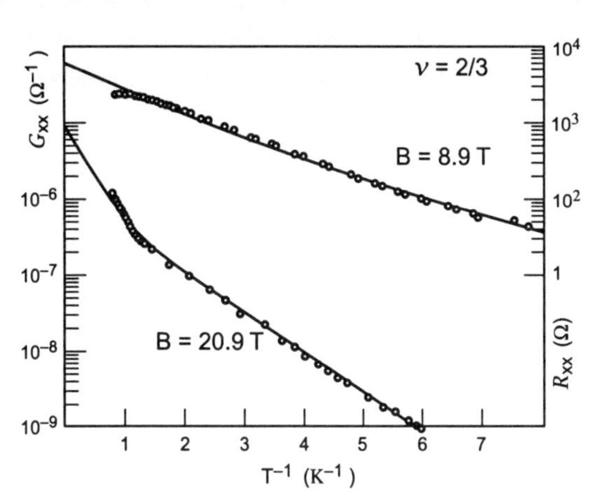

Fig. 12.6 Temperature dependence of the minimum of the diagonal conductance G_{xx} and resistance R_{xx} at $\nu = 2/3$. The data are taken from G.S. Boebinger et al.[70] Theoretical curves are given by the generalized the formula (12.4.9) with vortex excitation (12.2.3). See also Fig.12.2.

across the potential barrier with height A^* as in Fig.12.5. The transition rate is

$$R_{\uparrow}^{\text{tunnel}} = c\rho_0 e^{-S_{\text{tunnel}}/\hbar} \exp\left(-\frac{\Delta_{\text{gap}}}{k_B T}\right), \tag{12.4.7}$$

where S_{tunnel} is the Euclidean action for the tunneling process. It depends on the height A^* and the range r_0. It is obvious that the transition rate (12.4.7) dominates the rate (12.4.1) as $T \to 0$. The rate of recombination process is still given by (12.4.2), because of the plateau in the potential for $r > r_0$ in Fig.12.5. The detailed balance implies

$$R_{\uparrow} + R_{\uparrow}^{\text{tunnel}} = R_{\downarrow}, \tag{12.4.8}$$

with (12.4.1), (12.4.2) and (12.4.7), from which we obtain

$$n_{\text{qh}} = n_{\text{qe}} = n_0 \exp\left(-\frac{\Delta_{\text{gap}}}{2k_B T}\right)\left[1 + e^{-S_{\text{tunnel}}/\hbar} \exp\left(\frac{A^*}{k_B T}\right)\right]^{1/2}. \tag{12.4.9}$$

Hence, the Ohmic current is described by the formula

$$R_{xx} \propto \exp\left(-\frac{\Delta_{\text{gap}}}{2k_B T}\right)\left[1 + e^{-S_{\text{tunnel}}/\hbar} \exp\left(\frac{A^*}{k_B T}\right)\right]^{1/2}. \tag{12.4.10}$$

This formula contains two energy scales Δ_{gap} and A^*. Typical data due to Boebinger et al.[70] are fitted in Fig.12.6. In so doing we have determined Δ_{gap} by the theoretical formula (12.2.3), $\Delta_{\text{gap}} = \Delta_{\text{gap}}^{1/3}$ by using $\Gamma_{\text{offset}} = 6.8$ K. The theoretical curve (for $B = 8.9$ T) is obtained by using $\Delta_{\text{gap}} \simeq 1.7$ K, $A^* \simeq 0.69$ K and

$S_{\text{tunnel}}/\hbar \simeq 2.0$. The theoretical curve (for $B = 20.9$ T) is obtained by using $\Delta_{\text{gap}} \simeq 6.1$ K, $A^* \simeq 3.7$ K and $S_{\text{tunnel}}/\hbar \simeq 4.0$. The tunneling process makes an important contribution at strong magnetic field because A^* becomes larger.

CHAPTER 13

FIELD THEORY OF COMPOSITE PARTICLES

The composite-particle picture is essential for understanding various aspects of QH effects. A composite particle is obtained by attaching m quanta of the Chern-Simons flux to an electron: It is a *composite boson* when $m = 2p+1 =$ odd, and a *composite fermion* when $m = 2p =$ even. It acquires a physical reality by trading the Chern-Simons flux for the magnetic flux in the QH state. The $\nu = 1/(2p+1)$ fractional QH state may be viewed in the following two ways: (A) It is a Bose liquid of composite bosons, where they experience effectively no magnetic field and condense. (B) It is an integer QH state of composite fermions, where they undergo cyclotron motion in a reduced effective magnetic field corresponding to the effective filling factor $\nu_{\text{eff}} = 1$.

13.1 COMPOSITE PARTICLES

A new type of particles, called *composite particles*, emerge in the QH system. A composite particle is an electron bound to m quanta ($m\Phi_{\text{D}}$) of the Chern-Simons flux. The attached flux may alter the exchange statistics of electrons, as was reviewed in Chapter 8. It is a boson (*composite boson*) when $m = 2p+1 =$ odd, and a fermion (*composite fermion*) when $m = 2p =$ even. The concept of composite particles is the key to the fractional QH effect, just as the concept of the Cooper pairs is the key to superconductivity.

We represent an electron as a bound state of a composite particle and the Chern-Simons flux. This is possible in two dimensions (see Section 8.1). The system of electrons in the magnetic field turns out to be the system of composite particles in the magnetic flux and the Chern-Simons flux, where composite particles experience the effective magnetic field, $B_\perp^{\text{eff}} \equiv B_\perp - m\Phi_{\text{D}}\rho_0$.

Attaching an odd number of flux quanta ($m = 2p+1 =$ odd), we obtain composite bosons. At $\nu \equiv \Phi_{\text{D}}\rho_0/B_\perp = 1/(2p+1)$, the effective magnetic field vanishes,

$$B_\perp^{\text{eff}} \equiv B_\perp - (2p+1)\Phi_{\text{D}}\rho_0 = 0. \tag{13.1.1}$$

Composite bosons condense to generate the QH state in the zero effective magnetic

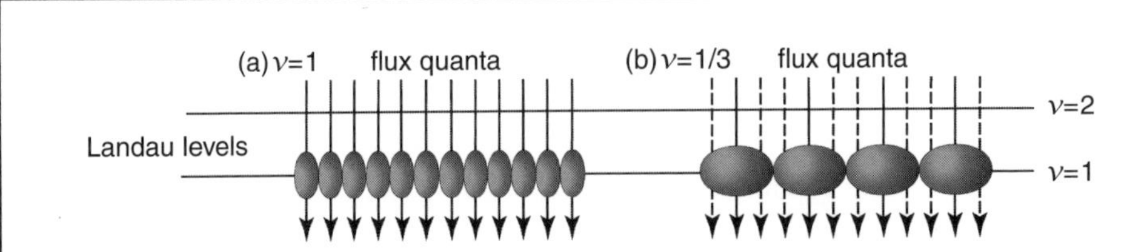

Fig. 13.1 (a) Electrons are bosonized at $\nu = 1$ by attaching one flux quantum to each electron. (b) Electrons are transformed into composite fermions at $\nu = 1/(2p+1)$, by attaching $2p$ flux quanta (dotted lines) to each electron. Here $p = 1$. A composite fermion behaves as a "fat electron" occupying $(2p+1)$ sites. Composite fermions fill the lowest Landau level, generating the $\nu_{\text{eff}} = 1$ integer QH state. Composite fermions are bosonized at $\nu_{\text{eff}} = 1$ by attaching one residual flux quantum (solid line) to each composite fermion.

field. This is the composite-boson theory[190,535,402,134,137,536] of the QH system.

Attaching an even number of flux quanta ($m = 2p$ = even), we obtain composite fermions. At $\nu \equiv \Phi_D\rho_0/B_\perp = 1/(2p+1)$, they experience a nonzero effective magnetic field,

$$B_\perp^{\text{eff}} \equiv B_\perp - 2p\Phi_D\rho_0 = \frac{1}{2p+1}B_\perp. \tag{13.1.2}$$

The effective filling factor of composite fermions is

$$\nu_{\text{eff}} = \frac{2\pi\hbar\rho_0}{eB_\perp^{\text{eff}}} = (2p+1)\nu. \tag{13.1.3}$$

Hence, the *fractional* QH state of electrons at $\nu = 1/(2p+1)$ is the *integer* QH state of composite fermions at $\nu_{\text{eff}} = 1$. This is the composite-fermion theory[241,316,256] of the QH system.

It is useful to bosonize electrons even at $\nu = 1$, since boson fields are easy to handle semiclassically [Fig.13.1(a)]. Indeed, this enables us to study vortex excitations as well as skyrmion excitations semiclassically. It is clear that composite bosons obey the Pauli exclusion principle at $\nu = 1$. The statistical transmutation does not change the exclusion property.[211]

There exist ample experimental evidences that composite fermions make cyclotron motion in the reduced effective magnetic field (13.1.2): See Section 17.1. They indicate that a composite fermion is a physical object and behaves as a "fat electron" occupying $(2p+1)$ sites [Fig.13.1(b)]. Composite fermions, also called superelectrons in reference,[515] have different exclusion statistics from electrons.

A composite fermion can be bosonized by attaching one residual flux [Fig.13.1(b)]. The composite boson thus obtained is the same with the one constructed by attaching all $(2p+1)$ flux quanta to an electron. Composite fermions and composite bosons obey the same exclusion statistics. They are two ways of representing one physical object, the composite particle.

We denote by $\varphi(\boldsymbol{x})$ a field operator describing the composite particle. The wave function of a generic N-body state $|\mathfrak{S}\rangle$ is defined by

$$\mathfrak{S}_{\rm CP}[\boldsymbol{x}] \equiv \langle 0|\varphi(\boldsymbol{x}_1)\cdots\varphi(\boldsymbol{x}_N)|\mathfrak{S}\rangle \tag{13.1.4}$$

in terms of *composite particles*, as it is defined by

$$\mathfrak{S}[\boldsymbol{x}] \equiv \langle 0|\psi(\boldsymbol{x}_1)\cdots\psi(\boldsymbol{x}_N)|\mathfrak{S}\rangle \tag{13.1.5}$$

in terms of *electrons*. The field operator for composite particles is by no means unique. With a judicious choice[402,398] we obtain a remarkable relation[147] between $\mathfrak{S}[\boldsymbol{x}]$ and $\mathfrak{S}_{\rm CP}[\boldsymbol{x}]$,

$$\mathfrak{S}[\boldsymbol{x}] = \mathfrak{S}_{\rm CP}[\boldsymbol{x}]\prod_{r<s}(z_r - z_s)^m \exp\left(-\frac{eB_\perp}{4\hbar}\sum_{k=1}^{N} r_k^2\right), \tag{13.1.6}$$

where m is an integer. When all electrons are in the lowest Landau level, the wave function $\mathfrak{S}(\boldsymbol{x})$ is given by (11.3.1), and hence $\mathfrak{S}_{\rm CP}[\boldsymbol{x}]$ is an analytic function, $\mathfrak{S}_{\rm CP}[\boldsymbol{x}] = \rho_0^{N/2}\omega[z]$.

When we take $m = 2p+1$ in (13.1.6), $\mathfrak{S}_{\rm CP}[\boldsymbol{x}]$ is a wave function of composite bosons. Let us denote it by $\mathfrak{S}_{\rm CB}[\boldsymbol{x}]$. The Laughlin state (11.3.2) is given simply by choosing

$$\mathfrak{S}_{\rm CB}[\boldsymbol{x}] = \rho_0^{N/2}\omega_{\rm LN}[z] = \rho_0^{N/2} \tag{13.1.7}$$

for composite bosons. The power of z_r in $\omega[z]$ is the angular momentum carried by the rth composite boson. Consequently, the Laughlin state (13.1.7) is the one where all composite bosons condense into the angular-momentum zero state, as anticipated. Similarly, a vortex state with $\omega[z] = \prod_r z_r$ is the one where all composite bosons condense into the angular-momentum one state. Composite bosons represent the physical degree of freedom describing solely the deviation $\omega[z]$ from the ground state at $\nu = 1/(2p+1)$ in the generic wave function (11.3.10).

When we take $m = 2p$ in (13.1.6), $\mathfrak{S}_{\rm CP}[\boldsymbol{x}]$ is a wave function of composite fermions. The Laughlin state (11.3.2) is given simply by choosing

$$\mathfrak{S}_{\rm CF}[\boldsymbol{x}] = \rho_0^{N/2}\omega_{\rm Jain}[z] = \rho_0^{N/2}\prod_{r<s}(z_r - z_s) \tag{13.1.8}$$

for composite fermions. It was interpreted by Jain[241] that this represents the integer QH state of composite fermions at $\nu_{\rm eff} = 1$.

However, it is not clear in the wave function (13.1.8) that composite fermions experience a nonzero effective magnetic field $B_\perp^{\rm eff}$ and constitutes the $\nu_{\rm eff} = 1$ integer QH state. It is possible to introduce another composite-fermion field operator

$\bar{\psi}(\mathbf{x})$, with which the wave function of the Laughlin state reads

$$\begin{aligned}\mathfrak{S}_{\rm CF}[\mathbf{x}] &\equiv \langle 0|\bar{\psi}(\mathbf{x}_1)\cdots\bar{\psi}(\mathbf{x}_N)|\mathfrak{S}\rangle \\ &= \rho_0^{N/2}\prod_{r<s}(z_r - z_s)\exp\left(-\frac{eB_\perp^{\rm eff}}{4\hbar}\sum_{k=1}^{N} r_k^2\right)\end{aligned} \tag{13.1.9}$$

instead of (13.1.8), together with an appropriate modification of the relation (13.1.6). This is the Laughlin wave function at $\nu_{\rm eff} = 1$. Since composite fermions fill the lowest Landau level, it turns out that each of them occupies $2p+1$ Landau sites at $\nu = 1/(2p+1)$, as illustrated in Fig.13.1(b).

13.2 Statistical Transmutation

Let $\psi(\mathbf{x})$ be the electron field. A composite-particle field $\phi(\mathbf{x})$ is defined by an operator phase transformation,

$$\phi(\mathbf{x}) = e^{-im\Theta(\mathbf{x})}\psi(\mathbf{x}). \tag{13.2.1}$$

The phase field $\Theta(\mathbf{x})$ is defined by

$$\Theta(\mathbf{x}) = \int d^2y\,\theta(\mathbf{x}-\mathbf{y})\rho(\mathbf{y}), \tag{13.2.2}$$

where m is an integer and $\theta(\mathbf{x}-\mathbf{y})$ is the angle made between the vector $\mathbf{x}-\mathbf{y}$ and the x axis. We call $\phi(\mathbf{x})$ the *bare* field, in contrast to the other operator, the *dressed* field $\varphi(\mathbf{x})$, to be introduced in Section 13.4.

The phase field (13.2.2) is a multivalued field,

$$\varepsilon_{jk}\partial_j\partial_k\Theta(\mathbf{x}) = 2\pi\rho(\mathbf{x}), \tag{13.2.3}$$

where we have used the relation $\varepsilon_{jk}\partial_j\partial_k\theta(\mathbf{x}) = 2\pi\delta(\mathbf{x})$: See Appendix E. Hence, under the phase transformation (13.2.1), the Coulomb term (10.1.7) and the Zeeman term (10.1.14) are invariant but the kinetic term (10.1.13) is affected. We study the kinetic term in Section 13.3.

We derive the commutation relations for the composite-particle field from those for the electron field. The analysis is essentially the same as the one presented in Section 8.5. Equation (13.2.1) is identical to (8.5.1), when we set $\alpha = m$ and regard the electron as an anyon. In the rest of this section we use the natural unit ($\hbar = c = 1$) for notational simplicity.

Using

$$[\Theta(\mathbf{y}),\psi(\mathbf{x})] = \int d^2z\;\theta(\mathbf{y}-\mathbf{z})[\rho(\mathbf{z}),\psi(\mathbf{x})] = -\theta(\mathbf{y}-\mathbf{x})\psi(\mathbf{x}), \tag{13.2.4}$$

we derive

$$e^{im\Theta(\mathbf{y})}\psi(\mathbf{x})e^{-im\Theta(\mathbf{y})} = e^{-im\theta(\mathbf{y}-\mathbf{x})}\psi(\mathbf{x}) \tag{13.2.5}$$

as in (8.5.6). With this formula we get

$$\begin{aligned}\phi(x)\phi(y) &= e^{-im\Theta(x)}\psi(x)e^{-im\Theta(y)}\psi(y) \\ &= e^{-im\theta(y-x)}e^{-im\Theta(x)}e^{-im\Theta(y)}\psi(x)\psi(y),\end{aligned} \tag{13.2.6}$$

and

$$\begin{aligned}\phi(y)\phi(x) &= e^{-im\theta(x-y)}e^{-im\Theta(y)}e^{-im\Theta(x)}\psi(y)\psi(x) \\ &= (-1)^{m+1}\phi(x)\phi(y).\end{aligned} \tag{13.2.7}$$

We have used $\psi(x)\psi(y) = -\psi(y)\psi(x)$ and $\theta(x-y) = \theta(y-x) + \pi$ in the last equality (see Appendix E). Hence,

$$[\phi(x), \phi(y)] = [\phi^\dagger(x), \phi^\dagger(y)] = 0 \qquad \text{for} \quad m = \text{odd}, \tag{13.2.8a}$$

and

$$\{\phi(x), \phi(y)\} = \{\phi^\dagger(x), \phi^\dagger(y)\} = 0 \qquad \text{for} \quad m = \text{even}. \tag{13.2.8b}$$

We also evaluate

$$\phi(x)\phi^\dagger(y) = e^{-im\Theta(x)}\psi(x)\psi^\dagger(y)e^{im\Theta(y)}. \tag{13.2.9}$$

We substitute $\psi(x)\psi^\dagger(y) = -\psi^\dagger(y)\psi(x) + \delta(x-y)$ into it,

$$\phi(x)\phi^\dagger(y) = -e^{-im\Theta(x)}\psi^\dagger(y)\psi(x)e^{im\Theta(y)} + \delta(x-y). \tag{13.2.10}$$

We use (13.2.5) to find

$$\phi(x)\phi^\dagger(y) = -e^{im\theta(y-x)}e^{-im\Theta(x)}\psi^\dagger(y)e^{im\Theta(y)}\psi(x) + \delta(x-y). \tag{13.2.11}$$

On the other hand,

$$\begin{aligned}\phi^\dagger(y)\phi(x) &= \psi^\dagger(y)e^{im\Theta(y)}e^{-im\Theta(x)}\psi(x) \\ &= e^{im\theta(x-y)}e^{-im\Theta(x)}\psi^\dagger(y)e^{im\Theta(y)}\psi(x),\end{aligned} \tag{13.2.12}$$

where use was made of the Hermitian conjugate of (13.2.5). Comparing these two equations we obtain

$$[\phi(x), \phi^\dagger(y)] = \delta(x-y) \qquad \text{for} \quad m = \text{odd}, \tag{13.2.13a}$$

and

$$\{\phi(x), \phi^\dagger(y)\} = \delta(x-y) \qquad \text{for} \quad m = \text{even}. \tag{13.2.13b}$$

These are the commutation relations of the composite-particle field.

We conclude that the composite-particle field ϕ is bosonic when m = odd, and that it is fermionic when m = even. Accordingly, we call ϕ the composite-boson field or the composite-fermion field. Although the field ϕ is bosonic with m odd, composite bosons are subject to the exclusion principle. This is seen by taking the limit $y \to x$ in (13.2.6). Though the factor $e^{-im\theta(x-y)}$ gets ill defined, it is bounded. Therefore, the exclusion principle $\psi(x)^2 = 0$ implies

$$\phi(x)^2 = 0. \tag{13.2.14}$$

This should be the case physically since they are constructed simply by attaching flux to electrons. Moreover, the statistical transformation does not change the spin, as we shall see in (15.4.4).

Wave Functions

We define the wave function of composite particles and relate it to the wave function of electrons. In terms of bare composite particles the wave function is defined by

$$\check{\mathfrak{S}}_{\text{CP}}[x] = \langle 0|\phi(x_1)\phi(x_2)\cdots\phi(x_N)|\mathfrak{S}\rangle. \tag{13.2.15}$$

The N-body state $|\mathfrak{S}\rangle$ associated with this wave function is given by

$$|\mathfrak{S}\rangle = \int [dx]\check{\mathfrak{S}}_{\text{CP}}[x]\phi^\dagger(x_1)\phi^\dagger(x_2)\cdots\phi^\dagger(x_N)|0\rangle, \tag{13.2.16}$$

as is verified based on the commutation relation (13.2.13). We are able to rewrite this as

$$|\mathfrak{S}\rangle = \int [dx]\mathfrak{S}[x]\psi^\dagger(x_1)\psi^\dagger(x_2)\cdots\psi^\dagger(x_N)|0\rangle \tag{13.2.17}$$

with

$$\mathfrak{S}[x] = e^{im\sum_{r<s}\theta(z_r-z_s)}\check{\mathfrak{S}}_{\text{CP}}[x]. \tag{13.2.18}$$

Equation (13.2.17) is derived from (13.2.16) as follows. We start with the Hermitian conjugate of (13.2.6),

$$\phi^\dagger(y)\phi^\dagger(x) = e^{im\theta(y-x)}\psi^\dagger(y)\psi^\dagger(x)e^{i\Theta(y)}e^{i\Theta(x)}. \tag{13.2.19}$$

Since the vacuum $|0\rangle$ contains no electrons, it satisfies $\rho(x)|0\rangle = 0$, or $\Theta(x)|0\rangle = 0$ due to (13.2.2). Hence, we find

$$\phi^\dagger(y)\phi^\dagger(x)|0\rangle = e^{im\theta(y-x)}\psi^\dagger(y)\psi^\dagger(x)|0\rangle. \tag{13.2.20}$$

In general we obtain

$$\phi^\dagger(x_1)\phi^\dagger(x_2)\cdots\phi^\dagger(x_N)|0\rangle = e^{im\sum_{r<s}\theta(z_r-z_s)}\psi^\dagger(x_1)\psi^\dagger(x_2)\cdots\psi^\dagger(x_N)|0\rangle. \tag{13.2.21}$$

This proves (13.2.17) with (13.2.18).

13.3 EFFECTIVE MAGNETIC FIELD

By substituting (13.2.1) into the kinetic Hamiltonian (10.6.1b) we obtain

$$H = \frac{1}{2M}\int d^2x\phi^\dagger(\mathbf{x})(\check{P}_x - i\check{P}_y)(\check{P}_x + i\check{P}_y)\phi(\mathbf{x}) + \frac{1}{2}N\hbar\omega_c. \tag{13.3.1}$$

This is the Hamiltonian for bare composite particles. The density operator is

$$\rho(\mathbf{x}) = \psi^\dagger(\mathbf{x})\psi(\mathbf{x}) = \phi^\dagger(\mathbf{x})\phi(\mathbf{x}). \tag{13.3.2}$$

The covariant momentum is

$$\check{P}_k \equiv P_k + eC_k = -i\hbar\frac{\partial}{\partial x^k} + e\mathcal{A}_k \tag{13.3.3}$$

with

$$C_k(\mathbf{x}) \equiv \frac{\hbar m}{e}\partial_k\Theta(\mathbf{x}) = m\frac{\hbar}{e}\int d^2y\partial_k\theta(\mathbf{x}-\mathbf{y})\rho(\mathbf{y}) \tag{13.3.4}$$

and

$$\mathcal{A}_j(\mathbf{x}) \equiv A_j^{\text{ext}}(\mathbf{x}) + C_j(\mathbf{x}). \tag{13.3.5}$$

The field $C_k(\mathbf{x})$ is an auxiliary field determined solely by the density $\rho(\mathbf{x})$. By comparing (13.3.4) with (8.4.12), it is identified with the Chern-Simons field.

The Chern-Simons field $C_k(\mathbf{x})$ makes the minimal coupling with the composite-particle field, and composite particles experience it as a "magnetic field". The Chern-Simons flux is an ensemble of infinitely thin solenoids,

$$\varepsilon_{jk}\partial_j C_k(\mathbf{x}) = \frac{\hbar m}{e}\varepsilon_{jk}\partial_j\partial_k\Theta(\mathbf{x}) = m\Phi_D\delta(\mathbf{x}), \tag{13.3.6}$$

where we have used (13.2.3); Φ_D is the Dirac flux unit.

The Lagrangian density is given by

$$\mathcal{L} = \phi^\dagger(i\hbar\partial_t - ceC_0 - ceA_0^{\text{ext}})\phi - \mathcal{H} + \mathcal{L}_{CS} \tag{13.3.7}$$

with

$$\mathcal{L}_{CS} = -\frac{e^2}{4m\pi\hbar}\varepsilon^{\mu\nu\lambda}C_\mu\partial_\nu C_\lambda. \tag{13.3.8}$$

It is easy to see that the Hamiltonian (13.3.1), the canonical momentum (13.3.3) and the commutation relation (13.2.13) follow from this Lagrangian by the standard procedure described in Part I. The constraint condition (13.3.6) for the Chern-Simons field is derived as an Euler-Lagrange equation. See Section 8.4 for details.

The effect of the operator phase transformation (13.2.1) is to attach m flux quanta to each electron. Composite particles experience the effective magnetic field $\mathcal{B}_{\text{eff}}(\mathbf{x})$ described by the potential $\mathcal{A}_j(\mathbf{x})$,

$$\mathcal{B}_{\text{eff}}(\mathbf{x}) = -\varepsilon_{ij}\partial_i\mathcal{A}_j(\mathbf{x}) = B_\perp - m\phi_D\rho(\mathbf{x}). \tag{13.3.9}$$

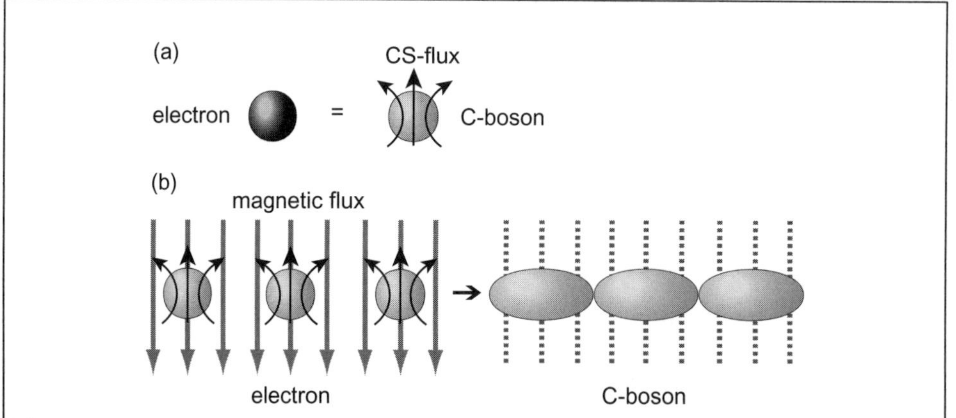

Fig. 13.2 (a) An electron is a bound state of a composite boson (C-boson) and the Chern-Simons flux (CS-flux). (b) The Chern-Simons flux is cancelled by the external magnetic flux, and composite bosons are left over to form the $\nu = 1/m$ QH state. Here, $m = 3$.

The effective magnetic flux is the sum of the real magnetic flux and the Chern-Simons flux. It follows from this equation that the effective magnetic field vanishes, $\langle \mathcal{B}_{\text{eff}} \rangle_g = 0$, on the homogeneous state, $\rho_0 = \langle \rho(\boldsymbol{x}) \rangle$, at the filling factor

$$\nu \equiv \frac{\rho_0 \phi_D}{B_\perp} = \frac{1}{m}. \tag{13.3.10}$$

It is equivalent to say that the Chern-Simons flux is cancelled by the magnetic flux on the homogeneous state [Fig.13.2]. The Chern-Simons flux is traded for the magnetic flux, and composite bosons acquire a physical reality. For odd m the homogeneous state is a Bose liquid upon which the QH effect is realized, while for even m it is a Fermi liquid. See Chapter 14 and 17 for more details.

Since $\mathcal{A}_j(\boldsymbol{x})$ satisfies the Coulomb gauge condition,

$$\partial_j \mathcal{A}_j(\boldsymbol{x}) = 0, \tag{13.3.11}$$

we may express it as

$$\mathcal{A}_j(\boldsymbol{x}) = -\frac{\hbar}{e} \varepsilon_{jk} \partial_k \mathcal{A}(\boldsymbol{x}) \tag{13.3.12}$$

in terms of a scalar field $\mathcal{A}(\boldsymbol{x})$. This is only possible in the planar geometry. The field $\mathcal{A}(\boldsymbol{x})$ is regarded as the scalar potential of the effective magnetic field,

$$\mathcal{B}_{\text{eff}}(\boldsymbol{x}) = -\frac{\hbar}{e} \nabla^2 \mathcal{A}(\boldsymbol{x}). \tag{13.3.13}$$

From (13.3.5) with (10.3.4) and (13.3.4), we obtain (see Appendix E)

$$\partial_j \mathcal{A}(\mathbf{x}) = m\varepsilon_{jk}\int d^2y \partial_k \theta(\mathbf{x}-\mathbf{y})\rho(\mathbf{y}) - \frac{1}{2\ell_B^2}x_j \tag{13.3.14a}$$

$$= m\partial_j \int d^2y \ln\left(\frac{|\mathbf{x}-\mathbf{y}|}{2\ell_B}\right)\rho(\mathbf{y}) - \partial_j |z|^2, \tag{13.3.14b}$$

and hence

$$\mathcal{A}(\mathbf{x}) = m\int d^2y \ln\left(\frac{|\mathbf{x}-\mathbf{y}|}{2\ell_B}\right)\rho(\mathbf{y}) - |z|^2 \tag{13.3.15}$$

up to an integration constant.

13.4 Dressed Composite Particles

A composite particle has been constructed by attaching m Dirac flux quanta to an electron by way of a singular gauge transformation. However, as we shall see in (13.5.9), its N-body wave function is more complicated than the Laughlin wave function itself. Furthermore, as we shall see in (14.1.10), it brings in singularities,

$$e^{im\Theta(\mathbf{x})} = \prod_r e^{im\theta(\mathbf{x}-\mathbf{x}_r)}, \tag{13.4.1}$$

to the ground-state wave function in the lowest order of the semiclassical analysis at $\nu = 1/m$ with m odd. We now introduce another composite-particle operator free from these difficulties.[402,398,439,145]

Since the key property of statistical transmutation is the multivalueness of the azimuthal angle $\theta(\mathbf{x})$, it is free to include any single-valued operator to define the composite-particle operator. We define[1]

$$\varphi(\mathbf{x}) = e^{-\mathcal{A}(\mathbf{x})}\phi(\mathbf{x}) = e^{-im\Theta(\mathbf{x})-\mathcal{A}(\mathbf{x})}\psi(\mathbf{x}), \tag{13.4.2}$$

in terms of the bare composite-particle field $\phi(\mathbf{x})$ and the auxiliary field $\mathcal{A}(\mathbf{x})$. In the basis where electron positions are diagonalized, we find

$$e^{\mathcal{A}(\mathbf{x})} \propto \prod_r |\mathbf{x}-\mathbf{x}_r|^m e^{-|z|^2}, \tag{13.4.3}$$

where we have used (13.3.15). It removes the singularities at $\mathbf{x} = \mathbf{x}_r$ in (13.4.1), mimicking precisely the repulsive effect induced by the Aharonov-Bohm phase due to the attached flux. We call $\varphi(\mathbf{x})$ the *dressed field* since it is constructed by dressing the bare operator $\phi(\mathbf{x})$ with a cloud of the effective magnetic flux described by $\mathcal{A}(\mathbf{x})$.

By substituting (13.4.2) into (10.6.1b) the kinetic Hamiltonian reads

$$H_K = \frac{1}{2M}\int d^2x \varphi^\ddagger(\mathbf{x})(\mathcal{P}_x - i\mathcal{P}_y)(\mathcal{P}_x + i\mathcal{P}_y)\varphi(\mathbf{x}) + \frac{1}{2}N\hbar\omega_c, \tag{13.4.4}$$

[1] This formula is good for the composite-boson field as it stands but needs a slight modification for the composite-fermion field [see (13.6.2)].

where we have defined

$$\varphi^\ddagger(x) \equiv \varphi^\dagger(x)e^{2\mathcal{A}(x)} = \psi^\dagger(x)e^{im\Theta(x)+\mathcal{A}(x)}. \tag{13.4.5}$$

The density operator is

$$\rho(x) = \psi^\dagger(x)\psi(x) = \varphi^\ddagger(x)\varphi(x). \tag{13.4.6}$$

The covariant momentum is

$$\mathcal{P}_k = -i\hbar\partial_k + eA_k^{\text{ext}}(x) + eC_k(x) - i\hbar\partial_k\mathcal{A}(x) \equiv -i\hbar\partial_k + e\tilde{\mathcal{A}}_k(x), \tag{13.4.7}$$

with

$$\tilde{\mathcal{A}}_\mu(x) = A_\mu^{\text{ext}}(x) + C_\mu(x) - i\frac{\hbar}{e}\partial_\mu\mathcal{A}(x), \tag{13.4.8}$$

in terms of the external electromagnetic potential $A_k^{\text{ext}}(x)$, the Chern-Simons field $C_k(x) = (m/e)\hbar\partial_k\Theta(x)$ and the auxiliary field $\mathcal{A}(x)$. It is easy to show

$$\mathcal{P}_x + i\mathcal{P}_y = -\frac{i\hbar}{\ell_B}\frac{\partial}{\partial z^*}, \tag{13.4.9}$$

by using (13.2.2) and (13.3.14a), which plays an important role in constructing the states in the lowest Landau level: See for instance (13.5.7).

Dressed composite particles experience the same effective magnetic field that bare composite particles do,

$$\mathcal{B}_{\text{eff}}(x) = -\varepsilon_{ij}\partial_i\tilde{\mathcal{A}}_j(x) = B_\perp - m\phi_D\rho(x). \tag{13.4.10}$$

They experience the same electric field that the electrons see,

$$\mathcal{E}_j(x) = c\partial_j\tilde{\mathcal{A}}_0(x) - \partial_t\tilde{\mathcal{A}}_j(x) = c\partial_j A_0^{\text{ext}}(x) - \partial_t A_j^{\text{ext}}(x) = E_j(x), \tag{13.4.11}$$

where we have used the facts that $\mathcal{A}(x)$ is a single-valued function and $\Theta(x)$ is a multivalued function in the xy plane [see (13.3.6)].

For later convenience we combine (13.3.9) and (13.3.13) to derive

$$\nabla^2\mathcal{A}(x) = 2\pi m\rho(x) - \frac{eB_\perp}{\hbar} = 2\pi m\left(\rho(x) - \frac{\rho_0}{\nu m}\right), \tag{13.4.12}$$

where ν is the filling factor, $\nu = 2\pi\hbar\rho_0/eB_\perp$. Thus, $\mathcal{A}(x)$ is expressible entirely in terms of the density fluctuation, $\Delta\rho(x) = \rho(x) - \rho_0$, at $\nu = 1/m$.

The Lagrangian density is

$$\begin{aligned}\mathcal{L} &= \psi^\dagger(i\hbar\partial_t - ecA_0^{\text{ext}})\psi - \mathcal{H} \\ &= \varphi^\dagger e^{2\mathcal{A}}(i\hbar\partial_t - ecA_0^{\text{ext}} - ecC_0 - \hbar\partial_t\mathcal{A})\varphi - \mathcal{H} + \mathcal{L}_{\text{CS}},\end{aligned} \tag{13.4.13}$$

with the Chern-Simons term (13.3.8). Because of the measure $e^{2\mathcal{A}(x)}$ in the Lagrangian density, the canonical conjugate of the field $\varphi(x)$ is not $i\varphi^\dagger(x)$ but $i\varphi^\ddagger(x)$. The equal-time canonical commutation relations should be

$$[\varphi(x), \varphi^\ddagger(y)]_\pm = \delta(x-y), \quad [\varphi(x), \varphi(y)]_\pm = [\varphi^\ddagger(x), \varphi^\ddagger(y)]_\pm = 0. \tag{13.4.14}$$

We can explicitly derive these relations from the commutation relations (13.2.13) for the bare composite-particle field, that is

$$[\phi(\mathbf{x}), \phi^\dagger(\mathbf{y})] = \delta(\mathbf{x}-\mathbf{y}), \quad [\phi(\mathbf{x}), \phi(\mathbf{y})] = [\phi^\dagger(\mathbf{x}), \phi^\dagger(\mathbf{y})] = 0 \tag{13.4.15a}$$

for m = odd, and

$$\{\phi(\mathbf{x}), \phi^\dagger(\mathbf{y})\} = \delta(\mathbf{x}-\mathbf{y}), \quad \{\phi(\mathbf{x}), \phi(\mathbf{y})\} = \{\phi^\dagger(\mathbf{x}), \phi^\dagger(\mathbf{y})\} = 0 \tag{13.4.15b}$$

for m = even. Using (13.4.2), (13.4.5) and (13.2.13a), we have

$$\begin{aligned}[\varphi(\mathbf{x}), \varphi^\ddagger(\mathbf{y})] &= e^{-\mathcal{A}(\mathbf{x})}\phi(\mathbf{x})\phi^\dagger(\mathbf{y})e^{\mathcal{A}(\mathbf{y})} - \phi^\dagger(\mathbf{y})e^{\mathcal{A}(\mathbf{y})}e^{-\mathcal{A}(\mathbf{x})}\phi(\mathbf{x}) \\ &= \delta(\mathbf{x}-\mathbf{y}) + e^{-\mathcal{A}(\mathbf{x})}\phi^\dagger(\mathbf{y})\phi(\mathbf{x})e^{\mathcal{A}(\mathbf{y})} - \phi^\dagger(\mathbf{y})e^{-\mathcal{A}(\mathbf{x})}e^{\mathcal{A}(\mathbf{y})}\phi(\mathbf{x}).\end{aligned} \tag{13.4.16}$$

We now use (13.3.15) to express $\mathcal{A}(\mathbf{y})$ explicitly in terms of the density $\rho(\mathbf{z})$. Because the commutation relation $[\rho(\mathbf{z}), \chi(\mathbf{x})] = i\delta(\mathbf{x}-\mathbf{z})$ holds between the density and phase, we obtain

$$\begin{aligned}\phi(\mathbf{x})e^{\mathcal{A}(\mathbf{y})} &= \left|\frac{\mathbf{x}-\mathbf{y}}{2\ell_B}\right|^m e^{\mathcal{A}(\mathbf{y})}\phi(\mathbf{x}), \\ e^{-\mathcal{A}(\mathbf{x})}\phi^\dagger(\mathbf{y}) &= \left|\frac{\mathbf{x}-\mathbf{y}}{2\ell_B}\right|^{-m} \phi^\dagger(\mathbf{y})e^{-\mathcal{A}(\mathbf{x})}.\end{aligned} \tag{13.4.17}$$

Hence, the first equation in (13.4.15a) follows. All other commutators are derived similarly.

Two comments are in order. The first comment is on the gauge invariance of the system. The original Hamiltonian (10.6.1b) is invariant under the gauge transformation. However, we have broken it apparently by choosing the symmetric gauge (10.3.4). In particular, the composite-particle theory is constructed based on this gauge choice. The symmetric-gauge choice breaks the translational symmetry. As will be discussed in Section 29.3, the system possesses a combined symmetry, called the magnetic translation, of the gauge transformation and the translation.

The second comment is on the unitarity of the system. It was claimed[398,439] that the unitarity was lost by the transformation (13.4.2) due to the factor $e^{-\mathcal{A}(\mathbf{x})}$. Indeed, the covariant momentum (13.4.7) has an unusual expression. However, any mathematical transformation does not break the unitarity: It just brings in a nontrivial measure to define the hermiticity. Thus, the canonical conjugate field of φ is not $i\varphi^\dagger$ but $i\varphi^\ddagger = i\varphi^\dagger e^{2\mathcal{A}}$. The hermiticity is to be defined together with the measure $e^{2\mathcal{A}(\mathbf{x})}$. The momentum $\mathcal{P}_j$ is Hermitian with this measure. We may check this explicitly for a one-body state. Using (13.4.7) and making a partial integration, we have

$$\begin{aligned}\langle\varphi_2|\mathcal{P}_j\varphi_1\rangle &\equiv \int d^2x\, e^{2\mathcal{A}(\mathbf{x})}\varphi_2^\dagger(\mathbf{x})\mathcal{P}_j\varphi_1(\mathbf{x}) \\ &= \int d^2x\, e^{2\mathcal{A}(\mathbf{x})}\mathcal{P}_j^\dagger\varphi_2^\dagger(\mathbf{x})\cdot\varphi_1(\mathbf{x}) = \langle\mathcal{P}_j\varphi_2|\varphi_1\rangle.\end{aligned} \tag{13.4.18}$$

This verifies the hermiticity of the covariant momentum (13.4.7). Furthermore, by the use of the relation (13.4.9) the Hamiltonian (13.4.4) is rewritten as

$$\begin{aligned} H_{\mathrm{K}} &= \frac{1}{2M}\int d^2x\ ((\mathcal{P}_x + i\mathcal{P}_y)\varphi(\boldsymbol{x}))^\dagger\, e^{2\mathcal{A}(\boldsymbol{x})}(\mathcal{P}_x + i\mathcal{P}_y)\varphi(\boldsymbol{x}) + \frac{1}{2}N\hbar\omega_{\mathrm{c}} \\ &= \frac{1}{2}\hbar\omega_{\mathrm{c}}\int d^2x\ \left(\frac{\partial\varphi(\boldsymbol{x})}{\partial z^*}\right)^\dagger e^{2\mathcal{A}(\boldsymbol{x})}\frac{\partial\varphi(\boldsymbol{x})}{\partial z^*} + \frac{1}{2}N\hbar\omega_{\mathrm{c}}. \end{aligned} \tag{13.4.19}$$

It is manifestly Hermitian.

Wave Functions

We define the wave function of composite particles and relate it to the wave function of electrons. In terms of dressed composite particles the N-body wave function is defined by

$$\mathfrak{S}_{\mathrm{CP}}[\boldsymbol{x}] = \langle 0|\varphi(\boldsymbol{x}_1)\varphi(\boldsymbol{x}_2)\cdots\varphi(\boldsymbol{x}_N)|\mathfrak{S}\rangle. \tag{13.4.20}$$

The N-body state $|\mathfrak{S}\rangle$ associated with this wave function is given by

$$|\mathfrak{S}\rangle = \int [d\boldsymbol{x}]\mathfrak{S}_{\mathrm{CP}}[\boldsymbol{x}]\varphi^\ddagger(\boldsymbol{x}_1)\varphi^\ddagger(\boldsymbol{x}_2)\cdots\varphi^\ddagger(\boldsymbol{x}_N)|0\rangle, \tag{13.4.21}$$

as is verified based on the commutation relation (13.4.15). We are able to rewrite this as

$$|\mathfrak{S}\rangle = \int [d\boldsymbol{x}]\mathfrak{S}[\boldsymbol{x}]\psi^\dagger(\boldsymbol{x}_1)\psi^\dagger(\boldsymbol{x}_2)\cdots\psi^\dagger(\boldsymbol{x}_N)|0\rangle \tag{13.4.22}$$

with

$$\mathfrak{S}[\boldsymbol{x}] = \mathfrak{S}_{\mathrm{CP}}[\boldsymbol{x}]\prod_{r<s}(z_r - z_s)^m e^{-\sum_{r=1}^N |z_r|^2}. \tag{13.4.23}$$

Equation (13.4.22) is derived from (13.4.21) as follows. The dressed composite-boson field is related to the electron field by (13.4.2), or

$$\varphi(\boldsymbol{x}) = \exp\left(-\int d^2y\ \ln(z - z')^m\rho(\boldsymbol{y}) + |z|^2\right)\psi(\boldsymbol{x}). \tag{13.4.24}$$

The conjugate field is given by (13.4.5), or

$$\varphi^\ddagger(\boldsymbol{x}) = \psi^\dagger(\boldsymbol{x})\exp\left(\int d^2y\ \ln(z - z')^m\rho(\boldsymbol{y}) - |z|^2\right). \tag{13.4.25}$$

The field operator $\varphi^\ddagger$ acts on the vacuum as

$$\varphi^\ddagger(\boldsymbol{x}_1)|0\rangle = e^{-|z_1|^2}\psi^\dagger(\boldsymbol{x}_1)|0\rangle, \tag{13.4.26}$$

since $\rho(\boldsymbol{x})|0\rangle = 0$. Next, it acts on the one-body state as

$$\varphi^\ddagger(\boldsymbol{x}_1)\varphi^\ddagger(\boldsymbol{x}_2)|0\rangle = (z_1 - z_2)^m e^{-|z_1|^2-|z_2|^2}\psi^\dagger(\boldsymbol{x}_1)\psi^\dagger(\boldsymbol{x}_2)|0\rangle, \tag{13.4.27}$$

since $\rho(\mathbf{x})\psi^\dagger(\mathbf{x}_2)|0\rangle = \delta(\mathbf{x}-\mathbf{x}_2)\psi^\dagger(\mathbf{x}_2)|0\rangle$. Repeating this procedure we find

$$\varphi^\ddagger(\mathbf{x}_1)\varphi^\ddagger(\mathbf{x}_2)\cdots\varphi^\ddagger(\mathbf{x}_N)|0\rangle$$
$$= \prod_{r<s}(z_r - z_s)^m e^{-\sum_r |z_r|^2}\psi^\dagger(\mathbf{x}_1)\psi^\dagger(\mathbf{x}_2)\cdots\psi^\dagger(\mathbf{x}_N)|0\rangle. \tag{13.4.28}$$

Substituting this into (13.4.21) we obtain (13.4.22) and (13.4.23).

13.5 Composite Particles in Lowest Landau Level

We construct the wave function for the state in the lowest Landau level by solving the LLL condition. Though all formulas are valid whether the attached flux number m is odd or even, actually they are useful only for composite bosons (m odd) as they stand. We shall present a slightly modified version in Section 13.6, which is useful for composite fermions (m even).

Bare Composite Particles

The kinetic Hamiltonian is given by (13.3.1) for bare composite particles. We define the relative coordinate $\check{\mathbf{R}}$ and the guiding center $\check{\mathbf{X}}$ of the composite particle by

$$\check{\mathbf{R}} = \frac{1}{eB}(-\check{P}_y, \check{P}_x), \qquad \check{\mathbf{X}} = \mathbf{x} - \check{\mathbf{R}} \tag{13.5.1}$$

in analogy to (10.2.7), where $\check{P}_k$ is the covariant momentum (13.3.3). Corresponding to (10.2.13a) and (10.2.13b), we introduce two sets of harmonic oscillators. We substitute the Chern-Simons field explicitly in the covariant momentum (13.3.3). The first set is the Landau-level ladder operator,

$$\check{a} \equiv \frac{\ell_B}{\sqrt{2}\hbar}(\check{P}_x + i\check{P}_y) = -\frac{i}{\sqrt{2}}\left(z + \frac{\partial}{\partial z^*} - \frac{m}{2}\sum_s \frac{1}{z^* - z_s^*}\right), \tag{13.5.2a}$$

$$\check{a}^\dagger \equiv \frac{\ell_B}{\sqrt{2}\hbar}(\check{P}_x - i\check{P}_y) = \frac{i}{\sqrt{2}}\left(z^* - \frac{\partial}{\partial z} - \frac{m}{2}\sum_s \frac{1}{z - z_s}\right). \tag{13.5.2b}$$

The second set is the angular-momentum ladder operators,

$$\check{b} \equiv \frac{1}{\sqrt{2}\ell_B}(\check{X} - i\check{Y}) = \frac{1}{\sqrt{2}}\left(z^* + \frac{\partial}{\partial z} + \frac{m}{2}\sum_s \frac{1}{z - z_s}\right), \tag{13.5.3a}$$

$$\check{b}^\dagger \equiv \frac{1}{\sqrt{2}\ell_B}(\check{X} + i\check{Y}) = \frac{1}{\sqrt{2}}\left(z - \frac{\partial}{\partial z^*} + \frac{m}{2}\sum_s \frac{1}{z^* - z_s^*}\right), \tag{13.5.3b}$$

which govern the physics in the lowest Landau level. These operators depend on the positions of composite particles present in the system. Note that

$$\check{a} - i\check{b}^\dagger = -\sqrt{2}iz, \qquad \check{a}^\dagger + i\check{b} = \sqrt{2}iz^* \tag{13.5.4}$$

as in (10.3.8). They satisfy

$$[\check{a},\check{a}^\dagger]=[\check{b},\check{b}^\dagger]=1,\qquad [\check{a},\check{b}]=[\check{a},\check{b}^\dagger]=0 \tag{13.5.5}$$

as in (10.2.14).

The Hamiltonian (13.3.1) is rewritten as

$$H=\hbar\omega_c\int d^2x\phi^\dagger(\mathbf{x})\check{a}^\dagger(\mathbf{x})\check{a}(\mathbf{x})\phi(\mathbf{x})+\frac{1}{2}N\hbar\omega_c. \tag{13.5.6}$$

It is clear from this Hamiltonian that the LLL condition is

$$\check{a}(\mathbf{x})\phi(\mathbf{x})|\mathfrak{S}\rangle=\frac{\ell_B}{\sqrt{2}\hbar}(\check{P}_x+i\check{P}_y)\phi(\mathbf{x})|\mathfrak{S}\rangle=0. \tag{13.5.7}$$

We rewrite this condition for the N-body wave function (13.2.15) as

$$\left(z_r+\frac{\partial}{\partial z_r^*}-\frac{m}{2}\sum_s\frac{1}{z_r^*-z_s^*}\right)\check{\mathfrak{S}}_{\text{CP}}[\mathbf{x}]=0 \tag{13.5.8}$$

for $r=1,2,\cdots,N$. The generic solution is

$$\check{\mathfrak{S}}_{\text{CP}}=\rho_0^{N/2}\omega[z]\prod_{r<s}|z_r-z_s|^m e^{-\sum_{r=1}^N|z_r|^2}, \tag{13.5.9}$$

where $\omega[z]\equiv\omega(z_1,\cdots,z_N)$ is a dimensionless arbitrary analytic function.

The N-body wave functions in terms of electrons and of composite particles are related by (13.2.18). Using (13.5.9) we obtain

$$\mathfrak{S}[\mathbf{x}]=\rho_0^{N/2}\omega[z]\prod_{r<s}(z_r-z_s)^m e^{-\sum_{r=1}^N|z_r|^2}. \tag{13.5.10}$$

This agrees with the generic N-body wave function (11.3.10) for electrons in the lowest Landau level.

Dressed Composite Particles

The kinetic Hamiltonian is (13.4.4) for dressed composite particles. Corresponding to (13.5.2) and (13.5.3), we define two sets of the harmonic oscillators. The first set is the Landau-level ladder operator,

$$\tilde{a}\equiv\frac{\ell_B}{\sqrt{2}\hbar}(\mathcal{P}_x+i\mathcal{P}_y)=-\frac{i}{\sqrt{2}}\frac{\partial}{\partial z^*}, \tag{13.5.11a}$$

$$\tilde{a}^\dagger\equiv\frac{\ell_B}{\sqrt{2}\hbar}(\mathcal{P}_x-i\mathcal{P}_y)=\frac{i}{\sqrt{2}}\left(2z^*-\frac{\partial}{\partial z}-m\sum_s\frac{1}{z-z_s}\right). \tag{13.5.11b}$$

The second set is the angular-momentum ladder operators,

$$\tilde{b}\equiv\frac{1}{\sqrt{2}\ell_B}(\tilde{X}-i\tilde{Y})=\frac{1}{\sqrt{2}}\left(\frac{\partial}{\partial z}+m\sum_s\frac{1}{z-z_s}\right), \tag{13.5.12a}$$

$$\tilde{b}^\dagger\equiv\frac{1}{\sqrt{2}\ell_B}(\tilde{X}+i\tilde{Y})=\frac{1}{\sqrt{2}}\left(2z-\frac{\partial}{\partial z^*}\right), \tag{13.5.12b}$$

where $(\widetilde{X}, \widetilde{Y})$ is the guiding center for the dressed composite particle. Note that

$$\widetilde{a} - i\widetilde{b}^\dagger = -\sqrt{2}iz, \qquad \widetilde{a}^\dagger + i\widetilde{b} = \sqrt{2}iz^* \tag{13.5.13}$$

as in (10.3.8). They satisfy

$$[\widetilde{a}, \widetilde{a}^\dagger] = [\widetilde{b}, \widetilde{b}^\dagger] = 1, \qquad [\widetilde{a}, \widetilde{b}] = [\widetilde{a}, \widetilde{b}^\dagger] = 0 \tag{13.5.14}$$

as in (10.2.14).

We rewrite the Hamiltonian (13.4.4) as

$$H_K = \hbar\omega_c \int d^2x \varphi^\ddagger(\boldsymbol{x})\widetilde{a}^\dagger(\boldsymbol{x})\widetilde{a}(\boldsymbol{x})\varphi(\boldsymbol{x}) + \frac{1}{2}N\hbar\omega_c. \tag{13.5.15}$$

It is clear from this Hamiltonian that the LLL condition[145] is

$$\widetilde{a}(\boldsymbol{x})\varphi(\boldsymbol{x})|\mathfrak{S}\rangle = \frac{\ell_B}{\sqrt{2}\hbar}(\mathcal{P}_x + i\mathcal{P}_y)\varphi(\boldsymbol{x})|\mathfrak{S}\rangle = -\frac{i}{\sqrt{2}}\frac{\partial}{\partial z^*}\varphi(\boldsymbol{x})|\mathfrak{S}\rangle = 0 \tag{13.5.16}$$

by using (13.4.9). Hence, the N-body wave function is an arbitrary analytic function, $\mathfrak{S}_{\text{CP}}[\boldsymbol{x}] = \rho_0^{N/2}\omega[z]$. When m = odd, it is totally *symmetric* in the N variables z_r. When m = even, it is totally *antisymmetric* in the N variables z_r.

The N-body wave functions in terms of electrons and of composite particles are related by (13.4.23), or

$$\mathfrak{S}[\boldsymbol{x}] = \rho_0^{N/2}\omega[z]\prod_{r<s}(z_r - z_s)^m e^{-\sum_{r=1}^N |z_r|^2}. \tag{13.5.17}$$

This is the generic N-body wave function (11.3.10) for electrons in the lowest Landau level.

When m = odd, the wave function $\mathfrak{S}_{\text{CF}}[\boldsymbol{x}] = \rho_0^{N/2}\omega[z]$ is symmetric in the N variables. The factorization holds, $\omega[z] = \prod_r \omega(z_r)$, for a wide class of states. Such a state is constructed from the vacuum by the Fock construction,

$$|\mathfrak{S}\rangle = \prod_{r=1}^N \left(\rho_0^{1/2}\int d^2x_r\, \omega(z_r)\varphi^\ddagger(\boldsymbol{x}_r)\right)|0\rangle. \tag{13.5.18}$$

The simplest example is given by the Laughlin state for which $\omega(z) = 1$, and the vortex state for which $\omega(z) = z$.

It is to be emphasized that the kinetic Hamiltonian (13.5.15) and the LLL condition (13.5.16) are exact formulas: No approximation has been used to derive them from the microscopic electron theory (10.1.12). Consequently, the Landau levels of composite particles agree with those of electrons, as follows from the Hamiltonian (13.4.19) and the commutation relation (13.5.14) with (13.5.11). It is controversial, but the spacing between Landau levels of composite particles is identical to that of electrons without the LLL projection: See Section 17.2 on this point. The lowest Landau level is filled up with composite particles [Fig.9.7(a) (page 170)], leading to the picture that one composite particle occupies m Landau sites at $\nu = 1/m$.

This property is valid both for composite fermions and composite bosons. See the succeeding section for more details on composite fermions.

13.6 Composite Fermions in Lowest Landau Level

We have defined the composite-particle field $\varphi(\boldsymbol{x})$ in section 13.4, by attaching m flux quanta to the electron field and then by dressing it with a cloud of the effective magnetic field $\mathcal{B}_{\text{eff}}$ that composite particles see. When we choose $m = 2p+1$, it is the composite-boson field and presents a microscopic description of the QH system at $\nu = 1/(2p+1)$ where $\langle \mathcal{B}_{\text{eff}} \rangle = 0$. When we choose $m = 2p$, it is the composite-fermion field. Composite fermions experience the effective magnetic field $\langle \mathcal{B}_{\text{eff}} \rangle = B_{\perp}^{\text{eff}}$ with

$$B_{\perp}^{\text{eff}} = B_{\perp} - 2p\Phi_{\text{D}}\rho_0, \tag{13.6.1}$$

and make cyclotron motion. Nevertheless, it is not precisely what we want. We reconsider the composite-fermion field at $\nu = 1/(2p+1)$.

We modify the definition of the composite-fermion field. The new composite-fermion field is

$$\bar{\psi}(\boldsymbol{x}) = e^{-i2p\Theta(\boldsymbol{x}) - \mathcal{A}_{\text{CS}}(\boldsymbol{x})}\psi(\boldsymbol{x}) \tag{13.6.2}$$

instead of (13.4.2), where

$$\Theta(\boldsymbol{x}) = \int d^2y \theta(\boldsymbol{x} - \boldsymbol{y})\rho(\boldsymbol{y}), \tag{13.6.3}$$

and

$$\mathcal{A}_{\text{CS}}(\boldsymbol{x}) = 2p \int d^2y \ln\left(\frac{|\boldsymbol{x} - \boldsymbol{y}|}{2\ell_B}\right)\rho(\boldsymbol{y}) - \frac{eB_{\perp}^{\text{CS}}}{4\hbar}|\boldsymbol{x}|^2, \tag{13.6.4}$$

instead of (13.3.15), with

$$B_{\perp}^{\text{CS}} = 2p\Phi_{\text{D}}\rho_0. \tag{13.6.5}$$

We have dressed the bare composite-fermion field only with a cloud of the field $B_{\perp}^{\text{CS}}$ due to the Chern-Simons flux. By substituting (13.6.2) into (10.6.1b) the kinetic Hamiltonian reads

$$H_{\text{K}} = \frac{1}{2M}\int d^2x \bar{\psi}^{\ddagger}(\boldsymbol{x})(\bar{\mathcal{P}}_x - i\bar{\mathcal{P}}_y)(\bar{\mathcal{P}}_x + i\bar{\mathcal{P}}_y)\bar{\psi}(\boldsymbol{x}) + \frac{1}{2}N\hbar\omega_{\text{c}}, \tag{13.6.6}$$

instead of (13.4.4). The covariant momentum is

$$\bar{\mathcal{P}}_\mu = -i\hbar\partial_\mu + eA_\mu^{\text{ext}}(\boldsymbol{x}) + 2p\hbar\partial_\mu\Theta(\boldsymbol{x}) - i\hbar\partial_\mu\mathcal{A}_{\text{CS}}(\boldsymbol{x}), \tag{13.6.7}$$

instead of (13.4.7). The new field $\bar{\psi}(\boldsymbol{x})$ satisfies the same anticommutation relations (13.4.15b) as the field $\varphi(\boldsymbol{x})$ satisfies.

The effective magnetic field is given by (13.3.9) since $\mathcal{A}_{\text{CS}}(\boldsymbol{x})$ does not contribute to it,

$$\mathcal{B}_{\text{eff}}(\boldsymbol{x}) = B_\perp - 2p\Phi_{\text{D}}\rho(\boldsymbol{x}). \tag{13.6.8}$$

The ground state is homogeneous, $\langle\rho(\boldsymbol{x})\rangle = \rho_0$, upon which we have $\langle\mathcal{B}_{\text{eff}}(\boldsymbol{x})\rangle = B_\perp^{\text{eff}}$. Both the field $\varphi(\boldsymbol{x})$ and the revised field $\bar{\psi}(\boldsymbol{x})$ experience the same effective magnetic field $B_\perp^{\text{eff}}$: There are no distinction between them on the mean-field level.

The covariant momentum (13.6.7) is explicitly written down as

$$\begin{aligned}\bar{\mathcal{P}}_x &= -i\hbar\partial_x + \frac{e}{2}(B_\perp y + iB_\perp^{\text{CS}}x) + 2p\hbar\partial_x\Theta - 2pi\hbar\partial_y\Theta,\\ \bar{\mathcal{P}}_y &= -i\hbar\partial_y + \frac{e}{2}(-B_\perp x + iB_\perp^{\text{CS}}y) + 2p\hbar\partial_y\Theta + 2pi\hbar\partial_x\Theta.\end{aligned} \tag{13.6.9}$$

The Landau-level ladder operators $(\bar{a},\bar{a}^\dagger)$ are

$$\bar{a} \equiv \frac{\ell_B}{\sqrt{2}\hbar}(\bar{\mathcal{P}}_x + i\bar{\mathcal{P}}_y) = -\frac{i}{\sqrt{2}}\left(\frac{\partial}{\partial z^*} + \frac{B_\perp^{\text{eff}}}{B_\perp}z\right), \tag{13.6.10a}$$

$$\bar{a}^\dagger \equiv \frac{\ell_B}{\sqrt{2}\hbar}(\bar{\mathcal{P}}_x - i\bar{\mathcal{P}}_y) = \frac{i}{\sqrt{2}}\left\{-\frac{\partial}{\partial z} - 2p\sum_s\frac{1}{z-z_s} + 2\left(1 - \frac{B_\perp^{\text{eff}}}{2B_\perp}\right)z^*\right\}, \tag{13.6.10b}$$

instead of (13.5.11). The angular-momentum ladder operators $(\bar{b},\bar{b}^\dagger)$ are given by way of the relations,

$$\bar{a} - i\bar{b}^\dagger = -\sqrt{2}iz, \qquad \bar{a}^\dagger + i\bar{b} = \sqrt{2}iz^*. \tag{13.6.11}$$

They satisfy the commutation relations

$$[\bar{a},\bar{a}^\dagger] = [\bar{b},\bar{b}^\dagger] = 1, \qquad [\bar{a},\bar{b}] = [\bar{a}^\dagger,\bar{b}] = 0. \tag{13.6.12}$$

The Hamiltonian (13.6.6) is rewritten as

$$H_{\text{K}} = \hbar\omega_c\int d^2x\bar{\psi}^\ddagger(\boldsymbol{x})\bar{a}^\dagger(\boldsymbol{x})\bar{a}(\boldsymbol{x})\bar{\psi}(\boldsymbol{x}) + \frac{1}{2}N\hbar\omega_c. \tag{13.6.13}$$

It follows from this Hamiltonian that the LLL condition is

$$\begin{aligned}\bar{a}(\boldsymbol{x})\bar{\psi}(\boldsymbol{x})|\mathfrak{S}\rangle &= \frac{\ell_B}{\sqrt{2}\hbar}(\bar{\mathcal{P}}_x + i\bar{\mathcal{P}}_y)\bar{\psi}(\boldsymbol{x})|\mathfrak{S}\rangle\\ &= -\frac{i}{\sqrt{2}}\left(\frac{\partial}{\partial z^*} + \frac{B_\perp^{\text{eff}}}{B_\perp}z\right)\bar{\psi}(\boldsymbol{x})|\mathfrak{S}\rangle = 0\end{aligned} \tag{13.6.14}$$

instead of (13.5.16). The N-body wave function is defined by

$$\mathfrak{S}_{\text{CF}}[\boldsymbol{x}] \equiv \langle 0|\bar{\psi}(\boldsymbol{x}_1)\bar{\psi}(\boldsymbol{x}_2)\cdots\bar{\psi}(\boldsymbol{x}_N)|\mathfrak{S}\rangle. \tag{13.6.15}$$

We solve (13.6.14) exactly as

$$\mathfrak{S}_{\text{CF}}[\boldsymbol{x}] = \rho_0^{N/2}\omega[z]\exp\left(-\frac{eB_\perp^{\text{eff}}}{4\hbar}\sum_{r=1}^N|\boldsymbol{x}_r|^2\right), \tag{13.6.16}$$

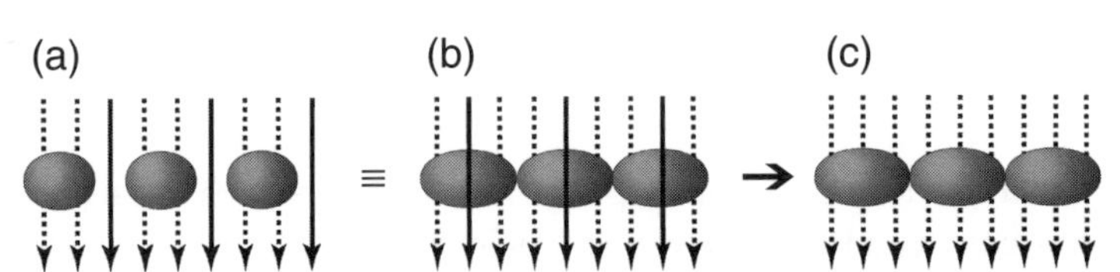

Fig. 13.3 The $\nu = 1/3$ QH state is illustrated in the composite-fermion and composite-boson theories. Arrows represent flux quanta. Their number is the number of Landau sites. (a) A composite fermion is obtained by attaching 2 flux quanta (dotted lines) to an electron. There is one effective magnetic flux (heavy line) left per one composite fermion at $\nu = 1/3$. (b) We may equivalently illustrate composite fermions as in this figure. The number of heavy lines (flux quanta left over) is the number of CF-Landau sites. (c) A composite fermion is converted into a composite boson when one flux is additionally attached.

where $\omega[z]$ is an arbitrary analytic function antisymmetric in all N variables z_r.

The lowest Landau level is filled up with composite fermions at $\nu_{\text{eff}} = 2\pi\rho_0(\ell_B^{\text{eff}})^2 = 1$. The composite-fermion wave function is

$$\mathfrak{S}_{\text{CF}}[\boldsymbol{x}] = \rho_0^{N/2}\prod_{r<s}(z_r - z_s)\exp\left(-\frac{eB_\perp^{\text{eff}}}{4\hbar}\sum_{r=1}^{N}|\boldsymbol{x}_r|^2\right). \tag{13.6.17}$$

This is the Slater determinant of one-body wave functions implied by (13.6.16) as in (11.2.2) for electrons: See also (17.2.3) in Section 17.2. The electron wave function $\mathfrak{S}[\boldsymbol{x}]$ is derived from the composite-fermion wave function $\mathfrak{S}_{\text{CF}}[\boldsymbol{x}]$ as before. Starting with (13.6.15) instead of (13.4.20), we may repeat the procedure from (13.4.21) to (13.4.28). Because of the difference between (13.6.4) and (13.3.15) we obtain

$$\mathfrak{S}[\boldsymbol{x}] = \mathfrak{S}_{\text{CF}}[\boldsymbol{x}]\prod_{r<s}(z_r - z_s)^{2p}\exp\left(-\frac{eB_\perp^{\text{CS}}}{4\hbar}\sum_{r=1}^{N}|\boldsymbol{x}_r|^2\right), \tag{13.6.18}$$

instead of (13.4.23), which yields

$$\mathfrak{S}[\boldsymbol{x}] = \rho_0^{N/2}\prod_{r<s}(z_r - z_s)^{2p+1}e^{-\sum_{r=1}^{N}|z_r|^2}. \tag{13.6.19}$$

This is precisely the Laughlin wave function for the fractional QH state at $\nu = 1/(2p+1)$. Thus, the Laughlin wave function of electrons at $\nu = 1/m$ is derived from that of composite fermions at $\nu_{\text{eff}} = 1$.

Composite fermions can be bosonized at $\nu_{\text{eff}} = 1$ just as electrons are bosonized at $\nu = 1$, as illustrated in fig.13.3. The relevant formula is (13.4.2) with $m = 1$ in the spin-frozen theory,

$$\varphi(\boldsymbol{x}) = e^{-i\Theta(\boldsymbol{x}) - \mathcal{A}_{\text{eff}}(\boldsymbol{x})}\bar{\psi}(\boldsymbol{x}), \tag{13.6.20}$$

where

$$\mathcal{A}_{\text{eff}}(\boldsymbol{x}) = \int d^2y \ln\left(\frac{|\boldsymbol{x}-\boldsymbol{y}|}{2\ell_B}\right)\rho(\boldsymbol{y}) - \frac{eB_\perp^{\text{eff}}}{4\hbar}|\boldsymbol{x}|^2, \tag{13.6.21}$$

since composite fermions experience the effective magnetic field $B_\perp^{\text{eff}}$ rather than the full external magnetic field $B_\perp$. Substituting (13.6.2) and (13.6.21) into (13.6.20), we find that the composite-boson field (13.6.20) is precisely the same one as (13.4.2) with $m = 2p+1$. The two theories are manifestly equivalent.

It follows from (13.6.16) that the one-body wave function of the composite fermion carrying angular momentum $n\hbar$ behaves as

$$\mathfrak{S}_{\text{CF}}^n(\boldsymbol{x}) \propto r^n \exp\left(-\frac{r^2}{4(\ell_B^{\text{eff}})^2}\right), \tag{13.6.22}$$

where ℓ_B^{eff} is the effective magnetic length

$$\ell_B^{\text{eff}} = \sqrt{\frac{\hbar}{e|B_\perp^{\text{eff}}|}} = \sqrt{2p+1}\,\ell_B \tag{13.6.23}$$

by the use of (13.1.3). The probability of finding the composite fermion at r is given by $|\mathfrak{S}_{\text{CF}}^n(\boldsymbol{x})|^2$, which has a sharp peak at $r = r_n$ with

$$r_n = \sqrt{2n}\,\ell_B^{\text{eff}}. \tag{13.6.24}$$

Compare this with (10.3.19) for electrons. The wave function (13.6.22) describes a composite fermion making cyclotron motion with magnetic length ℓ_B^{eff}, as was anticipated in (13.1.9). These states are represented by rings on a disc geometry [Fig.10.7 (page 190)]. The area of each ring is

$$\Delta S_{\text{CF}} = \pi r_{n+1}^2 - \pi r_n^2 = 2\pi\left(\ell_B^{\text{eff}}\right)^2 = 2\pi(2p+1)\ell_B^2, \tag{13.6.25}$$

where we have used (13.6.23). This is $(2p+1)$ times bigger than the area (10.2.24) of one Landau site. It determines the exclusion statistics of composite fermions.

Based on the Hamiltonian (13.6.13) and the LLL condition, we have reached the following conclusions: (A) The Landau levels of composite fermions agree with those of electrons, as follows from the Hamiltonian (13.6.13) and the commutation relation $[\bar{a}, \bar{a}^\dagger] = 1$; (B) One composite fermion behaves as a "fat electron" occupying $(2p+1)$ Landau sites at $\nu = 1/(2p+1)$: One CF-Landau site consists of $(2p+1)$ Landau sites. Composite fermions are identified with superelectrons in literature.[515] The property (A) is what we have already obtained in the last subsection of Section 13.5. The property (B) is consistent with the picture that one composite boson occupies $(2p+1)$ Landau sites [Fig.13.2]. See Section 17.2 for further properties of composite fermions.

CHAPTER 14

COMPOSITE BOSONS AND SEMICLASSICAL ANALYSIS

The fractional QH system is analyzed semiclassically in the composite-boson theory. Any perturbative excitations are dynamically frozen by a large cyclotron gap. The ground state is obtained by minimizing the Coulomb energy within the lowest Landau level. At the filling factor $\nu = 1/m$ with an odd integer m, the Laughlin wave function is derived to describe the ground state. Excitations are generated nonperturbatively by creating topological vortices. They are extended objects carrying electric charges.

14.1 GROUND STATE AND LAUGHLIN WAVE FUNCTION

We seek for the ground state and its wave function within a semiclassical approximation. The kinetic energy is a constant in the lowest Landau level. The Coulomb Hamiltonian is given classically by (10.1.10). It reads field-theoretically as

$$H_C = \frac{1}{2}\int d^2x d^2y \Delta\rho(\boldsymbol{x})V(\boldsymbol{x}-\boldsymbol{y})\Delta\rho(\boldsymbol{y}) \tag{14.1.1}$$

with

$$\Delta\rho(\boldsymbol{x}) = \psi^\dagger(\boldsymbol{x})\psi(\boldsymbol{x}) - \rho_0. \tag{14.1.2}$$

Here we have subtracted the homogeneous background charge from the electron density $\rho(\boldsymbol{x}) = \psi^\dagger(\boldsymbol{x})\psi(\boldsymbol{x})$. See Section 32.5 how the background charge arises.

The Coulomb Hamiltonian plays an essential role in all aspects of the QH effect. Even in integer QH states a large portion of the gap energy is the Coulomb energy. The Coulomb energy is minimized by the state such that $\langle\Delta\rho(\boldsymbol{x})\rangle = 0$, where the number density is homogeneous.

It is convenient to use the dressed composite boson, in terms of which we have

$$\langle\rho(\boldsymbol{x})\rangle = \langle\varphi^\ddagger(\boldsymbol{x})\varphi(\boldsymbol{x})\rangle = e^{2\langle\mathcal{A}(\boldsymbol{x})\rangle}\langle\varphi^\dagger(\boldsymbol{x})\varphi(\boldsymbol{x})\rangle = \rho_0. \tag{14.1.3}$$

More generally,

$$\rho_0^N = \langle \prod_{r=1}^N \rho(\mathbf{x}_r)\rangle = \langle \mathfrak{S}|\varphi^\ddagger(\mathbf{x}_N)\cdots\varphi^\ddagger(\mathbf{x}_2)\varphi^\ddagger(\mathbf{x}_1)\varphi(\mathbf{x}_1)\varphi(\mathbf{x}_2)\cdots\varphi(\mathbf{x}_N)|\mathfrak{S}\rangle$$
$$= \prod_{r=1}^N e^{2\langle \mathcal{A}(\mathbf{x}_r)\rangle} \langle \mathfrak{S}|\varphi^\dagger(\mathbf{x}_N)\cdots\varphi^\dagger(\mathbf{x}_2)\varphi^\dagger(\mathbf{x}_1)\varphi(\mathbf{x}_1)\varphi(\mathbf{x}_2)\cdots\varphi(\mathbf{x}_N)|\mathfrak{S}\rangle. \tag{14.1.4}$$

It is a nontrivial condition that the ground state is homogeneous, as we have seen in Section 13.3. We reexamine the essence. We insert a complete set $\sum |n\rangle\langle n| = 1$ between two operators $\varphi^\dagger(\mathbf{x}_1)$ and $\varphi(\mathbf{x}_1)$. When the state $|\mathfrak{S}\rangle$ contains N electrons, only the vacuum term $|0\rangle\langle 0|$ survives in the complete set because $\varphi(\mathbf{x}_r)$ decreases the electron number by one and because the vacuum state is unique,

$$\rho_0^N = \mathfrak{S}_{\rm CB}[\mathbf{x}]^*\mathfrak{S}_{\rm CB}[\mathbf{x}]\prod_{r=1}^N e^{2\langle \mathcal{A}(\mathbf{x}_r)\rangle}, \tag{14.1.5}$$

where $\mathfrak{S}_{\rm CB}[\mathbf{x}]$ is the N-body wave function (13.4.20). Hence,

$$\mathfrak{S}_{\rm CB}[\mathbf{x}] = \rho_0^{N/2}\prod_{r=1}^N e^{-\langle \mathcal{A}(\mathbf{x}_r)\rangle} \tag{14.1.6}$$

up to an irrelevant phase factor. The wave function $\mathfrak{S}_{\rm CB}[\mathbf{x}]$ should be analytic due to the LLL condition (13.5.16). This is only possible when $\langle \mathcal{A}(\mathbf{x})\rangle$ = constant since it is real. It implies that the effective magnetic flux vanishes and that the filling factor is given by $\nu = 1/m$: See (13.3.9), (13.3.10) and (13.3.12). Namely, the homogeneous ground state is realized only at the magic filling factor $\nu = 1/m$.

At $\nu = 1/m$ the ground-state wave function is given by

$$\mathfrak{S}_{\rm CB}[\mathbf{x}] = \rho_0^{N/2} \tag{14.1.7}$$

in terms of dressed composite bosons. We conclude that the ground state is a condensed phase of composite bosons. By substituting this into (13.4.23), the ground-state wave function reads

$$\mathfrak{S}_{\rm LN}[\mathbf{x}] = \rho_0^{N/2}\prod_{r<s}(z_r - z_s)^m e^{-\sum_{r=1}^N |z_r|^2} \tag{14.1.8}$$

in terms of electrons. Thus, we have derived the Laughlin wave function semiclassically.

For the sake of completeness we examine the same problem in terms of bare composite bosons. The Hamiltonian is (13.3.1). Repeating similar arguments, we derive the ground-state wave function to be

$$\check{\mathfrak{S}}_{\rm CB}[\mathbf{x}] = \rho_0^{N/2} \tag{14.1.9}$$

in terms of bare composite bosons. However, the state does not belong to the lowest Landau level since it violates the LLL condition (13.5.7). Furthermore, by substituting it into (13.2.18), the wave function reads

$$\mathfrak{S}[x] = \rho_0^{N/2} e^{im\sum_{r<s}\theta(z_r - z_s)} \tag{14.1.10}$$

in terms of electrons. This is too singular to be accepted, though the singularity is remediable when a higher order quantum correction is taken into account.[137]

14.2 Perturbative Excitations

We carry out a perturbative analysis around the homogeneous density distribution. We parametrize the bare composite-boson field $\phi(x)$ as

$$\phi(x) = e^{i\chi(x)}\sqrt{\rho_0 + \Delta\rho(x)}, \tag{14.2.1}$$

and hence the dressed composite-boson field as

$$\varphi(x) = e^{-\mathcal{A}(x)}\phi(x) = e^{-\mathcal{A}(x)}e^{i\chi(x)}\sqrt{\rho_0 + \Delta\rho(x)}, \tag{14.2.2}$$

where $\Delta\rho(x)$ and $\chi(x)$ denote the density fluctuation and the phase fluctuation. We then expand

$$\varphi(x) = \sqrt{\rho_0}\left(1 + \frac{\Delta\rho(x)}{2\rho_0} + i\chi(x) - \mathcal{A}(x)\right) + \cdots, \tag{14.2.3a}$$

$$\varphi^\dagger(x) = \sqrt{\rho_0}\left(1 + \frac{\Delta\rho(x)}{2\rho_0} - i\chi(x) + \mathcal{A}(x)\right) + \cdots, \tag{14.2.3b}$$

where only the first order terms are displayed.

We substitute this into the kinetic Hamiltonian (13.4.4), or

$$H_K = \frac{1}{2M}\int d^2x\, \varphi^\dagger(x)(\mathcal{P}_x - i\mathcal{P}_y)(\mathcal{P}_x + i\mathcal{P}_y)\varphi(x), \tag{14.2.4}$$

and calculate an excitation energy. It is convenient to introduce the Fourier components

$$\Delta\rho(x) = \int \frac{d^2k}{2\pi}\varrho_k e^{ikx}, \qquad \chi(x) = \int \frac{d^2k}{2\pi}\chi_k e^{ikx}. \tag{14.2.5}$$

They satisfy

$$\varrho_k^\dagger = \varrho_{-k}, \qquad \chi_k^\dagger = \chi_{-k}, \tag{14.2.6}$$

since the fields $\Delta\rho(x)$ and $\chi(x)$ are real. For the auxiliary field $\mathcal{A}(x)$ we use (13.4.12) to obtain

$$\mathcal{A}(x) = -2\pi m \int \frac{d^2k}{2\pi}\frac{1}{k^2}\varrho_k e^{ikx}. \tag{14.2.7}$$

The Fourier component $\varphi_{\boldsymbol{k}}$ of the composite-boson field is

$$\varphi_{\boldsymbol{k}} = i\chi_{\boldsymbol{k}} + \frac{1}{2\rho_0} G_{\boldsymbol{k}} \tag{14.2.8}$$

with

$$G_{\boldsymbol{k}} = \frac{1}{2\rho_0} + \frac{2\pi m}{\boldsymbol{k}^2} = \frac{1}{2\rho_0}\left(1 + \frac{2}{\ell_B^2 \boldsymbol{k}^2}\right). \tag{14.2.9}$$

Hence we find

$$\begin{aligned}(\mathcal{P}_x + i\mathcal{P}_y)\varphi(\boldsymbol{x}) &= \hbar \int \frac{d^2k}{2\pi} e^{i\boldsymbol{kx}}(k_x + ik_y)(G_{\boldsymbol{k}}\varrho_{\boldsymbol{k}} + i\chi_{\boldsymbol{k}}) \\ &= \hbar \int \frac{d^2k}{2\pi} e^{i\boldsymbol{kx}}(k_x + ik_y)\sqrt{2G_{\boldsymbol{k}}}\xi_{\boldsymbol{k}},\end{aligned} \tag{14.2.10}$$

where we have defined

$$\xi_{\boldsymbol{k}} = \frac{1}{\sqrt{2}}\left(\sqrt{G_{\boldsymbol{k}}}\varrho_{\boldsymbol{k}} + i\frac{1}{\sqrt{G_{\boldsymbol{k}}}}\chi_{-\boldsymbol{k}}^{\dagger}\right), \quad \xi_{\boldsymbol{k}}^{\dagger} = \frac{1}{\sqrt{2}}\left(\sqrt{G_{\boldsymbol{k}}}\varrho_{\boldsymbol{k}}^{\dagger} - i\frac{1}{\sqrt{G_{\boldsymbol{k}}}}\chi_{-\boldsymbol{k}}\right). \tag{14.2.11}$$

The density fluctuation and the phase obey the commutation relation

$$[\Delta\rho(t,\boldsymbol{x}), \chi(t,\boldsymbol{y})] = i\delta(\boldsymbol{x} - \boldsymbol{y}), \tag{14.2.12}$$

as implies

$$[\varrho_{\boldsymbol{l}}, \chi_{\boldsymbol{k}}] = i\delta(\boldsymbol{k} + \boldsymbol{l}). \tag{14.2.13}$$

The Fourier components of $\xi(\boldsymbol{x})$ and $\xi^{\dagger}(\boldsymbol{x})$ satisfy

$$[\xi_{\boldsymbol{k}}, \xi_{\boldsymbol{l}}^{\dagger}] = \delta(\boldsymbol{k} - \boldsymbol{l}). \tag{14.2.14}$$

It is easy to rewrite the kinetic Hamiltonian (14.2.4) with the aid of (14.2.10) as

$$H_{\text{K}} = \frac{\hbar^2\rho_0}{M}\int d^2k\, \boldsymbol{k}^2 G_{\boldsymbol{k}}\xi_{\boldsymbol{k}}^{\dagger}\xi_{\boldsymbol{k}} = \int d^2k\, E(\boldsymbol{k})\xi_{\boldsymbol{k}}^{\dagger}\xi_{\boldsymbol{k}} \tag{14.2.15}$$

with the dispersion relation

$$E(\boldsymbol{k}) = \frac{\hbar^2\boldsymbol{k}^2}{2M} + \hbar\omega_{\text{c}}. \tag{14.2.16}$$

The gap energy is the cyclotron energy $\hbar\omega_{\text{c}}$. It represents the Kohn mode[278] associated with the collective motion of the system.

Consequently there are no density fluctuations within the lowest Landau level: The U(1) field $\varphi(\boldsymbol{x})$ is dynamically frozen by a large cyclotron gap. This is one of the most characteristic features of the QH system.

14.3 Vortex Excitations

We proceed to study nonperturbative excitations at $\nu = 1/m$ within the lowest Landau level. The generic state $|\mathfrak{S}\rangle$ is described by the wave function (13.4.20),

$$\mathfrak{S}_{\text{CB}}[x] = \rho_0^{N/2}\omega_N[z] \equiv \rho_0^{N/2}\omega(z_1, \cdots, z_N), \tag{14.3.1}$$

which is an arbitrary analytic function in N variables. We now show that the wave function describes an ensemble of topological vortices having quantized electric charges and cores of size $\sim \ell_B$.

When the wave function is factorizable,

$$\langle 0|\varphi(\boldsymbol{x}_1)\varphi(\boldsymbol{x}_2)\cdots\varphi(\boldsymbol{x}_N)|\mathfrak{S}\rangle = \rho_0^{N/2}\omega_N[z] = \rho_0^{N/2}\prod_{r=1}^{N}\omega(z_r), \tag{14.3.2}$$

each factor is identified as

$$\langle\varphi(\boldsymbol{x})\rangle \equiv \lim_{N\to\infty}\langle\mathfrak{S}^N|\varphi(\boldsymbol{x})|\mathfrak{S}^{N+1}\rangle = \rho_0^{1/2}\omega(z), \tag{14.3.3}$$

where two states $|\mathfrak{S}^N\rangle$ and $|\mathfrak{S}^{N+1}\rangle$ are identical except the number of electrons. This is because the operator $\varphi(\boldsymbol{x})$ changes the electron number by one. Since $\phi^\dagger(\boldsymbol{x})\phi(\boldsymbol{x}) = \rho(\boldsymbol{x})$, we parametrize

$$\phi(\boldsymbol{x}) = e^{i\chi(\boldsymbol{x})}\sqrt{\rho(\boldsymbol{x})}. \tag{14.3.4}$$

Here, $\chi(\boldsymbol{x})$ is the phase field operator conjugate to the density operator $\rho(\boldsymbol{x})$,

$$[\chi(\boldsymbol{x}), \rho(\boldsymbol{y})] = \delta(\boldsymbol{x} - \boldsymbol{y}). \tag{14.3.5}$$

Thus, $e^{i\chi(\boldsymbol{x})}$ is the operator decreasing the electron number by one. Hence, from (13.4.2) and (14.3.3) we see

$$e^{i\chi_{\text{cl}}(\boldsymbol{x})}\langle\mathfrak{S}|e^{-\mathcal{A}(\boldsymbol{x})}\sqrt{\rho(\boldsymbol{x})}|\mathfrak{S}\rangle = \sqrt{\rho_0}\,\omega(z), \tag{14.3.6}$$

where $\chi_{\text{cl}}(\boldsymbol{x})$ is the c-number phase field which exists in general. Due to the incompressibility of the QH state, $|\mathfrak{S}\rangle$ is an eigenstate of the density operator,

$$\rho(\boldsymbol{x})|\mathfrak{S}\rangle = \rho_{\text{cl}}(\boldsymbol{x})|\mathfrak{S}\rangle, \qquad \mathcal{A}(\boldsymbol{x})|\mathfrak{S}\rangle = \mathcal{A}_{\text{cl}}(\boldsymbol{x})|\mathfrak{S}\rangle. \tag{14.3.7}$$

Consequently we have

$$e^{i\chi_{\text{cl}}(\boldsymbol{x})}e^{-\mathcal{A}_{\text{cl}}(\boldsymbol{x})}\sqrt{\rho_{\text{cl}}(\boldsymbol{x})} = \sqrt{\rho_0}\,\omega(z). \tag{14.3.8}$$

This is the LLL condition for vortex excitations.

We analyze the semiclassical LLL condition (14.3.8). We identify the imaginary parts in (14.3.8),

$$\omega(z) = |\omega(z)|e^{i\chi_{\text{cl}}(\boldsymbol{x})}. \tag{14.3.9}$$

The logarithm of (14.3.8) yields

$$\frac{1}{2}\ln\rho_{\text{cl}}(\boldsymbol{x}) - \mathcal{A}_{\text{cl}}(\boldsymbol{x}) = \ln[\sqrt{\rho_0}|\omega(z)|]. \tag{14.3.10}$$

At $\nu = 1/m$, we transform it into a differential equation,

$$\frac{1}{4\pi m}\nabla^2\ln\rho_{\text{cl}}(\boldsymbol{x}) - \rho_{\text{cl}}(\boldsymbol{x}) + \rho_0 = \frac{1}{m}J^0_{\text{vor}}(\boldsymbol{x}), \tag{14.3.11}$$

where use was made of (13.4.12), and

$$J^0_{\text{vor}}(\boldsymbol{x}) = \frac{1}{2\pi}\nabla^2\ln|\omega(z)|. \tag{14.3.12}$$

Solving (14.3.11) with (14.3.12) we determine the density modulation $\Delta\rho_{\text{cl}}(\boldsymbol{x})$ induced by the excitation $\omega(z)$. We call (14.3.11) the soliton equation, since it determines semiclassical properties of the soliton.

The source term $J^0_{\text{vor}}(\boldsymbol{x})$ is a topological charge. To see it we write down the Cauchy-Riemann equation (see Appendix E) for the analytic function $\ln\omega(z)$,

$$\partial_j\ln|\omega(z)| = \varepsilon_{jk}\partial_k\chi_{\text{cl}}(\boldsymbol{x}), \tag{14.3.13}$$

with $\chi_{\text{cl}}(\boldsymbol{x})$ the phase of $\omega(z)$, from which we obtain

$$J^0_{\text{vor}}(\boldsymbol{x}) = \frac{1}{2\pi}\nabla^2\ln|\omega(z)| = \frac{1}{2\pi}\varepsilon_{jk}\partial_j\partial_k\chi_{\text{cl}}(\boldsymbol{x}). \tag{14.3.14}$$

It is the time component of the topological current density

$$J^\mu_{\text{vor}}(\boldsymbol{x}) = \frac{1}{2\pi}\varepsilon^{\mu\nu\lambda}\partial_\nu\partial_\lambda\chi_{\text{cl}}(\boldsymbol{x}) \tag{14.3.15}$$

and represents the vorticity. Because the only scale the system has is the magnetic length ℓ_B, we expect that a topological soliton has a range of density modulation of order ℓ_B.

We may set $\omega(z) = z$ for the vortex sitting at the center of the system. The topological charge density (14.3.12) is

$$J^0_{\text{vor}}(\boldsymbol{x}) = \delta(\boldsymbol{x}). \tag{14.3.16}$$

The point $\boldsymbol{x} = 0$ is the branch singularity point of the classical phase field $\chi_{\text{cl}}(\boldsymbol{x})$, and gives the vortex center. Since the wave function $\omega(z) = z$ vanishes at the vortex center $z = 0$, we obtain the boundary condition

$$\rho_{\text{cl}}(\boldsymbol{x}) = 0 \qquad \text{at} \quad \boldsymbol{x} = 0 \tag{14.3.17}$$

from (14.3.8). Integrating (14.3.11) over the whole system, we obtain

$$N_{\text{vor}} = \int d^2x\ [\rho_{\text{cl}}(\boldsymbol{x}) - \rho_0] = -\frac{1}{m}. \tag{14.3.18}$$

The electron number is forced to decrease by a vortex excitation.

A generic vortex configuration is described by the analytic function of the type

$$\omega(z) = \prod_r (z - z_r)^{q_r} \tag{14.3.19}$$

with integer $q_r > 0$, for which we have

$$J^0_{\text{vor}}(\mathbf{x}) = \sum_r q_r \delta(\mathbf{x} - \mathbf{x}_r). \tag{14.3.20}$$

The topological charge density is a sum of isolated point distributions. We may associate the electron number $-q_r/m$ with each excitation. It is a hole made in the condensate of composite bosons [Fig.14.1(a)].

Comments are in order. The first comment is with respect to the electron number $1/m$ of one vortex. It is a fractional number for $m \neq 1$. Needless to say, it cannot exist by itself because the fundamental object is the electron. A pair of a vortex and an antivortex is excited thermally. An ensemble of m vortices may exist since its electron number is -1. When we analyze one vortex, it is implicit that there exist a partner far away from it in general [Fig.9.7 (page 170)]. However, an intriguing case is a finite QH droplet, where a single vortex may be generated together with a certain edge excitation: See Section 18.5.

The second comment is with respect to the electron number (14.3.18) of a vortex. It is tricky to derive it by a naive integration of the soliton equation (14.3.11) with (14.3.16). We give another derivation. The vortex is cylindrically symmetric. Because the density modulation $\Delta\rho_{\text{cl}}(\mathbf{x})$ is regular everywhere, the soliton equation (14.3.11) implies

$$\frac{1}{4\pi}\nabla^2 \ln \rho_{\text{cl}}(\mathbf{x}) = \delta(\mathbf{x}) \qquad \text{around} \quad \mathbf{x} = 0. \tag{14.3.21}$$

This is solved to yield

$$\rho_{\text{cl}}(\mathbf{x}) = ar^2 + \cdots \qquad \text{as} \quad r = |\mathbf{x}| \to 0, \tag{14.3.22}$$

where a is an integration constant. We now integrate the soliton equation (14.3.11) in the whole region excluding the origin,

$$\begin{aligned}\int d^2x \, [\rho_{\text{cl}}(\mathbf{x}) - \rho_0] &= \frac{1}{4\pi m}\int_{r>\varepsilon} d^2x \, \nabla^2 \ln \rho_{\text{cl}}(r) \\ &= \frac{1}{4\pi m}\lim_{\varepsilon\to 0}\int_\varepsilon^\infty 2\pi r \left[\frac{1}{r}\frac{d}{dr}\left(r\frac{d}{dr}\ln \rho_{\text{cl}}(r)\right)\right] dr \\ &= -\frac{1}{2m}\lim_{r\to 0} r\frac{d}{dr}\ln \rho_{\text{cl}}(r) = -\frac{1}{m},\end{aligned} \tag{14.3.23}$$

where use was made of (14.3.22) at $r = \varepsilon \to 0$. This gives (14.3.18).

Vortex Energy

The density modulation occurs around the zeros of the analytic function $\omega[z]$, generating topological excitations. It is governed by the soliton equation (14.3.11).

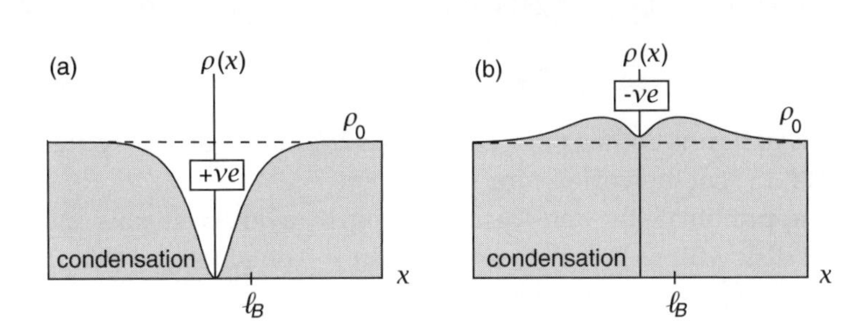

Fig. 14.1 (a) A vortex is a hole made in the condensate of composite bosons. (b) An antivortex is a lump made on the condensate of composite bosons.

We consider one isolated excitation and estimate its Coulomb energy,

$$\Delta_{\text{qh}}^{\text{vor}} = \frac{e^2}{2}\int d^2x d^2y \frac{\delta\rho_{\text{cl}}(\boldsymbol{x})\delta\rho_{\text{cl}}(\boldsymbol{y})}{4\pi\varepsilon|\boldsymbol{x}-\boldsymbol{y}|} = \frac{\alpha_{\text{qh}}^{\text{vor}}}{m^2}E_{\text{C}}^0, \tag{14.3.24}$$

where $\delta\rho_{\text{cl}}(\boldsymbol{x}) = \rho_{\text{cl}}(\boldsymbol{x}) - \rho_0$; $\alpha_{\text{qh}}^{\text{vor}}$ is a numerical constant, and $E_{\text{C}}^0 \equiv e^2/4\pi\varepsilon\ell_B$ is the Coulomb energy unit. The reason why the energy scales as in (14.3.24) reads as follows. We introduce a new function by

$$\rho_{\text{cl}}(\boldsymbol{x}) = \rho_0 e^{u(r)}. \tag{14.3.25}$$

The soliton equation (14.3.11) is reduced to

$$\frac{d^2u}{ds^2} + \frac{1}{s}\frac{du}{ds} + 1 = e^{u(s)}, \tag{14.3.26}$$

where $s = \sqrt{2}r/\ell_B$. This is a modified Liuville equation including a constant term. Its solution is a dimensionless function free from any parameter. Hence, the vortex energy (14.3.24) is proportional to $(\rho_0)^2(\ell_B)^{-3} = 1/(4\pi^2m^2\ell_B)$ with a numerical constant factor.

Although it is not possible to solve the soliton equation (14.3.11) analytically, an approximate solution is easy to obtain. In the regime where the modulation is small ($\Delta\rho_{\text{cl}}(\boldsymbol{x})/\rho_0 \ll 1$), we expand (14.3.11) as

$$\ell_B^2\nabla^2\rho_{\text{cl}}(\boldsymbol{x}) - 2\left(\rho_{\text{cl}}(\boldsymbol{x}) - \rho_0\right) = \frac{1}{m}\delta(\boldsymbol{x}) \qquad \text{at} \quad \boldsymbol{x} \neq \boldsymbol{x}_r \tag{14.3.27}$$

for $\nu = 2\pi\ell_B^2\rho_0$. Since it behaves as $\delta\rho_{\text{cl}} \propto e^{-\sqrt{2}r/\ell_B}$ asymptotically, it has a core with a radius $\sim \ell_B$, as is expected. Equation (14.3.27) has an exact solution,[294]

$$\delta\rho_{\text{cl}}(\boldsymbol{x}) = -\frac{1}{m\pi\ell_B^2}K_0(\sqrt{2}r/\ell_B) \tag{14.3.28}$$

with K_0 the modified Bessel function, where the normalization constant has been chosen to give the correct electron number (14.3.18). For this solution the energy is calculated and is given by (14.3.24) with

$$\alpha_{\text{qh}}^{\text{vor}} = \frac{\pi}{4\sqrt{2}} \simeq 0.56. \tag{14.3.29}$$

This approximation is rather poor since (14.3.28) does not satisfy the boundary condition (14.3.17) at the vortex core.

We next approximate the density modulation by a homogeneous one, $\Delta\rho_{\text{cl}}(\boldsymbol{x}) = -\rho_0$ inside a disk with radius $\sqrt{2}\ell_B$ and $\Delta\rho_{\text{cl}}(\boldsymbol{x}) = 0$ outside it, so that the topological charge is $2\pi\ell_B^2\rho_0 = \nu$. The energy is given by (14.3.24) with

$$\alpha_{\text{qh}}^{\text{vor}} = \frac{4\sqrt{2}}{3\pi} \simeq 0.60, \tag{14.3.30}$$

which is larger than (14.3.29).

We give a better approximation. The asymptotic solution of (14.3.27) is $\Delta\rho_{\text{cl}}(r) \simeq ce^{-\sqrt{2}r/\ell_B}$ with c an integration constant. Requiring the boundary condition (14.3.17) at $r = 0$, we extrapolate it into the vortex core,

$$\delta\rho_{\text{cl}}(r) \simeq -\rho_0(1 + b_1 r - b_2 r^2)e^{-\sqrt{2}r/\ell_B}, \tag{14.3.31}$$

keeping two adjustable parameters b_1 and b_2. Comparing this with (14.3.22) we fix the parameter b_1. We then require that the soliton has the correct electron number (14.3.18). As a result we obtain

$$\delta\rho_{\text{cl}}(r) \simeq -\rho_0\left(1 + \frac{\sqrt{2}r}{\ell_B} - \frac{r^2}{3\ell_B^2}\right)e^{-\sqrt{2}r/\ell_B}. \tag{14.3.32}$$

The number density $\rho_{\text{cl}}(r)$ is zero at the vortex center and monotonically increases toward the ground-state value $\rho_{\text{cl}}(r) = \rho_0$ for $r \gg \ell_B$, as illustrated in Fig.14.2. Substituting (14.3.32) into (14.3.24) we find

$$\alpha_{\text{qh}}^{\text{vor}} = \frac{5775\pi}{32768\sqrt{2}} \simeq 0.39. \tag{14.3.33}$$

This is a good approximation as checked by a numerical analysis.

To make a numerical analysis we use the modified Liuville equation (14.3.26). The behavior (14.3.22) around the vortex center ($s = 0$) is rewritten as

$$u(s) = 2\ln\frac{s}{s_0} + \cdots, \tag{14.3.34}$$

where s_0 is a constant. The boundary condition at infinity is

$$\lim_{s\to\infty} u(s) = 0, \tag{14.3.35}$$

so that the number density approaches the ground-state value. The method of a numerical computation reads as follows. Equation (14.3.26) is integrated numerically by assuming an appropriate initial data s_0 in the asymptotic solution at the vortex center, and it is examined whether or not an obtained solution satisfies the boundary condition (14.3.35) at infinity. This process is repeated by changing the

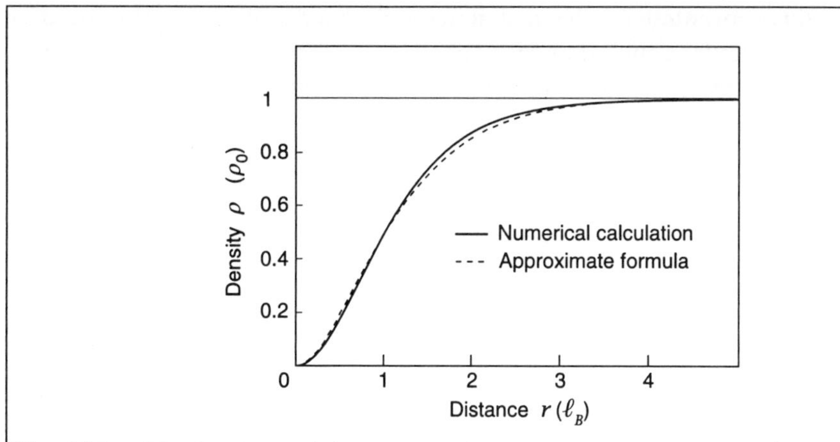

Fig. 14.2 The density modulation around a vortex with $q = 1$ is plotted. The solid curve is obtained by solving the differential equation (14.3.26) numerically. The dashed curve is drawn by using the approximate formula (14.3.32).

initial data. The numerical solution agrees with the approximation (14.3.31) quite well [Fig.14.2].

Antivortices

We have so far considered the case where the wave function is factorizable, $\omega[z] = \prod_r \omega(z_r)$. A vortex is described by the choice of (11.4.9),

$$\omega_{\text{qh}}[z] = \prod_r z_r, \tag{14.3.36}$$

which gives an angular momentum to each electron. Similarly, an antivortex is generated by decreasing an angular momentum of each electron. Naively, this is to multiply z_r^* to the rth electron. Though the wave function is factorizable, the resulting configuration does not stay within the lowest Landau level. After the LLL projection the wave function is given by (11.4.12),

$$\omega_{\text{qe}}[z] = \frac{\prod_r \frac{\partial}{\partial z_r} \prod_{r<s}(z_r - z_s)^m}{\prod_{r<s}(z_r - z_s)^m}. \tag{14.3.37}$$

The wave function is no longer factorizable though it is analytic. An antivortex is a lump made on the condensate of composite bosons [Fig.14.1(b)].

14.4 Field Theory of Vortex Solitons

We construct a local field theory of vortices[134] with vorticity $q = \pm 1$, considering the pointlike limit of vortices. The local field theory describes correctly the effects whose scale is larger than the scale ℓ_B of the vortex soliton. A pointlike vortex is a flux concentrated in a small domain. Such a flux is easily introduced by considering a singular gauge transformation (11.4.1). To create N_{vor} vortices at z_r, $r = 1, 2, \cdots, N_{\text{vor}}$, we perform a singular gauge transformation in the Hamiltonian (13.3.1),

$$\phi \to e^{if}\phi, \quad C_\mu \to C_\mu - \frac{\hbar}{e}\partial_\mu f, \tag{14.4.1}$$

where $f(\boldsymbol{x}) = q\sum_{r=1}^{N_{\text{vor}}} \theta(\boldsymbol{x} - \boldsymbol{x}_r)$. The constraint equation (13.3.6) is transformed into

$$2\pi m\hbar\rho(\boldsymbol{x}) = e\varepsilon_{ij}\partial_i C_j - 2\pi q e \rho_{\text{vor}}(\boldsymbol{x}), \tag{14.4.2}$$

where $\rho = \phi^\dagger\phi$ is the electron density, and

$$\rho_{\text{vor}}(\boldsymbol{x}) = \sum_{r=1}^{N_{\text{vor}}} \delta^2(\boldsymbol{x} - \boldsymbol{x}_r) \tag{14.4.3}$$

is the vortex density describing a set of pointlike vortices sitting at $x^k = x_r^k(t)$ in the 2-dimensional space at time t, where r labels the vortices.

Performing the singular gauge transformation to the Lagrangian density in (13.3.7), we obtain

$$\mathcal{L} \to \mathcal{L} + \Delta\mathcal{L}^{\text{vor}} \tag{14.4.4}$$

with

$$\Delta\mathcal{L}^{\text{vor}} = \frac{q}{m}C_\mu K^\mu + \frac{\hbar}{m}G, \tag{14.4.5}$$

which has been generated from the Chern-Simons term (13.3.8). Here, the term

$$K^\mu(\boldsymbol{x}) = \frac{1}{2\pi}\sum_{r=1}^{N_{\text{vor}}} \varepsilon^{\mu\nu\lambda}\partial_\nu\partial_\lambda\theta(\boldsymbol{x} - \boldsymbol{x}_r) = \sum_{r=1}^{N_{\text{vor}}} \dot{x}_r^\mu \delta^2(\boldsymbol{x} - \boldsymbol{x}_r) \tag{14.4.6}$$

represents world lines of the vortices; $x^\mu = (t, \boldsymbol{x})$, while the term

$$G(\boldsymbol{x}) = \frac{1}{4\pi}\sum_{r,s} \varepsilon^{\mu\nu\lambda}\partial_\mu\theta(\boldsymbol{x} - \boldsymbol{x}_r)\partial_\nu\partial_\lambda\theta(\boldsymbol{x} - \boldsymbol{x}_s) = \sum_s K^\mu\partial_\mu\theta(\boldsymbol{x} - \boldsymbol{x}_s) \tag{14.4.7}$$

describes the linking of world lines[83,392,497] of the vortices.

Integrating $\Delta\mathcal{L}^{\text{vor}}$ over the 2-dimensional space we get

$$\int d^2x\Delta\mathcal{L}^{\text{vor}} = \frac{q}{m}\sum_{r=1}^{N_{\text{vor}}} C_\mu\frac{dx_r^\mu}{dt} - \alpha_v\hbar\sum_{r<s}\frac{d}{dt}\theta(x_r - x_s) \tag{14.4.8}$$

with $\alpha_v = -1/m$. It describes how the vortices interact with the background Chern-Simons field C_μ. By comparing this with (8.3.2) in Part I, the coefficient α_v is the statistics parameter of the vortex. Since it is defined mod 2, the statistics parameter α_v is given actually by the generalized reciprocal relation

$$\alpha_v = -\frac{1}{\alpha} + 2p \tag{14.4.9}$$

with p being an integer.

The particle mechanics of the vortices is described by

$$\begin{aligned}\mathcal{L}^{\text{vor}} &= \sum_{r=1}^{N_{\text{vor}}} \frac{M_{\text{vor}}}{2}\left(\frac{dx_r^k}{dt}\right)^2 + \int d^2x \Delta\mathcal{L} \\ &= \sum_{r=1}^{N_{\text{vor}}} \left[\frac{M_{\text{vor}}}{2}\left(\frac{dx_r^k}{dt}\right)^2 + \frac{q}{m}C_\mu \frac{dz_r^\mu}{dt}\right] - \alpha_v \hbar \sum_{r<s} \frac{d}{dt}\theta(x_r - x_s),\end{aligned} \tag{14.4.10}$$

where the vortex mass M_{vor} is introduced for convenience: The kinetic term is quenched anyway since we are concerned about physics in the lowest Landau level. This describes the particle mechanics of vortices interacting with the background Chern-Simons field C_μ. Because the vortices live on the condensate of electrons, we substitute the mean-field value for C_k, that is, $C_k = eA_k$.

The particle-mechanical Hamiltonian is derived from this Lagrangian,

$$H^{\text{vor}} = \frac{1}{2M_{\text{vor}}} \sum_{r=1}^{N_{\text{vor}}} \left(p_r^k - \frac{qe}{m}A_k - eC_k^{(1)}(x_r)\right)^2 \tag{14.4.11}$$

with

$$C_k^{(1)}(\boldsymbol{x}) = \frac{\alpha_v \hbar}{e} \sum_s^{N_{\text{vor}}} \partial_k \theta(\boldsymbol{x} - \boldsymbol{x}_s). \tag{14.4.12}$$

This leads to the constraint equation

$$\varepsilon_{jk}\partial_j C_k^{(1)} = \alpha_v \Phi_{\text{D}} \rho_{\text{vor}} \tag{14.4.13}$$

for the Chern-Simons field $C_k^{(1)}$.

A quantum field theory of vortices is derived by following the scheme we have presented in the previous section. The Hamiltonian density is

$$H^{(1)} = \frac{1}{2M} \int d^2x |(D_x^{(1)} - iD_y^{(1)})\phi^{(1)}|^2 + H_{\text{int}}^{(1)},$$

where

$$iD_k^{(1)} = i\hbar\partial_k + eC_k^{(1)} + \frac{qe}{m}A_k, \tag{14.4.14}$$

and $H_{\text{int}}^{(1)}$ is the Coulomb interaction between the vortices. The electric charge e_{vor} of a vortex appears in the coefficient of the electromagnetic gauge potential A_k,

$$e_{\text{vor}} = \frac{qe}{m} = \pm\frac{e}{m}, \tag{14.4.15}$$

which agrees with (11.4.15).

14.5 HALDANE-HALPERIN HIERARCHY

As the filling factor deviates from $\nu = 1/m$ quasiparticles (vortices) are created as was discussed in Section 11.5. When the density becomes sufficiently large, we expect them to condense into a new phase by making quasiparticle-flux composites, just as electrons condense by making electron-flux composites into the QH state at $\nu = 1/m$. The condensate is a new QH state supporting a Hall plateau. This is the mechanism proposed by Haldane and Halperin[207,216] to account for QH states at $\nu = n/m$ with odd m. This Haldane-Halperin hierarchy is now known to be unphysical, as is commented at the end of this section. Nevertheless, it is worthwhile to present a field-theoretical formulation[137] for completeness.

One quasiparticle has the electron number $1/m$ and the electric charge (14.4.15),

$$e^{(1)} \equiv e_{\text{vor}} = \pm \frac{e}{m}. \tag{14.5.1}$$

The quasiparticle is neither a boson nor a fermion, but is an anyon. As was reviewed in Chapter 8, the anyon is said to have the statistical parameter α when the phase $e^{i\alpha\pi}$ is generated by interchanging two anyons, and it appears as a coefficient of the particle-mechanical Lagrangian (8.3.2). The statistical parameter $\alpha^{(1)}$ of the quasiparticle at $\nu = 1/m$ is given by (14.4.9), as follows from (14.4.8), or

$$\alpha^{(1)} = -\frac{1}{\alpha^{(0)}} + 2p^{(1)}, \tag{14.5.2}$$

with $\alpha^{(0)} = m$ and an arbitrary even integer $2p^{(1)}$.

The quasiparticle is a topological soliton whose size is $\sim \ell_B$. We may regard it as a pointlike object as far as the long-distance physics is concerned. The anyon statistics is taken care of by way of the Chern-Simons field $C_k^{(1)}(\boldsymbol{x})$ defined by (14.4.13), or

$$\varepsilon_{jk}\partial_j C_k^{(1)}(\boldsymbol{x}) = \alpha^{(1)}\Phi_{\text{D}}\rho^{(1)}(\boldsymbol{x}) \tag{14.5.3}$$

with $\alpha^{(1)} = \alpha_{\text{qh}}$ and $\rho^{(1)} = \rho_{\text{qh}}$. The Coulomb energy is given by

$$H_C = \frac{1}{2}\left(\frac{e^{(1)}}{e}\right)^2 \int d^2x d^2y\, V(\boldsymbol{x}-\boldsymbol{y})\rho^{(1)}(\boldsymbol{x})\rho^{(1)}(\boldsymbol{y}). \tag{14.5.4}$$

This is minimized by a homogeneous configuration of quasiparticles.

The effective magnetic field reads

$$\mathcal{B}_{\text{eff}}^{(1)}(\boldsymbol{x}) \equiv -\varepsilon_{jk}\partial_j \left\{A_k^{\perp} + C_k^{(1)}\right\} = B_\perp - \alpha^{(1)}\Phi_{\text{D}}\rho^{(1)}(\boldsymbol{x}). \tag{14.5.5}$$

We examine the condition that it vanishes, $\mathcal{B}_{\text{eff}}^{(1)}(\boldsymbol{x}) = 0$. First, it is necessary that the vortex density takes the value

$$\langle \rho^{(1)}(\boldsymbol{x})\rangle_g = \rho_0^{(1)} \equiv \frac{B_\perp}{\alpha^{(1)}\Phi_{\text{D}}} = \frac{eB}{2\pi\hbar m\alpha^{(1)}}. \tag{14.5.6}$$

Because one quasiparticle is made of $1/m$ of one electron in the QH state at $\nu = 1/m$, the vortex density $\rho_0^{(1)}$ amounts to the electron density $\Delta\rho \equiv \rho_0^{(1)}/m$: The total electron density is $\rho_0 + \Delta\rho$. Hence, the filling factor is

$$\nu = \nu^{(1)} \equiv \frac{2\pi(\rho_0 + \Delta\rho)\hbar}{eB} = \frac{1}{m}\left(1 + \frac{1}{m\alpha^{(1)}}\right) = \frac{1}{m - \dfrac{1}{2p^{(1)}}} = \frac{2p^{(1)}}{2mp^{(1)} - 1}. \tag{14.5.7}$$

At this filling factor, the effective magnetic field $\mathcal{B}_{\text{eff}}^{(1)}(\mathbf{x})$ vanishes, and hence we expect that quasiparticle-flux composites undergo a Bose condensation. It forms a QH state in the second generation, where the incompressibility follows precisely as before since all excitations are gapful. The Hall current gets contribution from the "electron" condensate and the "quasiparticle" condensate. The total electric charge density is $e(\rho_0 + \Delta\rho)$. Therefore, the Hall resistivity is given by

$$R_{xy} = -\frac{(\rho_0 + \Delta\rho)e}{B} = -\frac{\nu e^2}{2\pi\hbar} \tag{14.5.8}$$

at $\nu = \nu^{(1)}$ with (14.5.7). We call the QH state in the second generation the daughter state.

Excitations on this daughter state are new vortices, which we call daughter-quasiparticles. In the vicinity of $\nu = \nu^{(1)}$, daughter-quasiparticles are created but localized by impurities, as leads to a Hall plateau.

The electric charge of the daughter-quasiparticle is similarly discussed as that of the ordinary quasiparticle. One daughter-quasiparticle at $\nu = \nu^{(1)} = 2p^{(1)}/(2mp^{(1)} - 1)$ has the electric charge,

$$e^{(2)} = \pm\frac{1}{\alpha^{(0)}\alpha^{(1)}}e = \pm\frac{e}{2mp^{(1)} - 1}. \tag{14.5.9}$$

Its statistical parameter $\alpha^{(2)}$ is related to that of the parent quasiparticle by the reciprocal relation

$$\alpha^{(2)} = -\frac{1}{\alpha^{(1)}} + 2p^{(2)}, \tag{14.5.10}$$

where $2p^{(2)}$ is an even integer.

Daughter-quasiparticles undergo a Bose condensation and constitute a QH state in the third generation at the filling factor

$$\nu = \nu^{(2)} \equiv \frac{1}{\alpha^{(0)}}\left[1 + \frac{1}{\alpha^{(0)}\alpha^{(1)}}\left[1 + \frac{1}{\alpha^{(1)}\alpha^{(2)}}\right]\right] = \frac{1}{m - \dfrac{1}{2p^{(1)} - \dfrac{1}{2p^{(2)}}}}. \tag{14.5.11}$$

This step goes as far as we wish. At the kth step, the filling factor is

$$\begin{aligned}\nu = \nu^{(k)} &\equiv \frac{1}{\alpha^{(0)}}\left[1+\frac{1}{\alpha^{(0)}\alpha^{(1)}}\left[1+\frac{1}{\alpha^{(1)}\alpha^{(2)}}\left[1+\cdots\frac{1}{\alpha^{(k-1)}\alpha^{(k)}}\right]\cdots\right]\right] \\ &= \cfrac{1}{m-\cfrac{1}{2p^{(1)}-\cfrac{1}{2p^{(2)}-\cfrac{1}{\ddots\cfrac{1}{2p^{(k)}}}}}} \\ &\equiv \frac{n^{(k)}}{m^{(k)}},\end{aligned} \tag{14.5.12}$$

where $m^{(k)}$ is odd. The charge of one quasiparticle at $\nu = n^{(k)}/m^{(k)}$ is

$$e^{(k+1)} = \pm e\prod_{i=0}^{k}\frac{1}{\alpha^{(i)}} = \frac{e}{m^{(k)}}. \tag{14.5.13}$$

This is the field theoretical realization of the Haldane-Halperin hierarchy of QH states.

This hierarchy scheme has some unsatisfactory features theoretically and experimentally. A theoretical difficulty is that a large number of quasiparticles are required to create daughter states: For instance, $N/2$ quasiparticles are needed to build the $\nu = 2/5$ state on top of the $\nu = 1/3$ state. At such high densities quasiparticles significantly overlap, where the distance between them is of the same order as their size. Then, the quasiparticle description itself is questionable. In particular, we cannot justify the assumption that quasiparticles are pointlike objects carrying definite electric charges. Furthermore, it is even more questionable that excitations of this new state also make their own fractional QH state, and that this process can be repeated several times. A phenomenological difficulty is related to this point. From the reasoning we expect that the stability decreases substantially from parent to daughter. Experiments reveal an extraordinary stability of fractional QH states along the principal sequences, $\nu = n/(2pn \pm 1)$. For instance, the $\nu = 6/13$ QH state is well established experimentally [Fig.9.2 (page 165)], but it is necessary to repeat the hierarchical construction five times to reach this state. See Chapter 17 for a better hierarchy scheme.

Chapter 15

Quantum Hall Ferromagnets

Each Landau level contains two energy levels with up and down spins. These two energy levels are almost degenerate when the Zeeman effect is small enough. Nevertheless, the system becomes incompressible at $\nu = 1$ owing to a Coulomb gap generated by the exchange interaction. It is called a QH ferromagnet. The spin wave is a Goldstone mode. The Zeeman effect gives a gap energy to the Goldstone mode, and makes the coherence length finite.

15.1 Spin Coherence

We analyze the effect due to the spin degree of freedom. The Hamiltonian depends on the electron spin through the Zeeman energy $\Delta_Z = |g^* \mu_B B|$, where g^* is the gyromagnetic factor and μ_B the Bohr magneton. Each Landau level contains two energy levels with the one-body gap energy Δ_Z. If the Zeeman energy is sufficiently large, the spin degree of freedom is frozen and all previous arguments are valid as they stand.

However, the Zeeman energy $\Delta_Z = |g^* \mu_B B|$ is much smaller than the Coulomb energy of order $E_C^0 = e^2/4\pi\varepsilon\ell_B$. Their ratio is $\tilde{g} \equiv \Delta_Z/E_C^0 \simeq 0.02$ in typical samples at 10 Tesla. The gyromagnetic factor is 2 in the vacuum, but is determined in a semiconductor mostly by spin-orbit coupling[414,230] and therefore changes for example by the application of hydrostatic pressure to the system. Even the vanishing limit of the Zeeman energy ($g^* = 0$) is feasible.[338,300] In this case one Landau level contains two almost degenerate energy levels. What integer QH states would we expect? If the Coulomb interaction were absent, the first integer QH state would occur at $\nu = 2$, where the lowest Landau level is filled. However, there emerges a spin-polarized incompressible state at $\nu = 1$ with a Coulomb gap. The physical origin is understood as follows [Fig.15.1].

We explore the QH state at $\nu = 1$. One might think that the orientations of electron spins are random without the Zeeman effect ($g^* = 0$). This is not the case due to the exchange interaction as in ferromagnets (see Section 4.1). The exchange interaction emerges also in the QH system as if electrons were localized on lattice

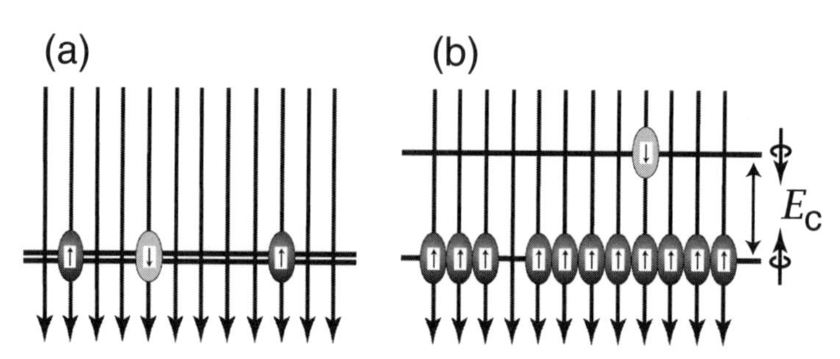

Fig. 15.1 (a) In the vanishing limit of the Zeeman energy ($g^* = 0$) one Landau level contains two degenerate energy levels. Electrons with opposite spins are accommodated within the same energy level. (b) At $\nu = 1$ all spins are spontaneously polarized into one direction even without the Zeeman effect. This direction is actually chosen by the external magnetic field however small it may be, and the up-spin level is filled up. Exciting one electron to the down-spin level implies a pair creation of an electron and a quasihole at a cost of a Coulomb gap energy.

points (Section 32.6). It is summarized by an effective Hamiltonian,

$$H_X = 2J_s \sum_a \int d^2x \, [\partial_k \mathcal{S}_a(\boldsymbol{x})]^2, \tag{15.1.1}$$

where $\mathcal{S}_a(\boldsymbol{x})$ is the spin field normalized as $\boldsymbol{\mathcal{S}}^2(\boldsymbol{x}) = 1/4$, and J_s is the spin stiffness,[254,351,144]

$$J_s = \frac{1}{16\sqrt{2\pi}} E_C^0. \tag{15.1.2}$$

The effective Hamiltonian is a nonlinear sigma model.

It is one of the main topics of Part IVto derive the effective Hamiltonian (15.1.1) from the microscopic theory. There exists an intriguing mechanism associated with "noncommutative geometry" for the emergence of the exchange energy in the QH system. We consider a local spin excitation, which in general increases the kinetic energy. However, because an increase of the kinetic energy brings the state out of the lowest Landau level, it is necessary to rotate merely the spin component belonging to the lowest Landau level. The relevant component is extracted self-consistently by making the LLL projection. It has a crucial consequence: After the LLL projection the spin density SU(2) operator and the number density U(1) operator become noncommutative, and hence a spin rotation increases the Coulomb energy,[351,144] as is described by the exchange energy (15.1.1).

The ground state is a coherent state owing to the exchange interaction (15.1.1). All spins are polarized spontaneously in the absence of the Zeeman effect, where the direction of the polarization is arbitrary. It means a spontaneous breaking[1] of

[1]See Appendix J for discussions on spontaneous symmetry breaking in the 2-dimensional space.

the SU(2) spin symmetry. The Goldstone mode is the spin wave. The system is referred to as a *QH ferromagnet*. (We may also call it the spin QH ferromagnet in contrast to the pseudospin QH ferromagnet in the bilayer system.)

The up-spin state is chosen by a nonzero Zeeman effect. Thus, the up-spin level is filled at $\nu = 1$. Exciting one electron to the down-spin level implies a pair creation of an electron and a quasihole at a cost of a Coulomb gap energy [Fig.15.1]. Hence, the system is incompressible. The same incompressibility follows also for the QH state at $\nu = 1/m$ (m odd) based on the composite-particle theory. A quasiparticle-pair creation will be investigated in Chapter 16.

The Zeeman effect makes the spin wave a gapful mode. The coherence length (spin correlation length) is given by

$$\xi_{\text{spin}} = 2\ell_B\sqrt{\frac{\pi J_s}{\Delta_Z}}. \tag{15.1.3}$$

When the exchange Coulomb energy is dominant, the coherence length is large ($J_s \gg \Delta_Z$) and the system is regarded as a QH ferromagnet. When the Zeeman energy is dominant ($\Delta_Z \gg J_s$), the coherence length is small and the spin degree of freedom is frozen. There exists no phase transition between the QH ferromagnet and the spin-frozen limit. In summary, the exchange interaction develops the spin coherence, though the coherence length is made finite by the Zeeman effect.

15.2 Spin Degree of Freedom

The Hamiltonian is given by

$$H = H_K + H_C + H_Z + \frac{N}{2}\hbar\omega_c \tag{15.2.1}$$

with the kinetic term H_K, the Coulomb term H_C and the Zeeman term H_Z,

$$H_K = \frac{1}{2M}\int d^2x\, \Psi^\dagger(\mathbf{x})(P_x - iP_y)(P_x + iP_y)\Psi(\mathbf{x}), \tag{15.2.2}$$

$$H_C = \frac{1}{2}\int d^2x d^2y\, V(\mathbf{x}-\mathbf{y})\rho(\mathbf{x})\rho(\mathbf{y}), \tag{15.2.3}$$

$$H_Z = -\Delta_Z \int d^2x\, \mathcal{S}_z(\mathbf{x}). \tag{15.2.4}$$

Here, P_k is the covariant momentum, $P_k = -i\hbar\partial_k + eA_k^{\text{ext}}$, with the external magnetic field taken in the symmetric gauge. Let us explain notations.

We have introduced the two-component electron field

$$\Psi = \begin{pmatrix} \psi^\uparrow \\ \psi^\downarrow \end{pmatrix} \tag{15.2.5}$$

with spin up and down components. The total number density is

$$\rho(\mathbf{x}) = \Psi^\dagger\Psi = \rho^\uparrow(\mathbf{x}) + \rho^\downarrow(\mathbf{x}) \tag{15.2.6}$$

with $\rho^\alpha(\boldsymbol{x}) \equiv \psi^\alpha(\boldsymbol{x})^\dagger \psi^\alpha(\boldsymbol{x})$. The spin operator is

$$
\begin{aligned}
\mathcal{S}_x(\boldsymbol{x}) &= \Psi^\dagger \frac{\tau_x}{2} \Psi = \frac{1}{2}(\psi^{\uparrow\dagger}\psi^{\downarrow} + \psi^{\downarrow\dagger}\psi^{\uparrow}),\\
\mathcal{S}_y(\boldsymbol{x}) &= \Psi^\dagger \frac{\tau_y}{2} \Psi = \frac{1}{2i}(\psi^{\uparrow\dagger}\psi^{\downarrow} - \psi^{\downarrow\dagger}\psi^{\uparrow}),\\
\mathcal{S}_z(\boldsymbol{x}) &= \Psi^\dagger \frac{\tau_z}{2} \Psi = \frac{1}{2}(\psi^{\uparrow\dagger}\psi^{\uparrow} - \psi^{\downarrow\dagger}\psi^{\downarrow}) = \frac{1}{2}(\rho^{\uparrow} - \rho^{\downarrow}),
\end{aligned} \tag{15.2.7}
$$

where τ_a is the Pauli matrix,

$$
\tau_x = \begin{pmatrix} 0 & 1 \\ 1 & 0 \end{pmatrix}, \qquad \tau_y = \begin{pmatrix} 0 & -i \\ i & 0 \end{pmatrix}, \qquad \tau_z = \begin{pmatrix} 1 & 0 \\ 0 & -1 \end{pmatrix}. \tag{15.2.8}
$$

The kinetic Hamiltonian has the global SU(2) spin symmetry in addition to the global U(1) phase symmetry. It determines the Landau-level spectrum.

The total number density $\rho(\boldsymbol{x})$ and the spin density $\mathcal{S}_a(\boldsymbol{x})$ satisfy the algebraic relations

$$
[\mathcal{S}_a(\boldsymbol{x}), \mathcal{S}_b(\boldsymbol{y})] = i\delta(\boldsymbol{x} - \boldsymbol{y})\varepsilon_{abc}\mathcal{S}_c(\boldsymbol{x}) \tag{15.2.9}
$$

and

$$
[\rho(\boldsymbol{x}), \mathcal{S}_a(\boldsymbol{y})] = 0. \tag{15.2.10}
$$

The number density operator generates a U(1) phase transformation,

$$
\Psi(\boldsymbol{x}) \quad \to \quad e^{-i\rho(f)}\Psi(\boldsymbol{x})e^{i\rho(f)} = \exp[if(\boldsymbol{x})]\Psi(\boldsymbol{x}) \tag{15.2.11}
$$

with

$$
\rho(f) = \int d^2x\, f(\boldsymbol{x})\rho(\boldsymbol{x}). \tag{15.2.12}
$$

The system has the U(1) gauge symmetry. The spin operator generates a local SU(2) transformation, $e^{-i\mathcal{S}(f)}$, with

$$
\mathcal{S}(f) = \sum_a \int d^2x\, f^a(\boldsymbol{x})\mathcal{S}_a(\boldsymbol{x}). \tag{15.2.13}
$$

It acts on the SU(2) field as

$$
\Psi(\boldsymbol{x}) \quad \to \quad e^{-i\mathcal{S}(f)}\Psi(\boldsymbol{x})e^{i\mathcal{S}(f)} = \exp\left[i\sum_a f^a(\boldsymbol{x})\frac{\tau_a}{2}\right]\Psi(\boldsymbol{x}). \tag{15.2.14}
$$

It generates a *spin texture* on the ground state $|\mathrm{g}\rangle$,

$$
|\mathfrak{S}\rangle = e^{i\mathcal{S}(f)}|\mathrm{g}\rangle. \tag{15.2.15}
$$

It is an excited state with spins rotated because the system does not possess the local SU(2) symmetry.

15.3 Composite Bosons and Spin-Charge Separation

We investigate the spin system based on the composite-boson theory at $\nu = 1/m$. We define the bare composite-boson field ϕ^α by an operator phase transformation,[144]

$$\phi^\alpha(\boldsymbol{x}) = e^{-im\Theta(\boldsymbol{x})}\psi^\alpha(\boldsymbol{x}), \tag{15.3.1}$$

where m is an odd integer and

$$\Theta(\boldsymbol{x}) = \int d^2y\theta(\boldsymbol{x}-\boldsymbol{y})\rho(\boldsymbol{y}). \tag{15.3.2}$$

The two-components $\phi^\alpha(\boldsymbol{x})$ share one common phase field $\Theta(\boldsymbol{x})$ in (15.3.1); otherwise, the global SU(2) symmetry of the kinetic Hamiltonian is explicitly broken in the composite-boson theory [see Section 23.1 and the comment under (23.1.21)]. The commutation relation is derived for $\phi^\alpha(\boldsymbol{x})$,

$$[\phi^\alpha(\boldsymbol{x}), \phi^{\beta\dagger}(\boldsymbol{y})] = \delta^{\alpha\beta}\delta(\boldsymbol{x}-\boldsymbol{y}) \tag{15.3.3}$$

precisely as (13.2.13a) is derived. See also (32.4.20) for this commutation relation.

We decompose the bare composite-boson field into the U(1) field $\phi(\boldsymbol{x})$ and the SU(2) field $n^\alpha(\boldsymbol{x})$,

$$\phi^\alpha(\boldsymbol{x}) = \phi(\boldsymbol{x})n^\alpha(\boldsymbol{x}), \qquad \phi(\boldsymbol{x}) = e^{i\chi(\boldsymbol{x})}\sqrt{\rho(\boldsymbol{x})}. \tag{15.3.4}$$

It is equivalent to the decomposition of the electron field into its U(1) and SU(2) parts,

$$\psi^\alpha(\boldsymbol{x}) = \psi(\boldsymbol{x})n^\alpha(\boldsymbol{x}), \quad \text{with} \quad \psi(\boldsymbol{x}) = e^{im\Theta(\boldsymbol{x})}\phi(\boldsymbol{x}). \tag{15.3.5}$$

Substituting this into the kinetic Hamiltonian (15.2.2), we find that the U(1) field $\psi(\boldsymbol{x})$ is the charge part since the electromagnetic field $A_k(\boldsymbol{x})$ couples only with it. On the other hand, the CP^1 field $n^\alpha(\boldsymbol{x})$ carries solely the spin degree of freedom, as we see in (15.4.5). Formula (15.3.5) is interpreted as a *spin-charge separation.*

Ground State

To discuss the ground state and its wave function, it is convenient to use the dressed composite boson field,

$$\varphi^\alpha(\boldsymbol{x}) = e^{-\mathcal{A}(\boldsymbol{x})}\phi^\alpha(\boldsymbol{x}) = e^{-im\Theta(\boldsymbol{x})-\mathcal{A}(\boldsymbol{x})}\psi^\alpha(\boldsymbol{x}). \tag{15.3.6}$$

The spin-charge separation reads

$$\varphi^\alpha(\boldsymbol{x}) = \varphi(\boldsymbol{x})n^\alpha(\boldsymbol{x}), \quad \text{with} \quad \varphi(\boldsymbol{x}) = e^{-\mathcal{A}(\boldsymbol{x})}\phi(\boldsymbol{x}). \tag{15.3.7}$$

We define the Chern-Simons field C_k by

$$C_k(\boldsymbol{x}) = \frac{\hbar m}{e}\partial_k\Theta(\boldsymbol{x}), \tag{15.3.8}$$

and the effective magnetic field $\mathcal{B}_{\text{eff}}(\boldsymbol{x})$ by

$$\mathcal{B}_{\text{eff}}(\boldsymbol{x}) = -\varepsilon_{ij}\partial_i \mathcal{A}_j(\boldsymbol{x}) = B_\perp - m\phi_{\text{D}}\rho(\boldsymbol{x}), \tag{15.3.9}$$

where

$$\mathcal{A}_j(\boldsymbol{x}) \equiv A_j^{\text{ext}}(\boldsymbol{x}) + C_j(\boldsymbol{x}) = -\frac{\hbar}{e}\varepsilon_{jk}\partial_k \mathcal{A}(\boldsymbol{x}) \tag{15.3.10}$$

with

$$\mathcal{A}(\boldsymbol{x}) = m\int d^2y \ln\left(\frac{|\boldsymbol{x}-\boldsymbol{y}|}{2\ell_B}\right)\rho(\boldsymbol{y}) - |z|^2. \tag{15.3.11}$$

It follows from (15.3.9) that the effective magnetic field vanishes at $\nu = 1/m$, $\langle \mathcal{B}_{\text{eff}}(\boldsymbol{x})\rangle = 0$, when the density is homogeneous, $\langle\rho(\boldsymbol{x})\rangle = \rho_0$, as in the spin-frozen theory.

Let us ignore the Zeeman effect. The ground state minimizes the Coulomb energy, which is achieved by a homogeneous electron density,

$$\langle\rho(\boldsymbol{x})\rangle = e^{2\langle\mathcal{A}(\boldsymbol{x})\rangle}\left(\langle\varphi^{\uparrow\dagger}(\boldsymbol{x})\varphi^{\uparrow}(\boldsymbol{x})\rangle + \langle\varphi^{\downarrow\dagger}(\boldsymbol{x})\varphi^{\downarrow}(\boldsymbol{x})\rangle\right) = \rho_0. \tag{15.3.12}$$

At $\nu = 1/m$, where $\langle\mathcal{A}(\boldsymbol{x})\rangle = 0$, we solve this as

$$\begin{aligned}\langle\varphi^{\uparrow\dagger}(\boldsymbol{x})\varphi^{\uparrow}(\boldsymbol{x})\rangle &= \frac{1}{2}\rho_0\left[1+\sqrt{1-\sigma_0^2}\right],\\ \langle\varphi^{\downarrow\dagger}(\boldsymbol{x})\varphi^{\downarrow}(\boldsymbol{x})\rangle &= \frac{1}{2}\rho_0\left[1-\sqrt{1-\sigma_0^2}\right],\end{aligned} \tag{15.3.13}$$

or

$$\langle\Phi(\boldsymbol{x})\rangle = \frac{\sqrt{\rho_0}}{2}\begin{pmatrix}\sqrt{1+\sigma_0}+\sqrt{1-\sigma_0}\\ \sqrt{1+\sigma_0}-\sqrt{1-\sigma_0}\end{pmatrix}. \tag{15.3.14}$$

Here, σ_0 is an arbitrary real number subject to $|\sigma_0| \le 1$. We have made this parametrization so that the up-spin polarized state is given by the choice of $\sigma_0 = 0$ [see also (15.4.6)]. The one-point function $\langle\Phi(\boldsymbol{x})\rangle$ is a constant vector pointing to an arbitrary direction in the SU(2) spin space, as implies a spontaneous breaking of the SU(2) symmetry. The ground state $|\text{g}\rangle$ is a coherent state due to the degeneracy.

Wave Functions

In terms of the dressed field the kinetic Hamiltonian term (15.2.2) reads

$$H_K = \frac{1}{2M}\int d^2x\, \Phi^{\ddagger}(\boldsymbol{x})(\mathcal{P}_x - i\mathcal{P}_y)(\mathcal{P}_x + i\mathcal{P}_y)\Phi(\boldsymbol{x}), \tag{15.3.15}$$

where

$$\Phi = \begin{pmatrix}\varphi^{\uparrow}\\ \varphi^{\downarrow}\end{pmatrix}, \qquad \Phi^{\dagger} = \left(\varphi^{\uparrow\dagger}, \varphi^{\downarrow\dagger}\right), \tag{15.3.16}$$

and $\Phi^{\ddagger}(\boldsymbol{x}) = \Phi^{\dagger}(\boldsymbol{x})e^{2\mathcal{A}(\boldsymbol{x})}$; $\mathcal{P}_k$ is the covariant momentum,

$$\mathcal{P}_j = -i\hbar\partial_j + e(\delta_{jk} - i\varepsilon_{jk})\left(A_k^{\text{ext}} + C_k\right). \tag{15.3.17}$$

The LLL condition follows from this kinetic Hamiltonian,

$$(\mathcal{P}_x + i\mathcal{P}_y)\Phi(\boldsymbol{x})|\mathfrak{S}\rangle = -\frac{i\hbar}{\ell_B}\frac{\partial}{\partial z^*}\Phi(\boldsymbol{x})|\mathfrak{S}\rangle = 0. \tag{15.3.18}$$

The N-body wave function is obtained by solving this condition,

$$\mathfrak{S}_{\rm CB}[\boldsymbol{x}] \equiv \langle 0|\Phi(\boldsymbol{x}_1)\Phi(\boldsymbol{x}_2)\cdots\Phi(\boldsymbol{x}_N)|\mathfrak{S}\rangle = \mathfrak{S}_{\rm CB}[z]. \tag{15.3.19}$$

It is analytic and totally symmetric in N variables z_r for odd integer m.

From (15.3.1) and (15.3.6) we obtain

$$\Phi(\boldsymbol{x}) = \exp\left(-\int d^2y\ \ln(z-z')^m\rho(\boldsymbol{y}) + |z|^2\right)\Psi(\boldsymbol{x}). \tag{15.3.20}$$

Repeating arguments from (13.4.24) to (13.4.28) we find

$$\begin{aligned}&\Phi^\ddagger(\boldsymbol{x}_1)\Phi^\ddagger(\boldsymbol{x}_2)\cdots\Phi^\ddagger(\boldsymbol{x}_N)|0\rangle\\ &\quad= \prod_{r<s}(z_r-z_s)^m e^{-\sum_r|z_r|^2}\Psi^\dagger(\boldsymbol{x}_1)\Psi^\dagger(\boldsymbol{x}_2)\cdots\Psi^\dagger(\boldsymbol{x}_N)|0\rangle.\end{aligned} \tag{15.3.21}$$

Therefore, with the composite-boson wave function (15.3.19), the state $|\mathfrak{S}\rangle$ is given by

$$|\mathfrak{S}\rangle = \int[d\boldsymbol{x}]\mathfrak{S}_{\rm CB}[z]\Phi^\ddagger(\boldsymbol{x}_1)\Phi^\ddagger(\boldsymbol{x}_2)\cdots\Phi^\ddagger(\boldsymbol{x}_N)|0\rangle \tag{15.3.22a}$$

$$= \int[d\boldsymbol{x}]\mathfrak{S}[\boldsymbol{x}]\Psi^\dagger(\boldsymbol{x}_1)\Psi^\dagger(\boldsymbol{x}_2)\cdots\Psi^\dagger(\boldsymbol{x}_N)|0\rangle, \tag{15.3.22b}$$

where

$$\mathfrak{S}[\boldsymbol{x}] = \mathfrak{S}_{\rm CB}[z]\mathfrak{S}_{\rm LN}[\boldsymbol{x}] \tag{15.3.23}$$

with $\mathfrak{S}_{\rm LN}[\boldsymbol{x}]$ the Laughlin function. The electron wave function is given by the Laughlin function multiplied by the analytic spinor factor $\mathfrak{S}_{\rm CB}[z]$ describing excitations confined to the lowest Landau level.

In the semiclassical analysis the boson field operator is approximated by a c-number field. It follows from (15.3.19) that the electron wave function reads

$$\mathfrak{S}[\boldsymbol{x}] = \prod_r\langle\Phi(\boldsymbol{x}_r)\rangle\mathfrak{S}_{\rm LN}[\boldsymbol{x}], \tag{15.3.24}$$

where $\langle\Phi(\boldsymbol{x})\rangle$ is analytic. The ground-state wave function is given by the choice of (15.3.14). In particular, we find

$$\mathfrak{S}_{\rm g}[\boldsymbol{x}] = \prod_j\begin{pmatrix}1\\0\end{pmatrix}_j\mathfrak{S}_{\rm LN}[\boldsymbol{x}] \tag{15.3.25}$$

for the up-spin polarized ground state.

15.4 Spin Field, Sigma Field and CP^1 Field

We have introduced a new field $n^\alpha(\boldsymbol{x})$ by way of the charge-spin separation (15.3.5). We explore its properties. By substituting (15.3.5) into the density operator $\rho(\boldsymbol{x})$, we find

$$\boldsymbol{n}^\dagger(\boldsymbol{x})\boldsymbol{n}(\boldsymbol{x}) = n^{\uparrow\dagger}(\boldsymbol{x})n^\uparrow(\boldsymbol{x}) + n^{\downarrow\dagger}(\boldsymbol{x})n^\downarrow(\boldsymbol{x}) = 1, \tag{15.4.1}$$

where

$$\boldsymbol{n}(\boldsymbol{x}) = \begin{pmatrix} n^\uparrow(\boldsymbol{x}) \\ n^\downarrow(\boldsymbol{x}) \end{pmatrix}. \tag{15.4.2}$$

We count the number of the real fields in the decomposition (15.3.4). The composite boson ϕ^α has four real fields in total, and the U(1) field ϕ has two real fields. Hence, the two-component complex field n^α must have only two real fields. Indeed, one real field is removed by the normalization condition (15.4.1) and another real field is removed by the local U(1) phase invariance[2]

$$n_a(x) \longrightarrow n_a(x)\exp[if(x)]. \tag{15.4.3}$$

Such a field is the CP^1 field reviewed in Section 7.8.

The spin density (15.2.7) is unaffected by the bosonization transformations (15.3.1) and (15.3.6). It is expressed as

$$S_a(\boldsymbol{x}) = \left(\phi^{\uparrow\dagger}, \phi^{\downarrow\dagger}\right)\frac{\tau_a}{2}\begin{pmatrix} \phi^\uparrow \\ \phi^\downarrow \end{pmatrix} = \Phi^\ddagger\frac{\tau_a}{2}\Phi = \rho(\boldsymbol{x})\mathcal{S}_a(\boldsymbol{x}), \tag{15.4.4}$$

where $\boldsymbol{\mathcal{S}}(\boldsymbol{x})$ is the normalized spin field (sigma field)

$$\mathcal{S}_a(\boldsymbol{x}) = \boldsymbol{n}^\dagger(\boldsymbol{x})\frac{\tau_a}{2}\boldsymbol{n}(\boldsymbol{x}), \tag{15.4.5}$$

satisfying $\boldsymbol{\mathcal{S}}^2(\boldsymbol{x}) = 1/4$: See the footnote in page 49 and (30.2.30) with respect to the normalization. It represents the spin of an individual electron. To see this let us consider a classical configuration of electrons with the density $\rho(\boldsymbol{x}) = \sum_i \delta(\boldsymbol{x} - \boldsymbol{x}_i)$, with which the total spin is $\boldsymbol{S} = \sum_i \boldsymbol{\mathcal{S}}(\boldsymbol{x}_i)$. Actually in the QH system an electron may be regarded almost localized within one Landau site. The normalized spin $\boldsymbol{\mathcal{S}}$ corresponds to the spin variable associated with each Landau site.

We parametrize the CP^1 field as

$$\boldsymbol{n}(\boldsymbol{x}) = \frac{1}{\sqrt{2}}T\begin{pmatrix} e^{i\vartheta(\boldsymbol{x})/2}\sqrt{1+\sigma(\boldsymbol{x})} \\ e^{-i\vartheta(\boldsymbol{x})/2}\sqrt{1-\sigma(\boldsymbol{x})} \end{pmatrix} \tag{15.4.6}$$

---

[2] The Coulomb term, the Zeeman term and the exchange term depend only on the density $\rho(\boldsymbol{x})$ and the spin $S_a(\boldsymbol{x})$. Hence, it is invariant under the transformation (15.4.3). The overall phase of the two-component complex field n^α is not dynamical: See (7.8.10).

in terms of two scalar fields $\sigma(\boldsymbol{x})$ and $\vartheta(\boldsymbol{x})$, where T is a unitary matrix,

$$T = \frac{1}{\sqrt{2}}\begin{pmatrix} 1 & 1 \\ 1 & -1 \end{pmatrix}, \qquad T^{\dagger}T = 1. \tag{15.4.7}$$

The spin field (15.4.5) reads

$$\mathcal{S}_x(\boldsymbol{x}) = \frac{1}{2}\sigma(\boldsymbol{x}), \quad \mathcal{S}_y(\boldsymbol{x}) = \frac{1}{2}\sqrt{1-\sigma^2(\boldsymbol{x})}\sin\vartheta(\boldsymbol{x}),$$
$$\mathcal{S}_z(\boldsymbol{x}) = \frac{1}{2}\sqrt{1-\sigma^2(\boldsymbol{x})}\cos\vartheta(\boldsymbol{x}). \tag{15.4.8}$$

Note that without the matrix T in (15.4.6) the ground state corresponds to $\langle\sigma\rangle = 1$ and $\langle\vartheta\rangle = 0$. We have introduced T so that the perturbative fluctuation around the ground state corresponds to the expansion around $\sigma = 0$ and $\vartheta = 0$.

The CP¹ field is also parametrized as

$$\boldsymbol{n}(\boldsymbol{x}) = \frac{1}{\sqrt{1+|eta^2(\boldsymbol{x})|}}\begin{pmatrix} 1 \\ eta(\boldsymbol{x}) \end{pmatrix}. \tag{15.4.9}$$

Up to the first order of field variables we may identify

$$eta(\boldsymbol{x}) = \frac{\sigma(\boldsymbol{x}) + i\vartheta(\boldsymbol{x})}{2}, \qquad eta^{\dagger}(\boldsymbol{x}) = \frac{\sigma(\boldsymbol{x}) - i\vartheta(\boldsymbol{x})}{2}, \tag{15.4.10}$$

and expand

$$\phi^{\uparrow}(\boldsymbol{x}) = \phi(\boldsymbol{x}) = \sqrt{\rho_0}\left[1 + \frac{1}{2\rho_0}\Delta\rho(\boldsymbol{x}) + i\chi(\boldsymbol{x})\right],$$
$$\phi^{\downarrow}(\boldsymbol{x}) = \phi(\boldsymbol{x})eta(\boldsymbol{x}) = \sqrt{\rho_0}eta(\boldsymbol{x}). \tag{15.4.11}$$

The composite-boson field satisfies the canonical commutation relation (15.3.3), or

$$[\phi^{\alpha}(\boldsymbol{x}), \phi^{\beta\dagger}(\boldsymbol{y})] = \delta^{\alpha\beta}\delta(\boldsymbol{x}-\boldsymbol{y}). \tag{15.4.12}$$

This is realized by

$$[\Delta\rho(\boldsymbol{x}), \chi(\boldsymbol{y})] = i\delta(\boldsymbol{x}-\boldsymbol{y}) \tag{15.4.13}$$

and

$$[eta(\boldsymbol{x}), eta^{\dagger}(\boldsymbol{y})] = \rho_0^{-1}\delta(\boldsymbol{x}-\boldsymbol{y}), \tag{15.4.14}$$

or

$$\frac{\rho_0}{2}[\sigma(\boldsymbol{x}), \vartheta(\boldsymbol{y})] = i\delta(\boldsymbol{x}-\boldsymbol{y}). \tag{15.4.15}$$

The U(1) and SU(2) parts constitute two independent canonical sets.

The ground state $|\mathrm{g}_0\rangle$ is given by the vacuum of $eta(\boldsymbol{x})$,

$$eta(\boldsymbol{x})|\mathrm{g}_0\rangle = 0, \tag{15.4.16}$$

from which it follows that

$$n(x)|g_0\rangle = \begin{pmatrix} 1 \\ 0 \end{pmatrix} |g_0\rangle. \tag{15.4.17}$$

As far as perturbative fluctuations are concerned within the lowest Landau level, we may set the U(1) field $\phi(x)$ to its ground-state value since it is dynamically frozen. In general, we have

$$\langle \mathfrak{S}|n(x)|\mathfrak{S}\rangle = \frac{1}{\sqrt{2}} T \begin{pmatrix} e^{i\vartheta_{\text{cl}}(x)/2}\sqrt{1+\sigma_{\text{cl}}(x)} \\ e^{-i\vartheta_{\text{cl}}(x)/2}\sqrt{1-\sigma_{\text{cl}}(x)} \end{pmatrix}, \tag{15.4.18}$$

where $\zeta_{\text{cl}}(x)$, $\sigma_{\text{cl}}(x)$ and $\vartheta_{\text{cl}}(x)$ are classical fields. The sigma field (15.4.8) reads

$$\mathcal{S}_x^{\text{cl}}(x) = \frac{1}{2}\sigma_{\text{cl}}(x), \quad \mathcal{S}_y^{\text{cl}}(x) = \frac{1}{2}\sqrt{1-\sigma_{\text{cl}}^2(x)}\sin\vartheta_{\text{cl}}(x),$$
$$\mathcal{S}_z^{\text{cl}}(x) = \frac{1}{2}\sqrt{1-\sigma_{\text{cl}}^2(x)}\cos\vartheta_{\text{cl}}(x), \tag{15.4.19}$$

where $\mathcal{S}_i^{\text{cl}}(x) \equiv \langle \mathfrak{S}|\mathcal{S}_i(x)|\mathfrak{S}\rangle$.

The spin texture is classified by the Pontryagin number (7.7.9),

$$Q_{\text{sky}} = \int d^2x\, J_{\text{sky}}^0(x), \tag{15.4.20}$$

where $J_{\text{sky}}^0(x)$ is the charge density of a topological current,

$$J_{\text{sky}}^\mu(x) = \frac{1}{\pi}\varepsilon_{abc}\varepsilon^{\mu\nu\lambda}\mathcal{S}_a^{\text{cl}}\partial_\nu\mathcal{S}_b^{\text{cl}}\partial_\lambda\mathcal{S}_c^{\text{cl}}. \tag{15.4.21}$$

It is an absolutely conserved charge [see (7.7.8)]. We may rewrite J_{sky}^0 as

$$J_{\text{sky}}^0(x) = \frac{1}{4\pi}\varepsilon_{ij}\partial_i\sigma_{\text{cl}}\partial_j\vartheta_{\text{cl}}. \tag{15.4.22}$$

A spin texture cannot be deformed into a trivial spin texture (the classical ground state) when it has a nonzero Pontryagin number. It is a coherent excitation called *skyrmions*, which we elucidate in Chapter 16.

It is necessary to examine the topological stability of a skyrmion since the basic symmetry of the spin system is SU(2). One might wonder why there are topological solitons in a simply connected manifold with $\pi_2(\text{SU}(2)) = 0$. This is because the U(1) field $\varphi(x)$ in the charge-spin separation (15.3.7) is dynamically frozen by a large cyclotron gap. Consequently, the dynamical symmetry group is given by CP^1, for which $\pi_2(\text{CP}^1) \neq 0$. It is possible to deform the skyrmion configuration continuously into the ground-state configuration within the SU(2) manifold since $\pi_2(\text{SU}(2)) = 0$. However, in so doing, the U(1) field $\varphi(x)$ is necessarily deformed and involves higher Landau levels, as costs a cyclotron energy $\hbar\omega_c$. Hence, the topological stability is valid rigorously only in the limit $\hbar\omega_c \to \infty$. One might then wonder why we may treat a skyrmion as a topologically stable object in actual systems with finite $\hbar\omega_c$. This is because we observe experimentally only quasiparticles activated thermally. Any virtual excitations with higher energy are practically suppressed in actual systems.

15.5 EFFECTIVE HAMILTONIAN

It is convenient to employ the effective Hamiltonian H_{eff} to explore the dynamical property of the coherent mode $eta(\mathbf{x})$ at $\nu = 1/m$. The effective Coulomb Hamiltonian consists of the direct term H_D and the exchange term H_X. The direct interaction describes the Coulomb energy (14.1.1) due to the density modulation,

$$H_D = \frac{e^2}{2}\int d^2x d^2y\, \Delta\rho(\mathbf{x})V(\mathbf{x}-\mathbf{y})\Delta\rho(\mathbf{y}), \tag{15.5.1}$$

while the exchange interaction describes a spin-spin interaction due to the spin modulation,

$$H_X = \frac{2}{m}J_s\sum_a\int d^2x\,[\partial_k\mathcal{S}_a(\mathbf{x})]^2, \tag{15.5.2}$$

where J_s is the spin stiffness (15.1.2). This exchange Hamiltonian is derived in Section 32.6: See (32.6.12).

The direct Coulomb energy is minimized by a homogeneous state, that is a QH state as in the spinless theory (see Section 14.1). The exchange energy (15.5.2) is minimized by any constant value of the spin field, $\mathcal{S}^g(\mathbf{x}) = \mathcal{S}_0$ =constant. It corresponds to $\sigma_g(\mathbf{x}) = \sigma_0$ and $\vartheta_g(\mathbf{x}) = \vartheta_0$ in (15.4.19),

$$\mathcal{S}_x^g(\mathbf{x}) = \frac{1}{2}\sigma_0, \quad \mathcal{S}_y^g(\mathbf{x}) = \frac{1}{2}\sqrt{1-\sigma_0^2}\sin\vartheta_0, \quad \mathcal{S}_z^g(\mathbf{x}) = \frac{1}{2}\sqrt{1-\sigma_0^2}\cos\vartheta_0. \tag{15.5.3}$$

Hence, there exists a degeneracy in the ground states as labeled by σ_0 and ϑ_0. The system is called *QH ferromagnet*, since the ground state is spontaneously magnetized [Fig.15.2]. When a continuous symmetry is spontaneously broken, there should arise a gapless mode known as the Goldstone mode, and quantum coherence develops spontaneously.[3]

The spin wave is described by the nonlinear sigma model (15.5.2). The excitation of the spin wave is exactly solved[254] at $\nu = 1$, where the spin stiffness (15.1.2) was first derived. The spin wave was also studied[364] numerically by an exact diagonalization of a few-electron system in the composite-fermion theory at $\nu = 1/m$.

Including the Zeeman interaction as a small perturbation on the coherent state, we obtain the effective Hamiltonian

$$H_{\text{eff}} = \frac{2}{m}J_s\sum_a\int d^2x\,[\partial_k\mathcal{S}_a(\mathbf{x})]^2 - \Delta_Z\rho_0\int d^2x\,\mathcal{S}_z(\mathbf{x}). \tag{15.5.4}$$

[3]See Appendix J for discussions on spontaneous symmetry breaking in the 2-dimensional space.

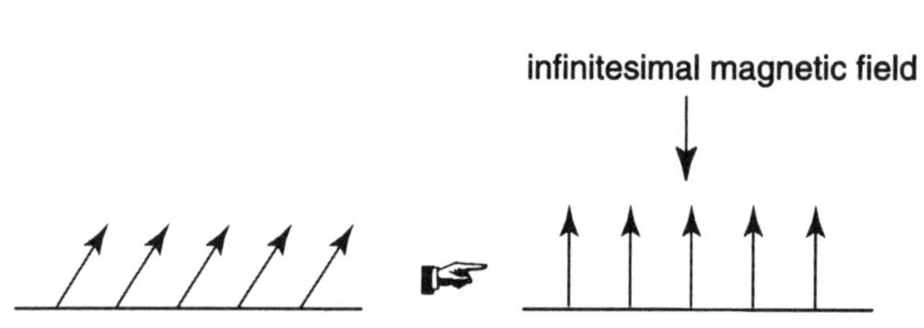

Fig. 15.2 The QH state is a ferromagnet in the absence of the Zeeman effect ($g^* = 0$), where all electron spins are spontaneously polarized into an arbitrary direction to minimize the exchange energy. The polarization axis is actually chosen by the Zeeman effect ($g^* \neq 0$) however small it may be.

It governs the spin wave around the spin-polarized ground state.

Goldstone Mode

The Goldstone mode describes the spin wave , or small fluctuations of the spin field around the ground state $|g_0\rangle$. By substituting the spin configuration (15.4.8) into the effective Hamiltonian (15.5.4), we find

$$\mathcal{H}_{\text{eff}} = \frac{J_s}{2m}(\partial_k \vartheta)^2 + \frac{J_s}{2m}(\partial_k \sigma)^2 + \frac{\Delta_Z \rho_0}{4}(\sigma^2 + \vartheta^2), \tag{15.5.5}$$

together with (15.4.15), or[4]

$$\frac{\rho_0}{2}[\sigma(\boldsymbol{x}), \vartheta(\boldsymbol{y})] = i\delta(\boldsymbol{x} - \boldsymbol{y}). \tag{15.5.6}$$

We rewrite (15.5.5) as

$$\mathcal{H}_{\text{eff}} = \frac{2J_s}{m\rho_0}\partial_k \zeta^\dagger(\boldsymbol{x})\partial_k \zeta(\boldsymbol{x}) + \Delta_Z \zeta^\dagger(\boldsymbol{x})\zeta(\boldsymbol{x}), \tag{15.5.7}$$

where $\zeta(\boldsymbol{x}) = \sqrt{\rho_0} eta(\boldsymbol{x})$ with (15.4.10). It describes the dynamics of the coherent mode.

In the momentum space the Hamiltonian reads

$$H_{\text{eff}} = \int d^2k\, E_k \zeta_k^\dagger \zeta_k \tag{15.5.8}$$

with

$$[\zeta_k, \zeta_l^\dagger] = \delta(\boldsymbol{k} - \boldsymbol{l}). \tag{15.5.9}$$

[4]The effective Hamiltonian (15.5.4) is derived from the noncommutative nature of the QH system in Chapter 32, where the Poisson bracket (32.4.20) holds for the classical field. This Poisson bracket corresponds to the canonical commutation relation (15.4.12), from which (15.4.15) follows.

The dispersion relation is

$$E_{\boldsymbol{k}} = \left(\frac{2J_s \boldsymbol{k}^2}{m\rho_0} + \Delta_Z \right). \tag{15.5.10}$$

In the vanishing limit of the Zeeman effect ($g^* = 0$), it reads

$$E_{\boldsymbol{k}} = \frac{2J_s}{m\rho_0}\boldsymbol{k}^2 = \frac{\sqrt{\pi}\ell_B^2}{4\sqrt{2}}\boldsymbol{k}^2 E_C^0, \tag{15.5.11}$$

as implies a Goldstone mode. When $g^* \neq 0$, it acquires a gap Δ_Z, and the eta mode is now a quasi-Goldstone mode.

The coherence length ξ_{spin} is derived from (15.5.7) at $\nu = 1/m$,

$$\xi_{\text{spin}} = \sqrt{\frac{2J_s}{m\Delta_Z\rho_0}} = \ell_B\sqrt{\frac{4\pi J_s}{\Delta_Z}}. \tag{15.5.12}$$

It is finite due to the Zeeman effect. We make the following observation based on this formula: (A) The coherence length is infinite in the absence of the Zeeman effect ($\Delta_Z = 0$); (B) The coherence length is zero and the spin degree of freedom is frozen in the absence of the exchange Coulomb energy ($J_s = 0$); (C) There exists no phase transition between the QH ferromagnet and the spin-frozen limit. The coherence length is estimated numerically as

$$\frac{\xi_{\text{spin}}}{\ell_B} \simeq \frac{7.3}{\sqrt{B_\perp}} \tag{15.5.13}$$

at $\nu = 1$ with the magnetic field $B_\perp$ in Tesla. In actual samples the ground state is a intralayer coherent state with a finite coherence length.

CHAPTER 16

SPIN TEXTURES

Charged excitations are skyrmions in the QH ferromagnet. A skyrmion is a spin texture, rotating several spins simultaneously, and carries electric charge $\pm e/m$ at $\nu = 1/m$. It is a coherent excitation spread over two energy levels. Its size is determined by a competition between the direct Coulomb energy and the Zeeman energy. We describe it as a topological soliton in the composite-boson theory.

16.1 SPIN EXCITATIONS

The QH ferromagnet is a macroscopic coherent state, where all spin components $\mathcal{S}_x(\boldsymbol{x})$, $\mathcal{S}_y(\boldsymbol{x})$ and $\mathcal{S}_z(\boldsymbol{x})$ are simultaneously measurable as classical fields. The presence of quantum coherence is confirmed experimentally by observing coherent excitations. The spin wave is a neutral perturbative excitation. What charged excitations do we expect? We may excite one electron in the $\nu = 1$ QH state [Fig.16.1(a)], where the excited-electron spin is reversed and there appears one hole. Each of them increases a Coulomb energy due to the electric charge $\pm e$ confined to a domain of area $2\pi\ell_B^2$. Actually, the Coulomb self-energy can be lowered by rotating[1] the spins of neighboring electrons coherently [Fig.16.1(b)] and thus modulating the charge density as well. When the charge is spread over a wider domain of radius $\sim 2\kappa\ell_B$, the Coulomb self-energy becomes smaller by $\sim 1/(2\kappa)$ times. On the other hand, the number of reversed spins is increased by $\sim 4\kappa^2$ times, as increases the Zeeman energy. The resulting object is identified with a pair of skyrmion and antiskyrmion reviewed in Section 7.7. The skyrmion scale κ is determined to minimize the total energy.

It is interesting that a skyrmion carries electric charge $\pm e/m$ at $\nu = 1/m$ though electrons are excited to the down-spin level. A fractional charge excitation is possible because it is a coherent excitation. A skyrmion is a topological soliton spread over two energy levels [Fig.16.1(c)]. Its importance was first recognized in the effective Chern-Simons gauge theory,[304,451] then studied in the Hartree-Fock approximation,[163,312,36,166] in small-size numerical analysis,[257,530] and also in a

[1]In the $\nu = 1$ QH ferromagnet, we shall argue in Section 31.2 that a skyrmion state is constructed from a hole state by making a $W_\infty(2)$ rotation, which is an SU(2) rotation in the noncommutative space.

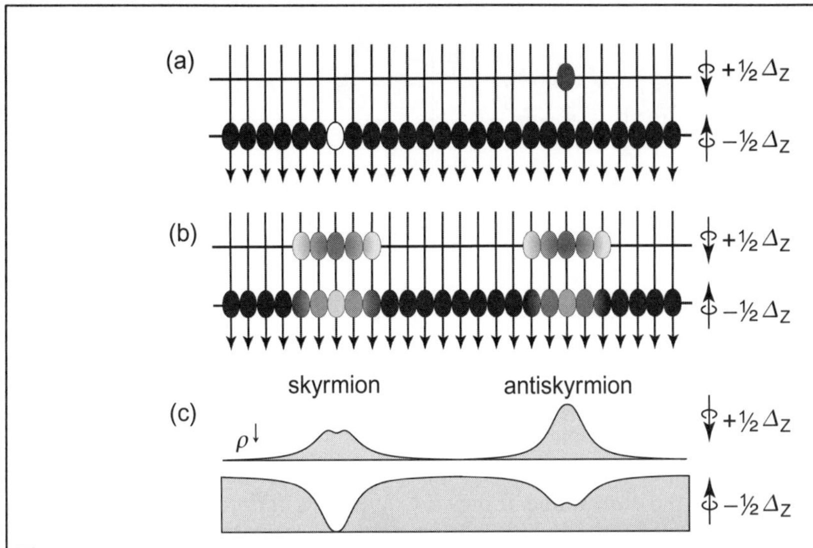

Fig. 16.1 (a) When one electron is excited at $\nu = 1$, its spin is reversed and there appears one hole. The Coulomb energy of each part is $\sim E_C^0$. (b) The Coulomb self-energy is reduced to $\sim (2\kappa)^{-1}E_C^0$ by exciting neighboring electrons coherently spread over a domain of size $\sim 2\kappa\ell_B$. The excited objects are known as a skyrmion and an antiskyrmion. (c) They are topological solitons in the composite-boson theory. A skyrmion may carry electric charge $\pm e/m$ at $\nu = 1/m$. The electron densities $\rho^\uparrow(\boldsymbol{x})$ and $\rho^\downarrow(\boldsymbol{x})$ are illustrated. The electron spin is reversed at the center of both the skyrmion and the antiskyrmion.

semiclassical field theory.[144,146,147] A microscopic formulation of skyrmions has revealed some new aspects,[163,479] which we shall review in Chapter 31.

Skyrmion excitations can be detected experimentally by measuring their spins. Recall the mechanism of the Hall plateau formation around the filling factor $\nu = 1/m$ (see Section 11.5). A plateau is generated around $\nu = 1/m$, because quasielectrons are created but trapped for $\nu > 1/m$ while quasiholes are created but trapped for $\nu < 1/m$, as in Fig.16.2. This mechanism works equally for skyrmion excitations as well as for vortex excitations. The number of quasiparticles at $\nu \neq 1/m$ is given by (11.5.3) with (11.5.2), or (11.5.5) with (11.5.4).

A crucial feature of a skyrmion is that it reverses spins more than a vortex or a hole excitation does. Skyrmions are detectable by measuring the number of reversed spins.

In the fractional QH state [Fig.16.2(a)], if quasiparticles are vortices, the spin polarization remains unchanged around $\nu = 1/m$ since their excitations reverse no spins; if they are skyrmions, the spin polarization is decreased rapidly by their excitations around $\nu = 1/m$.

In the integer QH state [Fig.16.2(b)], if quasiparticles are holes, no spin is reversed for $\nu < 1$, but one spin is reversed per one excitation for $\nu > 1$ since

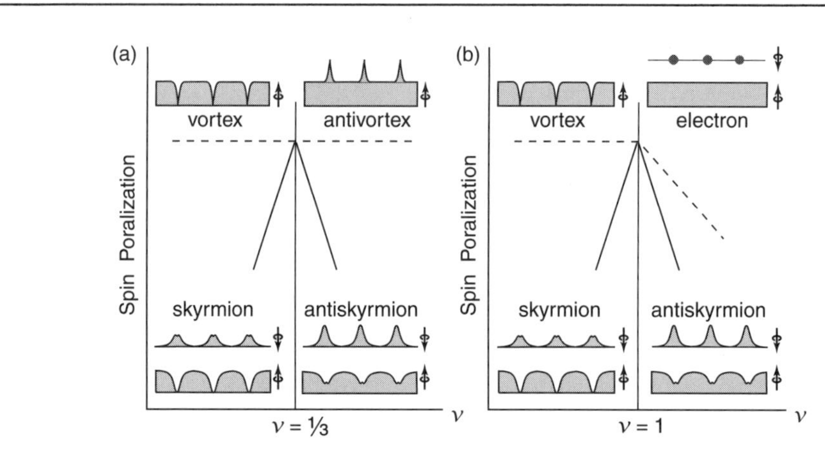

Fig. 16.2 (a) If quasiparticles are vortices, the spin polarization is not changed around $\nu = 1/3$ as indicated by the dashed line. If they are skyrmions, it decreases rapidly both for $\nu > 1/3$ and $\nu < 1/3$ as indicated by the solid lines. (b) If quasiparticles are vortices, the spin polarization is not changed for $\nu < 1$ but it decreases for $\nu > 1$. If they are skyrmions, it decreases rapidly both for $\nu > 1$ and $\nu < 1$. See Fig.16.3 for experimental data.

electrons are pushed up to a higher energy level; if they are skyrmions, spins are reversed both for $\nu > 1$ and $\nu < 1$ by their excitations. An experimental measurement[63] of the spin polarization has been made around $\nu = 1$, as is sketched in Fig.16.3. Here, the spin polarization is observed both for $\nu > 1$ and for $\nu < 1$. It is a clear signal of skyrmion excitations.

Skyrmion excitations are also observed by measuring the activation energy in magnetotransport experiments at[429] $\nu = 1$ and at[300] $\nu = 1/3$ by the use of the tilted-field method. Tilting the magnetic field $\boldsymbol{B}$ with respect to the sample is an indispensable method[157,468,226] to study the role of the spin degree of freedom, since it changes effectively the Zeeman gap. The skyrmion excitation energy E_{sky} consists of the Coulomb energy and the Zeeman energy,

$$E_{\text{sky}} = E_{\text{C}}^{\text{sky}} + N_{\text{spin}}\Delta_{\text{Z}}, \tag{16.1.1}$$

where both the Coulomb energy E_{C} and the skyrmion spin N_{spin} depend only on the perpendicular component $B_\perp$ of the applied field $\boldsymbol{B}$. Hence, the skyrmion spin is given by

$$N_{\text{spin}} = \frac{dE_{\text{sky}}}{d\Delta_{\text{Z}}}, \tag{16.1.2}$$

which is the number of flipped spins by one skyrmion excitation.

The skyrmion scale κ depends on the Zeeman effect. As the Zeeman effect becomes very large, the QH ferromagnet approaches the spin-frozen system. At

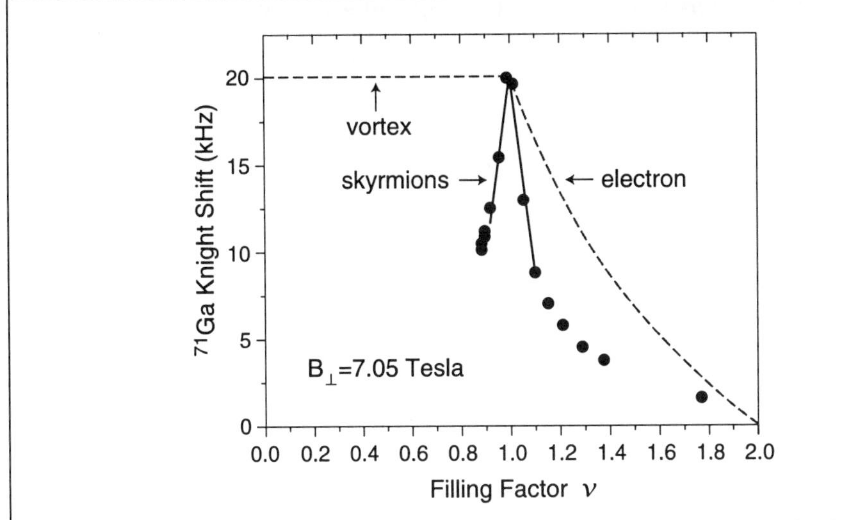

Fig. 16.3 The change of the spin polarization is found around $\nu = 1$ in the QH state by a measurement of the Knight shift. It is observed that a skyrmion and an antiskyrmion flip the same number of spins. The number of flipped spins is 3.6 per skyrmion at $B_\perp = 7.05$ Tesla. The data are taken from Barrett et al.[63] Skyrmions are clearly excited by comparing them with Fig.16.2.

$\nu = 1$, an antiskyrmion-skyrmion pair is reduced continuously to an electron-hole pair in this limit. On the contrary, at $\nu = 1/m$, $m \neq 1$, though a skyrmion is reduced to a vortex, there is a clear distinction between an antiskyrmion and an antivortex. An antiskyrmion has a predominant component in the down-spin level [Fig.16.1], but an antivortex is confined entirely within the up-spin level. It is a dynamical problem which solitons are excited.

16.2 Factorizable Skyrmions

We analyze topological excitations in the QH ferromagnet at $\nu = 1/m$ based on the composite-boson theory. See Chapter 31 for a microscopic analysis. The exchange interaction is represented by the Hamiltonian (15.1.1). It is the O(3) sigma model allowing skyrmions as topological excitations: See section 7.7. However, the Hamiltonian contains only the spin part (15.4.5) of the spin-charge separation (15.3.4). We investigate how the spin rotation affects the charge distribution.

The analysis of skyrmions is quite analogous to that of vortices given in Section 14.3. When the N-body wave function is factorizable,

$$\mathfrak{S}_{\text{CB}}[\boldsymbol{x}] = \langle 0|\Phi(\boldsymbol{x}_1)\Phi(\boldsymbol{x}_2)\cdots\Phi(\boldsymbol{x}_N)|\mathfrak{S}\rangle = \prod_r \langle\Phi(\boldsymbol{x}_r)\rangle, \tag{16.2.1}$$

the one-point function is analytic [see (14.3.3)] and given by

$$\langle \Phi(\boldsymbol{x}) \rangle = \rho_0^{1/2} \begin{pmatrix} \omega^{\uparrow}(z) \\ \omega^{\downarrow}(z) \end{pmatrix}. \tag{16.2.2}$$

From (15.3.6), (15.3.4) and (16.2.2) we derive the LLL condition for solitons,

$$e^{-\mathfrak{A}(\boldsymbol{x})}\sqrt{\rho(\boldsymbol{x})}n^{\alpha}(\boldsymbol{x}) = \rho_0^{1/2}\omega^{\alpha}(z), \tag{16.2.3}$$

where $n^{\alpha}(\boldsymbol{x})$ is the CP^1 field subject to $|n^{\uparrow}(\boldsymbol{x})|^2 + |n^{\downarrow}(\boldsymbol{x})|^2 = 1$. Here and hereafter, we assume various fields as classical fields since topological solitons are classical objects. The difference from the vortex case (14.3.8) is that we have the CP^1 field $n^{\alpha}(\boldsymbol{x})$ instead of the U(1) phase factor $e^{i\chi(\boldsymbol{x})}$.

We first solve the LLL condition (16.2.3) for the CP^1 field $n^{\alpha}(\boldsymbol{x})$ as

$$n^{\alpha}(\boldsymbol{x}) = \frac{\omega^{\alpha}(z)}{\sqrt{|\omega^{\uparrow}(z)|^2 + |\omega^{\downarrow}(z)|^2}}. \tag{16.2.4}$$

This is the most general form of the skyrmion configuration (7.8.46) given in Section 7.7. Skyrmions are topological solitons associated with the boundary condition (15.4.17) at infinity ($|z| \to \infty$).

We call the skyrmion described by (16.2.2) the *factorizable skyrmion* since the wave function is fatorizable as in (16.2.1). It is a strong requirement that the wave function is factorizable. Indeed, as we shall see in Section 31.7, the realistic skyrmion is not of this type. Nevertheless, we make an analysis based on it since it is simple and captures the basic nature of the charged excitation. Furthermore it is a very good approximation to the realistic skyrmion.

In order to identify the topological charge we substitute (16.2.4) into the LLL condition (16.2.3),

$$e^{-\mathfrak{A}(\boldsymbol{x})}\sqrt{\rho(\boldsymbol{x})} = \rho_0^{1/2}\sqrt{|\omega^{\uparrow}(z)|^2 + |\omega^{\downarrow}(z)|^2}. \tag{16.2.5}$$

Taking the logarithm and the derivative $\boldsymbol{\nabla}^2$, we obtain the soliton equation

$$\frac{1}{4\pi m}\boldsymbol{\nabla}^2 \ln \rho(\boldsymbol{x}) - \rho(\boldsymbol{x}) + \rho_0 = \frac{1}{m}J^0_{\text{sky}}(\boldsymbol{x}), \tag{16.2.6}$$

where use was made of (13.4.12) and

$$J^0_{\text{sky}}(\boldsymbol{x}) = \frac{1}{4\pi}\boldsymbol{\nabla}^2 \ln\left[|\omega^{\uparrow}(z)|^2 + |\omega^{\downarrow}(z)|^2\right]. \tag{16.2.7}$$

To see that this is a topological charge density, we write down the Cauchy-Riemann equation for the analytic function $\omega^{\alpha}(z)$ in (16.2.4),

$$0 = \frac{\partial \omega^{\alpha}(z)}{\partial z^*} = \frac{\partial n^{\alpha}(\boldsymbol{x})}{\partial z^*}\sqrt{|\omega^{\uparrow}(z)|^2 + |\omega^{\downarrow}(z)|^2} + n^{\alpha}(\boldsymbol{x})\frac{\partial}{\partial z^*}\sqrt{|\omega^{\uparrow}(z)|^2 + |\omega^{\downarrow}(z)|^2}. \tag{16.2.8}$$

We divide it by $\sqrt{|\omega^\uparrow(z)|^2 + |\omega^\downarrow(z)|^2}$, multiply it with $n^{\alpha*}(\boldsymbol{x})$ and take the sum over $\alpha = 1, 2$,

$$\frac{\partial}{\partial z^*} \ln \sqrt{|\omega^\uparrow(z)|^2 + |\omega^\downarrow(z)|^2} = -\sum_{\alpha=1}^{2} n^{\alpha*}(\boldsymbol{x}) \frac{\partial n^\alpha(\boldsymbol{x})}{\partial z^*}, \tag{16.2.9}$$

or

$$\partial_k \ln \sqrt{|\omega^\uparrow(z)|^2 + |\omega^\downarrow(z)|^2} = \varepsilon_{kj} K_j(\boldsymbol{x}), \tag{16.2.10}$$

where

$$K_k = -i \sum_\alpha n^{\alpha*} \partial_k n^\alpha \tag{16.2.11}$$

is the auxiliary field (7.8.8) associated with the CP^1 field. The source term (16.2.7) is rewritten as

$$J^0_{\text{sky}}(\boldsymbol{x}) = \frac{1}{4\pi} \nabla^2 \ln \left[|\omega^\uparrow(z)|^2 + |\omega^\downarrow(z)|^2 \right] = \frac{1}{2\pi} \varepsilon_{ij} \partial_i K_j(\boldsymbol{x}), \tag{16.2.12}$$

which is the time component of the topological current density

$$J^\mu_{\text{sky}}(\boldsymbol{x}) = \frac{1}{2\pi} \varepsilon^{\mu\nu\lambda} \partial_\nu K_\lambda(\boldsymbol{x}). \tag{16.2.13}$$

The topological charge is identical to the Pontryagin number (15.4.21): Compare (15.4.22) with (7.7.29), and (16.2.13) with (7.8.15).

The simplest skyrmion is described by the one-point function

$$\begin{pmatrix} \langle \varphi^\uparrow(\boldsymbol{x}) \rangle_{\text{sky}} \\ \langle \varphi^\downarrow(\boldsymbol{x}) \rangle_{\text{sky}} \end{pmatrix} = \sqrt{\rho_0} \begin{pmatrix} z \\ \kappa \end{pmatrix}, \tag{16.2.14}$$

or by the CP^1 field[2]

$$\begin{pmatrix} n^\uparrow(\boldsymbol{x}) \\ n^\downarrow(\boldsymbol{x}) \end{pmatrix} = \frac{1}{\sqrt{|z|^2 + \kappa^2}} \begin{pmatrix} z \\ \kappa \end{pmatrix}, \tag{16.2.15}$$

with an arbitrary constant κ. It represents a classical skyrmion with scale κ sitting at the origin of the system. The electron wave function is

$$\mathfrak{S}_{\text{sky}}[\boldsymbol{x}] = \prod_r \begin{pmatrix} z_r \\ \kappa \end{pmatrix} \mathfrak{S}_{\text{LN}}[\boldsymbol{x}]. \tag{16.2.16}$$

By the use of (16.2.14) the sigma field (15.4.5) is calculated as

$$\mathcal{S}_x^{\text{sky}} = \frac{1}{2}\sqrt{1 - \sigma_{\text{sky}}^2} \cos\theta, \quad \mathcal{S}_y^{\text{sky}} = -\frac{1}{2}\sqrt{1 - \sigma_{\text{sky}}^2} \sin\theta, \quad \mathcal{S}_z^{\text{sky}} = \frac{1}{2}\sigma_{\text{sky}}, \tag{16.2.17}$$

[2]In the $\nu = 1$ QH ferromagnet, we shall argue in Section 31.2 that a skyrmion state is constructed from a hole state by making a $W_\infty(2)$ rotation, which is an SU(2) rotation in the noncommutative space. In particular, as $\kappa \to 0$, the skyrmion state (16.2.15) is continuously reduced to the hole state created in the ground state where all up-spin states are filled up.

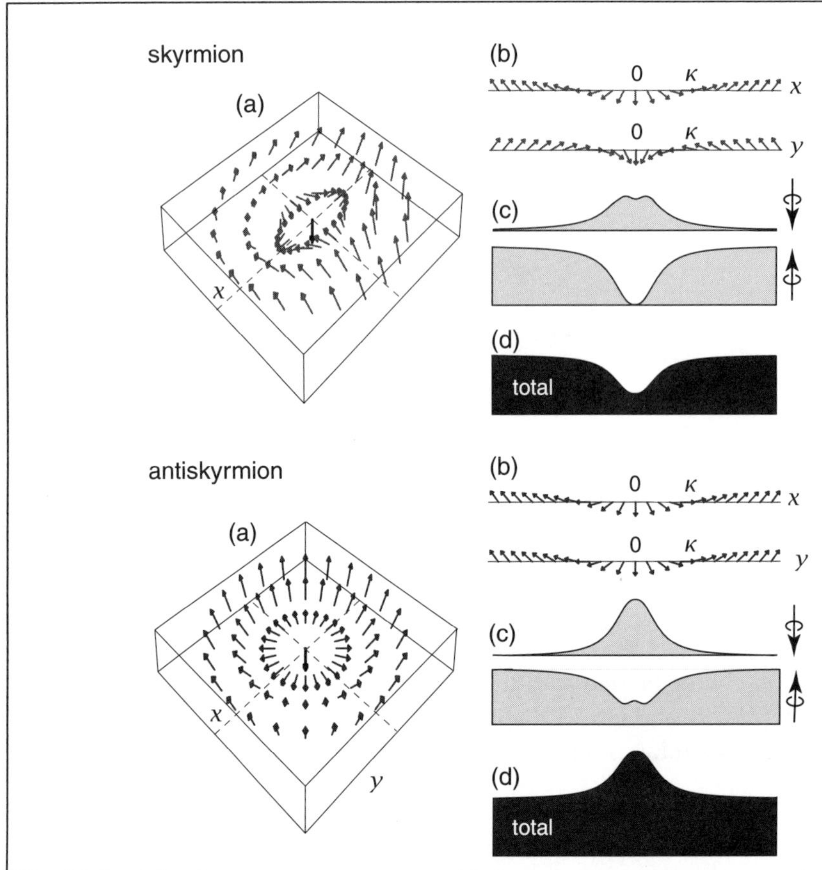

Fig. 16.4 The electron spin rotates smoothly in the presence of a skyrmion (16.2.18) or an antiskyrmion (16.2.29). It points downward at the center, and is in the xy plane at $r = 2\kappa\ell_B$, and points upward asymptotically. (a) An overall view of the normalized spin $s_a(\boldsymbol{x})$ at $r = 0$, $2\kappa\ell_B$, $4\kappa\ell_B$ and $6\kappa\ell_B$. (b) An overall view of the normalized spin $s_a(\boldsymbol{x})$ along the x axis for $-6\kappa\ell_B < x < 6\kappa\ell_B$, and along the y axis for $-6\kappa\ell_B < y < 6\kappa\ell_B$. (c) The up-spin and down-spin electron densities around a skyrmion and an antiskyrmion.

with

$$\sigma_{\text{sky}}(\boldsymbol{x}) = \frac{r^2 - (2\kappa\ell_B)^2}{r^2 + (2\kappa\ell_B)^2}. \tag{16.2.18}$$

The spin flips at the skyrmion center, $\boldsymbol{\mathcal{S}} = (0, 0, -1/2)$ at $r = 0$, while the spin-polarized ground state is approached away from it, $\boldsymbol{\mathcal{S}} = (0, 0, 1/2)$ for $r \gg \kappa\ell_B$. The spin polarization around a skyrmion is illustrated in Fig.16.4(a).

A skyrmion excitation induces the density modulation $\Delta\rho_{\text{sky}}(\boldsymbol{x})$ around the ground-state value ρ_0,

$$\rho_{\text{sky}}(\boldsymbol{x}) = \rho_0 + \Delta\rho_{\text{sky}}(\boldsymbol{x}). \tag{16.2.19}$$

The electron number associated with the skyrmion (16.2.14) is

$$N_{\text{sky}} = \int d^2x\, \Delta\rho_{\text{sky}}(\boldsymbol{x}) = -\frac{Q_{\text{sky}}}{m} = -\frac{1}{m}, \tag{16.2.20}$$

by integrating the soliton equation (16.2.6). A skyrmion excitation removes $1/m$ electron from the system.[3]

Here we discuss the relation between a skyrmion and a vortex in the fractional QH system. First of all, the skyrmion wave function (16.2.16) approaches the vortex wave function (11.4.10) in the spin-frozen limit as $\kappa \to 0$. Second, the soliton equation (16.2.6) for the skyrmion excitation is identical with the corresponding one (14.3.11) for the vortex excitation. The only difference is the explicit expression of the topological charge: It is the vorticity density for the vortex, and the Pontryagin number density for the skyrmion. Now, the Pontryagin number density (16.2.7) is calculated for a skyrmion with (16.2.14) as

$$J^0_{\text{sky}}(\boldsymbol{x}) = \frac{1}{\pi}\frac{(2\kappa\ell_B)^2}{[r^2 + (2\kappa\ell_B)^2]^2}, \tag{16.2.21}$$

which is reduced to

$$J^0_{\text{sky}}(\boldsymbol{x}) \to \delta(\boldsymbol{x}) \qquad \text{as} \quad \kappa \to 0. \tag{16.2.22}$$

This is the vorticity density (14.3.16). Hence, a skyrmion is reduced to a vortex in the small-skyrmion limit $\kappa \to 0$. We shall make an analysis on a skyrmion and a hole based on the microscopic theory in Section 31.4.

It is easy to obtain an approximate solution of the soliton equation in two limits, the large skyrmion limit ($\kappa \gg 1$) and the small skyrmion limit ($\kappa \ll 1$). First, in the large limit we may solve it iteratively as

$$\Delta\rho_{\text{sky}}(\boldsymbol{x}) = -\frac{1}{m}J^0_{\text{sky}}(\boldsymbol{x}) - \frac{1}{4\pi m^2\rho_0}\nabla^2 J^0_{\text{sky}}(\boldsymbol{x}) + \cdots, \tag{16.2.23}$$

which we approximate as $\Delta\rho_{\text{sky}}(\boldsymbol{x}) \simeq -m^{-1}J^0_{\text{sky}}(\boldsymbol{x})$, or

$$\Delta\rho_{\text{sky}}(\boldsymbol{x}) \simeq -\frac{1}{\pi m}\frac{(2\kappa\ell_B)^2}{[r^2 + (2\kappa\ell_B)^2]^2}, \qquad \text{for} \quad \kappa \gg 1. \tag{16.2.24}$$

On the other hand, in the small-size limit the skyrmion is reduced to the vortex in the fractional QH system,

$$\Delta\rho_{\text{sky}}(\boldsymbol{x}) \simeq -\rho_0\left(1 + \frac{\sqrt{2}r}{\ell_B} - \frac{r^2}{3\ell_B^2}\right)e^{-\sqrt{2}r/\ell_B}, \qquad \text{for} \quad \kappa \simeq 0, \tag{16.2.25}$$

and to the hole in the integer QH system [see (29.4.38)],

$$\Delta\rho_{\text{sky}}(\boldsymbol{x}) = -\rho_0 e^{-r^2/2\ell_B^2}. \tag{16.2.26}$$

[3] Needless to say, it is impossible to remove a fractional number of electrons from the system. Actually, a skyrmion-antiskyrmion pair is excited with the zero total electron number.

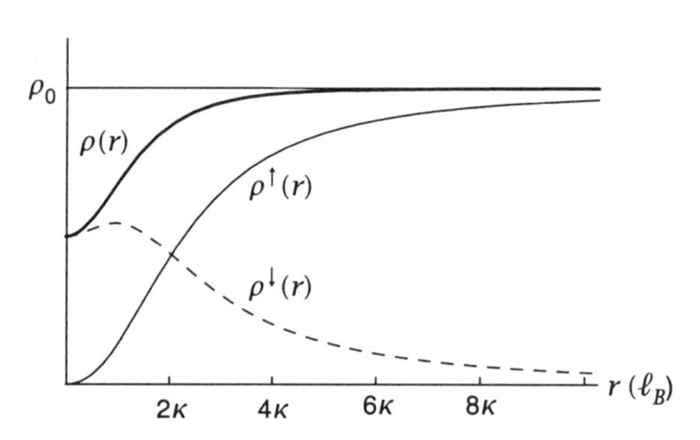

Fig. 16.5 Electron densities $\rho(\boldsymbol{x})$, $\rho^{\uparrow}(\boldsymbol{x})$ and $\rho^{\downarrow}(\boldsymbol{x})$ are modulated around a skyrmion as depicted by the thick curve, the thin curve and the dotted curve, respectively.

The density $\rho^{\alpha}_{\text{sky}}(\boldsymbol{x})$ for each spin component is given by

$$\rho^{\uparrow}_{\text{sky}}(\boldsymbol{x}) = \frac{1}{2}\rho_{\text{sky}}(\boldsymbol{x})[1 + \sigma_{\text{sky}}(\boldsymbol{x})] = \rho_{\text{sky}}(\boldsymbol{x})\frac{r^2}{r^2 + (2\kappa\ell_B)^2},$$
$$\rho^{\downarrow}_{\text{sky}}(\boldsymbol{x}) = \frac{1}{2}\rho_{\text{sky}}(\boldsymbol{x})[1 - \sigma_{\text{sky}}(\boldsymbol{x})] = \rho_{\text{sky}}(\boldsymbol{x})\frac{(2\kappa\ell_B)^2}{r^2 + (2\kappa\ell_B)^2}. \tag{16.2.27}$$

We have illustrated the electron densities $\rho_{\text{sky}}(r)$, $\rho^{\uparrow}_{\text{sky}}(\boldsymbol{x})$ and $\rho^{\downarrow}_{\text{sky}}(\boldsymbol{x})$ around a typical skyrmion in Fig.16.5.

Antiskyrmion

An antiskyrmion is a topological soliton possessing a negative Pontryagin number that approaches the same spin-polarized ground state asymptotically. Its classical configuration is given by reversing the sign of the spin component $\mathcal{S}^{\text{sky}}_y(\boldsymbol{x})$ in (16.2.17). The simplest antiskyrmion is given by (7.7.25) with $q = 1$, or

$$\mathcal{S}^{\text{antisky}}_x = \frac{1}{2}\sqrt{1 - \sigma^2_{\text{sky}}}\cos\theta = \mathcal{S}^{\text{sky}}_x,$$
$$\mathcal{S}^{\text{antisky}}_y = \frac{1}{2}\sqrt{1 - \sigma^2_{\text{sky}}}\sin\theta = -\mathcal{S}^{\text{sky}}_y, \tag{16.2.28}$$
$$\mathcal{S}^{\text{antisky}}_z = \frac{1}{2}\sigma_{\text{sky}} = \mathcal{S}^{\text{sky}}_z, \tag{16.2.29}$$

where σ_{sky} is given by (16.2.18). The spin flips at the antiskyrmion center, $\mathcal{S} = (0, 0, -1/2)$ at $r = 0$, while the spin-polarized ground state is approached away from it, $\mathcal{S} = (0, 0, 1/2)$ for $r \gg \kappa\ell_B$. The topological number density is

$$J^0_{\text{antisky}}(\boldsymbol{x}) = -\frac{1}{\pi}\frac{(2\kappa\ell_B)^2}{[r^2 + (2\kappa\ell_B)^2]^2}. \tag{16.2.30}$$

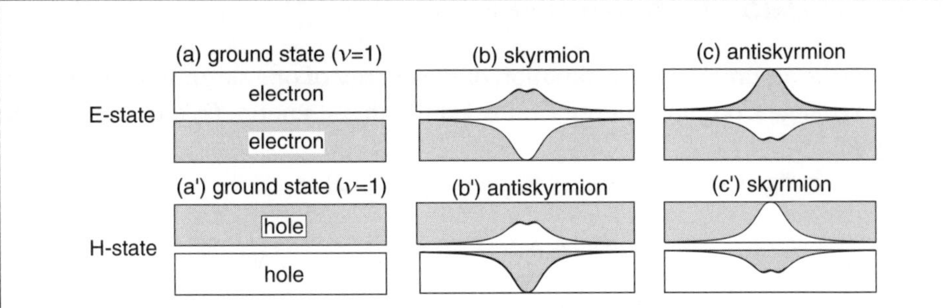

Fig. 16.6 (a) The QH ferromagnet at $\nu = 1$ is the state (E-state) where electron sites are half filled; (a′) It is equivalently viewed as the state (H-state) where hole sites are half filled. (b) Now, a skyrmion is a quasihole in the E-state; (b′) It is equivalently regarded as a quasielectron (antiskyrmion) in the H-state. (c) Similarly, an antiskyrmion is a quasielectron in the E-state; (c′) It is equivalently regarded as a quasihole (skyrmion) in the H-state.

The spin polarization around an antiskyrmion is illustrated in Fig.16.4(b).

An antiskyrmion induces the density modulation in the ground state,

$$\rho_{\text{antisky}}(\boldsymbol{x}) = \rho_0 + \Delta\rho_{\text{antisky}}(\boldsymbol{x}). \tag{16.2.31}$$

It is difficult to make a mean field analysis of $\Delta\rho_{\text{antisky}}(\boldsymbol{x})$ since the wave function of composite bosons is not factorizable into the product of one-body functions. Nevertheless, all the results obtained to a skyrmion apply equally to the case of an antiskyrmion because of an exact particle-hole symmetry that holds at $\nu = 1$. First, the QH ferromagnet is the state (E-state) where electron sites are half filled. It is equivalently viewed as the state (H-state) where hole sites are half filled. Now, an antiskyrmion is a quasiparticle in the E-state, but is equivalently regarded as a quasihole (skyrmion) in the H-state [Fig.16.6]. We can make a similar argument for an antiskyrmion at $\nu = 1/m$. Hence, from (16.2.21) and (16.2.30) we conclude

$$\Delta\rho_{\text{antisky}}(\boldsymbol{x}) = -\Delta\rho_{\text{sky}}(\boldsymbol{x}) \simeq \frac{1}{m} J^0_{\text{sky}}(\boldsymbol{x}). \tag{16.2.32}$$

The electron number associated with the antiskyrmion is

$$N_{\text{antisky}} = \int d^2x\, \Delta\rho_{\text{antisky}}(\boldsymbol{x}) = \frac{Q_{\text{sky}}}{m} = \frac{1}{m}. \tag{16.2.33}$$

An antiskyrmion adds $1/m$ electron to the system.

It is possible to discuss an antiskyrmion and a skyrmion precisely in the same way in the microscopic formalism, where we can show that the same number of spins are flipped around a skyrmion and an antiskyrmion: See Chapter 31. This is precisely what is observed experimentally at $\nu = 1$ [see Fig.16.3].

16.3 Skyrmion Excitation Energy

We make a semiclassical estimation of the energy of one skyrmion.[4] It consists of the exchange Coulomb energy E_X, the Coulomb self-energy E_D and the Zeeman energy E.

The exchange Coulomb energy is given by (15.5.2),

$$E_X = \frac{2}{m}J_s \int d^2x\, [\partial_k \mathcal{S}^{sky}(\boldsymbol{x})]^2 = \frac{4\pi}{m}J_s = \frac{1}{4m}\sqrt{\frac{\pi}{2}}E_C^0. \tag{16.3.1}$$

This is an exact result as far as the nonlinear sigma model is valid [see (7.7.12)]. However, the nonlinear sigma model is a poor approximation within the core of a skyrmion because the density modulation $\delta\rho_{sky}(\boldsymbol{x})$ due to the skyrmion excitation has been neglected, as discussed in Section 31.7: See Fig.31.1. To take the correction into account we set

$$E_X = \frac{\alpha}{4m}\sqrt{\frac{\pi}{2}}E_C^0, \tag{16.3.2}$$

by introducing a phenomenological parameter α. We expect $\alpha \simeq 1$.

The Coulomb self-energy is given by (15.5.1) together with the density modulation $\Delta\rho_{sky}(\boldsymbol{x})$,

$$E_D = \frac{1}{2}\int d^2x d^2y\, \Delta\rho_{sky}(\boldsymbol{x})V(\boldsymbol{x}-\boldsymbol{y})\Delta\rho_{sky}(\boldsymbol{y}). \tag{16.3.3}$$

On one hand, assuming a large skyrmion, we find

$$E_D^{sky} = \beta_C(\kappa) = \frac{1}{2m^2}\int dx\, x^2\, [K_1(x)]^2 = \frac{3\pi^2}{128\kappa m^2}E_C^0, \tag{16.3.4}$$

where we have used the skyrmion configuration (16.2.24): See (O.7) in Appendix. On the other hand, the Coulomb self-energy E_D of a hole at $\nu = 1$ is

$$E_D^{hole} = \frac{1}{2}\sqrt{\frac{\pi}{2}}E_C^0, \tag{16.3.5}$$

as we shall derive in Section 30.3: See (30.3.8) with (30.1.16b). We have illustrated these energies as functions of κ in Fig.16.7, where two curves meet at $\kappa \simeq 0.37$.

The Zeeman energy is given by (10.1.14),

$$E_Z = N_{spin}\Delta_Z. \tag{16.3.6}$$

It is proportional to the skyrmion spin N_{spin} given by

$$N_{spin} = -\int d^2x\, \left\{S_z^{sky}(\boldsymbol{x}) - \frac{1}{2}\rho_0\right\}. \tag{16.3.7}$$

[4] In computing the activation energy of a skyrmion or an antiskyrmion, one needs to know the chemical potential to remove or add an electron to the system. However, this problem does not exist as far as we consider a skyrmion-antiskyrmion pair because the number of electrons is unchanged. Thus, when we calculate the skyrmion energy, we actually consider the energy of a skyrmion in a well-separated skyrmion-antiskyrmion pair.

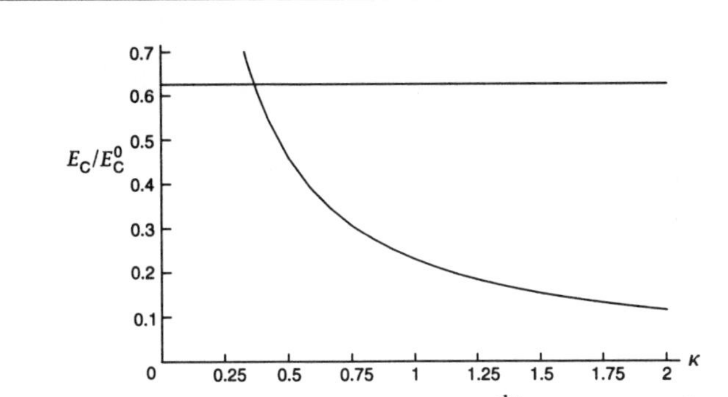

Fig. 16.7 The Coulomb energy of one skyrmion, E_C^{sky} in units of E_C^0, is illustrated as a function of the scale κ at $\nu = 1$. The curve represents an estimation based on the large-skyrmion formula (16.3.4), while the horizontal line is the hole excitation energy (16.3.5).

For the hole we have

$$N_{\text{spin}} = \frac{1}{2}. \tag{16.3.8}$$

To study a large skyrmion we rewrite it as

$$N_{\text{spin}} = \frac{1}{2}\int d^2x\, \rho_0 \left\{1 - \sigma_{\text{sky}}(\boldsymbol{x})\right\} - \frac{1}{2}\int d^2x\, \Delta\rho_{\text{sky}}(\boldsymbol{x})\sigma_{\text{sky}}(\boldsymbol{x}), \tag{16.3.9}$$

where we have used $S_z^{\text{sky}}(\boldsymbol{x}) = \rho_{\text{sky}}(\boldsymbol{x})\mathcal{S}_z^{\text{sky}}(x)$ according to (15.4.4) together with (16.2.17) and (16.2.19). The antiskyrmion spin is given by

$$N_{\text{spin}}^{\text{antisky}} = \frac{1}{2}\int d^2x\, \rho_0 \left\{1 - \sigma_{\text{sky}}(\boldsymbol{x})\right\} + \frac{1}{2}\int d^2x\, \Delta\rho_{\text{sky}}(\boldsymbol{x})\sigma_{\text{sky}}(\boldsymbol{x}), \tag{16.3.10}$$

as follows from (16.2.29) and (16.2.32).

We consider a pair of skyrmion and antiskyrmion, which is excited thermally. The second term in (16.3.9) is cancelled by the corresponding term in (16.3.10). We ignore it since it does not contribute to the Zeeman energy of a pair excitation. We focus on the first term in (16.3.9). It diverges logarithmically, which leads to an infinitely large Zeeman energy. We can show that a factorizable skyrmion (16.2.16) carries necessarily an infinitely large Zeeman energy because of its slow fall-off to the ground-state value, as implies that it cannot be created. A realistic skyrmion is such that its Zeeman energy is finite.[5] We shall construct explicitly such a skyrmion based on a microscopic theory in Section 31.7. Here, in order to reveal the essence, we simply cut off the divergence. We estimate the cut-off parameter as follows. First, the skyrmion excitation occurs within the coherent domain, only within which the spin can be modulated coherently. Second, the

[5]We have already remarked that the standard skyrmion configuration (16.2.17) is no longer a solution of the broken-O(3) sigma model (7.7.36).

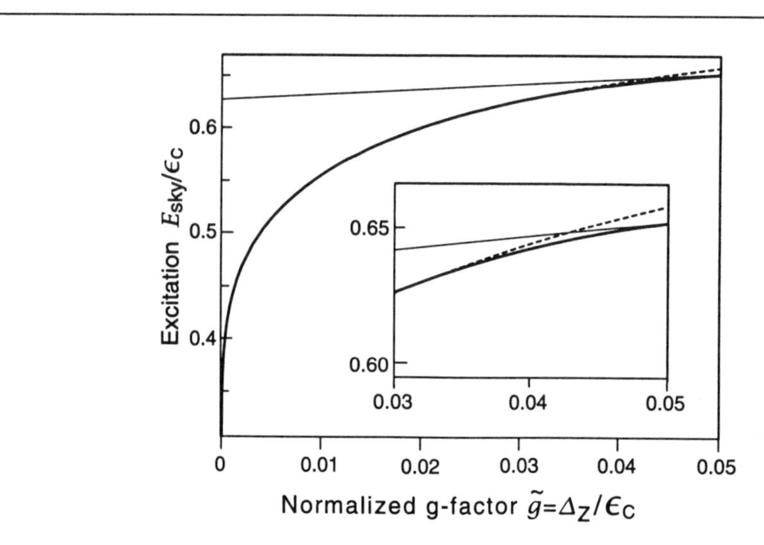

Fig. 16.8 The skrmion excitation energy is plotted as a function of the normalized Zeeman gap $\tilde{g} = \Delta_Z/\epsilon_C$. We have given the excitation energy calculated analytically based on the formula (16.3.16), which is represented by the dotted curve. The heavy solid curve is obtained by the numerical analysis made in Section 31.7: See Fig.31.2 there. The two curves are practically identical for $\tilde{g} < 0.038$. The thin line is the hole excitation energy. It is seen that a skyrmion is excited for $\tilde{g} < 0.045$ but a hole is excited for $\tilde{g} > 0.045$. The inset is an enlargement of the figure near $\tilde{g} = 0.04$, where three curves meet.

skyrmion size must be proportional to the scale parameter κ. Hence, we cut off the upper limit of the integration at $r \simeq \beta\kappa\xi_{\text{spin}}$ with a phenomenological parameter β,

$$N_{\text{spin}} \simeq \frac{1}{2}\int_{r<\kappa\xi_Z} d^2x\, \rho_0 \left\{1 - \sigma_{\text{sky}}(\boldsymbol{x})\right\} \simeq 2\kappa^2 \ln\left(\frac{\beta^2\xi_{\text{spin}}^2}{4\ell_B^2} + 1\right). \tag{16.3.11}$$

The coherence length is given by (15.5.12), or

$$\frac{\xi_{\text{spin}}^2}{\ell_B^2} = \frac{2J_s}{\Delta_Z\rho_0\ell_B^2} = \frac{\sqrt{2\pi}}{8\tilde{g}} \tag{16.3.12}$$

by the use of (9.1.10) and (15.1.2), where we have introduced the normalized Zeeman gap $\tilde{g}$ by

$$\tilde{g} \equiv \frac{\Delta_Z}{E_C^0}. \tag{16.3.13}$$

It reads $\tilde{g} \simeq 0.02$ and hence $\xi_{\text{spin}} \simeq 4\ell_B$ in typical samples at $B = 10$ Tesla.

The total excitation energy is given by $E_{sky} = E_X + E_D + E_Z$,

$$E_{sky} \simeq \left\{ \frac{1}{4m}\sqrt{\frac{\pi}{2}}\alpha + \frac{3\pi^2}{128\kappa m^2} + 2\kappa^2 \tilde{g} \ln\left(\frac{\beta^2\sqrt{2\pi}}{32\tilde{g}} + 1\right)\right\} E_C^0. \tag{16.3.14}$$

The Coulomb energy decreases while the Zeeman energy increases as the skyrmion scale κ becomes larger. The optimized scale κ is determined so as to minimize the total energy. The optimized scale is found to be

$$\kappa \simeq \frac{1}{2}\left(\frac{3\pi^2}{64m^2}\right)^{1/3} \left\{\tilde{g}\ln\left(\frac{\beta^2\sqrt{2\pi}}{32\tilde{g}} + 1\right)\right\}^{-1/3}, \tag{16.3.15}$$

with which the skyrmion energy is

$$E_{sky} \simeq \left\{ \frac{1}{4m}\sqrt{\frac{\pi}{2}}\alpha + \frac{9\pi^2}{256\kappa m^2}\right\} E_C^0. \tag{16.3.16}$$

We have derived the skyrmion excitation energy with the use of two phenomenological parameters α and β.

We are able to develop a microscopic theory of a skyrmion and make a reliable numerical estimation of the excitation energy, as will be described in Section 31.7. When we choose $\alpha \simeq 0.9$ and $\beta \simeq 1.478$, the agreement with the numerical result is remarkable at $\nu = 1$, as is seen in Fig.16.8. Consequently, though strictly speaking the factorizable skyrmion cannot be physical, it still presents a reasonable approximation with an appropriate cutoff of the divergent Zeeman energy.

16.4 Experimental Evidence

In this section we focus on the skyrmion at $\nu = 1$. A skyrmion excitation is characterized by a coherent excitation of spins. We would naively expect $N_{spin} = 1/2$ associated with either a hole or an electron excitation, where N_{spin} is the number of spins carried by one quasiparticle. Hence an evidence of the skyrmion excitation is given if $N_{spin} > 1/2$.

The number of flipped spins is given by the relation (16.1.2). We can confirm that the relation is preserved by the above minimization procedure as follows. The excitation energy (16.3.16) is the sum of the Coulomb and Zeeman energies, which depend on the scale parameter τ. The quantity to minimize with respect to τ is

$$E_{sky}(\tau) = E_C(\tau) + \Delta_Z N_{spin}(\tau), \tag{16.4.1}$$

with $E_C = E_X + E_D$. At the minimum we obtain

$$\frac{\partial E_C}{\partial \tau} + \Delta_Z \frac{\partial N_{spin}}{\partial \tau} = 0, \tag{16.4.2}$$

from which we solve out $\tau = f_0(\Delta_Z)$ and substituting it back into (16.4.2),

$$E_{\text{sky}}(\Delta_Z) = E_C[f_0(\Delta_Z)] + \Delta_Z N_{\text{spin}}[f_0(\Delta_Z)], \tag{16.4.3a}$$

$$N_{\text{spin}}(\Delta_Z) = N[f_0(\Delta_Z)]. \tag{16.4.3b}$$

Using (16.4.2) it is easy to verify the formula

$$N_{\text{spin}} = \frac{d\langle H\rangle_{\text{sky}}}{d\Delta_Z}, \tag{16.4.4}$$

which is (16.1.2).

To compare our theoretical result with experimental data, it is necessary to take into account two points so far neglected.

First, what we observe experimentally is the thermal activation energy of a skyrmion-antiskyrmion pair. But this activation takes place in the presence of charged impurities. The existence of charged impurities reduces the activation energy considerably. We include an offset parameter Γ_{offset} to treat this effect phenomenologically, as detailed in Section 12.1.

Second, we have so far assumed an ideal two-dimensional space for electrons. This is not the case. Electrons are confined within a quantum well of a finite width of order 200 Å. This will reduce the Coulomb energy considerably. It is quite difficult to make a rigorous analysis of the Coulomb energy in an actual quantum well. We simulate the effect by including the reduction factor γ.

We consider the excitation energy of a skyrmion-antiskyrmion pair since it is an observable quantity in magnetotransport experiments. It is simply twice of the skyrmion excitation energy. Taking into account these two points, instead of (16.4.1) we set the activation energy as

$$\Delta_{\text{gap}}(\tau) = 2\gamma E_C(\tau) + 2\Delta_Z N(\tau) - \Gamma_{\text{offset}}, \tag{16.4.5}$$

where $0 < \gamma < 1$. Repeating the same steps as for (16.4.3) we come to

$$E_\gamma(\Delta_Z) = \gamma E_{\text{sky}}\left(\frac{\Delta_Z}{\gamma}\right) - \Gamma_{\text{offset}}, \tag{16.4.6a}$$

$$N_\gamma(\Delta_Z) = N_{\text{spin}}\left(\frac{\Delta_Z}{\gamma}\right), \tag{16.4.6b}$$

with the use of $E_{\text{sky}}(\Delta_Z)$ and $N_{\text{spin}}(\Delta_Z)$ derived in (16.4.3). Thus the activation energy reads

$$E_{\text{sky}} \simeq \gamma\left\{\frac{1}{2}\sqrt{\frac{\pi}{2}}\alpha + \frac{9\pi^2}{128\kappa}\right\}E_C^0 - \Gamma_{\text{offset}} \tag{16.4.7}$$

together with

$$\kappa \simeq \frac{1}{2}\left(\frac{3\pi^2}{64}\right)^{1/3}\left\{\frac{\tilde{g}}{\gamma}\ln\left(\frac{\beta^2\gamma\sqrt{2\pi}}{32\tilde{g}} + 1\right)\right\}^{-1/3}. \tag{16.4.8}$$

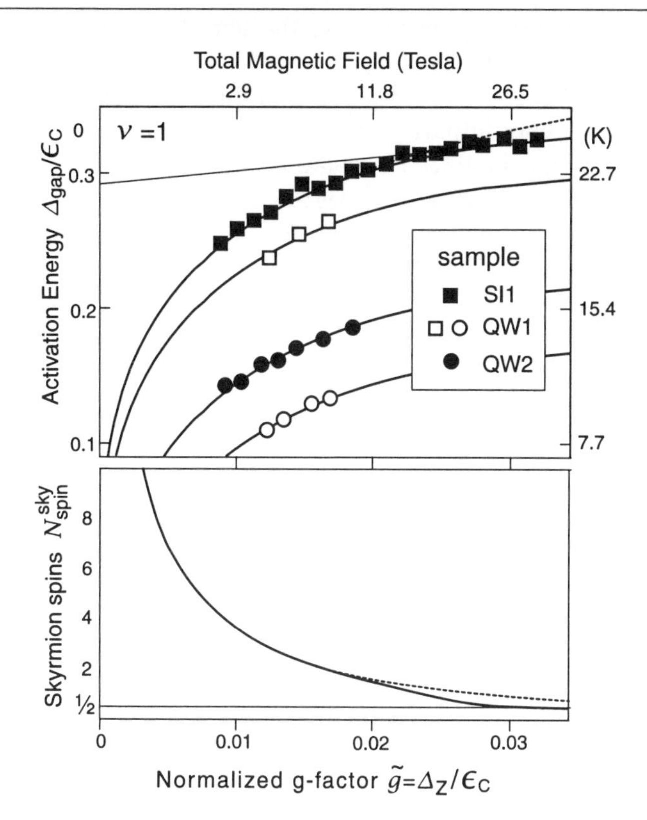

Fig. 16.9 A theoretical result on the activation energy Δ_{gap} of a skyrmion-antiskyrmion pair is compared with experimental data at $\nu = 1$. The data are taken from Schmeller et al.[429] By the dotted curve we give the activation energy calculated based on the formula (16.4.7). The thin line represents the hole excitation energy. The heavy solid curve is obtained by the numerical analysis made in Section 31.7. In these theoretical curves we have taken $\gamma = 0.56$, and $\Gamma_{offset} = 0.41$ for sample SI1. The offset Γ_{offset} increases as the mobility decreases. The skyrmion spin is one half of the slope of the activation-energy curve, $N_{spin} = \frac{1}{2}\partial\Delta_{gap}/\partial\tilde{g}$, where Δ_{gap} is taken in units of ϵ_C. The thin curve, the heavy curve and the dotted line correspond to those for the activation energy. The number N_{spin} depends sensitively on the normalized Zeeman gap $\tilde{g}$ for small $\tilde{g}$. It is seen that the hole excitation occurs for $\tilde{g} > 0.03$.

We have compared the theoretical result (16.4.7) with the experimental data by making appropriate choices of the reduction factor γ and the offset parameter Γ_{offset} in Fig.16.9. The best fit is obtained with $\alpha \simeq 0.9$, $\beta \simeq 1.478$, $\gamma \simeq 0.56$ and $\Gamma_{offset} = 0.41$ for sample SI1. The offset Γ_{offset} increases as the mobility decreases. We estimate $N_{sky}^{spin} \simeq 2.4$ at $B = 7.05$ Tesla, where $\kappa \simeq 1.0$ and $\tilde{g} \simeq 0.015$. The number $N_{sky}^{spin} \simeq 2.4$ is smaller than the one found in the experiment due to Barrett et al.[63], where

$N_{\rm sky}^{\rm spin} \simeq 3.5$. This simply implies that both the activation energy and the skyrmion spin will depend on samples.

As we have noticed, the above analysis of the skyrmion energy is made in an ideal thin layer without impurity effects and others. Furthermore, we have not taken into account the modification of the skyrmion wave function in a quantum well by tilting the field. Actual energies are determined by an unknown complicated Hamiltonian. In some samples we may encounter the following situation. The size of a skyrmion is fixed by this complicated Hamiltonian, where the change of the Zeeman energy due to the tilting is a very small effect. Then, we should treat it as a perturbation without changing the skyrmion size. In this case the number of reversed spins is a constant over a wide range of the normalized g-factor. Indeed, we have experiment data[342,289] indicating this feature. See Fig.26.17 (page 483) for an example, where $2N_{\rm spin} \simeq 7$ for $\tilde{g} < \tilde{g}_0$ and $2N_{\rm spin} \simeq 1$ for $\tilde{g} > \tilde{g}_0$ with a certain critical value $\tilde{g}_0$.

CHAPTER 17

HIERARCHY OF FRACTIONAL QH STATES

Composite fermions are obtained by attaching an even number of flux quanta to each electrons. They form Landau levels, fill a Fermi sea, execute semiclassical cyclotron orbits and show Schubnikov-de Haas oscillations in a reduced effective magnetic field. The fractional QH effect is interpreted as the integer QH effect of composite fermions.

17.1 JAIN HIERARCHY

The fractional QH state is realized as a result of composite-boson condensation at $\nu = 1/(2p+1)$, where p is an integer. We have invoked the Haldane-Halperin hierarchy to explain QH states at generic filling factors in Section 14.5. However, this hierarchy scheme is unnatural, as was commented there.

Experimentally, QH states are known to appear predominantly along the principal sequences [Fig.17.1],

$$\nu = \frac{n}{2pn \pm 1}, \tag{17.1.1}$$

where n is a positive integer. The sequences are derived naturally based on the composite-fermion theory.

Composite fermions are obtained by attaching $2p$ quanta of the Chern-Simons flux to each electron. The attached flux quanta amount to the averaged magnetic field,

$$B_\perp^{\text{CS}} = 2p\Phi_{\text{D}}\rho_0. \tag{17.1.2}$$

Composite fermions experience the effective magnetic field

$$B_\perp^{\text{eff}} = B_\perp - B_\perp^{\text{CS}}, \tag{17.1.3}$$

and make cyclotron motion in this reduced field. The effective filling factor for composite fermions is defined by

$$\nu_{\text{eff}} = \pm\frac{2\pi\hbar\rho_0}{eB_\perp^{\text{eff}}} > 0, \tag{17.1.4}$$

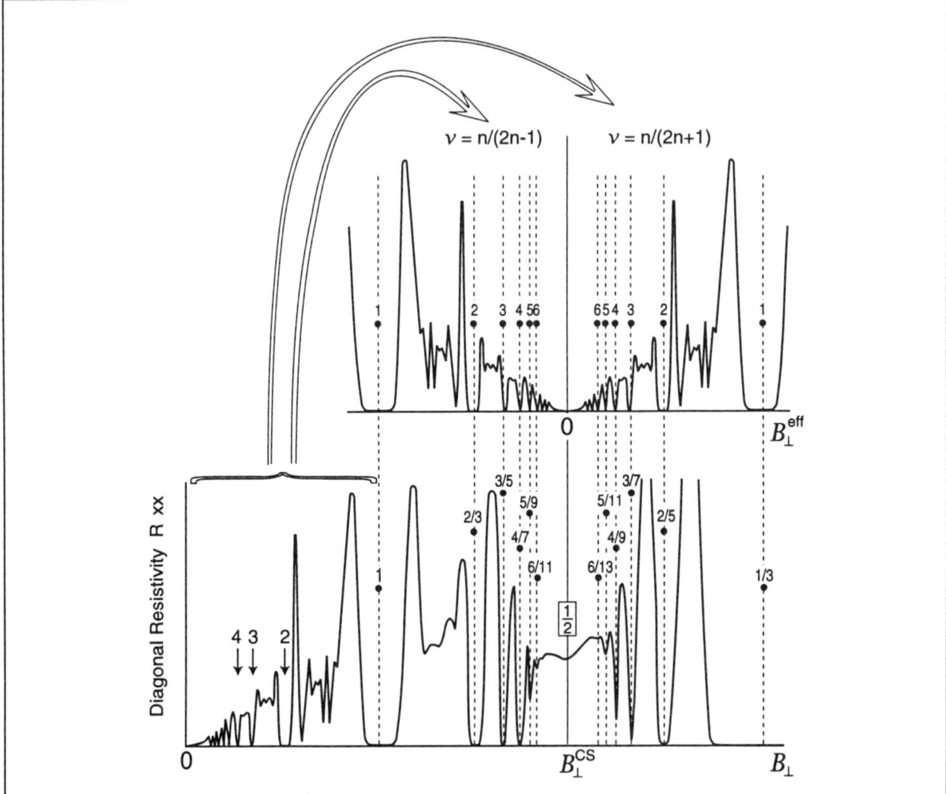

Fig. 17.1 The oscillation of R_{xx} around $\nu = 1/2$ ($B_\perp = B_\perp^{CS}$) is interpreted as the Schubnikov-de Haas oscillation of composite fermions around $B_\perp^{eff} = 0$. This figure is obtained from Fig.9.2 (page 165) by duplicating the pattern around $B_\perp = 0$ and moving it together with its reflection so that the point $B_\perp = 0$ coincides with $B_\perp = B_\perp^{CS}$.

and satisfies the relation

$$\pm \nu_{eff} B_\perp^{eff} = \nu B_\perp. \tag{17.1.5}$$

When $B_\perp^{eff} < 0$, the effective magnetic field is antiparallel to the external field. An identity follows from (17.1.3),

$$\frac{1}{\nu} \equiv \frac{eB_\perp}{2\pi\hbar\rho_0} = \pm\frac{eB_\perp^{eff}}{2\pi\hbar\rho_0} + \frac{eB_\perp^{CS}}{2\pi\hbar\rho_0} = \pm\frac{1}{\nu_{eff}} + \frac{1}{2p}, \tag{17.1.6}$$

or

$$\nu = \frac{\nu_{eff}}{(2p\nu_{eff} \pm 1)}. \tag{17.1.7}$$

The effective magnetic length is

$$\ell_B^{\text{eff}} = \sqrt{\frac{\hbar}{e|B_\perp^{\text{eff}}|}}. \tag{17.1.8}$$

The effective cyclotron energy $\hbar\omega_c^{\text{eff}}$ and the effective mass M_{eff} are defined phenomenologically by[218,118]

$$\Delta_{\text{gap}} \equiv \hbar\omega_c^{\text{eff}} = \frac{\hbar e|B_\perp^{\text{eff}}|}{M_{\text{eff}}}, \tag{17.1.9}$$

where Δ_{gap} is the gap energy of the fractional QH state.

How real is the effective magnetic field? An illuminating insight is gained by reviewing fractional QH states. The oscillations of the diagonal resistance R_{xx} around $\nu = \frac{1}{2}$ look very much like the Schubnikov-de Haas oscillations of composite fermions around $B_\perp^{\text{eff}} = 0$, as seen in Fig.17.1. Recall that the Schubnikov-de Haas oscillation occurs because the diagonal resistance R_{xx} depends on the density of states at the Fermi energy ε_{F}: The resistance R_{xx} vanishes ($R_{xx} = 0$) each time when one energy level is filled up. This suggests the following picture. (We consider the spin-frozen theory for simplicity.) Composite fermions make cyclotron motion in the effective magnetic field $B_\perp^{\text{eff}}$, and fill their own Landau levels called *CF-Landau levels*. Composite fermions must form an integer QH state at $\nu_{\text{eff}} = n$ when the nth CF-Landau level is filled up. Namely, the fractional QH state at $\nu = n/(2pn \pm 1)$ is interpreted as the integer QH state of composite fermions at $\nu_{\text{eff}} = n$. This is the magic filling factor (17.1.1) along the principal sequences.

The effective magnetic field vanishes at the limiting point ($\nu = 1/2p$) of the principal sequences (17.1.1),

$$|B_\perp^{\text{eff}}| = \frac{B_\perp}{2pn \pm 1} = \frac{2\pi\hbar\rho_0}{en} \to 0 \qquad \text{as} \quad n \to \infty, \tag{17.1.10}$$

where $B_\perp = B_\perp^{\text{CS}}$. Composite fermions experience no magnetic field at $\nu = 1/2p$. They would behave as if they were free electrons possessing a Fermi surface at $B_\perp^{\text{eff}} = 0$.[218]

The composite-fermion theory implies that the dynamics of charged particles is governed by the effective magnetic field $B_\perp^{\text{eff}}$ rather than the applied field $B_\perp$. The relevant number of the flux quanta is

$$N_\Phi^{\text{eff}} = \frac{B_\perp^{\text{eff}}}{\Phi_{\text{D}}}, \tag{17.1.11}$$

which gives the number of states of composite fermions in one energy level.

It is convenient to consider CF-Landau sites. The *CF-Landau site* is the quantum-mechanical state in the CF-Landau level. One CF-Landau site can accommodate one composite fermion in the spinless theory, and two composite fermions in the theory with spins. One CF-Landau site corresponds to m Landau sites at

$\nu = 1/m$: See Fig.13.3 (page 244) and the last subsection 13.6 (page 245) in Section 13.6.

Ample experimental data support the composite-fermion theory along the principal sequences. Surface acoustic wave experiments[502,503,504] provided the first evidence that something quite unusual was happening near $\nu = 1/2$. Geometric resonance experiments with antidot lattices[261,448] demonstrated the existence of the Fermi sea and the effective cyclotron motion. A magnetic focusing experiment[198] confirmed semiclassical cyclotron orbits of composite fermions near $\nu = 1/2$. The effective mass M_{eff} of composite fermions was estimated[118,119,299,333,106,370] by analyzing the activation energy and the Schubnikov-de Haas oscillations around $\nu = 1/2$. It was also estimated[370] by the use of the Aharonov-Bohm effect in antidot lattices [see (5.7.13)]. A review on experiments is found in literature.[507]

17.2 Landau Levels of Composite Fermions

The field theory of composite fermions was described in Section 13.6. From the Hamiltonian (13.6.13) and the LLL condition (13.6.14) we have derived the following results: The Landau levels of composite fermions agree with those of electrons, as follows from the Hamiltonian (13.6.13) and the commutation relation $[\bar{a}, \bar{a}^\dagger] = 1$. It is controversial, but the spacing between Landau levels of composite fermions is identical to that of electrons before the LLL projection is made.

The lowest CF-Landau level is filled up with composite fermions at $\nu_{\text{eff}} = 2\pi\rho_0(\ell_B^{\text{eff}})^2 = 1$. The composite-fermion wave function is given by (13.6.17),

$$\bar{\mathfrak{S}}_{\text{CF}}[\boldsymbol{x}] = \rho_0^{N/2} \prod_{r<s}(z_r - z_s) \exp\left(-\frac{eB_\perp^{\text{eff}}}{4\hbar}\sum_{r=1}^{N}|\boldsymbol{x}_r|^2\right). \tag{17.2.1}$$

We fill up the lowest n CF-Landau levels with composite fermions, as yields the integer QH state of composite fermions at $\nu_{\text{eff}} = n$. Let us focus on composite fermions in the jth Landau level, where the one-body wave function $\bar{\mathfrak{S}}_l^j(\boldsymbol{x})$ describes a composite fermion with angular momentum $l\hbar$ and is given by

$$\bar{\mathfrak{S}}_l^j(\boldsymbol{x}) \propto (\bar{a}^\dagger)^j(\bar{b}^\dagger)^{j+l} \exp\left(-\frac{eB_\perp^{\text{eff}}}{4\hbar}|\boldsymbol{x}|^2\right). \tag{17.2.2}$$

The Landau-level ladder operator $\bar{a}^\dagger$ and the angular-momentum ladder operator $\bar{b}^\dagger$ are defined by (13.6.10) and (13.6.11), respectively: Compare this with (10.3.17a) for electrons. The wave function is the Slater determinant,

$$\bar{\mathfrak{S}}_{\text{CF}}^j[\boldsymbol{x}] = \frac{1}{\sqrt{N!}}\begin{vmatrix} \bar{\mathfrak{S}}_0^j(\boldsymbol{x}_1) & \bar{\mathfrak{S}}_1^j(\boldsymbol{x}_1) & \cdots & \bar{\mathfrak{S}}_{N-1}^j(\boldsymbol{x}_1) \\ \bar{\mathfrak{S}}_0^j(\boldsymbol{x}_2) & \bar{\mathfrak{S}}_1^j(\boldsymbol{x}_2) & \cdots & \bar{\mathfrak{S}}_{N-2}^j(\boldsymbol{x}_2) \\ \vdots & \vdots & \ddots & \vdots \\ \bar{\mathfrak{S}}_0^j(\boldsymbol{x}_N) & \bar{\mathfrak{S}}_1^j(\boldsymbol{x}_N) & \cdots & \bar{\mathfrak{S}}_{N-1}^j(\boldsymbol{x}_N) \end{vmatrix}, \tag{17.2.3}$$

where $\bar{N} = N_\Phi/n$ is the number of composite fermions in each Landau level. The electron wave function $\mathfrak{S}^j[x]$ is given by the formula (13.6.18). The total wave function is a product of these wave functions together with a possible correlation between composite fermions belonging to different CF-Landau levels.

This wave function is quite complicated due to the term $\sum_s (z - z_s)^{-1}$ in the ladder operators (13.6.10). We can simplify them as

$$\bar{a} \equiv -\frac{i}{\sqrt{2}}\left(\frac{\partial}{\partial z^*} + \frac{B_\perp^{\text{eff}}}{B_\perp} z\right), \qquad \bar{a}^\dagger \equiv \frac{i}{\sqrt{2}}\left\{-\frac{\partial}{\partial z} + 2\left(1 - \frac{B_\perp^{\text{eff}}}{2B_\perp}\right) z^*\right\}, \tag{17.2.4a}$$

$$\bar{b} \equiv \frac{1}{\sqrt{2}}\left(\frac{\partial}{\partial z} + \frac{B_\perp^{\text{eff}}}{B_\perp} z^*\right), \qquad \bar{b}^\dagger \equiv -\frac{1}{\sqrt{2}}\left\{-\frac{\partial}{\partial z^*} - 2\left(1 - \frac{B_\perp^{\text{eff}}}{2B_\perp}\right) z\right\}, \tag{17.2.4b}$$

without changing the LLL condition (13.6.14) and the commutation relations (13.6.12), and hence the structure of the Landau levels. The modified one-body wave function (17.2.2) describes also the composite fermion with angular momentum $l\hbar$. The set of wave functions $\{\bar{\mathfrak{S}}_l(\boldsymbol{x}); l = 0, 1, \cdots, \bar{N} - 1\}$ is equivalent between the two choices of the ladder operators, because they describes the same QH droplet system with a finite area where all angular-momentum states are filled.

The wave function (17.2.3) involves composite fermions in higher Landau levels. At sufficiently low temperature, however, only the components belonging to the lowest Landau level are relevant. Such components are extracted by making the LLL projection.[1] But for the Coulomb interaction all energy levels of composite fermions would be degenerate within the lowest Landau level when the LLL projection is made. The Coulomb interaction produces an energy gap $\hbar\omega_c^{\text{eff}}$ between the two CF-Landau levels. Effects due to the Coulomb interaction are analyzed in Section 17.5.

Though the Hamiltonian (13.6.6) of composite fermions is considerably different from that of electrons, their wave functions are essentially the same except for the magnetic length. Hence, the cyclotron radius R_c^{eff} of the composite fermion is calculated in the same way as that of the electron. It is given by (10.2.29), or

$$R_c^{\text{eff}} = \sqrt{2\nu_{\text{eff}} + 1}\, \ell_B^{\text{eff}} \simeq k_F (\ell_B^{\text{eff}})^2 = \frac{\hbar k_F}{e B_\perp^{\text{eff}}}, \tag{17.2.5}$$

where $\hbar k_F$ is the Fermi momentum, $k_F = \sqrt{4\pi\rho_0}$. This result is unaffected by the LLL projection, since it is independent of the band mass M of electrons and thus of the cyclotron energy $\hbar\omega_c = \hbar e B_\perp/M$. Consequently, the semiclassical cyclotron orbit of a composite fermion has the radius R_c^{eff} around $\nu = 1/2$.

---

[1] It is known by small-size numerical analysis[245] that the wave function before the LLL projection is already a good approximation. The wave function has been numerically studied by making the LLL projection explicitly.[244]

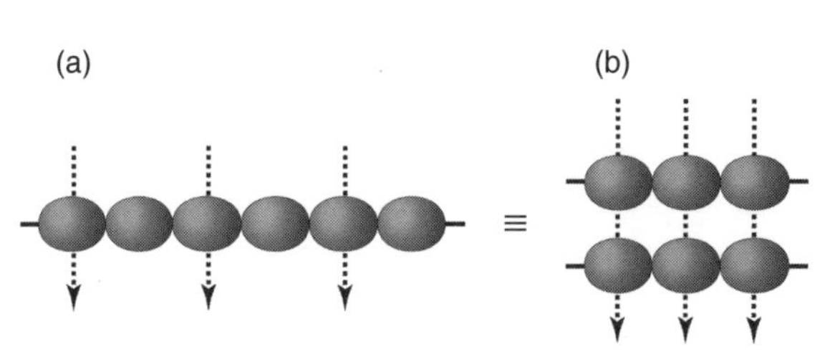

Fig. 17.2 The $\nu = 2$ QH state is illustrated. (a) There are two electrons per one flux quantum. (b) Electrons fill two Landau levels. We should consider that one flux exists per one electron in each Landau level. Electrons are bosonized in each Landau level.

17.3 Beyond Principal Sequences

The composite-fermion theory explains the fractional QH state at the principal sequences (17.1.1) quite well. However, there are many other fractions $\nu = n/m$ at which fractional QH states may exist. Indeed, FQH states have been discovered[525] along the sequence $\nu = (3n \pm 2)/(4n \pm 3)$.

Several microscopic realizations of fractional QH states are possible beyond the principal sequences,[531] but it is not clear which should be realized at a particular magnetic field. Here is one scenario. In the composite-fermion theory a state is mapped to another state by the antiparallel-flux-attachment transformation,[514] $\nu = \nu'/(2\nu' - 1)$, and the electron-hole symmetry transformation, $\nu'' = 2 - \nu'$. In this way, there exists a mapping from a sequence to the principal sequence,

$$\nu = \frac{3n \pm 2}{4n \pm 3} \qquad \Longrightarrow \qquad \nu'' = \frac{n}{2n \pm 1}. \tag{17.3.1}$$

For instance, the $\nu = 5/7$ state corresponds to the $\nu'' = 1/3$ state. If the mapping were correct, the $\nu = 5/7$ state would always be fully spin polarized as the $\nu = 1/3$ state is. However, a change of the spin polarization has been detected at $\nu = 5/7$ as a function of the Zeeman energy.[525] Thus, this scenario fails.

We wish to present a possible scenario valid for all fractional QH states.[4] In so doing we recall the microscopic formulation of the composite-fermion field given in Section 13.6. We have explicitly constructed the composite-fermion field at $\nu_{\text{eff}} = 1$ as in (13.6.2). How can we define the composite-fermion field at $\nu_{\text{eff}} = 2$, where two CF-Landau levels are filled?

We solve this problem by reconsidering how we can perform the statistical transmutation to electrons at $\nu = 2$ in the spinless theory. According to the standard procedure, we attach all flux quanta to electrons. As a result, one half of the electrons are bosonized and the other half are left as electrons [Fig.17.2(a)]. Should we conclude that these electrons are left in the zero effective magnetic fields? This is a nonsense view. We know that electrons fill two Landau levels. The correct view

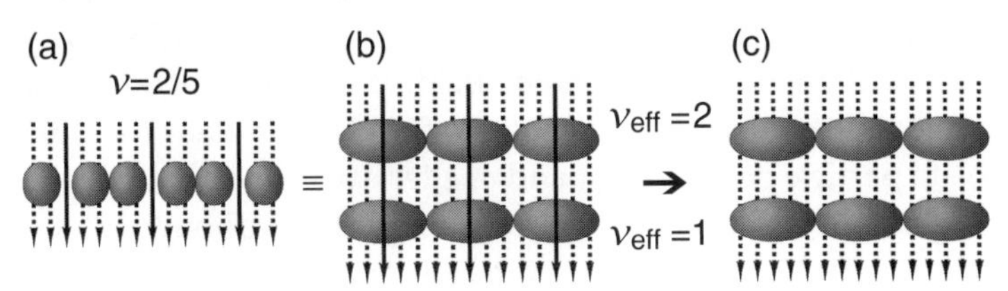

Fig. 17.3 The $\nu = 2/5$ QH state is illustrated in the composite-fermion and composite-boson theories. There are two electrons per five flux quanta. (a) A composite fermion is obtained by attaching two flux quanta (dotted lines) to an electron. There are two composite fermions ($\nu_{\text{eff}} = 2$) per one effective magnetic flux (heavy line). The number of heavy lines is the number of CF-Landau sites. (b) Composite fermions fill two CF-Landau levels, since there are two electrons per one CF-Landau sites. It is observed as if one composite fermion absorbed four flux quanta. (c) A composite fermion is converted into a composite boson when one flux is additionally attached in each CF-Landau level.

is that one flux quantum exists for each electron in each Landau level [Fig.17.2(b)]. We can now apply the statistical transmutation to electrons in each Landau level. Actually, to study essential physics in the $\nu = 2$ QH state, we are allowed to neglect all electrons in the lowest Landau level by regarding them inactive. Then, it is equivalent to a simple electron system at $\nu = 1$ with the electron density $\rho_0/2$. Bosonization is done as before by using the electron density $\rho_0/2$. The $\nu = n$ integer QH state is treated similarly by using the electron density ρ_0/n.

The same procedure can be taken for composite fermions at $\nu = n/m$ with $m = 2pn + 1$, by treating composite fermions simply as "fat electrons" governed by the Hamiltonian (13.6.13). We put composite fermions in their CF-Landau levels [Fig.17.3(b)]: n levels are filled up. The number of composite fermions in each CF-Landau level is N/n, when there are N electrons in the system. Hence, there are $n/\nu = m$ flux quanta per each electron in each CF-Landau level. In particular, if all composite fermions in the lower CF-Landau levels are regarded inactive, this is equivalent to a simple composite-fermion system at $\nu = 1/m$ with the electron density ρ_0/n. They must have been constructed from electrons by the statistical transmutation. The composite-fermion field operator is given by (13.6.2)~(13.6.4) with $2pn$ instead of $2p$ and $\rho(\mathbf{x})/n$ instead of $\rho(\mathbf{x})$. Namely, it is necessary to associate $2pn$ flux quanta, though an original composite fermion carries $2p$ flux quanta [Fig.17.3(a) and (b)]. We may as well bosonize them additionally [Fig.17.3(c)]. One composite particle occupies m Landau sites in each CF-Landau level at $\nu = n/m$.

It suggests a *heuristic hierarchy*:[4] For any odd m, the $\nu = n/m$ QH state consists of n filled CP-Landau levels, where each composite particle occupies one CP-Landau site in each CP-Landau level.[2] We have called them composite par-

[2] This is completely analogous to the following statement: The $\nu = n$ QH state consists of n filled Landau levels, where each electron occupies one Landau site in each Landau level.

ticles rather than composite fermions, since they are not necessarily the composite fermions in the Jain hierarchy. A natural question is why the CP-Landau levels are generated without an effective magnetic field. The key is the exclusion principle obeyed by composite particles. According to the Hamiltonian (13.5.15) or (13.6.13), one composite particle occupies m Landau sites at $\nu = n/m$, and the Landau-level structure is generated with the cyclotron energy $\hbar\omega_c$. Though all levels are collapsed to the lowest Landau level by the LLL projection, the energy gap $\hbar\omega_c^{\text{eff}}$ is actually produced between two successive CP-Landau levels due to Coulomb collective effects. Recall that we have already encountered such a Coulomb gap in the QH ferromagnet at $\nu = 1$ in the absence of the Zeeman effect (see Section 15.1). Consequently, it is possible that the exclusion statistics underlies the hierarchy of fractional QH states. There are two types of heuristic hierarchies in association with two types of sequences $\nu = n/(2pn \pm 1)$. The principal sequences are enhanced since Jain's composite fermions experience an effective magnetic field additionally. The heuristic hierarchy has unique predictions for the spin structure for all fractional QH states.

17.4 Spin Polarization

One might think that spins are always fully polarized in fractional QH states because all electrons are in the lowest Landau level under the influence of strong magnetic field. Early experiments revealed that this is not the case.[126,103,182,130] By tilting the magnetic field, we can change the Zeeman energy at a fixed filling factor and investigate the spin structure. Recently, as the Zeeman energy is increased by using the tilted-field method, fractional QH states were seen to disappear and then reappear,[120,121] indicating a transition between different spin-polarized states. It is possible even to measure the spin polarization and observe the spin transition directly by magnetoluminescence experiments.[287,288]

There had been numerous theoretical works[215,88,532,400,516,93,92,69] on the spin polarization of fractional QH states, before the composite-particle picture was presented. They are based on a trial wave function or on an exact numerical diagonalization of the model Hamiltonian of a small system. Though exact diagonalization studies can sometimes give quite accurate numbers, they are not so helpful in achieving an intuitive understanding of physics. Furthermore, they are limited to small systems: For example, even at $\nu = 2/5$ the biggest system studied contains only $N = 8$ electrons.[516,69]

The composite-fermion theory presents a systematic picture of the spin structure of fractional QH states in the principal sequences.[120,384] However, there are many other fractional QH states. We present a simple picture valid for all fractional QH states based on the heuristic hierarchy described in the previous section. The basic postulate is that the $\nu = n/m$ state is constructed when composite particles fill n CP-Landau levels. Due to the spin degree of freedom one CP-Landau level is

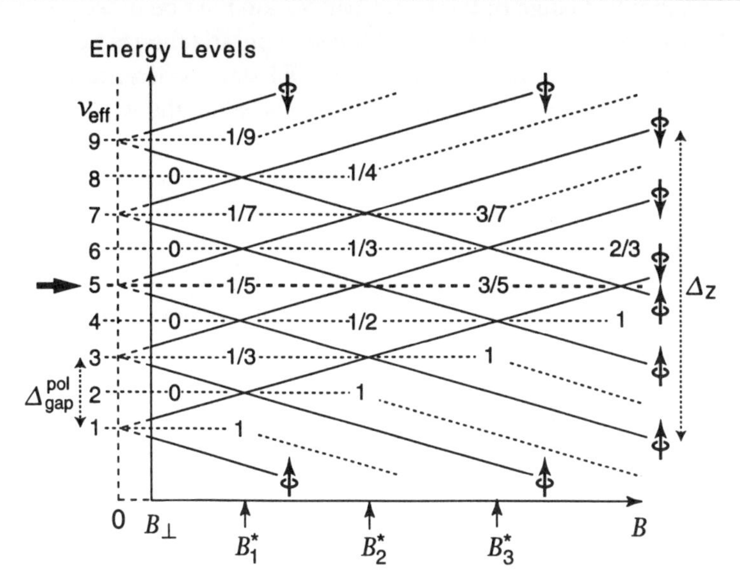

Fig. 17.4 Energy levels of composite fermions are illustrated schematically as a function of the total magnetic field B, $B \geq B_\perp$, by neglecting all effects from higher Landau levels. (Here $\nu = n/m$ with $m = 5$). One CP-Landau level is split into two energy levels, the up-spin and down-spin levels. They undergo level crossings, as the total magnetic field B is increased with the perpendicular component $B_\perp$ fixed. (The Zeeman gap Δ_Z changes linearly but the Coulomb gap $\Delta_{\rm gap}^{\rm pol}$ remains unchanged.) The dotted line represents the Fermi level. The $\nu = n/m$ QH state is realized when all levels are filled below the $\nu_{\rm eff} = n$ Fermi level. The number on the dotted line is the spin polarization $\gamma_e = (\rho_0^\uparrow - \rho_0^\downarrow)/(\rho_0^\uparrow + \rho_0^\downarrow)$. The symbol B_i^* represents a phase transition point. For instance, the $\nu = 3/5$ QH state is initially a partially polarized state ($\gamma_e = 1/3$), but as B increases it undergoes a phase transition to the fully polarized state ($\gamma_e = 1$) at B_2^*. Note that the $\nu = 5/5 = 1$ state is fully polarized always ($\nu_{\rm eff} = 5$ for $m = 5$): Nevertheless, we have given the numbers so that they are applicable to the $\nu = 5/m$ state with $m \geq 7$ with a trivial generalization.

split into two energy levels, the up-spin and down-spin levels [Fig.17.4]. Note that composite particles possess the same spins as the electrons. We consider the regime where the cyclotron energy $\hbar\omega_c$ is so large that all effects from higher Landau levels are negligible. There are two energy scales, the Coulomb energy $E_C^0 = e^2/4\pi\varepsilon\ell_B$ and the Zeeman energy $\Delta_Z = |g^*\mu_B B|$, which are independent parameters. We increase the Zeeman energy while keeping the Coulomb energy fixed by using the tilted-field method. The up-spin (down-spin) energy level goes down (up) linearly in B, and crosses the down-spin (up-spin) energy levels. As levels cross, the energy gap closes and opens in a zigzag fashion: The fractional QH state undergoes a phase transition between two different spin-polarized states. At a point of coincidence the gap vanishes, leading to a peak in the magnetoresistivity ρ_{xx}. This is the reason why a change of the spin polarization may be detected by magneto-transport experiments,[120,121,525] even if the spin polarization is not measured.

The spin polarization γ_e of the $\nu = n/m$ QH state is determined by counting the numbers of the up-spin and down-spin levels below the $\nu_{\text{eff}} = n$ Fermi level in Fig.17.4,

$$\begin{aligned}\gamma_e &= \frac{\rho_0^\uparrow - \rho_0^\downarrow}{\rho_0^\uparrow + \rho_0^\downarrow} \\ &= \frac{\text{Number of up-spin levels} - \text{Number of down-spin levels}}{\text{Number of up-spin levels} + \text{Number of down-spin levels}}. \end{aligned} \tag{17.4.1}$$

We list how the spin polarization γ_e changes as B increases with $B_\perp$ fixed. For the $\nu = n/m < 1$ state it changes as [Fig.17.4]:

$$\begin{aligned}
&\nu = 1/m(m \geq 3) : \gamma_e = 1 \\
&\nu = 2/m(m \geq 3) : \gamma_e = 0 \quad \to 1 \\
&\nu = 3/m(m \geq 5) : \gamma_e = 1/3 \to 1 \\
&\nu = 4/m(m \geq 5) : \gamma_e = 0 \quad \to 1/2 \to 1 \\
&\nu = 5/m(m \geq 7) : \gamma_e = 1/5 \to 3/5 \to 1 \\
&\nu = 6/m(m \geq 7) : \gamma_e = 0 \quad \to 1/3 \to 2/3 \to 1 \\
&\nu = 7/m(m \geq 9) : \gamma_e = 1/7 \to 3/7 \to 5/7 \to 1 \\
&\nu = 8/m(m \geq 9) : \gamma_e = 0 \quad \to 1/4 \to 1/2 \to 3/4 \to 1.
\end{aligned}$$

It is a characteristic feature that eventually the FQH state (at $\nu = n/m < 1$) gets fully polarized within the LLL. These polarization patterns are clearly seen at $\nu = 2/3, 3/5, 4/7, 2/5, 3/7, 4/9$ by magnetoluminescence experiments [Fig.17.5]. Although the polarization has not been measured, the existence of transitions has been detected by magnetotransport experiments[525] also at filling factors not in the principal sequences, such as at $\nu = 5/7, 8/11, 10/13, 7/9$ and $4/5$.

The polarization of the $\nu = n/m > 1$ state is controversial. In the standard composite-fermion theory the state at ν is related to the conjugate state at $\nu' = 2 - \nu$ by the electron-hole symmetry. Then, the polarization pattern must be the same between these two conjugate states, as was assumed to analyze the data in magnetic transport experiments.[525,120] However, the present scheme predicts the following polarization patterns [Fig.17.4]:

$$\begin{aligned}
&\nu = 4/3 : \gamma_e = 0 \quad \to 1/2 \\
&\nu = 5/3 : \gamma_e = 1/5 \\
&\nu = 6/5 : \gamma_e = 0 \quad \to 1/3 \to 2/3 \\
&\nu = 7/5 : \gamma_e = 1/7 \to 3/7 \\
&\nu = 8/5 : \gamma_e = 0 \quad \to 1/4 \\
&\nu = 9/5 : \gamma_e = 1/9.
\end{aligned}$$

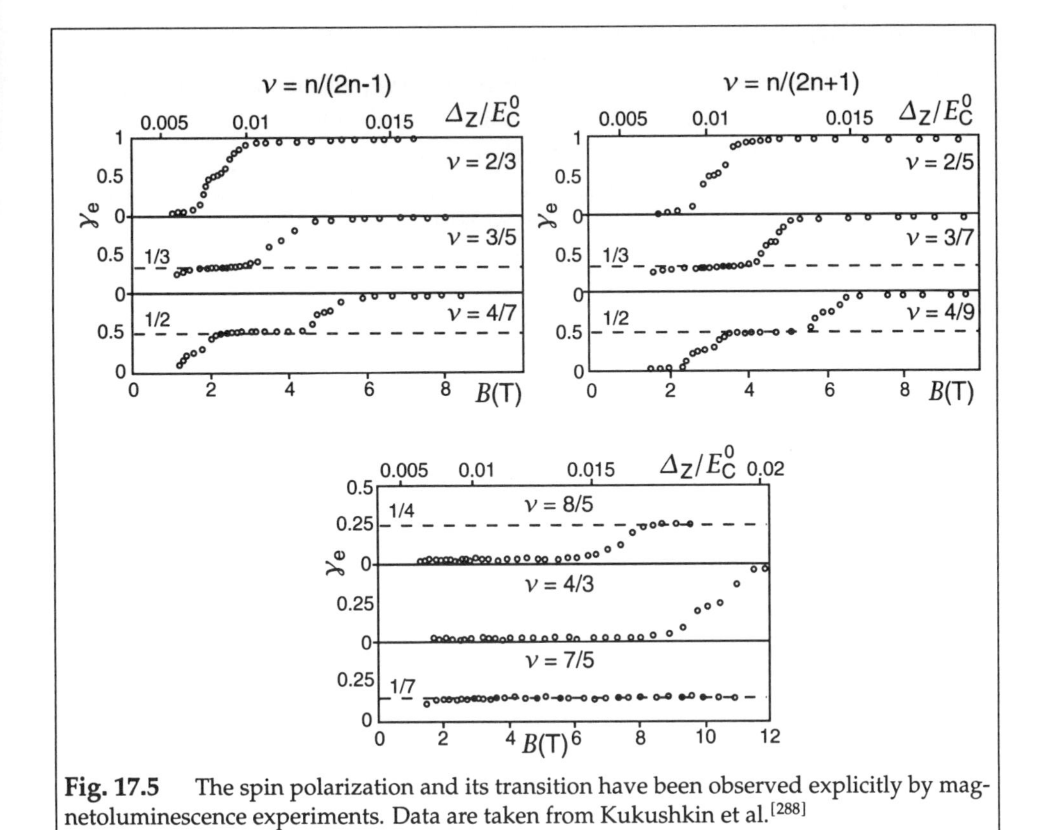

Fig. 17.5 The spin polarization and its transition have been observed explicitly by magnetoluminescence experiments. Data are taken from Kukushkin et al.[288]

It is a distinguished feature that the full polarization is never achieved without the effects from higher Landau levels. These polarization patterns are clearly observed at $\nu = 8/5$ and $4/3$, which are indeed very different from those at their electron-hole conjugate fillings $\nu = 2/5$ and $2/3$, as in Fig.17.5. With respect to the $\nu = 7/5$ state the measured polarization is almost constant in the whole range of measurement. Nevertheless, the value is consistent with the predicted one, $\gamma_e = 1/7$, for a small Zeeman splitting. We expect a spin transition to the state with $\gamma_e = 3/7$ for a large Zeeman splitting: The transition point would be $\Delta_Z/E_C^0 \simeq 0.023$ for the data given in Fig.17.5 due to Kukushkin et al.,[288] as will be argued below (17.5.10).

17.5 Gap Energies

The incompressibility of the fractional QH state at $\nu = n/m$ follows from the incompressibility of the integer QH state of composite particles. It is a result of a nonzero gap between the two successive CP-Landau levels. We present a possible scenario for the generation of gap energy based on the heuristic hierarchy.

Activation Gap

We first study the activation energy of a quasiparticle pair. It is a basic feature[294,463] that a quasiparticle possesses a fractional charge $\pm e/m$ at $\nu = n/m$. How do we reconcile it with the fact that composite particles carry charge $-e$. We have already seen that a topological soliton (skyrmion or vortex) possesses electric charges $\pm e/m$ at $\nu = 1/m$, though what are actually excited are composite bosons with charge $-e$. Indeed, a coherent excitation may carry a fractional charge.

It is a good approximation to consider merely thermal excitations from the occupied highest energy level to the unoccupied lowest energy level [Fig.17.6]. Though its detail depends on the spin polarizations of the two levels, the excitation energy is dominated by the exchange energy for a large m: The exchange energy is proportional to $1/m$ but the direct Coulomb energy to $1/m^2$ in the semiclassical analysis as in (16.3.1) and (16.3.4). It is approximated by

$$\Delta_{\text{pair}} \equiv \Delta_{\text{qh}} + \Delta_{\text{qe}} \simeq \frac{\alpha^{\text{act}}}{m} E_{\text{C}}^{0}, \tag{17.5.1}$$

where $\alpha^{\text{act}} = \sqrt{\pi/8} \simeq 0.63$ if the exchange energy is described by the simple sigma model [see (16.3.1)]. There are corrections in actual samples. At $\nu = 1$, to fit the experimental data in (16.4.7), we have set $\alpha^{\text{act}} = \alpha\gamma\sqrt{\pi/8}$ with $\alpha \simeq 0.9$ and $\gamma \simeq 0.56$, or

$$\alpha^{\text{act}} \simeq 0.32. \tag{17.5.2}$$

This value of α^{act} was derived at $\nu = 1/m$, but is valid also at $\nu = n/m$ when the lowest $n-1$ CP-Landau levels are regarded inactive.

To make a quantitative comparison of the activation energy with experimental data, we recall that the gap energy measured experimentally is reduced considerably by a sample-dependent offset, $\Delta_{\text{gap}} = \Delta_{\text{pair}} - \Gamma_{\text{offset}}$, as in (12.1.2). The main part of the offset is due to a Landau-level broadening by impurities in the bulk, and estimated as in (12.1.6). We expect

$$\Delta_{\text{gap}} \simeq \frac{\alpha^{\text{act}}}{m} E_{\text{C}}^{0} - \Gamma_{\text{offset}}^{\text{res}} \equiv \frac{\alpha_{\text{eff}}^{\text{act}}}{m} E_{\text{C}}^{0}, \tag{17.5.3}$$

with a sample-dependent parameter

$$\alpha_{\text{eff}}^{\text{act}} = \alpha^{\text{act}} - Z\frac{\ell_B}{d_{\text{imp}}}, \tag{17.5.4}$$

where Ze is the impurity charge, d_{imp} is the distance from impurities to the electron layer, and $\Gamma_{\text{offset}}^{\text{res}}$ is the residual offset.

We may define the effective mass M_{eff} (known as the activation mass) of composite particles phenomenologically by the formula (17.1.9). We equate the gap

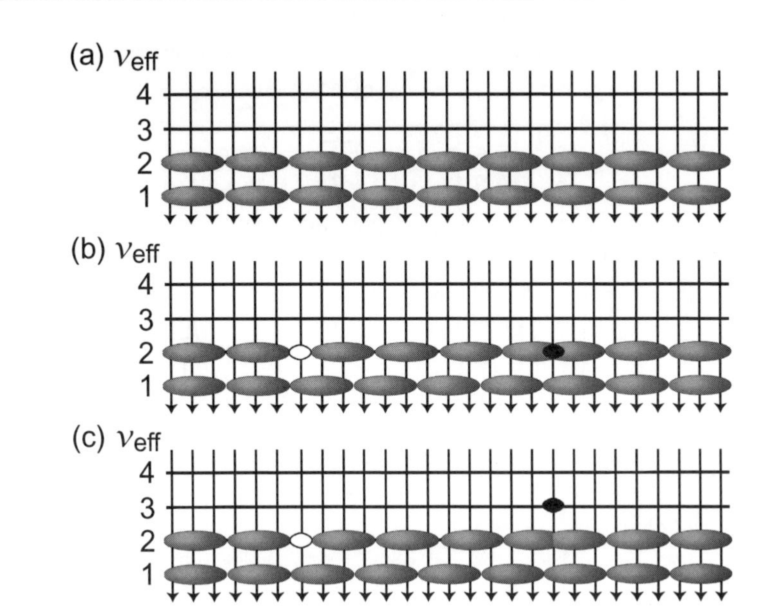

Fig. 17.6 (a) The ground state is illustrated at $\nu = 2/3$, where an oval represents a composite boson. (b) A would-be thermal excitation is illustrated at $\nu = 2/3$, where a "hole" represents a vortex and an "overlap" an antivortex [see Fig.9.7 (page 170)]. (c) A hole may evolve into a skyrmion and an antivortex into an antiskyrmion. They are coherent excitations carrying charge $\mp e/3$.

energy in (17.1.9) with (17.5.3),

$$\frac{\hbar e|B_{\perp}^{\text{eff}}|}{M_{\text{eff}}^{\text{act}}} = \Delta_{\text{gap}} = 51\alpha_{\text{eff}}^{\text{act}} \left(\frac{|B_{\perp}^{\text{eff}}|}{B_{\perp}} \right) \sqrt{B_{\perp}} \quad \text{K}, \tag{17.5.5}$$

where we have used the relation (17.1.5) and $\nu = \nu_{\text{eff}}/m$ to find $1/m = |B_{\perp}^{\text{eff}}|/B_{\perp}$; we have also used $E_{\text{C}}^{0} = 51\sqrt{B_{\perp}}$ K, with $B_{\perp}$ in Tesla and the energy in Kelvin. We compare the effective mass M_{eff} with the electron mass m_{e} in the vacuum. Since

$$\hbar\omega_{\text{c}}^{\text{vac}} = \frac{\hbar e B_{\perp}}{m_{\text{e}}} = 1.38 B_{\perp} \quad \text{K}, \tag{17.5.6}$$

we find

$$\frac{M_{\text{eff}}^{\text{act}}}{m_{\text{e}}} = \frac{0.027}{\alpha_{\text{eff}}^{\text{act}}} \sqrt{B_{\perp}}. \tag{17.5.7}$$

The parameter $\alpha_{\text{eff}}^{\text{act}}$ is determined by the observed activation mass $M_{\text{eff}}^{\text{act}}$. We sum-

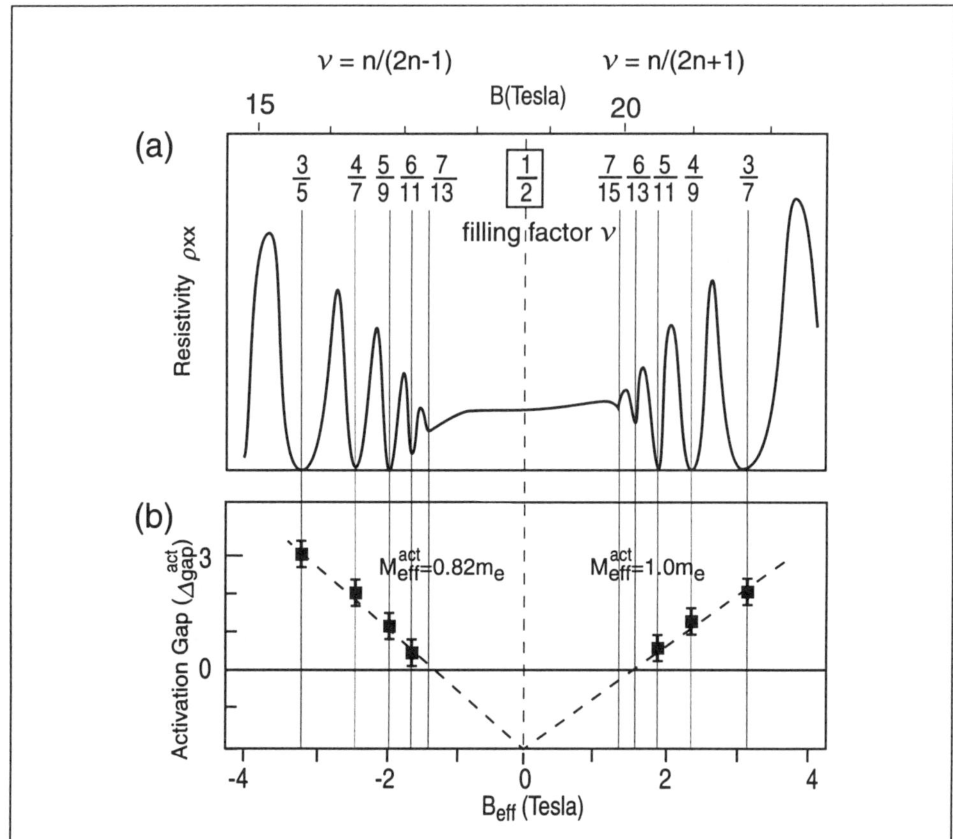

Fig. 17.7 (a) Diagonal resistance R_{xx} is illustrated around the point $\nu = 1/2$, where the effective magnetic field vanishes. Its characteristic behavior is understood as the Schubnikov-de Haas oscillations (integer QH effects) of composite fermions around $B_\perp^{\text{eff}} = 0$. (b) The activation mass $M_{\text{eff}}^{\text{act}}$ is found to decrease linearly as $B_{\text{eff}} \to 0$. The data are from Du et al.[119]

marize the experiments due to Du et al.[119] as follows [Fig.17.7],

$$M_{\text{eff}}^{\text{act}} \simeq 0.82m_e \quad \text{leads to} \quad \alpha_{\text{eff}}^{\text{act}} \simeq 0.10 \quad \text{for the sequence} \quad \nu = n/(2n-1),$$

$$M_{\text{eff}}^{\text{act}} \simeq 1.0m_e \quad \text{leads to} \quad \alpha_{\text{eff}}^{\text{act}} \simeq 0.12 \quad \text{for the sequence} \quad \nu = n/(2n+1),$$

for the measured range of $B_\perp$. These values of $\alpha_{\text{eff}}^{\text{act}}$ are consistent with $\alpha^{\text{act}} \simeq 0.32$, $d_{\text{imp}}/Z \simeq 300$ Å and $\Gamma_{\text{offset}}^{\text{res}} \simeq 2$ K in the formula (17.5.4). The activation mass depends on sample qualities sensitively.

Polarization Gap

The spin transition (change of polarization) occurs when the Zeeman energy exceeds the Coulomb energy between two successive CP-Landau levels. What is this Coulomb gap energy? There is an attempt to determine the gap energy $\Delta_{\text{gap}}^{\text{pol}}$ by a numerical analysis.[384] Here, we make a speculative argument. At $\nu = n/m$, n CP-Landau levels are completely filled. If an equal spacing is assumed for the energy levels of composite particles at the zero Zeeman coupling ($g^* = 0$), we would have

$$\Delta_{\text{gap}}^{\text{pol}} \simeq \frac{\alpha^{\text{pol}}}{m} E_{\text{C}}^0. \tag{17.5.8}$$

We treat α^{pol} as a phenomenological parameter depending on the sequence of fractional QH states. The exchange energy does not contribute to the polarization gap because the number of the filled energy levels is not changed by the level crossing (see Appendix I). Hence, we expect that the polarization gap will be much smaller than the activation gap, or $\alpha^{\text{pol}} \ll \alpha^{\text{act}}$. This is indeed the case experimentally, as we now see.

The energy of the up-spin level decreases while that of the down-spin level increases as the Zeeman effect increases. The level crossing occurs between two successive CF-Landau levels, leading to the change of polarization. It is clear in Fig.17.4 that the transition point B_j^* is determined by the condition $\Delta_{\text{Z}} = j\Delta_{\text{gap}}^{\text{pol}}$. This picture explains various experimental data[120,525,288] quite well. Here, we analyze the data given in Fig.17.5 due to Kukushkin et al.[288]

First, the full polarization is achieved at B_{n-1}^* in the $\nu = n/m \leq 1$ state [Fig.17.4], where $\Delta_{\text{Z}} = (n-1)\Delta_{\text{gap}}^{\text{pol}}$, or

$$\frac{\Delta_{\text{Z}}}{E_{\text{C}}^0} = \frac{n-1}{m}\alpha^{\text{pol}}. \tag{17.5.9}$$

The data are fitted reasonably well by choosing $\alpha^{\text{pol}} \simeq 0.028$ for the sequence $\nu = n/(2n-1)$, and $\alpha^{\text{pol}} \simeq 0.047$ for the sequence $\nu = n/(2n+1)$.

Second, the maximum polarization is achieved at B_{2m-n-1}^* in the $\nu = n/m > 1$ state [Fig.17.4], where $\Delta_{\text{Z}} = (2m-n-1)\Delta_{\text{gap}}^{\text{pol}}$, or

$$\frac{\Delta_{\text{Z}}}{E_{\text{C}}^0} = \frac{2m-n-1}{m}\alpha^{\text{pol}}. \tag{17.5.10}$$

We find $\alpha^{\text{pol}} \simeq 0.057$ for the sequence $\nu = 2 - n/(2n-1)$, and $\alpha^{\text{pol}} \simeq 0.080$ for the sequence $\nu = 2 - n/(2n+1)$. These numerical values may not be so reliable since there exists only one datum in each sequence,[288] that is, at $\nu = 4/3$ and $8/5$. Nevertheless, we use it to predict the spin transition at $\Delta_{\text{Z}}/E_{\text{C}}^0 \simeq 0.023$ in the $\nu = 7/5$ state [Fig.17.5].

The effective mass $M_{\text{eff}}^{\text{pol}}$ (known as polarization mass) is defined by the formula,

$$\frac{\hbar e|B_\perp^{\text{eff}}|}{M_{\text{eff}}^{\text{pol}}} = \Delta_{\text{gap}}^{\text{pol}} = 51\alpha^{\text{pol}}\left(\frac{|B_\perp^{\text{eff}}|}{B_\perp}\right)\sqrt{B_\perp} \quad \text{K}. \tag{17.5.11}$$

We find

$$\frac{M_{\text{eff}}^{\text{pol}}}{m_{\text{e}}} = \frac{0.027}{\alpha^{\text{pol}}}\sqrt{B_{\perp}}. \tag{17.5.12}$$

It follows that $M_{\text{eff}}^{\text{pol}}/M_{\text{eff}}^{\text{act}} = \alpha_{\text{eff}}^{\text{pol}}/\alpha^{\text{pol}} = 2 \sim 3$. It depends on sample qualities sensitively, since $M_{\text{eff}}^{\text{act}}$ does.

CHAPTER 18

EDGE EFFECTS

The QH state is an incompressible liquid confined within a finite domain. Electrons on the edge move as a gapless mode into one direction determined by the sign of the magnetic field. They form a chiral Tomonaga-Luttinger liquid in the fractional QH system. quasiparticles propagate freely along the edge, and they may hop from one edge to the other when the opposite edges are brought into close proximity. It provides us with a direct measurement of fractional charges.

18.1 EDGE CURRENTS AND BULK CURRENTS

The QH state is an incompressible liquid confined within a finite domain. Since the bulk is incompressible, the low-energy excitations are deformations of the boundary shape which preserve the area of the liquid. At the edge, the Landau levels are bent by the confining potential, cross the Fermi level, and form edge channels [Fig.18.1]. Edge excitations are gapless modes since an infinitesimal deformation of the boundary costs an infinitesimal energy. The importance of gapless edge modes was first elucidated by Halperin[214] and thereafter has attracted much attentions.

There exists a controversy over the Hall current whether it flows solely along the edge or the bulk. According to the edge transport picture the dominant contribution comes from gapless edge excitations both in integer QH states[478,322,462,240,79] and in fractional QH states.[96,66,324] Some experimental results[487,280,341,95,283,486] are interpreted to support the edge transport picture. The bulk transport picture has also experimental supports,[175,265,485] in which the Hall current was found to be proportional to the width of the sample. There is an attempt to relate the two pictures.[442,444] Readers are also referred to articles[48,281] surveying this problem.

Here, we remark that advanced technology makes it possible to measure directly the electric potential distribution on Hall bars.[40] Let us take a rectangular geometry, where the Hall current flows along the x axis [Fig.9.3]. The Hall current,

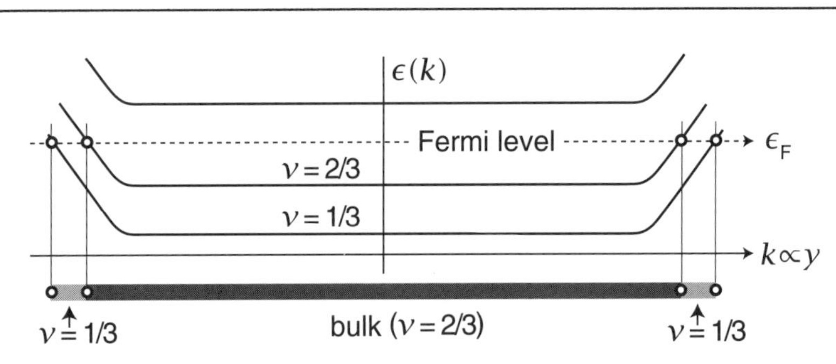

Fig. 18.1 Dispersion of energy levels is illustrated in a quantum Hall bar as a function of the one-dimensional momentum $\hbar k$. The horizontal axis is identified with the y axis, because the wave vector k is related to the electron position as $y = -\ell_B^2 k$ in the Landau gauge. Energy levels are bent by the confining potential. Gapless edge channels appear when the Fermi level crosses energy levels. In this example, there are two edge channels, as indicated by open circles. The $\nu = 1/3$ narrow domain is formed between the two channels along the edge, though the $\nu = 2/3$ QH state occupies the bulk. Shot-noise experiments[419] are carried out by using the $\nu = 1/3$ narrow domain of the QH state with the bulk filling factor $\nu = 2/3$, whose results are illustrated in Fig.18.3.

driven by the Lorentz force, is expressed by

$$\mathcal{J}_x(\boldsymbol{x}) = \frac{1}{R_{xy}} E_y(\boldsymbol{x}) = \frac{\nu e^2}{2\pi\hbar} \frac{\partial V(\boldsymbol{x})}{\partial y}, \tag{18.1.1}$$

where $V(\boldsymbol{x})$ is the electric potential. Hence, the local flow of the Hall current can be determined experimentally[40] by measuring the potential distribution $V(\boldsymbol{x})$ on the sample. This might settle down the controversy whether the current flows solely along the edge or the bulk.

18.2 Shot Noises of Fractional Charges

Our main concern is about edge excitations in the absence of the Hall current. The edge channel presents a concrete realization of the *Tomonaga-Luttinger liquid* made of interacting one-dimensional electron gas. The Tomonaga-Luttinger liquid has been known theoretically for a few decades[319,450,318] but inaccessible experimentally because of difficulty to fabricate clean one-dimensional quantum wires: The QH edge state now provides us with a unique laboratory for its study.

Electrons are confined within a sample by a confining electric field $\boldsymbol{E}$, whose magnitude $E = |\boldsymbol{E}|$ is assumed to be a constant along the edge. For simplicity we include no electric field driving the Hall current. Electrons experience the Lorentz forcein the presence of the electric ($\boldsymbol{E}$) and magnetic ($\boldsymbol{B}$) fields, as generates a persistent current flowing along the edge. The drift velocity $\boldsymbol{v}$ is determined by the

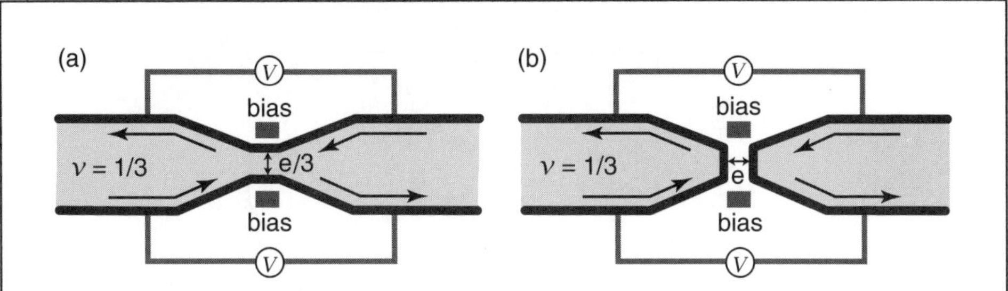

Fig. 18.2 Quasiparticles flow along the one-dimensional edges, as indicated arrows. (a) With weak pinch-off, fractional charged quasiparticles tunnel between the top and bottom edges through the QH liquid. (b) With strong pinch-off, electrons tunnel between two isolated QH liquids.

relation $\boldsymbol{E} = -\boldsymbol{v} \times \boldsymbol{B}$ according to (9.1.1). Its magnitude is

$$v = E/B_{\perp}. \tag{18.2.1}$$

The edge wave propagates with the drift velocity v in one direction determined by the sign of the magnetic field. Edge excitations have a definite *chirality*, that is, either the left-handedness or the right-handedness. The chiral dynamics of the edge mode was explored by Wen and others.[491,492,259,350]

We have seen that quasiparticles are gapful in the bulk. Moreover, they are pinned in place by impurities. On the contrary, they are gapless and propagate freely along the edge. A remarkable phenomenon becomes observable when they are allowed to tunnel between two edges.[259] The simplest setup is a point contact made in a QH bar, where the sample is pinched electrostatically by a bias voltage [Fig.18.2]. When the constriction is open, the two-terminal conductance is given by its quantized value. The conductance is reduced, as the channel is pinched off and the top and bottom edges are brought into close proximity, because charges begin to backscatter between the left-moving and right-moving edge channels [Fig.18.2(a)]. Ultimately, as the gate voltage is increased, the Hall bar is pinched off completely, and the weak tunneling current is carried by electrons [Fig.18.2(b)].

By tuning the bias voltage quasiparticles tunnel one by one between the two edges through the QH liquid [Fig.18.2(a)]. The tunneling current contributes to the backscattered current. Now, an electrical current has tiny fluctuations, or *shot noise*, intrinsic to the granularity of the current.

W. Schottky[430] predicted in 1918 that a vacuum tube would have two intrinsic sources of time-dependent current fluctuations; noise from the thermal agitation of electrons (thermal noise) and noise from the discreteness of the electric charge (shot noise). In a vacuum tube the cathode emits electrons randomly and independently. It is a Poisson process, where the mean of the squared fluctuation of the number of emission events is equal to the average count of electrons. The corresponding

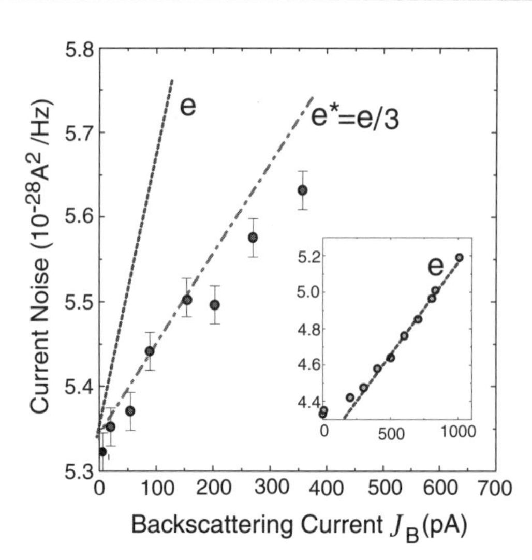

Fig. 18.3 Tunneling noise was measured in the fractional QH state with the bulk filling factor $\nu = 2/3$, and also in the integer QH state at the bulk filling factor $\nu = 4$ (given in the inset). See Fig.18.1 with respect to the bulk filling factor $\nu = 2/3$. The slopes are shown for quasiparticles with $e/3$ and electrons with charge e. Data are taken from Saminadayar et al.[419]

current noise S is given by the classical Schottky formula,[1]

$$S = 2qJ, \tag{18.2.2}$$

where J is the mean current and $q = e$ because the electron carries the charge $-e$ in the vacuum tube. Here the factor 2 appears since positive and negative frequencies contribute independently.

In the QH system at $\nu = 1/m$, since the current is carried by discrete quasiparticles, each of which has charge $q = e/m$ and travels independently of the others, the current noise is given by the classical Schottky formula (18.2.2)

$$S = 2\frac{e}{m}J_B, \tag{18.2.3}$$

where J_B is the backscattered current. It presents a direct measurement of the charge of quasiparticles. Observed charges were $e/3$ in the $\nu = 1/3$ QH state,[391,419] as is seen in Fig.18.3, and $e/5$ in the $\nu = 2/5$ QH state.[413] We remark that there is also a report[101] that the backscattered charges were $e/3$, $2e/5$ and $\sim 3e/7$

[1]Schottky proposed a measurement of the value of electric charge e by this formula. An interesting example of the shot noise occurs at a tunnel junction between a normal met al. and a superconductor, where we expect $q = 2e$ since charge is added to the superconductor in Cooper pairs. Such a characteristic noise has been measured recently.[309]

at $\nu = 1/3$, $2/5$ and $3/7$, respectively, at extremely low electron temperatures ($T < 10$mK). Namely, backscattering in this regime seems to be that of correlated quasiparticles.

18.3 Chiral Edge Excitations

We consider the QH system at $\nu = 1/m$ without the bulk Hall current in the spinless theory. We take a rectangular geometry [Fig.10.8 (page 192)] and analyze excitations along an edge taken as the x axis. The quantum-mechanical Hamiltonian is given by (10.7.10) in the Landau gauge,

$$H^{\text{QM}} = eyE, \tag{18.3.1}$$

where the kinetic term has been quenched. The field-theoretical Hamiltonian is constructed by the standard prescription (2.4.7),

$$H^{\text{FT}} = eE\langle\psi|y|\psi\rangle = eE\int d^2x\, y\rho(x,y). \tag{18.3.2}$$

We may use this to describe the edge located at $y = 0$, by demanding the electric field E to confine electrons inside the bulk ($y \leq 0$). We have neglected the Coulomb interaction between electrons.[350] A possible electron-electron contact interaction is harmless.[2] Furthermore, edge excitations are insensitive to impurities since all excitations propagate in one direction. Even if they are scattered by impurities, they are forced to move to the original direction again.[3]

The confining electric field is absent in the bulk away from the edge, where the energy level is flat and the Landau level is degenerate. The degeneracy of the Landau level is resolved by the confining electric field near the edge [Fig.18.4]. By neglecting the term which vanishes as $\hbar\omega_c \to \infty$ in the formula (10.7.15), the energy is given by the linear dispersion relation

$$\epsilon(k_x) = -eE\ell_B^2 k_x. \tag{18.3.3}$$

The edge excitation is a gapless mode, $\epsilon(k_x) \to 0$ as $k_x \to 0$. This is physically reasonable because an infinitesimal deformation of an incompressible droplet costs an infinitesimal energy.

The electron state having the momentum $\hbar k_x$ is represented by a plane wave in the x direction but localized in the y direction. It is localized at $y = \bar{y}_k$ as in

---

[2] The effect of contact interactions is to add the term $b\rho^2(x)$ to the Hamiltonian, where b is a numerical constant. Because the Hamiltonian (18.3.2) is equivalent to the hydrodynamical Hamiltonian $H = \pi m\hbar v \int dx\rho^2(x)$, as will be shown in (18.4.4), the term $b\rho^2(x)$ is simply absorbed into the velocity v. Contact interactions merely sift the edge velocity.

[3] The effect of impurities is simulated by a spatially dependent random edge potential $\lambda(x)\rho(x)$. We analyze the problem in the equivalent fermion theory (18.5.1) to be derived later. It is modified as $\mathcal{L} = i\hbar\psi^\dagger(x)\,(\partial_t - v\partial_x)\,\psi(x) + \lambda_{\text{imp}}(x)\psi^\dagger(x)\psi(x)$. We perform a local phase transformation, $\psi(x) \to \psi'(x) = e^{-i\delta(x)}\psi(x)$ with $\delta(x) = (\hbar v)^{-1}\int_{-\infty}^{x} dx'\ \lambda_{\text{imp}}(x')$. Then, the impurity term is eliminated. Hence, the effect of the impurity is merely to introduce a phase shift $\delta(x)$.

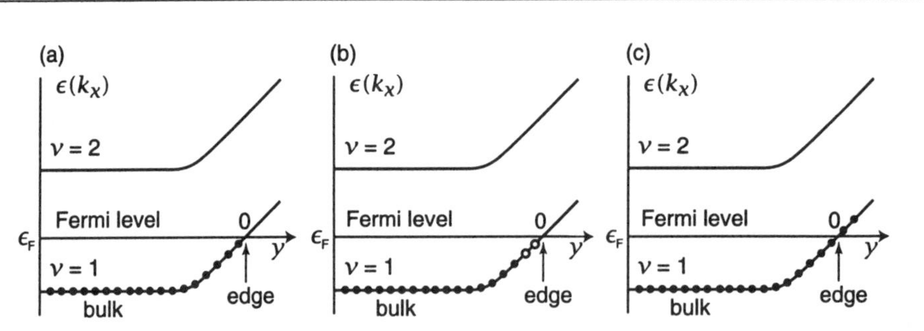

Fig. 18.4 The dispersion relation $\epsilon(k_x)$ is illustrated near the edge of the QH system at $\nu = 1$. The horizontal axis is identified with the y axis with $y = -\ell_B^2 k_x$ in the Landau gauge. The QH droplet has a sharp edge. All sites with $y < 0$ are filled while those with $y > 0$ are empty in the ground state [Fig.(a)]. Edge excitations imply that a few electrons are removed locally from sites with $y < 0$ [Fig.(b)] or added locally to sites with $y > 0$ [Fig.(c)]. See also Fig.18.5.

(10.7.13),

$$\bar{y}_k = -k_x \ell_B^2. \tag{18.3.4}$$

The dispersion relation (18.3.3) may be expressed also as

$$\epsilon(k_x) = -eE\ell_B^2 k_x = eE\bar{y}_k = -v\hbar k_x \tag{18.3.5}$$

by using of (18.3.4) and (18.2.1). In the box normalization the wave number is quantized,

$$k_x = \frac{2\pi n_x}{L_x}, \qquad n_x = 0, \pm 1, \pm 2, \cdots, \tag{18.3.6}$$

with L_x the length of the edge. The QH droplet has a sharp edge. In the ground state $|\text{g}\rangle$, all one-body states with $\bar{y}_k < 0$ ($k_x > 0$) are filled, while those with $\bar{y}_k > 0$ ($k_x < 0$) are empty. An electron may be added to (removed from) the system at $\bar{y} = 2\pi n_x \ell_B^2 / L_x$, where n_x is a small positive (negative) integer.

18.4 Chiral Tomonaga-Luttinger Liquid

We extract solely the edge part from the Hamiltonian (18.3.2), employing the hydrodynamical approach.[492] The QH liquid has a sharp boundary within which the electron density is constant [Fig.18.5],

$$\rho(x, y) = \Theta\left(h(x) - y\right) \rho_0. \tag{18.4.1}$$

Here, $y = h(x)$ parametrizes the edge. The one-dimensional density is defined by

$$\rho(x) = \int dy\, \rho(x, y) = \rho_0 h(x), \tag{18.4.2}$$

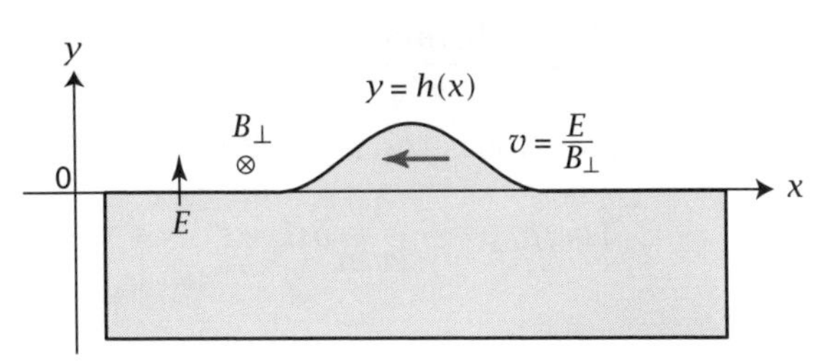

Fig. 18.5 The edge excitation of the QH droplet is nothing but a deformation of the boundary. Its shape is parametrized by a function $y = h(x)$. The electron density is constant within the droplet and zero outside it. By the Lorentz force a density wave propagates with velocity $v = E/B$, where E and $B_\perp$ are the confining electric field and the perpendicular magnetic field.

apart from an irrelevant constant arising from the ground-state contribution. The edge wave propagates with the drift velocity v to the left direction in the present setting [Fig.18.5]. The wave equation is

$$\partial_t \rho = v\partial_x \rho, \tag{18.4.3}$$

which is solved as $\rho(t,x) = \rho(x + vt)$. By using (18.4.1) and (18.4.2) in (18.3.2) we derive the hydrodynamical Hamiltonian,

$$\begin{aligned} H &= eE\rho_0 \int d^2x\, y\Theta\left(h(x) - y\right) \\ &= eE\rho_0 \int dx \int_{-\infty}^{h(x)} dy\, y = \pi m\hbar v \int dx\, \rho^2(x) \end{aligned} \tag{18.4.4}$$

at $\nu = 2\pi\rho_0\ell_B^2 = 1/m$, where we have discarded the ground-state energy. The wave equation (18.4.3) and the Hamiltonian (18.4.4) read

$$\dot{\rho}_k = ivk\rho_k, \tag{18.4.5}$$

$$H = 2\pi m\hbar v \int_{-\infty}^{0} dk\, \rho_{-k}\rho_k \tag{18.4.6}$$

in the momentum space.

We are able to construct the edge Lagrangian leading to the above Hamiltonian and wave equation,

$$L = 2\pi m\hbar v \int_{-\infty}^{0} dk \left(-\frac{i}{vk}\rho_{-k}\dot{\rho}_k + \rho_{-k}\rho_k\right). \tag{18.4.7}$$

Indeed, identifying ρ_k $(k < 0)$ as the "coordinate", the "momentum" is given by

$$\pi_k = \frac{\partial L}{\partial \dot{\rho}_k} = -\frac{2\pi i m\hbar}{k}\rho_{-k} \qquad \text{for} \quad k < 0. \tag{18.4.8}$$

The canonical commutation relation is $[\rho_k, \pi_\ell] = i\hbar\delta(k-\ell)$, or

$$[\rho_k, \rho_\ell] = -\frac{k}{2\pi m}\delta(k+\ell). \tag{18.4.9}$$

The Hamiltonian is

$$H = \pi_k\dot{\rho}_k - L = 2\pi m\hbar v\int_{-\infty}^{0} dk\, \rho_{-k}\rho_k, \tag{18.4.10}$$

and the Heisenberg equation of motion is

$$\dot{\rho}_k = [H, \rho_k] = ivk\rho_k. \tag{18.4.11}$$

They agree with (18.4.6) and (18.4.5). The algebraic relation (18.4.9) is known as the U(1) Kac-Moody algebra.[192] The Hamiltonian (18.4.6) together with this commutation relation (18.4.9) defines the chiral Tomonaga-Luttinger model.[477,319,337]

The annihilation and creation operators for edge excitations of left-going electrons are

$$a_k = \sqrt{\frac{2\pi m}{k}}\rho_k, \qquad a_k^\dagger = \sqrt{\frac{2\pi m}{k}}\rho_{-k} \qquad \text{for} \quad k < 0, \tag{18.4.12}$$

respectively. They obey

$$[a_k, a_\ell^\dagger] = \delta(k-\ell). \tag{18.4.13}$$

The ground state is annihilated by a_k,

$$a_k|\mathrm{g}\rangle = 0, \tag{18.4.14}$$

as agrees with the physical picture, where the electron density is zero outside the QH droplet ($y = -\ell_B^2 k > 0$). The Hamiltonian (18.4.6) is

$$H = \int_{-\infty}^{0} dk\, \epsilon(k) a_k^\dagger a_k \tag{18.4.15}$$

with the dispersion relation (18.3.5), where normal ordering has been made so that the filled Fermi sea has zero energy. The Hilbert space is constructed as a Fock space by operating $a_k^\dagger$ to the "vacuum" $|\mathrm{g}\rangle$. It presents a representation (the neutral sector) of the Kac-Moody algebra (29.8.8).

We have adopted the hydrodynamical approach to derive the edge Hamiltonian (18.4.6). We have then constructed the Lagrangian leading to this Hamiltonian: It is the edge Lagrangian (18.4.7). We may as well take the algebraic approach. As will be detailed in Chapter 29, the symmetry of the QH system is given by the W_∞ algebra,[236,84] and it yields the Kac-Moody algebra (18.4.9) when restricted to the edge. The spectrum of edge excitations must be a representation of this algebra.[86,85] We construct the Lagrangian yielding the Kac-Moody algebra as the canonical commutation relation: It is the edge Lagrangian (18.4.7). We have thus come to the same edge Lagrangian in both approaches.

Charged Excitations

After having analyzed neutral edge excitations, we proceed to discuss charged excitations.[492] They are electrons and quasielectrons. We represent their field operators in terms of the density field $\rho(x)$ with the aid of bosonization in the one-dimensional space. Bosonization has its historical roots in reference,[461] and was used to solve the Thirring model[473,431,331] and the Schwinger model.[433,317,87] It is applicable also to the present chiral model.[445,171]

Leaving a rigorous treatment of bosonization to Appendix N, here we summarize the result. Let us define an exponentiated field $\psi(x)$ by

$$\psi(x) = C\exp\left[2\pi im\int^x dx'\rho(x')\right] \equiv C\exp\left[2\pi im\phi(x)\right], \tag{18.4.16}$$

where we have defined the boson field $\phi(x)$ by

$$\rho(x) = \partial_x\phi(x), \tag{18.4.17}$$

and C is a regularization c-number factor [see (N.18) in Appendix]. In the real space the Kac-Moody algebra (18.4.9) reads

$$[\rho(x), \rho(x')] = \frac{i}{2\pi m}\partial_x\delta(x-x'). \tag{18.4.18}$$

Integrating this we find[4]

$$[\phi(x), \phi(x')] = -\frac{i}{4\pi m}\text{sgn}(x-x'). \tag{18.4.19}$$

Since the density $\rho(x)$ obeys the wave equation (18.4.3), both $\phi(x)$ and $\psi(x)$ obey the same equation,

$$\partial_t\phi(x) = v\partial_x\phi(x), \qquad \partial_t\psi(x) = v\partial_x\psi(x). \tag{18.4.20}$$

Using the commutation relation (18.4.19) we are able to prove the following formulas [see (N.21), (N.22) and (N.17) in Appendix],

$$\{\psi(x), \psi(x')\} = 0, \qquad \{\psi(x), \psi^\dagger(x')\} = 0 \quad \text{for} \quad x \neq x', \tag{18.4.21}$$

$$[\rho(x), \psi(x')] = -\delta(x-x')\psi(x'), \tag{18.4.22}$$

$$\rho(x) = \psi^\dagger(x)\psi(x). \tag{18.4.23}$$

Anticommutation relations (18.4.21) say that $\psi(x)$ is a fermion field. The physical meaning of (18.4.22) is that $\psi(x')$ annihilates one electron at position x' by decreasing the density $\rho(x)$ at $x = x'$ [see (2.3.7) in Part I]. Equations (18.4.22) and (18.4.23)

---

[4] The symbol sgn(x) stands for the sign function such that sgn$(x) = 1$ for $x > 0$ and sgn$(x) = -1$ for $x < 0$. Note that $\partial_x\text{sgn}(x) = 2\delta(x)$.

imply that $\{\psi(x), \psi^\dagger(x')\} = \delta(x - x')$. The operator $\psi(x)$ is identified with the one-dimensional electron field on the edge.

All N-point functions are calculable exactly. In particular, two-point functions read [see (N.36) and (N.37) in Appendix]

$$\langle g|\psi(t,x)\psi(0)|g\rangle = 0, \quad \langle g|\psi^\dagger(t,x)\psi(0)|g\rangle \propto \frac{1}{(x - vt)^m}. \tag{18.4.24}$$

The electron propagators are the standard ones in the integer QH state at $\nu = 1/m = 1$, where the edge state is described by noninteracting electrons and forms a Fermi liquid. However, it acquires a nontrivial exponent $m \neq 1$ in the fractional QH state at $\nu = 1/m$. It implies that electrons on the edge are strongly correlated. This type of an electron state has been named a chiral Tomonaga-Luttinger liquid.[491,206] The exponent m was successfully measured by a Maryland-IBM group[347,97] through the temperature dependence of tunneling conductance between two edges, which had been predicted[491,259,350] to have the form $\sigma \propto T^{2m-2}$.

The electron field (18.4.16) may be regarded as a composite of m basic fields $\psi_{\text{qe}}(x)$,

$$\psi_{\text{qe}}(x) = Ce^{2\pi i\phi(x)}. \tag{18.4.25}$$

Using the Kac-Moody algebra (18.4.18), we are able to prove [see (N.24) and (N.25) in Appendix]

$$\psi_{\text{qe}}(x)\psi_{\text{qe}}(x') = e^{\pi i/m}\psi_{\text{qe}}(x')\psi_{\text{qe}}(x), \tag{18.4.26}$$

$$[\rho(x), \psi_{\text{qe}}(x')] = -\frac{1}{m}\delta(x - x')\psi_{\text{qe}}(x'). \tag{18.4.27}$$

The relation (18.4.26) says that it is an anyon field [see (8.5.9) in Part I]. The physical meaning of (18.4.27) is that $\psi_{\text{qe}}(x')$ annihilates a $1/m$ electron at position x' by decreasing the density $\rho(x)$ at $x = x'$; the electric charge is $-e/m$. We identify it with the quasielectron field on edge.

18.5 ELECTRODYNAMICS ON EDGE

Edge electrons are governed by the Lagrangian density,

$$\mathcal{L} = i\hbar\psi^\dagger(x)\,(\partial_t - v\partial_x)\,\psi(x). \tag{18.5.1}$$

This is checked as follows. First, the canonical momentum is

$$\pi(x) = \frac{\delta L}{\delta\dot{\psi}(x)} = i\hbar\psi^\dagger(x), \tag{18.5.2}$$

and the canonical commutation relations are

$$\{\psi(x), \psi^\dagger(x')\} = \delta(x - x'), \qquad \{\psi(x), \psi(x')\} = 0. \tag{18.5.3}$$

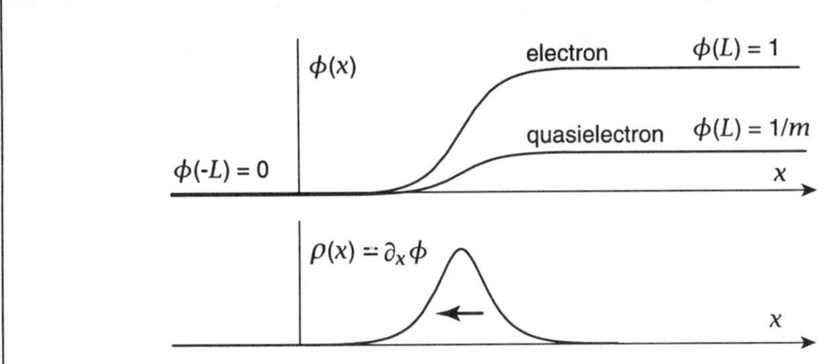

Fig. 18.6 The boson field $\phi(x)$ interpolates two asymptotic values, $\phi(L)$ and $\phi(-L) = 0$, whose difference is the electron number N_e. The electron is given by $\phi(L) = 1$, while the quasielectron by $\phi(L) = 1/m$ at $\nu = 1/m$. It is a topological number as in the sine-Gordon model [Fig.7.5]. The number density $\rho(x) = \partial_x \phi(x)$ is localized within a wave packet.

Second, the Hamiltonian density is

$$\mathcal{H} = \pi(x)\dot{\psi}(x) - \mathcal{L} = i\hbar v \psi^\dagger(x)\partial_x \psi(x). \tag{18.5.4}$$

Third, the Heisenberg equation of motion is

$$i\hbar \partial_t \psi(x) = [\psi(x), H] = i\hbar v \partial_x \psi(x), \tag{18.5.5}$$

which is the wave equation (18.4.20). These are the properties (18.4.20) ~ (18.4.23) required by the exponentiated field (18.4.16). The Lagrangian density (18.5.1) describes massless fermions possessing a definite chirality. They are Weyl fermions, as we have reviewed in Section 6.5: Compare (18.5.5) with (6.5.5b).

The Lagrangian density (18.5.1) is invariant under the global phase transformation,

$$\psi(x) \to e^{i\alpha}\psi(x), \tag{18.5.6}$$

where α is a real constant. According to the Nöther theorem in Section 3.5, there exits a conserved charge, that is the electron number,

$$N_e \equiv \int dx\, \rho(x) = \int_{-L}^{L} dx\, \partial_x \phi(x) = \phi(L) - \phi(-L). \tag{18.5.7}$$

The local "charge" and "current" are $\rho_e(x) = -e\rho(x)$ and $j_e(x) = ev\rho(x)$. The conservation law is nothing but the wave equation (18.4.3). The edge electron is a topological soliton [Fig.18.6], as is a reminiscence of the sine-Gordon soliton in Section 7.4. However, there is a difference: In the sine-Gordon model the topological charge is quantized due to the potential term, but in the present system it is quantized by the bulk effect, as we shall see in (18.5.15).

It is straightforward to incorporate the electromagnetic potential into the Lagrangian density (18.5.1) by making the minimal substitution (5.3.10),

$$\mathcal{L} = \psi^\dagger(x)\left(i\hbar\partial_t - i\hbar v\partial_x - ceA_0(x) + evA_x(x)\right)\psi(x). \tag{18.5.8}$$

The Lagrangian is invariant under the gauge transformation,

$$\psi(x) \to e^{if(x)}\psi(x), \qquad A_\mu(x) \to A_\mu(x) - \frac{\hbar}{e}\partial_\mu f(x). \tag{18.5.9}$$

The electric charge and current are

$$\rho_e(x) = \frac{\delta L}{c\delta A_0(x)} = -e\psi^\dagger(x)\psi(x), \tag{18.5.10a}$$

$$j_e(x) = \frac{\delta L}{\delta A_x(x)} = ev\psi^\dagger(x)\psi(x). \tag{18.5.10b}$$

One would naively anticipate that the electric charge density $\rho_e(x)$ is related to the electron number density $\rho(x)$ by the well-known relation, $\rho_e(x) = -e\rho(x)$ also in the presence of the electromagnetic potential.

This cannot be the case. The gauge transformation (18.5.9) translates the boson field $\phi(x)$ as

$$\phi(x) \to \phi(x) + \frac{1}{2\pi m}f(x), \tag{18.5.11}$$

according to the bosonization formula (18.4.16). Consequently, the electron number (18.4.17) is not gauge invariant,

$$\rho(x) \to \rho(x) + \frac{1}{2\pi m}\partial_x f(x). \tag{18.5.12}$$

Thus, the identification $\rho_e(x) = -e\rho(x)$ would imply that the electric charge were not gauge invariant.

On one hand, the electron number $\rho(x)$ is defined independently of the electromagnetic potential, because it is the Nöther charge associated with a global phase transformation (18.5.6). On the other hand, the electric charge $\rho_e(x)$ is to be defined in a gauge invariant manner. It is proved that the electron charge is (see Appendix N)

$$\rho_e(x) = -e\rho(x) - \frac{e^2}{2\pi m\hbar}A_x(x) = -e\partial_x\phi(x) - \frac{e^2}{2\pi m\hbar}A_x(x). \tag{18.5.13}$$

This is manifestly gauge invariant. Why do we obtain two different expressions, (18.4.17) for $\rho(x)$ and (18.5.13) for $\rho_e(x)$, from the same operator product $\psi^\dagger(x)\psi(x)$? This is because the operator product $\psi^\dagger(x)\psi(x')$ is ill defined at the same point ($x = x'$). It is necessary to make a short-distance regularization. A naive regularization breaks the gauge invariance and leads to (18.4.17), while a gauge-invariant regularization leads to (18.5.13). The two types of densities are produced from two ways of regularizations (N.15) and (N.41) in Appendix N.

It is impossible to make both the electron number and the electric charge simultaneously conserved. The conserved one is the electric charge associated with local gauge invariance. Namely, the conservation of the electron number (Nöther charge) is broken. This is an example of anomaly in the Nöether current, as was mentioned in Section 3.5: See the footnote in page 44.

As an important consequence, the electron number is intrinsically entangled with the electromagnetic potential on the edge. For instance, electrons are generated by fluctuations of the potential on the edge. It may sound strange, but is quite natural from a physical point of view. The edge theory describes electrons on the edge of the QH droplet, where the edge is determined by the Fermi level [Fig.18.4]. Fluctuations of the electromagnetic field lead to fluctuations of the Fermi level, as allows electrons to appear or disappear on the edge.

It has another important implication. Let us take a disk geometry, and insert one unit of magnetic flux at its center. As was discussed in Section 11.4, it increases the angular momenta of all electrons by one and pushes them outward, creating a quasihole at the origin. It affects the edge of the QH droplet. We are able to examine what happens based on the charge formula (18.5.13). The quasihole is a vortex, which changes the electromagnetic potential without affecting the electromagnetic fields far away from it. Since the electric charge is conserved, $\delta\rho_e = 0$, the electron number changes,

$$\delta\rho(x) = -\frac{e}{2\pi m\hbar}\delta A_x(x). \tag{18.5.14}$$

The change of the potential $\delta A_x(x)$ is such that its integration along the closed path is a unit flux carried by the vortex,

$$\delta Q \equiv \oint dx_k \delta\rho = -\frac{e}{2\pi m\hbar}\oint dx_k\ \delta A_k(\mathbf{x}) = -\frac{e\Phi_D}{2\pi m\hbar} = \frac{1}{m}. \tag{18.5.15}$$

The generated electric number is $1/m$, which is precisely the electron number of one quasielectron. This is the minimum charge allowed on the edge.

The same conclusion is reached based on the electron-number formula (18.4.17), which is not gauge invariant as in (18.5.12). A magnetic flux is inserted by a singular gauge transformation. It affects the electromagnetic potential on the edge according to (18.5.9) and (18.5.11) with $f = \theta(\mathbf{x})$, where $\theta(\mathbf{x})$ is the azimuthal angle. The electron number changes locally on the edge,

$$\delta\rho(x) = \frac{1}{2\pi m}\partial_x\theta(\mathbf{x}). \tag{18.5.16}$$

We integrate this along a closed path along the edge,

$$\delta Q \equiv \oint dx_k\ \delta\rho = \frac{1}{2\pi m}\oint dx_k\ \partial_k\theta(\mathbf{x}) = \frac{1}{m}. \tag{18.5.17}$$

The singular gauge transformation in the bulk generates a net electron number $1/m$ on the edge.

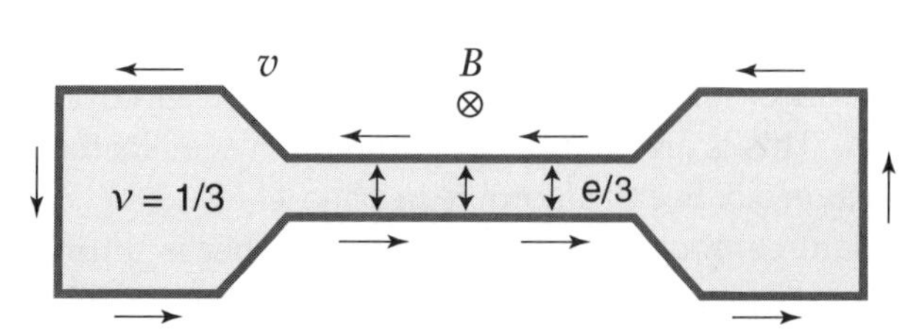

Fig. 18.7 Interedge tunneling becomes feasible when the opposite edge is brought into close proximity.

We may reexamine the problem based on the equivalent boson theory. The edge Lagrangian (18.4.7) is expressed as

$$\mathcal{L} = \pi\hbar m \left(2\partial_t\phi(x)\partial_x\phi(x) - v\partial_x\phi(x)\partial_x\phi(x)\right) \tag{18.5.18}$$

in the real space. This is known as the chiral boson Lagrangian.[171] To keep the local gauge invariance we incorporate the electromagnetic potential into the Lagrangian density (18.5.18) via the minimal substitution,

$$\partial_\mu\phi \to \partial_\mu\phi(x) + \frac{e}{2\pi m\hbar}A_\mu(x). \tag{18.5.19}$$

The Heisenberg equation reads

$$\partial_t\phi(x) - v\partial_x\phi(x) + \frac{e}{2\pi m\hbar}\left(cA_0(x) - vA_x(x)\right) = 0, \tag{18.5.20}$$

which is the conservation law of the electric charge (18.5.13),

$$\partial_t\rho_e - v\partial_x\rho_e = 0, \tag{18.5.21}$$

and implies the non-conservation of the electron number,

$$\partial_t\rho - v\partial_x\rho = -\frac{e}{2\pi m\hbar}\left(c\partial_x A_0(x) - v\partial_x A_x(x)\right). \tag{18.5.22}$$

A change of the electromagnetic fields $\boldsymbol{E}$ and $\boldsymbol{B}$ acts as a source of the electron number.

18.6 Edge Tunneling and Sine-Gordon Solitons

Edge excitations propagate to one direction. Backscattering occurs when the opposite edge is brought into close proximity and interedge tunneling becomes feasible [Fig.18.7]. We investigate edge excitations, allowing for interedge tunneling, based on the hydrodynamical Hamiltonian (18.4.4). When there are two edges, including both the left and right movers, we obtain the Tomonaga-Luttinger model[477,319,337]

$$H = \pi m\hbar v \int dx \left(\rho_R^2(x) + \rho_L^2(x)\right), \tag{18.6.1}$$

where the two edges are taken parallel to the x axis [Fig.18.7]. Note that the y coordinate does not appear since it has already been integrated out as in (18.4.4). Chiral boson fields $\phi_R(x)$ and $\phi_L(x)$ are introduced as in (18.4.17),

$$\rho_L(x) = \partial_x \phi_L(x), \qquad \rho_R(x) = \partial_x \phi_R(x). \tag{18.6.2}$$

They satisfy the algebraic relation (18.4.19) independently,

$$[\phi_L(x), \phi_L(x')] = -\frac{i}{4\pi m}\text{sgn}(x - x'), \tag{18.6.3a}$$

$$[\phi_R(x), \phi_R(x')] = \frac{i}{4\pi m}\text{sgn}(x - x'), \tag{18.6.3b}$$

$$[\phi_R(x), \phi_L(x')] = 0. \tag{18.6.3c}$$

Commutator (18.6.3a) has the opposite sign to (18.6.3b) since the direction of propagation is opposite.

The Tomonaga-Luttinger model (18.6.1) is equivalent to the massless Klein-Gordon model,

$$\mathcal{L}_0 = -\frac{\pi m\hbar v}{2}\partial_\mu \varphi(x)\partial^\mu \varphi(x), \tag{18.6.4}$$

where $\varphi(x)$ is the boson field defined by

$$\partial_x \varphi(x) = \rho_L(x) - \rho_R(x), \tag{18.6.5a}$$

$$\partial_t \varphi(x) = -v\left[\rho_L(x) + \rho_R(x)\right]. \tag{18.6.5b}$$

These two conditions are compatible because of the wave function (18.4.3), $\partial_t \rho_L = v\partial_x \rho_L$ and $\partial_t \rho_R = -v\partial_x \rho_R$. It follows that

$$\phi_L(x) \equiv \frac{1}{2}\left(\varphi(x) - \frac{1}{v}\int_{-\infty}^{x} dx'\, \dot{\varphi}(x')\right), \tag{18.6.6a}$$

$$\phi_R(x) \equiv \frac{1}{2}\left(-\varphi(x) - \frac{1}{v}\int_{-\infty}^{x} dx'\, \dot{\varphi}(x')\right). \tag{18.6.6b}$$

The equivalence is readily verified because the Hamiltonian associated with the Klein-Gordon model (18.6.4) agrees with (18.6.1), and because the canonical commutation relations are

$$[\varphi(x), \varphi(x')] = [\dot{\varphi}(x), \dot{\varphi}(x')] = 0,$$

$$[\varphi(x), \dot{\varphi}(x')] = \frac{iv}{\pi m}\delta(x - x'), \tag{18.6.7}$$

from which the commutation relations (18.6.3) follow.

Bosonization is done with respect to the left mover and the right mover independently. According to the bosonization formulas (18.4.16), we set

$$\psi_{\rm L}(x) = Ce^{2\pi i m\phi_{\rm L}(x)}, \qquad \psi_{\rm R}(x) = Ce^{2\pi i m\phi_{\rm R}(x)}. \tag{18.6.8}$$

Then,

$$\rho_{\rm L}(x) = \psi_{\rm L}^\dagger(x)\psi_{\rm L}(x) = \partial_x\phi_{\rm L}(x), \tag{18.6.9a}$$

$$\rho_{\rm R}(x) = \psi_{\rm R}^\dagger(x)\psi_{\rm R}(x) = \partial_x\phi_{\rm R}(x). \tag{18.6.9b}$$

The Tomonaga-Luttinger model is equivalent to a fermion theory,

$$\mathcal{L}_0 = i\hbar\psi_{\rm L}^\dagger(x)\,(\partial_t - v\partial_x)\,\psi_{\rm L}(x) + i\hbar\psi_{\rm R}^\dagger(x)\,(\partial_t + v\partial_x)\,\psi_{\rm R}(x). \tag{18.6.10}$$

It is rewritten as

$$\begin{aligned}\mathcal{L}_0 &= i\hbar v\overline{\Psi}(x)\gamma^\mu\partial_\mu\Psi(x)\\ &= \Psi^\dagger(x)\begin{pmatrix} i\hbar(\partial_t - v\partial_x) & 0 \\ 0 & i\hbar(\partial_t + v\partial_x)\end{pmatrix}\Psi(x),\end{aligned} \tag{18.6.11}$$

where $\Psi(x)$ is the two-component field $\overline{\Psi}(x) \equiv \Psi^\dagger\gamma^0$ and γ^μ is the two-dimensional Dirac matrix,

$$\Psi(x) = \begin{pmatrix}\psi_{\rm L}(x)\\ \psi_{\rm R}(x)\end{pmatrix}, \qquad \psi_{\rm L} = \frac{1-\gamma^5}{2}\Psi, \qquad \psi_{\rm R} = \frac{1+\gamma^5}{2}\Psi, \tag{18.6.12}$$

with

$$\gamma^0 = \begin{pmatrix}0&1\\1&0\end{pmatrix}, \qquad \gamma^1 = \begin{pmatrix}0&1\\-1&0\end{pmatrix}, \qquad \gamma^5 \equiv \gamma^0\gamma^1 = \begin{pmatrix}-1&0\\0&1\end{pmatrix}. \tag{18.6.13}$$

Bosonization formulas (18.6.8) are combined into

$$\Psi(x) = C\exp\left\{-2\pi i m\left(\gamma^5\varphi(x) + \frac{1}{v}\int_{-\infty}^x dx'\,\dot\varphi(x')\right)\right\}. \tag{18.6.14}$$

The system of independent left and right movers is described equivalently by the massless Klein-Gordon theory (18.6.4) or the massless Dirac theory (18.6.11).

The left-moving and right-moving quasielectron fields are given as a generalization of (18.4.25),

$$\psi_{\rm qeL}(x) = Ce^{2\pi i\phi_{\rm L}(x)}, \qquad \psi_{\rm qeR}(x) = Ce^{2\pi i\phi_{\rm R}(x)}. \tag{18.6.15}$$

The tunneling of a quasielectron is introduced phenomenologically by annihilating one right moving quasielectron and then creating one left moving quasielectron, and vice versa. The interaction Hamiltonian is

$$\mathcal{H}_{\rm int}^{\rm qe} = \frac{\lambda^{\rm qe}}{2}\left(\psi_{\rm qeL}^\dagger(x)\psi_{\rm qeR}(x) + \psi_{\rm qeR}^\dagger(x)\psi_{\rm qeL}(x)\right), \tag{18.6.16}$$

which reads

$$\mathcal{H}_{\text{int}}^{\text{qe}} = -\lambda_{\text{SG}}^{\text{qe}} \cos 2\pi\varphi(x) \tag{18.6.17}$$

by using (18.6.15) and (18.6.6). If tunneling is allowed to occur only at a point contact [Fig.18.2], say, at $x = 0$, the interaction term is given by

$$\mathcal{H}_{\text{int}}^{\text{qe}} = -\lambda_{\text{SG}}^{\text{qe}} \cos 2\pi\varphi(0). \tag{18.6.18}$$

This is a point-contact interaction, and is realized experimentally by pinching the sample electrostatically by a bias voltage [Fig.18.2]: The coupling constant $\lambda_{\text{SG}}^{\text{qe}}$ is tunable by the gate voltage.

A point contact plays the role of impurity. The Tomonaga-Luttinger model (18.6.1) with the point-contact interaction (18.6.18) is integrable.[186] In integral models, the scattering is completely elastic and factorizable. It means that the energy and momentum of quasiparticles are individually conserved, and that the multiparticle S matrix decomposes into a product of two-particle elements. Physically speaking, quasiparticles scatter off the point contact without particle production, i.e., one by one. Based on this fact various correlation functions are exactly calculable. We can prove the following:[160,161] quasiparticles tunnel at the point contact in the weak-coupling limit, while electrons tunnel in the strong-coupling limit. In the both limits the tunneling events happen independently, so the shot noise[430] is proportional to the charge of the carriers as in (18.2.3). Thus, the weak-backscattering limit is a direct signal of the fractional charge of quasiparticles,[260] and indeed fractional charges were unambiguously identified [391,419] by this method [Fig.18.3].

We investigate the case where interedge tunneling occurs in a wider region as in Fig.18.7. Including the tunneling term (18.6.17) into the massless Lagrangian (18.6.4), it becomes

$$\mathcal{L} = -\frac{\pi m\hbar v}{2}\partial_\mu\varphi(x)\partial^\mu\varphi(x) + \lambda_{\text{SG}}^{\text{qe}} \cos 2\pi\varphi(x). \tag{18.6.19}$$

For tunneling of electrons we would have

$$\mathcal{L} = -\frac{\pi m\hbar v}{2}\partial_\mu\varphi(x)\partial^\mu\varphi(x) + \lambda_{\text{SG}}^{\text{e}} \cos 2\pi m\varphi(x). \tag{18.6.20}$$

Let us scale $\phi(x) = \sqrt{\pi m}\varphi(x)$ and we have

$$\mathcal{L} = -\frac{\hbar v}{2}\partial_\mu\phi(x)\partial^\mu\phi(x) + \lambda_{\text{SG}} \cos \gamma\phi(x), \tag{18.6.21}$$

where

$$\gamma = \sqrt{4\pi/m} \qquad \text{for tunneling of quasielectrons} \tag{18.6.22a}$$

$$\gamma = \sqrt{4\pi m} \qquad \text{for tunneling of electrons.} \tag{18.6.22b}$$

Quantum-mechanical properties of the sine-Gordon model (18.6.21) has been explored extensively.[104,331]

Let us cite one important result.[104] The sine-Gordon model (18.6.21) is well defined quantum mechanically only when $\gamma < \sqrt{8\pi}$. Hence, it is well defined for quasielectron tunneling (18.6.22a) but ill defined for electron tunneling (18.6.22b) in the fractional QH state ($\nu = 1/m$ with $m > 3$). This would implies that electrons do not propagate along the edge in the fractional QH state: The edge modes must consist of quasielectrons and density waves.

The sine-Gordon model (18.6.21) is equivalent to the massive Thirring model[104,331] defined by

$$\mathcal{L} = i\overline{\Psi}_{\mathrm{T}}(x)\gamma^{\mu}\partial_{\mu}\Psi_{\mathrm{T}}(x) + \frac{\lambda}{2}\overline{\Psi}_{\mathrm{T}}(x)\Psi_{\mathrm{T}}(x) - \frac{1}{2}g\overline{\Psi}_{\mathrm{T}}(x)\gamma^{\mu}\Psi_{\mathrm{T}}(x)\overline{\Psi}_{\mathrm{T}}(x)\gamma_{\mu}\Psi_{\mathrm{T}}(x) \tag{18.6.23}$$

with

$$\Psi_{\mathrm{T}}(x) = \begin{pmatrix} \psi_{\mathrm{TL}}(x) \\ \psi_{\mathrm{TR}}(x) \end{pmatrix}. \tag{18.6.24}$$

The coupling constant g is given by

$$\frac{\gamma^2}{4\pi} = \frac{1}{1+g/\pi}. \tag{18.6.25}$$

The equivalence is proved based on the bosonization formula[331]

$$\psi_{\mathrm{TL}}(x) = Ce^{2\pi i m\phi_{\mathrm{TL}}(x)}, \qquad \psi_{\mathrm{TR}}(x) = Ce^{2\pi i m\phi_{\mathrm{TR}}(x)} \tag{18.6.26}$$

as in (18.6.8) but with

$$\phi_{\mathrm{TL}}(x) \equiv \frac{1}{2}\left(\frac{1}{m}\varphi(x) - \frac{1}{v}\int_{-\infty}^{x} dx'\, \dot{\varphi}(x')\right), \tag{18.6.27a}$$

$$\phi_{\mathrm{TR}}(x) \equiv \frac{1}{2}\left(-\frac{1}{m}\varphi(x) - \frac{1}{v}\int_{-\infty}^{x} dx'\, \dot{\varphi}(x')\right). \tag{18.6.27b}$$

Formula (18.6.27) is slightly different from the corresponding formula (18.6.6). They agree with each other only when $m = 1$, where the four-point interaction is absent ($g = 0$) and the massive Thirring field is reduced to the massive Dirac field. This is the case for the integer QH system ($\nu = 1$). In the fractional Hall system ($\nu = 1/m \neq 1$), edge electrons are described by the massive Thirring field with $g \neq 0$. quasielectrons and density fluctuations are gapful due to the tunneling effect.

CHAPTER 19

STRIPES AND BUBBLES IN HIGHER LANDAU LEVELS

Electrons occupy only the lowest Landau level in the limit of high magnetic field. With decreasing magnetic field, the number of degeneracy in each Landau level decreases and part of electrons occupy higher Landau levels. Since the Coulomb potential is not a monotonically decreasing function in higher Landau levels, we may expect charge density waves (CDWs) such as stripes and bubbles to emerge. In these CDW states electrons form clusters with other electrons, which contrasts to the Laughlin state realized in the lowest Landau level. The stripe phase has been observed experimentally by way of a strong anisotropy and nonlinearity in the diagonal resistivity R_{xx}.[1]

19.1 HIGHER LANDAU LEVELS

We have so far analyzed physics taking place mainly in the lowest Landau level. Electrons occupy higher Landau levels in weak magnetic fields. Since the filled Landau levels are inactive, it is a good approximation to consider only electrons in the highest Landau level partially occupied. We may think naively that their behavior is similar to the one in the partially filled lowest Landau level. However, this is not correct. Electrons have larger kinetic energy and their wave functions extend over space with larger cyclotron radius R_c as in (10.2.29). Short range Coulomb repulsions between electrons is effectively reduced in higher Landau level. This change modifies physics in each Landau level. In particular, the Laughlin state would not be realized in higher Landau levels, because it is stabilized by a strong short range repulsion. Indeed, no fractional QH effects have been observed experimentally for $\nu > 4$.

On the other hand, remarkable transport anomalies have been observed experimentally[313,122,108] on high quality samples for $\nu > 4$, especially when ν is near a half integer, namely, when $\nu_N \simeq \frac{1}{2}$, where ν_N is the filling of the Landau level partially occupied. Such anomalies include a strong anisotropy and nonlinearity in the diagonal resistivity R_{xx}: See Fig.9.10 (page 175). They reflect intriguing cor-

[1]This chapter was prepared by Naokazu Shibata, to whom the author is very grateful.

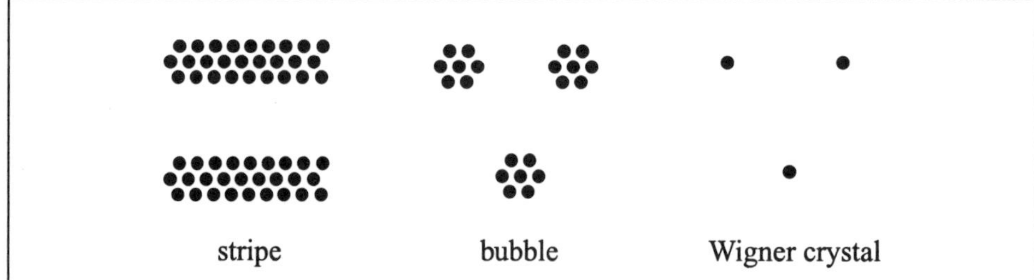

Fig. 19.1 Stripe and bubble states have been predicted in Hartree-Fock calculations. The dots represent electrons.

relation physics at work that is qualitatively different from fractional QH effects.

It was argued,[282,172,348,173] before the experimental discovery of the above mentioned transport anomalies, that the uniform electron liquid may be unstable against the formation of CDW in the $N = 2$ and higher Landau levels. The CDW states represent stripes and bubbles, as illustrated in Fig.19.1. Near half filling for the partially filled Landau level ($\nu_N \simeq \frac{1}{2}$), the CDW has been predicted to be a unidirectional *stripe phase* having a wavelength of order the cyclotron radius. It would give rise to transport anisotropy as the orientation of the stripe picks out a special direction in space. The formation of the CDW state has been confirmed to be stable against quantum fluctuations by numerical studies.[412,212,436]

These CDW states arise owing to an intriguing behavior of the Coulomb interaction in higher Landau levels. The Coulomb Hamiltonian is given by

$$H_{\rm C}^N = \pi \int d^2q\, V(\boldsymbol{q})\rho_N(-\boldsymbol{q})\rho_N(\boldsymbol{q}) \tag{19.1.1}$$

with

$$V(\boldsymbol{q}) = \frac{e^2}{4\pi\varepsilon|\boldsymbol{q}|}, \tag{19.1.2}$$

where $\rho_N(\boldsymbol{q})$ is the projected density describing solely electrons in the Nth Landau level. As we present a detailed analysis in Section 29.4, it is given by

$$\rho_N(\boldsymbol{q}) = F_N(\boldsymbol{q})\hat{\rho}(\boldsymbol{q}), \tag{19.1.3}$$

where $\hat{\rho}(\boldsymbol{q})$ is the bare density, and $F_N(\boldsymbol{q})$ is the Landau-level form factor,

$$F_N(\boldsymbol{q}) = L_N\left(\frac{\ell_B^2\boldsymbol{q}^2}{2}\right)e^{-\ell_B^2 q^2/4}, \tag{19.1.4}$$

with $L_N(x)$ the Laguerre polynomial: See (29.4.17). Hence, the Hamiltonian reads

$$H_{\rm C}^N = \pi \int d^2q\, V_N^{\rm eff}(\boldsymbol{q})\hat{\rho}(-\boldsymbol{q})\hat{\rho}(\boldsymbol{q}) \tag{19.1.5}$$

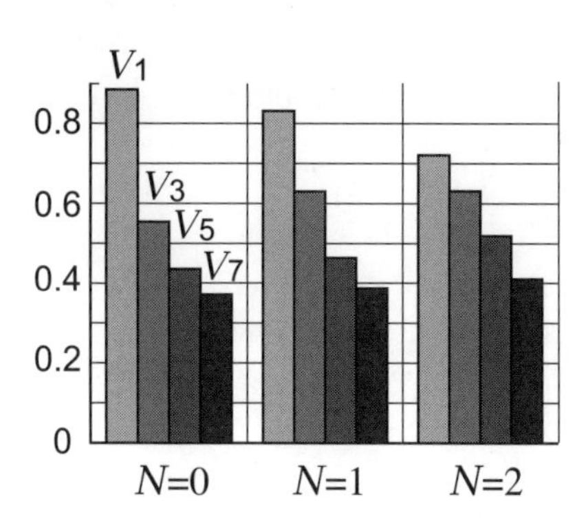

Fig. 19.2 Haldane's pseudopotential V_n is shown for the Nth Landau levels with $N = 0, 1$ and 2, where n is the relative angular momentum of two electrons.

with the effective Coulomb potential

$$V_N^{\text{eff}}(\boldsymbol{q}) = V(\boldsymbol{q})F_N^2(\boldsymbol{q}) = \frac{e^2}{4\pi\varepsilon|\boldsymbol{q}|}\left[L_N\left(\frac{\ell_B^2\boldsymbol{q}^2}{2}\right)\right]^2 e^{-\ell_B^2 q^2/2}. \tag{19.1.6}$$

The bare density$\hat{\rho}(\boldsymbol{q})$ is common to all Landau levels, and the form factor $F_N(\boldsymbol{q})$ characterizes each Landau level.

19.2 Haldane's Pseudopotentials

To see the interaction between electrons in each Landau level we expand the effective Coulomb potential (19.1.6) in terms of Laguerre polynomials,

$$V_N^{\text{eff}}(\boldsymbol{q}) = 2\ell_B^2 \sum_n V_n L_n(q^2\ell_B^2) e^{-q^2\ell_B^2/2}. \tag{19.2.1}$$

Then, setting $x = q^2\ell_B^2$, and using the normalization of the Laguerre polynomials,

$$\int_0^\infty dx\, e^{-x} L_m(x) L_n(x) = \delta_{mn}, \tag{19.2.2}$$

we obtain

$$\int_0^\infty dx\, V_N^{\text{eff}}(\boldsymbol{q}) L_n(x) e^{-x/2} = 2\ell_B^2 \sum_m V_m \int_0^\infty dx\, e^{-x} L_m(x) L_n(x) = 2\ell_B^2 V_n, \tag{19.2.3}$$

or

$$V_n = \int_0^\infty q dq\, V(\boldsymbol{q}) \left[L_N\left(\ell_B^2 q^2/2\right)\right]^2 L_n(q^2\ell_B^2) e^{-\ell_B^2 q^2}. \tag{19.2.4}$$

This is called the Haldane's pseudopotential parameter.[207] It represents the energy of a pair of electrons with relative angular momentum n.

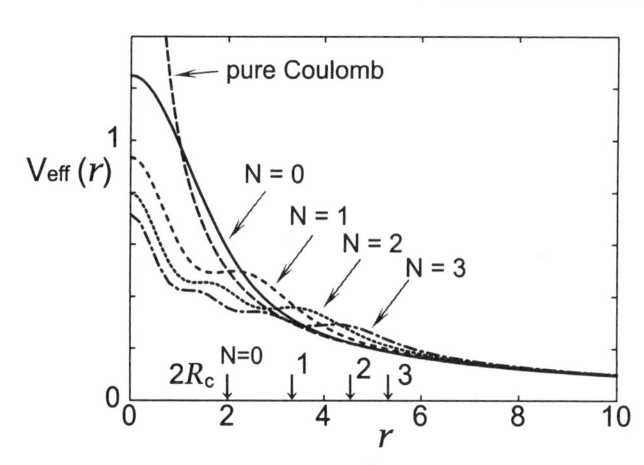

Fig. 19.3 The effective Coulomb potential $V_N^{\text{eff}}(\boldsymbol{x})$ is drawn for the Landau level index $N = 0, 1, 2, 3$. Non-monotonic structure appears below twice the classical cyclotron radius $2R_c = 2\sqrt{2N+1}$, as shown by the arrows. The magnetic length is set unity ($\ell_B = 1$).

Each term of the expansion (19.2.1) is regarded as the projection of the potential $V(\boldsymbol{q})$ to the electron pair state with the relative angular momentum $n\hbar$. Since the relative angular momentum is equal to or larger than $m\hbar$ for any pair of electrons in the Laughlin state at $\nu = 1/m$, the terms V_n with $n < m$ make no contribution to the Laughlin state in (19.2.1). Hence, if it were that $V_n = 0$ for $n \geq m$, the energy of the Laughlin state would remain zero while any other state would acquire an energy proportional to V_n. This energy gap stabilizes the Laughlin state. Thus the $\nu = \frac{1}{3}$ fractional QH state is stable if the term V_1 dominates all other terms.

The parameter V_m is shown for $N = 0, 1$ and 2 for the Coulomb potential in Fig.19.2. This figure demonstrates that V_1 is prominent in the lowest Landau level. This is a reason why the Laughlin state is realized in the lowest Landau level. In higher Landau levels, however, V_1 becomes smaller and almost comparable to other V_n. Then the energy gap between the Laughlin state and another state becomes small, and it is no more stable in higher Landau levels.

19.3 Effective Coulomb Interactions

In the lowest Landau level, the effective Coulomb potential reads

$$V_0^{\text{eff}}(\boldsymbol{q}) = \frac{e^2}{4\pi\varepsilon|\boldsymbol{q}|} e^{-\ell_B^2 q^2/2}, \tag{19.3.1}$$

or

$$V_0^{\text{eff}}(\boldsymbol{x}) = \frac{e^2\sqrt{2\pi}}{8\pi\varepsilon\ell_B} I_0(\boldsymbol{x}^2/4\ell_B^2) e^{-\boldsymbol{x}^2/4\ell_B^2}, \tag{19.3.2}$$

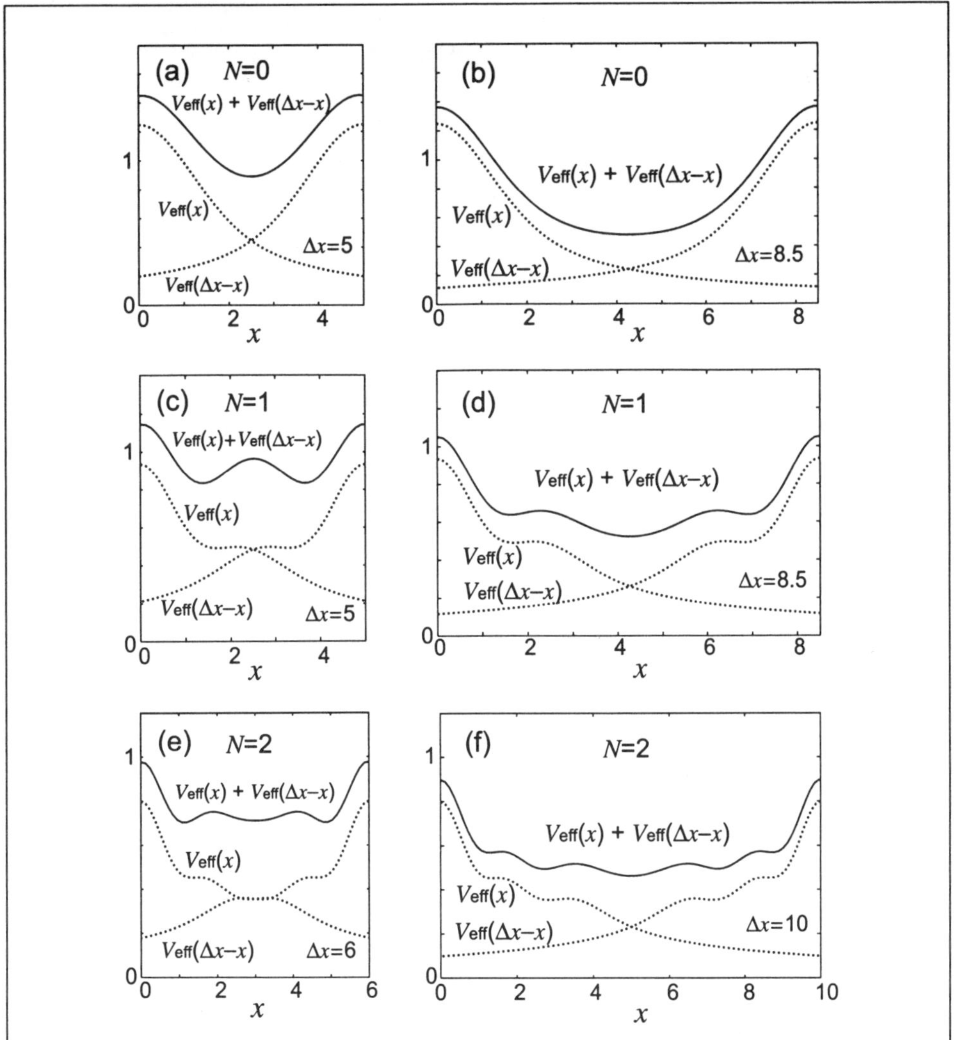

Fig. 19.4 The effective Coulomb interaction $\mathcal{V}_{\text{eff}}(x) = V_N^{\text{eff}}(x) + V_N^{\text{eff}}(\Delta x - x)$ is shown between a pair of electrons with the distance Δx in the Nth Landau level. There are several local minimum points in higher Landau levels. The magnetic length is set unity ($\ell_B = 1$).

with $I_0(x)$ the modified Bessel function.[2] It yields

$$V_0^{\text{eff}}(\boldsymbol{x}) \to \frac{e^2\sqrt{2\pi}}{8\pi\varepsilon\ell_B} = \sqrt{\frac{\pi}{2}}E_C^0 \tag{19.3.3}$$

in short distance ($r = |\boldsymbol{x}| \to 0$), and monotonically decreases with the increase in r.

Such a monotonic behavior is lost in higher Landau levels. Non-monotonic

[2]Its asymptotic behaviors are $I_0(x) \to 1$ as $x \to 0$, and $I_0(x) \to e^x/\sqrt{2\pi x}$ as $x \to \infty$.

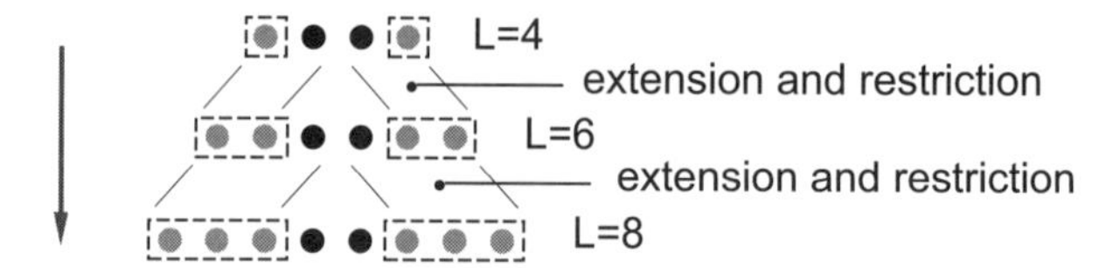

finite system algorithm of DMRG

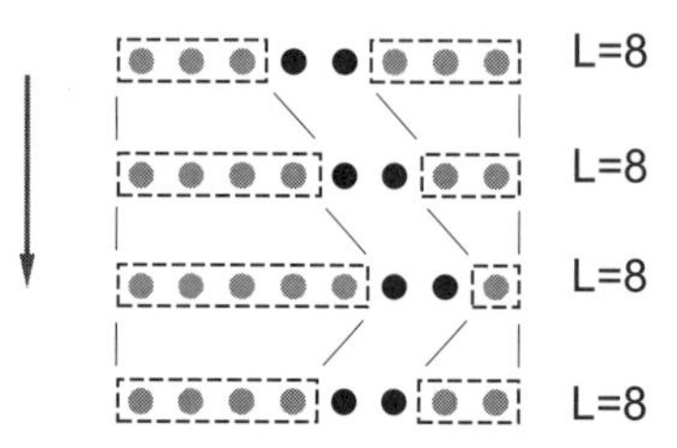

Fig. 19.5 Infinite and finite system algorithm of the DMRG are illustrated. Small circles represent local one-particle orbitals.

structure appears below twice the classical cyclotron radius $2R_c = 2\sqrt{2N+1}$, as shown by the arrows in Fig.19.3. It means an overlap between the two wave functions rapidly increases when two electrons approach twice the cyclotron radius. We expect it to produce a rich variety of CDW states.

The non-monotonic structure in $V_N^{\text{eff}}(\boldsymbol{x})$ explains why the clustering of electrons is realized in higher Landau levels.[438] When we fix the two electrons at $\boldsymbol{x}_1$ and $\boldsymbol{x}_2$, other electrons at $\boldsymbol{x}$ feel the potential

$$\mathcal{V}_{\text{eff}}(\boldsymbol{x}) = V_N^{\text{eff}}(\boldsymbol{x} - \boldsymbol{x}_1) + V_N^{\text{eff}}(\boldsymbol{x} - \boldsymbol{x}_2) \tag{19.3.4}$$

made by the two fixed electrons, which is shown in Fig.19.4 by the solid lines. In the figure, we have set two electrons at $(0,0)$ and $(\Delta x, 0)$ with $\Delta x = |\boldsymbol{x}_1 - \boldsymbol{x}_2|$ the distance between the two fixed electrons, and calculated the potential sum along the x-axis. In the lowest Landau level, we always find the potential minimum at the center of the two fixed electrons. This is due to the monotonic decrease in the effective Coulomb interaction $V_N^{\text{eff}}(\boldsymbol{x})$. Thus the electrons always keep away from the other electrons. In higher Landau levels, however, the effective Coulomb potential has potential minima near the two fixed electrons, which are made by the non-monotonic decrease in $V_N^{\text{eff}}(\boldsymbol{x})$. These potentials suggest that the electrons in higher Landau levels tend to form clusters when the mean distance between the electrons is not so long compared with the classical cyclotron radius R_c. This would lead to the formation of stripes and bubbles in higher Landau levels.

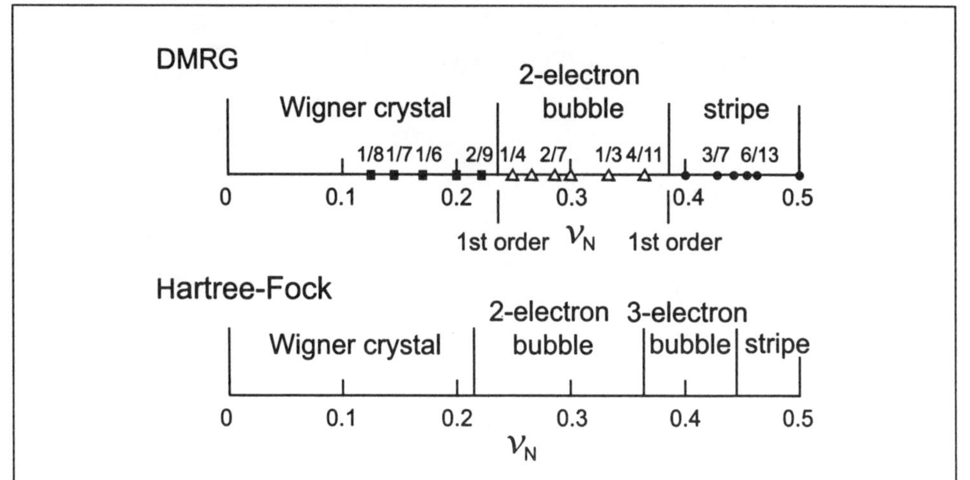

Fig. 19.6 Ground state phase diagram for $N = 2$ Landau level determined by (a) DMRG and (b) Hartree-Fock calculations. ν_N is the filling of $N = 2$ Landau level.

19.4 Density Matrix Renormalization Group Method

Mean field type analysis[282,172,173] based on Hartree-Fock theory neglects quantum fluctuations. This means even though Hartree-Fock theory predicts stripe and bubble states, there still exist possibility of liquid ground states.

The effect of quantum fluctuations has been extensively studied by exact diagonalization calculations[412]. This method enables us to obtain the exact ground state including effects of quantum fluctuations. However, the size of the system is limited up to about 10 electrons, which is too small to determine the electronic structure of CDW states and phase diagram for general filling ν.

To extend the limitations of exact diagonalizations the density matrix renormalization group (DMRG) method has been developed[509] and applied to quantum Hall systems.[436] In this method it is possible to calculate the ground state wave function of large systems with controlled accuracy. The algorithm of this method is summarized as follows: We start from a small system, i.e. a system consisting of only four single particle orbitals. We divide the system into two blocks, and add new orbitals at the end of two blocks to expand the blocks. We then calculate the ground state wave function of the system and obtain the density matrix of each expanded blocks. We then restrict the basis states in the expanded blocks by keeping only eigenstates of large eigenvalues of the density matrix. We add new orbitals between the blocks again and repeat the above procedure until we get desired size of system: See Fig.19.5. We then use the finite system algorithm of the DMRG to improve the ground state wave function. After several sweeps, it is possible to obtain the optimal ground state wave function within a restricted

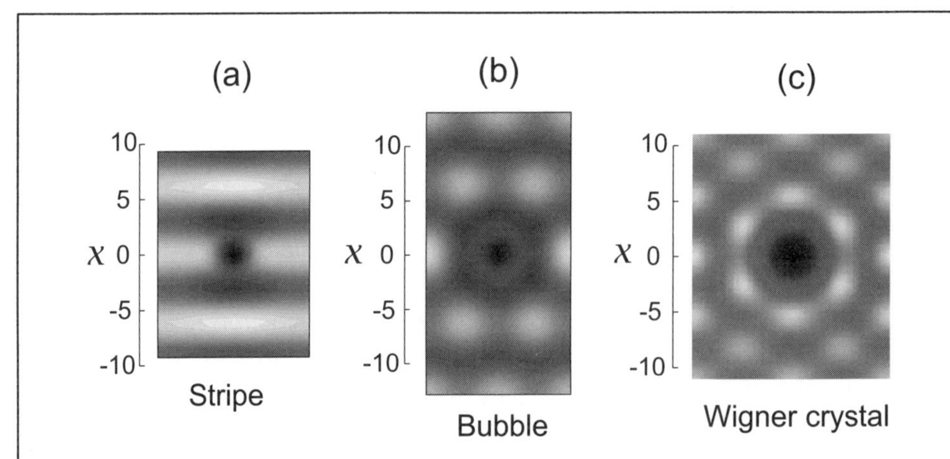

Fig. 19.7 The pair correlation functions $g(\boldsymbol{r})$ is illustrated. (a) 18 electrons in the unit cell at $\nu_N = 3/7$. (b) 16 electrons in the unit cell at $\nu_N = 8/27$. (c) 16 electrons in the unit cell at low fillings. White regions indicate that the electron density is high. The magnetic length is set unity ($\ell_B = 1$). See also Fig.8 in the frontispiece.

number of basis states. The error due to the truncation of basis states is estimated from the eigenvalues of the density matrix, and the accuracy of the wave function is systematically improved by increasing the number of basis states kept in the blocks.

19.5 Stripes, Bubbles and Wigner Crystal

The results obtained by DMRG for the $N = 2$ Landau level are summarized[436] in the ground state phase diagram shown in Fig.19.6. The DMRG calculation confirms the existence of stripe and bubble states predicted in Hartree-Fock calculations. The ground state phase diagram obtained by DMRG consists of three different CDW states; stripe state near half filling, bubble state around $\nu_N = 0.3$, and Wigner crystal below $\nu_N = 0.24$, where ν_N is the filling of the $N = 2$ Landau level.

The characteristic feature of each CDW state is shown in the pair correlation functions $g(\boldsymbol{x})$ in guiding center coordinates,[436,437]

$$g(\boldsymbol{x}) \equiv \frac{L_x L_y}{N_e(N_e - 1)} \langle\Psi| \sum_{i \neq j} \delta(\boldsymbol{x} + \boldsymbol{X}_i - \boldsymbol{X}_j)|\Psi\rangle, \tag{19.5.1}$$

where $|\Psi\rangle$ is the ground state and N_e is the number of the electrons in the unit cell. This is the correlation function of the center of the cyclotron motion. Figure 19.7(a) shows $g(\boldsymbol{x})$ at $\nu_N = 3/7$. This figure clearly shows that the stripe structure similar to that predicted in Hartree-Fock calculations is realized in the Landau level of $N = 2$. The stripe structure is obtained also for any fillings between $\nu_N = 1/2$

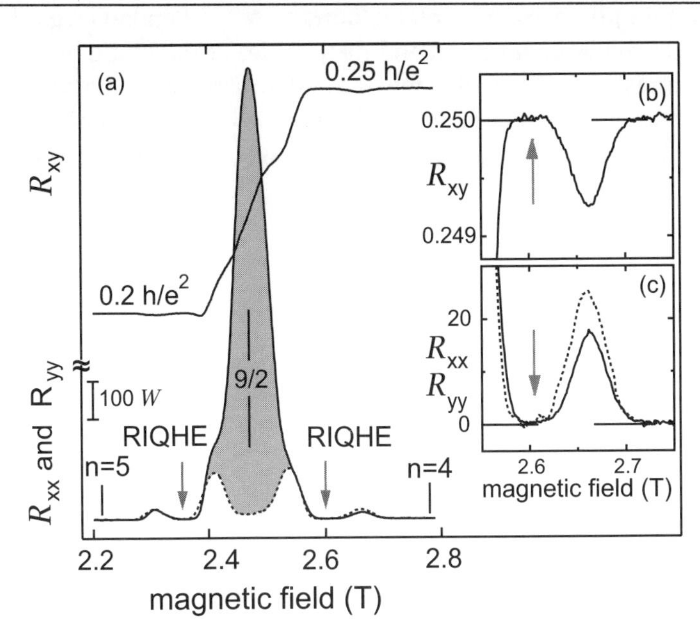

Fig. 19.8 (a) Diagonal resistance R_{xx}, R_{yy} and Hall resistance R_{xy} are shown in the QH systems, where the $N = 2$ Landau level is partially filled by electrons. There is a clear difference between R_{xx} and R_{yy} showing strong anisotropy around half filling of $N = 2$ Landau level ($\nu = 9/2$). Reentrant integer quantum Hall effect (RIQHE) state ($\nu \simeq 4\frac{1}{4}, 4\frac{3}{4}$) is expected to be two-electron bubble phase. Insets (b) and (c) magnify the RIQHE region. Data are taken from Lilly et al.[313]

and 0.38. This is consistent with experiments, where strong anisotropy in longitudinal resistance is observed in $N = 2$ Landau level around half filling as shown in Fig.19.8.

With decreasing ν_N the correlation function discontinuously changes at around $\nu_N = 0.38$ and the ground state is characterized by bubbles shown in Fig.19.7(b). This figure shows 8 bubbles in the unit cell on the triangular lattice. Since the number of electrons is 16 in the unit cell, two electrons are clustering together. This pairing of the two electrons makes ring structure in the correlation functions around the origin. These two-electron bubbles are obtained at intermediate fillings between $\nu_N = 0.38$ and 0.24. The similar two-electron bubbles are obtained in Hartree-Fock calculations. However, as shown in Fig.19.5(b), Hartree-Fock theory predicts also the three-electron bubbles, each of which contains three electrons. Since three-electron bubbles are not obtained in the DMRG calculation, the energy gain due to quantum fluctuations is relatively small for three-electron bubbles.

The absence of the three-electron bubble phase in $N = 2$ Landau level is consistent with experiments. The DMRG results suggest the two-electron bubble phase is the re-entrant phase experimentally found in $N = 2$ Landau level.[313,122,108] The coexistence of the Wigner crystal and the bubbles at the phase boundary around $\nu_N = 0.24$ brings finite dissipation into the system separating the two integer quantum Hall states as shown in Fig.19.8. This idea would be jeopardized if the three-electron bubble were realized, since then there would be another re-entrant phase.

CHAPTER 20

QUANTUM HALL EFFECTS IN GRAPHENE

A most recent topic is the discovery of an unconventional QH effect in graphene, where electrons are described by the Dirac equation. The key observation is that the intrinsic Zeeman effect causes a Landau-level mixing to create the SUSY energy spectrum with a 4-fold degeneracy. The filling factors form a series $\nu = \pm 2, \pm 6, \pm 10, \cdots$. When strong magnetic field is applied, the Coulomb interaction becomes important, and an excitonic gap opens by making a BCS-type condensation of electron-hole pairs at $\nu = 0$. It also gives birth to the Ising QH ferromagnet at $\nu = \pm 1$. According to the SUSY spectrum, a single energy level contains up-spin and down-spin electrons belonging to different Landau levels. Remarkable consequences follow from this fact that the effective Coulomb potential depends on spins and that the valley polarization occurs in the ground state at $\nu = \pm 4, \pm 8, \pm 12, \cdots$.[1]

20.1 UNCONVENTIONAL QH EFFECTS

Recent experimental developments have revealed unconventional QH effects in graphene.[374,375,538,376,539] The filling factors form a series [Fig.20.1(a)],

$$\nu = 0, \pm 1, \mathbf{\pm 2}, \pm 4, \mathbf{\pm 6}, \pm 8, \mathbf{\pm 10}, \pm 12, \cdots, \tag{20.1.1}$$

where the bold-face series had been predicted[47,203,390] before it was found experimentally[374,375,538,376], while the full series was discovered later[539] with larger magnetic field applied. Subsequently theoretical works[43,373,193,204,132] have been made to interpret the series. Furthermore, in bilayer graphene the series reads

$$\nu = 0, \pm 1, \pm 2, \pm 3, \mathbf{\pm 4}, \pm 6, \mathbf{\pm 8}, \pm 10, \mathbf{\pm 12}, \cdots, \tag{20.1.2}$$

where the bold-face series has been found experimentally[376] and studied theoretically,[340] while in trilayer graphene it reads

$$\nu = 0, \pm 1, \pm 2, \pm 3, \pm 4, \pm 5, \mathbf{\pm 6}, \pm 8, \mathbf{\pm 10}, \pm 12, \cdots, \tag{20.1.3}$$

[1]This chapter was prepared by Motohiko Ezawa, to whom the author is very grateful.

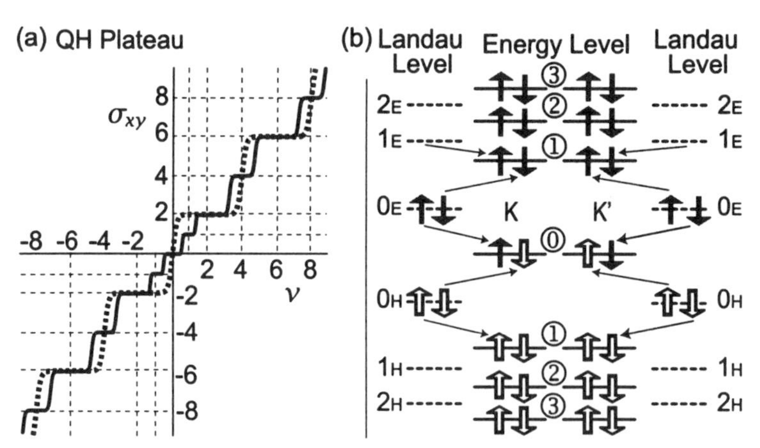

Fig. 20.1 (a) The QH conductivity σ_{xy} is illustrated in graphene. The dotted curve shows the sequence $\nu = \pm 2, \pm 6, \pm 10, \cdots$, while the solid curve the sequence $\nu = 0, \pm 1, \pm 2, \pm 4, \cdots$. (b) The energy level (number in circle) and the Landau level (number with index E or H) are illustrated for electrons and holes. The spin is indicated by a solid (open) arrow for electron and hole at the K and K′ points. This energy spectrum is a manifestation of SUSY. The Nth energy level contains up(down)-spin electrons from the Nth Landau level and down(up)-spin electrons from the (N-1)th Landau level at the K (K′) point for $N \geq 1$. The zero-energy level ($N = 0$) contains up(down)-spin electrons and down(up)-spin holes at the K (K′) point. See also Fig.9 in the frontispiece.

where the bold-face series has been predicted theoretically.[201] The full series is expected[132] to emerge when Coulomb interactions become important, i.e., when strong magnetic field is applied.[2] It is notable that the basic height in the Hall conductance is $4e^2/h$ for the bold-face series for all these systems, indicating the 4-fold degeneracy of the energy level except for the first step at the $\nu = 0$ point within noninteracting theory.

We make a detailed study of monolayer graphene. There are two inequivalent Brillouin zone corners, called the K and K′ points [Fig.20.2]. Conduction and valence bands in graphene form conically shaped valleys, touching at these points [Fig.20.3], where the two-dimensional energy dispersion relation is linear and *massless Dirac electrons* emerge[447,434,42]. Let us refer to the valley as the *Dirac valley*, and assign the valley index to the electron so that the electron at the K (K′) point carries the index $\tau = +\,(-)$.

A remarkable feature of the Dirac electron in magnetic field is that the intrinsic Zeeman energy is precisely one half of the cyclotron energy, as we have seen in Section 6.6. Then, two Landau levels mix to create one energy level [Fig.20.1(b)]. It has two important consequences leading to the SUSY energy spectrum; the emer-

[2]The Coulomb energy increases as $E_C^0 = e^2/4\pi\varepsilon\ell_B \propto \sqrt{B_\perp}$.

gence of the zero-energy state, and the degeneracy of the up-spin and down-spin states for each nonzero-energy level. Since this holds separately at the K and K′ points, each energy level has a 4-fold degeneracy [Fig.20.1(b)], and the noninteracting system has the SU(4) symmetry. The resulting series is $\nu = \pm 2, \pm 6, \pm 10, \cdots$, as is the bold-face series in (20.1.1) for monolayer graphene.

To study how the degeneracy is resolved by Coulomb interactions, it is necessary to treat the zero-energy state and nonzero-energy states separately.

The zero-energy state is distinctive, since it contains both electrons and holes. Electron-hole pairs form an excitonic condensation[270,271,202,204] due to attractive interaction, producing an excitonic gap. It is shown[132] that the ground state is a BCS-type state of electron-hole pairs at $\nu = 0$. As a consequence of excitonic condensation, the degeneracy of the zero-energy state is resolved into two subbands each of which contains either electrons or holes. The Coulomb Hamiltonian, projected to each of them, has the U(1) symmetry but is broken into the Z_2 symmetry. Hence the Ising QH ferromagnet[132] arises at $\nu = \pm 1$.

On the other hand, with respect to nonzero-energy states, since each energy level contains up-spin and down-spin electrons belonging to different Landau levels [Fig.20.1(b)], a remarkable result is obtained[132] that the effective Coulomb potential depends on the spin and the valley in each energy level. Namely, it depends not only on the spin of electrons but also on which valley electrons belongs to. This follows because wave functions are different for electrons belonging to different Landau levels. The Coulomb Hamiltonian, projected to a single energy level, possesses only the $U(1)\otimes U(1)\otimes Z_2$ symmetry. Consequently, the Coulomb interaction leads to the resolution of the 4-fold degeneracy into two 2-fold degeneracies, and a new series emerges at $\nu = \pm 4, \pm 8, \pm 12, \cdots$. The *valley polarization* occurs in the ground state[132] in such a way that up-spin electrons are more dense in the Dirac valley at the K point than at the K′ point. This occurs because the Coulomb energy is lower in higher Landau levels.

The degeneracy of the zero-energy state in noninteracting theory was originally derived from the arguments[375,538,376] on the Berry phase. The SUSY analysis can be generalized[132] to the system where the dispersion relation has the form $E(p) \propto p^{j_\uparrow + j_\downarrow}$ and the Berry phase is given by $\pi(j_\uparrow - j_\downarrow)$ in the absence of the magnetic field. The monolayer, bilayer and trilayer graphene correspond to $(j_\uparrow, j_\downarrow) = (1,0)$, $(2,0)$ and $(3,0)$, respectively. We show that the zero-energy states have the $j_\uparrow$-fold ($j_\downarrow$-fold) degeneracy for up-spin (down-spin) electrons at the K point. As a result the quantized values of the Hall conductivity become

$$\sigma_{xy} = \pm\left(n + \frac{j_\uparrow + j_\downarrow}{2}\right)\frac{4e^2}{h}, \qquad n = 0, 1, 2, \cdots, \tag{20.1.4}$$

within noninteracting theory. The degeneracy is resolved by Coulomb interactions as in monolayer graphene.

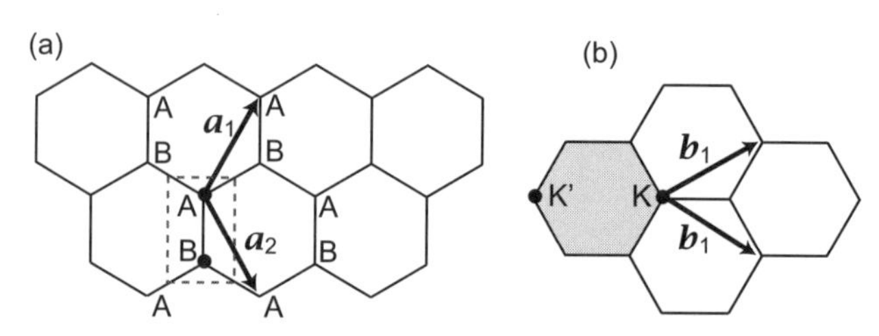

Fig. 20.2 (a) The two-dimensional honeycomb lattice is made of two triangular sublattices generated by the two basis vectors $\boldsymbol{a}_1$ and $\boldsymbol{a}_2$ from the base points A and B in primitive cell (dotted rectangle). (b) The reciplocal lattice is also a honeycomb lattice with the basis vectors $\boldsymbol{b}_1$ and $\boldsymbol{b}_2$. The first Brillouin zone is depicted in gray. The Fermi points are located at its corners. As two inequivalent points we take K and K′ as indicated.

20.2 Graphene and Dirac Electrons

We introduce basic notions of graphene, reviewing how Dirac electrons arise from the one-particle electronic states in the honeycomb lattice [Fig.20.2]. See also a review article due to Ando.[49] We take the basis vectors $\boldsymbol{a}_1$ and $\boldsymbol{a}_2$ as

$$\boldsymbol{a}_1 = \left(\frac{12}{2}, \frac{\sqrt{3}}{2}\right) a, \qquad \boldsymbol{a}_2 = \left(\frac{1}{2}, -\frac{\sqrt{3}}{2}\right) a, \tag{20.2.1}$$

with a the lattice constant. The honeycomb lattice has two different atoms per primitive cell, which we call the A and B sites. There are three B sites adjoint one A site, which are specified by the three vectors $\boldsymbol{r}_i$ with

$$\boldsymbol{r}_1 = \left(0, -\frac{1}{\sqrt{3}}\right) a, \quad \boldsymbol{r}_2 = \left(\frac{1}{2}, \frac{1}{2\sqrt{3}}\right) a, \quad \boldsymbol{r}_3 = \left(-\frac{1}{2}, \frac{1}{2\sqrt{3}}\right) a. \tag{20.2.2}$$

The A and B sites are generated as

$$A(n_1, n_2) = n_1\boldsymbol{a}_1 + n_2\boldsymbol{a}_2, \tag{20.2.3a}$$

$$B(n_1, n_2) = n_1\boldsymbol{a}_1 + n_2\boldsymbol{a}_2 + \boldsymbol{r}_1, \tag{20.2.3b}$$

where n_i are integers. The basis vectors $\boldsymbol{b}_i$ of the reciprocal lattice are given by solving the relations $\boldsymbol{b}_i \cdot \boldsymbol{a}_j = 2\pi\delta_{ij}$,

$$\boldsymbol{b}_1 = \left(1, \frac{1}{\sqrt{3}}\right) \frac{2\pi}{a}, \qquad \boldsymbol{b}_2 = \left(1, \frac{-1}{\sqrt{3}}\right) \frac{2\pi}{a}. \tag{20.2.4}$$

The Brillouin zone is a hexagon in the reciprocal lattice with opposite sides identified.

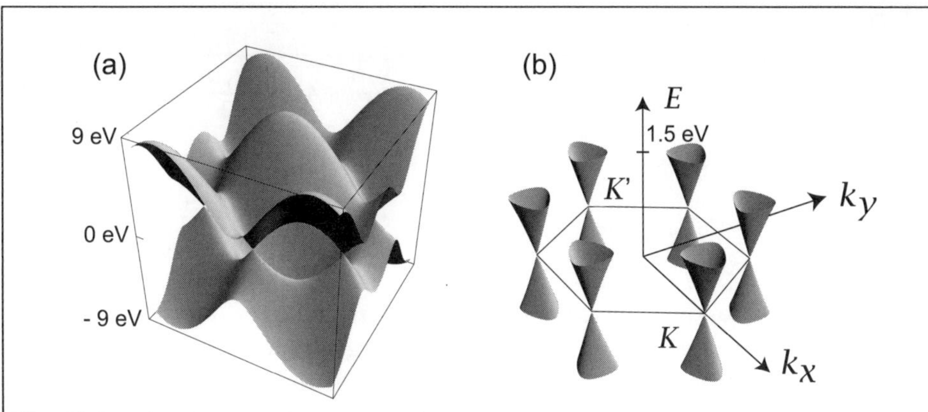

Fig. 20.3 (a) The overall structure of the dispersion relation (20.2.10) is illustrated. (b) Conduction and valence bands form conically shaped valleys near Brillouin zone corners, at which they touch. We refer to the valley as the Dirac valley. The dispersion relation becomes linear, and massless Dirac electrons emerge, within the Dirac valley. There are two inequivalent Brillouin zone corners (K and K′ points). The valence band is filled up with negative-energy electrons, as is a reminiscence of the Dirac sea.

In the tight-binding approximation electrons in graphene are described by the Hamiltonian,[434]

$$H = t \sum_{\sigma} \sum_{\boldsymbol{x},i} \left[c_{\mathrm{A}}^{\sigma\dagger}(\boldsymbol{x})\, c_{\mathrm{B}}^{\sigma}(\boldsymbol{x} + \boldsymbol{r}_i) + c_{\mathrm{B}}^{\sigma\dagger}(\boldsymbol{x} + \boldsymbol{r}_i)\, c_{\mathrm{A}}^{\sigma}(\boldsymbol{x}) \right], \tag{20.2.5}$$

where c_S^{σ} and $c_S^{\sigma\dagger}$ are the annihilation and creation operators for electrons with spin σ localized at the S site (S =A,B). We introduce the Fourier transform of the electron annihilation operator at the A and B sites,

$$c_A^{\sigma}(\boldsymbol{x}) = \int \frac{d^2k}{2\pi} e^{i\boldsymbol{k}\cdot\boldsymbol{x}} c_A^{\sigma}(\boldsymbol{k}), \qquad c_B^{\sigma}(\boldsymbol{x}) = \int \frac{d^2k}{2\pi} e^{i\boldsymbol{k}\cdot\boldsymbol{x}} c_B^{\sigma}(\boldsymbol{k}). \tag{20.2.6}$$

It is to be remarked that the Hamiltonian consists solely of the transfer term of an electron between neighboring A and B sites. Namely, the Hamiltonian has only off-diagonal components. In the momentum space it reads

$$H = t \sum_{\sigma} \int d^2k \left(c_{\mathrm{A}}^{\sigma\dagger}, c_{\mathrm{B}}^{\sigma\dagger} \right) \begin{pmatrix} 0 & f(\boldsymbol{k}) \\ f^*(\boldsymbol{k}) & 0 \end{pmatrix} \begin{pmatrix} c_{\mathrm{A}}^{\sigma} \\ c_{\mathrm{B}}^{\sigma} \end{pmatrix} \tag{20.2.7}$$

with

$$f(\boldsymbol{k}) = e^{ik_x a/\sqrt{3}} + 2e^{-ik_x a/2\sqrt{3}} \cos\left(k_y \frac{a}{2}\right). \tag{20.2.8}$$

It is diagonalized,

$$H = \sum_{\sigma} \int d^2k\, \mathcal{E}(\boldsymbol{k}) [C_{\vee}^{\sigma\dagger}(\boldsymbol{k}) C_{\vee}^{\sigma}(\boldsymbol{k}) - C_{\wedge}^{\sigma\dagger}(\boldsymbol{k}) C_{\wedge}^{\sigma}(\boldsymbol{k})], \tag{20.2.9}$$

where

$$\mathcal{E}(\boldsymbol{k}) = t\sqrt{1 + 4\cos\frac{\sqrt{3}k_x a}{2}\cos\frac{k_y a}{2} + 4\cos^2\frac{k_y a}{2}}, \tag{20.2.10}$$

and we have defined the operators $C_\vee^\sigma(\boldsymbol{k})$ and $C_\wedge^\sigma(\boldsymbol{k})$ by

$$C_\vee^\sigma(\boldsymbol{k}) = \frac{1}{\sqrt{2}}[c_A^\sigma(\boldsymbol{k}) + c_B^\sigma(\boldsymbol{k})], \qquad C_\wedge^\sigma(\boldsymbol{k}) = \frac{1}{\sqrt{2}}[c_A^\sigma(\boldsymbol{k}) - c_B^\sigma(\boldsymbol{k})]. \tag{20.2.11}$$

The operators $C_\vee^\sigma$ and $C_\wedge^\sigma$ describes electrons in the conduction band with positive energy and in the valence band with negative energy, respectively.

The band structure, as follows from (20.2.9), is illustrated in Fig.20.3. It is symmetric between positive-energy and negative-energy electrons. The ground state of graphene is half-filled and the Fermi energy is $\mathcal{E}(\boldsymbol{k}) = 0$. Namely, all negative-energy states are occupied, as is a reminiscence of the Dirac sea.

As we have stated, the Fermi level exists at $\mathcal{E}(\boldsymbol{k}) = 0$. Solving $\mathcal{E}(\boldsymbol{k}) = 0$ with the dispersion relation (20.2.10), we find that the Fermi point is only reached by six corners of the first Brillouin zone, among which there are only two inequivalent points due to the periodicity of the reciprocal lattice. We take them as

$$\boldsymbol{K} = \left(\frac{4\pi}{3a}, 0\right), \qquad \boldsymbol{K}' = \left(-\frac{4\pi}{3a}, 0\right). \tag{20.2.12}$$

It is customary to call them the K and K′ points [Fig.20.2(b) and Fig.20.3(b)]. They endow graphene with a two-component structure, the valley degree of freedom, corresponding to two independent Fermi points.

The band dispersion (20.2.10) is linear around each of these points in the continuum limit [Fig.20.3],

$$\mathcal{E}_\tau(\boldsymbol{k}) = \hbar v_F|\boldsymbol{k} - \tau\boldsymbol{K}| \qquad \text{for} \quad \boldsymbol{k} \simeq \tau\boldsymbol{K}, \tag{20.2.13}$$

where $\tau = \pm$ is the valley index, and v_F is the Fermi velocity,

$$v_F = \frac{\sqrt{3}}{2\hbar}ta \tag{20.2.14}$$

It is seen that the Fermi velocity v_F is substituted for the speed of light. This linear behavior has been confirmed experimentally[540] up to 3eV. The electron is a Weyl fermion due to this linear dispersion. Then the quantization axis of the spin should be taken along the momentum. We take the chirality as the spin index σ.

We explore low-energy physics near the Fermi points, $\boldsymbol{k} \simeq \tau\boldsymbol{K}$. It is clear from the dispersion relation illustrated in Fig.20.3 that the contribution to low-energy physics comes only from the regions around the two Fermi points, $\boldsymbol{k} \simeq \tau\boldsymbol{K}$. We make the change of variable as $\boldsymbol{k} = \boldsymbol{k}' + \tau\boldsymbol{K}$, where the variable $\boldsymbol{k}'$ represents a "small" variable, $|\boldsymbol{k}'| \ll |\boldsymbol{K}|$. We may approximate (20.2.6) as

$$c_S^\sigma(\boldsymbol{x}) \simeq \sum_{\tau=\pm} \int \frac{d^2k'}{2\pi} e^{i(\boldsymbol{k}'+\tau\boldsymbol{K})\cdot\boldsymbol{x}} c_S^\sigma(\boldsymbol{k}'+\tau\boldsymbol{K}) \equiv \sum_{\tau=\pm} e^{i\tau\boldsymbol{K}\cdot\boldsymbol{x}} \int \frac{d^2k'}{2\pi} e^{i\boldsymbol{k}'\cdot\boldsymbol{x}} c_{S\tau}^\sigma(\boldsymbol{k}'), \tag{20.2.15}$$

where we have set

$$c_{S\tau}^\sigma(\boldsymbol{k}') = c_S^\sigma(\boldsymbol{k}'+\tau\boldsymbol{K}). \tag{20.2.16}$$

Consequently the electron operator is a sum of the two terms coming from the two Dirac valleys,

$$c_S^\sigma(\boldsymbol{x}) \simeq e^{i\boldsymbol{K}\cdot\boldsymbol{x}} c_{S+}^\sigma(\boldsymbol{x}) + e^{-i\boldsymbol{K}\cdot\boldsymbol{x}} c_{S-}^\sigma(\boldsymbol{x}), \tag{20.2.17}$$

with the phase factors $e^{i\tau\boldsymbol{K}\cdot\boldsymbol{x}}$. The new operator $c_{S\tau}^\sigma(\boldsymbol{x})$ annihilates an electron located at the K or K′ point, which is described by the so-called envelope function.[49] We are able to assign the valley index $\tau = \pm$ to electrons in this way.

We come back to the Hamiltonian (20.2.9). The contribution comes from the momentum integration only around $\boldsymbol{k} = \boldsymbol{K}$ or $\boldsymbol{k} = -\boldsymbol{K}$. We can approximate the Hamiltonian as

$$H \simeq \sum_{\tau=\pm}\sum_{\sigma} \int_\tau d^2k\, \mathcal{E}_\tau(\boldsymbol{k})[C_\vee^{\sigma\dagger}(\boldsymbol{k})C_\vee^\sigma(\boldsymbol{k}) - C_\wedge^{\sigma\dagger}(\boldsymbol{k})C_\wedge^\sigma(\boldsymbol{k})], \tag{20.2.18}$$

where the integration $\int_\tau d^2k$ is made around $\boldsymbol{k} = \tau\boldsymbol{K}$. Here, making the change of variable $\boldsymbol{k} = \boldsymbol{k}' + \tau\boldsymbol{K}$, we obtain

$$H \simeq \sum_{\tau=\pm}\sum_{\sigma} \int d^2k'\, \mathcal{E}(\boldsymbol{k}')[C_{\vee\tau}^{\sigma\dagger}(\boldsymbol{k}')C_{\vee\tau}^\sigma(\boldsymbol{k}') - C_{\wedge\tau}^{\sigma\dagger}(\boldsymbol{k}')C_{\wedge\tau}^\sigma(\boldsymbol{k}')], \tag{20.2.19}$$

where $\mathcal{E}(\boldsymbol{k}') = \hbar v_{\rm F}|\boldsymbol{k}'|$, and we have set $C_{\vee\tau}^\sigma(\boldsymbol{k}') = C_\vee^\sigma(\boldsymbol{k}'+\tau\boldsymbol{K})$ and $C_{\wedge\tau}^\sigma(\boldsymbol{k}') = C_\wedge^\sigma(\boldsymbol{k}'+\tau\boldsymbol{K})$ in accord with (20.2.16). We introduce the electron and hole operators c_τ^σ and d_τ^σ by

$$c_\tau^\sigma(\boldsymbol{k}') = C_{\vee\tau}^\sigma(\boldsymbol{k}'), \qquad d_\tau^{\sigma\dagger}(\boldsymbol{k}') = C_{\wedge\tau}^\sigma(-\boldsymbol{k}'). \tag{20.2.20}$$

Note that σ stands for the chirality. We rewrite the Hamiltonian (20.2.19) as

$$H = \sum_{\tau=\pm}\sum_{\sigma} \int d^2k'\, \mathcal{E}(\boldsymbol{k}') \left[c_\tau^{\sigma\dagger}(\boldsymbol{k}')c_\tau^\sigma(\boldsymbol{k}') + d_\tau^{\sigma\dagger}(\boldsymbol{k}')d_\tau^\sigma(\boldsymbol{k}')\right], \tag{20.2.21}$$

together with

$$\mathcal{E}(\boldsymbol{k}') = \hbar v_{\rm F}|\boldsymbol{k}'|. \tag{20.2.22}$$

We use the variable $\boldsymbol{k}'$ to show that $\hbar\boldsymbol{k}'$ is the relative momentum of electrons, $|\boldsymbol{k}'| \ll |\boldsymbol{K}|$, measured from $\tau\boldsymbol{K}$ at the K or K′ point.

The Hamiltonian (20.2.21) contains a set of two independent electrons located at the K and K′ points, where the information on the locations of the K and K′ points is absent. The approximation is good for the analysis of the QH effect, where the fundamental length scale is given by the magnetic length ℓ_B. The distance between the K and K′ points is of the order $1/a$ in the momentum, which is much larger than $1/\ell_B$. Recall that $a \simeq 0.14$nm, while $\ell_B \simeq 2.56\text{nm}/\sqrt{|\boldsymbol{B}|(\text{Tesla})}$ with the magnetic field $\boldsymbol{B}$.

The Hamiltonian (20.2.21) is the second-quantized Hamiltonian for massless Dirac (Weyl) fermions, which we have reviewed in Section 6.5. To show this, we start with the Hamiltonian

$$H = \sum_{\tau=\pm} \int d^2x \, \Psi_\tau^\dagger(\boldsymbol{x}) H_{\text{D}}^\tau \Psi_\tau(\boldsymbol{x}), \tag{20.2.23}$$

where H_{D}^τ is the quantum mechanical Dirac Hamiltonian,

$$H_{\text{D}}^\tau = v_{\text{F}} \left(\tau \alpha_x p_x + \alpha_y p_y\right) = v_{\text{F}} \left(\tau \sigma_x p_x + \sigma_y p_y\right) \gamma_5, \tag{20.2.24}$$

with $p_i \equiv -i\hbar\partial_i$ and

$$\alpha_i = \begin{pmatrix} 0 & \sigma_i \\ \sigma_i & 0 \end{pmatrix}, \qquad \gamma_5 = \begin{pmatrix} 0 & 1 \\ 1 & 0 \end{pmatrix}. \tag{20.2.25}$$

Let us verify that (20.2.23) is equivalent to (20.2.21).

An orthonormal complete set of solutions of the Dirac Hamiltonian H_{D}^τ in (20.2.23) are explicitly given by (6.5.8) or

$$u_{\text{R}}^\tau(\boldsymbol{k}') = \frac{1}{\sqrt{2}} \left(1 + \frac{\tau\sigma_x k_i' + \sigma_y k_y'}{|\boldsymbol{k}'|}\right) \begin{pmatrix} 1 \\ 0 \end{pmatrix}_{\text{R}}, \tag{20.2.26a}$$

$$v_{\text{R}}^\tau(\boldsymbol{k}') = \frac{1}{\sqrt{2}} \left(1 - \frac{\tau\sigma_x k_i' + \sigma_y k_y'}{|\boldsymbol{k}'|}\right) \begin{pmatrix} 0 \\ 1 \end{pmatrix}_{\text{R}}, \tag{20.2.26b}$$

$$u_{\text{L}}^\tau(\boldsymbol{k}') = \frac{1}{\sqrt{2}} \left(1 - \frac{\tau\sigma_x k_i' + \sigma_y k_y'}{|\boldsymbol{k}'|}\right) \begin{pmatrix} 1 \\ 0 \end{pmatrix}_{\text{L}}, \tag{20.2.26c}$$

$$v_{\text{L}}^\tau(\boldsymbol{k}') = \frac{1}{\sqrt{2}} \left(1 + \frac{\tau\sigma_x k_i' + \sigma_y k_y'}{|\boldsymbol{k}'|}\right) \begin{pmatrix} 0 \\ 1 \end{pmatrix}_{\text{L}}. \tag{20.2.26d}$$

We expand the field operator in terms of these eigenfunctions,

$$\Psi_\tau(\boldsymbol{x}) = \Psi_{\text{e}\tau}(\boldsymbol{x}) + \Psi_{\text{h}\tau}(\boldsymbol{x}), \tag{20.2.27}$$

with

$$\Psi_{\text{e}\tau}(\boldsymbol{x}) = \sum_\sigma \int \frac{d^2k'}{2\pi} c_\tau^\sigma(\boldsymbol{k}') u_\sigma^\tau(\boldsymbol{k}') e^{i\boldsymbol{k}'\boldsymbol{x}}, \tag{20.2.28a}$$

$$\Psi_{\text{h}\tau}(\boldsymbol{x}) = \sum_\sigma \int \frac{d^2k'}{2\pi} d_\tau^{\sigma\dagger}(\boldsymbol{k}') v_\sigma^\tau(\boldsymbol{k}') e^{-i\boldsymbol{k}'\boldsymbol{x}}. \tag{20.2.28b}$$

where $\Psi_{\text{e}\tau}(\boldsymbol{x})$ and $\Psi_{\text{h}\tau}(\boldsymbol{x})$ describe electrons and holes, respectively. We can now derive (20.2.21) from (20.2.23) by substituting (20.2.27) into (20.2.23).

The functions $u_\sigma^\tau(\boldsymbol{k}')$ and $v_\sigma^\tau(\boldsymbol{k}')$ are the envelope functions and not the wave functions.[49] They are different by phase factors in accord with the formula

(20.2.17). Hence the field operator $\psi_\tau(\boldsymbol{x})$ of electrons and holes in graphene is given by

$$\psi_\tau(\boldsymbol{x}) = e^{i\tau K x}\Psi_\tau(\boldsymbol{x}). \tag{20.2.29}$$

The phase factor translates the origin of the momentum ($\boldsymbol{k} = \boldsymbol{k}' + \tau\boldsymbol{K}$).

We make a comment on the factor τ in front of α_x in (20.2.24). It has the standard expression of the Dirac Hamiltonian for $\tau = +$, that is, at the K point. The original Hamiltonian (20.2.5) for graphene has the mirror symmetry: The K point is transformed into the K′ point under the mirror reflection. Corresponding to this we have derived the mirror-reflected Dirac Hamiltonian (20.2.24) for $\tau = -$, that is, at the K′ point. We have shown that the low-energy dynamics of electrons at the K and K′ points is described by the Dirac Hamiltonian and the mirror-reflected Dirac Hamiltonian, respectively.

20.3 Dirac Hamiltonian and Supersymmetry

We apply the magnetic field to a graphene sheet taken on the xy plane. It is introduced to the Hamiltonian (20.2.23) by making the minimal substitution,

$$H = \sum_{\tau=\pm}\int d^2x\, \Psi_\tau^\dagger(\boldsymbol{x}) H_{\mathrm{D}}^\tau \Psi_\tau(\boldsymbol{x}), \tag{20.3.1}$$

where

$$H_{\mathrm{D}}^\tau = v_{\mathrm{F}}\left(\tau\sigma_x P_x + \sigma_y P_y\right)\gamma_5, \tag{20.3.2}$$

with the covariant momentum $P_i \equiv -i\hbar\partial_i + eA_i$. We assume a homogeneous magnetic field $\boldsymbol{B} = \nabla\times\boldsymbol{A} = (0,0,-B)$ with $B > 0$ along the z axis. The presence of the external magnetic field changes the mirror symmetry into the modified mirror symmetry: The modified mirror reflection not only transforms the K point into the K′ point but also reverses the direction of the magnetic field to maintain the symmetry of the system. This is most clearly seen in (6.6.8), as we discuss later. Consequently, the K and K′ points become physically distinguishable.

Landau Levels and Energy Spectrum

Electrons make cyclotron motion in magnetic field, and fill Landau levels successively. The helicity is no longer a good variable to describe the eigenstate of the Hamiltonian (20.3.2), since there exists a special direction for the spin, that is, the direction of magnetic field. The quantization axis of spin should be this direction. We solve the quantum mechanical problem in each Dirac valley,

$$H_{\mathrm{D}}^\tau \Psi_\tau(\boldsymbol{x}) = \mathcal{E}_\tau \Psi_\tau(\boldsymbol{x}), \tag{20.3.3}$$

and to expand the field operator in terms of the wave functions describing the new eigenstates in the presence of magnetic field.

As we have reviewed in Section 6.6, Dirac electrons in magnetic field are described by the SUSY quantum mechanics. Here we have two kinds of electrons carrying the valley index $\tau = \pm$. The Hamiltonian (20.3.2) is expressed as

$$H_{\text{D}}^{\tau} = \begin{pmatrix} 0 & Q_\tau \\ Q_\tau & 0 \end{pmatrix} \tag{20.3.4}$$

with

$$Q_\tau = v_{\text{F}} \left(\tau \sigma_x P_x + \sigma_y P_y \right) = v_{\text{F}} \begin{pmatrix} 0 & \tau P_x - iP_y \\ \tau P_x + iP_y & 0 \end{pmatrix}. \tag{20.3.5}$$

It is diagonalized as

$$H_{\text{D}}^{\tau} = \text{diag.} \left(\sqrt{Q_\tau Q_\tau}, -\sqrt{Q_\tau Q_\tau} \right). \tag{20.3.6}$$

The operators (20.3.5) are rewritten as

$$Q_+ = \hbar\omega_c \begin{pmatrix} 0 & a^\dagger \\ a & 0 \end{pmatrix}, \quad Q_- = \hbar\omega_c \begin{pmatrix} 0 & a \\ a^\dagger & 0 \end{pmatrix}, \tag{20.3.7}$$

with the use of the Landau-level ladder operator

$$a \equiv \frac{\ell_B}{\sqrt{2}\hbar}(P_x + iP_y), \quad a^\dagger \equiv \frac{\ell_B}{\sqrt{2}\hbar}(P_x - iP_y), \tag{20.3.8}$$

where $\omega_c = \sqrt{2}\hbar v_{\text{F}}/\ell_B$. Hence the diagonalized Hamiltonian reads

$$H_{\text{D}}^{+} = \hbar\omega_c \, \text{diag.} \left(\sqrt{a^\dagger a}, \sqrt{aa^\dagger}, -\sqrt{a^\dagger a}, -\sqrt{aa^\dagger} \right), \tag{20.3.9a}$$

$$H_{\text{D}}^{-} = \hbar\omega_c \, \text{diag.} \left(\sqrt{aa^\dagger}, \sqrt{a^\dagger a}, -\sqrt{aa^\dagger}, -\sqrt{a^\dagger a} \right). \tag{20.3.9b}$$

The energy spectrum of the Dirac Hamiltonian $H_{\text{D}}^{\pm}$ is found to be

$$\mathcal{E}_N^{+} = \hbar\omega_c \, \text{diag.} \left(\sqrt{N}, \sqrt{N+1}, -\sqrt{N}, -\sqrt{N+1} \right), \tag{20.3.10a}$$

$$\mathcal{E}_N^{-} = \hbar\omega_c \, \text{diag.} \left(\sqrt{N+1}, \sqrt{N}, -\sqrt{N+1}, -\sqrt{N} \right), \tag{20.3.10b}$$

by acting $H_{\text{D}}^{\pm}$ on the state $|N\rangle$ with $N = 0, 1, 2, 3, \cdots$.

Since a and $a^\dagger$ are the Landau-level ladder operators, N in (20.3.10) represents the Landau-level index. We have found that the energy of an electron in the Nth Landau level has the energy either $\hbar\omega_c\sqrt{N}$ or $\hbar\omega_c\sqrt{N+1}$. It is curious that the energy of an electron in the lowest Landau level ($N = 0$) can be zero though it undergoes cyclotron motion . We now solve this puzzle.

To reveal the intrinsic structure of the energy spectrum (20.3.10) we appeal to the SUSY quantum mechanics. We investigate the Pauli Hamiltonian

$$H_{\text{P}}^{\pm} \equiv Q_\pm Q_\pm = v_{\text{F}}^2 \left[(-i\hbar\nabla + eA)^2 \mp e\hbar\sigma_z B \right], \tag{20.3.11}$$

which is the building blocks of the Dirac Hamiltonian (20.3.6). Here, the direction of the magnetic field is found to be effectively opposite at the K and K′ points. As we have shown in Section 6.6, the spectrum of the relativistic Dirac Hamiltonian is mapped from that of the nonrelativistic Pauli Hamiltonian. Thus the eigenvalue $\mathcal{E}_\tau$ of the Dirac Hamiltonian $H_{\rm D}^\tau$ is constructed as $\mathcal{E}_\tau = \pm\sqrt{E_\tau}$ from the eigenvalue E_τ of the Pauli Hamiltonian $H_{\rm P}^\tau$. Furthermore, massless electrons in magnetic field are described as if they were nonrelativistic particles.

In the Pauli Hamiltonian (20.3.11), the first term is the kinetic term while the second term is the Zeeman term. The Landau level is created by electrons making cyclotron motion. According to the Pauli Hamiltonian (20.3.11), there exists the intrinsic Zeeman energy which is precisely one half of the cyclotron energy, and two Landau levels mix to create one energy level, as illustrated in Fig.20.1(b).

We consider the K point ($\tau = +$). It is obvious that the up-spin and down-spin states are eigenstates of the Pauli Hamiltonian (20.3.11) and hence those of the Dirac Hamiltonian (20.3.6). It is also obvious that the up-spin state has a lower energy than the down-spin state when they belong to the same Landau level. On the other hand, the direction of the spin is opposite at the K and K′ points since that of the magnetic field is opposite in the Pauli Hamiltonian (20.3.11). Hence, the spin index of the energy spectrum (20.3.10) in the Nth Landau level reads

$$\mathcal{E}_N^{+\uparrow} = \mathcal{E}_N^{-\downarrow} = \hbar\omega_c\sqrt{N}, \qquad \mathcal{E}_N^{+\downarrow} = \mathcal{E}_N^{-\uparrow} = \hbar\omega_c\sqrt{N+1}. \tag{20.3.12}$$

From these we can read the quantum numbers of the states in the Nth energy level as

$$\mathcal{E}_N^{+\uparrow} = \mathcal{E}_{N-1}^{+\downarrow} = \mathcal{E}_N^{-\downarrow} = \mathcal{E}_{N-1}^{-\uparrow} = \hbar\omega_c\sqrt{N}. \tag{20.3.13}$$

Namely, the Nth energy level contains up-spin electrons coming from the Nth Landau level and down-spin electrons coming from the (N-1)th Landau level at the K point, down-spin electrons coming from the Nth Landau level and up-spin electrons coming from the (N-1)th Landau level at the K′ point. A similar identification of spin is made for holes in the Nth energy level ($N \geq 1$). The zeroth energy level ($N = 0$) consists of the up-spin electron, the down-spin hole at the K point, and the down-spin electron, the up-spin hole at the K′ point coming from the lowest Landau levels for electrons and holes [Fig.20.1(b)].

There exists a degeneracy between the up-spin eigenstate and the down-spin eigenstate for nonzero-energy states [Fig.20.1(b)]. They make a supermultiplet. Furthermore there exist zero-energy states. Consequently, the SUSY is a good symmetry within noninteracting theory. Counting the states at the K and K′ points all together, one energy level has a 4-fold degeneracy. Each filled energy level contributes one conductance quantum $e^2/\hbar$ to the Hall conductivity. It follows that the resulting series is $\nu = \pm 2, \pm 6, \pm 10, \cdots$, which accounts for the bold-face series (20.1.1) in the monolayer graphene.

For long time the SUSY quantum mechanics (20.3.6) has served only as a toy model to all SUSY theories in modern physics. It is remarkable that the SUSY quantum mechanics is realized in graphene and furthermore its physical consequence is experimentally observed as the bold-face series (20.1.1) of the filling factors.

Field Operators

Thermal transitions between two energy levels are prohibited at sufficiently low temperature and in sufficiently strong magnetic field. Hence it is enough to consider the quantum-mechanical states in a single energy level. We focus on electrons in the Nth energy level. (Essentially the same analysis is applicable to holes.)

The quantum-mechanical states in the Nth Landau level are the Fock states $|N, n\rangle$ given by (10.2.17), or

$$|N, n\rangle = \frac{1}{\sqrt{N!n!}}(a^\dagger)^N (b^\dagger)^n |0\rangle \tag{20.3.14}$$

with $|0\rangle$ the Fock vacuum, $a|0\rangle = b|0\rangle = 0$. The Nth energy level ($N \neq 0$) contains electrons coming from the two Dirac valleys ($\tau = \pm$), and their field operators are expanded in terms of the eigenfunctions. Corresponding to $\mathcal{E}_N^{+\uparrow}$, $\mathcal{E}_{N-1}^{+\downarrow}$, $\mathcal{E}_N^{-\downarrow}$ and $\mathcal{E}_{N-1}^{-\uparrow}$ in (20.3.13), we have

$$\psi_{N+}^{\uparrow}(\boldsymbol{x}) = e^{iKx} \sum_n \langle \boldsymbol{x}|N, n\rangle c_+^{\uparrow}(N, n), \tag{20.3.15a}$$

$$\psi_{N+}^{\downarrow}(\boldsymbol{x}) = e^{iKx} \sum_n \langle \boldsymbol{x}|N-1, n\rangle c_+^{\downarrow}(N-1, n), \tag{20.3.15b}$$

$$\psi_{N-}^{\downarrow}(\boldsymbol{x}) = e^{-iKx} \sum_n \langle \boldsymbol{x}|N, n\rangle c_-^{\downarrow}(N, n), \tag{20.3.15c}$$

$$\psi_{N-}^{\uparrow}(\boldsymbol{x}) = e^{-iKx} \sum_n \langle \boldsymbol{x}|N-1, n\rangle c_-^{\uparrow}(N-1, n), \tag{20.3.15d}$$

respectively. Here, $c_\tau^\sigma(N, n)$ is the annihilation operator acting on the Fock state $|N, n\rangle$ at the τ point. In what follows we suppress the Landau-level index N in $c_\tau^\sigma(N, n)$ with the understanding that

$$c_+^{\uparrow}(n) = c_+^{\uparrow}(N, n), \quad c_-^{\downarrow}(n) = c_-^{\downarrow}(N, n), \quad c_+^{\downarrow}(n) = c_+^{\downarrow}(N-1, n), \quad c_-^{\uparrow}(n) = c_-^{\uparrow}(N-1, n). \tag{20.3.16}$$

See (20.2.29) with respect to the factor $e^{\pm iKx}$. The electron operators in the zeroth energy level are given by $\psi_{N+}^{\uparrow}(\boldsymbol{x})$ and $\psi_{N-}^{\downarrow}(\boldsymbol{x})$ with $N = 0$.

20.4 Effective Coulomb Interactions

We proceed to discuss the Coulomb interaction. It is described in the original graphene system by the Hamiltonian

$$H_C = \sum_{S,S'} \int d^2x d^2y \, V_{SS'}(\boldsymbol{x} - \boldsymbol{y}) \rho_S(\boldsymbol{x}) \rho_{S'}(\boldsymbol{y}), \tag{20.4.1}$$

where $\rho_S(\boldsymbol{x})$ is the density operators at the S site (S =A,B). It is the sum of those with spin σ [Fig.20.2(a)],

$$\rho_S(\boldsymbol{x}) = \rho_S^{\uparrow}(\boldsymbol{x}) + \rho_S^{\downarrow}(\boldsymbol{x}), \tag{20.4.2}$$

where

$$\rho_A^\sigma = \frac{1}{2}\left[\psi_{\rm e}^{\sigma\dagger}\psi_{\rm e}^\sigma - \psi_{\rm h}^{\sigma\dagger}\psi_{\rm h}^\sigma + \psi_{\rm e}^{\sigma\dagger}\psi_{\rm h}^{\sigma\dagger} + \psi_{\rm h}^\sigma\psi_{\rm e}^\sigma\right], \tag{20.4.3a}$$

$$\rho_B^\sigma = \frac{1}{2}\left[\psi_{\rm e}^{\sigma\dagger}\psi_{\rm e}^\sigma - \psi_{\rm h}^{\sigma\dagger}\psi_{\rm h}^\sigma - \psi_{\rm e}^{\sigma\dagger}\psi_{\rm h}^{\sigma\dagger} - \psi_{\rm h}^\sigma\psi_{\rm e}^\sigma\right]. \tag{20.4.3b}$$

Substituting (20.4.3) into (20.4.1) we obtain the Hamiltonian for electrons and holes. It is easy to see that there exists an attractive force between an electron and a hole, as expected.

As in the conventional QH effect we are extensively exploring, we make the basic assumption that the cyclotron energy is much larger than the Coulomb energy. We neglect the Landau-level mixing by the Coulomb interaction, and analyze it within one level of the SUSY energy spectrum [Fig.20.1(b)]. There exists a consistent formalism, known as the Landau-level projection, which we shall present details in Chapter 30. The projected theory presents not only a good approximation but also an essential way to reveal a new physics inherent to the QH system.

In this projected formalism we may treat the electron system, the hole system and the zero-energy system independently of each other. The hole system has the same structure as the electron system due to the electron-hole symmetry. On the other hand, the zero-energy system contains both electrons and holes. We study the electron system in this section, and the zero-energy system composed of electrons and holes in Section 20.5.

Landau-Level Projection

We are concerned about electrons in the Nth energy level in the presence of magnetic field. The total Coulomb Hamiltonian is given by (20.4.1) together with (20.4.3). We extract the terms involving only electrons from (20.4.3), $\rho_A^\sigma = \rho_B^\sigma = \frac{1}{2}\psi_{\rm e}^{\sigma\dagger}\psi_{\rm e}^\sigma$. The Hamiltonian for electrons reads

$$H = \frac{1}{2}\int\int d^2x d^2y\, V(\boldsymbol{x}-\boldsymbol{y})\rho(\boldsymbol{x})\rho(\boldsymbol{y}), \tag{20.4.4}$$

with $V(\boldsymbol{x}-\boldsymbol{y}) = V_{AA}(\boldsymbol{x}-\boldsymbol{y}) + V_{AB}(\boldsymbol{x}-\boldsymbol{y})$, and

$$\rho(\boldsymbol{x}) = \sum_\sigma \psi_{\rm e}^{\sigma\dagger}(\boldsymbol{x})\psi_{\rm e}^\sigma(\boldsymbol{x}) = \sum_{\sigma,\tau\tau'} \psi_\tau^{\sigma\dagger}(\boldsymbol{x})\psi_{\tau'}^\sigma(\boldsymbol{x}), \tag{20.4.5}$$

where $\psi_{\rm e}^\sigma(\boldsymbol{x}) = \sum_\tau \psi_\tau^\sigma(\boldsymbol{x})$ is the electron field operator. The Coulomb Hamiltonian (20.4.4) does not have the SU(2) invariance with respect to the valley degree of freedom, though the kinetic Hamiltonian (20.2.21) does. The origin of the non-invariance can be traced back to the mechanism how the valley index has been introduced. Namely, the mixing between the K and K′ points is suppressed due to the dispersion relation $E(\boldsymbol{k})$ given by (20.2.10) in the kinetic Hamiltonian, but this is not the case in the Coulomb Hamiltonian. Nevertheless, as we shall see in (20.4.15), such a mixing is exponentially small and actually negligible.

We consider electrons confined within the Nth energy level (20.3.13) in the presence of magnetic field ($N \geq 1$), assuming all lower levels are filled up. To investigate the Coulomb effect for these electrons, we make the projection of the Coulomb Hamiltonian (20.4.4). As we have noticed, the electron fields projected to the Nth energy level are given by (20.3.15). Hence, we construct the projected density operator $\rho_N(\boldsymbol{x})$ by

$$\rho_N(\boldsymbol{x}) = \sum_{\sigma\tau\tau'} \psi_{N\tau}^{\sigma\dagger}(\boldsymbol{x})\psi_{N\tau'}^{\sigma}(\boldsymbol{x}), \tag{20.4.6}$$

where $\psi_{N\tau}^{\sigma}(\boldsymbol{x})$ is the field operator given by (20.3.15). We examine each term closely. For instance,

$$\rho_{N+-}^{\uparrow\uparrow}(\boldsymbol{x}) \equiv \psi_{N+}^{\uparrow\dagger}(\boldsymbol{x})\,\psi_{N-}^{\uparrow}(\boldsymbol{x}) = \sum_{mn} e^{-2i\boldsymbol{K}\boldsymbol{x}}\langle N, m|\boldsymbol{x}\rangle\langle \boldsymbol{x}|N-1, n\rangle c_{+}^{\uparrow\dagger}(m)c_{-}^{\uparrow}(n). \tag{20.4.7}$$

Its Fourier transformation is

$$\rho_{N+-}^{\uparrow\uparrow}(\boldsymbol{q}) = \frac{1}{2\pi}\sum_{mn}\int d^2x\,\langle N, m|\boldsymbol{x}\rangle e^{-i(\boldsymbol{q}+2\boldsymbol{K})\boldsymbol{x}}\langle \boldsymbol{x}|N-1, n\rangle c_{+}^{\uparrow\dagger}(m)c_{-}^{\uparrow}(n). \tag{20.4.8}$$

As we describe details in Section 29.4, this can be factorized as

$$\rho_{N+-}^{\uparrow\uparrow}(\boldsymbol{q}) = \frac{1}{2\pi}\langle N|e^{-i(\boldsymbol{q}+2\boldsymbol{K})\boldsymbol{R}}|N-1\rangle \sum_{mn}\langle m|e^{-i[\boldsymbol{q}+2\boldsymbol{K}]\boldsymbol{X}}|n\rangle c_{+}^{\uparrow\dagger}(m)c_{-}^{\uparrow}(n), \tag{20.4.9}$$

where $\boldsymbol{R}$ and $\boldsymbol{X}$ are the relative coordinate and the guiding center, respectively.

Thus the projected density operator (20.4.6) is rewritten as

$$\rho_N(\boldsymbol{q}) = \sum_{\tau\tau'\sigma} F_{\tau\tau'}^{\sigma}(\boldsymbol{q})\,\hat{\mathcal{D}}_{\tau\tau'}^{\sigma\sigma}(\boldsymbol{q}), \tag{20.4.10}$$

where $\hat{\mathcal{D}}_{\tau\tau'}^{\sigma\sigma'}(\boldsymbol{q})$ is the bare density operator,

$$\hat{\mathcal{D}}_{\tau\tau'}^{\sigma\sigma'}(\boldsymbol{q}) = \frac{1}{2\pi}\sum_{mn}\langle m|e^{-i[\boldsymbol{q}+\tau\boldsymbol{K}-\tau'\boldsymbol{K}]\boldsymbol{X}}|n\rangle c_{\tau}^{\sigma\dagger}(m)c_{\tau'}^{\sigma'}(n), \tag{20.4.11}$$

and $F_{\tau\tau'}^{\sigma}(\boldsymbol{q})$ is the form factor,

$$F_{++}^{\uparrow}(\boldsymbol{q}) = F_{--}^{\downarrow}(\boldsymbol{q}) = F_N(\boldsymbol{q}), \tag{20.4.12a}$$

$$F_{++}^{\downarrow}(\boldsymbol{q}) = F_{--}^{\uparrow}(\boldsymbol{q}) = F_{N-1}(\boldsymbol{q}), \tag{20.4.12b}$$

$$F_{+-}^{\uparrow}(\boldsymbol{q}) = F_{-+}^{\downarrow}(\boldsymbol{q}) = G_N(\boldsymbol{q}), \tag{20.4.12c}$$

$$F_{-+}^{\uparrow}(\boldsymbol{q}) = F_{+-}^{\downarrow}(\boldsymbol{q}) = G_N^{*}(-\boldsymbol{q}), \tag{20.4.12d}$$

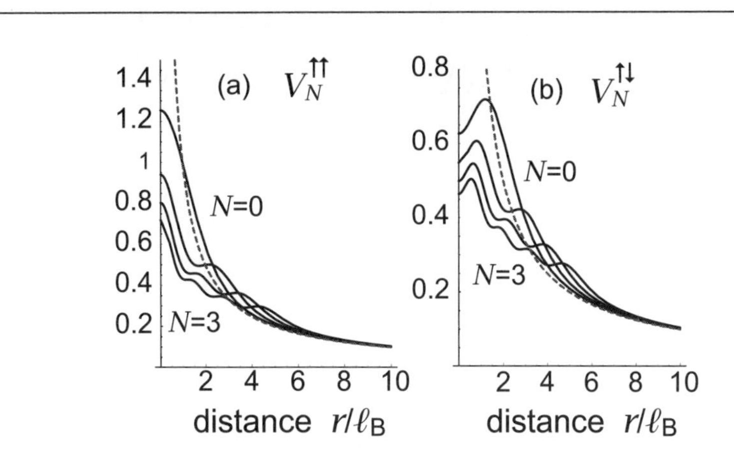

Fig. 20.4 We illustrate the spin dependence of the effective Coulomb potential. The vertical axis is the potential in unit of $e^2/4\pi\varepsilon\ell_B$. The horizontal axis is the distance r in unit of ℓ_B. The dotted curve represents the ordinary Coulomb potential without the form factor. (a) The effective Coulomb potential $V_N^{\uparrow\uparrow}(r)$ in the Nth energy level for $N = 0, 1, 2, 3$ from top to bottom. Note that $V_N^{\downarrow\downarrow}(r) = V_{N-1}^{\uparrow\uparrow}(r)$. It describes interactions between electrons with the same spin. (b) The effective Coulomb potential $V_N^{\uparrow\downarrow}(r)$ for $N = 0, 1, 2, 3$ from top to bottom. It describes interactions between electrons with the different spins.

with

$$F_N(\boldsymbol{q}) = \langle N|e^{-i\boldsymbol{q}\boldsymbol{R}}|N\rangle, \tag{20.4.13a}$$

$$G_N(\boldsymbol{q}) = \langle N|e^{-i(\boldsymbol{q}-\boldsymbol{K})\boldsymbol{R}}|N-1\rangle. \tag{20.4.13b}$$

Here we have used the relation, $\boldsymbol{q} = \boldsymbol{q} + 3\boldsymbol{K}$, due to the lattice structure [Fig.20.2(b)]. The form factors are explicitly given by using[45]

$$\langle N+M|e^{i\boldsymbol{q}\boldsymbol{R}}|N\rangle = \frac{\sqrt{N!}}{\sqrt{(N+M)!}}\left(\frac{\ell_B q}{\sqrt{2}}\right)^M L_N^M\left(\frac{\ell_B^2 q^2}{2}\right)e^{-\frac{1}{4}\ell_B^2 q^2} \tag{20.4.14}$$

for $M \geq 0$ in terms of associated Laguerre polynomials. It should be remarked that the bare density operator $\hat{\mathcal{D}}_{\tau\tau'}^{\sigma\sigma'}(\boldsymbol{q})$ involves only the guiding center coordinate $\boldsymbol{X}$, while the form factor $F_{\tau\tau'}^{\sigma}(\boldsymbol{q})$ involves only the relative coordinate $\boldsymbol{R}$.

A comment is in order on the form factors. Note that

$$G_N(\boldsymbol{q}) \simeq e^{-\frac{1}{2}\ell_B^2|\boldsymbol{K}|^2}F_N(\boldsymbol{q}). \tag{20.4.15}$$

Thus, $G_N(\boldsymbol{q})$ is exponentially smaller than $F_N(\boldsymbol{q})$. For instance, $F_{+-}^{\uparrow}(\boldsymbol{q}) \equiv G_N(\boldsymbol{q})$ represents the transfer of the up-spin electron ($\sigma = \uparrow$) from the K′ point ($\tau = -$) to the K point ($\tau = +$). Therefore, such a mixing between the K and K′ points is exponentially small and negligible.

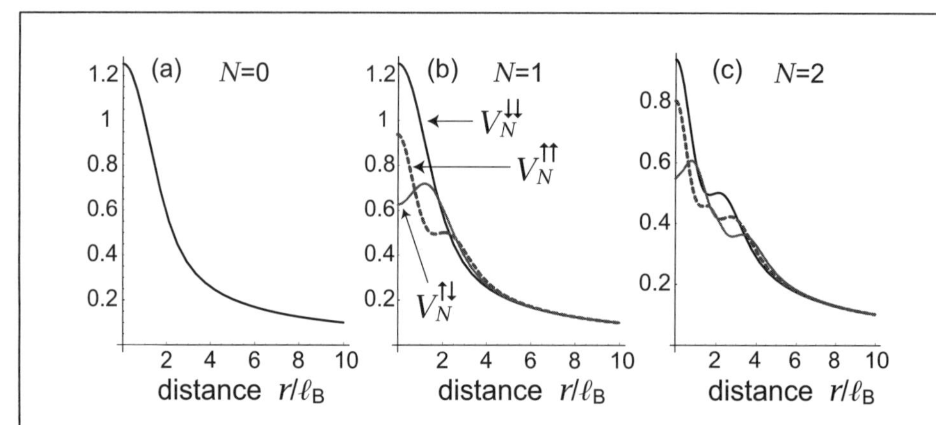

Fig. 20.5 We illustrate the spin dependence of the effective Coulomb potential $V_N^{\downarrow\downarrow}(r)$, $V_N^{\uparrow\uparrow}(r)$ and $V_N^{\uparrow\downarrow}(r)$ in the Nth energy level for (a) $N = 0$, (b) $N = 1$, and (c) $N = 2$. All effective potentials agree for the lowest energy level ($N = 0$). The vertical axis is the potential in unit of $e^2/4\pi\varepsilon\ell_B$.

The projected Coulomb Hamiltonian is constructed by substituting the projected density (20.4.10) into the Hamiltonian (20.4.4),

$$\begin{aligned} H_N =&\pi \int d^2q\, V(\boldsymbol{q})\rho_N(-\boldsymbol{q})\rho_N(\boldsymbol{q}) \\ =&\pi \sum_{\tau\tau'\sigma'} \sum_{\lambda\lambda'\sigma'} \int d^2q\, V^{\sigma\sigma'}_{N;\tau\tau'\lambda\lambda'}(\boldsymbol{q})\hat{\mathcal{D}}^{\sigma}_{\tau\tau'}(-\boldsymbol{q})\hat{\mathcal{D}}^{\sigma'}_{\lambda\lambda'}(\boldsymbol{q}), \end{aligned} \tag{20.4.16}$$

where $V^{\sigma\sigma'}_{N;\tau\tau'\lambda\lambda'}(\boldsymbol{q})$ is the effective Coulomb potential in the Nth energy level,

$$V^{\sigma\sigma'}_{N;\tau\tau'\lambda\lambda'}(\boldsymbol{q}) = V(\boldsymbol{q})F^{\sigma}_{\tau\tau'}(-\boldsymbol{q})F^{\sigma'}_{\lambda\lambda'}(\boldsymbol{q}) \tag{20.4.17}$$

with

$$V(\boldsymbol{q}) = \frac{e^2}{4\pi\varepsilon|\boldsymbol{q}|}. \tag{20.4.18}$$

It is remarkable that the effective Coulomb potential depends on the spin and the valley through the form factors characterizing Landau levels. The dependence arises since effective Coulomb potentials are different in different Landau levels.

The effective Coulomb potentials in the Nth energy level at the K point are

$$V_N^{\uparrow\uparrow}(\boldsymbol{q}) = V(\boldsymbol{q})F_N(-\boldsymbol{q})F_N(\boldsymbol{q}), \tag{20.4.19a}$$

$$V_N^{\downarrow\downarrow}(\boldsymbol{q}) = V(\boldsymbol{q})F_{N-1}(-\boldsymbol{q})F_{N-1}(\boldsymbol{q}), \tag{20.4.19b}$$

$$V_N^{\uparrow\downarrow}(\boldsymbol{q}) = V(\boldsymbol{q})F_N(-\boldsymbol{q})F_{N-1}(\boldsymbol{q}). \tag{20.4.19c}$$

The potential $V_N^{\sigma\sigma}(\boldsymbol{q})$ stands for the interaction between electrons with the same spin σ, $V_N^{\uparrow\downarrow}(\boldsymbol{q})$ for the interaction between electrons with the different spin. The spin direction is reversed at the K′ point. We have illustrated the effective Coulomb potentials in Figs.20.4 and 20.5.

We note that the effective potential $V_N^{\uparrow\uparrow}(\boldsymbol{q})$ for electrons with the same spin in the Nth energy level has precisely the same form as in the standard QH effect for electrons in the Nth Landau level: Compare Fig.20.4(a) with Fig.19.3. On the other hand the effective potential $V_N^{\uparrow\downarrow}(\boldsymbol{q})$ for electrons with the different spins are entirely new.

Symmetry of Projected Density

The symmetry group of the density operator is far from being trivial in graphene. In the rest of this section we investigate the projected density operator $\rho_N(\boldsymbol{q})$. We do this as a preparation for the study of the symmetry of the projected Coulomb Hamiltonian H_N, which is the key issue to determine how the Coulomb effect modifies the energy spectrum of the noninteracting theory. It is the subject of Section 20.6.

The symmetry group of the density operator $\rho_N(\boldsymbol{q})$ given by (20.4.10) reads as follows. We abbreviate its structure as

$$\begin{aligned}\rho_N \propto\; & F_N[c_+^{\uparrow\dagger}c_+^{\uparrow} + c_-^{\downarrow\dagger}c_-^{\downarrow}] + F_{N-1}[c_-^{\uparrow\dagger}c_-^{\uparrow} + c_+^{\downarrow\dagger}c_+^{\downarrow}] \\ & + G_N[c_+^{\uparrow\dagger}c_-^{\uparrow} + c_-^{\downarrow\dagger}c_+^{\downarrow}] + G_N^*[c_-^{\uparrow\dagger}c_+^{\uparrow} + c_+^{\downarrow\dagger}c_-^{\downarrow}].\end{aligned} \tag{20.4.20}$$

We make a global linear transformation of c_τ^σ,

$$c_{\tau'}^{\sigma'} = \sum_{\sigma\tau} f_{\tau'\tau}^{\sigma'\sigma} c_\tau^\sigma. \tag{20.4.21}$$

The first line of (20.4.20) is invariant under the SU(2) transformation of the following two vectors independently,

$$\begin{pmatrix} c_+^{\uparrow} \\ c_-^{\downarrow} \end{pmatrix} \to S_1 \begin{pmatrix} c_+^{\uparrow} \\ c_-^{\downarrow} \end{pmatrix}, \quad \begin{pmatrix} c_-^{\uparrow} \\ c_+^{\downarrow} \end{pmatrix} \to S_2 \begin{pmatrix} c_-^{\uparrow} \\ c_+^{\downarrow} \end{pmatrix}. \tag{20.4.22}$$

Note that the two electrons in the first set come from the Nth Landau level, while those in the second set from the (N-1)th Landau level. The symmetry SU(4) is broken to SU(2)×SU(2) since there are two types of electrons coming from different Landau levels.

The symmetry group of ρ_N is much smaller than this because there exist a mixing of these two Landau levels. The first line of (20.4.20) is invariant under four independent phase transformations, $c_\tau^\sigma \to \exp[f_\tau^\sigma]\, c_\tau^\sigma$, for the four electron fields. However, two of them are broken by the second line. The remaining invariant transformations are

$$\begin{pmatrix} c_+^{\uparrow} \\ c_-^{\uparrow} \end{pmatrix} \to e^{i\alpha} \begin{pmatrix} c_+^{\uparrow} \\ c_-^{\uparrow} \end{pmatrix}, \quad \begin{pmatrix} c_+^{\downarrow} \\ c_-^{\downarrow} \end{pmatrix} \to e^{i\beta} \begin{pmatrix} c_+^{\downarrow} \\ c_-^{\downarrow} \end{pmatrix}, \tag{20.4.23}$$

with two arbitrary constants α and β. Additionally there exist the Z_2 symmetry which exchanges the up-spin and down spin components with the different valley indices,

$$\begin{pmatrix} c_+^{\uparrow} \\ c_-^{\uparrow} \end{pmatrix} \longleftrightarrow \begin{pmatrix} c_-^{\downarrow} \\ c_+^{\downarrow} \end{pmatrix}, \tag{20.4.24}$$

as corresponds to the fact that the magnetic field is opposite at the K and K′ sites. This is the modified mirror symmetry present in the graphene system with external magnetic field, which we mentioned in a paragraph below (20.3.2). We conclude that the symmetry of the projected density operator is U(1)×U(1)×Z_2 for $N \geq 1$.

20.5 EXCITONIC CONDENSATION

It is necessary to pay a special attention to the zero-energy level ($N = 0$), since it contains both electrons and holes. Let us recapture the properties of the zero-energy level [Fig.20.6]. In the absence of the magnetic field the band structure is described by the Dirac valleys associated with the dispersion relation (20.2.13), as illustrated in Fig.20.6(a). When the magnetic field is applied, the band structure is changed to generate Landau levels [Fig.20.6(b)]. However, the Dirac electron is subject to the intrinsic Zeeman effect, which induces a Landau-level mixing, and the zero-energy state emerges for up-spin electrons as well as down-spin holes [Fig.20.6(c)]. There exists an attractive Coulomb force between an electron and a hole. Hence we expect them to make an excitonic condensation, producing a gap to the electron and hole states [Fig.20.6(d)].

The excitonic condensation[270,271,202,204] has been studied in graphene. A clear-cut approach[132] has been proposed to this problem on the analogy of the BCS theory. Our physical picture is summarized in Fig.20.6. The remarkable point is that the kinetic term is quenched in each Landau level, and hence it is also absent in the zero-energy level. This simplifies the analysis considerably.

The field operators present in the zero-energy level are $\psi_{\mathrm{e}+}^{\uparrow}(\boldsymbol{x})$, $\psi_{\mathrm{e}-}^{\downarrow}(\boldsymbol{x})$ for electrons and $\psi_{\mathrm{h}+}^{\downarrow}(\boldsymbol{x})$, $\psi_{\mathrm{h}-}^{\uparrow}(\boldsymbol{x})$ for holes coming from the $N = 0$ Landau levels [Fig.20.1(b)]. It is sufficient to investigate electron-hole pairs at the K and K′ points separately. An exciton composed of an electron and a hole belonging to different Dirac valleys is fragile, because their effective Coulomb potential involves the factor $e^{-\frac{1}{2}\ell_B^2|\boldsymbol{K}|^2}$ compared with the one within the K point: See (20.4.12) and (20.4.15). For definiteness, we consider the K point [Fig.20.6].

The effective Coulomb potential (20.4.17) is simple in the lowest Landau level,

$$V^{\text{eff}}(\boldsymbol{q}) = V(\boldsymbol{q})e^{-\ell_B^2 q^2/2}. \tag{20.5.1}$$

We consider the effective Hamiltonian together with this effective potential,

$$H^{\text{eff}} = -\int d^2x d^2y\, V^{\text{eff}}(\boldsymbol{x}-\boldsymbol{y})\psi_{\mathrm{e}}^{\uparrow\dagger}(\boldsymbol{x})\psi_{\mathrm{e}}^{\uparrow}(\boldsymbol{x})\psi_{\mathrm{h}}^{\downarrow\dagger}(\boldsymbol{y})\psi_{\mathrm{h}}^{\downarrow}(\boldsymbol{y}), \tag{20.5.2}$$

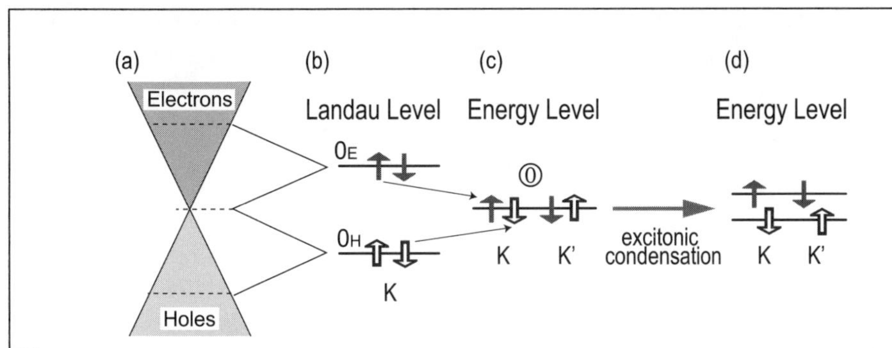

Fig. 20.6 (a) We illustrate electron states and hole states in the Dirac valley. (b) When the magnetic field is applied, the band structure is changed to generate Landau levels. (c) The intrinsic Zeeman effect induces a Landau-level mixing, and the zero-energy state with the up-spin electron and the down-spin hole emerges. (d) They make an excitonic condensation to form a BCS-type state, producing a gap to the electron and hole states.

where $\psi_{\text{e}}^{\uparrow}(\boldsymbol{x})$ and $\psi_{\text{h}}^{\downarrow}(\boldsymbol{x})$ are the up-spin electron field and the down-spin hole field, respectively. The Hamiltonian is rewritten as

$$H^{\text{eff}} = -\pi\int d^2q d^2k d^2k'\, V^{\text{eff}}(\boldsymbol{q})\psi_{\text{e}}^{\uparrow\dagger}(\boldsymbol{k}+\boldsymbol{q})\psi_{\text{h}}^{\downarrow\dagger}(\boldsymbol{k}')\psi_{\text{h}}^{\downarrow}(\boldsymbol{k}'-\boldsymbol{q})\psi_{\text{e}}^{\uparrow}(\boldsymbol{k}). \tag{20.5.3}$$

The gap equation is derived on the analogy of the analysis familiar in the BCS theory. We take the terms satisfying $\boldsymbol{q} = \boldsymbol{k}' - \boldsymbol{k}$ as the dominant ones, and approximate the Hamiltonian as

$$H^{\text{eff}} \simeq -\pi\int d^2k d^2k'\, V^{\text{eff}}(\boldsymbol{k}'-\boldsymbol{k})\psi_{\text{e}}^{\uparrow\dagger}(\boldsymbol{k}')\psi_{\text{h}}^{\downarrow\dagger}(\boldsymbol{k}')\psi_{\text{h}}^{\downarrow}(\boldsymbol{k})\psi_{\text{e}}^{\uparrow}(\boldsymbol{k}). \tag{20.5.4}$$

We define the singlet excitonic gap function by

$$\Delta(\boldsymbol{k}) = \int d^2k'\, V^{\text{eff}}(\boldsymbol{k}'-\boldsymbol{k})\langle\psi_{\text{e}}^{\uparrow\dagger}(\boldsymbol{k}')\psi_{\text{h}}^{\downarrow\dagger}(\boldsymbol{k}')\rangle, \tag{20.5.5}$$

which can be taken to be positive without loss of generality.

The mean-field Hamiltonian reads

$$H^{\text{eff}} \simeq -\pi\int d^2k\,[\Delta(\boldsymbol{k})\,\psi_{\text{h}}^{\downarrow}(\boldsymbol{k})\,\psi_{\text{e}}^{\uparrow}(\boldsymbol{k}) + \psi_{\text{e}}^{\uparrow\dagger}(\boldsymbol{k})\,\psi_{\text{h}}^{\downarrow\dagger}(\boldsymbol{k})\,\Delta^*(\boldsymbol{k})]. \tag{20.5.6}$$

By setting

$$\begin{aligned}
\Psi_1(\boldsymbol{k}) &= \frac{1}{\sqrt{2}}[\psi_{\text{e}}^{\uparrow}(\boldsymbol{k}) + \psi_{\text{h}}^{\downarrow\dagger}(\boldsymbol{k})],\\
\Psi_2(\boldsymbol{k}) &= \frac{1}{\sqrt{2}}[\psi_{\text{e}}^{\uparrow\dagger}(\boldsymbol{k}) - \psi_{\text{h}}^{\downarrow}(\boldsymbol{k})],
\end{aligned} \tag{20.5.7}$$

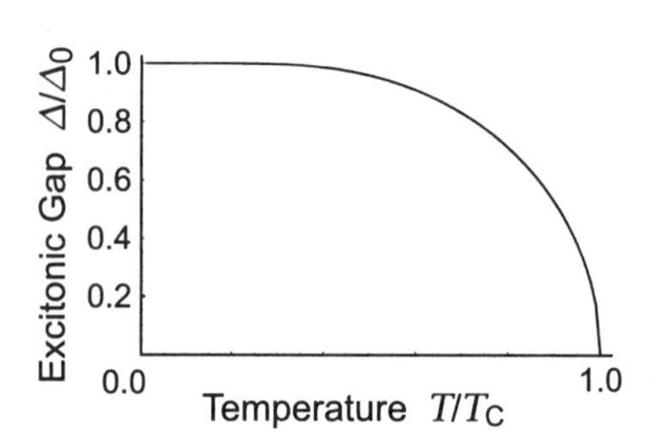

Fig. 20.7 We illustrate the temperature dependence of the excitonic gap Δ_T. The gap vanishes at the critical temperature T_C.

it is easy to diagonalize (20.5.6) as

$$H^{\text{eff}} \simeq \pi \int d^2k\,\Delta(\boldsymbol{k})\,[\Psi_1^\dagger(\boldsymbol{k})\,\Psi_1(\boldsymbol{k}) + \Psi_2^\dagger(\boldsymbol{k})\,\Psi_2(\boldsymbol{k})]. \tag{20.5.8}$$

Since Ψ_1 and Ψ_2 are free fields, the ground state is given by solving

$$\Psi_1(\boldsymbol{k})\,|\Phi_{\text{exc}}\rangle = \Psi_2(\boldsymbol{k})\,|\Phi_{\text{exc}}\rangle = 0, \tag{20.5.9}$$

or

$$|\Phi_{\text{exc}}\rangle = \prod_{\boldsymbol{k}} \frac{1}{\sqrt{2}}[1 + \psi_{\text{h}}^{\downarrow\dagger}(\boldsymbol{k})\,\psi_{\text{e}}^{\uparrow\dagger}(\boldsymbol{k})]|0\rangle. \tag{20.5.10}$$

This is a BCS-type state representing the condensation of electron-hole pairs.

Due to the Fermi statistics the thermodynamical average $\langle \Psi_i^\dagger(\boldsymbol{k})\,\Psi_j(\boldsymbol{k})\rangle$ is given by

$$\begin{aligned}\langle \Psi_1^\dagger(\boldsymbol{k})\,\Psi_1(\boldsymbol{k})\rangle &= \langle \Psi_2^\dagger(\boldsymbol{k})\,\Psi_2(\boldsymbol{k})\rangle = \frac{1}{1+e^{\Delta(\boldsymbol{k})/k_{\text{B}}T}},\\ \langle \Psi_2^\dagger(\boldsymbol{k})\,\Psi_1(\boldsymbol{k})\rangle &= \langle \Psi_1^\dagger(\boldsymbol{k})\,\Psi_2(\boldsymbol{k})\rangle = 0,\end{aligned} \tag{20.5.11}$$

where k_{B} is the Boltzmann factor. Combining these with (20.5.7) we obtain

$$\langle \psi_{\text{e}}^{\uparrow\dagger}(\boldsymbol{k})\,\psi_{\text{h}}^{\downarrow\dagger}(\boldsymbol{k})\rangle = \frac{1}{2}\tanh\frac{\Delta(\boldsymbol{k})}{2k_{\text{B}}T}. \tag{20.5.12}$$

It is straightforward to derive the gap equation by substituting this into (20.5.5),

$$\Delta(\boldsymbol{k}) = \frac{1}{2}\int d^2\boldsymbol{k}' V^{\text{eff}}(\boldsymbol{k}'-\boldsymbol{k})\tanh\frac{\Delta(\boldsymbol{k}')}{2k_{\text{B}}T}. \tag{20.5.13}$$

In the limit $T \to 0$, the gap $\Delta(\boldsymbol{k})$ is given by

$$\Delta(\boldsymbol{k})|_{T=0} = \frac{1}{2}\int d^2\boldsymbol{k}'\, V^{\text{eff}}(\boldsymbol{k}') = \pi\sqrt{\frac{\pi}{2}}\frac{e^2}{4\pi\varepsilon\ell_B} \equiv \Delta_0, \tag{20.5.14}$$

which is dispersionless (independent of $\boldsymbol{k}$). For finite temperature, assuming the gap is dispersionless, we obtain the relation

$$\frac{\Delta_T}{\Delta_0} = \tanh\frac{\Delta_T}{2k_B T}. \tag{20.5.15}$$

The critical temperature T_C at which $\Delta_T = 0$ is solved as

$$T_C = \frac{\Delta_0}{2k_B}. \tag{20.5.16}$$

We have illustrated the temperature dependence of the gap Δ_T in Fig.20.7.

According to the diagonalized Hamiltonian (20.5.8), the excitonic condensation provides them with the mass Δ_0, resolving the electron-hole degeneracy in the zero-energy state. Combining the results at the K and K′ points, the 4-fold degenerate levels split into two 2-fold degenerate levels [Fig.20.6(d)] with the gap energy (20.5.14). This leads to a new plateau at $\nu = 0$, as illustrated in Fig.20.8.

20.6 Valley Polarization

It is now possible to treat electrons and holes separately since the gap has opened between the electron and hole bands. Furthermore, it is enough to study only electrons due to the electron-hole symmetry. We shall show that the Coulomb effect modifies the energy spectrum of the noninteracting theory so that plateaux emerges at $\nu = \pm 1, \pm 2n$ for $n = 1, 2, 3, \cdots$ [Fig.20.8].

We express the Coulomb Hamiltonian (20.4.16) as

$$H_N = \sum V_{N;mnij}^{\sigma\sigma';\tau\tau'\lambda\lambda'} c_\tau^{\sigma\dagger}(m) c_{\tau'}^{\sigma}(n) c_\lambda^{\sigma'\dagger}(i) c_{\lambda'}^{\sigma'}(j), \tag{20.6.1}$$

where the summations over repeated indices, m, n, i, j, σ, σ', τ, τ', λ, λ' are understood, and

$$\begin{aligned} V_{N;mnij}^{\sigma\sigma';\tau\tau'\lambda\lambda'} =& \frac{1}{4\pi}\int d^2q\, V(\boldsymbol{q}) F_{\tau\tau'}^{\sigma}(-\boldsymbol{q}) F_{\lambda\lambda'}^{\sigma'}(\boldsymbol{q}) \\ &\times \langle m|e^{-i[-\boldsymbol{q}+\tau\boldsymbol{K}-\tau'\boldsymbol{K}]X}|n\rangle\langle i|e^{-i[\boldsymbol{q}+\lambda\boldsymbol{K}-\lambda'\boldsymbol{K}]X}|j\rangle. \end{aligned} \tag{20.6.2}$$

We introduce

$$V_{N;\mathrm{D}}^{\tau\tau'\lambda\lambda'} = \sum_{ij} V_{N;iijj}^{\uparrow\uparrow;\tau\tau'\lambda\lambda'}, \quad V_{N;\mathrm{X}}^{\tau\tau'\lambda\lambda'} = \sum_{ij} V_{N;ijji}^{\uparrow\uparrow;\tau\tau'\lambda\lambda'}. \tag{20.6.3}$$

For instance,

$$\begin{aligned} V_{N;\mathrm{D}}^{++++} &= \frac{N_\Phi}{2\ell_B^2} V(\boldsymbol{q}=0), \\ V_{N;\mathrm{X}}^{++++} &= \frac{N_\Phi}{4\pi}\int d^2q\, V(\boldsymbol{q}) F_{++}^{\uparrow}(-\boldsymbol{q}) F_{++}^{\uparrow}(\boldsymbol{q}), \end{aligned} \tag{20.6.4}$$

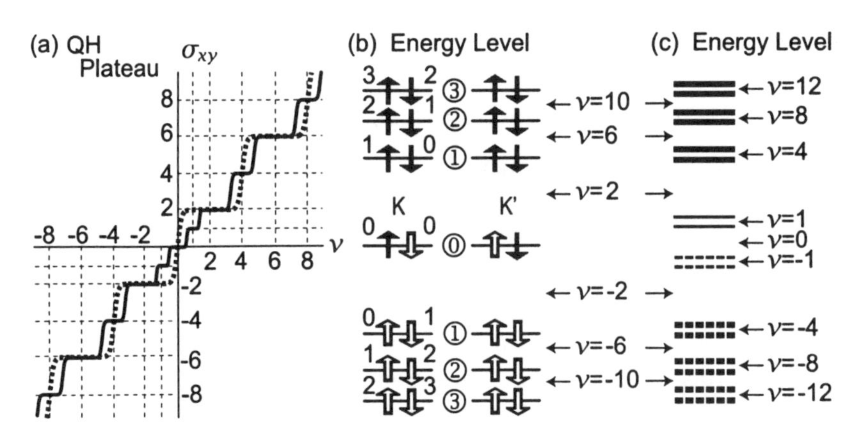

Fig. 20.8 (a) The QH conductivity is illustrated in graphene. The dotted curve shows the sequence $\nu = \pm 2, \pm 6, \pm 10, \cdots$, while the solid curve the sequence $\nu = 0, \pm 1, \pm 2, \pm 4, \pm 6, \pm 8, \cdots$. (b) The energy level within noninteracting theory, indicated by the number in circle. The spin is indicated by a solid (open) arrow for electron (hole) at the K and K′ points. The number attached to a solid (open) arrow shows from which Landau level the electron (hole) comes from. This energy spectrum is a manifestation of SUSY. (c) When Coulomb interactions are included, the zero-energy level splits into four nondegenerate subbands, while each nonzero-energy level splits into two 2-fold degenerate subbands with the exact U(1) symmetry.

since $F^{\sigma}_{\tau\tau'}(\boldsymbol{q} = 0) = 1$. They represent the direct and exchange Coulomb energies, respectively, proportional to the number of the Landau site N_Φ: See (30.1.16) for more details. It is easy to evaluate them numerically. See Fig.20.9(a) for $V^{++++}_{N;X}$ as a function of N.

We analyze the lowest energy level and higher energy levels separately.

Lowest Energy Level

We first study the lowest energy level, which contain two degenerate states [Fig.20.8(b)]. The degeneracy is resolved obviously if a strong extrinsic Zeeman effect exists additionally. We consider the case where the extrinsic Zeeman energy is absent at temperature $T = 0$, or is present but quite small compared with the thermal energy at $T \neq 0$. We ask whether the degeneracy is resolved even in such cases.

As a generic trial function we take

$$|\Phi_0\rangle = \prod_n \left(u c_+^{\downarrow\dagger}(n) + v c_-^{\downarrow\dagger}(n) \right) |0\rangle, \tag{20.6.5}$$

with $|u|^2 + |v|^2 = 1$. The state covers the entire SU(2) space of the lowest energy level for electrons, when the two parameters u and v are varied.

We calculate the Coulomb energy $\langle H \rangle_0 \equiv \langle \Phi_0 | H | \Phi_0 \rangle$ with (20.6.1). Most terms are SU(2) invariant, but there exists a noninvariant term,

$$\langle H_{\text{noninv}} \rangle_0 = 2\pi |uv|^2 V_{0;X}^{++++}. \tag{20.6.6}$$

Hence the SU(2) symmetry is broken explicitly into the U(1) symmetry by Coulomb interactions.

The energy is minimized either $u = 0$ or $v = 0$ due to the term (20.6.6), corresponding to the state

$$|\Phi_0^{\downarrow}\rangle = \prod_n c_-^{\downarrow\dagger}(n)\,|0\rangle, \qquad \text{or} \qquad |\Phi_0^{\uparrow}\rangle = \prod_n c_+^{\uparrow\dagger}(n)\,|0\rangle. \tag{20.6.7}$$

The ground state is either $|\Phi_0^{\downarrow}\rangle$ or $|\Phi_0^{\uparrow}\rangle$, though there exists still the Z_2 symmetry. The energy barrier between these two states is of the order of the Coulomb energy, which is much larger than the thermal energy.

We conclude as follows: The two levels split explicitly by an extrinsic Zeeman effect if exits. Then the first energy level is up-spin polarized, and the second energy level is down-spin polarized. Even without such an extrinsic Zeeman effect, driven by the Coulomb exchange interaction, the spontaneous breakdown of the Z_2 symmetry turns the system into a QH ferromagnet. It is reasonable to call it the Ising QH ferromagnet due to the Z_2 symmetry. In any case, a plateau emerges at $\nu = 1$, where the activation energy is of the order of the typical Coulomb energy as in the conventional QH effect.

*N*th Energy Level

We next study the Nth energy level with $N \geq 1$. It contains four degenerate states with the SU(4) symmetry in noninteracting theory. However, the projected density is invariant only under U(1)×U(1)×Z_2, as we have seen in (20.4.23). Hence we take a set of trial functions by requiring this symmetry,

$$|\Phi_N^{\uparrow}\rangle = \prod_n \left(\sin\alpha_N e^{i\theta^{\uparrow}} c_+^{\uparrow\dagger}(n) + \cos\alpha_N e^{-i\theta^{\uparrow}} c_-^{\uparrow\dagger}(n) \right) |0\rangle, \tag{20.6.8a}$$

$$|\Phi_N^{\downarrow}\rangle = \prod_n \left(\cos\alpha_N e^{-i\theta^{\downarrow}} c_+^{\downarrow\dagger}(n) + \sin\alpha_N e^{i\theta^{\downarrow}} c_-^{\downarrow\dagger}(n) \right) |0\rangle, \tag{20.6.8b}$$

where the phase factors $e^{i\theta^{\uparrow}}$ and $e^{i\theta^{\downarrow}}$ assure the U(1)×U(1) symmetry. These two states are degenerate, $\langle \Phi_N^{\uparrow} | H_N | \Phi_N^{\uparrow} \rangle = \langle \Phi_N^{\downarrow} | H_N | \Phi_N^{\downarrow} \rangle$, due to the Z_2 symmetry.

It is easy to determine the angle α_N by minimizing the Coulomb energy $\langle \Phi_N^{\sigma} | H_N | \Phi_N^{\sigma} \rangle$. After some calculations we find

$$\langle \Phi_N^{\uparrow} | H_N | \Phi_N^{\uparrow} \rangle = A |\sin\alpha_N|^4 + B |\cos\alpha_N|^4 + C |\sin\alpha_N \cos\alpha_N|^2, \tag{20.6.9}$$

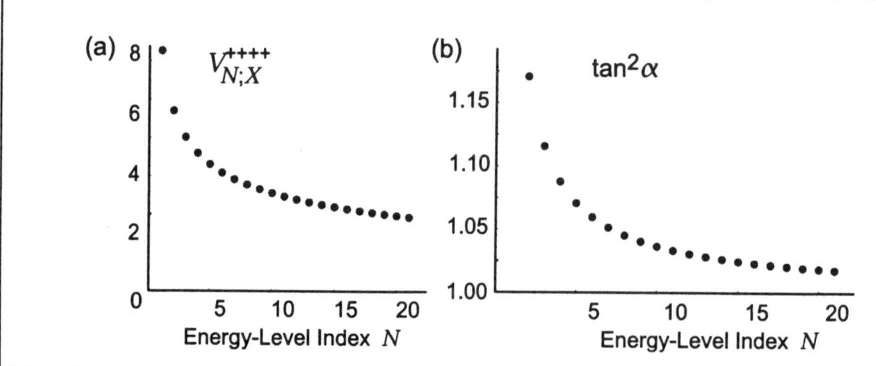

Fig. 20.9 (a) We show the exchange energy $V^{++++}_{N;X}$ of the Nth energy level as a function of N. The vertical axis is the energy in unit of $e^2/4\pi\varepsilon\ell_B$. Note that $V^{----}_{N;X} = V^{++++}_{N-1;X}$. (b) We show the function $\tan^2\alpha_N$ given by (20.6.13). It approaches 1 assymptotically, $\lim_{N\to\infty}\tan^2\alpha_N = 1$.

where

$$A = V^{++++}_{N;D} - V^{++++}_{N;X}, \qquad B = V^{----}_{N;D} - V^{----}_{N;X}, \tag{20.6.10a}$$

$$C = V^{++--}_{N;D} + V^{--++}_{N;D} - V^{+--+}_{N;X} - V^{-++-}_{N;X} \tag{20.6.10b}$$

with (20.6.3). Here, $V^{++++}_{N;D} = V^{----}_{N;D} = V^{++--}_{N;D} = V^{--++}_{N;D}$ since $F^{\sigma}_{\tau\tau'}(0) = 1$. Furthermore, because of (20.4.15), $V^{+--+}_{N;X}$ and $V^{-++-}_{N;X}$ are exponentially smaller in $(a/\ell_B)^2$ than $V^{++++}_{N;X}$ or $V^{----}_{N;X}$, and can be neglected. Hence

$$\tan^2\alpha_N = \frac{B - C/2}{A - C/2} \simeq \frac{V^{----}_{N;X}}{V^{++++}_{N;X}}. \tag{20.6.11}$$

Now it follows from (20.4.12) that $V^{----}_{N;X} = V^{++++}_{N-1;X}$. We also see [Fig.20.9(a)]

$$V^{++++}_{N-1;X} > V^{++++}_{N;X}, \tag{20.6.12}$$

implying that the Coulomb energy of an electron in higher Landau level is lower. It follows that

$$\tan^2\alpha_N \simeq \frac{V^{++++}_{N-1;X}}{V^{++++}_{N;X}} > 1. \tag{20.6.13}$$

We have depicted $\tan^2\alpha_N$ as a function of N in Fig.20.9(b).

We conclude that, since $|\sin\alpha_N| > |\cos\alpha_N|$, more electrons are present in the Dirac valley at the K point than at the K′ point in the state $|\Phi^{\uparrow}_N\rangle$. Similarly, more electrons are present at the K′ point than at the K point in the state $|\Phi^{\downarrow}_N\rangle$. Namely, the valley polarization occurs both in $|\Phi^{\uparrow}_N\rangle$ and $|\Phi^{\downarrow}_N\rangle$. Accordingly, the ground state is a valley polarized state. This can be understood physically as follows [20.1(b)]: Up-spin electrons in the K point belong to the (N-1)th Landau level but those in the K′ point belong to the Nth Landau level. Up-spin electrons are more dense at the K point than at the K′ point because the Coulomb energy is lower in higher Landau levels. A similar argument is made with respect to down-spin electrons. It should be remarked that the valley polarization is possible because the naive mirror symmetry is broken by the external magnetic field, and the K and K′ points are physically distinguishable. The valley polarization disappears as $N \to \infty$, since $V^{----}_{N+1;X} = V^{----}_{N;X}$ in the limit [Fig.20.9(b)].

An extrinsic Zeeman effect opens a gap between these two spin polarized states. Even without such an effect, driven by the Coulomb exchange interaction, the spontaneous breakdown of the Z_2 symmetry turns the system into a QH ferromagnet. Note that it has still the 2-fold degeneracy. In any case, a plateau emerges at $\nu = 4N$ [Fig.20.8(c)], where the activation energy is of the order of the typical Coulomb energy as in the conventional QH effect.

QH Plateaux

The symmetry group is SU(4) in the noninteracting theory of graphene, where gaps open at $\nu = \pm 2, \pm 6, \pm 10, \cdots$. This series of QH plateaux is the first experimental result[375] for QH effects in graphene. When Coulomb interactions are included, QH plateaux emerge at $\nu = 0, \pm 1, \pm 4, \pm 8, \cdots$ since the symmetry SU(4) is broken. This series has been observed[539] when larger magnetic field is applied.

There still exists the 2-fold degeneracy associated with the U(1) symmetry. The remaining problem is whether a further resolution of this degeneracy may occur. In Part III, we analyze a similar problem in the spin-frozen bilayer QH effect in GaAs semiconductor, where the pseudospin describes the layer degree of freedom. Let us summarize the key points. If the tunneling gap Δ_{SAS} is neglected, there is the pseudospin SU(2) symmetry in the noninteracting theory, which is broken to the U(1) symmetry explicitly by the capacitance energy. Then, the exchange Coulomb interaction generates the Goldstone mode associated with spontaneous breakdown of a continuous symmetry, and turns the bilayer system into the pseudospin QH ferromagnet.[351] Based on this analogy it might be suggested[373] that the remaining U(1) symmetry would also be broken spontaneously in graphene. If this were the case, all the degeneracy would be completely resolved.

We question[3] why the spontaneous breakdown of a continuous symmetry is possible in spite of the Mermin-Wagner theorem[345,346,232] in the conventional QH effect. This is because there exists actually the tunneling gap $\Delta_{SAS} \neq 0$ in the actual bilayer system, which gives a gap to the Goldstone mode. Namely, the pseudospin U(1) symmetry is broken explicitly by the tunneling gap in actual samples, and hence the Mermin-Wagner theorem is not applicable.

We now come back to the QH effect in graphene. First of all,. graphene is an

[3]Recall that, according to the Mermin-Wagner theorem, the spontaneous breakdown of a continuous symmetry is impossible at finite temperature in the two-dimensional space. See Appendix J for the underlying physics[105,320] behind the Mermin-Wagner theorem.

ideal two-dimensional system. On one hand, the spin U(1) symmetry may be broken by a small extrinsic Zeeman effect to produce the Ising QH ferromagnet at $\nu = \pm 1$. On the other hand, the U(1) symmetry with respect to the valley degree of freedom is exact. There is no external parameter which breaks it explicitly to make the Mermin-Wagner theorem inapplicable. It is concluded[132] that spontaneous symmetry breaking of the valley U(1) symmetry cannot occur in graphene, so that the plateaux at $\nu = \pm 3, \pm 5, \pm 7, \cdots$ will not emerge in monolayer graphene.

20.7 MULTILAYER GRAPHENE SYSTEMS

Bilayer Graphene (Bernal Stacking)

We proceed to generalize the above analysis to a bilayer graphene, which is a system made of two coupled hexagonal lattices according to the Bernal stacking. In the absence of magnetic field, the low-energy spectrum of the bilayer graphene is known[376,340] to be parabolic,

$$\mathcal{E}(k) \propto |\boldsymbol{k}|^2, \tag{20.7.1}$$

near the K and K′ points. In the presence of the magnetic field, we can reformulate the model Hamiltonian[340,201] as the generalized Dirac Hamiltonian defined by

$$H^{\pm} = \text{diag.}\left(\sqrt{Q_{\pm}Q_{\pm}}, -\sqrt{Q_{\pm}Q_{\pm}}\right), \tag{20.7.2}$$

together with

$$Q_{+} = \begin{pmatrix} 0 & A^{\dagger} \\ A & 0 \end{pmatrix}, \quad Q_{-} = \begin{pmatrix} 0 & A \\ A^{\dagger} & 0 \end{pmatrix}. \tag{20.7.3}$$

Here, $A = \hbar\omega_c a^2$, with a the Landau-level ladder operator. We also consider the generalized Pauli Hamiltonian

$$H_{\text{P}}^{+} \equiv Q_{+}Q_{+} = (\hbar\omega_c)^2 \begin{pmatrix} a^{2\dagger}a^2 & 0 \\ 0 & a^2a^{2\dagger} \end{pmatrix}, \tag{20.7.4a}$$

$$H_{\text{P}}^{-} \equiv Q_{-}Q_{-} = (\hbar\omega_c)^2 \begin{pmatrix} a^2a^{2\dagger} & 0 \\ 0 & a^{2\dagger}a^2 \end{pmatrix}. \tag{20.7.4b}$$

Since the commutation relation $[a, a^{\dagger}] = 1$ has not been used to derive this, we may switch off the magnetic field in this formula, and reproduce the energy spectrum (20.7.1). Hence, the bilayer graphene provides us with another model of the SUSY quantum mechanics [Fig.20.10]: See Section 6.6 for details.

By using the relation $a^{2\dagger}a^2 = a^{\dagger}a\,(a^{\dagger}a - 1)$, the energy spectrum is found to be

$$H^{\pm}|N\rangle = \text{diag.}\left(\mathcal{E}_N^{\pm\uparrow}, \mathcal{E}_N^{\pm\downarrow}, -\mathcal{E}_N^{\pm\uparrow}, -\mathcal{E}_N^{\pm\downarrow}\right)|N\rangle \tag{20.7.5}$$

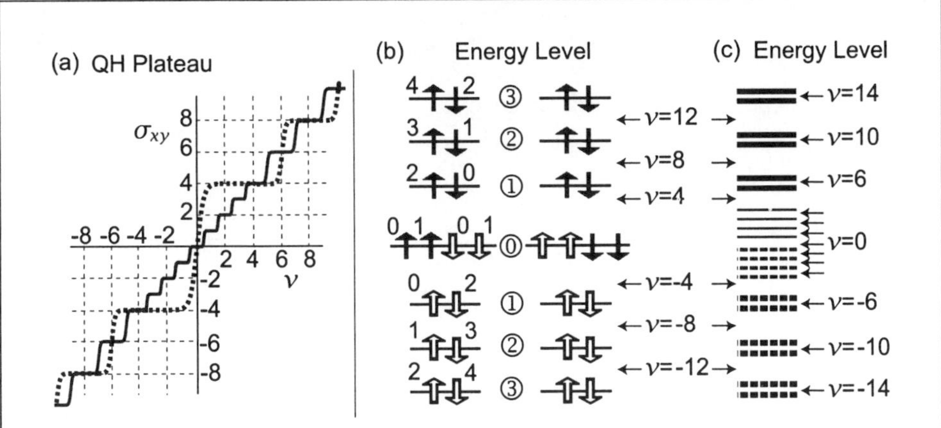

Fig. 20.10 (a) The QH conductivity is illustrated in bilayer graphene. The dotted curve shows the sequence $\nu = \pm 4, \pm 8, \pm 12, \cdots$, while the solid curve the sequence $\nu = 0, \pm 1, \pm 2, \pm 3, \pm 4, \pm 6, \pm 8, \cdots$. (b) The energy level within noninteracting theory is illustrated. The spin is indicated by a solid (open) arrow for electron (hole) at the K and K′ points. The number attached to a solid (open) arrow shows from which Landau level the electron (hole) comes from. This energy spectrum is a manifestation of SUSY. (c) When Coulomb interactions are included, the zero-energy level splits into eight nondegenerate subbands, while each nonzero-energy level splits into two 2-fold degenerate subbands with the exact U(1) symmetry.

where $|N\rangle = (N!)^{-1/2}(a^\dagger)^N|0\rangle$, and

$$\mathcal{E}_N^{+\uparrow} = \mathcal{E}_N^{-\downarrow} = \hbar\omega_c\sqrt{N(N-1)}, \tag{20.7.6a}$$

$$\mathcal{E}_N^{+\downarrow} = \mathcal{E}_N^{-\uparrow} = \hbar\omega_c\sqrt{(N+2)(N+1)} \tag{20.7.6b}$$

for $N = 0, 1, 2, 3, \cdots$, as illustrated in Fig.20.10. Here, N is the Landau-level index.

It is interesting that two Landau levels mix to create one nonzero-energy level, but that four Landau levels mix to create the zero-energy level. Thus there exists the 4-fold degeneracy in the nonzero-energy level but the 8-fold degeneracy in the zero-energy state [Fig.20.11], as results in the bold-face series (20.1.2). This agrees with the previous result.[376,340,201]

We include Coulomb interactions. The Nth energy level ($N \neq 0$) has the same structure as in the monolayer graphene system. Coulomb interactions make each energy level split into two subbands.

We discuss the zero-energy state in some details. There are electron-hole pairs coming from the $N = 0$ Landau level and the $N = 1$ Landau level for electrons and holes [Fig.20.10(b)]. At the K point there are four gap functions $\Delta_{00}(\boldsymbol{k})$, $\Delta_{11}(\boldsymbol{k})$,

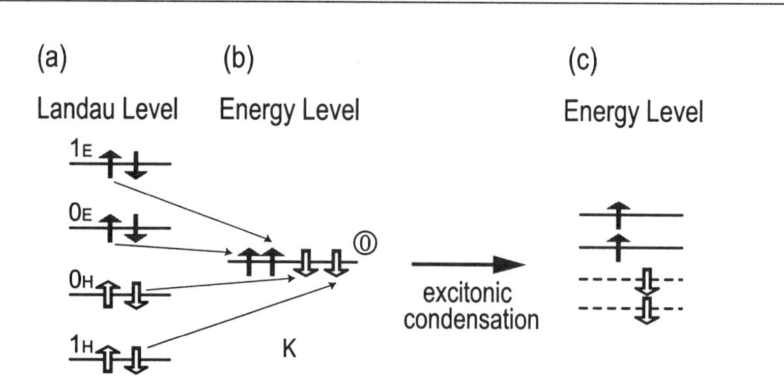

Fig. 20.11 In bilayer graphene, the zero-energy state contains up-spin electrons and down-spin holes coming from the zeroth Landau level and 1st Landau level. (a) The Nth Landau levels for electrons and holes, where $N = 0, 1$. (b) The intrinsic Zeeman effect induces a Landau-level mixing. (c) They make an excitonic condensation, producing gaps to electron and hole states.

$\Delta_{01}(\boldsymbol{k})$ and $\Delta_{10}(\boldsymbol{k})$,

$$\Delta_{NN'}(\boldsymbol{k}) = \int d^2k'\, V^{\text{eff}}(\boldsymbol{k}' - \boldsymbol{k})\langle \psi_{\text{e}}^{N\uparrow\dagger}(\boldsymbol{k}')\psi_{\text{h}}^{N'\downarrow\dagger}(\boldsymbol{k}')\rangle, \tag{20.7.7}$$

where $\psi_{\text{e}}^{N\uparrow\dagger}$ and $\psi_{\text{h}}^{N'\uparrow\dagger}$ are creation operators of electrons in the Nth Landau level and holes in the N'th Landau level, respectively. Repeating similar analysis we have made in Section 20.5, we solve the gap equations at $T = 0$ as

$$\Delta_{00}(\boldsymbol{k}) = \Delta_0, \qquad \Delta_{11}(\boldsymbol{k}) = \frac{3}{4}\Delta_0, \qquad \Delta_{01}(\boldsymbol{k}) = \Delta_{10}(\boldsymbol{k}) = \frac{1}{2}\Delta_0, \tag{20.7.8}$$

where Δ_0 is given by (20.5.14); $\Delta_0 = \pi\sqrt{\pi/2}(e^2/4\pi\varepsilon l_B)$. Thus, the 8-fold degenerate zero-energy level splits into four 2-fold degenerated subbands, producing plateaux at $\nu = \pm 2, \pm 4$. Furthermore we have Ising QH ferromagnets at $\nu = \pm 1, \pm 3$. Consequently, the full series (20.1.2) is obtained.

Trilayer Graphene (Rhombohedral Stacking)

The above analysis is applicable also to a trilayer graphene with the ABC stacking (rhombohedral stacking). In the absence of the magnetic field, the low-energy spectrum of trilayer graphene has been argued[201] to be cubic,

$$\mathcal{E}(k) \propto |\boldsymbol{k}|^3, \tag{20.7.9}$$

near the K and K′ points. In the presence of the magnetic field, we can reformulate the model Hamiltonian[201] as the generalized Dirac Hamiltonian as

$$H^{\pm} = \text{diag.}\left(\sqrt{Q_{\pm}Q_{\pm}}, -\sqrt{Q_{\pm}Q_{\pm}}\right), \tag{20.7.10}$$

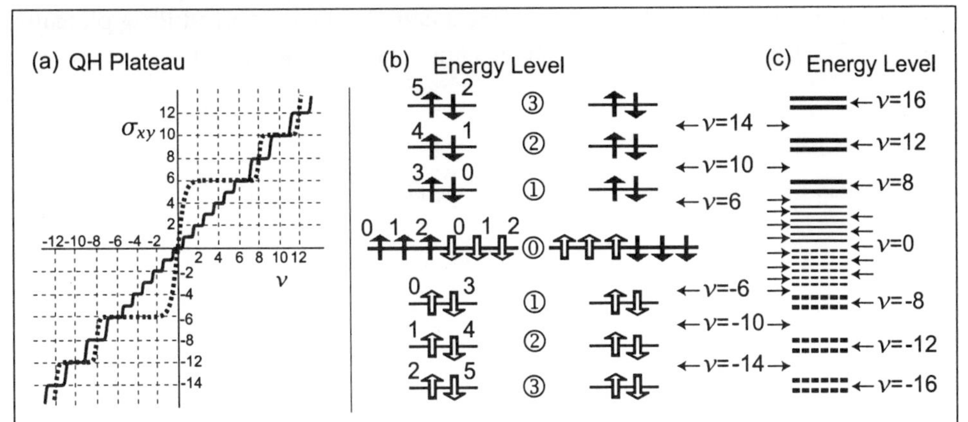

Fig. 20.12 (a) The QH conductivity is illustrated in trilayer graphene. The dotted curve shows the sequence $\nu = \pm 6, \pm 10, \pm 14, \cdots$, while the solid curve the sequence $\nu = 0, \pm 1, \pm 2, \pm 3, \pm 4, \pm 5, \pm 6, \pm 8, \pm 10, \cdots$. (b) The energy level within noninteracting theory. The spin is indicated by a solid (open) arrow for electron (hole) at the K and K′ points. The number attached to a solid (open) arrow shows from which Landau level the electron (hole) comes from. This energy spectrum is a manifestation of SUSY . (c) When Coulomb interactions are included, the zero-energy level splits into twelve nondegenerate subbands, while each nonzero-energy level splits into two 2-fold degenerate subbands with the exact U(1) symmetry.

together with

$$Q_+ = \begin{pmatrix} 0 & A^\dagger \\ A & 0 \end{pmatrix}, \quad Q_- = \begin{pmatrix} 0 & A \\ A^\dagger & 0 \end{pmatrix}. \tag{20.7.11}$$

Here, $A = \hbar\omega_c a^3$. Hence, the trilayer graphene provides us with another model of the SUSY quantum mechanics [Fig.20.12]: See Section 6.6 for details.

The Hamiltonian $H^\pm$ is diagonalized as in (20.7.5) with

$$\mathcal{E}_N^{+\uparrow} = \mathcal{E}_N^{-\downarrow} = \hbar\omega_c\sqrt{N(N-1)(N-2)} \tag{20.7.12a}$$

and

$$\mathcal{E}_N^{+\downarrow} = \mathcal{E}_N^{-\uparrow} = \hbar\omega_c\sqrt{(N+3)(N+2)(N+1)} \tag{20.7.12b}$$

for $N = 0, 1, 2, 3, \cdots$, as illustrated in Fig.20.10. There exists the 4-fold degeneracy in the nonzero-energy level but the 12-fold degeneracy in the zero-energy state, as results in the bold-face series (20.1.3). This agrees with the previous result.[201]

We briefly argue how the SUSY spectrum is modified by Coulomb interactions. The Nth energy level ($N \neq 0$) has the same structure as in the monolayer graphene system. Hence, each energy level splits into two subbands. There are electron-hole pairs in the zero-energy state, which make excitonic condensation by the same mechanism as in the monolayer and bilayer cases. Thus, the 12-fold degenerate zero-energy level splits into six 2-fold degenerated subbands, producing plateaux at $\nu = \pm 2, \pm 4, \pm 6$. Furthermore we have Ising QH ferromagnets at $\nu = \pm 1, \pm 3, \pm 5$. Consequently, the full series (20.1.3) is obtained.

General Case

We next investigate the generalized Dirac Hamiltonian as

$$H^{\pm} = \text{diag.}\left(\sqrt{Q_{\pm}Q_{\pm}}, -\sqrt{Q_{\pm}Q_{\pm}}\right), \tag{20.7.13}$$

together with

$$Q_{+} = \begin{pmatrix} 0 & A^{\dagger} \\ A & 0 \end{pmatrix}, \quad Q_{-} = \begin{pmatrix} 0 & A \\ A^{\dagger} & 0 \end{pmatrix}, \tag{20.7.14}$$

where A is given by

$$A = \hbar\omega_c a^{\dagger j_{\downarrow}} a^{j_{\uparrow}}, \qquad A^{\dagger} = \hbar\omega_c a^{\dagger j_{\uparrow}} a^{j_{\downarrow}}, \tag{20.7.15}$$

with $j_{\uparrow}$ and $j_{\downarrow}$ being integers such that $j_{\uparrow} > j_{\downarrow} \geq 0$. The monolayer graphene is given by $j_{\uparrow} = 1$, $j_{\downarrow} = 0$, the bilayer graphene by $j_{\uparrow} = 2$, $j_{\downarrow} = 0$, and the trilayer graphene by $j_{\uparrow} = 3$, $j_{\downarrow} = 0$. Though a physical system with $j_{\downarrow} \neq 0$ is yet to be found, we study the case since it presents a new insight on the relation between the degeneracy in the energy spectrum and the topological nature of graphene.

We explore the energy spectrum of the generalized Pauli Hamiltonian

$$H_P^{\pm} = Q_{\pm}Q_{\pm}. \tag{20.7.16}$$

Although the Hamiltonian is written in terms of A and $A^{\dagger}$, the basic physical variables are the Landau-level ladder operators a and $a^{\dagger}$. Thus, the eigenstate is given by

$$|N\rangle = \frac{1}{\sqrt{N!}}(a^{\dagger})^N|0\rangle. \tag{20.7.17}$$

We make an analysis at the K point. The zero-energy up-spin states are given by the condition $A|n\rangle = 0$. They are $|0\rangle, |1\rangle, \cdots, |j_{\uparrow} - 1\rangle$, which are degenerate in $|\psi_0^{+\uparrow}\rangle$. On the other hand the zero-energy down-spin states are determined by requiring $A^{\dagger}|n\rangle = 0$. If $j_{\downarrow} = 0$, there are no zero-energy states, and the supermultiplet is given by (6.6.17a) [Fig.6.3(a)]. If $j_{\downarrow} \neq 0$, they are $|0\rangle, |1\rangle, \cdots, |j_{\downarrow} - 1\rangle$, which are degenerate in $|\psi_0^{+\downarrow}\rangle$. Consequently, there exists the $(j_{\uparrow} + j_{\downarrow})$-fold degeneracy in the zero-energy state, corresponding to $j_{\uparrow}$ up-spin electrons and $j_{\downarrow}$ down-spin electrons [Fig.6.3(b)]. The similar analysis is made also at the K′ point.

The diagonal elements are easily calculated by evaluating $A^{\dagger}A|N\rangle$ and $AA^{\dagger}|N\rangle$. We derive (20.7.5) with

$$\mathcal{E}_N^{+\uparrow} = \mathcal{E}_N^{-\downarrow} = \hbar\omega_c\sqrt{\frac{N!\,(N - j_{\uparrow} + j_{\downarrow})!}{\{(N - j_{\uparrow})!\}^2}} \tag{20.7.18a}$$

for the bosonic sector, and

$$\mathcal{E}_N^{+\downarrow} = \mathcal{E}_N^{-\uparrow} = \hbar\omega_c \sqrt{\frac{N!\,(N + j_\uparrow - j_\downarrow)!}{\{(n - j_\downarrow)!\}^2}} \tag{20.7.18b}$$

for the fermionic sector, where $N = 0, 1, 2, \cdots$. Counting the contributions from the electron and hole sectors at the K and K′ point, the zero-energy states are $4(j_\uparrow + j_\downarrow)$-fold degenerate and all other states are 4-fold degenerate. This energy spectrum implies that the Hall conductivity is quantized as

$$\sigma_{xy} = \pm\left(n + \frac{j_\uparrow + j_\downarrow}{2}\right)\frac{4e^2}{h}, \qquad n = 0, 1, 2, \cdots. \tag{20.7.19}$$

When Coulomb interactions are included, we expect that the zero-energy level splits into nondegenerate subbands and that each nonzero-energy level splits into two 2-fold degenerate subbands with the exact U(1) symmetry.

20.8 Berry Phase and Index Theorem

We study generalized Dirac systems (20.7.13) with $A = \hbar\omega_c a^{\dagger j_\downarrow} a^{j_\uparrow}$ to derive a new insight[132] on the relation between the degeneracy of the zero-energy state, the Berry phase and the Witten index. The degeneracy of the zero-energy state was originally derived from the arguments[375,538,376] on the Berry phase in the noninteracting graphene.

We first derive the energy dispersion relation in the zero field ($B = 0$). Setting $P_i = p_i$ and parametrizing $\boldsymbol{p} = (p\cos\phi, p\sin\phi)$ in (20.7.15), we obtain

$$A = \mathcal{E}(p)\, e^{i(j_\uparrow - j_\downarrow)\phi}, \quad A^\dagger = \mathcal{E}(p)\, e^{-i(j_\uparrow - j_\downarrow)\phi}, \tag{20.8.1}$$

with

$$\mathcal{E}(p) \propto p^{j_\uparrow + j_\downarrow}. \tag{20.8.2}$$

The eigenvalues of the Dirac Hamiltonian (20.7.13) are $\pm\mathcal{E}(p)$ with the eigenstates

$$|\pm\rangle = \frac{1}{\sqrt{2}}\begin{pmatrix} 1 \\ \pm e^{i(j_\uparrow - j_\downarrow)\phi} \end{pmatrix} \tag{20.8.3}$$

at the K point.

The Berry phase is defined as a loop integration

$$\Gamma_{\rm B} = \oint \sum_k C_k dx_k \tag{20.8.4}$$

of the connection field

$$C_k = \langle n|\, i\frac{\partial}{\partial x_k}\, |n\rangle. \tag{20.8.5}$$

It becomes

$$\Gamma_{\mathrm{B}} = \pi\,(j_{\uparrow} - j_{\downarrow}) \tag{20.8.6}$$

for the eigenstates $|\pm\rangle$. Thus the indices $j_{\uparrow}$ and $j_{\downarrow}$ are fixed by the dispersion relation and the Berry phase in the zero field.

We now study the Witten index and the Atiyah-Singer index theorem. We focus on the K point. The Witten index[508] is given by

$$\Delta_{\mathrm{W}} = \dim[\ker H^{+\uparrow}] - \dim[\ker H^{+\downarrow}] = j_{\uparrow} - j_{\downarrow}, \tag{20.8.7}$$

where

$$j_{\uparrow} \equiv \dim[\ker H^{+\uparrow}], \qquad j_{\downarrow} \equiv \dim[\ker H^{+\downarrow}] \tag{20.8.8}$$

are the numbers of zero-energy states of $H^{+\uparrow}$ and $H^{+\downarrow}$, respectively. We have explicitly shown in the present model that the Witten index defined in the nonzero field ($B \neq 0$) is equal to the Berry phase defined in the zero field ($B = 0$), $\Gamma_{\mathrm{B}} = \pi\Delta_{\mathrm{W}}$. This is a general property, as we briefly sketch. The Witten index is equal to the Fredholm index[31]

$$\Delta_{\mathrm{F}} = \dim\,[\ker A] - \dim[\ker A^{\dagger}], \tag{20.8.9}$$

since

$$\ker H^{+\uparrow} = \ker A^{\dagger}A = \ker A, \qquad \ker H^{+\downarrow} = \ker AA^{\dagger} = \ker A^{\dagger}. \tag{20.8.10}$$

Now, according to the Atiyah-Singer index theorem, Δ_{F} is invariant as $B \to 0$, where Δ_{F} becomes the chiral anomaly.[31] The chiral anomaly is given by the Berry phase. Consequently it follows that $\Gamma_{\mathrm{B}} = \pi\Delta_{\mathrm{W}}$.

The Witten index, $j_{\uparrow} - j_{\downarrow}$, and the degeneracy of the zero-energy state, $j_{\uparrow} + j_{\downarrow}$, are different objects in generalized graphene systems. Nevertheless, they are identical in graphene where $j_{\downarrow} = 0$. Thus, by combining the contributions from the K and K′ points, the degeneracy $4j_{\uparrow}$ of the zero-energy state can be related to the Berry phase, as pointed out in literature.[375,538,376]

PART 3

BILAYER QUANTUM HALL SYSTEMS

CHAPTER 21

OVERVIEW OF BILAYER QH SYSTEMS

Bilayer systems exhibit a new variety of QH states. Various QH states are realized by controlling system parameters such as the strengths of the interlayer and intralayer Coulomb interactions, the tunneling interaction and the Zeeman effect. Moreover, electrons are transferable between the two layers by applying bias voltages. The interlayer exchange interaction develops the interlayer coherence spontaneously. In this section we overview diverse aspects of bilayer QH systems.

Bilayer QH Systems

A bilayer system is made by trapping electrons in two thin layers at the interface of semiconductors, where the structure introduces an additional degree of freedom in the third direction.[1] The Hall current is induced by the Lorentz force precisely as in the monolayer system,

$$\mathcal{J}_i = \frac{e^2\nu}{2\pi\hbar}\varepsilon_{ij}E_j, \tag{21.1.1}$$

where $\mathcal{J}_i$ is the sum of the currents flowing on the front and back layers, and $\nu = 2\pi\ell_B^2\rho_0$ is the filling factor with ρ_0 the total electron density. Hall plateaux develop in the Hall conductivity at "magic" values of the filling factor ν, where the system becomes incompressible. There exists a variety of QH states[71,465,127,359,298,220] depending on the relative strength of various interactions. The key interactions are the intralayer Coulomb interaction ($\sim e^2/4\pi\varepsilon\ell_B$), the interlayer Coulomb interaction ($\sim e^2/4\pi\varepsilon d$), the Zeeman interaction ($\sim \Delta_Z$) and the tunneling interaction ($\sim \Delta_{SAS}$). Here, d is the separation between the two layers [Fig.21.1]. Since their relative strength is only relevant, we measure the energy in units of the intralayer Coulomb energy $E_C^0 \equiv e^2/4\pi\varepsilon\ell_B$, or equivalently measure the length in units of the magnetic length ℓ_B.

By applying bias voltages electrons are transferable from one layer to the other [Fig.21.1(b)]. It is important that a certain bilayer QH state is stable against charge

[1] A generalization to N-layer system is straightforward,[149] where no new physics seems to appear.[141] However, the edge of a multilayer system with more than 100 layers may present a chiral torus akin to a carbon nanotube.[266]

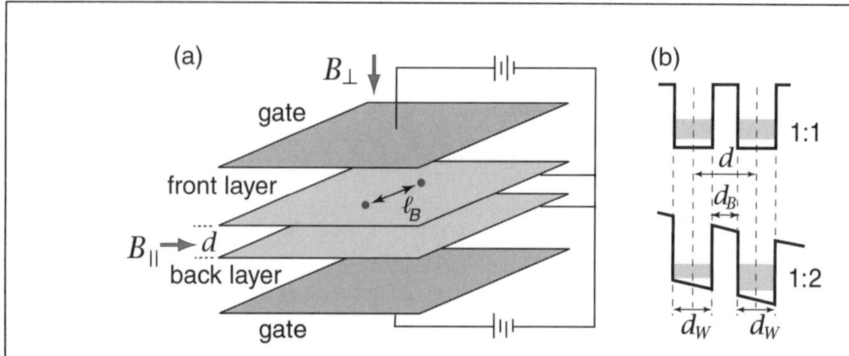

Fig. 21.1 (a) The bilayer system has four scale parameters, the magnetic length ℓ_B, the Zeeman gap Δ_Z, the interlayer distance d and the tunneling gap Δ_{SAS}. It is customary to take $d = d_B + d_W$ with the width d_W and the separation d_B of the two quantum wells. Here, $d_W \simeq$ 200 Å and $d_B \simeq 31$ Å in typical samples. The number density ρ^α in each quantum well is controlled by applying gate bias voltages. A parallel magnetic field $B_\parallel$ may be additionally applied to the system. (b) Two typical examples are given where $\rho^f : \rho^b = 1:1$ (balanced) and $1:2$ (imbalanced).

transfer.[423,424,357] A most prominent feature is that the interlayer coherence may develop spontaneously on charge-transferable states, as was first noticed by Ezawa and Iwazaki,[138,140] and independently by Wen and Zee.[493,494]

There exists a variety of bilayer QH states, where typical ones are listed in Fig.21.2. See early theoretical works[399,400,91,529,162,324,75,227,228] before 1993 and also a review article[128] for experiments before 1996.

We label the two layers by the index α = f, b, and call the α = f layer the front layer and the α = b layer the back layer. The bilayer QH system possesses the SU(2) *pseudospin* structure, where the up-pseudospin component ($\mathcal{P}_z = 1/2$) is the front-layer component while the down-pseudospin component ($\mathcal{P}_z = -1/2$) is the back-layer component. The pseudospin rotates when electrons are transferred from one layer to the other. When the number density is balanced between the two quantum wells, it is referred to as the balanced configuration and otherwise as the imbalanced configuration.

There are four types of electrons associated with the field operators $\psi^{f\uparrow}$, $\psi^{f\downarrow}$, $\psi^{b\uparrow}$ and $\psi^{b\downarrow}$. One Landau level contains four energy levels. The underlying group structure is enlarged to SU(4) in each Landau level. Let us call it the isospin SU(4) in contrast to the spin SU(2) and the pseudospin SU(2). In particular, the system enjoys the SU(4) isospin symmetry[2] in the degenerate limit of the four levels.[147] Then, the SU(4) isospin is spontaneously polarized on the ground state,[3] and the

[2] The dynamical symmetry is not SU(4) but CP^3 when the cyclotron gap is sufficiently large. See Section 15.4.

[3] See Section 4.1 and Appendix J on the spontaneous symmetry breaking in the 2-dimensional space.

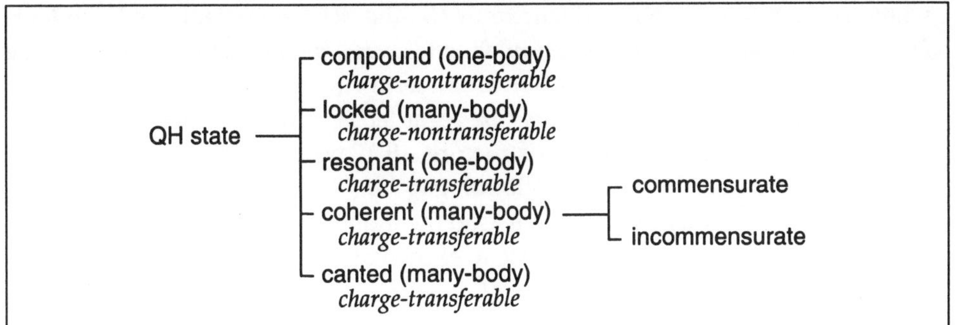

Fig. 21.2 Bilayer systems exhibit a rich family of QH states depending on the type of correlations between the two layers. Charges are not transferable between the two layers in the compound state and the locked state, but transferable in the resonant state and the coherent state. A canted antiferromagnetic state arises at $\nu = 2$ between the compound and resonant states by the SU(4) noninvariant part of the exchange interaction.

system is called the *SU(4) QH ferromagnet*.

The algebra SU(4) has the subalgebra $SU_{spin}(2) \times SU_{ppin}(2)$. We denote the 3 generators of the spin SU(2) algebra by τ_a^{spin}, and those of the pseudospin SU(2) generator by τ_a^{ppin}. The remaining 9 generators of the algebra SU(4) are given by $\tau_a^{spin}\tau_b^{ppin}$. The physical operators are constructed as bilinear combinations of $\psi(\boldsymbol{x})$ and $\psi^\dagger(\boldsymbol{x})$. They are 16 density operators

$$\begin{aligned} \rho(\boldsymbol{x}) &= \psi^\dagger(\boldsymbol{x})\psi(\boldsymbol{x}), \\ \mathcal{S}_a(\boldsymbol{x}) &= \frac{1}{2}\psi^\dagger(\boldsymbol{x})\tau_a^{spin}\psi(\boldsymbol{x}), \\ \mathcal{P}_a(\boldsymbol{x}) &= \frac{1}{2}\psi^\dagger(\boldsymbol{x})\tau_a^{ppin}\psi(\boldsymbol{x}), \\ \mathcal{R}_{ab}(\boldsymbol{x}) &= \frac{1}{2}\psi^\dagger(\boldsymbol{x})\tau_a^{spin}\tau_a^{ppin}\psi(\boldsymbol{x}), \end{aligned} \tag{21.1.2}$$

where $\mathcal{S}_a$ describes the total spin of the system, $2\mathcal{P}_z$ measures the electron-density difference between the two layers. The operator $\mathcal{R}_{ab}$ transforms as a spin under $SU_{spin}(2)$ and as a pseudospin under $SU_{ppin}(2)$.

Hamiltonian

The Hamiltonian is decomposed into the SU(4)-invariant part H_{inv} and the SU(4)-noninvariant part ΔH,

$$H = H_{inv} + \Delta H \tag{21.1.3}$$

with

$$H_{inv} = H_K + H_C^+, \qquad \Delta H = H_C^- + H_T + H_Z + H_{bias}, \tag{21.1.4}$$

where various terms are the kinetic term $H_{\rm K}$, the SU(4)-invariant Coulomb term $H_{\rm C}^{+}$, the capacitance term $H_{\rm C}^{-}$, the tunneling term $H_{\rm T}$, the Zeeman term $H_{\rm Z}$ and the bias-voltage term $H_{\rm bias}$. The SU(4)-invariant term $H_{\rm inv}$ dominates the system, and the SU(4)-noninvariant part ΔH may be regarded as a perturbation. The kinetic term $H_{\rm K}$ creates the Landau level structure. It plays no more roles if the level mixing is suppressed by a large cyclotron gap $\hbar\omega_{\rm c}$. The Coulomb term $H_{\rm C}^{+}$ leads to the incompressibility of the system as in the monolayer system.

The Coulomb interaction gives rise to exchange interactions, among which the SU(4)-invariant component is dominant. It reads

$$H_{\rm X}^{+} = J_s^{+} \int d^2x \left(\sum_a \left\{ [\partial_k \mathcal{S}_a(\boldsymbol{x})]^2 + [\partial_k \mathcal{P}_a(\boldsymbol{x})]^2 \right\} + \sum_{ab} [\partial_k \mathcal{R}_{ab}(\boldsymbol{x})]^2 \right), \tag{21.1.5}$$

for sufficiently smooth configuration, where $\mathcal{S}_a(\boldsymbol{x})$, $\mathcal{P}_a(\boldsymbol{x})$ and $\mathcal{R}_{ab}(\boldsymbol{x})$ are the normalized isospin densities (sigma fields),

$$S_a(\boldsymbol{x}) = \rho_\Phi \mathcal{S}_a(\boldsymbol{x}), \qquad P_a(\boldsymbol{x}) = \rho_\Phi \mathcal{P}_a(\boldsymbol{x}), \qquad R_{ab}(\boldsymbol{x}) = \rho_\Phi \mathcal{R}_{ab}(\boldsymbol{x}), \tag{21.1.6}$$

with ρ_Φ the number density of Landau sites, $\rho_\Phi = 1/2\pi\ell_B^2$. The driving force of the SU(4) quantum coherence is this Coulomb exchange interaction. Provided it is not too small, it is still good to regard the system as the SU(4) QH ferromagnet, even if the SU(4) symmetry is explicitly broken by the noninvariant interactions.

Pseudospin QH Ferromagnet

It is instructive to freeze the spin degree of freedom to elucidate the essence of the bilayer system. The Hamiltonian consists of the interlayer exchange interaction $H_{\rm X}^{\rm inter}$, the direct Coulomb interactions $H_{\rm D}^{\pm}$, the tunneling term $H_{\rm T}$ and the bias-voltage term $H_{\rm bias}$. The interlayer exchange interaction is

$$H_{\rm X}^{\rm inter} = 2 \int d^2x \left(J_s [\partial_k \mathcal{P}_x(\boldsymbol{x})]^2 + J_s [\partial_k \mathcal{P}_y(\boldsymbol{x})]^2 + J_s^d [\partial_k \mathcal{P}_z(\boldsymbol{x})]^2 \right), \tag{21.1.7}$$

which is an anisotropic O(3) nonlinear sigma model. Various isospin stiffness parameters are related by

$$J_s^{\pm} = \frac{1}{2}\left(J_s \pm J_s^d \right), \tag{21.1.8}$$

where $J_s^{\pm}$ is ascribable to the Coulomb interaction $H_{\rm C}^{\pm}$: See (A.14) in Appendix. The pseudospin field is parametrized as

$$\begin{aligned} &\mathcal{P}_x(\boldsymbol{x}) = \frac{1}{2}\sqrt{1-\sigma(\boldsymbol{x})^2}\cos\vartheta(\boldsymbol{x}), \qquad \mathcal{P}_y(\boldsymbol{x}) = -\frac{1}{2}\sqrt{1-\sigma(\boldsymbol{x})^2}\sin\vartheta(\boldsymbol{x}), \\ &\mathcal{P}_z(\boldsymbol{x}) = \frac{1}{2}\sigma(\boldsymbol{x}), \end{aligned} \tag{21.1.9}$$

where $\vartheta(\boldsymbol{x})$ is the interlayer phase field, and $\sigma(\boldsymbol{x})$ the interlayer imbalance field. The pair of them represents a canonical set of Goldstone modes.

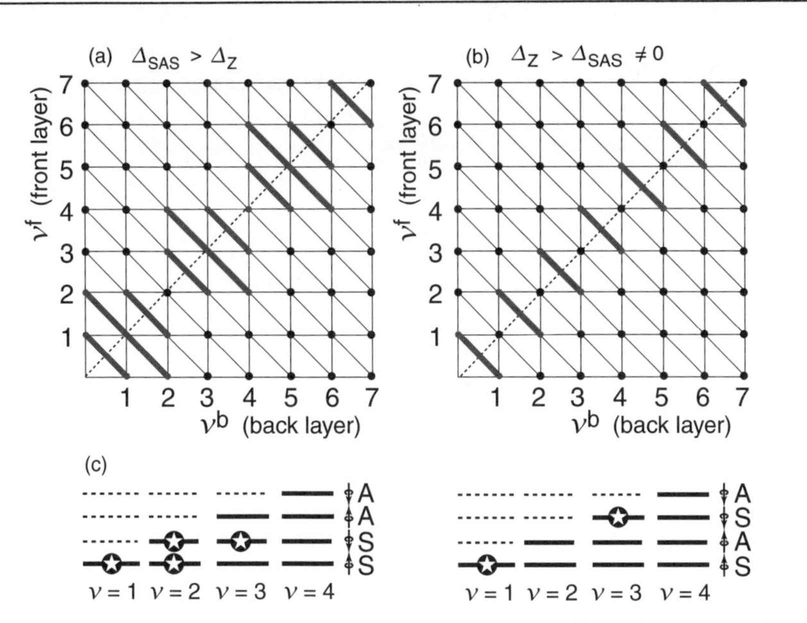

Fig. 21.3 Various bilayer QH states are realizable at integer filling factors. A solid circle on each lattice point (ν^b, ν^f) denotes a *compound state* at $\nu = \nu^b + \nu^f$ made of the $\nu = \nu^b$ and $\nu = \nu^f$ monolayer QH states. The tunneling gap collapses on the compound state. A heavy line represents a *charge-transferable state* (a) when $\Delta_{SAS} > \Delta_Z$ and (b) when $\Delta_Z > \Delta_{SAS} \neq 0$, though the phase-transition point $\Delta_{SAS} = \Delta_Z$ is actually modified by the exchange interaction effect. A charage-transferable state is a *resonant state* or *coherent state.* (c) Due to the tunneling gap one Landau level may be split into four energy levels at the balanced point, the up-spin symmetric state (↑S), the down-spin symmetric state (↓S), the up-spin antisymmetric state (↑A) and the down-spin antisymmetric state (↓A). A heavy line indicates a filled level. A heavy line with symbol ✪ indicates that it is a charge-transferable level. Charge transfer is not allowed if the state is a pseudospin singlet.

Resonant and Interlayer-Coherent States

The interlayer exchange energy (21.1.7) has an expression similar to the intralayer exchange energy. Furthermore, the sum of the tunneling term and the bias term looks very much like the Zeeman term,

$$H_{pZ} \equiv H_T + H_{bias} = -N_\Phi \boldsymbol{\mathcal{B}} \cdot \boldsymbol{\mathcal{P}}, \tag{21.1.10}$$

where $\boldsymbol{\mathcal{B}} = (\Delta_{SAS}, 0, \Delta_{bias})$ is the pseudomagnetic field with Δ_{SAS} the tunneling gap and Δ_{bias} the bias parameter to describe the imbalanced density configuration between the two layers, while

$$\mathcal{P}_a = \frac{1}{N_\Phi} \int d^2x \, \mathcal{P}_a(\boldsymbol{x}) \tag{21.1.11}$$

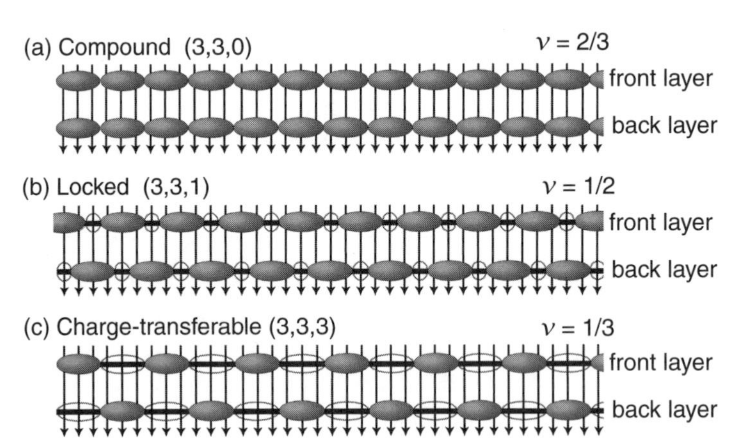

Fig. 21.4 Typical examples of fractional QH states. In these three figures the magnetic field is the same but the electron densities are different. (a) Compound state (3,3,0) at $\nu = 2/3$: The QH state develops almost independently in each layer. The density ratio is fixed between the two layers. (b) Bilayer-locked state (3,3,1) at $\nu = 1/2$: The Coulomb repulsive force repels electrons on the opposite layer and creates a "hole" as indicated by an open oval. The density ratio is fixed between the two layers. (c) Charge-transferable state at $\nu = 1/3$: The QH state is stabilized by the tunneling interaction. The density ratio can be tuned arbitrarily between the two layers. Electrons do not have home layers. The interlayer coherence is driven to develop on the state by the exchange Coulomb energy.

is the pseudospin per Landau site. A good analogy holds between the bilayer spinless system and the monolayer spin system provided the capacitance term is neglected. The analogy reads as follows:

(A) When the "pseudo-Zeeman effect " is large enough, the bilayer system is approximated by the pseudospin-frozen system. One Landau level is split into two energy levels for the symmetric and antisymmetric states at the balanced point. An integer QH state is realized when each level is filled up. Fractional QH states are generated when it is partially filled at $\nu = n/m$ with odd m. As the front and back bias voltages are applied ($\Delta_{\text{bias}} \neq 0$) they evolve into the bonding and antibonding states. The direction of the pseudospin is controlled by "tilting" the pseudomagnetic field, i.e., by applying the bias voltages. It implies physically that electrons are transferable between the two layers without breaking the QH system [Fig.21.3]. Such a state is referred to as the *resonant state* since the QH state is stabilized essentially by the tunneling gap.

(B) A new phase emerges when the interlayer exchange energy (21.1.7) is dominant over the pseudo-Zeeman energy. According to the Hamiltonian (21.1.7) the pseudospin wave is a low energy excitation, leading to a pseudospin coherence

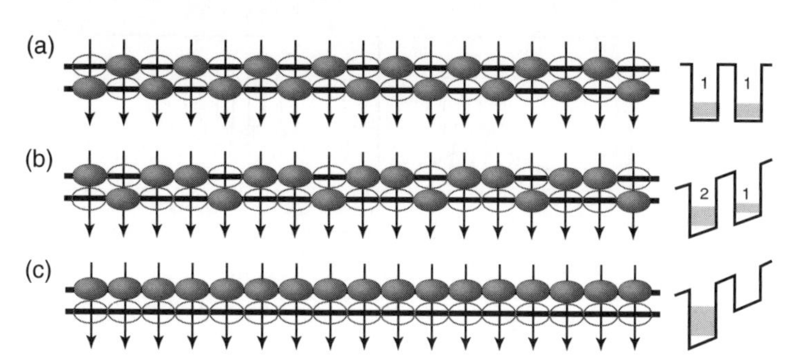

Fig. 21.5 The density ratio is freely changed at $\nu = 1$ by controlling the bias gate voltage. On average, (a) it is 1/1; (b) it is 2/1; (c) all electrons are transferred to the front layer. Electrons do not have home layers.

(interlayer coherence) with two coherence lengths

$$\xi_{\text{ppin}}^{\vartheta} = \ell_B \sqrt{\frac{4\pi J_s^d}{\Delta_{\text{SAS}}}}, \qquad \xi_{\text{ppin}}^{\sigma} = \ell_B \sqrt{\frac{4\pi J_s}{\epsilon_{\text{cap}} + \Delta_{\text{SAS}}}} \tag{21.1.12}$$

at $\nu = 1$, where ϵ_{cap} is a capacitance energy. The formula implies that the pseudospin degree of freedom would be frozen ($\xi_{\text{ppin}} = 0$) but for the interlayer exchange interaction ($J_s^d = 0$). The interlayer coherence is well developed for $J_s^d \gtrsim \Delta_{\text{SAS}}$. In particular, we find

$$\xi_{\text{ppin}}^{\vartheta} \to \infty, \qquad \text{as} \quad \Delta_{\text{SAS}} \to 0. \tag{21.1.13}$$

Such a state is referred to as the *interlayer-coherent state.* No phase transition exists between the resonant and interlayer-coherent states.

Charge transferability characterizes resonant and interlayer-coherent states even at higher filling factors.[357] It is not true, however, that all electrons are always transferable to one layer by applying bias voltages in these states. Here, the analogy between the spin polarization and the pseudospin polarization is useful. As we have seen in Section 17.4, the maximum polarization depends on which spin multiplet the QH state belongs to. Similarly, the maximum charge transferability depends on which pseudospin multiplet the bilayer QH state belongs to. In particular, no charge transfer is allowed in the pseudospin-singlet state. We have illustrated typical charge-transferable states for integer QH states in Fig.21.3. See Fig.14 and Fig.15 in the frontispiece for experimental data.

Compound States

The pseudomagnetic field is oriented to the z axis in the absence of the tunneling interaction ($\Delta_{\text{SAS}} \simeq 0$), $\boldsymbol{\mathcal{B}} = (0, 0, \Delta_{\text{bias}})$, and the Hamiltonian H contains only

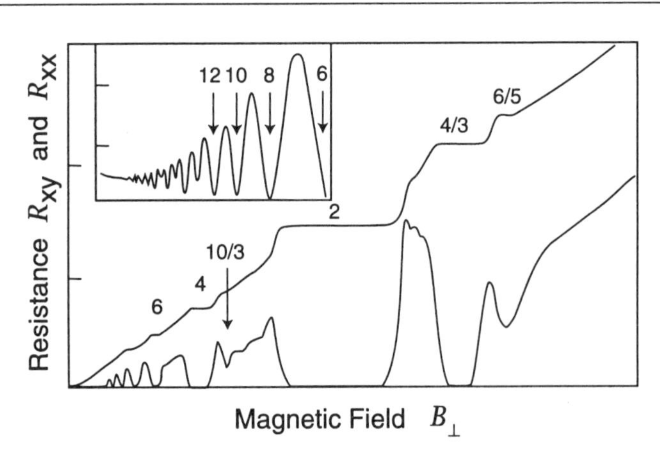

Fig. 21.6 All QH states here may be identified with compound states, where $\nu = 2\nu^{\mathrm{f}}$ at the balanced point. The data are taken from Hamilton et al.[220]

the pseudospin component $\mathcal{P}_z$. Then, the QH state is an eigenstate belonging to a discrete spectrum of the operator $\mathcal{P}_z$. We may regard it as a *compound state* of two monolayer QH states at $\nu = \nu^{\mathrm{b}}$ and $\nu = \nu^{\mathrm{f}}$, which we denote by $(\nu^{\mathrm{b}}, \nu^{\mathrm{f}})$, as illustrated in Fig.21.3 for integer QH states. It is realized at the filling factor $\nu = \nu^{\mathrm{b}} + \nu^{\mathrm{f}}$, where the density ratio is fixed as $\rho_0^{\mathrm{f}}/\rho_0^{\mathrm{b}} = \nu^{\mathrm{f}}/\nu^{\mathrm{b}}$. In particular, it follows that $\nu = 2\nu^{\mathrm{f}}$ at the balanced point [Fig.21.6].

Compound states are stable even in the presence of a tunneling interaction ($\Delta_{\mathrm{SAS}} \neq 0$). Though electrons in different layers are strongly correlated and they are no longer made of two independent monolayer states, they are still realized at $\nu = \nu^{\mathrm{b}}+\nu^{\mathrm{f}}$ and the density ratio is fixed as $\rho_0^{\mathrm{f}}/\rho_0^{\mathrm{b}} = \nu^{\mathrm{f}}/\nu^{\mathrm{b}}$. They are the pseudospin-singlet states, where charge is not transferable between the two layers. See Fig.14 (page xviii) and Fig.15 (page xix) in the frontispiece for experimental data.

Bilayer-Locked States

There exists another type of QH states peculiar to the bilayer system. Since electrons expel one another even if they belong to different layers, they create "holes" in the opposite layer as in Fig.21.4(b). Such a fractional QH state is described by the Halperin wave function[215] which is a product of two Laughlin wave functions with an interlayer correlation included,

$$\mathfrak{S}_{\mathrm{HN}}[\mathbf{x}^{\mathrm{f}}, \mathbf{x}^{\mathrm{b}}] = (\rho_0^{\mathrm{f}})^{N_{\mathrm{f}}/2}(\rho_0^{\mathrm{b}})^{N_{\mathrm{b}}/2} \prod_{r<s}(z_r^{\mathrm{f}} - z_s^{\mathrm{f}})^{m_{\mathrm{f}}} \prod_{p<q}(z_p^{\mathrm{b}} - z_q^{\mathrm{b}})^{m_{\mathrm{b}}} \times \prod_{r,p}(z_r^{\mathrm{f}} - z_p^{\mathrm{b}})^m e^{-\sum|z_r^{\mathrm{f}}|^2-\sum|z_p^{\mathrm{b}}|^2}, \tag{21.1.14}$$

where the term $(z_r^{\mathrm{f}} - z_p^{\mathrm{b}})^m$ describes the interlayer correlation with m being an

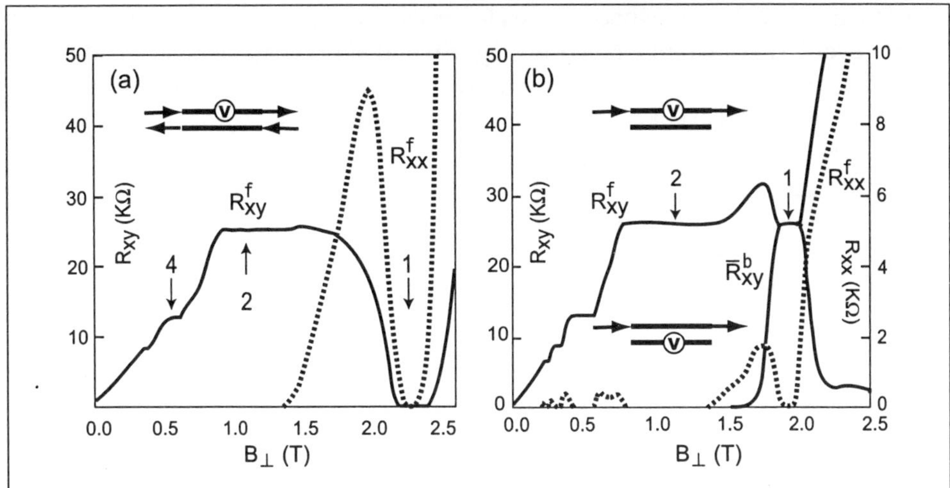

Fig. 21.7 By feeding different currents into the two layers, the Hall resitance R_{xy} (solid curve) and the diagonal resistance R_{xx} (dotted curve) are measured on one of the layers (indicated by V in a circle). (a) In the counterflow experiment the opposite amounts of currents are fed to the two layers, where $R^{f}_{xx} = R^{f}_{xy} = 0$ anomalously at $\nu = 1$. Data are taken from Tutuc et al.[481] (b) In the drag experiment the current is fed only to the front layer, where R^{f}_{xy} takes anomalously the same value at $\nu = 1$ and $\nu = 2$. It is also remarkable that $R^{f}_{xy} = \bar{R}^{b}_{xy}$ at $\nu = 1$, where $\bar{R}^{b}_{xy} \equiv E^{b}_{y}/\mathcal{J}^{f}_{x}$. Data are taken from Kellogg et al.[267] See also Fig.20 in the frontispiece.

integer. We must have $m_\alpha \geq m$ because the interlayer correlation is smaller than the intralayer correlation due to the interlayer separation.

The wave function (21.1.14) is self-explanatory. Two monolayer states are *locked* so that a pair of electron and quasihole is formed to lower the interlayer Coulomb energy [Fig.21.4(b)]. We call it the *bilayer-locked state*, and denote it by $(m_{\rm f}, m_{\rm b}, m)$. It emerges at the filling factor

$$\nu = \frac{m_{\rm f} + m_{\rm b} - 2m}{m_{\rm f} m_{\rm b} - m^2}, \tag{21.1.15}$$

where the density ratio is fixed as $\rho_0^{\rm f}/\rho_0^{\rm b} = (m_{\rm b} - m)/(m_{\rm f} - m)$. A typical QH state is given at $\nu = \frac{1}{2}$ by the choice of $m_{\rm f} = m_{\rm b} = 3$ and $m = 1$, which has been observed experimentally:[465,127] See Fig.23.1 (page 402) for experimental data. The bilayer QH state attracted the first attention because of the discovery of this new state.

Anomalous Hall Resistivity

As the unique signal of the QH effect, it is customary to regard the emergence of the plateau in the Hall resistance R_{xy} together with the vanishing diagonal resistance $R_{xx} = 0$. This is certainly the case in the monolayer QH system as well as in the standard bilayer QH system. However, recent experiments[268,481] have revealed an anomalous behavior of the Hall resistance in a counterflow geometry in the bilayer QH system that both the diagonal and Hall resistances vanish at the total bilayer filling factor $\nu = 1$ in samples with $\Delta_{SAS} \simeq 0$, as illustrated in Fig.21.7(a). Another anomalous Hall resistance has been reported in a drag experiment[267], as illustrated in Fig.21.7(b).

These new phenomena occur due to the interlayer coherence[153]. As is seen in (21.1.13), the interlayer phase field has an infinitely large coherence length for $\Delta_{SAS} = 0$. There flows the phase current $\propto \partial_i \vartheta(\boldsymbol{x})$, when there is an inhomogeneity in $\vartheta(\boldsymbol{x})$. The Hall current in each layer is given by

$$\mathcal{J}_x^{\mathrm{f}} = \frac{e}{\hbar}(1-\sigma_0^2)J_s^d \partial_x \vartheta + \frac{e^2 \ell_B^2 \rho_0}{2\hbar}(1+\sigma_0)E_y, \tag{21.1.16a}$$

$$\mathcal{J}_x^{\mathrm{b}} = -\frac{e}{\hbar}(1-\sigma_0^2)J_s^d \partial_x \vartheta + \frac{e^2 \ell_B^2 \rho_0}{2\hbar}(1-\sigma_0)E_y, \tag{21.1.16b}$$

which add up to (21.1.1). Here, the phase current arranges itself to minimize the total energy of the system, as results in the Hall resistivity

$$R_{xy}^{\mathrm{f}} \equiv \frac{E_y^{\mathrm{f}}}{\mathcal{J}_x^{\mathrm{f}}} = \frac{2\pi\hbar}{\nu e^2}\left(1 + \frac{\mathcal{J}_x^{\mathrm{b}}}{\mathcal{J}_x^{\mathrm{f}}}\right), \qquad R_{xy}^{\mathrm{b}} \equiv \frac{E_y^{\mathrm{b}}}{\mathcal{J}_x^{\mathrm{b}}} = \frac{2\pi\hbar}{\nu e^2}\left(1 + \frac{\mathcal{J}_x^{\mathrm{f}}}{\mathcal{J}_x^{\mathrm{b}}}\right) \tag{21.1.17}$$

in terms of the Hall currents. The Hall resistance vanishes in a counterflow geometry ($\mathcal{J}_x^{\mathrm{b}} = -\mathcal{J}_x^{\mathrm{f}}$). The anomalous Hall resistance in the drag experiment is also explained owing to the interlayer phase coherence.

Spin-Pseudospin Texture

Charged excitations are spin and/or pseudospin textures in the interlayer-coherent state. At the balanced point the lowest energy state is the symmetric up-spin state. We consider the $\nu = 1$ QH system. The simplest charged excitation is a hole created in the ground state. To lower the Coulomb energy, the hole may be transformed into a pseudospin texture, called a pseudospin-skyrmion or a *ppin-skyrmion* for brevity, by exciting the antisymmetric up-spin component coherently. This occurs precisely in the same mechanism as a hole transformed into a spin texture (skyrmion) in the monolayer QH ferromagnet. Since the bilayer QH system can be continuously brought into the monolayer QH system by controlling the bias voltage [Fig.21.5], a pseudospin texture is continuously transformed into a spin texture. In this process a pseudospin texture at the balanced point evolves into an SU(4) isospin texture and then regressed to a spin texture in the monolayer limit.[150] The corresponding continuous transformation of the activation energy has been observed experimentally by Sawada et al.[426,472]

Activation Energy Anomaly

The $\nu = 1$ bilayer QH state is a spin-doublet and pseudospin-doublet state, where both spins and pseudospins are spontaneously polarized in the SU(4) in-

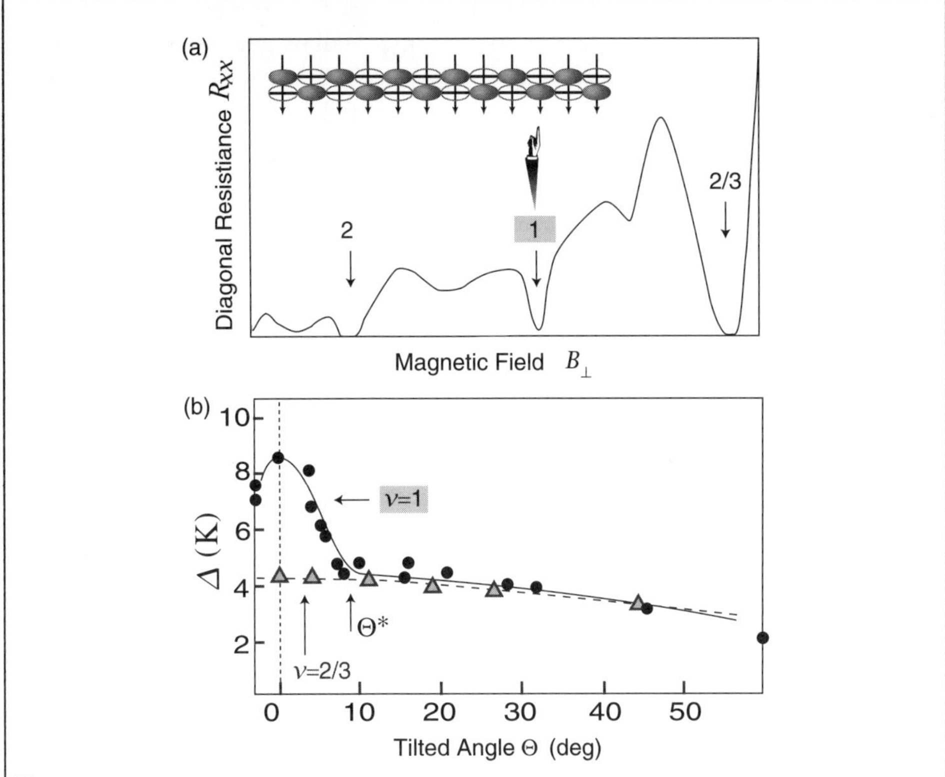

Fig. 21.8 (a) Three well developed QH states are observed in the diagonal resistance R_{xx}. (b) Activation energies at $\nu = 1$ and $\nu = 2/3$ show entirely different behaviors when the sample is tilted with an angle Θ. The one at $\nu = 1$ exhibits a rapid decrease up to a critical angle Θ^*, and then decreases very smoothly. This rapid decrease, known as the activation enery anomaly, is an indication of the interlayer coherence. The data are taken from Murphy et al.[359]

variant limit. Because it is a spin ferromagnet, we expect naively that the activation energyincreases by the Zeeman effect by tilting samples [see Section 12.2]. On the contrary, an entirely opposite behavior was observed by Murphy et al.,[359] where the activation energy decreases rapidly by tilting samples [Fig.21.8]. This unexpected behavior is called the activation energy anomaly. The activation anomaly occurs[351,150] due to the loss of the exchange energy of a pseudospin texture: It is the signal of the phase transition between the commensurate (C) phase and the incommensurate (IC) phase. The system is in the C phase for $B_\parallel < B_\parallel^*$ and in the IC phase for $B_\parallel > B_\parallel^*$, with a certain critical field $B_\parallel^*$.[519]

Soliton Lattice and Anomalous Diagonal Resistivity

In the C phase the parallel magnetic field $B_\parallel$ penetrates between the two lay-

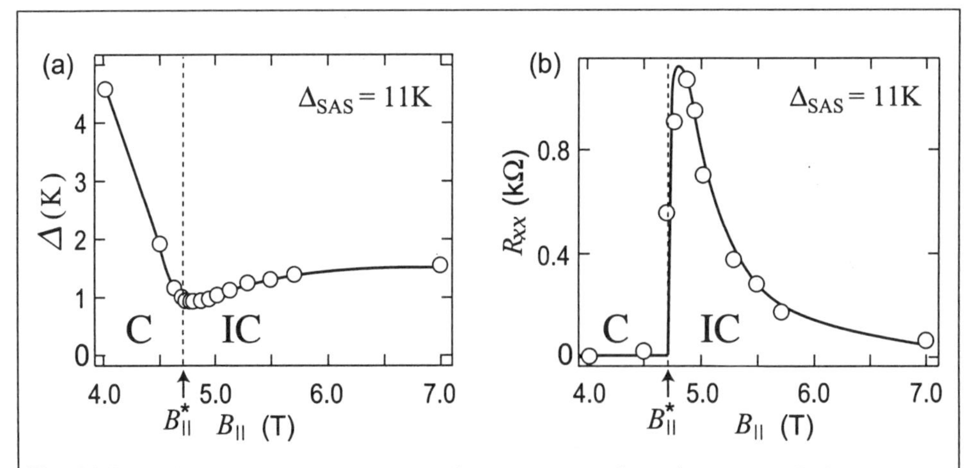

Fig. 21.9 (a) The activation energy Δ shows an anomalous decrease as $B_{\parallel}$ increases towards the C-IC transition point $B_{\parallel}^{*}$. (b) The diagonal resistance R_{xx} yields an anomolous peak in $B_{\parallel}$ when it crosses $B_{\parallel}^{*}$, though there exists a well-developed Hall plateau. It is interpreted that the anomalous resistance arises due to the backscattering of electrons by solitons. Soltions are placed periodically at T = 0K, but the lattice structure is distorted by thermal fluctuations as the temperature increases. The data are taken from the doctor thesis (March, 2005) of D. Terasawa. See also Fig.22 in the frontispiece and a similar work due to Fukuda et al.[181]

ers homogeneously. However, it penetrates as sine-Gordon solitons[139] in the IC phase. As $B_{\parallel}^{*}$ increases, the number of solitons increases rapidly and they create a soliton lattice[4,222] for $B_{\parallel} > B_{\parallel}^{*}$, with $B_{\parallel}^{*}$ the C-IC phase transition point. At the zero temperature the soliton lattice is a perfectly periodic system, where electrons propagate as a Bloch wave without reflection. However, the periodicity is broken at finite temperature, as would lead to a nontrivial transmission coefficient. Hence, we expect an anomalous increase of the diagonal resistance R_{xx} due to the backscattering of electrons against a thermally fluctuating soliton lattice in the QH regime with a well-developed Hall plateau. This will the interpretation of the recent experimental result[181] on the anomalous increase of R_{xx} just beyond the phase transition point $B_{\parallel}^{*}$ as shown in Fig.21.9.

Canted Antiferromagnetic States

According to the one-body picture we expect to have two phases at $\nu = 2$ depending on the relative strength between the Zeeman gap Δ_Z and the tunneling gap Δ_{SAS}. One is the spin-ferromagnet and pseudospin-singlet phase (abridged as the spin phase) for $\Delta_Z > \Delta_{SAS}$; the other is the spin-singlet and pseudospin-ferromagnet phase (abridged as the ppin phase) for $\Delta_Z < \Delta_{SAS}$. They are distinguishable by the charge-transferable property. Between these two phases, driven

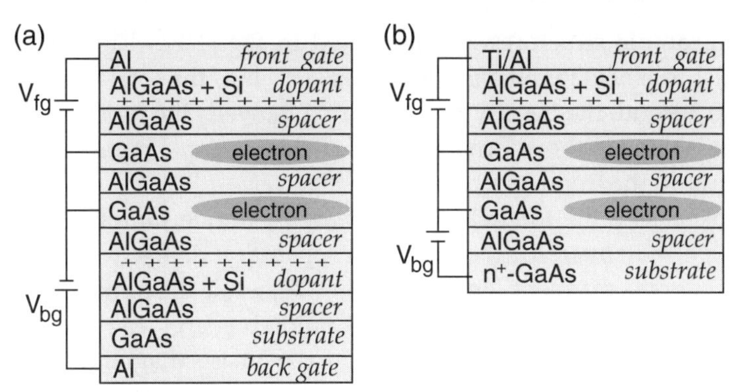

Fig. 21.10 Typical sample structures are illustrated. The tunneling gap Δ_{SAS} between the two quantum wells is tuned by the amount of aluminum in the spacer between the two electron layers. Electron density in each layer is controlled by applying the front or back gate voltage. (a) Electrons are provided by two remote δ-doped donor layers separated by AlGaAs spacers. Typical bias voltages are $V_{fg} \simeq -0.4$ V and $V_{bg} \simeq -40$ V. This type of sample is used in the experiments.[423,424] (b) Electrons in the front layer are provided by a remote δ-doped donor layer separated by an AlGaAs spacer, while those in the back layer are induced from the substrate.[358] Typical bias voltages are $V_{fg} \simeq -0.4$ V and $V_{bg} \simeq 0.8$ V. It is easy to control the density difference.

by interlayer correlations, a novel canted antiferromagnetic phase (abridged as the canted phase) has been predicted to emerge.[537,420] The first experimental indication of a phase transition at $\nu = 2$ was revealed by inelastic light scattering spectroscopy[388]. Subsequently, capacitance spectroscopy[269] as well as magnetotransport measurements[423,424,183,180] followed to establish these three phases.

Bilayer QH Junctions

Driven by the interlayer exchange interaction, the interlayer coherence develops on charge-transferable states. Charge transfer occurs coherently between the two layers. The interlayer coherence is affected by the parallel magnetic field, as is well known in the physics of superconductivity.[32,13] It brings about new features not present in the intralayer coherence (spin coherence).

(A) The bilayer QH system closely resembles the superconductor Josephson junction,[140] though the Meissner effect does not exit.[519,143]

(C) The pseudospin wave is quantized as *plasmons*,[142] just as the spin wave is quantized as magnons. Indeed, the interlayer coherence supports a resonating excitation of charge transfer.

(D) The Hall current may tunnel as a dc Josephson current between the two layers,[140] which seems to be verified experimentally[454] though still somewhat controversial.[61,457,174]

Experimental Approaches

Typical sample structures are illustrated in Fig.21.10. Three standard techniques exist to elucidate bilayer QH states: (I) First, the total electron density is varied. (II) Second, the density ratio is varied between the two layers. (III) Third, the sample is tilted in the external magnetic field. We explain physics behind them, using figures in the frontispiece.

- **(I)** By decreasing the total electron density ρ_0 in one sample at a fixed filling factor $\nu = 2\pi\hbar\rho_0/eB_\perp$, the interlayer separation d effectively decreases with respect to the magnetic length, $d/\ell_B \propto \sqrt{\rho_0}$, while the tunneling energy effectively increases with respect to the Coulomb energy E_C^0, $\Delta_{SAS}/E_C^0 \propto 1/\sqrt{\rho_0}$. The QH state moves along a line in the phase diagram, as indicated in Fig.16(a) for $\nu = 1$ (page xx) and in Fig.17(a) for $\nu = 2$ (page xxi).
- **(II)** The density difference is controlled between the two quantum wells by applying bias voltages. As the number density is imbalanced, the compound and locked states become unstable, but charge-transferable states (coherent and resonant states) remain stable. In the data [Fig.11(a) (page xvi)], the $\nu = 1$ state is stable as the density imbalance σ increases, but the $\nu = 2$ and $\nu = 2/3$ states become unstable off the balanced point ($\sigma = 0$). See also Figs.16 (page xx) and 17 (page xxi) for experimental data at $\nu = 1$ and $\nu = 2$. Further, a quasiparticle is an SU(4) skyrmions at $\nu = 1$, which may be continuously deformed from a pseudospin SU(2) texture at the balanced point to a spin SU(2) texture at the monolayer limit by controlling bias voltages
- **(III)** The tilted-field method is the standard way to enhance the Zeeman effect at a fixed filling factor. It has additional effects in the bilayer system. On one hand, the parallel magnetic field $B_\parallel$ penetrates between the two layers and reduces the one-body tunneling energy.[234] On the other hand, it affects the interlayer coherence, and decreases the activation energy anomalously.[359] For instance, see Fig.16 (page xx) for experimental data at $\nu = 1$. Since the $\nu = 1$ state is a ferromagnet [Fig.14 (page xviii)], as the sample is tilted, the activation energy is naively expected to increase as in the monolayer QH ferromagnet. On the contrary, it decreases anomalously [Fig.21.8] toward the C-IC phase transition point. The penetrated field $B_\parallel$ forms a soliton lattice in the IC phase [Fig.22 (page xxiv)].

CHAPTER 22

SU(2) PSEUDOSPIN STRUCTURE

The SU(2) pseudospin structure is introduced into the bilayer system by assigning "up" and "down" pseudospins to the electrons on the front and back layers, respectively. The electron configuration is locally described by the total number density $\rho(\boldsymbol{x})$ and the pseudospin density $\boldsymbol{P}(\boldsymbol{x})$. The bilayer QH state belongs to a pseudospin multiplet, where the pseudospin direction is controlled by applying bias voltages. The state belongs to a discrete spectrum or a continuous spectrum.

22.1 BILAYER PLANAR ELECTRONS

We study bilayer electron systems in the spin-frozen theory. The spin degree of freedom is recovered in Sections 22.7. It is instructive to consider a double-well potential made of two δ-functions,

$$V(z) = -V_0 \left\{ \delta\left(z - \frac{1}{2}d\right) + \delta\left(z + \frac{1}{2}d\right) \right\}, \tag{22.1.1}$$

as a generalization of the δ-function potential (10.1.3) for the monolayer system, where d is the interlayer distance. The wave function of one electron is factorized, $\mathfrak{S}(\boldsymbol{x}, z) = \mathfrak{S}(\boldsymbol{x})\mathfrak{S}_0(z)$, as in (10.1.1).

When the two layers are sufficiently separated, an electron is localized either in the front or back layer as in (10.1.4). The wave functions are

$$\mathfrak{S}_0^{\mathrm{f}}(z) = \sqrt{\beta} e^{-\beta|z-d/2|}, \qquad \mathfrak{S}_0^{\mathrm{b}}(z) = \sqrt{\beta} e^{-\beta|z+d/2|}, \tag{22.1.2}$$

where $\beta = MV_0/\hbar^2$ with the electron mass M.

The kinetic term H_{K} is

$$H_{\mathrm{K}} = \frac{1}{2M} \int d^2x dz\, \psi^\dagger(\boldsymbol{x}, z)[(P_x - iP_y)(P_x + iP_y) + (P_z)^2]\psi(\boldsymbol{x}, z) + \frac{N}{2}\hbar\omega_{\mathrm{c}}, \tag{22.1.3}$$

where[1] $P_k = -i\hbar\partial_k + eA_k^{\mathrm{ext}}$. Since electrons are confined within the front or back

[1]We assume there is no parallel magnetic field. The parallel magnetic field induces various novel phenomena, which we discuss in Section 25.6 and Chapter 27.

layers, we may approximate this as

$$H_{\rm K} = \frac{1}{2M} \sum_{\alpha={\rm f,b}} \int d^2x \psi_\alpha^\dagger(\boldsymbol{x})(P_x - iP_y)(P_x + iP_y)\psi_\alpha(\boldsymbol{x}) + \frac{N}{2}\hbar\omega_{\rm c}, \tag{22.1.4}$$

where the field $\psi_\alpha(\boldsymbol{x})$ describes electrons at the layer $\alpha(= {\rm f,b})$. The kinetic term $H_{\rm K}$ creates Landau levels.

The LLL condition reads

$$(P_x + iP_y)\psi_\alpha(\boldsymbol{x})|\mathfrak{S}\rangle = 0, \tag{22.1.5}$$

corresponding to (10.6.5) in the monolayer system. The projected field operator is given by

$$\psi_\alpha(\boldsymbol{x}) \equiv \sum_{n=0}^{\infty} c_\alpha(n)\mathfrak{S}_n(\boldsymbol{x}), \tag{22.1.6}$$

corresponding to (10.6.22) in the monolayer system. Here, $c_\alpha(n)$ is the annihilation operator of the electron with wave function $\mathfrak{S}_n(\boldsymbol{x}) = \langle \boldsymbol{x}|n\rangle$ at Landau site $|n\rangle$ in the front-layer for α =f or in the back-layer for α =b, obeying

$$\{c_\alpha(n), c_\beta^\dagger(m)\} = \delta_{nm}\delta_{\alpha\beta}, \quad \{c_\alpha(n), c_\beta(m)\} = \{c_\alpha^\dagger(n), c_\beta^\dagger(m)\} = 0. \tag{22.1.7}$$

The analysis is considerably complicated in terms of the projected operator (22.1.6). Here we employ an effective theory derived from the microscopic formalism with the use of the projected operator presented in Part IV.

The Coulomb interaction term $H_{\rm C}$ is given by (10.1.7), or

$$H_{\rm C} = \frac{1}{2}\int d^2x dz d^2x' dz' \frac{\rho_e(\boldsymbol{x},z)\rho_e(\boldsymbol{x}',z')}{4\pi\varepsilon\sqrt{(\boldsymbol{x}-\boldsymbol{x}')^2 + (z-z')^2}}, \tag{22.1.8}$$

where $\rho_e(\boldsymbol{x},z)$ is the electric charge density. Because electrons are confined within the front and back layers, we have

$$H_{\rm C} = \frac{1}{2}\sum_{\alpha,\beta}\int d^2x d^2y\, V^{\alpha\beta}(\boldsymbol{x}-\boldsymbol{y})\rho^\alpha(\boldsymbol{x})\rho^\beta(\boldsymbol{y}) \tag{22.1.9}$$

with $\rho^\alpha(\boldsymbol{x}) \equiv \psi_\alpha^\dagger(\boldsymbol{x})\psi_\alpha(\boldsymbol{x})$ and

$$V^{\alpha\beta}(\boldsymbol{x}) = \frac{e^2}{4\pi\varepsilon}\frac{1}{\sqrt{|\boldsymbol{x}|^2 + d_{\alpha\beta}^2}}, \tag{22.1.10}$$

where $d_{\rm ff} = d_{\rm bb} = 0$ and $d_{\rm fb} = d_{\rm bf} = d$. For later convenience we define

$$V^\pm(\boldsymbol{x}) = \frac{1}{2}\left(V^{\rm ff}(\boldsymbol{x}) \pm V^{\rm fb}(\boldsymbol{x})\right). \tag{22.1.11}$$

Their Fourier transforms read

$$V^{\rm ff}(\boldsymbol{q}) = V^{\rm bb}(\boldsymbol{q}) = \frac{e^2}{4\pi\varepsilon|\boldsymbol{q}|}, \qquad V^{\rm fb}(\boldsymbol{q}) = V^{\rm bf}(\boldsymbol{q}) = \frac{e^2}{4\pi\varepsilon|\boldsymbol{q}|}e^{-|\boldsymbol{q}|d}, \tag{22.1.12}$$

and

$$V^{\pm}(\boldsymbol{q}) = \frac{e^2}{8\pi\varepsilon|\boldsymbol{q}|}\left(1 \pm e^{-|\boldsymbol{q}|d}\right). \tag{22.1.13}$$

In the QH system electrons are provided from dopants in the remote donor layers [Fig.21.10]. When their contribution is taken into account, as is shown in Section 32.5, the Coulomb Hamiltonian becomes

$$H_C = \frac{1}{2}\sum_{\alpha,\beta}\int d^2x d^2y\, V^{\alpha\beta}(\boldsymbol{x}-\boldsymbol{y})\Delta\rho^{\alpha}(\boldsymbol{x})\Delta\rho^{\beta}(\boldsymbol{y}), \tag{22.1.14}$$

where

$$\Delta\rho^{\alpha}(\boldsymbol{x}) = \rho^{\alpha}(\boldsymbol{x}) - \rho_0^{\alpha} \tag{22.1.15}$$

is the modulation of the electron density around its mean value ρ_0^{α} in each layer.

22.2 Pseudospins

It is convenient to introduce the concept of pseudospins. We assign "up" and "down" pseudospins ($\mathscr{P}_z = \pm\frac{1}{2}$) to the electrons belonging to the front and back layers, respectively, and construct a two-component electron field Ψ,

$$\Psi = \begin{pmatrix} \psi_f \\ \psi_b \end{pmatrix}. \tag{22.2.1}$$

The pseudospin density operator is defined by

$$\mathcal{P}_x(\boldsymbol{x}) = \Psi^{\dagger}\frac{\tau_x}{2}\Psi = \frac{1}{2}(\psi_f^{\dagger}\psi_b + \psi_b^{\dagger}\psi_f), \tag{22.2.2a}$$

$$\mathcal{P}_y(\boldsymbol{x}) = \Psi^{\dagger}\frac{\tau_y}{2}\Psi = \frac{1}{2i}(\psi_f^{\dagger}\psi_b - \psi_b^{\dagger}\psi_f), \tag{22.2.2b}$$

$$\mathcal{P}_z(\boldsymbol{x}) = \Psi^{\dagger}\frac{\tau_z}{2}\Psi = \frac{1}{2}(\psi_f^{\dagger}\psi_f - \psi_b^{\dagger}\psi_b) = \frac{1}{2}(\rho^f - \rho^b) \tag{22.2.2c}$$

in analogy to the spin density operator (15.2.7). They are rewritten as

$$\mathcal{P}_x(\boldsymbol{x}) = \frac{1}{2}(\psi_S^{\dagger}\psi_S - \psi_A^{\dagger}\psi_A), \tag{22.2.3a}$$

$$\mathcal{P}_y(\boldsymbol{x}) = \frac{i}{2}(\psi_S^{\dagger}\psi_A - \psi_A^{\dagger}\psi_S), \tag{22.2.3b}$$

$$\mathcal{P}_z(\boldsymbol{x}) = \frac{1}{2}(\psi_S^{\dagger}\psi_A + \psi_A^{\dagger}\psi_S) \tag{22.2.3c}$$

in terms of the symmetric and antisymmetric field operators defined by

$$\psi_S = \frac{1}{\sqrt{2}}(\psi_f + \psi_b), \qquad \psi_A = \frac{1}{\sqrt{2}}(\psi_f - \psi_b). \tag{22.2.4}$$

The physical meaning of the component $\mathcal{P}_z$ is the density difference (22.2.2c) between the two layers. To see the physical meaning of P_x and P_y we define

$$\mathcal{P}^+(\boldsymbol{x}) \equiv \mathcal{P}_x(\boldsymbol{x}) + i\mathcal{P}_y(\boldsymbol{x}) = \psi_\mathrm{f}^\dagger \psi_\mathrm{b},$$
$$\mathcal{P}^-(\boldsymbol{x}) \equiv \mathcal{P}_x(\boldsymbol{x}) - i\mathcal{P}_y(\boldsymbol{x}) = \psi_\mathrm{b}^\dagger \psi_\mathrm{f}. \tag{22.2.5}$$

The operator $\mathcal{P}^+(\boldsymbol{x})$ $[\mathcal{P}^-(\boldsymbol{x})]$ transfers one electron at $\boldsymbol{x}$ from the back (front) layer to the front (back) layer.

The physical variables are the pseudospin density $\mathcal{P}_a(\boldsymbol{x})$ as well as the total number density

$$\rho(\boldsymbol{x}) = \Psi^\dagger\Psi = \psi_\mathrm{f}^\dagger\psi_\mathrm{f} + \psi_\mathrm{b}^\dagger\psi_\mathrm{b} = \rho^\mathrm{f}(\boldsymbol{x}) + \rho^\mathrm{b}(\boldsymbol{x}). \tag{22.2.6}$$

They satisfy the commutation relations

$$[\mathcal{P}_a(\boldsymbol{x}), \mathcal{P}_b(\boldsymbol{y})] = i\delta(\boldsymbol{x}-\boldsymbol{y})\varepsilon_{abc}\mathcal{P}_c(\boldsymbol{x}) \tag{22.2.7}$$

and

$$[\rho(\boldsymbol{x}), \mathcal{P}_a(\boldsymbol{y})] = 0. \tag{22.2.8}$$

The algebra is U(1)⊗SU(2).

We can define the normalized pseudospin field $\mathscr{P}_a$ in the present effective theory by

$$\mathcal{P}_a(\boldsymbol{x}) = \nu^{-1}\rho(\boldsymbol{x})\mathscr{P}_a(\boldsymbol{x}), \tag{22.2.9}$$

since $\rho(\boldsymbol{x})$ and $\mathcal{P}_a(\boldsymbol{y})$ commute. This is not the case in the microscopic theory for electrons in the lowest Landau level, as we extensively study in Part IV. However, to analyze the Goldstone mode we can impose the incompressibility condition, $\rho(\boldsymbol{x}) = \rho_0$, with which we have

$$\mathcal{P}_a(\boldsymbol{x}) = \nu^{-1}\rho_0\mathscr{P}_a(\boldsymbol{x}) = \rho_\Phi\mathscr{P}_a(\boldsymbol{x}). \tag{22.2.10}$$

The density modulation occurs only in association with quasiparticle excitations, for which (22.2.9) is a good approximation.

The Coulomb term (22.1.14) is decomposed into two terms,

$$H_\mathrm{C} = H_\mathrm{C}^+ + H_\mathrm{C}^-, \tag{22.2.11}$$

where

$$H_\mathrm{C}^+ = \frac{1}{2}\int d^2x d^2y\, V^+(\boldsymbol{x}-\boldsymbol{y})\Delta\rho(\boldsymbol{x})\Delta\rho(\boldsymbol{y}), \tag{22.2.12a}$$
$$H_\mathrm{C}^- = 2\int d^2x d^2y\, V^-(\boldsymbol{x}-\boldsymbol{y})\Delta\mathcal{P}_z(\boldsymbol{x})\Delta\mathcal{P}_z(\boldsymbol{y}). \tag{22.2.12b}$$

The Coulomb term H_C^+ is invariant under the global SU(2) transformation. The Coulomb term H_C^- describes the capacitance energy of the bilayer system, as we shall see in Section 22.5. It breaks the global SU(2) invariance. The potential V^-

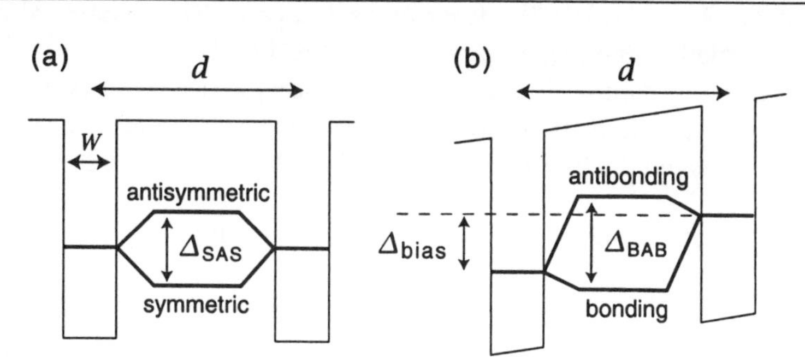

Fig. 22.1 One-body energy levels are depicted in a double quantum well structure. The δ-potential limit (22.1.1) is obtained by letting the width $W \to 0$ with the area of the potential well fixed finite. (a) The tunneling interaction generates the gap energy Δ_{SAS} between the symmetric state and the antisymmetric state. (b) The gap energy is increased as $\Delta_{BAB} = \sqrt{\Delta_{SAS}^2 + \Delta_{bias}^2}$ when the bias parameter Δ_{bias} is increased. The symmetric and antisymmetric states are turned into the bonding and antibonding states for $\Delta_{bias} \neq 0$.

becomes strongest ($V^- = V^+$) when there is no interlayer Coulomb interaction ($d \to \infty$), while it vanishes ($V^- = 0$) when the interlayer Coulomb interaction is maximum ($d \to 0$).

22.3 Tunneling Interaction

When the two layers are placed close enough, the wave functions of two electrons belonging to different layers have a nontrivial overlap. The Schrödinger equation is easily solved provided the Coulomb interaction is neglected. There arise two energy levels [Fig.22.1(a)]. The ground state is a symmetric state well approximated by

$$\mathfrak{S}_0^{S}(z) = \frac{1}{\sqrt{2}}\left\{\mathfrak{S}_0^{f}(z) + \mathfrak{S}_0^{b}(z)\right\}, \tag{22.3.1}$$

while the excited state is an antisymmetric state well approximated by

$$\mathfrak{S}_0^{A}(z) = \frac{1}{\sqrt{2}}\left\{\mathfrak{S}_0^{f}(z) - \mathfrak{S}_0^{b}(z)\right\}, \tag{22.3.2}$$

where $\mathfrak{S}_0^{\alpha}(z)$ is given by (22.1.2). The excitation gap is calculated as

$$\Delta_{SAS} \simeq \langle \mathfrak{S}_0^{A}|V|\mathfrak{S}_0^{A}\rangle - \langle \mathfrak{S}_0^{S}|V|\mathfrak{S}_0^{S}\rangle = \frac{MV_0^2}{\hbar^2}\exp\left[-\frac{Md}{\hbar^2}|V_0|\right] \tag{22.3.3}$$

with (22.1.1). An electron is described by the wave function $\mathfrak{S}(\boldsymbol{x})\mathfrak{S}_0^{S}(z)$ or $\mathfrak{S}(\boldsymbol{x})\mathfrak{S}_0^{A}(z)$ in the symmetric or antisymmetric state, or the wave function

$\mathfrak{S}(\boldsymbol{x})\mathfrak{S}_0^{\rm f}(z)$ or $\mathfrak{S}(\boldsymbol{x})\mathfrak{S}_0^{\rm b}(z)$ in the front or back layers. The dynamical degree of freedom is carried by the planar wave function $\mathfrak{S}(\boldsymbol{x})$.

Electrons in the symmetric and antisymmetric states are described by the symmetric and antisymmetric fields (22.2.4), respectively. The existence of the gap energy $\Delta_{\rm SAS}$ between the two states is summarized into the tunneling term

$$H_{\rm T} = \frac{1}{2}\Delta_{\rm SAS}\int d^2x\, (\psi_{\rm A}^\dagger\psi_{\rm A} - \psi_{\rm S}^\dagger\psi_{\rm S}). \tag{22.3.4}$$

This is also written as

$$H_{\rm T} = -\frac{1}{2}\Delta_{\rm SAS}\int d^2x\, (\psi_{\rm f}^\dagger\psi_{\rm b} + \psi_{\rm b}^\dagger\psi_{\rm f}) = -\Delta_{\rm SAS}\int d^2x\, \mathcal{P}_x(\boldsymbol{x}) \tag{22.3.5}$$

in terms of the pseudospin operator.

22.4 Imbalanced Configuration

Applying gate bias voltages to the front and back layers, we can control the electron number densities in the two layers independently [Fig.21.10]. To create an imbalanced configuration we introduce the bias term[2],

$$H_{\rm bias} = -\frac{1}{2}\Delta_{\rm bias}\left(N^{\rm f} - N^{\rm b}\right) = -\Delta_{\rm bias}\int d^2x\, \mathcal{P}_z(\boldsymbol{x}). \tag{22.4.1}$$

The sum of the tunneling and bias terms is

$$H_{\rm pZ} \equiv H_{\rm T} + H_{\rm bias} = -\int d^2x\, [\Delta_{\rm SAS}\mathcal{P}_x(\boldsymbol{x}) + \Delta_{\rm bias}\mathcal{P}_z(\boldsymbol{x})]. \tag{22.4.2}$$

The zero-momentum component is

$$H_{\rm pZ}^0 = -N_\Phi\,(\Delta_{\rm SAS}\mathscr{P}_x + \Delta_{\rm bias}\mathscr{P}_z) = -N_\Phi\boldsymbol{\mathscr{B}}\cdot\boldsymbol{\mathscr{P}}, \tag{22.4.3}$$

with $\boldsymbol{\mathscr{B}} = (\Delta_{\rm SAS}, 0, \Delta_{\rm bias})$ and (21.1.11) for $\boldsymbol{\mathscr{P}}$. The Hamiltonian $H_{\rm pZ}$ describes the "pseudo-Zeeman effect "with $\boldsymbol{\mathscr{B}}$ the pseudomagnetic field. The pseudomagnetic field $\boldsymbol{\mathscr{B}}$ has the magnitude

$$\Delta_{\rm BAB} \equiv \sqrt{\Delta_{\rm SAS}^2 + \Delta_{\rm bias}^2}. \tag{22.4.4}$$

The pseudomagnetic field originally points to the x axis when $\Delta_{\rm bias} = 0$ but is "tilted" by increasing the bias parameter ($\Delta_{\rm bias} \neq 0$). It affects the pseudospin orientation via the pseudo-Zeeman term (22.4.3).

The pseudo-Zeeman term $H_{\rm pZ}$ is expressed as

$$H_{\rm pZ} = \frac{1}{2}\int d^2x\, (\psi_{\rm f}^\dagger, \psi_{\rm b}^\dagger)\begin{pmatrix} -\Delta_{\rm bias} & \Delta_{\rm SAS} \\ \Delta_{\rm SAS} & \Delta_{\rm bias}\end{pmatrix}\begin{pmatrix}\psi_{\rm f} \\ \psi_{\rm b}\end{pmatrix}. \tag{22.4.5}$$

[2]The bias parameter $\Delta_{\rm bias}$ is determined by the gate voltages applied to the front and back layers. However, it has nothing to do with the potential difference between the front and back layers. See Section 32.5 for the precise mechanism how the bias term $H_{\rm bias}$ arises.

The matrix is diagonalized

$$T_\sigma^\dagger \begin{pmatrix} -\Delta_{\text{bias}} & \Delta_{\text{SAS}} \\ \Delta_{\text{SAS}} & \Delta_{\text{bias}} \end{pmatrix} T_\sigma = \begin{pmatrix} -\Delta_{\text{BAB}} & 0 \\ 0 & \Delta_{\text{BAB}} \end{pmatrix}, \tag{22.4.6}$$

with the use of a unitary matrix

$$T_\sigma = \frac{1}{\sqrt{2}} \begin{pmatrix} \sqrt{1+\sigma_0} & \sqrt{1-\sigma_0} \\ \sqrt{1-\sigma_0} & -\sqrt{1+\sigma_0} \end{pmatrix}, \tag{22.4.7}$$

where σ_0 is given by

$$\Delta_{\text{bias}} = \frac{\sigma_0}{\sqrt{1-\sigma_0^2}} \Delta_{\text{SAS}}. \tag{22.4.8}$$

Substituting (22.4.8) into (22.4.4) we find

$$\Delta_{\text{BAB}} = \frac{1}{\sqrt{1-\sigma_0^2}} \Delta_{\text{SAS}}. \tag{22.4.9}$$

The pseudo-Zeeman term (22.4.5) is diagonalized,

$$H_{\text{pZ}} = \frac{1}{2} \Delta_{\text{BAB}} \int d^2x \, \left\{ \psi_{\text{A}}^\dagger(\boldsymbol{x}) \psi_{\text{A}}(\boldsymbol{x}) - \psi_{\text{B}}^\dagger(\boldsymbol{x}) \psi_{\text{B}}(\boldsymbol{x}) \right\}, \tag{22.4.10}$$

where we have introduced the bonding and antibonding states.

$$\psi_{\text{B}}(\boldsymbol{x}) = \sqrt{\frac{1+\sigma_0}{2}} \psi_{\text{f}}(\boldsymbol{x}) + \sqrt{\frac{1-\sigma_0}{2}} \psi_{\text{b}}(\boldsymbol{x}), \tag{22.4.11a}$$

$$\psi_{\text{A}}(\boldsymbol{x}) = \sqrt{\frac{1-\sigma_0}{2}} \psi_{\text{f}}(\boldsymbol{x}) - \sqrt{\frac{1+\sigma_0}{2}} \psi_{\text{b}}(\boldsymbol{x}). \tag{22.4.11b}$$

There arises a one-body energy gap Δ_{BAB} between the bonding and antibonding states [Fig.22.1 (b)]. The bonding and antibonding states (22.4.11) regress to the symmetric and antisymmetric fields (22.2.4) at $\sigma_0 = 0$.

On the other hand, with the use of (22.4.8), the pseudo-Zeeman term (22.4.2) is given by

$$H_{\text{pZ}} = -\Delta_{\text{SAS}} \int d^2x \left[\mathcal{P}_x(\boldsymbol{x}) + \frac{\sigma_0}{\sqrt{1-\sigma_0^2}} \mathcal{P}_z(\boldsymbol{x}) \right] \tag{22.4.12}$$

as a function of the parameter σ_0. This is a more useful formula than (22.4.2) since the parameter σ_0 is more physical than the parameter Δ_{bias}. See also Section 24.2.

The physical meaning of the parameter σ_0 reads as follows. The ground state is the bonding state, which electrons begin to fill first. We evaluate the electron numbers in the front and back layers when all electrons are in the bonding state. It follows from (22.4.11) that

$$\langle \rho^{\text{f}} \rangle = \frac{1+\sigma_0}{2} \langle \rho \rangle, \qquad \langle \rho^{\text{b}} \rangle = \frac{1-\sigma_0}{2} \langle \rho \rangle, \tag{22.4.13}$$

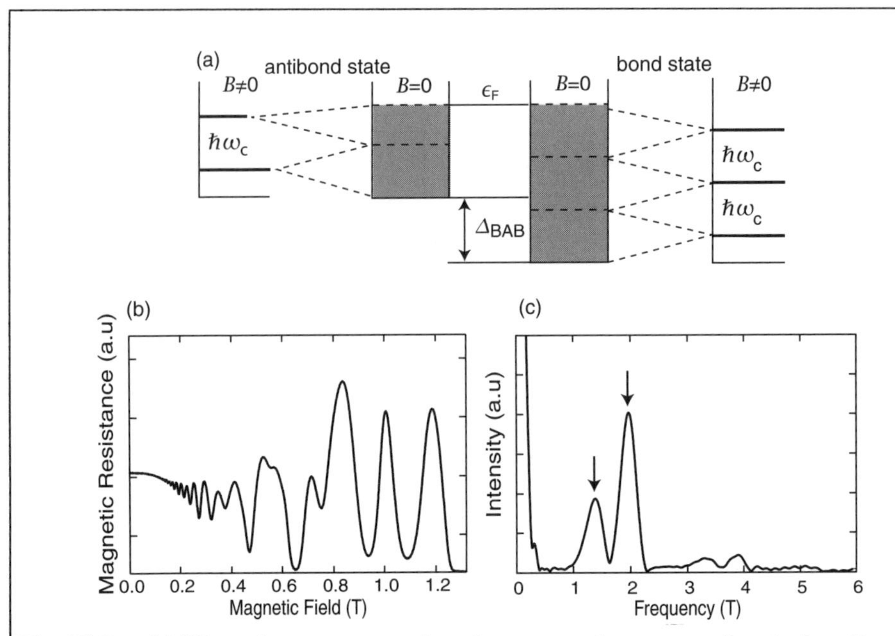

Fig. 22.2 (a) When electrons are put into the system, they occupy first the bonding state until the Fermi energy reaches the gap energy $\Delta_{\rm BAB}$, and then the bonding and antibonding states equally. The number density in each state is experimentally measured as follows. (b) We measure the diagonal resistance R_{xx} in a weak magnetic field B by changing B and keeping the density ρ_0 fixed, as exhibits Schubnikov-de Haas oscillations. (c) The electron densities in the two states are determined separately by making Fourier analysis of the oscillations. A Fourier analysis yields a double periodic behavior in the resistance corresponding to the two layers. Now, in each layer the resistance exhibits a periodic behavior as a function of $1/B$ (the Schubnikov-de Haas effect): Indeed, it vanishes ($R_{xx} = 0$) each time when one Landau level is filled up ($\nu \equiv 2\pi\rho_0\ell_B^2 =$ integer). The density ρ_0 is determined from the periodicity $\Delta(1/B) = e/(2\pi\hbar\rho_0)$.

or

$$\sigma_0 = \frac{\langle\rho^{\rm f}\rangle - \langle\rho^{\rm b}\rangle}{\langle\rho^{\rm f}\rangle + \langle\rho^{\rm b}\rangle}. \tag{22.4.14}$$

In general we define

$$\sigma_0 = \frac{\rho_0^{\rm f} - \rho_0^{\rm b}}{\rho_0^{\rm f} + \rho_0^{\rm b}} = \frac{N^{\rm f} - N^{\rm b}}{N^{\rm f} + N^{\rm b}}, \tag{22.4.15}$$

where $N^{\rm f}$ and $N^{\rm b}$ are the electron numbers in the front and back layers. Namely, it represents the normalized density difference between the two layer. We call it the imbalance parameter. It is also expressed as

$$\sigma_0 = 2\nu^{-1}\mathcal{P}_z, \tag{22.4.16}$$

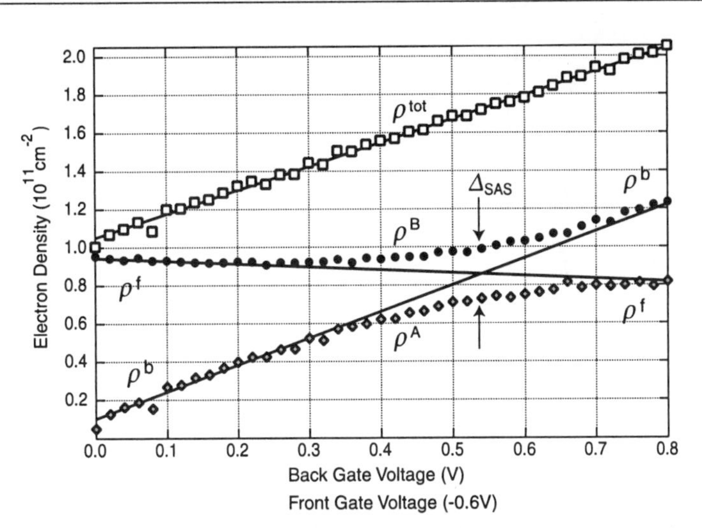

Fig. 22.3 Electron densities in the bonding and antibonding states are determined by making Fourier analysis of Schubnikov-de Haas oscillations, as explained in the caption of Fig.22.2. The tunneling gap is given by $\Delta_{SAS} = (\pi\hbar^2/M)\delta\rho_0$, where $\delta\rho_0$ is the minimum value of the difference $\rho_0^B - \rho_0^A$. It reads that $\Delta_{SAS} \simeq 11$ K. The data are taken from Muraki et al.[358]

where $\mathcal{P}_z$ is the pseudospin per Landau site (21.1.11).

We describe how to measure experimentally the tunneling gap Δ_{SAS}. Let us put electrons into the system. Electrons occupy first the bonding state until the Fermi energy reaches this gap, and then occupy the bonding and antibonding states equally [Fig.22.2(a)]. See Fig.10.2 (page 180) and Fig.10.3 (page 182) to recall how electrons occupy quantum states in the planar system when $B = 0$ and $B \neq 0$. Since the density of states is $D(\varepsilon) = M/(\pi\hbar^2)$ when $B = 0$, the gap energy is given by

$$\Delta_{BAB} = \sqrt{\Delta_{SAS}^2 + \Delta_{bias}^2} = \frac{\pi\hbar^2}{M}\left(\rho_0^B - \rho_0^A\right) \tag{22.4.17}$$

in terms of the density difference, where the spin degree of freedom has been taken into account: See (10.1.22). On the other hand, the carrier densities ρ_0^α are determined from the Fourier transform of Schubnikov-de Haas oscillations in a weak magnetic field [Fig.22.2(b)]. In particular, Δ_{SAS} is given by Δ_{BAB} at $V_{bias} = 0$, or

$$\Delta_{SAS} = \frac{\pi\hbar^2}{M}\delta\rho_0, \tag{22.4.18}$$

where $\delta\rho_0$ is the minimum value of the difference between ρ_0^B and ρ_0^A by scanning the bias voltage [Fig.22.3].

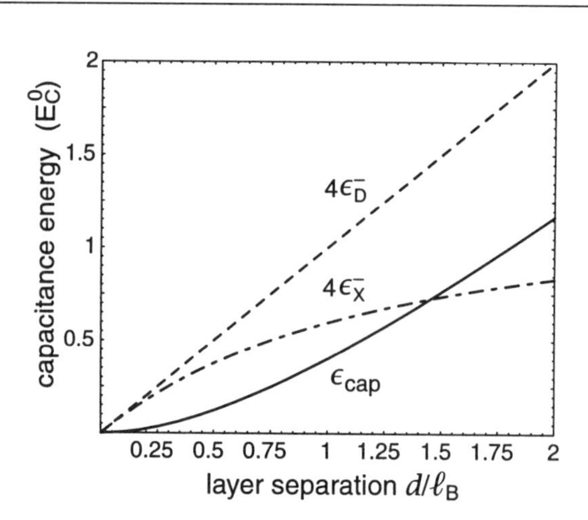

Fig. 22.4 The capacitance energy (ϵ_{cap}) consists of two contributions from the direct energy($\epsilon^0_{cap} = 4\epsilon^-_D$) and from the exchange energy ($4\epsilon^-_X$), $\epsilon_{cap} = 4(\epsilon^-_D - \epsilon^-_X)$. The former ($4\epsilon^-_D$) represents the capacitance energy of the condenser made of two planes with separation d as in (22.5.4). The interlayer coherence decreases this energy considerably for $d/\ell_B < 0.5$.

22.5 Capacitance Energy

The capacitance energy arises from the Coulomb interaction due to a density modulation around the ground state,

$$H_{cap} = \frac{\epsilon_{cap}}{\rho_\Phi} \int d^2x\, \Delta\mathcal{P}_z(\boldsymbol{x})\Delta\mathcal{P}_z(\boldsymbol{x}). \tag{22.5.1}$$

The parameter ϵ_{cap} would be estimated as follows. For a sufficiently smooth pseudospin configuration we make a derivative expansion by setting $\boldsymbol{y} = \boldsymbol{x} - \boldsymbol{z}$ in the Coulomb term (22.2.12b). Keeping the lowest-order term we obtain

$$H^-_C \simeq 2\int d^2y\, V^-(\boldsymbol{z}) \int d^2x\, \Delta\mathcal{P}_z(\boldsymbol{x})\Delta\mathcal{P}_z(\boldsymbol{x}), \tag{22.5.2}$$

where the equality holds when $\Delta P_z(\boldsymbol{x})$ is a constant.

Comparing this with (22.5.1), we would expect

$$\epsilon_{cap} = \epsilon^0_{cap} \equiv 2\rho_\Phi \int d^2z\, V^-(\boldsymbol{z}) = \frac{d}{\ell_B} E^0_C = \frac{e^2\rho_\Phi d}{2\varepsilon}, \tag{22.5.3}$$

and

$$H_{cap} \simeq \frac{\epsilon^0_{cap} S}{\rho_\Phi} (\Delta\mathcal{P}_z)^2 = \frac{Q^2}{2}\frac{d}{\varepsilon S}, \tag{22.5.4}$$

where $Q = e(\Delta\mathcal{P}_z)S$. This is the standard expression for the capacitance energy of the condenser made of two planes with separation d, with capacitance $C = \varepsilon S/d$.

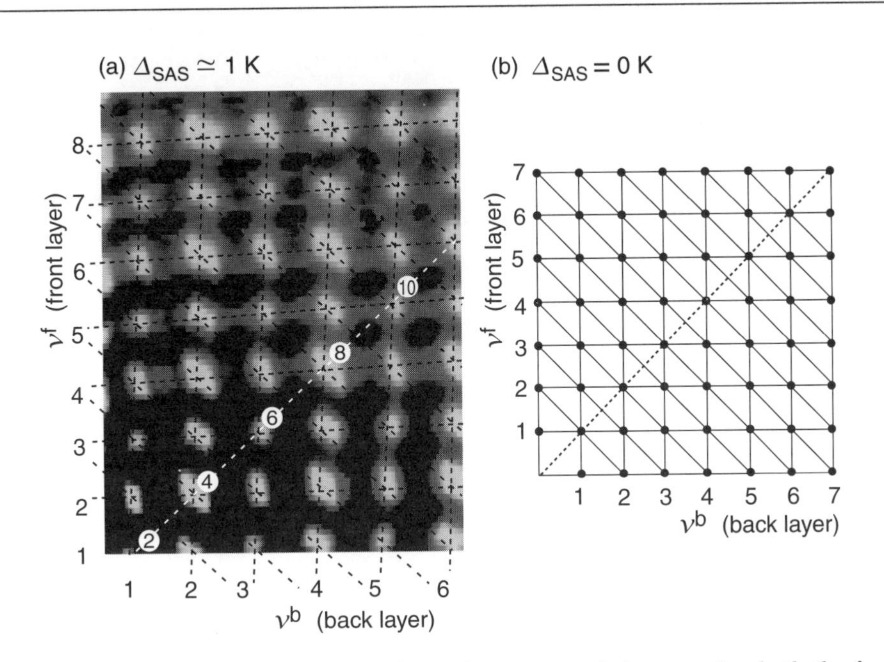

Fig. 22.5 The diagonal resistance is shown in a gray scale by scanning both the front and back gate voltages. The bright regions correspond to small resistance, indicating QH states in a sample with $\Delta_{SAS} \simeq 1$ K and $d = 23.1$ nm. (a) QH states are observed only on discrete lattice points (ν^b, ν^f), with integers ν^b and ν^f. They represent the filling factors in the back and front layers. In particular, no odd-integer QH states are seen at the balanced point for $\nu \geq 2$, i.e., along a white dashed line. The data are taken from Muraki et al.[357] (b) A schematic diagram is drawn for compound states (ν^b, ν^f) at $\Delta_{SAS} = 0$ K. See Fig.14 (page xviii) and Fig.15 (page xix) in the frontispiece for more details.

Actually, since the separation of the front and back layers is so small that the exchange Coulomb interaction becomes important. Taking account of it we obtain [see (30.5.14) and (32.5.17)]

$$\epsilon_{\text{cap}} = \epsilon^0_{\text{cap}} - 4\epsilon^-_X = \left\{ \frac{d}{\ell_B} - \sqrt{\frac{\pi}{2}} \left[1 - e^{d^2/2\ell_B^2} \text{erfc}\left(d/\sqrt{2}\ell_B \right) \right] \right\} E^0_C. \tag{22.5.5}$$

The interlayer coherence decreases the capacitance energy considerably for small separation ($d/\ell_B < 0.5$), as illustrated in Fig.22.4.

22.6 Compound States

There exists a trivial class of bilayer QH states when the tunneling interaction is small. They are compound states $|\nu^b, \nu^f\rangle$ made of the two monolayer states at $\nu = \nu^b$ and ν^f. When $\Delta_{SAS} = 0$ the ground state is a discrete eigenstate of the

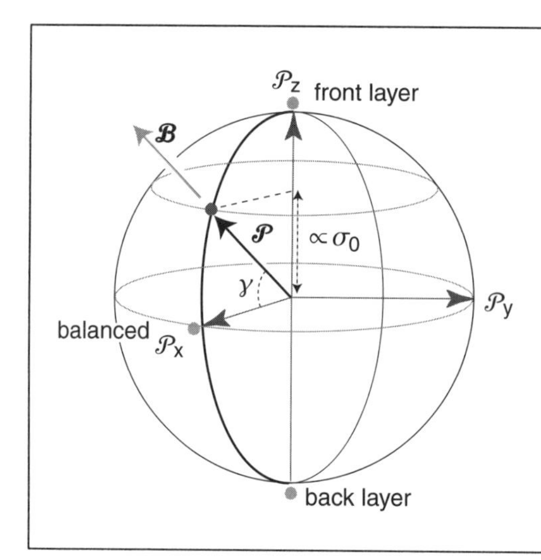

Fig. 22.6 The pseudomagnetic field $\boldsymbol{\mathcal{B}}$ aligns the pseudospin $\boldsymbol{\mathcal{P}}$; $\boldsymbol{\mathcal{P}} = (\mathcal{P}_x, \mathcal{P}_y, \mathcal{P}_z)$ and $\boldsymbol{\mathcal{B}} = (\Delta_{\text{SAS}}, 0, \Delta_{\text{bias}})$. The imbalance parameter is parametrized as $\sigma_0 = \sin\gamma$, where γ is the angle between the pseudospin $\boldsymbol{\mathcal{P}}$ and the x axis. A rotation of the pseudospin $\boldsymbol{\mathcal{P}}$ is induced by tilting the pseudomagnetic field $\boldsymbol{\mathcal{P}}$. This is the process of a charge transfer between the two layers by changing the bias voltage.

operator $\mathcal{P}_z$,

$$\mathcal{P}_z|\nu^{\text{b}}, \nu^{\text{f}}\rangle = \frac{1}{2}(\nu^{\text{f}} - \nu^{\text{b}})|\nu^{\text{b}}, \nu^{\text{f}}\rangle. \tag{22.6.1}$$

It was experimentally confirmed[357] in a sample with very small tunneling gap that QH states are observed precisely in discrete spectra of the pseudospin component $\mathcal{P}_z$ [see Fig.22.5].

Compound states are stable even in the presence of the tunneling interaction unless there exist other stabler states at the same filling factor. See Fig.15 (page xix) in the frontispiece for experimental data, where many compound states are seen though charge-transferable states are prominent at certain filling factors. In particular, a competition between a compound state (ferromagnet) and a charge-transferable state (spin singlet) has been studied carefully at $\nu = 2$ by magnetotransport experiments,[423,424] optical experiments[388] and capacitance spectroscopy.[269] See Fig.17 (page xxi) for experimental data. Detailed discussions on the $\nu = 2$ QH states are given in Chapter 26.

22.7 Charge-Transferable States

We recover the spin degree of freedom in this and the next sections. The $\nu = 1$ QH state is obtained by filling the up-spin bonding state. In general, integer QH states are constructed by filling one-body states successively as in the monolayer system. Fractional QH states are also constructed. The QH state at $\nu = 1$ is given by operating the bonding field operators (22.4.11) to the vacuum,

$$|g_{\text{B}}\rangle = \int [d\boldsymbol{x}]\mathfrak{S}_{\text{LN}}[\boldsymbol{x}]\psi^{B\uparrow\dagger}(\boldsymbol{x}_1)\psi^{B\uparrow\dagger}(\boldsymbol{x}_2)\cdots\psi^{B\uparrow\dagger}(\boldsymbol{x}_N)|0\rangle, \tag{22.7.1}$$

where $\mathfrak{S}_{\rm LN}[x]$ is the Laughlin wave function familiar in the monolayer system. When $\Delta_{\rm SAS} > \Delta_{\rm Z}$, the $\nu = 2$ level consists of the down-spin bonding state, the $\nu = 3$ level consists of the up-spin antibonding state, and the $\nu = 4$ level consists of the down-spin antibonding state [Fig.21.3 (page 373)].

The parameter σ_0 in the bonding and antibonding operators (22.4.11) represents the imbalance between the electron densities in the front and back layers as in (22.4.16). In the bonding state $|g_{\rm B}\rangle$, all electrons are moved to the back layer when $\sigma_0 = -1$, and to the front layer when $\sigma_0 = 1$, which is achieved by making $V_{\rm bias} \to \mp\infty$. Namely, the QH state at $\nu = 1$ is a charge-transferable state. It is interpreted that, as the pseudomagnetic field is tilted, the pseudospin rotates to minimize the pseudo-Zeeman energy [Fig.22.6]. The charge transfer between the two layers is a rotation of the pseudospin.

We classify all charge-transferable states. We investigate the ground state at an integer filling factor $\nu = n$. (The analysis is applicable to the fractional QH state at $\nu = n/m$ by considering composite particles.) It is convenient to use the pseudospin operator $\mathcal{P}$ acting on electrons in one Landau site. At $\nu = n$ it is decomposed [Fig.22.7] as $\mathcal{P} = \mathcal{P}_1 + \mathcal{P}_2 + \cdots + \mathcal{P}_n$, where $\mathcal{P}_i$ is the pseudospin in the ith energy level.

The ground state $|\jmath\rangle_{\rm g}$ is a simultaneous eigenstate of H_0 and $\mathcal{P}^2$,

$$H^0_{\rm pZ}|\jmath\rangle_{\rm g} = E_{\rm g}|\jmath\rangle_{\rm g}, \qquad \mathcal{P}^2|\jmath\rangle_{\rm g} = \jmath(\jmath+1)|\jmath\rangle_{\rm g}. \tag{22.7.2}$$

It is a linear superposition of the $(2\jmath+1)$ eigenstates $|\jmath, m\rangle$ of $\mathcal{P}_{\rm z}$,

$$|\jmath\rangle_{\rm g} = \sum_{m=-\jmath}^{\jmath} C_m(\sigma_0)|\jmath, m\rangle, \tag{22.7.3}$$

where

$$\mathcal{P}^2|\jmath, m\rangle = \jmath(\jmath+1)|\jmath, m\rangle, \qquad \mathcal{P}_{\rm z}|\jmath, m\rangle = m|\jmath, m\rangle, \tag{22.7.4}$$

with $\jmath = 0, \frac{1}{2}, 1, \frac{3}{2}, \cdots$, and $m = -\jmath, -\jmath+1, \cdots, \jmath-1, \jmath$. The ground state $|\jmath\rangle_{\rm g}$ moves from the state $|\jmath, -\jmath\rangle$ to $|\jmath, \jmath\rangle$ within the pseudospin $(2\jmath+1)$-multiplet, as σ_0 moves from -1 to 1, or $V_{\rm bias}$ from $-\infty$ to $+\infty$. The maximum number difference occurs when $\mathcal{P}_{\rm z} = \pm\jmath$, where the change of the $\mathcal{P}_{\rm z}$ component is

$$\Delta\mathcal{P}_{\rm z} = \jmath - (-\jmath) = 2\jmath. \tag{22.7.5}$$

This is the maximum charge transferability of the state $|\jmath\rangle_{\rm g}$.

Let us study the system explicitly at a few low filling factors. In the $\nu = 1$ state, where N_Φ is equal to the number N of electrons, the pseudospin $\mathcal{P}$ is identified with that of one electron. The QH state belongs to a pseudospin doublet $|\frac{1}{2}\rangle\!\rangle$: Two independent states are $|\frac{1}{2}, \frac{1}{2}\rangle$ and $|\frac{1}{2}, -\frac{1}{2}\rangle$, which correspond to the QH states where all electrons are transferred to the front and the back layers, respectively. The maximum charge transferability is $\Delta\mathcal{P}_{\rm z} = 1$.

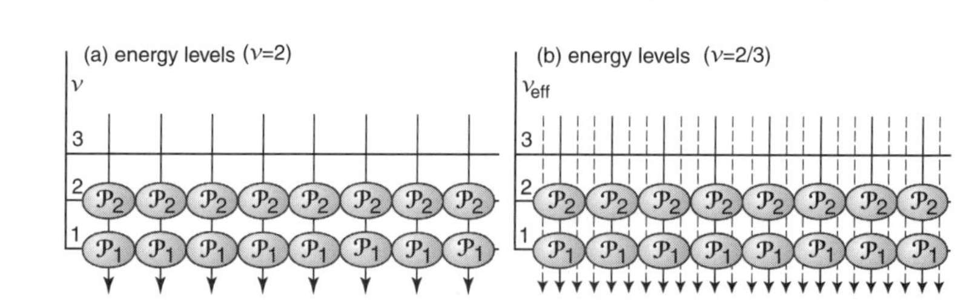

Fig. 22.7 (a) The pseudospin of the $\nu = 2$ QH state is characterized by the sum of the pseudospins of two electrons, $\mathcal{P} = \mathcal{P}_1 + \mathcal{P}_2$. (b) The pseudospin of the $\nu = 2/3$ QH state is also characterized by the sum of the pseudospins of two composite particles, $\mathcal{P} = \mathcal{P}_1 + \mathcal{P}_2$.

In the $\nu = 2$ state there are two electrons in one Landau site [Fig.22.7]. (At $\nu = 2/m$ there are two composite particles in one CF-Landau site.) The relevant pseudospin is $\mathcal{P} = \mathcal{P}_1 + \mathcal{P}_2$. Thus, the problem is reduced to the composition of two pseudospins,

$$\frac{1}{2} \otimes \frac{1}{2} = 0 \oplus 1. \tag{22.7.6}$$

The QH state belongs to either a pseudospin singlet $|0\rangle\!\rangle$ or a pseudospin triplet $|1\rangle\!\rangle$. The charge transfer is impossible in the singlet state where $\Delta\mathcal{P}_z = 0$, but is allowed in the triplet state with $\Delta\mathcal{P}_z = 2$. The singlet state is identified with the compound state. All electrons are transferred to one of the layers in the triplet state $|1, \pm 1\rangle$. See Chapter 26 for more details.

In the $\nu = 3$ state ($\nu = 3/m$ state) the problem is reduced to the composition of three pseudospins, $\mathcal{P} = \mathcal{P}_1 + \mathcal{P}_2 + \mathcal{P}_3$, or

$$\frac{1}{2} \otimes \frac{1}{2} \otimes \frac{1}{2} = \frac{1}{2} \oplus \frac{1}{2} \oplus \frac{3}{2}. \tag{22.7.7}$$

The QH state belongs to either a doublet $|\frac{1}{2}\rangle\!\rangle$ or a quartet $|\frac{3}{2}\rangle\!\rangle$. Only the doublet (partially spin polarized) is possible within the lowest Landau level [Fig.22.8], because three bonding states are necessary to construct a quartet. The charge transfer occurs in the doublet with $\Delta\mathcal{P}_z = 1$. The quartet is realizable as a charge-transferable state in samples with large Δ_{SAS}, as will be explained soon.

The $\nu = 4$ state ($\nu = 4/m$ state) belongs to either the singlet $|0\rangle\!\rangle$, the triplet $|1\rangle\!\rangle$ or the quintet $|2\rangle\!\rangle$. Only the singlet (spin unpolarized) is possible within the lowest Landau level [Fig.22.8], where no charge transfer is allowed. The $\nu = n$ state (the $\nu = n/m$ state) is determined in general by examining the composition of n pseudospins.

To see what is the ground state, we take a fixed real magnetic field at zero bias voltage, and examine how the energy levels evolve as the tunneling gap Δ_{SAS} increases from $\Delta_{SAS} = 0$. The energy of the symmetric state (antisymmetric state)

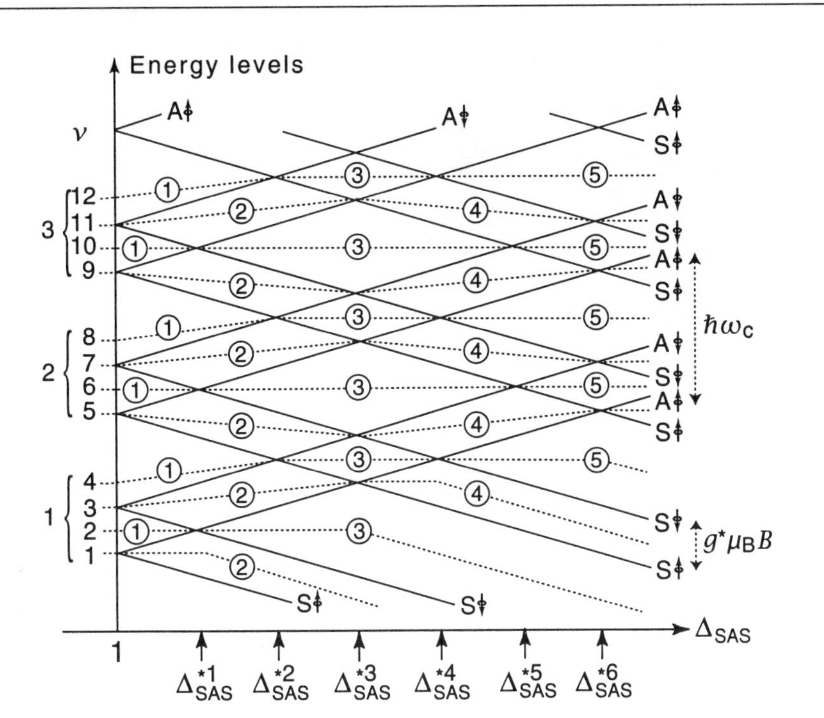

Fig. 22.8 Energy levels are given at the balance point as a function of the tunneling gap Δ_{SAS}, where the magnetic field is fixed. A level crossing occurs at $\Delta_{\text{SAS}} = \Delta_{\text{SAS}}^{*}$ because the energy of the symmetric (antisymmetric) state decreases (increases) by $\frac{1}{2}\Delta_{\text{SAS}}$. The number k in circle represents the pseudospin multiplicity of the multiplet $|\mathfrak{j}\rangle\!\rangle$ with $k = 2j+1$. The dotted line represents the Fermi level. For instance, the $\nu = 4$ state is realized when all levels are filled below the $\nu = 4$ Fermi level.

decreases (increases) by $\frac{1}{2}\Delta_{\text{SAS}}$, as illustrated in Fig.22.8. For instance, a level crossing takes place between the $\nu = 2$ state and the $\nu = 3$ state at $\Delta_{\text{SAS}} = \Delta_{\text{SAS}}^{*1}$, where the tunneling gap becomes larger than the Zeeman gap and a phase transition occurs. It is easy to read all phase-transition points from the figure,

$$\begin{aligned} &\Delta_{\text{SAS}}^{*1} = \Delta_{\text{Z}}, \qquad &&\Delta_{\text{SAS}}^{*2} = \hbar\omega_c - \Delta_{\text{Z}}, \qquad &&\Delta_{\text{SAS}}^{*3} = \hbar\omega_c, \\ &\Delta_{\text{SAS}}^{*4} = \hbar\omega_c + \Delta_{\text{Z}}, \qquad &&\Delta_{\text{SAS}}^{*5} = 2\hbar\omega_c - \Delta_{\text{Z}}, \qquad &&\Delta_{\text{SAS}}^{*6} = 2\hbar\omega_c, \end{aligned} \tag{22.7.8}$$

and so on, where $\Delta_{\text{Z}} = |g^{*}\mu_B B|$. The pseudospin $(2\mathfrak{j}+1)$ multiplet $|\mathfrak{j}\rangle\!\rangle$ appears as indicated in the same figure, with the maximum charge transferability $\Delta\mathcal{P}_z = 2j$.

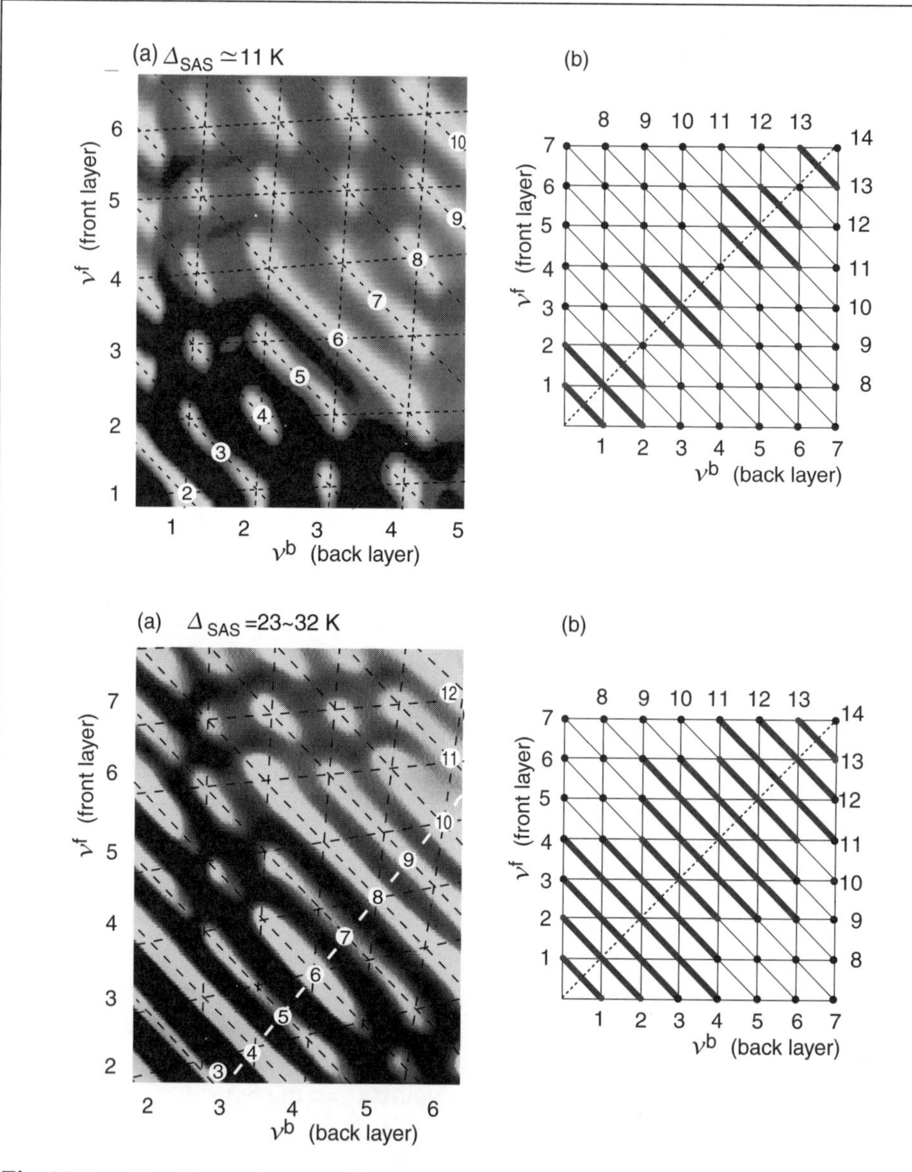

Fig. 22.9 The diagonal resistance is shown in a gray scale by scanning both the front and back gate voltages. The bright regions correspond to small resistance, indicating QH states (a) in a sample with $\Delta_{SAS} \simeq 11$ K and $d = 23.1$ nm, (b) in a sample with $\Delta_{SAS} \simeq 23 \sim 32$ K and $d = 23.1$ nm. QH states are observed not only on discrete lattice points (ν^b, ν^f), but also on elongated regions connecting two lattice points. The data are taken from an unpublished work carried out by Sawada's group. A schematic diagram is drawn for compound states (ν^b, ν^f) and charge-transferable states $|\jmath\rangle_g$ (a′) with a sequence of multiplicity pattern $\{2, 3, 2, 1\}$, (b′) with a sequence of multiplicity pattern $\{2345; 4345\}$. See also Fig.14 (page xviii) and Fig.15 (page xix) in the frontispiece.

We list the sequences of the multiplicity for some values of Δ_{SAS} [Fig.22.8]:

$$
\begin{aligned}
\Delta_{\text{SAS}} < \Delta_{\text{SAS}}^{*1} \quad &: \{2,1,2,1;2,1,2,1;\cdots\}, \\
\Delta_{\text{SAS}}^{*1} < \Delta_{\text{SAS}} < \Delta_{\text{SAS}}^{*2} &: \{2,3,2,1;2,3,2,1;\cdots\}, \\
\Delta_{\text{SAS}}^{*2} < \Delta_{\text{SAS}} < \Delta_{\text{SAS}}^{*3} &: \{2,3,2,3;2,3,2,3;\cdots\}, \\
\Delta_{\text{SAS}}^{*3} < \Delta_{\text{SAS}} < \Delta_{\text{SAS}}^{*4} &: \{2,3,4,3;4,3,4,3;\cdots\}, \\
\Delta_{\text{SAS}}^{*4} < \Delta_{\text{SAS}} < \Delta_{\text{SAS}}^{*5} &: \{2,3,4,5;4,3,4,5;\cdots\}, \\
\Delta_{\text{SAS}}^{*5} < \Delta_{\text{SAS}} < \Delta_{\text{SAS}}^{*6} &: \{2,3,4,5;4,5,4,5;\cdots\}.
\end{aligned}
$$

Various sequences have been observed experimentally.[357] See experimental results demonstrating the second sequence in Fig.22.9(a) and the fifth sequence in Fig.22.9(b).

CHAPTER 23

BILAYER-LOCKED STATES

The composite-boson theory reveals a rich phase structure present in the bilayer QH system. In defining the composite-boson field there is a freedom to introduce a set of three integers (m_f, m_b, m), characterizing electron correlations and indexing bilayer QH liquids. In this chapter we discuss the (m_f, m_b, m) phase with $m_f m_b \neq m^2$. A typical one is the $\nu = \frac{1}{2}$ state with $m_f = m_b = 3$ and $m = 1$.

23.1 COMPOSITE-BOSON FIELD

We have so far explored integer QH states. Though fractional QH states were mentioned, they are the states which are regarded as integer QH states of composite particles. The composite-boson theory reveals the existence of new fractional QH states in the bilayer system.[138,140]

We define the bare composite-boson field ϕ^α by[138]

$$\phi^\alpha(\mathbf{x}) = e^{i\Theta^\alpha(\mathbf{x})}\psi^\alpha(\mathbf{x}), \qquad \alpha = \mathrm{f}, \mathrm{b}. \tag{23.1.1}$$

The most general condition for the phase field Θ^α is

$$\begin{aligned}\Theta^{\mathrm{f}}(\mathbf{x}) &= m_{\mathrm{f}}\int d^2y\, \theta(\mathbf{x}-\mathbf{y})\rho^{\mathrm{f}}(\mathbf{y}) + m\int d^2y\, \theta(\mathbf{x}-\mathbf{y})\rho^{\mathrm{b}}(\mathbf{y}),\\ \Theta^{\mathrm{b}}(\mathbf{x}) &= m\int d^2y\, \theta(\mathbf{x}-\mathbf{y})\rho^{\mathrm{f}}(\mathbf{y}) + m_{\mathrm{b}}\int d^2y\, \theta(\mathbf{x}-\mathbf{y})\rho^{\mathrm{b}}(\mathbf{y}),\end{aligned} \tag{23.1.2}$$

where m_α and m are integers and $\theta(\mathbf{x}-\mathbf{y})$ is the angle made between the vector $\mathbf{x}-\mathbf{y}$ and the x axis. Here, m_α is associated with the statistics between composite particles within layer α, while m with the relative statistics between composite particles on the different layers. When m_α is odd (even), composite particles are bosons (fermions). We choose m_α odd.

With the singular phase transformation the kinetic term in the Hamiltonian (22.1.4) is transformed into

$$H_{\mathrm{K}} = \frac{1}{2M}\sum_{\alpha=1}^{2}\int d^2x\phi^{\alpha\dagger}(\mathbf{x})(\check{P}_x^\alpha - i\check{P}_y^\alpha)(\check{P}_x^\alpha + i\check{P}_y^\alpha)\phi^\alpha(\mathbf{x}), \tag{23.1.3}$$

while the Coulomb term is unchanged with $\rho^\alpha = \psi^{\alpha\dagger}\psi^\alpha = \phi^{\alpha\dagger}\phi^\alpha$. The covariant momentum is

$$\mathcal{P}_k^\alpha = -i\hbar\partial_k + e\mathcal{A}_k^\alpha \tag{23.1.4}$$

with

$$\mathcal{A}_j^\alpha(\boldsymbol{x}) \equiv A_j^{\rm ext}(\boldsymbol{x}) + C_j^\alpha(\boldsymbol{x}), \tag{23.1.5}$$

where $C_j^\alpha(\boldsymbol{x})$ is the Chern-Simons field,

$$C_j^\alpha(\boldsymbol{x}) = \frac{\hbar}{e}\partial_j\Theta^\alpha(\boldsymbol{x}). \tag{23.1.6}$$

It follows from (23.1.2) and (23.1.6) that

$$\begin{aligned}\varepsilon_{jk}\partial_j C_k^{\rm f}(\boldsymbol{x}) &= \Phi_{\rm D}\left(m_{\rm f}\rho^{\rm f}(\boldsymbol{x}) + m\rho^{\rm b}(\boldsymbol{x})\right),\\ \varepsilon_{jk}\partial_j C_k^{\rm b}(\boldsymbol{x}) &= \Phi_{\rm D}\left(m\rho^{\rm f}(\boldsymbol{x}) + m_{\rm b}\rho^{\rm b}(\boldsymbol{x})\right),\end{aligned} \tag{23.1.7}$$

where $\Phi_{\rm D} \equiv 2\pi\hbar/e$ is the Dirac flux unit. This is the constraint equation which determines the Chern-Simons field $C_k^\alpha(\boldsymbol{x})$ in terms of the number density $\rho^\alpha(\boldsymbol{x})$.

The effect of the phase transformation (23.1.1) is to attach the Chern-Simons flux to an electron in each layer according to the rule (23.1.7). The attached flux gives rise to a relative angular momentum and keeps other electrons away. The repulsive range for a pair of electrons in the layer α is proportional to m_α, while the one for a pair of electrons belonging to the different layers is proportional to m. This is intuitively clear in the Halperin wave function (21.1.14), which will be derived semiclassically in Section 23.2. It is necessary that $m_\alpha \geq m \geq 0$ in general, since the interlayer repulsive force is supplied by the interlayer Coulomb interaction. A special limit is given by $m = 0$ with no interlayer correlation. The other special limit is given by $m_{\rm f} = m_{\rm b} = m$ with equal interlayer and intralayer correlations. Physics is very different between the two phases, the $(m_{\rm f}, m_{\rm b}, m)$ phase with $m_\alpha > m$ and the (m, m, m) phase, because the two constraint equations (23.1.7) are reduced to a single equation in the (m, m, m) phase. We discuss the $(m_{\rm f}, m_{\rm b}, m)$ phase in this chapter and the (m, m, m) phase in Chapter 25.

The effective magnetic field that composite particles experience is given by

$$\begin{aligned}\mathcal{B}_{\rm eff}^{\rm f} &\equiv -\varepsilon_{ij}\partial_i\mathcal{A}_j^{\rm f}(\boldsymbol{x}) = B_\perp - \Phi_{\rm D}\left(m_{\rm f}\rho^{\rm f}(\boldsymbol{x}) + m\rho^{\rm b}(\boldsymbol{x})\right),\\ \mathcal{B}_{\rm eff}^{\rm b} &\equiv -\varepsilon_{ij}\partial_i\mathcal{A}_j^{\rm b}(\boldsymbol{x}) = B_\perp - \Phi_{\rm D}\left(m\rho^{\rm f}(\boldsymbol{x}) + m_{\rm b}\rho^{\rm b}(\boldsymbol{x})\right).\end{aligned} \tag{23.1.8}$$

As in the monolayer case, we require the effective magnetic field to vanish on the homogeneous state: See Section 23.3. For the state we obtain

$$\begin{aligned}m_{\rm f}\rho_0^{\rm f} + m\rho_0^{\rm b} &= \frac{B_\perp}{\Phi_{\rm D}},\\ m\rho_0^{\rm f} + m_{\rm f}\rho_0^{\rm b} &= \frac{B_\perp}{\Phi_{\rm D}}.\end{aligned} \tag{23.1.9}$$

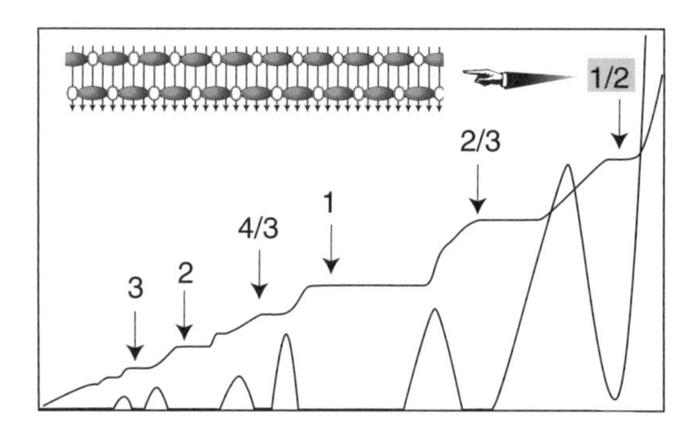

Fig. 23.1 The QH state at $\nu = 1/2$ is a genuine bilayer QH state. It is identified with the bilayer-locked state. The data are taken from Suen et al.[465]

We may solve this as

$$\rho_0^{\mathrm{f}} = \frac{m_{\mathrm{b}} - m}{m_{\mathrm{f}} m_{\mathrm{b}} - m^2} \frac{B_\perp}{\Phi_{\mathrm{D}}} = \frac{m_{\mathrm{b}} - m}{m_{\mathrm{f}} + m_{\mathrm{b}} - 2m} \rho_0,$$
$$\rho_0^{\mathrm{b}} = \frac{m_{\mathrm{f}} - m}{m_{\mathrm{f}} m_{\mathrm{b}} - m^2} \frac{B_\perp}{\Phi_{\mathrm{D}}} = \frac{m_{\mathrm{f}} - m}{m_{\mathrm{f}} + m_{\mathrm{b}} - 2m} \rho_0. \tag{23.1.10}$$

The density ratio is quantized as

$$\frac{\rho_0^{\mathrm{f}}}{\rho_0^{\mathrm{b}}} = \frac{m_{\mathrm{b}} - m}{m_{\mathrm{f}} - m}. \tag{23.1.11}$$

The state is realized at the filling factor

$$\nu \equiv \frac{\Phi_{\mathrm{D}} \rho_0}{B} = \frac{m_{\mathrm{f}} + m_{\mathrm{b}} - 2m}{m_{\mathrm{f}} m_{\mathrm{b}} - m^2}, \tag{23.1.12}$$

where $\rho_0 = \rho_0^{\mathrm{f}} + \rho_0^{\mathrm{b}}$ is the total number density.

It is required that $m_{\mathrm{f}} = m_{\mathrm{b}}$ in the balanced configuration, $\rho_0^{\mathrm{f}} = \rho_0^{\mathrm{b}}$, which is realized at the filling factor

$$\nu = \frac{2}{m_{\mathrm{f}} + m}, \tag{23.1.13}$$

as follows from (23.1.9). The simplest example is given by the $\nu = \frac{1}{2}$ state with $m_{\mathrm{f}} = 3$ and $m = 1$. Such a state has been experimentally observed.[465,127] Since it has an even denominator, there exists no doubt that the state is a genuine bilayer QH state [see Fig.23.1]. The $\nu = \frac{1}{2}$ state was also studied[411,529,227] based on numerical calculations for few-particle systems.

Since $\mathcal{A}_j^\alpha(x)$ satisfies the Coulomb gauge condition, we may introduce a scalar field $\mathcal{A}^\alpha$ obeying

$$\partial_i \mathcal{A}^\alpha = \frac{e}{\hbar}\varepsilon_{ij}\mathcal{A}_j^\alpha. \tag{23.1.14}$$

It is explicitly given by

$$\mathcal{A}^{\rm f}(x) = \int d^2y \ln\left(\frac{|x-y|}{2\ell_B}\right)\left\{m_{\rm f}\rho^{\rm f}(y) + m\rho^{\rm b}(y)\right\} - |z|^2,$$
$$\mathcal{A}^{\rm b}(x) = \int d^2y \ln\left(\frac{|x-y|}{2\ell_B}\right)\left\{m\rho^{\rm f}(y) + m_{\rm b}\rho^{\rm b}(y)\right\} - |z|^2. \tag{23.1.15}$$

We proceed to define the dressed composite-boson field,

$$\varphi^\alpha(x) = e^{-\mathcal{A}^\alpha(x)}\phi^\alpha(x), \tag{23.1.16}$$

whose canonical conjugate field is

$$\varphi^{\alpha\ddagger}(x) = \phi^{\alpha\dagger}(x)e^{\mathcal{A}^\alpha(x)}, \tag{23.1.17}$$

as follows from the Lagrangian density similar to (13.4.13) in the monolayer system. The density operator is given by $\rho^\alpha = \phi^{\alpha\dagger}\phi^\alpha = \varphi^{\alpha\ddagger}\varphi^\alpha$. When m^α is an odd (even) integer, the canonical commutation (anticommutation) relations are derived for these fields as in the monolayer case (see Section 13.4).

The kinetic part of the Hamiltonian for the dressed composite-boson field is

$$H_{\rm K} = \frac{1}{2M}\sum_\alpha \int d^2x \varphi^{\alpha\ddagger}(x)(\mathcal{P}_x^\alpha - i\mathcal{P}_y^\alpha)(\mathcal{P}_x^\alpha + i\mathcal{P}_y^\alpha)\varphi^\alpha(x), \tag{23.1.18}$$

where the covariant momentum is

$$\mathcal{P}_j^\alpha = \check{P}_j^\alpha - i\partial_j\mathcal{A}^\alpha(x) = -i\hbar\partial_j + e(\delta_{jk} - i\varepsilon_{jk})\mathcal{A}_k^\alpha(x). \tag{23.1.19}$$

It is easy to see

$$\mathcal{P}_x^\alpha + i\mathcal{P}_y^\alpha = -\frac{i\hbar}{\ell_B}\frac{\partial}{\partial z^{\alpha *}}. \tag{23.1.20}$$

The Hamiltonian is simple in terms of the dressed field,

$$H_{\rm K} = \frac{1}{2}\hbar\omega_c\sum_\alpha \int d^2x \left(\frac{\partial}{\partial z^*}\varphi^\alpha(x)\right)^\dagger e^{2\mathcal{A}^\alpha(x)}\left(\frac{\partial}{\partial z^*}\varphi^\alpha(x)\right). \tag{23.1.21}$$

The dressed field $\varphi^\alpha(x)$ is a nonlocal operator in terms of the density fluctuation $\Delta\rho(x)$ as in the monolayer theory: See Section 23.3 for more details.

We mention an important feature revealed in the composite-boson theory. Although the electron kinetic Hamiltonian (22.1.4) is invariant under the global SU(2) transformation, the composite-boson Hamiltonians (23.1.3) and (23.1.18) are not so due to the field $\mathcal{A}_k^\alpha(x)$ in the covariant momentum (23.1.4) unless $m_{\rm f} = m_{\rm b} = m$. The existence of the $(m_{\rm f}, m_{\rm b}, m)$ phases is hidden in the original electron theory, where these phases must be realized by way of spontaneous breaking of the SU(2) symmetry. On the contrary, in the composite-boson theory each phase is described by its own Lagrangian containing the symmetry-breaking parameters explicitly.

23.2 Wave Functions

We derive the wave function in the (m_f, m_b, m) phase. Since electrons in the different layers are distinguishable, we label them by the layer index α. The LLL condition reads

$$(\mathcal{P}_x^\alpha + i\mathcal{P}_y^\alpha)\varphi^\alpha(\boldsymbol{x}^\alpha)|\mathfrak{S}\rangle = -\frac{i\hbar}{\ell_B}\frac{\partial}{\partial z^{\alpha *}}\varphi^\alpha(\boldsymbol{x}^\alpha)|\mathfrak{S}\rangle = 0. \tag{23.2.1}$$

It implies that the wave function for dressed composite bosons is an arbitrary analytic function,

$$\mathfrak{S}_{\rm CB}[z^{\rm f}, z^{\rm b}] = \langle 0|\varphi^{\rm f}(\boldsymbol{x}_1^{\rm f})\cdots\varphi^{\rm f}(\boldsymbol{x}_{N_{\rm f}}^{\rm f})\varphi^{\rm b}(\boldsymbol{x}_1^{\rm b})\cdots\varphi^{\rm b}(\boldsymbol{x}_{N_{\rm b}}^{\rm b})|\mathfrak{S}\rangle, \tag{23.2.2}$$

where N_α is the number of electrons in the α-layer.

The N-body state $|\mathfrak{S}\rangle$ is given by

$$|\mathfrak{S}\rangle = \int[dx]\mathfrak{S}_{\rm CB}[z^{\rm f}, z^{\rm b}]\varphi^{\rm f\ddagger}(\boldsymbol{x}_1^{\rm f})\cdots\varphi^{\alpha\ddagger}(\boldsymbol{x}_r^\alpha)\cdots\varphi^{\rm b\ddagger}(\boldsymbol{x}_N^{\rm b})|0\rangle, \tag{23.2.3}$$

where $\int[dx] = \prod_{r,p}\int d^2x_r^{\rm f}d^2x_p^{\rm b}$. We are able to rewrite this as

$$|\mathfrak{S}\rangle = \int[dx]\mathfrak{S}[\boldsymbol{x}^{\rm f}, \boldsymbol{x}^{\rm b}]\psi^{\rm f\dagger}(\boldsymbol{x}_1^{\rm f})\cdots\psi^{\alpha\dagger}(\boldsymbol{x}_r^\alpha)\cdots\psi^{\rm b\dagger}(\boldsymbol{x}_N^{\rm b})|0\rangle \tag{23.2.4}$$

in terms of the electron field, where

$$\mathfrak{S}[\boldsymbol{x}^{\rm f}, \boldsymbol{x}^{\rm b}] = \mathfrak{S}_{\rm CB}[z^{\rm f}, z^{\rm b}]\mathfrak{S}_{\rm HN}[\boldsymbol{x}^{\rm f}, \boldsymbol{x}^{\rm b}] \tag{23.2.5}$$

with the Halperin wave function

$$\mathfrak{S}_{\rm HN}[\boldsymbol{x}^{\rm f}, \boldsymbol{x}^{\rm b}] = (\rho_0^{\rm f})^{N_{\rm f}/2}(\rho_0^{\rm b})^{N_{\rm b}/2}\prod_{r<s}(z_r^{\rm f} - z_s^{\rm f})^{m_{\rm f}}\prod_{p<q}(z_p^{\rm b} - z_q^{\rm b})^{m_{\rm b}} \\ \times\prod_{r,p}(z_r^{\rm f} - z_p^{\rm b})^m e^{-\sum|z_r^{\rm f}|^2-\sum|z_p^{\rm b}|^2}. \tag{23.2.6}$$

To derive it we may repeat the arguments from (13.4.26) to (13.4.28) made in the monolayer case. We start with (23.1.17), or

$$\varphi^{\alpha\ddagger}(\boldsymbol{x}) = \psi^{\alpha\dagger}(\boldsymbol{x})e^{\mathcal{B}^\alpha(\boldsymbol{x})}, \tag{23.2.7}$$

where $\mathcal{B}^\alpha(\boldsymbol{x}) = \mathcal{A}^\alpha(\boldsymbol{x}) + i\Theta^\alpha(\boldsymbol{x})$, or

$$\mathcal{B}^{\rm f}(\boldsymbol{x}) = \int d^2x'\ln(z - z')\left\{m_{\rm f}\rho^{\rm f}(\boldsymbol{y}) + m\rho^{\rm b}(\boldsymbol{y})\right\} - |z|^2,$$

$$\mathcal{B}^{\rm b}(\boldsymbol{x}) = \int d^2x'\ln(z - z')\left\{m\rho^{\rm f}(\boldsymbol{y}) + m_{\rm b}\rho^{\rm b}(\boldsymbol{y})\right\} - |z|^2. \tag{23.2.8}$$

First, the operator $\varphi^{f\ddagger}(x_1^f)$ acts on the vacuum as

$$\varphi^{f\ddagger}(x_1^f)|0\rangle = e^{-|z_1^f|^2}\psi^{f\dagger}(x_1^f)|0\rangle, \tag{23.2.9}$$

since $\rho^\alpha(x)|0\rangle = 0$. Operating $\varphi^{\alpha\ddagger}(x_2)$ to this state, we find

$$\begin{aligned}
\varphi^{f\ddagger}(x_2^f)\varphi^{f\ddagger}(x_1^f)|0\rangle &= (z_2^f - z_1^f)^{m_f} e^{-|z_2^f|^2-|z_1^f|^2}\psi^{f\dagger}(x_2^f)\psi^{f\dagger}(x_1^f)|0\rangle, \\
\varphi^{b\ddagger}(x_2^b)\varphi^{f\ddagger}(x_1^f)|0\rangle &= (z_2^b - z_1^f)^{m} e^{-|z_2^b|^2-|z_1^f|^2}\psi^{b\dagger}(x_2^b)\psi^{f\dagger}(x_1^f)|0\rangle,
\end{aligned} \tag{23.2.10}$$

since $\rho^\alpha(x)\psi^{f\dagger}(x_1)|0\rangle = \delta^{\alpha f}\delta(x - x_1)\psi^{f\dagger}(x_1)|0\rangle$. In general we obtain

$$\begin{aligned}
&\varphi^{f\ddagger}(x_1^f)\cdots\varphi^{\alpha\ddagger}(x_r^\alpha)\cdots\varphi^{b\ddagger}(x_N^b)|0\rangle \\
&= \prod(z_r^f - z_s^f)^{m_f}\prod(z_p^b - z_q^b)^{m_b}\prod(z_r^f - z_p^b)^{m} e^{-\sum|z_r^f|^2-\sum|z_p^b|^2} \\
&\quad\times \psi^{f\dagger}(x_1^f)\cdots\psi^{\alpha\dagger}(x_r^\alpha)\cdots\varphi^{b\dagger}(x_N^b)|0\rangle.
\end{aligned} \tag{23.2.11}$$

This proves the equality between (23.2.3) and (23.2.4).

23.3 Ground State

We determine the ground state semiclassically precisely as in the monolayer system. It is the state minimizing the total energy $\langle H\rangle$, and realized when the number density is homogeneous,

$$\langle\rho^\alpha(x)\rangle = \langle\varphi^\ddagger(x)\varphi(x)\rangle = e^{2\langle\mathcal{A}^\alpha(x)\rangle}\langle\varphi^{\alpha\dagger}(x)\varphi^\alpha(x)\rangle = \rho_0^\alpha. \tag{23.3.1}$$

The wave function $\Omega[z^f, z^b]$ factorizes in the semiclassical approximation,

$$\Omega[z^f, z^b] = (\rho_0^f)^{N_f/2}(\rho_0^b)^{N_b/2}\prod_{r=1} e^{-\langle\mathcal{A}^f(x_r^f)\rangle}\prod_{p=1} e^{-\langle\mathcal{A}^b(x_p^b)\rangle}. \tag{23.3.2}$$

The analyticity of the wave function implies $\langle\mathcal{A}^\alpha(x)\rangle = 0$, or

$$\langle\mathcal{A}_j^\alpha(x)\rangle = A_j^{ext}(x) + \langle C_j^\alpha(x)\rangle = 0. \tag{23.3.3}$$

We have already shown that this occurs only at the filling factor (23.1.12), or

$$\nu = \frac{m_f + m_b - 2m}{m_f m_b - m^2}. \tag{23.3.4}$$

The density ratio is quantized as in (23.1.11).

The semiclassical ground state is given by

$$\mathfrak{S}_{CB}[z^f, z^b] = \text{constant} \tag{23.3.5}$$

in terms of composite bosons. Hence, from (23.2.5) the ground-state wave function is given by the Halperin wave function (23.2.6) in terms of electrons.

We next consider whether there are any perturbative fluctuation within the lowest Landau level. Any state in the lowest Landau level is given by (23.2.3), or

$$|\mathfrak{S}\rangle = \int [dx]\mathfrak{S}_{\mathrm{CB}}[z^{\mathrm{f}}, z^{\mathrm{b}}]\varphi^{\mathrm{f}\ddagger}(\boldsymbol{x}_1^{\mathrm{f}})\cdots\varphi^{\mathrm{f}\ddagger}(\boldsymbol{x}_1^{\mathrm{f}})\varphi^{\mathrm{b}\ddagger}(\boldsymbol{x}_1^{\mathrm{b}})\cdots\varphi^{\mathrm{b}\ddagger}(\boldsymbol{x}_{N_{\mathrm{b}}}^{\mathrm{b}})|0\rangle. \tag{23.3.6}$$

The state $|\mathfrak{S}\rangle$ contains only nonlocal excitations as quasiparticles, because $\varphi^{\alpha\ddagger}(\boldsymbol{x})$ is a nonlocal field in terms of the density fluctuation $\Delta\rho(\boldsymbol{x})$. They are topological excitations, as is seen in the next section. There are no perturbative fluctuation confined to the lowest Landau level, as in the spinless monolayer QH system. Topological solitons carry quantized electric charges, and require a Coulomb energy $\sim E_{\mathrm{C}}^0$ for their activation. The bilayer-locked $(m_{\mathrm{f}}, m_{\mathrm{b}}, m)$ state is incompressible precisely by the same reasoning that is given in the monolayer system: The ground state is unique without gapless mode. In Chapter 25 we shall study the special case $m_{\mathrm{f}} = m_{\mathrm{b}} = m$, where the interlayer coherence develops spontaneously.

23.4 Vortex Excitations

We have found that the ground state is given by the trivial solution (23.3.5) in the composite-boson theory. The generic state $|\mathfrak{S}\rangle$ is described by an analytic function in terms of composite bosons. When it factorizes,

$$\mathfrak{S}_{\mathrm{CB}}[z^{\mathrm{f}}, z^{\mathrm{b}}] = (\rho_0^{\mathrm{f}})^{N_{\mathrm{f}}/2}(\rho_0^{\mathrm{b}})^{N_{\mathrm{b}}/2}\prod_r^{N_{\mathrm{f}}}\omega^{\mathrm{f}}(z_r^{\mathrm{f}})\prod_r^{N_{\mathrm{b}}}\omega^{\mathrm{b}}(z_r^{\mathrm{b}}), \tag{23.4.1}$$

we identify

$$\langle\varphi^{\alpha}(\boldsymbol{x})\rangle = \sqrt{\rho_0^{\alpha}}\,\omega^{\alpha}(z). \tag{23.4.2}$$

We make a semiclassical analysis.

All fields are classical fields hereafter in this section. From (23.1.16) and (23.4.2) we set

$$\varphi^{\alpha}(\boldsymbol{x}) = e^{-\mathcal{A}^{\alpha}(\boldsymbol{x})}e^{-\chi^{\alpha}(\boldsymbol{x})}\sqrt{\rho_0^{\alpha} + \Delta\rho^{\alpha}(\boldsymbol{x})} = \sqrt{\rho_0^{\alpha}}\,\omega^{\alpha}(z). \tag{23.4.3}$$

Since the auxiliary field $\mathcal{A}^{\alpha}(\boldsymbol{x})$ is given by (23.1.15), or

$$\begin{aligned}\boldsymbol{\nabla}^2\mathcal{A}^{\mathrm{f}}(\boldsymbol{x}) &= 2\pi\left\{m_{\mathrm{f}}\Delta\rho^{\mathrm{f}}(\boldsymbol{x}) + m\Delta\rho^{\mathrm{b}}(\boldsymbol{x})\right\},\\ \boldsymbol{\nabla}^2\mathcal{A}^{\mathrm{b}}(\boldsymbol{x}) &= 2\pi\left\{m\Delta\rho^{\mathrm{f}}(\boldsymbol{x}) + m_{\mathrm{b}}\Delta\rho^{\mathrm{b}}(\boldsymbol{x})\right\}\end{aligned} \tag{23.4.4}$$

at the filling factor (23.1.12), we obtain the soliton equation

$$\begin{aligned}\frac{1}{4\pi}\boldsymbol{\nabla}^2\ln\left(1 + \frac{\Delta\rho^{\mathrm{f}}(\boldsymbol{x})}{\rho_0^{\mathrm{f}}}\right) &= m_{\mathrm{f}}\Delta\rho^{\mathrm{f}}(\boldsymbol{x}) + m\Delta\rho^{\mathrm{b}}(\boldsymbol{x}) + J_{\mathrm{vor}}^{0\mathrm{f}}(\boldsymbol{x}),\\ \frac{1}{4\pi}\boldsymbol{\nabla}^2\ln\left(1 + \frac{\Delta\rho^{\mathrm{b}}(\boldsymbol{x})}{\rho_0^{\mathrm{b}}}\right) &= m\Delta\rho^{\mathrm{f}}(\boldsymbol{x}) + m_{\mathrm{b}}\Delta\rho^{\mathrm{b}}(\boldsymbol{x}) + J_{\mathrm{vor}}^{0\mathrm{b}}(\boldsymbol{x})\end{aligned} \tag{23.4.5}$$

with the topological charge density

$$J^{0\alpha}_{\rm vor}(\boldsymbol{x}) = \frac{\nu}{2\pi}\nabla^2 \ln[|\omega^\alpha(z)|]. \tag{23.4.6}$$

The soliton equation (23.4.5) is a generalization of (14.3.11) in the monolayer spinless system.

For the analytic function of the type

$$\omega^\alpha(z) = \prod_r (z - z^\alpha_r)^{q^\alpha_r}, \tag{23.4.7}$$

we have

$$J^{0\alpha}_{\rm vor}(\boldsymbol{x}) = \sum_r q^\alpha_r \delta(\boldsymbol{x} - \boldsymbol{x}^\alpha_r). \tag{23.4.8}$$

Here, the point $\boldsymbol{x} = \boldsymbol{x}^\alpha_r$ is the branch singularity point of the phase field $\chi^\alpha(\boldsymbol{x})$, defining the center of a vortex, which we call the $(q^{\rm f}, q^{\rm b})$-vortex. Although (23.4.7) is valid only for quasiholes ($q^\alpha_r > 0$), we would well approximate the topological charge of quasielectrons ($q^\alpha_r < 0$) by (23.4.8).

Integrating (23.4.5) over the whole system, we obtain

$$\begin{aligned} m_{\rm f} N^{\rm f}_{\rm vor} + m N^{\rm b}_{\rm vor} &= -\sum_r q^{\rm f}_r, \\ m N^{\rm f}_{\rm vor} + m_{\rm b} N^{\rm b}_{\rm vor} &= -\sum_r q^{\rm b}_r, \end{aligned} \tag{23.4.9}$$

where

$$N^\alpha_{\rm vor} = \int d^2x\, \Delta\rho^\alpha(\boldsymbol{x}) \tag{23.4.10}$$

is the electron number associated with the excitation. We solve the electron number of the $(q^{\rm f}, q^{\rm b})$-vortex as

$$\begin{aligned} N^{\rm f}_{\rm vor}(q^{\rm f}, q^{\rm b}) &= \frac{-m_{\rm b} q^{\rm f} + m q^{\rm b}}{m_{\rm f} m_{\rm b} - m^2}, \\ N^{\rm b}_{\rm vor}(q^{\rm f}, q^{\rm b}) &= \frac{-m_{\rm f} q^{\rm b} + m q^{\rm f}}{m_{\rm f} m_{\rm b} - m^2}. \end{aligned} \tag{23.4.11}$$

For instance, the (1,0)-vortex has the electric charge

$$-eN^{\rm f}_{\rm vor}(1,0) = \frac{m_{\rm b}}{m_{\rm f} m_{\rm b} - m^2} e, \quad -eN^{\rm b}_{\rm vor}(1,0) = -\frac{m}{m_{\rm f} m_{\rm b} - m^2} e \tag{23.4.12}$$

in each layer. It has a positive charge on the front layer (α = f) and a negative charge on the back layer (α = b). The total charge is

$$-e[N^{\rm f}_{\rm vor}(1,0) + N^{\rm b}_{\rm vor}(1,0)] = \frac{m_{\rm b} - m}{m_{\rm f} m_{\rm b} - m^2} e, \tag{23.4.13}$$

which is positive definite since $m_\alpha > m$. The $(0,1)$-vortex is given by interchanging $m_{\rm f}$ and $m_{\rm b}$ in the above formulas.

The (1,-1)-vortex has the electric charge

$$-eN_{\text{vor}}^{\text{f}}(1,-1) = \frac{m_{\text{b}} + m}{m_{\text{f}}m_{\text{b}} - m^2}e, \quad -eN_{\text{vor}}^{\text{b}}(1,-1) = -\frac{m_{\text{f}} + m}{m_{\text{f}}m_{\text{b}} - m^2}e \tag{23.4.14}$$

in each layer. The total charge is

$$-e[N_{\text{vor}}^{\text{f}}(1,-1) + N_{\text{vor}}^{\text{b}}(1,-1)] = \frac{m_{\text{b}} - m_{\text{f}}}{m_{\text{f}}m_{\text{b}} - m^2}e, \tag{23.4.15}$$

which vanishes when $m_{\text{f}} = m_{\text{b}}$. It is not a charged excitation in the balanced configuration. Because of the charge relation

$$\Delta N_{\text{vor}}^{\alpha}(1,-1) = \Delta N_{\text{vor}}^{\alpha}(1,0) + \Delta N_{\text{vor}}^{\alpha}(0,-1), \tag{23.4.16}$$

it is regarded as an excitonic bound state of the (1,0)-vortex and the (0,-1) vortex.

CHAPTER 24

INTERLAYER COHERENCE

The interlayer coherence is studied in the spin-frozen bilayer QH system. Each Landau level contains two energy levels with up and down pseudospins. These two levels are degenerate in the vanishing limit of the tunneling interaction, where pseudospins are spontaneously polarized to minimize the interlayer exchange energy. The pseudospin wave is a Goldstone mode. It affects the Hall current in an intriguing way such that the Hall plateau disappears though the QH state is well developed in a counterflow transport setting.

24.1 PSEUDOSPIN FERROMAGNET

A prominent feature intrinsic to the bilayer QH system is a spontaneous development of the interlayer coherence . For instance, there exists a unique QH state at $\nu = 1$, which is a charge-transferable state upon which the interlayer coherence develops (see Fig.24.1). To elucidate its essence, in this chapter we freeze the spin degree of freedom. As far as a sufficiently smooth pseudospin configuration is concerned, we can approximate the Coulomb Hamiltonian H_{C} by the sum of the interlayer exchange term $H_{\text{X}}^{\text{inter}}$ and the capacitance energy H_{cap},

$$H_{\text{C}} = H_{\text{X}}^{\text{inter}} + H_{\text{cap}}, \tag{24.1.1}$$

where H_{cap} is given by (22.5.1). The interlayer exchange interaction is described by an anisotropic O(3) nonlinear sigma model

$$H_{\text{X}}^{\text{inter}} = 2J_s^d \sum_{a=x,y} \int d^2x\, [\partial_k \mathcal{P}_a(\mathbf{x})]^2 + 2J_s \int d^2x\, [\partial_k \mathcal{P}_z(\mathbf{x})]^2, \tag{24.1.2}$$

where $\boldsymbol{\mathcal{P}}(\mathbf{x})$ is the normalized SU(2) pseudospin field (22.2.9); the stiffness parameters J_s and J_s^d are given by (A.6) and (A.15). We derive the nonlinear sigma model $H_{\text{X}}^{\text{inter}}$ and the capacitance energy H_{cap} from the Coulomb interaction in Chapter 32.6: See (32.6.28) and (32.6.29).

By including the pseudo-Zeeman term H_{pZ} given by (22.4.12), the relevant

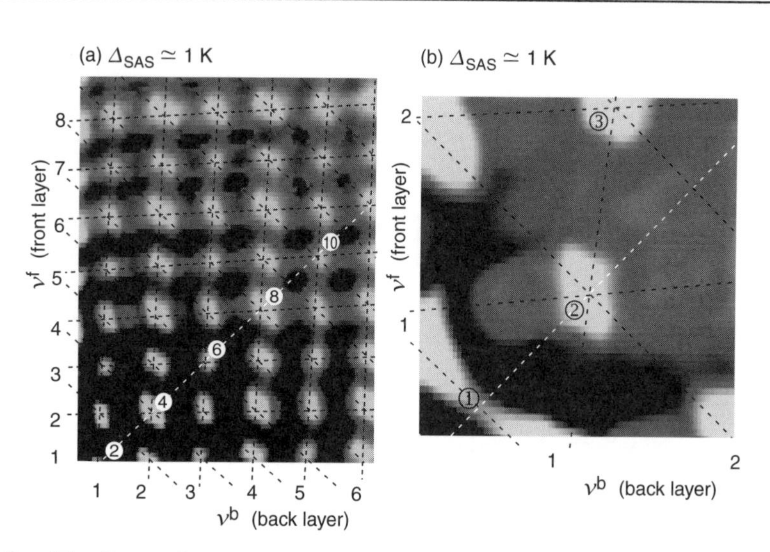

Fig. 24.1 The diagonal resistance is shown in a gray scale by scanning both the front and back gate voltages. The bright regions correspond to small resistance, indicating QH states. (a) In a sample with $\Delta_{SAS} \simeq 1$ K and $d = 23.1$ nm, no odd-integer QH states are seen at the balanced point for $\nu > 2$, i.e., along a white dashed line. Only compound states are seen. [The figure (a) is the same one as Fig.22.5(a)]. (b) In the same sample, the $\nu = 1$ state is elongated, indicating the development of the interlayer coherence at $\nu = 1$. See Fig.14 (page xviii) and Fig.15 (page xix) in the frontispiece for more details.

Hamiltonian is

$$H = H_C + H_{pZ} = H_X^{inter} + H_{cap} + H_{pZ}. \tag{24.1.3}$$

A close analogy holds between the spin-frozen bilayer system and the monolayer spin system. The pseudo-Zeeman term plays the role of the Zeeman term. The interlayer exchange interaction supports a pseudospin wave as a low energy excitation, leading to the interlayer coherence. Just as the monolayer spin system is a spin ferromagnet, the spin-frozen bilayer system is a pseudospin ferromagnet. A new feature is that the pseudospin wave carries the electric current, as yields anomalous QH currents in some experimental settings such as the counterflow transport[268,481] and the drag experiment:[267] See Section 24.4.

In the interlayer-coherent state, the number difference is subject to a large uncertainty proportional to $N_0^{1/2}$, though the total electron number N_0 is fixed. The phase difference, being the conjugate variable, is subject only to a small uncertainty proportional to $N_0^{-1/2}$. Consequently, there are practically no uncertainties both in the normalized density difference $\sigma(\boldsymbol{x})$ and the phase difference $\vartheta(\boldsymbol{x})$ in thermodynamical limit ($N_0 \to \infty$): They are macroscopic observables. The macroscopic coherence between the two layers is the basic ingredient of the interlayer coherence.

Composite Bosons

The composite-boson theory provides us with a convenient way to formulate the interlayer coherence. In the (m, m, m) phase, since the two phase fields $\Theta^\alpha(\boldsymbol{x})$ collapse into one in (23.1.2), all formulas become identical mathematically to those in the monolayer QH ferromagnet.

We convert the electron field into the bare and dressed composite-boson fields,

$$\psi^\alpha(\boldsymbol{x}) = e^{i\Theta(\boldsymbol{x})}\phi^\alpha(\boldsymbol{x}) = e^{i\Theta(\boldsymbol{x})+\mathcal{A}(\boldsymbol{x})}\varphi^\alpha(\boldsymbol{x}), \tag{24.1.4}$$

where

$$\Theta(\boldsymbol{x}) = m\int d^2y\,\theta(\boldsymbol{x}-\boldsymbol{y})\rho(\boldsymbol{y}), \tag{24.1.5}$$

$$\mathcal{A}(\boldsymbol{x}) = m\int d^2y\,\ln\left(\frac{|\boldsymbol{x}-\boldsymbol{y}|}{2\ell_B}\right)\rho(\boldsymbol{y}) - |z|^2 \tag{24.1.6}$$

with the use of the total number density

$$\rho(\boldsymbol{x}) = \sum_\alpha \psi^{\alpha\dagger}(\boldsymbol{x})\psi^\alpha(\boldsymbol{x}) = \sum_\alpha \phi^{\alpha\dagger}(\boldsymbol{x})\phi^\alpha(\boldsymbol{x}) = \sum_\alpha \varphi^{\alpha\dagger}(\boldsymbol{x})e^{2\mathcal{A}(\boldsymbol{x})}\varphi^\alpha(\boldsymbol{x}). \tag{24.1.7}$$

The canonical commutation relation is given by (15.3.3), or

$$[\phi^\alpha(\boldsymbol{x}), \phi^{\beta\dagger}(\boldsymbol{y})] = \delta^{\alpha\beta}\delta(\boldsymbol{x}-\boldsymbol{y}) \tag{24.1.8}$$

for the bare composite-boson field $\phi^\alpha(\boldsymbol{x})$.

The kinetic Hamiltonian is

$$H_{\text{K}} = \frac{1}{2M}\sum_\alpha\int d^2x\,\Phi^\ddagger(\boldsymbol{x})(\mathcal{P}_x - i\mathcal{P}_y)(\mathcal{P}_x + i\mathcal{P}_y)\Phi(\boldsymbol{x}) \tag{24.1.9}$$

with

$$\Phi(\boldsymbol{x}) = \begin{pmatrix}\varphi^{\text{f}}(\boldsymbol{x})\\ \varphi^{\text{b}}(\boldsymbol{x})\end{pmatrix}, \qquad \Phi^\ddagger(\boldsymbol{x}) = \left(\varphi^{\text{f}\dagger}(\boldsymbol{x}), \varphi^{\text{b}\dagger}(\boldsymbol{x})\right)e^{2\mathcal{A}(\boldsymbol{x})}. \tag{24.1.10}$$

The covariant momentum is common to both of the components,

$$\mathcal{P}_j = -i\hbar\partial_j + \hbar(\varepsilon_{jk} + \delta_{jk})\partial_k\mathcal{A}(\boldsymbol{x}). \tag{24.1.11}$$

The Coulomb interactions (22.2.12a), (22.2.12b) and the tunneling interaction (22.3.5) remain as they stand with $\rho(\boldsymbol{x}) = \Phi^\ddagger(\boldsymbol{x})\Phi(\boldsymbol{x})$ and $P^a(\boldsymbol{x}) = \frac{1}{2}\Phi^\ddagger(\boldsymbol{x})\tau^a\Phi(\boldsymbol{x})$.

The ground state $|\text{g}\rangle$ is a homogeneous configuration of electrons. It is a coherent state, upon which the composite-boson field is a constant, $\Phi(\boldsymbol{x})|\text{g}\rangle = \Phi^{\text{g}}(\boldsymbol{x})|\text{g}\rangle$,

$$\Phi^{\text{g}}(\boldsymbol{x}) = \begin{pmatrix}\varphi^{\text{g}}_{\text{f}}(\boldsymbol{x})\\ \varphi^{\text{g}}_{\text{b}}(\boldsymbol{x})\end{pmatrix} = \sqrt{\frac{\rho_0}{2}}\begin{pmatrix}e^{i\vartheta_0/2}\sqrt{1+\sigma_0}\\ e^{-i\vartheta_0/2}\sqrt{1-\sigma_0}\end{pmatrix}, \tag{24.1.12}$$

and the effective magnetic field vanishes

$$\mathcal{B}_{\text{eff}}^{\text{g}}(\boldsymbol{x}) = -\frac{\hbar}{e}\nabla^2\mathcal{A}^{\text{g}}(\boldsymbol{x}) = B_\perp - m\phi_{\text{D}}\rho_0 = 0. \tag{24.1.13}$$

The state is realizable only at the filling factor

$$\nu \equiv \frac{2\pi\hbar\rho_0}{eB} = \frac{1}{m} \tag{24.1.14}$$

as in the monolayer system.

CP¹ Field

We study the bilayer QH system at $\nu = 1$. The bare composite-boson field ϕ^α is decomposed into the U(1) field ϕ and the CP^1 field n^α as in the monolayer QH ferromagnet,

$$\phi^\alpha(\boldsymbol{x}) = \phi(\boldsymbol{x})n^\alpha(\boldsymbol{x}), \qquad \phi(\boldsymbol{x}) = e^{i\chi(\boldsymbol{x})}\sqrt{\rho(\boldsymbol{x})}. \tag{24.1.15}$$

The electron field is decomposed into its U(1) and SU(2) parts,

$$\psi^\alpha(\boldsymbol{x}) = \psi(\boldsymbol{x})n^\alpha(\boldsymbol{x}), \tag{24.1.16}$$

where

$$\psi(\boldsymbol{x}) = e^{i\Theta(\boldsymbol{x})}e^{i\chi(\boldsymbol{x})}\sqrt{\rho(\boldsymbol{x})}. \tag{24.1.17}$$

It is interpreted as a charge-pseudospin separation. The field $n^\alpha(\boldsymbol{x})$ is subject to the constraint

$$\boldsymbol{n}^\dagger(\boldsymbol{x})\boldsymbol{n}(\boldsymbol{x}) = n^{\text{f}\dagger}(\boldsymbol{x})n^{\text{f}}(\boldsymbol{x}) + n^{\text{b}\dagger}(\boldsymbol{x})n^{\text{b}}(\boldsymbol{x}) = 1. \tag{24.1.18}$$

The normalized pseudospin field (22.2.9) is expressed as

$$\mathcal{P}_a(\boldsymbol{x}) = \boldsymbol{n}^\dagger(\boldsymbol{x})\frac{\tau_a}{2}\boldsymbol{n}(\boldsymbol{x}), \tag{24.1.19}$$

and satisfies $\boldsymbol{\mathcal{P}}^2(\boldsymbol{x}) = 1/4$.

The overall phase of the CP^1 field is unphysical: See a comment below (15.4.5). Parametrizing it as

$$\boldsymbol{n}(\boldsymbol{x}) = \frac{1}{\sqrt{2}}\begin{pmatrix} e^{i\vartheta(\boldsymbol{x})/2}\sqrt{1+\sigma(\boldsymbol{x})} \\ e^{-i\vartheta(\boldsymbol{x})/2}\sqrt{1-\sigma(\boldsymbol{x})} \end{pmatrix}, \tag{24.1.20}$$

we have

$$\mathcal{P}_x(\boldsymbol{x}) = \frac{1}{2}\sqrt{1-\sigma(\boldsymbol{x})^2}\cos\vartheta(\boldsymbol{x}), \qquad \mathcal{P}_y(\boldsymbol{x}) = -\frac{1}{2}\sqrt{1-\sigma(\boldsymbol{x})^2}\sin\vartheta(\boldsymbol{x}),$$
$$\mathcal{P}_z(\boldsymbol{x}) = \frac{1}{2}\sigma(\boldsymbol{x}), \tag{24.1.21}$$

and

$$\mathcal{P}^\pm(\boldsymbol{x}) \equiv \mathcal{P}_x(\boldsymbol{x}) \pm i\mathcal{P}_y = \frac{1}{2}\sqrt{1-\sigma(\boldsymbol{x})^2}e^{\mp i\vartheta(\boldsymbol{x})}. \tag{24.1.22}$$

Recall that operator $P^+(\mathbf{x})$ $[P^-(\mathbf{x})]$ transfers one electron at $\mathbf{x}$ from the back (front) layer to the front (back) layer [see (22.2.5)]. Hence, the charge transfer is caused by the phase operator $e^{\mp i\vartheta(\mathbf{x})}$.

The pseudospin field $\mathcal{P}_a^{\rm g}(\mathbf{x}) = \langle {\rm g}|\mathcal{P}_a|{\rm g}\rangle$ in the ground state reads

$$\mathcal{P}_x^{\rm g}(\mathbf{x}) = \frac{1}{2}\sqrt{1-\sigma_0^2}\cos\vartheta_0, \qquad \mathcal{P}_y^{\rm g}(\mathbf{x}) = -\frac{1}{2}\sqrt{1-\sigma_0^2}\sin\vartheta_0,$$
$$\mathcal{P}_z^{\rm g}(\mathbf{x}) = \frac{1}{2}\sigma_0. \tag{24.1.23}$$

All these states are degenerate in the absence of the SU(2)-violating term in the Hamiltonian (24.1.3), leading to a spontaneous breaking of the SU(2) pseudospin symmetry and the birth to the Goldstone mode.

The tunneling energy $\langle H_{\rm T}\rangle \propto -\mathcal{P}_x^{\rm cl}$ is minimized by the choice of $\vartheta_0 = 0$ in (24.1.23),[1]

$$\mathcal{P}_x^{\rm g} = \frac{1}{2}\sqrt{1-\sigma_0^2}, \qquad \mathcal{P}_y^{\rm g} = 0, \qquad \mathcal{P}_z^{\rm g} = \frac{1}{2}\sigma_0, \tag{24.1.24}$$

which characterizes the actual ground state.

24.2 Effective Hamiltonian

The effective Hamiltonian for the Goldstone mode is given by (24.1.3). It consists of the exchange term (24.1.2), the capacitance term (22.5.1) and pseudo-Zeeman terms (22.4.2),

$$H_{\rm eff} = \sum_{a=x,y}\int d^2x\,\left\{2J_s^d[\partial_k\mathcal{P}_a(\mathbf{x})]^2 + 2J_s[\partial_k\mathcal{P}_z(\mathbf{x})]^2 + \epsilon_{\rm cap}\rho_0\,[\Delta\mathcal{P}_z(\mathbf{x})]^2\right\}$$
$$-\int d^2x\,[\Delta_{\rm SAS}\mathcal{P}_x(\mathbf{x}) + \Delta_{\rm bias}\mathcal{P}_z(\mathbf{x})]. \tag{24.2.1}$$

With the use of the parametrization (24.1.20), the Hamiltonian density reads

$$\mathcal{H}_{\rm eff} = \frac{J_s^d}{2}(1-\sigma^2)(\partial_k\vartheta)^2 + \frac{1}{2}\left(J_s + \frac{\sigma^2}{1-\sigma^2}J_s^d\right)(\partial_k\sigma)^2$$
$$+\frac{\epsilon_{\rm cap}\rho_0}{4}(\sigma-\sigma_0)^2 - \frac{\rho_0}{2}\left(\Delta_{\rm SAS}\sqrt{1-\sigma^2}\cos\vartheta + \Delta_{\rm bias}\sigma\right). \tag{24.2.2}$$

By minimizing the Hamiltonian (24.2.2) for the ground state we reproduce (22.4.8), or

$$\Delta_{\rm bias} = \frac{\sigma_0}{\sqrt{1-\sigma_0^2}}\Delta_{\rm SAS}. \tag{24.2.3}$$

---

[1] The phase difference becomes nonzero, $\vartheta_0 \neq 0$, in the presence of a constant tunneling current between the two layers: See (27.8.11).

Hence the Hamiltonian density (24.2.2) reads

$$\mathcal{H}_{\text{eff}} = \frac{J_s^d}{2}(1-\sigma^2)(\partial_k \vartheta)^2 + \frac{1}{2}\left(J_s + \frac{\sigma^2}{1-\sigma^2}J_s^d\right)(\partial_k \sigma)^2$$
$$+ \frac{\epsilon_{\text{cap}}\rho_0}{4}(\sigma - \sigma_0)^2 - \frac{\Delta_{\text{SAS}}\rho_0}{2}\left(\sqrt{1-\sigma^2}\cos\vartheta + \frac{\sigma_0}{\sqrt{1-\sigma_0^2}}\sigma\right). \tag{24.2.4}$$

The canonical commutation relation is derived from (24.1.8) as

$$\frac{\rho_0}{2}[\sigma(\boldsymbol{x}), \vartheta(\boldsymbol{y})] = i\delta(\boldsymbol{x} - \boldsymbol{y}), \tag{24.2.5}$$

just (15.4.15) is derived from (15.4.12): See the footnote in page 272.

We calculate the Heisenberg equations of motion

$$i\hbar\partial_t \vartheta = [\vartheta, \int d^2x\, \mathcal{H}_{\text{eff}}], \qquad i\hbar\partial_t \sigma = [\sigma, \int d^2x\, \mathcal{H}_{\text{eff}}]. \tag{24.2.6}$$

They yield

$$\hbar\partial_t \vartheta = \frac{2}{\rho_0}\partial_k(J_s^\sigma \partial_k \sigma) + \frac{2J_s^d}{\rho_0}\sigma\left[(\partial_k\vartheta)^2 - \frac{1}{(1-\sigma^2)^2}(\partial_k\sigma)^2\right]$$
$$- \epsilon_{\text{cap}}(\sigma - \sigma_0) - \frac{\sigma\cos(\vartheta - \delta_{\text{m}}x)}{\sqrt{1-\sigma^2}}\Delta_{\text{SAS}} + \frac{\sigma_0}{\sqrt{1-\sigma_0^2}}\Delta_{\text{SAS}}, \tag{24.2.7a}$$

$$\hbar\partial_t \sigma = -\frac{2}{\rho_0}\partial_k\left(J_s^\vartheta \partial_k \vartheta\right) + \Delta_{\text{SAS}}\sqrt{1-\sigma^2}\sin(\vartheta - \delta_{\text{m}}x), \tag{24.2.7b}$$

where

$$J_s^\vartheta = (1-\sigma^2)J_s^d, \qquad J_s^\sigma = \left(J_s + \frac{\sigma^2}{1-\sigma^2}J_s^d\right). \tag{24.2.8}$$

The physical meaning of these equations reads as follows.

To investigate the physical meaning of (24.2.7a) let us make a thought experiment to move electrons from the back layer to the front layer at the position $\boldsymbol{x}$. The charge densities on the layers are

$$\rho_e^{\text{f}} = -e\rho_0(1+\sigma)/2, \qquad \rho_e^{\text{b}} = -e\rho_0(1-\sigma)/2. \tag{24.2.9}$$

The charge transfer $\delta\rho_e$ makes a work δE if the voltage difference V_{junc} is present in the junction, $\delta E = V_{\text{junc}}\delta\rho_e$. Here, $\delta\rho_e \equiv \delta\rho_e^{\text{f}} = -e\sigma(\boldsymbol{x})\rho_0/2$, and δE is the change of the energy. Consequently,

$$eV_{\text{junc}} = -\frac{2}{\rho_0}\frac{\delta\int d^2x\, \mathcal{H}_{\text{eff}}(\boldsymbol{x})}{\delta\sigma(\boldsymbol{x})} = \hbar\partial_t\vartheta(\boldsymbol{x}), \tag{24.2.10}$$

which is equal to (24.2.7a). It tells that there exists no voltage difference ($V_{\text{junc}} = 0$) between the two layers in the static configuration ($\partial_t\vartheta = 0$).

To investigate the physical meaning of (24.2.7b), we rewrite it in terms of the charge density on each layer. It follows from (24.2.9) that

$$\partial_t \rho_e^{\mathrm{f}} = -\frac{e\rho_0}{2}\partial_t \sigma = \frac{e}{\hbar}\partial_k\left(J_s^{\vartheta}\partial_k\vartheta\right) - J_{\mathrm{T}}\sqrt{1-\sigma^2}\sin\vartheta,$$
$$\partial_t \rho_e^{\mathrm{b}} = +\frac{e\rho_0}{2}\partial_t \sigma = -\frac{e}{\hbar}\partial_k\left(J_s^{\vartheta}\partial_k\vartheta\right) + J_{\mathrm{T}}\sqrt{1-\sigma^2}\sin\vartheta, \tag{24.2.11}$$

where

$$J_{\mathrm{T}} \equiv \frac{e\Delta_{\mathrm{SAS}}\rho_0}{2\hbar}. \tag{24.2.12}$$

The time derivative of the charge is associated with the current via the continuity equation,

$$\partial_t \rho_e^{\mathrm{f}} = \partial_k \mathcal{J}_k^{\mathrm{f}} - \mathcal{J}_z, \qquad \partial_t \rho_e^{\mathrm{b}} = \partial_k \mathcal{J}_k^{\mathrm{b}} + \mathcal{J}_z. \tag{24.2.13}$$

We thus identify

$$\mathcal{J}_k^{\mathrm{f}}(\boldsymbol{x}) = -\mathcal{J}_k^{\mathrm{b}}(\boldsymbol{x}) = \frac{e}{\hbar}J_s^{\vartheta}\partial_k\vartheta = \frac{eJ_s^d}{\hbar}(1-\sigma^2)\partial_k\vartheta \tag{24.2.14}$$

as the phase currents on the two layers, and

$$\mathcal{J}_z(\boldsymbol{x}) = J_{\mathrm{T}}\sqrt{1-\sigma^2}\sin\vartheta \tag{24.2.15}$$

as the tunneling current between the two layers. We conclude that the electric current flows in each layer to reduce the inhomogeneity in the interlayer phase difference $\vartheta(\boldsymbol{x})$. The phase current (24.2.14) will be derived based on the microscopic formalism in Section 32.9: See (32.9.22).

24.3 Pseudospin Waves

Small fluctuations around the ground state describe pseudospin waves, which are Goldstone modes governed by the effective Hamiltonian (24.2.4). Here we make an analysis at the balanced point, where $\sigma_0 = \langle\sigma\rangle = 0$ and $\langle\vartheta\rangle = 0$. We carry out the analysis in imbalanced configuration ($\sigma_0 \neq 0$) in Section 25.3, including the spin degree of freedom into the bilayer system: See (25.3.18).

Expanding the Hamiltonian (24.2.4) to the second order in $\sigma(\boldsymbol{x})$ and $\vartheta(\boldsymbol{x})$, we obtain

$$\mathcal{H}_{\mathrm{eff}} = \frac{J_s^d}{2}(\nabla\vartheta)^2 + \frac{\Delta_{\mathrm{SAS}}\rho_0}{4}\vartheta^2 + \frac{J_s}{2}(\nabla\sigma)^2 + \frac{\rho_0}{4}\left(\epsilon_{\mathrm{cap}} + \Delta_{\mathrm{SAS}}\right)\sigma^2, \tag{24.3.1}$$

or

$$\mathcal{H}_{\mathrm{eff}}(\boldsymbol{k}) = \left(\frac{J_s^d}{2}\boldsymbol{k}^2 + \frac{\Delta_{\mathrm{SAS}}\rho_0}{4}\right)\vartheta_{\boldsymbol{k}}^{\dagger}\vartheta_{\boldsymbol{k}} + \left(\frac{J_s}{2}\boldsymbol{k}^2 + \frac{\Delta_{\mathrm{SAS}}\rho_0}{4} + \frac{\epsilon_{\mathrm{cap}}\rho_0}{4}\right)\sigma_{\boldsymbol{k}}^{\dagger}\sigma_{\boldsymbol{k}}$$
$$= \frac{1}{2M_{\boldsymbol{k}}}\vartheta_{\boldsymbol{k}}^{\dagger}\vartheta_{\boldsymbol{k}} + \frac{M_{\boldsymbol{k}}}{8}E_{\boldsymbol{k}}^2\rho_0^2\sigma_{\boldsymbol{k}}^{\dagger}\sigma_{\boldsymbol{k}} \tag{24.3.2}$$

in the momentum space, where

$$M_{\boldsymbol{k}} = \frac{2}{2J_s^d \boldsymbol{k}^2 + \Delta_{\rm SAS}\rho_0}, \tag{24.3.3}$$

$$E_{\boldsymbol{k}}^2 = \left(\frac{2J_s^d \boldsymbol{k}^2}{\rho_0} + \Delta_{\rm SAS}\right)\left(\frac{2J_s \boldsymbol{k}^2}{\rho_0} + \epsilon_{\rm cap} + \Delta_{\rm SAS}\right). \tag{24.3.4}$$

The Hamiltonian describes a collection of harmonic oscillators with the commutation relation (see Section 4.5),

$$\frac{\rho_0}{2}[\sigma_{\boldsymbol{k}}, \vartheta_{\boldsymbol{l}}] = i\delta(\boldsymbol{k}+\boldsymbol{l}). \tag{24.3.5}$$

It is diagonalized,

$$H_{\rm eff} = \int d^2k\, E_{\boldsymbol{k}} \zeta_{\boldsymbol{k}}^\dagger \zeta_{\boldsymbol{k}}, \tag{24.3.6}$$

where

$$\zeta_{\boldsymbol{k}} = \frac{1}{2}\left(\sqrt{G_{\boldsymbol{k}}\rho_0^2}\sigma_{\boldsymbol{k}} + i\frac{1}{\sqrt{G_{\boldsymbol{k}}}}\vartheta_{\boldsymbol{k}}\right), \quad \zeta_{\boldsymbol{k}}^\dagger = \frac{1}{2}\left(\sqrt{G_{\boldsymbol{k}}\rho_0^2}\sigma_{\boldsymbol{k}}^\dagger - i\frac{1}{\sqrt{G_{\boldsymbol{k}}}}\vartheta_{\boldsymbol{k}}^\dagger\right) \tag{24.3.7}$$

with $G_{\boldsymbol{k}} = M_{\boldsymbol{k}}E_{\boldsymbol{k}}/2$. The dispersion relation $E_{\boldsymbol{k}}$ is given by (24.3.4).

When the tunneling interaction is extremely small ($\Delta_{\rm SAS} \to 0$), the dispersion relation yields

$$E_{\boldsymbol{k}} = |\boldsymbol{k}|\sqrt{\frac{2\epsilon_{\rm cap}J_s^d}{\rho_0} + \frac{4J_sJ_s^d}{\rho_0^2}\boldsymbol{k}^2}. \tag{24.3.8}$$

The ζ mode is a Goldstone mode exhibiting a linear dispersion relation as $|\boldsymbol{k}| \to 0$. It reflects the rotational symmetry of the Hamiltonian (24.2.1) with $\Delta_{\rm SAS} = 0$ around the pseudospin z axis. Here we remark that the Hamiltonian density (24.3.2) is identical to the Hamiltonian density (4.5.32). The capacitance term ($\propto \epsilon_{\rm cap}$) in (24.3.2) corresponds to the term ($\propto g$) in (4.5.32). Consequently, the capacitance energy makes the ground state a squeezed coherent state.[367] The squeezing factor is given by (4.5.35), which is finite except in the limit $\boldsymbol{k} \to 0$,

$$e^{\tau_{\boldsymbol{k}}} = \sqrt{\rho_0 G_{\boldsymbol{k}}} \propto \left(\frac{1}{\boldsymbol{k}^2}\right)^{1/4} \quad \text{as} \quad \boldsymbol{k} \to 0. \tag{24.3.9}$$

The phase and density uncertainties are given by (4.5.36),

$$\langle\Delta\vartheta_{\boldsymbol{k}}\rangle = \frac{2\pi}{L}\sqrt{M_{\boldsymbol{k}}}\langle\Delta p\rangle = \frac{2\pi}{vL}e^{2\tau_{\boldsymbol{k}}}, \qquad \langle\Delta\sigma_{\boldsymbol{k}}\rangle = \frac{2\pi}{L}\frac{2\langle\Delta q\rangle}{\hbar v^2\sqrt{M_{\boldsymbol{k}}}} = \frac{(2\pi)}{vL}e^{-2\tau_{\boldsymbol{k}}}, \tag{24.3.10}$$

where L is the size of the system. In thermodynamical limit ($L \to \infty$) there are no uncertainties for $\boldsymbol{k} \neq 0$. However, the phase uncertainty is infinite ($\langle\Delta\vartheta_{\boldsymbol{k}}\rangle \to \infty$) in the ground state ($\boldsymbol{k} \to 0$) due to (24.3.9). This is because we have set $\Delta_{\rm SAS} = 0$.

In the presence of the tunneling interaction, it is no longer a Goldstone mode but a quasi-Goldstone mode with a finite gap,

$$E_{\text{P}} = \sqrt{\epsilon_{\text{cap}}\Delta_{\text{SAS}} + \Delta_{\text{SAS}}^2}, \tag{24.3.11}$$

as follows from (24.3.4). The ground state is still a squeezed state. Nevertheless, owing to the tunneling gap Δ_{SAS}, the squeezing factor (24.3.9) remains finite as $\boldsymbol{k} \to 0$ and hence the uncertainties vanish for all value of $\boldsymbol{k}$ in thermodynamical limit ($L \to \infty$).

The QH liquid is a superfluid because of a linear dispersion relation (24.3.8) near $\boldsymbol{k} = 0$ for $\Delta_{\text{SAS}} = 0$ or because of the lack of gapless modes for $\Delta_{\text{SAS}} \neq 0$. The pseudospin wave is quantized as plasmons, and $\omega_{\text{P}} = E_{\text{P}}/\hbar$ turns out to be the frequency of the plasmon zero mode (see Section 27.7).

It is instructive to regard (24.3.1) as a classical Hamiltonian. Then, the pseudospin wave is characterized by two coherence lengths,

$$\xi_{\text{ppin}}^{\vartheta} = 2\ell_B\sqrt{\frac{\pi J_s^d}{\Delta_{\text{SAS}}}}, \qquad \xi_{\text{ppin}}^{\sigma} = 2\ell_B\sqrt{\frac{\pi J_s}{\epsilon_{\text{cap}} + \Delta_{\text{SAS}}}}, \tag{24.3.12}$$

with respect to the ϑ and σ modes, and their dispersion relations are

$$E_{\boldsymbol{k}}^{\vartheta} = \frac{2J_s^d\boldsymbol{k}^2}{\rho_0} + \Delta_{\text{SAS}}, \qquad E_{\boldsymbol{k}}^{\sigma} = \frac{2J_s\boldsymbol{k}^2}{\rho_0} + \epsilon_{\text{cap}} + \Delta_{\text{SAS}}. \tag{24.3.13}$$

The energy (24.3.4) is the geometric average, $E_{\boldsymbol{k}} = \sqrt{E_{\boldsymbol{k}}^{\vartheta}E_{\boldsymbol{k}}^{\sigma}}$, of two energies $E_{\boldsymbol{k}}^{\vartheta}$ and $E_{\boldsymbol{k}}^{\sigma}$. We conclude that the interlayer coherence develops spontaneously provided $J_s^d \gtrsim \Delta_{\text{SAS}}$. In particular, the ϑ mode is gapless for $\Delta_{\text{SAS}} = 0$, though the σ mode is gapful due to the capacitance term ϵ_{cap}. Hence, the pseudospin $\mathscr{P}$ is most likely confined to the pseudospin xy plane at the balanced point: It is said that the pseudospin system has an "easy-plane" anisotropy that gives the itinerant ferromagnet with the xy symmetry.

24.4 Anomalous Bilayer QH Currents

It is customary to regard the emergence of the Hall plateau together with the vanishing diagonal resistance as the unique signal of the QH effect. This is certainly the case in the monolayer QH system as well as in the standard bilayer QH system. However, recent experiments[268,481] have revealed an anomalous behavior of the Hall resistance in a counterflow geometry in the bilayer QH system that both the diagonal and Hall resistances vanish at the total bilayer filling factor $\nu = 1$ in samples with $\Delta_{\text{SAS}} \simeq 0$. Another anomalous Hall resistance has been reported in a drag experiment.[267] We explain[153] these experimental data based on the interlayer coherence.

Let us first review the QH current in the monolayer system with homogeneous electron density ρ_0. The electric field $\boldsymbol{E}$ drives the Hall current $\mathcal{J}_i$ into the direction perpendicular to it,

$$\mathcal{J}_i = \frac{e^2\ell_B^2}{\hbar}\varepsilon_{ij}E_j\rho_0. \tag{24.4.1}$$

We apply the electric field so that the current flows into the x direction, as implies

$$E_x = 0. \tag{24.4.2}$$

Hence the diagonal resistance vanishes,

$$R_{xx} \equiv \frac{E_x}{\mathcal{J}_x} = 0, \tag{24.4.3}$$

and the Hall current is given by

$$\mathcal{J}_x = \frac{e^2\ell_B^2}{\hbar}E_y\rho_0. \tag{24.4.4}$$

The signals of the QH effect consist of the dissipationless current (24.4.3) and the development of the Hall plateau at a magic filling factor ν.

What occurs in actual systems is as follows. We feed the current $\mathcal{J}_x$ into the x direction. Due to the Lorentz force an electric field E_y that makes the given amount of current $\mathcal{J}_x$ flow into the x direction [Fig.24.2]. The relation between the current and the electric field is fixed kinematically by the formula (24.4.4) in the monolayer system.

We proceed to study the QH current in the bilayer system with $\Delta_{\text{SAS}} = 0$. As we have seen in the previous section, the interlayer phase field $\vartheta(\boldsymbol{x})$ is gapless while the imbalance field $\sigma(\boldsymbol{x})$ is gapful at the balanced point ($\sigma_0 = 0$). This property remains true in imbalanced configuration ($\sigma_0 \neq 0$): See (25.3.19). Consequently, the excitation in $\sigma(\boldsymbol{x})$ is suppressed at sufficiently low energy, and we can assume homogeneous electron densities $\rho_0^{\text{f}} = (1+\sigma_0)\rho_0/2$ and $\rho_0^{\text{b}} = (1-\sigma_0)\rho_0/2$ in the two layers. We imagine the electric fields E_j^{f} and E_j^{b} driving the Hall currents to flow into the x direction [Fig.24.2]. Naively we would expect the Hall current (24.4.4) in each layer,

$$\mathcal{J}_i^{\text{f}} = \frac{e^2\ell_B^2}{\hbar}\varepsilon_{ij}E_j^{\text{f}}\rho_0^{\text{f}}, \qquad \mathcal{J}_i^{\text{b}} = \frac{e^2\ell_B^2}{\hbar}\varepsilon_{ij}E_j^{\text{b}}\rho_0^{\text{b}}. \tag{24.4.5}$$

However, as we have argued in (24.2.14), the Goldstone mode carries the electric current as well. The basic formula for the current is the sum of the phase current (24.2.14) and the standard QH current (24.4.5),

$$\mathcal{J}_i^{\text{f}} = \frac{e}{\hbar}(1-\sigma_0^2)J_s^d\partial_i\vartheta + \frac{e^2\ell_B^2}{\hbar}\varepsilon_{ij}E_j^{\text{f}}\rho_0^{\text{f}}, \tag{24.4.6a}$$

$$\mathcal{J}_i^{\text{b}} = -\frac{e}{\hbar}(1-\sigma_0^2)J_s^d\partial_i\vartheta + \frac{e^2\ell_B^2}{\hbar}\varepsilon_{ij}E_j^{\text{b}}\rho_0^{\text{b}}. \tag{24.4.6b}$$

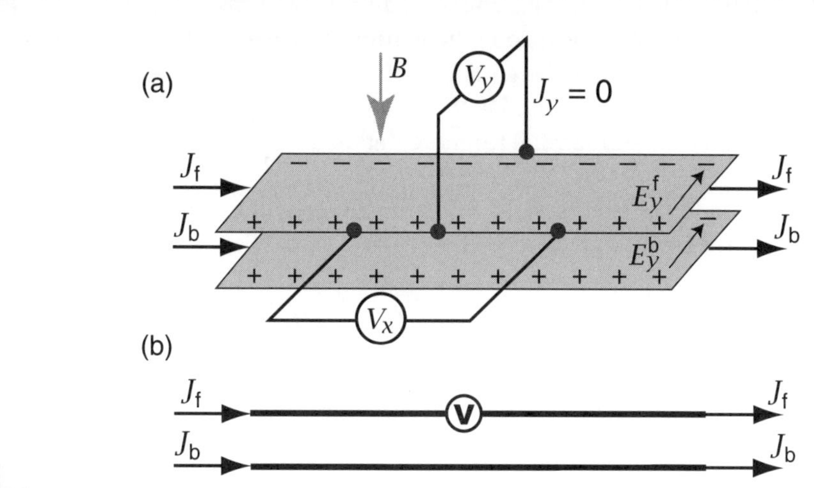

Fig. 24.2 (a) Hall currents are injected to the front and back layers independently. The Hall and diagonal resistances are measured only in one of the layers. (b) A simplified picture is given to represent the same measurement, where the symbol V in a circle indicates that the measurement is done on this layer.

We shall derive these formulas based on the microscopic formalism in Section 32.9: See (32.9.22).

Since our system is assumed to be homogeneous in the y direction, the variables depend only on x. Thus,

$$E_x^f = E_x^b = 0. \qquad \partial_y \vartheta = 0, \tag{24.4.7}$$

and

$$R_{xx}^f \equiv \frac{E_x^f}{\mathcal{J}_x^f} = 0, \qquad R_{xx}^b \equiv \frac{E_x^b}{\mathcal{J}_x^b} = 0, \tag{24.4.8}$$

for $\mathcal{J}_x^f \neq 0$ and $\mathcal{J}_x^b \neq 0$. The Hall current is given by

$$\mathcal{J}_x^f = \frac{e}{\hbar}(1 - \sigma_0^2) J_s^d \partial_x \vartheta + \frac{e^2 \ell_B^2 \rho_0}{2\hbar}(1 + \sigma_0) E_y^f, \tag{24.4.9a}$$

$$\mathcal{J}_x^b = -\frac{e}{\hbar}(1 - \sigma_0^2) J_s^d \partial_x \vartheta + \frac{e^2 \ell_B^2 \rho_0}{2\hbar}(1 - \sigma_0) E_y^b. \tag{24.4.9b}$$

Consequently the relation between the current and the electric field is not fixed kinematically in the presence of the interlayer phase difference ϑ.

Any set of E_y^f, E_y^b and ϑ seems to yield the given amounts of currents $\mathcal{J}_x^f$ and $\mathcal{J}_x^b$ provided they satisfy (24.4.9). In the actual system the unique set of them is realized: It is the one that minimizes the energy of the system. It is a dynamical problem how E_y^f and E_y^b are determined in the bilayer system.

A bilayer system consists of the two layers and the volume between them. We have so far neglected the electric field in the volume between the two layers. However, it is necessary to analyze the equation of motion also for the electric field. Equivalently it is necessary to minimize the Coulomb energy stored in the volume between the two layers. See (32.9.23) for more detailed discussions on this point in Section 32.9.

The energy due to the electric field $\boldsymbol{E}(\boldsymbol{x}, z)$ is given by the sum of the Maxwell term and the source term,

$$H_{\text{Maxwell}} = \frac{\varepsilon}{2}\int d^2x dz\, \boldsymbol{E}^2(\boldsymbol{x}, z) + ec\int d^2x dz\, A_0(\boldsymbol{x}, z)\rho(\boldsymbol{x}, z), \tag{24.4.10}$$

where

$$E_x = c\partial_x A_0, \qquad E_y = c\partial_y A_0, \qquad E_z = \frac{cA_0^{\text{f}} - cA_0^{\text{b}}}{d}, \tag{24.4.11}$$

with d the interlayer separation.

When there are constant fields E_y^{f} and E_y^{b} into the y axis on the layers, the electric field E_z between the two layers is given by

$$E_z = \frac{cA_0^{\text{f}} - cA_0^{\text{b}}}{d} = \frac{E_y^{\text{f}} - E_y^{\text{b}}}{d}y. \tag{24.4.12}$$

We carry out the integration over z and then over the plane in (24.4.10),

$$\begin{aligned} H_{\text{Maxwell}} =& \frac{\varepsilon d_w}{2}\int d^2x \left((E_y^{\text{f}})^2 + (E_y^{\text{b}})^2\right) + \frac{\varepsilon d}{2}\int d^2x\, E_z^2 \\ =& \frac{\varepsilon d_w L^2}{2}\left((E_y^{\text{f}})^2 + (E_y^{\text{b}})^2\right) + \frac{\varepsilon L^4}{24d}(E_y^{\text{f}} - E_y^{\text{b}})^2, \end{aligned} \tag{24.4.13}$$

where d_w is the thickness of the layer, and L is the size of the sample. Note that there is no contribution from the source term,

$$\int d^2x dz\, A_0(\boldsymbol{x}, z)\rho(\boldsymbol{x}, z) = d_w \sum_{\alpha=\text{f,b}} \rho^\alpha E_y^\alpha \int_{-L/2}^{L/2} dx \int_{-L/2}^{L/2} dy\, y = 0 \tag{24.4.14}$$

due to the parity. The energy density is given by

$$\mathcal{H}_{\text{Maxwell}} = \frac{\varepsilon d_w}{2}\left((E_y^{\text{f}})^2 + (E_y^{\text{b}})^2\right) + \frac{\varepsilon L^2}{24d}(E_y^{\text{f}} - E_y^{\text{b}})^2. \tag{24.4.15}$$

The important observation is that, unless $E_y^{\text{f}} = E_y^{\text{b}}$, the second term diverges in the large limit of the sample size L. The order of the sample size is $L \simeq 1$ mm, while the typical size parameter is $\ell_B \simeq d \simeq d_B \simeq 10$ nm, as implies $L/\ell_B \simeq 10^6$. It is a good approximation to take the limit $L \to \infty$. We then find

$$E_y^{\text{f}} = E_y^{\text{b}} \tag{24.4.16}$$

to make the energy density finite.

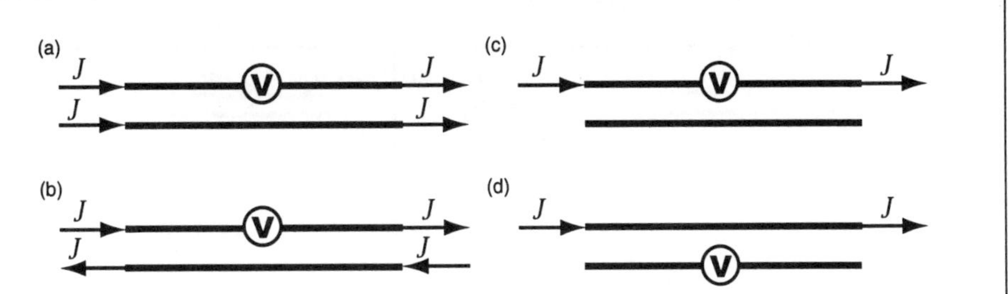

Fig. 24.3 (a) The same amount of current flows on both layers in the same direction. (b) The same amount of current flows on both layers in the opposite directions. (c) and (d) The current flows only on the front layer. In these experiments the diagonal and Hall resistances are measured at one of the layers indicated by the symbol V in a circle. The tunneling interaction is assumed negligible.

We rewrite (24.4.9) as

$$E_y^{\mathrm{f}} = \frac{2\hbar}{e^2\ell_B^2\rho_0}\left[\frac{\mathcal{J}_x^{\mathrm{f}}}{1+\sigma_0} - \frac{e}{\hbar}(1-\sigma_0)J_s^d\partial_x\vartheta\right], \tag{24.4.17a}$$

$$E_y^{\mathrm{b}} = \frac{2\hbar}{e^2\ell_B^2\rho_0}\left[\frac{\mathcal{J}_x^{\mathrm{b}}}{1-\sigma_0} + \frac{e}{\hbar}(1+\sigma_0)J_s^d\partial_x\vartheta\right]. \tag{24.4.17b}$$

The condition $E_y^{\mathrm{f}} = E_y^{\mathrm{b}}$ requires

$$E_y^{\mathrm{f}} - E_y^{\mathrm{b}} = \frac{2\hbar}{e^2\ell_B^2\rho_0}\left(\frac{\mathcal{J}_x^{\mathrm{f}}}{1+\sigma_0} - \frac{\mathcal{J}_x^{\mathrm{b}}}{1-\sigma_0} - \frac{2eJ_s^d}{\hbar}\partial_x\vartheta\right) = 0, \tag{24.4.18}$$

or

$$\partial_x\vartheta = \frac{\hbar}{2eJ_s^d}\left(\frac{\mathcal{J}_x^{\mathrm{f}}}{1+\sigma_0} - \frac{\mathcal{J}_x^{\mathrm{b}}}{1-\sigma_0}\right). \tag{24.4.19}$$

Substituting (24.4.19) into (24.4.17) we obtain

$$E_y^{\mathrm{f}} = E_y^{\mathrm{b}} = \frac{\hbar}{e^2\ell_B^2\rho_0}\left(\mathcal{J}_x^{\mathrm{f}} + \mathcal{J}_x^{\mathrm{b}}\right). \tag{24.4.20}$$

We conclude that the Hall resistance is given by

$$R_{xy}^{\mathrm{f}} \equiv \frac{E_y^{\mathrm{f}}}{\mathcal{J}_x^{\mathrm{f}}} = \frac{2\pi\hbar}{\nu e^2}\left(1 + \frac{\mathcal{J}_x^{\mathrm{b}}}{\mathcal{J}_x^{\mathrm{f}}}\right), \qquad R_{xy}^{\mathrm{b}} \equiv \frac{E_y^{\mathrm{b}}}{\mathcal{J}_x^{\mathrm{b}}} = \frac{2\pi\hbar}{\nu e^2}\left(1 + \frac{\mathcal{J}_x^{\mathrm{f}}}{\mathcal{J}_x^{\mathrm{b}}}\right) \tag{24.4.21}$$

in terms of the Hall currents. We note that both the diagonal and Hall resistances are independent of the imbalance parameter σ_0.

We apply these formulas to analyze typical bilayer Hall currents illustrated in Fig.24.3.

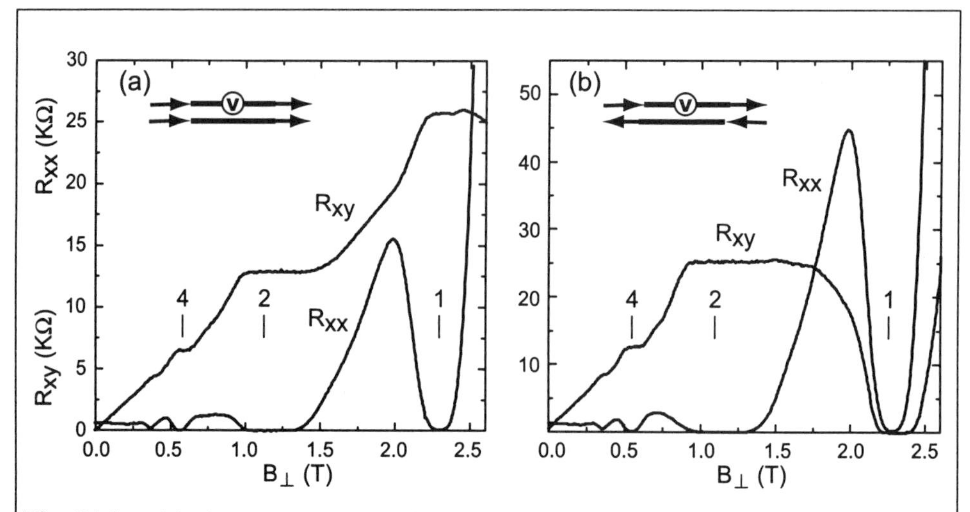

Fig. 24.4 (a) The same amounts of currents are feeded into the same direction in the front and back layers. The diagonal and Hall resistances are measured at the front layers. Their behaviors are the standard ones as the QH effect. (b) The same amounts of currents are feeded into the opposite directions in the front and back layers. The diagonal and Hall resistances are measured at the front layers. The behavior of the diagonal resistance is the standard one as the QH effect, but the behavior of the Hall resistance is anomalous at $\nu = 1$, where the interlayer coherence develops. The data are taken from Tutuc et al.[481]

Experiment (a)

The same amounts of currents are fed to the two layers in the experiment [Fig.24.3(a)]. Since $\mathcal{J}_x^{\rm f} = \mathcal{J}_x^{\rm b}$, we obtain from (24.4.19) that

$$\vartheta = \text{constant}, \tag{24.4.22}$$

and

$$R_{xy}^{\rm f} \equiv \frac{E_y^{\rm f}}{\mathcal{J}_x^{\rm f}} = \frac{4\pi\hbar}{\nu e^2}, \qquad R_{xy}^{\rm b} \equiv \frac{E_y^{\rm b}}{\mathcal{J}_x^{\rm b}} = \frac{4\pi\hbar}{\nu e^2}, \tag{24.4.23}$$

as has been measured experimentally.[268,481] This leads to the standard result of the bilayer Hall current,

$$R_{xy} \equiv \frac{E_y}{\mathcal{J}_x} = \left(\frac{1}{R_{xy}^{\rm f}} + \frac{1}{R_{xy}^{\rm b}}\right)^{-1} = \frac{2\pi\hbar}{\nu e^2} \tag{24.4.24}$$

with $E_y = E_y^{\rm f} = E_y^{\rm b}$ and $\mathcal{J}_x = \mathcal{J}_x^{\rm f} + \mathcal{J}_x^{\rm b}$, where the Hall resistance is measured as a total contribution from the front and back layers.

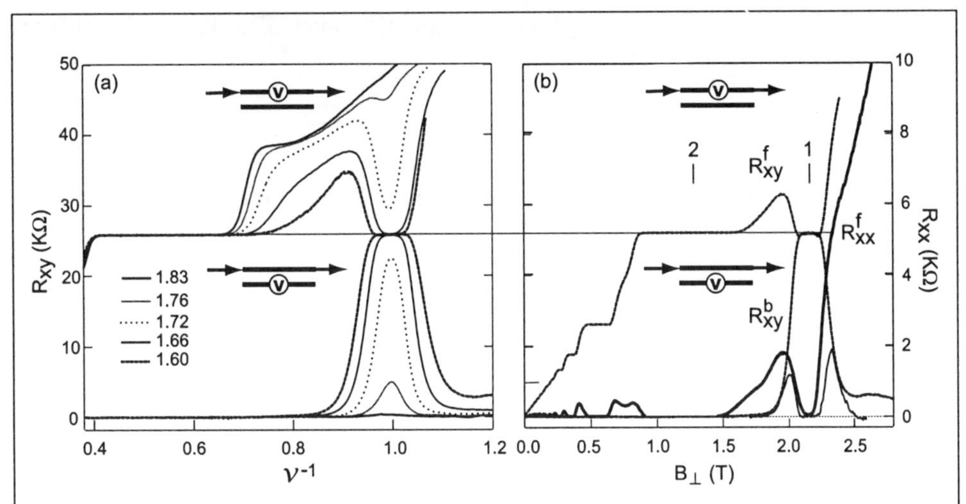

Fig. 24.5 The current flows only on the front layer. The diagonal and Hall resistances are measured at one of the layers indicated by solid circles. (a) At $\nu = 1$, the Hall plateau and hence the interlayer coherence develop better as the effective layer separation $d/\ell_B \propto \sqrt{\rho_0}$ becomes smaller. (b) The diagonal resistance vanishes at $\nu = 1$, where the QH state emerges. Layer densities are $\rho_f = \rho_b = 2.6 \times 10^{10}$ cm^{-2}, giving $d/\ell_B = 1.6$ at $\nu = 1$. The data are taken from Kellogg et al.[267]

Counterflow Experiment (b)

The counterflow experiment [Fig.24.3(b)] is most interesting, where the same amounts of currents are fed to the two layers in the opposite direction. Since $\mathcal{J}_x^f = -\mathcal{J}_x^b$, we obtain from (24.4.19) that

$$\vartheta = \frac{\hbar \mathcal{J}_x^f}{e J_s^d} x + \text{constant}, \tag{24.4.25}$$

and

$$R_{xy}^f \equiv \frac{E_y^f}{\mathcal{J}_x^f} = 0, \qquad R_{xy}^b \equiv \frac{E_y^b}{\mathcal{J}_x^b} = 0. \tag{24.4.26a}$$

The result is remarkable since it is against the naive picture of the QH effect. Recall that the essential signal of the QH effect is considered to be the development of the plateau. The vanishing of the Hall resistance in the QH regime is a new phenomenon. This anomalous behavior has been observed experimentally by Kellogg et al.[268] and Tutuc et al.[481] at $\nu = 1$, as illustrated in Fig.24.4.

Drag Current (c) & (d)

The drag experiment [Figs.24.3(c) and (d)] is also very interesting, where the current is fed only to the front layer. The Hall resistance is measure in the front layer [Fig.(c)] and also in the back layer [Figs.(d)]. Since $\mathcal{J}_x^{\rm b} = 0$, we obtain from (24.4.19) that

$$\vartheta = \frac{\hbar \mathcal{J}_x^{\rm f}}{2eJ_s^d} x + \text{constant}. \tag{24.4.27}$$

In the drag experiment the definition (24.4.21) for $R_{xy}^{\rm b}$ becomes meaningless since $\mathcal{J}_x^{\rm b} = 0$. We adopt the definition

$$\mathcal{R}_{xy}^{\rm b} \equiv \frac{E_y^{\rm b}}{\mathcal{J}_x^{\rm f}} \tag{24.4.28}$$

with the use of $\mathcal{J}_x^{\rm f}$. Then, we find

$$R_{xy}^{\rm f} \equiv \frac{E_y^{\rm f}}{\mathcal{J}_x^{\rm f}} = \frac{2\pi\hbar}{\nu e^2}, \qquad \mathcal{R}_{xy}^{\rm b} \equiv \frac{E_y^{\rm b}}{\mathcal{J}_x^{\rm f}} = \frac{2\pi\hbar}{\nu e^2}. \tag{24.4.29a}$$

In particular, we have

$$R_{xy}^{\rm f} = \mathcal{R}_{xy}^{\rm b} = \frac{E_y^{\rm f}}{\mathcal{J}_x^{\rm f}} = \frac{2\pi\hbar}{e^2} \qquad \text{at} \quad \nu = 1. \tag{24.4.30}$$

On the other hand, if there is no interlayer coherence, at the balance point, the Hall current in the front layer is

$$\mathcal{J}_i^{\rm f} = \frac{\nu e^2}{4\pi\hbar} \varepsilon_{ij} E_j^{\rm f}, \tag{24.4.31}$$

and

$$R_{xy}^{\rm f} = \frac{E_y^{\rm f}}{\mathcal{J}_x^{\rm f}} = \frac{4\pi\hbar}{\nu e^2}. \tag{24.4.32}$$

In particular we have

$$R_{xy}^{\rm f} = \frac{E_y^{\rm f}}{\mathcal{J}_x^{\rm f}} = \frac{2\pi\hbar}{e^2} \qquad \text{at} \quad \nu = 2. \tag{24.4.33}$$

It is prominent that from (24.4.30) and (24.4.33) the Hall resistance is the same at $\nu = 1$ and $\nu = 2$. These theoretical results have been confirmed by the drag experimental data due to Kellogg et al.[267], as illustrated in Fig.24.5.

24.5 Pseudospin Texture

Charged excitations are CP^1-skyrmions, as in spin ferromagnets. Indeed, the LLL condition follows from the kinetic Hamiltonian (24.1.9),

$$(\mathcal{P}_x + i\mathcal{P}_y)\,\Phi(\boldsymbol{x})|\mathfrak{S}\rangle = -\frac{i\hbar}{\ell_B}\frac{\partial}{\partial z^*}\Phi(\boldsymbol{x})|\mathfrak{S}\rangle = 0, \tag{24.5.1}$$

where $\Phi(x)$ is the two-component dressed composite-boson field (24.1.10). Hence the N-body wave function is analytic and symmetric in N variables,

$$\mathfrak{S}_{\rm CB}[z] = \langle 0|\Phi(x_1)\cdots\Phi(x_N)|\mathfrak{S}\rangle. \tag{24.5.2}$$

The nontrivial simplest analytic function is $\mathfrak{S}_{\rm CB}[z] = \prod_r \mathfrak{S}(z_r)$ with[2]

$$\tilde{\mathfrak{S}}(z) = \begin{pmatrix} \langle \varphi^{\rm B}(x)\rangle \\ \langle \varphi^{\rm A}(x)\rangle \end{pmatrix} = \sqrt{\rho_0}\begin{pmatrix} z \\ \kappa \end{pmatrix} \tag{24.5.3}$$

in terms of the bonding-antibonding state, and

$$\mathfrak{S}(z) = \begin{pmatrix} \langle \varphi^{\rm B}(x)\rangle \\ \langle \varphi^{\rm A}(x)\rangle \end{pmatrix} = \sqrt{\frac{\rho_0}{2}}\begin{pmatrix} z\sqrt{1+\sigma_0} + \kappa\sqrt{1-\sigma_0} \\ z\sqrt{1-\sigma_0} - \kappa\sqrt{1+\sigma_0} \end{pmatrix} \tag{24.5.4}$$

in terms of the layer fields. The CP^1 field (24.1.15) is

$$\boldsymbol{n}_{\rm sky}(x) = \begin{pmatrix} n^{\rm f}(x) \\ n^{\rm b}(x) \end{pmatrix} = \frac{1}{\sqrt{2(z^2+\kappa^2)}}\begin{pmatrix} z\sqrt{1+\sigma_0} + \kappa\sqrt{1-\sigma_0} \\ z\sqrt{1-\sigma_0} - \kappa\sqrt{1+\sigma_0} \end{pmatrix}. \tag{24.5.5}$$

It describes a CP^1-skyrmion. The electron wave function is

$$\mathfrak{S}_{\rm sky}[x] = \prod_r \begin{pmatrix} z_r\sqrt{1+\sigma_0} + \kappa\sqrt{1-\sigma_0} \\ z_r\sqrt{1-\sigma_0} - \kappa\sqrt{1+\sigma_0} \end{pmatrix} \mathfrak{S}_{\rm LN}[x]. \tag{24.5.6}$$

We call it the ppin-skyrmion since it reverses only pseudospins; on the other hand we call it the spin-skyrmion when the skyrmion reverses only spins.

The pseudospin associated with the skyrmion texture is

$$\mathcal{P}_a^{\rm sky}(x) = \rho_{\rm sky}(x)\mathcal{P}_a^{\rm sky}(x) \quad \text{with} \quad \mathcal{P}_a^{\rm sky}(x) = \boldsymbol{n}_{\rm sky}^\dagger(x)\frac{\tau_a}{2}\boldsymbol{n}_{\rm sky}(x). \tag{24.5.7}$$

The normalized pseudospin is calculated explicitly as

$$2\mathcal{P}_x^{\rm sky}(x) = \sqrt{1-\sigma_0^2}\,\sigma_{\rm sky}(x) - \sigma_0\sqrt{1-\sigma_{\rm sky}^2(x)}\cos\theta, \tag{24.5.8a}$$

$$2\mathcal{P}_y^{\rm sky}(x) = \sqrt{1-\sigma_{\rm sky}^2(x)}\sin\theta, \tag{24.5.8b}$$

$$2\mathcal{P}_z^{\rm sky}(x) = \sigma_0\sigma_{\rm sky}(x) + \sqrt{1-\sigma_0^2}\sqrt{1-\sigma_{\rm sky}^2(x)}\cos\theta, \tag{24.5.8c}$$

where the function $\sigma_{\rm sky}(x)$ is given by (16.2.18),

$$\sigma_{\rm sky}(x) = \frac{r^2 - 4(\kappa\ell_B)^2}{r^2 + 4(\kappa\ell_B)^2}. \tag{24.5.9}$$

---

[2]In the $\nu = 1$ QH ferromagnet, we shall argue in Section 31.2 that a skyrmion state is constructed from a hole state by making a $W_\infty(2)$ rotation, which is an SU(2) rotation in the noncommutative space. In particular, as $\kappa \to 0$, the skyrmion state (24.5.3) is continuously reduced to the hole state created in the ground state where all bonding states are filled up.

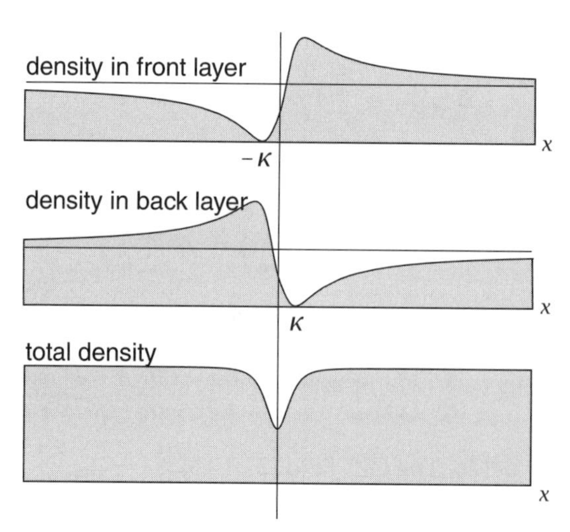

Fig. 24.6 A ppin-skyrmion excitation (24.5.5) modulates the electron density $\rho^{\alpha}(\boldsymbol{x})$ in each layer, which is depicted along the x axis in the balanced configuration. The density $\rho^{\mathrm{f}}(\boldsymbol{x})$ vanishes at $x = -\kappa$ in the front layer, while $\rho^{\mathrm{b}}(\boldsymbol{x})$ vanishes at $x = \kappa$ in the back layer, representing a vortex formation in each layer. However, there appears a density excess on the opposite layer to compensate for a charge defect induced by the vortex. A ppin-skyrmion is a kind of an excitonic excitation across the layers, carrying a total charge e/m. The total charge density ρ_{sky} is cylindrical symmetric around $x = y = 0$.

The density modulation $\Delta\rho_{\text{sky}}(\boldsymbol{x}) = \rho_{\text{sky}}(\boldsymbol{x}) - \rho_0$ is determined by the soliton equation, and given by

$$\Delta\rho_{\text{sky}}(\boldsymbol{x}) \simeq -\frac{1}{\pi}\frac{4(\kappa\ell_B)^2}{[r^2 + 4(\kappa\ell_B)^2]^2} \tag{24.5.10}$$

for a large skyrmion ($\kappa \gg 1$) at $\nu = 1/m$, as in spin ferromagnets [see (16.2.24)].

The skyrmion induces a cylindrical-asymmetric density imbalance between the two layers,

$$\Delta\mathcal{P}_z(\boldsymbol{x}) = \rho_{\text{sky}}(\boldsymbol{x})\mathcal{P}_z^{\text{sky}}(\boldsymbol{x}) - \frac{\rho_0\sigma_0}{2}. \tag{24.5.11}$$

The total charge imbalance due to the excitation is

$$q_{\mathrm{f}}^e - q_{\mathrm{b}}^e = -2e\int d^2x\,\Delta\mathcal{P}_z(\boldsymbol{x}) \equiv 2e\sigma_0 N_{\text{ppin}}, \tag{24.5.12}$$

where N_{ppin} is the number of pseudospin flips,

$$N_{\text{ppin}} = -\int d^2x\,\left\{\mathcal{P}_z^{\text{sky}}(\boldsymbol{x}) - \frac{1}{2}\rho_0\right\}. \tag{24.5.13}$$

In the vortex limit we have

$$N_{\text{ppin}} = \frac{1}{2}. \tag{24.5.14}$$

For a large skyrmion we have

$$N_{\text{ppin}} = \frac{1}{2} \int d^2x \left[\rho_0 - \rho_{\text{sky}}(\boldsymbol{x}) \sigma_{\text{sky}}(\boldsymbol{x}) \right]. \tag{24.5.15}$$

It diverges with the use of (24.5.9). We have already encountered such a divergence for a skyrmion in monolayer QH ferromagnets: See (16.3.10). As we have noticed there, this is a property of the factorizable skyrmion (24.5.6), and there is no divergence in a realistic skyrmion. Here, in order to reveal the essence, we simply cut off the divergence as in (16.3.11) for the number of flipped spins in monolayer QH ferromagnets,

$$N_{\text{ppin}} = \rho_0 \int d^2x \frac{4(\kappa_p \ell_B)^2}{r^2 + 4(\kappa \ell_B)^2} \simeq 2\kappa^2 \ln \left(\frac{\xi^2}{4\ell_B^2} + 1 \right), \tag{24.5.16}$$

where ξ is the coherence length.

We remark a feature specific to the skyrmion in the pseudospin space. The CP^1 field is given by (24.5.5). It represents a pair of vortices whose centers are located at

$$z = -\kappa \sqrt{\frac{1 - \sigma_0}{1 + \sigma_0}} \qquad \text{and} \qquad z = \kappa \sqrt{\frac{1 + \sigma_0}{1 - \sigma_0}} \tag{24.5.17}$$

in the front and back layers, respectively. Indeed, the densities vanish at these points [Fig.24.6], as implies that there are charge defects on the layers. However, lumps are created precisely at these points in the opposite layers to compensate for the defects, so that the constraint, $\rho^{\text{f}}(\boldsymbol{x}) + \rho^{\text{b}}(\boldsymbol{x}) = \rho_{\text{sky}}(\boldsymbol{x})$, is obeyed. The total number density $\rho_{\text{sky}}(\boldsymbol{x})$ is cylindrically symmetric as in (25.4.11).

We estimate the excitation energy of one ppin-skyrmion in Section 25.5.

CHAPTER 25

SU(4) QUANTUM HALL FERROMAGNETS

The bilayer QH system has four energy levels in the lowest Landau level, corresponding to the layer and spin degrees of freedom. We investigate the system in the regime where all four levels are nearly degenerate and equally active, where the underlying group structure is SU(4). At $\nu = 1$ the QH state is a charge-transferable state, and the SU(4) isospin coherence develops spontaneously. Quasiparticles are SU(4)-skyrmions. One quasiparticle consists of a pair of charged excitations in the front and back layers. The SU(4)-skyrmion evolves continuously from the ppin-skyrmion limit into the spin-skyrmion limit as the system is transformed from the balanced point to the monolayer point by controlling the bias voltage.

25.1 SU(4) ISOSPIN STRUCTURE

An electron has two types of indices, the layer index (f, b) and the spin index $(\uparrow, \downarrow)$. They are combined into the SU(4) isospin index, $\alpha = \mathrm{f}\uparrow, \mathrm{f}\downarrow, \mathrm{b}\uparrow, \mathrm{b}\downarrow$. Since the electron field has four components, the bilayer system possesses the underlying algebra SU(4). The generators are represented by Hermitian, traceless, 4×4 matrices. There are $(4^2 - 1)$ independent ones. The standard basis is customarily written as[23] $\{\lambda_A; A = 1, 2, \cdots, 15\}$ with the normalization

$$\mathrm{Tr}(\lambda_A \lambda_B) = 2\delta_{AB}. \tag{25.1.1}$$

They are constructed by generalizing Pauli matrices τ_a of SU(2) to SU(3), and then from SU(3) to SU(4), as reviewed in Appendix D. For our purpose it is better to use another basis. Embedding SU(2)⊗SU(2) into SU(4) we define the spin matrices by

$$\tau_x^{\text{spin}} = \begin{pmatrix} \tau_x & 0 \\ 0 & \tau_x \end{pmatrix}, \quad \tau_y^{\text{spin}} = \begin{pmatrix} \tau_y & 0 \\ 0 & \tau_y \end{pmatrix}, \quad \tau_z^{\text{spin}} = \begin{pmatrix} \tau_z & 0 \\ 0 & \tau_z \end{pmatrix}, \tag{25.1.2}$$

and the pseudospin matrices by

$$\tau_x^{\text{ppin}} = \begin{pmatrix} 0 & \mathbf{1}_2 \\ \mathbf{1}_2 & 0 \end{pmatrix}, \quad \tau_y^{\text{ppin}} = \begin{pmatrix} 0 & -i\mathbf{1}_2 \\ i\mathbf{1}_2 & 0 \end{pmatrix}, \quad \tau_z^{\text{ppin}} = \begin{pmatrix} \mathbf{1}_2 & 0 \\ 0 & -\mathbf{1}_2 \end{pmatrix}, \tag{25.1.3}$$

where $\mathbf{1}_2$ is the unit matrix in two dimensions. Nine remaining matrices are simple products of the spin and pseudospin matrices:

$$\tau_a^{\text{spin}}\tau_x^{\text{ppin}} = \begin{pmatrix} 0 & \tau_a \\ \tau_a & 0 \end{pmatrix}, \quad \tau_a^{\text{spin}}\tau_y^{\text{ppin}} = \begin{pmatrix} 0 & -i\tau_a \\ i\tau_a & 0 \end{pmatrix}, \quad \tau_a^{\text{spin}}\tau_z^{\text{ppin}} = \begin{pmatrix} \tau_a & 0 \\ 0 & -\tau_a \end{pmatrix}. \tag{25.1.4}$$

Let us denote them $T_{a0} \equiv \tau_a^{\text{spin}}$, $T_{0a} \equiv \tau_a^{\text{ppin}}$, $T_{ab} \equiv \tau_a^{\text{spin}}\tau_b^{\text{ppin}}$. They satisfy the normalization condition

$$\text{Tr}(T_{ab}T_{cd}) = 4\delta_{ac}\delta_{bd}, \tag{25.1.5}$$

which is different from (25.1.1). We denote the spin SU(2) as $\text{SU}_{\text{spin}}(2)$, and the pseudospin SU(2) as $\text{SU}_{\text{ppin}}(2)$. See Appendix D for more details with respect to generating matrices of SU(4), where the relation between the two sets of matrices λ_A and T_{ab} is given.

All the physical operators required for the description of the system are constructed as bilinear combinations of $\psi(\boldsymbol{x})$ and $\psi^\dagger(\boldsymbol{x})$. They are 16 density operators

$$\begin{aligned}
\rho(\boldsymbol{x}) &= \psi^\dagger(\boldsymbol{x})\psi(\boldsymbol{x}), \\
\mathcal{S}_a(\boldsymbol{x}) &= \frac{1}{2}\psi^\dagger(\boldsymbol{x})\tau_a^{\text{spin}}\psi(\boldsymbol{x}), \\
\mathcal{P}_a(\boldsymbol{x}) &= \frac{1}{2}\psi^\dagger(\boldsymbol{x})\tau_a^{\text{ppin}}\psi(\boldsymbol{x}), \\
\mathcal{R}_{ab}(\boldsymbol{x}) &= \frac{1}{2}\psi^\dagger(\boldsymbol{x})\tau_a^{\text{spin}}\tau_a^{\text{ppin}}\psi(\boldsymbol{x}),
\end{aligned} \tag{25.1.6}$$

where $\mathcal{S}_a$ describes the total spin, $2\mathcal{P}_z$ measures the electron-density difference between the two layers. The operator $\mathcal{R}_{ab}$ transforms as a spin under $\text{SU}_{\text{spin}}(2)$ and as a pseudospin under $\text{SU}_{\text{ppin}}(2)$. To see this we perform an $\text{SU}_{\text{spin}}(2)$ transformation,

$$\psi \to \exp(i\omega_a\tau_a^{\text{spin}})\psi, \tag{25.1.7}$$

under which

$$\mathcal{S}_a \to \Sigma_{ab}\mathcal{S}_b, \quad \mathcal{P}_a \to \mathcal{P}_a, \quad \mathcal{R}_{ab} \to \Sigma_{ac}\mathcal{R}_{cb}, \tag{25.1.8}$$

where

$$\Sigma_{ab}\tau_b^{\text{spin}} = \exp(-i\omega_a\tau_a^{\text{spin}})\tau_a^{\text{spin}}\exp(i\omega_a\tau_a^{\text{spin}}). \tag{25.1.9}$$

An $\text{SU}_{\text{ppin}}(2)$ transformation is similarly treated.

The Landau level structure is generated by the kinetic Hamiltonian,

$$H_{\text{K}} = \frac{1}{2M}\sum_\alpha \int d^2x\psi^{\alpha\dagger}(\boldsymbol{x})(P_x^2 + P_y^2)\psi^\alpha(\boldsymbol{x}), \tag{25.1.10}$$

where the sum runs over $\alpha = \text{f}\uparrow, \text{f}\downarrow, \text{b}\uparrow, \text{b}\downarrow$. The kinetic Hamiltonian is invariant under the global SU(4) transformation. The Coulomb interaction is decomposed into the SU(4)-invariant and SU(4)-noninvariant terms as in (22.2.11),

$$H_C^+ = \frac{1}{2}\int d^2x d^2y\, V^+(\boldsymbol{x}-\boldsymbol{y})\Delta\rho(\boldsymbol{x})\Delta\rho(\boldsymbol{y}), \tag{25.1.11a}$$

$$H_C^- = 2\int d^2x d^2y\, V^-(\boldsymbol{x}-\boldsymbol{y})\Delta\mathcal{P}_z(\boldsymbol{x})\Delta\mathcal{P}_z(\boldsymbol{y}), \tag{25.1.11b}$$

with

$$V^\pm(\boldsymbol{x}) = \frac{e^2}{8\pi\varepsilon}\left(\frac{1}{|\boldsymbol{x}|} \pm \frac{1}{\sqrt{|\boldsymbol{x}|^2+d^2}}\right), \tag{25.1.12}$$

and $2\mathcal{P}_z = \rho^{f\uparrow} + \rho^{f\downarrow} - \rho^{b\uparrow} - \rho^{b\downarrow}$. The other interaction terms are the Zeeman and pseudo-Zeeman terms,

$$\mathcal{H}_Z = -\Delta_Z \mathcal{S}_z(\boldsymbol{x}), \quad \mathcal{H}_{pZ} = -\Delta_{SAS}\mathcal{P}_x(\boldsymbol{x}) - \Delta_{bias}\mathcal{P}_z(\boldsymbol{x}), \tag{25.1.13}$$

with $\mathcal{S}_a(\boldsymbol{x}) = \mathcal{S}_a^f(\boldsymbol{x}) + \mathcal{S}_a^b(\boldsymbol{x})$ and $\mathcal{P}_a(\boldsymbol{x}) = \mathcal{P}_a^\uparrow(\boldsymbol{x}) + \mathcal{P}_a^\downarrow(\boldsymbol{x})$. The bias term plays a role of creating an imbalanced configuration with σ_0 as in (22.4.8), or

$$\Delta_{bias} = \frac{\sigma_0}{\sqrt{1-\sigma_0^2}}\Delta_{SAS}. \tag{25.1.14}$$

The total Hamiltonian is

$$H = H_K + H_C^+ + H_C^- + H_Z + H_{pZ}. \tag{25.1.15}$$

The lowest Landau level contains four energy levels.

To explore the one-body ground state at $\nu = 1$, we study the zero-momentum Hamiltonian,

$$H_{ZpZ}^0 = \mathcal{H}_Z + \mathcal{H}_{pZ} = -N_\Phi\left(\Delta_{SAS}\mathcal{P}_x + \Delta_{bias}\mathcal{P}_z + \Delta_Z\mathcal{S}_z\right). \tag{25.1.16}$$

It is diagonalized as [Fig.25.1]

$$H_0 = -\frac{1}{2}\begin{pmatrix} \Delta_{BAB}+\Delta_Z & 0 & 0 & 0 \\ 0 & \Delta_{BAB}-\Delta_Z & 0 & 0 \\ 0 & 0 & -\Delta_{BAB}+\Delta_Z & 0 \\ 0 & 0 & 0 & -\Delta_{BAB}-\Delta_Z \end{pmatrix} \tag{25.1.17}$$

with the eigenstates $|B\uparrow\rangle$, $|B\downarrow\rangle$, $|A\uparrow\rangle$ and $|A\downarrow\rangle$, where

$$\Delta_{BAB} = \frac{\Delta_{SAS}}{\sqrt{1-\sigma_0^2}} \tag{25.1.18}$$

in the imbalanced configuration at σ_0. The one-body ground state is the up-spin bonding state. The $\nu = 1$ QH state is generated when all up-spin bonding states are filled up. It is the spin ferromagnet and pseudospin ferromagnet state.

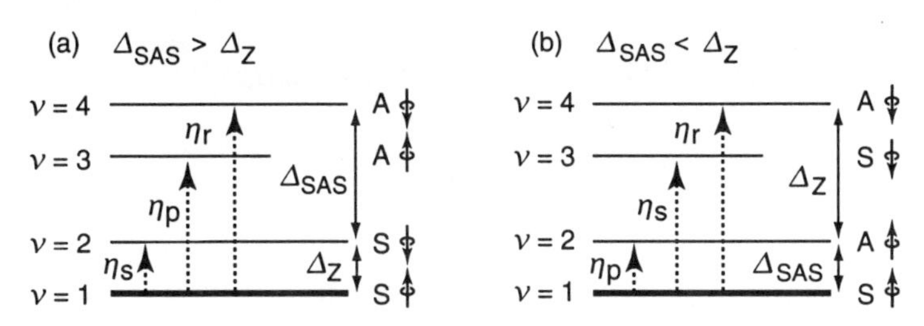

Fig. 25.1 The lowest Landau level contains four energy levels corresponding to the two layers and the two spin states. They are represented (a) for $\Delta_{\text{SAS}} > \Delta_{\text{Z}}$ and (b) for $\Delta_{\text{SAS}} < \Delta_{\text{Z}}$. The lowest-energy level consists of up-spin symmetric states in the balanced configuration, and is filled at $\nu = 1$. It is the spin ferromagnet and pseudospin ferromagnet state. Small fluctuations are Goldstone modes η_{s}, η_{p} and η_{r}.

25.2 SU(4) Isospin Fields

The dynamical field is the CP^3 field at $\nu = 1$. It is introduced precisely as the CP^1 field in the monolayer system. All formulas from (24.1.4) to (24.1.17) hold as they stand when the index α runs over the SU(4) index. Indeed we make a charge-isospin separation of the electron field,

$$\psi^\alpha(\boldsymbol{x}) = \psi(\boldsymbol{x}) n^\alpha(\boldsymbol{x}), \tag{25.2.1}$$

where the U(1) field $\psi(\boldsymbol{x})$ is defined by (24.1.17). Here, $n^\alpha(\boldsymbol{x})$ is the CP^3 field subject to the normalization condition

$$\sum_{\alpha=1}^{4} n^{\alpha\dagger}(\boldsymbol{x}) n^\alpha(\boldsymbol{x}) = 1. \tag{25.2.2}$$

The U(1) field $\psi(\boldsymbol{x})$ carries the charge and the CP^3 field $n^\alpha(\boldsymbol{x})$ carries the isospin. Each component has the density $\rho^\alpha(\boldsymbol{x}) = \rho(\boldsymbol{x}) n^{\alpha\dagger}(\boldsymbol{x}) n^\alpha(\boldsymbol{x})$. The U(1) field $\psi(\boldsymbol{x})$ is dynamically frozen by a large cyclotron gap as in the monolayer QH ferromagnet. The CP^3 field contains 3 independent complex fields, which are the dynamical fields in the bilayer QH system at $\nu = 1$.

The SU(4) isospin field and its normalized one are given by

$$\mathcal{I}_A(\boldsymbol{x}) = \psi^\dagger(\boldsymbol{x}) \frac{\lambda_A}{2} \psi(\boldsymbol{x}), \qquad \mathscr{I}_A(\boldsymbol{x}) = \boldsymbol{n}^\dagger(\boldsymbol{x}) \frac{\lambda_A}{2} \boldsymbol{n}(\boldsymbol{x}) \tag{25.2.3}$$

in the standard basis, where $\mathscr{I}(\boldsymbol{x})$ is normalized as $\mathscr{I}(\boldsymbol{x})\mathscr{I}(\boldsymbol{x}) = 3/8$ at $\nu = 1$ in accord with (25.1.1): See also (30.2.32a). Note that $\mathcal{I}_A(\boldsymbol{x}) = \rho(\boldsymbol{x}) \mathscr{I}_A(\boldsymbol{x})$. It is also given by

$$\mathcal{S}_a(\boldsymbol{x}) = \boldsymbol{n}^\dagger(\boldsymbol{x})\frac{\tau_a^{\text{spin}}}{2}\boldsymbol{n}(\boldsymbol{x}), \quad \mathcal{P}_a(\boldsymbol{x}) = \boldsymbol{n}^\dagger(\boldsymbol{x})\frac{\tau_a^{\text{ppin}}}{2}\boldsymbol{n}(\boldsymbol{x}),$$
$$\mathcal{R}_{ab}(\boldsymbol{x}) = \boldsymbol{n}^\dagger(\boldsymbol{x})\frac{\tau_a^{\text{spin}}\tau_b^{\text{ppin}}}{2}\boldsymbol{n}(\boldsymbol{x}), \tag{25.2.4}$$

with

$$\boldsymbol{n}^T = \left(n^{\text{f}\uparrow}, n^{\text{f}\downarrow}, n^{\text{b}\uparrow}, n^{\text{b}\downarrow}\right). \tag{25.2.5}$$

The normalization

$$\sum_a \mathcal{S}_a(\boldsymbol{x})\mathcal{S}_a(\boldsymbol{x}) + \sum_a \mathcal{P}_a(\boldsymbol{x})\mathcal{P}_a(\boldsymbol{x}) + \sum_{ab} \mathcal{R}_{ab}(\boldsymbol{x})\mathcal{R}_{ab}(\boldsymbol{x}) = 3/4 \tag{25.2.6}$$

holds at $\nu = 1$ in accord with (25.1.5).

The kinetic Hamiltonian (25.1.10) is a constant in the lowest Landau level. On the other hand, as we see in Section 32.6, from the exchange interaction the SU(4)-invariant Coulomb Hamiltonian (25.1.11a) yields the SU(4) nonlinear sigma model,

$$H_{\text{X}}^{+} = 2J_s^{+}\sum_{A=1}^{15}\int d^2x\, [\partial_k \mathcal{I}_A(\boldsymbol{x})]^2 \tag{25.2.7}$$

in the standard basis [see (32.6.12)], where $J_s^{\pm} = \frac{1}{2}(J_s \pm J_s^d)$ with (A.15). By including SU(4)-noninvariant terms, the total effective Coulomb Hamiltonian is found to be

$$\mathcal{H}_{\text{C}}^{\text{eff}} = \frac{4\epsilon_{\text{D}}^{-}}{\rho_\Phi}\mathcal{P}_z^2 - \frac{2\epsilon_{\text{X}}^{-}}{\rho_\Phi}\left(\sum_a \mathcal{S}_a^2 + \mathcal{P}_z^2 + \sum_a \mathcal{R}_{az}^2\right)$$
$$+ J_s^d\left(\sum_a [\partial_k \mathcal{S}_a]^2 + \sum_a [\partial_k \mathcal{P}_a]^2 + \sum_{ab} [\partial_k \mathcal{R}_{ab}]^2\right)$$
$$+ 2J_s^{-}\left(\sum_a [\partial_k \mathcal{S}_a]^2 + [\partial_k \mathcal{P}_z]^2 + \sum_a [\partial_k \mathcal{R}_{az}]^2\right) \tag{25.2.8}$$

in the physical basis [see (32.6.35)].

We examine two important limits. When all electrons are moved to the front layer,

$$\left(n^{\text{f}\uparrow}, n^{\text{f}\downarrow}, n^{\text{b}\uparrow}, n^{\text{b}\downarrow}\right) = \left(n^{\uparrow}, n^{\downarrow}, 0, 0\right), \tag{25.2.9}$$

it follows that $\mathcal{P}_a = \frac{1}{2}\delta_{az}$ and $\mathcal{R}_{ab} = \mathcal{S}_a\delta_{bz}$. The Hamiltonian (25.2.8) is reduced to

$$\mathcal{H}_{\text{X}} = 2J_s \sum_{a=xyz} [\partial_k \mathcal{S}_a]^2, \tag{25.2.10}$$

which is the spin-ferromagnet Hamiltonian (15.1.1). Similarly, when the spin degree of freedom is frozen,

$$\left(n^{\text{f}\uparrow}, n^{\text{f}\downarrow}, n^{\text{b}\uparrow}, n^{\text{b}\downarrow}\right) = \left(n^{\text{f}}, 0, n^{\text{b}}, 0\right), \tag{25.2.11}$$

it follows that $\mathcal{S}_a = \frac{1}{2}\delta_{az}$ and $\mathcal{R}_{ab} = \delta_{az}\mathcal{P}_b$. The Hamiltonian is reduced to

$$\mathcal{H}_X = 2J_s^d \sum_{a=xy} [\partial_k \mathcal{P}_a]^2 + 2J_s [\partial_k \mathcal{P}_z]^2, \tag{25.2.12}$$

which is the pseudospin-ferromagnet Hamiltonian (24.1.2).

25.3 SU(4) Isospin Waves

The SU(4)-invariant Coulomb energy H_D^+ dominates the bilayer system unless $d \gg \ell_B$. It is reasonable to start with the SU(4)-invariant limit, where the Hamiltonian consists solely of the Coulomb term (25.1.11a). All isospins are spontaneously polarized to lower the exchange energy (25.2.7), which arises from the Coulomb interaction. There are four degenerate states, any one of which can be chosen as the ground state. It implies the spontaneous breaking of the SU(4) symmetry, giving rise to Goldstone modes and topological solitons.

We choose the ground state at the imbalanced configuration σ_0 as

$$(n^{B\uparrow}, n^{B\downarrow}, n^{A\uparrow}, n^{A\downarrow})_g = (1, 0, 0, 0) \tag{25.3.1}$$

in the bonding-antibonding representation, which reads

$$(n^{f\uparrow}, n^{f\downarrow}, n^{b\uparrow}, n^{b\downarrow}) = \left(\sqrt{\frac{1+\sigma_0}{2}}, 0, \sqrt{\frac{1-\sigma_0}{2}}, 0\right) \tag{25.3.2}$$

in the layer representation. The ground-state values of the isospin fields are

$$\begin{aligned} \mathcal{S}_a^g &= \frac{1}{2}\delta_{az}, \qquad \mathcal{P}_a^g = \frac{1}{2}(\sqrt{1-\sigma_0^2}\,\delta_{ax} + \sigma_0 \delta_{az}), \\ \mathcal{R}_{ab}^g &= \frac{1}{2}\delta_{az}(\sqrt{1-\sigma_0^2}\,\delta_{bx} + \sigma_0 \delta_{bz}). \end{aligned} \tag{25.3.3}$$

The residual symmetry keeping the ground state invariant is U(3). Thus, the pattern of the symmetry breaking is SU(4)⟶U(3). The target space is the coset space

$$CP^3 = SU(4)/U(3) = U(4)/U(1) \otimes U(3), \tag{25.3.4}$$

which is the complex-projective (CP) space. Goldstone bosons are described by the CP^3 field.

We analyze small fluctuations around the ground state at $\nu = 1/m \le 1$. We parametrizes the bonding-antibonding stateas

$$n^{B\uparrow} = \sqrt{1 - \eta_s^\dagger \eta_s - \eta_p^\dagger \eta_p - \eta_r^\dagger \eta_r}, \quad n^{B\downarrow} = \eta_s, \quad n^{A\uparrow} = \eta_p, \quad n^{A\downarrow} = \eta_r, \tag{25.3.5}$$

with $[\eta_i(\boldsymbol{x}), \eta_j^\dagger(\boldsymbol{y})] = \rho_0^{-1}\delta_{ij}\delta(\boldsymbol{x}-\boldsymbol{y})$, where the mode $\eta_s(\boldsymbol{x})$ describes the spin wave, $\eta_p(\boldsymbol{x})$ the pseudospin wave discussed in Section 24.3, and $\eta_r(\boldsymbol{x})$ the additional wave connecting the ground state to the highest level in the lowest Landau level [Fig.25.1]. The layer field reads

$$\begin{pmatrix} n^{f\uparrow} \\ n^{f\downarrow} \\ n^{b\uparrow} \\ n^{b\downarrow} \end{pmatrix} = \frac{1}{\sqrt{2}} \begin{pmatrix} \sqrt{1+\sigma_0} & 0 & \sqrt{1-\sigma_0} & 0 \\ 0 & \sqrt{1+\sigma_0} & 0 & \sqrt{1-\sigma_0} \\ \sqrt{1-\sigma_0} & 0 & -\sqrt{1+\sigma_0} & 0 \\ 0 & \sqrt{1-\sigma_0} & 0 & -\sqrt{1+\sigma_0} \end{pmatrix} \begin{pmatrix} n^{B\uparrow} \\ n^{B\downarrow} \\ n^{A\uparrow} \\ n^{A\uparrow} \end{pmatrix}. \tag{25.3.6}$$

Expanding

$$(n^{B\uparrow}, n^{B\downarrow}, n^{A\uparrow}, n^{A\downarrow}) = \left(1, \eta_s, \eta_p, \eta_r\right) + \cdots, \tag{25.3.7}$$

for small fluctuations around the ground state, we obtain

$$\begin{aligned} n^{f\uparrow} &\simeq \sqrt{1+\sigma_0} + \eta_p\sqrt{1-\sigma_0}, \qquad & n^{f\downarrow} &\simeq \eta_s\sqrt{1+\sigma_0} + \eta_r\sqrt{1-\sigma_0}, \\ n^{b\uparrow} &\simeq \sqrt{1-\sigma_0} - \eta_p\sqrt{1+\sigma_0}, \qquad & n^{b\downarrow} &\simeq \eta_s\sqrt{1-\sigma_0} - \eta_r\sqrt{1+\sigma_0} \end{aligned} \tag{25.3.8}$$

in the linear approximation.

We express the exchange interaction in terms of fluctuation fields. The SU(4)-invariant part (25.2.7) is rewritten as

$$H_X^+ = 2J_s^+ \sum_{\alpha=1}^{4} \int d^2x \left(\partial_j n^{\alpha\dagger} + iK_j n^{\alpha\dagger}\right)\left(\partial^j n^\alpha + iK_j n^\alpha\right) \tag{25.3.9}$$

with the auxiliary field $K_j = -i\sum_\alpha n^{\alpha\dagger}\partial_j n^\alpha$ in terms of the CP^3 field: See Section 7.8. This Hamiltonian is characterized by the U(1) gauge symmetry (7.8.9),

$$n_a(x) \longrightarrow n_a(x)\exp[if(x)], \qquad K_j(x) \longrightarrow K_j(x) + \partial_j f(x). \tag{25.3.10}$$

We expand (25.3.9) up to the second order,

$$\mathcal{H}_X^+ = 2J_s^+ \sum_{i=1}^{3} \partial_k \eta_i^\dagger(\boldsymbol{x})\partial_k \eta_i(\boldsymbol{x}) = \frac{J_s^+}{2}\sum_{i=1}^{3}\left\{(\partial_k\sigma_i)^2 + (\partial_k\vartheta_i)^2\right\}, \tag{25.3.11}$$

where we have set

$$\eta_i(\boldsymbol{x}) = \frac{\sigma_i(\boldsymbol{x}) + i\vartheta_i(\boldsymbol{x})}{2}, \qquad eta_i^\dagger(\boldsymbol{x}) = \frac{\sigma_i(\boldsymbol{x}) - i\vartheta_i(\boldsymbol{x})}{2}. \tag{25.3.12}$$

Here, $\rho_0\sigma_i(\boldsymbol{x})$ is the number density excited from the ground state to the i-th level designated by (25.3.8), and $\vartheta_i(\boldsymbol{x})$ is the conjugate phase field.

To derive the full effective Hamiltonian, we express the isospin field (25.2.4) in terms of the CP^3 field (25.3.8),

$$\begin{aligned}
2\mathcal{S}_a &= (\quad \sigma_s, \quad \vartheta_s, 1 - 2|\eta_s|^2 - 2|\eta_r|^2), \\
2\mathcal{P}_a &= (\quad f_-(s,p,r), -\vartheta_p, \quad g_+(s,p,r)), \\
2\mathcal{R}_{xa} &= (\sqrt{1-\sigma_0^2}\sigma_s - \sigma_0\sigma_r, -\vartheta_r, \sigma_0\sigma_s + \sqrt{1-\sigma_0^2}\sigma_r), \\
2\mathcal{R}_{ya} &= (\sqrt{1-\sigma_0^2}\vartheta_s - \sigma_0\vartheta_r, \quad \sigma_r, \sigma_0\vartheta_s + \sqrt{1-\sigma_0^2}\vartheta_r), \\
2\mathcal{R}_{ya} &= (\quad f_+(r,p,s), -\vartheta_p, \quad g_-(r,p,s)),
\end{aligned} \tag{25.3.13}$$

with

$$\begin{aligned}
f_\pm(s,p,r) &= \sqrt{1-\sigma_0^2} - \sigma_0\sigma_p - 2\sqrt{1-\sigma_0^2}(|\eta_p|^2 + |\eta_r|^2) \pm \sigma_0(\eta_s^*\eta_r + \eta_r^*\eta_s), \\
g_\pm(s,p,r) &= \sigma_0 + \sqrt{1-\sigma_0^2}\sigma_p - 2\sigma_0(|\eta_p|^2 + |\eta_r|^2) \pm \sqrt{1-\sigma_0^2}(\eta_s^*\eta_r + \eta_r^*\eta_s).
\end{aligned} \tag{25.3.14}$$

Substitute them into (25.2.8), we obtain the effective Coulomb Hamiltonian,

$$\begin{aligned}
\mathcal{H}_C^{\text{eff}} =& \frac{1}{2}[(1-\sigma_0^2)J_s + \sigma_0^2 J_s^d](\partial_k\sigma_p)^2 + \frac{1}{2}J_s^d(\partial_k\vartheta_p)^2 + \frac{\rho_0}{4}\epsilon_{\text{cap}}(1-\sigma_0^2)\sigma_p^2 \\
&+ \frac{1}{2}J_s[(\partial_k\sigma_s)^2 + (\partial_k\vartheta_s)^2 + (\partial_k\sigma_r)^2 + (\partial_k\vartheta_r)^2] \\
&- \frac{1}{2}J_s^-[(1-\sigma_0^2)\{(\partial_k\sigma_s)^2 + (\partial_k\vartheta_s)^2\} + (1+\sigma_0^2)\{(\partial_k\sigma_r)^2 + (\partial_k\vartheta_r)^2\}] \\
&+ J_s^-\sigma_0\sqrt{1-\sigma_0^2}(\partial_k\sigma_s\partial_k\sigma_r + \partial_k\vartheta_2\partial_k\vartheta_r),
\end{aligned} \tag{25.3.15}$$

where we have neglected irrelevant constant terms. The two modes η_s and η_r are coupled by the SU(4)-noninvariant exchange interaction in imbalanced configuration.

Goldstone modes are made gapful by the Zeeman and pseudo-Zeeman terms,

$$\begin{aligned}
\mathcal{H}_{\text{ZpZ}}^{\text{eff}} &= -\rho_0[\Delta_Z\mathcal{S}_z + \Delta_{\text{SAS}}\mathcal{P}_x + \Delta_{\text{bias}}\mathcal{P}_z] \\
&= \frac{\rho_0}{4}\Delta_Z(\sigma_s^2 + \vartheta_s^2 + \sigma_r^2 + \vartheta_r^2) + \frac{\rho_0}{4}\frac{\Delta_{\text{SAS}}}{\sqrt{1-\sigma_0^2}}(\sigma_p^2 + \vartheta_p^2 + \sigma_r^2 + \vartheta_r^2),
\end{aligned} \tag{25.3.16}$$

where Δ_{bias} is given by (25.1.14), or

$$\Delta_{\text{bias}} = \frac{\sigma_0}{\sqrt{1-\sigma_0^2}}\Delta_{\text{SAS}}. \tag{25.3.17}$$

We combine the Coulomb term $\mathcal{H}_C^{\text{eff}}$ and the Zeeman-pseudo-Zeeman term $\mathcal{H}_{\text{ZpZ}}^{\text{eff}}$ to derive the total effective Hamiltonian.

The pseudospin mode is decoupled from the others, and the effective Hamiltonian reads

$$\mathcal{H}_{\mathrm{ppin}} = \frac{(1-\sigma_0^2)J_s + \sigma_0^2 J_s^d}{2}(\partial_k \sigma_{\mathrm{p}})^2 + \frac{\rho_0}{4}\left[\epsilon_{\mathrm{cap}}(1-\sigma_0^2) + \frac{\Delta_{\mathrm{SAS}}}{\sqrt{1-\sigma_0^2}}\right]\sigma_{\mathrm{p}}^2$$
$$+ \frac{1}{2}J_s^d(\partial_k \vartheta_{\mathrm{p}})^2 + \frac{\rho_0}{4}\frac{\Delta_{\mathrm{SAS}}}{\sqrt{1-\sigma_0^2}}\vartheta_{\mathrm{p}}^2, \tag{25.3.18}$$

as is reduced to the effective Hamiltonian (24.3.1) at the balanced point in the pseudospin ferromagnet. The coherence lengths of the interlayer phase field ϑ_{p} and the imbalance field σ_{p} are

$$\xi_{\mathrm{ppin}}^{\vartheta} = 2\ell_B\sqrt{\frac{\pi\sqrt{1-\sigma_0^2}J_s^d}{\Delta_{\mathrm{SAS}}}},$$
$$\xi_{\mathrm{ppin}}^{\sigma} = 2\ell_B\sqrt{\frac{\pi\left[(1-\sigma_0^2)J_s + \sigma_0^2 J_s^d\right]}{\epsilon_{\mathrm{cap}}(1-\sigma_0^2) + \Delta_{\mathrm{SAS}}/\sqrt{1-\sigma_0^2}}}. \tag{25.3.19}$$

The ϑ_{p} mode is gapless even in imbalanced configuration for $\Delta_{\mathrm{SAS}} = 0$, though the σ mode is gapful due to the capacitance term ϵ_{cap}.

To study the other two modes η_{s} and η_{r}, we change the variables as

$$\eta_s = \sqrt{\frac{1+\sigma_0}{2}}\eta_1 + \sqrt{\frac{1-\sigma_0}{2}}\eta_2, \qquad \eta_r = \sqrt{\frac{1-\sigma_0}{2}}\eta_1 - \sqrt{\frac{1+\sigma_0}{2}}\eta_2. \tag{25.3.20}$$

The effective Hamiltonian reads

$$\mathcal{H}_{\mathrm{mix}} = \frac{J_s^+ + \sigma_0 J_s^-}{2}[(\partial_k\sigma_1)^2 + (\partial_k\vartheta_1)^2] + \frac{\rho_0}{4}\left(\Delta_{\mathrm{Z}} + \frac{1}{2}\Delta_{\mathrm{SAS}}\sqrt{\frac{1-\sigma_0}{1+\sigma_0}}\right)[\sigma_1^2 + \vartheta_1^2]$$
$$+ \frac{J_s^+ - \sigma_0 J_s^-}{2}[(\partial_k\sigma_2)^2 + (\partial_k\vartheta_2)^2] + \frac{\rho_0}{4}\left(\Delta_{\mathrm{Z}} + \frac{1}{2}\Delta_{\mathrm{SAS}}\sqrt{\frac{1+\sigma_0}{1-\sigma_0}}\right)[\sigma_2^2 + \vartheta_2^2]$$
$$+ \frac{\rho_0}{4}\Delta_{\mathrm{SAS}}(\sigma_1\sigma_2 + \vartheta_1\vartheta_2). \tag{25.3.21}$$

The two modes η_1 and η_2 are decoupled for $\Delta_{\mathrm{SAS}} = 0$. There exist no gapless modes in this Hamiltonian provided $\Delta_{\mathrm{Z}} \neq 0$.

25.4 SU(4) Isospin Textures

Charged excitations are investigated as before [see Sections 16.3 and 24.5]. The LLL condition reads

$$(\mathcal{P}_x + i\mathcal{P}_y)\,\varphi^{\alpha}(\mathbf{x})|\mathfrak{S}\rangle = -\frac{i\hbar}{\ell_B}\frac{\partial}{\partial z^*}\varphi^{\alpha}(\mathbf{x})|\mathfrak{S}\rangle = 0. \tag{25.4.1}$$

The N-body wave function is analytic and symmetric in N variables,

$$\mathfrak{S}_{\rm CB}[z] = \langle 0|\Phi(\boldsymbol{x}_1)\cdots\Phi(\boldsymbol{x}_N)|\mathfrak{S}\rangle. \tag{25.4.2}$$

The wave function for electrons is given by

$$\mathfrak{S}[\boldsymbol{x}] = \mathfrak{S}_{\rm CB}[z]\mathfrak{S}_{\rm LN}[\boldsymbol{x}] \tag{25.4.3}$$

as in (15.3.23) for the monolayer system, where $\mathfrak{S}_{\rm LN}[\boldsymbol{x}]$ is the Laughlin wave function.

The analysis is simple when the analytic function is factorizable, $\mathfrak{S}_{\rm CB}[z] = \prod_r \mathfrak{S}(z_r)$. It follows that $\mathfrak{S}(z) = \langle\Phi(\boldsymbol{x})\rangle$ and that $\langle \boldsymbol{n}(\boldsymbol{x})\rangle = \mathfrak{S}(z)/|\mathfrak{S}(z)|$, or

$$n^\alpha(\boldsymbol{x}) = \frac{\omega^\alpha(z)}{\sqrt{\sum_{\alpha=1}^4 |\omega^\alpha(z)|^2}} \tag{25.4.4}$$

with $\omega^\alpha(z) = \sqrt{\rho_0}\langle\varphi^\alpha(\boldsymbol{x})\rangle$. The nontrivial simplest analytic function is

$$\widetilde{\mathfrak{S}}(z) = \begin{pmatrix} \langle\varphi^{\rm B\uparrow}(\boldsymbol{x})\rangle \\ \langle\varphi^{\rm B\downarrow}(\boldsymbol{x})\rangle \\ \langle\varphi^{\rm A\uparrow}(\boldsymbol{x})\rangle \\ \langle\varphi^{\rm A\downarrow}(\boldsymbol{x})\rangle \end{pmatrix} = \sqrt{\rho_0}\begin{pmatrix} z \\ \kappa_s \\ \kappa_p \\ \kappa_r \end{pmatrix} \tag{25.4.5}$$

in terms of the bonding-antibonding state. The CP^3 field reads[1]

$$\widetilde{\boldsymbol{n}}(\boldsymbol{x}) = \begin{pmatrix} n^{\rm B\uparrow}(\boldsymbol{x}) \\ n^{\rm B\downarrow}(\boldsymbol{x}) \\ n^{\rm A\uparrow}(\boldsymbol{x}) \\ n^{\rm A\downarrow}(\boldsymbol{x}) \end{pmatrix} = \frac{1}{\sqrt{z^2+\kappa^2}}\begin{pmatrix} z \\ \kappa_s \\ \kappa_p \\ \kappa_r \end{pmatrix} \tag{25.4.6}$$

with $\kappa^2 = \kappa_s^2 + \kappa_p^2 + \kappa_r^2$, or

$$\boldsymbol{n}(\boldsymbol{x}) = \begin{pmatrix} n^{\rm f\uparrow}(\boldsymbol{x}) \\ n^{\rm f\downarrow}(\boldsymbol{x}) \\ n^{\rm b\uparrow}(\boldsymbol{x}) \\ n^{\rm b\downarrow}(\boldsymbol{x}) \end{pmatrix}_{\rm sky} = \frac{1}{\sqrt{2(z^2+\kappa^2)}}\begin{pmatrix} z\sqrt{1+\sigma_0} + \kappa_p\sqrt{1-\sigma_0} \\ \kappa_s\sqrt{1+\sigma_0} + \kappa_r\sqrt{1-\sigma_0} \\ z\sqrt{1-\sigma_0} - \kappa_p\sqrt{1+\sigma_0} \\ \kappa_s\sqrt{1-\sigma_0} - \kappa_r\sqrt{1+\sigma_0} \end{pmatrix} \tag{25.4.7}$$

in terms of the layer field. It turns out to describe a SU(4)-skyrmion.

The SU(4)-skyrmion emerges as a topological soliton according to the homotopy theorem $\pi_2(\mathrm{CP}^3) = \mathbb{Z}$ in an incompressible fluid (see Section 15.4), where $\mathrm{CP}^3 = \mathrm{SU}(4)/[\mathrm{U}(1)\otimes\mathrm{SU}(3)]$. The topological charge density is

$$J^0_{\rm sky}(\boldsymbol{x}) = \frac{1}{4\pi}\boldsymbol{\nabla}^2 \ln\left[\sum_{\alpha=1}^4 |\omega^\alpha(z)|^2\right]. \tag{25.4.8}$$

[1] In the $\nu = 1$ QH ferromagnet, we shall argue in Section 31.2 that a skyrmion state is constructed from a hole state by making a $W_\infty(4)$ rotation, which is an SU(4) rotation in the noncommutative space. In particular, as $\kappa \to 0$, the skyrmion state (25.4.6) is continuously reduced to the hole state created in the ground state where all bonding up-spin states are filled up.

It is calculated with (25.4.6) as

$$J^0_{\text{sky}}(\boldsymbol{x}) = \frac{1}{\pi}\frac{4(\kappa\ell_B)^2}{[r^2+4(\kappa\ell_B)^2]^2}. \tag{25.4.9}$$

The topological charge is one, $\int d^2x\, J^0_{\text{sky}}(\boldsymbol{x}) = 1$. The soliton equation,

$$\frac{1}{4\pi m}\nabla^2 \ln\rho(\boldsymbol{x}) - \rho(\boldsymbol{x}) + \rho_0 = \frac{1}{m}J^0_{\text{sky}}(\boldsymbol{x}), \tag{25.4.10}$$

follows from the LLL condition (25.4.1). It is solved as $\rho_{\text{sky}}(\boldsymbol{x}) = \rho_0 + \delta\rho_{\text{sky}}(\boldsymbol{x})$, where

$$\Delta\rho_{\text{sky}}(\boldsymbol{x}) \simeq -\frac{1}{\pi m}\frac{4(\kappa\ell_B)^2}{[r^2+4(\kappa\ell_B)^2]^2} \tag{25.4.11}$$

for a large skyrmion, and

$$\Delta\rho_{\text{sky}}(\boldsymbol{x}) \simeq -\rho_0\left(1+\frac{\sqrt{2}r}{\ell_B}-\frac{r^2}{3\ell_B^2}\right)e^{-\sqrt{2}r/\ell_B} \tag{25.4.12}$$

in the small-skyrmion limit ($\kappa \to 0$), where $J^0_{\text{sky}}(\boldsymbol{x}) \to \delta(\boldsymbol{x})$. They have the same expressions as in the case of the monolayer spin-skyrmion [see Section 16.2].

It is seen in (25.4.6) that three parameters κ_i represent how electrons are excited in various energy levels [Fig.25.1]. They are to be determined to minimize the skyrmion excitation energy. In a special limit only the spin or the pseudospin component is excited. In this case we refer to it as the spin-skyrmion or the ppin-skyrmion.

By setting $\kappa = \kappa_s$, $\kappa_p = \kappa_r = 0$, the spin-skyrmion limit is

$$\tilde{\boldsymbol{n}}^{\text{spin}}_{\text{sky}}(\boldsymbol{x}) = \begin{pmatrix} n^{\text{B}\uparrow}(\boldsymbol{x}) \\ n^{\text{B}\downarrow}(\boldsymbol{x}) \\ n^{\text{A}\uparrow}(\boldsymbol{x}) \\ n^{\text{A}\downarrow}(\boldsymbol{x}) \end{pmatrix}^{\text{spin}}_{\text{sky}} = \frac{1}{\sqrt{z^2+\kappa^2}}\begin{pmatrix} z \\ \kappa \\ 0 \\ 0 \end{pmatrix} \tag{25.4.13}$$

in terms of the bonding-antibonding state, or

$$\boldsymbol{n}^{\text{spin}}_{\text{sky}}(\boldsymbol{x}) = \begin{pmatrix} n^{\text{f}\uparrow}(\boldsymbol{x}) \\ n^{\text{f}\downarrow}(\boldsymbol{x}) \\ n^{\text{b}\uparrow}(\boldsymbol{x}) \\ n^{\text{b}\downarrow}(\boldsymbol{x}) \end{pmatrix}^{\text{spin}}_{\text{sky}} = \frac{1}{\sqrt{2(z^2+\kappa^2)}}\begin{pmatrix} z\sqrt{1+\sigma_0} \\ \kappa\sqrt{1+\sigma_0} \\ z\sqrt{1-\sigma_0} \\ \kappa\sqrt{1-\sigma_0} \end{pmatrix} \tag{25.4.14}$$

in terms of the layer field. The number density in each layer is given by

$$\begin{aligned} \rho^{\text{f}}(\boldsymbol{x}) &= \rho^{\text{f}\uparrow}(\boldsymbol{x}) + \rho^{\text{f}\downarrow}(\boldsymbol{x}) = \frac{1+\sigma_0}{2}\rho_{\text{sky}}(\boldsymbol{x}), \\ \rho^{\text{b}}(\boldsymbol{x}) &= \rho^{\text{b}\uparrow}(\boldsymbol{x}) + \rho^{\text{b}\downarrow}(\boldsymbol{x}) = \frac{1-\sigma_0}{2}\rho_{\text{sky}}(\boldsymbol{x}). \end{aligned} \tag{25.4.15}$$

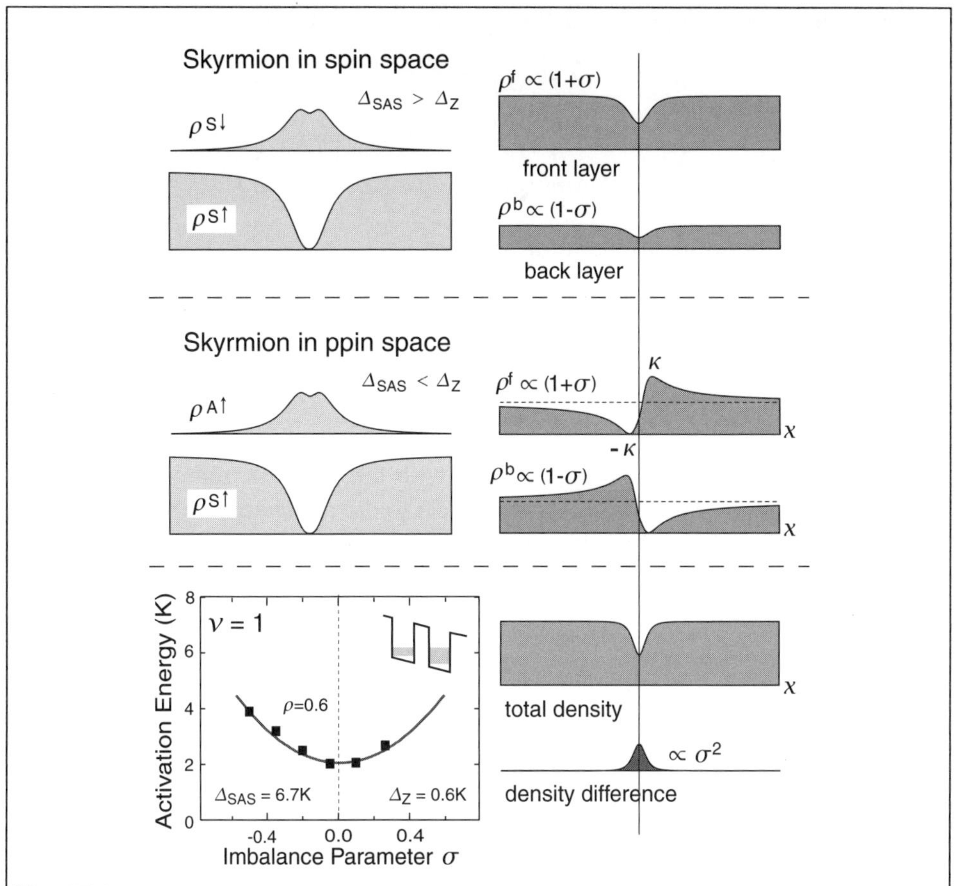

Fig. 25.2 (a) A spin-skyrmion (spin-antiskyrmion) modifies the number density in the up-spin and down-spin bonding states. It has the charge $\pm e/m$ at $\nu = 1/m$. (b) The same skyrmion modulates the number densities on both layers. It has the charge $\pm(1+\sigma_0)e/2m$ on the front layer, and $\pm(1-\sigma_0)e/2m$ on the back layer. Hence, the capacitance energy is proportional to $(\sigma_0 e/m)^2$.

One spin-skyrmion (as well as spin-antiskyrmion) represents a hole in the up-spin bonding state and a lump in the down-spin bonding state [Fig.25.2(a)]. It implies a pair of holes for a spin-skyrmion and a pair of lumps for a spin-antiskyrmion in two layers [Fig.25.2(b)]. There arises a local density difference. The net charge imbalance

$$q_f^e - q_b^e = -2e \int d^2x\, \Delta P_z(\boldsymbol{x}) = \frac{e\sigma_0}{m} \tag{25.4.16}$$

is associated with a pair of holes for a spin-skyrmion.

By setting $\kappa = \kappa_p$, $\kappa_s = \kappa_r = 0$, the ppin-skyrmion limit is

$$\tilde{\boldsymbol{n}}_{\text{sky}}^{\text{ppin}}(\boldsymbol{x}) = \begin{pmatrix} n^{\text{B}\uparrow}(\boldsymbol{x}) \\ n^{\text{B}\downarrow}(\boldsymbol{x}) \\ n^{\text{A}\uparrow}(\boldsymbol{x}) \\ n^{\text{A}\downarrow}(\boldsymbol{x}) \end{pmatrix}_{\text{sky}}^{\text{ppin}} = \frac{1}{\sqrt{z^2+\kappa^2}} \begin{pmatrix} z \\ 0 \\ \kappa \\ 0 \end{pmatrix} \tag{25.4.17}$$

in terms of the bonding-antibonding state, or

$$\boldsymbol{n}_{\text{sky}}^{\text{ppin}}(\boldsymbol{x}) = \begin{pmatrix} n^{\text{f}\uparrow}(\boldsymbol{x}) \\ n^{\text{f}\downarrow}(\boldsymbol{x}) \\ n^{\text{b}\uparrow}(\boldsymbol{x}) \\ n^{\text{b}\downarrow}(\boldsymbol{x}) \end{pmatrix}_{\text{sky}}^{\text{ppin}} = \frac{1}{\sqrt{2(z^2+\kappa^2)}} \begin{pmatrix} z\sqrt{1+\sigma_0}+\kappa\sqrt{1-\sigma_0} \\ 0 \\ z\sqrt{1-\sigma_0}-\kappa\sqrt{1+\sigma_0} \\ 0 \end{pmatrix} \tag{25.4.18}$$

in terms of the layer field. It is identified with the ppin-skyrmion (24.5.5) in the spin-frozen system. See (24.5.12) with respect to the net charge imbalance, as corresponds to (25.4.16) for the spin-skyrmion.

25.5 Excitation Energy of SU(4) Skyrmions

The SU(4)-skyrmion is parametrized by three shape parameters κ_s, κ_p and κ_r. Its special limit is the spin-skyrmion and the ppin-skyrmion. It is a dynamical problem what shape of skyrmion is activated thermally. Physically it is important to know how one skyrmion evolves continuously from the balanced point ($\sigma_0 = 0$) to the monolayer point ($\sigma_0 = 1$) by controlling the bias voltage. To study these problems we estimate the excitation energy of a pair of SU(4)-skyrmion and antiskyrmion. We make a further study of SU(4) skyrmion by controlling parallel magnetic field in Section 25.6.

The antiskyrmion configuration is related to the skyrmion configuration by

$$\begin{aligned} &\mathcal{S}_x^{\text{antisky}}(\boldsymbol{x}) = \mathcal{S}_x^{\text{sky}}(\boldsymbol{x}), \quad &&\mathcal{S}_y^{\text{antisky}}(\boldsymbol{x}) = -\mathcal{S}_y^{\text{sky}}(\boldsymbol{x}), \quad &&\mathcal{S}_z^{\text{antisky}}(\boldsymbol{x}) = \mathcal{S}_z^{\text{sky}}(\boldsymbol{x}), \\ &\mathcal{P}_x^{\text{antisky}}(\boldsymbol{x}) = \mathcal{P}_x^{\text{sky}}(\boldsymbol{x}), \quad &&\mathcal{P}_y^{\text{antisky}}(\boldsymbol{x}) = -\mathcal{P}_y^{\text{sky}}(\boldsymbol{x}), \quad &&\mathcal{P}_z^{\text{antisky}}(\boldsymbol{x}) = \mathcal{P}_z^{\text{sky}}(\boldsymbol{x}). \end{aligned} \tag{25.5.1}$$

A skyrmion (antiskyrmion) induces the modulation of the spin and the pseudospin,

$$\Delta S_a^{\text{(anti)sky}}(\boldsymbol{x}) = \rho_{\text{(anti)sky}}(\boldsymbol{x})\mathcal{S}_a^{\text{(anti)sky}}(\boldsymbol{x}) - \rho_0\mathcal{S}_a^{\text{g}}, \tag{25.5.2a}$$

$$\Delta P_a^{\text{(anti)sky}}(\boldsymbol{x}) = \rho_{\text{(anti)sky}}(\boldsymbol{x})\mathcal{P}_a^{\text{(anti)sky}}(\boldsymbol{x}) - \rho_0\mathcal{P}_a^{\text{g}}, \tag{25.5.2b}$$

where $\mathcal{S}_a^{\text{g}}$ and $\mathcal{P}_a^{\text{g}}$ are the ground-state values (25.3.3), and

$$\rho_{\text{sky}}(\boldsymbol{x}) = \rho_0 + \Delta\rho_{\text{sky}}(\boldsymbol{x}), \qquad \rho_{\text{antisky}}(\boldsymbol{x}) = \rho_0 - \Delta\rho_{\text{sky}}(\boldsymbol{x}) \tag{25.5.3}$$

with (25.4.11).

It is the energy of a skyrmion-antiskyrmion pair,

$$\Delta_{\text{pair}} = \frac{1}{2}(E_{\text{sky}} + E_{\text{antisky}}), \tag{25.5.4}$$

that is observed experimentally. We estimate the energy of one skyrmion.[2] For simplicity we set $\kappa_r = 0$ in the SU(4)-skyrmion (25.4.6). Though this approximation reduces the validity of some of our results, we use it to reveal an essential physics of SU(4)-skyrmions, because otherwise various formulas become too complicated to handle with. By an essential physics we mean a continuous transformation of the SU(4)-skyrmion from the ppin-skyrmion limit to the spin-skyrmion limit.

Namely we study a skyrmion parametrized by two scales κ_s and κ_p with $\kappa^2 = \kappa_s^2 + \kappa_p^2$. We calculate the SU(4) generators (25.2.4) for this field configuration,

$$\begin{aligned}
\mathcal{S}_x^{\text{sky}} &= \frac{\alpha_s x}{r^2+\alpha^2}, \quad \mathcal{S}_y^{\text{sky}} = \frac{\alpha_s y}{r^2+\alpha^2}, \quad \mathcal{S}_z^{\text{sky}} = \frac{1}{2} - \frac{\alpha_s^2}{r^2+\alpha^2}, \\
\mathcal{P}_x^{\text{sky}} &= \frac{\cos\beta}{2} - \frac{\alpha_p x\sin\beta + \alpha_p^2\cos\beta}{r^2+\alpha^2}, \quad \mathcal{P}_y^{\text{sky}} = -\frac{\alpha_p y}{r^2+\alpha^2}, \\
\mathcal{P}_z^{\text{sky}} &= \frac{\sin\beta}{2} + \frac{\alpha_p x\cos\beta - \alpha_p^2\sin\beta}{r^2+\alpha^2},
\end{aligned} \tag{25.5.5}$$

and

$$\begin{aligned}
\mathcal{R}_{xx}^{\text{sky}} &= \mathcal{S}_x^{\text{sky}}\cos\beta + \mathcal{R}_{yy}^{\text{sky}}\sin\beta, \quad \mathcal{R}_{xy}^{\text{sky}} = 0, \quad \mathcal{R}_{xz}^{\text{sky}} = \mathcal{S}_x^{\text{sky}}\sin\beta - \mathcal{R}_{yy}^{\text{sky}}\cos\beta, \\
\mathcal{R}_{yx}^{\text{sky}} &= \mathcal{S}_y^{\text{sky}}\cos\beta, \quad \mathcal{R}_{yy}^{\text{sky}} = -\frac{\alpha_s\alpha_p}{r^2+\alpha^2}, \quad \mathcal{R}_{yz}^{\text{sky}} = \mathcal{S}_y^{\text{sky}}\sin\beta, \\
\mathcal{R}_{zx}^{\text{sky}} &= \mathcal{P}_x^{\text{sky}} + \left(\mathcal{S}_z^{\text{sky}} - \frac{1}{2}\right)\cos\beta, \quad \mathcal{R}_{zy}^{\text{sky}} = \mathcal{P}_y^{\text{sky}}, \quad \mathcal{R}_{zz}^{\text{sky}} = \mathcal{P}_z^{\text{sky}} + \left(\mathcal{S}_z^{\text{sky}} - \frac{1}{2}\right)\sin\beta,
\end{aligned} \tag{25.5.6}$$

where $\alpha_s = 2\kappa_s\ell_B$, $\alpha_p = 2\kappa_p\ell_B$, $\alpha^2 = 4(\kappa\ell_B)^2$, $\cos\beta = \sqrt{1-\sigma_0^2}$ and $\sin\beta = \sigma_0$.

The skyrmion energy consists of the Coulomb energy, the Zeeman energy and the pseudo-Zeeman energy. The Coulomb energy is decomposed into the self-energy (32.6.23), the capacitance energy (32.6.31) and the exchange energy (32.6.33), as we shall see in Section 32.6.

The dominant one is the self-energy due to the SU(4)-invariant Coulomb interaction,

$$E_{\text{self}} = \frac{1}{2}\int d^2x d^2y\, \Delta\rho_{\text{sky}}(\boldsymbol{x})V^+(\boldsymbol{x}-\boldsymbol{y})\Delta\rho_{\text{sky}}(\boldsymbol{y}). \tag{25.5.7}$$

After a straightforward calculation we find [see (O.10) in Appendix]

$$E_{\text{self}}(\kappa) = \frac{1}{8\kappa}E_C^0\int z^2[K_1(z)]^2\left(1+e^{-\frac{d}{2\ell}\frac{z}{\kappa}}\right)dz. \tag{25.5.8}$$

[2]See a footnote in page 284.

It depends only on the total skyrmion scale $\kappa = \sqrt{\kappa_s^2 + \kappa_p^2}$, since the Coulomb term H_D^+ depends only on the total density $\delta\rho_{\text{sky}}(\boldsymbol{x})$.

The capacitance energy arises from the SU(4)-noninvariant Coulomb interaction,

$$H_{\text{cap}} = \frac{\epsilon_{\text{cap}}}{\rho_\Phi} \int d^2x \, \Delta\mathcal{P}_z^{\text{sky}}(\boldsymbol{x}) \Delta\mathcal{P}_z^{\text{sky}}(\boldsymbol{x}) \tag{25.5.9}$$

with (22.5.5). It is calculated as in (O.14) in Appendix O. The leading term is

$$E_{\text{cap}} \simeq \frac{1}{2}(1 - \sigma_0^2)\epsilon_{\text{cap}} N_{\text{ppin}}(\kappa_p), \tag{25.5.10}$$

where $N_{\text{ppin}}(\kappa_p)$ is the number of flipped pseudospins to be defined by (25.5.22).

The exchange energy is produced by the SU(4)-invariant and SU(4)-noninvariant Coulomb interactions. It is calculated based on (25.2.8),[3]

$$E_X^{\text{SU}(4)} = 4\pi \left[J_s^+ - \frac{1}{3} J_s^- \left\{ \frac{\kappa_p^2}{\kappa^2} - \frac{\kappa_s^2}{\kappa^2} \left(1 + 2\frac{\kappa_s^2}{\kappa^2} \right) \sigma_0^2 \right\} \right]. \tag{25.5.11}$$

It contains scale parameters explicitly via the SU(4)-noninvariant term. The SU(4)-invariant part of (25.2.8) is the SU(4) sigma model (25.2.7) yielding a topological invariant value $4\pi J_s^+$ as in (7.8.25). In the spin-skyrmion limit it is reduced to

$$E_X^{\text{spin}} = 4\pi \left(J_s^+ + J_s^- \sigma_0^2 \right) = 2\pi \left[J_s + J_s^d + (J_s - J_s^d)\sigma_0^2 \right], \tag{25.5.12}$$

which agrees with the previous formula (16.3.1) in the monolayer limit ($\sigma_0 \to 1$). In the ppin skyrmion limit it is reduced to

$$E_X^{\text{ppin}} = 4\pi (J_s^+ - \frac{1}{3} J_s^-) = \frac{4\pi}{3} \left(J_s + 2J_s^d \right), \tag{25.5.13}$$

which is independent of the imbalance parameter σ_0.

The Zeeman energy of one skyrmion is given by

$$E_Z = -\Delta_Z \int d^2x \, \Delta S_Z(\boldsymbol{x}) \equiv N_{\text{spin}}(\kappa_s)\Delta_Z, \tag{25.5.14}$$

where N_{spin} is the number of flipped spins,

$$N_{\text{spin}}(\kappa_s) = -\int d^2x \, \Delta\mathcal{S}_z(\boldsymbol{x}). \tag{25.5.15}$$

Note that

$$\Delta\mathcal{S}_z^{\text{sky}}(\boldsymbol{x}) + \Delta\mathcal{S}_z^{\text{antisky}}(\boldsymbol{x}) = 2\rho_0 \left(\mathcal{S}_a^{\text{sky}}(\boldsymbol{x}) - \mathcal{S}_a^{\text{g}} \right) \tag{25.5.16}$$

[3]The effective Hamiltonian (25.2.8) is obtained by setting $\rho(\boldsymbol{x}) = \rho_0$ in (32.6.33). It is necessary to calculate the exchange energy based on (32.6.33) instead of (25.2.8) since the density is not uniform around a skyrmion. Nevertheless, this is a good approximation for a large skyrmion. See Appendix O, where the exchange energy is calculated by the use of (32.6.33).

with (25.5.5). By neglecting the term which is cancelled out in a skyrmion-antiskyrmion pair excitation, we obtain

$$N_{\text{spin}}(\kappa_s) = \rho_0 \int d^2x \frac{4(\kappa_s \ell_B)^2}{r^2 + 4(\kappa \ell_B)^2} \simeq \kappa_s^2 N_\xi \tag{25.5.17}$$

with

$$N_\xi = 2\ln\left(1 + \frac{\xi^2}{4\ell_B^2}\right), \tag{25.5.18}$$

where the divergence has been cut off at $r \simeq \kappa\xi$ with a typical coherence length ξ as in the case of a monolayer skyrmion: See (16.3.9)~(16.3.11).

The pseudo-Zeeman energy (22.4.12) is given by

$$E_{\text{pZ}} = -\Delta_{\text{SAS}} \int d^2x \left[\Delta\mathcal{P}_x(\boldsymbol{x}) + \frac{\sigma_0}{\sqrt{1-\sigma_0^2}}\Delta\mathcal{P}_z(\boldsymbol{x})\right]. \tag{25.5.19}$$

The pseudospin-density modulation is given by (25.5.2). Note that

$$\Delta\mathcal{P}_a^{\text{sky}}(\boldsymbol{x}) + \Delta\mathcal{P}_a^{\text{antisky}}(\boldsymbol{x}) = 2\rho_0\left(\mathscr{P}_a^{\text{sky}}(\boldsymbol{x}) - \mathscr{P}_a^{\text{g}}\right) \qquad \text{for} \quad a = x, z \tag{25.5.20}$$

with (25.5.5). From (25.5.19) we extract the terms which are not cancelled out in a skyrmion-antiskyrmion pair excitation. It is equal to

$$E_{\text{pZ}} = \frac{N_{\text{ppin}}(\kappa_p)}{\sqrt{1-\sigma_0^2}}\Delta_{\text{SAS}}, \tag{25.5.21}$$

where N_{ppin} is the number of flipped pseudospins,

$$N_{\text{ppin}}(\kappa_p) = \rho_0 \int d^2x \frac{4(\kappa_p \ell_B)^2}{r^2 + 4(\kappa \ell_B)^2} \simeq \kappa_p^2 N_\xi \tag{25.5.22}$$

with (25.5.18): We have cut off the divergence as in (25.5.17).

The total excitation energy is

$$E_{\text{sky}} = E_{\text{X}}^{\text{SU(4)}} + E_{\text{self}} + E_{\text{cap}} + E_{\text{Z}} + E_{\text{pZ}} \tag{25.5.23}$$

with (25.5.11), (25.5.8), (25.5.10), (25.5.14) and (25.5.21). Skyrmion parameters κ_s and κ_p with $\kappa^2 = \kappa_s^2 + \kappa_p^2$ are to be determined by minimizing the excitation energy (25.5.23).

According to an examination of (25.5.23) presented at the end of this section, provided

$$\frac{4\pi J_s^-}{3} > \left(\frac{1}{2}\epsilon_{\text{cap}} - \Delta_{\text{Z}} + \Delta_{\text{SAS}}\right)\kappa^2 N_\xi, \tag{25.5.24}$$

ppin-skyrmions are excited at the balanced point ($\sigma_0 = 0$): See (25.5.28). Then, it evolves continuously into a spin-skyrmion at the monolayer point ($\sigma_0 = 1$) via a generic skyrmion ($\kappa_s\kappa_p \neq 0$). As we have stated, to simplify calculations we have set $\kappa_r = 0$ in the SU(4)-skyrmion (25.4.6). When we allow a skyrmion to be excited into the r-level ($\kappa_r \neq 0$), a genuine SU(4)-skyrmions ($\kappa_s \kappa_p \kappa_r \neq 0$) would be excited except at the monolayer point, as illustrated in Fig.25.3.

We examine the condition (25.5.24) numerically. When we adopt typical values of sample parameters ($\rho_0 = 1.2 \times 10^{11}/\text{cm}^2$ and $d = 231\text{Å}$), we find the capacitance (22.5.5) to be $\epsilon_{\text{cap}} \simeq 132\text{K}$ while the exchange-energy difference to be $4\pi J_s^-/3 \simeq 5.4\text{K}$. The condition is hardly satisfied. Here, we question the validity of the standard identification of the layer separation, $d = d_B + d_W$, where $d_W \simeq 200\text{Å}$ is the width of a quantum well and $d_B \simeq 31\text{Å}$ is the separation of the two quantum wells. In this identification the electron cloud is assumed to be localized in the center of each quantum well. However, it is a dynamical problem. Let us minimize the ppin-skyrmion energy as a function of d. The minimum is found to be achieved at $d < d_B$. Namely, the energy increases monotonically for $d > d_B$. Then, it would be reasonable to use $d = d_B$ as the layer separation to estimate the energy of the ppin-skyrmion. When we choose $d \simeq 31\text{Å}$ we find $\epsilon_{\text{cap}} \simeq 4.5\text{K}$. Then, the condition (25.5.24) is satisfied in some parameter regions, where ppin-skyrmions are excited.

However, the condition (25.5.24) cannot be taken literally, because it rules out excitations of large ppin-skyrmions against experimental indications. It seems that the exchange energy has been underestimated by making the derivative expansion. It is necessary to go beyond the present approximation to fully understand the problem.

We proceed to estimate the energies of a spin-skyrmion and a ppin skyrmion in all range of σ_0, because their estimation is much more reliable than that of a generic SU(4)-skyrmion. We use sample parameters $\rho_0 = 1.2 \times 10^{11}/\text{cm}^2$, $d = 31\text{Å}$ and $\Delta_{\text{SAS}} = 1\text{K}$, and normalize the energy to the experimental data[472] at the monolayer point for a spin-skyrmion and at the balanced point for a spin-skyrmion. Recall that the absolute value of the activation energy cannot be determined theoretically because they depends essentially on samples: See Section 12.1.

First, we calculate the energy of a spin-skyrmion as a function of σ_0 by setting $\kappa = \kappa_s$ and $\kappa_p = 0$ in (25.5.23). The skyrmion scale κ, to be determined by minimizing this, depends on σ_0 very weakly. We obtain $N_{\text{spin}}(\kappa) \simeq 3.5$ as in the monolayer QH ferromagnet. We have depicted the excitation energy $E_{\text{sky}}^{\text{spin}}$ in Fig.25.3, where it is normalized to the data at the monolayer point.

Next, we calculate the energy of a ppin-skyrmion as a function of σ_0 by setting $\kappa = \kappa_p$ and $\kappa_s = 0$ in (25.5.23). There is an important remark. The effective tunneling gap diverges as $\sigma_0 \to 1$ unless $\Delta_{\text{SAS}} = 0$. Furthermore, the minimum pseudospin flip is $N_{\text{ppin}} = 1$ for a skyrmion-antiskyrmion pair. Hence, the pair excitation energy (25.5.4) diverges as

$$\Delta_{\text{pair}} \to \frac{\Delta_{\text{SAS}}}{\sqrt{1 - \sigma_0^2}}, \tag{25.5.25}$$

and ppin-skyrmions are not excited at the monolayer point ($\sigma_0 \to 1$). We have

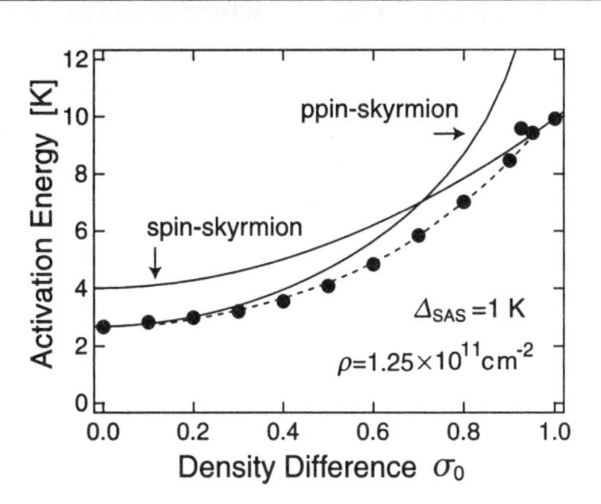

Fig. 25.3 The skyrmion energy is estimated as a function of the imbalanced parameter σ_0 for a sample ($\Delta_{SAS} = 1$ K, $\rho_0 = 1.25 \times 10^{11}/\text{cm}^2$). The experimental data are taken from Terasawa et al. [472]. Solid lines show the energies of a ppin-skyrmion and a spin-skyrmion, normalized to the data at the balanced point and the monolayer point, respectively. The experimental data would be explained by the excitation of a genuine SU(4)-skyrmion.

depicted the excitation energy $E_{\text{sky}}^{\text{ppin}}$ in Fig.25.3, where it is normalized to the data at the balanced point. In so doing we have used the full expression (O.14) for the capacitance energy E_{cap} since other terms are as important as (25.5.10) for a ppin-skyrmion of an ordinary size.

Continuous Transformation

Based on the skyrmion-energy formula (25.5.23) we verify that, if a ppin-skyrmion ($\kappa_s = 0$, $\kappa_p = \kappa$) is excited at the balanced point, it evolves continuously into a spin-skyrmion ($\kappa_s = \kappa$, $\kappa_p = 0$) via a generic skyrmion ($\kappa_s\kappa_p \neq 0$) as the imbalance parameter σ_0 increases.

We summarize the total skyrmion energy (25.5.23) as a function of $\kappa = \sqrt{\kappa_s^2 + \kappa_p^2}$, $z = (\kappa_s/\kappa)^2$ and σ_0 in the following form,

$$E_{\text{sky}}(\kappa, z; \sigma_0) = \frac{4\pi J_s^+ \sigma_0^2}{3} z^2 + \kappa^2 A(\kappa; \sigma_0) z + B(\kappa; \sigma_0), \tag{25.5.26}$$

where

$$A(\kappa; \sigma_0) = \frac{4\pi J_s^-}{3\kappa^2}(1 + \sigma_0^2) - \frac{1 - \sigma_0^2}{2}\epsilon_{\text{cap}} N_\xi + \Delta_Z N_\xi - \frac{\Delta_{\text{SAS}}}{\sqrt{1 - \sigma_0^2}} N_\xi \tag{25.5.27}$$

with (25.5.18). The explicit expression of $B(\kappa; \sigma_0)$ is not necessary. The variable z is limited in the range $0 \le z \le 1$.

We wish to minimize $E_{\text{sky}}(\kappa, z; \sigma_0)$ with respect to z at fixed values of κ and σ_0.

We start with the balanced point ($\sigma_0 = 0$), where $E_{\text{sky}}(\kappa, z; \sigma_0)$ is a linear function of z. Let us assume

$$A(\kappa; \sigma_0) > 0 \qquad \text{at} \quad \sigma_0 = 0. \tag{25.5.28}$$

Then the energy is minimized at $z = 0$ or $\kappa_s = 0$, where ppin-skyrmions are excited. We now increases σ_0 from $\sigma_0 = 0$. As far as $A(\kappa; \sigma_0) > 0$, ppin-skyrmions have the lowest energy. Due to the Δ_{SAS} term, $A(\kappa; \sigma_0)$ decreases and vanishes at $\sigma_0 = \sigma_{\text{p}}$,

$$A(\kappa; \sigma_0) = 0 \qquad \text{at} \quad \sigma_0 = \sigma_{\text{p}}. \tag{25.5.29}$$

Then it becomes negative for $\sigma_0 > \sigma_{\text{p}}$, and the energy is minimized at $z = z_{\text{min}}$ with

$$z_{\text{min}} = \frac{3}{8\pi J_s^+ \sigma_0^2} |A(\kappa; \sigma_0)|. \tag{25.5.30}$$

Because z_{min} increases continuously from $z_{\text{min}} = 0$ at $\sigma_0 = \sigma_{\text{p}}$, the spin component ($\kappa_s$) is excited gradually to form a generic skyrmion ($\kappa_s \kappa_p \neq 0$). The point z_{min} increases and achieve at $z_{\text{min}} = 1$ at $\sigma_0 = \sigma_{\text{s}}$,

$$|A(\kappa; \sigma_0)| = \frac{8\pi J_s^+ \sigma_0^2}{3} \qquad \text{at} \quad \sigma_0 = \sigma_{\text{s}}, \tag{25.5.31}$$

where the pseudospin component (κ_p) vanishes gradually.

We conclude that, if $A(\kappa; \sigma_0) > 0$ at $\sigma_0 = 0$, ppin-skyrmions are excited for $0 \le \sigma_0 < \sigma_{\text{p}}$, generic skyrmions are excited for $\sigma_{\text{p}} < \sigma_0 < \sigma_{\text{s}}$, and finally spin-skyrmions are excited for $\sigma_{\text{s}} < \sigma_0 \le 1$. The transition occurs continuously as illustrated in Fig.25.3, with the critical points σ_{p} and σ_{s} being fixed by (25.5.28) and (25.5.31). On the other hand, generic skyrmions are excited at the balanced point if $A(\kappa; \sigma_0) < 0$ and $z_{\text{min}} < 1$ at $\sigma_0 = 0$, and spin skyrmions are excited at the balanced point if $A(\kappa; \sigma_0) < 0$ and $z_{\text{min}} \ge 1$ at $\sigma_0 = 0$.

25.6 Activation Energy Anomaly

We have studied how one skyrmion evolves continuously from the balanced point ($\sigma_0 = 0$) to the monolayer point ($\sigma_0 = 1$) by changing its shape. It is important how to distinguish various shapes of skyrmions experimentally. As is well known, as the sample is tilted, the activation energy of a spin-skyrmion increases due to the Zeeman energy. On the contrary, as we now show,[150] the activation energy of a ppin-skyrmion decreases due to the loss of the exchange energy. Thus, the tilted-field method provides us with a remarkable experimental method[359] to reveal the existence of various shapes of a skyrmion in bilayer QH systems.

As the sample is tilted, the parallel magnetic field $B_\parallel$ is penetrated between the two layers [Fig.25.4]. We shall discuss in detail how it affects the interlayer coherence between the two layers in Chapter 27. In particular we shall show in Section 27.4 that the ground state is in the commensurate (C) phase for $B_\parallel < B_\parallel^*$

and in the incommensurate (IC) phase for $B_\parallel > B_\parallel^*$, where the phase transition point is given by (27.5.13), or

$$B_\parallel^* = \frac{\hbar}{\pi e d}\sqrt{\frac{\rho_0 \Delta_{\text{SAS}}}{J_s^d}}. \tag{25.6.1}$$

It is convenient to use the variable

$$\theta(x) = \vartheta(x) - \delta_{\text{m}} x \tag{25.6.2}$$

instead of the interlayer phase field $\vartheta(x)$, where

$$\delta_{\text{m}} = \frac{edB_\parallel}{\hbar} \tag{25.6.3}$$

is called the misfit parameter.

In the C phase the ground state is given by

$$\theta(x) = 0, \tag{25.6.4}$$

where the tunneling energy is minimized rather than the Coulomb energy: See (27.4.10). By changing the variable from $\vartheta(x)$ to $\theta(x)$, the exchange energy for an excitation around $\theta = 0$ is described by (27.4.3), or

$$E_{\text{X}}^{\text{com}}(B_\parallel) = 2J_s^d \sum_{a=x,y} \int d^2x\, [\nabla \mathcal{P}_a^{\text{sky}}(\boldsymbol{x}; B_\parallel)]^2 + 2J_s \int d^2x\, [\nabla \mathcal{P}_z^{\text{sky}}(\boldsymbol{x}; B_\parallel)]^2, \tag{25.6.5}$$

where the pseudospin field is

$$\mathcal{P}_x^{\text{sky}}(\boldsymbol{x}; B_\parallel) = \frac{1}{2}\sqrt{1-\sigma(\boldsymbol{x})^2}\cos[\theta(\boldsymbol{x}) + \delta_{\text{m}} x],$$

$$\mathcal{P}_y^{\text{sky}}(\boldsymbol{x}; B_\parallel) = -\frac{1}{2}\sqrt{1-\sigma(\boldsymbol{x})^2}\sin[\theta(\boldsymbol{x}) + \delta_{\text{m}} x], \quad \mathcal{P}_z^{\text{sky}}(\boldsymbol{x}; B_\parallel) = \frac{1}{2}\sigma(\boldsymbol{x}), \tag{25.6.6}$$

and the tunneling term remains as it is,

$$H_{\text{T}} = -\Delta_{\text{SAS}}\rho_0 \int d^2x\, \mathcal{P}_x(\boldsymbol{x}). \tag{25.6.7}$$

The above pseudospin field is constructed by parametrizing the CP^1 field as

$$\boldsymbol{n}(\boldsymbol{x}; B_\parallel) = \frac{1}{\sqrt{2}}\begin{pmatrix} e^{+i(\theta+\delta_{\text{m}}x)/2}\sqrt{1+\sigma} \\ e^{-i(\theta+\delta_{\text{m}}x)/2}\sqrt{1-\sigma} \end{pmatrix} = \exp\left(-\frac{i}{2}\delta_{\text{m}}\tau_z^{\text{ppin}} x\right)\boldsymbol{n}(\boldsymbol{x}; 0). \tag{25.6.8}$$

Accordingly, the skyrmion configuration acquires different phase factors between the two layers, where $\boldsymbol{n}(\boldsymbol{x}; 0)$ is the configuration (25.4.7) in the absence of the parallel magnetic field. Various isospin fields are given by (25.2.4) with this CP^3 field. It is important that the parallel magnetic field only affects the exchange energy in the commensurate phase.

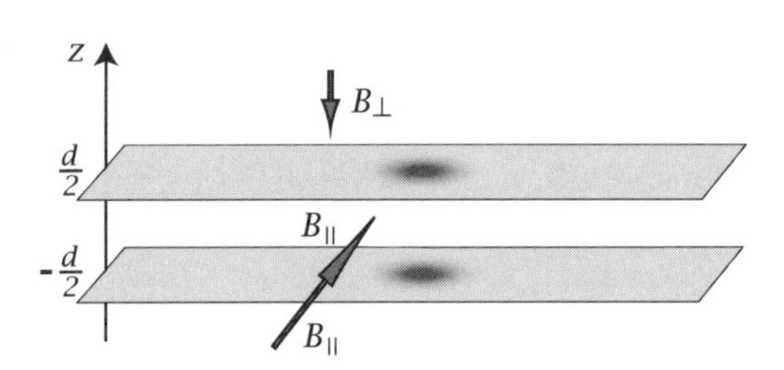

Fig. 25.4 We take the two layers parallel to the xy plane at $z = \pm d/2$. A ppin-skyrmion is excited across the layers. The activation energy decreases due to the loss of the exchange energy as the parallel magnetic field $B_\parallel$ penetrates between the layers. The loss of the exchange energy increases as $B_\parallel$ increases in the C phase ($B_\parallel < B_\parallel^*$), but it stops in the IC phase ($B_\parallel > B_\parallel^*$). See Sections 27.4 and 27.5 for details on the C-IC phase transition.

We first consider the balanced configuration ($\sigma_0 = 0$), where we assume ppin-skyrmions are excited [Fig.25.3]. The excitation energy at $B_\parallel = 0$ is given by,

$$E_{\text{sky}}^{\text{ppin}}(0) = E_{\text{X}}^{\text{com}}(0) + E_{\text{self}} + E_{\text{cap}} + N_{\text{ppin}}\Delta_{\text{SAS}}, \tag{25.6.9}$$

where $E_{\text{X}}^{\text{com}}(0)$ is the exchange energy (25.5.13) in the absence of the parallel magnetic field. We may rewrite the exchange Hamiltonian (25.6.5) as

$$E_{\text{X}}^{\text{ppin}}(B_\parallel) = E_{\text{X}}^{\text{com}}(0) + 2J_s^d\delta_m^2 \int d^2x \left[\frac{1}{4} - \left(\mathcal{P}_z^{\text{sky}}(\boldsymbol{x})\right)^2\right], \tag{25.6.10}$$

where $\mathcal{P}_z^{\text{sky}}(\boldsymbol{x})$ is the pseudospin field in the absence of the parallel magnetic field,

$$\mathcal{P}_z^{\text{sky}}(\boldsymbol{x}) = \frac{2\kappa_p\ell_B x}{r^2 + 4(\kappa_p\ell_B)^2}. \tag{25.6.11}$$

The pseudospin configuration (25.6.6) may describe the ground state as well by setting $\sigma(\boldsymbol{x}) = \vartheta(\boldsymbol{x}) = 0$, where (25.6.10) yields the ground-state energy $\frac{1}{2}J_s^d\delta_m^2\int d^2x$ induced by the parallel magnetic field: see also (27.5.3). Subtracting it we deduce the $B_\parallel$ dependence of the skyrmion excitation energy,

$$E_{\text{X}}^{\text{ppin}}(B_\parallel) = E_{\text{X}}^{\text{com}}(0) + \Delta E_{\text{X}}^{\text{ppin}}(B_\parallel) \tag{25.6.12}$$

with

$$\Delta E_{\text{X}}^{\text{ppin}}(B_\parallel) = -2J_s^d\delta_m^2 \int d^2x \left(\mathcal{P}_z^{\text{sky}}(\boldsymbol{x})\right)^2. \tag{25.6.13}$$

It is proportional to the capacitance energy (32.6.31). Thus the leading order term is

$$\Delta E_{\text{X}}^{\text{ppin}}(B_\parallel) \simeq -\frac{2\pi d^2 J_s^d}{\ell_B^2} N_{\text{ppin}} \tan^2\Theta, \tag{25.6.14}$$

where $\tan\Theta = B_{\parallel}/B_{\perp}$ and N_{ppin} is the number of pseudospins flipped around the skyrmion. Thus the excitation energy decreases as the tilting angle Θ increases. The rate of the decrease depends on the number N_{ppin} of flipped pseudospins and the amount of the penetrated magnetic field $B_{\parallel}$.

On the other hand, in the IC phase the ground state is given by

$$\vartheta(x) = 0, \tag{25.6.15}$$

where the Coulomb energy is minimized: See (27.4.21). The exchange energy no longer decreases even if $B_{\parallel}$ is decreased more. The activation energy becomes flat for $\Theta > \Theta^*$ with $\tan\Theta^* = B_{\parallel}^*/B_{\perp}$.

Hence, from (25.6.9) and (25.6.14) the excitation energy is

$$E_{\text{sky}}^{\text{ppin}}(B_{\parallel}) = E_{\text{X}}^{\text{ppin}} + E_{\text{self}} + E_{\text{cap}} + N_{\text{ppin}}(\kappa_p)\Delta_{\text{SAS}}^{\Theta}, \tag{25.6.16}$$

where we have absorbed $\Delta E_{\text{X}}^{\text{ppin}}(B_{\parallel})$ into the effective tunneling gap,

$$\Delta_{\text{SAS}}^{\Theta} = \begin{cases} \Delta_{\text{SAS}} - (2\pi d^2 J_s^d/\ell_B^2)\tan^2\Theta & \text{for} \quad \Theta < \Theta^* \\ \Delta_{\text{SAS}} - (2\pi d^2 J_s^d/\ell_B^2)\tan^2\Theta^* & \text{for} \quad \Theta > \Theta^* \end{cases}. \tag{25.6.17}$$

At each angle Θ, in principle, it is necessary to minimize the total excitation energy (25.6.16) and determine the scale parameter κ. However, since the energy change (25.6.17) is quite small compared with the total energy, we may treat it as a small perturbation. Namely, we may assume that the flipped pseudospin number N_{ppin} is a constant independent of Θ. This reminds us of experimental data,[342,289] where the flipped spin number N_{spin} is nearly constant by tilting the sample in the excitation of spin-skyrmions.

We have fitted the data due to Murphy et al.[359] in Fig.25.5 by assuming an appropriate flipped pseudospin number N_{ppin} per one skyrmion-antiskyrmion pair. We summarize the result in a table.

sample	A	B	C	D
Δ_{SAS}	0.81	0.86	4.43	8.53
N_{ppin}	18	24	2.1	1.5

(25.6.18)

It takes a large value in samples with $\Delta_{\text{SAS}} < 1$ K but take a small value in samples with $\Delta_{\text{SAS}} > 4$ K. Recall spin-skyrmion excitations in the monolayer QH system, where the flipped spin number remains small when the Zeeman energy is moderate but becomes quite large when the Zeeman energy is almost zero.

We proceed to analyze skyrmion excitations in imbalanced configuration, where the exchange Hamiltonian is given by (25.2.8). The exchange-energy loss $\Delta E_{\text{X}}^{\text{SU(4)}}(B_{\parallel})$ is again proportional to the capacitance energy, and the leading order term is

$$\Delta E_{\text{X}}(B_{\parallel}) \simeq -\frac{2\pi d^2 J_s^d}{\ell_B^2}(1-\sigma_0^2)N_{\text{ppin}}\tan^2\Theta. \tag{25.6.19}$$

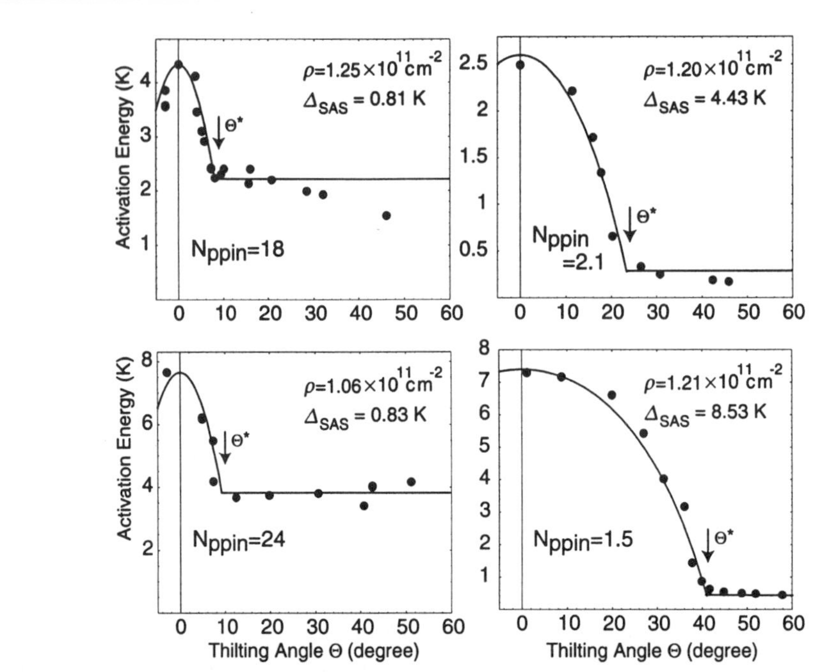

Fig. 25.5 The activation energy at $\nu = 1$ is plotted as a function of the tilting angle Θ in several samples with different tunneling gaps Δ_{SAS}. It shows a rapid decrease towards the critical angle Θ^*, and then becomes almost flat. The data are taken from Murphy et al.[359] They are well fitted by the theoretical formula (25.6.16), where the activation energy decreases in the C phase and becomes flat in the IC phase. To fit the data, we have adjusted the activation energy at $\Theta = 0$ with the experimental value, and assumed that the number of flipped pseudospins N_{ppin} is constant for all values of the tilting angle.

It is reduced to (25.6.14) in the balanced point.

Hence, in the C phase ($\Theta < \Theta_\sigma^*$) the excitation energy turns out to be

$$E_{\text{sky}} = E_{\text{X}}^{\text{SU(4)}} + E_{\text{self}} + E_{\text{cap}} + N_{\text{spin}}(\kappa_s)\sqrt{1 + \tan^2\Theta}\Delta_{\text{Z}}^0 + N_{\text{ppin}}(\kappa_p)\Delta_{\text{BAB}}^{\Theta} \tag{25.6.20}$$

with

$$\Delta_{\text{BAB}}^{\Theta} = \frac{1}{\sqrt{1-\sigma_0^2}}\Delta_{\text{SAS}} - \frac{2\pi d^2 J_s^d}{\ell_B^2}(1-\sigma_0^2)\tan^2\Theta. \tag{25.6.21}$$

In the IC phase ($\Theta > \Theta_\sigma^*$) it is given by this formula by replacing $\tan^2\Theta$ with $\tan^2\Theta_\sigma^*$. Note that the C-IC transition point increases slowly as in (27.5.41), or

$$\Theta_\sigma^* = (1-\sigma_0^2)^{-1/4}\Theta^*, \tag{25.6.22}$$

as the imbalance parameter σ_0 increases. As we have noticed in Section 25.5, as the

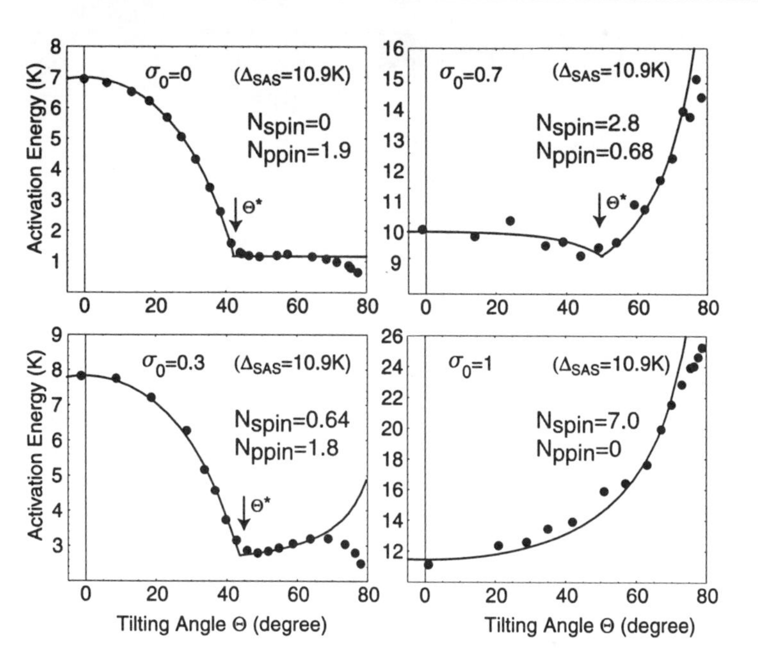

Fig. 25.6 The activation energy at $\nu = 1$ is plotted as a function of the tilting angle Θ in onel sample ($\Delta_{SAS} = 11$ K, $\rho_0 = 0.6 \times 10^{11}$ /cm^2) with different imbalance parameter σ_0. It shows a decrease towards the critical angle Θ^*, and then begin to increase for $\sigma_0 = 0.3$ and $\sigma_0 = 0.7$. The data are taken from Sawada et al.[426,472] To fit the data by the theoretical formula (25.6.20), we have adjusted the activation energy at $\Theta = 0$ with the experimental value, and assumed that the flipped spin number N_{ppin} and the flipped pseudospin number N_{ppin} are constant for all values of the tilting angle.

imbalance parameter σ_0 increases and approaches the monolayer limit ($\sigma_0 = 1$), the pseudospin component is hardly excited since it costs a very large energy. The pseudospin component is entirely suppressed in the monolayer limit.

Experiments have been carried out by Sawada et al.[426,472] in a bilayer sample (Fig.25.6), where the activation energy was measured at $\nu = 1$ by controlling both the tilting angle Θ and the imbalance parameter σ_0. Their data are interpreted based on the theoretical result (25.6.20) as follows. We focus on the behavior of the activation energy E_{sky} by changing the tilting angle at each fixed imbalance parameter. The relevant term in (25.6.20) is

$$\Delta E_{\text{sky}}(\Theta) = N_{\text{spin}}(\kappa_s)\sqrt{1+\tan^2\Theta}\Delta_Z^0 - N_{\text{ppin}}(\kappa_p)\frac{2\pi d^2 J_s^d(1-\sigma_0^2)}{\ell_B^2}\tan^2\Theta \quad (25.6.23)$$

for $\Theta < \Theta_\sigma^*$. By adjusting the theoretical curve to the data at the point $\Theta = 0$, we fit the data by this curve throughout the observed range of the tilting angle Θ. As

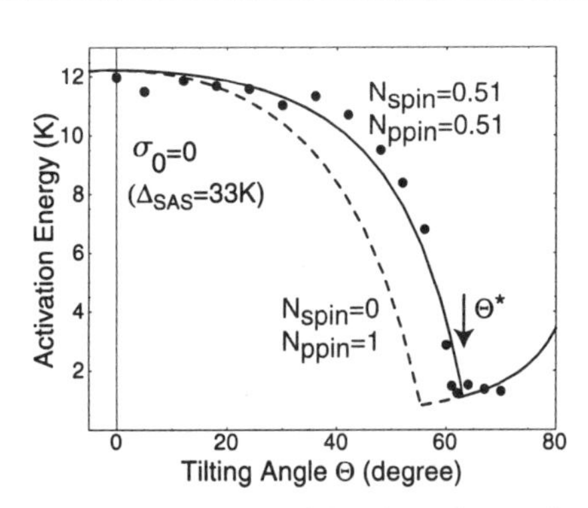

Fig. 25.7 The activation energy at $\nu = 1$ is plotted as a function of Θ in a sample ($\Delta_{SAS} = 33$K, $\rho_0 = 1.0 \times 10^{11}/\text{cm}^2$) . The data are taken from Terasawa et al.[472]. To fit the data by the theoretical formula, we have adjusted the activation energy at $\Theta = 0$ with the experimental value. We have also assumed that N_{spin} and N_{ppin} are constant for all values of the tilting angle. A better fitting is obtained as indicated by a solid line when both spins and speudospins are excited.

seen in Fig.25.6, the fitting is quite good when we assume constant values of N_{spin} and N_{ppin} throughout the range of Θ. A deviation of the theoretical curve from the data for large tilting angles $\Theta > 70°$ would be due to effects not taken into account in the above analysis. We summarize the numbers N_{spin} and N_{ppin} per one skyrmion-antiskyrmion pair determined by this fitting in a table.

σ_0	0	0.3	0.6	0.7	1
N_{spin}	0	0.32	1.2	1.4	3.5
N_{ppin}	0.9	0.8	0.7	0.34	0

(25.6.24)

It is found that N_{spin} increases and N_{ppin} decreases as the sample is tilted. In particular, spins and pseudospins are flipped simultaneously unless $\sigma_0 = 0$ or $\sigma_0 = 1$. Consequently, the SU(4)-skyrmion evolves *continuously* from the ppin-skyrmion limit to the spin-skyrmion limit as σ_0 increases from 0 to 1. It never makes a jump from a ppin-skyrmion to a spin-skyrmion at a certain point.

It is interesting to study the same problem in a sample having a very large tunneling gap. Terasawa et al.[472] have measured the activation energy by controlling the tilting angle Θ at the balanced point in a sample with $\Delta_{SAS} \simeq 33$K. We have fitted their data [Fig.25.7] by the ppin-excitation formula (25.6.16) and by the generic formula (25.6.20). It is difficult to fit the data if pure pseudospin excitations are assumed since it is required that $N_{ppin} \geq 1$ per one pair. A better fitting is obtained if spins and pseudospins are excited simultaneously since N_{ppin} may take a smaller value than 1. Such a simultaneous excitation is allowed at the balanced point as explained in the last paragraph of Section 25.5.

In passing we comment on the mechanism[351,520,404] proposed to explain the activation energy anomaly based on the exchange-energy loss of bimeron excitations. A bimeron has the same quantum numbers as a skyrmion, and it can be viewed as a deformed skyrmion with two meron cores with a string between them. The bimeron excitation energy consists of the core energy, the string energy and the Coulomb repulsive energy between the two cores. It is argued that the parallel magnetic field decreases the string tension and hence the bimeron activation energy. Clearly the mechanism works well only when the string length is much larger than the core size. A microscopic calculation has already revealed[77] that the meron core size is large enough to invalidate the naive picture. Furthermore, the skyrmion is almost as small as the hole itself in samples with large tunneling gap. On the contrary, in the present mechanism the decrease of the exchange energy follows simply from the phase difference induced by the parallel magnetic field between the wave functions associated with the two layers, and it is valid even for small skyrmions.

CHAPTER 26

BILAYER QUANTUM HALL SYSTEMS AT $\nu = 2$

At the filling factor $\nu = 2$ the bilayer QH system has three phases, the ferromagnetic phase (spin phase), the spin singlet phase (ppin phase) and the canted antiferromagnetic phase (canted phase). The dynamical field is the Grassmannian $G_{4,2}$ field carrying four complex degrees of freedom. In each of the spin and ppin phases there are four complex soft waves (Goldstone modes) and one kind of skyrmion excitations ($G_{4,2}$ skyrmions) flipping either spins or pseudospins coherently. An intriguing feature is that a quasiparticle is a $G_{4,2}$-skyrmion essentially consisting of two SU(4)-skyrmions and thus possesses charge $2e$.

26.1 SPIN PHASE, PPIN PHASE AND CANTED PHASE

There are four energy levels $|S\uparrow\rangle$, $|S\downarrow\rangle$, $|A\uparrow\rangle$ and $|A\downarrow\rangle$ within the lowest Landau level of the bilayer QH system [Fig.26.1], where S and A stand for the symmetric and antisymmetric states. As we have seen, the ground state is rather trivial at $\nu = 1$: It is uniquely given by the spin ferromagnet and the pseudospin ferromagnet state $|S\uparrow\rangle$ [Fig.26.1]. On the contrary, it is quite nontrivial at $\nu = 2$, since there are several ways to fill in two energy levels.

Since there are two electrons in one Landau site, the isospin state is classified according to the group-theoretical composition rule,

$$\mathbf{4} \otimes \mathbf{4} = \mathbf{10} \oplus \mathbf{6}. \tag{26.1.1}$$

The **10**-dimensional irreducible representation is a symmetric state, while the **6**-dimensional irreducible representation is an antisymmetric state. Two electrons in one Landau site must form an antisymmetric state due to the Pauli exclusion principle. Hence, we choose the antisymmetric **6**-dimensional irreducible representation.

In the language of the subgroup $SU_{spin}(2) \otimes SU_{ppin}(2)$, the **6**-dimensional irreducible representation of the group SU(4) is divided into two different irreducible representations,

$$\mathbf{6} = (\mathbf{3}, \mathbf{1}) + (\mathbf{1}, \mathbf{3}), \tag{26.1.2}$$

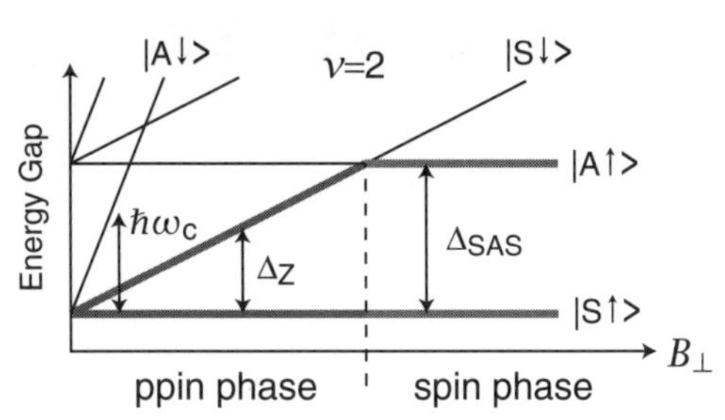

Fig. 26.1 The energy level is depicted in the noninteracting bilayer electron system. The lowest-Landau level contains four states, $|\mathrm{S}\uparrow\rangle$, $|\mathrm{S}\downarrow\rangle$, $|\mathrm{A}\uparrow\rangle$ and $|\mathrm{A}\downarrow\rangle$, in the lowest Landau level. A phase transition occurs naively from the ppin phase to the spin phase as the perpendicular magnetic field $B_\perp$ increases.

where **3** is the symmetric representation of SU(2), and **1** is the antisymmetric representation of SU(2). Consequently, they takes six states at each Landau site, which are grouped into two sectors, i.e., the $(\mathbf{3},\mathbf{1})$ sector and the $(\mathbf{1},\mathbf{3})$ sector. We call the $(\mathbf{3},\mathbf{1})$ sector the spin sector and the $(\mathbf{1},\mathbf{3})$ sector the ppin sector. It is to be emphasized that the spin and the pseudospin are equal partners.

The spin sector $(\mathbf{3},\mathbf{1})$ consists of spin-triplet pseudospin-singlet states,

$$|\mathcal{S}_\uparrow\rangle \equiv c^\dagger_{\mathrm{f}\uparrow}c^\dagger_{\mathrm{b}\uparrow}|0\rangle, \quad |\mathcal{S}_0\rangle \equiv \frac{c^\dagger_{\mathrm{f}\uparrow}c^\dagger_{\mathrm{b}\downarrow} + c^\dagger_{\mathrm{f}\downarrow}c^\dagger_{\mathrm{b}\uparrow}}{\sqrt{2}}|0\rangle, \quad |\mathcal{S}_\downarrow\rangle \equiv c^\dagger_{\mathrm{f}\downarrow}c^\dagger_{\mathrm{b}\downarrow}|0\rangle, \tag{26.1.3a}$$

while the ppin sector $(\mathbf{1},\mathbf{3})$ consists of spin-singlet pseudospin-triplet states,

$$|\mathcal{P}_\mathrm{f}\rangle \equiv c^\dagger_{\mathrm{f}\uparrow}c^\dagger_{\mathrm{f}\downarrow}|0\rangle, \quad |\mathcal{P}_0\rangle \equiv \frac{c^\dagger_{\mathrm{f}\uparrow}c^\dagger_{\mathrm{b}\downarrow} - c^\dagger_{\mathrm{f}\downarrow}c^\dagger_{\mathrm{b}\uparrow}}{\sqrt{2}}|0\rangle, \quad |\mathcal{P}_\mathrm{b}\rangle \equiv c^\dagger_{\mathrm{b}\uparrow}c^\dagger_{\mathrm{b}\downarrow}|0\rangle \tag{26.1.3b}$$

at each Landau site.

All these six states are degenerate in the SU(4)-invariant limit. We may choose any superposition of them as the ground state. The degeneracy is resolved by SU(4)-noninvariant interactions. We determine the ground state by minimizing the energy $\langle \mathrm{g}|H|\mathrm{g}\rangle$, where

$$H = H_\mathrm{C}^+ + H_\mathrm{C}^- + H_\mathrm{ZpZ}. \tag{26.1.4}$$

We first study the balanced point. It is clear that in the spin sector the lowest energy state is

$$|\mathrm{g}_\mathrm{spin}\rangle = |\mathrm{A}\uparrow;\mathrm{S}\uparrow\rangle = |\mathcal{S}_\uparrow\rangle \tag{26.1.5}$$

due to the Zeeman interaction, while in the ppin sector it is

$$|\mathrm{g}_\mathrm{ppin}\rangle = |\mathrm{S}\downarrow,\mathrm{S}\uparrow\rangle = \frac{1}{2}|\mathcal{P}_\mathrm{f}\rangle + \frac{1}{\sqrt{2}}|\mathcal{P}_0\rangle + |\mathcal{P}_\mathrm{b}\rangle \tag{26.1.6}$$

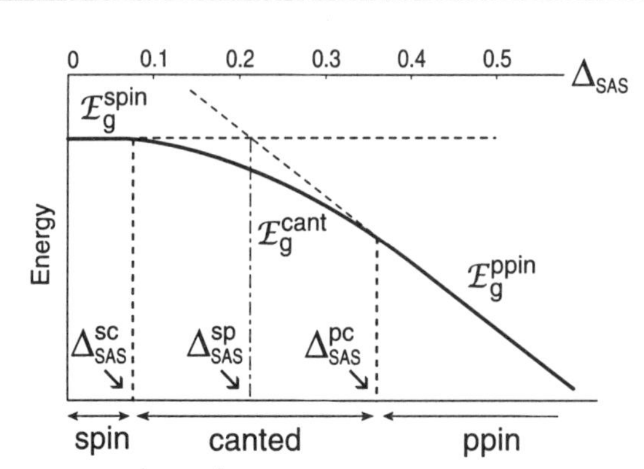

Fig. 26.2 The energies $\mathcal{E}_g^{spin}$, $\mathcal{E}_g^{ppin}$ and $\mathcal{E}_g^{cant}$ for the ground state in the spin, ppin and canted phases are depicted. A phase transition occurs along the heavy curve from the spin phase to the canted phase, and then to the ppin phase, as Δ_{SAS} increases.

due to the tunneling interaction [Fig.26.1]. We expect the realization of two phases[1] by the competition between the Zeeman and tunneling energies: One is the *spin phase* with the ground state $|g_{spin}\rangle$ belonging to the spin-triplet and pseudospin-singlet; the other is the *ppin phase* with the ground state $|g_{ppin}\rangle$ belonging to the spin-singlet and pseudospin-triplet.

Indeed, the ground-state energy $\mathcal{E}_g = \langle g|H|g\rangle / N_\Phi$ per one Landau site is calculated as

$$\mathcal{E}_g^{spin} = -2\epsilon_X^+ - 2\epsilon_X^- - \Delta_Z, \\ \mathcal{E}_g^{ppin} = -2\epsilon_X^+ - \Delta_{SAS} \tag{26.1.7}$$

in each phase, where $\epsilon_X^\pm$ are the exchange energies,

$$\epsilon_X^\pm = \frac{1}{4}\sqrt{\frac{\pi}{2}}\left[1 \pm e^{\frac{1}{2}(d/\ell)^2}\text{erfc}\left(\frac{d}{\sqrt{2}\ell}\right)\right]E_C^0. \tag{26.1.8}$$

Due to the crossing of the energy levels as Δ_{SAS} increases, a phase transition would occur suddenly from the spin phase to the ppin phase together with the would-be phase transition point [Fig.26.2]

$$\Delta_{SAS}^{sp} = 2\epsilon_X^- + \Delta_Z. \tag{26.1.9}$$

Actually, there arises a mixing of states which has a lower energy, as turns the level crossing into the level anticrossing [Fig.26.2]. A new phase emerges between the spin and ppin phases. It is called the *canted phase*[537] by the reason as we describe

[1]In the standard literature the spin-ferromagnet and pseudospin-singlet phase is called the *ferromagnet phase*, while the spin-singlet and pseudospin-ferromagnet phase is called the *singlet phase*. Here, we call them the *spin phase* and the *ppin phase* to respect their equal partnership.

later: See Fig.26.4. Now phase transitions occur smoothly from the spin phase to the ppin phase via the canted phase as Δ_{SAS} increases.

26.2 GROUND-STATE ENERGY

To explore the phase space we investigate[151] the ground-state energy E_g for a general homogeneous state $|g\rangle$. As we shall derive in Section 30.7 [see (30.7.27)], it is given by $E_g = \mathcal{E}_g N_\Phi$ with

$$\mathcal{E}_g = 4\epsilon_D^- \mathcal{P}_z^2 - 2\epsilon_X^- \left(\mathcal{S}_a^2 + \mathcal{P}_z^2 + \mathcal{R}_{az}^2\right) - \Delta_Z \mathcal{S}_z - \Delta_{SAS}\mathcal{P}_x - \Delta_{bias}\mathcal{P}_z, \tag{26.2.1}$$

together with the kinematical constraint (30.7.20), or

$$\boldsymbol{\mathcal{S}}^2 + \boldsymbol{\mathcal{P}}^2 + \boldsymbol{\mathcal{R}}^2 = 1, \tag{26.2.2}$$

where $\mathcal{S}_a$, $\mathcal{P}_a$ and $\mathcal{R}_{ab}$ are the isospins (25.1.6) per one Landau site; $\epsilon_X^\pm$ is the exchange energy (26.1.8); $4\epsilon_D^-$ is the naive capacitance energy (22.5.4); Δ_{bias} is the bias parameter (22.4.8) determined in terms of the imbalance parameter σ_0.

There are 15 isospin components, $\mathcal{S}_a$, $\mathcal{P}_a$ and $\mathcal{R}_{ab}$, but they are not all independent. We may choose $\mathcal{S}_a$ and $\mathcal{P}_a$ as 6 independent variables. We can show that $\mathcal{R}_{ab}$ are written in terms of these 6 variables and 3 extra variables [see (30.7.30)],

$$\begin{aligned}\mathcal{R}_{ab} =& \frac{\boldsymbol{\mathcal{S}}^2\mathcal{P}_a - (\boldsymbol{\mathcal{S}}\boldsymbol{\mathcal{P}})\mathcal{S}_a}{\mathcal{S}\mathcal{Q}} \frac{\boldsymbol{\mathcal{P}}^2\mathcal{S}_b - (\boldsymbol{\mathcal{S}}\boldsymbol{\mathcal{P}})\mathcal{P}_b}{\mathcal{P}\mathcal{Q}}\mathcal{R}_{PS} + \frac{\boldsymbol{\mathcal{S}}^2\mathcal{P}_a - (\boldsymbol{\mathcal{S}}\boldsymbol{\mathcal{P}})\mathcal{S}_a}{\mathcal{S}\mathcal{Q}}\frac{\mathcal{Q}_b}{\mathcal{Q}}\mathcal{R}_{PQ} \\ &+ \frac{\mathcal{Q}_a}{\mathcal{Q}}\frac{\boldsymbol{\mathcal{P}}^2\mathcal{S}_b - (\boldsymbol{\mathcal{S}}\boldsymbol{\mathcal{P}})\mathcal{P}_b}{\mathcal{P}\mathcal{Q}}\mathcal{R}_{QS} + \frac{\mathcal{Q}_a}{\mathcal{Q}}\frac{\mathcal{Q}_b}{\mathcal{Q}}\mathcal{R}_{QQ},\end{aligned} \tag{26.2.3}$$

where $\mathcal{Q}_a \equiv \varepsilon_{abc}\mathcal{S}_b\mathcal{P}_c$, $\mathcal{S} = |\boldsymbol{\mathcal{S}}|$, $\mathcal{P} = |\boldsymbol{\mathcal{P}}|$ and $\mathcal{Q} = |\boldsymbol{\mathcal{Q}}|$, while $\mathcal{R}_{PS}$, $\mathcal{R}_{PQ}$, $\mathcal{R}_{QS}$ and $\mathcal{R}_{QQ}$ are parametrized as

$$\mathcal{R}_{PS} + i\mathcal{R}_{QS} = e^{i\omega}\left(-i\lambda + \frac{1}{\xi}\mathcal{S}\right), \qquad \mathcal{R}_{PQ} + i\mathcal{R}_{QQ} = -ie^{i\omega}\xi\mathcal{P}. \tag{26.2.4}$$

There are 9 independent variables, ξ, λ, ω together with $\mathcal{S}_a$, $\mathcal{P}_a$.[2]

Let us show that $\mathcal{S}_x = \mathcal{S}_y = \mathcal{P}_y = 0$ for the ground state. First of all $\mathcal{S}_a$ and $\mathcal{R}_{az}$ rotate as vectors under the $SU_{spin}(2)$ transformation. So, if we have any configuration with nonvanishing $\mathcal{S}_x$ and $\mathcal{S}_y$, we can perform an $SU_{spin}(2)$ rotation to increase $\mathcal{S}_z$, without affecting $\mathcal{S}_a^2$ and $\mathcal{R}_{az}^2$, as far as possible until $\mathcal{S}_x = \mathcal{S}_y = 0$. This decreases the energy (26.2.1). Similarly, performing an $SU_{ppin}(2)$ rotation in the pseudospin xy plane, we can lower the energy via the tunneling term by increasing $\mathcal{P}_x$, without affecting $\mathcal{P}_z$ and $\mathcal{R}_{az}^2$, as far as possible until $\mathcal{P}_y = 0$.

---

[2] On the ground state one variable is fixed via (26.2.8). Thus, there are 8 real independent variables. They agree with the number of Goldstone modes discussed in Section 26.6.

Substituting $\mathcal{S}_x = \mathcal{S}_y = \mathcal{P}_y = 0$ into $\mathcal{R}_{ab}$ we come to

$$\sum_{ab} \mathcal{R}_{ab}^2 = \frac{\mathcal{S}_z^2}{\xi^2} + \lambda^2 + \xi^2 \left(\mathcal{P}_x^2 + \mathcal{P}_z^2\right), \qquad \sum_a \mathcal{R}_{az}^2 = \left(\frac{\mathcal{S}_z^2}{\xi^2} + \lambda^2\right) \frac{\mathcal{P}_x^2}{\mathcal{P}_x^2 + \mathcal{P}_z^2}. \tag{26.2.5}$$

Hence the energy (26.2.1) yields

$$\begin{aligned}\mathcal{E}_g = & -2\epsilon_X^- \mathcal{S}_z^2 - 2\epsilon_X^- \frac{\mathcal{P}_x^2}{\mathcal{P}_x^2 + \mathcal{P}_z^2} \frac{\mathcal{S}_z^2}{\xi^2} + \epsilon_{\text{cap}} \mathcal{P}_z^2 - 2\epsilon_X^- \frac{\lambda^2 \mathcal{P}_x^2}{\mathcal{P}_x^2 + \mathcal{P}_z^2} \\ & - \Delta_Z \mathcal{S}_z - \Delta_{\text{SAS}} \mathcal{P}_x - \Delta_{\text{bias}} \mathcal{P}_z,\end{aligned} \tag{26.2.6}$$

where we have set

$$\epsilon_{\text{cap}} \equiv 4\epsilon_D^- - 2\epsilon_X^-. \tag{26.2.7}$$

Here $\epsilon_{\text{cap}} \mathcal{P}_z^2$ is the capacitance energy per one Landau site.[3]

We minimize the energy density under the kinematical condition (26.2.2), which reads

$$\mathcal{S}^2 + \mathcal{P}^2 + \mathcal{R}^2 = \left(1 + \frac{1}{\xi^2}\right) \mathcal{S}_z^2 + \left(\mathcal{P}_x^2 + \mathcal{P}_z^2\right)\left(1 + \xi^2\right) + \lambda^2 = 1. \tag{26.2.8}$$

We eliminate λ^2 in (26.2.6) by using (26.2.8),

$$\begin{aligned}\mathcal{E}_g = & 2\epsilon_X^- \frac{\mathcal{P}_z^2}{\mathcal{P}_x^2 + \mathcal{P}_z^2} \left(1 - \mathcal{S}_z^2\right) + 2\epsilon_X^- \left(1 + \xi^2\right) \mathcal{P}_x^2 + \epsilon_{\text{cap}} \mathcal{P}_z^2 - 2\epsilon_X^- \\ & - \Delta_Z \mathcal{S}_z - \Delta_{\text{SAS}} \mathcal{P}_x - \Delta_{\text{bias}} \mathcal{P}_z.\end{aligned} \tag{26.2.9}$$

It is clear that we can decrease the energy by increasing $\mathcal{S}_z$ without affecting other terms in (26.2.9). This is achieved at by decreasing λ^2 until $\lambda^2 = 0$ in (26.2.8), which yields

$$\mathcal{S}_z^2 + \xi^2 \left(\mathcal{P}_x^2 + \mathcal{P}_z^2\right) = \frac{\xi^2}{1 + \xi^2}. \tag{26.2.10}$$

We solve this as

$$\mathcal{S}_z = \frac{\xi}{\sqrt{1 + \xi^2}} \sqrt{1 - \alpha^2}, \quad \mathcal{P}_x = \frac{1}{\sqrt{1 + \xi^2}} \alpha \sqrt{1 - \beta^2}, \quad \mathcal{P}_z = \frac{1}{\sqrt{1 + \xi^2}} \alpha\beta \tag{26.2.11}$$

in terms of two parameters $|\alpha| \leqslant 1$ and $|\beta| \leqslant 1$.

Substituting these into (26.2.9) we obtain

$$\begin{aligned}\mathcal{E}_g = & 2\epsilon_X^- \alpha^2 + \left[2\epsilon_X^- + 4\left(\epsilon_D^- - \epsilon_X^-\right)\alpha^2\right] \frac{\beta^2}{1 + \xi^2} - 2\epsilon_X^- \\ & - \frac{\Delta_Z \xi}{\sqrt{1 + \xi^2}} \sqrt{1 - \alpha^2} - \frac{\Delta_{\text{SAS}}}{\sqrt{1 + \xi^2}} \alpha \sqrt{1 - \beta^2} - \frac{\Delta_{\text{bias}}}{\sqrt{1 + \xi^2}} \alpha\beta.\end{aligned} \tag{26.2.12}$$

[3]The capacitance parameter (26.2.7) is different from the $\nu = 1$ QH system, where $\epsilon_{\text{cap}}^{\nu=1} \equiv 4\epsilon_D^- - 4\epsilon_X^-$ as in (22.5.5). Note that the energy formula (26.2.1) holds as it stands also at $\nu = 1$. The difference arises because $\sum_a \mathcal{R}_{az}^2 = \mathcal{P}_z^2$ in (26.2.1) at $\nu = 1$ but it is given by (26.2.5) at $\nu = 2$.

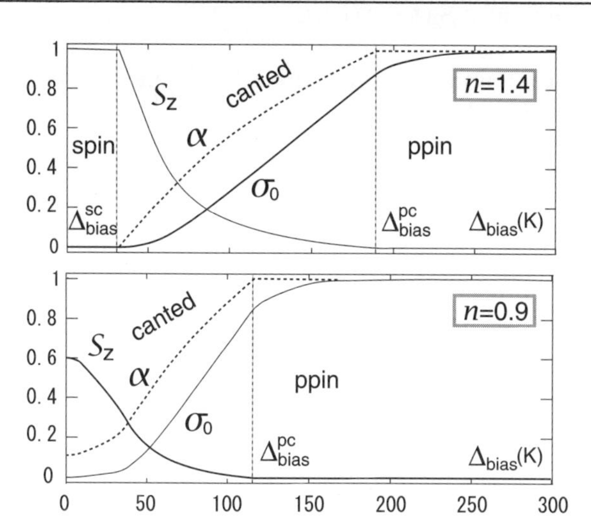

Fig. 26.3 The spin component $\mathcal{S}_z$ and the imbalance parameter $\sigma_0 \equiv \mathcal{P}_z$ are illustrated together with α as functions of Δ_{bias} for typical sample parameters (d = 23nm, Δ_{SAS} = 6.7K and $\rho_0 = n \times 10^{11}/\text{cm}^2$). The spin phase ($\alpha = 0$), the canted phase ($0 < \alpha < 1$) and the ppin phase ($\alpha = 1$) are realized for $\Delta_{\text{bias}} < \Delta_{\text{bias}}^{\text{sc}}$, $\Delta_{\text{bias}}^{\text{sc}} < \Delta_{\text{bias}} < \Delta_{\text{bias}}^{\text{pc}}$ and $\Delta_{\text{bias}}^{\text{pc}} < \Delta_{\text{bias}}$, respectively. There are 3 phases for n = 1.4 but only 2 phases for n = 0.9.

To minimize this with respect to α, β and ξ, we write down the equations $\partial_\alpha E = \partial_\beta E = \partial_\xi E = 0$. We rearrange them as

$$\Delta_Z^2 = \frac{\Delta_{\text{SAS}}^2}{1-\beta^2} - \frac{4\epsilon_X^- \left(\Delta_0^2 - \beta^2 \Delta_{\text{SAS}}^2\right)}{\Delta_0\sqrt{1-\beta^2}}, \tag{26.2.13a}$$

$$\frac{\Delta_{\text{bias}}}{\beta\Delta_{\text{SAS}}} = \frac{4\left(\epsilon_X^- + 2\alpha^2\left(\epsilon_D^- - \epsilon_X^-\right)\right)}{\Delta_0} + \frac{1}{\sqrt{1-\beta^2}}, \tag{26.2.13b}$$

$$\xi = \frac{\Delta_Z}{\Delta_{\text{SAS}}}\frac{\sqrt{1-\alpha^2}}{\alpha}\sqrt{1-\beta^2}, \tag{26.2.13c}$$

where

$$\Delta_0 \equiv \sqrt{\Delta_{\text{SAS}}^2\alpha^2 + \Delta_Z^2\left(1-\alpha^2\right)\left(1-\beta^2\right)}. \tag{26.2.14}$$

The ground state is determined by these equations. The parameters α and β are solved out from (26.2.13a) and (26.2.13b) in terms of the sample parameters [Fig.26.3]. Then, ξ is given by (26.2.13c) in terms of them.

26.3 Ground-State Structure

We discuss the ground-state structure as a function of parameters α and β. Substituting (26.2.13c) into (26.2.11) we get

$$\mathcal{S}_z = \frac{\Delta_Z}{\Delta_0}\left(1-\alpha^2\right)\sqrt{1-\beta^2}, \quad \mathcal{P}_x = \frac{\Delta_{SAS}}{\Delta_0}\alpha^2\sqrt{1-\beta^2}, \quad \mathcal{P}_z = \frac{\Delta_{SAS}}{\Delta_0}\alpha^2\beta. \tag{26.3.1}$$

Recall that $\mathcal{S}_x = \mathcal{S}_y = 0$ and $\mathcal{P}_y = 0$. Using (26.2.4) in (26.2.3), we perform all the necessary substitutions and obtain

$$\begin{aligned}
\mathcal{R}_{xx} + i\mathcal{R}_{yx} &= -\frac{\Delta_{SAS}}{\Delta_0}\alpha\sqrt{1-\alpha^2}\beta e^{i\omega}, \\
\mathcal{R}_{yy} + i\mathcal{R}_{xy} &= +\frac{\Delta_Z}{\Delta_0}\alpha\sqrt{1-\alpha^2}\sqrt{1-\beta^2}e^{i\omega}, \\
\mathcal{R}_{xz} + i\mathcal{R}_{yz} &= +\frac{\Delta_{SAS}}{\Delta_0}\alpha\sqrt{1-\alpha^2}\sqrt{1-\beta^2}e^{i\omega},
\end{aligned} \tag{26.3.2}$$

while the rest components vanish, $\mathcal{R}_{za} = 0$. We have expressed all the isospin components $\mathcal{S}_a$, $\mathcal{P}_a$ and $\mathcal{R}_{ab}$ in terms of three variables α, β and ω.

Using $2\mathcal{S}_a^{\rm f} = \mathcal{S}_a + \mathcal{R}_{az}$ and $2\mathcal{S}_a^{\rm b} = \mathcal{S}_a - \mathcal{R}_{az}$ we find

$$\begin{aligned}
\mathcal{S}_x^{\rm f} &= -\mathcal{S}_x^{\rm b} = \frac{1}{2}\frac{\Delta_{SAS}}{\Delta_0}\alpha\sqrt{1-\alpha^2}\sqrt{1-\beta^2}\cos\omega, \\
\mathcal{S}_y^{\rm f} &= -\mathcal{S}_y^{\rm b} = \frac{1}{2}\frac{\Delta_{SAS}}{\Delta_0}\alpha\sqrt{1-\alpha^2}\sqrt{1-\beta^2}\sin\omega, \\
\mathcal{S}_z^{\rm f} &= \mathcal{S}_z^{\rm b} = \frac{1}{2}\mathcal{S}_a.
\end{aligned} \tag{26.3.3}$$

We see that ω describes the orientation of $\mathcal{S}_a^{\rm f}$ and $\mathcal{S}_a^{\rm b}$ in the xy plane. Since the energy does not depend on ω, it is the zero-energy mode associated with the rotational invariance in the xy plane.

The ground state $|g\rangle$ can be reconstructed from (26.3.1) and (26.3.2). For the sake of simplicity we take $\omega = \frac{\pi}{4}$ and come to

$$|g\rangle = \gamma_\uparrow^{\rm s}|\mathcal{S}_\uparrow\rangle + \gamma_0^{\rm s}|\mathcal{S}_0\rangle + \gamma_\downarrow^{\rm s}|\mathcal{S}_\downarrow\rangle + \gamma_{\rm f}^{\rm p}|\mathcal{P}_{\rm f}\rangle + \gamma_0^{\rm p}|\mathcal{P}_0\rangle + \gamma_{\rm b}^{\rm p}|\mathcal{P}_{\rm b}\rangle, \tag{26.3.4}$$

where the 6 states $|\mathcal{S}_\uparrow\rangle$ etc. are given by (26.1.3), and[4]

$$\begin{aligned}
&\gamma_\uparrow^{\rm s} = \frac{1+i}{2\sqrt{2}}\sqrt{1-\alpha^2}\left(1+\frac{\Delta_Z}{\Delta_0}\sqrt{1-\beta^2}\right), \quad \gamma_0^{\rm s} = 0, \quad \gamma_\downarrow^{\rm s} = \frac{1-i}{2\sqrt{2}}\sqrt{1-\alpha^2}\left(1-\frac{\Delta_Z}{\Delta_0}\sqrt{1-\beta^2}\right), \\
&\gamma_{\rm f}^{\rm p} = -\frac{i\alpha}{2}\left(1+\frac{\Delta_{SAS}}{\Delta_0}\beta\right), \quad \gamma_0^{\rm p} = -\frac{i\alpha}{\sqrt{2}}\frac{\Delta_{SAS}}{\Delta_0}\sqrt{1-\beta^2}, \quad \gamma_{\rm b}^{\rm p} = -\frac{i\alpha}{2}\left(1-\frac{\Delta_{SAS}}{\Delta_0}\beta\right).
\end{aligned} \tag{26.3.5}$$

All quantities are parametrized by α and β. When we solve (26.2.13a) and (26.2.13b) for them, we find $\alpha < 0$ for $\Delta_{\rm bias} < \Delta_{\rm bias}^{\rm sc}$, and $\alpha > 1$ for $\Delta_{\rm bias} > \Delta_{\rm bias}^{\rm pc}$

[4]These coefficients are derived in Appendix C of reference.[151]

with certain values of $\Delta_{\text{bias}}^{\text{sc}}$ and $\Delta_{\text{bias}}^{\text{pc}}$: see (26.4.5). Accordingly the ground-state energy (26.2.12) is minimized by $\alpha = 0$ for $\Delta_{\text{bias}} < \Delta_{\text{bias}}^{\text{sc}}$, and $\alpha = 1$ for $\Delta_{\text{bias}} > \Delta_{\text{bias}}^{\text{pc}}$. Then, the spin component $\mathcal{S}_z$ and the density imbalance $\mathcal{P}_z$ are calculated from (26.3.1). In Fig.26.3 we illustrate $\mathcal{S}_z$ and $\sigma_0 \equiv \mathcal{P}_z$ together with α as functions of Δ_{bias} for typical sample parameters; d = 23nm, Δ_{SAS} = 6.7K and $\rho_0 = n \times 10^{11}/\text{cm}^2$ with n = 1.4 and = 0.9.

First, when $\alpha = 0$, it follows that $\mathcal{S}_z = 1$ and $\mathcal{P}_z = 0$ since $\Delta_0 = \Delta_Z\sqrt{1-\beta^2}$. Note that β disappears from all formulas in (26.3.5). The spin phase is characterized by the fact that the isospin is fully polarized into the spin direction with $\mathcal{S}_z = 1$ and all others being zero. The spins in both layers point to the positive z axis due to the Zeeman effect. Substituting $\alpha = 0$ and $\Delta_0 \equiv \Delta_Z\sqrt{1-\beta^2}$ into (26.3.5) we find the ground state to be

$$|g_{\text{spin}}\rangle = |\mathcal{S}_\uparrow\rangle = c_{\text{f}\uparrow}^\dagger c_{\text{b}\uparrow}^\dagger |0\rangle \tag{26.3.6}$$

at each Landau site, as we have noticed in (26.1.5). A prominent feature is that, even if the bias voltage is applied, no charge transfer occurs between the two layers ($\sigma_0 = 0$) as far as $\Delta_{\text{bias}} < \Delta_{\text{bias}}^{\text{sc}}$, where the system is in the spin phase [Fig.26.3]. However, see a comment at the end of Section 26.4 on this point in actual samples with impurities.

Second, when $\alpha = 1$, it follows that $\mathcal{S}_z = 0$ and $\mathcal{P}_z \neq 0$ (actually $\mathcal{P}_x^2 + \mathcal{P}_z^2 = 1$). The ppin phase is characterized by the fact that the isospin is fully polarized into the pseudospin direction with

$$\mathcal{P}_x = \sqrt{1-\beta^2}, \qquad \mathcal{P}_z = \beta, \tag{26.3.7}$$

and all others being zero. Because $\mathcal{P}_z$ represents the density difference between the two layers, β is identified with the imbalance parameter σ_0. Substituting $\alpha = 1$ into (26.3.5) we find the ground state to be

$$\begin{aligned} |g_{\text{ppin}}\rangle &= \frac{1}{2}(1+\sigma_0)|\mathcal{P}_\text{f}\rangle + \frac{1}{\sqrt{2}}\sqrt{1-\sigma_0^2}|\mathcal{P}_0\rangle + \frac{1}{2}(1-\sigma_0)|\mathcal{P}_\text{b}\rangle \\ &= \frac{1}{2}\left(\sqrt{1+\sigma_0}c_{\text{f}\uparrow}^\dagger + \sqrt{1-\sigma_0}c_{\text{b}\uparrow}^\dagger\right)\left(\sqrt{1+\sigma_0}c_{\text{f}\downarrow}^\dagger + \sqrt{1-\sigma_0}c_{\text{b}\downarrow}^\dagger\right)|0\rangle \end{aligned} \tag{26.3.8}$$

at each Landau site, as we have noticed in (26.1.6). It is found that all electrons are in the front layer when $\sigma_0 = 1$, and in the back layer when $\sigma_0 = -1$.

For intermediate values of α ($0 < \alpha < 1$) none of the spin and pseudospin vanish, where we may control the density imbalance by applying a bias voltage as in the ppin phase. The ground state $|g_{\text{cant}}\rangle$ is given by (26.3.4) with (26.3.5). All states except $|\mathcal{S}_0\rangle$ contribute to form the ground state. This is so even in the balanced configuration ($\beta = 0$). It follows from (26.3.1) and (26.3.3) that, as the system goes away from the spin phase, the spins begin to cant and make antiferromagnetic correlations between the two layers [Fig.26.4]. Hence, it is called the canted antiferromagnetic phase.[537]

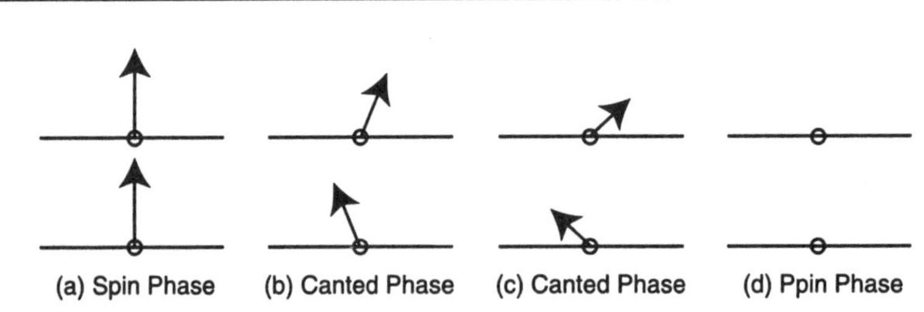

Fig. 26.4 The directions and magnitudes of the spin are shown in the front and back layers in various phases. In the spin phase the spins in the both layers point to the positive z-axis. When the state crosses from the spin phase into the canted phase, the spin directions begin to cant and make antiferromagnetic correlations between the two layers. The magnitude of the spin decreases in each layer, and finally it vanishes as the state crosses into the ppin phase. Note that the equal number of up-spin and down-spin electrons are present so that the magnitude can vanish in each layer in the ppin phase: See (26.1.6).

We conclude that there are three phases in general; the spin phase, the canted phase and the ppin phase. They are characterized as

spin	canted	ppin
$\mathcal{S}^2 = 1$	$\mathcal{S}^2 \neq 0$	$\mathcal{S}^2 = 0$
$\mathcal{P}^2 = 0$	$\mathcal{P}^2 \neq 0$	$\mathcal{P}^2 = 1$

(26.3.9)

The order parameters are $\mathcal{S}^2$ and $\mathcal{P}^2$.

Let us study the system at the balanced point ($\beta = 0$) more in detail. The spin and pseudospin are

$$\mathcal{S}_z = \frac{\Delta_Z}{\sqrt{\Delta_{SAS}^2\alpha^2 + \Delta_Z^2(1-\alpha^2)}}\left(1-\alpha^2\right), \qquad \mathcal{P}_x = \frac{\Delta_{SAS}}{\sqrt{\Delta_{SAS}^2\alpha^2 + \Delta_Z^2(1-\alpha^2)}}\alpha^2, \tag{26.3.10}$$

where α^2 is easily obtained from (26.2.13a) with $\beta = 0$,

$$\alpha^2 = \frac{\Delta_{SAS}^2 - \Delta_Z^2}{4(2\epsilon_X^-)^2} - \frac{\Delta_Z^2}{\Delta_{SAS}^2 - \Delta_Z^2}. \tag{26.3.11}$$

The ground-state energy in each phase is given by

$$\begin{aligned}
\mathcal{E}_g^{\text{spin}} &= -2\epsilon_X^- - \Delta_Z, \\
\mathcal{E}_g^{\text{cant}} &= -2\epsilon_X^-\left[\frac{1}{4}\left(\frac{\Delta_{SAS}}{2\epsilon_X^-}\right)^2 - \frac{1}{4}\left(\frac{\Delta_Z}{2\epsilon_X^-}\right)^2 + \frac{\Delta_{SAS}^2}{\Delta_{SAS}^2 - \Delta_Z^2}\right], \\
\mathcal{E}_g^{\text{ppin}} &= -\Delta_{SAS}.
\end{aligned} \tag{26.3.12}$$

We have depicted them as a function of Δ_{SAS} in Fig.26.2. The spin-canted phased transition points are

$$\Delta_{SAS}^{sc} = \sqrt{\Delta_Z^2 + 4\epsilon_X^- \Delta_Z} \tag{26.3.13}$$

from $\mathcal{E}_g^{spin} = \mathcal{E}_g^{cant}$, while the ppin-canted transition points are

$$\Delta_{SAS}^{pc} = 2\epsilon_X^- + \sqrt{\Delta_Z^2 + (2\epsilon_X^-)^2} \tag{26.3.14}$$

from $\mathcal{E}_g^{ppin} = \mathcal{E}_g^{cant}$.

We briefly discuss effects due to higher order corrections.[151] We can easily check that the fully spin polarized state (26.3.6) is an exact eigenstate of the total Hamiltonian. Hence the mean-field equations (26.2.13a) ~ (26.2.13c) are exact ones in the spin phase, implying that the spin-canted phase boundary is not affected by higher order perturbations. On the contrary, the fully pseudospin polarized state (26.3.8) cannot be an eigenstate of the total Hamiltonian. It is well known that the second order correction always lower the energy obtained in the first order perturbation. It can be shown that the ppin-canted phase boundary is considerably modified by higher order corrections. These features are precisely what are found in the exact diagonalization of a few electron system:[428] See Fig.26.5.

26.4 PHASE DIAGRAMS

The phase diagram has been studied numerically within the Hartree-Fock approximation.[537,420,78,328] We present analytic formulas[151] for it by searching for the boundaries separating the canted phase from the spin and ppin phases. These can be extracted from (26.2.13a) and (26.2.13b).

Along the spin-canted boundary we have $\alpha = 0$. Substituting this value into (26.2.13b) we get

$$\frac{\beta}{\sqrt{1-\beta^2}} = \frac{\Delta_{bias}}{\Delta_{SAS}} \frac{\Delta_Z}{\Delta_Z + 4\epsilon_X^-}. \tag{26.4.1}$$

We solve this for β and substitute it into (26.2.13a),

$$\Delta_{SAS}^2 = \Delta_Z^2 + 4\epsilon_X^- \Delta_Z - \frac{\Delta_Z \Delta_{bias}^2}{\Delta_Z + 4\epsilon_X^-}. \tag{26.4.2}$$

This determines the spin-canted boundary in the Δ_{SAS}-Δ_Z plane for typical values of Δ_{bias}.

Along the ppin-canted boundary we have $\alpha = 1$. Substituting this value into (26.2.13a) and (26.2.13b) we get

$$\Delta_{SAS} = \sqrt{1-\beta^2}\left[\frac{\Delta_{bias}}{\beta} - 2\epsilon_{cap}\right] \tag{26.4.3}$$

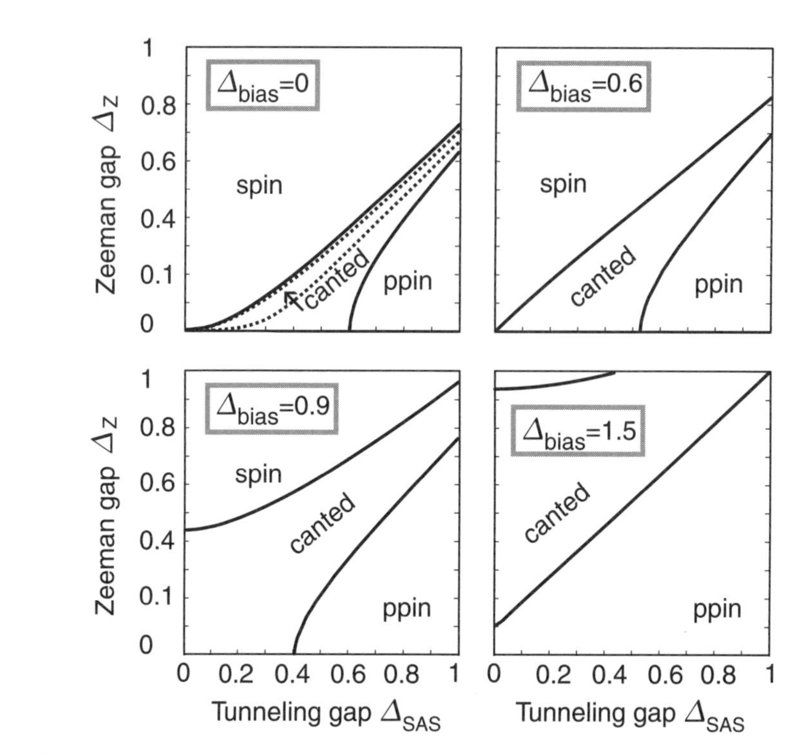

Fig. 26.5 The phase diagram is given in the Δ_{SAS}-Δ_Z plane by changing the bias parameter Δ_{bias}. Here we have set $d = \ell_B$, and taken the Coulomb energy unit E_C^0 for the tunneling gap (horizontal axis) and the Zeeman gap (vertical axis). It is observed that both the canted and ppin phases are stabilarized even for $\Delta_{SAS} = 0$ by applying the bias voltage. The dotted curves in the top left panel represent the exact diagonalization result due to Schliemann et al.[428] for the 12 electron system. It is observed that the ppin-canted phase boundary is modified considerably by higher oder quantum corrections.

and

$$\Delta_Z^2 = \left(\frac{\Delta_{bias}}{\beta} - 2\varepsilon_{cap}\right)\left(\frac{\Delta_{bias}}{\beta} - 8\varepsilon_D^- + 4\beta^2\varepsilon_X^-\right) \tag{26.4.4}$$

after some manipulation. These two equations give a parametric representation of the ppin-canted boundary in terms of β in the Δ_{SAS}-Δ_Z plane for typical values of Δ_{bias}.

In drawing the phase diagram in the Δ_{SAS}-Δ_Z plane we need to fix the layer separation d. We have examined the case $d = \ell_B$ and presented the phase diagram in Fig.26.5. The present results agree with results obtained numerically in references[78,328] for imbalanced configurations.

These phase diagrams are, however, not so useful to study experimental data,

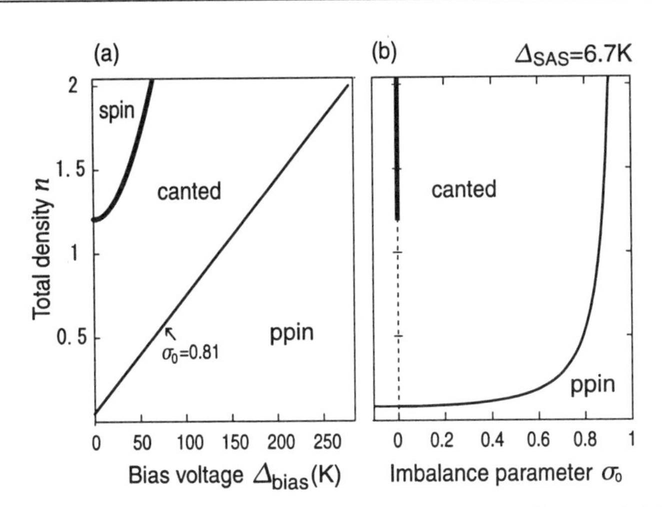

Fig. 26.6 The phase diagrams are given (a) in the Δ_{bias}-ρ_0 plane and (b) in the σ_0-ρ_0 plane for the sample with $d = 23$nm and $\Delta_{SAS} = 6.7$K, where $\rho_0 = n \times 10^{11}/\text{cm}^{-2}$. The two curves stand for the spin-canted phase boundary and the ppin-canted phase boundary in the lowest order of perturbation. The spin phase is to be realized only in the balance configuration but for impurity.

since we need many samples with different d to realize, e.g., $d = \ell_B$, but this is impossible. It is more interesting to see experimentally[423] how the phase transition occurs by controlling the total density ρ_0 and also the imbalance parameter σ_0 in a single sample with fixed values of d and Δ_{SAS}. Here we wish to describe new aspects of the phase diagram[151] from this point of view.

We start with the spin phase in the balanced configuration with $\Delta_{bias} = 0$. (The spin phase realizes provided the electron density is more than a certain critical value.) When the bias voltage is applied ($\Delta_{bias} \neq 0$) the parameter β becomes nonzero according to (26.4.1). However, no charge imbalance is induced between the two layers ($\sigma_0 = 0$) as far as $\alpha = 0$. A charge imbalance occurs only above a certain critical value Δ_{bias}^{sc} of the bias voltage as in Fig.26.3, which is given by solving (26.4.2) as

$$\left(\Delta_{bias}^{sc}\right)^2 = (\Delta_Z + 4\epsilon_X^-)^2 - \left(1 + 4\frac{\epsilon_X^-}{\Delta_Z}\right)\Delta_{SAS}^2. \tag{26.4.5}$$

This gives the spin-canted phase boundary in the Δ_{bias}-ρ_0 plane [Fig.26.6(a)]. For $\Delta_{bias} > \Delta_{bias}^{sc}$ the system is driven into the canted phase with $\alpha \neq 0$. As the bias voltage increases above a second critical value Δ_{bias}^{pc}, the system turns into the ppin phase with $\alpha = 1$ as in Fig.26.3. The critical value Δ_{bias}^{pc} is obtained by eliminating β in (26.4.3) and (26.4.4). This gives the ppin-canted phase boundary in the Δ_{bias}-ρ_0

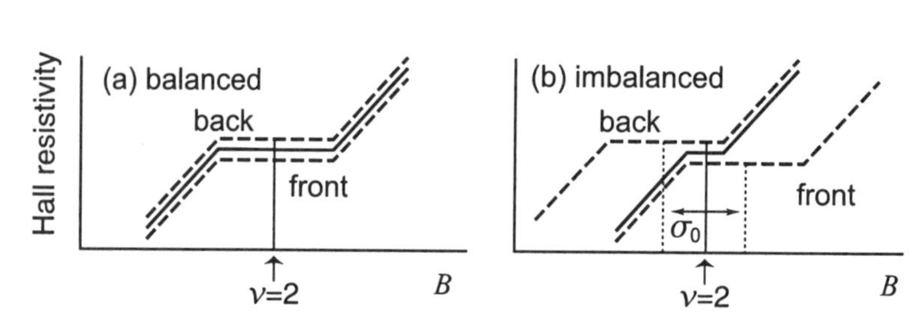

Fig. 26.7 The ground state of the spin phase is a compound state consisting of two monolayer $\nu = 1$ QH states in the front and back layers. (a) At the balanced point, the $\nu = 1$ Hall plateau (dashed lines) develops due to impurity effects in each layer, and so does the $\nu = 2$ Hall plateau (thick line). (b) At the imbalanced point ($\sigma_0 > 0$), the electron number is more (less) than the flux number in the front (back) layer. In the front (back) layer with more (less) electrons, the Hall plateux (dashed line) would be generated with its center being located at a higher (lower) value of the magnetic flux B, as indicated by a dotted vertical line. Then the $\nu = 2$ Hall plateau (thick line) is generated in the region where the two $\nu = 1$ Hall plateaux (dashed lines) in the front and back layers overlap.

plane [Fig.26.6(a)]. We remark that, as the electron density decreases, the critical point $\Delta_{\text{bias}}^{\text{sc}}$ decreases and eventually becomes zero so that the spin phase disappears at all: The critical density is $\rho_0 = 1.19 \times 10^{11}/\text{cm}^{-2}$ in the case of Fig.26.6. To compare the experimental data it is more convenient to use the phase diagram in the σ_0-ρ_0 plane [Fig.26.6(b)], which is constructed from the relation between the bias voltage and the imbalance parameter implied by (26.4.3) and (26.4.4).

We have argued that the charge transfer does not occur in the spin phase. Namely the spin phase is realized only at the balanced point. However, all the analysis has been done in ideal samples without impurities. Recall that the Hall plateau is generated as an impurity effect [Fig.11.4]. In this circumstance the spin phase arises even in imbalanced configuration, as illustrated in Fig.26.7.

26.5 Experimental Data

The first experimental indication of a variety of phases in the $\nu = 2$ bilayer QH system was given by inelastic light scattering spectroscopy by Pelligrini et al.[388] They also observed mode softening signals indicating second-order phase transitions. Capacitance spectroscopy[269] as well as magnetotransport measurements[423,424,290,183,180] were also performed to study various phases at $\nu = 2$.

Here, based on the present theoretical results, we wish to interpret mainly the experimental data due to Fukuda et al.[180] obtained from a sample with $\Delta_{\text{SAS}} =$

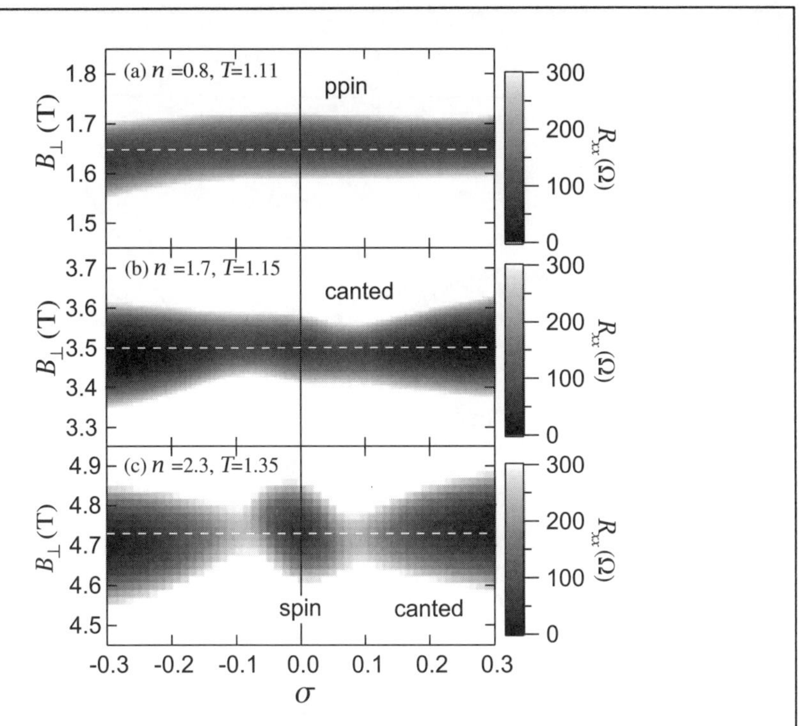

Fig. 26.8 Image plots of the magnetoresistance R_{xx} are given in the σ_0-$B_\perp$ plane at typical values of n. The black region represents the $\nu = 2$ QH state. The values of the density n and the temperature (K) are shown in the figure. The solid line indicates the balanced point ($\sigma_0 = 0$). The white dashed line is located just on the $\nu = 2$ filling. The corresponding activation-energy data are given in Fig.26.11. The sample parameters are $d = 23$nm and $\Delta_{SAS} = 11$K. Data are taken from Fukuda et al.[180]

11K, since they yield dramatic evidences for phase transitions. In this section we set $\rho_0 = n \times 10^{11}/\text{cm}^2$, and use n to represent the total density.

We start with presenting image plots of the magnetoresistance R_{xx} by changing the perpendicular magnetic field $B_\perp$ and the density imbalance σ_0. We display three typical patterns of the magnetoresistance R_{xx} corresponding to three values of the electron density n in Fig.26.8. The black region represents the well-developed $\nu = 2$ QH state. We remark the following characteristic features of the image pattern as $|\sigma_0|$ is increased: (a) At $n = 0.8$, the width of the stable region is almost constant; (b) At $n = 1.7$, it becomes slightly narrower and then wider; (c) At $n = 2.3$, it becomes drastically narrower and then wider. To investigate their intrinsic property the measurement of the activation energy is important.

We first present the measured activation energy as a function of n at the balanced point ($\sigma_0 = 0$) in Fig.26.9. In the beginning, the activation energy Δ_{gap} gradually increases for the increment of n from 0.6 to 1.1. After crossing the max-

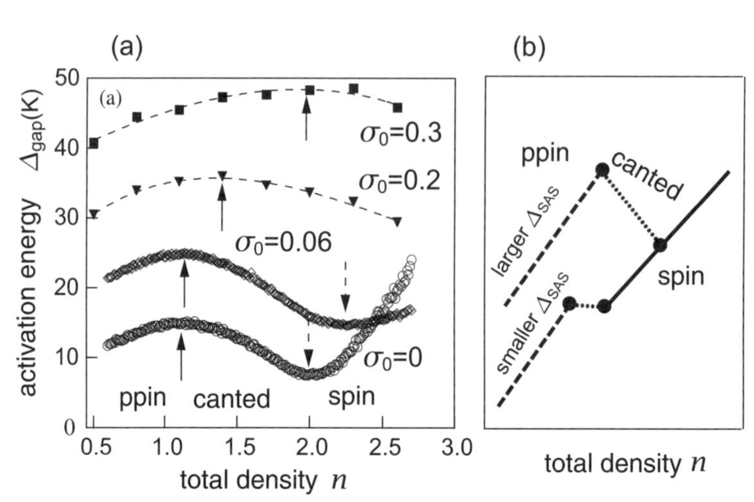

Fig. 26.9 (a) The activation energy Δ_{gap} is measured by changing the total density $\rho_0 = n \times 10^{11}/\text{cm}^2$ for several imbalanced points σ_0 in a sample with $d = 23$nm and $\Delta_{SAS} = 11$K. The Δ_{gap} at $\sigma = 0$ has a correct scale, while each trace for $\sigma \neq 0$ is shifted by 10K. Solid (dashed) arrows indicate the local maximum (minimum) of Δ_{gap}, separating three different phases (ppin, canted and spin) as indicated. Data are taken from Fukuda et al.[180] (b) The dependence of Δ_{gap} on n is illustrated schematically in each phase. The major part of Δ_{gap} comes from the Coulomb energy, which increases as n increases since the magnetic length ℓ_B decreases. Pseudoparticles are spin-skyrmions (ppin-skyrmions) in the spin (ppin) phase. The Δ_{gap} of the spin-skyrmion (heavy line) is a function of ℓ_B and Δ_Z. The Δ_{gap} of of the ppin-skyrmion (two dashed lines) is a function of ℓ_B and Δ_{SAS}, where Δ_{gap} is larger when Δ_{SAS} is larger. Since the ground state changes continuously from the $SU_{ppin}(2)$ space to the $SU_{spin}(2)$ space via the SU(4) space, the pseudoparticle transforms continuously from the ppin-skyrmion to the spin-skyrmion via the SU(4)-skyrmion. Thus, the activation energy changes smoothly from that of the ppin-skyrmion to that of the spin-skyrmion via the canted region (two dotted lines). Two typical examples are given, as explains the present data due to Fukuda et al.[180] with $\Delta_{SAS} = 11$K, where Δ_{gap} decreases, and the data due to Sawada et al.[423] with $\Delta_{SAS} = 6.7$K, where Δ_{gap} is almost constant in the canted phase.

imum of Δ_{gap} at $n = 1.1$, it decreases anomalously to the minimum point of Δ_{gap} at $n = 2.1$, and finally increases steeply. This figure indicates that there are three phases in the balanced configuration. We have argued theoretically that the phase transition is smooth. Thus the exact phase transition point is not clear due to a smooth change of Δ_{gap}, but it would exist around the n which gives the local maximum or minimum. These three phases must be the ppin phase, the canted phase and the spin phase according to the theoretical analysis we have developed. The activation energy Δ_{gap} is the skyrmion excitation energy in each phase. It is illustrated in Fig.26.9(b) how the activation energy Δ_{gap} is expected to change from the ppin phase to the spin phase via the canted phase.

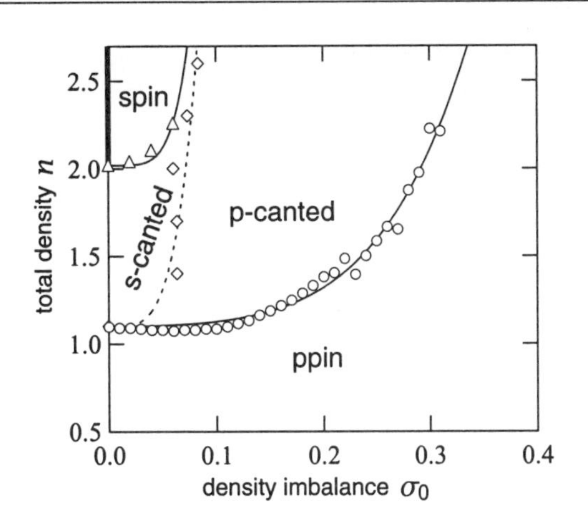

Fig. 26.10 The phase diagram in the σ_0-ρ_0 plane is constructed experimentally for a sample with Δ_{SAS} = 11K. The heavy solid line just on the vertical axis for $n > 2.021 \times 10^{11}\text{cm}^{-2}$ represents the spin phase that is only stable at the balanced point but for an impurity effect. Open triangles (circles) are the densities n that give the local minima (minima) of activation energy at fixed σ_0. Open diamonds are the imbalance parameters σ_0 that give the minima of the activation energy in Fig.26.11, which stand for the boundary between the s-canted and p-canted regions. Data are taken from Fukuda et al.[180]

We repeat the same analysis at various imbalance points σ_0 and extract these two types of phase transition points [Fig.26.9]. We mark the $\sigma_0 - \rho_0$ phase diagram with open circles and triangles representing the maxima and minima, respectively [Fig.26.10]. The spin phase is found to occur not only at the balanced point but also in imbalanced configuration. This is interpreted as in Fig.26.7.

We next present the measured activation energy as a function of σ_0 at several values of n in Fig.26.11 in order to investigate quantitative features of the image plots given in Fig.26.8. In Fig.26.11(a) the activation energy Δ_{gap} is almost constant or gradually increases as the density imbalance is increased, indicating that the state is robust against the density imbalance. It is consistent that all these states are in the ppin phase for all values of $\sigma_0 < 0.4$. In Fig.26.11(b), the behavior of Δ_{gap} is slightly different from the one in in Fig.26.11(a). It decreases slightly and then increases more rapidly as σ_0 increases. We interpret this behavior to mean that the canted phase is composed of two regions [Fig.26.4], where Δ_{gap} decreases in the s-canted region and it increases in the f-canted region. Finally, in Fig.26.11(c), Δ_{gap} steeply decreases for the initial increment of σ, followed by the steep increasing of Δ_{gap} above the critical point. The initial decrease of Δ_{gap} would be due to the feebleness of the spin phase against the density imbalance. A further decrease would be explained by the instability of the spin-component in the canted phase. The steep increase would be due to the ppin-component in the canted phase which

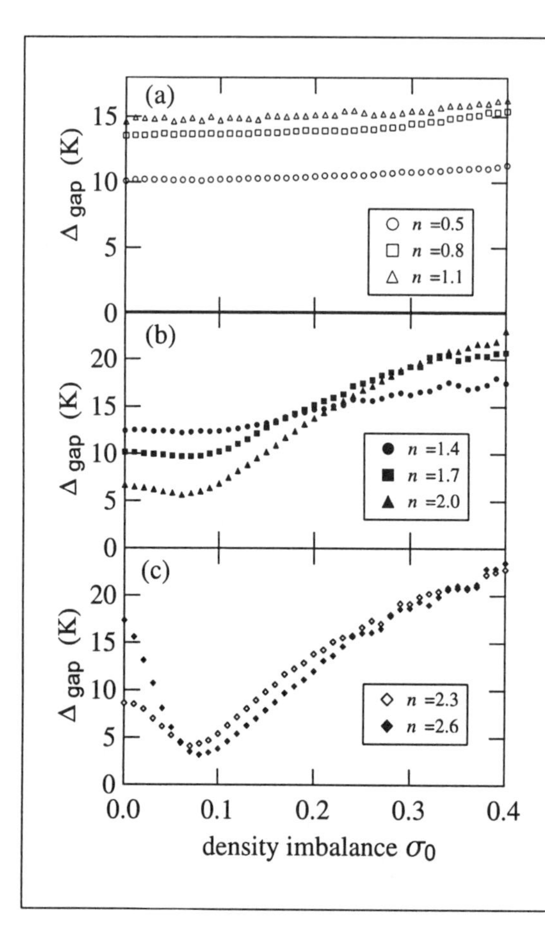

Fig. 26.11 The activation energy is plotted as a function of σ_0 for several values of n. At the balance point, (a) the states are in the ppin phase, (b) they are in the canted phase, and (c) they are in the spin phase. The data are taken from Fukuda et al.[180]

increases the interlayer correlation. The minimum point would separate the s-canted and p-canted regions, as indicated by open diamonds in the σ_0-ρ_0 phase diagram [Fig.26.10].

We proceed to study the ρ_0-Δ_{SAS} phase diagram [Fig.26.12], where the theoretically estimated phase boundaries are shown by a dotted line and a dashed line. We have also displayed the phase boundaries determined based on an exact-diagonalization analysis[180] by triangles with solid lines. The convergency for the canted-ppin phase boundary is quite good for a system containing more than 6 particles to be diagonalized. On the other hand, the calculated points on the spin-canted boundary converges only gradually against the number of particles N (see inset in Fig.26.12), and we have displayed the asymptotic values for $N \to \infty$. The agreement of the spin–canted boundary between the HF approximation and the exact-diagonalization analysis is extremely good, as is expected. We have plotted the experimental data taken from Sawada et al.[423] and Fukuda et al.[180] The agreement with the exact-diagonalization analysis is reasonably well but not per-

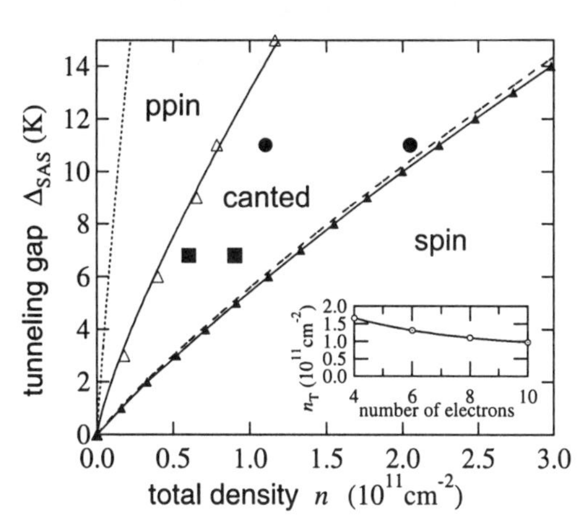

Fig. 26.12 The phase diagram in the ρ_0-Δ_{SAS} plane is given at the balanced point. Solid circles and solid squares are the experimental data representing phase transition points, which are taken from Fukuda et at.[180] and Sawada et al.[423], respectively. Dotted line and dashed line are the estimated phase boundaries by the Hartree-Fock approximation. Open triangles and solid triangles are the phase boundary points derived by an exact diagonalization method.[180]

fect. The disagreement would be due to an impurity effect and a finite width of quantum wells, which are neglected in theoretical calculations.

26.6 SU(4) Breaking and Grassmannian Fields

Isospin states in the $\nu = 2$ bilayer QH system belong to the **6**-dimensional irreducible representation of SU(4) as in (26.1.1). We restrict the Hamiltonian to this representation. In the SU(4)-invariant limit the Hamiltonian consists of the direct Coulomb term H_{D}^{+} and the exchange interaction given by

$$\mathcal{H}_{\text{X}}^{+} = 2J_s^{+}\partial_k \mathfrak{I}(\boldsymbol{x})\cdot\partial_k \mathfrak{I}(\boldsymbol{x}), \tag{26.6.1}$$

as derived in Section 32.6: See (32.6.12). All isospins are spontaneously polarized to lower the exchange energy. There are six degenerate states any one of which can be chosen as the ground state. It implies a spontaneous breaking of the SU(4) symmetry, giving rise to Goldstone modes and topological solitons.

Actually, the ground state is chosen explicitly by SU(4)-noninvariant interactions, as described in the previous section. Nevertheless, we regard the system as an isospin ferromagnet, where Goldstone modes are made gapful. When the explicit breaking is small enough, the pattern of the spontaneous symmetry breaking provides us with an essential information on perturbative and nonperturbative excitations.

SU(4) Spontaneous Symmetry Breaking

We first study the spontaneous symmetry breaking in the spin phase.[225] The ground state is a spin triplet and a ppin singlet. Let the spins be polarized to the z direction. The unbroken symmetry is such one that keeps the ground state invariant. One unbroken symmetry is the rotation about the spin-z axis, which is generated by the generator $\mathcal{S}_z$. The ground state is also a ppin SU(2) singlet. Thus, the rotation in the ppin space generated by $\mathcal{P}_a$ or $\mathcal{R}_{za}$ keeps the ground state invariant. These seven transformations generated by $\mathcal{S}_z, \mathcal{P}_a$ and $\mathcal{R}_{za}$ exhaust the unbroken symmetry, which is U(1)⊗SU(2)⊗SU(2). They form the algebra,

$$[\mathcal{S}_z, \mathcal{J}_a] = 0, \qquad [\mathcal{S}_z, \mathcal{K}_b] = 0, \qquad [\mathcal{J}_a, \mathcal{J}_b] = i\epsilon_{abc}\mathcal{J}_c,$$
$$[\mathcal{K}_a, \mathcal{K}_b] = i\epsilon_{abc}\mathcal{K}_c, \qquad [\mathcal{J}_a, \mathcal{K}_b] = 0, \tag{26.6.2}$$

where $\mathcal{J}_a$ and $\mathcal{K}_a$ are defined by

$$\mathcal{J}_a = \frac{1}{2}(\mathcal{P}_a + \mathcal{R}_{za}), \qquad \mathcal{K}_a = \frac{1}{2}(\mathcal{P}_a - \mathcal{R}_{za}). \tag{26.6.3}$$

The pattern of the symmetry breaking is

$$\mathrm{SU(4)} \to \mathrm{U(1)} \otimes \mathrm{SU(2)} \otimes \mathrm{SU(2)}, \tag{26.6.4}$$

where the U(1) transformation is generated by $\mathcal{S}_z$.

The target space is given by the coset space

$$G_{4,2} = \frac{\mathrm{SU(4)}}{\mathrm{U(1)} \otimes \mathrm{SU(2)} \otimes \mathrm{SU(2)}}, \tag{26.6.5}$$

which is an example ($N = 4$, $k = 2$) of the complex Grassmannian manifold,

$$G_{N,k} = \frac{\mathrm{U(N)}}{\mathrm{U(k)} \otimes \mathrm{U(N\text{-}k)}} = \frac{\mathrm{SU(N)}}{\mathrm{U(1)} \otimes \mathrm{SU(k)} \otimes \mathrm{SU(N\text{-}k)}}, \tag{26.6.6}$$

whose dimension is $U^2 - k^2 - (N-k)^2 = 2k(N-k)$. Since the dimension of the manifold $G_{4,2}$ is 8, there emerge eight Goldstone modes associated with the broken eight generators, $\mathcal{S}_x$, $\mathcal{S}_y$, $\mathcal{R}_{xa}$ and $\mathcal{R}_{ya}$. The Grassmannian manifold $G_{N,k}$ allows topological solitons called $G_{N,k}$ solitons, as we shall see in Section 26.7. The nontrivial mapping characterizing the soliton is the U(1) group generated by $\mathcal{S}_z$. The $G_{4,1}$ skyrmion is the same object as the SU(4)-skyrmion.

Similarly we can study the spontaneous symmetry breaking in the ppin phase. The ground state is a ppin triplet and a spin singlet. For simplicity we consider the case, $d = 0$, where all ppins are polarized to the x direction by the tunneling interaction. The pattern of the symmetry breaking is the same as in the spin phase, as should be the case. The only difference is that the nontrivial mapping is the U(1) group generated by $\mathcal{P}_x$. The seven residual symmetries are generated by $\mathcal{P}_x$, $\mathcal{S}_a$ and $\mathcal{R}_{ax}$, while the eight broken symmetries are generated by $\mathcal{P}_y$, $\mathcal{P}_z$, $\mathcal{R}_{ay}$ and $\mathcal{R}_{az}$.

Grassmannian Fields

There are fifteen generators in the nonlinear sigma model (26.6.1), but only eight of them are independent fields. To elucidate them we introduce the CP^3 field $n^\sigma(x)$ by (25.2.1). There are two electrons per one Landau site at $\nu = 2$. We introduce two CP^3 fields $\boldsymbol{n}_1(\boldsymbol{x})$ and $\boldsymbol{n}_2(\boldsymbol{x})$ to describe them. Due to the Pauli exclusion principle they should be orthogonal one to another. Hence, we require

$$\boldsymbol{n}_i^\dagger(\boldsymbol{x}) \cdot \boldsymbol{n}_j(\boldsymbol{x}) = \delta_{ij}. \tag{26.6.7}$$

Using a set of two CP^3 fields subject to this normalization condition we introduce a 4×2 matrix field

$$Z(\boldsymbol{x}) = (\boldsymbol{n}_1, \boldsymbol{n}_2), \tag{26.6.8}$$

obeying

$$Z^\dagger Z = 1. \tag{26.6.9}$$

Though we have introduced two fields $\boldsymbol{n}_1(\boldsymbol{x})$ and $\boldsymbol{n}_2(\boldsymbol{x})$, we cannot distinguish them quantum mechanically since they describe two electrons in the same Landau site. Namely, two fields $Z(\boldsymbol{x})$ and $Z'(\boldsymbol{x})$ are indistinguishable physically when they are related by a local U(2) transformation $U(\boldsymbol{x})$,

$$Z'(\boldsymbol{x}) = Z(\boldsymbol{x})U(\boldsymbol{x}). \tag{26.6.10}$$

By identifying these two fields $Z(\boldsymbol{x})$ and $Z'(\boldsymbol{x})$, the 4×2 matrix field $Z(\boldsymbol{x})$ takes values on the Grassmann manifold $G_{4,2}$ defined by (26.6.5). The field $Z(\boldsymbol{x})$ is no longer a set of two independent CP^3 fields. It is a new object, called the Grassmannian field, carrying eight real degrees of freedom, as mentioned just below (26.6.6).

We have argued that the dynamical fields are given by the Grassmannian field $Z(\boldsymbol{x})$ in the bilayer QH system at $\nu = 2$.

In the SU(4)-invariant limit the effective Hamiltonian is given by the SU(4) nonlinear sigma model (26.6.1). By way of the relation

$$\mathcal{I}_a = \frac{1}{2}\mathrm{Tr}[Z^\dagger \lambda_a Z] = \frac{1}{2}\boldsymbol{n}_1^\dagger \lambda_a \boldsymbol{n}_1 + \frac{1}{2}\boldsymbol{n}_2^\dagger \lambda_a \boldsymbol{n}_2, \tag{26.6.11}$$

we are able to rewrite it into the well-known Hamiltonian[329] for the Grassmannian field,

$$\mathcal{H}_X = 2J_s^+ \mathrm{Tr}\left[(\partial_j Z - iZK_j)^\dagger)(\partial_j Z - iZK_j)\right], \tag{26.6.12}$$

where

$$K_\mu(\boldsymbol{x}) = -iZ^\dagger(\boldsymbol{x})\partial_\mu Z(\boldsymbol{x}). \tag{26.6.13}$$

This Hamiltonian has the local U(2) gauge symmetry,

$$Z(\boldsymbol{x}) \to Z(\boldsymbol{x})U(\boldsymbol{x}), \tag{26.6.14a}$$

$$K_\mu(\boldsymbol{x}) \to U(\boldsymbol{x})^\dagger K_\mu(\boldsymbol{x})U(\boldsymbol{x}) - iU(\boldsymbol{x})^\dagger \partial_\mu U(\boldsymbol{x}). \tag{26.6.14b}$$

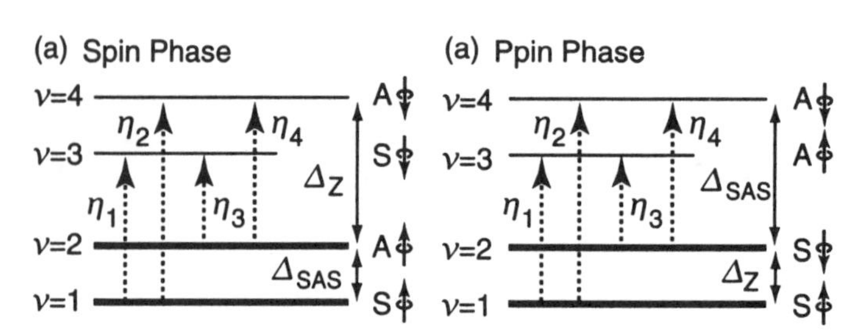

Fig. 26.13 The lowest two energy levels are occupied in the ground state of the spin phase (a) and the ppin phase (b) at $\nu = 2$. Small fluctuations are Goldstone modes η_1, η_2, η_3 and η_4.

The gauge field K_μ is not a dynamical field since it is an auxiliary field given by (26.6.13).

The lowest-energy one-body electron state is the up-spin symmetric state. The second lowest energy state is either the up-spin antisymmetric state or the down-spin symmetric state (Fig.26.13). It is convenient to use the CP^3 field whose components are taken in the symmetric-antisymmetric basis,

$$n^{S\alpha}(\boldsymbol{x}) = \frac{1}{\sqrt{2}}\left(n^{f\alpha}(\boldsymbol{x}) + n^{b\alpha}(\boldsymbol{x})\right), \qquad n^{A\alpha}(\boldsymbol{x}) = \frac{1}{\sqrt{2}}\left(n^{f\alpha}(\boldsymbol{x}) - n^{b\alpha}(\boldsymbol{x})\right), \quad (26.6.15)$$

where $\alpha = \uparrow, \downarrow$.

We first study the spin phase, where it is convenient to take the components of the CP^3 field as [Fig.26.13(a)]

$$\boldsymbol{n} = (n^{S\uparrow}, n^{A\uparrow}, n^{S\downarrow}, n^{A\downarrow}). \quad (26.6.16)$$

In the ground state all up-spin states are occupied,

$$\boldsymbol{n}_1 = (1,0,0,0), \qquad \boldsymbol{n}_2 = (0,1,0,0). \quad (26.6.17)$$

Accordingly the ground state is represented by the $G_{4,2}$ field as

$$Z_g = \begin{pmatrix} 1 & 0 \\ 0 & 1 \\ 0 & 0 \\ 0 & 0 \end{pmatrix} U(\boldsymbol{x}). \quad (26.6.18)$$

Up to the second order of fluctuation fields, we may expand (26.6.8) as

[Fig.26.13(a)]

$$Z = \begin{pmatrix} 1 - \frac{1}{2}|\eta_1|^2 - \frac{1}{2}|\eta_2|^2 & -\frac{1}{2}\eta_1^* \eta_3 - \frac{1}{2}\eta_2^* \eta_4 \\ -\frac{1}{2}\eta_3^* \eta_1 - \frac{1}{2}\eta_4^* \eta_2 & 1 - \frac{1}{2}|\eta_3|^2 - \frac{1}{2}|\eta_4|^2 \\ \eta_1 & \eta_3 \\ \eta_2 & \eta_4 \end{pmatrix} U(\boldsymbol{x}), \qquad (26.6.19)$$

where η_k are the four complex Goldstone modes accompanied with the spontaneous SU(4) breaking (26.6.4). The Goldstone modes η_1 and η_4 are pure spin and pseudospin waves, while η_2 and η_3 are waves mixing spins and pseudospins.

We can similarly analyze the ppin phase, where it is convenient to take the components of the CP^3 field as [Fig.26.13(b)]

$$\boldsymbol{n} = (n^{S\uparrow}, n^{S\downarrow}, n^{A\uparrow}, n^{A\downarrow}). \qquad (26.6.20)$$

In the ground state all up-spin states are occupied,

$$\boldsymbol{n}_1 = (1, 0, 0, 0), \quad \boldsymbol{n}_2 = (0, 1, 0, 0). \qquad (26.6.21)$$

The $G_{4,2}$ field is given by (26.6.18), and the eight Goldstone modes are parameterized as in (26.6.19) with this choice of the components of the CP^3 field.

26.7 Grassmannian $G_{4,2}$ Solitons

The Grassmannian field $Z(x)$ takes values on the Grassmannian manifold $G_{4,2}$. The existence of topological solitons ($G_{4,2}$-skyrmions) follows from the homotopy theorem,

$$\pi_2(G_{N,k}) = \mathbb{Z}, \qquad (26.7.1)$$

as we have discussed in Chapter 7. The topological charge is defined[329] as a gauge invariant by

$$Q = \frac{i}{2\pi} \int d^2x\, \epsilon_{jk} \mathrm{Tr}\left[(\partial_j Z - iZK_j)^\dagger (\partial_k Z - iZK_k) \right]. \qquad (26.7.2)$$

It is a topological invariant since it is equal to

$$Q = \frac{i}{2\pi} \int d^2x\, \epsilon_{jk} \mathrm{Tr}\left[(\partial_j Z)^\dagger (\partial_k Z) \right]. \qquad (26.7.3)$$

We rewrite it with the use of (26.6.8) as

$$Q = \frac{i}{2\pi} \int d^2x\, \epsilon_{jk} \left[(\partial_j \boldsymbol{n}_1)^\dagger \cdot (\partial_k \boldsymbol{n}_1) + (\partial_j \boldsymbol{n}_2)^\dagger \cdot (\partial_k \boldsymbol{n}_2) \right]. \qquad (26.7.4)$$

It is the sum of the topological charges associated with the CP^3 fields $\boldsymbol{n}_1$ and $\boldsymbol{n}_2$ [see (7.8.16).] Hence, the $G_{4,2}$-skyrmion consists of two CP^3-skyrmions,

$$n_k^\sigma(\boldsymbol{x}) = \frac{1}{\sqrt{\sum |\omega_k^\sigma(z)|}} \omega_k^\sigma(z), \qquad (26.7.5)$$

excited in the front and back layers in the spin phase. Here, $\omega_k^\sigma(z)$ are arbitrary analytic functions if they describe two independent CP[3]-skyrmions.

We recapitulate the LLL condition for soliton states to derive a condition on the pair of CP[3] fields (26.7.5). We start with the requirement that the kinetic energy is quenched,

$$\frac{\partial}{\partial z^*}\varphi^\sigma(\boldsymbol{x})|\Phi\rangle = 0, \tag{26.7.6}$$

where $\varphi^\sigma(\boldsymbol{x})$ is the dressed CB field parametrized as

$$\varphi^\sigma(\boldsymbol{x}) = e^{i\chi(\boldsymbol{x})}e^{\mathcal{A}(\boldsymbol{x})}\sqrt{\rho(\boldsymbol{x})}n^\sigma(\boldsymbol{x}). \tag{26.7.7}$$

The auxiliary field $\mathcal{A}(\boldsymbol{x})$ is given by (24.1.6), and hence it satisfies

$$\boldsymbol{\nabla}^2\mathcal{A}(\boldsymbol{x}) = 2\pi\left(\rho(\boldsymbol{x}) - \rho_0\right). \tag{26.7.8}$$

Let $|\Phi\rangle$ be an eigenstate of the density operator $\rho(\boldsymbol{x})$ and a coherent state of the CP[3] field $\boldsymbol{n}(\boldsymbol{x})$,

$$\rho(\boldsymbol{x})|\Phi\rangle = \rho^{\text{cl}}(\boldsymbol{x})|\Phi\rangle, \qquad \boldsymbol{n}(\boldsymbol{x})|\Phi\rangle = \boldsymbol{n}^{\text{cl}}(\boldsymbol{x})|\Phi\rangle. \tag{26.7.9}$$

For the state the LLL condition (26.7.6) implies

$$e^{-\mathcal{A}^{\text{cl}}(\boldsymbol{x})}\sqrt{\rho^{\text{cl}}(\boldsymbol{x})}n^{\text{cl}(\sigma)}(\boldsymbol{x}) = \omega^\sigma(z), \tag{26.7.10}$$

where $\omega^\sigma(z)$ is an analytic function, and $\mathcal{A}^{\text{cl}}(\boldsymbol{x})$, $\rho^{\text{cl}}(\boldsymbol{x})$ and $n^{\text{cl}(\sigma)}(\boldsymbol{x})$ are classical fields. The argument is precisely the same one as given for CP[1]-skyrmions in Section 16.2. The point is that it is valid at any filling factor.

At $\nu = 2$ the state $|\Phi\rangle$ contains two electrons per one Landau site. Thus, we have

$$\boldsymbol{n}(\boldsymbol{x})|\Phi\rangle = \boldsymbol{n}^{\text{cl}}(\boldsymbol{x})|\Phi\rangle = \left[\boldsymbol{n}_1^{\text{cl}}(\boldsymbol{x}) + \boldsymbol{n}_2^{\text{cl}}(\boldsymbol{x})\right]|\Phi\rangle, \tag{26.7.11}$$

together with $\boldsymbol{n}_i^{\text{cl}}(\boldsymbol{x})\cdot\boldsymbol{n}_j^{\text{cl}}(\boldsymbol{x}) = \delta_{ij}$ and

$$\rho(\boldsymbol{x})|\Phi\rangle = \rho^{\text{cl}}(\boldsymbol{x})|\Phi\rangle = [\rho_1^{\text{cl}}(\boldsymbol{x}) + \rho_2^{\text{cl}}(\boldsymbol{x})]|\Phi\rangle. \tag{26.7.12}$$

We may solve (26.7.10) as

$$n^{\text{cl}(\sigma)}(\boldsymbol{x}) = \frac{\sqrt{2}}{\sqrt{\sum|\omega^\sigma(x)|^2}}\omega^\sigma(z). \tag{26.7.13}$$

Comparing this with (26.7.5) we conclude

$$\sum_\sigma|\omega_1^\sigma(z)|^2 = \sum_\sigma|\omega_2^\sigma(z)|^2 \tag{26.7.14}$$

and $\omega^\sigma(z) = \omega_1^\sigma(z) + \omega_2^\sigma(z)$. Therefore, for each component the LLL condition (26.7.7) holds and we obtain from (26.7.8) that

$$\frac{1}{4\pi}\boldsymbol{\nabla}^2\ln\rho^{\text{cl}}(\boldsymbol{x}) - \rho^{\text{cl}}(\boldsymbol{x}) + \rho_0 = j_{\text{sky}}^0(\boldsymbol{x}) \tag{26.7.15}$$

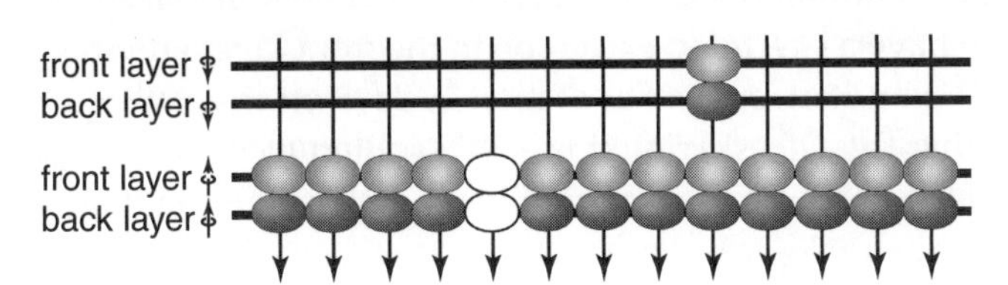

Fig. 26.14 A quasiparticle is a biskyrmion consisting of a pair of two skyrmions in a genuin bilayer QH state at $\nu = 2$. We give an example of a vortex-pair excitations (zero-size skyrmions).

with

$$j^0_{\text{sky}}(\boldsymbol{x}) = \frac{1}{4\pi}\nabla^2 \ln \sum_\sigma |\omega_1^\sigma(z)|^2 = \frac{1}{4\pi}\nabla^2 \ln \sum_\sigma |\omega_2^\sigma(z)|^2. \tag{26.7.16}$$

It is easy to see that the topological charge (26.7.4) is given by $Q = 2\int d^2x\, j^0_{\text{sky}}(\boldsymbol{x})$.

The $G_{4,2}$-skyrmion has a general expression,

$$Z_{\text{sky}} = \frac{1}{\sqrt{\sum_\sigma |\omega_1^\sigma(z)|^2}} \begin{pmatrix} \omega_1^{f\uparrow}(z) & \omega_2^{f\uparrow}(z) \\ \omega_1^{f\downarrow}(z) & \omega_2^{f\downarrow}(z) \\ \omega_1^{b\uparrow}(z) & \omega_2^{b\uparrow}(z) \\ \omega_1^{b\downarrow}(z) & \omega_2^{b\downarrow}(z) \end{pmatrix}, \tag{26.7.17}$$

together with the condition (26.7.14). The simplest $G_{4,2}$-skyrmion in the spin sector is given by

$$Z^{\text{spin}}_{\text{sky}} = \frac{1}{\sqrt{|z|^2+\kappa^2}} \begin{pmatrix} \omega_1^{S\uparrow}(z) & \omega_2^{S\uparrow}(z) \\ \omega_1^{A\uparrow}(z) & \omega_2^{A\uparrow}(z) \\ \omega_1^{S\downarrow}(z) & \omega_2^{S\downarrow}(z) \\ \omega_1^{A\downarrow}(z) & \omega_2^{A\downarrow}(z) \end{pmatrix} = \frac{1}{\sqrt{|z|^2+\kappa^2}} \begin{pmatrix} z & 0 \\ 0 & z \\ \kappa & 0 \\ 0 & k \end{pmatrix}, \tag{26.7.18}$$

which describes a pair of CP^3-skyrmions excited in both layers. Similarly the simplest $G_{4,2}$-skyrmion in the ppin sector is given by

$$Z^{\text{spin}}_{\text{sky}} = \frac{1}{\sqrt{|z|^2+\kappa^2}} \begin{pmatrix} \omega_1^{S\uparrow}(z) & \omega_2^{S\uparrow}(z) \\ \omega_1^{S\downarrow}(z) & \omega_2^{S\downarrow}(z) \\ \omega_1^{A\uparrow}(z) & \omega_2^{A\uparrow}(z) \\ \omega_1^{A\downarrow}(z) & \omega_2^{A\downarrow}(z) \end{pmatrix} = \frac{1}{\sqrt{|z|^2+\kappa^2}} \begin{pmatrix} z & 0 \\ 0 & z \\ \kappa & 0 \\ 0 & k \end{pmatrix} \tag{26.7.19}$$

in the balanced configuration. We call them biskyrmions [Fig.26.14].

We emphasize this peculiar situation by recalling the formula (26.7.7), which introduces the CP^3 field $\boldsymbol{n}(\boldsymbol{x})$. It is essential that the total electron density $\rho(\boldsymbol{x})$ is common to all the four components; otherwise the SU(4) symmetry is explicitly broken by hand. Even if we try to excite a skyrmion only in the front layer, the density modulation associated with it affects equally electrons in the back layer as far as the LLL condition (26.7.7) is respected. In a genuine bilayer QH system it is impossible to have a skyrmion excitation in the front layer and the ground state in the back layer simultaneously: See section 26.8 for more details on this point.

We note that the $G_{4,2}$-skyrmion is a BPS soliton of the exchange Hamiltonian (26.6.12). Indeed, the following inequality[329] holds between the exchange energy (26.6.12) and the topological charge (26.7.2),

$$E_X \geqq 4\pi J^+ Q, \tag{26.7.20}$$

where the equality is achieved by the $G_{4,2}$-skyrmion (26.7.17). As in the $\nu = 1$ case, it is shown that the wave function for the biskyrmion state (26.7.17) is given by

$$\Psi[x] = \prod_{j=1}^{N} \begin{pmatrix} \omega_1^{f\uparrow}(z_j) & \omega_2^{f\uparrow}(z_j) \\ \omega_1^{f\downarrow}(z_j) & \omega_2^{f\downarrow}(z_j) \\ \omega_1^{b\uparrow}(z_j) & \omega_2^{b\uparrow}(z_j) \\ \omega_1^{b\downarrow}(z_j) & \omega_2^{b\downarrow}(z_j) \end{pmatrix} \prod_{r<s}(z_r - z_s) e^{-\sum_{r=1}^{N}|z_r|^2/4\pi\ell_B^2}, \tag{26.7.21}$$

where $[x]$ stands for $(x_1, x_2, \cdots, x_N)$.

Spin-1 Sigma Models for Biskyrmions

We have argued that charged excitations are Grassmannian $G_{4,2}$-skyrmions in the $\nu = 2$ bilayer QH system. We now represent them in terms of the SU(4) sigma field by way of relation (26.6.11).

We first treat the spin phase, where the biskyrmion is described by (26.7.18). We calculate the SU(4) generators $\mathcal{I}_a$ by using this biskyrmion configuration in (26.6.11), or alternatively $\mathcal{S}_a$, $\mathcal{P}_a$ and $\mathcal{R}_{ab}$ by way of (25.2.4). We find that $\mathcal{P}_a = \mathcal{R}_{ab} = 0$ and that $\mathcal{S}_a$ are given by the well-known formula of the O(3)-skyrmion,

$$\mathcal{S}_x = \frac{2\kappa|z|}{|z|^2 + \kappa^2}\cos\theta, \quad \mathcal{S}_y = -\frac{2\kappa|z|}{|z|^2 + \kappa^2}\sin\theta, \quad \mathcal{S}_z = \frac{|z|^2 - \kappa^2}{|z|^2 + \kappa^2}, \tag{26.7.22}$$

normalized as $\sum \mathcal{S}_a(x)\mathcal{S}_a(x) = 1$. This biskyrmion configuration is purely spin-like and expanded by three states (26.1.3a). We call it the spin biskyrmion.

It is instructive to reformulate it in terms of Schwinger bosons. We define the Schwinger-boson operator s_a by

$$s_\uparrow(x) = n^{f\uparrow}(x)n^{b\uparrow}(x), \qquad s_0(x) = \frac{1}{\sqrt{2}}\left(n^{f\uparrow}(x)n^{b\downarrow}(x) + n^{f\downarrow}(x)n^{b\uparrow}(x)\right),$$
$$s_\downarrow(x) = n^{f\downarrow}(x)n^{b\downarrow}(x). \tag{26.7.23}$$

We evaluate them on biskyrmion configuration (26.7.18),

$$s(x) = (s_\uparrow, s_0, s_\downarrow) = \frac{1}{|z|^2 + \kappa^2}(z^2, \sqrt{2}z\kappa, \kappa^2). \tag{26.7.24}$$

The spin-1 field is given by

$$\mathcal{S}_a(x) = s^\dagger(x)\hat{\tau}_a s(x), \tag{26.7.25}$$

where $\hat{\tau}_a$ are the SU(2) generators in the adjoint representation,

$$\hat{\boldsymbol{\tau}} \equiv \left\{ \frac{1}{\sqrt{2}} \begin{pmatrix} 0 & 1 & 0 \\ 1 & 0 & 1 \\ 0 & 1 & 0 \end{pmatrix}, \frac{1}{\sqrt{2}} \begin{pmatrix} 0 & -i & 0 \\ i & 0 & -i \\ 0 & i & 0 \end{pmatrix}, \begin{pmatrix} 1 & 0 & 0 \\ 0 & 0 & 0 \\ 0 & 0 & -1 \end{pmatrix} \right\}. \tag{26.7.26}$$

It is easy to see that (26.7.25) with (26.7.23) reproduces skyrmion configuration (26.7.22). Hence, the Grassmannian $G_{4,2}$-skyrmion in the spin phase is equal to the O(3)-skyrmion in the spin sector.

We reduce the exchange Hamiltonian (25.2.8) to the spin sector, where $\mathcal{P}_a = \mathcal{R}_{ab} = 0$. Hence we find

$$\mathcal{H}^{\text{eff}} = J_s \partial_k \boldsymbol{\mathcal{S}}(\boldsymbol{x}) \cdot \partial_k \boldsymbol{\mathcal{S}}(\boldsymbol{x}) + \mathcal{H}_{\text{self}} - \Delta_Z \rho_\Phi \mathcal{S}_z(\boldsymbol{x}), \tag{26.7.27}$$

where $\mathcal{H}_D$ stands for the Coulomb self-energy. The O(3)-skyrmion (26.7.22) is the BPS soliton of the O(3)-nonlinear-sigma-model part of this Hamiltonian. The total Hamiltonian (26.7.27) is quite similar to the one in the monolayer QH system at $\nu = 1$ with the replacement of the spin-$\frac{1}{2}$ field by the spin-1 field.

The excitation energy of the biskyrmion is easily calculable based on (26.7.27) as in the $\nu = 1$ monolayer QH system: See Section 16.3. The skyrmion scale κ is determined to optimize the Coulomb energy and the Zeeman energy. It is to be remarked that the spin biskyrmion is insensible to the interlayer stiffness J^d.

We may similarly discuss the ppin phase in the balanced configuration, where only $\mathcal{P}_a$ is nonvanishing for the biskyrmion configuration. The biskyrmion turns out to be the O(3)-skyrmion,

$$\mathcal{P}_x = \frac{2\kappa|z|}{|z|^2 + \kappa^2} \cos\theta, \quad \mathcal{P}_y = -\frac{2\kappa|z|}{|z|^2 + \kappa^2} \sin\theta,$$
$$\mathcal{P}_z = \frac{|z|^2 - \kappa^2}{|z|^2 + \kappa^2}. \tag{26.7.28}$$

The effective Hamiltonian (25.2.8) is restricted to the ppin sector, which reads

$$\mathcal{H}^{\text{eff}} = J_s^d \{\partial_k \mathcal{P}_x(\boldsymbol{x}) \cdot \partial_k \mathcal{P}_x(\boldsymbol{x}) + \partial_k \mathcal{P}_y(\boldsymbol{x}) \cdot \partial_k \mathcal{P}_y(\boldsymbol{x})\} + J_s \partial_k \mathcal{P}_z(\boldsymbol{x}) \cdot \partial_k \mathcal{P}_z(\boldsymbol{x})$$
$$+ \mathcal{H}_{\text{self}} + \mathcal{H}_{\text{cap}} - \Delta_{\text{SAS}} \rho_\Phi \mathcal{P}_x(\boldsymbol{x}), \tag{26.7.29}$$

where $\mathcal{H}_{\text{self}}$ and $\mathcal{H}_{\text{cap}}$ stand for the Coulomb self-energy and the capacitance energy. The total Hamiltonian is quite similar to that in the spin-frozen bilayer QH system at $\nu = 1$ with the replacement of the ppin-$\frac{1}{2}$ field with the ppin-1 field.

26.8 Genuine Bilayer versus Two-Monolayer Systems

It is intriguing that a quasiparticle is a biskyrmion at $\nu = 2$. However, it is clear intuitively that a charged excitation is a simple skyrmion excited in one of the two layers even at $\nu = 2$ if the two layers are sufficiently separated [Figs.26.15(a) and 26.16(a)]. Let us first study what is the criterion for it to be a genuine bilayer system

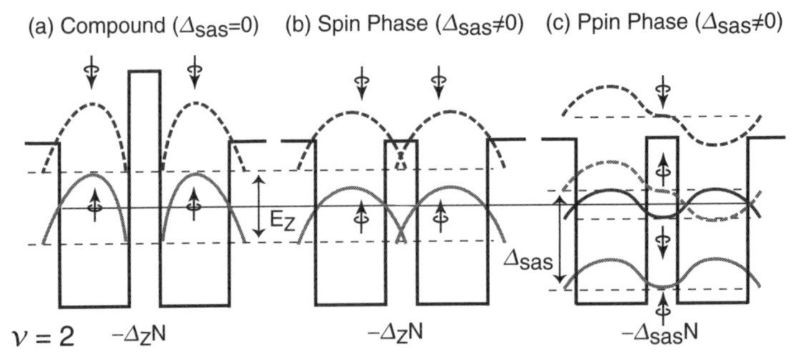

Fig. 26.15 One-body energy levels and wave functions are illustrated schematically in a double quantum well structure at $\nu = 2$, by neglecting Coulomb interactions. The levels are determined by the competition between the Zeeman (Δ_Z) and tunneling (Δ_{SAS}) gap energies. (a) When the interlayer barrier is high enough ($\Delta_{SAS} = 0$), the one-body wave function is confined within each quantum well, and a simple skyrmion will be excited in one of the two layers. (b) When the interlayer barrier is not so high ($\Delta_{SAS} \neq 0$), the wave functions overlap between the two quantum wells. Then, correlated compound states are realized to form the spin phase for $\Delta_Z > \Delta_{SAS}$. (c) On the other hand, resonant states are realized to form the ppin phase for $\Delta_{SAS} > \Delta_Z$ on the symmetric and antisymmetric states. A biskyrmion is expected to emerge in these cases (b) and (c).

or a set of two monolayer systems at $\nu = 1$. We examine the local symmetry of the direct interactions (25.1.11) and (25.1.13), which only involves $\rho(\boldsymbol{x})$, $\mathcal{S}_z(\boldsymbol{x})$, $\mathcal{P}_x(\boldsymbol{x})$ and $\mathcal{P}_z(\boldsymbol{x})$. It is given by a direct product of two U(1) symmetries, $\mathrm{U}^{\uparrow}(1)\otimes\mathrm{U}^{\downarrow}(1)$, because these operators are invariant under two local phase transformations,

$$\begin{pmatrix} \psi^{\mathrm{f}\uparrow}(\boldsymbol{x}) \\ \psi^{\mathrm{b}\uparrow}(\boldsymbol{x}) \end{pmatrix} \longrightarrow e^{i\alpha(\boldsymbol{x})} \begin{pmatrix} \psi^{\mathrm{f}\uparrow}(\boldsymbol{x}) \\ \psi^{\mathrm{b}\uparrow}(\boldsymbol{x}) \end{pmatrix},$$

$$\begin{pmatrix} \psi^{\mathrm{f}\downarrow}(\boldsymbol{x}) \\ \psi^{\mathrm{b}\downarrow}(\boldsymbol{x}) \end{pmatrix} \longrightarrow e^{i\beta(\boldsymbol{x})} \begin{pmatrix} \psi^{\mathrm{f}\downarrow}(\boldsymbol{x}) \\ \psi^{\mathrm{b}\downarrow}(\boldsymbol{x}) \end{pmatrix}. \tag{26.8.1}$$

The exchange interaction breaks it into a single U(1) symmetry,

$$\psi^{\sigma}(\boldsymbol{x}) \longrightarrow e^{i\alpha(\boldsymbol{x})}\psi^{\sigma}(\boldsymbol{x}). \tag{26.8.2}$$

Note that this is the case even if $J_s^d = 0$ provided $J_s \neq 0$. It is the exact local symmetry of the total Hamiltonian. Corresponding to this U(1) symmetry, we have introduced the normalized CB field $n^{\sigma}(\boldsymbol{x})$ in (26.7.7), or

$$\phi^{\sigma}(\boldsymbol{x}) = e^{i\chi(\boldsymbol{x})}\sqrt{\rho(\boldsymbol{x})}n^{\sigma}(\boldsymbol{x}). \tag{26.8.3}$$

It is the CP^3 field containing 3 independent complex fields, because one real field is eliminated by the constraint (25.2.2) and furthermore the U(1) phase field is not dynamical due to the local U(1) symmetry (26.8.2). At $\nu = 2$ we introduce a set of

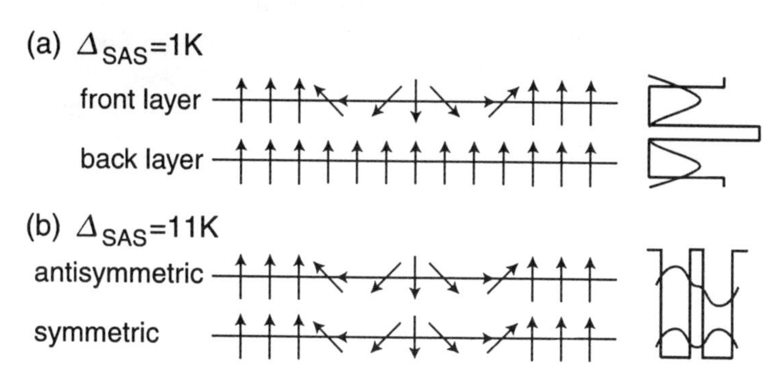

Fig. 26.16 Schematic diagram of spin flip is illustrated in the compound $\nu = 2$ state. An arrow represents the direction of a spin. The overlap of the wave functions is given on the right side. (a) In the sample with a small tunneling gap (Δ_{SAS} = 1 K), spin excitations are identical to those in the monolayer $\nu = 1$ QH state. (b) In the samples with a large tunneling gap (Δ_{SAS} = 11 K), spin excitations in one of the layers affect those in the other layer: A skyrmion-skyrmion pair is excited.

two CP^3 fields for two electrons in one Landau site with the U(2) local symmetry (26.6.10). The set turns out to be a Grassmannian $G_{4,2}$ field with four independent complex fields. Topological solitons are biskyrmions [Figs.26.15(b,c) and 26.16(b)].

We next consider a system where the two layers are separated sufficiently so that there are no interlayer exchange interaction ($J_s^d = 0$) nor the tunneling interaction ($\Delta_{SAS} = 0$). Then, the total Hamiltonian is invariant under two local transformations, $U^f(1)$ and $U^b(1)$, which act on electrons on the two layers independently,

$$\begin{pmatrix} \psi^{f\uparrow}(x) \\ \psi^{f\downarrow}(x) \end{pmatrix} \longrightarrow e^{i\alpha(x)} \begin{pmatrix} \psi^{f\uparrow}(x) \\ \psi^{f\downarrow}(x) \end{pmatrix},$$

$$\begin{pmatrix} \psi^{b\uparrow}(x) \\ \psi^{b\downarrow}(x) \end{pmatrix} \longrightarrow e^{i\beta(x)} \begin{pmatrix} \psi^{b\uparrow}(x) \\ \psi^{b\downarrow}(x) \end{pmatrix}. \tag{26.8.4}$$

Corresponding to these two U(1) symmetries, we should introduced two normalized CB fields by

$$\phi^{f\alpha}(x) = e^{i\chi^f(x)}\sqrt{\rho^f(x)}n^{f\alpha}(x), \quad \phi^{b\alpha}(x) = e^{i\chi^b(x)}\sqrt{\rho^b(x)}n^{b\alpha}(x), \tag{26.8.5}$$

where $\alpha = \uparrow, \downarrow$ and

$$\sum_{\alpha=\uparrow\downarrow} n^{f\alpha\dagger}(x)n^{f\alpha}(x) = \sum_{\alpha=\uparrow\downarrow} n^{b\alpha\dagger}(x)n^{b\alpha}(x) = 1. \tag{26.8.6}$$

We have a set of two CP^1 fields, each of which is the dynamical field for each layer at $\nu = 2$. Topological solitons are simple skyrmions [Figs.26.15(a) and 26.16(a)].

It is interesting to consider a case without the interlayer exchange interaction ($J_s^d \simeq 0$) but with a nonnegligible tunneling interaction ($\Delta_{SAS} \neq 0$). The basic field is the CP^3 field at $\nu = 1$ because the Hamiltonian possesses only the local U(1) symmetry (26.8.2). Hence, we have the $G_{4,2}$ field at $\nu = 2$. It is a genuine bilayer QH system and topological solitons are biskyrmions.

We do a thinking experiment to make it convincing that a biskyrmion is excited even for $J_s^d \simeq 0$ provided $\Delta_{SAS} \neq 0$. We start with the SU(4)-invariant limit of the exchange interaction ($J_s^d \simeq J_s$), where a biskyrmion must be excited. We question what would happen as J_s^d is decreased. As far as Δ_{SAS} is significant, two electrons in one Landau site are indistinguishable, and hence we need the Grassmannian field $Z(\boldsymbol{x})$ to describe the system. Furthermore, the spin biskyrmion (26.7.22) is insensible to the interlayer stiffness J_s^d because it is governed by the Hamiltonian (26.7.27). Hence, nothing would happen for the spin biskyrmion as $J_s^d \to 0$. We stress that the existence of topological solitons is the property of the Grassmannian manifold and not the property of the Hamiltonian (26.6.12). When $J_s^d \simeq 0$ the Grassmannian soliton is not the BPS state of the Hamiltonian, but its very existence is guaranteed by the homotopy theorem (26.7.1).

26.9 Experimental Indication of Biskyrmions

Topological solitons arise as quasiparticles (charged excitations) in the spin and ppin phases. They are studied experimentally by measuring the activation energy.

Spin Biskyrmions

In the spin phase only the spin degree of freedom is activated. By measuring the activation energy as a function of the tilting angle Θ, we can tell how many spins are flipped by one skyrmion as in the monolayer QH system. Kumada et al.[289] made a careful measurement of activation energy by using two bilayer samples with a large tunneling gap ($\Delta_{SAS} \simeq 11$ K) and a small tunneling gap ($\Delta_{SAS} \simeq 1$ K). These two samples have precisely the same sample parameters except for the tunneling gap, where $J_s^d/J_s = 0.15$ with use of (A.15). Thus, J_s^d is quite small compared with J_s. They have also measured activation energy in the monolayer limit of the same samples. (The monolayer state is constructed by emptying the back layer by tuning the bias voltage in the bilayer sample. The total electron density in the bilayer system is controlled so that it is precisely twice of that in the monolayer system.) They have found 7 flipped spins in the 1K-sample while 14 flipped spins in the 11K-sample when the tilting angle is small [Fig.26.17]. When the tilting angle becomes large, the number of flipped spins makes a transition from 14 to 7 in the 11K-sample. This is understood as follows. As the sample is tilted, the tunneling gap decrease as [see (27.2.16)]

$$\Delta_{SAS}(\Theta) = \exp\{-(d/2\ell_B)^2 \tan^2 \Theta\}\Delta_{SAS}. \tag{26.9.1}$$

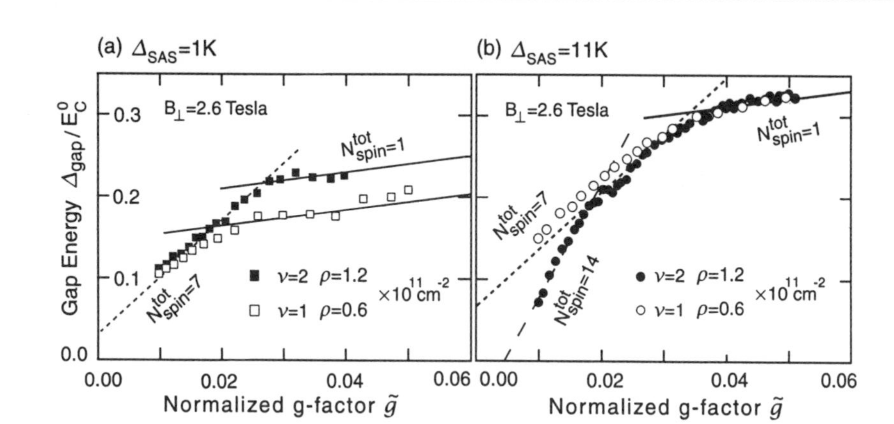

Fig. 26.17 The activation energy is measured by tilting the sample in the bilayer $\nu = 2$ compound state (solid marks) and the monolayer $\nu = 1$ state (open marks). The monolayer state is constructed by emptying the back layer by tuning the bias voltage in the sample of the type in Fig.21.10(b). The total electron density in the bilayer system is controlled so that it is precisely twice of that in the monolayer system. The vertical axis is the activation energy Δ_{gap} in units of E_C^0. The horizontal axis is the normalized g-factor $\tilde{g}$. (a) In the sample with small tunneling gap ($\Delta_{SAS} \simeq 1$ K), the reversed-spin number reads as follows: $N_{spin}^{tot} \simeq 7$ both for the bilayer and monolayer states for small $\tilde{g}$; $N_{spin}^{tot} \simeq 1$ both for the bilayer and monolayer states for large $\tilde{g}$. (b) In the sample with large tunneling gap ($\Delta_{SAS} \simeq 11$ K), the reversed-spin number reads as follows: $N_{spin}^{tot} \simeq 14$ for the bilayer state but $N_{spin}^{tot} \simeq 7$ for the monolayer states for small $\tilde{g}$; $N_{spin}^{tot} \simeq 7$ both for the bilayer and monolayer states for middle $\tilde{g}$; $N_{spin}^{tot} \simeq 1$ both for the bilayer and monolayer states for large $\tilde{g}$. The data are taken from Kumada et al.[289]

In Fig.26.17 the transition occurs at $\Theta = 60°$, where $\Delta_{SAS} \simeq 2$K. It is small enough compared with other interactions so that it is regarded as a two-monolayer system. They have also confirmed 7 flipped spins in the monolayer limit of both samples. (Only one spin is flipped in all cases for a sufficiently large tilting angle, where a vortex is excited in one of the layers since the tunneling gap becomes so small and the Zeeman effect becomes so large.) These facts are consistent with our conclusion that biskyrmions (simple skyrmion) are excited in a sample with a large (negligible) tunneling gap.

Ppin Biskyrmions

In the ppin phase only the pseudospin degree of freedom is activated. As we discuss in Section 25.6, it is possible to tell how many pseudospins are flipped by one skyrmion by tilting the sample. However, it is impossible to compare the result so far with the corresponding one in the bilayer system at $\nu = 1$. Let us use another method.[147] We remark that one biskyrmion carries the electric charge $2e$. We are able to tell such excitations by measuring the activation energy as a function of the

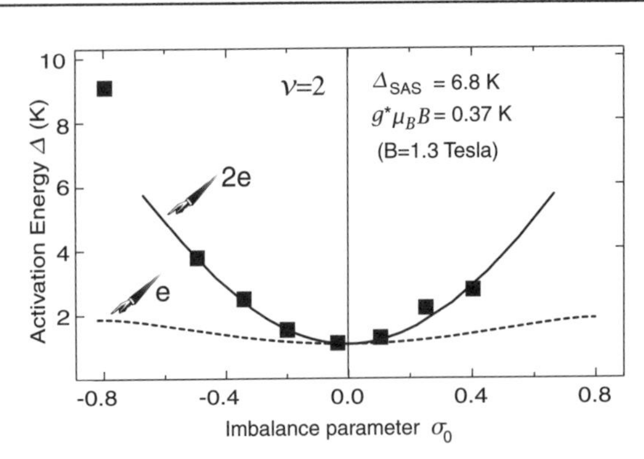

Fig. 26.18 The excitation energy of one ppin-skyrmion is calculated in the small-size limit at $\nu = 2$ as a function of the imbalance parameter σ_0. We have taken $\ell_B = 22.8$ nm, $d = 23$ nm and $\Delta_{SAS} = 6.8$ K, to compare the theoretical curve with the data taken from Sawada et al.[423] The curve is adjusted to the observed data at the balanced point ($\sigma_0 = 0$) by assuming a phenomenological offset Γ_{offset}. The CP^3-skyrmion (vortex) carries the electric charge $2e$. The dotted curve represents a theoretical estimation if the skyrmion carried the charge e.

imbalance parameter σ_0. The estimation given in Section 25.5 is valid as it is by replacing the skyrmion charge e by $2e$.

As the tunneling gap Δ_{SAS} increases, the coherence length becomes short and skyrmions are approximated by vortices. The key point is not the size of the coherence length but the fact that two electrons behave as a pair. The excitation energy is estimated numerically, and compared with experimental data in Fig.26.18, where $\Delta_{SAS} \simeq 6.8$ K. A reasonable agreement is obtained[147] only when one skyrmion (vortex) has the electric charge $2e$ and not e. If the charge were just e, the theoretical curve does not fit the data at all, as is depicted by a dotted line in Fig.26.18, because the capacitance and tunneling energies are cancelled almost completely.

CHAPTER 27

BILAYER QUANTUM-HALL JUNCTION

The bilayer QH system closely resembles the superconductor Josephson junction in the presence of the interlayer phase coherence. Various Josephson-like phenomena are expected to occur except for the Meissner effect. Electrons tunnel between the two layers coherently. Plasmon excitations are detectable by applying microwave to the system. Furthermore, the Hall current flows as the dc Josephson current between the two layers. The parallel magnetic field $B_{\|}$ penetrates between the two layers as sine-Gordon solitons, creating a soliton lattice for $B_{\|} > B_{\|}^*$ with $B_{\|}^*$ a certain transition point. Electrons are backscattered by a soliton lattice fluctuating thermally, as induces an anomalous peak of the diagonal resistivity R_{xx} in the QH regime with a well-depeloped Hall plateau.

27.1 JOSEPHSON-LIKE PHENOMENA

A most intriguing property of quantum coherence is the Josephson-like phenomenon predicted[143,142] in the bilayer QH system. The interlayer coherence develops necessarily at $\nu = 1$, where both the density difference $\sigma(\boldsymbol{x})$ and the phase field $\vartheta(\boldsymbol{x})$ are macroscopic observables. They are controllable electromagnetically because the phase difference is coupled with the electromagnetic potential. The bilayer QH system is reminiscent of the superconductor Josephson junction. We neglect the spin degree of freedom since it is not essential.

First of all, the tunneling current $\mathcal{J}_z$ may flow as the dc Josephson current between the two layers,

$$\mathcal{J}_z = J_{\rm T} \sin \vartheta, \tag{27.1.1}$$

where $J_{\rm T} \equiv e\Delta_{\rm SAS}\rho_0/2\hbar$.

Quantum mechanically, the phase difference ϑ oscillates around its mean value with frequency $\omega_{\rm P}$, which reads

$$\hbar\omega_{\rm P} = \sqrt{\epsilon_{\rm cap}\Delta_{\rm SAS} + \Delta_{\rm SAS}^2} \tag{27.1.2}$$

at the balance point. The corresponding oscillation mode is the Anderson plasmon in superconductor Josephson junction.[32,13] Here, it is also a plasmon mode, accompanied with the oscillation of the charging ($e\sigma\rho_0$) between the two layers. Plasmon excitations are detectable by applying microwave to the system.[142]

We anticipate rich physics by applying the parallel magnetic field $B_\parallel$ between the two layers. When a uniform parallel field $B_\parallel$ is applied into the y direction, its effect is to change locally the phase difference $\vartheta(x)$,

$$\vartheta(x) \longrightarrow \theta(x) \equiv \vartheta(x) - \delta_{\text{m}} x, \qquad \delta_{\text{m}} \equiv \frac{edB_\parallel}{\hbar} = \frac{d}{\ell_B^2} \tan\Theta \tag{27.1.3}$$

with $\tan\Theta = B_\parallel / B_\perp$. First of all, when $B_\parallel$ is sufficiently small, the ground-state field configuration is such that $\theta(\boldsymbol{x}) = 0$, or

$$\vartheta(x) = \delta_{\text{m}} x. \tag{27.1.4}$$

This minimizes the tunneling energy. Since the interlayer phase $\vartheta(\boldsymbol{x})$ is commensurate to the applied parallel magnetic field $B_\parallel$, this regime is called the *commensurate phase* (C phase). The shift of the ground state induces effectively a local rotation of the pseudospin,

$$\begin{aligned}
&\mathcal{P}_x(\boldsymbol{x}; B_\parallel) = \frac{1}{2}\sqrt{1 - \sigma(\boldsymbol{x})^2} \cos\left[\theta(x) + \delta_{\text{m}} x\right], \\
&\mathcal{P}_y(\boldsymbol{x}; B_\parallel) = -\frac{1}{2}\sqrt{1 - \sigma(\boldsymbol{x})^2} \sin\left[\theta(x) + \delta_{\text{m}} x\right], \quad \mathcal{P}_z(\boldsymbol{x}; B_\parallel) = \frac{1}{2}\sigma(x).
\end{aligned} \tag{27.1.5}$$

As $B_\parallel$ becomes larger, the pseudospin rotates rapidly. It increases the exchange energy of the ground state proportional to $(\nabla\mathcal{P})^2$. When $B_\parallel$ is large enough, it is energitically favored to minimize the exchange energy, leading to $\nabla\mathcal{P} = 0$. The regime is called the *incommensurate phase* (IC phase). A phase transition occurs by the competition between the tunneling energy and the exchange energy.[519] Actually this C-IC phase transition was first discovered experimentally[359] by the anomalous behavior of the activation energy as a function of $B_\parallel$, as already explained in Section 25.6: See Fig.25.5.

In the C phase, the parallel field penetrates between the two layers homogeneously, where a topological soliton is the sine-Gordon soliton.[139] When a soliton is created at $x = x_0$, the phase $\theta(x)$ jumps by $\pm 2\pi$ around $x = x_0$: See Fig.7.5.

In the IC phase, the parallel field penetrates between the two layers as sine-Gordon solitons on top of the homogeneous field $B_\parallel$. The ground state is no longer a homogeneous but modulated configuration in $\theta(x)$. Solitons are generated at random in the vicinity of $B_\parallel^*$. However, the soliton density becomes dense rapidly, and they form a *soliton lattice*[4,222] in terms of the variable $\theta(x)$. This is analogous to the well-known soliton-lattice state in long Josephson junctions.[301,164] The magnitude $|\theta(x)|$ is a monotonically increasing function of x. On the other hand, the interlayer phase field $\vartheta(x)$ is a periodic function in x, and its magnitude $|\vartheta(x)|$ decreases as $B_\parallel$ increases, and eventually it approaches

$$\vartheta(x) = 0 \tag{27.1.6}$$

in the asymptotic regime.

At the zero temperature the soliton lattice is a perfectly periodic system, where electrons propagate as a Bloch wave without reflection. However, the periodicity is broken at finite temperature, as would lead to a nontrivial transmission coefficient. Hence, there exists an intriguing possibility to observe a soliton lattice experimentally by way of an anomalous increase of the diagonal resistance R_{xx} in the QH state with a well-developed Hall plateau. See the experimental data to support this picture due to Fukuda et al.[181] in Fig.22 (page xxiv) in the frontispiece.

In spite of these similarities there exists an essential difference between the superconductor Josephson junction[32,13] and the bilayer QH junction. On one hand, the superconductor Josephson junction consists of two superconductors; thus, the phase fields are well defined dynamical variables separately in two bulks. On the other hand, in the QH-state junction the phase difference is only a dynamical variable: The phase sum is ill defined because the total electron number is fixed in the incompressible liquid. This difference leads to the fact that the Meissner effect does not exist in the bilayer QH state.[143]

27.2 Parallel Magnetic Field

We introduce the electromagnetic field into the bilayer Hamiltonian additionally to the perpendicular magnetic field $B_\perp$. The bilayer system is set parallel to the xy plane with the two layers at $z = \pm d/2$. For definiteness we apply a constant magnetic field $B_\parallel$ in the y direction and regard the system uniform in the y and z directions [Fig.27.1]. This allows us to set $B_x = 0$ and $B_y = B_\parallel$. To proceed further we make a gauge choice to represent $\boldsymbol{B} = \nabla \times \boldsymbol{A} = (0, B_\parallel, -B_\perp)$.

In the Landau gauge together with $A_z = 0$, the electromagnetic potential is given by

$$\boldsymbol{A} = (B_\perp y + B_\parallel z, 0, 0)\,. \tag{27.2.1}$$

The kinetic Hamiltonian is

$$H_K = \sum_{\alpha=\mathrm{f,b}} \int d^2x \left[\frac{1}{2M} \psi^{\alpha\dagger}(\boldsymbol{x})(P_x^\alpha - iP_y^\alpha)(P_x^\alpha + iP_y^\alpha)\psi^\alpha(\boldsymbol{x}) \right]. \tag{27.2.2}$$

The covariant momentum (10.2.3) and the guiding center (10.2.4) read in each layer as

$$P_x^\alpha = -i\hbar\frac{\partial}{\partial x} + \frac{\hbar}{\ell_B^2}\left(y \pm \frac{B_\parallel d}{2B_\perp}\right), \qquad P_y^\alpha = -i\hbar\frac{\partial}{\partial y}, \tag{27.2.3a}$$

$$X^\alpha = x - i\ell_B^2\frac{\partial}{\partial y}, \qquad Y^\alpha = i\ell_B^2\frac{\partial}{\partial x} \mp \frac{B_\parallel d}{2B_\perp}, \tag{27.2.3b}$$

where we have set $z = \pm d/2$ with $\pm$ for $\alpha =$f and b.

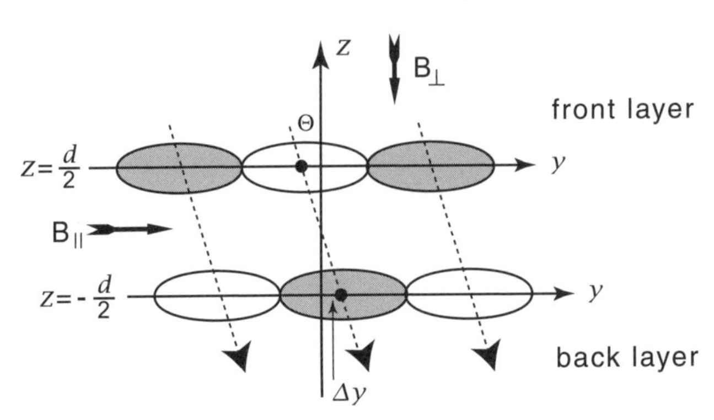

Fig. 27.1 We consider a compound state made of two monolayer QH states. Ovals represent Landau sites in the front and back layers. They are pierced by magnetic flux lines (dotted lines). In the absence of the parallel magnetic field ($B_\parallel = 0$), the term $\Delta_{\text{SAS}}\psi^{\text{f}\dagger}(\boldsymbol{x})\psi^{\text{b}}(\boldsymbol{x})$ describes the one-body tunneling probability by eliminating an electron at $\boldsymbol{x}$ in the back layer and creating it at $\boldsymbol{x}$ in the front layer. In the presence of the parallel magnetic into the y-direction ($B_\parallel \neq 0$), as the flux line is tilted, the corresponding Landau sites are shifted in the front and back layers by $\mp\Delta y$, where $\Delta y = d\tan\Theta/2$ with $\tan\Theta = B_\parallel/B_\perp$. The guiding center tunnels along the flux line with an electron making a cyclotron motion around it. Then, the one-body tunneling probability turns out to be $\Delta_{\text{SAS}}(\Theta)\psi^{\text{f}\dagger}(\boldsymbol{x})\psi^{\text{b}}(\boldsymbol{x})$, where $\Delta_{\text{SAS}}(\Theta) = \exp[-(\Delta y/\ell_B)^2]\Delta_{\text{SAS}} = \exp[-(d\tan\Theta/2\ell_B)^2]\Delta_{\text{SAS}}$.

Hereafter we introduce the parameter

$$\delta_{\text{m}} = \frac{edB_\parallel}{\hbar} = \frac{d}{\ell_B^2}\tan\Theta \tag{27.2.4}$$

with $\tan\Theta = B_\parallel/B_\perp$. The Landau-level ladder operator (10.4.4) stands as it is,

$$a^\alpha = \frac{1}{\sqrt{2}}\left(\ell_B\frac{\partial}{\partial y} + \frac{y - Y^\alpha}{\ell_B}\right), \quad a^{\alpha\dagger} = -\frac{1}{\sqrt{2}}\left(\ell_B\frac{\partial}{\partial y} - \frac{y - Y^\alpha}{\ell_B}\right). \tag{27.2.5}$$

The LLL condition, $a^\alpha|\mathfrak{S}^\alpha\rangle = 0$, yields

$$\left(\ell_B\frac{\partial}{\partial y} - i\ell_B\frac{\partial}{\partial x} + \frac{y}{\ell_B} \pm \frac{1}{2}\ell_B\delta_m\right)\mathfrak{S}_k^\alpha(\boldsymbol{x}; B_\parallel) = 0 \tag{27.2.6}$$

for the wave function $\mathfrak{S}^\alpha(\boldsymbol{x}) = \langle\boldsymbol{x}|\mathfrak{S}^\alpha\rangle$ in each layer.

A general solution of (27.2.6) reads

$$\mathfrak{S}_k^\alpha(\boldsymbol{x}; B_\parallel) = \mathcal{N}\exp\left(\beta_1 x + \beta_2 y + \frac{1}{2\ell_B^2}y^2\right), \tag{27.2.7}$$

where

$$\beta_1 = ik \mp \frac{\lambda}{2}\delta_m, \qquad \beta_2 = -k \mp \frac{1-\lambda}{2}\delta_m \tag{27.2.8}$$

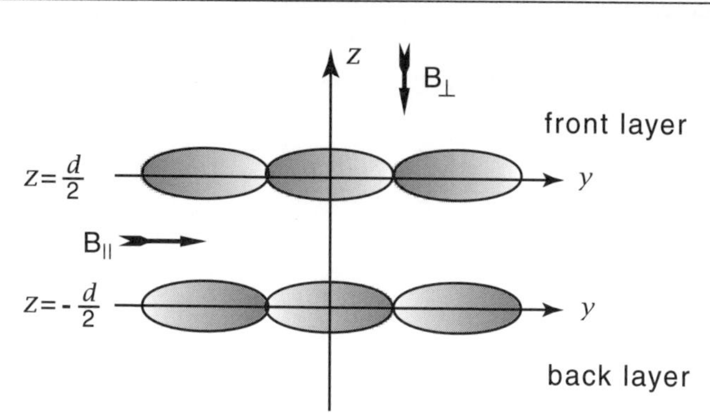

Fig. 27.2 We consider an interlayer-coherent state. Ovals represent Landau sites in the front and back layers. Even in the presence of the parallel magnetic into the y-direction ($B_\parallel \neq 0$), where the flux line is tilted, the corresponding Landau sites are not shifted in the front and back layers. Instead, the phases are shifted by $\mp(i/2)\delta_m x$.

with λ being an arbitrary constant, and $\mathcal{N}$ the normalization factor.

In the absence of the parallel magnetic field ($\delta_m = 0$), we obtain the same wave function (10.4.7) in the both layers,

$$\mathfrak{S}_k(\boldsymbol{x}) = \frac{1}{\sqrt{\pi^{1/2}\ell_B}} \exp(ixk) \exp\left[-\frac{1}{2\ell_B^2}(y + k\ell_B^2)^2\right]. \tag{27.2.9}$$

In what follows we study two important cases for $\delta_m \neq 0$.

Compound System

In the first case, we choose $\lambda = 0$, or

$$\beta_1 = ik, \qquad \beta_2 = -k \mp \frac{1}{2}\delta_m. \tag{27.2.10}$$

It gives an orthogonal set of functions,

$$\begin{aligned} \mathfrak{S}_k^\alpha(\boldsymbol{x}; B_\parallel) &= \frac{1}{\sqrt{\pi^{1/2}\ell_B}} \exp(ixk) \exp\left[-\frac{1}{2\ell_B^2}\left(y + k\ell_B^2 \pm \frac{1}{2}\delta_\mathrm{m}\ell_B^2\right)^2\right] \\ &= \exp\left[\mp(y + k\ell_B^2)\frac{\delta_\mathrm{m}}{2} - \frac{1}{8}\delta_\mathrm{m}^2\ell_B^2\right]\mathfrak{S}_k(\boldsymbol{x}), \end{aligned} \tag{27.2.11}$$

where $\mathfrak{S}_k(\boldsymbol{x})$ is the wave function (27.2.9) in the absence of the parallel magnetic field. They satisfy

$$\int d^2x\, \mathfrak{S}_k^{\alpha *}(\mathbf{x};B_\parallel)\mathfrak{S}_\ell^{\alpha}(\mathbf{x};B_\parallel) = 2\pi\delta(k-\ell), \tag{27.2.12a}$$

$$\int d^2x\, \mathfrak{S}_k^{\mathrm{f}*}(\mathbf{x};B_\parallel)\mathfrak{S}_\ell^{\mathrm{b}}(\mathbf{x};B_\parallel) = 2\pi e^{-\delta_{\mathrm{m}}\ell_B^2/4}\delta(k-\ell). \tag{27.2.12b}$$

By applying the guiding center (27.2.3b) to the wave function (27.2.11), it is found to be an eigenfunction of Y^α such that

$$Y^\alpha\mathfrak{S}_k^{\alpha}(\mathbf{x};B_\parallel) = \ell_B^2\left(-k\mp\frac{1}{2}\delta_{\mathrm{m}}\right)\mathfrak{S}_k^{\alpha}(\mathbf{x};B_\parallel). \tag{27.2.13}$$

The characteristic features are that the same plane wave $\exp(ixk)$ is taken in each layer, and that the guiding centers Y^α are shifted by $\mp\delta_{\mathrm{m}}\ell_B^2/2$ in the front and back layers [Fig.27.1].

We expand the LLL-projected field operator in terms of $\mathfrak{S}_k^{\alpha}(\mathbf{x};B_\parallel)$ as

$$\psi^\alpha(\mathbf{x};B_\parallel) \equiv \int \frac{dk}{\sqrt{2\pi}} c^\alpha(k)\mathfrak{S}_k^{\alpha}(\mathbf{x};B_\parallel) \tag{27.2.14}$$

together with $\{c^\alpha(k), c^{\beta\dagger}(\ell)\} = 2\pi\delta_{\alpha\beta}\delta(k-\ell)$. The tunneling interaction (22.3.5) is given by

$$\begin{aligned} H_{\mathrm{T}} &= -\frac{1}{2}\Delta_{\mathrm{SAS}}\int d^2x\left[\psi^{\mathrm{f}\dagger}(\mathbf{x};B_\parallel)\psi^{\mathrm{b}}(\mathbf{x};B_\parallel) + \psi^{\mathrm{b}\dagger}(\mathbf{x};B_\parallel)\psi^{\mathrm{f}}(\mathbf{x};B_\parallel)\right] \\ &= -\frac{1}{2}\Delta_{\mathrm{SAS}}e^{-\delta_{\mathrm{m}}\ell_B^2/4}\int d^2x\left[\psi^{\mathrm{f}\dagger}(\mathbf{x};0)\psi^{\mathrm{b}}(\mathbf{x};0) + \psi^{\mathrm{b}\dagger}(\mathbf{x};0)\psi^{\mathrm{f}}(\mathbf{x};0)\right] \\ &= -\Delta_{\mathrm{SAS}}(\Theta)\int d^2x\, P_x(\mathbf{x}), \end{aligned} \tag{27.2.15}$$

where use was made of (27.2.12b), and

$$\Delta_{\mathrm{SAS}}(\Theta) = \exp\left[-\left(\frac{d}{2\ell_B}\right)^2\tan^2\Theta\right]\Delta_{\mathrm{SAS}} \tag{27.2.16}$$

with $\tan\Theta = B_\parallel/B_\perp$. The tunneling gap $\Delta_{\mathrm{SAS}}(\Theta)$ decreases exponentially as $B_\parallel$ increases,[234] as illustrated in Fig.27.1.

The ground state energy due to the tunneling term is

$$\langle H_{\mathrm{T}}\rangle_{\mathrm{compound}} = -\frac{1}{2}\Delta_{\mathrm{SAS}}(\Theta)N_{\mathrm{e}} \tag{27.2.17}$$

at the balance point with N_{e} the number of electrons. It is a function of Θ.

Coherent System

In the second case, we choose $\lambda = 1$, or

$$\beta_1 = ik \mp \frac{1}{2}\delta_m, \qquad \beta_2 = -k. \tag{27.2.18}$$

It gives an orthogonal set of functions,

$$\begin{aligned}\mathfrak{S}_k^\alpha(\boldsymbol{x};B_\parallel) &= \frac{1}{\sqrt{\pi^{1/2}\ell_B}}\exp[i(k\mp\delta_{\rm m})x]\exp\left[-\frac{1}{2\ell_B^2}\left(y+k\ell_B^2\right)^2\right] \\ &= \exp\left[\mp\frac{i}{2}\delta_{\rm m}x\right]\mathfrak{S}_k(\boldsymbol{x}),\end{aligned}\tag{27.2.19}$$

where $\mathfrak{S}_k(\boldsymbol{x})$ is the wave function (27.2.9) in the absence of the parallel magnetic field. They satisfy

$$\int d^2x\, \mathfrak{S}_k^{\alpha *}(\boldsymbol{x};B_\parallel)\mathfrak{S}_\ell^\alpha(\boldsymbol{x};B_\parallel) = 2\pi\delta(k-\ell\mp\delta_{\rm m}),\tag{27.2.20a}$$

$$\int d^2x\, \mathfrak{S}_k^{{\rm f}*}(\boldsymbol{x};B_\parallel)\mathfrak{S}_\ell^{\rm b}(\boldsymbol{x};B_\parallel) = 2\pi\delta(k-\ell).\tag{27.2.20b}$$

By applying the guiding center (27.2.3b) to the wave function (27.2.11), it is found to be an eigenfunction of Y^α such that

$$Y^\alpha\mathfrak{S}_k^\alpha(\boldsymbol{x};B_\parallel) = -k\ell_B^2\mathfrak{S}_k^\alpha(\boldsymbol{x};B_\parallel).\tag{27.2.21}$$

Expanding the LLL-projected field operator in terms of $\mathfrak{S}_k^\alpha(\boldsymbol{x};B_\parallel)$, we find

$$\psi^\alpha(\boldsymbol{x};B_\parallel) = \exp\left(\mp\frac{i}{2}\delta_{\rm m}x\right)\psi^\alpha(\boldsymbol{x})\tag{27.2.22}$$

with

$$\psi^\alpha(\boldsymbol{x}) = \int\frac{dk}{\sqrt{2\pi}}c^\alpha(k)\mathfrak{S}_k(\boldsymbol{x}).\tag{27.2.23}$$

This is to be contrasted with (27.2.14). The essential feature is that the field operators are different only by a phase factor between the two layers [Fig.27.2]. Such a choice is possible only when there exists a dynamical degree of freedom associated with the interlayer phase difference, namely, only when the interlayer coherence develops.

Though we have used the Landau gauge to derive (27.2.22), it is valid in any gauge. The tunneling interaction (22.3.5) is now given by

$$\begin{aligned}H_{\rm T} &= -\frac{1}{2}\Delta_{\rm SAS}\int d^2x\left[\psi^{{\rm f}\dagger}(\boldsymbol{x};B_\parallel)\psi^{\rm b}(\boldsymbol{x};B_\parallel)+\psi^{{\rm b}\dagger}(\boldsymbol{x};B_\parallel)\psi^{\rm f}(\boldsymbol{x};B_\parallel)\right] \\ &= -\Delta_{\rm SAS}\int d^2x\,\left[P_x(\boldsymbol{x})\cos\delta_{\rm m}x - P_y(\boldsymbol{x})\sin\delta_{\rm m}x\right].\end{aligned}$$

This yields

$$H_{\rm T} = -\frac{1}{2}\Delta_{\rm SAS}\int d^2x\,\rho(\boldsymbol{x})\sqrt{1-\sigma(\boldsymbol{x})^2}\cos\left[\vartheta(\boldsymbol{x})-\delta_{\rm m}x\right],\tag{27.2.24}$$

with the use of the parametrization (24.1.21) for the pseudospin field. It is minimized by[1]

$$\vartheta(\mathbf{x}) = \delta_{\mathrm{m}} x. \tag{27.2.25}$$

The ground state energy due to the tunneling term is

$$\langle H_{\mathrm{T}} \rangle_{\mathrm{coherent}} = -\frac{1}{2}\Delta_{\mathrm{SAS}} N_{\mathrm{e}} \tag{27.2.26}$$

at the balance point with N_{e} the number of electrons.

We have studied two special choices of the wave function in the lowest Landau level. They describe physically different systems. Indeed, the ground-state energies are different. From (27.2.17) and (27.2.26) we find

$$\langle H_{\mathrm{T}} \rangle_{\mathrm{coherent}} \leq \langle H_{\mathrm{T}} \rangle_{\mathrm{compound}}, \tag{27.2.27}$$

where the equality holds in the absence of the parallel magnetic field. There is any other choice of λ in (27.2.8), where the wave function is a mixture of these two ones. However, we can show

$$\langle H_{\mathrm{T}} \rangle_{\mathrm{coherent}} \leq \langle H_{\mathrm{T}} \rangle_{\mathrm{other}} \leq \langle H_{\mathrm{T}} \rangle_{\mathrm{compound}}. \tag{27.2.28}$$

Consequently, in the presence of the interlayer coherence, the state (27.2.19) is chosen dynamically and the system is described by the effective Hamiltonian (27.2.24). On the other hand, in the absence of the interlayer coherence it is necessary to adopt the Hamiltonian (27.2.15) to describe the bilayer system.

A comment is in order on the gauge invariance. The definition of the tunneling interaction needs a nontrivial modification in general since it is a nonlocal interaction: It annihilates an electron on the back layer and creates an electron on the front layer, and vice versa. There is a standard way to make such a term gauge invariant. Let $(\mathbf{x}, z)$ be a point in the 3-dimensional space. We take a path $\mathcal{P}$ from the point $(\mathbf{x}^{\mathrm{f}}, z^{\mathrm{f}})$ to another point $(\mathbf{x}^{\mathrm{b}}, z^{\mathrm{b}})$ in the 3-dimensional space [Fig.27.3], along which we make a line integration of the gauge potential $A_k(\mathbf{x}, z)$. The following quantity is gauge invariant,

$$\psi^{\mathrm{f}\dagger}(\mathbf{x}^{\mathrm{f}}, z^{\mathrm{f}}; B_{\parallel}) \exp\left[\frac{ie}{\hbar}\int_{(\mathbf{x}^{\mathrm{b}}, z^{\mathrm{b}})}^{(\mathbf{x}^{\mathrm{f}}, z^{\mathrm{f}})} dx_k A_k(\mathbf{x}, z)\right] \psi^{\mathrm{b}}(\mathbf{x}^{\mathrm{b}}, z^{\mathrm{b}}; B_{\parallel}). \tag{27.2.29}$$

It is convenient to take the shortest path $\mathcal{P}$ connecting the two points $(\mathbf{x}, z^{\mathrm{f}})$ and $(\mathbf{x}, z^{\mathrm{b}})$ by a perpendicular line segment. Furthermore, for a sufficiently small layer distance d, the z-dependence of the component $A_z(\mathbf{x}, z)$ may be neglected,

$$\exp\left[\frac{ie}{\hbar}\int_{(\mathbf{x}, z^{\mathrm{b}})}^{(\mathbf{x}, z^{\mathrm{f}})} dx_k A_k(\mathbf{x}, z)\right] = \exp\left[\frac{ie}{\hbar}\int_{z^{\mathrm{b}}}^{z^{\mathrm{f}}} dz A_z(\mathbf{x}, z)\right] = \exp\left[\frac{ied}{\hbar} A_z(\mathbf{x})\right] \tag{27.2.30}$$

[1] As we shall argue in Section 27.4, this is the ground-state configuration in the commensurate phase: See (27.4.12).

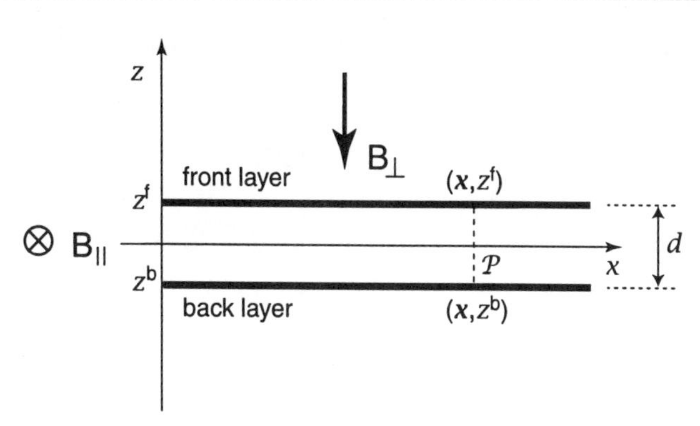

Fig. 27.3 We take the xy plane parallel to the bilayer system. The front layer is placed at $z = z^{\text{f}}$ and the back layer at $z = z^{\text{b}}$. The tunneling interaction annihilates an electron at $(\boldsymbol{x}, z^{\text{b}})$ on the back layer and creates an electron at $(\boldsymbol{x}, z^{\text{f}})$ on the front layer, and vice versa. We apply the parallel magnetic field $B_{\parallel}$ along the y axis. The parallel magnetic field is partially screened between the two layers in the interlayer-coherent state.

with $d = z^{\text{f}} - z^{\text{b}}$. The gauge-invariant tunneling interaction reads

$$H_{\text{T}} = -\frac{1}{2}\Delta_{\text{SAS}} \int d^2x \, (\psi^{\text{f}\dagger} e^{-iedA_z/\hbar} \psi^{\text{b}} + \psi^{\text{b}\dagger} e^{iedA_z/\hbar} \psi^{\text{f}}), \tag{27.2.31}$$

which is reduced to (27.2.24) in the gauge $A_z = 0$.

27.3 Effective Hamiltonian

After the LLL projection the Hamiltonian consists of the Coulomb and tunneling terms. The Coulomb term involves only the total density $\rho(\boldsymbol{x})$ and the density difference $P_z(\boldsymbol{x})$. It is not affected by the parallel magnetic field $B_{\parallel}$ since it is invariant under local gauge transformations. On the other hand, the tunneling term depends on $B_{\parallel}$ as in (27.2.24). Hence the effective Hamiltonian is given by modifying only the tunneling term in (24.2.4),

$$\begin{aligned}\mathcal{H}_{\text{eff}} =& \frac{J_s^{\vartheta}}{2}(\partial_k\vartheta)^2 + \frac{J_s^{\sigma}}{2}(\partial_k\sigma)^2 + \frac{\epsilon_{\text{cap}}\rho_0}{4}(\sigma - \sigma_0)^2 \\ &- \frac{\Delta_{\text{SAS}}\rho_0}{2}\sqrt{1-\sigma^2}\cos(\vartheta - \delta_{\text{m}}x) - \frac{\Delta_{\text{SAS}}\rho_0}{2}\frac{\sigma_0\sigma}{\sqrt{1-\sigma_0^2}},\end{aligned} \tag{27.3.1}$$

with

$$J_s^{\vartheta} = (1-\sigma^2)J_s^d, \qquad J_s^{\sigma} = \left(J_s + \frac{\sigma^2}{1-\sigma^2}J_s^d\right). \tag{27.3.2}$$

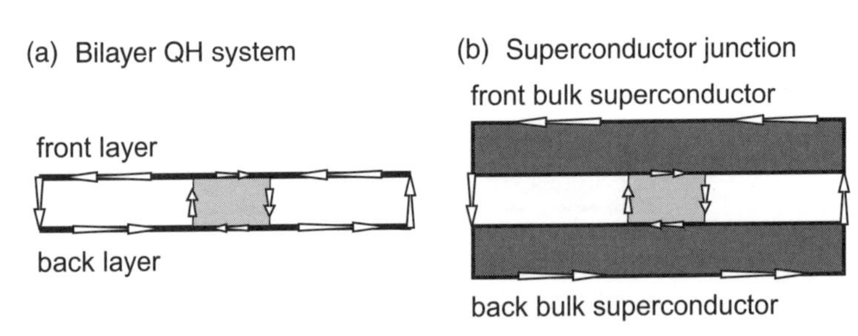

Fig. 27.4 (a) The magnetic field is $B_\parallel$ just outside the junction in bilayer QH states, as implies that the Meissner effect does not exist ($B_y \simeq B_\parallel$) between the junction. (b) It is $B_y = 0$ just outside the Josephson junction in superconductors due to the bulk Meissner effect, as implies that the Meissner effect does exist ($B_y = 0$) between the junction. In the incommensurate phase ($B_\parallel > B_\parallel^*$) magnetic flux penetrates into the junction and forms a soliton in addition to the homogeneous one $B_\parallel$. The screening current $\mathcal{J}_x^\alpha$ flows as indicated.

The equations of motion (24.2.7) are modified as

$$\hbar\partial_t\vartheta = \frac{2}{\rho_0}\partial_k\left(J_s^\sigma\partial_k\sigma\right) + \frac{2J_s^d}{\rho_0}\sigma\left[(\partial_k\vartheta)^2 - \frac{1}{(1-\sigma^2)^2}(\partial_k\sigma)^2\right]$$
$$- \epsilon_{\text{cap}}(\sigma-\sigma_0) - \frac{\sigma\cos(\vartheta-\delta_{\text{m}}x)}{\sqrt{1-\sigma^2}}\Delta_{\text{SAS}} + \frac{\sigma_0}{\sqrt{1-\sigma_0^2}}\Delta_{\text{SAS}}, \tag{27.3.3a}$$

$$\hbar\partial_t\sigma = -\frac{2}{\rho_0}\partial_k\left(J_s^\vartheta\partial_k\vartheta\right) + \Delta_{\text{SAS}}\sqrt{1-\sigma^2}\sin(\vartheta-\delta_{\text{m}}x)\,. \tag{27.3.3b}$$

They govern the dynamics of the interlayer coherence in the presence of the parallel magnetic field.

The phase current (24.2.14) reads

$$\mathcal{J}_k^{\text{f}}(\mathbf{x}) = -\mathcal{J}_k^{\text{b}}(\mathbf{x}) = \frac{e}{\hbar}J_s^\vartheta\partial_k\vartheta = \frac{eJ_s^d}{\hbar}(1-\sigma^2)\partial_k\vartheta, \tag{27.3.4}$$

while the tunneling current (24.2.15) reads

$$\mathcal{J}_z(\mathbf{x}) = J_{\text{T}}\sqrt{1-\sigma^2}\sin(\vartheta-\delta_{\text{m}}x)\,, \tag{27.3.5}$$

by making the same argument given in Section 24.2.

Josephson Junction versus Quantum-Hall Junction

The QH junction has a striking resemblance to the Josephson junction, except that the Meissner effect does not exist.[519,143] We trace back the origin of this difference to the fact that the superconductor Josephson junction is sandwiched between two superconducting bulks but the QH junction is not [Fig.27.4].

We study the static Maxwell equation in the 3-dimensional space. By assuming homogeneity in the y direction, the nontrivial Maxwell equations are

$$\partial_x B_y = \mu_m \mathcal{J}_z, \tag{27.3.6a}$$

$$\partial_z B_y = -\mu_m \mathcal{J}_x. \tag{27.3.6b}$$

It is necessary to pay attention to the fact that the current $\mathcal{J}_x$ flows only on the two layers,

$$\mathcal{J}_x(x,z) = \mathcal{J}_x^f(x)\delta(z - z^f) + \mathcal{J}_x^b(x)\delta(z - z^b), \tag{27.3.7}$$

where $\mathcal{J}_x^\alpha(x)$ is given by (27.3.4). Taking account of the boundary condition that the parallel magnetic field just outside the junction is given by $B_\parallel$, we integrate the Maxwell equation (27.3.6b) across one of the layers as

$$B_\parallel - B_y(x) = \mu_m \int dz\, \mathcal{J}_x(x) = \mu_m \mathcal{J}_x^f(x) = -\mu_m \mathcal{J}_x^b(x), \tag{27.3.8}$$

or

$$B_y(x) = -\frac{e\mu_m J_s^\vartheta}{\hbar}\partial_x \vartheta + B_\parallel. \tag{27.3.9}$$

This is the magnetic field present between the two layers. It can be confirmed based on the Maxwell equation (27.3.6a). Indeed, substituting the current (27.3.5) and the magnetic field (27.3.9) into the Maxwell equation (27.3.6a), we reproduce the equation of motion (27.3.3b) for static field ($\partial_t \sigma = 0$).

27.4 Commensurate-Incommensurate Phase Transition

Though the interlayer phase field is described by $\vartheta(\boldsymbol{x})$, it is convenient to introduce a new variable

$$\theta(\boldsymbol{x}) = \vartheta(\boldsymbol{x}) - \delta_m x, \tag{27.4.1}$$

and rewrite the effective Hamiltonian (27.3.1) as

$$\begin{aligned}\mathcal{H}_{\text{eff}} =& \frac{J_s^\vartheta}{2}\left[(\partial_x\theta + \delta_m)^2 + (\partial_y\theta)^2\right] + \frac{J_s^\sigma}{2}(\nabla\sigma)^2 + \frac{\epsilon_{\text{cap}}\rho_0}{4}(\sigma - \sigma_0)^2 \\ &- \frac{\Delta_{\text{SAS}}\rho_0}{2}\left(\sqrt{1-\sigma^2}\cos\theta + \frac{\sigma_0\sigma}{\sqrt{1-\sigma_0^2}}\right).\end{aligned} \tag{27.4.2}$$

This is derived from the modified exchange interaction

$$E_X^{\text{ppin}}(B_\parallel) = 2J_s^d \sum_{a=x,y} \int d^2x\, [\nabla \mathcal{P}_a(\boldsymbol{x}; B_\parallel)]^2 + 2J_s \int d^2x\, [\nabla \mathcal{P}_z(\boldsymbol{x}; B_\parallel)]^2 \tag{27.4.3}$$

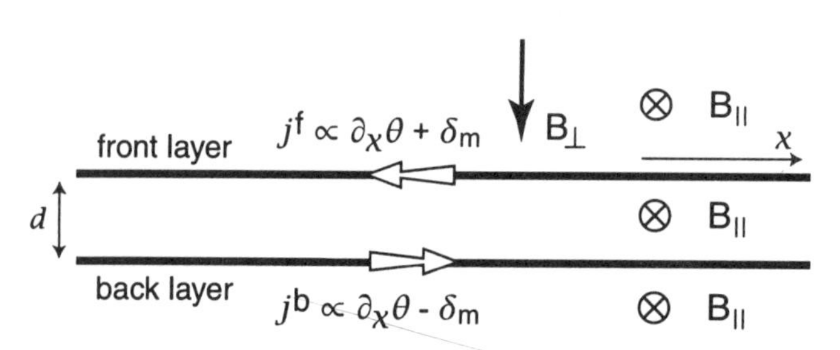

Fig. 27.5 The external parallel magnetic field $B_\parallel$ induces phase currents to screen it on the two layers. In the ground state of the commensurate phase the phase currents are $\pm eJ_s^d\delta_{\rm m}/\hbar$.

together with the unmodified tunneling term, where

$$\mathcal{P}_x(\boldsymbol{x};B_\parallel) = \frac{1}{2}\sqrt{1-\sigma(\boldsymbol{x})^2}\cos[\theta+\delta_{\rm m}x],$$

$$\mathcal{P}_y(\boldsymbol{x};B_\parallel) = -\frac{1}{2}\sqrt{1-\sigma(\boldsymbol{x})^2}\sin[\theta+\delta_{\rm m}x], \qquad \mathcal{P}_z(\boldsymbol{x};B_\parallel) = \frac{1}{2}\sigma(\boldsymbol{x}). \tag{27.4.4}$$

In terms of the new variable it appears as if the parallel magnetic field would rotate the pseudospin and affect the exchange interaction without changing the tunneling interaction.

The pseudospin field (27.4.4) is constructed by parametrizing the CP^1 field as

$$\boldsymbol{n} = \frac{1}{\sqrt{2}}\begin{pmatrix} e^{+i(\theta+\delta_{\rm m}x)/2}\sqrt{1+\sigma} \\ e^{-i(\theta+\delta_{\rm m}x)/2}\sqrt{1-\sigma}\end{pmatrix}. \tag{27.4.5}$$

The phase currents (27.3.4) read [Fig.27.5]

$$\mathcal{J}_x^{\rm f}(\boldsymbol{x}) = -\mathcal{J}_x^{\rm b}(\boldsymbol{x}) = \frac{eJ_s^d}{\hbar}(1-\sigma^2)\partial_k\vartheta = \frac{eJ_s^\vartheta}{\hbar}(\partial_x\theta+\delta_{\rm m}), \tag{27.4.6}$$

$$\mathcal{J}_z(\boldsymbol{x}) = J_{\rm T}\sqrt{1-\sigma^2}\sin(\vartheta-\delta_{\rm m}x) = J_{\rm T}\sqrt{1-\sigma^2}\sin\theta. \tag{27.4.7}$$

The choice of variable θ corresponds to the gauge with $A_z = \hbar\delta_{\rm m}x/ed$, where the tunneling term is independent of $\delta_{\rm m}x$.

The equations of motion (27.3.3) read

$$\hbar\partial_t\theta = \frac{2}{\rho_0}\partial_k(J_s^\sigma\partial_k\sigma) + \frac{2J_s^d}{\rho_0}\sigma\left[(\partial_k\theta)^2 - \frac{1}{(1-\sigma^2)^2}(\partial_k\sigma)^2\right]$$
$$- \epsilon_{\rm cap}(\sigma-\sigma_0) - \frac{\sigma\cos\theta}{\sqrt{1-\sigma^2}}\Delta_{\rm SAS} + \frac{\sigma_0}{\sqrt{1-\sigma_0^2}}\Delta_{\rm SAS}, \tag{27.4.8a}$$

$$\hbar\partial_t\sigma = -\frac{2}{\rho_0}\partial_k\left(J_s^\vartheta\partial_k\theta\right) + \Delta_{\rm SAS}\sqrt{1-\sigma^2}\sin\theta. \tag{27.4.8b}$$

We search for the classical ground state by solving the static field equations with $\partial_t \sigma = \partial_t \theta = 0$. A trivial set of solutions is given by $\theta = \theta_0$ with

$$\theta_0 = 2\pi n, \qquad n = 0, \pm 1, \pm 2, \cdots, \tag{27.4.9}$$

and $\sigma = \sigma_0$. In what follows we mainly use the variable $\theta(\boldsymbol{x})$ instead of the original phase field $\vartheta(\boldsymbol{x})$.

Commensurate Phase

The configuration (27.4.9) presents the classical ground state in the absence of the parallel magnetic field. Provided $\delta_{\rm m}$ is sufficiently small, we expect the ground states are still given by (27.4.9). We may set $\theta_0 = 0$ without loss of generality. Namely, the ground state is described by

$$\theta(x) = 0. \tag{27.4.10}$$

The energy density of the ground state is

$$\mathcal{E}_{\rm C} = \frac{1}{2}\left[J_s^\vartheta \delta_{\rm m}^2 - \Delta_{\rm SAS}\rho_0\sqrt{1-\sigma_0^2} - \Delta_{\rm SAS}\rho_0 \frac{\sigma_0^2}{\sqrt{1-\sigma_0^2}} \right]. \tag{27.4.11}$$

We remark that the tunneling energy is minimized by this configuration. The interlayer phase field $\vartheta(x)$ takes the value

$$\vartheta(x) = \delta_{\rm m} x = \frac{ed}{\hbar} B_{\parallel} x \tag{27.4.12}$$

on the ground state. Since the phase is commensurate to the applied parallel magnetic field $B_{\parallel}$, it is called the *commensurate phase*[519] or the C phase [Fig.27.6].

The CP^1 field reads

$$\boldsymbol{n} = \frac{1}{\sqrt{2}} \begin{pmatrix} e^{+i\delta_{\rm m}x/2}\sqrt{1+\sigma_0} \\ e^{-i\delta_{\rm m}x/2}\sqrt{1-\sigma_0} \end{pmatrix}. \tag{27.4.13}$$

The phase currents (27.4.6) flow on the both layers in the opposite directions [Fig.27.5],

$$\mathcal{J}_x^{\rm f} = -\mathcal{J}_x^{\rm b} = \delta_{\rm m} \frac{e J_s^d}{\hbar}(1-\sigma_0^2) = \frac{e^2 J_s^d d}{\hbar^2}(1-\sigma_0^2) B_{\parallel}, \tag{27.4.14}$$

while the tunneling current (27.4.7) vanishes. The magnitude of the phase current increases as $B_{\parallel}$ increases.

We estimate numerically the screening current (27.4.14) and the screening effect (27.3.9) on the parallel magnetic field for typical sample parameters (A.16) given in Appendix. We find

$$\mathcal{J}_x^{\rm f} = -\mathcal{J}_x^{\rm b} = \frac{e^2 J_s^d d}{\hbar^2}(1-\sigma^2) B_{\parallel} \simeq 0.24(1-\sigma_0^2) B_{\parallel}, \tag{27.4.15}$$

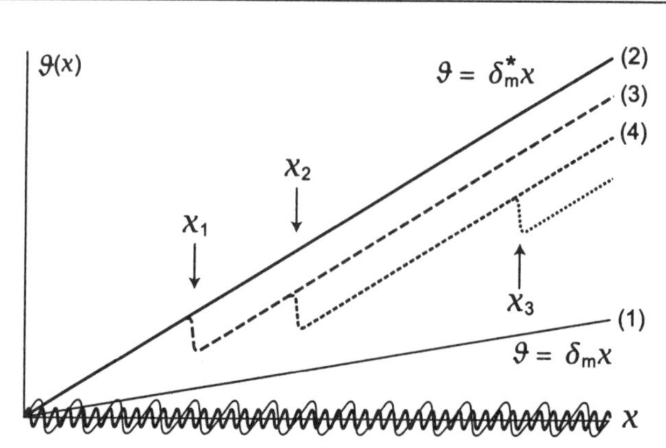

Fig. 27.6 The interlayer phase field $\vartheta(x)$ is illustrated schematically for various values of δ_m. We can see how it changes from the C phase to the IC phase. In the C phase ($\delta_m < \delta_m^*$), the phase field increases as $\vartheta(x) = \delta_m x$ [line (1)]. This bahavior continues until $\vartheta(x) = \delta_m^* x$ [line (2)], where solitons begin to emerge. One soliton is created [line (3)] when the magnetic flux is increased just by one flux unit at $\delta_m = \delta_m^*$: We take the position of the soliton to be $x = x_1$. Another soliton is created [line (4)] at $x = x_2$ when it is increased by another flux unit. The positions x_1 and x_2 are independent. When sufficiently many solitons are created, they are periodically spaced, as represented by wavy lines around the x axis: This is the soliton-lattice regime. The amplitude $|\vartheta(x)|$ decreases as δ_m increases, and the phase $\vartheta(x)$ approaches the asymptotic value $\vartheta(x) = 0$ deep inside the IC phase.

and

$$B_y(x) = B_\parallel - \frac{e\mu_m J_s^d}{\hbar}\delta_m = \left(1 - \frac{e^2\mu_m J_s^d d}{\hbar^2}\right) B_\parallel. \tag{27.4.16}$$

The screening current is quite large,

$$\mathcal{J}_x^{\rm f} = -\mathcal{J}_x^{\rm b} \simeq 0.38\,{\rm A/m} \tag{27.4.17}$$

at $B_\parallel = 1.6\,$Tesla and $\sigma = 0$. Note that the measurement of the Hall current is done around $\sim 10^{-4}\,$A/m. On the other hand, the screening effect is

$$\frac{e^2\mu_m J_s^d d}{\hbar^2} \simeq 2.7 \times 10^{-7}. \tag{27.4.18}$$

It is negligible though the screening current is quite large.

Incommensurate Phase

There exists another phase in the model (27.3.1) or (27.4.2). For a large misfit δ_m or $B_\parallel$, the exchange energy $J_s^\vartheta \delta_m^2/2$ increases and becomes dominant in the ground-state energy (27.4.11). We should minimize the exchange term in the effective Hamiltonian, by choosing

$$\theta(x) = -\delta_{\rm m} x + \vartheta_0 = -\frac{edB_\parallel x}{\hbar} + \vartheta_0 \tag{27.4.19}$$

in (27.4.2). The CP^1 field reads

$$\boldsymbol{n} = \frac{1}{\sqrt{2}} \begin{pmatrix} e^{+i\vartheta_0/2}\sqrt{1+\sigma_0} \\ e^{-i\vartheta_0/2}\sqrt{1-\sigma_0} \end{pmatrix}. \tag{27.4.20}$$

The original interlayer phase field $\vartheta(x)$ takes a constant value ϑ_0 on the ground state, which we may choose $\vartheta_0 = 0$ without loss of generality,

$$\vartheta(x) = 0. \tag{27.4.21}$$

It is notable that there exists no phase current, $\mathcal{J}_x^{\rm f} = -\mathcal{J}_x^{\rm b} = 0$, and there exists no screening of the parallel magnetic field, $B_y(x) = B_\parallel$.

Since the interlayer phase is not commensurate to the applied parallel magnetic field $B_\parallel$, it is called the *incommensurate phase*[519] or the IC phase. This presents only a gross idea of the IC phase. In reality, the IC phase is much more complicated [Fig.27.6], as we shall see more in detail in the succeeding section: The phase field takes the value (27.4.21) deep inside the IC phase.

Nevertheless, it is worthwhile to estimate the phase transition point. The energy density of the state is

$$\mathcal{E}_{\rm IC} = -\frac{1}{2}\Delta_{\rm SAS}\rho_0 \frac{\sigma_0^2}{\sqrt{1-\sigma_0^2}} - \frac{1}{2L}\Delta_{\rm SAS}\rho_0\sqrt{1-\sigma_0^2}\int_0^L dx \cos(\vartheta_0 - \delta_{\rm m} x), \tag{27.4.22}$$

where L is the size of the sample. The last term is negligible for $\delta_{\rm m} L \gg 1$. Comparing this with (27.4.11) we find the phase transition point $B_\parallel^{\rm CIC}$ at $\mathcal{E}_{\rm C} = \mathcal{E}_{\rm IC}$, which yields $J_s^\vartheta \delta_{\rm m}^2 = \Delta_{\rm SAS}\rho_0\sqrt{1-\sigma_0^2}$, or

$$B_\parallel^{\rm CIC} = (1-\sigma_0^2)^{-1/4}\frac{\hbar}{ed}\sqrt{\frac{\rho_0\Delta_{\rm SAS}}{J_s^d}}. \tag{27.4.23}$$

This gives a rough estimation of the C-IC phase transition point. See (27.5.41) for the correct one.

27.5 Soliton Lattice

The phase transition from the C phase to the IC phase requires further elucidation. We continue to use the variable $\theta(x)$ to discuss topological solitons. We restrict the analysis to the balanced configuration ($\sigma_0 = 0$) by setting $V_{\rm bias} = 0$. For a later convenience, we derive the time-dependent sine-Gordon equation from (27.4.8a) and (27.4.8b),

$$\frac{\lambda_J^2}{v_{\rm F}^2}\partial_t^2\theta + \lambda_J^2\partial_x^2\theta - \sin\theta = 0, \tag{27.5.1}$$

where v_F stands for the velocity of the pseudospin wave, and[2]

$$\lambda_J = \sqrt{\frac{2J_s^d}{\Delta_{SAS}\rho_0}}, \qquad v_F^2 = \frac{2J_s^d \epsilon_{cap}}{\rho_0 \hbar^2}. \tag{27.5.2}$$

The effective Hamiltonian (27.4.2) is reduced to

$$\mathcal{H}_{eff}(x) \to \mathcal{E}_{mSG}(x) = \frac{J_s^d}{2}\left[\frac{1}{v_F^2}(\partial_t\theta)^2 + (\partial_x\theta + \delta_m)^2 + \frac{2}{\lambda_J^2}(1-\cos\theta)\right] - \frac{1}{2}J_s^d\delta_m^2, \tag{27.5.3}$$

where we have subtracted the ground-state energy. We later use the time-dependent equation to derive the mass of the soliton: See (27.6.7).

Hereafter we study the static system. The system (27.5.3) defines a modified sine-Gordon model,[3] where δ_m is known as the misfit parameter. Since the Euler-Lagrange equation is the sine-Gordon equation (27.5.1), the system must contain sine-Gordon solitons, as reviewed in Section 7.4. On the other hand, when $|\theta| \ll 1$ we may approximate $\sin\theta \simeq \theta$, and its solution is

$$\theta(x) \simeq \theta_0 e^{-x\lambda_J} \qquad \text{for} \quad x \geq 0. \tag{27.5.4}$$

Thus, λ_J corresponds to the penetration depth in the Josephson junction.

In the absence of the misfit ($\delta_m = 0$), the model (27.5.3) is reduced to the sine-Gordon model (7.4.4),

$$\mathcal{E}_{SG}(x) = \frac{J_s^d}{2}\left[\frac{1}{v_F^2}(\partial_t\theta)^2 + (\partial_x\theta)^2 + \frac{2}{\lambda_J^2}(1-\cos\theta)\right]. \tag{27.5.5}$$

The classical ground states are given by (27.4.9), which are all degenerate.

We study soliton excitations in the Hamiltonian density (27.5.3). Their existence follows from the degenerate set (27.4.9) of the ground states. A soliton emerges when, say, $\theta = 0$ as $x \to -\infty$ and $\theta = 2\pi q$ as $x \to \infty$: See Fig.7.5 (page 120). According to the formula (27.3.9) one soliton carries the magnetic flux

$$\begin{aligned}\Phi_{SG} &= \int_{-d/2}^{d/2} dz \int dx\, [B_y(x) - \text{background}] \\ &= -\frac{e\mu_m J_s^d}{\hbar}\int_{-d/2}^{d/2} dz \int dx\, \partial_x\theta = \frac{2\pi\hbar}{e}\frac{e^2\mu_m J_s^d d}{\hbar^2} q.\end{aligned} \tag{27.5.6}$$

The magnetic flux is quantized, but not in units of the Dirac flux unit. See also (27.4.18). This is not surprising since there exists no Meissner effect.

---

[2]The penetration depth is estimated numerically as $\lambda_J \simeq 2.5\times10^{-8}$m for typical sample parameters (A.16) in Appendix A.

[3] The model with a cosine potential was originally introduced by Y.I. Frenkel and T. Kontorowa,[177] and has been studied in various contexts by various authors. For instance, the Hamiltonian (27.5.3) describes the electron distribution in the crystal, where the number density is commensurate to the lattice when there is no misfit: Here, the system is said to be in the commensurate phase. This model is also called the Pokrovsky-Talapov model according to the article[393] in some references. A concise review is given by P. Bak.[58]

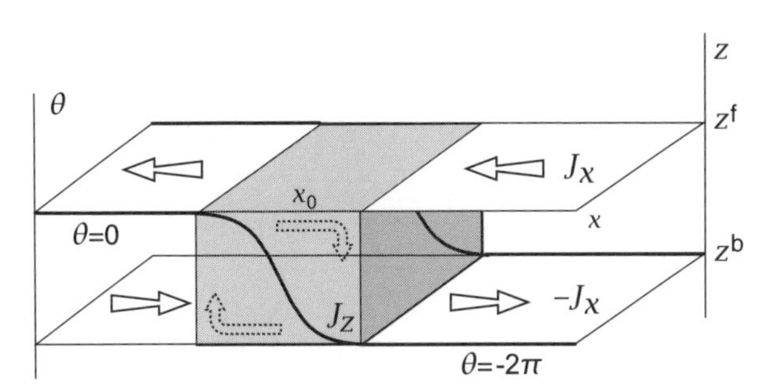

Fig. 27.7 We illustrate the penetration of a soliton between the two layers in terms of the variable $\theta(\boldsymbol{x}) \equiv \vartheta(\boldsymbol{x}) - \delta_{\text{m}}x$. One soliton separates two commensurate regions (white areas with $\theta = 0$ and $\theta = -2\pi$) in the QH junction. When $B_{\parallel} < B_{\parallel}^*$ it is a finite-energy topological soliton in the C phase. When $B_{\parallel} > B_{\parallel}^*$ its presence lowers the energy. The shaded area represents a region occupied by one soliton where $\partial_x\theta \neq 0$: Screening currents on the two layers become smaller and allow a flux penetration, while tunneling current flows between the two layers and squeezes the flux to form a sine-Gordon soliton.

We rewrite the effective Hamiltonian (27.5.3) as

$$\mathcal{E}_{\text{mSG}}(x) = \mathcal{E}_{\text{SG}}(x) + J_s^d \delta_{\text{m}} \partial_x \theta(x) \tag{27.5.7}$$

with (27.5.5). Since a soliton of the modified sine-Gordon system $\mathcal{E}_{\text{mSG}}$ is a soliton of the sine-Gordon system $\mathcal{E}_{\text{SG}}(x)$, we may relate the excitation energy E_{mSG} in the modified sine-Gordon system to the excitation energy E_{SG} in the sine-Gordon system. We integrate this equation for a soliton configuration,

$$E_{\text{mSG}} \equiv \int dx\, \mathcal{E}_{\text{mSG}}(x) = E_{\text{SG}} - 2\pi q J_s^d \delta_{\text{m}}, \tag{27.5.8}$$

where

$$E_{\text{SG}} = \frac{8J_s^d}{\lambda_J}, \tag{27.5.9}$$

and

$$q = -\frac{1}{2\pi} \int dx\, \partial_x \theta(x) \tag{27.5.10}$$

is the topological charge. Here we make a convention[4] to call it a soliton for $q > 0$ and an antisoliton for $q < 0$.

We consider one soliton ($q = 1$) or antisoliton ($q = -1$) excitation. Its field configuration is given in terms of the variable $\theta(x)$ as in (7.4.8), or

$$\theta(x) = -4q \arctan \exp\left(\frac{x - x_0}{\lambda_J}\right). \tag{27.5.11}$$

[4]Note that this convention is opposite to the one adopted in the original version of this book.[4]

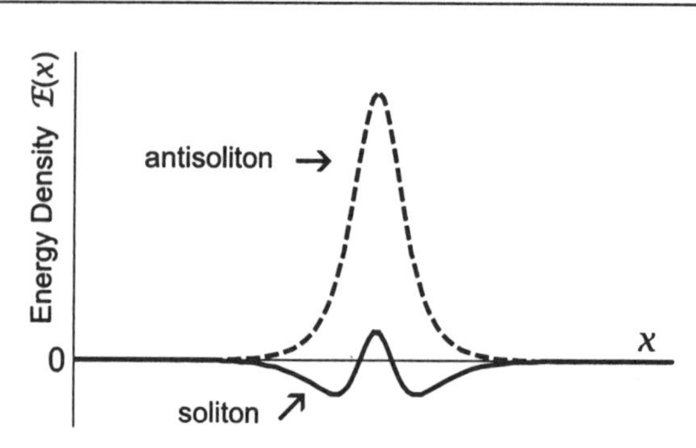

Fig. 27.8 We illustrate the energy density $\mathcal{E}(x)$ around a soliton ($q = 1$) by a solid line and an antisoliton ($q = -1$) by a dotted line. It is intriguing that there exist negative energy regions inside a soliton. In the IC phase, the totol energy $E = \int dx\, \mathcal{E}(x)$ is negative for a soliton, and hence solitons are spontaneously created.

We illustrate the energy density (27.5.3) around a soliton ($q = 1$) and an antisoliton ($q = -1$) in Fig.27.8. It is intriguing that there exist negative energy regions inside a soliton, by taking the energy of the homogeneous configuration in θ to be zero. The total energy (27.5.8) reads

$$E_{\mathrm{mSG}} = 2\pi J_s^d \left(\frac{4}{\pi\lambda_J} - q\delta_{\mathrm{m}}\right) = \frac{2\pi J_s^d ed}{\hbar}\left(B_{\parallel}^* - qB_{\parallel}\right) = E_{\mathrm{SG}}\left(1 - q\frac{\delta_{\mathrm{m}}}{\delta_{\mathrm{m}}^*}\right), \tag{27.5.12}$$

where we have set[5]

$$\delta_{\mathrm{m}}^* = \frac{4}{\pi\lambda_J}, \qquad B_{\parallel}^* = \frac{4\hbar}{\pi ed\lambda_J} = \frac{\hbar}{\pi ed}\sqrt{\frac{\rho_0\Delta_{\mathrm{SAS}}}{J_s^d}}. \tag{27.5.13}$$

It costs necessarily a positive energy ($E_{\mathrm{mSG}} > 0$) to excite an antisoliton ($q < 0$) in (27.5.12), but this is not always the case for a soliton ($q > 0$). Indeed, when $B_{\parallel} > B_{\parallel}^*$, it costs a negative energy ($E_{\mathrm{mSG}} < 0$) to excite a soliton: One soliton is generated spontaneously within the junction to lower the energy of the system. Hence, the C-IC phase transition occurs at $B_{\parallel}^*$ given by (27.5.13).

The penetrated soliton forms a wall centered at $x = x_0$, separating two commensurate regions, one with $\theta = 0$ and the other with $\theta = -2\pi$ [Fig.27.7]. The CP^1 field is

$$\boldsymbol{n} = \frac{1}{\sqrt{2}}\begin{pmatrix} e^{+i[\theta(x)+\delta_{\mathrm{m}}^* x]/2}\sqrt{1+\sigma_0} \\ e^{-i[\theta(x)+\delta_{\mathrm{m}}^* x]/2}\sqrt{1-\sigma_0} \end{pmatrix} \tag{27.5.14}$$

with (27.5.11), where $\theta(x) \simeq 0$ for $x \ll x_0$ and $\theta(x) \simeq -2\pi$ for $x \gg x_0$. The ground state is no longer a homogeneous but modulated configuration in $\theta(x)$.

[5]The critical value is estimated numerically as $B_{\parallel}^* \simeq 1.6$ Tesla for typical sample parameters (A.16) in Appendix A.

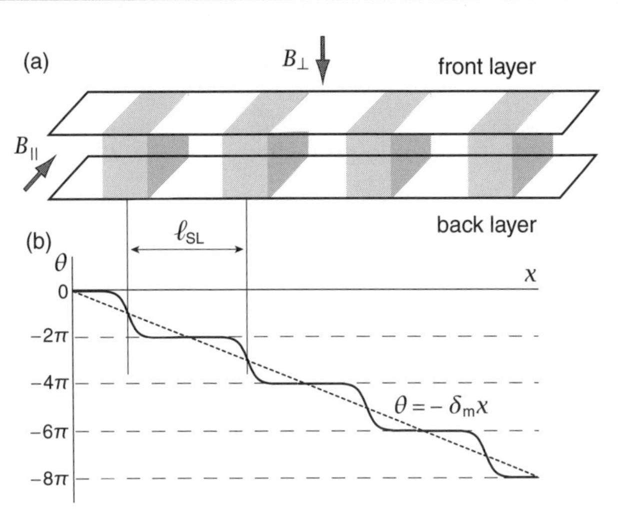

Fig. 27.9 We illustrate a soliton lattice in terms of the variable $\theta(x) = \vartheta(x) - \delta_m x$. (a) Solitons are periodically created in the soliton-lattice regime. The shaded regions represent the area where $\partial_x \theta \neq 0$. The lattice spacing ℓ_{SL} is given by (27.5.27). (b) The phase $\theta(x)$ shows a staircase for $\delta_m \simeq \delta_m^*$, and appoaches a dashed straight line as δ_m increases. The dashed straight line represents the asymptotic ground state (27.4.19) deep inside the IC phase. They are represented by a wavy line around the x axis and the x axis itself, respectively, in term of the interlayer phase field $\vartheta(x)$ in Fig.27.6.

As the external parallel magnetic field increases, more solitons are generated. They are created at random as far as the soliton density is low, as illustrated in Fig.27.6. When a certain density is achieved they are regularly spaced and form a soliton lattice [Fig.27.9].

The soliton lattice is obtained as a periodic solution of the sine-Gordon equation (27.5.1) in terms of the variable $\theta(x)$. It is given by (see Appendix L)

$$\cos \frac{\theta_0 - \theta(x)}{2} = \mathrm{cn}\left(\frac{x}{\kappa \lambda_J} \middle| \kappa^2 \right), \tag{27.5.15}$$

where θ_0 and κ are integration constants, and $\mathrm{cn}(x|\kappa^2)$ is the Jacobian elliptic function with $0 < \kappa^2 < 1$. We may choose $\theta_0 = 0$ without loss of generality. The Jacobian elliptic function $\mathrm{cn}(x|\kappa^2)$ is a periodic function with the periodicity $4K(\kappa^2)$, where $K(\kappa^2)$ is the complete elliptic integral of the first kind and given by (27.5.23a).

To grasp the gist of the solution it is convenient to use the differentiated form,

$$\frac{d\theta(x)}{dx} = -\frac{2}{\kappa \lambda_J} \sqrt{1 - \kappa^2 \cos^2(\theta/2)} \leq 0. \tag{27.5.16}$$

Thus, $\theta(x)$ is a monotonically decreasing function of x. Its minimum value is

$$\frac{d\theta(x)}{dx} = -\frac{2}{\kappa\lambda_J}, \tag{27.5.17}$$

which occurs at $\theta(x) = \pi - 2n\pi$, while its maximum value is

$$\frac{d\theta(x)}{dx} = -\frac{2}{\kappa\lambda_J}\sqrt{1-\kappa^2}, \tag{27.5.18}$$

which occurs at $\theta(x) = -2n\pi$. The minimum and maximum values become identical as $\kappa \to 0$, which implies the solution is reduced to a straight line,

$$\theta(x) = -\frac{2}{\kappa\lambda_J}x. \tag{27.5.19}$$

We have illustrated $\theta(x)$ in Fig.27.9(b) for a typical value of κ.

We proceed to determine the integration constant κ in (27.5.15) by minimizing the energy of the system for a given value of δ_{m}. We rewrite the energy (27.5.8) for a soliton lattice as

$$E_{\text{SL}} = J_s^d \int dx \left[\frac{1}{2}(\partial_x\theta)^2 + \frac{1}{\lambda_J^2}(1-\cos\theta)\right] - 2\pi J_s^d \delta_{\text{m}} N_{\text{SL}}, \tag{27.5.20}$$

where $N_{\text{SL}} = -\frac{1}{2\pi}\int dx\, \partial_x\theta(x)$ is the number of solitons in the lattice. Using the relation (27.5.16), or

$$dx = -\frac{\kappa\lambda_J}{\sqrt{1-\kappa^2\cos^2(\theta/2)}}d(\theta/2), \tag{27.5.21}$$

we change the integration variable from x to θ in (27.5.20) and obtain

$$E_{\text{SL}}(\kappa,\delta_{\text{m}}) = \frac{4J_s^d L}{\lambda_J^2}\left[\frac{1}{2} + \frac{1}{\kappa^2}\frac{E(\kappa^2)}{K(\kappa^2)} - \frac{1}{2\kappa^2} - \frac{1}{\kappa K(\kappa^2)}\frac{\delta_{\text{m}}}{\delta_{\text{m}}^*}\right], \tag{27.5.22}$$

with

$$K(\kappa^2) = \int_0^{\pi/2} \frac{d\theta}{\sqrt{1-\kappa^2\cos^2\theta}} = \frac{\pi}{2}F\left(\frac{1}{2},\frac{1}{2};1;\kappa^2\right), \tag{27.5.23a}$$

$$E(\kappa^2) = \int_0^{\pi/2} d\theta\sqrt{1-\kappa^2\cos^2\theta} = \frac{\pi}{2}F\left(-\frac{1}{2},\frac{1}{2};1;\kappa^2\right), \tag{27.5.23b}$$

where $F(a,b;c;z)$ is the hypergeometric function. We minimize the energy (27.5.22). Using the relations

$$\frac{dK(x)}{dx} = \frac{E(x)}{2x(1-x)} - \frac{K(x)}{2x}, \qquad \frac{dE(x)}{dx} = \frac{E(x)-K(x)}{2x}, \tag{27.5.24}$$

we obtain

$$\frac{dE_{\text{SL}}(\kappa,\delta_{\text{m}})}{d\kappa} = \frac{E(\kappa^2)}{K^2(\kappa^2)}\frac{\epsilon_{\text{m}}\kappa - E(\kappa^2)}{\kappa^3(1-\kappa^2)} \tag{27.5.25}$$

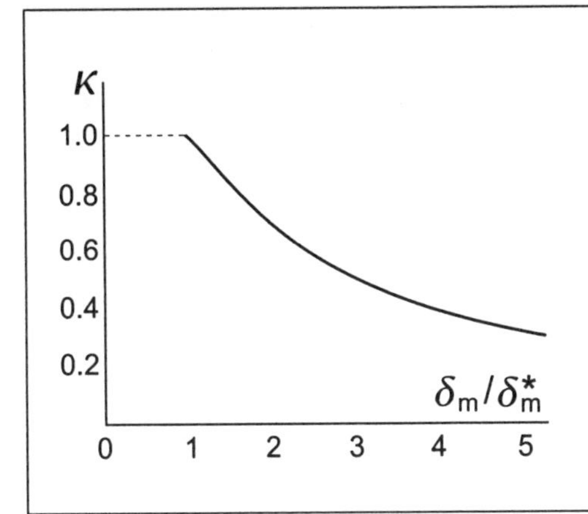

Fig. 27.10 The parameter $\kappa = \bar{\kappa}(\delta_{\rm m})$ in the soliton lattice solution (27.5.15) is determined by minimizing the energy (27.5.22) of the soliton lattice system.

with $\epsilon_{\rm m} \equiv \delta_{\rm m}/\delta^*_{\rm m}$. Hence, $\kappa = \bar{\kappa}(\delta_m)$ is determined by solving

$$\frac{E(\bar{\kappa}^2)}{\bar{\kappa}} = \frac{\delta_{\rm m}}{\delta^*_{\rm m}}. \tag{27.5.26}$$

The function (27.5.26) can be determined numerically as in Fig.27.10.

It is easy to see in (27.5.22) that $E_{\rm SL}(\kappa, \delta_{\rm m}) > 0$ for $\delta_{\rm m} < \delta^*_{\rm m}$, where no soliton lattice is formed. On the other hand, for $\delta_m > \delta^*_{\rm m}$, $E_{\rm SL}(\kappa, \delta_{\rm m})$ takes one unique minimum value at $\kappa = \bar{\kappa}(\delta_m)$ given by (27.5.26), and the minimum value is negative, $E_{\rm SL}(\bar{\kappa}, \delta_{\rm m}) < 0$: Hence, a soliton lattice is formed.

Since the distance $\ell_{\rm SL}$ between two neighboring solitons is related to the periodicity as

$$\frac{\ell_{\rm SL}}{\kappa\lambda_J} = 2K(\kappa^2), \tag{27.5.27}$$

the density of solitons is given by

$$\rho_{\rm SL}(\kappa) = \frac{1}{\ell_{\rm SL}} = \frac{1}{2\kappa\lambda_J K(\kappa^2)} = \frac{\pi\delta^*_{\rm m}}{8\kappa K(\kappa^2)}. \tag{27.5.28}$$

Thus the soliton number is

$$N_{\rm SL}(\kappa) = L\rho_{\rm SL}(\kappa) = \frac{L}{2\kappa\lambda_J K(\kappa^2)}, \tag{27.5.29}$$

with L the size of the sample. The soliton density $\rho_{\rm SL}$, given by (27.5.28) with (27.5.26) as a function of $\delta_{\rm m}$, is illustrated in Fig.27.11(a). The density $\rho_{\rm SL}$ is zero in the C phase ($\delta_{\rm m} < \delta^*_{\rm m}$), and it increases rapidly for $\delta_{\rm m} \geq \delta^*_{\rm m}$, and approaches an asymptotic line for $\delta_{\rm m} \gtrsim 1.2\delta^*_{\rm m}$.

It is instructive to calculate the energy per one soliton at $\delta_{\rm m}$ from (27.5.29) and (27.5.22),

$$\frac{E_{\rm SL}(\bar{\kappa}, \delta_{\rm m})}{N_{\rm SL}(\bar{\kappa})} = E_{\rm SG}\frac{K(\bar{\kappa}^2)}{2\bar{\kappa}}[\bar{\kappa}^2 - 1], \tag{27.5.30}$$

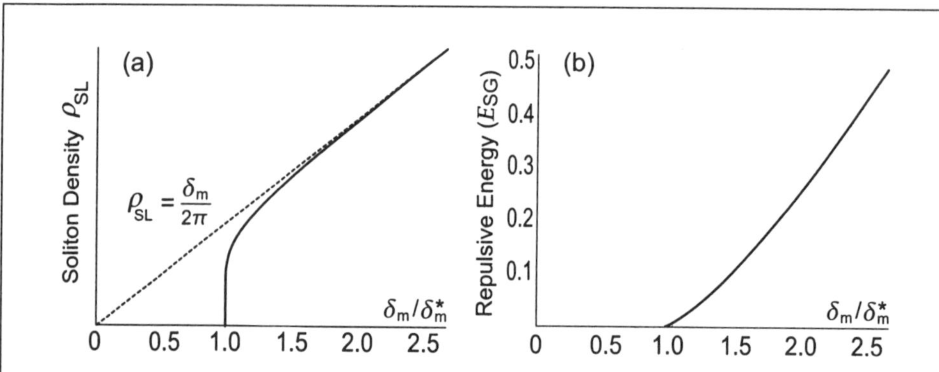

Fig. 27.11 (a) The soliton density $\rho_{\rm SL}$ is illustrated as a function of the misfit $\delta_{\rm m}$. No soliton exists below the phase-transition point $\delta_{\rm m}^*$. Above the point the density increases rapidly as described by (27.5.38), and approaches the dashed straight line representing the ground state (27.4.19) deep inside the IC phase. (b) The repulsive energy (27.5.31) per one soliton in the soliton lattice is illustrated. It increases almost lenearly as a function of $\delta_{\rm m}/\delta_{\rm m}^*$ for $\delta_{\rm m}/\delta_{\rm m}^* \geq 1$.

where use was made of (27.5.26). We compare this with the energy (27.5.12) of a single soliton,

$$\frac{E_{\rm SL}(\bar{\kappa},\delta_{\rm m})}{N_{\rm SL}(\bar{\kappa})} - E_{\rm mSG} = E_{\rm SG}\left(\frac{K(\bar{\kappa}^2)}{2\bar{\kappa}}[\bar{\kappa}^2-1] + \frac{E(\bar{\kappa}^2)}{\bar{\kappa}} - 1\right) > 0. \tag{27.5.31}$$

It represents the repulsive energy between two solitons, which increases almost linearly as a function of $\delta_{\rm m}$ for $\delta_{\rm m} > \delta_{\rm m}^*$ as in Fig.27.11(b).

The typical behaviors of $\rho_{\rm SL}$ as a function of $\delta_{\rm m}$ are understood as follows. We first examine the regime $\delta_{\rm m} \gg \delta_{\rm m}^*$, or $\kappa \to 0$, where

$$K(\kappa^2) \to \frac{\pi}{2}, \qquad E(\kappa^2) \to \frac{\pi}{2}. \tag{27.5.32}$$

From (27.5.26) we find

$$\bar{\kappa} \simeq \frac{\pi}{2}\frac{\delta_{\rm m}^*}{\delta_{\rm m}} = \frac{\pi}{2}\frac{B_\parallel^*}{B_\parallel}. \tag{27.5.33}$$

Now the soliton density (27.5.28) is

$$\rho_{\rm SL} = \frac{\pi\delta_{\rm m}^*}{8\bar{\kappa}K(\bar{\kappa}^2)} \simeq \frac{\delta_{\rm m}^*}{4\bar{\kappa}} \simeq \frac{\delta_{\rm m}}{2\pi}, \tag{27.5.34}$$

as is the dashed straight line in Fig.27.11. Note that the straight line (27.5.19) reads

$$\theta(x) = -\frac{2}{\bar{\kappa}\lambda_J}x = -\frac{\pi\delta_{\rm m}^*}{2\bar{\kappa}}x \simeq -\delta_{\rm m}x, \tag{27.5.35}$$

which agrees with (27.4.19), and describes the ground state deep inside the IC phase.

We next examine the regime $\delta_{\rm m} \simeq \delta_{\rm m}^*$, or $\kappa \to 1$, where we may approximate

$$K(\kappa^2) \to -\frac{1}{2}\ln(1-\kappa^2), \qquad K'(\kappa^2) \to \frac{1}{2}\frac{1}{1-\kappa^2},$$
$$E(\kappa^2) \to 1 - \frac{1}{4}(1-\kappa^2)\ln(1-\kappa^2), \qquad E'(\kappa^2) \to \frac{1}{4}\ln(1-\kappa^2). \tag{27.5.36}$$

Using them in (27.5.26), we find

$$\bar{\kappa} \simeq \frac{\delta_{\rm m}^*}{\delta_{\rm m}}. \tag{27.5.37}$$

Hence the soliton density (27.5.28) is

$$\rho_{\rm SL} = \frac{\pi\delta_{\rm m}^*}{8\bar{\kappa}K(\bar{\kappa}^2)} \simeq -\frac{\pi\delta_{\rm m}}{4\ln[1-(\delta_{\rm m}^*/\delta_{\rm m})^2]} = -\frac{\delta_{\rm m}/\delta_{\rm m}^*}{\lambda_J\ln[1-(\delta_{\rm m}^*/\delta_{\rm m})^2]} \tag{27.5.38}$$

near the phase transition point, $\delta_{\rm m} \gtrsim \delta_{\rm m}^*$, which describes the rapid increase in Fig.27.11.

We next analyze the classical ground state in unbalanced configuration. The system is in the C phase for small parallel magnetic field. We search for a soliton solution. Though it is difficult to solve the equations of motion (27.4.8) explicitly since they are coupled equations in σ and θ, the existence of topological solitons is guaranteed by the existence of the degenerate set (27.4.9) of the ground states in θ. Hence, there exists the soliton-lattice regime, where topological solitons are spontaneously generated as in the balanced configuration.

To determine the phase transition point we consider a single soliton excitation, which is surely akin to a sine-Gordon soliton. It appears when the phase field θ deviates from one of the ground states, say, $\theta = 0$ as $x \to -\infty$ and approaches another ground state $\theta = 2\pi q$ as $x \to \infty$: See Fig.7.5 (page 120). A new aspect is that $\sigma = \sigma_0$ is violated inside the soliton. Nevertheless, we might approximate $\sigma = \sigma_0$ even inside the soliton for samples with $\epsilon_{\rm cap} \gg \Delta_{\rm SAS}$. Then, (27.4.8b) yields the sine-Gordon equation

$$(\lambda_J^\sigma)^2\partial_x^2\theta - \sin\theta = 0, \tag{27.5.39}$$

but with the penetration depth

$$\lambda_J^\sigma = (1-\sigma_0^2)^{1/4}\sqrt{\frac{2J_s^d}{\Delta_{\rm SAS}\rho_0}}. \tag{27.5.40}$$

The phase transition point (27.5.13) becomes

$$B_\parallel^* = \frac{4\hbar}{\pi e d\lambda_J^\sigma} = (1-\sigma_0^2)^{-1/4}\frac{\hbar}{\pi e d}\sqrt{\frac{\rho_0\Delta_{\rm SAS}}{J_s^d}}, \tag{27.5.41}$$

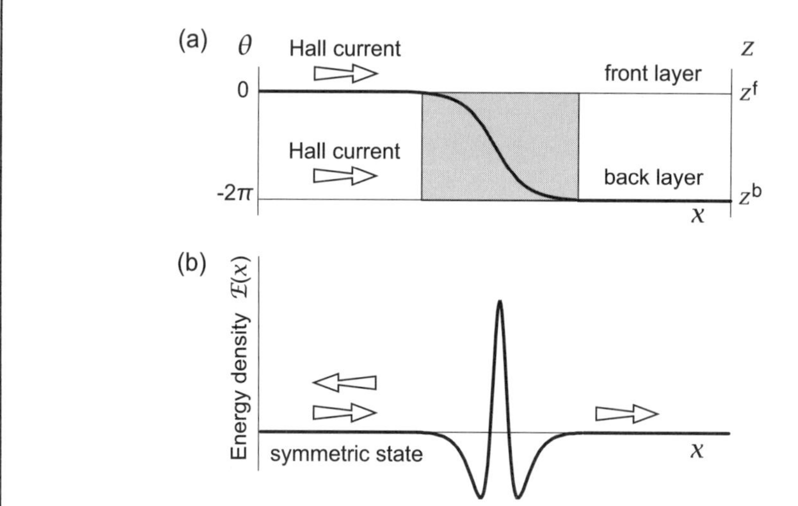

Fig. 27.12 (a) The Hall current flows in the presence of a single soliton. The phase field takes $\theta = 0$ in the left-hand side but takes $\theta = -2\pi$ in the right-hand side of the soliton. (b) An electron is in the symmetric state in the ground state, but deviates from it inside the soliton. The energy density $\mathcal{E}(x)$ is calculated around a soliton based on (27.5.3). It acts as a potential barrier for the electron. It would backscatter the electron coming in and going away in the symmetric state, and the transmission coefficient $\mathcal{T}$ is less than one, $\mathcal{T} < 1$.

which increases very slowly towards the monolayer point ($\sigma_0 = 1$), though the formula would be reliable only near $\sigma_0 = 0$.

27.6 Anomalous Diagonal Resistivity

We have presented a detailed structure of the C-IC phase transition in the previous section, from which we obtain the following picture. The applied parallel magnetic field $B_\parallel$ penetrates between the two layers homogeneously in the C phase. There exists a phase transition point $B_\parallel^*$, at which it begins to penetrate as quantized solitons on top of the homogeneous field $B_\parallel$. A few solitons are created at random very near the phase transition point, and more solitons are generated as $B_\parallel$ increases. They are created at random when their mean separation ℓ_{SL} is much more than the penetration depth λ_J. We call it the random-soliton regime. The number of solitons increases quite rapidly beyond $B_\parallel^*$, and it soon becomes proportional to the applied parallel magnetic field $B_\parallel$, as illustrated in Fig.27.11. They are periodically spaced to make a soliton lattice.[4,222] We call it the soliton-lattice regime. The interlayer phase field $\vartheta(x)$ behaves as follows [Fig.27.6]:

- It is commensurate to $B_\parallel$, $\vartheta(x) = \delta_{\mathrm{m}} x$, in the C phase ($B_\parallel < B_\parallel^*$).
- It behaves in a complicated way in the random-soliton regime of the IC phase.

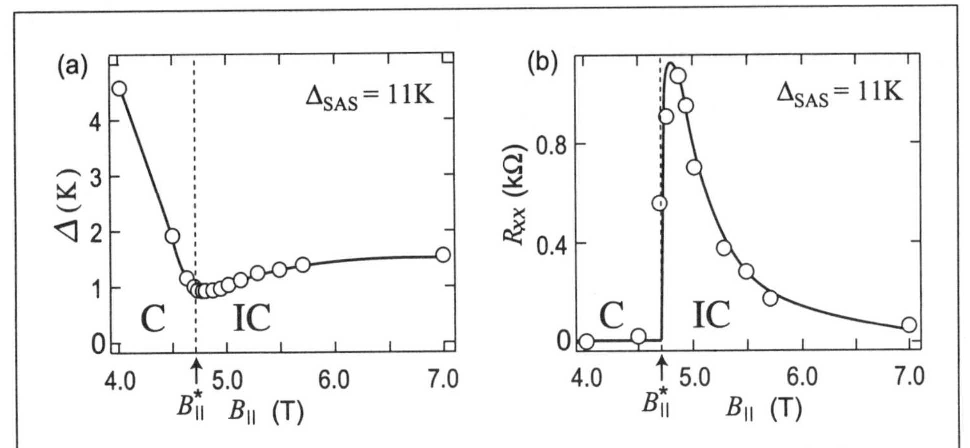

Fig. 27.13 (a) The activation energy Δ is plotted against $B_{\parallel}$ around the C-IC transition point $B_{\parallel}^*$. (b) The diagonal resistance R_{xx} is plotted against $B_{\parallel}$ around the C-IC transition point. It yields an anomalous peak in the vicinity of $B_{\parallel}^*$ in the IC phase. The data are taken from the doctor thesis (March, 2005) of D. Terasawa, where $\rho_0 = 0.86 \times 10^{11}\text{cm}^{-2}$ and $T = 60$mK.

- It becomes a periodic function in the soliton-lattice regime of the IC phase.
- It eventually becomes homogeneous, $\vartheta(x) = 0$, in the asymptotic IC regime.

We now discuss[154] how to detect solitons. Indeed, there exists an experimental report[181] that the soliton lattice has been observed by way of an anomalous increase of the diagonal resistance R_{xx} in the QH state with a well-developed Hall plateau [Figs.27.13 and 27.14]. See also Fig.22 (page xxiv) in the frontispiece.

A soliton is a topological excitation flipping the pseudospin. Let us focus on an electron in the QH current [Fig.27.12]. An electron is initially in the symmetric state, and moves to a higher energy state within a soliton since the antisymmetric component is mixed. The energy density $\mathcal{E}(x)$ is given by the formula (27.5.3). It is reasonable to simulate this as a one-dimensional potential problem with potential $\mathcal{E}(x)$, where a point particle makes a ballistic scattering against a potential barrier [Fig.27.12(b)]. Let $\mathcal{T}$ and $\mathcal{R}$ be the corresponding transmission and reflection coefficients with $\mathcal{T}+\mathcal{R} = 1$ and $0 < \mathcal{R} \ll 1$. When there are N solitons placed at random [Fig.27.15(a)], the transmission coefficient increases as $\mathcal{T}^N = (1 - \mathcal{R})^N \simeq 1 - N\mathcal{R}$. According to the 4-terminal Landauer formula[291] we would conclude an increase of the diagonal resistance as

$$R_{xx} = \frac{\pi\hbar}{e^2}\frac{1 - \mathcal{T}^N}{\mathcal{T}^N} \simeq N\frac{\pi\hbar}{e^2}\mathcal{R} \equiv NR_{\text{SG}}. \tag{27.6.1}$$

Thus one soliton gives a contribution $R_{\text{SG}} = \pi\hbar\mathcal{R}/e^2$ to the resistance. Such an increase of R_{xx} occurs in the random-soliton regime even at the zero-temperature limit, though its experimental observation is practically impossible since the num-

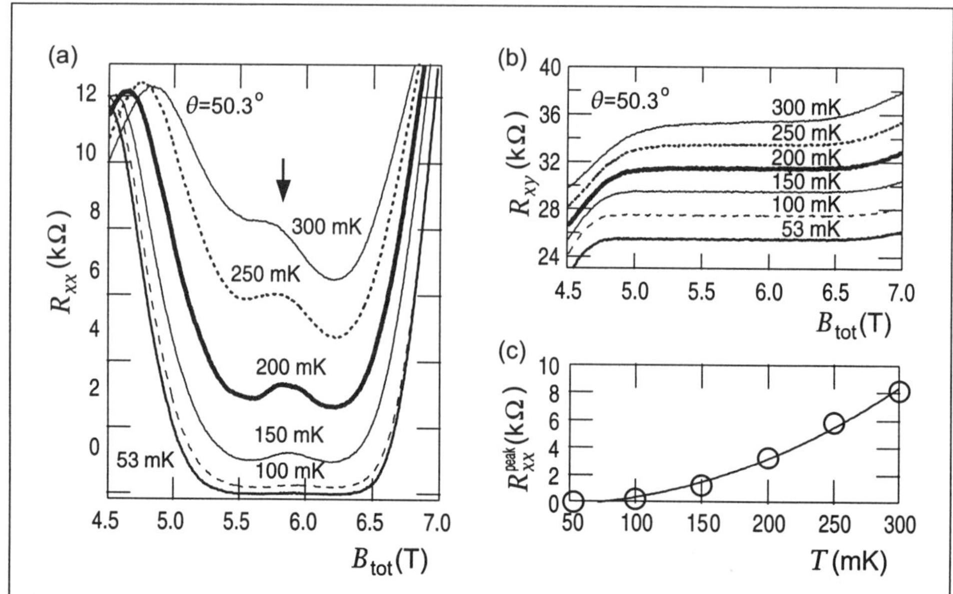

Fig. 27.14 Diagonal and Hall resistances are measured by changing the total magnetic field B_{tot} in a tilted sample at 50.3° at various temperature T. (a) A local maximum is clearly observed in R_{xx} for $T > 150$K. (b) Nevertheless, the Hall plateau develops well in this range, implying the dissipation occurs in the QH state for $T > 150$K. Hall resistance is offset by 2kΩ for ease of visibility. (c) The peak value R_{xx}^{peak} decreases as the temperature T decreases. Data are taken from Fukuda et al.[181], where $\rho_0 = 0.96 \times 10^{11}\text{cm}^{-2}$. See also Fig.22 in the frontispiece.

ber of solitons increases so rapidly as to form a soliton lattice just beyond the phase transition point $B_{\parallel}^*$ [Fig.27.11].

We now argue the mechanism how the dissipation ($R_{xx} \neq 0$) occurs in the soliton-lattice regime [Fig.27.15(b)]. We start with an analysis at the zero temperature, where the soliton lattice makes a perfect periodic potential for electrons. The wave function of electrons spreads over the whole system and makes a Bloch state. There is no dissipation since there is no backscattering of electrons in a perfect periodic system. This agrees with the experimental fact that $R_{xx} = 0$ at sufficiently low temperature [Fig.27.14(c)]. We next consider the system at a higher temperature. Since the soliton lattice fluctuates thermally, the perfect periodicity is lost. Backscattering must occur by individual solitons fluctuating thermally. If we may model a fluctuation around its mean place by a harmonic oscillator, after quantization it is characterized by a certain gap energy Δ_{mSG}. The fluctuation probability is the thermo-active type, and the dissipation would occur according to the formula

$$R_{xx}(\delta_{\text{m}}, T) = L\rho_{\text{SL}}(\delta_{\text{m}}) \exp\left[-\frac{\Delta_{\text{mSG}}(\delta_{\text{m}})}{k_B T}\right] R_{\text{SG}} + R_0(T), \qquad (27.6.2)$$

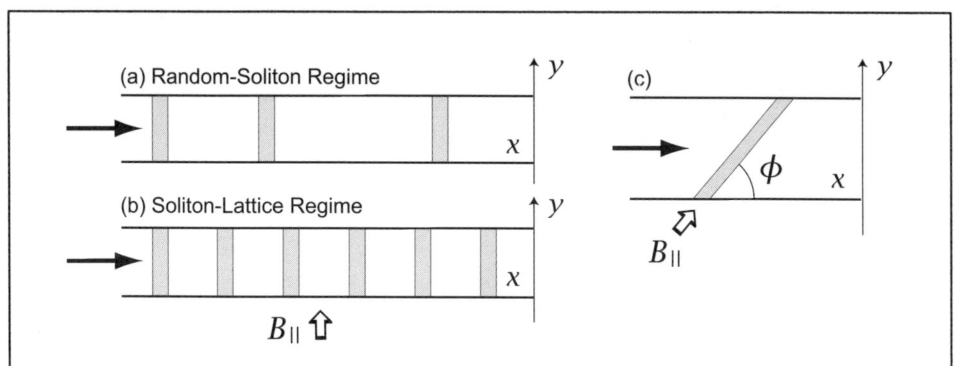

Fig. 27.15 An electron (arrow) travels along the x axis in the presence of solitons (gray regions). (a) An electron collides against solitons placed parallel to the y axis at random. Each soliton gives a contribution R_{SL} to the diagonal resistance R_{xx}. (b) An electron travels over solitons placed periodically as a Bloch wave. There arises no resistivity at the zero-temperature. However, a resistance arises at finite temperature since the periodicity is broken by thermal fluctuations. (c) The resistance R_{xx} decreases as the angle ϕ increases, where ϕ is the angle between the parallel magnetic field $B_{\|}$ and the QH current.

where we have set $N = \rho_{SL}L$ with L being the size of the system, and $R_0(T)$ is a residual resistance. It arises from the excitation of quasiparticles (skyrmions). As we have argued in Section 25.6, it is almost insensitive to the parallel magnetic field δ_m in the IC phase.

We study how $\Delta_{mSG}(\delta_m)$ depends on δ_m. In the IC phase, as we have argued, the soliton density $\rho_{SL}(\delta_m)$ is determined by the competition between a negative creation energy of solitons and a positive repulsive energy between solitons. It proliferates rapidly until the distance between solitons becomes comparable to the width of a soliton [Fig.27.11(a)]. This is the reason why there arises an anomalous increase of the diagonal resistance R_{xx} just beyond the C-IC phase transition in the experimental data [Fig.27.13(b)]. As the soliton density becomes dense, the rigidity of the soliton lattice, and hence $\Delta_{mSG}(\delta_m)$, increases because the repulsive interaction increases [Fig.27.11(b)]. Consequently, the diagonal resistance R_{xx} decreases and approaches the residual value $R_0(T)$ as $B_{\|}$ increases [Fig.27.13(b)].

It is possible to confirm the present picture of the electron-soliton scattering by measuring R_{xx} in a sample where an angle ϕ is made between the direction of the applied parallel magnetic field and the direction of the Hall current [Fig.27.15(c)]. In this case the resistance should behave as

$$R_{xx}(\delta_m, T, \phi) = \frac{1}{2}(1 - \cos\phi)L\rho_{SL}(\delta_m)\exp\left[-\frac{\Delta_{mSG}(\delta_m)}{k_BT}\right]R_{SG} + R_0(T). \quad (27.6.3)$$

It takes a maximum value at $\phi = 90°$ and a minimum value at $\phi = 0°$. This is precisely what is observed experimentally [Fig.27.16] by controlling the direction of the applied parallel magnetic field against the direction of the Hall current.

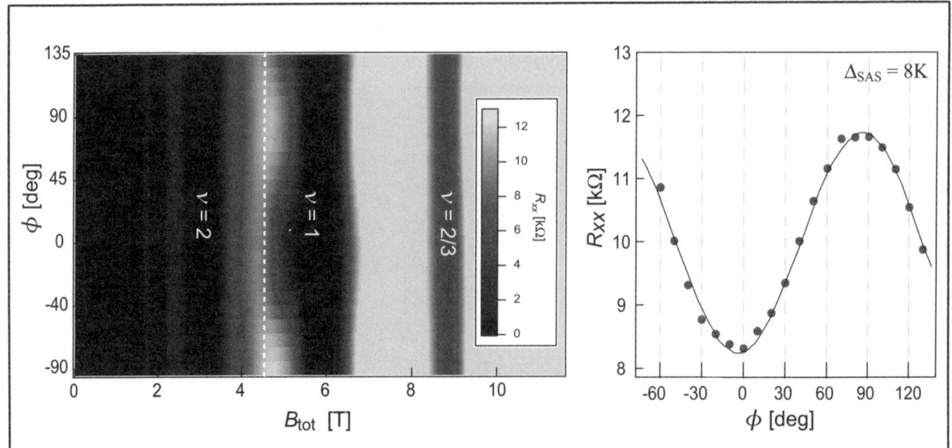

Fig. 27.16 (a) The image plot of R_{xx} is given against the total magnetic field B_{tot} and the angle ϕ, where ϕ is the angle between the directions of the parallel magnetic field and the Hall current. The experiments are carried out at $T = 0.07$K, $\rho_0 = 1.0 \times 10^{11}\text{cm}^{-2}$, $\Theta = 45.0°$ and $\mathcal{J} = 10$nA. The QH states develop in dark regions. (b) The measured values of R_{xx} are plotted against the angle ϕ in the $\nu = 1$ QH state. The dependence is well fitted by $R_{xx}(\phi) = R_0 + \frac{1}{2}(1 - \cos\phi)N_{\text{eff}}R_{\text{SG}}$, where R_{SG} is the resistance due to one electron-soliton scattering, N_{eff} is the effective number of scatterers in the soliton lattice, and R_0 is a residual resistance. The data are taken from the doctor thesis (March, 2006) of M. Morino. See also Fig.22 in the frontispiece taken from a similar work due to Fukuda et al.[181]

It should be noticed that the Hall plateau develops [Fig.27.14(b)] even in the presence of the anomalous peak in R_{xx} since electron-soliton scatterings does not affect the incompressibility of the system.

We make a rough estimation[154] of $\Delta_{\text{mSG}}(\delta_{\text{m}})$. By simulating the fluctuation of a soliton by the harmonic oscillator,

$$H = -\frac{\hbar^2}{2M(\delta_{\text{m}})}\left(\frac{\partial}{\partial x}\right)^2 + \frac{1}{2}k(\delta_{\text{m}})x^2, \tag{27.6.4}$$

the gap is given by

$$\Delta_{\text{mSG}}(\delta_{\text{m}}) = \hbar\sqrt{\frac{k(\delta_{\text{m}})}{M(\delta_{\text{m}})}}. \tag{27.6.5}$$

Here, x is the coordinate of one soliton about its mean point taken at the origin.

The mass M of one soliton is derived from the effective Hamiltonian (27.5.3) as

$$\mathcal{E}_{\text{mSG}}(x) = J_s^d \int dx \left[\frac{1}{2v_{\text{F}}^2}\partial_t^2\theta\right] + E_{\text{mSG}} = \frac{J_s^d v^2}{2v_{\text{F}}^2}\int dx \left[\partial_x^2\theta\right] + E_{\text{mSG}} \equiv \frac{N_{\text{SL}}M}{2}v^2 + E_{\text{mSG}}, \tag{27.6.6}$$

by considering a moving soliton solution of the form $\theta(x - vt)$ with v the velocity.

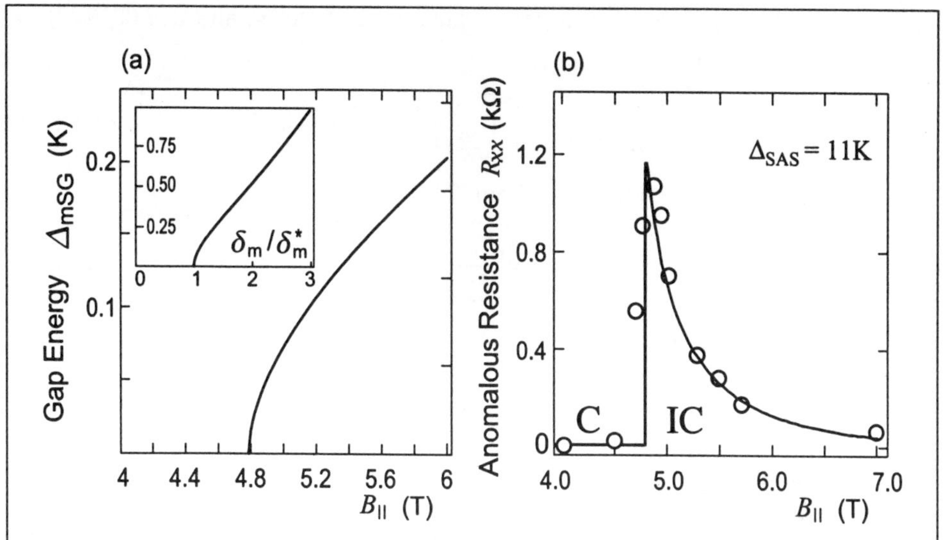

Fig. 27.17 (a) The gap energy (27.6.13) is illustrated as a function of $B_{\|}$ near the phase transition point (Inset: as a function of the parameter δ_m for $0 < \delta_m < 3\delta_m^*$). (b) The experimental data for R_{xx} in Fig.27.13(b), where $B_{\|}$ is varied at a fixed temperature $T \simeq 60$mK, are fitted by the formula (27.6.2) together with (27.6.13). The used valules of parameters are $\Delta_{SAS} = 11$K, $d = 23.1$nm, $\rho_0 = 0.86 \times 10^{15}/\text{m}^{-2}$ and $L = 180\mu$m, where $B_{\|}^* = 4.78$ Tesla from (27.5.13).

Hence,

$$M(\delta_m) = \frac{J_s^d}{v_F^2 N_{SL}} \int dx \left[\partial_x^2 \theta\right] = \frac{8J_s^d}{\lambda_J v_F^2} \frac{E(\bar{\kappa}^2)}{\bar{\kappa}}, \tag{27.6.7}$$

where we have used the static soliton-lattice solution (27.5.15).

To estimate the elasticity $k(\delta_m)$, we enquire how rigid the soliton lattice is. Quantitatively it is given by the energy increase of the soliton lattice when we change the lattice spacing ℓ_{SL} by x. It is given per soliton by

$$\mathcal{E}(x) = \frac{1}{2}\left[\frac{E_{SL}(\bar{\ell}_{SL}+x)}{\rho_{SL}(\bar{\ell}_{SL}+x)} + \frac{E_{SL}(\bar{\ell}_{SL}-x)}{\rho_{SL}(\bar{\ell}_{SL}-x)}\right], \tag{27.6.8}$$

where $E_{SL}(\bar{\ell}_{SL}) = E_{SL}(\bar{\kappa}, \delta_m)$ together with (27.5.27) and $\kappa = \bar{\kappa}$. Expanding this in powers of x, we come to

$$\mathcal{E}(x) = \mathcal{E}(0) + \frac{1}{2}\frac{E_{SL}''(\bar{\ell}_{SL})}{\rho_{SL}(\bar{\ell}_{SL})}x^2 + O(x^4). \tag{27.6.9}$$

We identify this energy increase with the "elasticity" in the Hamiltonian (27.6.4),

$$k(\delta_m) = \frac{1}{\rho_{SL}(\bar{\ell}_{SL})}\frac{d^2}{d\ell_{SL}^2}E_{SL}(\ell_{SL}). \tag{27.6.10}$$

To calculate $E''_{\text{SL}}(\ell_{\text{SL}})$, we change the variable from ℓ_{SL} to κ, and use (27.5.27) together with (27.5.24),

$$\frac{d^2}{d\ell_{\text{SL}}^2}E_{\text{SL}}(\ell_{\text{SL}}) = \frac{1-\kappa^2}{2\lambda_J E(\kappa^2)}\frac{d}{d\kappa}\left[\frac{1-\kappa^2}{2\lambda_J E(\kappa^2)}\frac{d}{d\kappa}E_{\text{SL}}(\kappa,\delta_{\text{m}})\right]. \tag{27.6.11}$$

Evaluating this at $\kappa = \bar{\kappa}$, we obtain from (27.6.10) that

$$k(\delta_{\text{m}}) = \frac{2J_s^d}{\lambda_J^3}\frac{1-\bar{\kappa}^2}{\bar{\kappa}^4}\frac{\delta_{\text{m}}^*}{\delta_{\text{m}}}. \tag{27.6.12}$$

The gap energy (27.6.5) reads

$$\Delta_{\text{mSG}}(\delta_{\text{m}}) = \frac{\hbar v_{\text{F}}}{2\lambda_J}\frac{\sqrt{1-\bar{\kappa}^2}}{\bar{\kappa}^2}\frac{\delta_{\text{m}}^*}{\delta_{\text{m}}}. \tag{27.6.13}$$

We illustrate the gap energy (27.6.13) and the diagonal resistivity (27.6.2) together with the experimental data in Fig.27.17. The theoretical curve explains quite well the anomalous behavior of R_{xx}, provided that we take phenomenological values $v_{\text{F}} = 900\text{m/sec}$ and $\mathcal{R} = 5\times 10^{-5}$.

27.7 Plasmon Excitations

Low-lying collective excitations are oscillation modes of the charging $e\sigma\rho_0$, which are induced by the coherent tunneling of electrons between the two layers. The charging $e\sigma\rho_0$ generates the electric field, which acts to decrease the charging itself. Thus, the excitations look like plasma oscillations in electron gas. It is convenient to use the variable $\theta(x)$ rather than the interlayer phase field $\vartheta(x)$ also in this section.

Commensurate Phase

We analyze plasmon excitations in the commensurate phase, where the ground state is given by (27.4.9) or $\theta = 0$. Small fluctuations around the balanced point ($\sigma = 0$) are described by assuming $|\sigma| \ll 1$ and $|\theta| \ll \pi$. It follows from (27.4.8) that

$$\hbar\partial_t\sigma = -\frac{2\gamma J_s^d}{\rho_0}\nabla^2\theta + \Delta_{\text{SAS}}\theta, \tag{27.7.1a}$$

$$\hbar\partial_t\theta = \frac{2J_s}{\rho_0}\nabla^2\sigma - \left(\epsilon_{\text{cap}} + \Delta_{\text{SAS}}\right)\sigma, \tag{27.7.1b}$$

where only the linear terms in σ and θ are taken. We search for solutions of the type,

$$\sigma = \sigma_1 e^{ikx}\sin(\omega_k t), \tag{27.7.2a}$$

$$\theta = \theta_1 e^{ikx}\cos(\omega_k t), \tag{27.7.2b}$$

where σ_1 and θ_1 are integration constants.

Let us apply an oscillating bias voltage around $V_{\text{bias}} = 0$,

$$V_{\text{bias}}(t) = V_0 \sin(\omega_{\text{ext}} t). \tag{27.7.3}$$

The equation for the zero mode is derived from (27.7.1),

$$\hbar^2 \partial_t^2 \sigma + \hbar^2 \omega_{\text{P}}^2 \sigma = e\Delta_{\text{SAS}} V_0 \sin(\omega_{\text{ext}} t), \tag{27.7.4}$$

where ω_{P} is the plasmon frequency $\omega_{\boldsymbol{k}}$ at $\boldsymbol{k} = 0$, and given by

$$\hbar\omega_{\text{P}} = \sqrt{\epsilon_{\text{cap}}\Delta_{\text{SAS}} + \Delta_{\text{SAS}}^2}. \tag{27.7.5}$$

It is a well-known equation of the forced oscillation, whose solution is

$$\sigma(t) = \frac{e\Delta_{\text{SAS}} V_0}{\hbar^2(\omega_{\text{P}}^2 - \omega_{\text{ext}}^2)} \sin(\omega_{\text{ext}} t). \tag{27.7.6}$$

The phase difference oscillates,

$$\theta(t) = \frac{eV_0\omega_{\text{ext}}}{\hbar(\omega_{\text{P}}^2 - \omega_{\text{ext}}^2)} \cos(\omega_{\text{ext}} t). \tag{27.7.7}$$

Accordingly, the tunneling current (27.3.5) oscillates,

$$\mathcal{J}_z = J_{\text{T}} \frac{eV_0\omega_{\text{ext}}}{\hbar(\omega_{\text{P}}^2 - \omega_{\text{ext}}^2)} \cos(\omega_{\text{ext}} t). \tag{27.7.8}$$

Because a charge transfer occurs in a nonzero voltage difference,

$$eV_{\text{junc}} = \hbar\partial_t\theta \neq 0, \tag{27.7.9}$$

it makes a work by emitting a microwave with the frequency ω_{ext}. The intensity diverges when the frequency ω_{ext} agrees with the plasmon frequency ω_{P}. Consequently, the plasmon frequency ω_{P} is measurable by the microwave emitted from the sample. Actually, it is impractical to change the bias voltage periodically. It is practical to excite plasmons by applying microwave to the sample and to detect a sudden increase of temperature due to the microwave radiation.

Incommensurate Phase

We analyze plasmon excitations in the incommensurate phase, where the ground state is approximated by (27.4.19) or $\theta = \vartheta_0 - \delta_{\text{m}} x$. Small fluctuations around the balanced point ($\sigma = 0$) are studied by setting

$$\theta = \vartheta_0 - \delta_{\text{m}} x + \xi \tag{27.7.10}$$

in (27.3.3), and by assuming $|\sigma| \ll 1$ and $|\xi| \ll \pi$. The plasmon zero mode obeys

$$\hbar^2 \partial_t^2 \sigma + \Delta_{\text{SAS}} \cos(\vartheta_0 - \delta_{\text{m}} x) \left\{ \frac{\epsilon_{\text{cap}}}{2} + \Delta_{\text{SAS}} \cos(\vartheta_0 - \delta_{\text{m}} x) \right\} \sigma = \frac{\sigma_0}{\sqrt{1-\sigma_0^2}} \Delta_{\text{SAS}}^2 \cos(\vartheta_0 - \delta_{\text{m}} x). \tag{27.7.11}$$

We average this equation over the size L of the junction,[111] using

$$\langle \cos(\vartheta_0 - \delta_{\text{m}} x) \rangle = \frac{1}{L} \int_{-L/2}^{L/2} dx \cos(\vartheta_0 - \delta_{\text{m}} x) = \cos \vartheta_0 \frac{\sin(\pi \Phi_{\parallel}/\Phi_{\text{D}})}{\pi \Phi_{\parallel}/\Phi_{\text{D}}}, \tag{27.7.12a}$$

$$\langle \cos^2(\vartheta_0 - \delta_{\text{m}} x) \rangle = \frac{1}{L} \int_{-L/2}^{L/2} dx \cos^2(\vartheta_0 - \delta_{\text{m}} x) = \frac{1}{2} + O(L^{-1}), \tag{27.7.12b}$$

where

$$\frac{L\delta_{\text{m}}}{2} = \frac{LedB_{\parallel}}{2\hbar} = \frac{\pi \Phi_{\parallel}}{\Phi_{\text{D}}} \tag{27.7.13}$$

with (27.2.4). Here, $\Phi_{\parallel} \equiv LdB_{\parallel}$ is the total parallel magnetic field penetrated into the junction, and $\Phi_{\text{D}} \equiv 2\pi\hbar/e$ is the Dirac flux unit. The resulting equation is

$$\hbar^2 \partial_t^2 \sigma + \hbar^2 \omega_{\text{IP}}^2 \sigma = e\Delta_{\text{SAS}} V_0 \left| \frac{\sin(\pi \Phi_{\parallel}/\Phi_{\text{D}})}{\pi \Phi_{\parallel}/\Phi_{\text{D}}} \right| \sin(\omega_{\text{ext}} t) \tag{27.7.14}$$

with

$$\hbar \omega_{\text{IP}} = \sqrt{\frac{e^2 \Delta_{\text{SAS}} \rho_0}{2C} \left| \frac{\sin(\pi \Phi_{\parallel}/\Phi_{\text{D}})}{\pi \Phi_{\parallel}/\Phi_{\text{D}}} \right| + \frac{1}{2} \Delta_{\text{SAS}}^2}. \tag{27.7.15}$$

The plasmon frequency ω_{IP} depends on the total magnetic flux $\Phi_{\parallel}$ penetrated into the junction. As in the commensurate phase, the tunneling current emits a microwave with the frequency ω_{ext}, whose intensity diverges when the frequency ω_{ext} agrees with the plasmon frequency ω_{IP}. It is detectable experimentally. The commensurate and incommensurate transition is also checked by the transition of the plasmon mode from (27.7.5) to (27.7.15).

27.8 Josephson-Like Effects

Because of a similarity to the superconductor Josephson junction, it is natural to anticipate the Josephson-like effect in the interlayer-coherent state.[140,494] The Josephson-like effect implies a phenomenon where the electric current flows between the two layers as a dissipationless current. We analyze this problem affirmatively. We comment that there is a counter-argument[351,326] made on this problem but their argument is not correct. Indeed, they argue the absence of the dc Josephson effect without feeding any current into the system. As we show, when a current is fed in an appropriate way, the Hall current flows as a tunneling current at zero interlayer voltage ($V_{\text{junc}} = 0$).

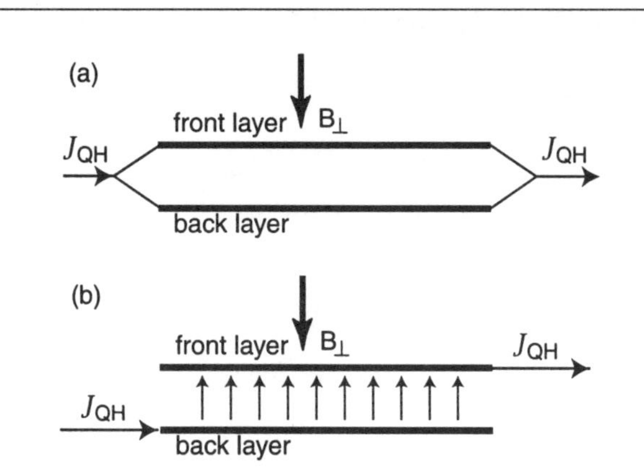

Fig. 27.18 (a) In standard samples the Hall current flows on both layers without tunneling current (σ_0 = constant, $\vartheta = 0$). (b) It is possible to feed the Hall current to one of the layers and extract it from the other layer. The tunneling current may flow as a supercurrent (σ_0 = constant, ϑ = constant).

In the standard sample of bilayer QH systems, a current $\mathcal{J}_{\text{QH}}$ is fed to both layers and extracted from them simultaneously [Fig.27.18(a)]. There is no tunneling current in the ground state.

We consider a sample where a current $\mathcal{J}_{\text{QH}}$ is fed to one layer and extracted from the other layer [Fig.27.18(b)]. The current (27.4.6) in each layer reads

$$\mathcal{J}_x^{\text{f}}(\boldsymbol{x}) = \frac{eJ_s^d}{\hbar}\partial_x\left(\theta + \frac{edB_\parallel}{\hbar}x\right) + \mathcal{J}_{\text{QH}}\left(\frac{1}{2} - \frac{x}{L}\right), \tag{27.8.1a}$$

$$\mathcal{J}_x^{\text{b}}(\boldsymbol{x}) = -\frac{eJ_s^d}{\hbar}\partial_x\left(\theta + \frac{edB_\parallel}{\hbar}x\right) + \mathcal{J}_{\text{QH}}\left(\frac{1}{2} + \frac{x}{L}\right), \tag{27.8.1b}$$

where the edges of the sample are placed at $x = -L/2$ and $x = L/2$. The Hall current is given by the total current $\mathcal{J}_{\text{QH}} = \mathcal{J}_x^{\text{f}} + \mathcal{J}_x^{\text{b}}$, but it now tunnels between the two layers.

The effective Hamiltonian leading to such currents is given by

$$\mathcal{H}_{\text{eff}}^{\text{JJ}} = \mathcal{H}_{\text{eff}} + \frac{\mathcal{J}_{\text{QH}}}{2}\left(A_x^{\text{f}} + A_x^{\text{b}}\right) - \frac{x\mathcal{J}_{\text{QH}}}{L}\Delta A_x. \tag{27.8.2}$$

The Heisenberg equations of motion are calculated,

$$i\hbar\partial_t\vartheta = [\vartheta, \int d^2x\, \mathcal{H}_{\text{eff}}^{\text{JJ}}], \qquad i\hbar\partial_t\sigma = [\sigma, \int d^2x\, \mathcal{H}_{\text{eff}}^{\text{JJ}}] \tag{27.8.3}$$

based on the commutation relation (15.4.15),

$$\frac{\rho_0}{2}[\sigma(\boldsymbol{x}), \vartheta(\boldsymbol{y})] = i\delta(\boldsymbol{x} - \boldsymbol{y}). \tag{27.8.4}$$

The result is a simple modification of (27.4.8). The Hall current $\mathcal{J}_{QH}$ is controlled externally.

The interlayer voltage is given by

$$eV_{\text{junc}} = -\frac{2}{\rho_0}\frac{\delta \int d^2x\, \mathcal{H}_{\text{eff}}^{\text{JJ}}(\boldsymbol{x})}{\delta\sigma(\boldsymbol{x})} = \hbar\partial_t\vartheta(\boldsymbol{x}), \tag{27.8.5}$$

or

$$eV_{\text{junc}} = \hbar\partial_t\theta(\boldsymbol{x}), \tag{27.8.6}$$

which does not vanish in general. It can be measured directly by attaching probes to the two layers.

In the C phase, the ground state is described by a static configuration

$$\boldsymbol{n}_{\text{g}} = \frac{1}{\sqrt{2}}\begin{pmatrix} e^{i(\theta_0+\delta_{\text{m}}x)/2}\sqrt{1+\sigma_0} \\ e^{-i(\theta_0+\delta_{\text{m}}x)/2}\sqrt{1-\sigma_0} \end{pmatrix} \tag{27.8.7}$$

in terms of the CP^1 field. The pseudospin field reads

$$\mathcal{P}_x = \frac{1}{2}\sqrt{1-\sigma_0^2}\cos(\theta_0+\delta_{\text{m}}x), \quad \mathcal{P}_y = -\frac{1}{2}\sqrt{1-\sigma_0^2}\sin(\theta_0+\delta_{\text{m}}x), \quad \mathcal{P}_z = \frac{1}{2}\sigma_0, \tag{27.8.8}$$

while the tunneling current is given by

$$\mathcal{J}_z(\boldsymbol{x}) = J_{\text{T}}\sqrt{1-\sigma_0^2}\sin\theta_0. \tag{27.8.9}$$

The parameters σ_0 and θ_0 are given by solving

$$\mathcal{J}_z \equiv J_{\text{T}}\sqrt{1-\sigma_0^2}\sin\theta_0 = \frac{1}{L}\mathcal{J}_{\text{QH}}, \tag{27.8.10a}$$

$$eV_{\text{junc}} \equiv \Delta_{\text{SAS}}\frac{\sigma_0}{\sqrt{1-\sigma_0^2}}(1-\cos\theta_0) = 0 \tag{27.8.10b}$$

in terms of the Hall current $\mathcal{J}_{\text{QH}}$ and the bias voltage V_{bias}. Here, $\mathcal{J}_{\text{T}} \equiv ed\Delta_{\text{SAS}}\rho_0/2\hbar$. In particular, when $V_{\text{bias}} = 0$, the balanced configuration ($\sigma_0 = 0$) is realized with $V_{\text{junc}} = 0$, and we obtain

$$\mathcal{J}_z = \frac{1}{L}\mathcal{J}_{\text{QH}} = J_{\text{T}}\sin\theta_0. \tag{27.8.11}$$

When the Hall current is less than the critical current, $|\mathcal{J}_{\text{QH}}| \leq LJ_{\text{T}}$, the phase difference θ_0 is fixed by this formula. Because charge transfers occur in the absence of the voltage difference ($V_{\text{junc}} = 0$), the tunneling current $\mathcal{J}_z$ is a dissipationless current.

In the IC phase, the ground state is described by a static configuration

$$\boldsymbol{n}_{\text{g}} = \frac{1}{\sqrt{2}}\begin{pmatrix} e^{+i\vartheta_0/2}\sqrt{1+\sigma_0} \\ e^{-i\vartheta_0/2}\sqrt{1-\sigma_0} \end{pmatrix} \tag{27.8.12}$$

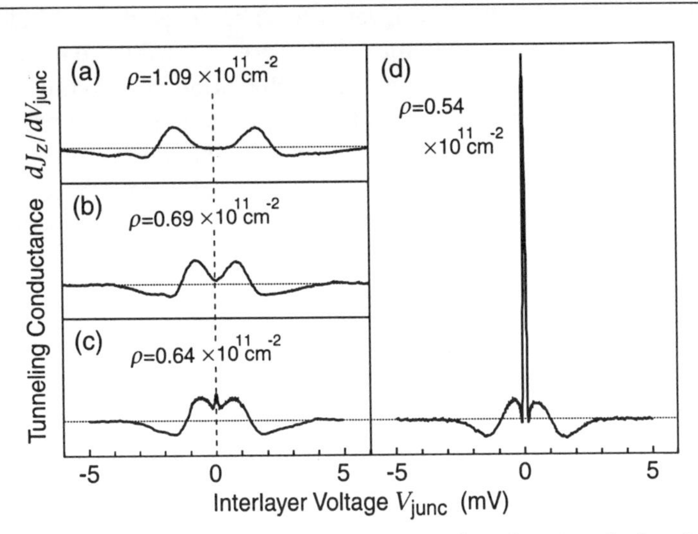

Fig. 27.19 The tunneling conductance $dJ_z/dV_{\rm junc}$ is plotted against the interlayer voltage $V_{\rm junc}$. In samples (a)~(c) with high densities the bilayer system is not a QH system, where the electron tunneling is suppressed. In sample (d) with low density the interlayer coherence develops between the two layers, where the electron tunneling occurs at zero interlayer voltage ($V_{\rm junc} = 0$) as the dc Josephson effect. The data are taken from Spielman et al.[454] Sample parameters are $d \simeq 27.9$ nm and $\Delta_{\rm SAS} \simeq 90\ \mu$K.

in terms of the CP^1 field. The pseudospin field reads

$$\mathcal{P}_x = \frac{1}{2}\sqrt{1-\sigma_0^2}\cos\vartheta_0, \quad \mathcal{P}_y = -\frac{1}{2}\sqrt{1-\sigma_0^2}\sin\vartheta_0, \quad \mathcal{P}_z = \frac{1}{2}\sigma_0, \tag{27.8.13}$$

while the tunneling current is given by

$$\mathcal{J}_z(\boldsymbol{x}) = J_{\rm T}\sqrt{1-\sigma_0^2}\sin(\vartheta_0 - \delta_{\rm m}x)\,. \tag{27.8.14}$$

We take the average of them over the size L of the sample as in (27.7.12). In particular, in the balanced configuration ($\sigma_0 = 0$) we obtain,

$$\mathcal{J}_z = \frac{1}{L}\mathcal{J}_{\rm QH} = J_{\rm T}\langle\sin(\vartheta_0 - \delta_{\rm m}x)\rangle = J_{\rm T}\frac{\sin(\pi\Phi_\parallel/\Phi_{\rm D})}{\pi\Phi_\parallel/\Phi_{\rm D}}\sin\vartheta_0. \tag{27.8.15}$$

The maximum Hall current shows a Fraunhofer pattern: It decreases and then increases as the sample is tilted.

The existence of the Josephson effect is detectable by measuring the tunneling current-voltage (IV) characteristics. We cite a recent experimental result[454] in the $\nu = 1$ bilayer QH system, where the tunneling conductance exhibits a sharp peak at zero interlayer voltage ($V_{\rm junc} = 0$) indicating the dc Josephson effect [Fig.27.19]. It is important that the phenomenon occurs only when the total density is low enough.

It agrees with the fact that the interlayer coherence develops only in low-density samples: See Fig.16(a) (page xx) in the frontispiece. According to the above analysis we expect an infinite zero-voltage peak in the differential conductance dI/dV. Nevertheless, a surprising feature is that the peak maintains an appreciable width signaling the presence of dissipation. A microscopic understanding of the mechanism that cuts off the divergence in the differential conductance $dI/dV|_{V=0}$ is presently lacking, though there are some phenomenological discussions.[61,457,174]

PART 4

MICROSCOPIC THEORY

CHAPTER 28

OVERVIEW OF MICROSCOPIC THEORY

Excitations across Landau levels are suppressed when the cyclotron energy is large enough. A self-consistent theory without these excitations is constructed by making the Landau-level projection. An electron confined to a single Landau level is described by the guiding center (X, Y) subject to noncommutativity, $[X, Y] = -i\ell_B^2$. Thus the QH system provides us with a realistic world of noncommutative geometry. It presents not only a good approximation but also an essential way to reveal a new physics inherent to the QH system. In formulating a microscopic theory of the integer QH system, we derive some key formulas already used in the semiclassical analysis. In this section we overview a microscopic theory of the QH effects based on noncommutative geometry. The complex number is defined by $z = (x + iy)/\ell_B$ throughout this Part.

Lowest-Landau-Level Projection

The coordinate $\boldsymbol{x} = (x, y)$ of the electron making cyclotron motion is decomposed into the guiding center $\boldsymbol{X} = (X, Y)$ and the relative coordinate $\boldsymbol{R} = (R_x, R_y)$, $\boldsymbol{x} = \boldsymbol{X} + \boldsymbol{R}$, where $R_x = -P_y/eB_\perp$ and $R_y = P_x/eB_\perp$ with $\boldsymbol{P} = (P_x, P_y)$ the covariant momentum. The commutation relation,

$$[X, Y] = -i\ell_B^2, \quad [P_x, P_y] = i\frac{\hbar^2}{\ell_B^2},$$
$$[X, P_x] = [X, P_y] = [Y, P_x] = [Y, P_y] = 0, \tag{28.1.1}$$

implies that the guiding center and the relative coordinate are independent variables. The kinetic Hamiltonian,

$$H_K = \frac{\boldsymbol{P}^2}{2M} = \frac{1}{2M}(P_x - iP_y)(P_x + iP_y) + \frac{1}{2}\hbar\omega_c, \tag{28.1.2}$$

creates Landau levels with gap energy $\hbar\omega_c$. When it is large enough, excitations across Landau levels are suppressed. It is a good approximation to prohibit all such excitations by requiring electron confinement to a single Landau level.

We have already required the LLL condition on the state,

$$(P_x + iP_y)\psi(\boldsymbol{x})|\mathfrak{S}\rangle = 0, \tag{28.1.3}$$

and determined the Hilbert space $\mathbb{H}_{\text{LLL}}$. A semiclassical theory of topological solitons has been developed based on this condition. However, the requirement of the LLL condition on the state is not enough. Even if we start with a state in the lowest Landau level, a potential term kicks it out into higher Landau levels. It is necessary to make the LLL projection of the potential term additionally, that is to say, to freeze out the relative coordinate in the potential term. This turns out to reveal a new physics associated with noncommutative geometry in QH systems.

The importance of the LLL projection was first pointed out by Girvin et al.,[187,189] and subsequently the algebraic structure was elucidated.[236,84,418,443] The internal symmetry was included to study the spin coherence and the pseudospin coherence.[144,351] The electron system with the SU(N) symmetry is characterized by the W_∞(N) algebra,[149] which is the SU(N) extension of the W_∞ algebra. The microscopic understanding of QH systems is fully obtained based on noncommutative geometry.[152] In this Part we explore the Landau level projection to a single Landau level, assuming that all the lower Landau levels are filled up.

Noncommutative Geometry

We explore the physics of electrons confined to a single Landau level, where the electron position is specified solely by the guiding center $\boldsymbol{X} = (X, Y)$, whose X and Y components are noncommutative,

$$[X, Y] = -i\ell_B^2. \tag{28.1.4}$$

The holomorphic basis is constructed by introducing the operators

$$b = \frac{1}{\sqrt{2}\ell_B}(X - iY), \qquad b^\dagger = \frac{1}{\sqrt{2}\ell_B}(X + iY), \tag{28.1.5}$$

obeying $[b, b^\dagger] = 1$, together with the Fock states

$$|n\rangle = \frac{1}{\sqrt{n!}}(b^\dagger)^n|0\rangle, \qquad n = 0, 1, 2, \cdots, \qquad b|0\rangle = 0. \tag{28.1.6}$$

They are identified with Landau sites.

In this noncommutative plane a classical quantity $f(\boldsymbol{x})$ depending on the electron position $\boldsymbol{x}$ becomes an operator given by the Weyl ordering,

$$f(\boldsymbol{x}) \Rightarrow W[f] = \frac{1}{(2\pi)^2}\int d^2q d^2x \, e^{iq(x-X)} f(\boldsymbol{x}). \tag{28.1.7}$$

The function $f(\boldsymbol{x})$ is called the symbol of $W[f]$. A product of two Weyl operators $W[f]W[g]$ turns out to be a Weyl operator, whose symbol is denoted as $f(\boldsymbol{x}) \star g(\boldsymbol{x})$.

It is called the star product of $f(\boldsymbol{x})$ and $g(\boldsymbol{x})$. Thus,

$$f(\boldsymbol{x}) \star g(\boldsymbol{x}) \Rightarrow W[f]W[g]. \tag{28.1.8}$$

The QH system provides us with an ideal 2-dimensional world with the built-in *noncommutative geometry*.[16]

Applying the Weyl ordering to the plane wave we find

$$e^{i\boldsymbol{p}\boldsymbol{x}} \Rightarrow e^{i\boldsymbol{p}\boldsymbol{X}} \tag{28.1.9}$$

The plane wave $e^{i\boldsymbol{p}\boldsymbol{x}}$ generates the translation in the ordinary space. The translation turns out to be non-Abelian within the lowest Landau level, and is called the *magnetic translation*,

$$e^{i\boldsymbol{p}\boldsymbol{X}}e^{i\boldsymbol{q}\boldsymbol{X}} = e^{i(\boldsymbol{p}+\boldsymbol{q})\boldsymbol{X}} \exp\left[\frac{i}{2}\ell_B^2 \boldsymbol{p}\wedge\boldsymbol{q}\right], \tag{28.1.10}$$

where $\boldsymbol{p}\wedge\boldsymbol{q} = p_x q_y - p_y q_x$. It yields an algebraic relation,

$$[e^{i\boldsymbol{p}\boldsymbol{X}}, e^{i\boldsymbol{q}\boldsymbol{X}}] = 2ie^{i(\boldsymbol{p}+\boldsymbol{q})\boldsymbol{X}} \exp\left(i\ell_B^2 \frac{\boldsymbol{p}\wedge\boldsymbol{q}}{2}\right), \tag{28.1.11}$$

isomorphic to the W_∞ algebra.[236,84]

In ordinary physics the dynamics arises from the kinetic Hamiltonian, which is just a c-number constant for electrons confined to a single Landau level. It is quite complicated to construct a conventional Lagrangian for the QH system.[459] This is because the dynamics arises from the very nature of noncommutative geometry. The Heisenberg equation of motion is to be calculated by the use of the commutation relation (28.1.4),

$$i\hbar\frac{d}{dt}O = [O, H], \tag{28.1.12}$$

where O is the projected operator and H is the projected Hamiltonian composed solely of the potential terms.

Projected Density Operator

We consider electrons confined to the Nth Landau level, where Fock states are given by $|N, n\rangle$ [see (10.2.17)]. The unique physical variable is the electron density $\rho_N(\boldsymbol{x})$ projected to the Nth Landau level. In the spinless monolayer theory it is given by[1]

$$\rho_N(\boldsymbol{x}) = \psi^\dagger(\boldsymbol{x})\psi(\boldsymbol{x}) \tag{28.1.13}$$

where $\psi(\boldsymbol{x})$ is the field operator describing electrons in the Nth Landau level, and given by

$$\psi(\boldsymbol{x}) = \sum_n \langle \boldsymbol{x}|N, n\rangle c(n) \tag{28.1.14}$$

[1]In Section 10.6, we have used the notation $\psi_N(\boldsymbol{x})$ and $c_N(\boldsymbol{x})$ for operators in the Nth Landau level: See (10.6.15). Here we omit the index N since we only treat the projected quantities.

with $c(n)$ the annihilation operator acting on the Landau site $|N,n\rangle$, $\{c(n), c^\dagger(m)\} = \delta_{mn}$. In the momentum space the *projected density* is given by

$$\rho_N(\boldsymbol{q}) \equiv \int \frac{d^2x}{2\pi} e^{-i\boldsymbol{p}\boldsymbol{x}} \rho(\boldsymbol{x}) = F_N(\boldsymbol{p})\hat{\rho}(\boldsymbol{p}), \tag{28.1.15}$$

where we have defined the *bare density*

$$\hat{\rho}(\boldsymbol{p}) = \frac{1}{2\pi} \sum_{mn} \langle m|e^{-i\boldsymbol{p}\boldsymbol{X}}|n\rangle c^\dagger(m)c(n), \tag{28.1.16}$$

and the *Landau-level form factor*

$$F_N(\boldsymbol{p}) = \langle N|e^{-i\boldsymbol{p}\boldsymbol{R}}|N\rangle \tag{28.1.17}$$

with $\boldsymbol{R}$ the relative coordinate. It reads

$$F_0(\boldsymbol{p}) = e^{-\ell_B^2 \boldsymbol{p}^2/4} \tag{28.1.18}$$

for the lowest Landau level ($N = 0$). It is convenient to construct a formalism based on the bare density $\hat{\rho}(\boldsymbol{p})$ rather than the projected density $\rho_N(\boldsymbol{q})$, though $\rho_N(\boldsymbol{x})$ is the physical density. The difference between $\rho_N(\boldsymbol{p})$ and $\hat{\rho}(\boldsymbol{p})$ is negligible for sufficiently smooth field configurations since $F_N(\boldsymbol{p}) \to 1$ as $\boldsymbol{p} \to 0$.

It follows from (28.1.11) that the bare density satisfies the algebraic relation,[189]

$$[\hat{\rho}(\boldsymbol{p}), \hat{\rho}(\boldsymbol{q})] = \frac{i}{\pi} \hat{\rho}(\boldsymbol{p}+\boldsymbol{q}) \sin\left(\ell_B^2 \frac{\boldsymbol{p}\wedge\boldsymbol{q}}{2}\right). \tag{28.1.19}$$

It determines both the kinematical and dynamical properties of the spinless QH system completely.

W_∞(N) Density Algebra

In the SU(N) theory, the electron field $\Psi(\boldsymbol{x})$ has N isospin components and is given by

$$\psi_\mu(\boldsymbol{x}) = \sum_n \langle \boldsymbol{x}|N,n\rangle c_\mu(n) \tag{28.1.20}$$

in the Nth Landau level,[2] with $\{c_\mu(n), c_\nu^\dagger(m)\} = \delta_{mn}\delta_{\mu\nu}$. The physical variables are the electron density ρ_N and the isospin field $\mathcal{J}_{N;A}$ projected to the Nth Landau level,

$$\rho_N(\boldsymbol{x}) = \Psi^\dagger(\boldsymbol{x})\Psi(\boldsymbol{x}), \qquad \mathcal{J}_{N;A}(\boldsymbol{x}) = \frac{1}{2}\Psi^\dagger(\boldsymbol{x})\lambda_A\Psi(\boldsymbol{x}). \tag{28.1.21}$$

They are summarized into

$$\mathcal{D}_{N;\mu\nu} = \frac{1}{N}\delta_{\mu\nu}\rho_N + (\lambda_A)_{\mu\nu}\mathcal{J}_{N;A}, \tag{28.1.22}$$

where λ_A are the generating matrices of SU(N).

---

[2]No confusion should be made on the index N, though we use the same N for the group index and the Landau level index.

The bare densities $\hat{\rho}$ and $\hat{\mathfrak{I}}_A$ are defined similarly as in (28.1.16) and summarized into the density matrix $\hat{\mathcal{D}}_{\mu\nu}$ as

$$\hat{\mathcal{D}}_{\mu\nu} = \frac{1}{N}\delta_{\mu\nu}\hat{\rho} + (\lambda_A)_{\mu\nu}\hat{\mathfrak{I}}_A. \tag{28.1.23}$$

It is related to the projected density as

$$\mathcal{D}_{N;\mu\nu}(\boldsymbol{q}) = F_N(\boldsymbol{p})\hat{\mathcal{D}}_{\mu\nu}(\boldsymbol{p}), \tag{28.1.24}$$

where $F_N(\boldsymbol{p})$ is the Landau-level form factor (28.1.17).

The density algebra (28.1.19) is enlarged into[151]

$$2\pi[\hat{\mathcal{D}}_{\mu\nu}(\boldsymbol{p}), \hat{\mathcal{D}}_{\sigma\tau}(\boldsymbol{q})] = \delta_{\mu\tau}e^{+\frac{i}{2}\ell_B^2 \boldsymbol{p}\wedge\boldsymbol{q}}\hat{\mathcal{D}}_{\sigma\nu}(\boldsymbol{p}+\boldsymbol{q}) - \delta_{\sigma\nu}e^{-\frac{i}{2}\ell_B^2 \boldsymbol{p}\wedge\boldsymbol{q}}\hat{\mathcal{D}}_{\mu\tau}(\boldsymbol{p}+\boldsymbol{q}), \tag{28.1.25}$$

or[144]

$$[\hat{\rho}(\boldsymbol{p}), \hat{\rho}(\boldsymbol{q})] = \frac{i}{\pi}\hat{\rho}(\boldsymbol{p}+\boldsymbol{q})\sin\left(\ell_B^2\frac{\boldsymbol{p}\wedge\boldsymbol{q}}{2}\right), \tag{28.1.26a}$$

$$[\hat{\mathfrak{I}}_A(\boldsymbol{p}), \hat{\rho}(\boldsymbol{q})] = \frac{i}{\pi}\hat{\mathfrak{I}}_A(\boldsymbol{p}+\boldsymbol{q})\sin\left(\ell_B^2\frac{\boldsymbol{p}\wedge\boldsymbol{q}}{2}\right), \tag{28.1.26b}$$

$$\begin{aligned}[\hat{\mathfrak{I}}_A(\boldsymbol{p}), \hat{\mathfrak{I}}_B(\boldsymbol{q})] = {} & \frac{i}{2\pi}f_{ABC}\hat{\mathfrak{I}}_C(\boldsymbol{p}+\boldsymbol{q})\cos\left(\ell_B^2\frac{\boldsymbol{p}\wedge\boldsymbol{q}}{2}\right) \\ & + \frac{i}{2\pi}d_{ABC}\hat{\mathfrak{I}}_C(\boldsymbol{p}+\boldsymbol{q})\sin\left(\ell_B^2\frac{\boldsymbol{p}\wedge\boldsymbol{q}}{2}\right) \\ & + \frac{i}{2\pi N}\delta_{AB}\hat{\rho}(\boldsymbol{p}+\boldsymbol{q})\sin\left(\ell_B^2\frac{\boldsymbol{p}\wedge\boldsymbol{q}}{2}\right),\end{aligned} \tag{28.1.26c}$$

where f_{ABC} is the structure constants characterizing the SU(N) algebra: See Appendix D. It is referred to as the $W_\infty(N)$ algebra.[149] The isospin field I_A and the density operator ρ become noncommutative within the lowest Landau level. Consequently, the isospin rotation modulates the electron number density.

Noncommutative Constraint

What is observed experimentally is the classical field $\hat{\mathcal{D}}^{\text{cl}}_{\mu\nu}(\boldsymbol{x})$, which is the expectation value of $\hat{\mathcal{D}}_{\mu\nu}(\boldsymbol{x})$ by a Fock state,

$$\hat{\mathcal{D}}^{\text{cl}}_{\mu\nu}(\boldsymbol{x}) = \langle\hat{\mathcal{D}}_{\mu\nu}(\boldsymbol{x})\rangle. \tag{28.1.27}$$

For a wide class of Fock states, noncommutativity manifests itself as a constraint on $\hat{\mathcal{D}}^{\text{cl}}_{\mu\nu}(\boldsymbol{x})$,

$$\sum_{\lambda=1}^{N}\hat{\mathcal{D}}^{\text{cl}}_{\mu\lambda}(\boldsymbol{x})\star\hat{\mathcal{D}}^{\text{cl}}_{\lambda\nu}(\boldsymbol{x}) = \frac{1}{2\pi\ell_B^2}\hat{\mathcal{D}}^{\text{cl}}_{\mu\nu}(\boldsymbol{x}). \tag{28.1.28}$$

To describe SU(N) excitations we resolve this as

$$\hat{\mathcal{D}}^{\text{cl}}_{\mu\nu}(\boldsymbol{x}) = \frac{1}{2\pi\ell_B^2}n_\mu(\boldsymbol{x})\star n^*_\nu(\boldsymbol{x}) \tag{28.1.29}$$

with the aid of the noncommutative CP^{N-1} field $n_\mu(\boldsymbol{x})$ and its complex conjugate $n^*_\mu(\boldsymbol{x})$ subject to the noncommutative normalization condition

$$\sum_{\mu=1}^{N} n^*_\mu(\boldsymbol{x}) \star n_\mu(\boldsymbol{x}) = 1. \tag{28.1.30}$$

Note that the CP^{N-1} field is introduced as a classical field from the beginning in the microscopic theory.

Incompressibility Condition

The symmetry of the QH system is represented by the density algebra. Consequently, the excitation spectrum must be a representation of this algebra. As we have argued in the semiclassical framework, there exist no gapless density fluctuations in QH systems. Accordingly, the ground state is annihilated by all the generators $\hat{\rho}(\boldsymbol{q})$ except for the zero mode,

$$\hat{\rho}(\boldsymbol{q})|\mathrm{g}\rangle_{\mathrm{bulk}} = 2\pi\rho_0\delta(\boldsymbol{q})|\mathrm{g}\rangle_{\mathrm{bulk}}. \tag{28.1.31}$$

It summarizes algebraically the incompressibility of the QH state.[84]

The incompressibility condition is a very special condition that never holds in the presence of both the kinetic and Coulomb energy terms since they do not commute with one another. The system is solely governed by the Coulomb term in each Landau level since the kinetic term is quenched. The ground state is given just by minimizing the Coulomb energy. Only in this circumstance the ground state can be an eigenstate of the density operator as in (28.1.31).

Edge States

Edge excitations must also provide a representation of the density algebra (28.1.19). This is a powerful requirement to categorize all possible edge excitations.[86] As far as sufficiently smooth edge excitations are concerned, the density algebra (28.1.19) is reduced to the Kac-Moody algebra

$$[\hat{\rho}_k, \hat{\rho}_l] = -\frac{k}{2\pi m}\delta(k+l), \tag{28.1.32}$$

where $\hat{\rho}_k$ is the density of edge electrons with the index k the one-dimensional wave vector along the edge. The edge ground state is characterized by

$$\hat{\rho}_k|\mathrm{g}\rangle_{\mathrm{edge}} = 0 \qquad \text{for} \quad k > 0, \tag{28.1.33}$$

while neutral edge excitations are generated by operating $\hat{\rho}_{-k}$ $(k > 0)$ to the ground state $|\mathrm{g}\rangle_{\mathrm{edge}}$. Nonperturbative excitations are given by gapless edge electrons and they are quasiparticles. They constitute charged sectors of the Kac-Moody algebra.

Topological Density

The topological charge density is defined by the formula[152]

$$J_0(\mathbf{x}) = \frac{1}{2\pi\ell_B^2}\sum_\mu [n_\mu^*(\mathbf{x})\star n_\mu(\mathbf{x}) - n_\mu(\mathbf{x})\star n_\mu^*(\mathbf{x})] \tag{28.1.34}$$

in terms of the the noncommutative CP^{N-1} field. It is a remarkable property of a noncommutative soliton that the topological charge density is essentially the electron number density excited around it,

$$J_0(\mathbf{x}) = -\Delta\rho^{\text{cl}}(\mathbf{x}). \tag{28.1.35}$$

Hence a soliton carries necessarily the electron number

$$\Delta N_{\text{e}}^{\text{cl}} = -Q_{\text{sky}}, \tag{28.1.36}$$

where $\Delta N_{\text{e}}^{\text{cl}} = \int d^2x\, \Delta\rho^{\text{cl}}(\mathbf{x})$ and $Q_{\text{sky}} = \int d^2x\, J_0(\mathbf{x})$.

Microscopic Skyrmions

A charged excitation is a skyrmion in QH ferromagnets. Though the skyrmion is originally introduced as a classical solution in the nonlinear sigma model, as reviewed in Sections 7.7 and 7.8, it is possible to consider a quantum mechanical state in QH ferromagnets.[163,327,36]

The simplest skyrmion carries the topological charge $Q_{\text{sky}} = 1$. The skyrmion state $|\mathfrak{S}_{\text{sky}}\rangle$ is constructed[152] as a $W_\infty(N)$-rotated state of a hole state $|\text{h}\rangle$,

$$|\mathfrak{S}_{\text{sky}}\rangle = e^{iW}|\text{h}\rangle, \tag{28.1.37}$$

where W is an element of the $W_\infty(N)$ algebra. The corresponding CP^{N-1} field is given by

$$\mathfrak{n}_\mu = \sum_{n=0}^{\infty} [u_\mu(n)|n\rangle\langle n| + v_\mu(n)|n+1\rangle\langle n|], \tag{28.1.38}$$

where $\mathfrak{n}_\mu$ is the Weyl operator whose symbol is $n_\mu(x)$, $\mathfrak{n}_\mu = W[n_\mu]$, while $u_\mu(n)$ and $v_\mu(n)$ are infinitely many parameters characterizing the skyrmion.

These parameters are determined once the Hamiltonian is given. An analytic solution is obtained for the hardcore model,[308] where

$$u^2(n) = \frac{\omega^2}{n+1+\omega^2}, \quad v^2(n) = \frac{n+1}{n+1+\omega^2}. \tag{28.1.39}$$

This is the condition for the skyrmion to be factorizable. In this model the skyrmion state is constructed analytically as an eigenstate of the Hamiltonian. The noncommutative CP^{N-1} model[307] is also studied. The realistic case with the Coulomb interaction can be discussed by carrying out a numerical analysis.[479]

Projected Coulomb Hamiltonian

We are mainly concerned about physics of electrons confined within the lowest Landau level ($N = 0$) and governed mainly by Coulomb interactions. The Coulomb Hamiltonian is given in terms of the projected density $\rho_N(\boldsymbol{q})$, which we rewrite in terms of the bare density $\hat{\rho}(\boldsymbol{q})$. It is given by

$$H_\mathrm{C} = \pi \int d^2q\, V_\mathrm{D}(\boldsymbol{q})\hat{\rho}(-\boldsymbol{q})\hat{\rho}(\boldsymbol{q}), \tag{28.1.40}$$

where $V_\mathrm{D}(\boldsymbol{q})$ is the effective potential in the lowest Landau level,

$$V_\mathrm{D}(\boldsymbol{q}) \equiv V_0^\mathrm{eff}(\boldsymbol{q}) = V(\boldsymbol{q})F_0(-\boldsymbol{q})F_0(\boldsymbol{q}), \tag{28.1.41}$$

with $F_0(\boldsymbol{q})$ the Landau-level form factor (28.1.17) for the lowest Landau level. It is given by

$$V_\mathrm{D}(\boldsymbol{q}) = \frac{e^2}{4\pi\varepsilon|\boldsymbol{q}|}e^{-\ell_B^2 q^2/2}, \tag{28.1.42}$$

or

$$V_\mathrm{D}(\boldsymbol{x}) = \frac{e^2\sqrt{2\pi}}{8\pi\varepsilon\ell_B}I_0(\boldsymbol{x}^2/4\ell_B^2)e^{-\boldsymbol{x}^2/4\ell_B^2} \tag{28.1.43}$$

in the real space, where $I_0(x)$ is the modified Bessel function. It approaches the ordinary Coulomb potential at large distance, as expected,

$$V_\mathrm{D}(\boldsymbol{x}) \to V(\boldsymbol{x}) = \frac{e^2}{4\pi\varepsilon|\boldsymbol{x}|} \quad \text{as} \quad |\boldsymbol{x}| \to \infty, \tag{28.1.44}$$

but at short distance it does not diverge in contrast to the ordinary Coulomb potential,

$$V_\mathrm{D}(\boldsymbol{x}) \to \frac{e^2\sqrt{2\pi}}{8\pi\varepsilon\ell_B} = \sqrt{\frac{\pi}{2}}E_\mathrm{C}^0 \quad \text{as} \quad |\boldsymbol{x}| \to 0. \tag{28.1.45}$$

This is physically reasonable because an electron cannot be localized to a point within the lowest Landau level.

Exchange Interaction

The Coulomb Hamiltonian (28.1.40) is rewritten into an entirely different form,

$$H_\mathrm{X} = -\pi \int d^2p\, V_\mathrm{X}(\boldsymbol{p}) \left[\sum_{A=1}^{N^2-1} \hat{\mathfrak{I}}_A(-\boldsymbol{p})\hat{\mathfrak{I}}_A(\boldsymbol{p}) + \frac{1}{2N}\hat{\rho}(-\boldsymbol{p})\hat{\rho}(\boldsymbol{p}) \right], \tag{28.1.46}$$

where the exchange potential $V_\mathrm{X}(\boldsymbol{p})$ is defined by

$$V_\mathrm{X}(\boldsymbol{p}) = \frac{\ell_B^2}{\pi}\int d^2k\, e^{-i\ell_B^2 \boldsymbol{p}\wedge\boldsymbol{k}}V_\mathrm{D}(\boldsymbol{k}). \tag{28.1.47}$$

It has a typical property of the exchange interaction with a short-range potential $V_X(\boldsymbol{x})$. Two Hamiltonians (28.1.40) and (28.1.46) are equivalent as the microscopic Hamiltonian.

Decomposition Formula

Though these two Hamiltonians are equivalent as quantum mechanical ones, $H_C = H_X$, they are not when they are regarded as the corresponding classical Hamiltonians. For instance, let us consider two well-separated charged excitations. There is a long-range Coulomb interaction $V_D(\boldsymbol{x})$ between them, but not a short-range exchange interaction $V_X(\boldsymbol{x})$. Hence, $H_D^{cl} \neq H_X^{cl}$. They describe the direct Coulomb energy and the exchange Coulomb energy of charged objects, respectively. The total energy is simply the sum of them,[151]

$$\langle H_C \rangle = H_D^{cl} + H_X^{cl}, \tag{28.1.48}$$

which we call the decomposition formula. Here H_D^{cl} and H_X^{cl} are the Hamiltonians of the direct-interaction form (28.1.40) and of the exchange-interaction form (28.1.46), where the density operators $\hat{\rho}(\boldsymbol{x})$ and $\hat{\mathfrak{I}}_A(\boldsymbol{x})$ are replaced by the classical ones $\hat{\rho}^{cl}(\boldsymbol{x})$ and $\hat{\mathfrak{I}}_A^{cl}(\boldsymbol{x})$.

Spontaneous Symmetry Breaking

Since the Coulomb Hamiltonian (28.1.40) does not involve isospin variables, it seems that the energy of a state is independent of isospin orientations. This is not the case because of the decomposition formula (28.1.48), according to which the energy consists of the direct and exchange ones. Though the direct energy does not involves isospins just as the original Hamiltonian (28.1.40), the exchange energy (28.1.46) does. The exchange energy is minimized when all spins are polarized into one arbitrary direction, leading to a spontaneous breaking of the SU(N) symmetry.

Classical Equation of Motion

The Heisenberg equation of motion (28.1.12) for the density $\hat{\mathfrak{D}}_{\mu\nu}(\boldsymbol{x})$ is

$$i\hbar \frac{d}{dt} \hat{\mathfrak{D}}_{\mu\nu}(\boldsymbol{x}) = [\hat{\mathfrak{D}}_{\mu\nu}(\boldsymbol{x}), H]. \tag{28.1.49}$$

The total Hamiltonian H consists of the Coulomb term H_C and the rest term H_{rest},

$$H = H_C + H_{rest}, \tag{28.1.50}$$

where H_{rest} stands for the Zeeman term in the monolayer system and additionally the tunneling and bias terms in the bilayer system.

The classical equation of motion is derived by taking the expectation value of the Heisenberg equation of motion,

$$i\hbar \frac{d}{dt} \hat{\mathfrak{D}}_{\mu\nu}^{cl}(\boldsymbol{x}) = \langle \hat{\mathfrak{D}}_{\mu\nu}(\boldsymbol{p}) H \rangle - \langle H \hat{\mathfrak{D}}_{\mu\nu}(\boldsymbol{p}) \rangle. \tag{28.1.51}$$

We can verify that

$$\frac{d}{dt}\hat{\mathcal{D}}^{\text{cl}}_{\mu\nu}(\boldsymbol{x}) = [\hat{\mathcal{D}}^{\text{cl}}_{\mu\nu}(\boldsymbol{x}), H^{\text{cl}}]_{\text{PB}}, \tag{28.1.52}$$

where $H^{\text{cl}} = \langle H \rangle$ is the classical Hamiltonian, provided the classical density is endowed with the Poisson structure

$$2\pi i\hbar[\hat{\mathcal{D}}^{\text{cl}}_{\mu\nu}(\boldsymbol{p}), \hat{\mathcal{D}}^{\text{cl}}_{\sigma\tau}(\boldsymbol{q})]_{\text{PB}} = \delta_{\mu\tau}e^{+\frac{i}{2}\ell_B^2\boldsymbol{p}\wedge\boldsymbol{q}}\hat{\mathcal{D}}^{\text{cl}}_{\sigma\tau}(\boldsymbol{p}+\boldsymbol{q}) - \delta_{\sigma\nu}e^{-\frac{i}{2}\ell_B^2\boldsymbol{p}\wedge\boldsymbol{q}}\hat{\mathcal{D}}^{\text{cl}}_{\mu\tau}(\boldsymbol{p}+\boldsymbol{q}), \tag{28.1.53}$$

as corresponds to the $W_\infty(N)$ algebra (28.1.25). The classical Coulomb energy consists of the direct and exchange energies, $\langle H_{\text{C}} \rangle = H^{\text{cl}}_{\text{D}} + H^{\text{cl}}_{\text{X}}$, and we obtain

$$H^{\text{cl}} = H^{\text{cl}}_{\text{D}} + H^{\text{cl}}_{\text{X}} + H^{\text{cl}}_{\text{rest}} \tag{28.1.54}$$

from the total Hamiltonian (28.1.50).

Effective Hamiltonian

We study the classical Hamiltonian H^{cl} for sufficiently smooth field configurations. Though we have so far considered the SU(N) field, we are mainly concerned about the spin SU(2) field $\mathcal{S}_a(\boldsymbol{x})$ and the pseudospin SU(2) field $\mathcal{P}_a(\boldsymbol{x})$.

We introduce new notations,

$$\mathcal{D}_{\mu\nu}(\boldsymbol{q}) = \frac{1}{\rho_\Phi}\hat{\mathcal{D}}^{\text{cl}}_{\mu\nu}(\boldsymbol{q}), \qquad \boldsymbol{\mathcal{S}}(\boldsymbol{q}) = \frac{1}{\rho_\Phi}\hat{\boldsymbol{\mathcal{S}}}^{\text{cl}}(\boldsymbol{q}), \qquad \boldsymbol{\mathcal{P}}(\boldsymbol{q}) = \frac{1}{\rho_\Phi}\hat{\boldsymbol{\mathcal{P}}}^{\text{cl}}(\boldsymbol{q}). \tag{28.1.55}$$

Making the derivative expansion of the classical Hamiltonian H^{cl}, we derive the effective Hamiltonian appropriate for the description of low-energy phenomena.

In the instance of the monolayer system the Coulomb exchange interaction H^{cl}_{X} yields the SU(2) nonlinear sigma model[144,351]

$$H_{\text{eff}} = 2J_s \int d^2x\, [\partial_k \boldsymbol{\mathcal{S}}(\boldsymbol{x})]^2. \tag{28.1.56}$$

We have used it to analyze spin waves as perturbative excitations. We have also applied it to study skyrmion excitations.

In the instance of the spinless bilayer system the Coulomb exchange interaction H^{cl}_{X} yields the anisotropic O(3) nonlinear sigma model

$$H_{\text{eff}} = 2J_s^d \sum_{a=x,y} \int d^2x\, [\partial_k \mathcal{P}_a(\boldsymbol{x})]^2 + 2J_s \int d^2x\, [\partial_k \mathcal{P}_z(\boldsymbol{x})]^2. \tag{28.1.57}$$

We have used it to analyze pseudospin waves as well as skyrmion excitations. We have also applied it to study Josephson-like phenomena. These are typical examples of quantum coherent phenomena originating in noncommutative geometry.

Electric Currents

The electric current on the QH fluid is controversial. The Nöther current is given by $J_k = (e/M)\psi^\dagger P_k \psi$ in the ordinary theory, but it vanishes identically on the state $|\mathfrak{S}\rangle$ obeying the LLL condition (28.1.3). It seems that no current could flow on the QH fluid. This is not surprising because the kinetic Hamiltonian is quenched in each Landau level. The current flows due to the noncommutative nature of the coordinate.

The Nöther current is introduced to guarantee the conservation of the electric charge. In noncommutative space the continuity equation is given by

$$-e\frac{d}{dt}\hat{\rho}(\boldsymbol{x}) = \partial_i \hat{\mathcal{J}}_i(\boldsymbol{x}), \tag{28.1.58}$$

where $-e\hat{\rho}(\boldsymbol{x})$ is the electric charge density. Using the Heisenberg equation of motion (28.1.12), we find

$$\partial_i \hat{\mathcal{J}}_i(\boldsymbol{x}) = -e\frac{1}{i\hbar}[\hat{\rho}(\boldsymbol{x}), H]. \tag{28.1.59}$$

Hence the current is constructed by integrating this equation, after calculating the right-hand side with the use of the $W_\infty(N)$ algebra (28.1.25).

The observable quantity is the classical current, which we denote as

$$\mathcal{J}_i(\boldsymbol{x}) = \langle \hat{\mathcal{J}}_i(\boldsymbol{x}) \rangle. \tag{28.1.60}$$

We take the expectation value of the continuity equation (28.1.59) and use the classical equation of motion (28.1.52),

$$\partial_i \mathcal{J}_i(\boldsymbol{x}) = -e[\hat{\rho}^{\text{cl}}(\boldsymbol{x}), H^{\text{cl}}]_{\text{PB}}, \tag{28.1.61}$$

where H^{cl} is given by (28.1.54) and the Poisson structure is defined by (28.1.53). This leads to the standard formula for the Hall current in the monolayer QH system,

$$\mathcal{J}_i(\boldsymbol{x}) = \frac{e^2\ell_B^2}{\hbar}\varepsilon_{ij}E_j\rho_0. \tag{28.1.62}$$

On the other hand, there arises the phase current in addition to the standard Hall current in the bilayer QH system with interlayer phase coherence,

$$\mathcal{J}_i^{\text{f}}(\boldsymbol{x}) = \frac{4J_s^d}{\hbar}(\partial_i\mathcal{P}_x \cdot \mathcal{P}_y - \partial_i\mathcal{P}_y \cdot \mathcal{P}_x) + \frac{e^2\ell_B^2}{\hbar}\varepsilon_{ij}E_j\rho_0^{\text{f}}, \tag{28.1.63a}$$

$$\mathcal{J}_i^{\text{b}}(\boldsymbol{x}) = -\frac{4J_s^d}{\hbar}(\partial_i\mathcal{P}_x \cdot \mathcal{P}_y - \partial_i\mathcal{P}_y \cdot \mathcal{P}_x) + \frac{e^2\ell_B^2}{\hbar}\varepsilon_{ij}E_j\rho_0^{\text{b}}, \tag{28.1.63b}$$

where $\mathcal{P}_a$ is the normalized pseudospin field. It also leads to the tunneling current between the two layers,

$$\mathcal{J}_z(\boldsymbol{x}) = -\frac{ed\rho_0}{\hbar}\Delta_{\text{SAS}}\mathcal{P}_y(\boldsymbol{x}). \tag{28.1.64}$$

The tunneling current is anticipated to be the Josephson current of single electrons.[140]

CHAPTER 29

NONCOMMUTATIVE GEOMETRY

The QH system provides us with a simple world of noncommutative geometry. In such a world it is the Weyl operator that is fundamental. When the electron possesses the SU(N) isospin index, the physical variables are the electron number density and the isospin SU(N) density. They form an SU(N) extension of the W_∞ algebra, in which they are entangled in an intrinsic way. Due to this entanglement a topological soliton induces an excitation of the electron number density around it. It carries the electric charge proportional to its topological charge.

29.1 NONCOMMUTATIVE COORDINATE

We recapitulate the electron system in a single Landau level in the context of noncommutative geometry. The position of an electron confined to a single Landau level is described by the guiding center $\boldsymbol{X} = (X, Y)$. Since X and Y do not commute,

$$[X, Y] = -i\ell_B^2, \tag{29.1.1}$$

the QH system provides us with a simple example of noncommutative geometry.

To represent the noncommutative algebra (29.1.1) it is most convenient to use the holomorphic basis. It is constructed by introducing the operators

$$b = \frac{1}{\sqrt{2}\ell_B}(X - iY), \qquad b^\dagger = \frac{1}{\sqrt{2}\ell_B}(X + iY), \tag{29.1.2}$$

obeying $[b, b^\dagger] = 1$, together with the Fock states

$$|n\rangle = \frac{1}{\sqrt{n!}}(b^\dagger)^n|0\rangle, \qquad n = 0, 1, 2, \cdots, \qquad b|0\rangle = 0. \tag{29.1.3}$$

The completeness condition

$$\sum_n |n\rangle\langle n| = 1 \tag{29.1.4}$$

holds within the Fock space. The Fock state $|n\rangle$ is identified with the Landau site in the symmetric gauge.

The coherent-state basis is also useful. It is given by a set of states

$$|\hat{x}\rangle = \sqrt{\frac{1}{2\pi\ell_B^2}} e^{-|z|^2/4} e^{z^* b^\dagger/\sqrt{2}} |0\rangle$$
$$= \sqrt{\frac{1}{2\pi\ell_B^2}} \sum_{n=0}^{\infty} \sqrt{\frac{1}{2^n n!}} \left(\frac{x - iy}{\ell_B}\right)^n e^{-r^2/4\ell_B^2} |n\rangle. \tag{29.1.5}$$

It is a coherent state of the annihilation operator b,

$$b|\hat{x}\rangle = \frac{(x - iy)}{\sqrt{2}\ell_B} |\hat{x}\rangle, \tag{29.1.6}$$

where $z = (x + iy)/\ell_B$ and $r^2 = x^2 + y^2$. According to (29.1.5), the state $|\hat{x}\rangle$ is the projection of the eigenstate $|\boldsymbol{x}\rangle$ of the position operator $\boldsymbol{x}$,

$$|\boldsymbol{x}\rangle \to |\hat{x}\rangle \equiv \sum_{n=0}^{\infty} |n\rangle\langle n|\boldsymbol{x}\rangle. \tag{29.1.7}$$

The completeness condition

$$\int d^2x \, |\hat{x}\rangle\langle\hat{x}| = 1 \tag{29.1.8}$$

follows from (29.1.4). The wave function is derived from (29.1.5) as

$$\langle\boldsymbol{x}|n\rangle = \langle\hat{x}|n\rangle = \sqrt{\frac{1}{2^{n+1}\pi\ell_B^2 n!}} \left(\frac{x + iy}{\ell_B}\right)^n e^{-r^2/4\ell_B^2}. \tag{29.1.9}$$

We also note

$$\langle\hat{x}|\hat{y}\rangle = \frac{1}{2\pi\ell_B^2} \exp\left(i\frac{\boldsymbol{x}\wedge\boldsymbol{y}}{2\ell_B^2}\right) \exp\left(-\frac{|\boldsymbol{x} - \boldsymbol{y}|^2}{4\ell_B^2}\right). \tag{29.1.10}$$

We have explained these formulas in Section 10.6.

29.2 Weyl Operator and Symbol

There exists a procedure (the Weyl prescription[495,496,510]) to construct a non-commutative theory with the coordinate $\boldsymbol{X} = (X, Y)$ from a commutative theory with the coordinate $\boldsymbol{x} = (x, y)$. From an arbitrary function $f(\boldsymbol{x})$, the *Weyl operator* $W[f]$ is defined by

$$W[f] = \frac{1}{(2\pi)^2} \int d^2q d^2x \, e^{-i\boldsymbol{q}(\boldsymbol{x}-\boldsymbol{X})} f(\boldsymbol{x}) = \frac{1}{2\pi} \int d^2q \, e^{i\boldsymbol{q}\boldsymbol{X}} f(\boldsymbol{q}), \tag{29.2.1}$$

where $f(\boldsymbol{q})$ is the Fourier transform of $f(\boldsymbol{x})$. The inversion formula is

$$f(\boldsymbol{q}) = \ell_B^2 \sum_{n=0}^{\infty} \langle n|e^{-i\boldsymbol{q}\boldsymbol{X}} W[f]|n\rangle, \tag{29.2.2}$$

as we shall see later in (29.2.9).

Taking the plane wave $f(\boldsymbol{x}) = e^{i\boldsymbol{p}\boldsymbol{x}}$ in (29.2.1), we find

$$W[e^{i\boldsymbol{p}\boldsymbol{x}}] = \frac{1}{(2\pi)^2}\int d^2q d^2x\, e^{i(\boldsymbol{p}-\boldsymbol{q})\boldsymbol{x}} e^{i\boldsymbol{q}\boldsymbol{X}} = e^{i\boldsymbol{p}\boldsymbol{X}}. \tag{29.2.3}$$

The Weyl operator $e^{i\boldsymbol{p}\boldsymbol{X}}$ generates the projective translation group,

$$e^{i\boldsymbol{p}\boldsymbol{X}} e^{i\boldsymbol{q}\boldsymbol{X}} = e^{i(\boldsymbol{p}+\boldsymbol{q})\boldsymbol{X}} \exp\left[\frac{i}{2}\ell_B^2 \boldsymbol{p}\wedge\boldsymbol{q}\right], \tag{29.2.4}$$

where $\boldsymbol{p}\wedge\boldsymbol{q} \equiv \varepsilon_{ij}p_i q_j = p_x q_y - p_y q_x$. It is also called the magnetic translation group, about which we shall explain in Section 29.3.

There holds a basic relation

$$\mathrm{Tr}\left(e^{i\boldsymbol{p}\boldsymbol{X}}\right) \equiv \sum_{n=0}^{\infty} \langle n|e^{i\boldsymbol{p}\boldsymbol{X}}|n\rangle = \frac{2\pi}{\ell_B^2}\delta(\boldsymbol{p}), \tag{29.2.5}$$

which we shall drive later [see (29.2.31)]. Here, "Tr" is defined by

$$\mathrm{Tr}\left(W[f]\right) = \sum_n \langle n|W[f]|n\rangle \tag{29.2.6}$$

for any Weyl operator $W[f]$ in the holomorphic basis (29.1.3).

We derive the inversion formula of the Weyl operator. We calculate

$$\int d^2p\, e^{i\boldsymbol{p}\boldsymbol{x}}\mathrm{Tr}\left(e^{-i\boldsymbol{p}\boldsymbol{X}}W[f]\right) = \frac{1}{(2\pi)^2}\int d^2p d^2q d^2y\, \mathrm{Tr}\left(e^{-i\boldsymbol{p}\boldsymbol{X}}e^{i\boldsymbol{q}\boldsymbol{X}}\right) e^{i\boldsymbol{p}\boldsymbol{x}-i\boldsymbol{q}\boldsymbol{y}} f(\boldsymbol{y}), \tag{29.2.7}$$

where we have substituted (29.2.1). Then, we use (29.2.4) and (29.2.5) to find

$$\begin{aligned}
&\int d^2p\, e^{i\boldsymbol{p}\boldsymbol{x}}\mathrm{Tr}\left(e^{-i\boldsymbol{p}\boldsymbol{X}}W[f]\right) \\
&= \frac{1}{(2\pi)^2}\int d^2p d^2q d^2y\, \exp\left[-\frac{i}{2}\ell_B^2\boldsymbol{p}\wedge\boldsymbol{q}\right]\mathrm{Tr}\left(e^{i(\boldsymbol{q}-\boldsymbol{p})\boldsymbol{X}}\right) e^{i\boldsymbol{p}\boldsymbol{x}-i\boldsymbol{q}\boldsymbol{y}} f(\boldsymbol{y}) \\
&= \frac{1}{2\pi\ell_B^2}\int d^2p d^2y\, e^{i\boldsymbol{p}(\boldsymbol{x}-\boldsymbol{y})} f(\boldsymbol{y}) = \frac{1}{\ell_B^2}\int d^2p\, e^{i\boldsymbol{p}\boldsymbol{x}} f(\boldsymbol{p}) = \frac{2\pi}{\ell_B^2} f(\boldsymbol{x}).
\end{aligned} \tag{29.2.8}$$

Hence

$$f(\boldsymbol{x}) = \frac{\ell_B^2}{2\pi}\int d^2p\, e^{i\boldsymbol{p}\boldsymbol{x}}\mathrm{Tr}\left(e^{-i\boldsymbol{p}\boldsymbol{X}}W[f]\right), \tag{29.2.9}$$

and

$$f(\boldsymbol{p}) = \ell_B^2 \mathrm{Tr}\left(e^{-i\boldsymbol{p}\boldsymbol{X}}W[f]\right). \tag{29.2.10}$$

This is the inversion relation. It follows from (29.2.9) that

$$\int d^2x\, f(\boldsymbol{x}) = 2\pi\ell_B^2 \mathrm{Tr}\left(W[f]\right). \tag{29.2.11}$$

There is a one-to-one mapping between a function f and a Weyl operator $W[f]$. The c-number function f is called the *symbol* of $W[f]$.

Moyal Product

There is a standard way of representing a product of two Weyl operators by a symbol. We denotes the symbol of $W[f]W[g]$ by $f(\boldsymbol{x})\star g(\boldsymbol{x})$,

$$W[h] = W[f]W[g] \qquad \Longleftrightarrow \qquad h(\boldsymbol{x}) = f(\boldsymbol{x})\star g(\boldsymbol{x}). \tag{29.2.12}$$

The function $h(\boldsymbol{x})$ is called the Moyal product or the star product[356,170,460] of $f(\boldsymbol{x})$ and $g(\boldsymbol{x})$. It is explicitly constructed as follows.

The magnetic translation (29.2.4) implies that the Moyal product of two plane waves is a plane wave. Since $e^{i\boldsymbol{p}\boldsymbol{x}}\star e^{i\boldsymbol{q}\boldsymbol{x}}$ is the symbol of $e^{i\boldsymbol{p}\boldsymbol{X}}e^{i\boldsymbol{q}\boldsymbol{X}}$ by definition, applying the inversion formula (29.2.9) we obtain

$$e^{i\boldsymbol{p}\boldsymbol{x}}\star e^{i\boldsymbol{q}\boldsymbol{x}} = \frac{\ell_B^2}{2\pi}\int d^2k\, e^{i\boldsymbol{k}\boldsymbol{x}}\mathrm{Tr}\left(e^{-i\boldsymbol{k}\boldsymbol{X}}e^{i\boldsymbol{p}\boldsymbol{X}}e^{i\boldsymbol{q}\boldsymbol{X}}\right) = \exp\left(\frac{i}{2}\ell_B^2\boldsymbol{p}\wedge\boldsymbol{q}\right)e^{i(\boldsymbol{p}+\boldsymbol{q})\boldsymbol{x}}, \tag{29.2.13}$$

where use was made of (29.2.4). In general we have

$$\begin{aligned} f(\boldsymbol{x})\star g(\boldsymbol{x}) &= \frac{1}{(2\pi)^2}\int d^2p d^2q\, f(\boldsymbol{p})g(\boldsymbol{q})e^{i\boldsymbol{p}\boldsymbol{x}}\star e^{i\boldsymbol{q}\boldsymbol{x}} \\ &= \frac{1}{(2\pi)^2}\int d^2p d^2q\, \exp\left(\frac{i}{2}\ell_B^2\boldsymbol{p}\wedge\boldsymbol{q}\right)e^{i(\boldsymbol{p}+\boldsymbol{q})\boldsymbol{x}}f(\boldsymbol{p})g(\boldsymbol{q}). \end{aligned} \tag{29.2.14}$$

We may rewrite this as

$$\begin{aligned} f(\boldsymbol{x})\star g(\boldsymbol{x}) &= \frac{1}{(2\pi)^2}\lim_{\boldsymbol{y}\to\boldsymbol{x}}\int d^2p d^2q\, \exp\left(\frac{i}{2}\ell_B^2\boldsymbol{p}\wedge\boldsymbol{q}\right)e^{i(\boldsymbol{p}\boldsymbol{x}+\boldsymbol{q}\boldsymbol{y})}f(\boldsymbol{p})g(\boldsymbol{q}) \\ &= \lim_{\boldsymbol{y}\to\boldsymbol{x}}\exp\left(-\frac{i}{2}\ell_B^2\nabla_x\wedge\nabla_y\right)f(\boldsymbol{x})g(\boldsymbol{y}), \end{aligned} \tag{29.2.15}$$

where $\nabla_x\wedge\nabla_y \equiv \varepsilon_{ij}\partial_i^x\partial_j^y$. This defines the Moyal product $f\star g$ explicitly. It follows from (29.2.14) that

$$\int d^2x\, f(\boldsymbol{x})\star g(\boldsymbol{x}) = \int d^2p d^2q\, \delta(\boldsymbol{p}+\boldsymbol{q})\exp\left(\frac{i}{2}\ell_B^2\boldsymbol{p}\wedge\boldsymbol{q}\right)f(\boldsymbol{p})g(\boldsymbol{q}) = \int d^2x\, f(\boldsymbol{x})g(\boldsymbol{x}). \tag{29.2.16}$$

Namely the integral of the Moyal product of two functions is equal to that of the ordinary product.

Matrix Representation

Using the completeness condition (29.1.4) we can expand a Weyl operator $W[f]$ in terms of the basis $|m\rangle\langle n|$,

$$W[f] = \sum_{mn} f_{mn}|m\rangle\langle n|, \tag{29.2.17}$$

where

$$f_{mn} = \langle m|W[f]|n\rangle = \frac{1}{2\pi}\int d^2q\, \langle m|e^{i\boldsymbol{q}\boldsymbol{X}}|n\rangle f(\boldsymbol{q}). \tag{29.2.18}$$

Thus the operator $W[f]$ may be represented by a matrix f_{mn}. The product of two Weyl operators $W[f]$ and $W[g]$ corresponds to a matrix multiplication,

$$W[h] = W[f]W[g] \quad \Longleftrightarrow \quad h_{mn} = \sum_j f_{mj} g_{jn}, \tag{29.2.19}$$

where $f_{mn} = \langle m|W[f]|n\rangle$, $g_{mn} = \langle m|W[g]|n\rangle$ and $h_{mn} = \langle m|W[h]|n\rangle$.

The inversion formula of $W[f]$ follows from (29.2.10) as

$$f(\boldsymbol{q}) = \ell_B^2 \sum_n \langle n|e^{-iqX} W[f]|n\rangle = \ell_B^2 \sum_{mn} \langle n|e^{-iqX}|m\rangle f_{mn}, \tag{29.2.20}$$

which enables us to construct $f(\boldsymbol{q})$ from the matrix f_{mn}.

Comparing the coefficients of f_{mn} in (29.2.17) and (29.2.20), we establish the correspondence for the basis $|m\rangle\langle n|$,

$$W(\Xi_{mn}) \equiv |m\rangle\langle n| \quad \Longleftrightarrow \quad \Xi_{mn}(\boldsymbol{q}) \equiv \ell_B^2 \langle n|e^{-iqX}|m\rangle. \tag{29.2.21}$$

The symbol reads[224]

$$\Xi_{mn}(\boldsymbol{x}) = \frac{\ell_B^2}{2\pi} \int d^2q \, e^{iqx} \langle n|e^{-iqX}|m\rangle = 2^{\frac{m-n}{2}+1} z^{m-n} e^{-|z|^2} \frac{(-1)^n \sqrt{n!}}{\sqrt{m!}} L_n^{m-n}(2|z|^2) \tag{29.2.22}$$

in the real space.

Derivative

We show that the Weyl operator of the derivative $\partial_i f$ is

$$W[\partial_i f] = -\frac{i}{\ell_B^2} \varepsilon_{ij} [X_j, W[f]]. \tag{29.2.23}$$

According to the definition (29.2.1), we have

$$W[\partial_i f] = \frac{1}{(2\pi)^2} \int d^2q d^2x \, e^{-iq(x-X)} \partial_i f(\boldsymbol{x}) = \frac{i}{(2\pi)^2} \int d^2q \, q_i e^{iqX} f(\boldsymbol{q}). \tag{29.2.24}$$

Using the Hausdorff formula together with the algebra (29.1.1) we obtain

$$e^{iqX} = e^{iq_xX+iq_yY} = e^{-[iq_xX, iq_yY]/2} e^{iq_xX} e^{iq_yY} = e^{-i\ell_B^2 q_x q_y/2} e^{iq_xX} e^{iq_yY}. \tag{29.2.25}$$

Thus

$$\partial_{q_x} e^{iqX} = \left(-\frac{i}{2}\ell_B^2 q_y + iX\right) e^{iqX}. \tag{29.2.26}$$

Similarly, from

$$e^{iqX} = e^{i\ell_B^2 q_x q_y/2} e^{iq_yY} e^{iq_xX}, \tag{29.2.27}$$

we find

$$\partial_{q_x} e^{iqX} = e^{iqX} \left(\frac{i}{2}\ell_B^2 q_y + iX\right). \tag{29.2.28}$$

Equating (29.2.26) and (29.2.28) we find

$$q_i e^{iqX} = -\frac{1}{\ell_B^2}\varepsilon_{ij}[X_j, e^{iqX}], \tag{29.2.29}$$

where we have combined a similar formula arising from $\partial_{q_y} e^{iqX}$. Substituting this into (29.2.24) we obtain

$$W[\partial_i f] = -\frac{i}{(2\pi)^2 \ell_B^2}\varepsilon_{ij}\int d^2q\,[X_j, e^{iqX}]f(\boldsymbol{q}) = -\frac{i}{\ell_B^2}\varepsilon_{ij}[X_j, W[f]]. \tag{29.2.30}$$

This is the formula (29.2.23) for the derivative.

Some Basic Formulas

We present two formulas,

$$\mathrm{Tr}\left(e^{ipX}\right) \equiv \sum_{n=0}^{\infty}\langle n|e^{ipX}|n\rangle = \frac{2\pi}{\ell_B^2}\delta(\boldsymbol{p}), \tag{29.2.31}$$

$$\int d^2p\,\langle m|e^{-ipX}|n\rangle\langle i|e^{ipX}|j\rangle = \frac{2\pi}{\ell_B^2}\delta_{ni}\delta_{mj}, \tag{29.2.32}$$

which play basic roles in noncommutative geometry.

A simplest proof of (29.2.31) is given in a representation of the noncommutativity, $[X, Y] = -i\ell_B^2$, in which the coordinate X is diagonalized,

$$X|x\rangle = x|x\rangle. \tag{29.2.33}$$

The noncommutativity is represented by setting

$$Y = i\ell_B^2\frac{\partial}{\partial X}. \tag{29.2.34}$$

The merit of this representation is that the orthonormal completeness condition holds within the LLL,

$$\langle x'|x\rangle = \delta(x - x'), \qquad \int dx\,|x\rangle\langle x| = 1. \tag{29.2.35}$$

Since Y is a shifting operator it is easy to show

$$e^{ipX}|x\rangle = \exp[-\frac{i}{2}\ell_B^2 p_x p_y]\exp[ip_x x]\,|x + \ell_B^2 p_y\rangle. \tag{29.2.36}$$

Hence,

$$\langle x'|e^{ipX}|x\rangle = \exp[ip_x x + \frac{i}{2}\ell_B^2 p_x p_y]\delta(x - x' + \ell_B^2 p_y) \tag{29.2.37a}$$

$$= \exp[ip_x x' - \frac{i}{2}\ell_B^2 p_x p_y]\delta(x - x' + \ell_B^2 p_y). \tag{29.2.37b}$$

We set $x' = x$ and integrate over it,

$$\int dx \, \langle x|e^{ipX}|x\rangle = \frac{2\pi}{\ell_B^2}\delta(\boldsymbol{p}), \tag{29.2.38}$$

from which (29.2.31) follows with the use of the orthonormal completeness condition (29.2.35).

To derive (29.2.32) we next study

$$A_{mnij} \equiv \int d^2p \, \langle m|e^{-ipX}|n\rangle\langle i|e^{ipX}|j\rangle. \tag{29.2.39}$$

Substituting $\int dx \, |x\rangle\langle x| = 1$, we find

$$A_{mnij} = \int dx_m dx_n dx_i dx_j \int d^2p \, \langle m|x_m\rangle\langle x_n|n\rangle\langle i|x_i\rangle\langle x_j|j\rangle\langle x_m|e^{-ipX}|x_n\rangle\langle x_i|e^{ipX}|x_j\rangle. \tag{29.2.40}$$

We use (29.2.37a) and (29.2.37b),

$$\begin{aligned} A_{mnij} &= \int dx_m dx_n dx_i dx_j \int dp_x dp_y \, \langle m|x_m\rangle\langle x_n|n\rangle\langle i|x_i\rangle\langle x_j|j\rangle \\ &\quad \times \exp\left[ip_x(x_j - x_m)\right]\delta(x_m - x_n + \ell_B^2 p_y)\delta(x_i - x_j - \ell_B^2 p_y) \\ &= \frac{2\pi}{\ell_B^2}\int dx_m dx_n \langle m|x_m\rangle\langle x_m|j\rangle\langle i|x_n\rangle\langle x_n|n\rangle = \frac{2\pi}{\ell_B^2}\delta_{ni}\delta_{mj}, \end{aligned} \tag{29.2.41}$$

which is (29.2.32).

It is instructive to present another proof of (29.2.31) by using the representation in accord with (10.3.5b),

$$X = \frac{1}{2}x - i\ell_B^2\frac{\partial}{\partial y}, \qquad Y = \frac{1}{2}y + i\ell_B^2\frac{\partial}{\partial x}. \tag{29.2.42}$$

Here, both X and Y are shifting operators,

$$e^{ipX}|x, y\rangle = e^{ipx/2}|x + \ell_B^2 p_y, y - \ell_B^2 p_x\rangle, \tag{29.2.43}$$

where we have written $|\hat{\boldsymbol{x}}\rangle \equiv |x, y\rangle$ for clarity. Thus,

$$\langle x, y|e^{ipX}|x', y'\rangle = e^{ipx/2}\langle x, y|x' + \ell_B^2 p_y, y' - \ell_B^2 p_x\rangle, \tag{29.2.44}$$

or

$$\langle \boldsymbol{x}|e^{ipX}|\boldsymbol{x}'\rangle = \frac{1}{2\pi\ell_B^2}e^{ipx}\exp\left(i\frac{\boldsymbol{x}\wedge\boldsymbol{x}'}{2\ell_B^2}\right)\exp\left(-\frac{|x - x' + \ell_B^2 p_y|^2 + |y - y' - \ell_B^2 p_x|^2}{4\ell_B^2}\right), \tag{29.2.45}$$

where the use was made of (29.1.10). In particular, the diagonal element is

$$\langle \boldsymbol{x}|e^{ipX}|\boldsymbol{x}\rangle = \frac{1}{2\pi\ell_B^2}e^{ipx}e^{-\ell_B^2 p^2/4}, \tag{29.2.46}$$

from which (29.2.31) follows trivially. We can prove (29.2.32) also in this representation though slightly complicated.

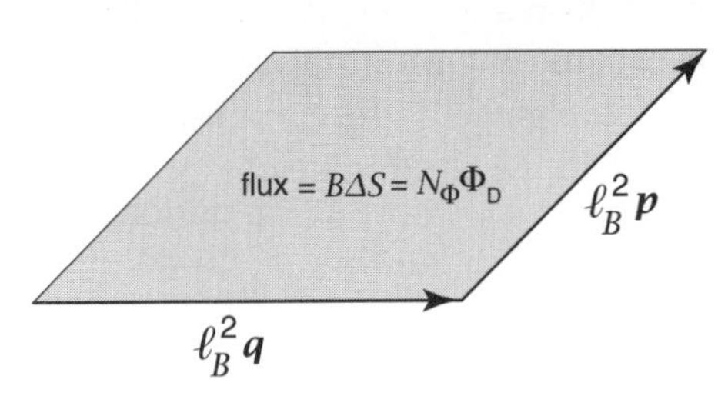

Fig. 29.1 The operator e^{-ipX} translates the coordinate x_j by $\ell_B^2 \varepsilon_{jk} p_k$. Two successive operations not only translate a point but also generate a phase factor as in (29.3.2). Four successive operations on the left hand side of (29.3.12) draw a parallelogram spanned two vectors $\ell_B^2 q$ and $\ell_B^2 p$. They leave the Aharonov-Bohm phase as on the right hand side of (29.3.12).

29.3 Magnetic Translation

The translation symmetry is a basic symmetry, whose generator is the plane wave e^{ipx}. Translations form an Abelian group,

$$e^{ipx} e^{iqx} = e^{i(p+q)x}. \tag{29.3.1}$$

The guiding center X replaces the coordinate x in the lowest Landau level. The Weyl operator e^{iqX} satisfies the magnetic translational group (29.2.4), or

$$e^{ipX} e^{iqX} = e^{i(p+q)X} \exp\left(i\ell_B^2 \frac{p \wedge q}{2} \right), \tag{29.3.2}$$

from which we find

$$[e^{ipX}, e^{iqX}] = 2ie^{i(p+q)X} \sin\left(i\ell_B^2 \frac{p \wedge q}{2} \right). \tag{29.3.3}$$

This gives the basic algebraic structure of the QH system.

A physical meaning of these formulas becomes clear when they are acted on the wave function. It follows from (29.2.43) that

$$e^{ipX} \mathfrak{S}(x, y) = e^{ipx/2} \mathfrak{S}(x - \ell_B^2 p_y, y + \ell_B^2 p_x). \tag{29.3.4}$$

Thus the magnetic translation is accompanied by a phase factor: The operator e^{ipX} not only translates the x_j coordinate as $x_j \to x_j - \ell_B^2 \varepsilon_{jk} p_k$, but also produces a phase $e^{ipx/2}$. A successive operation yields

$$e^{ipX} e^{iqX} \mathfrak{S}(x) = \exp\left(\frac{i}{2} \ell_B^2 \varepsilon_{jk} p_j q_k \right) e^{i(p+q)X} \mathfrak{S}(x), \tag{29.3.5}$$

which is identical to (29.3.2).

We may associate the origin of the phase factor with a gauge choice we have made. In a homogeneous magnetic field there is a translation invariance physically, but a choice of gauge breaks it superficially. This is the reason why a translation is accompanied by a compensating gauge transformation. Let us show this explicitly by taking the symmetric gauge (10.3.4). When we make a translation of the coordinate, $\boldsymbol{x} \to \boldsymbol{x}' = \boldsymbol{x} - \boldsymbol{a}$, in the electron field,

$$\psi(\boldsymbol{x}) \to \psi(\boldsymbol{x}') = \psi(\boldsymbol{x} - \boldsymbol{a}), \tag{29.3.6}$$

the Hamiltonian (10.6.1a) changes as

$$H \to H_{\boldsymbol{a}} \equiv \frac{1}{2M}\int d^2x\, \psi^\dagger(\boldsymbol{x}')(P_x^2 + P_y^2)\psi(\boldsymbol{x}') + H_{\mathrm{C}}, \tag{29.3.7}$$

where

$$P_j = -i\hbar\frac{\partial}{\partial x^j} + \frac{eB}{2}\varepsilon_{jk}x_k = P_j' + \frac{eB}{2}\varepsilon_{jk}a_k \tag{29.3.8}$$

with

$$P_j' = -i\hbar\frac{\partial}{\partial x'^j} + \frac{eB}{2}\varepsilon_{jk}x_k'. \tag{29.3.9}$$

It is not a symmetry transformation ($H_{\boldsymbol{a}} \neq H$) because $P_j \neq P_j'$. The symmetry breaking occurs since the external potential is fixed by making a gauge choice. Instead of (29.3.6), we consider a transformation

$$\psi(\boldsymbol{x}) \to \psi'(\boldsymbol{x}') \equiv \exp\left(-i\frac{1}{2\ell_B^2}\varepsilon_{jk}a_k x_j'\right)\psi(\boldsymbol{x}'), \tag{29.3.10}$$

where the translation is supplemented by a phase transformation. It follows that

$$H \to H_{\boldsymbol{a}}' = \frac{1}{2M}\int d^2x\, \psi^\dagger(\boldsymbol{x}')(P'^2_x + P'^2_y)\psi'(\boldsymbol{x}') + H_{\mathrm{C}}. \tag{29.3.11}$$

The Hamiltonian is invariant ($H_{\boldsymbol{a}}' = H$) under the transformation (29.3.10). The change (29.3.10) of the field operator corresponds to the change (29.3.4) of the wave function with $a_j = \ell_B^2\varepsilon_{jk}p_k$.

The translational symmetry implies that two points on the plane are physically identical. This does not imply that the symmetry group is Abelian. The translational symmetry holds even in the presence of a uniform magnetic field. When we move a charged particle along a closed path in a magnetic field, it acquires an Aharonov-Bohm phase, as reviewed in Section 5.7. It follows from (29.3.3) that

$$e^{ipX}e^{iqX}e^{-ipX}e^{-iqX} = \exp[i\ell_B^2(\boldsymbol{p}\wedge\boldsymbol{q})]. \tag{29.3.12}$$

Each operator on the left hand side generates a translation, and their net effect is to make a parallelogram spanned by two vectors $\ell_B^2\boldsymbol{p}$ and $\ell_B^2\boldsymbol{q}$, as in Fig.29.1. The phase on the right hand side reads

$$\ell_B^2(\boldsymbol{p}\wedge\boldsymbol{q}) = \frac{S}{\ell_B^2} = 2\pi N_\Phi \tag{29.3.13}$$

with

$$N_\Phi = \frac{S}{2\pi \ell_B^2} = \frac{B}{\Phi_D} S, \tag{29.3.14}$$

where Φ_D is the Dirac flux unit and $S = \ell_B^4 (\boldsymbol{p} \wedge \boldsymbol{q})$ is the area spanned by the two vectors $\ell_B^2 \boldsymbol{p}$ and $\ell_B^2 \boldsymbol{q}$. This phase counts the number N_Φ of the Dirac flux quanta passing through the parallelogram. Formula (29.3.12) describes precisely the Aharonov-Bohm phase.

29.4 Density Operators

In ordinary theory the density operator is defined by

$$\rho(\boldsymbol{x}) = \psi^\dagger(\boldsymbol{x})\psi(\boldsymbol{x}). \tag{29.4.1}$$

We consider electrons confined to the Nth Landau level, where Fock states are given by (10.2.17), or

$$|N, n\rangle = \sqrt{\frac{1}{N! n!}} (a^\dagger)^N (b^\dagger)^n |0\rangle. \tag{29.4.2}$$

The projected field operator is given by (10.6.15), or[1]

$$\psi(\boldsymbol{x}) = \sum_n \langle \boldsymbol{x} | N, n \rangle c(n), \tag{29.4.3}$$

with $c(n)$ the annihilation operator of electrons acting on the Landau site $|N, n\rangle$, $\{c(n), c^\dagger(m)\} = \delta_{nm}$. Using (29.4.3) in the density operator $\rho(\boldsymbol{x})$, we obtain

$$\rho_N(\boldsymbol{x}) = \psi^\dagger(\boldsymbol{x})\psi(\boldsymbol{x}) = \sum_{mn} \langle N, m | \boldsymbol{x} \rangle \langle \boldsymbol{x} | N, n \rangle c^\dagger(m) c(n), \tag{29.4.4}$$

which we call the *projected density operator*. Its Fourier transformation reads

$$\rho_N(\boldsymbol{q}) = \frac{1}{2\pi} \sum_{mn} \int d^2x \, \langle N, m | \boldsymbol{x} \rangle e^{-i\boldsymbol{q}\boldsymbol{x}} \langle \boldsymbol{x} | N, n \rangle c^\dagger(m) c(n) \tag{29.4.5a}$$

$$= \frac{1}{2\pi} \sum_{mn} \langle N | e^{-i\boldsymbol{q}\boldsymbol{R}} | N \rangle \langle m | e^{-i\boldsymbol{q}\boldsymbol{X}} | n \rangle c^\dagger(m) c(n), \tag{29.4.5b}$$

as we shall soon verify: See (29.4.12). Hence we may express

$$\rho_N(\boldsymbol{q}) = F_N(\boldsymbol{q}) \hat{\rho}(\boldsymbol{q}), \tag{29.4.6}$$

where

$$F_N(\boldsymbol{q}) = \langle N | e^{-i\boldsymbol{q}\boldsymbol{R}} | N \rangle, \tag{29.4.7}$$

[1] In Section 10.6, we have used the notation $\psi_N(\boldsymbol{x})$ and $c_N(\boldsymbol{x})$ for operators in the Nth Landau level. Here we omit the index N since we only treat the projected quantities to the Nth Landau level.

and

$$\hat{\rho}(\boldsymbol{q}) = \frac{1}{2\pi}\sum_{mn}\langle m|e^{-i\boldsymbol{q}\boldsymbol{X}}|n\rangle c^\dagger(m)c(n) = \frac{1}{2\pi}\sum_{mn}\langle m|e^{-i\boldsymbol{q}\boldsymbol{X}}|n\rangle \rho(n,m) \tag{29.4.8}$$

with

$$\rho(n,m) = c^\dagger(m)c(n). \tag{29.4.9}$$

We call $F_N(\boldsymbol{q})$ the *Landau-level form factor*, and $\hat{\rho}(\boldsymbol{q})$ the *bare density*. We note that

$$\int d^2x\, \rho_N(\boldsymbol{x}) = \int d^2x\, \hat{\rho}(\boldsymbol{x}) = \sum_m \rho(m,m), \tag{29.4.10}$$

with the use of $F_N(\boldsymbol{q}=0) = 1$. It counts the number of electrons in the Nth Landau level.

We now derive (29.4.5b). Since the state $|\boldsymbol{x}\rangle$ is an eigenstate of the position operator $\boldsymbol{x}_{\text{op}}$, $\boldsymbol{x}_{\text{op}}|\boldsymbol{x}\rangle = \boldsymbol{x}|\boldsymbol{x}\rangle$, we obtain

$$e^{-i\boldsymbol{q}\boldsymbol{x}_{\text{op}}} = \int d^2x\, e^{-i\boldsymbol{q}\boldsymbol{x}_{\text{op}}}|\boldsymbol{x}\rangle\langle\boldsymbol{x}| = \int d^2x\, |\boldsymbol{x}\rangle e^{-i\boldsymbol{q}\boldsymbol{x}}\langle\boldsymbol{x}|, \tag{29.4.11}$$

where the use of made of the completeness condition $\int d^2x\, |\boldsymbol{x}\rangle\langle\boldsymbol{x}| = 1$. Hence

$$\int d^2x\, \langle N,m|\boldsymbol{x}\rangle e^{-i\boldsymbol{q}\boldsymbol{x}}\langle\boldsymbol{x}|N,n\rangle = \langle N,m|e^{-i\boldsymbol{q}\boldsymbol{x}_{\text{op}}}|N,n\rangle. \tag{29.4.12}$$

We decompose $\boldsymbol{x}_{\text{op}}$ into the guiding center $\boldsymbol{X}$ and the relative coordinate $\boldsymbol{R}$, $\boldsymbol{x}_{\text{op}} = \boldsymbol{X} + \boldsymbol{R}$, as in (10.2.4) with (10.2.7). Since $\boldsymbol{X}$ and $\boldsymbol{R}$ commutes, (29.4.12) becomes

$$\langle N,m|e^{-i\boldsymbol{q}\boldsymbol{x}_{\text{op}}}|N,n\rangle = \langle N|e^{-i\boldsymbol{q}\boldsymbol{R}}|N\rangle\langle m|e^{-i\boldsymbol{q}\boldsymbol{X}}|n\rangle. \tag{29.4.13}$$

Substituting (29.4.12) and (29.4.13) into (29.4.5a), we obtain (29.4.5b).

We calculate the Landau-level form factor $F_N(\boldsymbol{q})$. We write

$$\boldsymbol{q}\boldsymbol{R} = -i\frac{\ell_B}{\sqrt{2}}(qa^\dagger - q^*a), \tag{29.4.14}$$

where $q = q_x + iq_y$, and a is the Landau-level ladder operator. We use the Hausdorff formula $e^{A+B} = e^A e^B e^{[B,A]/2}$, which holds when $[A,B]$ is a c-number, to find that

$$F_N(\boldsymbol{q}) = \frac{1}{N!}\langle 0|a^N e^{-\ell_B q a^\dagger/\sqrt{2}} e^{\ell_B q^* a/\sqrt{2}}(a^\dagger)^N|0\rangle e^{-\ell_B^2 q^2/4}. \tag{29.4.15}$$

For the lowest Landau level ($N = 0$) we find

$$F_0(\boldsymbol{q}) = e^{-\ell_B^2 q^2/4}. \tag{29.4.16}$$

In general it follows that

$$F_N(\boldsymbol{q}) = L_N\left(\frac{\ell_B^2 \boldsymbol{q}^2}{2}\right) e^{-\ell_B^2 q^2/4}, \tag{29.4.17}$$

where $L_N(x)$ is the Laguerre polynomial.

We are mainly concerned about electrons in the lowest Landau level, where

$$\psi(\boldsymbol{x}) = \sum_n \langle \boldsymbol{x}|n\rangle c(n). \tag{29.4.18}$$

The physical density is related to the bare density via

$$\rho(\boldsymbol{q}) = e^{-\frac{1}{4}\ell_B^2 q^2}\hat{\rho}(\boldsymbol{q}), \tag{29.4.19}$$

where we have omitted the Landau level index $N = 0$. In the real space the relation reads

$$\rho(\boldsymbol{x}) = e^{\ell_B^2 \nabla^2/4}\hat{\rho}(\boldsymbol{x}) = \frac{1}{\pi\ell_B^2}\int d^2y\, e^{-|\boldsymbol{x}-\boldsymbol{y}|^2/\ell_B^2}\hat{\rho}(\boldsymbol{y}). \tag{29.4.20}$$

It implies that the physical density $\rho(\boldsymbol{x})$ is not localizable to a point even if $\hat{\rho}(\boldsymbol{x})$ is localized.

It is convenient to construct the theory based on the bare density $\hat{\rho}(\boldsymbol{x})$ since it is prescribed solely by the noncommutative geometry and common to all Landau levels. We can always come back to the physical density with the help of the formula (29.4.6). Furthermore, $\hat{\rho}(\boldsymbol{x}) \to \rho_N(\boldsymbol{x})$ for sufficiently smooth configuration, where $F_N(\boldsymbol{q}) \to 1$ since we may take the limit $\boldsymbol{q} \to 0$.

Electron and Hole

We wish to explore the difference between the physical and bare densities in an instance of a single electron and a single hole. We define the classical density $\hat{\rho}^{\text{cl}}$ of the bare density (29.4.8) by the expectation value,

$$\hat{\rho}^{\text{cl}}(\boldsymbol{q}) = \frac{1}{2\pi}\sum_{mn}\langle n|e^{-i\boldsymbol{q}\boldsymbol{X}}|m\rangle \rho^{\text{cl}}(m,n) \tag{29.4.21}$$

with

$$\rho^{\text{cl}}(m,n) = \langle c^\dagger(n)c(m)\rangle. \tag{29.4.22}$$

Due to the correspondence (29.2.21) the Weyl operator and the symbol read

$$W[\hat{\rho}^{\text{cl}}] = \frac{1}{2\pi\ell_B^2}\sum_{mn}\rho^{\text{cl}}(m,n)|m\rangle\langle n|, \tag{29.4.23}$$

$$\hat{\rho}^{\text{cl}}(\boldsymbol{x}) = \frac{1}{2\pi\ell_B^2}\sum_{mn}\rho^{\text{cl}}(m,n)\Xi_{mn}(\boldsymbol{x}), \tag{29.4.24}$$

where $\Xi_{mn}(\boldsymbol{x})$ is given by (29.2.22). The symbol is the bare density $\hat{\rho}^{\text{cl}}(\boldsymbol{x})$, which is calculable once $\rho^{\text{cl}}(m,n)$ is given.

We consider a single electron placed in the Landau site $|0\rangle$ in the lowest Landau level. It is given by

$$|\text{e}\rangle = c^\dagger(0)|0\rangle, \tag{29.4.25}$$

for which

$$\rho_{\text{e}}^{\text{cl}}(m,n) = \langle \text{e}|c^\dagger(m)c(n)|\text{e}\rangle = \delta_{m0}\delta_{n0}. \tag{29.4.26}$$

The Weyl operator (29.4.23) and the symbol (29.4.24) are calculated,

$$W[\hat{\rho}_{\text{e}}^{\text{cl}}] = \frac{1}{2\pi\ell_B^2}\sum_{mn}\delta_{m0}\delta_{n0}|m\rangle\langle n| = \frac{1}{2\pi\ell_B^2}|0\rangle\langle 0|, \tag{29.4.27}$$

$$\hat{\rho}_{\text{e}}^{\text{cl}}(\boldsymbol{x}) = \frac{1}{2\pi\ell_B^2}\Xi_{00}(\boldsymbol{x}) = \frac{1}{\pi\ell_B^2}e^{-r^2/\ell_B^2}, \tag{29.4.28}$$

for one electron with the angular-momentum zero.

We recall that we have already constructed the wave function of one electron as (10.3.14), or

$$\mathfrak{S}_0(\boldsymbol{x}) = \frac{1}{\sqrt{2\pi\ell_B^2}}\exp\left(-\frac{r^2}{4\ell_B^2}\right). \tag{29.4.29}$$

The density is

$$\rho_{\text{e}}^{\text{cl}}(\boldsymbol{x}) = |\mathfrak{S}_0(\boldsymbol{x})|^2 = \frac{1}{2\pi\ell_B^2}e^{-r^2/2\ell_B^2}, \tag{29.4.30}$$

which is half of $\hat{\rho}_{\text{e}}^{\text{cl}}(\boldsymbol{x})$. In the momentum space we find

$$\hat{\rho}_{\text{e}}^{\text{cl}}(\boldsymbol{q}) = \frac{1}{\pi\ell_B^2}\int\frac{d^2x}{2\pi}e^{-iqx}e^{-r^2/\ell_B^2} = \frac{1}{2\pi}e^{-\ell_B^2q^2/4}, \tag{29.4.31a}$$

$$\rho_{\text{e}}^{\text{cl}}(\boldsymbol{q}) = \frac{1}{2\pi\ell_B^2}\int\frac{d^2x}{2\pi}e^{-iqx}e^{-r^2/2\ell_B^2} = \frac{1}{2\pi}e^{-\ell_B^2q^2/2}. \tag{29.4.31b}$$

There holds a relation,

$$\rho_{\text{e}}^{\text{cl}}(\boldsymbol{q}) = e^{-\ell_B^2q^2/4}\hat{\rho}_{\text{e}}^{\text{cl}}(\boldsymbol{q}). \tag{29.4.32}$$

This is precisely the relation (29.4.19) between the physical and bare densities.

We next study a hole excitation placed in the angular-momentum zero state in the ground state of the QH state at $\nu = 1$, where $\rho_0 = 1/2\pi\ell_B^2$. It is given by

$$|\text{h}\rangle = \prod_{n\geq 1}c^\dagger(n)|0\rangle, \tag{29.4.33}$$

for which

$$\langle \text{h}|c^\dagger(m)c(n)|\text{h}\rangle = \delta_{mn} - \delta_{m0}\delta_{n0}. \tag{29.4.34}$$

The Weyl operator (29.4.23) and the symbol (29.4.24) are calculated,

$$W[\hat{\rho}_{\text{h}}^{\text{cl}}] = \rho_0\sum_{mn}(\delta_{mn} - \delta_{m0}\delta_{n0})\,|m\rangle\langle n| = \rho_0\,(1 - |0\rangle\langle 0|)\,, \tag{29.4.35}$$

$$\hat{\rho}_{\text{h}}^{\text{cl}}(\boldsymbol{x}) = \rho_0\left(1 - 2e^{-r^2/\ell_B^2}\right). \tag{29.4.36}$$

The hole density becomes negative at the origin,

$$\lim_{x\to 0}\hat{\rho}^{\rm h}(x) = -\rho_0. \tag{29.4.37}$$

There is nothing wrong with this mathematically since the electron cannot be localized within the LLL. Nevertheless, this represents an unphysical nature of the bare density. On the other hand, the physical density is defined by (29.4.20), and reads

$$\rho_{\rm h}^{\rm cl}(x) = \rho_0\left(1 - e^{-r^2/2\ell_B^2}\right). \tag{29.4.38}$$

The hole density satisfies

$$\lim_{x\to 0}\rho_{\rm h}^{\rm cl}(x) = 0. \tag{29.4.39}$$

This behavior is what we expect for the physical density at the origin.

Density Algebra

We evaluate the commutation relation of the density operator $\hat{\rho}(\boldsymbol{p})$,

$$[\hat{\rho}(\boldsymbol{p}),\hat{\rho}(\boldsymbol{q})] = \frac{1}{2\pi}\sum_{mn}^{\infty}\sum_{ij}^{\infty}\langle n|e^{-i\boldsymbol{p}X}|m\rangle\langle j|e^{-i\boldsymbol{q}X}|i\rangle[\rho(m,n),\rho(i,j)]. \tag{29.4.40}$$

Here, it is trivial to derive the algebra of $\rho(m,n) = c^\dagger(n)c(m)$,

$$[\rho(m,n),\rho(i,j)] = \delta_{mj}\rho(i,n) - \delta_{in}\rho(m,j), \tag{29.4.41}$$

based on the anticommutation relation of $c(m)$. Using it we find

$$[\hat{\rho}(\boldsymbol{p}),\hat{\rho}(\boldsymbol{q})] = \left(\frac{1}{2\pi}\right)^2\sum_{mn}^{\infty}\langle m|[e^{-i\boldsymbol{p}X},e^{-i\boldsymbol{q}X}]|n\rangle\rho(n,m), \tag{29.4.42}$$

which yields

$$[\hat{\rho}(\boldsymbol{p}),\hat{\rho}(\boldsymbol{q})] = \frac{i}{\pi}\hat{\rho}(\boldsymbol{p}+\boldsymbol{q})\sin\left(\ell_B^2\frac{\boldsymbol{p}\wedge\boldsymbol{q}}{2}\right), \tag{29.4.43}$$

owing to the magnetic translation algebra (29.3.2). The density algebra (29.4.41) or (29.4.43) is isomorphic to the W_∞ algebra; hence it is also called the W_∞ density algebra: See Appendix P on this point.

The W_∞ algebra is far from being trivial as the commutation relation for the quantized field. We may introduce a redundant variable to reduce the dynamical system to the standard one. Let us introduce a complex bosonic field $\phi(x)$ by

$$\hat{\rho}(x) = \phi(x)\star\phi^*(x), \tag{29.4.44}$$

or

$$\hat{\rho}(\boldsymbol{p}) = \frac{1}{2\pi}\int d^2q\, e^{-\frac{i}{2}\ell_B^2\boldsymbol{p}\wedge\boldsymbol{q}}\phi(\boldsymbol{q})\phi^*(\boldsymbol{q}-\boldsymbol{p}). \tag{29.4.45}$$

We calculate

$$[\hat{\rho}(\boldsymbol{p}),\hat{\rho}(\boldsymbol{q})]$$
$$=\frac{1}{4\pi^2}\int d^2p'd^2q'\, e^{-\frac{i}{2}\ell_B^2(\boldsymbol{p}\wedge\boldsymbol{p}'+\boldsymbol{q}\wedge\boldsymbol{q}')}\phi(\boldsymbol{p}')[\phi^*(\boldsymbol{p}'-\boldsymbol{p}),\phi(\boldsymbol{q}')]\phi^*(\boldsymbol{q}'-\boldsymbol{q})$$
$$+\frac{1}{4\pi^2}\int d^2p'd^2q'\, e^{-\frac{i}{2}\ell_B^2(\boldsymbol{p}\wedge\boldsymbol{p}'+\boldsymbol{q}\wedge\boldsymbol{q}')}\phi(\boldsymbol{q}')[\phi^*(\boldsymbol{p}'),\phi(\boldsymbol{q}'-\boldsymbol{q})]\phi^*(\boldsymbol{p}'-\boldsymbol{p}). \tag{29.4.46}$$

The W_∞ algebra (29.4.43) implies

$$[\phi(\boldsymbol{p}),\phi^\dagger(\boldsymbol{q})]=\delta(\boldsymbol{p}-\boldsymbol{q}),\quad \text{or} \quad [\phi(\boldsymbol{x}),\phi^\dagger(\boldsymbol{y})]=\delta(\boldsymbol{x}-\boldsymbol{y}). \tag{29.4.47}$$

It agrees with the commutation relation (13.2.13b) of the composite boson.

Smeared Density Operator

We define a smeared density operator by

$$\hat{\rho}(f)=\int d^2x f(\boldsymbol{x})\hat{\rho}(\boldsymbol{x})=\int d^2q\, f(-\boldsymbol{q})\hat{\rho}(\boldsymbol{q}), \tag{29.4.48}$$

where $f(\boldsymbol{x})$ is a c-number function interpreted as a "wave packet". We study the commutator of two such operators,

$$[\hat{\rho}(f),\hat{\rho}(g)]=\int d^2qd^2p\, f(-\boldsymbol{p})g(-\boldsymbol{q})[\hat{\rho}(\boldsymbol{p}),\hat{\rho}(\boldsymbol{q})]. \tag{29.4.49}$$

With the use of the density algebra (29.4.43) we find[236]

$$[\hat{\rho}(f),\hat{\rho}(g)]=\hat{\rho}(h), \tag{29.4.50}$$

where the Fourier transform of $h(\boldsymbol{x})$ is given by

$$h(-\boldsymbol{k})=\frac{i}{\pi}\int d^2qd^2p\,\delta(\boldsymbol{k}-\boldsymbol{p}-\boldsymbol{q})\sin\left(\ell_B^2\frac{\boldsymbol{p}\wedge\boldsymbol{q}}{2}\right)f(-\boldsymbol{p})g(-\boldsymbol{q}). \tag{29.4.51}$$

It is the Moyal bracket of two functions $f(\boldsymbol{x})$ and $g(\boldsymbol{x})$,

$$h(\boldsymbol{x})=[\![f(\boldsymbol{x}),g(\boldsymbol{x})]\!]\equiv f\star g(\boldsymbol{x})-g\star f(\boldsymbol{x}) \tag{29.4.52}$$

with the Moyal product (29.2.15).

A special wave packet is the localized one, $\delta_{\boldsymbol{z}}(\boldsymbol{x})=\delta(\boldsymbol{x}-\boldsymbol{z})$ as $f(\boldsymbol{x})$, whose Fourier transformation is $f(\boldsymbol{q})=e^{-i\boldsymbol{z}\boldsymbol{q}}/2\pi$. In this case the smeared density (29.4.48) is reduced to (29.4.8), $\hat{\rho}(f)\to\hat{\rho}(\boldsymbol{z})$. The commutator (29.4.50) reads

$$[\hat{\rho}(\boldsymbol{x}),\hat{\rho}(\boldsymbol{y})]=\int d^2z\,[\![\delta_{\boldsymbol{x}}(\boldsymbol{z}),\delta_{\boldsymbol{y}}(\boldsymbol{z})]\!]\,\hat{\rho}(\boldsymbol{z}), \tag{29.4.53}$$

where $[\![\delta_x(z), \delta_y(z)]\!]$ is the Moyal bracket,

$$[\![\delta_x(z), \delta_y(z)]\!] = \delta_x(z) \star \delta_y(z) - \delta_y(z) \star \delta_x(z)$$
$$= \frac{2i}{(2\pi)^4} \int d^2p d^2q\, e^{i\boldsymbol{p}(\boldsymbol{x}-\boldsymbol{z})+i\boldsymbol{q}(\boldsymbol{y}-\boldsymbol{z})} \sin \ell_B^2 \frac{\boldsymbol{p} \wedge \boldsymbol{q}}{2}. \quad (29.4.54)$$

The commutator (29.4.53) vanishes in the commutative limit $\ell_B \to 0$, as is consistent with the fact that $\hat{\rho}(\boldsymbol{x})$ coincides with the ordinary density $\rho(\boldsymbol{x})$ in the commutative limit.

The Moyal bracket is expanded as

$$[\![f(\boldsymbol{x}), g(\boldsymbol{x})]\!] = -2i \lim_{\boldsymbol{y}\to\boldsymbol{x}} \sin\left(\ell_B^2 \frac{\nabla_x \wedge \nabla_y}{2}\right) f(\boldsymbol{x}) g(\boldsymbol{y})$$
$$= -i\ell_B^2 \left(\partial_x f(\boldsymbol{x}) \partial_y g(\boldsymbol{x}) - \partial_y f(\boldsymbol{x}) \partial_x g(\boldsymbol{x})\right) + O(\ell_B^4). \quad (29.4.55)$$

The first term is the Poisson bracket up to a trivial factor. The Moyal bracket was introduced[356] originally to make an alternative formulation of quantum mechanics in terms of statistical distributions on phase space. It has been shown[170,460] that the Moyal bracket is the only possible deformation of the Poisson bracket; here the deformation parameter is the magnetic length ℓ_B. The Moyal bracket is reduced to the Poisson bracket in the limit $\ell_B \to 0$.

As we shall see later [see (30.2.2)], the ground state satisfies the incompressibility condition,

$$\hat{\rho}(\boldsymbol{q})|\mathrm{g}\rangle = 2\pi\rho_0\delta(\boldsymbol{q})|\mathrm{g}\rangle. \quad (29.4.56)$$

Due to the condition the local phase transformation acts on this state as a c-number,

$$e^{i\hat{\rho}(f)}|\mathrm{g}\rangle = e^{i\alpha}|\mathrm{g}\rangle, \quad (29.4.57)$$

where $\hat{\rho}(f)$ is the smeared density operator (29.4.48) and $\alpha = \rho_0 \int d^2x\, f(\boldsymbol{x})$. Since the c-number phase factor is irrelevant, it means that the ground state is invariant under the W_∞ transformation.

29.5 SU(N)-Extended W∞ Algebra

We generalize the scheme so that the electron field $\Psi(\boldsymbol{x})$ possesses N isospin components. We have $N = 2$ in the monolayer QH ferromagnet and $N = 4$ in the bilayer QH ferromagnet. The physical variables are the number density and the isospin density projected to the Nth Landau level[2],

$$\rho_N(\boldsymbol{x}) = \Psi^\dagger(\boldsymbol{x})\Psi(\boldsymbol{x}), \qquad \mathcal{I}_{N;A}(\boldsymbol{x}) = \frac{1}{2}\Psi^\dagger(\boldsymbol{x})\lambda_A\Psi(\boldsymbol{x}), \quad (29.5.1)$$

[2]There should be no confusion with respect to the index N, though we use it as the group index and the Landau level index.

which are summarized into the density matrix,

$$\mathcal{D}_{N;\mu\nu}(\boldsymbol{x}) = \frac{1}{N}\delta_{\mu\nu}\rho_N(\boldsymbol{x}) + (\lambda_A)_{\mu\nu}\mathfrak{I}_{\mu\nu N;A}(\boldsymbol{x}). \tag{29.5.2}$$

Here λ_A are the generating matrices of SU(N): See Appendix D.

We introduce the bare densities as in the spinless theory [see (29.4.8)]

$$\hat{\rho}(\boldsymbol{q}) = \frac{1}{2\pi}\sum_{mn}^{\infty}\langle n|e^{-i\boldsymbol{q}\boldsymbol{X}}|m\rangle\rho(m,n), \quad \hat{\mathfrak{I}}_A(\boldsymbol{q}) = \frac{1}{2\pi}\sum_{mn}^{\infty}\langle n|e^{-i\boldsymbol{q}\boldsymbol{X}}|m\rangle I_A(m,n), \tag{29.5.3}$$

where

$$\rho(m,n) = \sum_{\sigma} c_\sigma^\dagger(n)c_\sigma(m), \quad I_A(m,n) = \frac{1}{2}\sum_{\sigma\tau} c_\sigma^\dagger(n)(\lambda_A)_{\sigma\tau}c_\tau(m). \tag{29.5.4}$$

In terms of the density matrix operator we have

$$\hat{\mathcal{D}}_{\mu\nu}(\boldsymbol{q}) = \frac{1}{2\pi}\sum_{mn}\langle n|e^{-i\boldsymbol{q}\boldsymbol{X}}|m\rangle D_{\mu\nu}(m,n), \tag{29.5.5}$$

together with

$$D_{\mu\nu}(m,n) \equiv c_\nu^\dagger(n)c_\mu(m) = \frac{1}{N}\delta_{\mu\nu}\rho(m,n) + (\lambda_A)_{\mu\nu}I_A(m,n). \tag{29.5.6}$$

Similarly, $\hat{\mathcal{D}}_{\mu\nu}(\boldsymbol{q})$ is related to $\hat{\rho}(\boldsymbol{q})$ and $\hat{\mathfrak{I}}_A(\boldsymbol{q})$ in the same way as in (29.5.2).

The bare densities are related to the physical densities as in (29.4.6),

$$\rho_N(\boldsymbol{q}) = F_N(\boldsymbol{q})\hat{\rho}(\boldsymbol{q}), \quad \mathfrak{I}_{N;A}(\boldsymbol{x}) = F_N(\boldsymbol{q})\hat{\mathfrak{I}}_A(\boldsymbol{q}), \tag{29.5.7}$$

with the use of the Landau-level form factor $F_N(\boldsymbol{q})$ given by (29.4.7).

It is straightforward to verify that

$$[D_{\mu\nu}(m,n), D_{\sigma\tau}(s,t)] = \delta_{\mu\tau}\delta_{mt}D_{\sigma\nu}(s,n) - \delta_{\sigma\nu}\delta_{sn}D_{\mu\tau}(m,t) \tag{29.5.8}$$

based on the anticommutation relation (10.6.17) of $c_\sigma(m)$. From this formula together with the magnetic translation algebra (29.3.3), we derive the algebraic relation among $\hat{\mathcal{D}}_{\mu\nu}(\boldsymbol{p})$,

$$2\pi[\hat{\mathcal{D}}_{\mu\nu}(\boldsymbol{p}), \hat{\mathcal{D}}_{\sigma\tau}(\boldsymbol{q})] = \delta_{\mu\tau}e^{+\frac{i}{2}\ell_B^2\boldsymbol{p}\wedge\boldsymbol{q}}\hat{\mathcal{D}}_{\sigma\nu}(\boldsymbol{p}+\boldsymbol{q}) - \delta_{\sigma\nu}e^{-\frac{i}{2}\ell_B^2\boldsymbol{p}\wedge\boldsymbol{q}}\hat{\mathcal{D}}_{\mu\tau}(\boldsymbol{p}+\boldsymbol{q}). \tag{29.5.9}$$

They are the generalization of (29.4.41) and (29.4.43) in the spinless theory. We call this the W_∞(N) algebra[149] since it is the SU(N) extension of W_∞.

The W_∞(N) algebra is far from being trivial as the commutation relation for the quantized field. Just as we did for the W_∞ algebra in (29.4.44), we introduce the N-component field $\phi_\mu(\boldsymbol{x})$ and resolve the density matrix as

$$\hat{\mathcal{D}}_{\mu\nu}(\boldsymbol{x}) = \phi_\mu(\boldsymbol{x}) \star \phi_\nu^\dagger(\boldsymbol{x}). \tag{29.5.10}$$

By calculating $[\hat{\mathcal{D}}_{\mu\nu}(\boldsymbol{p}), \hat{\mathcal{D}}_{\sigma\tau}(\boldsymbol{q})]$, we find the $W_\infty(N)$ algebra (29.5.9) is reproduced by requiring the standard commutation relation,

$$[\phi_\alpha(\boldsymbol{x}), \phi_\beta^\dagger(\boldsymbol{y})] = \delta_{\alpha\beta}\delta(\boldsymbol{x}-\boldsymbol{y}). \tag{29.5.11}$$

This agrees with the commutation relation (15.3.3) of the N-component composite boson field.

We may rewrite the density algebra (29.5.8) and (29.5.9) in terms of the physical densities,

$$[\rho(m,n), \rho(i,j)] = \delta_{mj}\rho(i,n) - \delta_{in}\rho(m,j), \tag{29.5.12a}$$

$$[\rho(m,n), I_A(i,j)] = \delta_{mj}I_A(i,n) - \delta_{in}I_A(m,j), \tag{29.5.12b}$$

$$\begin{aligned}[I_A(m,n), I_B(i,j)] = &\frac{i}{2}f_{ABC}\left[\delta_{mj}I_C(i,n) + \delta_{in}I_C(m,j)\right] \\ &+ \frac{1}{2}d_{ABC}\left[\delta_{mj}I_C(m,j) - \delta_{in}I_C(m,j)\right] \\ &+ \frac{1}{2N}\delta_{AB}\left[\delta_{mj}\rho(i,n) - \delta_{in}\rho(m,j)\right],\end{aligned} \tag{29.5.12c}$$

and

$$[\hat{\rho}(\boldsymbol{p}), \hat{\rho}(\boldsymbol{q})] = \frac{i}{\pi}\hat{\rho}(\boldsymbol{p}+\boldsymbol{q})\sin\left(\ell_B^2\frac{\boldsymbol{p}\wedge\boldsymbol{q}}{2}\right), \tag{29.5.13a}$$

$$[\hat{\mathfrak{J}}_A(\boldsymbol{p}), \hat{\rho}(\boldsymbol{q})] = \frac{i}{\pi}\hat{\mathfrak{J}}_A(\boldsymbol{p}+\boldsymbol{q})\sin\left(\ell_B^2\frac{\boldsymbol{p}\wedge\boldsymbol{q}}{2}\right), \tag{29.5.13b}$$

$$\begin{aligned}[\hat{\mathfrak{J}}_A(\boldsymbol{p}), \hat{\mathfrak{J}}_B(\boldsymbol{q})] = &\frac{i}{2\pi}f_{ABC}\hat{\mathfrak{J}}_C(\boldsymbol{p}+\boldsymbol{q})\cos\left(\ell_B^2\frac{\boldsymbol{p}\wedge\boldsymbol{q}}{2}\right) \\ &+ \frac{i}{2\pi}d_{ABC}\hat{\mathfrak{J}}_C(\boldsymbol{p}+\boldsymbol{q})\sin\left(\ell_B^2\frac{\boldsymbol{p}\wedge\boldsymbol{q}}{2}\right) \\ &+ \frac{i}{2\pi N}\delta_{AB}\hat{\rho}(\boldsymbol{p}+\boldsymbol{q})\sin\left(\ell_B^2\frac{\boldsymbol{p}\wedge\boldsymbol{q}}{2}\right).\end{aligned} \tag{29.5.13c}$$

It is reduced to the standard U(1)⊗SU(N) algebra for the $\boldsymbol{q}=0$ component,

$$[\hat{\rho}(0), \hat{\rho}(0)] = 0, \quad [\hat{\mathfrak{J}}_A(0), \hat{\rho}(0)] = 0, \quad [\hat{\mathfrak{J}}_A(0), \hat{\mathfrak{J}}_B(0)] = \frac{i}{2\pi}f_{ABC}\hat{\mathfrak{J}}_C(0). \tag{29.5.14}$$

The entanglement between the U(1) and SU(N) fields is removed on the ground state.

We introduce the Fourier components,

$$\hat{\rho}(\boldsymbol{x}) \equiv \int \frac{d^2p}{2\pi}\, e^{i\boldsymbol{px}}\hat{\rho}(\boldsymbol{p}), \qquad \hat{\mathfrak{J}}_A(\boldsymbol{x}) \equiv \int \frac{d^2p}{2\pi}\, e^{i\boldsymbol{px}}\hat{\mathfrak{J}}_A(\boldsymbol{p}). \tag{29.5.15}$$

Their commutators are calculated as

$$\begin{aligned}
[\hat{\rho}(\boldsymbol{x}), \hat{\rho}(\boldsymbol{y})] &= \int d^2z \, [\![\delta_{\boldsymbol{x}}(\boldsymbol{z}), \delta_{\boldsymbol{y}}(\boldsymbol{z})]\!] \, \hat{\rho}(\boldsymbol{z}), \\
[\hat{\mathfrak{I}}_A(\boldsymbol{x}), \hat{\rho}(\boldsymbol{y})] &= \int d^2z \, [\![\delta_{\boldsymbol{x}}(\boldsymbol{z}), \delta_{\boldsymbol{y}}(\boldsymbol{z})]\!] \, \hat{\mathfrak{I}}_A(\boldsymbol{z}), \\
[\hat{\mathfrak{I}}_A(\boldsymbol{x}), \hat{\mathfrak{I}}_B(\boldsymbol{y})] &= \frac{i}{2} f_{ABC} \int d^2z \, \{\!\{\delta_{\boldsymbol{x}}(\boldsymbol{z}), \delta_{\boldsymbol{y}}(\boldsymbol{z})\}\!\} \, \hat{\mathfrak{I}}_C(\boldsymbol{z}) \\
&\quad + \frac{1}{2} d_{ABC} \int d^2z \, [\![\delta_{\boldsymbol{x}}(\boldsymbol{z}), \delta_{\boldsymbol{y}}(\boldsymbol{z})]\!] \, \hat{\mathfrak{I}}_C(\boldsymbol{z}) \\
&\quad + \frac{1}{2N} \delta_{AB} \int d^2z \, [\![\delta_{\boldsymbol{x}}(\boldsymbol{z}), \delta_{\boldsymbol{y}}(\boldsymbol{z})]\!] \, \hat{\rho}(\boldsymbol{z}),
\end{aligned} \tag{29.5.16}$$

where $[\![\delta_{\boldsymbol{x}}(\boldsymbol{z}), \delta_{\boldsymbol{y}}(\boldsymbol{z})]\!]$ is the Moyal bracket (29.4.54), or

$$\begin{aligned}
[\![\delta_{\boldsymbol{x}}(\boldsymbol{z}), \delta_{\boldsymbol{y}}(\boldsymbol{z})]\!] &= \delta_{\boldsymbol{x}}(\boldsymbol{z}) \star \delta_{\boldsymbol{y}}(\boldsymbol{z}) - \delta_{\boldsymbol{y}}(\boldsymbol{z}) \star \delta_{\boldsymbol{x}}(\boldsymbol{z}) \\
&= \frac{2i}{(2\pi)^4} \int d^2p d^2q \, e^{i\boldsymbol{p}(\boldsymbol{x}-\boldsymbol{z})+i\boldsymbol{q}(\boldsymbol{y}-\boldsymbol{z})} \sin \frac{\boldsymbol{p} \wedge \boldsymbol{q}}{2\kappa_B^2},
\end{aligned} \tag{29.5.17}$$

and $\{\!\{\delta_{\boldsymbol{x}}(\boldsymbol{z}), \delta_{\boldsymbol{y}}(\boldsymbol{z})\}\!\}$ is the Moyal antibracket,

$$\begin{aligned}
\{\!\{\delta_{\boldsymbol{x}}(\boldsymbol{z}), \delta_{\boldsymbol{y}}(\boldsymbol{z})\}\!\} &= \delta_{\boldsymbol{x}}(\boldsymbol{z}) \star \delta_{\boldsymbol{y}}(\boldsymbol{z}) + \delta_{\boldsymbol{y}}(\boldsymbol{z}) \star \delta_{\boldsymbol{x}}(\boldsymbol{z}) \\
&= \frac{2}{(2\pi)^4} \int d^2p d^2q \, e^{i\boldsymbol{p}(\boldsymbol{x}-\boldsymbol{z})+i\boldsymbol{q}(\boldsymbol{y}-\boldsymbol{z})} \cos \frac{\boldsymbol{p} \wedge \boldsymbol{q}}{2\kappa_B^2}.
\end{aligned} \tag{29.5.18}$$

In the commutative limit ($\ell_B \to 0$) we have

$$\begin{aligned}
&[\![\delta_{\boldsymbol{x}}(\boldsymbol{z}), \delta_{\boldsymbol{y}}(\boldsymbol{z})]\!] \to 0, \\
&\{\!\{\delta_{\boldsymbol{x}}(\boldsymbol{z}), \delta_{\boldsymbol{y}}(\boldsymbol{z})\}\!\} \to 2\delta(\boldsymbol{x} - \boldsymbol{z})\delta(\boldsymbol{y} - \boldsymbol{z}),
\end{aligned} \tag{29.5.19}$$

and the density algebra (29.5.16) regresses to the standard algebra,

$$\begin{aligned}
&[\hat{\rho}(\boldsymbol{x}), \hat{\rho}(\boldsymbol{y})] = 0, \\
&[\hat{\mathfrak{I}}_A(\boldsymbol{x}), \hat{\rho}(\boldsymbol{y})] = 0, \\
&[\hat{\mathfrak{I}}_A(\boldsymbol{x}), \hat{\mathfrak{I}}_B(\boldsymbol{y})] = i f_{ABC} \hat{\mathfrak{I}}_C(\boldsymbol{x}) \delta(\boldsymbol{x} - \boldsymbol{y}).
\end{aligned} \tag{29.5.20}$$

The same algebra holds for the physical densities, since the bare and physical densities coincide with each other in the commutative limit.

29.6 Classical Fields

What we observe experimentally is the expectation value of a quantum operator. It is a classical object. quasiparticles are skyrmions in QH ferromagnets, which are described by classical fields.

We explore the kinematics and the dynamics[3] of the classical field $\hat{\mathfrak{D}}^{\text{cl}}_{\mu\nu}(\boldsymbol{q}) =$

---

[3]We analyze the equation of motion of the classical field $\hat{\mathfrak{D}}^{\text{cl}}_{\mu\nu}(\boldsymbol{q})$ in Section 32.4.

$\langle\mathfrak{S}|\hat{\mathcal{D}}_{\mu\nu}(\boldsymbol{q})|\mathfrak{S}\rangle$ constructed over a wide class of Fock states. We consider the states presented as

$$|\mathfrak{S}\rangle = e^{iW}|\mathfrak{S}_0\rangle, \tag{29.6.1}$$

where W is an arbitrary element of the $\mathrm{W}_\infty(\mathrm{N})$ algebra, and $|\mathfrak{S}_0\rangle$ is a state of the form

$$|\mathfrak{S}_0\rangle = \prod_{\mu n}\left[c_\mu^\dagger(n)\right]^{\nu_\mu(n)}|0\rangle, \tag{29.6.2}$$

where $\nu_\mu(n)$ may take the value either 0 or 1 depending whether the isospin state μ at a site n is occupied or not. The class of states (29.6.1) is quite general though it may not embrace all possible ones. Nevertheless all physically relevant states at integer filling factors seem to fall in this category. Indeed, as far as we know, perturbative excitations are isospin waves and nonperturbative excitations are skyrmions in QH systems. They belong surely to this category.

From (29.5.5) the classical field is given by

$$\hat{\mathcal{D}}_{\mu\nu}^{\mathrm{cl}}(\boldsymbol{q}) = \frac{1}{2\pi}\sum_{mn}\langle n|e^{-iqX}|m\rangle D_{\mu\nu}^{\mathrm{cl}}(m,n), \tag{29.6.3}$$

where

$$D_{\mu\nu}^{\mathrm{cl}}(m,n) = \langle\mathfrak{S}|c_\nu^\dagger(n)c_\mu(m)|\mathfrak{S}\rangle. \tag{29.6.4}$$

The Weyl operator is

$$W[\hat{\mathcal{D}}_{\mu\nu}^{\mathrm{cl}}] = \frac{1}{2\pi\ell_B^2}\sum_{mn}D_{\mu\nu}^{\mathrm{cl}}(m,n)|m\rangle\langle n|. \tag{29.6.5}$$

They are the generalization of (29.4.21), (29.4.22) and (29.4.23) to the system with internal symmetry.

As shown in a later section, there holds the relation [see (32.2.13)],

$$\sum_k\sum_\kappa D_{\mu\kappa}^{\mathrm{cl}}(m,k)D_{\kappa\nu}^{\mathrm{cl}}(k,n) = D_{\mu\nu}^{\mathrm{cl}}(m,n). \tag{29.6.6}$$

It reads

$$\sum_\kappa W[\hat{\mathcal{D}}_{\mu\kappa}^{\mathrm{cl}}]W[\hat{\mathcal{D}}_{\kappa\nu}^{\mathrm{cl}}] = \frac{1}{2\pi\ell_B^2}W[\hat{\mathcal{D}}_{\mu\nu}^{\mathrm{cl}}] \tag{29.6.7}$$

in terms of the Weyl operator, and

$$\sum_\kappa \hat{\mathcal{D}}_{\mu\kappa}^{\mathrm{cl}}(\boldsymbol{x})\star\hat{\mathcal{D}}_{\kappa\nu}^{\mathrm{cl}}(\boldsymbol{x}) = \frac{1}{2\pi\ell_B^2}\hat{\mathcal{D}}_{\mu\nu}^{\mathrm{cl}}(\boldsymbol{x}) \tag{29.6.8}$$

in terms of the symbol, which we call the noncommutative constraint. Using

$$\hat{\mathcal{D}}_{\mu\nu}^{\mathrm{cl}}(\boldsymbol{x}) = \frac{1}{N}\delta_{\mu\nu}\hat{\rho}^{\mathrm{cl}}(\boldsymbol{x}) + (\lambda_A)_{\mu\nu}\hat{\mathfrak{I}}_A^{\mathrm{cl}}(\boldsymbol{x}), \tag{29.6.9}$$

we may rewrite (29.6.8) as

$$\mathfrak{I}_A^{\rm cl}(\boldsymbol{x})\star\mathfrak{I}_A^{\rm cl}(\boldsymbol{x})+\frac{1}{2N}\hat{\rho}^{\rm cl}(\boldsymbol{x})\star\hat{\rho}^{\rm cl}(\boldsymbol{x})=\frac{1}{4\pi\ell_B^2}\hat{\rho}^{\rm cl}(\boldsymbol{x}), \tag{29.6.10}$$

and

$$\frac{1}{2}(if_{ABC}+d_{ABC})\mathfrak{I}_B^{\rm cl}(\boldsymbol{x})\star\mathfrak{I}_C^{\rm cl}(\boldsymbol{x})=\frac{1}{4\pi\ell_B^2}\mathfrak{I}_A^{\rm cl}(\boldsymbol{x})-\frac{1}{2N}\hat{\rho}^{\rm cl}(\boldsymbol{x})\star\mathfrak{I}_A^{\rm cl}(\boldsymbol{x})-\frac{1}{2N}\mathfrak{I}_A^{\rm cl}(\boldsymbol{x})\star\hat{\rho}^{\rm cl}(\boldsymbol{x}). \tag{29.6.11}$$

These relations are the manifestation of the microscopic noncommutativity at the level of the classical fields.

Spin-Frozen QH System

The noncommutative constraint (29.6.8) reads

$$\hat{\rho}^{\rm cl}(\boldsymbol{x})\star\hat{\rho}^{\rm cl}(\boldsymbol{x})=\frac{1}{2\pi\ell_B^2}\hat{\rho}^{\rm cl}(\boldsymbol{x}) \tag{29.6.12}$$

in the instance of the spin-frozen system. Since the star product is reduced to the ordinary product for a constant configuration, if we require the boundary condition $\lim_{r\to\infty}\hat{\rho}^{\rm cl}(\boldsymbol{x})=$const, only the following two cases are possible:

$$\text{(a)}\qquad \lim_{r\to\infty}\hat{\rho}^{\rm cl}(\boldsymbol{x})=0, \tag{29.6.13a}$$

$$\text{(b)}\qquad \lim_{r\to\infty}\hat{\rho}^{\rm cl}(\boldsymbol{x})=\frac{1}{2\pi\ell_B^2}. \tag{29.6.13b}$$

Obviously the QH system corresponds to the case (b).

We try to resolve the noncommutative constraint (29.6.12) by setting

$$\hat{\rho}^{\rm cl}(\boldsymbol{x})=\frac{1}{2\pi\ell_B^2}\phi^{\rm cl}(\boldsymbol{x})\star\phi^{\rm cl*}(\boldsymbol{x}). \tag{29.6.14}$$

By substituting this into (29.6.12), the consistency condition is found to be

$$\phi^{\rm cl*}(\boldsymbol{x})\star\phi^{\rm cl}(\boldsymbol{x})=1. \tag{29.6.15}$$

This is quite nontrivial, since we would conclude $\hat{\rho}^{\rm cl}(\boldsymbol{x})=1/2\pi\ell_B^2$ in the commutative limit, where $\phi^{\rm cl}(\boldsymbol{x})$ and $\phi^{\rm cl*}(\boldsymbol{x})$ do commute. Nevertheless, we can show that the resolution (29.6.14) is always possible together with (29.6.15) for an arbitrary function $\hat{\rho}^{\rm cl}(\boldsymbol{x})$. In so doing we examine explicitly what are the densities $\hat{\rho}^{\rm cl}(\boldsymbol{x})$ subject to the noncommutative constraint (29.6.12).

The condition (29.6.12) is rewritten in terms of the Weyl operator as

$$W[\hat{\rho}^{\rm cl}]W[\hat{\rho}^{\rm cl}]=\frac{1}{2\pi\ell_B^2}W[\hat{\rho}^{\rm cl}], \tag{29.6.16}$$

where $W[\hat{\rho}^{\text{cl}}]$ is expanded as in (29.4.23), or

$$W[\hat{\rho}^{\text{cl}}] = \frac{1}{2\pi\ell_B^2}\sum_{mn=0}^{\infty}\rho^{\text{cl}}(m,n)|n\rangle\langle m|. \tag{29.6.17}$$

Since $\rho^{\text{cl}}(m,n)$ is a Hermitian matrix, it can be brought into a diagonal matrix by a unitary transformation[4] U,

$$UW[\hat{\rho}^{\text{cl}}]U^\dagger = \frac{1}{2\pi\ell_B^2}\sum_{n=0}^{\infty}\lambda_n|n\rangle\langle n|. \tag{29.6.18}$$

Substituting this into (29.6.16) we find

$$\lambda_n = 0, \qquad \text{or} \qquad \lambda_n = 1. \tag{29.6.19}$$

The value $\lambda_n = 1$ ($\lambda_n = 0$) implies that there is one (no) electron in the nth Landau site. A multi-hole state is given by taking a finite number of $\lambda_n = 0$ in (29.6.18). If there are k zeros, (29.6.18) represents a QH state with k holes. This is the physical meaning of the state subject to the the noncommutative constraint (29.6.12).

We may take the diagonal matrix such that the first k sites are empty,

$$\begin{aligned} UW[\hat{\rho}^{\text{cl}}]U^\dagger &= \frac{1}{2\pi\ell_B^2}\Big(\sum_{n=0}^{\infty}|n+k\rangle\langle n+k|\Big) \\ &= \frac{1}{2\pi\ell_B^2}\Big(\sum_{n=0}^{\infty}|n+k\rangle\langle n|\Big)\Big(\sum_{m=0}^{\infty}|m\rangle\langle m+k|\Big). \end{aligned} \tag{29.6.20}$$

Consequently, we can express $W[\hat{\rho}^{\text{cl}}]$ as a product of two Weyl operators,

$$W[\hat{\rho}^{\text{cl}}] = \frac{1}{2\pi\ell_B^2}W[\phi^{\text{cl}}]W[\phi^{\text{cl}}]^\dagger, \tag{29.6.21}$$

as corresponds to (29.6.14), where

$$W[\phi^{\text{cl}}] = U^\dagger\Big(\sum_{n=0}^{\infty}|n+k\rangle\langle n|\Big). \tag{29.6.22}$$

It follows that

$$\begin{aligned} W[\phi^{\text{cl}}]^\dagger W[\phi^{\text{cl}}] &= \Big(\sum_{m=0}^{\infty}|m\rangle\langle m+k|\Big)UU^\dagger\Big(\sum_{n=0}^{\infty}|n+k\rangle\langle n|\Big) \\ &= 1, \end{aligned} \tag{29.6.23}$$

as corresponds to (29.6.15).

[4]The two states whose densities are $W[\hat{\rho}^{\text{cl}}]$ and $UW[\hat{\rho}^{\text{cl}}]U^\dagger$ are physically different. The state $UW[\hat{\rho}^{\text{cl}}]U^\dagger$ given by (29.6.18) describes a cylindrical symmetric state, but $W[\hat{\rho}^{\text{cl}}]$ does not in general.

Noncommutative CP^{N-1} Fields

We proceed to analyze the noncommutative constraint (29.6.8) in the N-component theory. We first study the $\nu = 1$ QH system. It is easy to see that the noncommutative constraint (29.6.8) is resolved as

$$\hat{\mathcal{D}}^{\text{cl}}_{\mu\nu}(\boldsymbol{x}) = \frac{1}{2\pi\ell_B^2}\phi^{\text{cl}}_\mu(\boldsymbol{x}) \star \phi^{\text{cl}*}_\nu(\boldsymbol{x}), \tag{29.6.24}$$

by introducing an N-component complex field $\phi^{\text{cl}}_\mu(\boldsymbol{x})$, provided it satisfies the noncommutative normalization condition

$$\sum_{\mu=1}^{N} \phi^{\text{cl}*}_\mu(\boldsymbol{x}) \star \phi^{\text{cl}}_\mu(\boldsymbol{x}) = 1. \tag{29.6.25}$$

We decompose it into the U(1) part $\phi^{\text{cl}}(\boldsymbol{x})$ and the SU(N) part $n_\mu(\boldsymbol{x})$ by setting

$$\phi^{\text{cl}}_\mu(\boldsymbol{x}) = \phi^{\text{cl}}(\boldsymbol{x}) \star n_\mu(\boldsymbol{x}), \tag{29.6.26}$$

where $\phi^{\text{cl}}(\boldsymbol{x})$ satisfies the noncommutative normalization condition (29.6.15). The SU(N) field $n_\mu(\boldsymbol{x})$ contains $N-1$ complex fields. From (29.6.25) and (29.6.26) we find

$$\sum_{\mu=1}^{N} n^*_\mu(\boldsymbol{x}) \star n_\mu(\boldsymbol{x}) = 1. \tag{29.6.27}$$

It is called the noncommutative CP^{N-1} field.

To explore SU(N) excitations it is enough to set $\phi^{\text{cl}}(\boldsymbol{x}) = 1$. Then, (29.6.24) is reduced to

$$\hat{\mathcal{D}}^{\text{cl}}_{\mu\nu}(\boldsymbol{x}) = \frac{1}{2\pi\ell_B^2} n_\mu(\boldsymbol{x}) \star n^*_\nu(\boldsymbol{x}), \tag{29.6.28}$$

or equivalently

$$\hat{\rho}^{\text{cl}}(\boldsymbol{x}) = \frac{1}{2\pi\ell_B^2}\sum_\mu n_\mu(\boldsymbol{x}) \star n^*_\mu(\boldsymbol{x}), \tag{29.6.29}$$

$$\hat{\mathfrak{I}}^{\text{cl}}_A(\boldsymbol{x}) = \frac{1}{4\pi\ell_B^2}\sum_{\mu\nu} (\lambda_A)_{\mu\nu}\, n_\mu(\boldsymbol{x}) \star n^*_\nu(\boldsymbol{x}). \tag{29.6.30}$$

We only study this type of excitations in what follows.

Noncommutative Grassmannian Fields

To analyze the $\nu = k$ QH system we need the Grassmannian $G_{N,k}$ field as described in Section 26.6 for the commutative theory. Equivalently we introduce k CP^{N-1} fields $\mathrm{n}^i_\mu(\boldsymbol{x})$, $i = 1, 2, \cdots k$, satisfying

$$\sum_{\mu=1}^{N} n^{i*}_\mu(\boldsymbol{x}) \star n^j_\mu(\boldsymbol{x}) = \delta_{ij}, \tag{29.6.31}$$

as corresponds to the normalization condition (26.6.7) in the commutative theory. We may resolve the noncommutative condition (29.6.8) by

$$\hat{\mathfrak{D}}^{\text{cl}}_{\mu\nu}(\boldsymbol{x}) = \frac{1}{2\pi\ell_B^2}\sum_{i=1}^{k} n^i_\mu(\boldsymbol{x}) \star n^{i*}_\nu(\boldsymbol{x}), \tag{29.6.32}$$

or

$$\hat{\rho}^{\text{cl}}(\boldsymbol{x}) = \frac{1}{2\pi\ell_B^2}\sum_{i,\mu} n^i_\mu(\boldsymbol{x}) \star n^{i*}_\mu(\boldsymbol{x}), \tag{29.6.33}$$

$$\hat{\mathfrak{I}}^{\text{cl}}_A(\boldsymbol{x}) = \frac{1}{4\pi\ell_B^2}\sum_{i,\mu\nu} (\lambda_A)_{\mu\nu}\, n^i_\mu(\boldsymbol{x}) \star n^{i*}_\nu(\boldsymbol{x}). \tag{29.6.34}$$

The Grassmannian $G_{N,k}$ field is defined as a collection of these fields, $Z(\boldsymbol{x}) = (n^1_\mu, n^2_\mu, \cdots, n^k_\mu)$.

Commutative Limit

Taking the commutative limit is a subtle process. In the spin-frozen system, both the constraint (29.6.12) and the normalization condition (29.6.15) become valid only for the ground state. Namely, it is impossible to analyze electron or hole excitation in the commutative effective theory. Similarly the U(1) part is frozen to the ground-state value in the multi-component system. However, the SU(N) part remains dynamical. Namely, the noncommutative CP^{N-1} field together with the noncommutative normalization condition (29.6.27) is reduced to the ordinary CP^{N-1} field together with the ordinary normalization condition. This is the reason why we are able to analyze isospin waves (Goldstone modes) and skyrmion excitations in the commutative effective theory.

29.7 Topological Charge Density

The topological charge in the commutative CP^{N-1} theory is given by (7.8.16), or

$$Q_{\text{sky}} = \frac{1}{2\pi i}\epsilon_{ij}\sum_\mu \int d^2x\, \partial_i n^*_\mu(\boldsymbol{x})\partial_j n_\mu. \tag{29.7.1}$$

This is generalized into the topological charge in the noncommutative plane as

$$Q_{\text{sky}} = \frac{1}{2\pi\ell_B^2}\sum_\mu \int d^2x\, [\![n^*_\mu(\boldsymbol{x}), n_\mu(\boldsymbol{x})]\!]. \tag{29.7.2}$$

Since the Moyal bracket is expanded as in (29.8.1), or

$$[\![n^*_\mu(\boldsymbol{x}), n_\mu(\boldsymbol{x})]\!] = -i\ell_B^2\epsilon_{jk}\partial_j n^*_\mu(\boldsymbol{x})\partial_k n_\mu(\boldsymbol{x}) + O(\ell_B^4), \tag{29.7.3}$$

the quantity (29.7.2) is reduced to the topological charge (29.7.1) in the commutative limit ($\ell_B \to 0$).

We now show that (29.7.2) meets all standard requirements as the topological charge. The topological charge density is

$$J^0_{\text{sky}}(\boldsymbol{x}) = \frac{1}{2\pi\ell_B^2}\sum_\mu [\![n^*_\mu(\boldsymbol{x}), n_\mu(\boldsymbol{x})]\!]. \tag{29.7.4}$$

First, due to an important property of the star product,

$$f(\boldsymbol{x})\star g(\boldsymbol{x}) = f(\boldsymbol{x})g(\boldsymbol{x}) + \partial_k\Lambda_k(\boldsymbol{x}), \tag{29.7.5}$$

the topological charge density is of the form $J^0_{\text{sky}}(\boldsymbol{x}) = \partial_k\Lambda_k(\boldsymbol{x})$ with a certain function $\Lambda_k(\boldsymbol{x})$. Thus (29.7.2) reads

$$Q_{\text{sky}} = \int d^2x\,\partial_k\Lambda_k(\boldsymbol{x}) = \oint dx_k\,\Lambda_k(\boldsymbol{x}), \tag{29.7.6}$$

where the contour integration is taken along a circle of an infinite radius. At the spatial infinity, the normalization (29.6.27) turns into a commutative condition (since the derivatives must vanish), and the asymptotic behavior is given by $n_\mu(\boldsymbol{x}) \sim r^0$, yielding $\Lambda_k \sim r^{-1}$. Thus, $Q \sim \oint d\theta$ with θ the azimuthal angle. The integration over the angle gives the number of windings. This is the standard way how the number of windings appear.

We next examine the continuity condition,

$$\partial_0 J^0_{\text{sky}}(\boldsymbol{x}) = \frac{1}{2\pi\ell_B^2}\sum_\mu [\![\partial_0 n^*_\mu(\boldsymbol{x}), n_\mu(\boldsymbol{x})]\!] + \frac{1}{2\pi\ell_B^2}\sum_\mu [\![n^*_\mu(\boldsymbol{x}), \partial_0 n_\mu(\boldsymbol{x})]\!] \equiv -\partial_k J^k_{\text{sky}}(\boldsymbol{x}), \tag{29.7.7}$$

where the existence of J^k_{sky} is guaranteed by the property (29.7.5). Based on dimensional considerations, the quantity $J^k_{\text{sky}}(\boldsymbol{x})$ behaves as $J^k_{\text{sky}}(\boldsymbol{x}) \sim r^{-2}$, implying the vanishing of the corresponding surface integral, or the conservation of the topological charge, $\partial_0 Q_{\text{sky}} = 0$.

Finally we relate the topological charge to the electron number of the excitation. The electron number of the $W_\infty(N)$-rotated state (29.6.1) is given by

$$\langle \mathfrak{S}|N|\mathfrak{S}\rangle = \langle \mathfrak{S}_0|e^{-iW}N_{\text{e}}e^{+iW}|\mathfrak{S}_0\rangle = \langle \mathfrak{S}_0|N_{\text{e}}|\mathfrak{S}_0\rangle, \tag{29.7.8}$$

where we have used the fact that the total electron number

$$N_{\text{e}} = \int d^2x\,\rho(\boldsymbol{x}) = \int d^2x\,\hat\rho(\boldsymbol{x}) = \sum_n\sum_\mu c^\dagger_\mu(n)c_\mu(n) \tag{29.7.9}$$

is a Casimir operator. The electron number of the state $|\mathfrak{S}\rangle$ is the same as that of the state $|\mathfrak{S}_0\rangle$. We set

$$\Delta N^{\text{cl}}_{\text{e}} = \langle \mathfrak{S}|N_{\text{e}}|\mathfrak{S}\rangle - \langle \text{g}|N_{\text{e}}|\text{g}\rangle = \int d^2x\,\Delta\hat\rho^{\text{cl}}(\boldsymbol{x}). \tag{29.7.10}$$

This is the electron number carried by the excitation described by the state $|\mathfrak{S}\rangle$.

We change the order of multiplications in (29.6.29) to derive

$$\hat{\rho}^{\text{cl}}(\mathbf{x}) = \rho_0 + \frac{1}{2\pi\ell_B^2}\sum_\mu [\![n_\mu(\mathbf{x}), n_\mu^*(\mathbf{x})]\!], \tag{29.7.11}$$

with the use of the constraint (29.6.31), where $\rho_0 = \nu/2\pi\ell_B^2$. The second term is the density excitation,

$$\Delta\hat{\rho}^{\text{cl}}(\mathbf{x}) = \frac{1}{2\pi\ell_B^2}\sum_\mu [\![n_\mu(\mathbf{x}), n_\mu^*(\mathbf{x})]\!]. \tag{29.7.12}$$

Comparing this with (29.7.4) we find

$$J_{\text{sky}}^0(\mathbf{x}) = -\Delta\hat{\rho}^{\text{cl}}(\mathbf{x}), \tag{29.7.13}$$

or

$$Q_{\text{sky}} = -\int d^2x\, \Delta\hat{\rho}^{\text{cl}}(\mathbf{x}) = -\Delta N_{\text{e}}^{\text{cl}}. \tag{29.7.14}$$

This is an integer, since $\langle\mathfrak{S}_0|N_{\text{e}}|\mathfrak{S}_0\rangle$ is an integer. Hence, the topological charge is quantized.

There are several different ways to define the topological charge Q. Needless to say, they are all equivalent. Nevertheless the topological charge densities are different, and in general they have no physical meaning. Among them the topological charge density (29.7.4) has a privileged role that it is essentially the electron density modulation induced by the topological soliton. As a consequence, the topological soliton carries necessarily the electric charge. This is a very peculiar phenomenon that occurs in the noncommutative theory.

The generalization to the noncommutative $\text{G}_{N,k}$ theory is straightforward. The topological charge density is

$$J_{\text{sky}}^0(\mathbf{x}) = \frac{1}{2\pi\ell_B^2}\sum_{r,\mu} [\![n_\mu^{r*}(\mathbf{x}), n_\mu^r(\mathbf{x})]\!], \tag{29.7.15}$$

which is equal to the density modulation as in (29.7.13). All the above arguments in this section hold as they stand.

29.8 Kac-Moody Algebra on Edges

We have already seen in Chapter 18 that rich excitations arise on the edge of the QH system. Any deformation of the incompressible droplet changes necessarily the shape of the QH system, which is an edge excitation. Edge excitations must also be a representation of the W_∞ algebra.[236,85] It has a strong power to determine all possible edge excitations[86] with the use of the representation theory of the algebra.[253,57]

Here, we study only sufficiently smooth excitations on the scale of the magnetic length ℓ_B. Then, the Moyal bracket (29.4.52) is approximated by the Poisson

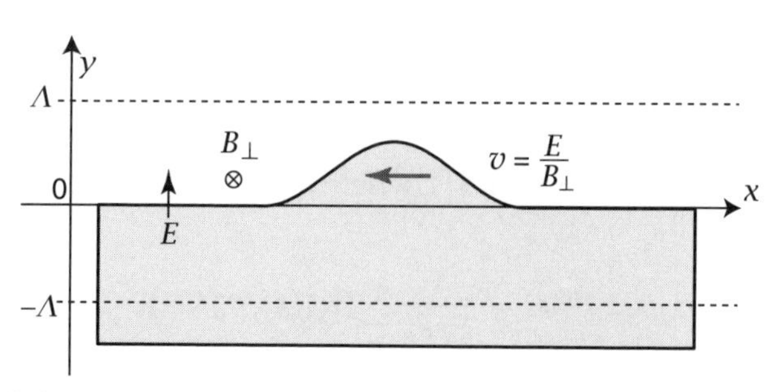

Fig. 29.2 A deformation of the incompressible droplet induces an edge excitation. The electron density is constant within the droplet and zero outside it. The height of an edge excitation is quite small, $\sim 2\pi\ell_B^2/L$, in a macroscopic system with size L.

bracket as in (29.4.55), or

$$h(\boldsymbol{x}) \equiv [\![f(\boldsymbol{x}), g(\boldsymbol{x})]\!] \simeq -i\ell_B^2 \left(\partial_x f \partial_y g - \partial_y f \partial_x g\right). \tag{29.8.1}$$

Let us smear the density operator by the following functions,[459]

$$\begin{aligned} f(\boldsymbol{x}) &= \frac{1}{\sqrt{2\pi}} e^{-ikx} \Theta(\Lambda - y)\Theta(\Lambda + y), \\ g(\boldsymbol{x}) &= \frac{1}{\sqrt{2\pi}} e^{-i\ell x} \Theta(\Lambda - y)\Theta(\Lambda + y), \end{aligned} \tag{29.8.2}$$

where $\Theta(\Lambda - y)$ is the step function, $\Theta(\Lambda - y) = 0$ for $\Lambda < y$ and $\Theta(\Lambda - y) = 1$ for $\Lambda > y$; Λ is taken so that any deformation is within the range $|y| < \Lambda$. We can take a very small quantity, $\Lambda \simeq 2\pi\ell_B^2/L$, in a macroscopic system with size L, because the height of the excitation is given by (18.3.4) with (18.3.6). With the choice of (29.8.2) the smeared density operator (29.4.48) is

$$\hat{\rho}(f) = \frac{1}{\sqrt{2\pi}} \int dx\, e^{-ikx} \int_{-\Lambda}^{\Lambda} dy\, \hat{\rho}(x, y) \equiv \hat{\rho}(k). \tag{29.8.3}$$

It depends only on the momentum component in the x direction. The Moyal bracket (29.8.1) yields

$$h(\boldsymbol{x}) \simeq \frac{-\ell_B^2}{2\pi}(k - \ell)\left\{\delta(\Lambda + y) - \delta(\Lambda - y)\right\} e^{-i(k+\ell)x} \Theta(\Lambda - y)\Theta(\Lambda + y). \tag{29.8.4}$$

Hence, we have

$$\hat{\rho}(h) = \int d^2x\, h(\boldsymbol{x})\hat{\rho}(\boldsymbol{x}) = -k\rho_0\ell_B^2\delta(k + l), \tag{29.8.5}$$

where we have used $\Theta(0) = 1/2$ and set

$$\delta(\Lambda - y)\hat{\rho}(\boldsymbol{x}) = 0, \qquad \delta(\Lambda + y)\hat{\rho}(\boldsymbol{x}) = \delta(\Lambda + y)\rho_0, \tag{29.8.6}$$

since no electrons exist at $y = \Lambda$, while the electron density is constant deep inside the bulk ($y = -\Lambda$). We are interested only in the dynamics of electrons on the edge [Fig.29.2].

We study the algebra for the smeared density (29.4.50),

$$[\hat{\rho}(f), \hat{\rho}(g)] = \hat{\rho}(h). \tag{29.8.7}$$

Using (29.8.4) and (29.8.5) we find

$$[\hat{\rho}(k), \hat{\rho}(l)] = -\frac{k}{2\pi m}\delta(k+l) \tag{29.8.8}$$

at $\nu = 2\pi\rho_0\ell_B^2 = 1/m$. This is the U(1) Kac-Moody algebra. It is the basic algebraic structure of the edge state. We construct the Lagrangian yielding this algebra as the canonical commutation relation, which is the chiral boson Lagrangian (18.4.7) we have already encountered in the hydrodynamical approach. The Hamiltonian is given by (18.4.6),

$$H = 2\pi m\hbar v \int_{-\infty}^{0} dk\, \rho(-k)\rho(k). \tag{29.8.9}$$

Here, ρ_k and ρ_{-k} are essentially the annihilation and creation operators. The edge ground state is defined by (18.4.14), or

$$\hat{\rho}(k)|\mathrm{g}\rangle_{\mathrm{edge}} = 0 \qquad \text{for} \quad k < 0. \tag{29.8.10}$$

The neutral sector is created by operating the density operator $\hat{\rho}(k)$ ($k > 0$) to the ground state $|\mathrm{g}\rangle$. Charged sectors consist of gapless edge electrons and quasiparticles, as was studied in Chapter 18. The ground-state condition (29.8.10) for the edge state is to be contrasted with the incompressibility condition (30.2.2) for the bulk state.

CHAPTER 30

LANDAU LEVEL PROJECTION

A self-consistent theory of the QH system is constructed by making the Landau-level projection. Almost all essential properties of QH systems are determined by the Coulomb interaction. We evaluate the energies of the ground state and an electron-hole pair in the monolayer SU(N) QH ferromagnet. We also analyze the bilayer QH system at $\nu = 1$ and $\nu = 2$. The ground state is determined as an eigenstate of the SU(4)-invariant Coulomb potential. All other interactions are regarded small and treated perturbatively.

30.1 PROJECTED COULOMB INTERACTIONS

The kinetic Hamiltonian is irrelevant as far as we are concerned about physics taking place in a single Landau level. The physical variable is solely given by the density operator. The Coulomb potential in the SU(N) QH system is given by

$$H_C = \frac{1}{2}\int d^2x d^2y\, V(\boldsymbol{x}-\boldsymbol{y})\Delta\rho(\boldsymbol{x})\Delta\rho(\boldsymbol{y}) = \pi\int d^2q\, V(\boldsymbol{q})\Delta\rho(-\boldsymbol{q})\Delta\rho(\boldsymbol{q}), \tag{30.1.1}$$

where $\Delta\rho(\boldsymbol{x})$ is the density excitation operator,

$$\Delta\rho(\boldsymbol{x}) = \rho(\boldsymbol{x}) - \rho_0. \tag{30.1.2}$$

It is necessary to project the density operator $\rho(\boldsymbol{q})$ to the Nth Landau level, assuming all the lower Landau levels are filled up, in order to prohibit electrons from escaping to a higher Landau level. Note that we should keep $V(\boldsymbol{q})$ as it is since it is just a c-number.

We have already studied the problem of the projection in Section 29.4, where the projected density operator is given by (29.4.6), or

$$\rho_N(\boldsymbol{q}) = F_N(\boldsymbol{q})\hat{\rho}(\boldsymbol{q}) \tag{30.1.3}$$

with $F_N(\boldsymbol{q})$ the Landau-level form factor (29.4.7) and $\hat{\rho}(\boldsymbol{q})$ the bare density (29.4.8). The projected Coulomb Hamiltonian is given by using the projected density in

(30.1.1), or

$$H_C^N = \pi \int d^2q\, V(\boldsymbol{q})\Delta\rho_N(-\boldsymbol{q})\Delta\rho_N(\boldsymbol{q}) = \pi \int d^2q\, V_N^{\text{eff}}(\boldsymbol{q})\Delta\hat{\rho}(-\boldsymbol{q})\hat{\rho}(\boldsymbol{q}), \tag{30.1.4}$$

with

$$V_N^{\text{eff}}(\boldsymbol{q}) = V(\boldsymbol{q})F_N(-\boldsymbol{q})F_N(\boldsymbol{q}). \tag{30.1.5}$$

This is the effective Coulomb potential in the Nth Landau level: See (19.1.6).

The potential $V_N^{\text{eff}}(\boldsymbol{x})$ approaches the ordinary Coulomb potential at large distance, since $F_N(\boldsymbol{q}) \to 1$ as $\boldsymbol{q} \to 0$,

$$V_N^{\text{eff}}(\boldsymbol{x}) \to V(\boldsymbol{x}) = \frac{e^2}{4\pi\varepsilon|\boldsymbol{x}|} \quad \text{as} \quad |\boldsymbol{x}| \to \infty. \tag{30.1.6}$$

However, at short distance it does not diverge in contrast to the ordinary Coulomb potential because of the factor $e^{-\ell_B^2 q^2/4}$ in the Landau-level form factor $F_N(\boldsymbol{q})$,

$$V_N^{\text{eff}}(\boldsymbol{x}) \to \text{constant}, \qquad \text{as} \quad |\boldsymbol{x}| \to 0. \tag{30.1.7}$$

This is reasonable because a real electron cannot be localized to a point within the lowest Landau level. See some profiles of the effective Coulomb potential $V_N^{\text{eff}}(\boldsymbol{x})$ in Fig.20.4.

In this Part we are mainly concerned about the physics taking place in the lowest Landau level ($N = 0$). In this case we set

$$V_D(\boldsymbol{q}) \equiv V_0^{\text{eff}}(\boldsymbol{q}) = V(\boldsymbol{q})e^{-\ell_B^2 q^2/2} = \frac{e^2}{4\pi\varepsilon|\boldsymbol{q}|}e^{-\ell_B^2 q^2/2}. \tag{30.1.8}$$

In the real space it is

$$V_D(\boldsymbol{x}) = \frac{e^2\sqrt{2\pi}}{8\pi\varepsilon\ell_B} I_0(x^2/4\ell_B^2)e^{-x^2/4\ell_B^2}, \tag{30.1.9}$$

with $I_0(x)$ the modified Bessel function,[1] and yields

$$V_D(\boldsymbol{x}) \to \frac{e^2\sqrt{2\pi}}{8\pi\varepsilon\ell_B} = \sqrt{\frac{\pi}{2}} E_C^0 \tag{30.1.10}$$

in short distance ($|\boldsymbol{x}| \to 0$). We also set $H_C = H_C^{N=0}$ for notational simplicity.

We rewrite the projected Coulomb Hamiltonian (30.1.4) in the Landau-site representation,

$$\begin{aligned} H_C &= \sum_{mnij} V_{mnij}[\rho(n,m) - \nu\delta_{mn}][\rho(j,i) - \nu\delta_{ij}] \\ &= \sum_{mnij} V_{mnij} \sum_{\sigma,\tau} c_\sigma^\dagger(m)c_\tau^\dagger(i)c_\tau(j)c_\sigma(n) + N_e\epsilon_X - \nu\,(N_e + \delta N_e)\,\epsilon_D, \end{aligned} \tag{30.1.11}$$

[1] Its asymptotic behaviors are $I_0(x) \to 1$ as $x \to 0$, and $I_0(x) \to e^x/\sqrt{2\pi x}$ as $x \to \infty$.

with

$$N_e = \int d^2x\, \rho(\boldsymbol{x}) = 2\pi \lim_{\boldsymbol{q}\to 0} \rho(\boldsymbol{q}) = \sum_n c^\dagger(n)c(n), \tag{30.1.12a}$$

$$\delta N_e = \int d^2x\, \Delta\rho(\boldsymbol{x}) = \int d^2x\, \rho(\boldsymbol{x}) - \rho_0 S = N_e - \nu N_\Phi. \tag{30.1.12b}$$

The numerical constant V_{mnij} is given by

$$V_{mnij} = \frac{1}{4\pi}\int d^2k\, V_D(\boldsymbol{k})\langle m|e^{iX\boldsymbol{k}}|n\rangle\langle i|e^{-iX\boldsymbol{k}}|j\rangle, \tag{30.1.13}$$

which satisfies $V_{mnij} = V_{ijmn} = V_{nmji}$. The angular momentum conservation forces V_{mnij} to vanish unless $m + i = n + j$, or

$$V_{mnjj} = V_{nnjj}\delta_{mn}, \qquad V_{mnnj} = V_{jnnj}\delta_{mj} \tag{30.1.14}$$

In particular,

$$V_{nnjj} = \frac{1}{4\pi}\int d^2k\, V_D(\boldsymbol{k})\langle n|e^{iX\boldsymbol{k}}|n\rangle\langle j|e^{-iX\boldsymbol{k}}|j\rangle \tag{30.1.15a}$$

is the direct integral, while

$$V_{njjn} = \frac{1}{4\pi}\int d^2k\, V_D(\boldsymbol{k})\langle n|e^{iX\boldsymbol{k}}|j\rangle\langle j|e^{-iX\boldsymbol{k}}|n\rangle \tag{30.1.15b}$$

is the exchange integral ($n \neq j$). The parameters ϵ_D and ϵ_X in (30.1.11) are defined by

$$\epsilon_D \equiv \sum_j V_{nnjj} = \frac{1}{4\pi\ell_B^2}\int d^2x\, V_D(\boldsymbol{x}) = \frac{1}{2\ell_B^2}V_D(\boldsymbol{q}=0), \tag{30.1.16a}$$

$$\epsilon_X \equiv \sum_j V_{njjn} = \frac{1}{4\pi}\int d^2k\, V_D(\boldsymbol{k}) = \frac{1}{4\ell_B^2}V_X(\boldsymbol{q}=0), \tag{30.1.16b}$$

as follows from (30.1.15) with the aid of (29.2.31), where $V_X(\boldsymbol{q})$ is the exchange potential defined later by (32.1.12): The summations $\sum_j V_{nnjj}$ and $\sum_j V_{njjn}$ are taken over j with n fixed arbitrarily. In particular,

$$\epsilon_X = \frac{1}{4\pi}\int d^2k\, V(\boldsymbol{k})e^{-\ell_B^2 k^2/2} = \frac{1}{2}\sqrt{\frac{\pi}{2}}E_C^0 \tag{30.1.17}$$

for the lowest Landau level. The operator N_e and δN_e represents the total electron number and the electron number brought by the excitation, respectively. The total electron number is fixed to the ground-state value, $N_e = \nu N_\Phi$, in the analysis of a Goldstone mode and also of a pair excitation of quasiparticle and quasihole. The quantities ϵ_D and ϵ_X are the direct and exchange energy parameters, respectively.[2]

[2]Strictly speaking, the term V_{nnnn} common in ϵ_D and ϵ_X is to be interpreted as the self-energy: See (30.3.9).

30.2 Monolayer QH System

We analyze the ground state in the N-component QH system at integer filling factor ν. The Coulomb Hamiltonian is invariant under the global SU(N) transformation, and given by (30.1.11) in the Landau-site representation. The ground state is given by a homogeneous distribution of electrons satisfying,

$$\rho(n,m)|g\rangle = \sum_{\sigma} c_{\sigma}^{\dagger}(m)c_{\sigma}(n)|g\rangle = \nu\delta_{mn}|g\rangle. \tag{30.2.1}$$

In the momentum space, with the use of (29.2.31), it reads

$$\hat{\rho}(\boldsymbol{q})|g\rangle = \frac{1}{2\pi}\sum_{mn}\langle m|e^{-iqX}|n\rangle\rho(n,m)|g\rangle = 2\pi\rho_0\delta(\boldsymbol{q})|g\rangle, \tag{30.2.2}$$

as is the incompressibility condition (28.1.31). It then follows that

$$H_C|g\rangle = 0. \tag{30.2.3}$$

The Coulomb Hamiltonian is normalized to be zero on the ground state.

SU(2) Spin Ferromagnet

We consider the $\nu = 1$ QH system with spin. The Hamiltonian is given by

$$H = H_C - \Delta_Z\int d^2x\, \mathcal{S}_a(\boldsymbol{x}) \tag{30.2.4}$$

with the spin density

$$\mathcal{S}_a(\boldsymbol{x}) = \frac{1}{2}\Psi^{\dagger}(\boldsymbol{x})\tau_a\Psi(\boldsymbol{x}). \tag{30.2.5}$$

Within the lowest Landau level it looks as

$$\int d^2x\, \mathcal{S}_a(\boldsymbol{x}) = 2\pi\lim_{\boldsymbol{q}\to 0}\mathcal{S}_a(\boldsymbol{q}) = \sum_n S_a(n,n), \tag{30.2.6}$$

with

$$S_a(m,n) = \frac{1}{2}c^{\dagger}(n)\tau_a c(m) \tag{30.2.7}$$

as in (30.1.12a).

Without the Zeeman effect ($\Delta_Z = 0$), the ground state is degenerate with respect to the spin degree of freedom. The ground state, satisfying the homogeneity condition (30.2.1), is expressed as

$$|g\rangle = \prod_n\left(e^{i\vartheta_s/2}\sqrt{\frac{1+\sigma_s}{2}}c_{\uparrow}^{\dagger}(n) + e^{-i\vartheta_s/2}\sqrt{\frac{1-\sigma_s}{2}}c_{\downarrow}^{\dagger}(n)\right)|0\rangle \tag{30.2.8}$$

with the use of two parameters σ_s and ϑ_s. The ground-state energy is independent of σ_s and ϑ_s,

$$\langle g|H_C|g\rangle = 0. \tag{30.2.9}$$

Due to its degeneracy it is a coherent state such that

$$\langle \mathrm{g}|S_a(m,n)|\mathrm{g}\rangle = \mathcal{S}_a^{\mathrm{g}}\delta_{mn} \tag{30.2.10}$$

or

$$\mathcal{S}_a^{\mathrm{g}}(\boldsymbol{x}) \equiv \frac{1}{\rho_\Phi}\langle \mathrm{g}|\hat{\mathcal{S}}_a(\boldsymbol{x})|\mathrm{g}\rangle = \mathcal{S}_a^{\mathrm{g}} \tag{30.2.11}$$

with

$$\mathcal{S}_x^{\mathrm{g}} = \frac{1}{2}\sqrt{1-\sigma_{\mathrm{s}}^2}\sin\vartheta_{\mathrm{s}}, \qquad \mathcal{S}_y^{\mathrm{g}} = \frac{1}{2}\sqrt{1-\sigma_{\mathrm{s}}^2}\cos\vartheta_{\mathrm{s}}, \qquad \mathcal{S}_z^{\mathrm{g}} = \frac{1}{2}\sigma_{\mathrm{s}}. \tag{30.2.12}$$

It is the QH ferromagnet, where all spins are spontaneously polarized into an arbitrary direction labeled by σ_{s} and ϑ_{s}. Without loss of generality we may choose $\sigma_{\mathrm{s}} = 1$ and $\vartheta_{\mathrm{s}} = 0$, as is realized by the ground state

$$|\mathrm{g}\rangle = \prod_n c_\uparrow^\dagger(n)|0\rangle. \tag{30.2.13}$$

We can prove[479] that any locally spin rotated state has a higher energy than the spin polarized state (30.2.13).

With the Zeeman effect ($\Delta_Z \neq 0$), the Hamiltonian (30.2.4) yields

$$\langle \mathrm{g}|H|\mathrm{g}\rangle = -\frac{1}{2}\sigma_{\mathrm{s}}\Delta_Z N_\Phi. \tag{30.2.14}$$

The ground state is the minimum-energy state, which is given (30.2.13) with $\sigma_{\mathrm{s}} = 1$. The spin configuration is

$$\langle \mathrm{g}|S_a(m,n)|\mathrm{g}\rangle = \frac{1}{2}\delta_{az}\delta_{mn}, \tag{30.2.15}$$

or

$$\mathcal{S}_a^{\mathrm{g}}(\boldsymbol{q}) = \frac{1}{\rho_\Phi}\langle \mathrm{g}|\hat{\mathcal{S}}_a(\boldsymbol{q})|\mathrm{g}\rangle = \pi\delta_{az}\delta(\boldsymbol{q}). \tag{30.2.16}$$

The ground state is an eigenstate of the total Hamiltonian,

$$H|\mathrm{g}\rangle = -\frac{1}{2}\sigma_{\mathrm{s}}\Delta_Z N_\Phi|\mathrm{g}\rangle, \tag{30.2.17}$$

since

$$S_z(n,n)|\mathrm{g}\rangle = \frac{1}{2}|\mathrm{g}\rangle. \tag{30.2.18}$$

The SU(2) symmetry is broken explicitly by the Zeeman effect.

Nevertheless, driven by the exchange interaction, the spin wave arises as a low-energy excitation with the correlation length ξ_{spin}: See Section 15.4. The spin wave should be regarded as a Goldstone mode as far as the correlation length ξ_{spin} is larger than the magnetic length ℓ_B.

SU(N) QH Systems

We generalize the SU(2) spin system to the SU(N) isospin system. Since there are N isospin states in one Landau site, we may consider the integer QH state at $\nu \leq N-1$. The generalization is trivial at $\nu = 1$, where the ground state is the SU(N) QH ferromagnet with the isospin spontaneously polarized into an arbitrary SU(N) direction. The case $\nu \geq 2$ is similarly discussed. See the bilayer SU(4) system in Section 30.7 for details.

There are many ways to fill in ν levels. The ground state is chosen by imposing the homogeneity condition (30.2.1),

$$\rho(m,n)|\mathrm{g}\rangle = \sum_{\sigma} c_{\sigma}^{\dagger}(n)c_{\sigma}(m)|\mathrm{g}\rangle = \nu\delta_{mn}|\mathrm{g}\rangle. \tag{30.2.19}$$

An example is given by

$$|\mathrm{g}\rangle = \prod_{n} c_1^{\dagger}(n)c_2^{\dagger}(n)\cdots c_{\nu}^{\dagger}(n)|0\rangle. \tag{30.2.20}$$

For the isospin density I_A we find

$$\langle \mathrm{g}|I_A(m,n)|\mathrm{g}\rangle = \frac{1}{2}\delta_{mn}\sum_{i=1}^{\nu}(\lambda_A)_{ii} \equiv \delta_{mn}\mathcal{I}[A], \tag{30.2.21}$$

where λ_A is the standard generator in SU(N): See Appendix D. In the momentum space they read

$$\hat{\rho}(\boldsymbol{q})|\mathrm{g}\rangle = \frac{1}{2\pi}\sum_{mn}\langle n|e^{-iqX}|m\rangle\rho(m,n)|\mathrm{g}\rangle = 2\pi\rho_0\delta(\boldsymbol{q})|\mathrm{g}\rangle, \tag{30.2.22}$$

and

$$\langle \mathrm{g}|\hat{\mathcal{I}}_A(\boldsymbol{q})|\mathrm{g}\rangle = \frac{1}{2\pi}\sum_{mn}\langle n|e^{-iqX}|m\rangle\langle \mathrm{g}|I_A(m,n)|\mathrm{g}\rangle = \pi\rho_{\Phi}\mathcal{I}[A]\delta(\boldsymbol{q}), \tag{30.2.23}$$

where use was made of (29.2.31).

We study $\mathcal{I}[A] = \frac{1}{2}\sum_{i=1}^{\nu}(\lambda_A)_{ii}$. Only the elements $\mathcal{I}[3], \mathcal{I}[8], \cdots, \mathcal{I}[N^2-1]$ may have nonvanishing values, since the diagonal elements of λ_A are involved. Even among them there are vanishing ones. The nonvanishing ones are given by

$$\begin{aligned}
\mathcal{I}[(\nu+1)^2-1] &= \frac{\nu}{\sqrt{2\nu(\nu+1)}}, \\
\mathcal{I}[(\nu+2)^2-1] &= \frac{\nu}{\sqrt{2(\nu+1)(\nu+2)}}, \qquad \cdots, \\
\mathcal{I}[N^2-1] &= \frac{\nu}{\sqrt{2(N-1)N}}.
\end{aligned} \tag{30.2.24}$$

Thus,

$$\sum_{A=1}^{N^2-1}\mathcal{I}[A]\mathcal{I}[A] = \frac{\nu(N-\nu)}{2N}. \tag{30.2.25}$$

From (30.2.23) we have

$$\langle \mathrm{g}|\hat{\mathcal{I}}_A(\boldsymbol{x})|\mathrm{g}\rangle = \rho_\Phi \mathcal{I}_A^{\mathrm{g}} = \rho_\Phi \mathcal{I}[A] \tag{30.2.26}$$

for the ground state (30.2.20). By setting

$$\hat{\rho}^{\mathrm{cl}}(\boldsymbol{x}) = \nu\rho_\Phi, \quad \hat{\mathcal{I}}_A^{\mathrm{cl}}(\boldsymbol{x}) = \rho_\Phi \mathcal{I}_A^{\mathrm{g}}, \tag{30.2.27}$$

the relation

$$\sum_{A=1}^{N^2-1} \mathcal{I}_A^{\mathrm{g}} \mathcal{I}_A^{\mathrm{g}} = \frac{\nu(N-\nu)}{2N} \tag{30.2.28}$$

follows for the ground state. This is derived also from (29.6.1).

The low-lying excitation is the isospin wave, which is a perturbative excitation without the electron density modulation. We substitute

$$\hat{\rho}^{\mathrm{cl}}(\boldsymbol{x}) = \nu\rho_\Phi, \quad \hat{\mathcal{I}}_A^{\mathrm{cl}}(\boldsymbol{x}) = \rho_\Phi \mathcal{I}_A^{\mathrm{cl}}(\boldsymbol{x}) \tag{30.2.29}$$

into (29.6.1) and find

$$\sum_{A=1}^{N^2-1} \mathcal{I}_A^{\mathrm{cl}}(\boldsymbol{x}) \star \mathcal{I}_A^{\mathrm{cl}}(\boldsymbol{x}) = \frac{\nu(N-\nu)}{2N}. \tag{30.2.30}$$

It is reduced to

$$\sum_{A=1}^{N^2-1} \mathcal{I}_A^{\mathrm{cl}}(\boldsymbol{x}) \mathcal{I}_A^{\mathrm{cl}}(\boldsymbol{x}) = \frac{\nu(N-\nu)}{2N} \tag{30.2.31}$$

in the commutative limit. For the instance of the SU(4) QH ferromagnet the normalized isospin field obeys

$$\sum_{A=1}^{15} \mathcal{I}_A^{\mathrm{cl}}(\boldsymbol{x}) \mathcal{I}_A^{\mathrm{cl}}(\boldsymbol{x}) = \frac{3}{8} \quad \text{at} \quad \nu = 1, \tag{30.2.32a}$$

$$\sum_{A=1}^{15} \mathcal{I}_A^{\mathrm{cl}}(\boldsymbol{x}) \mathcal{I}_A^{\mathrm{cl}}(\boldsymbol{x}) = \frac{1}{2} \quad \text{at} \quad \nu = 2 \tag{30.2.32b}$$

in the commutative limit $\ell_B \to 0$. Isospin waves are Goldstone modes. There are $N^2 - 1$ real components in $\mathcal{I}_A^{\mathrm{cl}}(\boldsymbol{x})$, but not all of them are independent. The independent number of real components is only $2(N-1)$ at $\nu = 1$. Later we shall derive the effective Hamiltonian for the Goldstone modes in terms of the CP^{N-1} field: See (32.7.18).

30.3 Electron-Hole Pair Excitations

We study an electron-hole state[3] in the SU(N) QH ferromagnet at $\nu = 1$,

$$|e_j; h_k\rangle = c_\downarrow^\dagger(j) c_\uparrow(k)|g\rangle, \tag{30.3.1}$$

where one electron is removed from the ground state $|g\rangle = \prod_n c_\uparrow^\dagger(n)|0\rangle$ at site k and added at site j. The total electron number is unchanged.

In the simplest example of the SU(2) QH ferromagnet, the electron density and the spin density are

$$\begin{aligned}
\rho^{\text{cl}}(m,n) =& \langle e_j; h_k|\rho(m,n)|e_j; h_k\rangle = \delta_{mn} - \delta_{mk}\delta_{nk} + \delta_{mj}\delta_{nj}, \\
S_z^{\text{cl}}(m,n) =& \langle e_j; h_k|S_z(m,n)|e_j; h_k\rangle = \frac{1}{2}\left(\delta_{mn} - \delta_{mk}\delta_{nk} - \delta_{mj}\delta_{nj}\right), \\
S_x^{\text{cl}}(m,n) =& S_y^{\text{cl}}(m,n) = 0.
\end{aligned} \tag{30.3.2}$$

With these values the following relation is readily checked,

$$\sum_a \sum_i S_a^{\text{cl}}(m,i) S_a^{\text{cl}}(i,n) + \frac{1}{4}\sum_i \rho^{\text{cl}}(m,i)\rho^{\text{cl}}(i,n) = \frac{1}{2}\rho^{\text{cl}}(m,n). \tag{30.3.3}$$

In the case of the SU(N) theory we have

$$\sum_A \sum_i I_A^{\text{cl}}(m,i) I_A^{\text{cl}}(i,n) + \frac{1}{2N}\sum_i \rho^{\text{cl}}(m,i)\rho^{\text{cl}}(i,n) = \frac{1}{2}\rho^{\text{cl}}(m,n). \tag{30.3.4}$$

In the real space it reads

$$\sum_A \hat{\mathfrak{I}}_A^{\text{cl}}(\boldsymbol{x}) \star \hat{\mathfrak{I}}_A^{\text{cl}}(\boldsymbol{x}) + \frac{1}{2N}\hat{\rho}^{\text{cl}}(\boldsymbol{x}) \star \hat{\rho}^{\text{cl}}(\boldsymbol{x}) = \frac{1}{4\pi\ell_B^2}\hat{\rho}^{\text{cl}}(\boldsymbol{x}), \tag{30.3.5}$$

where $\star$ stands for the star product. This is a special example of the formula (29.6.10) valid for a wide class of states.

We next calculate the excitation energy. It is straightforward to derive

$$\begin{aligned}
\Delta E\left(e_j; h_k\right) \equiv& \langle e_j; h_k|H_C|e_j; h_k\rangle - \langle g|H_C|g\rangle \\
=& \sum_n V_{jnnj} + \sum_n V_{knnk} - 2V_{kkjj}.
\end{aligned} \tag{30.3.6}$$

The energy is invariant under the exchange of electron and hole. This is the pair excitation energy. We may rewrite this as

$$\Delta E\left(e_j; h_k\right) = 2\epsilon_X - 2V_{kkjj}, \tag{30.3.7}$$

where we have used (30.1.16b). The term $2V_{kkjj}$ represents the direct Coulomb energy between the hole at site k and the electron at site j.

---

[3]Here, the indices ↑ and ↓ represent any two different isospin labels in the SU(N) QH ferromagnet. In the spin SU(2) case they represent the up-spin and down-spin lavels.

When they are separated far enough, we may set $V_{kkjj} = 0$. Indeed, as we have argued in Section 12.4, thermal spin fluctuations create well-separated pairs. Hence the pair creation energy is[254]

$$\Delta E_{\infty}(\mathrm{e};\mathrm{h}) = 2\epsilon_{\mathrm{X}}. \tag{30.3.8}$$

This agrees with the energy associated with two Landau sites: See (30.1.16b).

We decompose ϵ_{X} as

$$\epsilon_{\mathrm{X}} = E_{\mathrm{D}} + E_{\mathrm{X}} \tag{30.3.9}$$

with

$$E_{\mathrm{D}} = V_{kkkk}, \qquad E_{\mathrm{X}} = \sum_{n\neq k} V_{knnk}. \tag{30.3.10}$$

We may interpret E_{D} and E_{X} as the self-energy and the genuine exchange energy.

In the pair excitation energy (30.3.7), V_{jjkk} is a decreasing function of the distance between the two sites j and k. The minimum excitation energy occurs when the two cites coincide,

$$\Delta E(\mathrm{e}_k;\mathrm{h}_k) = 2\sum_{n\neq k} V_{knnk} = 2E_{\mathrm{X}}. \tag{30.3.11}$$

Hence, when we flip one spin, it costs only the genuine exchange energy.

30.4 Electron Excitation and Hole Excitation

We investigate the excitation of a hole and an electron independently in the SU(2) QH ferromagnet at $\nu = 1$,

$$|\mathrm{h}_j\rangle = c_{\uparrow}(j)|\mathrm{g}\rangle = \prod_{n\neq j} c_{\uparrow}^{\dagger}(n)|0\rangle, \tag{30.4.1a}$$

$$|\mathrm{e}_j\rangle = c_{\downarrow}^{\dagger}(j)|\mathrm{g}\rangle = c_{\downarrow}^{\dagger}(j)\prod_{n} c_{\uparrow}^{\dagger}(n)|0\rangle. \tag{30.4.1b}$$

Their excitation energies are

$$\Delta E(\mathrm{h}) \equiv \langle \mathrm{h}_j|\hat{H}_{\mathrm{C}}|\mathrm{h}_j\rangle - \langle \mathrm{g}|\hat{H}_{\mathrm{C}}|\mathrm{g}\rangle = 2\sum_{n\neq k}(V_{knnk} - V_{kknn}) = 2E_{\mathrm{X}} - 2\sum_{n\neq k} V_{kknn}, \tag{30.4.2a}$$

$$\Delta E(\mathrm{e}) \equiv \langle \mathrm{e}_j|\hat{H}_{\mathrm{C}}|\mathrm{e}_j\rangle - \langle \mathrm{g}|\hat{H}_{\mathrm{C}}|\mathrm{g}\rangle = 2\sum_{n} V_{jjnn} = 2E_{\mathrm{D}} + 2\sum_{n\neq k} V_{kknn}. \tag{30.4.2b}$$

The pair excitation energy (30.3.8) is the sum of these two energies, as expected. The exchange energy is needed to remove an electron from the ground state, but not to add an electron to the ground state: See also Section 4.1.

In spite of this fact it is more convenient to decompose their energies as

$$\Delta E(\mathrm{h}) = \epsilon_{\mathrm{X}} - 2E_{\mathrm{extra}}, \qquad \Delta E(\mathrm{e}) = \epsilon_{\mathrm{X}} + 2E_{\mathrm{extra}} \tag{30.4.3}$$

with

$$E_{\text{extra}} = 2\sum_n V_{jjnn} - \sum_n V_{jnnj}. \tag{30.4.4}$$

Then we can treat the electron and the hole in a symmetric way. Recall that we have made this type of decomposition to evaluate the excitation energy of skyrmions.

We use the von-Neumann-lattice representation to evaluate the self-energy E_{D}. In this representation the Landau site $|n\rangle$ is indexed by a complex number,

$$n = n_x + in_y, \tag{30.4.5}$$

where n_x, n_y =integers for the square lattice, and given by

$$|n\rangle = \exp\left[-\frac{1}{2}\pi|n|^2 + \sqrt{\pi}nb^\dagger\right]|0\rangle. \tag{30.4.6}$$

The Coulomb energy is given formally by the same equation as (30.1.13). The self-energy reads

$$E_{\text{D}} = V_{nnnn} = \frac{1}{4\pi}\int d^2k\, e^{-\ell_B^2 k^2/2} V(\mathbf{k})|\langle n|e^{ikX}|n\rangle|^2. \tag{30.4.7}$$

Here, using

$$X = \frac{\ell_B}{\sqrt{2}}(b + b^\dagger), \qquad Y = \frac{i\ell_B}{\sqrt{2}}(b - b^\dagger), \tag{30.4.8}$$

we have

$$\langle n|e^{ikX}|n\rangle = e^{-\ell_B^2 k^2/4} e^{i\sqrt{\pi}\ell_B(k_x n_x + k_y n_y)}, \tag{30.4.9}$$

and hence

$$E_{\text{D}} = \frac{1}{4\pi}\int d^2k\, e^{-\ell_B^2 k^2} V(\mathbf{k}) = \frac{1}{\sqrt{2}}\epsilon_{\text{X}}. \tag{30.4.10}$$

The exchange energy is given by

$$E_{\text{X}} = \epsilon_{\text{X}} - E_{\text{D}} = \left(\frac{\sqrt{2}-1}{\sqrt{2}}\right)\epsilon_{\text{X}}. \tag{30.4.11}$$

We have obtained these for the square lattice but it is independent of the type of the von Neumann lattice.

We comment that the self-energy E_{D} is a function of the site in the symmetric-gauge representation,

$$E_{\text{D}}(n) = V_{nnnn} = \frac{1}{4\pi}\int d^2k\, e^{-\ell_B^2 k^2/2} V(\mathbf{k})|\langle n|e^{ikX}|n\rangle|^2 = \frac{E_{\text{C}}^0}{\sqrt{2}}\int_0^\infty dx\, e^{-2x^2}\left[L_n(x^2)\right]^2. \tag{30.4.12}$$

This is reasonable. The Landau site with the angular-momentum index n is a ring with its width $\ell_B/\sqrt{2n}$ for $n \neq 0$. The Coulomb energy $E_{\text{D}}(n)$ becomes smaller for larger n since the electric charge $-e$ is spread over a larger ring,

$$E_D(n) = V_{nnnn} \simeq \frac{1}{n}\ln n \qquad \text{as} \quad n \to \infty. \tag{30.4.13}$$

The largest value of the self-energy is given by

$$E_D(0) = V_{0000} = \frac{E_C^0}{\sqrt{2}}\int_0^\infty dx\, e^{-2x^2} = \frac{1}{\sqrt{2}}\epsilon_X, \tag{30.4.14}$$

as agrees with (30.4.10), for which the exchange energy takes the minimum value,

$$E_X(0) = \epsilon_X - E_D(0) = \left(\frac{\sqrt{2}-1}{\sqrt{2}}\right)\epsilon_X, \tag{30.4.15}$$

as agrees with (30.4.11). Thus, when one spin flips in a single site, it must occur in the zero-angular-momentum site ($n = 0$), since the excitation energy $2E_X$ is minimum. What would happen when two electrons flip their spins. Do the second one occur at $n = 1$ with a larger excitation energy? This cannot be the case physically because there are no special points in the plane. This seemingly curious phenomenon is an artifact of the gauge choice. Indeed, various Landau sites are not equivalent in the symmetric-gauge representation, where a genuine local state is given only by the zero-angular-momentum site, and all other states are made of rings. Thus, when we estimate the excitation energy of an electron or a hole in the symmetric gauge we should put it in the angular-momentum zero state.

The classical densities associated with one-hole and one-electron excitations (30.4.1) are given by

$$\rho_h^{cl}(n,m) = \sum_\sigma \langle h|c_\sigma^\dagger(m)c_\sigma(n)|h\rangle = \delta_{mn} - \delta_{m0}\delta_{n0}, \tag{30.4.16a}$$

$$\rho_e^{cl}(n,m) = \sum_\sigma \langle e|c_\sigma^\dagger(m)c_\sigma(n)|e\rangle = \delta_{mn} + \delta_{m0}\delta_{n0}, \tag{30.4.16b}$$

where one hole or one electron is placed in the angular-momentum zero state in the symmetric gauge. We can construct the Weyl operators $W[\hat\rho_h^{cl}]$, $W[\hat\rho_e^{cl}]$ and the symbols $\hat\rho_h^{cl}(\boldsymbol{x})$, $\hat\rho_e^{cl}(\boldsymbol{x})$ according to (29.4.23) and (29.4.24). Explicitly we obtain

$$\hat\rho_h^{cl}(\boldsymbol{x}) = \frac{1}{2\pi\ell_B^2}[1 - \Xi_{00}(\boldsymbol{x})] = \rho_0\left(1 - 2e^{-r^2/\ell_B^2}\right), \tag{30.4.17a}$$

$$\hat\rho_e^{cl}(\boldsymbol{x}) = \frac{1}{2\pi\ell_B^2}[1 + \Xi_{00}(\boldsymbol{x})] = \rho_0\left(1 + 2e^{-r^2/\ell_B^2}\right). \tag{30.4.17b}$$

The spin densities are

$$\hat{\mathcal{S}}_x^{cl}(\boldsymbol{x}) = \hat{\mathcal{S}}_y^{cl}(\boldsymbol{x}) = 0, \qquad \hat{\mathcal{S}}_z^{cl}(\boldsymbol{x}) = \frac{1}{2}\rho_0\left(1 - 2e^{-r^2/\ell_B^2}\right). \tag{30.4.18}$$

The spin texture of the hole state is trivial.

30.5 BILAYER SYSTEM WITHOUT SPIN ($\nu = 1$)

We proceed to discuss the bilayer system at $\nu = 1$. Making the LLL projection of the Coulomb Hamiltonian (22.2.11) we obtain

$$H_C = \pi \int d^2p \left[V_D^+(\boldsymbol{p})\Delta\hat{\rho}(-\boldsymbol{p})\Delta\hat{\rho}(\boldsymbol{p}) + 4V_D^-(\boldsymbol{p})\Delta\hat{\mathcal{P}}_z(-\boldsymbol{p})\Delta\hat{\mathcal{P}}_z(\boldsymbol{p}) \right], \tag{30.5.1}$$

where the potential is

$$V_D^\pm(\boldsymbol{p}) = \frac{e^2}{8\pi\varepsilon|\boldsymbol{p}|}\left(1 \pm e^{-|\boldsymbol{p}|d}\right) e^{-\frac{1}{2}\ell_B^2 p^2}. \tag{30.5.2}$$

The Fourier transformation is

$$V_D^\pm(\boldsymbol{x}) = \frac{e^2\sqrt{2\pi}}{16\pi\varepsilon\ell_B} e^{-|\boldsymbol{x}|^2/4\ell_B^2} I_0(|\boldsymbol{x}|^2/4\ell_B^2) \pm \frac{e^2}{8\pi\varepsilon}\int_0^\infty dk\, e^{-\frac{1}{2}\ell_B^2 k^2 - dk} J_0(k|\boldsymbol{x}|). \tag{30.5.3}$$

The potential $V_D^\pm(\boldsymbol{x})$ approaches the standard one $V^\pm(\boldsymbol{x})$ asymptotically ($|\boldsymbol{x}| \gg \ell_B$) but is modified considerably at the short distance ($|\boldsymbol{x}| \ll \ell_B$).

The Coulomb Hamiltonian is decomposed into two terms,

$$H_C^+ = \pi \int d^2q\, V_D^+(\boldsymbol{p})\Delta\hat{\rho}(-\boldsymbol{p})\Delta\hat{\rho}(\boldsymbol{p}), \tag{30.5.4a}$$

$$H_C^- = 4\pi \int d^2q\, V_D^-(\boldsymbol{p})\Delta\hat{\mathcal{P}}_z(-\boldsymbol{p})\Delta\hat{\mathcal{P}}_z(\boldsymbol{p}). \tag{30.5.4b}$$

The Landau-site Hamiltonians are

$$H_C^+ = \sum_{mnij} V_{mnij}^+ [\rho(n,m) - \nu\delta_{mn}][\rho(j,i) - \nu\delta_{ij}], \tag{30.5.5a}$$

$$H_C^- = 4V_{mnij}^- \left[P_z(n,m) - \frac{1}{2}\nu\sigma_0\delta_{mn} \right] \left[P_z(j,i) - \frac{1}{2}\nu\sigma_0\delta_{ij} \right], \tag{30.5.5b}$$

with

$$V_{mnij}^\pm = \frac{1}{4\pi}\int d^2k\, V_D^\pm(\boldsymbol{k})\langle m|e^{iX\boldsymbol{k}}|n\rangle\langle i|e^{-iX\boldsymbol{k}}|j\rangle. \tag{30.5.6}$$

The tunneling and bias terms are combined into the pseudo-Zeeman term,

$$H_{pZ} = \sum_m [-\Delta_{SAS} P_x(m,m) + \Delta_{bias} P_z(m,m)], \tag{30.5.7}$$

where

$$P_a(n,m) = \frac{1}{2}\sum_{\alpha\beta} c_\alpha^\dagger(m)(\tau_a)_{\alpha\beta} c_\beta(n) \tag{30.5.8}$$

and

$$\Delta_{bias} = \frac{\sigma_0}{\sqrt{1-\sigma_0^2}}\Delta_{SAS}. \tag{30.5.9}$$

The bias term represents the energy increase due to an imbalanced density configuration between the two layers.

In this section we study the spinless bilayer QH system at $\nu = 1$. The SU(2)-invariant Hamiltonian (30.5.4a) is rewritten just as in (30.1.11),

$$H_C^+ = \sum_{mnij} V_{mnij}^+ \sum_{\sigma,\tau} c_\sigma^\dagger(m) c_\tau^\dagger(i) c_\tau(j) c_\sigma(n) + N_e \epsilon_X^+ - (N_e + \delta N_e)\, \epsilon_D^+. \tag{30.5.10}$$

Similarly the SU(2)-noninvariant Hamiltonian (30.5.4b) is rewritten as

$$H_C^- = \sum_{mnij} V_{mnij}^- \sum_{\sigma,\tau} \xi_\sigma \xi_\tau c_\sigma^\dagger(m) c_\tau^\dagger(i) c_\tau(j) c_\sigma(n)$$
$$+ N_e \epsilon_X^- - 4\sigma_0 \epsilon_D^- \sum_m P_z(m,m) + \sigma_0^2 (N_e - \delta N_e)\, \epsilon_D^-, \tag{30.5.11}$$

where $\xi_f = 1$, $\xi_b = -1$. Here

$$\epsilon_D^\pm \equiv \sum_j V_{nnjj}^\pm = \frac{1}{4\pi \ell_B^2} \int d^2x\, V_D^\pm(\boldsymbol{x}) = \frac{1}{2\ell_B^2} V_D^\pm(\boldsymbol{q} = 0), \tag{30.5.12a}$$

$$\epsilon_X^\pm \equiv \sum_j V_{njjn}^\pm = \frac{1}{4\pi} \int d^2k\, V_D^\pm(\boldsymbol{k}) = \frac{1}{4\ell_B^2} V_X^\pm(\boldsymbol{q} = 0). \tag{30.5.12b}$$

They are calculated as

$$\epsilon_D^- = \sum_j V_{nnjj}^- = \sqrt{\frac{1}{2\pi}} \frac{d}{\ell_B} \epsilon_X, \tag{30.5.13a}$$

$$\epsilon_X^\pm = \frac{1}{2}\left[1 \pm e^{d^2/2\ell_B^2} \text{erfc}\left(d/\sqrt{2}\ell_B\right)\right] \epsilon_X. \tag{30.5.13b}$$

We also define

$$\epsilon_{\text{cap}} = \frac{4}{N_\Phi} \sum_{jn} \left(V_{nnjj}^- - V_{jnnj}^-\right) \equiv 4(\epsilon_D^- - \epsilon_X^-). \tag{30.5.14}$$

Note that ϵ_X^+ is reduced to ϵ_X given by (30.1.16b) in the limit $d \to 0$. See (32.6.15) and (32.6.25) for more details on these parameters.

Ground State

We consider a homogeneous state satisfying the homogeneity condition $\rho(m,n)|g\rangle = \delta_{mn}|g\rangle$. It is parametrized as

$$|g\rangle = \prod_n \left(e^{i\vartheta_0/2} \sqrt{\frac{1+\sigma_0}{2}} c_f^\dagger(n) + e^{-i\vartheta_0/2} \sqrt{\frac{1-\sigma_0}{2}} c_b^\dagger(n) \right) |0\rangle, \tag{30.5.15}$$

where σ_0 and ϑ_0 are constants: σ_0 is the imbalance parameter,

$$\sigma_0 \equiv \frac{\rho_f - \rho_b}{\rho_f + \rho_b} = \frac{2}{\nu} \langle g|P_z(n,n)|g\rangle = 2\mathcal{P}_z. \tag{30.5.16}$$

It should be noted that the state $|g\rangle$ is not an eigenstate of the total Hamiltonian. Thus, the present scheme is justified in the regime where the SU(2)-invariant

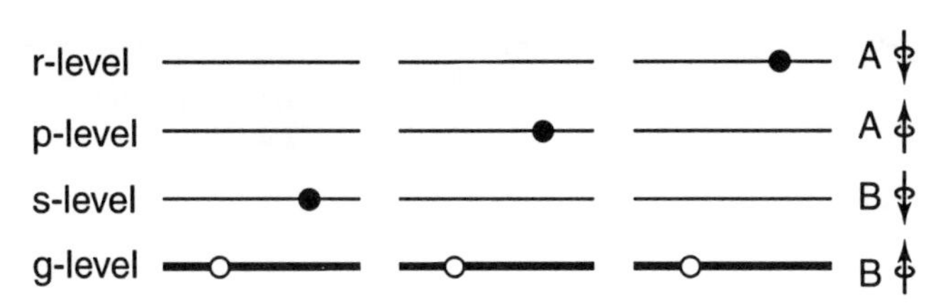

Fig. 30.1 The lowest Landau level contains four energy levels corresponding to the two layers and the two spin states. (We call them the g-level, s-level, p-level and r-level.) At $\nu = 1$ the ground state is the up-spin bonding state. An electron may be moved to any one of the other three levels to form an electron-hole pair excitation.

Coulomb term H_C^+ dominates all other interactions. Then we may treat the SU(2)-invariant term H_C^+ as the unperturbed Hamiltonian and all the SU(2)-noninvariant terms as perturbation. The unperturbed ground state is (30.5.15).

30.6 Bilayer System with Spin ($\nu = 1$)

We include the spin degree of freedom to the bilayer system[150] at $\nu = 1$. The spin and pseudospin operators are

$$S_a(n,m) = \frac{1}{2}\sum_{\mu\nu} c_\mu^\dagger(m)(\tau_a^{\text{spin}})_{\mu\nu}c_\tau(n), \quad P_a(n,m) = \frac{1}{2}\sum_{\mu\nu} c_\mu^\dagger(m)(\tau_a^{\text{ppin}})_{\mu\nu}c_\nu(n). \tag{30.6.1}$$

The Coulomb Hamiltonian consists of the SU(4)-invariant term and the SU(4)-noninvariant term, which are given by the same formulas as (30.5.5a) and (30.5.5b) with (30.6.1), respectively. The Zeeman-pseudo-Zeeman term is

$$H_{\text{ZpZ}} = \sum_m [-\Delta_{\text{SAS}}P_x(m,m) + \Delta_{\text{bias}}P_z(m,m) - \Delta_Z S_z(m,m)]. \tag{30.6.2}$$

The total Hamiltonian is $H = H_C^+ + H_C^- + H_{\text{ZpZ}}$.

Ground State

We search for the ground state. There are four states in each Landau site [Fig.30.1], and a homogeneous state is given by

$$|\text{g}\rangle = \sum_n\sum_\mu g_\mu c_\mu^\dagger(n)|0\rangle, \tag{30.6.3}$$

where g_μ are site-independent constants subject to the normalization condition $\sum g_\mu^* g_\mu = 1$. It satisfies the homogeneity condition

$$\rho(m,n)|\text{g}\rangle = \delta_{mn}|\text{g}\rangle. \tag{30.6.4}$$

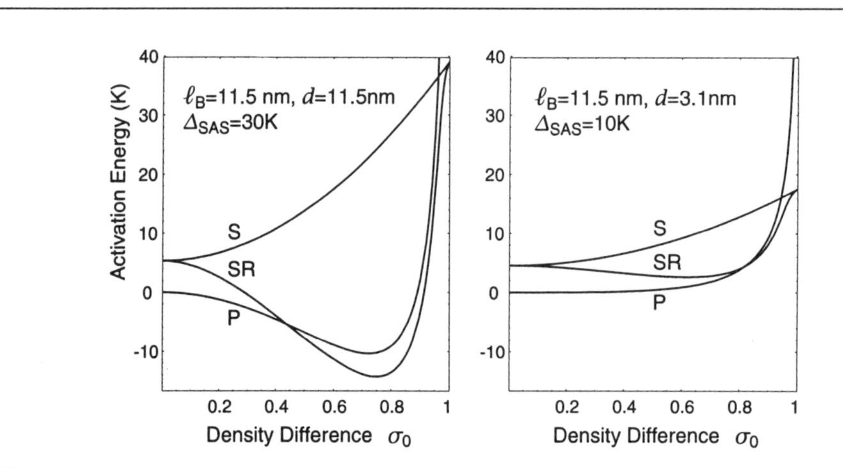

Fig. 30.2 The energy of an electron-hole pair state is depicted based on the formulas (30.3.8) and (30.6.17) with typical sample parameters as indicated. Three curves corresponds to the state $|\psi_s\rangle$, $|\psi_{sr}\rangle$ and $|\psi_p\rangle$. We have normalized the activation energy of the pseudospin excitation to zero at $\sigma_0 = 0$.

In the regime where the SU(4)-invariant Coulomb term H_C^+ dominates all other interactions, we may treat the SU(4)-invariant term H_C^+ as the unperturbed Hamiltonian and all the SU(4)-noninvariant terms as perturbation.

It follows trivially that

$$H_C^+|g\rangle = 0, \tag{30.6.5}$$

where $|g\rangle$ is given by (30.6.3) with arbitrary constants g_μ. One of them is removed due to the normalization condition and one of them is unphysical since it represents the overall phase. Thus, there are 6 real independent parameters, upon which the ground-state energy does not depend. They are the Goldstone modes. They get gapful by the SU(4)-noninvariant interactions.

Electron-Hole Pair Energy

There are four energy levels within the lowest Landau level [Fig.30.1], consisting of the up-spin bonding state $|B\uparrow\rangle$, the down-spin bonding state $|B\downarrow\rangle$, the up-spin antibonding state $|A\uparrow\rangle$ and the down-spin antibonding state $|A\downarrow\rangle$. They are constructed as

$$|B\uparrow\rangle = \prod_n B_\uparrow^\dagger(n)|0\rangle, \quad |A\uparrow\rangle = \prod_n A_\uparrow^\dagger(n)|0\rangle, \tag{30.6.6}$$

where

$$\begin{pmatrix} B_\uparrow(n) \\ A_\uparrow(n) \end{pmatrix} = \frac{1}{\sqrt{2}} \begin{pmatrix} \sqrt{1+\sigma_0} & \sqrt{1-\sigma_0} \\ \sqrt{1-\sigma_0} & -\sqrt{1+\sigma_0} \end{pmatrix} \begin{pmatrix} c_{f\uparrow}(n) \\ c_{b\uparrow}(n) \end{pmatrix}. \tag{30.6.7}$$

and a similar formula for $|\mathrm{B}\downarrow\rangle$ and $|\mathrm{A}\downarrow\rangle$.

One electron may be added into the down-spin bonding state, the up-spin antibonding state or the down-spin antibonding state [Fig.30.1]. These electron-hole states are

$$|\psi_{\mathrm{s}}\rangle = |e_j^{\mathrm{B}\downarrow}; h_k^{\mathrm{B}\uparrow}\rangle = B_\downarrow^\dagger(j)B_\uparrow(k)|\mathrm{B}\uparrow\rangle, \qquad |\psi_{\mathrm{p}}\rangle = |e_j^{\mathrm{A}\uparrow}; h_k^{\mathrm{B}\uparrow}\rangle = A_\uparrow^\dagger(j)B_\uparrow(k)|\mathrm{B}\uparrow\rangle,$$
$$|\psi_{\mathrm{r}}\rangle = |e_j^{\mathrm{A}\downarrow}; h_k^{\mathrm{B}\uparrow}\rangle = A_\downarrow^\dagger(j)B_\uparrow(k)|\mathrm{A}\uparrow\rangle. \tag{30.6.8}$$

The energy matrix looks as

$$\begin{pmatrix} \langle\psi_{\mathrm{s}}|H|\psi_{\mathrm{s}}\rangle & \langle\psi_{\mathrm{s}}|H|\psi_{\mathrm{r}}\rangle & 0 \\ \langle\psi_{\mathrm{r}}|H|\psi_{\mathrm{s}}\rangle & \langle\psi_{\mathrm{r}}|H|\psi_{\mathrm{r}}\rangle & 0 \\ 0 & 0 & \langle\psi_{\mathrm{p}}|H|\psi_{\mathrm{p}}\rangle \end{pmatrix}, \tag{30.6.9}$$

where

$$\begin{aligned}
\langle\psi_{\mathrm{s}}|H|\psi_{\mathrm{s}}\rangle &= E_0 + \Delta_{\mathrm{Z}}, \\
\langle\psi_{\mathrm{r}}|H|\psi_{\mathrm{r}}\rangle &= E_0 - 4\sigma_0^2\left(\epsilon_{\mathrm{D}}^- - V_{jjkk}^-\right) + \Delta_{\mathrm{Z}} + \Delta_{\mathrm{BAB}}, \\
\langle\psi_{\mathrm{p}}|H|\psi_{\mathrm{p}}\rangle &= E_0 - 4\sigma_0^2\left(\epsilon_{\mathrm{D}}^- - V_{jjkk}^-\right) - 2\left(1-\sigma_0^2\right)\left(\epsilon_{\mathrm{X}}^- - V_{jkkj}^-\right) + \Delta_{\mathrm{BAB}}, \\
\langle\psi_{\mathrm{r}}|H|\psi_{\mathrm{s}}\rangle &= 2\sigma_0\sqrt{1-\sigma_0^2}\left(\epsilon_{\mathrm{D}}^- - V_{jjkk}^-\right),
\end{aligned} \tag{30.6.10}$$

with

$$E_0 = \langle \mathrm{g}|H|\mathrm{g}\rangle + 2\left(\epsilon_{\mathrm{X}}^+ - V_{jjkk}^+\right) + 2\sigma_0^2\left(\epsilon_{\mathrm{X}}^- - V_{jjkk}^-\right), \tag{30.6.11}$$

and

$$\Delta_{\mathrm{BAB}} = \frac{1}{\sqrt{1-\sigma_0^2}}\Delta_{\mathrm{SAS}}. \tag{30.6.12}$$

We may set $V_{jjkk}^\pm = V_{jkkj}^- = 0$ for a well separated electron-hole pair.

The matrix (30.6.9) is diagonal at the balanced point. The minimum eigenvalue is the spin-excitation energy

$$\langle\psi_{\mathrm{s}}|H|\psi_{\mathrm{s}}\rangle = \langle \mathrm{g}|H|\mathrm{g}\rangle + 2\epsilon_{\mathrm{X}}^+ + \Delta_{\mathrm{Z}}, \tag{30.6.13}$$

or the ppin excitation energy

$$\langle\psi_{\mathrm{p}}|H|\psi_{\mathrm{p}}\rangle = \langle \mathrm{g}|H|\mathrm{g}\rangle + 2\epsilon_{\mathrm{X}}^+ - 2\epsilon_{\mathrm{X}}^- + \Delta_{\mathrm{SAS}}. \tag{30.6.14}$$

The pseudospin excitation occurs when

$$2\epsilon_{\mathrm{X}}^- > \Delta_{\mathrm{SAS}} - \Delta_{\mathrm{Z}}. \tag{30.6.15}$$

We remark that $2\epsilon_{\mathrm{X}}^- \simeq 47$K in a typical sample with $d \simeq 231$ nm.

The matrix (30.6.9) is diagonal also at the monolayer point, where the spin excitation always occurs with the excitation energy

$$\langle\psi_s|H|\psi_s\rangle = \langle g|H|g\rangle + 2\,(\epsilon_X^+ + \epsilon_X^-) + \Delta_Z, \tag{30.6.16}$$

since the energies of the other two modes diverge.

The two excitation states $|\psi_s\rangle$ and $|\psi_r\rangle$ mix to make a new state, $|\psi_{sr}\rangle$, to lower the excitation energy except for $\sigma_0 = 0$ and $\sigma_0 = 1$. After diagonalization it is

$$\begin{aligned}\langle\psi_{sr}|H|\psi_{sr}\rangle =&\langle g|H|g\rangle + 2\epsilon_X^+ - \frac{1}{2}\sigma_0^2\epsilon_{cap} + \Delta_Z + \frac{1}{2}\Delta_{BAB}\\ &- \frac{1}{2}\sqrt{(\Delta_{BAB} - 4\sigma_0^2\epsilon_D^-)^2 + 16\sigma_0^2(1-\sigma_0^2)(\epsilon_D^-)^2}.\end{aligned} \tag{30.6.17}$$

Thus, when the condition (30.6.15) is satisfied, an electron is excited to the $|\psi_p\rangle$ state at the balanced point, and it makes a sudden transition to the $|\psi_{sr}\rangle$ state, and finally transforms smoothly into the $|\psi_s\rangle$ state, as depicted in Fig.30.2. The sudden transition would be smoothed out in a skyrmion-antiskyrmion excitation involving several electrons simultaneously.

30.7 BILAYER SYSTEM WITH SPINS ($\nu = 2$)

We proceed to analyze the bilayer QH system at $\nu = 2$. The total Hamiltonian consists of the Coulomb Hamiltonian (30.5.5) and the Zeeman-pseudo-Zeeman term (30.6.2),

$$H = H_C^+ + H_C^- + H_{ZpZ}. \tag{30.7.1}$$

The Hamiltonian is identical to the one at $\nu = 1$. Nevertheless, the ground-state structure is not so simple as at $\nu = 1$. We have already shown the existence of the spin phase, the canted phase and the ppin phase in Chapter 26 based on the energy formula (26.2.1), which we shall derive in this section.[4]

At $\nu = 2$ there are two electrons in one Landau site. We consider a creation operator of a pair of electrons at site n,

$$G^\dagger(n) = \frac{1}{2}\sum_{\mu\nu} g_{\mu\nu}c_\mu^\dagger(n)c_\nu^\dagger(n), \tag{30.7.2}$$

where $g_{\mu\nu}$ is an antisymmetric complex matrix, $g_{\mu\nu} = -g_{\nu\mu}$. With site-independent constants $g_{\mu\nu}$, the state

$$|g\rangle = \prod_n G^\dagger(n)|0\rangle \tag{30.7.3}$$

[4]The final equation for the ground-state energy is (30.7.26). The Coulomb energy part of this formula is also derived by setting

$$\hat{\mathcal{S}}_a^{cl} = \rho_\Phi\mathcal{S}_a,\quad \hat{\mathcal{P}}_z^{cl} = \rho_\Phi\mathcal{P}_a,\quad \hat{\mathcal{R}}_{ab}^{cl} = \rho_\Phi\mathcal{R}_{ab},$$

and assuming constants values for $\mathcal{S}_a$, $\mathcal{P}_a$ and $\mathcal{R}_{ab}$ in the classical Hamiltonian (32.6.30).

satisfies the homogeneity condition

$$\rho(m,n)|\mathrm{g}\rangle = 2\delta_{mn}|\mathrm{g}\rangle. \tag{30.7.4}$$

The normalization of the state, $\langle \mathrm{g}|\mathrm{g}\rangle = 1$, leads to

$$\mathrm{Tr}\left(gg^\dagger\right) = 2. \tag{30.7.5}$$

Since the matrix $g_{\mu\nu}$ contains 6 independent complex parameters, there are 6 independent homogeneous states. They span the **6**-dimensional irreducible representation of SU(4)⊗SU(4).

It is hard to diagonalize the total Hamiltonian. We investigate the regime where the SU(4)-invariant Coulomb term H_{C}^{+} dominates all other interactions. Thus we start with the ground state of the Hamiltonian H_{C}^{+}, and include the SU(4)-noninvariant terms as small perturbation. It is easy to see that $|\mathrm{g}\rangle$ is an eigenstate of the Hamiltonian H_{C}^{+},

$$H_{\mathrm{C}}^{+}|\mathrm{g}\rangle = 0. \tag{30.7.6}$$

The unperturbed system realized in the limits $d \to 0$, $\Delta_{\mathrm{Z}} \to 0$, $\Delta_{\mathrm{SAS}} \to 0$ and $\Delta_{\mathrm{bias}} \to 0$.

We introduce the expectation values of isospin operators,

$$\begin{aligned}
\mathcal{S}_a &= \langle \mathrm{g}|S_a(n,n)|\mathrm{g}\rangle = \frac{1}{2}\mathrm{Tr}\left(\tau_a^{\mathrm{spin}} gg^\dagger\right),\\
\mathcal{P}_a &= \langle \mathrm{g}|P_a(n,n)|\mathrm{g}\rangle = \frac{1}{2}\mathrm{Tr}\left(\tau_a^{\mathrm{ppin}} gg^\dagger\right),\\
\mathcal{R}_{ab} &= \langle \mathrm{g}|R_{ab}(n,n)|\mathrm{g}\rangle = \frac{1}{2}\mathrm{Tr}\left(\tau_a^{\mathrm{spin}}\tau_b^{\mathrm{ppin}} gg^\dagger\right).
\end{aligned} \tag{30.7.7}$$

They are the total spin per one Landau site, and so on. At $\nu = 2$ the imbalance parameter σ_0 is defined by

$$\sigma_0 \equiv \frac{\rho_{\mathrm{f}} - \rho_{\mathrm{b}}}{\rho_{\mathrm{f}} + \rho_{\mathrm{b}}} = \frac{2}{\nu}\langle \mathrm{g}|P_z(n,n)|\mathrm{g}\rangle = \mathcal{P}_z. \tag{30.7.8}$$

We wish to represent various quantities in terms of these 15 physical densities.

Ground-State Condition

We investigate the ground-state condition(30.7.6) by studying the incompressibility condition (30.7.4). It holds trivially for $m = n$. The condition $\rho(m,n)|\mathrm{g}\rangle = 0$ for $m \neq n$ yields

$$\sum_{\alpha\beta\mu\nu} g_{\alpha\beta}g_{\mu\nu}c_\alpha^\dagger(m)c_\beta^\dagger(m)c_\mu^\dagger(m)c_\nu^\dagger(n)|0\rangle = 0, \tag{30.7.9}$$

which in turn leads to

$$\sum_{\alpha\beta\mu\nu} \epsilon_{\alpha\beta\mu\nu}g_{\alpha\beta}g_{\mu\nu} = 0, \tag{30.7.10}$$

where $\epsilon_{\alpha\beta\mu\nu}$ is the totally antisymmetric tensor. We now express this condition in terms of the physical densities (30.7.7).

The basis of the algebra SU(4) is spanned by 15 Hermitian traceless matrices τ_a^{spin}, τ_a^{ppin} and $\tau_a^{\text{spin}}\tau_b^{\text{ppin}}$ as given in Appendix D. A basis in the space of 4×4 antisymmetric matrices, comprised of 6 matrices, can be chosen as

$$\tau_z^{\text{spin}}\tau_y^{\text{ppin}} = \begin{pmatrix} 0 & 0 & -i & 0 \\ 0 & 0 & 0 & +i \\ +i & 0 & 0 & 0 \\ 0 & -i & 0 & 0 \end{pmatrix}, \quad \tau_y^{\text{ppin}} = \begin{pmatrix} 0 & 0 & -i & 0 \\ 0 & 0 & 0 & -i \\ +i & 0 & 0 & 0 \\ 0 & +i & 0 & 0 \end{pmatrix},$$

$$\tau_x^{\text{spin}}\tau_y^{\text{ppin}} = \begin{pmatrix} 0 & 0 & 0 & -i \\ 0 & 0 & -i & 0 \\ 0 & +i & 0 & 0 \\ +i & 0 & 0 & 0 \end{pmatrix}, \quad \tau_y^{\text{spin}}\tau_z^{\text{ppin}} = \begin{pmatrix} 0 & -i & 0 & 0 \\ +i & 0 & 0 & 0 \\ 0 & 0 & 0 & +i \\ 0 & 0 & -i & 0 \end{pmatrix},$$

$$\tau_y^{\text{spin}} = \begin{pmatrix} 0 & -i & 0 & 0 \\ +i & 0 & 0 & 0 \\ 0 & 0 & 0 & -i \\ 0 & 0 & +i & 0 \end{pmatrix}, \quad \tau_y^{\text{spin}}\tau_x^{\text{ppin}} = \begin{pmatrix} 0 & 0 & 0 & -i \\ 0 & 0 & +i & 0 \\ 0 & -i & 0 & 0 \\ +i & 0 & 0 & 0 \end{pmatrix}. \tag{30.7.11}$$

Hence, introducing four tree-dimensional vectors $\boldsymbol{A}$, $\boldsymbol{B}$, $\boldsymbol{C}$ and $\boldsymbol{D}$, we may parameterize the matrix $g_{\mu\nu}$ as

$$\begin{aligned} 2g =& (A_x + iB_x)\tau_z^{\text{spin}}\tau_y^{\text{ppin}} + (C_x + iD_x)\tau_y^{\text{spin}}\tau_z^{\text{ppin}} \\ &+ (A_y + iB_y)(-i\tau_y^{\text{ppin}}) + (C_y + iD_y)(-i\tau_y^{\text{spin}}) \\ &+ (A_z + iB_z)(-\tau_x^{\text{spin}}\tau_y^{\text{ppin}}) + (C_z + iD_z)(-\tau_y^{\text{spin}}\tau_x^{\text{ppin}}) \\ =& -i\left[(\boldsymbol{A} + i\boldsymbol{B})\boldsymbol{\tau}^{\text{spin}} + (\boldsymbol{C} + i\boldsymbol{D})\boldsymbol{\tau}^{\text{ppin}}\right]\tau_x^{\text{spin}}\tau_y^{\text{ppin}}, \end{aligned} \tag{30.7.12}$$

where we have used relations (D.12). We come to

$$\begin{aligned} gg^\dagger =& \frac{1}{4}(\boldsymbol{A}^2 + \boldsymbol{B}^2 + \boldsymbol{C}^2 + \boldsymbol{D}^2) \\ &+ \frac{1}{2}\tau_a^{\text{spin}}\varepsilon_{abc}A_bB_c + \frac{1}{2}\tau_a^{\text{ppin}}\varepsilon_{abc}C_bD_c + \frac{1}{2}\tau_a^{\text{spin}}\tau_b^{\text{ppin}}(A_aC_b + B_aD_b). \end{aligned} \tag{30.7.13}$$

On the other hand, (30.7.7) implies that $gg^\dagger$ is expressed as

$$gg^\dagger = \frac{1}{2} + \frac{1}{2}\left(\tau_a^{\text{spin}}\mathcal{S}_a + \tau_a^{\text{ppin}}\mathcal{P}_a + \tau_a^{\text{spin}}\tau_b^{\text{ppin}}\mathcal{R}_{ab}\right), \tag{30.7.14}$$

where we have used the normalization condition (30.7.5). Comparing these two equations we get

$$\boldsymbol{A}^2 + \boldsymbol{B}^2 + \boldsymbol{C}^2 + \boldsymbol{D}^2 = 2, \tag{30.7.15}$$

and

$$\mathcal{S}_a = \varepsilon_{abc} A_b B_c, \qquad \mathcal{P}_a = \varepsilon_{abc} C_b D_c, \qquad \mathcal{R}_{ab} = A_a C_b + B_a D_b, \tag{30.7.16}$$

from which we derive the constraints on $\mathcal{S}_a$, $\mathcal{P}_a$ and $\mathcal{R}_{ab}$,

$$\mathcal{S}_a \mathcal{R}_{ab} = 0, \qquad \mathcal{R}_{ab} \mathcal{P}_b = 0, \qquad \mathcal{S}_a \mathcal{P}_b - \varepsilon_{acd} \varepsilon_{bhe} \mathcal{R}_{ch} \mathcal{R}_{de} = 0. \tag{30.7.17}$$

The last constraint in (30.7.17) is verified based on the identity

$$\varepsilon_{acd} \varepsilon_{bhe} = \delta_{ab} \left(\delta_{ch} \delta_{de} - \delta_{ce} \delta_{dh} \right) - \delta_{ah} \left(\delta_{cb} \delta_{de} - \delta_{ce} \delta_{bd} \right) + \delta_{ae} \left(\delta_{cb} \delta_{dh} - \delta_{ch} \delta_{bd} \right), \tag{30.7.18}$$

while the first two constraints in (30.7.17) follows trivially from the first two equations in (30.7.16).

We substitute the expansion (30.7.12) into (30.7.10), and use (30.7.15), to obtain

$$\boldsymbol{A}^2 + \boldsymbol{D}^2 = \boldsymbol{B}^2 + \boldsymbol{C}^2 = 1, \quad \boldsymbol{AB} = \boldsymbol{CD}. \tag{30.7.19}$$

Using (30.7.16) we express $\boldsymbol{\mathcal{S}}^2 + \boldsymbol{\mathcal{P}}^2 + \boldsymbol{\mathcal{R}}^2$ in terms of $\boldsymbol{A}$, $\boldsymbol{B}$, $\boldsymbol{C}$ and $\boldsymbol{D}$. We then use (30.7.19) to derive

$$\boldsymbol{\mathcal{S}}^2 + \boldsymbol{\mathcal{P}}^2 + \boldsymbol{\mathcal{R}}^2 = 1. \tag{30.7.20}$$

This is the ground-state condition in terms of physical variables.

Ground-State Energy

We include the SU(4)-noninvariant interactions as small perturbation. The first order perturbation is to diagonalize the full Hamiltonian H within this subspace. Equivalently, the ground state is determined by minimizing the energy

$$E_{\mathrm{g}} \equiv \mathcal{E}_{\mathrm{g}} N_\Phi = \langle \mathrm{g} | H | \mathrm{g} \rangle \tag{30.7.21}$$

within this parameter space.

It is straightforward to show that

$$\langle \mathrm{g} | c_\mu^\dagger(m) c_\nu(n) | \mathrm{g} \rangle = \delta_{mn} (g g^\dagger)_{\nu\mu}, \tag{30.7.22}$$

and

$$\langle \mathrm{g} | c_\mu^\dagger(m) c_\sigma^\dagger(i) c_\tau(j) c_\nu(n) | \mathrm{g} \rangle = \begin{cases} +\delta_{mn} g_{\mu\sigma}^\dagger g_{\nu\tau} & \text{for} \quad m = i, j = n \\ +\delta_{mn} \delta_{ij} (g g^\dagger)_{\tau\sigma} (g g^\dagger)_{\nu\mu} & \text{for} \quad m > i, j < n \\ -\delta_{mj} \delta_{in} (g g^\dagger)_{\nu\sigma} (g g^\dagger)_{\tau\mu} & \text{for} \quad m > i, j > n \\ -\delta_{mj} \delta_{in} (g g^\dagger)_{\nu\sigma} (g g^\dagger)_{\tau\mu} & \text{for} \quad m < i, j < n \\ +\delta_{mn} \delta_{ij} (g g^\dagger)_{\tau\sigma} (g g^\dagger)_{\nu\mu} & \text{for} \quad m < i, j > n \end{cases}. \tag{30.7.23}$$

Thus

$$\frac{1}{N_\phi}\sum V^{\pm}_{mnij}\langle \mathrm{g}|c^\dagger_\mu(m)c^\dagger_\sigma(i)c_\tau(j)c_\nu(n)|\mathrm{g}\rangle$$
$$=(gg^\dagger)_{\tau\sigma}(gg^\dagger)_{\nu\mu}\varepsilon^{\pm}_{\mathrm{D}}-(gg^\dagger)_{\nu\sigma}(gg^\dagger)_{\tau\mu}\varepsilon^{\pm}_{\mathrm{X}}. \tag{30.7.24}$$

By using these formulas together with (30.7.14) the ground-state energy per Landau site is calculated as

$$\begin{aligned}\mathcal{E}_{\mathrm{g}}=&\varepsilon^-_{\mathrm{D}}\left[\mathrm{Tr}\left(\tau^{\mathrm{ppin}}_z gg^\dagger\right)\right]^2-\varepsilon^-_{\mathrm{X}}\mathrm{Tr}\left(\tau^{\mathrm{ppin}}_z gg^\dagger\tau^{\mathrm{ppin}}_z gg^\dagger\right)-\varepsilon^+_{\mathrm{X}}\mathrm{Tr}\left(gg^\dagger gg^\dagger\right)+2\epsilon^+_{\mathrm{X}}\\&-\frac{1}{2}\Delta_{\mathrm{Z}}\mathrm{Tr}\left(\tau^{\mathrm{spin}}_z gg^\dagger\right)-\frac{1}{2}\Delta_{\mathrm{SAS}}\mathrm{Tr}\left(\tau^{\mathrm{ppin}}_x gg^\dagger\right)-\frac{1}{2}\Delta_{\mathrm{bias}}\mathrm{Tr}\left(\tau^{\mathrm{ppin}}_z gg^\dagger\right),\end{aligned} \tag{30.7.25}$$

or

$$\begin{aligned}\mathcal{E}_{\mathrm{g}}=&4\varepsilon^-_{\mathrm{D}}\mathcal{P}^2_z-(\varepsilon^+_{\mathrm{X}}-\varepsilon^-_{\mathrm{X}})\left(\boldsymbol{\mathcal{S}}^2+\boldsymbol{\mathcal{P}}^2+\boldsymbol{\mathcal{R}}^2\right)-2\varepsilon^-_{\mathrm{X}}\left(\mathcal{S}^2_a+\mathcal{P}^2_z+\mathcal{R}^2_{az}\right)+\varepsilon^+_{\mathrm{X}}-\varepsilon^-_{\mathrm{X}}\\&-\Delta_{\mathrm{Z}}\mathcal{S}_z-\Delta_{\mathrm{SAS}}\mathcal{P}_x-\Delta_{\mathrm{bias}}\mathcal{P}_z.\end{aligned} \tag{30.7.26}$$

Here we can neglect constant terms without loss of generality. Since $\boldsymbol{\mathcal{S}}^2+\boldsymbol{\mathcal{P}}^2+\boldsymbol{\mathcal{R}}^2=1$ due to (30.7.20), we may set

$$\mathcal{E}_{\mathrm{g}}=4\varepsilon^-_{\mathrm{D}}\mathcal{P}^2_z-2\varepsilon^-_{\mathrm{X}}\left(\mathcal{S}^2_a+\mathcal{P}^2_z+\mathcal{R}^2_{az}\right)-\Delta_{\mathrm{Z}}\mathcal{S}_z-\Delta_{\mathrm{SAS}}\mathcal{P}_x-\Delta_{\mathrm{bias}}\mathcal{P}_z. \tag{30.7.27}$$

This is the energy formula (26.2.1), based upon which we have shown the existence of the spin phase, the ppin phase and the canted phase at $\nu=2$ in Chapter 26.

Independent Degree of Freedom

To determine the ground state it is necessary to minimize the energy (30.7.27) by varying all independent variables. There are 15 physical densities. We may choose the spin $\mathcal{S}_a$ and the pseudospin $\mathcal{P}_a$ as independent variables. The problem is to extract the independent degree of freedom from the isospin component $\mathcal{R}_{ab}$. For this purpose we construct $\mathcal{R}_{ab}$ satisfying (30.7.17). To satisfy the constraint $\mathcal{S}_a\mathcal{R}_{ab}=0$, we use the two normalized vectors

$$\frac{\boldsymbol{\mathcal{S}}^2\mathcal{P}_a-(\boldsymbol{\mathcal{S}}\boldsymbol{\mathcal{P}})\mathcal{S}_a}{\mathcal{S}\mathcal{Q}},\qquad\frac{\mathcal{Q}_a}{\mathcal{Q}}, \tag{30.7.28}$$

which are orthogonal to $\mathcal{S}_a$, and orthogonal one to another. To satisfy the constraint $\mathcal{R}_{ab}\mathcal{P}_b=0$, we use the two normalized vectors

$$\frac{\boldsymbol{\mathcal{P}}^2\mathcal{S}_b-(\boldsymbol{\mathcal{S}}\boldsymbol{\mathcal{P}})\mathcal{P}_b}{\mathcal{P}\mathcal{Q}},\qquad\frac{\mathcal{Q}_b}{\mathcal{Q}}, \tag{30.7.29}$$

which are orthogonal to $\mathcal{P}_b$, and orthogonal one to another. Hence we are able to expand $\mathcal{R}_{ab}$ in terms of 4 tensors made out of these vectors with 4 coefficients $\mathcal{R}_{PS}$,

$\mathcal{R}_{PQ}$, $\mathcal{R}_{QS}$ and $\mathcal{R}_{QQ}$ as

$$\mathcal{R}_{ab} = \frac{\mathcal{S}^2\mathcal{P}_a - (\mathcal{S}\mathcal{P})\mathcal{S}_a}{\mathcal{S}\mathcal{Q}} \frac{\mathcal{P}^2\mathcal{S}_b - (\mathcal{S}\mathcal{P})\mathcal{P}_b}{\mathcal{P}\mathcal{Q}} \mathcal{R}_{PS} + \frac{\mathcal{S}^2\mathcal{P}_a - (\mathcal{S}\mathcal{P})\mathcal{S}_a}{\mathcal{S}\mathcal{Q}} \frac{\mathcal{Q}_b}{\mathcal{Q}} \mathcal{R}_{PQ}$$
$$+ \frac{\mathcal{Q}_a}{\mathcal{Q}} \frac{\mathcal{P}^2\mathcal{S}_b - (\mathcal{S}\mathcal{P})\mathcal{P}_b}{\mathcal{P}\mathcal{Q}} \mathcal{R}_{QS} + \frac{\mathcal{Q}_a}{\mathcal{Q}} \frac{\mathcal{Q}_b}{\mathcal{Q}} \mathcal{R}_{QQ}. \tag{30.7.30}$$

This is the formula (26.2.3) we have already used. We now substitute (30.7.30) into the constraint (30.7.17) to find that

$$\mathcal{R}_{QS}\mathcal{R}_{PQ} - \mathcal{R}_{PS}\mathcal{R}_{QQ} = \mathcal{S}\mathcal{P}. \tag{30.7.31}$$

Consequently three variables are independent among $\mathcal{R}_{PS}$, $\mathcal{R}_{PQ}$, $\mathcal{R}_{QS}$ and $\mathcal{R}_{QQ}$. They are parametrized as

$$\mathcal{R}_{PS} + i\mathcal{R}_{QS} = e^{i\omega}\left(-i\lambda + \frac{1}{\xi}\mathcal{S}\right), \qquad \mathcal{R}_{PQ} + i\mathcal{R}_{QQ} = -ie^{i\omega}\xi\mathcal{P}. \tag{30.7.32}$$

There are 9 independent variables, ξ, λ, ω together with $\mathcal{S}_a$, $\mathcal{P}_a$, among which one variable is fixed by the ground-state condition (30.7.20). Consequently there are 8 independent variables parametrizing the SU(4)-invariant ground state. They are the Goldstone modes in the SU(4)-invariant system.

CHAPTER 31

NONCOMMUTATIVE SOLITONS

The algebraic structure of the bare density is the $W_\infty(N)$ algebra in a multicomponent QH system, where the electron density is intrinsically entangled with the isospin density. Accordingly a topological soliton modulates not only the isospin texture but also the number density of electrons. A new feature of a topological soliton is that it carries the electric charge as well. Namely a topological soliton with the topological charge Q_{sky} carries the electric charge eQ_{sky}. We study skyrmions in a microscopic theory of the SU(N) QH ferromagnet, and then extend the scheme to the SU(4) bilayer QH ferromagnet.

31.1 TOPOLOGICAL CHARGE AND ELECTRIC CHARGE

We investigate the monolayer QH system with the SU(N) isospin symmetry, where the algebraic structure of the bare density is the $W_\infty(N)$ algebra (29.5.12c). We consider a wide class of states presented in (29.6.1), or

$$|\mathfrak{S}\rangle = e^{iW}|\mathfrak{S}_0\rangle \quad \text{with} \quad |\mathfrak{S}_0\rangle = \prod_{\mu n}\left[c_\mu^\dagger(n)\right]^{\nu_\mu(n)}|0\rangle. \tag{31.1.1}$$

In this section we shall derive a theorem that, if the state $|\mathfrak{S}\rangle$ describe a topological soliton with the topological charge Q_{sky}, it carries the electric charge eQ_{sky}.

For definiteness we first give concrete examples in the SU(2) QH system. The hole excited state $|h\rangle$ and the electron excited state $|e\rangle$ are given by (30.4.1). We consider their $W_\infty(2)$-rotated states,

$$|\mathfrak{S}_{sky}^-\rangle = e^{iW^-}|h\rangle = e^{iW^-}c_\uparrow(0)|g\rangle, \tag{31.1.2a}$$

$$|\mathfrak{S}_{sky}^+\rangle = e^{iW^+}|e\rangle = e^{iW^+}c_\downarrow^\dagger(0)|g\rangle, \tag{31.1.2b}$$

where $W^\pm$ are arbitrary elements of the $W_\infty(2)$ algebra. For instance the $W_\infty(2)$ transformation e^{iW^-} modulates the spin texture around the hole. It modulates simultaneously the electron density around the hole due to the $W_\infty(2)$-algebraic entanglement, and decreases the Coulomb energy. The element W^- is to be determined by requiring the excitation energy to be minimized. As we shall show explicitly in Section 31.2, $|\mathfrak{S}_{\text{sky}}^-\rangle$ and $|\mathfrak{S}_{\text{sky}}^+\rangle$ describe the skyrmion and antiskyrmion states. Similarly,

$$|\mathfrak{S}_{\text{sky}}^-;k\rangle = e^{iW^-}c_\uparrow(0)\cdots c_\uparrow(k-1)|g\rangle, \tag{31.1.3a}$$

$$|\mathfrak{S}_{\text{sky}}^+;k\rangle = e^{iW^+}c_\downarrow^\dagger(0)\cdots c_\downarrow^\dagger(k-1)|g\rangle \tag{31.1.3b}$$

describe a multi-skyrmion and a multi-antiskyrmion as $W_\infty(2)$-rotated states of a multi-hole state and a multi-electron state.

To discuss topological issues, it is convenient to employ the CP^{N-1} field, which is a N-component complex field $n_\mu(x)$ subject to the normalization condition (29.6.27). To describe the $\nu = k$ QH system, we need the Grassmannian $G_{N,k}$ field, or equivalently k CP^{N-1} fields subject to the normalization condition (29.6.31).

The density excitation is given by (29.7.12), or

$$\Delta\hat{\rho}^{\text{cl}}(x) = \frac{1}{2\pi\ell_B^2}\sum_{r,\mu}[\![n_\mu^r(x), n_\mu^{r*}(x)]\!]. \tag{31.1.4}$$

This is nothing but the topological charge density (29.7.13),

$$J_{\text{sky}}^0(x) = -\Delta\hat{\rho}^{\text{cl}}(x) = \frac{1}{2\pi\ell_B^2}\sum_{r,\mu}[\![n_\mu^{r*}(x), n_\mu^{r}(x)]\!]. \tag{31.1.5}$$

Consequently,

$$Q_{\text{sky}} \equiv \frac{1}{2\pi\ell_B^2}\int d^2x \sum_{r,\mu}[\![n_\mu^r(x), n_\mu^{r*}(x)]\!] = -\Delta N_e^{\text{cl}}. \tag{31.1.6}$$

It is intriguing that a skyrmion possessing the topological charge Q_{sky} carries the electric charge $-eN_e^{\text{cl}} = eQ_{\text{sky}}$.

31.2 Microscopic Skyrmion States

We study the SU(2) QH system. We have argued in the semiclassical approach that the skyrmion is described by the CP^1 field (16.2.15), or

$$\begin{pmatrix} n^\uparrow(x) \\ n^\downarrow(x)\end{pmatrix} = \frac{1}{\sqrt{|z|^2+\lambda^2}}\begin{pmatrix} z \\ \lambda\end{pmatrix}. \tag{31.2.1}$$

We have called it the factorizable skyrmion since the wave function is factorizable as in (16.2.16). The associated nonlinear sigma field (normalized spin field) is (16.2.17), or

$$\mathcal{S}_x(x) = \frac{\lambda x}{r^2+\lambda^2},\quad \mathcal{S}_y(x) = \frac{\mp\lambda y}{r^2+\lambda^2},\quad \mathcal{S}_z(x) = \frac{1}{2}\frac{r^2-\lambda^2}{r^2+\lambda^2} \tag{31.2.2}$$

for a skyrmion (−) and an antiskyrmion (+), with $\lambda = 2\kappa\ell_B$.

We construct a microscopic skyrmion state $|\mathfrak{S}\rangle$ whose classical spin field $\langle\mathfrak{S}|S_a(\boldsymbol{x})|\mathfrak{S}\rangle$ has this tensorial structure. We recall the formula (29.6.3), or

$$\hat{\rho}^{\text{cl}}(\boldsymbol{k}) = \frac{1}{2\pi}\sum_{mn}^{\infty}\langle m|e^{-i\boldsymbol{k}\boldsymbol{X}}|n\rangle\hat{\rho}^{\text{cl}}(n,m),$$

$$\hat{\mathcal{S}}_a^{\text{cl}}(\boldsymbol{k}) = \frac{1}{2\pi}\sum_{mn}^{\infty}\langle m|e^{-i\boldsymbol{k}\boldsymbol{X}}|n\rangle\hat{\mathcal{S}}_a^{\text{cl}}(n,m), \tag{31.2.3}$$

where we can show[1]

$$\langle m|e^{-i\boldsymbol{k}\boldsymbol{X}}|m+n\rangle = \frac{(-i)^n\sqrt{m!}}{\sqrt{(m+n)!}}\left(\frac{(k_x+ik_y)\,\ell_B}{\sqrt{2}}\right)^n e^{-\boldsymbol{k}^2\ell_B^2/4}L_m^n\left(\frac{\boldsymbol{k}^2\ell_B^2}{2}\right). \tag{31.2.4}$$

Hence it is necessary that $\langle\mathfrak{S}|S_{x,y}(m,n)|\mathfrak{S}\rangle \propto \delta_{m,n\pm1}$ and $\langle\mathfrak{S}|S_z(m,n)|\mathfrak{S}\rangle \propto \delta_{m,n}$ for the state $|\mathfrak{S}\rangle$ to have the tensorial structure (31.2.2). Namely only the nearest neighboring sites are mixed in the skyrmion state $|\mathfrak{S}\rangle$.

We consider the W_∞(N)-rotated state (31.1.2a) of a hole. The simplest W_∞(N) rotation mixing only the nearest neighboring sites is given by the choice of $W^- = \sum_{n=0}^{\infty}W_n^-$ with

$$iW_n^- = \alpha_n\left[c_\downarrow^\dagger(n)c_\uparrow(n+1) - c_\uparrow^\dagger(n+1)c_\downarrow(n)\right], \tag{31.2.5}$$

where α_n is a real parameter. Note that W_n^- is a Hermitian operator belonging to the $\text{W}_\infty(2)$ algebra. Since $[W_n^-, W_m^-] = 0$, when we set

$$\xi^\dagger(n) \equiv e^{+iW^-}c_\uparrow^\dagger(n+1)e^{-iW^-}, \tag{31.2.6}$$

we find

$$\xi^\dagger(n) \equiv e^{+iW_n^-}c_\uparrow^\dagger(n+1)e^{-iW_n^-}. \tag{31.2.7}$$

We calculate $\xi^\dagger(n)$ using the standard technique of deriving the differential equation with respect to α_n. Since it satisfies

$$\frac{d^2\xi^\dagger(n)}{d\alpha_n^2} = -\xi^\dagger(n), \tag{31.2.8}$$

we integrate it as

$$\xi^\dagger(n) = u_-(n)c_\downarrow^\dagger(n)_n + v_-(n)c_\uparrow^\dagger(n+1), \tag{31.2.9}$$

where

$$u_-(n) = \sin\alpha_n, \quad \text{and} \quad v_-(n) = \cos\alpha_n. \tag{31.2.10}$$

[1]This formula is obtained by calculating

$$\langle m|e^{-i\boldsymbol{k}\boldsymbol{X}}|n\rangle = \int d^2x\,\langle m|e^{-i\boldsymbol{k}\boldsymbol{X}}|\boldsymbol{x}\rangle\langle\boldsymbol{x}|n\rangle = e^{\boldsymbol{k}^2\ell_B^2/4}\int d^2x\,e^{-i\boldsymbol{k}\boldsymbol{x}}\langle m|\boldsymbol{x}\rangle\langle\boldsymbol{x}|n\rangle,$$

together with the wave function (29.1.9). See also (29.2.22).

Hence the $W_\infty(N)$-rotated state (31.1.2a) is given by

$$|\mathfrak{S}^-_{\text{sky}}\rangle = e^{iW^-}|\text{h}\rangle = e^{iW^-}\prod_{n=0}^{\infty} c_\uparrow(n+1)|0\rangle = \prod_{n=0}^{\infty}\xi^\dagger(n)|0\rangle, \tag{31.2.11}$$

with the use of $e^{-iW^-}|0\rangle = |0\rangle$. It is a generic skyrmion state involving infinitely many parameters $u_-(n)$ and $v_-(n)$.

It is also easy to analyze the $W_\infty(N)$-rotated state (31.1.2b) of an electron-excited state. By choosing $W^+ = \sum_{n=0}^{\infty} W_n^+$ with

$$iW_n^+ = \beta_n\left[c_\downarrow^\dagger(n+1)c_\uparrow(n) - c_\uparrow^\dagger(n)c_\downarrow(n+1)\right], \tag{31.2.12}$$

we get

$$|\mathfrak{S}^+_{\text{sky}}\rangle = e^{iW^+}|\text{e}\rangle = e^{iW^+}c_\downarrow^\dagger(0)\prod_{n=0}^{\infty} c_\uparrow(n)|0\rangle = c_\downarrow^\dagger(0)\prod_{n=0}^{\infty}\zeta^\dagger(n)|0\rangle, \tag{31.2.13}$$

where

$$\zeta^\dagger(n) = u_+(n)c_\uparrow^\dagger(n) + v_+(n)c_\downarrow^\dagger(n+1), \tag{31.2.14}$$

and

$$v_+(n) = \sin\beta_n, \quad \text{and} \quad u_+(n) = \cos\beta_n. \tag{31.2.15}$$

The state (31.2.13) is a generic antiskyrmion state involving infinitely many parameters $u_+(n)$ and $v_+(n)$.

The operators $\xi(m)$ and $\zeta(n)$ satisfy the standard canonical commutation relations,

$$\begin{aligned} \{\xi(m),\xi^\dagger(n)\} &= \delta_{mn}, \quad \{\xi(m),\xi(n)\} = 0, \\ \{\zeta(m),\zeta^\dagger(n)\} &= \delta_{mn}, \quad \{\zeta(m),\zeta(n)\} = 0, \end{aligned} \tag{31.2.16}$$

based on the relation

$$u_\pm^2(n) + v_\pm^2(n) = 1, \tag{31.2.17}$$

as follows from (31.2.10) and (31.2.15). Since these states should approach the ground state asymptotically it is necessary that

$$\begin{aligned} \lim_{n\to\infty} u_-(n) &= 0, \quad \lim_{n\to\infty} v_-(n) = 1, \\ \lim_{n\to\infty} u_+(n) &= 1, \quad \lim_{n\to\infty} v_+(n) = 0. \end{aligned} \tag{31.2.18}$$

For a later convenience we define

$$\begin{aligned} u_-(-1) &= 1, \quad v_-(-1) = 0, \\ u_+(-1) &= 0, \quad v_+(-1) = 1. \end{aligned} \tag{31.2.19}$$

Recall that $|\mathfrak{S}_{\text{sky}}^-\rangle$ describes one hole excited state in (30.4.1) when $u_-(n) = 0$ and $v_-(n) = 1$ for all n, while $|\mathfrak{S}_{\text{sky}}^+\rangle$ describes one electron excited state in (30.4.1) when $u_+(n) = 1$ and $v_+(n) = 0$ for all n.

In what follows we calculate the classical density (31.2.3) explicitly. For instance the basic relations are

$$
\begin{aligned}
c_\uparrow(n)|\mathfrak{S}_{\text{sky}}^-\rangle =& v_-(n-1)\xi(n-1)|\mathfrak{S}_{\text{sky}}^-\rangle, \\
c_\downarrow(n)|\mathfrak{S}_{\text{sky}}^-\rangle =& u_-(n)\xi(n)|\mathfrak{S}_{\text{sky}}^-\rangle.
\end{aligned} \tag{31.2.20}
$$

They reduce the action of c-operators to that of ξ's, thus allowing to carry out exact calculus. We then employ $\langle\mathfrak{S}_{\text{sky}}^-|\xi^\dagger(m)\xi(n)|\mathfrak{S}_{\text{sky}}^-\rangle = \delta_{mn}$ together with (31.2.20) and its conjugate. In this way we come to

$$
\begin{aligned}
\hat{\rho}^\pm(n,n) &= u_\pm^2(n) + v_\pm^2(n-1), \\
\hat{\mathcal{S}}_z^\pm(n,n) &= \pm\frac{1}{2}\left[u_\pm^2(n) - v_\pm^2(n-1)\right], \\
\hat{\mathcal{S}}_x^\pm(n+1,n) = \hat{\mathcal{S}}_x^\pm(n,n+1) &= \frac{1}{2}u_\pm(n)v_\pm(n), \\
\hat{\mathcal{S}}_y^\pm(n+1,n) = -\hat{\mathcal{S}}_y^{\text{cl}}(n,n+1) &= \frac{i}{2}u_\pm(n)v_\pm(n),
\end{aligned} \tag{31.2.21}
$$

where we have omitted the superscript "cl". All other components, $\hat{\rho}^\pm(n,m)$, etc., vanish. By substituting these into the physical density (30.1.3) together with (29.1.9), we obtain

$$
\begin{aligned}
\frac{\rho^\pm(\boldsymbol{x})}{\rho_0} =& e^{-|z|^2/2}\sum_{n=0}\left[u_\pm^2(n) + v_\pm^2(n-1)\right]\left(\frac{|z|^2}{2}\right)^n, \\
\frac{\mathcal{S}_z^\pm(\boldsymbol{x})}{\rho_0} =& \pm\frac{1}{2}e^{-|z|^2/2}\sum_{n=0}\frac{u_\pm^2(n) - v_\pm^2(n-1)}{n!}\left(\frac{|z|^2}{2}\right)^n, \\
\frac{\mathcal{S}_x^\pm(\boldsymbol{x})}{\rho_0} =& \frac{x}{\sqrt{2}\ell_B}e^{-|z|^2/2}\sum_{n=0}\frac{u_\pm(n)v_\pm(n)}{n!\sqrt{n+1}}\left(\frac{|z|^2}{2}\right)^n, \\
\frac{\mathcal{S}_y^\pm(\boldsymbol{x})}{\rho_0} =& \pm\frac{y}{\sqrt{2}\ell_B}e^{-|z|^2/2}\sum_{n=0}\frac{u_\pm(n)v_\pm(n)}{n!\sqrt{n+1}}\left(\frac{|z|^2}{2}\right)^n.
\end{aligned} \tag{31.2.22}
$$

We have calculated the classical densities of the skyrmion and antiskyrmion states characterized by infinitely many variables $u_\pm(n)$ and $v_\pm(n)$. They are determined by minimizing the energies of these states, which we shall carry out in later sections.

The multi-skyrmion and multi-antiskyrmion states are analyzed[479] in the same way. They are given by (31.1.3) together with

$$iW^- = \sum_n \alpha_n \left[c_\downarrow^\dagger(n)c_\uparrow(n+k) - c_\uparrow^\dagger(n+k)c_\downarrow(n)\right], \tag{31.2.23a}$$

$$iW^+ = \sum_n \beta_n \left[c_\downarrow^\dagger(n+k)c_\uparrow(n) - c_\uparrow^\dagger(n)c_\downarrow(n+k)\right]. \tag{31.2.23b}$$

We obtain

$$|\mathfrak{S}_{\text{sky}}^-; k\rangle = \prod_{n=0} \xi_k^\dagger(n)|0\rangle \tag{31.2.24a}$$

$$|\mathfrak{S}_{\text{sky}}^+; k\rangle = c_\downarrow^\dagger(0)\cdots c_\downarrow^\dagger(k-1)\prod_{n=0} \zeta_k^\dagger(n)|0\rangle \tag{31.2.24b}$$

with

$$\xi_k^\dagger(n) = u_-(n)c_\downarrow^\dagger(n) + v_-(n)c_\uparrow^\dagger(n+k), \tag{31.2.25a}$$

$$\zeta_k^\dagger(n) = u_+(n)c_\uparrow^\dagger(n) + v_+(n)c_\downarrow^\dagger(n+k). \tag{31.2.25b}$$

The electron numbers are evidently $\delta N_e^{\text{cl}-} = -k$ and $\delta N_e^{\text{cl}+} = k$ for the skyrmion state and the antiskyrmion state, respectively.

31.3 Noncommutative CP¹ Skyrmion

We have constructed the skyrmion state yielding the same tensorial structure as the standard configuration (31.2.2). We proceed to reformulate it in terms of the noncommutative CP¹ field for the skyrmion case $|\mathfrak{S}_{\text{sky}}^-; k\rangle$. For notational simplicity, we set $u(n) \equiv u_-(n)$, $v(n) \equiv v_-(n)$ and

$$\begin{pmatrix} u_\uparrow \\ u_\downarrow \end{pmatrix} = \begin{pmatrix} 0 \\ u \end{pmatrix}, \quad \begin{pmatrix} v_\uparrow \\ v_\downarrow \end{pmatrix} = \begin{pmatrix} v \\ 0 \end{pmatrix}, \tag{31.3.1}$$

so that (31.2.25a) reads

$$\xi_k^\dagger(n) = u_\mu(n)c_\mu^\dagger(n) + v_\mu(n)c_\mu^\dagger(n+k). \tag{31.3.2}$$

It is convenient to use the density matrix (29.5.6), or

$$D_{\mu\nu}^{\text{cl}}(m,n) \equiv \langle\mathfrak{S}_{\text{sky}}^-|c_\nu^\dagger(n)c_\mu(m)|\mathfrak{S}_{\text{sky}}^-\rangle. \tag{31.3.3}$$

We calculate this with the aid of (31.2.20),

$$\begin{aligned} D_{\mu\nu}^{\text{cl}}(n,n) &= u_\mu(n)u_\nu^*(n) + v_\mu(n-k)v_\nu^*(n-k), \\ D_{\mu\nu}^{\text{cl}}(n,n+k) &= u_\mu(n)v_\nu^*(n), \\ D_{\mu\nu}^{\text{cl}}(n+k,n) &= v_\mu(n)u_\nu^*(n). \end{aligned} \tag{31.3.4}$$

All other matrix elements vanish. The formula (31.3.4) is equivalent to (31.2.21).

The Weyl operator for the density matrix is constructed by using (29.2.17),

$$2\pi\ell_B^2 W\left[\hat{\mathcal{D}}^{\text{cl}}_{\mu\nu}\right] = u_\mu(n)u_\nu^*(n)|n\rangle\langle n| + v_\mu(n-k)v_\nu^*(n-k)|n\rangle\langle n| + u_\mu(n)v_\nu^*(n)|n\rangle\langle n+k| + v_\mu(n)u_\nu^*(n)|n+k\rangle\langle n|. \tag{31.3.5}$$

Corresponding to (29.6.28) we have the decomposition

$$W\left[\hat{\mathcal{D}}^{\text{cl}}_{\mu\nu}\right] = \frac{1}{2\pi\ell_B^2}\mathfrak{n}_\mu \mathfrak{n}_\nu^*, \tag{31.3.6}$$

where $\mathfrak{n}_\mu \equiv W[n_\mu]$ is the Weyl operator of the CP^1 field $n_\mu(\boldsymbol{x})$. This is uniquely solved as

$$\mathfrak{n}_\mu = u_\mu(n)|n\rangle\langle n| + v_\mu(n)|n+k\rangle\langle n|, \tag{31.3.7}$$

or

$$\mathfrak{n}_\uparrow = \sum_n v(n)|n+k\rangle\langle n|, \qquad \mathfrak{n}_\downarrow = \sum_n u(n)|n\rangle\langle n|. \tag{31.3.8}$$

Due to the correspondence (29.2.21) the symbol reads

$$n_\uparrow(\boldsymbol{x}) = \sum_n v(n)\Xi_{n+k,n}(\boldsymbol{x}), \qquad n_\downarrow(\boldsymbol{x}) = \sum_n u(n)\Xi_{n,n}(\boldsymbol{x}), \tag{31.3.9}$$

where $\Xi_{m,n}(\boldsymbol{x})$ is given by (29.2.22). This is the CP^1 multi-skyrmion in the noncommutative plane.

To relate this to the standard expression (31.2.1), we make some algebraic works. From (31.3.8) we find

$$(b^\dagger)^k \mathfrak{n}_\downarrow = \sum_n u(n)(b^\dagger)^k|n\rangle\langle n| = \sum_n u(n)\sqrt{\frac{(n+k)!}{n!}}|n+k\rangle\langle n|, \tag{31.3.10}$$

which has the same operator structure as $\mathfrak{n}_\uparrow$. Hence, if we require

$$u(n)\sqrt{\frac{(n+k)!}{n!}} = \omega^k v(n) \tag{31.3.11}$$

with ω being a complex constant, we obtain

$$(b^\dagger)^k \mathfrak{n}_\downarrow = \omega^k \mathfrak{n}_\uparrow, \tag{31.3.12}$$

or[2]

$$z^k \star n_\downarrow(\boldsymbol{x}) = (\sqrt{2}\omega)^k n_\uparrow(\boldsymbol{x}), \tag{31.3.13}$$

where we have used $z^k \star z = z^{k+1}$. Thus,

$$\begin{pmatrix} n_\uparrow(\boldsymbol{x}) \\ n_\downarrow(\boldsymbol{x}) \end{pmatrix} = \frac{1}{\lambda^k} \begin{pmatrix} z^k \\ \lambda^k \end{pmatrix} \star n_\downarrow(\boldsymbol{x}) \tag{31.3.14}$$

with $\lambda = \sqrt{2}\omega\ell_B$. We have constructed the noncommutative generalization of the CP^1-skyrmion (31.2.1) by requiring the condition (31.3.11) on the generic skyrmion state (31.2.11) with (31.1.2b).

Topological Charge

It follows from the definition of the topological charge that the skyrmion state (31.2.24) carrying the electric number $\delta N_{\mathrm{e}}^{\mathrm{cl}} = -k$ carries also the topological charge $Q = k$ as in (31.1.6). We confirm this by an explicit calculation.

The topological charge (31.1.6) is expressed as

$$Q = \mathrm{Tr}([\mathfrak{n}_\mu^\dagger, \mathfrak{n}_\mu]) \tag{31.3.15}$$

in the Weyl form. Here, we cannot use $\mathrm{Tr}([A,B]) = \mathrm{Tr}(AB) - \mathrm{Tr}(BA) = 0$, which is valid only if $\mathrm{Tr}(AB)$ and $\mathrm{Tr}(BA)$ are separately well defined. This is not the case here, and the trace operation must be carried out after the commutator is calculated.

Using (31.3.7) we obtain

$$[\mathfrak{n}_\mu^\dagger, \mathfrak{n}_\mu] = \sum_{n=0}^{\infty} \bar{v}_\mu(n) v_\mu(n) \left[|n\rangle\langle n| - |n+k\rangle\langle n+k| \right]. \tag{31.3.16}$$

This is still inappropriate for calculating the trace, since the separate pieces are divergent due to the behavior $|v_\mu(n)| \to 1$ as $n \to \infty$ as implied by the boundary condition (31.2.18). We rewrite the formula in terms of $u_\mu(n)$ instead of $v_\mu(n)$, and obtain

$$[\mathfrak{n}_\mu^\dagger, \mathfrak{n}_\mu] = \sum_{n=0}^{k-1} |n\rangle\langle n| - \sum_{n=0}^{\infty} \bar{u}_\mu(n) u_\mu(n) |n\rangle\langle n| + \sum_{n=0}^{\infty} \bar{u}_\mu(n) u_\mu(n) |n+k\rangle\langle n+k|. \tag{31.3.17}$$

---

[2]The factor $\sqrt{2}$ in (31.3.13) is understand from (29.1.2). It may be instructive to verify (31.3.13) by explicit calculation. For simplicity we give the case of $k = 1$.

$$\begin{aligned} z \star \hat{n}_\downarrow(x) =& z\hat{n}_\downarrow(x) - \frac{\ell_B}{2}\left(\partial_x + i\partial_y\right)\hat{n}_\downarrow(x) = 4ze^{-|z|^2} \sum_{n=0}^{\infty} (-1)^n u_-(n) \left[L_n - L_n'\right] \\ =& 4ze^{-|z|^2} \sum_{n=0}^{\infty} (-1)^n u_-(n) L_n^1\left(2|z|^2\right), \end{aligned}$$

where we have used $L_n' - L_n = L_n^1$.

Each piece is well defined since $u_\mu(n) \to 0$ as $n \to \infty$, and the traces can be calculated separately, which eventually leads to $Q = k$.

31.4 Hole and Skyrmion

We have introduced a skyrmion state as a $W_\infty(N)$-rotated state of a hole state. A skyrmion state is reduced continuously to the hole state by controlling various parameters. The skyrmion carries the topological charge and the electron number. The hole carries the electron number. Does the hole carry also the topological charge? This is surely not the case in the commutative theory.

We investigate this problem in order to understand the difference of the noncommutative theory from the commutative one. The hole state is given by

$$|\mathrm{h}\rangle = c_\uparrow(0)|\mathrm{g}\rangle. \tag{31.4.1}$$

As derived in Section 30.4, the bare densities are

$$\hat{\rho}_{\mathrm{h}}^{\mathrm{cl}}(\boldsymbol{x}) = \rho_0\left(1 - 2e^{-r^2/\ell_B^2}\right), \tag{31.4.2a}$$

$$\hat{\mathcal{S}}_x^{\mathrm{cl}}(\boldsymbol{x}) = \hat{\mathcal{S}}_y^{\mathrm{cl}}(\boldsymbol{x}) = 0, \quad \hat{\mathcal{S}}_z^{\mathrm{cl}}(\boldsymbol{x}) = \frac{1}{2}\hat{\rho}_{\mathrm{h}}^{\mathrm{cl}}(\boldsymbol{x}). \tag{31.4.2b}$$

They represent literally a hole of the size ℓ_B. The spin texture is trivial.

In the commutative theory the topological number is given by (29.7.1), or

$$Q_{\mathrm{sky}} = \frac{1}{2\pi i}\epsilon_{kl}\sum_s \int d^2x\, (\partial_k \bar{n}_\mu)(\partial_l n_\mu). \tag{31.4.3}$$

It is equivalent to the Pontryagin number (15.4.21) in the commutative theory,

$$Q_{\mathrm{P}} = \frac{1}{\pi}\int d^2x\, \varepsilon_{abc}\varepsilon_{ij}\mathcal{S}_a\partial_i\mathcal{S}_b\partial_j\mathcal{S}_c, \tag{31.4.4}$$

where $\mathcal{S}_a$ is the normalized spin density, $\mathcal{S}_a \equiv \frac{1}{2}\bar{n}_\mu(\sigma_a)_{\mu\nu}n_\nu$. We would obviously conclude $Q_{\mathrm{P}} = 0$ for the trivial spin texture such as (31.4.2b). On the other hand, the CP^1 field giving the trivial spin texture (31.4.2b) is

$$n_\uparrow(\boldsymbol{x}) = e^{i\vartheta}, \qquad n_\downarrow(\boldsymbol{x}) = 0. \tag{31.4.5}$$

Though it carries a winding number except at the origin, since it is ill-defined there, we cannot calculate the topological number based on the formula (31.4.3).

However, this argument is not applicable to the noncommutative theory. According to the argument presented in the previous subsection, its topological charge is given by

$$Q = \mathrm{Tr}([\mathrm{n}_\mu^\dagger, \mathrm{n}_\mu]) = \mathrm{Tr}(|0\rangle\langle 0|) = 1, \tag{31.4.6}$$

since $u_\mu(n) = 0$ for all n in (31.3.17). There is no mystery here. Let us explain this by calculating the topological number explicitly in the real space.

The spin density is related to the CP^1 field via (29.6.34), or

$$\mathcal{S}_a^{cl}(x) = \frac{1}{4\pi\ell_B^2}\sum_{i,\mu\nu}(\tau_a)_{\mu\nu}\, n_\mu^i(x) \star n_\nu^{i*}(x). \tag{31.4.7}$$

The CP^1 field which gives the trivial spin texture (31.4.2b) is highly nontrivial,

$$n_\uparrow(x) = 2^{\frac{3}{2}} z e^{-\bar{z}z} \sum_{n=0}^{\infty} \frac{(-1)^n}{\sqrt{n+1}} L_n^1(2\bar{z}z), \tag{31.4.8a}$$

$$n_\downarrow(x) = 0. \tag{31.4.8b}$$

It is well-defined everywhere. Only the asymptotic behavior contributes to the topological number (29.7.2) in the noncommutative formulation, or equivalently to (31.4.3) in the commutative formulation. Any finite order terms in n vanish in the limit $\bar{z}z \to \infty$ due to the term $e^{-\bar{z}z}$ in (31.4.8a). Thus, we may use

$$\frac{1}{\sqrt{(n+1)}} \simeq \frac{\Gamma(n+\frac{3}{2})}{\Gamma(n+2)} \qquad \text{for} \quad n \gg 1. \tag{31.4.9}$$

We then use

$$\sum_{n=0}^{\infty} \frac{\Gamma(N+a)}{\Gamma(N+c)} t^n L_n^{(c-1)}(x) = \frac{\Gamma(a)}{\Gamma(c)} \frac{1}{(1-t)^a} M\left(a; c; \frac{xt}{t-1}\right), \tag{31.4.10}$$

where $M(a;b;x)$ is the Kummer function. Taking $a = \frac{3}{2}$, $c = 2$ and $t = -1$, we obtain

$$n_\uparrow(x) \to z e^{-\bar{z}z} \frac{\Gamma(\frac{3}{2})}{\Gamma(2)} M\left(\frac{3}{2}; 2; \bar{z}z\right) \to e^{i\vartheta}. \tag{31.4.11}$$

It carries a proper winding number. The topological number is clearly $Q = 1$, as is consistent with (31.4.6).

Hence we conclude that a hole is a kind of skyrmion: Its spin texture is trivial but the associated CP^{N-1} field is highly nontrivial and carries a proper winding number in the noncommutative theory. It is the essential feature of the noncommutative theory that the singularity (31.4.5) is regulated over the region of area θ. This is the reason why the noncommutative CP^1 field for the hole state is well defined everywhere and carries a winding number.

31.5 Skyrmion Wave Functions

We derive the wave function of the skyrmion state (31.2.11). Recall that the wave function of a many-body state $|\mathfrak{S}\rangle$ is given by

$$\mathfrak{S}_{\mu_1\mu_2\cdots\mu_N}[x] = \langle 0|\psi_{\mu_1}(x_1)\psi_{\mu_2}(x_2)\cdots\psi_{\mu_N}(x_N)|\mathfrak{S}\rangle, \tag{31.5.1}$$

where $\psi_\mu(\boldsymbol{x})$ is given by (29.4.18) with (29.1.9), or

$$\psi_\mu(\boldsymbol{x}) = \rho_0^{1/2} e^{-|z|^2/4} \sum_{n=0} \alpha(n) z^n c_\mu(n) \tag{31.5.2}$$

with

$$\alpha(n) = \sqrt{\frac{1}{2^n n!}}. \tag{31.5.3}$$

We find

$$\mathfrak{S}_{\mathrm{g}}[\boldsymbol{x}] = \prod_r \begin{pmatrix} 1 \\ 0 \end{pmatrix} \mathfrak{S}_{\mathrm{LN}}[\boldsymbol{x}] \tag{31.5.4}$$

for the ground state (30.2.13), and

$$\mathfrak{S}_{\mathrm{h}}[\boldsymbol{x}] = \prod_r \begin{pmatrix} z_r \\ 0 \end{pmatrix} \mathfrak{S}_{\mathrm{LN}}[\boldsymbol{x}] \tag{31.5.5}$$

for the hole state (30.4.1a), where $\mathfrak{S}_{\mathrm{LN}}[\boldsymbol{x}]$ is the Slater determinant, of the one-body wave functions,

$$\mathfrak{S}_{\mathrm{LN}}[\boldsymbol{x}] = \rho_0^{N/2} \begin{vmatrix} 1 & z_1 & \cdots & z_1^{N-1} \\ 1 & z_2 & \cdots & z_2^{N-1} \\ \vdots & \vdots & \ddots & \vdots \\ 1 & z_N & \cdots & z_N^{N-1} \end{vmatrix} e^{-\sum_{r=1}^N |z_r|^2/4}, \tag{31.5.6}$$

known as the Laughlin wave function.

The skyrmion state is given by $|\mathfrak{S}\rangle = \prod_{n=0}^\infty \xi^\dagger(n)|0\rangle$ with (31.2.9), or

$$\xi^\dagger(n) = u_-(n) c_\downarrow^\dagger(n)_n + v_-(n) c_\uparrow^\dagger(n+1). \tag{31.5.7}$$

We first study the component of the skyrmion wave function with all spins up, to which only the term $v_-(n) c_\uparrow^\dagger(n+1)$ in $\xi^\dagger(n)$,

$$\mathfrak{S}^{\mathrm{sky}}_{\uparrow\uparrow\cdots\uparrow}[\boldsymbol{x}] = \prod_{n=0}^{N-1} v_-(n) \langle 0| \prod_{i=1}^N \psi_\uparrow(\boldsymbol{x}_i) \prod_{n=1}^N c_\uparrow^\dagger(n)|0\rangle. \tag{31.5.8}$$

Apart from the factor $\prod_{n=0}^{N-1} v_-(n)$ this is nothing but the wave function of a hole. Thus

$$\mathfrak{S}^{\mathrm{sky}}_{\uparrow\uparrow\cdots\uparrow}[\boldsymbol{x}] = C_N \prod_{n=1}^N z_n \mathfrak{S}_{\mathrm{LN}}[\boldsymbol{x}], \tag{31.5.9}$$

where $C_N = \prod_{n=1}^N \alpha(n) v_-(n-1)$. Next we consider the component with all spins down, which arises only from the term $u_-(n) c_\downarrow^\dagger(n)$ in $\xi^\dagger(n)$,

$$\mathfrak{S}^{\mathrm{sky}}_{\downarrow\downarrow\cdots\downarrow}[\boldsymbol{x}] = \prod_{n=1}^N u_-(n) \langle 0| \prod_{i=1}^N \psi_\downarrow(\boldsymbol{x}_i) \prod_{n=1}^N c_\downarrow^\dagger(n)|0\rangle = C_N \prod_{n=0}^{N-1} \beta_n \mathfrak{S}_{\mathrm{LN}}[\boldsymbol{x}]. \tag{31.5.10}$$

Here we have set

$$\beta_n = \frac{\alpha(n)u_-(n)}{\alpha(n+1)v_-(n)} = \frac{u_-(n)}{v_-(n)}\sqrt{2(n+1)} \tag{31.5.11}$$

with (31.5.3). Comparing (31.5.9) and (31.5.10) we remark that the wave function $\mathfrak{S}^{\text{sky}}_{\downarrow\downarrow\cdots\downarrow}[x]$ is obtained by replacing z_n with β_{n-1} within the factor $\prod z_n$ of the wave function $\mathfrak{S}_{\uparrow\uparrow\cdots\uparrow}[x]$ for all n. The wave function with mixed spin components is similarly derived, where z_n is replaced with β_{n-1} for certain indices n within the factor $\prod z_n$. In general we derive

$$\mathfrak{S}^{\text{sky}}_{\mu_1\mu_2\cdots\mu_N}[x] = C_N e^{-\sum_{r=1}^{N-1}|z_r|^2/4} \times \begin{vmatrix} \begin{pmatrix} z_1 \\ \beta_0 \end{pmatrix}_{\mu_1} & z_1\begin{pmatrix} z_1 \\ \beta_1 \end{pmatrix}_{\mu_1} & \cdots & z_1^{N-1}\begin{pmatrix} z_1 \\ \beta_{N-1} \end{pmatrix}_{\mu_1} \\ \begin{pmatrix} z_2 \\ \beta_0 \end{pmatrix}_{\mu_2} & z_2\begin{pmatrix} z_2 \\ \beta_1 \end{pmatrix}_{\mu_2} & \cdots & z_2^{N-1}\begin{pmatrix} z_2 \\ \beta_{N-1} \end{pmatrix}_{\mu_2} \\ \vdots & \vdots & \ddots & \vdots \\ \begin{pmatrix} z_N \\ \beta_0 \end{pmatrix}_{\mu_N} & z_N\begin{pmatrix} z_N \\ \beta_1 \end{pmatrix}_{\mu_N} & \cdots & z_N^{N-1}\begin{pmatrix} z_N \\ \beta_{N-1} \end{pmatrix}_{\mu_N} \end{vmatrix} \tag{31.5.12}$$

for the wave function of the skyrmion.

It is notable that, when all β_n are equal ($\beta_n = \sqrt{2}\omega$), this is reduced to the wave function of the factorizable skyrmion (16.2.16) derived in the semiclassical approximation,

$$\mathfrak{S}_{\text{sky}}[x] = \prod_r \begin{pmatrix} z_r \\ \sqrt{2}\omega \end{pmatrix} \mathfrak{S}_{\text{LN}}[x]. \tag{31.5.13}$$

Setting $\beta_n = \sqrt{2}\omega$ in (31.5.11), the condition is reduced to (31.3.11). Thus the generic skyrmion (31.2.11) becomes the factorizable skyrmion together with the condition (31.3.11).

31.6 Hardcore Interaction

We have presented the noncommutative generalization of the factorizable skyrmion. As an explicit example we investigate the system governed by the hard-core interaction,[308,323]

$$V(x-y) = \ell_B \delta^2(x-y). \tag{31.6.1}$$

The Hamiltonian (15.2.3) reads

$$H_{\text{hc}} = \frac{1}{2}\int d^2x\, \Delta\rho(x)\Delta\rho(x). \tag{31.6.2}$$

All previous formulas hold with

$$V_{mnij} = \frac{1}{8\pi^2}\int d^2k\, e^{-\ell_B^2 k^2/2}\langle m|e^{iXk}|n\rangle\langle i|e^{-iXk}|j\rangle$$
$$= \frac{1}{8\pi\ell_B}\frac{\sqrt{(m+i)!(n+j)!}}{\sqrt{m!i!n!j!}}\frac{\delta_{m+i,n+j}}{\sqrt{2^{m+i+n+j}}}, \tag{31.6.3}$$

and

$$\epsilon_{\mathrm{D}} = \epsilon_{\mathrm{X}} = \frac{1}{4\pi\ell_B}. \tag{31.6.4}$$

The Hamiltonian (31.6.2) is rewritten into the normal ordered form,

$$H_{\mathrm{hc}} = \int d^2x\, \psi_\downarrow^\dagger(\boldsymbol{x})\psi_\uparrow^\dagger(\boldsymbol{x})\psi_\downarrow(\boldsymbol{x})\psi_\uparrow(\boldsymbol{x}) - \frac{1}{4\pi\ell_B}\int d^2x\, \Delta\rho(\boldsymbol{x}). \tag{31.6.5}$$

We examine if the skyrmion state can be as an eigenstate of the Hamiltonian. We treat the skyrmion state (31.2.11) explicitly, but similar formulas follow also for the antiskyrmion state (31.2.13).

Defining the state

$$|\mathfrak{S}'\rangle = \psi_\downarrow(\boldsymbol{x})\psi_\uparrow(\boldsymbol{x})|\mathfrak{S}_{\mathrm{sky}}^-\rangle = \sum_{mn}\varphi_m(\boldsymbol{x})\varphi_n(\boldsymbol{x})c_\downarrow(m)c_\uparrow(n)|\mathfrak{S}_{\mathrm{sky}}^-\rangle, \tag{31.6.6}$$

we have

$$\langle\mathfrak{S}_{\mathrm{sky}}^-|H_{\mathrm{hc}}|\mathfrak{S}_{\mathrm{sky}}^-\rangle = \int d^2x\, \langle\mathfrak{S}'|\mathfrak{S}'\rangle + \frac{1}{4\pi\ell_B}. \tag{31.6.7}$$

Using (31.2.20) we come to

$$|\mathfrak{S}'\rangle = \sum_{m=0}\sum_{n=0}\varphi_m(\boldsymbol{x})\varphi_{n+1}(\boldsymbol{x})u_-(m)v_-(n)\xi(m)\xi(n)|\mathfrak{S}_{\mathrm{sky}}^-\rangle. \tag{31.6.8}$$

Using (31.6.3) we obtain

$$\int d^2x\, \langle\mathfrak{S}'|\mathfrak{S}'\rangle = 2\sum_{k,l=0} v_-(k)u_-(l)v_-(l)u_-(k)V_{k+1,k,l,l+1}$$
$$= \frac{1}{8\pi\ell_B}\sum_{k,l=0}\frac{(k+l+1)!}{k!l!2^{k+l+1}}\left[\frac{v_-(k)u_-(l)}{\sqrt{k+1}} - \frac{v_-(l)u_-(k)}{\sqrt{l+1}}\right]^2 \geqslant 0. \tag{31.6.9}$$

The skyrmion energy is minimized as

$$E_{\mathrm{sky}} = \frac{1}{4\pi\ell_B} = \epsilon_{\mathrm{X}}, \tag{31.6.10}$$

when the equality is achieved, or

$$\frac{v_-(k)u_-(l)}{\sqrt{k+1}} = \frac{v_-(l)u_-(k)}{\sqrt{l+1}}. \tag{31.6.11}$$

Due to the normalization (31.2.17) it gives

$$u_-^2(n) = \frac{\omega^2}{n+1+\omega^2}, \quad v_-^2(n) = \frac{n+1}{n+1+\omega^2}. \tag{31.6.12}$$

This is the condition (31.3.11) that the skyrmion is factorizable. It follows $|\mathfrak{S}'\rangle = 0$ when the equality holds in (31.6.9). It implies that $|\mathfrak{S}_{\text{sky}}^-\rangle$ is an eigenstate of the Hamiltonian, $H_{\text{hc}}|\mathfrak{S}_{\text{sky}}^-\rangle = \epsilon_X|\mathfrak{S}_{\text{sky}}^-\rangle$.

Similarly, when

$$u_+^2(n) = \frac{n+1}{n+1+\omega^2}, \quad v_+^2(n) = \frac{\omega^2}{n+1+\omega^2}, \tag{31.6.13}$$

the antiskyrmion state is an eigenstate of the Hamiltonian, $H_{\text{hc}}|\mathfrak{S}_{\text{sky}}^+\rangle = \epsilon_X|\mathfrak{S}_{\text{sky}}^+\rangle$.

We may summarize them as

$$H_{\text{hc}}|\mathfrak{S}_{\text{sky}}^\pm\rangle = \frac{1}{4\pi\ell_B}\int d^2x\, \Delta\rho(\boldsymbol{x})|\mathfrak{S}_{\text{sky}}^\pm\rangle = E_{\text{hc}}|\mathfrak{S}_{\text{sky}}^\pm\rangle \quad \text{with} \quad E_{\text{hc}} = \frac{1}{4\pi\ell_B}|\delta N^{\text{cl}}|. \tag{31.6.14}$$

The electron number of the skyrmion or antiskyrmion state is different from that of the ground state by its topological charge, $Q_{\text{sky}} = -\Delta N_{\text{e}}^{\text{cl}}$. Each of them is the lowest energy state in each topological sector $Q_{\text{sky}} = \pm 1$. In conclusion, we have determined the $W_\infty(2)$-rotated states (31.1.2) by requiring the energy minimization, where the skyrmion energy is independent of the scale ω and degenerates with the hole-excited and electron-excited state.

We write down the physical densities (31.2.22) with the use of (31.6.12) and (31.6.13). They are expressed in terms of the Kummer function $M(a;b;x)$,

$$M(a;a+1;x) = a\sum_{n=0}^{\infty}\frac{x^n}{(n+a)n!}, \tag{31.6.15}$$

as

$$\begin{aligned}
\frac{\Delta\rho^\pm(\boldsymbol{x})}{\rho_0} &= \pm e^{-\frac{1}{2}|z|^2}M(\omega^2;\omega^2+1;|z|^2/2) \mp \frac{\omega^2}{\omega^2+1}e^{-\frac{1}{2}z^2}M(\omega^2+1;\omega^2+2;|z|^2/2),\\
\frac{\mathcal{S}_z^\pm(\boldsymbol{x})}{\rho_0} &= \frac{1}{2} - \frac{1}{2}e^{-\frac{1}{2}|z|^2}M(\omega^2;\omega^2+1;|z|^2/2)\\
&\qquad - \frac{1}{2}\frac{\omega^2}{\omega^2+1}e^{-\frac{1}{2}|z|^2}M(\omega^2+1;\omega^2+2;|z|^2/2),\\
\frac{\mathcal{S}_x^\pm(\boldsymbol{x})}{\rho_0} &= \frac{1}{\sqrt{2}}\frac{\omega x/\ell_B}{\omega^2+1}e^{-\frac{1}{2}|z|^2}M(\omega^2+1;\omega^2+2;|z|^2/2),\\
\frac{\mathcal{S}_y^\pm(\boldsymbol{x})}{\rho_0} &= \frac{\pm 1}{\sqrt{2}}\frac{\omega y/\ell_B}{\omega^2+1}e^{-\frac{1}{2}|z|^2}M(\omega^2+1;\omega^2+2;|z|^2/2),
\end{aligned} \tag{31.6.16}$$

with $|z|^2 = r^2/\ell_B^2$, where we have used the formula (13.4.3) of reference[12] to derive $\rho(\boldsymbol{x})$. We have illustrated the density $\rho^-(r)$ for typical values of the parameters ω in Fig.31.1.

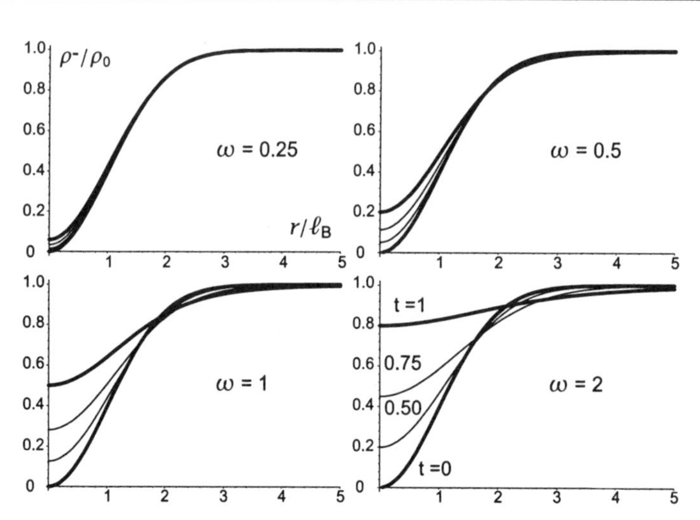

Fig. 31.1 The density modulation (ρ^-/ρ_0) accompanied with the skyrmion excitation is plotted as a function of the radius (r/ℓ_B). The heavy curve $(t = 1)$ is for the hardcore model based on (31.6.16) for the choice of $\omega = 0.25, 0.5, 1, 2$, while the heavy curve $(t = 0)$ represents the density of a hole. The thin curves represent the interpolating formula (31.7.6) for $t = 0.50, 0.75$, where the parameter t is defined by (31.7.3).

Using the relations

$$
\begin{aligned}
M(1;b;x) &= 1 + \frac{x}{b}M(1;b+1;x), \\
M(a;b;z) &= e^z M(b-a:b:-z),
\end{aligned}
\tag{31.6.17}
$$

we are able to represent the densities in terms of a single function,

$$
\begin{aligned}
\frac{\Delta\rho^\pm(\boldsymbol{x})}{\rho_0} &= \mp\left[\Omega(r)-1\right], &\qquad \frac{\mathcal{S}_z^\pm(\boldsymbol{x})}{\rho_0} &= \frac{1}{2}\frac{r^2-\lambda^2}{r^2+\lambda^2}\Omega(r), \\
\frac{\mathcal{S}_x^\pm(\boldsymbol{x})}{\rho_0} &= \frac{\lambda x}{r^2+\lambda^2}\Omega(r), &\qquad \frac{\mathcal{S}_y^\pm(\boldsymbol{x})}{\rho_0} &= \pm\frac{\lambda y}{r^2+\lambda^2}\Omega(r),
\end{aligned}
\tag{31.6.18}
$$

where we have set $\lambda = \sqrt{2}\omega\ell_B$ and

$$
\Omega(r) = \frac{r^2+\lambda^2}{2\ell_B^2+\lambda^2}M(1;\omega^2+2,-|z|^2/2).
\tag{31.6.19}
$$

It follows that

$$
S_a^\pm(\boldsymbol{x}) = \rho^-(\boldsymbol{x})\mathcal{S}_a^\pm(\boldsymbol{x})
\tag{31.6.20}
$$

with (31.2.2). This agrees with the formula (15.4.4) derived based on the composite-boson theory, whose physical meaning is the spin-charge separation.

We explore some properties of the density modulation. On one hand, the skyrmion is reduced to a hole for $\lambda = 0$, where $\Omega(r)$ is given by

$$\Omega(r) = 1 - e^{-r^2/2\ell_B^2}. \tag{31.6.21}$$

It vanishes at the origin, $\Omega(r) = 0$ at $r = 0$.

On the other hand, $\Omega(r) > 0$ for all $\lambda \neq 0$ with

$$\Omega(0) = \frac{\lambda^2}{\lambda^2 + 2\ell_B^2}. \tag{31.6.22}$$

We can expand $\Omega(r)$ in a power series of ℓ_B^2 as follows. Let us set

$$\Omega(r) = 1 - \frac{2\ell_B^2}{\lambda^2 + 2\ell_B^2} f(r^2). \tag{31.6.23}$$

We can show that $f(x)$ is the Kummer function of the form

$$f(x) = M(2; \frac{\lambda^2}{2\ell_B^2} + 2, -\frac{x}{\ell_B^2}) \tag{31.6.24}$$

with $x = r^2$, and satisfies the Kummer equation,

$$2\ell_B^2 x \frac{d^2 f}{dx^2} + (\lambda^2 + x + 4\ell_B^2)\frac{df}{dx} + 2f = 0. \tag{31.6.25}$$

This can be solved by expanding f in a power series of ℓ_B^2, $f = \sum_{n=0} f_n \ell_B^{2n}$. In particular, the lowest-order term is given by setting $\ell_B^2 = 0$, whose result is

$$f_0(x) = \frac{\lambda^4}{(x + \lambda^2)^2}. \tag{31.6.26}$$

Hence we have

$$\Omega(r) = 1 - \frac{2\lambda^2 \ell_B^2}{(r^2 + \lambda^2)^2} + O(\ell_B^4), \tag{31.6.27}$$

or

$$\frac{\Delta\rho^{\pm}(\boldsymbol{x})}{\rho_0} = \pm \frac{2\lambda^2 \ell_B^2}{(r^2 + \lambda^2)^2} + O(\ell_B^4), \tag{31.6.28a}$$

$$\frac{\mathcal{S}_z^{\pm}(\boldsymbol{x})}{\rho_0} = \frac{1}{2} - \frac{\lambda^2}{r^2 + \lambda^2} + O(\ell_B^2). \tag{31.6.28b}$$

The density $\rho(\boldsymbol{x})$ approaches the ground-state value ρ_0 only polynomially unless $\lambda = 0$.

Finally we examine what happens when the Zeeman interaction is taken into account. The number of spins flipped around a skyrmion is given by

$$N_{\text{spin}} = \int d^2x \left\{ \mathcal{S}_z^{\text{cl}}(\boldsymbol{x}) - \frac{1}{2}\rho_0 \right\}. \tag{31.6.29}$$

Substituting (31.6.18) into this, unless $\lambda = 0$, we find $N_{\rm spin}^{\rm sky}$ to diverge logarithmically due to the asymptotic behavior,

$$\lim_{r\to\infty} \mathcal{S}_z^{\rm cl}(\boldsymbol{x}) = \frac{\rho_0}{2}\left(1 - 2\frac{\lambda^2}{r^2}\right). \tag{31.6.30}$$

The Zeeman energy $H_{\rm Z}^{\rm sky} = -\Delta_{\rm Z} N_{\rm spin}$ is divergent, except for the hole, from the infrared contribution however small the Zeeman effect is. The factorizable skyrmion (31.6.20) is no longer valid. There exists a skyrmion state which has a finite Zeeman energy: See an example of (31.7.3) we use for the Coulomb interaction. Nevertheless, we can show that the hole state has the lowest energy. The reason reads as follows. The factorizable skyrmion is an eigenstate of the hardcore Hamiltonian, $H_{\rm hc}|\mathfrak{S}_{\rm sky}^{\pm}\rangle = E_{\rm hc}|\mathfrak{S}_{\rm sky}^{\pm}\rangle$ with $E_{\rm hc} = |\delta N^{\rm cl}|/4\pi\ell_B$, as in (31.6.14). Accordingly any spin texture $|\mathfrak{S}\rangle$ possessing the same electron number $\delta N^{\rm cl}$ has a higher energy, $\langle\mathfrak{S}|H_{\rm hc}|\mathfrak{S}\rangle \geq E_{\rm hc}$. Furthermore its Zeeman energy is larger than that of the hole, $\langle\mathfrak{S}|H_{\rm Z}|\mathfrak{S}\rangle \geq \frac{1}{2}\Delta_{\rm Z}$. Hence,

$$\langle\mathfrak{S}|H_{\rm hc} + H_{\rm Z}|\mathfrak{S}\rangle \geq E_{\rm hc} + \frac{1}{2}\Delta_{\rm Z}, \tag{31.6.31}$$

where the equality holds for the hole state. Consequently there are no skyrmions in the presence of the external magnetic field in the system with the hardcore interaction.

31.7 Coulomb Interaction

We next investigate the realistic system (15.2.3) governed by the Coulomb Hamiltonian $H_{\rm C}$ together with the potential

$$V(\boldsymbol{x}-\boldsymbol{y}) = \frac{e^2}{4\pi\varepsilon|\boldsymbol{x}-\boldsymbol{y}|}. \tag{31.7.1}$$

It is hard to construct the skyrmion state as an eigenstate of $H_{\rm C}$. We are satisfied to estimate the excitation energy of a skyrmion by minimizing the expectation value of the Hamiltonian.

Before so doing, it is instructive to estimate the Coulomb energy (31.7.9) by using the factorizable skyrmion (31.6.18) obtained in the hardcore model. The result is very different from the one in the hardcore model. The Coulomb energy is a monotonically decreasing function of the scale parameter λ, and

$$\lim_{\lambda\to\infty}\langle H_{\rm C}\rangle = 8\pi J_s = \epsilon_{\rm X}. \tag{31.7.2}$$

Consequently, an infinitely large skyrmion is necessarily excited. However, when the Zeeman interaction is introduced, the Zeeman energy of the skyrmion diverges due to the asymptotic behavior (31.6.30) as in the case of the hardcore interaction.

Namely the factorizable skyrmion cannot describe a quasiparticle in the realistic system.

It is necessary to consider a skyrmion not factorizable as in (31.6.18). Making a slight generalization of the parameter (31.6.12) we search for a skyrmion possessing a finite energy even in the presence of the Zeeman effect. We propose an ansatz,

$$u_-^2(n) = v_+^2(n) = \frac{\omega^2 t^{2n+2}}{n+1+\omega^2}. \tag{31.7.3}$$

The parameter t presents a smooth interpolation between the hole ($t = 0$) and the factorizable skyrmion ($t = 1$). This ansatz automatically satisfies the condition (31.2.19) for $n = -1$.

Substituting (31.7.3) into (31.2.22) we obtain

$$\begin{aligned}
\frac{\Delta\rho^\pm(\boldsymbol{x})}{\rho_0} &= \pm e^{-\frac{1}{2}|z|^2} M(\omega^2; \omega^2+1; t^{2|z|2}/2) \mp \frac{t^2\omega^2}{\omega^2+1} e^{-\frac{1}{2}|z|^2} M(\omega^2+1; \omega^2+2; t^{2|z|2}/2),\\
\frac{\mathcal{S}_z^\pm(\boldsymbol{x})}{\rho_0} &= \frac{1}{2} - \frac{1}{2} e^{-\frac{1}{2}|z|^2} M(\omega^2; \omega^2+1; t^{2|z|2}/2)\\
&\quad - \frac{1}{2}\frac{t^2\omega^2}{\omega^2+1} e^{-\frac{1}{2}|z|^2} M(\omega^2+1; \omega^2+2; t^{2|z|2}/2),\\
\frac{\mathcal{S}_x^\pm(\boldsymbol{x})}{\rho_0} &= \frac{t\omega x/\ell_B}{\sqrt{2}} e^{-\frac{1}{2}|z|^2} \sum_n \frac{1}{n!}\frac{\vartheta_n(w,t)}{n+1+\omega^2}\left(\frac{t|z|^2}{2}\right)^n,\\
\frac{\mathcal{S}_y^\pm(\boldsymbol{x})}{\rho_0} &= \frac{\pm t\omega y/\ell_B}{\sqrt{2}} e^{-\frac{1}{2}|z|^2} \sum_n \frac{1}{n!}\frac{\vartheta_n(w,t)}{n+1+\omega^2}\left(\frac{t|z|^2}{2}\right)^n,
\end{aligned} \tag{31.7.4}$$

where

$$\vartheta_n(w,t) = \sqrt{1 + \frac{1-t^{2n+2}}{n+1}\omega^2}. \tag{31.7.5}$$

Using (31.6.17), we can rewrite these as

$$\begin{aligned}
\frac{\Delta\rho^\pm(\boldsymbol{x})}{\rho_0} &= \pm e^{-\frac{1}{2}(1-t^2)|z|^2} M(1; \omega^2+1; -t^2|z|^2/2)\\
&\quad \mp \frac{t^2\omega^2}{\omega^2+1} e^{-\frac{1}{2}(1-t^2)|z|^2} M(1; \omega^2+2; -t^2|z|^2/2),\\
\frac{\mathcal{S}_z^\pm(\boldsymbol{x})}{\rho_0} &= \frac{1}{2} - \frac{1}{2} e^{-\frac{1}{2}(1-t^2)|z|^2} M(1; \omega^2+1; -t^2|z|^2/2)\\
&\quad - \frac{1}{2}\frac{t^2\omega^2}{\omega^2+1} e^{-\frac{1}{2}(1-t^2)|z|^2} M(1; \omega^2+2; -t^2|z|^2/2).
\end{aligned} \tag{31.7.6}$$

We have illustrated the density $\rho^-(r)$ for typical values of the parameters ω and t in Fig.31.1.

It is instructive to expand them in power series of ℓ_B^2. Setting

$$(1-t^2)|z|^2 = (1-t^2)r^2/\ell_B^2 = r^2/\left(\beta^2 + \ell_B^2\right), \tag{31.7.7}$$

and writing down similar equations to (31.6.25), we obtain the lowest order term as

$$\frac{\Delta\rho^{\pm}(\boldsymbol{x})}{\rho_0} = \pm\left[\frac{2\lambda^2\ell_B^2}{(r^2+\lambda^2)^2} + \frac{2\lambda^2\ell_B^2}{(r^2+\lambda^2)\,\beta^2} + O(\ell_B^4)\right] e^{-r^2/\beta^2}, \tag{31.7.8a}$$

$$\frac{\mathcal{S}_z^{\pm}(\boldsymbol{x})}{\rho_0} = \frac{1}{2} - \left[\frac{\lambda^2}{r^2+\lambda^2} + O(\ell_B^2)\right] e^{-r^2/\beta^2}. \tag{31.7.8b}$$

The Zeeman energy remains finite due to a rapid decrease to the ground-state value.

Excitation Energy

In calculating the expectation value of the Coulomb Hamiltonian we appeal to the decomposition formula,

$$\langle H_{\mathrm{C}}\rangle = H_{\mathrm{D}}^{\mathrm{cl}} + H_{\mathrm{X}}^{\mathrm{cl}}, \tag{31.7.9}$$

dictating that the Coulomb energy consists of the direct energy $H_{\mathrm{D}}^{\mathrm{cl}}$ and the exchange energy $H_{\mathrm{X}}^{\mathrm{cl}}$. We have already used it on a case-by-case basis in our previous sections. We shall derive the formula in Section 32.2. Here the direct and exchange energies read

$$H_{\mathrm{D}}^{\mathrm{cl}} = \pi\int d^2k\, V(\boldsymbol{k}) e^{-\frac{1}{2}k^2}\Delta\hat{\rho}^{\mathrm{cl}}(-\boldsymbol{k})\Delta\hat{\rho}^{\mathrm{cl}}(\boldsymbol{k}) \tag{31.7.10a}$$

$$H_{\mathrm{X}}^{\mathrm{cl}} = -\pi\int d^2k\, \Delta V_{\mathrm{X}}(\boldsymbol{k})\left[\hat{\mathcal{S}}_a^{\mathrm{cl}}(-\boldsymbol{k})\hat{\mathcal{S}}_a^{\mathrm{cl}}(\boldsymbol{k}) + \frac{1}{4}\hat{\rho}^{\mathrm{cl}}(-\boldsymbol{k})\hat{\rho}^{\mathrm{cl}}(\boldsymbol{k})\right], \tag{31.7.10b}$$

where $\Delta V_{\mathrm{X}}(\boldsymbol{k}) = V_{\mathrm{X}}(\boldsymbol{k}) - V_{\mathrm{X}}(0)$ with

$$V_{\mathrm{X}}(\boldsymbol{k}) \equiv \frac{1}{\pi}\int d^2k\, e^{-i\boldsymbol{k}\wedge\boldsymbol{k}'} e^{-\frac{1}{2}k'^2} V(\boldsymbol{k}') = 4\varepsilon_{\mathrm{X}} e^{-\frac{1}{4}k^2} I_0\left(\frac{k^2}{4}\right), \tag{31.7.11}$$

and $I_0(z)$ is the modified Bessel function.

In order to calculate the Coulomb energy based on the decomposition formula (31.7.10) we express the bare densities as

$$\Delta\hat{\rho}_{\pm}(\boldsymbol{k}) = \pm\frac{1}{2\pi}\varpi(\boldsymbol{k}), \tag{31.7.12a}$$

$$\hat{\mathcal{S}}_{x,y}^{\pm}(\boldsymbol{k}) = \frac{1}{4\pi i}\left\{\frac{k_x}{k}, \pm\frac{k_y}{k}\right\}\xi(\boldsymbol{k}), \qquad \hat{\mathcal{S}}_z^{\pm}(\boldsymbol{k}) = \frac{1}{2}\delta(\boldsymbol{k}) - \frac{1}{4\pi}\sigma(\boldsymbol{k}), \tag{31.7.12b}$$

where we have introduced the notations

$$\varpi(k) = e^{-\frac{1}{4}\ell_B^2 k^2} \sum_{n=0}^{\infty} \left[\frac{\omega^2 t^{2n}}{n+\omega^2} - \frac{\omega^2 t^{2n+2}}{n+1+\omega^2} \right] L_n\left(\frac{\ell_B^2 k^2}{2}\right), \tag{31.7.13a}$$

$$\xi(k) = \sqrt{2}\omega t k e^{-\frac{1}{4}\ell_B^2 k^2} \sum_{n=0}^{\infty} \frac{\vartheta_n(\omega,t) t^n}{n+1+\omega^2} L_n^{(1)}\left(\frac{\ell_B^2 k^2}{2}\right), \tag{31.7.13b}$$

$$\sigma(k) = e^{-\frac{1}{4}\ell_B^2 k^2} \sum_{n=0}^{\infty} \left[\frac{\omega^2 t^{2n}}{n+\omega^2} + \frac{\omega^2 t^{2n+2}}{n+1+\omega^2} \right] L_n\left(\frac{\ell_B^2 k^2}{2}\right), \tag{31.7.13c}$$

with $\vartheta_n(\omega, t)$ given by (31.7.5). Substituting these into (31.7.10) and using

$$\int_0^\infty dx\, x^\alpha e^{-x} L_m^\alpha(x) L_n^\alpha(x) = \frac{\Gamma(n+\alpha+1)}{n!}\delta_{mn},$$

$$\int_0^\infty k dk \left[\xi^2(k) + \sigma^2(k) + \varpi^2(k)\right] = 2 + 4a^2 \sum_{n=1}^{\infty} \frac{t^{2n}}{n+a^2}, \tag{31.7.14}$$

we get

$$\begin{aligned}\langle H_{\rm C}\rangle_{\rm sky} =& \frac{1}{2}\varepsilon_{\rm C} \int_0^\infty dk\, e^{-\frac{1}{2}\ell_B^2 k^2} \varpi^2(k) \\ &+ \frac{1}{2}\varepsilon_{\rm X} \int_0^\infty k dk \left[1 - e^{-\frac{1}{4}\ell_B^2 k^2} I_0\left(\frac{\ell_B^2 k^2}{4}\right)\right] \left[\xi^2(k) + \sigma^2(k) + \varpi^2(k)\right].\end{aligned} \tag{31.7.15}$$

The Zeeman energy is expressed as

$$\langle H_{\rm Z}\rangle_{\rm sky} = -\Delta_{\rm Z} N_{\rm spin} = -\Delta_{\rm Z}\sigma(k=0). \tag{31.7.16}$$

It is easy to check the convergence of these integrals.

We are able to determine these parameters so as to minimize the sum of the Coulomb and Zeeman energies, $\langle H\rangle_{\rm sky} = \langle H_{\rm C}\rangle_{\rm sky} + \langle H_{\rm Z}\rangle_{\rm sky}$, as a function of the Zeeman gap $\Delta_{\rm Z}$. When the Zeeman gap is infinitesimal, an infinitely large skyrmion is excited to decreases the direct Coulomb energy and we have

$$\langle H_{\rm D}\rangle_{\rm sky} = 0, \qquad \langle H_{\rm X}\rangle_{\rm sky} = \frac{1}{2}\epsilon_{\rm X}. \tag{31.7.17}$$

On the other hand, when the Zeeman gap is very large, an infinitesimal skyrmion (hole) is excited and we have

$$\langle H_{\rm D}\rangle_{\rm sky} = \frac{1}{\sqrt{2}}\epsilon_{\rm X}, \qquad \langle H_{\rm X}\rangle_{\rm sky} = \left(\frac{\sqrt{2}-1}{\sqrt{2}}\right)\epsilon_{\rm X}. \tag{31.7.18}$$

In general, it is necessary to calculate $\langle H\rangle_{\rm sky}$ numerically as a function of $\Delta_{\rm Z}$. For this purpose we calculate $\langle H\rangle_{\rm sky}$ numerically as a function of ω and t for a given value of $\Delta_{\rm Z}$, and we determine the values of ω and t which minimizes $\langle H\rangle_{\rm sky}$ at each $\Delta_{\rm Z}$. In this way we obtain the skyrmion excitation energy $\langle H\rangle_{\rm sky}$ as a function

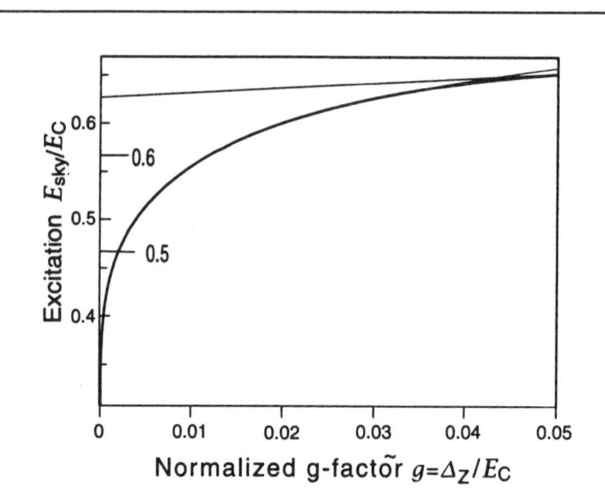

Fig. 31.2 The skrmion excitation energy is plotted as a function of the normalized Zeeman gap $\tilde{g} = \Delta_Z/\epsilon_C$. The heavy solid curve is obtained by the numerical analysis based on (31.7.15) and (31.7.16). The thin line is for the hole excitation energy.

of Δ_Z. We have plotted the excitation energy $\langle H\rangle_{\text{sky}}$ as a function of the normalized Zeeman gap $\tilde{g} = \Delta_Z/E_C^0$ in Fig.31.2. We have already pointed out that this theoretical result explains the experimental data quite well in Fig.16.8 of Section 16.3.

Tilted Field Method

We have calculated the excitation energy $\langle H\rangle_{\text{sky}}$ as a function of the Zeeman gap $\Delta_Z = |g|\mu_B B_\perp$ in the $\nu = 1$ system. We review how to get experimental data to compare this result. The problem is that the magnetic flux $B_\perp$ is fixed as a function of the density ρ_0 through the relation $\nu = 2\pi\hbar\rho_0/eB_\perp$. However, it is not practical to change the density ρ_0 in order to change the Zeeman gap Δ_Z. There exists a remarkable way known as the tilted field method.

By introducing the parallel magnetic field $B_\parallel$ the Hamiltonian becomes

$$H_\vartheta = H_C(B_\perp) - |g^*|\mu_B \int d^2x\, [B_\perp \mathcal{S}_z + B_\parallel \mathcal{S}_x]. \tag{31.7.19}$$

We define new electron annihilation operators by

$$c_\uparrow(n) = \tilde{c}_\uparrow(n)\cos\frac{\vartheta}{2} - \tilde{c}_\downarrow(n)\sin\frac{\vartheta}{2}, \tag{31.7.20a}$$

$$c_\downarrow(n) = \tilde{c}_\uparrow(n)\sin\frac{\vartheta}{2} + \tilde{c}_\downarrow(n)\cos\frac{\vartheta}{2}, \tag{31.7.20b}$$

where $B_\parallel = B_\perp \tan\vartheta$. We rewrite H_ϑ as

$$H_\vartheta = H_C(B_\perp) - \frac{\Delta_Z}{\cos\vartheta}\sum_n \tilde{\mathcal{S}}_z(n,n), \tag{31.7.21}$$

where

$$\tilde{\mathcal{S}}_z(m,n) = \frac{1}{2}\left[\tilde{c}_\uparrow^\dagger(m)\tilde{c}_\uparrow(n) - \tilde{c}_\downarrow^\dagger(m)\tilde{c}_\downarrow(n)\right]. \tag{31.7.22}$$

Therefore, all the scheme developed in previous sections remains valid by constructing the ground state and the skyrmion state with $\tilde{c}_\mu^\dagger(k)$ instead of $c_\mu^\dagger(k)$. The result is to replace the Zeeman gap Δ_Z with $\Delta_Z/\cos\vartheta$ in the excitation energy formula. Thus the Zeeman gap is effectively controlled by changing the parallel magnetic field $B_\parallel$. In actual experiments this is achieved by tilting the sample with the angle ϑ.

31.8 SU(4) Skyrmions

The generalization to the SU(4) bilayer system is straightforward. The ground state is given by the up-spin bonding state. The skyrmion state is given by

$$|\mathfrak{S}_{\text{sky}}\rangle = \prod_{n=0} \xi^\dagger(n)|0\rangle, \tag{31.8.1}$$

where

$$\xi^\dagger(n) = u_{\text{r}}(n)A_\downarrow^\dagger(n) + u_{\text{p}}(n)A_\uparrow^\dagger(n) + u_{\text{s}}(n)B_\downarrow^\dagger(n) + v(n)B_\uparrow^\dagger(n+1) \tag{31.8.2}$$

with constraint, $u_{\text{r}}^2(n) + u_{\text{p}}^2(n) + u_{\text{s}}^2(n) + v^2(n) = 1$. The hole state $B_\uparrow(0)|\text{B}\uparrow\rangle$ is given by setting $v(n) = 1$ and $u_{\text{s}}(n) = u_{\text{p}}(n) = u_{\text{r}}(n) = 0$ for all n.

There are three types of antiskyrmions,

$$\begin{aligned}
|\mathfrak{S}_{\text{asky}}^{\text{s}}\rangle &= \prod_{n=0} \zeta_s^\dagger(n)B_\downarrow^\dagger(0)|0\rangle, \\
|\mathfrak{S}_{\text{asky}}^{\text{p}}\rangle &= \prod_{n=0} \zeta_p^\dagger(n)A_\uparrow^\dagger(0)|0\rangle, \\
|\mathfrak{S}_{\text{asky}}^{\text{r}}\rangle &= \prod_{n=0} \zeta_r^\dagger(n)A_\downarrow^\dagger(0)|0\rangle,
\end{aligned} \tag{31.8.3}$$

with

$$\begin{aligned}
\zeta_s^\dagger(n) =& u_{\text{r}}(n)A_\downarrow^\dagger(n) + u_{\text{p}}(n)A_\uparrow^\dagger(n) - u_{\text{s}}(n)B_\downarrow^\dagger(n+1) + v(n)B_\uparrow^\dagger(n+1), \\
\zeta_p^\dagger(n) =& u_{\text{r}}(n)A_\downarrow^\dagger(n) - u_{\text{p}}(n+1)A_\uparrow^\dagger(n) + u_{\text{s}}(n)B_\downarrow^\dagger(n) + v(n)B_\uparrow^\dagger(n+1), \\
\zeta_r^\dagger(n) =& -u_{\text{r}}(n)A_\downarrow^\dagger(n+1) + u_{\text{p}}(n)A_\uparrow^\dagger(n) + u_{\text{s}}(n)B_\downarrow^\dagger(n) + v(n)B_\uparrow^\dagger(n+1).
\end{aligned} \tag{31.8.4}$$

They are reduced to three different electron excited states $B_\downarrow^\dagger(0)|\text{B}\uparrow\rangle$, $A_\uparrow^\dagger(0)|\text{B}\uparrow\rangle$ and $A_\downarrow^\dagger(0)|\text{B}\uparrow\rangle$ when we set $v(n) = 1$ and $u_{\text{s}}(n) = u_{\text{p}}(n) = u_{\text{r}}(n) = 0$ for all n: See Fig.30.1.

There are important SU(2) limits of SU(4)-skyrmions. When we set $u_{\text{p}}(n) = u_{\text{r}}(n) = 0$, $|\mathfrak{S}_{\text{sky}}^{\text{s}}\rangle$ and $|\mathfrak{S}_{\text{asky}}^{\text{s}}\rangle$ describe a skyrmion and an antiskyrmion where only spins are excited. We call them the spin-skyrmion and the spin-antiskyrmion.

Similarly, when we set $u_s(n) = u_r(n) = 0$, $|\mathfrak{S}^{\mathrm{P}}_{\mathrm{sky}}\rangle$ and $|\mathfrak{S}^{\mathrm{P}}_{\mathrm{asky}}\rangle$ describe a skyrmion and an antiskyrmion where only pseudospins are excited. We call them the ppin-skyrmion and the ppin-antiskyrmion.

It is a dynamical problem which skyrmion-antiskymion pairs are excited thermally. As we have shown, as far as electron-hole pairs are concerned, only pseudospins are excited at the balanced point unless the tunneling gap is too large, while only spins are excited at the monolayer point. This is the case also for skyrmion-antiskyrmion pair excitations. However, in general, all components are excited to lower the total energy, which leads to genuine SU(4)-skyrmions. Contrary to the case of the electron-hole limit, the transition from the ppin-skyrmion at the balanced point ($\sigma_0 = 0$) to the spin-skyrmion at the monolayer point ($\sigma_0 = 1$) will occur continuously via a genuine SU(4)-skyrmion since the matrix elements of the total Hamiltonian H between various skyrmion states are nonvanishing.

CHAPTER 32

EXCHANGE INTERACTIONS AND EFFECTIVE THEORY

The Coulomb Hamiltonian H_C depends only on the electron density. However, it is rewritten into an entirely different expression H_X which involves explicitly spin variables. The energy of a state $\langle H_C \rangle$ consists of two parts, one representing the direct energy H_D^{cl} and the other the exchange energy H_X^{cl}. They are obtained from H_C and H_X by replacing the density operators by the corresponding classical densities. We can verify the spontaneous symmetry breaking in the SU(N) QH system based on this decomposition formula. The dynamical behavior of the classical density is governed by the Hamiltonian $H_C^{cl} = H_D^{cl} + H_X^{cl}$. We then make the derivative expansion and derive the effective Hamiltonian to describe the quantum coherence such as spin waves and skyrmions. The exchange interaction is well described by the nonlinear sigma model.

32.1 EXCHANGE HAMILTONIAN

We analyze the Coulomb interaction described by

$$H_C = \pi \int d^2p\, V_D(\boldsymbol{p})\hat{\rho}(-\boldsymbol{p})\hat{\rho}(\boldsymbol{p}) = \frac{1}{2}\int d^2x d^2y\, \hat{\rho}(\boldsymbol{x})V_D(\boldsymbol{x}-\boldsymbol{y})\hat{\rho}(\boldsymbol{y}). \tag{32.1.1}$$

In classical theory the Coulomb energy $E_C = \langle H_C \rangle$ is simply given by replacing $\rho(\boldsymbol{x})$ with the classical density $\rho^{cl}(\boldsymbol{x}) = \langle \rho(\boldsymbol{x}) \rangle$, but this is not the case in quantum theory. The exchange Coulomb interaction emerges as an important contribution from the exchange integral over wave functions. We present a systematic way of analyzing the exchange interaction in QH systems. See also Appendix I for the physical meaning of the exchange energy.

Monolayer SU(N) QH Systems

We start with the monolayer QH system with SU(N) symmetry. The Hamiltonian (32.1.1) yields the Landau-site Hamiltonian given by

$$H_C = \sum_{mnij} V_{mnij} \sum_{\sigma,\tau} c_\sigma^\dagger(m)c_\sigma(n)c_\tau^\dagger(i)c_\tau(j) = \sum_{mnij} V_{mnij}\rho(n,m)\rho(j,i), \tag{32.1.2}$$

where

$$\rho(n,m) = \sum_\sigma c_\sigma^\dagger(m)c_\sigma(n). \tag{32.1.3}$$

We use an algebraic relation,

$$\delta_{\alpha\beta}\delta_{\sigma\tau} = \frac{1}{2}\sum_A^{N^2-1}\lambda_{\alpha\tau}^A\lambda_{\sigma\beta}^A + \frac{1}{N}\delta_{\alpha\tau}\delta_{\sigma\beta}, \tag{32.1.4}$$

to which we multiply $c_\sigma^\dagger(m)c_\tau(n)c_\alpha^\dagger(i)c_\beta(j)$ and obtain

$$\begin{aligned}\rho(n,m)\rho(j,i) = &-2[I_A(j,m)I_A(n,i) + \frac{1}{2N}\rho(j,m)\rho(n,i)] \\ &+ \frac{2N^2-1}{N}\rho(n,m)\delta_{ij} + \rho(j,m)\delta_{in},\end{aligned} \tag{32.1.5}$$

where

$$I_A(n,m) = \frac{1}{2}\sum_{\sigma\tau} c_\sigma^\dagger(m)(\lambda_A)_{\sigma\tau}c_\tau(n). \tag{32.1.6}$$

Recall that

$$\sum_{mnij} V_{mnij}\rho(n,m)\delta_{ij} = \sum_{nj} V_{nnjj}\rho(n,n) = \epsilon_D N_e, \tag{32.1.7a}$$

$$\sum_{mnij} V_{mnij}\rho(j,m)\delta_{in} = \sum_{nj} V_{jnnj}\rho(j,j) = \epsilon_X N_e, \tag{32.1.7b}$$

due to (30.1.14) and (30.1.16), where $N_e = \sum_n \rho(n,n)$ is the total electron number. We are analyzing the system where the total electron number is fixed. We introduce a new symbol H_X for the Hamiltonian,

$$H_X = -2\sum_{mnij} V_{mnij}[I_A(j,m)I_A(n,i) + \frac{1}{2N}\rho(j,m)\rho(n,i)]. \tag{32.1.8}$$

It is equivalent to the original Hamiltonian (32.1.2) except for the constant term $(\epsilon_D + \epsilon_X)N_e$.

To express H_X in the momentum space, we use the inversion formula (29.2.18),

$$\rho(n,m) = \ell_B^2\int d^2q\,\langle n|e^{iqX}|m\rangle\hat{\rho}(\boldsymbol{q}), \quad I_A(n,m) = \ell_B^2\int d^2q\,\langle n|e^{iqX}|m\rangle\hat{\jmath}_A(\boldsymbol{q}). \tag{32.1.9}$$

We also need a relation,

$$\sum_n \langle n|e^{-iXk}e^{iXp}e^{iXk}e^{iXq}|n\rangle = \frac{2\pi}{\ell_B^2}\delta(\boldsymbol{p}+\boldsymbol{q})\exp\left(i\ell_B^2\boldsymbol{p}\wedge\boldsymbol{k}\right), \tag{32.1.10}$$

which follows from (29.2.4) and (29.2.31). We substitute (32.1.9) and (30.1.13) into (32.1.8), and use (32.1.10) to obtain

$$H_X = -\pi \int d^2p\, V_X(\boldsymbol{p}) \left[\hat{\boldsymbol{I}}(-\boldsymbol{p})\hat{\boldsymbol{I}}(\boldsymbol{p}) + \frac{1}{2N}\hat{\rho}(-\boldsymbol{p})\hat{\rho}(\boldsymbol{p})\right]$$
$$= -\frac{1}{2}\int d^2x d^2y\, V_X(\boldsymbol{x}-\boldsymbol{y}) \left[\hat{\boldsymbol{I}}(\boldsymbol{x})\hat{\boldsymbol{I}}(\boldsymbol{y}) + \frac{1}{2N}\hat{\rho}(\boldsymbol{x})\hat{\rho}(\boldsymbol{y})\right], \tag{32.1.11}$$

where

$$V_X(\boldsymbol{p}) = \frac{\ell_B^2}{\pi}\int d^2k\, e^{-i\ell_B^2 \boldsymbol{p}\wedge\boldsymbol{k}} V_D(\boldsymbol{k}), \qquad V_X(\boldsymbol{x}) = \frac{1}{2\pi}\int d^2p\, e^{i\boldsymbol{p}\boldsymbol{x}} V_X(\boldsymbol{p}). \tag{32.1.12}$$

In particular, in the lowest Landau level they read

$$V_X(\boldsymbol{p}) = \frac{\sqrt{2\pi}e^2\ell_B}{4\pi\varepsilon} I_0(\ell_B^2\boldsymbol{p}^2/4)e^{-\ell_B^2\boldsymbol{p}^2/4}, \qquad V_X(\boldsymbol{x}) = 2V(\boldsymbol{x})e^{-\boldsymbol{x}^2/2\ell_B^2}. \tag{32.1.13}$$

We call H_X the Hamiltonian in the exchange-interaction form. The two Hamiltonians H_C and H_X are equivalent, $H_C = H_X$, though they look very different.

Bilayer QH System without Spin

In the bilayer spinless QH system the Coulomb Hamiltonian is given by

$$H_C = \pi\int d^2p\, V_D^+(\boldsymbol{p})\hat{\rho}(-\boldsymbol{p})\hat{\rho}(\boldsymbol{p}) + 4\pi\int d^2p\, V_D^-(\boldsymbol{p})\hat{\mathcal{P}}_z(-\boldsymbol{p})\hat{\mathcal{P}}_z(\boldsymbol{p}), \tag{32.1.14}$$

or

$$H_C = \sum_{mnij}\left[V_{mnij}^+\rho(n,m)\rho(j,i) + 4V_{mnij}^- P_z(n,m)P_z(j,i)\right]. \tag{32.1.15}$$

We transform this into the exchange-interaction form. The SU(2)-invariant part is transformed into

$$H_X^+ = -2\sum_{mnij} V_{mnij}^+\left[\sum_{a=xyz} P_a(j,m)P_a(n,i) + \frac{1}{4}\rho(j,m)\rho(n,i)\right], \tag{32.1.16}$$

as in (32.1.8) with $N = 2$. The SU(2)-noninvariant part is transformed into

$$H_X^- = -2\sum_{mnij} V_{mnij}^-\left[\sum_{a=xyz} \xi_a P_a(j,m)P_a(n,i) + \frac{1}{4}\rho(j,m)\rho(n,i)\right], \tag{32.1.17}$$

where $\xi_x = \xi_y = -1$ and $\xi_z = 1$. Combining these we obtain

$$H_X = -2\sum_{mnij}\Big[V_{mnij}^d \sum_{a=xy} P_a(j,m)P_a(n,i) + V_{mnij}P_z(j,m)P_z(n,i)$$
$$+ \frac{1}{4}V_{mnij}\rho(j,m)\rho(n,i)\Big], \tag{32.1.18}$$

where

$$V_{mnij} = V_{mnij}^+ + V_{mnij}^-, \qquad V_{mnij}^d = V_{mnij}^+ - V_{mnij}^-. \tag{32.1.19}$$

The Hamiltonian is expressed in the momentum space as

$$H_X = -\pi \int d^2p \Big[V_X^d(\boldsymbol{p}) \sum_{a=xy} \hat{\mathcal{P}}_a(-\boldsymbol{p})\hat{\mathcal{P}}_a(\boldsymbol{p}) + V_X(\boldsymbol{p})\hat{\mathcal{P}}_z(-\boldsymbol{p})\hat{\mathcal{P}}_z(\boldsymbol{p}) + \frac{1}{4}V_X(\boldsymbol{p})\hat{\rho}(-\boldsymbol{p})\hat{\rho}(\boldsymbol{p}) \Big], \tag{32.1.20}$$

where

$$V_X^{\pm}(\boldsymbol{p}) = \frac{\ell_B^2}{\pi}\int d^2k\, e^{-i\ell_B^2 \boldsymbol{p}\wedge \boldsymbol{k}} V_D^{\pm}(\boldsymbol{k}), \qquad V_X^{\pm}(\boldsymbol{x}) = \frac{1}{2\pi}\int d^2p\, e^{i\boldsymbol{p}\boldsymbol{x}} V_X^{\pm}(\boldsymbol{p}),$$

and

$$V_X = V_X^+ + V_X^-, \qquad V_X^d = V_X^+ - V_X^-. \tag{32.1.21}$$

In particular, in the lowest Landau level they read

$$\begin{aligned} V_X^{\pm}(\boldsymbol{p}) &= \frac{\sqrt{2\pi}e^2\ell_B}{8\pi\varepsilon} I_0(\ell_B^2 \boldsymbol{p}^2/4) e^{-\ell_B^2\boldsymbol{p}^2/4} \pm \frac{e^2\ell_B^2}{4\pi\varepsilon}\int_0^\infty dk\, e^{-\frac{1}{2}\ell_B^2 k^2 - dk} J_0(\ell_B^2 |\boldsymbol{p}| k), \\ V_X^{\pm}(\boldsymbol{x}) &= V(\boldsymbol{x})\left(1 \pm e^{-|\boldsymbol{x}|d/\ell_B^2}\right) e^{-\boldsymbol{x}^2/2\ell_B^2}. \end{aligned} \tag{32.1.22}$$

The two Hamiltonians (32.1.14) and (32.1.20) are equivalent, $H_C = H_X$.

Bilayer QH System with Spin

The Coulomb Hamiltonian in the bilayer QH system is given by

$$H_C = \pi \int d^2q\, V_D^+(\boldsymbol{p})\hat{\rho}(-\boldsymbol{p})\hat{\rho}(\boldsymbol{p}) + 4\pi \int d^2q\, V_D^-(\boldsymbol{p})\hat{\mathcal{P}}_z(-\boldsymbol{p})\hat{\mathcal{P}}_z(\boldsymbol{p}), \tag{32.1.23}$$

or

$$H_C = \sum_{mnij} \left[V_{mnij}^+ \rho(n,m)\rho(j,i) + 4V_{mnij}^- P_z(n,m)P_z(j,i) \right]. \tag{32.1.24}$$

The SU(4)-invariant part is transformed trivially into

$$H_X^+ = -2\sum_{mnij} V_{mnij}^+ \left[\sum_{A=1}^{15} I_A(j,m)I_A(n,i) + \frac{1}{8}\rho(j,m)\rho(n,i) \right], \tag{32.1.25}$$

as in (32.1.8) with $N = 2$. To deal with the SU(4)-noninvariant term we use the formula

$$4P_z(n,m)P_z(j,i) = -2\sum_{AB} M_{AB} I_A(j,m) I_B(n,i) - \frac{1}{4}\rho(j,m)\rho(n,i), \tag{32.1.26}$$

where

$$M_{AB} = \frac{1}{2}\mathrm{Tr}(\lambda_A \tau_z^{\mathrm{ppin}} \lambda_B \tau_z^{\mathrm{ppin}}) = \xi_A \delta_{AB} \tag{32.1.27}$$

with

$$\xi_{1,2,3,8,13,14,15} = +1, \qquad \xi_{4,5,6,7,9,10,11,12} = -1. \tag{32.1.28}$$

Thus it is transformed into

$$H_X^- = -2\sum_{mnij} V_{mnij}^- \left[\sum_{A=1}^{15} \xi_A I_A(j,m) I_A(n,i) + \frac{1}{8}\rho(j,m)\rho(n,i) \right]. \tag{32.1.29}$$

From (32.1.25) and (32.1.29) we obtain the Landau-site Hamiltonian

$$H_X = -2\sum_{mnij} \left[\sum_{A=1}^{15} \left(V_{mnij}^+ + V_{mnij}^- \xi_A\right) I_A(j,m) I_A(n,i) + \frac{1}{8} V_{mnij}\rho(j,m)\rho(n,i) \right] \tag{32.1.30}$$

in the exchange-interaction form.

We change the SU(4) basis from λ_A to τ_a^{spin}, τ_a^{ppin} and $\tau_a^{\text{spin}}\tau_b^{\text{ppin}}$ by way of the formula (D.13) in Appendix, which transforms variable $I_A(n,m)$ to a set of variables,

$$S_a(n,m) = \frac{1}{2}\sum_{\sigma\tau} c_\sigma^\dagger(m)(\tau_a^{\text{spin}})_{\sigma\tau} c_\tau(n), \quad P_a(n,m) = \frac{1}{2}\sum_{\sigma\tau} c_\sigma^\dagger(m)(\tau_a^{\text{ppin}})_{\sigma\tau} c_\tau(n),$$

$$R_{ab}(n,m) = \frac{1}{2}\sum_{\sigma\tau} c_\sigma^\dagger(m)(\tau_a^{\text{spin}}\tau_b^{\text{ppin}})_{\sigma\tau} c_\tau(n). \tag{32.1.31}$$

It is convenient to rewrite (32.1.30) as

$$\begin{aligned} H_X = &-2\sum_{mnij}\sum_{A=1}^{15} \left[V_{mnij}^d I_A(j,m) I_A(n,i) + V_{mnij}^- \left(1+\xi_A\right) I_A(j,m) I_A(n,i) \right] \\ &- \frac{1}{4}\sum_{mnij} V_{mnij}\rho(j,m)\rho(n,i), \end{aligned} \tag{32.1.32}$$

where $V_{mnij}^\pm = (V_{mnij} \pm V_{mnij}^d)/2$, from which we find

$$\begin{aligned} H_X = &-\sum_{mnij} V_{mnij}^d \left[S_a(j,m)S_a(n,i) + P_a(j,m)P_a(n,i) + R_{ab}(j,m)R_{ab}(n,i) \right] \\ &-2\sum_{mnij} V_{mnij}^- \left[S_a(j,m)S_a(n,i) + P_z(j,m)P_z(n,i) + R_{az}(j,m)R_{az}(n,i) \right] \\ &-\frac{1}{4}\sum_{mnij} V_{mnij}\rho(j,m)\rho(n,i). \end{aligned} \tag{32.1.33}$$

In the momentum space the Hamiltonian reads

$$\begin{aligned} H_X = &-\frac{\pi}{2}\int d^2p\, V_X^d(\boldsymbol{p}) \left[\hat{\mathcal{S}}_a(-\boldsymbol{p})\hat{\mathcal{S}}_a(\boldsymbol{p}) + \hat{\mathcal{P}}_a(-\boldsymbol{p})\hat{\mathcal{P}}_a(\boldsymbol{p}) + \hat{\mathcal{R}}_{ab}(-\boldsymbol{p})\hat{\mathcal{R}}_{ab}(\boldsymbol{p}) \right] \\ &-\pi\int d^2p\, V_X^-(\boldsymbol{p}) \left[\hat{\mathcal{S}}_a(-\boldsymbol{p})\hat{\mathcal{S}}_a(\boldsymbol{p}) + \hat{\mathcal{P}}_z(-\boldsymbol{p})\hat{\mathcal{P}}_z(\boldsymbol{p}) + \hat{\mathcal{R}}_{az}(-\boldsymbol{p})\hat{\mathcal{R}}_{az}(\boldsymbol{p}) \right] \\ &-\frac{\pi}{8}\int d^2p\, V_X(\boldsymbol{p})\rho(-\boldsymbol{p})\rho(\boldsymbol{p}), \end{aligned} \tag{32.1.34}$$

where $V_X^d(\boldsymbol{p})$, $V_X(\boldsymbol{p})$ and $V_X^d(\boldsymbol{p})$ are given by the same formulas as in the bilayer spinless system. The two Hamiltonians (32.1.23) and (32.1.34) are equivalent, $H_C = H_X$.

32.2 Decomposition Formula

We have demonstrated that the Coulomb Hamiltonian possesses two entirely different forms, the direct-interaction form H_C and the exchange-interaction form H_X. For instance, in the monolayer SU(N) QH system H_C given by (32.1.1) and H_X given by (32.1.11). They are equivalent and hence we should adopt only one of them as the Hamiltonian of the system. However, this is not the case for the corresponding classical quantity, the Coulomb energy. For instance, let us consider two well-separated charged excitations. There operates a direct Coulomb interaction between them since the potential $V_D(\boldsymbol{x})$ is long-ranged, but not an exchange interaction since $V_X(\boldsymbol{x})$ is short-ranged. We expect that the total energy is simply the sum of them, or

$$\langle H_C \rangle = H_D^{\text{cl}} + H_X^{\text{cl}}, \tag{32.2.1}$$

where H_D^{cl} and H_X^{cl} are obtained by replacing $\hat{\rho}$ and $\hat{\mathfrak{I}}$ with $\hat{\rho}^{\text{cl}}$ and $\hat{\mathfrak{I}}^{\text{cl}}$ in H_C and H_X, respectively. It can be proved that the decomposition formula (32.2.1) holds up to a trivial numerical constant with the use of (32.1.1) and (32.1.11) for H_D^{cl} and H_X^{cl}, respectively.

To make precise the decomposition formula, we take the Hamiltonian by subtracting the background charge as in (30.1.11), which is

$$H_C = \pi \int d^2k\, V_D(\boldsymbol{k}) \Delta\hat{\rho}(-\boldsymbol{k}) \Delta\hat{\rho}(\boldsymbol{k}) \tag{32.2.2}$$

in the momentum space. Then the classical energy consists of the direct energy and the exchange energy as

$$H_D^{\text{cl}} = \pi \int d^2k\, V_D(\boldsymbol{k}) \Delta\hat{\rho}^{\text{cl}}(-\boldsymbol{k}) \Delta\hat{\rho}^{\text{cl}}(\boldsymbol{k}), \tag{32.2.3a}$$

$$H_X^{\text{cl}} = -\pi \int d^2k\, [V_X(\boldsymbol{k}) - V_X(0)] \left[\hat{\mathfrak{I}}^{\text{cl}}(-\boldsymbol{k}) \hat{\mathfrak{I}}^{\text{cl}}(\boldsymbol{k}) + \frac{1}{2N} \hat{\rho}^{\text{cl}}(-\boldsymbol{k}) \hat{\rho}^{\text{cl}}(\boldsymbol{k}) \right]. \tag{32.2.3b}$$

We now prove the decomposition formula (32.2.1) with the use of these two classical Hamiltonians for H_D^{cl} and H_X^{cl}. We evaluate the energy of a generic state $|\mathfrak{S}\rangle$ in the class of states (29.6.1),

$$\langle H_C \rangle = \langle \mathfrak{S} | H_C | \mathfrak{S} \rangle. \tag{32.2.4}$$

For the sake of accuracy we first consider a system with a finite number of Landau sites ($m = 0, 1, \ldots, N_\Phi - 1$) and take the limit $N_\Phi \to \infty$ in final expressions. It is convenient to combine the isospin index ($\mu = 1, 2, \ldots, N$) and the site index into a multi-index $M \equiv (\mu, m)$, where the multi-index runs over the values $M = 1, 2, \ldots, NN_\Phi$.

The $W_\infty(N)$ algebra is identical to the algebra $U(NN_\Phi)$ in the limit $N_\Phi \to \infty$, and the transformation rules for the fermion operators appear as

$$e^{-iW} c_M e^{iW} = (U)_{MM'} c_{M'},$$
$$e^{-iW} c_M^\dagger e^{iW} = c_{M'}^\dagger (U^\dagger)_{M'M}, \tag{32.2.5}$$

where U is an $(NN_\Phi) \times (NN_\Phi)$ unitary matrix, $UU^\dagger = U^\dagger U = \mathbb{I}_{(NN_\Phi)\times(NN_\Phi)}$. Here and hereafter the repeated index implies the summation over it.

We first calculate the two-point averages by the state $|\mathfrak{S}\rangle = e^{iW}|\mathfrak{S}_0\rangle$ with

$$|\mathfrak{S}_0\rangle = \prod_{J=1}^{NN_\Phi} \left[c_J^\dagger\right]^{\nu_J} |0\rangle. \tag{32.2.6}$$

We get

$$\langle\mathfrak{S}|c_M^\dagger c_N|\mathfrak{S}\rangle = (U^\dagger)_{KM}(U)_{NL}\langle\mathfrak{S}_0|c_K^\dagger c_L|\mathfrak{S}_0\rangle \tag{32.2.7}$$

and

$$\langle\mathfrak{S}_0|c_K^\dagger c_L|\mathfrak{S}_0\rangle = \nu_K\delta_{KL}, \tag{32.2.8}$$

which eventually leads to

$$\langle\mathfrak{S}|c_M^\dagger c_N|\mathfrak{S}\rangle = \nu_K(U)_{NK}(U^\dagger)_{KM}. \tag{32.2.9}$$

We consider the quantity

$$\langle\mathfrak{S}|c_M^\dagger c_K|\mathfrak{S}\rangle\langle\mathfrak{S}|c_K^\dagger c_N|\mathfrak{S}\rangle = \nu_K(U)_{NK}(U^\dagger)_{KM}, \tag{32.2.10}$$

where we have used (32.2.9) and $\nu_K^2 = \nu_K$ since $\nu_K = 0$ or 1. Comparing this with (32.2.9) we find

$$\langle\mathfrak{S}|c_M^\dagger c_K|\mathfrak{S}\rangle\langle\mathfrak{S}|c_K^\dagger c_N|\mathfrak{S}\rangle = \langle\mathfrak{S}|c_M^\dagger c_N|\mathfrak{S}\rangle, \tag{32.2.11}$$

which is

$$\langle\mathfrak{S}|c_\mu^\dagger(m)c_\kappa(k)|\mathfrak{S}\rangle\langle\mathfrak{S}|c_\kappa^\dagger(k)c_\nu(n)|\mathfrak{S}\rangle = \langle\mathfrak{S}|c_\mu^\dagger(m)c_\nu(n)|\mathfrak{S}\rangle, \tag{32.2.12}$$

or

$$\sum_k\sum_\kappa \hat{\mathfrak{D}}^{\text{cl}}_{\nu\kappa}(n,k)\hat{\mathfrak{D}}^{\text{cl}}_{\kappa\mu}(k,m) = \hat{\mathfrak{D}}^{\text{cl}}_{\nu\mu}(n,m), \tag{32.2.13}$$

where $\hat{\mathfrak{D}}^{\text{cl}}_{\nu\mu}(n,m) = \langle\mathfrak{S}|c_\mu^\dagger(m)c_\nu(n)|\mathfrak{S}\rangle$. Thus we have derived the formula (29.6.6) used in the previous section.

We carry out analogous manipulations in four-point averages . We get

$$\langle\mathfrak{S}|c_M^\dagger c_S^\dagger c_T c_N|\mathfrak{S}\rangle = (U^\dagger)_{KM}(U^\dagger)_{IS}(U)_{TJ}(U)_{NL}\langle\mathfrak{S}_0|c_K^\dagger c_I^\dagger c_J c_L|\mathfrak{S}_0\rangle. \tag{32.2.14}$$

We can use

$$\langle\mathfrak{S}_0|c_K^\dagger c_I^\dagger c_J c_L|\mathfrak{S}_0\rangle = \nu_J\nu_L(\delta_{IJ}\delta_{KL} - \delta_{IL}\delta_{KJ}) \tag{32.2.15}$$

for the state $|\mathfrak{S}_0\rangle$. Substituting (32.2.15) into (32.2.14) and accounting (32.2.9) we summarize as

$$\langle\mathfrak{S}|c_M^\dagger c_S^\dagger c_T c_N|\mathfrak{S}\rangle = \langle\mathfrak{S}|c_M^\dagger c_N|\mathfrak{S}\rangle\langle\mathfrak{S}|c_S^\dagger c_T|\mathfrak{S}\rangle - \langle\mathfrak{S}|c_M^\dagger c_T|\mathfrak{S}\rangle\langle\mathfrak{S}|c_S^\dagger c_N|\mathfrak{S}\rangle. \quad (32.2.16)$$

This yields

$$\begin{aligned}\langle\mathfrak{S}|c_\sigma^\dagger(m)c_\tau^\dagger(i)c_\tau(j)c_\sigma(n)|\mathfrak{S}\rangle =&\langle\mathfrak{S}|c_\sigma^\dagger(m)c_\sigma(n)|\mathfrak{S}\rangle\langle\mathfrak{S}|c_\tau^\dagger(i)c_\tau(j)|\mathfrak{S}\rangle\\ &- \langle\mathfrak{S}|c_\sigma^\dagger(m)c_\tau(j)|\mathfrak{S}\rangle\langle\mathfrak{S}|c_\tau^\dagger(i)c_\sigma(n)|\mathfrak{S}\rangle, \quad (32.2.17)\end{aligned}$$

where

$$\langle\mathfrak{S}|c_\mu^\dagger(m)c_\nu(n)|\mathfrak{S}\rangle = \frac{\delta_{\nu\mu}}{N}\rho^{\text{cl}}(n,m) + (\lambda_A)_{\nu\mu}I_A^{\text{cl}}(n,m), \quad (32.2.18)$$

as follows from (29.5.6).

The Coulomb Hamiltonian (32.2.2) is given in the Landau site representation as in (30.1.11), or

$$H_{\text{C}} = \sum_{mnij} V_{mnij}\sum_{\sigma,\tau} c_\sigma^\dagger(m)c_\tau^\dagger(i)c_\tau(j)c_\sigma(n) + N_{\text{e}}\epsilon_{\text{X}} - \nu\,(N_{\text{e}} + \Delta N_{\text{e}})\,\epsilon_{\text{D}}, \quad (32.2.19)$$

to which we substitute (32.2.17) with four of (32.2.18). In this way we come to the decomposition formula,

$$H_{\text{C}}^{\text{cl}} \equiv \langle H_{\text{C}}\rangle = H_{\text{D}}^{\text{cl}} + H_{\text{X}}^{\text{cl}}, \quad (32.2.20)$$

with

$$H_{\text{D}}^{\text{cl}} = V_{mnij}\rho^{\text{cl}}(n,m)\rho^{\text{cl}}(j,i) - \nu(N_{\text{e}}^{\text{cl}} + \Delta N_{\text{e}}^{\text{cl}})\epsilon_{\text{D}}, \quad (32.2.21\text{a})$$

$$H_{\text{X}}^{\text{cl}} = -\,2V_{mnij}I_A^{\text{cl}}(j,m)I_A^{\text{cl}}(n,i) - \frac{1}{N}V_{mnij}\rho^{\text{cl}}(j,m)\rho^{\text{cl}}(n,i) + N_{\text{e}}^{\text{cl}}\epsilon_{\text{X}}, \quad (32.2.21\text{b})$$

by separating the direct and exchange terms.

The direct energy H_{D}^{cl} is rewritten as

$$H_{\text{D}}^{\text{cl}} = \pi\int d^2k\, V_{\text{D}}(\boldsymbol{k})\hat\rho^{\text{cl}}(-\boldsymbol{k})\hat\rho^{\text{cl}}(\boldsymbol{k}) - \nu(N_{\text{e}}^{\text{cl}} + \Delta N_{\text{e}}^{\text{cl}})\epsilon_{\text{D}}. \quad (32.2.22)$$

Absorbing the constant term we obtain (32.2.3a).

The exchange energy H_{X}^{cl} is rewritten as

$$H_{\text{X}}^{\text{cl}} = -\pi\int d^2k\, V_{\text{X}}(\boldsymbol{k})\left[\hat{\mathfrak{I}}^{\text{cl}}(-\boldsymbol{k})\hat{\mathfrak{I}}^{\text{cl}}(\boldsymbol{k}) + \frac{1}{2N}\hat\rho^{\text{cl}}(-\boldsymbol{k})\hat\rho^{\text{cl}}(\boldsymbol{k})\right] + N_{\text{e}}^{\text{cl}}\epsilon_{\text{X}}. \quad (32.2.23)$$

It is less trivial to absorb the constant term. We note that

$$(\nu N_\Phi + \Delta N_{\text{e}}^{\text{cl}})\epsilon_{\text{X}} = \frac{1}{4\ell_B^2}V_{\text{X}}(0)\int d^2x\,\hat\rho^{\text{cl}}(\boldsymbol{x}), \quad (32.2.24)$$

by using $V_X(0) = 4\ell_B^2\epsilon_X$. As we have pointed out, from (32.2.13) we obtain (29.6.10), or

$$\frac{1}{4\pi\ell_B^2}\int d^2x\,\hat{\rho}^{cl}(x) = \int d^2x\left[\mathfrak{J}_A^{cl}(x)\mathfrak{J}_A^{cl}(x) + \frac{1}{2N}\hat{\rho}^{cl}(x)\hat{\rho}^{cl}(x)\right]. \tag{32.2.25}$$

Hence,

$$N_e^{cl}\epsilon_X = \pi V_X(0)\int d^2k\left[\mathfrak{J}^{cl}(-k)\mathfrak{J}^{cl}(k) + \frac{1}{2N}\hat{\rho}^{cl}(-k)\hat{\rho}^{cl}(k)\right]. \tag{32.2.26}$$

We obtain (32.2.3b) from (32.2.23) and (32.2.26).

32.3 Spontaneous Symmetry Breaking

We analyze the problem of spontaneous symmetry breaking due to a repulsive interaction between electrons in QH system. The Hamiltonian is given by (32.2.2), or

$$H_C = \pi\int d^2k\,V_D(k)\Delta\hat{\rho}(-k)\Delta\hat{\rho}(k). \tag{32.3.1}$$

We assume that every Landau site is occupied by an equal number of electrons. The problem is to show that the spin polarized state is the lowest-energy state though this Hamiltonian involves no spin variables.

The state $|\mathfrak{S}\rangle$ in problem belongs to the class of states (29.6.1), for which the decomposition formula (32.2.1) is valid. According to the decomposition formula the energy of the Coulomb Hamiltonian, $H_C^{cl} \equiv \langle\mathfrak{S}|H_C|\mathfrak{S}\rangle$, is a sum of the direct and exchange energies, or

$$H_C^{cl} = H_D^{cl} + H_X^{cl} \tag{32.3.2}$$

with

$$H_D^{cl} = \pi\int d^2k\,V_D(k)\left|\Delta\hat{\rho}^{cl}(k)\right|^2, \tag{32.3.3a}$$

$$H_X^{cl} = \pi\int d^2k\,\Delta V_X(k)\left[\left|\mathfrak{J}^{cl}(k)\right|^2 + \frac{1}{4N}\left|\hat{\rho}^{cl}(k)\right|^2\right], \tag{32.3.3b}$$

where $V(k) > 0$ and $\Delta V_X(k) \equiv V_X(0) - V_X(k) > 0$. Here, $\hat{\rho}^{cl}(k) = \langle\mathfrak{S}|\hat{\rho}(k)|\mathfrak{S}\rangle$ and $\mathfrak{J}^{cl}(k) = \langle\mathfrak{S}|\mathfrak{J}(k)|\mathfrak{S}\rangle$. The key observation is that, though the Hamiltonian (32.3.1) involves no spin variables, the energy of a state does. It is important that both the energies are positive semidefinite. The direct energy H_D^{cl} is insensitive to spin orientations, and it vanishes for the homogeneous electron distribution since $\Delta\hat{\rho}^{cl}(k) = 0$. The exchange energy H_X^{cl} depends on spin orientations. The spin texture is homogeneous when the spin is completely polarized, where $\mathfrak{J}^{cl}(k) \propto \delta(k)$. Furthermore, $\hat{\rho}^{cl}(k) \propto \delta(k)$ due to the homogeneous electron distribution. For such a spin orientation the exchange energy also vanishes since $\Delta V_X(k) = V_X(0) - V_X(k) = 0$ in (32.3.3b). On the other hand, $H_X^{cl} > 0$ if the spin is not polarized completely since $\hat{\mathfrak{I}}^{\text{cl}}(\boldsymbol{k})$ contains nonzero momentum components. Consequently the spin-polarized state has the lowest energy, which is zero. Hence the exchange interaction is the driving force of spontaneous symmetry breaking.

32.4 Classical Equations of Motion

We wish to derive the equation of motion for the classical fields $\hat{\rho}^{\text{cl}}(\boldsymbol{x})$ and $\mathfrak{I}^{\text{cl}}(\boldsymbol{x})$, or collectively for $\hat{\mathfrak{D}}^{\text{cl}}_{\mu\nu}(\boldsymbol{x})$. We shall verify that the classical equation of motion

$$\frac{d}{dt}\hat{\mathfrak{D}}^{\text{cl}}_{\mu\nu}(\boldsymbol{x}) = [\hat{\mathfrak{D}}^{\text{cl}}_{\mu\nu}(\boldsymbol{x}), H^{\text{cl}}]_{\text{PB}} \tag{32.4.1}$$

holds, where $H^{\text{cl}} = \langle H_{\text{C}} \rangle$ and $[A, B]_{\text{PB}}$ is an appropriately defined Poisson bracket.

We start with the quantum equation of motion for the density matrix $D_{\mu\nu}(k, l)$,

$$i\hbar\frac{d}{dt}D_{\mu\nu}(k,l) = [D_{\mu\nu}(k,l), H_{\text{C}}]. \tag{32.4.2}$$

The Coulomb Hamiltonian H_{C} is given by (32.1.2) in the monolayer system. In the case of the bilayer system it is given by (32.1.24), which consists of the SU(4)-invariant term H_{C}^{+} and the noninvariant term H_{C}^{-},

$$H_{\text{C}} = H_{\text{C}}^{+} + H_{\text{C}}^{-}. \tag{32.4.3}$$

They are given by

$$H_{\text{C}}^{\pm} = \Upsilon_{\mu\nu}\Upsilon_{\sigma\tau}V_{mnij}D_{\nu\mu}(n,m)D_{\tau\sigma}(j,i), \tag{32.4.4}$$

where Υ is a matrix with $\Upsilon^2 = \mathbb{I}$; $\Upsilon = \mathbb{I}$ for the SU(4)-invariant Hamiltonian H_{C}^{+} and $\Upsilon = \text{diag.}(+1, +1, -1, -1)$ for the SU(4)-noninvariant Hamiltonian H_{C}^{-}. In what follows we consider the case of the bilayer system. The case of the SU(N) monolayer system is simply given by the choice of $H_{\text{C}} = H_{\text{C}}^{+}$ with $\Upsilon = \mathbb{I}$ in the N-dimensional space.

With the aid of (29.5.8) we calculate the commutator (32.4.2) explicitly,

$$\begin{aligned} i\hbar\frac{d}{dt}D_{\gamma\alpha}(k,l) =& \Upsilon_{\mu\nu}\Upsilon_{\gamma\tau}V_{mnkj}D_{\nu\mu}(n,m)D_{\tau\alpha}(j,l) - \Upsilon_{\mu\nu}\Upsilon_{\sigma\alpha}V_{mnil}D_{\nu\mu}(n,m)D_{\gamma\sigma}(k,i) \\ &+ \Upsilon_{\gamma\nu}\Upsilon_{\sigma\tau}V_{knij}D_{\nu\alpha}(n,l)D_{\tau\sigma}(j,i) - \Upsilon_{\mu\alpha}\Upsilon_{\sigma\tau}V_{mlij}D_{\gamma\mu}(k,m)D_{\tau\sigma}(j,i). \end{aligned} \tag{32.4.5}$$

We then take the expectation value of this equation with respect to the Fock state (29.6.1). In so doing we use the relation

$$\begin{aligned} \langle\mathfrak{S}|D_{\nu\mu}(n,m)D_{\tau\sigma}(j,i)|\mathfrak{S}\rangle =& D^{\text{cl}}_{\nu\mu}(n,m)D^{\text{cl}}_{\tau\sigma}(j,i) - D^{\text{cl}}_{\tau\mu}(j,m)D^{\text{cl}}_{\nu\sigma}(n,i) \\ &+ \delta_{\nu\sigma}\delta_{ni}D^{\text{cl}}_{\tau\mu}(j,m), \end{aligned} \tag{32.4.6}$$

and obtain

$$\begin{aligned} i\hbar\frac{d}{dt}D^{\text{cl}}_{\gamma\alpha}(l,k) \\ =&2V_{mnlj}\text{Tr}[D^{\text{cl}}(n,m)\Upsilon][\Upsilon D^{\text{cl}}(j,k)]_{\gamma\alpha} - 2V_{mnik}\text{Tr}[\Upsilon D^{\text{cl}}(n,m)][D^{\text{cl}}(l,i)\Upsilon]_{\gamma\alpha} \\ &- 2V_{mnlj}[\Upsilon D^{\text{cl}}(j,m)\Upsilon D^{\text{cl}}(n,k)]_{\gamma\alpha} + 2V_{mnik}[D^{\text{cl}}(l,m)\Upsilon D^{\text{cl}}(n,i)\Upsilon]_{\gamma\alpha}, \end{aligned} \tag{32.4.7}$$

where we have used used $\Upsilon^2 = \mathbb{I}$, and $\text{Tr}[D^{\text{cl}}(n,m)\Upsilon] = [D^{\text{cl}}(n,m)\Upsilon]_{\mu\mu}$. Recall that $D^{\text{cl}}_{\nu\mu}(n,m) = \langle\mathfrak{S}|D_{\nu\mu}(n,m)|\mathfrak{S}\rangle$.

We wish to summarize the equation of motion (32.4.7) in a compact form. The expectation value of the Hamiltonian (32.4.4) consists of the direct and exchange energies,

$$H^{\text{cl}\pm} \equiv \langle\mathfrak{S}|H^{\pm}_{\text{C}}|\mathfrak{S}\rangle = H^{\text{cl}\pm}_{\text{D}} + H^{\text{cl}\pm}_{\text{X}}, \tag{32.4.8}$$

with

$$H^{\text{cl}\pm}_{\text{D}} = V_{mnij}\text{Tr}[D^{\text{cl}}(n,m)\Upsilon]\text{Tr}[\Upsilon D^{\text{cl}}(j,i)], \tag{32.4.9}$$

$$H^{\text{cl}\pm}_{\text{X}} = -V_{mnij}\text{Tr}[\Upsilon D^{\text{cl}}(n,i)\Upsilon D^{\text{cl}}(j,m)] + \epsilon_{\text{X}}\text{Tr}[D^{\text{cl}}(n,n)]. \tag{32.4.10}$$

Note that $H^{\text{cl}+}_{\text{D}}$ and $H^{\text{cl}+}_{\text{X}}$ agrees with (32.2.21a) and (32.2.21b), respectively. A remarkable observation is that we can summarize the equation of motion (32.4.7) into

$$\frac{d}{dt}D^{\text{cl}}_{\mu\nu}(m,n) = [D^{\text{cl}}_{\mu\nu}(m,n), H^{\text{cl}\pm}]_{\text{PB}}, \tag{32.4.11}$$

where we have defined the Poisson structure by

$$i\hbar[D^{\text{cl}}_{\mu\nu}(m,n), D^{\text{cl}}_{\sigma\tau}(i,j)]_{\text{PB}} = \delta_{\mu\tau}\delta_{mj}D^{\text{cl}}_{\sigma\nu}(i,n) - \delta_{\sigma\nu}\delta_{in}D^{\text{cl}}_{\mu\tau}(m,j). \tag{32.4.12}$$

Consequently we have

$$\frac{d}{dt}D^{\text{cl}}_{\mu\nu}(m,n) = [D^{\text{cl}}_{\mu\nu}(m,n), H^{\text{cl}}]_{\text{PB}} \tag{32.4.13}$$

together with

$$H^{\text{cl}} = H^{\text{cl}+} + H^{\text{cl}-}. \tag{32.4.14}$$

This produces the classical equation of motion (32.4.2) in the real space.

The Poisson structure (32.4.12) is precisely the same as the algebra (29.5.8). To see its significance more in details, combining it with the magnetic-translation group property (29.2.4), we rewrite it in the momentum space,

$$2\pi i\hbar[\hat{\mathfrak{D}}^{\text{cl}}_{\mu\nu}(\boldsymbol{p}), \hat{\mathfrak{D}}^{\text{cl}}_{\sigma\tau}(\boldsymbol{q})]_{\text{PB}} = \delta_{\mu\tau}e^{+\frac{i}{2}\ell_B^2\boldsymbol{p}\wedge\boldsymbol{q}}\hat{\mathfrak{D}}^{\text{cl}}_{\sigma\nu}(\boldsymbol{p}+\boldsymbol{q}) - \delta_{\sigma\nu}e^{-\frac{i}{2}\ell_B^2\boldsymbol{p}\wedge\boldsymbol{q}}\hat{\mathfrak{D}}^{\text{cl}}_{\mu\tau}(\boldsymbol{p}+\boldsymbol{q}). \tag{32.4.15}$$

It corresponds to the $W_\infty(N)$ algebra (29.5.9), indicating that the classical density $\hat{\mathfrak{D}}^{\text{cl}}_{\mu\nu}$ should obey the $W_\infty(N)$ algebra as well.

We have previously noticed in (29.6.24) that the classical density $\hat{\mathcal{D}}^{\text{cl}}_{\mu\nu}$ is resolved in terms of the N-component bosonic field,

$$\hat{\mathcal{D}}^{\text{cl}}_{\mu\nu}(\boldsymbol{x}) = \frac{1}{2\pi\ell_B^2}\phi^{\text{cl}}_{\mu}(\boldsymbol{x}) \star \phi^{\text{cl}*}_{\nu}(\boldsymbol{x}), \tag{32.4.16}$$

or

$$\hat{\mathcal{D}}^{\text{cl}}_{\mu\nu}(\boldsymbol{p}) = \frac{1}{2\pi\ell_B^2}\int d^2q\, e^{-\frac{i}{2}\ell_B^2 \boldsymbol{p}\wedge\boldsymbol{q}}\phi^{\text{cl}}_{\mu}(\boldsymbol{q})\phi^{\text{cl}*}_{\nu}(\boldsymbol{q}-\boldsymbol{p}). \tag{32.4.17}$$

It follows that

$$\begin{aligned}
&2\pi[\hat{\mathcal{D}}^{\text{cl}}_{\mu\nu}(\boldsymbol{p}),\hat{\mathcal{D}}^{\text{cl}}_{\sigma\tau}(\boldsymbol{q})]_{\text{PB}}\\
=&\frac{1}{4\pi^2\ell_B^4}\int d^2p'd^2q'\, e^{-\frac{i}{2}\ell_B^2(\boldsymbol{p}\wedge\boldsymbol{p}'+\boldsymbol{q}\wedge\boldsymbol{q}')}\phi^{\text{cl}}_{\mu}(\boldsymbol{p}')[\phi^{\text{cl}*}_{\nu}(\boldsymbol{p}'-\boldsymbol{p}),\phi^{\text{cl}}_{\sigma}(\boldsymbol{q}')]_{\text{PB}}\phi^{\text{cl}*}_{\tau}(\boldsymbol{q}'-\boldsymbol{q})\\
&+\frac{1}{4\pi^2\ell_B^4}\int d^2p'd^2q'\, e^{-\frac{i}{2}\ell_B^2(\boldsymbol{p}\wedge\boldsymbol{p}'+\boldsymbol{q}\wedge\boldsymbol{q}')}\phi^{\text{cl}}_{\sigma}(\boldsymbol{q}')[\phi^{\text{cl}*}_{\mu}(\boldsymbol{p}'),\phi^{\text{cl}}_{\tau}(\boldsymbol{q}'-\boldsymbol{q})]_{\text{PB}}\phi^{\text{cl}*}_{\nu}(\boldsymbol{p}'-\boldsymbol{p}).
\end{aligned} \tag{32.4.18}$$

We can easily verify that this is reduced to the Poisson structure (32.4.15) with the use of (32.4.17), provided

$$i\hbar[\phi^{\text{cl}}_{\mu}(\boldsymbol{p}),\phi^{\text{cl}*}_{\nu}(\boldsymbol{q})]_{\text{PB}} = \delta_{\mu\nu}(\boldsymbol{q}-\boldsymbol{p}), \tag{32.4.19}$$

or

$$i\hbar[\phi^{\text{cl}}_{\mu}(\boldsymbol{x}),\phi^{\text{cl}*}_{\nu}(\boldsymbol{y})]_{\text{PB}} = \delta_{\mu\nu}\delta(\boldsymbol{x}-\boldsymbol{y}). \tag{32.4.20}$$

It simply implies that $\phi^{\text{cl}}_{\mu}(\boldsymbol{x})$ and $\phi^{\text{cl}*}_{\mu}(\boldsymbol{x})$ are the canonical conjugate variables in the classical theory. The field $\phi^{\text{cl}}_{\mu}(\boldsymbol{x})$ corresponds to the composite-boson field (15.3.1) together with the canonical commutation relation (15.3.3) in the continuum limit.

32.5 Four-Layer Condenser Model

In this section we explain the microscopic origin of the background terms in the Coulomb Hamiltonian (22.2.12), and also the origin of the bias term (22.4.1) with (22.4.8).

Typical sample structures of the bilayer system are illustrated in Fig.21.10 (page 381). Electrons are supplied from dopants separated from the quantum wells by spacers. When one layer is sufficiently separated from another, such a structure is modeled by a 4-layer condenser [Fig.32.1], by identifying the gate-layer with the dopant-layer for simplicity.

We put charges in these layers as shown in Fig.32.1: The hole density in the front (back) gate is $\rho_0(1\pm\bar{\sigma}_0)/2$, while the electron density in the front (back) layer

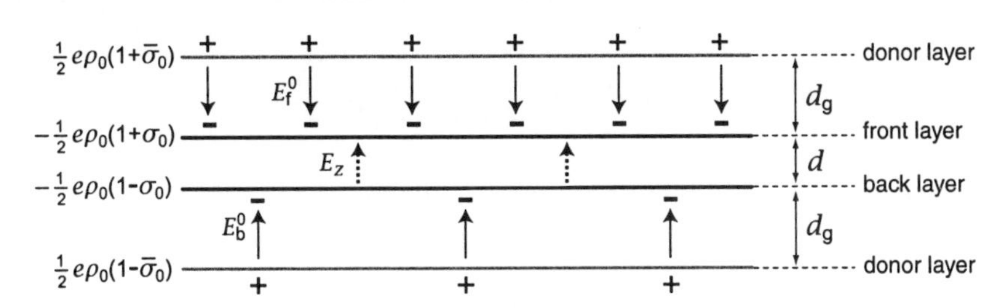

Fig. 32.1 A schematic picture of the bilayer QH system is given at $\sigma_0 = 1/3$. The electron density ρ_0^{f} (ρ_0^{b}) in the front (back) layer is independently controlled by the front (back) gate bias voltage. Electrons are supplied from dopants separated from the quantum wells [Fig.21.10 (page 381)] by controlling the gate bias voltages. All electric fields are homogeneous in the ground state.

is $\rho_0(1 \pm \sigma_0)/2$. The electric fields E_α^0 and the gate voltages V_α^{gate} between the layers are given by

$$E_{\text{f}}^0 = \frac{1}{2\varepsilon} e\rho_0(1 + \bar{\sigma}_0) = \frac{1}{d_{\text{g}}} V_{\text{f}}^{\text{gate}}, \tag{32.5.1a}$$

$$E_{\text{b}}^0 = \frac{1}{2\varepsilon} e\rho_0(1 - \bar{\sigma}_0) = \frac{1}{d_{\text{g}}} V_{\text{b}}^{\text{gate}}. \tag{32.5.1b}$$

The total density ρ_0 and the hole imbalance $\bar{\sigma}_0$ are controlled by applying the front and back gate voltages $V_{\text{f}}^{\text{gate}}$ and $V_{\text{b}}^{\text{gate}}$. The electric field between the front and back layers is given by

$$E_z^0 = \frac{1}{2\varepsilon} e\rho_0(\sigma_0 - \bar{\sigma}_0) \tag{32.5.2}$$

in the classical 4-layer condenser.

However, our system is not a classical condenser. The separation of the front and back layers is taken so small that the exchange Coulomb interaction becomes important and furthermore the tunneling occurs in general. Charge distributions would be determined so as to minimize the total Hamiltonian. The total Hamiltonian consists of the direct Coulomb term H_{D}, the exchange Coulomb term H_{X} and the tunneling term H_{T}.

We are interested in the dynamics of electrons in the presence of a homogeneous hole distribution in the donor layer. We take the charge density of the 4-layer capacitance as

$$\begin{aligned}\rho_e^{\text{cl}}(\boldsymbol{x}) =& \frac{1}{2} e\rho_0(1 + \bar{\sigma}_0)\delta(z - d/2 - d_{\text{g}}) + \frac{1}{2} e\rho_0(1 - \bar{\sigma}_0)\delta(z + d/2 + d_{\text{g}}) \\ & - e\rho^{\text{cl(f)}}(\boldsymbol{x})\delta(z - d/2) - e\rho^{\text{cl(b)}}(\boldsymbol{x})\delta(z + d/2),\end{aligned} \tag{32.5.3}$$

and substitute it into the direct Coulomb Hamiltonian,

$$H_{\rm D}^{\rm cl} = \frac{1}{2}\int d^2x dz d^2x' dz' \frac{\rho_e^{\rm cl}(\boldsymbol{x}, z)\rho_e^{\rm cl}(\boldsymbol{x}', z')}{4\pi\varepsilon\sqrt{(\boldsymbol{x}-\boldsymbol{x}')^2 + (z-z')^2}}. \tag{32.5.4}$$

Integrating it over the variables z and z', we obtain

$$\begin{aligned} H_{\rm D}^{\rm cl} =& \frac{1}{2}\int d^2xd^2x'\, V^+(\boldsymbol{x}-\boldsymbol{x}')\rho^{\rm cl}(\boldsymbol{x})\rho^{\rm cl}(\boldsymbol{x}') + 2\int d^2xd^2x'\, V^-(\boldsymbol{x}-\boldsymbol{x}')\mathcal{P}_z^{\rm cl}(\boldsymbol{x})\mathcal{P}_z^{\rm cl}(\boldsymbol{x}') \\ &- \frac{e^2\rho_0}{8\pi\varepsilon}\int d^2xd^2x' \left[\frac{\rho^{\rm cl}(\boldsymbol{x}) + 2\bar{\sigma}_0\mathcal{P}_z^{\rm cl}(\boldsymbol{x})}{\sqrt{\boldsymbol{x}'^2 + d_{\rm g}^2}} + \frac{\rho^{\rm cl}(\boldsymbol{x}) - 2\bar{\sigma}_0\mathcal{P}_z^{\rm cl}(\boldsymbol{x})}{\sqrt{\boldsymbol{x}'^2 + (d+d_{\rm g})^2}}\right] \\ &+ \frac{e^2\rho_0^2}{16\pi\varepsilon}\int d^2xd^2x' \left[\frac{1+\bar{\sigma}_0^2}{\sqrt{\boldsymbol{x}'^2}} + \frac{1-\bar{\sigma}_0^2}{\sqrt{\boldsymbol{x}'^2 + (d+2d_{\rm g})^2}}\right]. \end{aligned} \tag{32.5.5}$$

Here we use $\int d^2x\, \hat{\rho}^{\rm cl}(\boldsymbol{x}) = \rho_0 S$ and

$$\int d^2x \frac{1}{\sqrt{\boldsymbol{x}^2 + a^2}} = 2\pi \lim_{R\to\infty}\int_0^R rdr \frac{1}{\sqrt{r^2+a^2}} = 2\pi \lim_{R\to\infty}(R-a), \tag{32.5.6}$$

to summarize the direct Coulomb Hamiltonian as

$$\begin{aligned} H_{\rm D} =& \frac{1}{2}\int d^2xd^2x'\, V^+(\boldsymbol{x}-\boldsymbol{x}')\left[\rho^{\rm cl}(\boldsymbol{x}) - \rho_0\right]\left[\rho^{\rm cl}(\boldsymbol{x}') - \rho_0\right] \\ &+ 2\int d^2xd^2x'\, V^-(\boldsymbol{x}-\boldsymbol{x}')[\mathcal{P}_z^{\rm cl}(\boldsymbol{x}) - \frac{1}{2}\rho_0\bar{\sigma}_0][\mathcal{P}_z^{\rm cl}(\boldsymbol{x}') - \frac{1}{2}\rho_0\bar{\sigma}_0] \\ &+ \frac{e^2\rho_0^2 d_{\rm g}}{4\varepsilon}(1+\bar{\sigma}_0^2)S, \end{aligned} \tag{32.5.7}$$

where ρ_0 and $\bar{\sigma}_0$ are given by (32.5.1). It is to be noticed that the charge of holes appears as the background charge for the capacitive energy.

Taking care of the Landau-level projection we have

$$\begin{aligned} H_{\rm D} =& \frac{1}{2}\int d^2xd^2x'\, V_{\rm D}^+(\boldsymbol{x}-\boldsymbol{x}')\left[\hat{\rho}^{\rm cl}(\boldsymbol{x}) - \rho_0\right]\left[\hat{\rho}^{\rm cl}(\boldsymbol{x}') - \rho_0\right] \\ &+ 2\int d^2xd^2x'\, V_{\rm D}^-(\boldsymbol{x}-\boldsymbol{x}')[\hat{\mathcal{P}}_z^{\rm cl}(\boldsymbol{x}) - \frac{1}{2}\rho_0\bar{\sigma}_0][\hat{\mathcal{P}}_z^{\rm cl}(\boldsymbol{x}') - \frac{1}{2}\rho_0\bar{\sigma}_0], \end{aligned} \tag{32.5.8}$$

where an irrelevant constant term has been ignored. We include the exchange Coulomb interaction and the tunneling interaction, to which the electric charges on the donor layers do not contribute since they are far apart,

$$H_X = -\frac{1}{2}\int d^2x\, V_X^+(\mathbf{x}-\mathbf{x}')\Big[\hat{\mathcal{P}}_a^{cl}(\mathbf{x})\hat{\mathcal{P}}_a^{cl}(\mathbf{x}') + \frac{1}{4}\hat{\rho}^{cl}(\mathbf{x})\hat{\rho}^{cl}(\mathbf{x}')\Big]$$
$$-\frac{1}{2}\int d^2x\, V_X^-(\mathbf{x}-\mathbf{x}')\Big[2\hat{\mathcal{P}}_z^{cl}(\mathbf{x})\hat{\mathcal{P}}_z^{cl}(\mathbf{x}') - \sum_{a=xyz}\hat{\mathcal{P}}_a^{cl}(\mathbf{x})\hat{\mathcal{P}}_a^{cl}(\mathbf{x}') + \frac{1}{4}\hat{\rho}^{cl}(\mathbf{x})\hat{\rho}^{cl}(\mathbf{x}')\Big], \tag{32.5.9}$$

$$H_T = -\Delta_{SAS}\int d^2x\, \hat{\mathcal{P}}_x^{cl}(\mathbf{x}). \tag{32.5.10}$$

The total Hamiltonian is

$$H^{cl} = H_D^{cl} + H_X^{cl} + H_T^{cl} \tag{32.5.11}$$

with (32.5.5).

For the ground state,

$$\hat{\rho}^{cl}(\mathbf{x}) = \rho_0, \quad \hat{\mathcal{P}}_x^{cl}(\mathbf{x}) = \frac{1}{2}\rho_0\sqrt{1-\sigma_0}, \quad \hat{\mathcal{P}}_y^{cl}(\mathbf{x}) = 0, \quad \hat{\mathcal{P}}_z^{cl}(\mathbf{x}) = \frac{1}{2}\rho_0\sigma_0 \tag{32.5.12}$$

the total Hamiltonian (32.5.11) is reduced to

$$H^{cl} = \frac{1}{2}\rho_0\int d^2x\left[\frac{\epsilon_{cap}^D}{2}(\sigma_0-\bar{\sigma}_0)^2 - \frac{\epsilon_{cap}^X}{2}\sigma_0^2 - \Delta_{SAS}\sqrt{1-\sigma_0^2}\right], \tag{32.5.13}$$

apart from an irrelevant constant, with

$$\epsilon_{cap}^D \equiv 2\rho_0\int d^2x\, V_D^-(\mathbf{x}) = \frac{e^2\rho_0 d}{2\varepsilon} = 4\nu\epsilon_D^-, \tag{32.5.14}$$

$$\epsilon_{cap}^X \equiv \rho_0\int d^2x\, V_X^-(\mathbf{x}) = 4\nu\epsilon_X^-. \tag{32.5.15}$$

Minimizing this energy with respect to σ_0, we obtain

$$\epsilon_{cap}\sigma_0 + \frac{\sigma_0}{\sqrt{1-\sigma_0^2}}\Delta_{SAS} = \epsilon_{cap}^D\bar{\sigma}_0, \tag{32.5.16}$$

with

$$\epsilon_{cap} = \epsilon_{cap}^D - \epsilon_{cap}^X. \tag{32.5.17}$$

The electron imbalance σ_0 is related to the hole imbalance $\bar{\sigma}_0$ by (32.5.16).

We can eliminate $\bar{\sigma}_0$ from the Hamiltonian (32.5.11) with the use of (32.5.16). We extract the relevant term from the second line of (32.5.8), which reads

$$-\epsilon_{cap}^D\bar{\sigma}_0\int d^2x\, \hat{\mathcal{P}}_z^{cl}(\mathbf{x}) = -\sigma_0[\epsilon_{cap}^D - \epsilon_{cap}^X]\int d^2x\, \hat{\mathcal{P}}_z^{cl}(\mathbf{x}) - \frac{\sigma_0}{\sqrt{1-\sigma_0^2}}\Delta_{SAS}\int d^2x\, \hat{\mathcal{P}}_z^{cl}(\mathbf{x}). \tag{32.5.18}$$

We absorb the term containing the factor ϵ_{cap}^D into the second line of (32.5.8), and the term containing ϵ_{cap}^X into the second line of (32.5.9). We treat the last term as a new term,

$$H_{\text{bias}}^{\text{cl}} = -\frac{\sigma_0}{\sqrt{1-\sigma_0^2}}\Delta_{\text{SAS}}\int d^2x\, \hat{\mathcal{P}}_z^{\text{cl}}(\boldsymbol{x}) \equiv -\Delta_{\text{bias}}\int d^2x\, \hat{\mathcal{P}}_z^{\text{cl}}(\boldsymbol{x}). \tag{32.5.19}$$

This is the origin of the bias term (22.4.1) with (22.4.8). In this way the total Hamiltonian (32.5.11) is rewritten as

$$\begin{aligned} H^{\text{cl}} =& \frac{1}{2}\int d^2x d^2x'\, V_{\text{D}}^{+}(\boldsymbol{x}-\boldsymbol{x}')\left[\hat{\rho}^{\text{cl}}(\boldsymbol{x})-\rho_0\right]\left[\hat{\rho}^{\text{cl}}(\boldsymbol{x}')-\rho_0\right] \\ &+\int d^2x d^2x'\, [2V_{\text{D}}^{-}(\boldsymbol{x}-\boldsymbol{x}')-V_{\text{X}}^{-}(\boldsymbol{x}-\boldsymbol{x}')]\left[\hat{\mathcal{P}}_z^{\text{cl}}(\boldsymbol{x})-\frac{1}{2}\rho_0\sigma_0\right]\left[\hat{\mathcal{P}}_z^{\text{cl}}(\boldsymbol{x}')-\frac{1}{2}\rho_0\sigma_0\right] \\ &-\frac{1}{2}\sum_{a=xyz}\int d^2x\, [V_{\text{X}}^{+}(\boldsymbol{x}-\boldsymbol{x}')-V_{\text{X}}^{-}(\boldsymbol{x}-\boldsymbol{x}')]\, \hat{\mathcal{P}}_a^{\text{cl}}(\boldsymbol{x})\hat{\mathcal{P}}_a^{\text{cl}}(\boldsymbol{x}') \\ &-\frac{1}{4}\int d^2x\, [V_{\text{X}}^{+}(\boldsymbol{x}-\boldsymbol{x}')+V_{\text{X}}^{-}(\boldsymbol{x}-\boldsymbol{x}')]\, \hat{\rho}^{\text{cl}}(\boldsymbol{x})\hat{\rho}^{\text{cl}}(\boldsymbol{x}') \\ &+H_{\text{bias}}^{\text{cl}}+H_{\text{T}}^{\text{cl}}, \end{aligned} \tag{32.5.20}$$

where H_{T} is the tunneling term given by (32.5.10). The second line represents the capasitance energy.

A comment is in order. When we neglect the exchange and tunneling interactions ($\epsilon_{\text{cap}}^{\text{X}} = \Delta_{\text{SAS}} = 0$), the Hamiltonian (32.5.13) is reduced to

$$H = \frac{\varepsilon d}{2}\int d^2x\, (E_z^0)^2, \tag{32.5.21}$$

where E_z^0 is given by (32.5.2). This is the well-known result for the classical condenser.

32.6 Derivative Expansion and Effective Theory

There are two types of Coulomb interactions, the direct one and the exchange one. They are manifest in the classical Hamiltonian (32.2.20). To analyze small isospin fluctuations such as Goldstone modes and sufficiently smooth isospin textures, we make the derivative expansion and derive the effective Hamiltonian.

Monolayer SU(N) QH Systems

We discuss the monolayer QH system with SU(N) symmetry. The direct energy represents the well-known Coulomb interaction due to the charge density $-e\rho(\boldsymbol{x})$. We may set $\rho(\boldsymbol{x}) = \rho_0$ for the study of Goldstone modes since the QH state is robust against density fluctuations. We rewrite the exchange Hamiltonian (32.1.11) as

$$\hat{H}_{\text{X}} = -\frac{1}{2}\int d^2x d^2z\, V_{\text{X}}(\boldsymbol{z})\left[\hat{\mathfrak{I}}^{\text{cl}}(\boldsymbol{x}+\boldsymbol{z})\hat{\mathfrak{I}}^{\text{cl}}(\boldsymbol{x})+\frac{1}{2N}\hat{\rho}^{\text{cl}}(\boldsymbol{x}+\boldsymbol{z})\hat{\rho}^{\text{cl}}(\boldsymbol{x})\right]. \tag{32.6.1}$$

Since $V_X(z)$ is short ranged, it is a good approximation to make the Taylor expansion of $\mathfrak{I}^{cl}(x+z)$. The lowest order term is

$$\hat{H}_X^{(0)} = -\frac{1}{2}\int d^2z\, V_X(z)\int d^2x\left[\mathfrak{I}^{cl}(x)\mathfrak{I}^{cl}(x) + \frac{1}{2N}\hat{\rho}^{cl}(x)\hat{\rho}^{cl}(x)\right]. \quad (32.6.2)$$

Here we introduce the isospin $\mathcal{I}(x)$ per one Landau site,

$$\mathfrak{I}^{cl}(x) = \rho_\Phi \mathcal{I}(x), \quad (32.6.3)$$

which is subject to the normalization (30.2.30), or

$$\sum_{A=1}^{N^2-1}\mathcal{I}_A(x)\mathcal{I}_A(x) = \frac{\nu(N-\nu)}{2N}. \quad (32.6.4)$$

We can neglect the lowest order term $\hat{H}_X^{(0)}$ since it is just a constant.

The nontrivial first order term is

$$H_X^{\text{eff}} = \frac{1}{4}\int d^2z\, z_i z_j\, V_X(z)\int d^2x\left[\partial_i\mathfrak{I}^{cl}(x)\partial_j\mathfrak{I}^{cl}(x) + \frac{1}{2N}\partial_i\hat{\rho}^{cl}(x)\partial_j\hat{\rho}^{cl}(x)\right]. \quad (32.6.5)$$

Equivalently, we work in the momentum space, where

$$H_X^{\text{eff}} = -\pi\int d^2p\, V_X(p)\left[\mathfrak{I}^{cl}(-p)\mathfrak{I}^{cl}(p) + \frac{1}{2N}\hat{\rho}^{cl}(-p)\hat{\rho}^{cl}(p)\right] \quad (32.6.6)$$

with the potential (32.1.12). We expand $V_X(p)$ as

$$V_X(p) = V_X(0) - \frac{2J_s}{\pi\rho_\Phi^2}p^2 + O(p^4), \quad (32.6.7)$$

where

$$V_X(0) = 4\ell_B^2\epsilon_X \quad (32.6.8)$$

together with (30.1.16b) for ϵ_X, and

$$J_s = \frac{\ell_B^2}{16\pi}\int dk\, k^3 V_D(k). \quad (32.6.9)$$

The parameter J_s is called the spin stiffness. It reads

$$J_s = \frac{1}{8\pi}\epsilon_X = \frac{1}{16\pi}\sqrt{\frac{\pi}{2}}E_C^0 \quad (32.6.10)$$

for the lowest Landau level. Substituting this into (32.6.6) and taking the Fourier transformation, we obtain

$$H_X^{\text{eff}} = \frac{2J_s}{\rho_\Phi^2}\int d^2x\left[\partial_k\mathfrak{I}^{cl}(x)\partial_k\mathfrak{I}^{cl}(x) + \frac{1}{2N}\partial_k\hat{\rho}^{cl}(x)\partial_k\hat{\rho}^{cl}(x)\right] \quad (32.6.11)$$

as the first nontrivial term. Consequently, we have derived the SU(N) nonlinear sigma model,

$$H_X^{\rm eff} = 2J_s \int d^2x\, \partial_k \mathfrak{I}(x) \partial_k \mathfrak{I}(x), \tag{32.6.12}$$

together with the isospin stiffness (32.6.9), where $\mathfrak{I}(x)$ is the normalized isospin field. It represents the exchange interaction between isospins located in Landau sites: See (4.1.5b). It plays a role of the kinetic Hamiltonian for the isospin dynamics of electrons confined to the lowest Landau level.

Bilayer QH System without Spin

We next derive the effective Hamiltonian in the bilayer QH system. In the case of the spin-frozen system, the exchange Hamiltonian is given by (32.1.18),

$$\begin{aligned} H_X = -\pi \int d^2p \Big[V_X^d(p) \sum_{a=xy} \hat{\mathcal{P}}_a^{\rm cl}(-p)\hat{\mathcal{P}}_a^{\rm cl}(p) + V_X(p)\hat{\mathcal{P}}_z^{\rm cl}(-p)\hat{\mathcal{P}}_z^{\rm cl}(p) \\ + \frac{1}{4} V_X(p)\hat{\rho}^{\rm cl}(-p)\hat{\rho}^{\rm cl}(p) \Big]. \end{aligned} \tag{32.6.13}$$

The Coulomb potentials is expanded as

$$V_X^{\pm}(p) = V_X^{\pm}(0) - \frac{2J_s^{\pm}}{\pi\rho_\Phi^2} p^2 + O(p^4), \tag{32.6.14}$$

where[1]

$$V_X^{\pm}(0) = 2\ell_B^2 \left[1 \pm e^{d^2/2\ell_B^2} {\rm erfc}\left(d/\sqrt{2}\ell_B\right)\right] \epsilon_X = 4\ell_B^2 \epsilon_X^{\pm}, \tag{32.6.15}$$

$$J_s^{\pm} = \frac{1}{2} J_s \left[1 \pm \left\{ -\sqrt{\frac{2}{\pi}} \frac{d}{\ell_B} + \left(1 + \frac{d^2}{\ell_B^2}\right) e^{d^2/2\ell_B^2} {\rm erfc}\left(d/\sqrt{2}\ell_B\right) \right\}\right] \tag{32.6.16}$$

for the lowest Landau level. The lowest order terms are

$$\begin{aligned} &V_X^d(0) \sum_{a=xy} \hat{\mathcal{P}}_a^{\rm cl}(-p)\hat{\mathcal{P}}_a^{\rm cl}(p) + V_X(0)\hat{\mathcal{P}}_z^{\rm cl}(-p)\hat{\mathcal{P}}_z^{\rm cl}(p) \\ &= V_X^d(0) \sum_{a=xyz} \hat{\mathcal{P}}_a^{\rm cl}(-p)\hat{\mathcal{P}}_a^{\rm cl}(p) + 2V_X^-(0)\hat{\mathcal{P}}_z^{\rm cl}(-p)\hat{\mathcal{P}}_z^{\rm cl}(p), \end{aligned} \tag{32.6.17}$$

where $\sum_{a=xyz} \hat{\mathcal{P}}_a^{\rm cl}(-p)\hat{\mathcal{P}}_a^{\rm cl}(p)$ is a constant, and $2V_X^-(0) = 8\ell_B^2\epsilon_X^-$. Thus we find

$$\mathcal{H}_X^{\rm eff} = \mathcal{H}_X^{\rm inter} + \mathcal{H}_X^{\rm cap}, \tag{32.6.18}$$

---

[1]We use $J_0(z) = 1 - (z/2)^2 + O(z^4)$, and $I_0(z) = 1 + (z/2)^2 + O(z^4)$. The error function is defined by ${\rm erf}(x) = \frac{2}{\sqrt{\pi}}\int_0^x e^{-t^2}dt$, or ${\rm erfc}(x) = \frac{2}{\sqrt{\pi}}\int_x^\infty e^{-t^2}dt = 1 - {\rm erf}(x)$. It is expanded around $x = 0$ as ${\rm erf}(x) = \frac{2}{\sqrt{\pi}} e^{-x^2}(x + \frac{2}{3}x^3 + \cdots)$.

where

$$\mathcal{H}_{\mathrm{X}}^{\mathrm{inter}} = \frac{2J_s^d}{\rho_\Phi^2}\left(\partial_k\hat{\mathcal{P}}_x^{\mathrm{cl}}\partial_k\hat{\mathcal{P}}_x^{\mathrm{cl}} + \partial_k\hat{\mathcal{P}}_y^{\mathrm{cl}}\partial_k\hat{\mathcal{P}}_y^{\mathrm{cl}}\right) + \frac{2J_s}{\rho_\Phi^2}\partial_k\hat{\mathcal{P}}_z^{\mathrm{cl}}\partial_k\hat{\mathcal{P}}_z^{\mathrm{cl}}, \tag{32.6.19}$$

$$\mathcal{H}_{\mathrm{X}}^{\mathrm{cap}} = -\frac{4\epsilon_{\mathrm{X}}^-}{\rho_\Phi}\hat{\mathcal{P}}_z^{\mathrm{cl}}(\boldsymbol{x})\hat{\mathcal{P}}_z^{\mathrm{cl}}(\boldsymbol{x}), \tag{32.6.20}$$

and

$$J_s \equiv J_s^+ + J_s^-, \qquad J_s^d \equiv J_s^+ - J_s^-. \tag{32.6.21}$$

Here, $\mathcal{H}_{\mathrm{X}}^{\mathrm{cap}}$ and $\mathcal{H}_{\mathrm{X}}^{\mathrm{inter}}$ are the zeroth and first order terms in $\boldsymbol{p}^2$, respectively.

The direct Hamiltonian is given by (30.5.1),

$$H_{\mathrm{D}} = \pi\int d^2p\left[V_{\mathrm{D}}^+(\boldsymbol{p})\hat{\rho}^{\mathrm{cl}}(-\boldsymbol{p})\hat{\rho}^{\mathrm{cl}}(\boldsymbol{p}) + 4V_{\mathrm{D}}^-(\boldsymbol{p})\hat{\mathcal{P}}_z^{\mathrm{cl}}(-\boldsymbol{p})\hat{\mathcal{P}}_z^{\mathrm{cl}}(\boldsymbol{p})\right]. \tag{32.6.22}$$

We first extract the self-energy as

$$H_{\mathrm{self}} = \pi\int d^2p\, V_{\mathrm{D}}^+(\boldsymbol{p})\hat{\rho}^{\mathrm{cl}}(-\boldsymbol{p})\hat{\rho}^{\mathrm{cl}}(\boldsymbol{p}) = \frac{1}{2}\int d^2x d^2y\, \hat{\rho}^{\mathrm{cl}}(\boldsymbol{x})V_{\mathrm{D}}^+(\boldsymbol{x}-\boldsymbol{y})\hat{\rho}^{\mathrm{cl}}(\boldsymbol{y}). \tag{32.6.23}$$

We then extract the contribution to the capacitance energy,

$$H_{\mathrm{D}}^{\mathrm{cap}} = 4\pi V_{\mathrm{D}}^-(0)\int d^2p\,\hat{\mathcal{P}}_z^{\mathrm{cl}}(-\boldsymbol{p})\hat{\mathcal{P}}_z^{\mathrm{cl}}(\boldsymbol{p}) = \frac{4\epsilon_{\mathrm{D}}^-}{\rho_\Phi}\int d^2x\,\hat{\mathcal{P}}_z^{\mathrm{cl}}(\boldsymbol{x})\hat{\mathcal{P}}_z^{\mathrm{cl}}(\boldsymbol{x}) \tag{32.6.24}$$

with (30.5.12b), or

$$4\epsilon_{\mathrm{D}}^- = \frac{2V_{\mathrm{D}}^-(0)}{\ell_B^2} = \frac{d}{\ell_B}E_{\mathrm{C}}^0. \tag{32.6.25}$$

Indeed, this is the standard capacitance energy per unit area for a condenser made of two planes with separation d: See (22.5.4). Combining (32.6.20) and (32.6.24) we find

$$H_{\mathrm{cap}} = \frac{\epsilon_{\mathrm{cap}}}{\rho_\Phi}\int d^2x\,\hat{\mathcal{P}}_z(\boldsymbol{x})\hat{\mathcal{P}}_z(\boldsymbol{x}) \tag{32.6.26}$$

with (30.5.14) for the parameter ϵ_{cap}. The standard capacitance energy is decreased considerably by the exchange effect [see Fig.22.4].

When we may set $\hat{\rho}^{\mathrm{cl}}(\boldsymbol{x}) = \rho_0$ and $\hat{\mathcal{P}}_a^{\mathrm{cl}}(\boldsymbol{x}) = \rho_\Phi\mathcal{P}_a(\boldsymbol{x})$, the effective Hamiltonian is summarized as

$$\mathcal{H}_{\mathrm{C}}^{\mathrm{eff}} = \mathcal{H}_{\mathrm{X}}^{\mathrm{inter}} + \mathcal{H}_{\mathrm{cap}} \tag{32.6.27}$$

with

$$\mathcal{H}_{\mathrm{X}}^{\mathrm{inter}} = 2J_s^d\left(\partial_k\mathcal{P}_x\partial_k\mathcal{P}_x + \partial_k\mathcal{P}_y\partial_k\mathcal{P}_y\right) + 2J_s\partial_k\mathcal{P}_z\partial_k\mathcal{P}_z, \tag{32.6.28}$$

$$\mathcal{H}_{\mathrm{cap}} = \epsilon_{\mathrm{cap}}\rho_\Phi\mathcal{P}_z(\boldsymbol{x})\mathcal{P}_z(\boldsymbol{x}). \tag{32.6.29}$$

It describes Goldstone modes associated with quantum coherence in spin-frozen bilayer QH systems.

Bilayer QH System with Spin

We now include the spin degree of freedom to the bilayer system. The direct and exchange Hamiltonians are given by (32.1.23) and (32.1.34), respectively. By combining them the classical Hamiltonian reads

$$\begin{aligned} H^{\text{cl}} =& \pi \int d^2p\, V_{\text{D}}^{+}(\boldsymbol{p})\hat{\rho}^{\text{cl}}(-\boldsymbol{p})\hat{\rho}^{\text{cl}}(\boldsymbol{p}) + 4\pi \int d^2p\, V_{\text{D}}^{-}(\boldsymbol{p})\hat{\mathcal{P}}_z^{\text{cl}}(-\boldsymbol{p})\hat{\mathcal{P}}_z^{\text{cl}}(\boldsymbol{p}) \\ & - \frac{\pi}{2}\int d^2p\, V_{\text{X}}^{d}(\boldsymbol{p}) \left[\hat{\mathcal{S}}_a^{\text{cl}}(-\boldsymbol{p})\hat{\mathcal{S}}_a^{\text{cl}}(\boldsymbol{p}) + \hat{\mathcal{P}}_a^{\text{cl}}(-\boldsymbol{p})\hat{\mathcal{P}}_a^{\text{cl}}(\boldsymbol{p}) + \hat{\mathcal{R}}_{ab}^{\text{cl}}(-\boldsymbol{p})\hat{\mathcal{R}}_{ab}^{\text{cl}}(\boldsymbol{p})\right] \\ & - \pi \int d^2p\, V_{\text{X}}^{-}(\boldsymbol{p}) \left[\hat{\mathcal{S}}_a^{\text{cl}}(-\boldsymbol{p})\hat{\mathcal{S}}_a^{\text{cl}}(\boldsymbol{p}) + \hat{\mathcal{P}}_z^{\text{cl}}(-\boldsymbol{p})\hat{\mathcal{P}}_z^{\text{cl}}(\boldsymbol{p}) + \hat{\mathcal{R}}_{az}^{\text{cl}}(-\boldsymbol{p})\hat{\mathcal{R}}_{az}^{\text{cl}}(\boldsymbol{p})\right] \\ & - \frac{\pi}{8}\int d^2p\, V_{\text{X}}(\boldsymbol{p})\rho^{\text{cl}}(-\boldsymbol{p})\rho^{\text{cl}}(\boldsymbol{p}). \end{aligned} \tag{32.6.30}$$

The analysis proceeds almost identically to the bilayer spinless system. The self-energy term H_{self} is given by (32.6.23), while the lowest order term in $\boldsymbol{p}^2$ yields the capacitance term,

$$\mathcal{H}_{\text{cap}}^{\text{cl}} = \frac{\epsilon_{\text{cap}}}{\rho_\Phi}\hat{\mathcal{P}}_z^{\text{cl}}(\boldsymbol{x})\hat{\mathcal{P}}_z^{\text{cl}}(\boldsymbol{x}), \tag{32.6.31}$$

with

$$\epsilon_{\text{cap}} = \begin{cases} 4\epsilon_{\text{D}}^{-} - 4\epsilon_{\text{X}}^{-} \text{ at } \nu = 1 \\ 4\epsilon_{\text{D}}^{-} - 2\epsilon_{\text{X}}^{-} \text{ at } \nu = 2 \end{cases}. \tag{32.6.32}$$

See (30.5.14) and (26.2.7). The first order term in $\boldsymbol{p}^2$ yields a nonlinear sigma model,

$$\begin{aligned} \mathcal{H}_{\text{X}}^{\text{SU(4)}} =& \frac{J_s^d}{\rho_\Phi^2}\left(\partial_k\hat{\mathcal{S}}_a^{\text{cl}}\partial_k\hat{\mathcal{S}}_a^{\text{cl}} + \partial_k\hat{\mathcal{P}}_a^{\text{cl}}\partial_k\hat{\mathcal{P}}_a^{\text{cl}} + \partial_k\hat{\mathcal{R}}_{ab}^{\text{cl}}\partial_k\hat{\mathcal{R}}_{ab}^{\text{cl}}\right) \\ & + \frac{2J_s^-}{\rho_\Phi^2}\left(\partial_k\hat{\mathcal{S}}_a^{\text{cl}}\partial_k\hat{\mathcal{S}}_a^{\text{cl}} + \partial_k\hat{\mathcal{P}}_z^{\text{cl}}\partial_k\hat{\mathcal{P}}_z^{\text{cl}} + \partial_k\hat{\mathcal{R}}_{az}^{\text{cl}}\partial_k\hat{\mathcal{R}}_{az}^{\text{cl}}\right) \end{aligned} \tag{32.6.33}$$

with (32.6.16).

We have calculated the skyrmion energy based on the effective Hamiltonian $H_{\text{C}}^{\text{eff}} = H_{\text{self}} + H_{\text{X}}^{\text{SU(4)}} + H_{\text{cap}}$ in Section 25.5. We may set $\hat{\rho}^{\text{cl}}(\boldsymbol{x}) = \rho_0$, $\hat{\mathcal{S}}_a^{\text{cl}}(\boldsymbol{x}) = \rho_\Phi\mathcal{S}_a(\boldsymbol{x})$ and $\hat{\mathcal{P}}_a^{\text{cl}}(\boldsymbol{x}) = \rho_\Phi\mathcal{P}_a(\boldsymbol{x})$ for the study of Goldstone mode,

$$\mathcal{H}_{\text{cap}} = \epsilon_{\text{cap}}\rho_\Phi\mathcal{P}_z(\boldsymbol{x})\mathcal{P}_z(\boldsymbol{x}), \tag{32.6.34}$$

$$\begin{aligned} \mathcal{H}_{\text{X}}^{\text{SU(4)}} =& J_s^d\left(\sum \partial_k\mathcal{S}_a\partial_k\mathcal{S}_a + \partial_k\mathcal{P}_a\partial_k\mathcal{P}_a + \partial_k\mathcal{R}_{ab}\partial_k\mathcal{R}_{ab}\right) \\ & + 2J_s^-\left(\sum \partial_k\mathcal{S}_a\partial_k\mathcal{S}_a + \partial_k\mathcal{P}_z\partial_k\mathcal{P}_z + \partial_k\mathcal{R}_{az}\partial_k\mathcal{R}_{az}\right). \end{aligned} \tag{32.6.35}$$

We have analyzed the Goldstone modes based on the effective Hamiltonian $H_{\text{C}}^{\text{eff}} = H_{\text{X}}^{\text{SU(4)}} + H_{\text{cap}}$ in Section 25.3.

32.7 NONCOMMUTATIVE CP^{N-1} MODEL

We next derive the effective Hamiltonian in terms of the noncommutative CP^{N-1} field. For simplicity we take the monolayer system. We evaluate the energy based on the decomposition rule (32.2.1), where

$$H_{\text{D}}^{\text{cl}} = V_{mnij}D_{\mu\mu}^{\text{cl}}(n,m)D_{\nu\nu}^{\text{cl}}(j,i) - \nu(N_{\text{e}}^{\text{cl}} + \delta N_{\text{e}}^{\text{cl}})\epsilon_{\text{D}}, \tag{32.7.1a}$$

$$H_{\text{X}}^{\text{cl}} = -V_{mnij}D_{\mu\nu}^{\text{cl}}(j,m)D_{\nu\mu}^{\text{cl}}(n,i) + N_{\text{e}}^{\text{cl}}\epsilon_{\text{X}}. \tag{32.7.1b}$$

They are equivalent to (32.2.21).

For definiteness we consider the $\nu = 1$ case though the generalization to an arbitrary integer ν is straightforward with the use of the Grassmannian field $G_{N,\nu}$. We make the use of (31.3.6), or

$$W[\hat{\mathcal{D}}_{\mu\nu}^{\text{cl}}] = \frac{1}{2\pi\ell_B^2}\mathfrak{n}_\mu\mathfrak{n}_\nu^\dagger, \tag{32.7.2}$$

which is equivalent to

$$D_{\mu\nu}^{\text{cl}}(m,n) = \langle m|\mathfrak{n}_\mu\mathfrak{n}_\nu^\dagger|n\rangle, \tag{32.7.3}$$

and (29.7.12) reads

$$W\left[\Delta\hat{\rho}^{\text{cl}}\right] = \frac{1}{2\pi\ell_B^2}\left[\mathfrak{n}_\mu^\dagger, \mathfrak{n}_\mu\right]. \tag{32.7.4}$$

Substituting (32.7.3) into (32.7.1) we come to

$$H_{\text{D}}^{\text{cl}} = \int\frac{d^2k}{4\pi}e^{-\frac{1}{2}\ell_B^2k^2}V(\boldsymbol{k})\text{Tr}\left(\mathfrak{n}_\mu\mathfrak{n}_\mu^\dagger e^{ikX}\right)\text{Tr}\left(\mathfrak{n}_\nu\mathfrak{n}_\nu^\dagger e^{-ikX}\right) - (N_{\text{e}}^{\text{cl}} + \Delta N_{\text{e}}^{\text{cl}})\epsilon_D, \tag{32.7.5a}$$

$$H_{\text{X}}^{\text{cl}} = -\int\frac{d^2k}{4\pi}e^{-\frac{1}{2}\ell_B^2k^2}V(\boldsymbol{k})\text{Tr}\left(\mathfrak{n}_\nu\mathfrak{n}_\mu^\dagger e^{ikX}\mathfrak{n}_\nu\mathfrak{n}_\mu^\dagger e^{-ikX}\right) + N_{\text{e}}^{\text{cl}}\epsilon_{\text{X}}, \tag{32.7.5b}$$

where we have taken (30.1.13) into account.

We first study the direct energy H_{D}^{cl}. Since

$$\text{Tr}\left(\mathfrak{n}_\mu^\dagger\mathfrak{n}_\mu e^{ikX}\right) = \text{Tr}\left(e^{ikX} + \left[\mathfrak{n}_\mu^\dagger, \mathfrak{n}_\mu\right]e^{ikX}\right) = \frac{2\pi}{\ell_B^2}\delta(\boldsymbol{k}) + \text{Tr}\left(2\pi\ell_B^2W\left[\Delta\hat{\rho}^{\text{cl}}\right]e^{ikX}\right) \tag{32.7.6}$$

due to (29.2.31) and (32.7.4), we rewrite the direct energy as

$$\begin{aligned} H_{\text{D}}^{\text{cl}} &= \int\frac{d^2k}{4\pi}V_{\text{D}}(\boldsymbol{k})\text{Tr}\left(2\pi\ell_B^2W\left[\Delta\hat{\rho}^{\text{cl}}\right]e^{ikX}\right)\text{Tr}\left(2\pi\ell_B^2W\left[\Delta\hat{\rho}^{\text{cl}}\right]e^{-ikX}\right) \\ &= \pi\int d^2k\, V_{\text{D}}(\boldsymbol{k})\Delta\hat{\rho}^{\text{cl}}(-\boldsymbol{k})\,\Delta\hat{\rho}^{\text{cl}}(\boldsymbol{k}), \end{aligned} \tag{32.7.7}$$

where we have (29.2.10). It represents simply the classical energy of a charge distribution.

We proceed to discuss the exchange energy H_X^{cl}. To bring it to a more reasonable form, we use

$$N_e^{cl} = \mathrm{Tr}\left(\mathfrak{n}_\mu \mathfrak{n}_\mu^\dagger\right), \tag{32.7.8}$$

and rewrite the exchange energy as

$$H_X^{cl} = \int \frac{d^2k}{4\pi} e^{-\frac{1}{2}\ell_B^2 k^2} V(\boldsymbol{k}) \mathrm{Tr}\left(\mathfrak{n}_\mu \mathfrak{n}_\mu^\dagger - \mathfrak{n}_\mu \mathfrak{n}_\nu^\dagger e^{i\boldsymbol{k}\boldsymbol{X}} \mathfrak{n}_\nu \mathfrak{n}_\mu^\dagger e^{-i\boldsymbol{k}\boldsymbol{X}}\right), \tag{32.7.9}$$

where the cancellation of divergences occurs at $\boldsymbol{k} = 0$. We may change the order of operators under the trace simultaneously in both terms so that the cancellation at $\boldsymbol{k} = 0$ is maintained. We then use the normalization $\mathfrak{n}_\mu^\dagger \mathfrak{n}_\mu = 1$ and summarize the result as

$$H_X^{cl} = \int \frac{d^2k}{4\pi} e^{-\frac{1}{2}\ell_B^2 k^2} V(\boldsymbol{k}) \mathrm{Tr}\left(\mathbb{I} - \mathfrak{n}_\mu^\dagger e^{i\boldsymbol{k}\boldsymbol{X}} \mathfrak{n}_\mu \cdot \mathfrak{n}_\nu^\dagger e^{-i\boldsymbol{k}\boldsymbol{X}} \mathfrak{n}_\nu\right). \tag{32.7.10}$$

As far as we are concerned about sufficient smooth field configurations, making the derivative expansion we may take the lowest-order term. Here we note that

$$\mathbb{I} - \mathfrak{n}_\mu^\dagger e^{i\boldsymbol{k}\boldsymbol{X}} \mathfrak{n}_\mu \cdot \mathfrak{n}_\nu^\dagger e^{-i\boldsymbol{k}\boldsymbol{X}} \mathfrak{n}_\nu = [\mathfrak{n}_\mu^\dagger, e^{i\boldsymbol{k}\boldsymbol{X}}][e^{-i\boldsymbol{k}\boldsymbol{X}}, \mathfrak{n}_\mu] - \mathfrak{n}_\mu^\dagger [e^{i\boldsymbol{k}\boldsymbol{X}}, \mathfrak{n}_\mu] \mathfrak{n}_\nu^\dagger [e^{-i\boldsymbol{k}\boldsymbol{X}}, \mathfrak{n}_\nu]. \tag{32.7.11}$$

Because of the relation (29.2.23) or

$$\partial_i \mathfrak{n}_\mu = -\frac{i}{\ell_B^2} \epsilon_{ij} \left[X_j, \mathfrak{n}_\mu\right], \tag{32.7.12}$$

the derivative expansion corresponds to the expansion in $\boldsymbol{X}$,

$$e^{i\boldsymbol{k}\boldsymbol{X}} = \mathbb{I} + ik_i X_i - \frac{1}{2} k_i k_j X_i X_j + O(X^3). \tag{32.7.13}$$

Substituting this into (32.7.10) and assuming the rotational invariance of $V(\boldsymbol{k})$ we integrate over the angle in the momentum space. In such a way we come to

$$\begin{aligned} H_X^{cl} =& \frac{4\pi J_s}{\ell_B^2} \mathrm{Tr}\left(\mathfrak{n}_\mu^\dagger X_m X_m \mathfrak{n}_\mu - \mathfrak{n}_\mu^\dagger X_m \mathfrak{n}_\mu \mathfrak{n}_\nu^\dagger X_m \mathfrak{n}_\nu\right) + O(X^4) \\ =& \frac{4\pi J_s}{\ell_B^2} \mathrm{Tr}\left([\mathfrak{n}_\mu^\dagger, X_m][X_m, \mathfrak{n}_\mu] - \mathfrak{n}_\mu^\dagger [X_m, \mathfrak{n}_\mu] \mathfrak{n}_\nu^\dagger [X_m, \mathfrak{n}_\nu]\right) + O(X^4), \end{aligned} \tag{32.7.14}$$

where the constant J_s is given by

$$J_s \equiv \frac{\ell_B^2}{16\pi} \int dk\, k^3 e^{-\frac{1}{2}\ell_B^2 k^2} V(k). \tag{32.7.15}$$

We eventually come to

$$H_X^{cl} = 4\pi J_s \ell_B^2 \mathrm{Tr}\left[(\partial_m \mathfrak{n}_\mu^\dagger)(\partial_m \mathfrak{n}_\mu) + (\mathfrak{n}_\mu^\dagger \partial_m \mathfrak{n}_\mu)(\mathfrak{n}_\nu^\dagger \partial_m \mathfrak{n}_\nu)\right] \tag{32.7.16}$$

up to the lowest-order term. It is known as the noncommutative CP^{N-1} model.[307] It is interesting that the noncommutative CP^{N-1} model is derived from the four-fermion interaction Hamiltonian irrespective of the explicit form of the potential $V(\boldsymbol{x})$ as the lowest-order term in the derivative expansion.

We may discuss the Goldstone mode (isospin wave) based on the effective Hamiltonian. Small fluctuations of the CP^{N-1} field occur around the ground state without the density modulation ($\delta\rho_{\rm e}^{\rm cl}(\boldsymbol{x}) = 0$), thus minimizing the direct energy as $E_{\rm D} = 0$. The condition reads

$$n_\mu(\boldsymbol{x}) \star n_\mu^*(\boldsymbol{x}) = n_\mu^*(\boldsymbol{x}) \star n_\mu(\boldsymbol{x}) = 1. \tag{32.7.17}$$

The Goldstone modes are described by the noncommutative CP^{N-1} model (32.7.16), which is the effective Hamiltonian. Furthermore, sufficiently smooth isospin waves are described in its commutative limit,

$$E_{\rm X} = 2J_s \int d^2x \left[\left(\partial_m n_\mu^\dagger\right)\left(\partial_m n_\mu\right) + \left(n_\mu^\dagger \partial_m n_\mu\right)\left(n_\nu^\dagger \partial_m n_\nu\right)\right], \tag{32.7.18}$$

which is the ordinary CP^{N-1} model. There are $N-1$ complex Goldstone modes.

Skyrmions in Noncommutative CP^{N-1} Model

It is intriguing to analyze the problem of skyrmions in the noncommutative CP^{N-1} model (32.7.16). It is convenient to introduce the notation

$$\begin{aligned} \mathcal{D}_z \mathfrak{n}_\mu &\equiv \mathcal{D}_x \mathfrak{n}_\mu - i\mathcal{D}_y \mathfrak{n}_\mu, \\ \mathcal{D}_{\bar z} \mathfrak{n}_\mu &\equiv \mathcal{D}_x n_\mu + i\mathcal{D}_y \mathfrak{n}_\mu, \end{aligned} \tag{32.7.19}$$

where

$$\mathcal{D}_i \mathfrak{n}_\mu \equiv -\frac{i}{\ell_B^2}\epsilon_{ij}\left([X_j, \mathfrak{n}_\mu] - \mathfrak{n}_\mu \mathfrak{n}_\gamma [X_j, \mathfrak{n}_\gamma]\right), \tag{32.7.20}$$

and rewrite the noncommutative CP^{N-1} model (32.7.16) as

$$E_{\rm X} = 2\pi J_s \ell_B^2 {\rm Tr}\left[(\mathcal{D}_z \mathfrak{n}_\mu)^\dagger(\mathcal{D}_z \mathfrak{n}_\mu) + (\mathcal{D}_{\bar z}\mathfrak{n}_\mu)^\dagger(\mathcal{D}_{\bar z}\mathfrak{n}_\mu)\right]. \tag{32.7.21}$$

We have defined the topological charge Q by the formula (29.7.1). However, there are other equivalent ways of defining it, and here we use the representation

$$Q_{\rm sky} = \frac{1}{2}\ell_B^2 {\rm Tr}\left[(\mathcal{D}_z \mathfrak{n}_\mu)^\dagger(\mathcal{D}_z \mathfrak{n}_\mu) - (\mathcal{D}_{\bar z}\mathfrak{n}_\mu)^\dagger(\mathcal{D}_{\bar z}\mathfrak{n}_\mu)\right]. \tag{32.7.22}$$

The equivalence is readily proved because the topological charge densities associated with (29.7.1) and (7.8.16) are different only by a total derivative term that does not contribute to the topological charge.

It follows from (32.7.21) and (32.7.22) that

$$E_{\rm X} \geq 4\pi J_s |Q_{\rm sky}|. \tag{32.7.23}$$

The energy bound becomes saturated for

$$\mathcal{D}_{\bar{z}}\mathfrak{n}_\mu = 0 \qquad \text{(skyrmions)} \tag{32.7.24}$$

or

$$\mathcal{D}_{z}\mathfrak{n}_\mu = 0 \qquad \text{(antiskyrmions).} \tag{32.7.25}$$

We consider the case of skyrmions in what follows. The BPS equation (32.7.24) takes the form

$$[b^\dagger, \mathfrak{n}_\mu] - \mathfrak{n}_\mu \mathfrak{n}_\gamma [b^\dagger, \mathfrak{n}_\gamma] = 0. \tag{32.7.26}$$

We study explicitly the case of SU(2) and look for the solution to (32.7.26) in the form (31.3.7), or

$$\mathfrak{n}_\uparrow = \sum_{m=0}^{\infty} v(m)|m+k\rangle\langle m|, \quad \mathfrak{n}_\downarrow = \sum_{m=0}^{\infty} u(m)|m\rangle\langle m|, \tag{32.7.27}$$

where $|u(k)|^2 + |v(k)|^2 = 1$. This describes a multi-skyrmion carrying the topological charge $Q = k$. We get

$$\begin{aligned}
[b^\dagger, \mathfrak{n}_\uparrow] - \mathfrak{n}_\uparrow \mathfrak{n}_\mu [b^\dagger, \mathfrak{n}_\mu] &= \sum_{m=0}^{\infty} f_\uparrow(m)|m+k+1\rangle\langle m|, \\
[b^\dagger, \mathfrak{n}_\downarrow] - \mathfrak{n}_\downarrow \mathfrak{n}_\mu [b^\dagger, \mathfrak{n}_\mu] &= \sum_{m=0}^{\infty} f_\downarrow(m)|m+1\rangle\langle m|,
\end{aligned} \tag{32.7.28}$$

with

$$\begin{aligned}
f_\uparrow(m) =& v(m)\bar{u}(m+1)u(m+1)\sqrt{m+k+1} - v(m+1)\bar{u}(m+1)u(m)\sqrt{m+1}, \\
f_\downarrow(m) =& u(m)\bar{v}(m+1)v(m+1)\sqrt{m+1} - u(m+1)\bar{v}(m+1)v(m)\sqrt{m+k+1}.
\end{aligned} \tag{32.7.29}$$

The equation (32.7.26) holds if $f_\uparrow(m) = 0$ and $f_\downarrow(m) = 0$. They lead to a single equation, which can be summarized as

$$\frac{v(m)}{u(m)} = \frac{\sqrt{m+k}}{\sqrt{m}} \frac{v(m-1)}{u(m-1)}. \tag{32.7.30}$$

This recurrence relation is solved as

$$\frac{v(m)}{u(m)} = \frac{\sqrt{(m+k)!}}{\sqrt{m!}} \frac{1}{\omega^k}, \tag{32.7.31}$$

where we have introduced the complex parameter ω by

$$\omega^k \equiv \frac{u(0)}{v(0)} k!. \tag{32.7.32}$$

The parameters (32.7.31) are found to be the same as (31.3.11).

We conclude that the skyrmions are identical both in the hardcore model and the noncommutative CP^{N-1} model. Let us examine this equivalence more in details in the hardcore model, where the lowest-order term in the derivative expansion is the noncommutative CP^{N-1} model (32.7.21). We note that the spin stiffness J_s and the exchange energy parameter ϵ_X are defined by (32.7.15) and (30.1.16b), respectively, and for the hardcore model there holds the relation

$$\epsilon_X = 4\pi J_s. \tag{32.7.33}$$

On one hand, in the noncommutative CP^{N-1} model the skyrmion energy with $|Q_{sky}| = 1$ is given by

$$H_X^{cl} = 4\pi J_s, \tag{32.7.34}$$

as follows from (32.7.23). On the other hand, in the hardcore interaction model the skyrmion energy is

$$E_{sky} = \epsilon_X, \tag{32.7.35}$$

as in (31.6.10). The two energies (32.7.34) and (32.7.35) are identical. Namely, the skyrmion energy coming from the lowest-order term saturates the total energy. It implies that the direct energy and the higher order exchange energy have precisely cancelled out each other. Such a cancellation is possible in the hardcore model since both of them are short ranged.

In the case of the realistic theory of Coulomb interactions, where

$$\epsilon_X = 8\pi J_s, \tag{32.7.36}$$

the skyrmion energy depends on the size. It is $E_{sky} = \frac{1}{2}\epsilon_X$ for an infinitely large skyrmion, where the direct energy vanishes. It agrees with (32.7.34), implying that all the higher order exchange energy in (32.7.14) vanishes in this case.

32.8 Equations of Motion and Hall Currents

The kinetic energy is just a constant for electrons in the lowest Landau level. The dynamics arises not due to the kinetic Hamiltonian but due to the noncommutativity of the coordinates of the guiding center.

The dynamical variables are the electron number density $\hat{\rho}(\boldsymbol{x})$ and the three components $\hat{S}_a(\boldsymbol{x})$ of the spin field in the monolayer QH ferromagnet. The Heisenberg equations of motion are

$$i\hbar\frac{d\hat{\rho}}{dt} = [\hat{\rho}, H], \qquad i\hbar\frac{d\hat{S}_a}{dt} = [\hat{S}_a, H], \tag{32.8.1}$$

where $H = H_C + H_Z + H_E$ with

$$H_C = \pi \int d^2q\, V_D(\boldsymbol{q})\hat{\rho}(-\boldsymbol{q})\hat{\rho}(\boldsymbol{q}), \tag{32.8.2}$$

$$H_Z = -2\pi\Delta_Z\hat{S}_z(0). \tag{32.8.3}$$

$$H_E = ec \int d^2q\, e^{-q^2\ell_B^2/4} A_0(-\boldsymbol{q})\hat{\rho}(\boldsymbol{q}). \tag{32.8.4}$$

We have introduced the scalar potential to produce an electric field on the layer, $E_i(\boldsymbol{x}) = ec\partial_i A_0(\boldsymbol{x})$. It is straightforward to calculate the Heisenberg equations of motion (32.8.1) with the use of the W_∞ algebra (29.5.13).

Here we calculate the Hall current with the use of the equation of motion. The electric current is originally introduced so that the electric charge conserves,

$$\frac{d}{dt}\hat{\rho}_e(\boldsymbol{x}) = \partial_i \hat{\mathcal{J}}_i(\boldsymbol{x}), \tag{32.8.5}$$

with $\hat{\rho}_e(\boldsymbol{x}) = -e\hat{\rho}(\boldsymbol{x})$ the electric charge density. Thus,

$$\partial_i \hat{\mathcal{J}}_i(\boldsymbol{x}) = -e\frac{d}{dt}\hat{\rho}(\boldsymbol{x}) = -\frac{e}{i\hbar}[\hat{\rho}(\boldsymbol{x}), H]. \tag{32.8.6}$$

Since the observable quantity is the classical current

$$\mathcal{J}_i(\boldsymbol{x}) = \langle \hat{\mathcal{J}}_i(\boldsymbol{x})\rangle, \tag{32.8.7}$$

we take the expectation value of (32.8.6) and use the classical equation of motion (28.1.52),

$$\partial_i \mathcal{J}_i(\boldsymbol{x}) = -e\frac{d}{dt}\hat{\rho}^{\text{cl}}(\boldsymbol{x}) = -e[\hat{\rho}^{\text{cl}}(\boldsymbol{x}), H^{\text{cl}}]_{\text{PB}}. \tag{32.8.8}$$

The current $\mathcal{J}_i(\boldsymbol{x})$ is constructed from (32.8.8) by integrating it.

The Hamiltonian is $H^{\text{cl}} = H_D^{\text{cl}} + H_X^{\text{cl}} + H_Z^{\text{cl}} + H_E^{\text{cl}}$, with

$$H_D^{\text{cl}} = \pi \int d^2p \left[V_D(\boldsymbol{p})\hat{\rho}^{\text{cl}}(-\boldsymbol{p})\hat{\rho}^{\text{cl}}(\boldsymbol{p})\right], \tag{32.8.9a}$$

$$H_X^{\text{cl}} = -\pi \int d^2p \left[V_X(\boldsymbol{p}) \sum_{a=xyz} \hat{S}_a^{\text{cl}}(-\boldsymbol{p})\hat{S}_a^{\text{cl}}(\boldsymbol{p}) + \frac{1}{4}V_X(\boldsymbol{p})\hat{\rho}^{\text{cl}}(-\boldsymbol{p})\hat{\rho}^{\text{cl}}(\boldsymbol{p})\right], \tag{32.8.9b}$$

$$H_Z^{\text{cl}} = -2\pi\Delta_{\text{SAS}}\hat{S}_Z^{\text{cl}}(0), \tag{32.8.9c}$$

$$H_E^{\text{cl}} = ec \int d^2q\, e^{-q^2\ell_B^2/4} A_0(-\boldsymbol{q})\hat{\rho}^{\text{cl}}(\boldsymbol{q}). \tag{32.8.9d}$$

The basic Poisson structure is given by (32.4.12) or (32.4.15), or

$$[\hat{\rho}^{\rm cl}(\boldsymbol{p}), \hat{\rho}^{\rm cl}(\boldsymbol{q})]_{\rm PB} = \frac{1}{\pi\hbar}\hat{\rho}^{\rm cl}(\boldsymbol{p}+\boldsymbol{q}) \sin\left(\ell_B^2 \frac{\boldsymbol{p}\wedge\boldsymbol{q}}{2}\right), \tag{32.8.10a}$$

$$[\hat{S}_a^{\rm cl}(\boldsymbol{p}), \hat{\rho}^{\rm cl}(\boldsymbol{q})]_{\rm PB} = \frac{1}{\pi\hbar}\hat{S}_a^{\rm cl}(\boldsymbol{p}+\boldsymbol{q}) \sin\left(\ell_B^2 \frac{\boldsymbol{p}\wedge\boldsymbol{q}}{2}\right), \tag{32.8.10b}$$

$$[\hat{S}_a^{\rm cl}(\boldsymbol{p}), \hat{S}_b^{\rm cl}(\boldsymbol{q})]_{\rm PB} = \frac{1}{2\pi\hbar}\varepsilon_{abc}\hat{S}_c^{\rm cl}(\boldsymbol{p}+\boldsymbol{q}) \cos\left(\ell_B^2 \frac{\boldsymbol{p}\wedge\boldsymbol{q}}{2}\right) + \frac{1}{4\pi\hbar}\delta_{ab}\hat{\rho}^{\rm cl}(\boldsymbol{p}+\boldsymbol{q}) \sin\left(\ell_B^2 \frac{\boldsymbol{p}\wedge\boldsymbol{q}}{2}\right), \tag{32.8.10c}$$

in the instance of the SU(2) spin.

We calculate explicitly the Poisson bracket in (32.8.8). We then estimate it by using the incompressibility condition

$$\hat{\rho}^{\rm cl}(\boldsymbol{k}) = 2\pi\rho_0\delta^2(\boldsymbol{k}), \tag{32.8.11}$$

since we are interested in the current in the QH state. We first demonstrate that there are no contributions from the Coulomb and Zeeman interactions to the current.

We examine the contribution from $H_{\rm D}^{\rm cl}$. The relevant Poisson bracket is

$$[\hat{\rho}^{\rm cl}(\boldsymbol{k}), H_{\rm D}^{\rm cl}]_{\rm PB} = \frac{2}{\hbar}\int d^2q\, V_{\rm D}(\boldsymbol{q})\{\hat{\rho}^{\rm cl}(\boldsymbol{q}), \hat{\rho}^{\rm cl}(\boldsymbol{k}-\boldsymbol{q})\} \sin\left(\ell_B^2 \frac{\boldsymbol{k}\wedge\boldsymbol{q}}{2}\right). \tag{32.8.12}$$

This vanishes with the use of the incompressibility condition (32.8.11).

The Poisson bracket with $H_{\rm X}^{\rm cl}$ is

$$[\hat{\rho}^{\rm cl}(\boldsymbol{k}), H_{\rm X}^{\rm cl}]_{\rm PB} = -\frac{1}{2\hbar}\int d^2q\, V_{\rm X}(\boldsymbol{q})\hat{\rho}^{\rm cl}(-\boldsymbol{q})\hat{\rho}^{\rm cl}(\boldsymbol{k}+\boldsymbol{q}) \sin\left(\ell_B^2 \frac{\boldsymbol{k}\wedge\boldsymbol{q}}{2}\right) - \frac{2}{\hbar}\int d^2q\, V_{\rm X}(\boldsymbol{q})\hat{S}_a(-\boldsymbol{q})\hat{S}_a(\boldsymbol{k}+\boldsymbol{q}) \sin\left(\ell_B^2 \frac{\boldsymbol{k}\wedge\boldsymbol{q}}{2}\right). \tag{32.8.13}$$

The first term involving $\hat{\rho}^{\rm cl}(-\boldsymbol{q})\hat{\rho}^{\rm cl}(\boldsymbol{k}+\boldsymbol{q})$ vanishes by the incompressibility condition (32.8.11). Since we are concerned about a homogeneous flow of electrons, taking the nontrivial lowest order term in the derivative expansion of potential $V_{\rm X}(\boldsymbol{k})$, we find

$$\int d^2q\, V_{\rm X}(\boldsymbol{k}')\hat{S}_z^{\rm cl}(-\boldsymbol{q})\hat{S}_z^{\rm cl}(\boldsymbol{k}+\boldsymbol{q}) \sin\left(\frac{1}{2}\ell_B^2 \boldsymbol{k}\wedge\boldsymbol{q}\right) \simeq V_{\rm X}(0)\int d^2q\, \hat{S}_z^{\rm cl}(-\boldsymbol{q})\hat{S}_z^{\rm cl}(\boldsymbol{k}+\boldsymbol{q}) \sin\left(\frac{1}{2}\ell_B^2 \boldsymbol{k}\wedge\boldsymbol{q}\right) = 0. \tag{32.8.14}$$

This is zero because of the relation

$$\int d^2q f(-\boldsymbol{q})g(\boldsymbol{k}+\boldsymbol{q}) \sin(\boldsymbol{k}\wedge\boldsymbol{q}) = -\int d^2q f(\boldsymbol{k}+\boldsymbol{q})g(-\boldsymbol{q}) \sin(\boldsymbol{k}\wedge\boldsymbol{q}), \tag{32.8.15}$$

which holds for any two functions f and g. There is no contribution from $H_{\rm X}^{\rm cl}$.

No contribution arises also from $H_{\rm Z}^{\rm cl}$ since $[\rho(\boldsymbol{k}), H_{\rm Z}^{\rm cl}]_{\rm PB} = 0$.

We finally calculate the contribution from the electric-field term H_{E}. The Poisson bracket is

$$[\hat{\rho}^{\mathrm{cl}}(\boldsymbol{k}), H_{\mathrm{E}}^{\mathrm{cl}}]_{\mathrm{PB}} = \frac{\pi e c}{\hbar}\int d^2q\, e^{-q^2\ell_B^2/4}A_0(\boldsymbol{q})\hat{\rho}^{\mathrm{cl}}(\boldsymbol{k}-\boldsymbol{q})\sin\left(\ell_B^2\frac{\boldsymbol{k}\wedge\boldsymbol{q}}{2}\right). \tag{32.8.16}$$

Expanding

$$\sin\ell_B^2\frac{\boldsymbol{k}\wedge\boldsymbol{q}}{2} = \frac{\varepsilon_{ij}k_iq_j}{2}\ell_B^2\left[1-\frac{1}{3!}\left(\frac{\varepsilon_{ij}k_iq_j}{2}\ell_B^2\right)^2+\cdots\right], \tag{32.8.17}$$

we obtain

$$\mathcal{J}_i(\boldsymbol{k}) = i\frac{e^2\ell_B^2c}{2\pi\hbar}\varepsilon_{ij}\int d^2q\, e^{-k^2\ell_B^2/4}q_jA_0(\boldsymbol{q})\hat{\rho}^{\mathrm{cl}}(\boldsymbol{k}-\boldsymbol{q})\left[1-\frac{1}{3!}\left(\frac{\varepsilon_{ij}k_iq_j}{2}\ell_B^2\right)^2+\cdots\right]. \tag{32.8.18}$$

In a constant electric field E_j such that

$$q_icA_0(\boldsymbol{q}) = -2\pi iE_j\delta(\boldsymbol{q}), \tag{32.8.19}$$

we find

$$\mathcal{J}_i(\boldsymbol{k}) = \frac{e^2}{\hbar}\ell_B^2\varepsilon_{ij}E_je^{-k^2\ell_B^2/4}\hat{\rho}(\boldsymbol{k}-\boldsymbol{q}). \tag{32.8.20}$$

It reads

$$\mathcal{J}_i(\boldsymbol{k}) = \frac{2\pi e^2\ell_B^2}{\hbar}\varepsilon_{ij}E_j\rho_0\delta(\boldsymbol{k}) \tag{32.8.21}$$

on the incompressible state (32.8.11), or

$$\mathcal{J}_i(\boldsymbol{x}) = \frac{e^2\ell_B^2}{\hbar}\varepsilon_{ij}E_j\rho_0. \tag{32.8.22}$$

This is the standard formula for the Hall current.

32.9 Hall Currents in Pseudospin QH Ferromagnet

We proceed to study the electric currents in the bilayer system. Though the actual system has the SU(4) structure, we consider the spin-frozen system since the spin does not affect the current.

The dynamical variables are the electron number density $\hat{\rho}(\boldsymbol{x})$ and the three components $\hat{\mathcal{P}}_a(\boldsymbol{x})$ of the pseudospin field,

$$i\hbar\frac{d\hat{\rho}}{dt} = [\hat{\rho}, H], \qquad i\hbar\frac{d\hat{\mathcal{P}}_a}{dt} = [\hat{\mathcal{P}}_a, H]. \tag{32.9.1}$$

Since we are interested in the classical current, we may alternatively analyze the classical equations of motion (32.4.1),

$$\frac{d\hat{\rho}^{\mathrm{cl}}}{dt} = [\hat{\rho}^{\mathrm{cl}}, H^{\mathrm{cl}}]_{\mathrm{PB}}, \qquad \frac{d\hat{\mathcal{P}}_a^{\mathrm{cl}}}{dt} = [\hat{\mathcal{P}}_a^{\mathrm{cl}}, H^{\mathrm{cl}}]_{\mathrm{PB}}, \tag{32.9.2}$$

with the use of the Poisson bracket.

The Hamiltonian is given by

$$H^{\text{cl}} = H_{\text{D}}^{\text{cl}} + H_{\text{X}}^{\text{cl}} + H_{\text{T}}^{\text{cl}} + H_{\text{bias}}^{\text{cl}} + H_{\text{E}}^{\text{cl}}, \tag{32.9.3}$$

where H_{D}^{cl} is given by (32.5.5), H_{X}^{cl} by (32.5.9), H_{T}^{cl} by (32.5.10) and $H_{\text{bias}}^{\text{cl}}$ by (32.5.19). In the momentum space they are

$$H_{\text{D}}^{\text{cl}} = \pi \int d^2p \left[V_{\text{D}}^{+}(\boldsymbol{p})\hat{\rho}^{\text{cl}}(-\boldsymbol{p})\hat{\rho}^{\text{cl}}(\boldsymbol{p}) + 4V_{\text{D}}^{-}(\boldsymbol{p})\hat{\mathcal{P}}_z^{\text{cl}}(-\boldsymbol{p})\hat{\mathcal{P}}_z^{\text{cl}}(\boldsymbol{p})\right] - 2\pi\bar{\sigma}_0\epsilon_{\text{cap}}^{\text{D}}\hat{\mathcal{P}}_z^{\text{cl}}(0), \tag{32.9.4a}$$

$$H_{\text{X}}^{\text{cl}} = -\pi \int d^2p \Big[V_{\text{X}}^{d}(\boldsymbol{p}) \sum_{a=xy} \hat{\mathcal{P}}_a^{\text{cl}}(-\boldsymbol{p})\hat{\mathcal{P}}_a^{\text{cl}}(\boldsymbol{p}) + 2V_{\text{X}}^{-}(\boldsymbol{p})\hat{\mathcal{P}}_z^{\text{cl}}(-\boldsymbol{p})\hat{\mathcal{P}}_z^{\text{cl}}(\boldsymbol{p}) + \frac{1}{4}V_{\text{X}}(\boldsymbol{p})\hat{\rho}^{\text{cl}}(-\boldsymbol{p})\hat{\rho}^{\text{cl}}(\boldsymbol{p})\Big], \tag{32.9.4b}$$

$$H_{\text{T}}^{\text{cl}} = -2\pi\Delta_{\text{SAS}}\hat{\mathcal{P}}_x^{\text{cl}}(0), \tag{32.9.4c}$$

$$H_{\text{bias}}^{\text{cl}} = -2\pi\frac{\sigma_0}{\sqrt{1-\sigma_0^2}}\Delta_{\text{SAS}}\hat{\mathcal{P}}_z^{\text{cl}}(0). \tag{32.9.4d}$$

On the other hand, H_{E}^{cl} is the term coupled to the external electric source,

$$H_{\text{E}}^{\text{cl}} = ec \int d^2q\, e^{-q^2\ell_B^2/4}\left[A_0^{\text{f}}(-\boldsymbol{q})\hat{\rho}^{\text{f}}(\boldsymbol{q}) + A_0^{\text{b}}(-\boldsymbol{q})\hat{\rho}^{\text{b}}(\boldsymbol{q})\right]. \tag{32.9.4e}$$

We demonstrate that there arises the phase current from the exchange interaction H_{X}^{cl} in addition to the standard Hall current from the electric-field term H_{E}^{cl}. This is a novel feature in the bilayer QH system.

Poisson Brackets

The basic Poisson structure reads

$$[\hat{\rho}^{\text{cl}}(\boldsymbol{p}), \hat{\rho}^{\text{cl}}(\boldsymbol{q})]_{\text{PB}} = \frac{1}{\pi\hbar}\hat{\rho}^{\text{cl}}(\boldsymbol{p}+\boldsymbol{q})\sin\left(\ell_B^2\frac{\boldsymbol{p}\wedge\boldsymbol{q}}{2}\right), \tag{32.9.5a}$$

$$[\hat{\mathcal{P}}_a^{\text{cl}}(\boldsymbol{p}), \hat{\rho}^{\text{cl}}(\boldsymbol{q})]_{\text{PB}} = \frac{1}{\pi\hbar}\hat{\mathcal{P}}_a^{\text{cl}}(\boldsymbol{p}+\boldsymbol{q})\sin\left(\ell_B^2\frac{\boldsymbol{p}\wedge\boldsymbol{q}}{2}\right), \tag{32.9.5b}$$

$$[\hat{\mathcal{P}}_a^{\text{cl}}(\boldsymbol{p}), \hat{\mathcal{P}}_b^{\text{cl}}(\boldsymbol{q})]_{\text{PB}} = \frac{1}{2\pi\hbar}\varepsilon_{abc}\hat{\mathcal{P}}_c^{\text{cl}}(\boldsymbol{p}+\boldsymbol{q})\cos\left(\ell_B^2\frac{\boldsymbol{p}\wedge\boldsymbol{q}}{2}\right) + \frac{1}{4\pi\hbar}\delta_{ab}\hat{\rho}^{\text{cl}}(\boldsymbol{p}+\boldsymbol{q})\sin\left(\ell_B^2\frac{\boldsymbol{p}\wedge\boldsymbol{q}}{2}\right). \tag{32.9.5c}$$

It is straightforward to evaluate various Poisson brackets based on these formulas.

We list the results involving $H_{\rm D}^{\rm cl}$,

$$\hbar[\hat{\rho}^{\rm cl}(\boldsymbol{k}), H_{\rm D}^{\rm cl}]_{\rm PB} = 2\int d^2k'\, V_{\rm D}^{+}(\boldsymbol{k}')\hat{\rho}^{\rm cl}(-\boldsymbol{k}')\hat{\rho}^{\rm cl}(\boldsymbol{k}+\boldsymbol{k}')\sin\left(\frac{1}{2}\ell_B^2\boldsymbol{k}\wedge\boldsymbol{k}'\right) + 8\int d^2k'\, V_{\rm D}^{-}(\boldsymbol{k}')\hat{\mathfrak{P}}_z^{\rm cl}(-\boldsymbol{k}')\hat{\mathfrak{P}}_z^{\rm cl}(\boldsymbol{k}+\boldsymbol{k}')\sin\left(\frac{1}{2}\ell_B^2\boldsymbol{k}\wedge\boldsymbol{k}'\right), \tag{32.9.6a}$$

$$\hbar[\hat{\mathfrak{P}}_a^{\rm cl}(\boldsymbol{k}), H_{\rm D}^{\rm cl}]_{\rm PB} = 2\int d^2k'\, V_{\rm D}^{+}(\boldsymbol{k}')\hat{\rho}^{\rm cl}(-\boldsymbol{k}')\hat{\mathfrak{P}}_a^{\rm cl}(\boldsymbol{k}+\boldsymbol{k}')\sin\left(\frac{1}{2}\ell_B^2\boldsymbol{k}\wedge\boldsymbol{k}'\right) - 4\epsilon_{ab}\int d^2k'\, V_{\rm D}^{-}(\boldsymbol{k}')\hat{\mathfrak{P}}_z^{\rm cl}(-\boldsymbol{k}')\hat{\mathfrak{P}}_b^{\rm cl}(\boldsymbol{k}+\boldsymbol{k}')\cos\left(\frac{1}{2}\ell_B^2\boldsymbol{k}\wedge\boldsymbol{k}'\right) + \bar{\sigma}_0\epsilon_{\rm cap}^{\rm D}\epsilon_{ab}\hat{\mathfrak{P}}_b^{\rm cl}(\boldsymbol{k}), \tag{32.9.6b}$$

$$\hbar[\hat{\mathfrak{P}}_z^{\rm cl}(\boldsymbol{k}), H_{\rm D}^{\rm cl}]_{\rm PB} = 2\int d^2k'\, V_{\rm D}^{+}(\boldsymbol{k}')\hat{\rho}^{\rm cl}(-\boldsymbol{k}')\hat{\mathfrak{P}}_z^{\rm cl}(\boldsymbol{k}+\boldsymbol{k}')\sin\left(\frac{1}{2}\ell_B^2\boldsymbol{k}\wedge\boldsymbol{k}'\right) + 2\int d^2k'\, V_{\rm D}^{-}(\boldsymbol{k}')\hat{\mathfrak{P}}_z^{\rm cl}(-\boldsymbol{k}')\hat{\rho}^{\rm cl}(\boldsymbol{k}+\boldsymbol{k}')\sin\left(\frac{1}{2}\ell_B^2\boldsymbol{k}\wedge\boldsymbol{k}'\right). \tag{32.9.6c}$$

The results involving $H_{\rm X}^{\rm cl}$ read

$$\hbar[\hat{\rho}^{\rm cl}(\boldsymbol{k}), H_{\rm X}^{\rm cl}]_{\rm PB} = -\int d^2k'\, V_{\rm X}(\boldsymbol{k}')\left[\frac{1}{2}\hat{\rho}^{\rm cl}(-\boldsymbol{k}')\hat{\rho}^{\rm cl}(\boldsymbol{k}+\boldsymbol{k}') + 2\hat{\mathfrak{P}}_a^{\rm cl}(-\boldsymbol{k}')\hat{\mathfrak{P}}_a^{\rm cl}(\boldsymbol{k}+\boldsymbol{k}')\right]\sin\left(\frac{1}{2}\ell_B^2\boldsymbol{k}\wedge\boldsymbol{k}'\right) + 4\int d^2k'\, V_{\rm X}^{-}(\boldsymbol{k}')\hat{\mathfrak{P}}_a^{\rm cl}(-\boldsymbol{k}')\hat{\mathfrak{P}}_a^{\rm cl}(\boldsymbol{k}+\boldsymbol{k}')\sin\left(\frac{1}{2}\ell_B^2\boldsymbol{k}\wedge\boldsymbol{k}'\right), \tag{32.9.7a}$$

$$\begin{aligned}\hbar[\hat{\mathfrak{P}}_a^{\rm cl}(\boldsymbol{k}), H_{\rm X}^{\rm cl}]_{\rm PB} =& -\frac{1}{2}\int d^2k'\, V_{\rm X}^{+}(\boldsymbol{k}')\left[\hat{\rho}^{\rm cl}(-\boldsymbol{k}')\hat{\mathfrak{P}}_a^{\rm cl}(\boldsymbol{k}+\boldsymbol{k}') + \hat{\mathfrak{P}}_a^{\rm cl}(-\boldsymbol{k}')\hat{\rho}^{\rm cl}(\boldsymbol{k}+\boldsymbol{k}')\right]\sin\left(\frac{1}{2}\ell_B^2\boldsymbol{k}\wedge\boldsymbol{k}'\right)\\ &- \epsilon_{ab}\int d^2k'\, V_{\rm X}^{+}(\boldsymbol{k}')\left[\hat{\mathfrak{P}}_b^{\rm cl}(-\boldsymbol{k}')\hat{\mathfrak{P}}_z^{\rm cl}(\boldsymbol{k}+\boldsymbol{k}') - \hat{\mathfrak{P}}_z^{\rm cl}(-\boldsymbol{k}')\hat{\mathfrak{P}}_b^{\rm cl}(\boldsymbol{k}+\boldsymbol{k}')\right]\cos\left(\frac{1}{2}\ell_B^2\boldsymbol{k}\wedge\boldsymbol{k}'\right)\\ &- \frac{1}{2}\int d^2k'\, V_{\rm X}^{-}(\boldsymbol{k}')\left[\hat{\rho}^{\rm cl}(-\boldsymbol{k}')\hat{\mathfrak{P}}_a^{\rm cl}(\boldsymbol{k}+\boldsymbol{k}') - \hat{\mathfrak{P}}_a^{\rm cl}(-\boldsymbol{k}')\hat{\rho}^{\rm cl}(\boldsymbol{k}+\boldsymbol{k}')\right]\sin\left(\frac{1}{2}\ell_B^2\boldsymbol{k}\wedge\boldsymbol{k}'\right)\\ &+ \epsilon_{ab}\int d^2k'\, V_{\rm X}^{-}(\boldsymbol{k}')\left[\hat{\mathfrak{P}}_b^{\rm cl}(-\boldsymbol{k}')\hat{\mathfrak{P}}_z^{\rm cl}(\boldsymbol{k}+\boldsymbol{k}') + \hat{\mathfrak{P}}_z^{\rm cl}(-\boldsymbol{k}')\hat{\mathfrak{P}}_b^{\rm cl}(\boldsymbol{k}+\boldsymbol{k}')\right]\cos\left(\frac{1}{2}\ell_B^2\boldsymbol{k}\wedge\boldsymbol{k}'\right),\end{aligned} \tag{32.9.7b}$$

$$\hbar[\hat{\mathfrak{P}}_z^{\rm cl}(\boldsymbol{k}), H_{\rm X}^{\rm cl}]_{\rm PB} = -\frac{1}{2}\int d^2k'\, V_{\rm X}(\boldsymbol{k}')\left[\hat{\rho}^{\rm cl}(-\boldsymbol{k}')\hat{\mathfrak{P}}_z^{\rm cl}(\boldsymbol{k}+\boldsymbol{k}') + \hat{\mathfrak{P}}_z^{\rm cl}(-\boldsymbol{k}')\hat{\rho}^{\rm cl}(\boldsymbol{k}+\boldsymbol{k}')\right]\sin\left(\frac{1}{2}\ell_B^2\boldsymbol{k}\wedge\boldsymbol{k}'\right) - \epsilon_{ab}\int d^2k'\, V_{\rm X}^{d}(\boldsymbol{k}')\hat{\mathfrak{P}}_a^{\rm cl}(-\boldsymbol{k}')\hat{\mathfrak{P}}_b^{\rm cl}(\boldsymbol{k}+\boldsymbol{k}')\cos\left(\frac{1}{2}\ell_B^2\boldsymbol{k}\wedge\boldsymbol{k}'\right). \tag{32.9.7c}$$

The results involving $H_{\rm T}^{\rm cl}$ are

$$\hbar[\hat{\rho}^{\rm cl}(\boldsymbol{k}), H_{\rm T}^{\rm cl}]_{\rm PB} = 0, \tag{32.9.8a}$$

$$\hbar[\hat{\mathcal{P}}_x^{\rm cl}(\boldsymbol{k}), H_{\rm T}^{\rm cl}]_{\rm PB} = 0, \tag{32.9.8b}$$

$$\hbar[\hat{\mathcal{P}}_y^{\rm cl}(\boldsymbol{k}), H_{\rm T}^{\rm cl}]_{\rm PB} = \Delta_{\rm SAS}\hat{\mathcal{P}}_z^{\rm cl}(\boldsymbol{k}), \tag{32.9.8c}$$

$$\hbar[\hat{\mathcal{P}}_z^{\rm cl}(\boldsymbol{k}), H_{\rm T}^{\rm cl}]_{\rm PB} = -\Delta_{\rm SAS}\hat{\mathcal{P}}_y^{\rm cl}(\boldsymbol{k}). \tag{32.9.8d}$$

The results involving $H_{\rm bias}^{\rm cl}$ are

$$\hbar[\hat{\rho}^{\rm cl}(\boldsymbol{k}), H_{\rm bias}^{\rm cl}]_{\rm PB} = 0, \tag{32.9.9a}$$

$$\hbar[\hat{\mathcal{P}}_x^{\rm cl}(\boldsymbol{k}), H_{\rm bias}^{\rm cl}]_{\rm PB} = \frac{\sigma_0}{\sqrt{1-\sigma_0^2}}\Delta_{\rm SAS}\hat{\mathcal{P}}_y^{\rm cl}(\boldsymbol{k}), \tag{32.9.9b}$$

$$\hbar[\hat{\mathcal{P}}_y^{\rm cl}(\boldsymbol{k}), H_{\rm bias}^{\rm cl}]_{\rm PB} = -\frac{\sigma_0}{\sqrt{1-\sigma_0^2}}\Delta_{\rm SAS}\hat{\mathcal{P}}_x^{\rm cl}(\boldsymbol{k}), \tag{32.9.9c}$$

$$\hbar[\hat{\mathcal{P}}_z^{\rm cl}(\boldsymbol{k}), H_{\rm bias}^{\rm cl}]_{\rm PB} = 0. \tag{32.9.9d}$$

Finally we give the results involving $H_{\rm E}^{\rm cl}$,

$$\begin{aligned}\hbar[\hat{\rho}^{\rm cl}(\boldsymbol{k}), H_{\rm E}^{\rm cl}]_{\rm PB} =& \frac{ec}{2\pi}\int d^2k'\, e^{-\frac{1}{4}\ell_B^2 k'^2} A_0^+(-\boldsymbol{k}')\hat{\rho}^{\rm cl}(\boldsymbol{k}+\boldsymbol{k}')\sin\left(\frac{1}{2}\ell_B^2\boldsymbol{k}\wedge\boldsymbol{k}'\right)\\ &+\frac{ec}{\pi}\int d^2k'\, e^{-\frac{1}{4}\ell_B^2 k'^2} A_0^-(-\boldsymbol{k}')\hat{\mathcal{P}}_z^{\rm cl}(\boldsymbol{k}+\boldsymbol{k}')\sin\left(\frac{1}{2}\ell_B^2\boldsymbol{k}\wedge\boldsymbol{k}'\right),\end{aligned} \tag{32.9.10a}$$

$$\begin{aligned}\hbar[\hat{\mathcal{P}}_a^{\rm cl}(\boldsymbol{k}), H_{\rm E}^{\rm cl}]_{\rm PB} =& \frac{ec}{2\pi}\int d^2k'\, e^{-\frac{1}{4}\ell_B^2 k'^2} A_0^+(-\boldsymbol{k}')\hat{\mathcal{P}}_a^{\rm cl}(\boldsymbol{k}+\boldsymbol{k}')\sin\left(\frac{1}{2}\ell_B^2\boldsymbol{k}\wedge\boldsymbol{k}'\right)\\ &-\frac{ec}{2\pi}\epsilon_{ab}\int d^2k'\, e^{-\frac{1}{4}\ell_B^2 k'^2} A_0^-(-\boldsymbol{k}')\hat{\mathcal{P}}_b^{\rm cl}(\boldsymbol{k}+\boldsymbol{k}')\cos\left(\frac{1}{2}\ell_B^2\boldsymbol{k}\wedge\boldsymbol{k}'\right),\end{aligned} \tag{32.9.10b}$$

$$\begin{aligned}\hbar[\hat{\mathcal{P}}_z^{\rm cl}(\boldsymbol{k}), H_{\rm E}^{\rm cl}]_{\rm PB} =& \frac{ec}{2\pi}\int d^2k' \int e^{-\frac{1}{4}\ell_B^2 k'^2} A_0^+(-\boldsymbol{k}')\hat{\mathcal{P}}_z^{\rm cl}(\boldsymbol{k}+\boldsymbol{k}')\sin\left(\frac{1}{2}\ell_B^2\boldsymbol{k}\wedge\boldsymbol{k}'\right)\\ &+\frac{ec}{4\pi}\int d^2k' \int e^{-\frac{1}{4}\ell_B^2 k'^2} A_0^-(-\boldsymbol{k}')\hat{\rho}^{\rm cl}(\boldsymbol{k}+\boldsymbol{k}')\sin\left(\frac{1}{2}\ell_B^2\boldsymbol{k}\wedge\boldsymbol{k}'\right)dk',\end{aligned} \tag{32.9.10c}$$

where $A_0^{\pm} = A_0^{\rm f} \pm A_0^{\rm b}$. The equations of motion (32.9.2) are calculable with the use of these Poisson brackets.

Hall Currents

The currents flow in the front and back layers. The current is defined so that the electric charge conserves in each layer,

$$\frac{d\hat{\rho}_e^{\text{cl(f)}}}{dt} = \partial_i \mathcal{J}_i^{\text{f}}(\boldsymbol{x}) - \frac{1}{d}\mathcal{J}_z(\boldsymbol{x}), \qquad \frac{d\hat{\rho}_e^{\text{cl(b)}}}{dt} = \partial_i \mathcal{J}_i^{\text{b}}(\boldsymbol{x}) + \frac{1}{d}\mathcal{J}_z(\boldsymbol{x}), \tag{32.9.11}$$

where $\hat{J}_z(\boldsymbol{x})$ is the tunneling current.

The tunneling term H_{T}^{cl} contributes only to the tunneling current. Thus, it is given by

$$\frac{1}{d}\mathcal{J}_z(\boldsymbol{x}) = -\frac{1}{2}\frac{d}{dt}[\hat{\rho}_e^{\text{cl(f)}} - \hat{\rho}_e^{\text{cl(b)}}]\Big|_{H_{\text{T}}} = e[\hat{\mathcal{P}}_z^{\text{cl}}, H_{\text{T}}^{\text{cl}}]_{\text{PB}}, \tag{32.9.12}$$

which reads

$$\mathcal{J}_z(\boldsymbol{x}) = -\frac{ed}{\hbar}\Delta_{\text{SAS}}\hat{\mathcal{P}}_y^{\text{cl}}(\boldsymbol{x}), \tag{32.9.13}$$

where we have made use of (32.9.8a).

The Hall current is given by

$$\begin{aligned}\partial_i \mathcal{J}_i^{\alpha}(\boldsymbol{x}) &= \frac{d\hat{\rho}_e^{\text{cl}(\alpha)}}{dt}\Big|_{H_{\text{D}}+H_{\text{X}}+H_{\text{bias}}+H_{\text{E}}} \\ &= -e[\hat{\rho}^{\text{cl}(\alpha)}, H_{\text{D}}^{\text{cl}}]_{\text{PB}} - e[\hat{\rho}^{\text{cl}(\alpha)}, H_{\text{X}}^{\text{cl}}]_{\text{PB}} - e[\hat{\rho}^{\text{cl}(\alpha)}, H_{\text{bias}}^{\text{cl}}]_{\text{PB}} - e[\hat{\rho}^{\text{cl}(\alpha)}, H_{\text{E}}^{\text{cl}}]_{\text{PB}}\end{aligned} \tag{32.9.14}$$

for α =f,b. We express $\hat{\rho}^{\text{cl}(\alpha)}$ in terms of $\hat{\rho}$ and $\hat{\mathcal{P}}_z$ as

$$\hat{\rho}_e^{\text{cl(f)}} = -e\left(\frac{1}{2}\hat{\rho}^{\text{cl}} + \hat{\mathcal{P}}_z^{\text{cl}}\right), \qquad \hat{\rho}_e^{\text{cl(b)}} = -e\left(\frac{1}{2}\hat{\rho}^{\text{cl}} - \hat{\mathcal{P}}_z^{\text{cl}}\right), \tag{32.9.15}$$

and use the Poisson brackets calculated before. Since we are concerned about a homogeneous flow of electrons, we take nontrivial lowest order terms in the derivative expansion of potentials $V_{\text{D}}^{\pm}(\boldsymbol{k})$ and $V_{\text{X}}^{\pm}(\boldsymbol{k})$. We also take the nontrivial lowest order terms in

$$e^{-\frac{1}{4}\ell_B^2 k'^2}\sin\left(\frac{1}{2}\ell_B^2 \boldsymbol{k}\wedge\boldsymbol{k}'\right) \simeq \frac{1}{2}\ell_B^2 \boldsymbol{k}\wedge\boldsymbol{k}', \tag{32.9.16a}$$

$$e^{-\frac{1}{4}\ell_B^2 k'^2}\cos\left(\frac{1}{2}\ell_B^2 \boldsymbol{k}\wedge\boldsymbol{k}'\right) \simeq 1 - \frac{1}{4}\ell_B^2 \boldsymbol{k}'^2. \tag{32.9.16b}$$

Finally we use the incompressibility condition (32.8.11), or $\hat{\rho}^{\text{cl}}(\boldsymbol{k}) = 2\pi\rho_0\delta^2(\boldsymbol{k})$.

First, the Poisson bracket $[\hat{\rho}^{\alpha}, H_{\text{D}}]_{\text{PB}}$ vanishes when the incompressibility condition is used precisely as in the monolayer case.

In evaluating the contribution from the exchange Coulomb term H_{X}^{cl}, we note that $[\hat{\rho}^{\text{cl}}, H_{\text{X}}^{\text{cl}}]_{\text{PB}} = 0$ and that

$$\hbar[\hat{\mathcal{P}}_z^{\text{cl}}(\boldsymbol{k}), H_{\text{X}}^{\text{cl}}]_{\text{PB}} = -\epsilon_{ab}\int d^2k'\, V_{\text{X}}^{d}(\boldsymbol{k}')\hat{\mathcal{P}}_a^{\text{cl}}(-\boldsymbol{k}')\hat{\mathcal{P}}_b^{\text{cl}}(\boldsymbol{k}+\boldsymbol{k}'). \tag{32.9.17}$$

Using also the expansion (32.6.14), as the contribution from H_X we find

$$\mathcal{J}_i^{f(X)}(\boldsymbol{x}) = -\mathcal{J}_i^{b(X)}(\boldsymbol{x}) = \frac{4eJ_s^d}{\hbar\rho_0^2}(\partial_i\hat{\mathcal{P}}_x^{cl}\cdot\hat{\mathcal{P}}_y^{cl} - \partial_i\hat{\mathcal{P}}_y^{cl}\cdot\hat{\mathcal{P}}_x^{cl}). \tag{32.9.18}$$

This leads to the phase current in (32.9.22).

There is no contribution from H_{bias} precisely in the same reason as for H_D.

Finally, with respect to the contribution from the electric-field term H_E, from (32.9.10a) and (32.9.10c) we obtain

$$\hbar[\hat{\rho}^{cl(\alpha)}(\boldsymbol{k}), H_E^{cl}]_{PB} = \pi ec\int d^2q\, e^{-q^2\ell_B^2/4}A_0^\alpha(\boldsymbol{q})\hat{\rho}^\alpha(\boldsymbol{k}-\boldsymbol{q})\sin\left(\ell_B^2\frac{\boldsymbol{k}\wedge\boldsymbol{q}}{2}\right), \tag{32.9.19}$$

as corresponds to the monolayer formula (32.8.16). Hence, we obtain the standard formula for the Hall current in each layer,

$$\mathcal{J}_i^{\alpha(E)}(\boldsymbol{x}) = \frac{e^2\ell_B^2}{\hbar}\varepsilon_{ij}E_j\rho_0^\alpha, \tag{32.9.20}$$

on the incompressible ground state.

The Goldstone mode is described by a pair of the phase field $\vartheta(\boldsymbol{x})$ and the imbalance field $\sigma(\boldsymbol{x})$,

$$\hat{\mathcal{P}}_x^{cl}(\boldsymbol{x}) = \frac{1}{2}\rho_0\sqrt{1-\sigma(\boldsymbol{x})^2}\cos\vartheta(\boldsymbol{x}), \qquad \hat{\mathcal{P}}_y^{cl}(\boldsymbol{x}) = -\frac{1}{2}\rho_0\sqrt{1-\sigma(\boldsymbol{x})^2}\sin\vartheta(\boldsymbol{x}),$$
$$\hat{\mathcal{P}}_z^{cl}(\boldsymbol{x}) = \frac{1}{2}\rho_0\sigma(\boldsymbol{x}). \tag{32.9.21}$$

The Hall current is given by the sum of (32.9.20) and (32.9.18),

$$\mathcal{J}_i^f(\boldsymbol{x}) = \frac{eJ_s^d}{\hbar}(1-\sigma^2)\partial_i\vartheta + \frac{e^2\ell_B^2}{\hbar}\varepsilon_{ij}E_j^f\rho^f, \tag{32.9.22a}$$

$$\mathcal{J}_i^b(\boldsymbol{x}) = -\frac{eJ_s^d}{\hbar}(1-\sigma^2)\partial_i\vartheta + \frac{e^2\ell_B^2}{\hbar}\varepsilon_{ij}E_j^b\rho^b. \tag{32.9.22b}$$

It is prominent that the Goldstone mode affects the Hall current. We have already seen that it leads to anomalous Hall resistivity in Section 24.4.

We make a comment on the electric-field Hamiltonian (32.9.4e). A bilayer system consists of the two layers and the volume between them. Electrons are present only on the two planes. (The tunneling term has been introduced just to guarantee the charge conservation.) On the other hand, there exists the electric field also in the volume between the two layers. The energy due to the electric field $\boldsymbol{E}(\boldsymbol{x}, z)$ between the two layers is given by the sum of the Maxwell term and the source term,

$$H_E = \frac{\varepsilon}{2}\int d^2x dz\, \boldsymbol{E}^2(\boldsymbol{x}, z) + ec\int d^2x dz\, A_0(\boldsymbol{x}, z)\rho(\boldsymbol{x}, z), \tag{32.9.23}$$

where

$$E_x = c\partial_x A_0, \qquad E_y = c\partial_y A_0, \qquad E_z = \frac{cA_0^f - cA_0^b}{d}. \tag{32.9.24}$$

The reason why we have adopted (32.9.4e) rather than (32.9.23) is because they are equivalent in the equation of motion (32.9.2) since the electron density $\hat{\rho}^{\text{cl}}(\mathbf{x}, z)$ and the pseudospin density $\hat{\mathcal{P}}_a^{\text{cl}}(\mathbf{x}, z)$ are nonzero only on the two layers. However, it is necessary to use (32.9.23) when the equation of motion for the electric field is needed, as we did to determine the Hall current explicitly in the bilayer system: See (24.4.10).

Goldstone Mode

We have used the equations of motion for $\hat{\rho}$ and $\hat{\mathcal{P}}_z$ to define the Hall currents as well as the tunneling current. The remaining two equations of motion for $\hat{\mathcal{P}}_x$ and $\hat{\mathcal{P}}_y$ describe the dynamics of the Goldstone mode.

Making the derivative expansion and taking the nontrivial lowest order terms as before, we evaluate $d\hat{\mathcal{P}}_a/dt$ on the impressible ground state,

$$\frac{d}{dt}\hat{\mathcal{P}}_a^{\text{cl}} = \frac{16\pi^2\ell_B^4 J_s}{\hbar}\epsilon_{ab}(\hat{\mathcal{P}}_b^{\text{cl}}\nabla^2\hat{\mathcal{P}}_z^{\text{cl}}) - \frac{16\pi^2\ell_B^4 J_s^d}{\hbar}\epsilon_{ab}(\hat{\mathcal{P}}_z^{\text{cl}}\nabla^2\hat{\mathcal{P}}_b^{\text{cl}}) - \frac{4\pi\ell_B^2\epsilon_{\text{cap}}}{\hbar}\epsilon_{ab}\hat{\mathcal{P}}_b^{\text{cl}}\hat{\mathcal{P}}_z^{\text{cl}} + \frac{\epsilon_{\text{cap}}^{\text{D}}\bar{\sigma}_0}{\hbar}\epsilon_{ab}\hat{\mathcal{P}}_b^{\text{cl}} + \frac{\Delta_{\text{SAS}}}{\hbar}\delta_{ay}\hat{\mathcal{P}}_z^{\text{cl}} + \frac{e\ell_B^2}{2\hbar}\epsilon_{ij}(E_i^{\text{f}} + E_i^{\text{b}})\partial_j\hat{\mathcal{P}}_a^{\text{cl}}. \tag{32.9.25}$$

We combine them with $\hat{\mathcal{P}}_a^{\text{cl}}$ and then with $\epsilon_{ab}\hat{\mathcal{P}}_b^{\text{cl}}$. The first one is

$$\hat{\mathcal{P}}_a^{\text{cl}}\frac{d}{dt}\hat{\mathcal{P}}_a^{\text{cl}} = -\frac{16\pi^2\ell_B^4 J_s^d}{\hbar}\epsilon_{ab}\hat{\mathcal{P}}_z^{\text{cl}}(\hat{\mathcal{P}}_a^{\text{cl}}\nabla_b^2\hat{\mathcal{P}}^{\text{cl}}) + \frac{\Delta_{\text{SAS}}}{\hbar}\hat{\mathcal{P}}_y^{\text{cl}}\hat{\mathcal{P}}_z^{\text{cl}} + \frac{e\ell_B^2}{2\hbar}\epsilon_{ij}(E_i^{\text{f}} + E_i^{\text{b}})(\hat{\mathcal{P}}_a^{\text{cl}}\partial_j\hat{\mathcal{P}}_a^{\text{cl}}). \tag{32.9.26}$$

The second one is

$$\begin{aligned}\hbar\epsilon_{ab}\hat{\mathcal{P}}_a^{\text{cl}}\frac{d}{dt}\hat{\mathcal{P}}_b^{\text{cl}} =& -16\pi^2\ell_B^4 J_s(\hat{\mathcal{P}}_a^{\text{cl}}\hat{\mathcal{P}}_a^{\text{cl}}\nabla^2\hat{\mathcal{P}}_z^{\text{cl}}) + 16\pi^2\ell_B^4 J_s^d\hat{\mathcal{P}}_z^{\text{cl}}(\hat{\mathcal{P}}_a^{\text{cl}}\nabla^2\hat{\mathcal{P}}_a^{\text{cl}}) \\ &+ 4\pi\ell_B^2\epsilon_{\text{cap}}\hat{\mathcal{P}}_a^{\text{cl}}\hat{\mathcal{P}}_a^{\text{cl}}\hat{\mathcal{P}}_z^{\text{cl}} - \epsilon_{\text{cap}}^{\text{D}}\bar{\sigma}_0\hat{\mathcal{P}}_a^{\text{cl}}\hat{\mathcal{P}}_a^{\text{cl}} + \Delta_{\text{SAS}}\hat{\mathcal{P}}_x^{\text{cl}}\hat{\mathcal{P}}_z^{\text{cl}} \\ &+ \frac{e\ell_B^2}{2}\epsilon_{ij}(E_i^{\text{f}} + E_i^{\text{b}})\epsilon_{ab}\hat{\mathcal{P}}_a^{\text{cl}}\partial_j\hat{\mathcal{P}}_b^{\text{cl}}.\end{aligned} \tag{32.9.27}$$

We substitute the parametrization (32.9.21) of the pseudospin field into them,

$$\hbar\partial_t\sigma = -\frac{2}{\rho_0}J_s^d\partial_i[(1-\sigma^2)(\partial_i\vartheta)] + \Delta_{\text{SAS}}\sqrt{1-\sigma^2}\sin\vartheta + \frac{1}{2}e\ell_B^2\epsilon_{ij}(E_i^{\text{f}}+E_i^{\text{b}})\partial_j\sigma, \tag{32.9.28}$$

and

$$\hbar\partial_t\vartheta = \frac{2J_s}{\rho_0}(\nabla^2\sigma) + \frac{2J_s^d}{\rho_0}\sigma\left[\frac{(\partial_i\sigma)^2}{1-\sigma^2} + \frac{(\sigma\nabla^2\sigma)}{1-\sigma^2} + \frac{\sigma^2}{(1-\sigma^2)^2}(\partial_i\sigma)^2 + (\partial_i\vartheta)^2\right] - \epsilon_{\text{cap}}\sigma + \epsilon_{\text{cap}}^{\text{D}}\bar{\sigma}_0 - \frac{\sigma}{\sqrt{1-\sigma^2}}\Delta_{\text{SAS}}\cos\vartheta + \frac{e\ell_B^2}{2}\epsilon_{ij}(E_i^{\text{f}} + E_i^{\text{b}})(\partial_j\vartheta). \tag{32.9.29}$$

A comment is in order. Substituting the parametrization into $d\hat{\mathcal{P}}_z^{\text{cl}}/dt = [\hat{\mathcal{P}}_z^{\text{cl}}, H^{\text{cl}}]_{\text{PB}}$, we find an equation for σ and ϑ, but it is identical to (32.9.28). This is consistent with the fact that $\hat{\mathcal{P}}_a^{\text{cl}}\hat{\mathcal{P}}_a^{\text{cl}} = \rho_0^2/4$ is a constant of motion, $[\hat{\mathcal{P}}_z^{\text{cl}}\hat{\mathcal{P}}_z^{\text{cl}}, H^{\text{cl}}]_{\text{PB}} = 0$.

The equations (32.9.28) and (32.9.29) are written as

$$\hbar\partial_t\sigma(\boldsymbol{x}) = +\frac{2}{\rho_0}\frac{\delta\mathcal{H}^{\text{eff}}}{\delta\vartheta(\boldsymbol{x})}, \qquad \hbar\partial_t\vartheta(\boldsymbol{x}) = -\frac{2}{\rho_0}\frac{\delta\mathcal{H}^{\text{eff}}}{\delta\sigma(\boldsymbol{x})}, \tag{32.9.30}$$

where

$$\mathcal{H}^{\text{eff}} = \frac{1}{2}J_s^d(1-\sigma^2)(\partial_n\vartheta)^2 + \frac{1}{2}\left(J_s + \frac{\sigma^2}{1-\sigma^2}J_s^d\right)(\partial_n\sigma)^2 + \frac{\epsilon_{\text{cap}}\rho_0}{4}\sigma^2 - \frac{1}{2}\rho_0\epsilon_{\text{cap}}^{\text{D}}\bar{\sigma}_0\sigma$$
$$- \frac{\Delta_{\text{SAS}}\rho_0}{2}\sqrt{1-\sigma^2}\cos\vartheta - \frac{1}{4}\rho_0 e\ell_B^2 c(A_0^{\text{f}} + A_0^{\text{b}})\epsilon_{ij}(\partial_i\sigma)(\partial_j\vartheta). \tag{32.9.31}$$

We can eliminate the parameter $\bar{\sigma}_0$ with the use of the relation (32.5.16), or

$$\epsilon_{\text{cap}}^{\text{D}}\bar{\sigma}_0 = \epsilon_{\text{cap}}\sigma_0 + \frac{\sigma_0}{\sqrt{1-\sigma_0^2}}\Delta_{\text{SAS}}. \tag{32.9.32}$$

We then have

$$\mathcal{H}^{\text{eff}} = \frac{1}{2}J_s^d(1-\sigma^2)(\partial_n\vartheta)^2 + \frac{1}{2}\left(J_s + \frac{\sigma^2}{1-\sigma^2}J_s^d\right)(\partial_n\sigma)^2 + \frac{\epsilon_{\text{cap}}\rho_0}{4}(\sigma-\sigma_0)^2$$
$$- \frac{\Delta_{\text{SAS}}\rho_0}{2}\left(\sqrt{1-\sigma^2}\cos\vartheta + \frac{\sigma_0}{\sqrt{1-\sigma_0^2}}\sigma\right) - \frac{1}{4}\rho_0 e\ell_B^2 c(A_0^{\text{f}} + A_0^{\text{b}})\epsilon_{ij}(\partial_i\sigma)(\partial_j\vartheta). \tag{32.9.33}$$

It agrees with the effective Hamiltonian (32.6.27) derived previously but for the external electric potential ($A_0^{\text{f}} = A_0^{\text{b}} = 0$): See also (24.2.4).

APPENDICES

A ENERGY SCALES

We list typical energy scales together with their symbols we use in this book. We start with a remark that electrons in states near the conduction band minimum of a semiconductor behave like free electrons though band effects renormalize the electron mass and the gyromagnetic factor.[14] The band mass M in a GaAs semiconductor is considerably smaller than the electron mass m_e in the vacuum; $M \simeq 0.067m_e$. The gyromagnetic factor is $g^* = 2$ in the vacuum, but $g^* \simeq -0.44$ in a GaAs semiconductor.

In magnetic field $\boldsymbol{B} = (0, 0, -B_\perp)$ an electron executes cyclotron motion with magnetic length ℓ_B determined by the perpendicular component $B_\perp$,

$$\ell_B = \sqrt{\frac{\hbar}{eB_\perp}} \simeq \frac{256}{\sqrt{B_\perp(\text{Tesla})}} \quad \text{Å}. \tag{A.1}$$

The Coulomb energy associated with this length scale is

$$E_C^0 = \frac{e^2}{4\pi\varepsilon\ell_B} \simeq \frac{13000}{\ell_B(\text{Å})} \quad \text{K} \simeq 50.8\sqrt{B_\perp(\text{Tesla})} \quad \text{K}, \tag{A.2}$$

where we have used $\varepsilon \simeq 12.9\varepsilon_0$ for the background dielectric constant in GaAs semiconductors. Electrons fill Landau levels successively from the lowest Landau level. At the filling factor ν the magnetic length is

$$\ell_B = \sqrt{\frac{\nu}{2\pi\rho_0}} = 126 \times \sqrt{\frac{\nu}{n}} \quad \text{Å}, \tag{A.3}$$

when the electron density of the sample is $\rho_0 = n \times 10^{15}\text{m}^{-2}$. The gap energy between the two Landau levels is the cyclotron energy

$$\hbar\omega_c = \frac{\hbar e B_\perp}{M} \simeq 20.0 B_\perp(\text{Tesla}) \quad \text{K}, \tag{A.4}$$

where M is the band electron mass. Thus, the basic quantities are given in terms of the electron density.

Monolayer QH System

An electron has the up-spin and down-spin states. One Landau level is split into two levels with the Zeeman gap energy,

$$\Delta_Z \equiv |g^* \mu_B B| \simeq 0.296 B(\text{Tesla}) \quad \text{K}, \tag{A.5}$$

where μ_B is the Bohr magneton and $B = |\boldsymbol{B}|$. When $B = B_\perp$, $\Delta_Z = 2(e\hbar/2m_e) = \hbar\omega_c$ in the Dirac theory, but $\Delta_Z \simeq \hbar\omega_c/66$ in a GaAs semiconductor.

When two electrons get close enough an exchange integral of their wave functions gives rise to an exchange energy,

$$J_s = \frac{1}{16\sqrt{2\pi}} E_C^0 \simeq 1.27\sqrt{B_\perp(\text{Tesla})} \quad \text{K}, \tag{A.6}$$

as in the ferromagnet: See Section 4.1. This gives the spin stiffness J_s in the nonlinear sigma model (9.1.15).

The simplest excitation is an electron-hole pair, whose Coulomb energy is given by $\Delta_{\text{pair}} = 2\epsilon_X$ with

$$\epsilon_X = \frac{1}{2}\sqrt{\frac{\pi}{2}} E_C^0 \simeq 31.8\sqrt{B_\perp(\text{Tesla})} \quad \text{K}. \tag{A.7}$$

It gives the scale of the activation energy. The formulas for (A.6) and (A.7) are derived in (32.6.9) and (32.1.12), respectively.

Bilayer QH System

There exists an additional Coulomb energy scale associated with the layer separation d. The basic Coulomb potentials are

$$V^{\text{ff}}(\boldsymbol{q}) = V^{\text{bb}}(\boldsymbol{q}) = \frac{e^2}{4\pi\varepsilon|\boldsymbol{q}|}, \qquad V^{\text{fb}}(\boldsymbol{q}) = V^{\text{bf}}(\boldsymbol{q}) = \frac{e^2}{4\pi\varepsilon|\boldsymbol{q}|} e^{-|\boldsymbol{q}|d}, \tag{A.8}$$

for electrons in the same layers and in the different layers. We introduce

$$V^{\pm}(\boldsymbol{q}) = \frac{e^2}{8\pi\varepsilon|\boldsymbol{q}|}\left(1 \pm e^{-|\boldsymbol{q}|d}\right). \tag{A.9}$$

They are modified into

$$V_D(\boldsymbol{k}) = \frac{e^2}{4\pi\varepsilon|\boldsymbol{q}|} e^{-\ell_B^2 q^2/2}, \qquad V_D^{\pm}(\boldsymbol{q}) = \frac{e^2}{8\pi\varepsilon|\boldsymbol{q}|}\left(1 \pm e^{-|\boldsymbol{q}|d}\right) e^{-\ell_B^2 q^2/2} \tag{A.10}$$

for electrons confined to the lowest Landau level.

The exchange interaction is described by the short-range potentials,

$$V_X(\boldsymbol{p}) = \frac{\ell_B^2}{\pi}\int d^2k\, e^{-i\ell_B^2 \boldsymbol{p}\wedge\boldsymbol{k}} V_D(\boldsymbol{k}) = \frac{\sqrt{2\pi}e^2\ell_B}{4\pi\varepsilon} I_0(\ell_B^2 \boldsymbol{p}^2/4) e^{-\ell_B^2 \boldsymbol{p}^2/4}, \tag{A.11a}$$

$$V_X^{\pm}(\boldsymbol{p}) = \frac{\sqrt{2\pi}e^2\ell_B}{8\pi\varepsilon} I_0(\ell_B^2 \boldsymbol{p}^2/4) e^{-\ell_B^2 \boldsymbol{p}^2/4} \pm \frac{e^2\ell_B^2}{4\pi\varepsilon}\int_0^\infty dk\, e^{-\frac{1}{2}\ell_B^2 k^2 - dk} J_0(\ell_B^2|\boldsymbol{p}|k). \tag{A.11b}$$

We also introduce

$$V_X = V_X^+ + V_X^-, \qquad V_X^d = V_X^+ - V_X^-. \tag{A.12}$$

Corresponding to (A.7) and (A.6) we have

$$\epsilon_X^\pm = \frac{1}{2}\left[1 \pm e^{d^2/2\ell_B^2}\text{erfc}\left(d/\sqrt{2}\ell_B\right)\right]\epsilon_X, \tag{A.13}$$

and

$$J_s^\pm = \frac{1}{2}\left(J_s \pm J_s^d\right), \tag{A.14}$$

respectively, with

$$\frac{J_s^d}{J_s} = -\sqrt{\frac{2}{\pi}}\frac{d}{\ell_B} + \left(1 + \frac{d^2}{\ell_B^2}\right)e^{d^2/2\ell_B^2}\text{erfc}\left(d/\sqrt{2}\ell_B\right). \tag{A.15}$$

See (30.5.12b) and (32.6.16) for $\epsilon_X^\pm$ and $J_s^\pm$, respectively.

We estimate numerically various quantities at $\nu = 1$ for typical sample parameters,

$$\rho_0 \simeq 1.26 \times 10^{15}\ \text{m}^{-2}, \quad d \simeq 2.1 \times 10^{-8}\ \text{m}, \quad \Delta_{\text{SAS}} \simeq 0.8\ \text{K}, \tag{A.16}$$

for which

$$\ell_B \simeq 1.12 \times 10^{-8}\ \text{m}, \quad J_s \simeq 2.90\ \text{K}, \quad J_s^d \simeq 0.32\ \text{K}. \tag{A.17}$$

We have used

$$\begin{aligned} e &= 1.6022 \times 10^{-19}, \quad \mu_{\text{m}} \simeq 4\pi \times 10^{-7}, \\ \hbar &= 1.0546 \times 10^{-34}, \quad \text{K} = 1.3807 \times 10^{-23} \end{aligned} \tag{A.18}$$

in the MKSA unit.

B Hausdorff Formulas

We list several useful formulas related to the Hausdorff Formula.

Formula A

For arbitrary operators A and B the following formula holds,

$$e^{\lambda A}Be^{-\lambda A} = \sum \frac{\lambda^n}{n!}(\text{ad}_A)^n \cdot B = B + \lambda[A, B] + \frac{\lambda^2}{2}[A, [A, B]] + \cdots. \tag{B.1}$$

Proof: Setting

$$f(\lambda) = e^{\lambda A}Be^{-\lambda A}, \tag{B.2}$$

we find

$$\frac{d}{d\lambda}f(\lambda) = [A, f(\lambda)]. \tag{B.3}$$

Hence,

$$\left(\frac{d}{d\lambda}\right)^n f(\lambda) = \left(\frac{d}{d\lambda}\right)^{n-1} [A, f(\lambda)] = \left(\frac{d}{d\lambda}\right)^{n-2} [A, [A, f(\lambda)]] = \cdots$$
$$= [A, [A, \cdots [A, f(\lambda)] \cdots]] \equiv (\mathrm{ad}_A)^n \cdot f(\lambda). \tag{B.4}$$

Now, the Taylor expansion of $f(\lambda)$ yields

$$f(\lambda) = \sum_{n=0}^{\infty} \frac{\lambda^n}{n!} f^{(n)}(0) = \sum_{n=0}^{\infty} \frac{\lambda^n}{n!} (\mathrm{ad}_A)^n \cdot f(0) = \sum_{n=0}^{\infty} \frac{\lambda^n}{n!} (\mathrm{ad}_A)^n \cdot B. \tag{B.5}$$

Formula B

For arbitrary operators A and B the following formula holds,

$$e^A e^B e^{-A} = \exp\left(e^A B e^{-A}\right). \tag{B.6}$$

Proof: Making the power expansion we obtain

$$\exp\left(e^A B e^{-A}\right) = \sum_{n=0}^{\infty} \frac{1}{n!} \left(e^A B e^{-A}\right)^n = e^A \left(\sum_{n=0}^{\infty} \frac{1}{n!} B^n\right) e^{-A} = e^A e^B e^{-A}. \tag{B.7}$$

Formula C

If $[A, B]$ is a c-number the following formula holds,

$$e^A e^B = e^{[A,B]} e^B e^A = e^{[A,B]/2} e^{A+B}. \tag{B.8}$$

Proof: Setting

$$f(\lambda) = e^{\lambda A} e^{\lambda B} e^{-\frac{\lambda^2}{2}[A,B]}, \tag{B.9}$$

we find

$$\frac{d}{d\lambda} f(\lambda) = \left(A + e^{\lambda A} B e^{-\lambda A} - \lambda [A, B]\right) f(\lambda) = (A + B) f(\lambda), \tag{B.10}$$

where we have used (B.1) to expand $e^{\lambda A} B e^{-\lambda A}$. Its integration yields

$$f(\lambda) = e^{\lambda(A+B)}. \tag{B.11}$$

From (B.9) and (B.11), setting $\lambda = 1$, we obtain

$$e^A e^B e^{-[A,B]/2} = e^{A+B}, \tag{B.12}$$

which is equivalent to (B.8).

C GROUP SU(2) AND PAULI MATRICES

We examine some properties of the Pauli matrix τ_a,

$$\tau_x = \begin{pmatrix} 0 & 1 \\ 1 & 0 \end{pmatrix}, \qquad \tau_y = \begin{pmatrix} 0 & -i \\ i & 0 \end{pmatrix}, \qquad \tau_z = \begin{pmatrix} 1 & 0 \\ 0 & -1 \end{pmatrix}. \tag{C.1}$$

It is characterized by the properties

$$\tau_a \tau_b = i\varepsilon_{abc}\tau_c, \quad \text{and} \quad (\tau_x)^2 = (\tau_y)^2 = (\tau_z)^2 = 1. \tag{C.2}$$

For any operator D satisfying $D^2 = 1$ there is a formula

$$e^{i\lambda D} = \cos\lambda + iD\sin\lambda. \tag{C.3}$$

Therefore,

$$e^{-i\lambda\tau_a}\tau_b e^{i\lambda\tau_a} = \tau_b(\cos^2\lambda - \sin^2\lambda) + \varepsilon_{abc}\tau_c \sin 2\lambda, \tag{C.4}$$

where $a \neq b$. Choosing $\lambda = \pi/4$ we find

$$e^{-i\pi\tau_a/4}\tau_b e^{i\pi\tau_a/4} = \varepsilon_{abc}\tau_c, \qquad (a \neq b), \tag{C.5}$$

and choosing $\lambda = \pi/2$ we find

$$e^{-i\pi\tau_a/2}\tau_b e^{i\pi\tau_a/2} = -\tau_b, \qquad (a \neq b). \tag{C.6}$$

Using them it is easy to show that

$$T^{-1}\tau_x T = \tau_z, \quad T^{-1}\tau_y T = \tau_x, \quad T^{-1}\tau_z T = \tau_y, \tag{C.7}$$

when we choose

$$T = \exp\left(i\frac{\pi}{4}\tau_z\right)\exp\left(i\frac{\pi}{4}\tau_x\right)\exp\left(i\frac{\pi}{2}\tau_z\right). \tag{C.8}$$

The operator T makes a cyclic rotation of the spin operator S_a.

D GROUPS SU(N) AND SU(2N)

The special unitary group SU(N) has $(N^2 - 1)$ generators. According to the standard notation from elementary particle physics,[23] we denote them as λ_A, $A = 1, 2, \cdots, N^2 - 1$, and normalize them as

$$\text{Tr}(\lambda_A \lambda_B) = 2\delta_{AB}. \tag{D.1}$$

They are characterized by

$$[\lambda_A, \lambda_B] = 2if_{ABC}\lambda_C, \qquad \{\lambda_A, \lambda_B\} = \frac{4}{N} + 2d_{ABC}\lambda_C, \tag{D.2}$$

where f_{ABC} and d_{ABC} are the structure constants of SU(N). We have $\lambda_A = \tau_A$ (the Pauli matrix) with $f_{ABC} = \varepsilon_{ABC}$ and $d_{ABC} = 0$ in the case of SU(2).

Group SU(3)

Because SU(2) is a subgroup of SU(3), three generators are chosen to be Pauli matrices by extending them to three dimensions:

$$\lambda_1 = \begin{pmatrix} 0 & 1 & 0 \\ 1 & 0 & 0 \\ 0 & 0 & 0 \end{pmatrix}, \quad \lambda_2 = \begin{pmatrix} 0 & -i & 0 \\ i & 0 & 0 \\ 0 & 0 & 0 \end{pmatrix}, \quad \lambda_3 = \begin{pmatrix} 1 & 0 & 0 \\ 0 & -1 & 0 \\ 0 & 0 & 0 \end{pmatrix}. \tag{D.3}$$

We construct λ_4, λ_6 from λ_1, and λ_5, λ_7 from λ_2, by shifting the nonzero elements:

$$\begin{aligned} \lambda_1 = \begin{pmatrix} 0 & 1 & 0 \\ 1 & 0 & 0 \\ 0 & 0 & 0 \end{pmatrix} \quad &\Rightarrow \quad \lambda_4 = \begin{pmatrix} 0 & 0 & 1 \\ 0 & 0 & 0 \\ 1 & 0 & 0 \end{pmatrix}, \quad \lambda_6 = \begin{pmatrix} 0 & 0 & 0 \\ 0 & 0 & 1 \\ 0 & 1 & 0 \end{pmatrix}, \\ \lambda_2 = \begin{pmatrix} 0 & -i & 0 \\ i & 0 & 0 \\ 0 & 0 & 0 \end{pmatrix} \quad &\Rightarrow \quad \lambda_5 = \begin{pmatrix} 0 & 0 & -i \\ 0 & 0 & 0 \\ i & 0 & 0 \end{pmatrix}, \quad \lambda_7 = \begin{pmatrix} 0 & 0 & 0 \\ 0 & 0 & -i \\ 0 & i & 0 \end{pmatrix}. \end{aligned} \tag{D.4}$$

Finally, λ_8 is a traceless diagonal matrix:

$$\lambda_8 = \frac{1}{\sqrt{3}} \begin{pmatrix} 1 & 0 & 0 \\ 0 & 1 & 0 \\ 0 & 0 & -2 \end{pmatrix}. \tag{D.5}$$

We can check their algebraic structure to satisfy (D.2).

Group SU(4)

Because SU(3) is a subgroup of SU(4), eight generators are chosen to be the generators of SU(3) by extending them to four dimensions:

$$\lambda_A = \begin{pmatrix} \lambda_A & 0 \\ 0 & 0 \end{pmatrix} \qquad \text{for} \quad A = 1, 2, \cdots, 8. \tag{D.6}$$

We construct λ_9, λ_{11}, λ_{13} from λ_1, and λ_{10}, λ_{12}, λ_{14} from λ_2, by shifting the nonzero elements:

$$\begin{aligned} \lambda_9 &= \begin{pmatrix} 0 & 0 & 0 & 1 \\ 0 & 0 & 0 & 0 \\ 0 & 0 & 0 & 0 \\ 1 & 0 & 0 & 0 \end{pmatrix}, \quad \lambda_{11} = \begin{pmatrix} 0 & 0 & 0 & 0 \\ 0 & 0 & 0 & 1 \\ 0 & 0 & 0 & 0 \\ 0 & 1 & 0 & 0 \end{pmatrix}, \quad \lambda_{13} = \begin{pmatrix} 0 & 0 & 0 & 0 \\ 0 & 0 & 0 & 0 \\ 0 & 0 & 0 & 1 \\ 0 & 0 & 1 & 0 \end{pmatrix}, \\ \lambda_{10} &= \begin{pmatrix} 0 & 0 & 0 & i \\ 0 & 0 & 0 & 0 \\ 0 & 0 & 0 & 0 \\ -i & 0 & 0 & 0 \end{pmatrix}, \quad \lambda_{12} = \begin{pmatrix} 0 & 0 & 0 & 0 \\ 0 & 0 & 0 & i \\ 0 & 0 & 0 & 0 \\ 0 & -i & 0 & 0 \end{pmatrix}, \quad \lambda_{14} = \begin{pmatrix} 0 & 0 & 0 & 0 \\ 0 & 0 & 0 & 0 \\ 0 & 0 & 0 & i \\ 0 & 0 & -i & 0 \end{pmatrix}. \end{aligned} \tag{D.7}$$

Finally, λ_{15} is a traceless diagonal matrix analogous to (D.5):

$$\lambda_{15} = \frac{1}{\sqrt{6}} \begin{pmatrix} 1 & 0 & 0 & 0 \\ 0 & 1 & 0 & 0 \\ 0 & 0 & 1 & 0 \\ 0 & 0 & 0 & -3 \end{pmatrix}. \tag{D.8}$$

The standard matrix representation of SU(N) is similarly constructed.

Another representation is more useful in the bilayer SU(4) systems to emphasize the spin SU(2) symmetry in the front and back layers. We embed SU(2)⊗SU(2) into SU(4) with the four-component electron field $\boldsymbol{\psi} = (\psi^{\text{f}\uparrow}, \psi^{\text{f}\downarrow}, \psi^{\text{b}\uparrow}, \psi^{\text{b}\downarrow})$. The spin SU(2) is naively embedded,

$$\tau_x^{\text{spin}} = \begin{pmatrix} 0 & 1 & 0 & 0 \\ 1 & 0 & 0 & 0 \\ 0 & 0 & 0 & 1 \\ 0 & 0 & 1 & 0 \end{pmatrix}, \quad \tau_y^{\text{spin}} = \begin{pmatrix} 0 & -i & 0 & 0 \\ i & 0 & 0 & 0 \\ 0 & 0 & 0 & -i \\ 0 & 0 & i & 0 \end{pmatrix}, \quad \tau_z^{\text{spin}} = \begin{pmatrix} 1 & 0 & 0 & 0 \\ 0 & -1 & 0 & 0 \\ 0 & 0 & 1 & 0 \\ 0 & 0 & 0 & -1 \end{pmatrix}. \tag{D.9}$$

The pseudospin is the symmetry of the front and back layers. By shifting the nonzero elements in this representation, we define

$$\tau_x^{\text{ppin}} = \begin{pmatrix} 0 & 0 & 1 & 0 \\ 0 & 0 & 0 & 1 \\ 1 & 0 & 0 & 0 \\ 0 & 1 & 0 & 0 \end{pmatrix}, \quad \tau_y^{\text{ppin}} = \begin{pmatrix} 0 & 0 & -i & 0 \\ 0 & 0 & 0 & -i \\ i & 0 & 0 & 0 \\ 0 & i & 0 & 0 \end{pmatrix}, \quad \tau_z^{\text{ppin}} = \begin{pmatrix} 1 & 0 & 0 & 0 \\ 0 & 1 & 0 & 0 \\ 0 & 0 & -1 & 0 \\ 0 & 0 & 0 & -1 \end{pmatrix}. \tag{D.10}$$

The remaining nine matrices are given by $\tau_a^{\text{spin}} \tau_b^{\text{ppin}}$:

$$\tau_x^{\text{spin}}\tau_x^{\text{ppin}} = \begin{pmatrix} 0 & 0 & 0 & 1 \\ 0 & 0 & 1 & 0 \\ 0 & 1 & 0 & 0 \\ 1 & 0 & 0 & 0 \end{pmatrix}, \quad \tau_x^{\text{spin}}\tau_y^{\text{ppin}} = \begin{pmatrix} 0 & 0 & 0 & -i \\ 0 & 0 & -i & 0 \\ 0 & i & 0 & 0 \\ i & 0 & 0 & 0 \end{pmatrix}, \quad \tau_x^{\text{spin}}\tau_z^{\text{ppin}} = \begin{pmatrix} 0 & 1 & 0 & 0 \\ 1 & 0 & 0 & 0 \\ 0 & 0 & 0 & -1 \\ 0 & 0 & -1 & 0 \end{pmatrix},$$

$$\tau_y^{\text{spin}}\tau_x^{\text{ppin}} = \begin{pmatrix} 0 & 0 & 0 & -i \\ 0 & 0 & i & 0 \\ 0 & -i & 0 & 0 \\ i & 0 & 0 & 0 \end{pmatrix}, \quad \tau_y^{\text{spin}}\tau_y^{\text{ppin}} = \begin{pmatrix} 0 & 0 & 0 & -1 \\ 0 & 0 & 1 & 0 \\ 0 & 1 & 0 & 0 \\ -1 & 0 & 0 & 0 \end{pmatrix}, \quad \tau_y^{\text{spin}}\tau_z^{\text{ppin}} = \begin{pmatrix} 0 & -i & 0 & 0 \\ i & 0 & 0 & 0 \\ 0 & 0 & 0 & i \\ 0 & 0 & -i & 0 \end{pmatrix},$$

$$\tau_z^{\text{spin}}\tau_x^{\text{ppin}} = \begin{pmatrix} 0 & 0 & 1 & 0 \\ 0 & 0 & 0 & -1 \\ 1 & 0 & 0 & 0 \\ 0 & -1 & 0 & 0 \end{pmatrix}, \quad \tau_z^{\text{spin}}\tau_y^{\text{ppin}} = \begin{pmatrix} 0 & 0 & -i & 0 \\ 0 & 0 & 0 & i \\ i & 0 & 0 & 0 \\ 0 & -i & 0 & 0 \end{pmatrix}, \quad \tau_z^{\text{spin}}\tau_z^{\text{ppin}} = \begin{pmatrix} 1 & 0 & 0 & 0 \\ 0 & -1 & 0 & 0 \\ 0 & 0 & -1 & 0 \\ 0 & 0 & 0 & 1 \end{pmatrix}. \tag{D.11}$$

There hold relations

$$\begin{aligned}\tau_a^{\text{spin}}\tau_b^{\text{spin}} &= \delta_{ab} + i\varepsilon_{abc}\tau_c^{\text{spin}},\\ \tau_a^{\text{ppin}}\tau_b^{\text{ppin}} &= \delta_{ab} + i\varepsilon_{abc}\tau_c^{\text{ppin}},\\ \tau_a^{\text{spin}}\tau_b^{\text{ppin}} &= \tau_b^{\text{ppin}}\tau_a^{\text{spin}}.\end{aligned} \tag{D.12}$$

They are related with the standard SU(4) generators as

$$\begin{aligned}
\tau_x^{\text{spin}} &= \lambda_1 + \lambda_{13}, \quad \tau_y^{\text{spin}} = \lambda_2 + \lambda_{14}, \quad \tau_z^{\text{spin}} = \lambda_3 - \frac{1}{\sqrt{3}}\lambda_8 + \frac{\sqrt{6}}{3}\lambda_{15},\\
\tau_x^{\text{ppin}} &= \lambda_4 + \lambda_{11}, \quad \tau_y^{\text{ppin}} = \lambda_5 + \lambda_{12}, \quad \tau_z^{\text{ppin}} = \frac{2}{\sqrt{3}}\lambda_8 + \frac{2}{\sqrt{6}}\lambda_{15},\\
\tau_x^{\text{spin}}\tau_x^{\text{ppin}} &= \lambda_6 + \lambda_9, \quad \tau_x^{\text{spin}}\tau_y^{\text{ppin}} = \lambda_{10} + \lambda_7, \quad \tau_x^{\text{spin}}\tau_z^{\text{ppin}} = \lambda_1 - \lambda_{13},\\
\tau_y^{\text{spin}}\tau_x^{\text{ppin}} &= \lambda_{10} - \lambda_7, \quad \tau_y^{\text{spin}}\tau_y^{\text{ppin}} = \lambda_6 - \lambda_9, \quad \tau_y^{\text{spin}}\tau_z^{\text{ppin}} = \lambda_2 - \lambda_{14},\\
\tau_z^{\text{spin}}\tau_x^{\text{ppin}} &= \lambda_4 - \lambda_{11}, \quad \tau_z^{\text{spin}}\tau_y^{\text{ppin}} = \lambda_5 - \lambda_{12}, \quad \tau_z^{\text{spin}}\tau_z^{\text{ppin}} = \lambda_3 + \frac{1}{\sqrt{3}}\lambda_8 - \frac{\sqrt{6}}{3}\lambda_{15}.
\end{aligned} \tag{D.13}$$

The normalization condition reads

$$\text{Tr}(T_{ab}T_{cd}) = 4\delta_{ac}\delta_{bd}, \tag{D.14}$$

where $T_{a0} \equiv \tau_a^{\text{spin}}$, $T_{0a} \equiv \tau_a^{\text{ppin}}$, $T_{ab} \equiv \tau_a^{\text{spin}}\tau_b^{\text{ppin}}$. It is different from the standard one (D.1). A similar construction of SU(2N) generators is made by embedding SU(2)×SU(2)× · · · ×SU(2), which is useful to analyze the N-layer system with spins.

E Cauchy-Riemann Equations

An arbitrary function $f(\boldsymbol{x})$ of $\boldsymbol{x} = (x, y)$ may be written in terms of the complex number z and its conjugate number z^*, $f(\boldsymbol{x}) = f(z, z^*)$. The analytic function $\omega(z)$ depends only on the complex number. Separating it into the real and imaginary parts, $\omega(z) = \omega_{\Re}(\boldsymbol{x}) + i\omega_{\Im}(\boldsymbol{x})$, we find

$$\begin{aligned}\frac{\partial\omega(z)}{\partial z^*} &= \left(\frac{\partial}{\partial x} + i\frac{\partial}{\partial y}\right)(\omega_{\Re} + i\omega_{\Im})\\ &= (\partial_x\omega_{\Re} - \partial_y\omega_{\Im}) + i\,(\partial_y\omega_{\Re} + \partial_x\omega_{\Im}) = 0,\end{aligned} \tag{E.1}$$

or

$$\partial_j\omega_{\Re} = \varepsilon_{jk}\partial_k\omega_{\Im}. \tag{E.2}$$

We repeat the same step for the decomposition $\omega(z) = |\omega|e^{i\chi}$,

$$\frac{\partial\omega(z)}{\partial z^*} = \left(\frac{\partial}{\partial x} + i\frac{\partial}{\partial y}\right)|\omega|e^{i\chi}$$
$$= \omega(z)\left\{(\partial_x \ln|\omega| - \partial_y\chi) + i\,(\partial_y \ln|\omega| + \partial_x\chi)\right\} = 0, \tag{E.3}$$

or

$$\partial_j \ln|\omega(z)| = \varepsilon_{jk}\partial_k\chi(\boldsymbol{x}). \tag{E.4}$$

Applying this equation for the analytic function $\omega(z) = z = re^{i\theta}$, we find that

$$\partial_j \ln r = \varepsilon_{jk}\partial_k\theta, \tag{E.5}$$

or

$$\partial_x \ln r = \partial_y\theta, \qquad \partial_y \ln r = -\partial_x\theta. \tag{E.6}$$

It follows from (E.5) that

$$\varepsilon_{jk}\partial_j\partial_k\theta(\boldsymbol{x}) = \partial_j\partial_j \ln r = \nabla^2 \ln r = 2\pi\delta(\boldsymbol{x}). \tag{E.7}$$

The last equation gives the Green function of the Poisson equation in the 2-dimensional space.

F Green Function

We study the Green function of the Poisson equation in the 3-dimensional space. It satisfies

$$\triangle G(\boldsymbol{x},\boldsymbol{y}) = \delta(\boldsymbol{x}-\boldsymbol{y}). \tag{F.1}$$

To solve it, we introduce a function

$$F(\boldsymbol{x},\epsilon) = -\frac{1}{4\pi}\Delta\frac{1}{|\boldsymbol{x}|+\epsilon}. \tag{F.2}$$

By a direct calculation in the spherical coordinate we find

$$F(\boldsymbol{x},\epsilon) = \frac{\epsilon}{2\pi r(r+\epsilon)^3}, \tag{F.3}$$

where $r = |\boldsymbol{x}|$. On one hand, we obtain

$$\lim_{\epsilon\to+0} F(\boldsymbol{x},\epsilon) = 0 \qquad \text{for} \quad r \neq 0. \tag{F.4}$$

On the other hand, integrating it, we obtain

$$\int d^3x\, F(\boldsymbol{x},\epsilon) = 2\epsilon\int_0^\infty dr\frac{r}{(r+\epsilon)^3} = 1. \tag{F.5}$$

These two equations imply

$$\lim_{\epsilon\to+0} F(\boldsymbol{x},\epsilon) = \delta(\boldsymbol{x}). \tag{F.6}$$

Consequently, (F.1) is solved by

$$G(\boldsymbol{x},\boldsymbol{y}) = \lim_{\epsilon\to+0} F(\boldsymbol{x}-\boldsymbol{y},\epsilon) = -\frac{1}{4\pi}\Delta\frac{1}{|\boldsymbol{x}-\boldsymbol{y}|}. \tag{F.7}$$

This is the Green function of the Poisson equation.

G Bogoliubov Transformation

The Bogoliubov transformation relates two sets of annihilation and creation operators, $(a, a^\dagger)$ and $(b, b^\dagger)$. The most general expression is

$$b = a\cosh\vartheta + a^\dagger e^{i\beta}\sinh\vartheta, \qquad b^\dagger = a^\dagger\cosh\vartheta + ae^{-i\beta}\sinh\vartheta, \tag{G.1}$$

where ϑ and β are real constants. It is generated as

$$b = e^{-iG}ae^{iG}, \qquad b^\dagger = e^{-iG}a^\dagger e^{iG} \tag{G.2}$$

with

$$G = \frac{i}{2}\vartheta\left(e^{-i\beta}aa - e^{i\beta}a^\dagger a^\dagger\right). \tag{G.3}$$

We may verify it by a direct expansion of (G.2),

$$\begin{aligned} b = e^{-iG}ae^{iG} &= a + i[a,G] + \frac{i^2}{2!}[[a,G],G]] + \cdots \\ &= a + \vartheta e^{i\beta}a^\dagger + \frac{1}{2!}\vartheta a + \frac{1}{3!}\vartheta^3 e^{i\beta}a^\dagger + \cdots \\ &= a\cosh\vartheta + a^\dagger e^{i\beta}\sinh\vartheta. \end{aligned} \tag{G.4}$$

Similarly we have

$$e^{iG}ae^{-iG} = a\cosh\vartheta - a^\dagger e^{i\beta}\sinh\vartheta. \tag{G.5}$$

The vacuum state $|0\rangle$ of the a-mode is transformed into

$$|0\rangle\!\rangle \equiv e^{-iG}|0\rangle. \tag{G.6}$$

It is the vacuum of the b-mode,

$$b|0\rangle\!\rangle = e^{-iG}ae^{iG}e^{-iG}|0\rangle = e^{-iG}a|0\rangle = 0. \tag{G.7}$$

Hence,

$$(a\cosh\vartheta + a^\dagger\sinh\vartheta)|0\rangle\!\rangle = 0, \tag{G.8}$$

or

$$\left(\frac{\partial}{\partial a^\dagger} + a^\dagger e^{i\beta}\tanh\vartheta\right)|0\rangle\!\rangle = 0. \tag{G.9}$$

Solving this we obtain

$$|0\rangle\!\rangle \equiv e^{-iG}|0\rangle = C\exp\left(-\frac{1}{2}a^\dagger a^\dagger e^{i\beta}\tanh\vartheta\right)|0\rangle. \tag{G.10}$$

The normalization factor is given by

$$C(\vartheta) = \langle 0|e^{-iG}|0\rangle. \tag{G.11}$$

To calculate it, we take the derivative with respect to ϑ of $C(\vartheta)$,

$$\begin{aligned}\frac{dC(\vartheta)}{d\vartheta} &= \frac{1}{2}e^{i\beta}\langle 0|aae^{-iG}|0\rangle = \frac{1}{2}e^{i\beta}\langle 0|e^{-iG}e^{iG}ae^{-iG}e^{iG}ae^{-iG}|0\rangle \\ &= \frac{1}{2}e^{i\beta}\langle 0|e^{-iG}(a\cosh\vartheta - a^\dagger\sinh\vartheta)(a\cosh\vartheta - a^\dagger\sinh\vartheta)|0\rangle \\ &= \frac{1}{2}e^{i\beta}\left(\cosh\vartheta\sinh\vartheta\langle 0|e^{-iG}|0\rangle + \sinh^2\vartheta\langle 0|e^{-iG}a^\dagger a^\dagger|0\rangle\right),\end{aligned} \tag{G.12}$$

where we have used $\langle 0|a^\dagger = 0$, $a|0\rangle = 0$ and (G.5). We also have

$$\frac{dC(\vartheta)}{d\vartheta} = -\frac{1}{2}e^{-i\beta}\langle 0|e^{-iG}a^\dagger a^\dagger|0\rangle. \tag{G.13}$$

Combining (G.12) and (G.13) we derive a differential equation,

$$\frac{dC(\vartheta)}{d\vartheta} = -\frac{1}{2}C(\vartheta)\frac{e^{i\beta}\cosh\vartheta\sinh\vartheta}{1-e^{i\beta}\sinh^2\vartheta}. \tag{G.14}$$

When $\beta = 0$ we have

$$\frac{dC(\vartheta)}{d\vartheta} = -\frac{1}{2}C(\vartheta)\tanh\vartheta. \tag{G.15}$$

Solving this with the boundary condition $C(0) = 1$, we obtain

$$C = \frac{1}{\cosh\vartheta} = \exp(-\ln\cosh\vartheta). \tag{G.16}$$

This is the normalization factor in (G.10).

We next study the Bogoliubov transformation between $(\eta_k, \eta_k^\dagger)$ and $(\xi_k, \xi_k^\dagger)$,

$$\begin{aligned}\xi_k &= \eta_k\cosh\vartheta_k + \eta_{-k}^\dagger\sinh\vartheta_k, \\ \xi_k^\dagger &= \eta_k^\dagger\cosh\vartheta_k + \eta_{-k}\sinh\vartheta_k,\end{aligned} \tag{G.17}$$

as corresponds (G.1), but we have set $e^{i\beta_k} = 1$ for simplicity. The generator is

$$G = \frac{i}{2}\int d^2k\,\vartheta_k(\eta_k\eta_{-k} - \eta_k^\dagger\eta_{-k}^\dagger). \tag{G.18}$$

It is convenient to use the box normalization,

$$G = \frac{i}{2}\sum_{\boldsymbol{k}}\vartheta_k(\eta_k\eta_{-k} - \eta_k^\dagger\eta_{-k}^\dagger) = i\sum_{\boldsymbol{k}}{}'\,\vartheta_k(\eta_k\eta_{-k} - \eta_k^\dagger\eta_{-k}^\dagger), \tag{G.19}$$

where $\sum_{\boldsymbol{k}}'$ means the summation over $\boldsymbol{k} = (k_x, k_y, k_z)$ with $k_x \geq 0$.

All previous formulas are valid for each component

$$G_k = i\vartheta_k(\eta_k\eta_{-k} - \eta_k^\dagger\eta_{-k}^\dagger), \tag{G.20}$$

where η_k and η_{-k} are independent operators. We may write $\eta_k = \partial/\partial\eta_k^\dagger$. Then, the ground-state condition (4.5.24) with (4.5.15) reads

$$\left(\frac{\partial}{\partial\eta_k^\dagger} + \eta_{-k}^\dagger \tanh\vartheta_k\right)|g\rangle = 0. \tag{G.21}$$

This is solved as

$$\begin{aligned}|g\rangle &= C{\sum_k}' \exp\left(-\eta_k^\dagger\eta_{-k}^\dagger \tanh\vartheta_k\right)|0\rangle \\ &= C\exp\left(-\frac{1}{2}\int dk\eta_k^\dagger\eta_{-k}^\dagger \tanh\vartheta_k\right)|0\rangle\end{aligned} \tag{G.22}$$

with C a normalization factor, where $|0\rangle$ is the state obeying $\partial/\partial\eta_k^\dagger|0\rangle = \eta_k|0\rangle = 0$, i.e., the Fock vacuum defined by (4.5.11). We have found that the ground state $|g\rangle$ is a Bose condensate of $\eta_k\eta_{-k}$ pairs.

The normalization factor is explicitly given by

$$\begin{aligned}C &= \langle 0|g\rangle = \langle 0|e^{-iG}|0\rangle \\ &= \exp\left(-\frac{1}{4}\sum_k \ln\cosh\vartheta_k\right) = \exp\left(-\frac{V}{4(2\pi)^3}\int d^2k\ \ln\cosh\vartheta_k\right),\end{aligned} \tag{G.23}$$

where V is the volume of the system. The factor C is a finite constant only in a finite system. In this case two states $|0\rangle$ and $|g\rangle$ belong to the same Hilbert space. However, when the system has an infinitely large volume ($V \to \infty$), they belong to different Hilbert spaces since $C = 0$. It is said that these two Hilbert spaces are unitarity inequivalent. See a textbook[34] for more details.

H Energy-Momentum Tensor

The energy-momentum tensor (3.3.24) is derived in association with the Nöther theorem. The system is invariant under transformation $x_\mu \to x'_\mu$, when the action is invariant:

$$S = \frac{1}{c}\int d^4x \mathcal{L}[\phi(x), \partial_\nu\phi(x)] = \frac{1}{c}\int d^4x' \mathcal{L}[\phi(x'), \partial_\nu\phi(x')]. \tag{H.1}$$

Since the system does not contain a special point fixed externally, it is invariant under space-time translation $x_\mu \to x' \equiv x_\mu + \delta x_\mu$. We first examine the Lagrangian density,

$$\begin{aligned}\mathcal{L}[\phi(x'),\partial_\nu\phi(x')] &= \mathcal{L}[\phi(x+\delta x),\partial_\nu\phi(x+\delta x)] \\ &= \mathcal{L}[\phi(x),\partial_\nu\phi(x)] + \left(\frac{\partial\mathcal{L}}{\partial\phi}\partial^\mu\phi + \frac{\partial\mathcal{L}}{\partial(\partial_\nu\phi)}\partial^\mu\partial_\nu\phi\right)\delta x_\mu \\ &= \mathcal{L}[\phi(x),\partial_\nu\phi(x)] + \partial_\nu\left(\frac{\partial\mathcal{L}}{\partial(\partial_\nu\phi)}\partial^\mu\phi\right)\delta x_\mu, \end{aligned} \tag{H.2}$$

which holds up to the linear term in δx_μ. In the last equality we have used the Euler-Lagrange equation (3.5.6). The change of variable yields the Jacobian factor:[1]

$$d^4x' = \left|\frac{\partial x'^\mu}{\partial x^\nu}\right| d^4x = \det\left(\delta^\mu_\nu + \partial_\nu\delta x^\mu\right) d^4x = (1+\partial_\mu\delta x^\mu)d^4x. \tag{H.3}$$

Combining (H.2) and (H.3) we obtain

$$\begin{aligned}\int d^4x'\mathcal{L}[\phi(x'),\partial_\nu\phi(x')] &= \int d^4x(1+\partial_\mu\delta x^\mu)\left[\mathcal{L}(x) + \partial_\nu\left(\frac{\partial\mathcal{L}}{\partial(\partial_\nu\phi)}\partial^\mu\phi\right)\delta x_\mu\right] \\ &= \int d^4x\mathcal{L}[\phi(x),\partial_\nu\phi(x)] - \int d^4x\partial_\nu\Theta^{\nu\mu}\delta x_\mu \end{aligned} \tag{H.4}$$

up to the linear term in δx_μ, where $\Theta^{\nu\mu}(x)$ is the energy-momentum tensor,

$$\Theta^{\mu\nu} \equiv -\frac{\partial\mathcal{L}}{\partial(\partial_\mu\phi)}\partial^\nu\phi + g^{\mu\nu}\mathcal{L}. \tag{H.5}$$

The invariance of the action (H.1) implies that

$$\partial_\nu\Theta^{\nu\mu}(x) = 0. \tag{H.6}$$

The energy and momentum density is defined by $\Theta^{\mu 0}(x)$ as

$$\mathcal{H}(x) = \Theta^{00}(x), \qquad \mathcal{P}^k(x) = \frac{1}{c}\Theta^{k0}(x) \tag{H.7}$$

in consistent with the energy-momentum vector (3.3.7). Hence, the energy-momentum conservation is a result of the translational invariance of the system.

I Exchange Interaction

The ferromagnetic order is governed by the Hamiltonian (4.1.1) describing the spin-spin interaction. We explain the mechanism how such an interaction arises from the microscopic Coulomb interaction. See also Chapter 32 in text.

Two Electron Spins

[1]The Jacobian factor is obtained more easily from

$$\int d^4x' f(x) = \int d^4(x+\delta x) f(x) = \int d^4x f(x-\delta x) = \int d^4x\left[f(x) - \partial_\mu f(x)\delta x^\mu\right] = \int d^4x(1+\partial_\mu\delta x^\mu)f(x),$$

which holds for an arbitrary function $f(x)$.

We first consider two electrons localized at lattice sites i and j. The wave function consists of the coordinate part and the spin part. The spin part is either a spin singlet $|SS\rangle$,

$$|SS\rangle = \frac{1}{\sqrt{2}}(|i\uparrow; j\downarrow\rangle - |i\downarrow; j\uparrow\rangle), \tag{I.1}$$

or a spin triplet $|ST\rangle_i$,

$$|ST\rangle_+ = |i\uparrow; j\uparrow\rangle, \tag{I.2a}$$

$$|ST\rangle_0 = \frac{1}{\sqrt{2}}(|i\uparrow; j\downarrow\rangle + |i\downarrow; j\uparrow\rangle), \tag{I.2b}$$

$$|ST\rangle_- = |i\downarrow; j\downarrow\rangle. \tag{I.2c}$$

The spin singlet $|SS\rangle$ is antisymmetric while the spin triplet $|ST\rangle$ is symmetric under the exchange of two electrons. The Fermi statistics requires that the whole wave function is antisymmetric,

$$\mathfrak{S}_{SS}(\boldsymbol{x}, \boldsymbol{y}) = \frac{1}{\sqrt{2}}\left(\mathfrak{S}_i(\boldsymbol{x})\mathfrak{S}_j(\boldsymbol{y}) + \mathfrak{S}_i(\boldsymbol{y})\mathfrak{S}_j(\boldsymbol{x})\right)|SS\rangle, \tag{I.3a}$$

$$\mathfrak{S}_{ST}(\boldsymbol{x}, \boldsymbol{y}) = \frac{1}{\sqrt{2}}\left(\mathfrak{S}_i(\boldsymbol{x})\mathfrak{S}_j(\boldsymbol{y}) - \mathfrak{S}_i(\boldsymbol{y})\mathfrak{S}_j(\boldsymbol{x})\right)|ST\rangle. \tag{I.3b}$$

The Coulomb energies of these two states are

$$\langle\mathfrak{S}_{SS}|H_C|\mathfrak{S}_{SS}\rangle = 2U_{ij} + 2J_{ij}, \qquad \langle\mathfrak{S}_{ST}|H_C|\mathfrak{S}_{ST}\rangle = 2U_{ij} - 2J_{ij}, \tag{I.4}$$

where

$$U_{ij} = \frac{1}{2}\int d^m x d^m y\, \mathfrak{S}_i^*(\boldsymbol{x})\mathfrak{S}_i(\boldsymbol{x})V(\boldsymbol{x}-\boldsymbol{y})\mathfrak{S}_j^*(\boldsymbol{y})\mathfrak{S}_j(\boldsymbol{y}), \tag{I.5a}$$

$$J_{ij} = \frac{1}{2}\int d^m x d^m y\, \mathfrak{S}_i^*(\boldsymbol{x})\mathfrak{S}_j(\boldsymbol{x})V(\boldsymbol{x}-\boldsymbol{y})\mathfrak{S}_j^*(\boldsymbol{y})\mathfrak{S}_i(\boldsymbol{y}) \tag{I.5b}$$

in the m-dimensional space. The term U_{ij} represents the direct Coulomb energy between two charge distributions $\rho_i(\boldsymbol{x}) = \mathfrak{S}_i^*(\boldsymbol{x})\mathfrak{S}_i(\boldsymbol{x})$ and $\rho_j(\boldsymbol{y}) = \mathfrak{S}_j^*(\boldsymbol{y})\mathfrak{S}_j(\boldsymbol{y})$ localized on lattice points i and j. The term J_{ij} is the exchange Coulomb energy, which has no classical counterpart. The exchange energy J_{ij} vanishes when there is no overlap between the two wave functions $\mathfrak{S}_i(\boldsymbol{x})$ and $\mathfrak{S}_j(\boldsymbol{x})$.

We can rewrite the Coulomb energy (I.4) as

$$\langle\mathfrak{S}_{SS}|H_C^{\text{eff}}|\mathfrak{S}_{SS}\rangle = 2U_{ij} + 2J_{ij}, \qquad \langle\mathfrak{S}_{ST}|H_C^{\text{eff}}|\mathfrak{S}_{ST}\rangle = 2U_{ij} - 2J_{ij}, \tag{I.6}$$

where H_C^{eff} is the effective Hamiltonian representing the spin-spin interaction,

$$H_C^{\text{eff}} = 2U_{ij} - J_{ij}[1 + 4\boldsymbol{S}(i)\cdot\boldsymbol{S}(j)]. \tag{I.7}$$

Indeed, by using the relation

$$2\boldsymbol{S}(i)\cdot\boldsymbol{S}(j) = [\boldsymbol{S}(i) + \boldsymbol{S}(j)]^2 - \boldsymbol{S}(i)^2 - \boldsymbol{S}(j)^2 = s(s+1) - \frac{1}{2}\cdot\frac{3}{2} - \frac{1}{2}\cdot\frac{3}{2}, \tag{I.8}$$

we obtain (I.6) from (I.7) since $S(i) \cdot S(j) = -\frac{3}{4}$ for the spin singlet ($s = 0$), and $S(i) \cdot S(j) = \frac{1}{4}$ for the spin triplet ($s = 1$). The Coulomb energy (I.4) thus depends on the relative spin orientation.

Spin-Spin Interaction

After having studied the physical meaning of the exchange energy, we derive the spin-spin interaction (4.1.1) in text systematically. The field-theoretical Coulomb term in the m-dimensional space is

$$H_C = \frac{1}{2}\sum_{\sigma\tau}\int d^m x d^m y\, \psi_\sigma^\dagger(x)\psi_\tau^\dagger(y)V(x-y)\psi_\tau(y)\psi_\sigma(x). \tag{I.9}$$

We take a complete set of one-body wave functions $\mathfrak{S}_i(x)$ labeled by i. (The index i may label lattice sites.) We expand the electron field $\psi_\sigma(x)$ as

$$\psi_\sigma(x) = \sum_i c_\sigma(i)\mathfrak{S}_i(x), \tag{I.10}$$

where $c_\sigma(i)$ is the annihilation operator of the up-spin electron ($\sigma = \uparrow$) or down-spin electron ($\sigma = \downarrow$) at the site i,

$$\{c_\sigma(i), c_\tau^\dagger(j)\} = \delta_{ij}\delta_{\sigma\tau}, \qquad \{c_\sigma(i), c_\tau^\dagger(j)\} = \{c_\sigma^\dagger(i), c_\tau^\dagger(j)\} = 0. \tag{I.11}$$

Substituting (I.10) into (I.9), we find

$$H_C = \sum_{ijmn}\sum_{\sigma\tau} V_{ijmn} c_\sigma^\dagger(i)c_\tau^\dagger(m)c_\tau(n)c_\sigma(j) \tag{I.12}$$

with

$$V_{ijmn} = \frac{1}{2}\int d^m x d^m y\, \mathfrak{S}_i^*(x)\mathfrak{S}_j(x)V(x-y)\mathfrak{S}_m^*(y)\mathfrak{S}_n(y). \tag{I.13}$$

We are concerned with the expectation value $\langle H_C \rangle$, to which only the two types of terms contribute: (A) $m = n$ and $i = j$; (B) $m = j$ and $n = i$,

$$H_C^{\text{eff}} = \sum_{ij} U_{ij}\rho(i)\rho(j) + \sum_{ij}\sum_{\sigma\tau} J_{ij} c_\sigma^\dagger(i)c_\tau^\dagger(j)c_\tau(i)c_\sigma(j). \tag{I.14}$$

Here, integrals $U_{ij} \equiv V_{iijj}$ and $J_{ij} \equiv V_{ijji}$ are the direct and exchange energies (I.5) associated with two electrons localized at sites i and j, while $\rho(i) \equiv \sum_\sigma c_\sigma^\dagger(i)c_\sigma(i)$ is the electron number at site i.

At each site the spin S is defined by

$$S^a = (c_\uparrow^\dagger, c_\downarrow^\dagger)\frac{\tau^a}{2}\begin{pmatrix} c_\uparrow \\ c_\downarrow \end{pmatrix} \tag{I.15}$$

with τ^a the Pauli matrix. As is easily proved by a direct calculation, there exists the algebraic relation

$$\sum_{\sigma\tau} c_\sigma^\dagger(i)c_\tau^\dagger(j)c_\tau(i)c_\sigma(j) = -2S(i)\cdot S(j) - \frac{1}{2}\rho(i)\rho(j). \tag{I.16}$$

We extract the exchange interaction from the Hamiltonian (I.14),

$$H_X^{\text{eff}} = -2\sum_{ij} J_{ij}[S(i)\cdot S(j) + \frac{1}{4}\rho(i)\rho(j)]. \qquad \text{(I.17)}$$

When $J_{ij} > 0$, all spins are polarized, $S(i)\cdot S(j) = 1/4$. When all lattice points are occupied, $\rho(i) = 1$, its ground-state contribution is

$$\langle H_X^{\text{eff}}\rangle_g = -\sum_{ij} J_{ij}\rho(i)\rho(j) = -\frac{1}{2}\epsilon_X N, \qquad \text{(I.18)}$$

where N is the number of electrons, and $\epsilon_X = 2\sum_j J_{ij}$. Equation (I.17) is the basic formula (4.1.1) for the spin-spin interaction in text, with $\sum_{ij} = 2\sum_{\langle ij\rangle}$.

Spinless Electrons

The exchange interaction is present also in the spinless electron theory. The field operator is expanded as in (I.10),

$$\psi(x) = \sum_i c(i)\mathfrak{S}_i(x). \qquad \text{(I.19)}$$

The Hamiltonian reads

$$H_C = \sum_{ijmn} V_{ijmn} c^\dagger(i)c^\dagger(m)c(n)c(j). \qquad \text{(I.20)}$$

It is equivalent to

$$H_C^{\text{eff}} = 2\sum_{ij}\left(U_{ij} - J_{ij}\right)\rho(i)\rho(j) \qquad \text{(I.21)}$$

in the sense that $\langle H_C^{\text{eff}}\rangle = \langle H_C\rangle$. It agrees with the ground-state energy of the ferromagnet. The ground-state energy is identical whether electrons are spinless or have spins. It implies that the exchange energy arises solely from the Pauli exclusion principle.

By the use of the relations

$$\langle\psi^\dagger(x)\psi(x)\rangle = \sum_{ij}\mathfrak{S}_i^*(x)\mathfrak{S}_j(x)\langle c_i^\dagger c_j\rangle = \sum_i \mathfrak{S}_i^*(x)\mathfrak{S}_i(x)\rho(i),$$

$$\langle\psi^\dagger(x)\psi(y)\rangle = \sum_{ij}\mathfrak{S}_i^*(x)\mathfrak{S}_j(y)\langle c_i^\dagger c_j\rangle = \sum_i \mathfrak{S}_i^*(x)\mathfrak{S}_i(y)\rho(i) \qquad \text{(I.22)}$$

with $\rho(i) = c^\dagger(i)c(i)$, we may rewrite the Coulomb energy (I.21) as

$$H_C^{\text{eff}} = \frac{1}{2}\int d^m x d^m y\, \langle\psi^\dagger(x)\psi(x)\rangle V(|x-y|)\langle\psi^\dagger(y)\psi(y)\rangle$$

$$-\frac{1}{2}\int d^m x d^m y\, \langle\psi^\dagger(x)\psi(y)\rangle V(|x-y|)\langle\psi^\dagger(y)\psi(x)\rangle. \qquad \text{(I.23)}$$

It shows that the effective Hamiltonian consists of the direct-interaction and exchange-interaction terms.

J MERMIN-WAGNER THEOREM

Spontaneous symmetry breaking (SSB) is a central issue of modern physics, and plays an important role in QH ferromagnets. We have reviewed it in Chapter 4, but we have not paid attention to subtle problems encountered in lower dimensions. They are summarized as the Mermin-Wagner theorem, which is the connection between the impossibility of broken continuous symmetry and dimensionality for both zero and finite temperatures.[345,346,232] We summarize it as follows.

space dimensionality	$T > 0$	$T = 0$
$D = 1$	no	no
$D = 2$	no	yes
$D > 2$	yes	yes

(J.1)

We present a pedagogical argument[105,320] to bring out simply and more explicitly the physics behind the rigorously proven results on the Mermin-Wagner theorem.

As we have shown in Section 4.7, SSB of a continuous symmetry is accompanied necessarily by a gapless mode. It is described by a massless scalar boson (massless real Klein-Gordon field), since there exists no massless Schrödinger field. Thus we start with a discussion if the massless scalar field is well defined in lower dimensions.

The Hamiltonian is

$$\mathcal{H} = \frac{1}{2}[\nabla\eta(\boldsymbol{x})]^2. \tag{J.2}$$

The classical ground state is given by an arbitrary constant, $\langle\eta(\boldsymbol{x})\rangle = c$. It is related to the symmetry that the Hamiltonian is invariant under a continuous translation of the field,

$$\eta(\boldsymbol{x}) \to \eta(\boldsymbol{x}) + c. \tag{J.3}$$

We may take the classical ground state to be $\langle\eta(\boldsymbol{x})\rangle = 0$ without loss of generality.

We first restrict the system to a finite volume L^D. To realize the classical ground state $\eta = 0$, we add a symmetry breaking term to the Hamiltonian (J.2),

$$\mathcal{H}' = \frac{1}{2}\lambda^2\eta^2(\boldsymbol{x}), \tag{J.4}$$

where λ is a small, real and nonzero constant. Thus we consider the Hamiltonian

$$\mathcal{H}_\lambda = \frac{1}{2}[\nabla\eta(\boldsymbol{x})]^2 + \frac{1}{2}\lambda^2\eta^2(\boldsymbol{x}). \tag{J.5}$$

The problem is if all N-point functions are well-defined in the limit $\lambda \to 0$.

This is highly a nontrivial problem due to quantum fluctuations around the classical ground state. The quantum ground state has a wave function that spreads to other field configurations as well. The spread is harmless if it is within a finite range due to the gap term in (J.5). Though the gap disappears in the limit $\lambda \to 0$, the spread can be harmless depending on the dimensionality of the space.

The free scalar field is expanded as in (3.3.26) [see also (3.2.21)], or

$$\eta(x) = \frac{1}{L^{D/2}} \sum_{k} \frac{1}{\sqrt{2\omega_k}} \left(a_k e^{i(kx-\omega_k t)} + a_k^\dagger e^{-i(kx-\omega_k t)} \right), \tag{J.6}$$

with $[a_k, a_{k'}^\dagger] = \delta_{kk'}$ and $\omega_k = (k^2 + \lambda^2)^{1/2}$, where we have taken the natural unit ($c = \hbar = 1$). The ground state is defined by

$$a_k|0\rangle = 0. \tag{J.7}$$

The ground-state wave function is that of the harmonic oscillators corresponding to all modes $\boldsymbol{k}$. The quantum ground state has all the infinitely many $\boldsymbol{k}$ modes present. The effect of this spread on fluctuations of $\eta(\boldsymbol{x})$ can be estimated from the second moment,[2]

$$\lim_{L\to\infty} \langle 0|\eta(x)\eta(0)|0\rangle = \frac{1}{(2\pi)^D} \int_0^\Lambda d^D k \frac{e^{-ikx}}{2(k^2 + \lambda^2)^{1/2}} \equiv C_\lambda(x). \tag{J.8}$$

It is clear that, as $\lambda \to 0$, this is infrared convergent only for $D \geq 2$. Consequently the massless scalar field is ill defined in one space dimension ($D = 1$).

It should be emphasized that, though $\eta(\boldsymbol{x})$ itself is ill defined, its exponentiation $e^{\eta(x)}$ is well defined even at $D = 1$. To show this, we separate $\eta(\boldsymbol{x})$ into the annihilation operator $\eta_+(\boldsymbol{x})$ and the creation operator $\eta_-(\boldsymbol{x})$ as in (3.3.28). Their commutation relation is

$$[\eta_+(x), \eta_-(y)] = C_\lambda(x - y). \tag{J.9}$$

We now find

$$\langle 0|e^{\eta(x)}|0\rangle = e^{-C_\lambda(0)/2}\langle 0|e^{\eta_-(x)}e^{\eta_+(x)}|0\rangle = e^{-C_\lambda(0)/2} \to 0, \tag{J.10}$$

as $\lambda \to 0$. We interpret that the infrared divergence in (J.8) produces infinitely large fluctuations in $\eta(\boldsymbol{x})$ so that $\langle 0|e^{\eta(x)}|0\rangle$ vanishes. Furthermore we can verify

$$\langle 0|e^{\eta(x)}e^{\eta(y)}|0\rangle = e^{-C_\lambda(x)-C_\lambda(y)} \to 0. \tag{J.11}$$

On the other hand,

$$\langle 0|e^{-\eta(x)}e^{\eta(y)}|0\rangle = e^{C_\lambda(x)-C_\lambda(y)} = \exp\left[\frac{1}{2\pi}\int_0^\Lambda dk \frac{e^{-ikx} - e^{-iky}}{2(k^2 + \lambda^2)^{1/2}}\right]. \tag{J.12}$$

[2]The formula has an ultraviolet divergence in the momentum integral. This would be related to our approximation to use the simplified Hamiltonian (J.2). These ultraviolet divergences presumably do not appear in the the realistic interacting model of SSB. However, our results are independent of these technical complications in the ultraviolet range, since the restoration or otherwise of symmetry is an infrared phenomenon. We therefore assume an ultraviolet cutoff Λ when necessary in what follows.

This yields a well-defined nonzero quantity in the limit $\lambda \to 0$. Similarly we can show that all N-point functions of $e^{\pm\eta(\boldsymbol{x})}$ are well defined. See also Appendix N. Thus, $\cos\eta(\boldsymbol{x})$ and $\sin\eta(\boldsymbol{x})$ are well-defined field operators even if $\eta(\boldsymbol{x})$ is not.

The above analysis is limited to the case at the zero temperature ($T = 0$). At finite temperatures ($T > 0$), we expect the spread in configurations of $\eta(\boldsymbol{x})$ due to quantum effects to be further compounded by the demands of entropy. In other words, a system at finite T contains a statistical mixture of not only the ground state but also of excited states as well. More precisely, at finite T, the vacuum expectation value of any operator A is replaced by that of the thermodynamic expectation value,

$$\langle 0|A|0\rangle_T \equiv \frac{\mathrm{Tr}[e^{-\beta H}A]}{\mathrm{Tr}[e^{-\beta H}]}, \tag{J.13}$$

where $\beta = 1/k_B T$ with k_B the Boltzmann constant. Hence, in the limit $\lambda \to 0$ and $L \to \infty$, we have

$$\langle 0|a_{\boldsymbol{k}}^\dagger a_{\boldsymbol{k}'}|0\rangle_T = \frac{\delta(\boldsymbol{k}-\boldsymbol{k}')}{e^{\beta|\boldsymbol{k}|}-1}, \tag{J.14}$$

and

$$\langle 0|\eta(\boldsymbol{x})\eta(0)|0\rangle = \frac{1}{(2\pi)^D}\int_0^\Lambda d^D k\, \frac{e^{-ikx}}{2|\boldsymbol{k}|}\left(1+\frac{2}{e^{\beta|\boldsymbol{k}|}-1}\right). \tag{J.15}$$

In the limit $T \to 0$, or $\beta \to \infty$, only the first term survives, which is the infrared-divergent term (J.8) for $D < 2$. On the other hand, the second term has a stronger infrared behavior, $\propto 1/|\boldsymbol{k}|^2$ as $|\boldsymbol{k}| \to 0$. Hence, this is infrared divergent for $D \le 2$, producing infinitely large fluctuations in $\eta(\boldsymbol{x})$ at finite T. Consequently the massless scalar field is ill defined in two space dimensions ($D = 2$) at finite temperature. Though $\eta(\boldsymbol{x})$ itself is ill defined, its exponentiation $e^{\eta(\boldsymbol{x})}$ is well defined even at $D = 2$, by making a similar analysis similar to the $D = 1$ case.

We proceed to analyze the realistic model describing ferromagnets,

$$\mathcal{H} = 2J_s \sum_{a=xyz} [\nabla \mathcal{S}_a(\boldsymbol{x})]^2, \tag{J.16}$$

where the spin components are subject to

$$\sum_{a=xyz} [\mathcal{S}_a(\boldsymbol{x})]^2 = \frac{1}{4}. \tag{J.17}$$

The classical energy is minimized by any spin-polarized configuration,

$$\langle \mathcal{S}_x(\boldsymbol{x})\rangle = \frac{1}{2}\sin\theta_0, \qquad \langle \mathcal{S}_y(\boldsymbol{x})\rangle = \frac{1}{2}\cos\theta_0 \sin\vartheta_0,$$
$$\langle \mathcal{S}_z(\boldsymbol{x})\rangle = \frac{1}{2}\cos\theta_0\cos\vartheta_0, \tag{J.18}$$

where θ_0 and ϑ_0 are arbitrary constants. Without loss of generality we may choose $\theta_0 = \vartheta_0 = 0$, or

$$\langle \mathcal{S}_a(\boldsymbol{x}) \rangle = \frac{1}{2}(0, 0, 1). \tag{J.19}$$

The symmetry is broken spontaneously by the ground state.

We examine the condition on SSB at $T = 0$ more precisely. To realize this specific ground state we add a symmetry breaking term,

$$\mathcal{H}' = -\lambda^2 \mathcal{S}_z(\boldsymbol{x}), \tag{J.20}$$

which is nothing but the Zeeman term: See (15.5.4). The condition on SSB is

$$\langle \mathcal{S}_a \rangle = \lim_{\lambda \to 0} \lim_{L \to \infty} \langle 0 | \mathcal{S}_a(\boldsymbol{x}) | 0 \rangle = \frac{1}{2}\delta_{az}. \tag{J.21}$$

There would be no SSB if quantum fluctuations around the classical ground state are so large that all memory of the classically preferred direction is lost as $\lambda \to 0$.

It is convenient to parametrize the spin field as

$$\mathcal{S}_x(\boldsymbol{x}) = \frac{1}{2}\sin\theta(\boldsymbol{x}), \qquad \mathcal{S}_y(\boldsymbol{x}) = \frac{1}{2}\cos\theta(\boldsymbol{x})\sin\vartheta(\boldsymbol{x}),$$
$$\mathcal{S}_z(\boldsymbol{x}) = \frac{1}{2}\cos\theta(\boldsymbol{x})\cos\vartheta(\boldsymbol{x}), \tag{J.22}$$

and rewrite the Hamiltonian as

$$\mathcal{H} = \frac{J_s}{2}\left[\cos^2\theta(\boldsymbol{x})\,(\nabla\vartheta)^2 + (\nabla\theta)^2\right]. \tag{J.23}$$

The classical ground state is given by $\sigma(\boldsymbol{x}) = \sigma_0$ and $\vartheta(\boldsymbol{x}) = \vartheta_0$. Without loss of generality we may choose $\sigma_0 = 0$ and $\vartheta_0 = 0$ in accord with (J.19).

By including the symmetry breaking term (J.20), the total Hamiltonian becomes

$$\mathcal{H} = \frac{J_s}{2}(\nabla\theta)^2 + \frac{J_s}{2}(\nabla\vartheta)^2 + \frac{\lambda^2}{4}\theta^2(\boldsymbol{x}) + \frac{\lambda^2}{4}\vartheta^2(\boldsymbol{x}), \tag{J.24}$$

up to the second order in quantum fields. We define

$$\zeta_{\boldsymbol{k}} = \frac{\theta_{\boldsymbol{k}} + i\vartheta_{\boldsymbol{k}}}{2}, \qquad \zeta_{\boldsymbol{k}}^\dagger = \frac{\theta_{\boldsymbol{k}} - i\vartheta_{\boldsymbol{k}}}{2} \tag{J.25}$$

in the momentum space. Then, the Hamiltonian is diagonalized as in (15.5.8), or

$$H_{\text{eff}} = \int d^2k\, E_{\boldsymbol{k}} \zeta_{\boldsymbol{k}}^\dagger \zeta_{\boldsymbol{k}} \tag{J.26}$$

with

$$E_{\boldsymbol{k}} = 2J_s \boldsymbol{k}^2 + \lambda^2, \tag{J.27}$$

and

$$[\zeta_{\boldsymbol{k}}, \zeta_{\boldsymbol{l}}^\dagger] = \delta(\boldsymbol{k} - \boldsymbol{l}). \tag{J.28}$$

Comparing (J.26) with (3.3.34b), we can identify ζ_k and $\zeta_k^\dagger$ as the annihilation and creation operators of the real scalar field (3.3.26), or

$$\zeta(x) = \frac{1}{L^{D/2}} \sum_k \frac{1}{\sqrt{2E_k}} \left(\zeta_k e^{ikx} + \zeta_k^\dagger e^{-ikx} \right). \tag{J.29}$$

Note that

$$\zeta_k e^{ikx} + \zeta_k^\dagger e^{-ikx} = \theta_k \cos \boldsymbol{kx} - \vartheta_k \sin \boldsymbol{kx}. \tag{J.30}$$

The scalar field $\zeta(x)$ is massless in the limit $\lambda \to 0$ because of the dispersion relation (J.27).

We consider the theory in one space dimension for $T \geq 0$, or in two space dimensions for $T > 0$. Since $\eta(x)$ is massless, both $\theta(\boldsymbol{x})$ and $\vartheta(\boldsymbol{x})$ are ill defined. Nevertheless, $e^{\pm i\theta(\boldsymbol{x})}$ and $e^{\pm i\vartheta(\boldsymbol{x})}$ are well defined, and so are the spin components (J.22). Quantum fluctuations in $\theta(\boldsymbol{x})$ and $\vartheta(\boldsymbol{x})$ become infinitely large, and we have

$$\langle \cos\theta(\boldsymbol{x})\rangle = \langle \sin\theta(\boldsymbol{x})\rangle = \langle \cos\vartheta(\boldsymbol{x})\rangle = \langle \sin\vartheta(\boldsymbol{x})\rangle = 0, \tag{J.31}$$

or

$$\langle \mathcal{S}_x(\boldsymbol{x})\rangle = \langle \mathcal{S}_y(\boldsymbol{x})\rangle = \langle \mathcal{S}_z(\boldsymbol{x})\rangle = 0. \tag{J.32}$$

Quantum fluctuations have wiped out the spin polarization (J.19) initially imposed. Consequently, SSB does not occur.

Finally we make discussions on the relevance of the theorem to the QH system. Since it is a 2-dimensional system, strictly speaking, there exists no SSB for $T > 0$. Nevertheless, it makes sense to use the concept of SSB even at $T > 0$. In the QH system, SSB occurs at $T = 0$, because the spin is polarized owing to the Coulomb exchange interaction. The ground state is a ferromagnet. The Coulomb exchange interaction remains to be important for $T > 0$. Quantum-statistical effects destroy SSB only in the limit $\lambda \to 0$, that is, in the unphysical limit of the zero Zeeman coupling. Then, it is natural to regard the ground state as a ferromagnet even $T > 0$, though a small Zeeman effect is necessary to stabilize it. Furthermore, there are great merits in using the concept, since SSB implies various physical phenomena accompanying it such as Bose condensation, quantum coherence and topological excitations. Indeed, these important phenomena accompanying SSB continue to exist as T departs from $T = 0$ in the QH system. This is the reason why we use the concept of SSB even at $T > 0$ in the QH system.

K Lorentz Transformation

To discuss the Dirac equation we have defined the quantity

$$1, \qquad \gamma_\mu, \qquad \sigma_{\mu\nu} \equiv \frac{i}{2}[\gamma_\mu, \gamma_\nu], \qquad \gamma_\mu\gamma_5, \qquad \gamma_5, \tag{K.1}$$

and

$$M_{\mu\nu} \equiv \frac{1}{2}\sigma_{\mu\nu} = \frac{i}{4}[\gamma_\mu, \gamma_\nu]. \tag{K.2}$$

The following commutation relations are explicitly checked to hold[3],

$$[M_{\mu\nu}, 1] = [M_{\mu\nu}, \gamma_5] = 0, \tag{K.3a}$$

$$[M_{\mu\nu}, \gamma_\lambda] = i(g_{\nu\lambda}\gamma_\mu - g_{\mu\lambda}\gamma_\nu), \tag{K.3b}$$

$$[M_{\mu\nu}, \gamma_\lambda\gamma_5] = i(g_{\nu\lambda}\gamma_\mu\gamma_5 - g_{\mu\lambda}\gamma_\nu\gamma_5), \tag{K.3c}$$

$$[M_{\mu\nu}, M_{\lambda\rho}] = i(-g_{\nu\lambda}M_{\mu\rho} + g_{\mu\lambda}M_{\nu\rho} - g_{\nu\rho}M_{\lambda\mu} + g_{\mu\rho}M_{\lambda\nu}). \tag{K.3d}$$

We first explain (K.3d). As we have noticed in (6.1.15), M_{ij} is the generator of the rotation in the $x_i x_j$-plane. For instance, a rotation around the z-axis in the xy plane by angle ω_{xy} is given by

$$S = e^{-i\omega_{xy}M_{xy}}. \tag{K.4}$$

Similarly, a rotation in the Minkowsky $\mu\nu$ plane by angle $\omega_{\mu\nu}$ is given by $S(\omega) = e^{-i\omega^{\mu\nu}M_{\mu\nu}}$ for fixed μ and ν. An arbitrary Lorentz transformation is given by

$$S(\omega) = e^{-\frac{i}{2}\omega^{\mu\nu}M_{\mu\nu}} \tag{K.5}$$

with $\omega^{\mu\nu} = -\omega^{\nu\mu}$, where the sum is taken over μ and ν. Its inverse is

$$S(\omega)^{-1} = e^{\frac{i}{2}\omega^{\mu\nu}M_{\mu\nu}} = \gamma^0 S^\dagger(\omega)\gamma^0, \tag{K.6}$$

since $\gamma^0\sigma^\dagger_{\mu\nu}\gamma^0 = \sigma_{\mu\nu}$, as follows from (6.1.9). Thus, $M_{\mu\nu}$ is a generator of the Lorentz transformation.

We next explain (K.3b). We calculate $S(\omega)^{-1}\gamma^\mu S(\omega)$ for an infinitesimal value of $\omega^{\mu\nu}$,

$$S(\omega)^{-1}\gamma^\mu S(\omega) \simeq \left(1 + \frac{i}{2}\omega^{\mu\nu}M_{\mu\nu}\right)\gamma^\mu\left(1 - \frac{i}{2}\omega^{\lambda\sigma}M_{\lambda\sigma}\right) \simeq \gamma^\mu + \frac{i}{2}\omega_{\alpha\beta}[M_{\alpha\beta}, \gamma^\mu]. \tag{K.7}$$

We use (K.3b) to show

$$S(\omega)^{-1}\gamma^\mu S(\omega) \simeq \gamma^\mu + \omega^{\mu\nu}\gamma_\nu = (\delta^{\mu\nu} + \omega^{\mu\nu})\gamma_\nu = a^\mu{}_\nu\gamma^\nu, \tag{K.8}$$

where we have set

$$a^\mu{}_\nu = \delta^\mu{}_\nu + \omega^\mu{}_\nu \qquad \text{with} \, |\omega^\mu{}_\nu| \ll 1. \tag{K.9}$$

We can actually prove that the formula

$$S(\omega)^{-1}\gamma^\mu S(\omega) = a^\mu{}_\nu\gamma^\nu \tag{K.10}$$

---

[3]To verify them it is easier to use

$$M_{\mu\nu} = \frac{i}{4}(\gamma_\mu\gamma_\nu + \gamma_\nu\gamma_\mu) = \frac{i}{2}\gamma_\mu\gamma_\nu - \frac{i}{2}g_{\mu\nu},$$

since $g_{\mu\nu}$ does not contribute to the commutator.

holds for any value of $\omega^{\mu\nu}$, with a certain $a^{\mu}{}_{\nu}$ which agrees with (K.9) for $|\omega^{\mu}{}_{\nu}| \ll 1$. It implies that γ^{μ} transforms as a vector under the Lorentz transformation,

$$x^{\mu} \to x'^{\mu} = a^{\mu}{}_{\nu} x^{\nu}. \tag{K.11}$$

Similarly $\gamma^{\mu}\gamma_5$ transforms as a vector.

Relativistic Covariance

We proceed to study how the Dirac equation transforms under the Lorentz transformation (K.11). We shall show that the Lorentz invariance requires that the field $\psi(x)$ transform as

$$\psi(x) \to \psi'(x') = S\psi(x), \tag{K.12}$$

where S transforms γ^{μ} according to (K.10).

To move to a new Lorentz frame, we rewrite the Dirac equation as

$$(-i\hbar\gamma^{\mu}\partial_{\mu} + mc)\psi(x) = (-i\hbar a^{\nu}{}_{\mu}\gamma^{\mu}\partial'_{\nu} + mc)\psi(x) = (-i\hbar S^{-1}\gamma^{\nu} S\partial'_{\nu} + mc)\psi(x) = 0, \tag{K.13}$$

where we have used

$$\partial_{\mu} = \frac{\partial}{\partial x^{\mu}} = \frac{\partial x'^{\nu}}{\partial x^{\mu}}\frac{\partial}{\partial x'^{\nu}} = a^{\nu}{}_{\mu}\frac{\partial}{\partial x'^{\nu}} = a^{\nu}{}_{\mu}\partial'_{\nu} \tag{K.14}$$

and (K.10). Substituting $S^{-1}S = 1$, we find

$$(-i\hbar\gamma^{\mu}\partial_{\mu} + mc)\psi(x) = S^{-1}(-i\hbar\gamma^{\mu}\partial'_{\mu} + mc)S\psi(x) = 0. \tag{K.15}$$

Hence, if we assume (K.12), we recover the Dirac equation in the new frame,

$$(-i\hbar\gamma^{\mu}\partial'_{\mu} + mc)\psi'(x') = 0. \tag{K.16}$$

Note that the γ-matrix remains as it is. It is just a c-number matrix given explicitly by (6.1.10) in the standard representation.

Finally we analyze the Lorentz property of the current. Under the Lorentz transformation, the conjugate of the Dirac field (K.12) transforms as

$$\bar{\psi}(x) \to \bar{\psi}'(x') = \psi'^{\dagger}(x')\gamma^{0} = \psi^{\dagger}(x)S^{\dagger}\gamma^{0} = \bar{\psi}(x)S^{-1}, \tag{K.17}$$

with the use of (K.6). Thus, $\bar{\psi}(x)\psi(x)$ and $\bar{\psi}(x)\gamma_5\psi(x)$ transform as scalars. On the other hand, since

$$j^{\mu}(x) \to j'^{\mu}(x') = \bar{\psi}'(x')\gamma^{\mu}\psi'(x') = \bar{\psi}(x)S^{-1}\gamma^{\mu}S\psi(x) = a^{\mu}_{\nu} j^{\nu}(x), \tag{K.18}$$

with the use of (K.10), j^{μ} transforms as a vector. It is clear that $j_5^{\mu} = \bar{\psi}\gamma^{\mu}\gamma_5\psi$ transforms also as a vector. Similarly, we can show that $\bar{\psi}\sigma^{\mu\nu}\psi$ transforms as a tensor.

Both $\bar{\psi}(x)\psi(x)$ and $\bar{\psi}(x)\gamma_5\psi(x)$ transform as scalars. What is a difference between them? So far we have considered the Lorentz transformation of the form

(K.5), which can be constructed by a successive operation of an infinitesimal transformation. It is called the proper Lorentz transformation. There are other types of Lorentz transformations. For instance we consider the inversion of the space coordinate,

$$x^\mu = (x^0, x^i) \to x'^\mu = (x^0, -x^i). \tag{K.19}$$

The corresponding matrix is

$$S = \gamma^0. \tag{K.20}$$

Indeed,

$$S^{-1}\gamma^\mu S = \gamma^0\gamma^\mu\gamma^0 = (\gamma^0, -\gamma^i). \tag{K.21}$$

Under this transformation we obtain

$$\bar{\psi}(x)\psi(x) \to \bar{\psi}'(x')\psi'(x') = \bar{\psi}(x)S^{-1}S\psi(x) = \bar{\psi}(x)\psi(x), \tag{K.22}$$

but

$$\bar{\psi}(x)\gamma_5\psi(x) \to \bar{\psi}'(x')\gamma_5\psi'(x') = \bar{\psi}(x)S^{-1}\gamma_5 S\psi(x) = -\bar{\psi}(x)\gamma_5\psi(x). \tag{K.23}$$

Similarly,

$$j^\mu(x) \to j'^\mu(x') = \bar{\psi}(x)\gamma^0\gamma^\mu\gamma^0\psi(x) = \left(j^0, -j^i\right), \tag{K.24}$$

in accord with (K.19), but

$$j_5^\mu(x) \to j_5'^\mu(x') = \bar{\psi}(x)\gamma^0\gamma^\mu\gamma_5\gamma^0\psi(x) = \left(-j_5^0, j_5^i\right). \tag{K.25}$$

This is why $\bar{\psi}(x)\gamma_5\psi(x)$ is called a pseudo-scalar, and $\bar{\psi}(x)\gamma^\mu\gamma_5\psi(x)$ a pseudo-vector.

L One-Dimensional Soliton Solutions

We solve explicitly one-dimensional soliton solutions described by the following static energy density,

$$\mathcal{H} = \frac{c\hbar}{2}\left(\frac{d\phi}{dx}\right)^2 + U(\phi). \tag{L.1}$$

The static Euler-Lagrange equation is

$$c\hbar\frac{d^2\phi}{dx^2} - \frac{dU}{d\phi} = 0. \tag{L.2}$$

Multiplying $d\phi/dx$ to this equation, we find

$$\frac{c\hbar}{2}\frac{d}{dx}\left(\frac{d\phi}{dx}\right)^2 - \frac{dU}{dx} = 0. \tag{L.3}$$

Integrating this once, we have

$$\sqrt{c\hbar}\frac{d\phi}{dx} = \pm\sqrt{2U(\phi) - 2u_0}, \tag{L.4}$$

where u_0 is an integration constant. We integrate this as

$$x - x_0 = \pm\sqrt{c\hbar}\int_{\phi(x_0)}^{\phi(x)} \frac{d\eta}{\sqrt{2U(\eta) - 2u_0}}. \tag{L.5}$$

The energy density (L.1) is given by

$$\mathcal{H} = 2U(\phi) - u_0 \tag{L.6}$$

by the use of (L.4).

Kinks

We consider the kink system (7.3.3), where

$$U(\phi) = \frac{g}{4}(\phi^2 - v^2)^2. \tag{L.7}$$

To derive one-soliton solutions, we impose the boundary condition that $\phi \to \pm v$ as $x \to \pm\infty$. It requires $u_0 = 0$ in (L.5). Equation (L.5) reads

$$x - x_0 = \pm\sqrt{\frac{2c\hbar}{g}}\int_{\phi(x_0)}^{\phi(x)} \frac{d\eta}{\sqrt{\eta^2 - v^2}} = \pm\frac{1}{v}\sqrt{\frac{2c\hbar}{g}}\tan\frac{\phi(x) - \phi(x_0)}{v}. \tag{L.8}$$

We obtain the kink solution

$$\phi(x) - \phi(x_0) = \pm v \tanh\left(\frac{x - x_0}{2\xi}\right), \tag{L.9}$$

with $\xi = (1/v)(\sqrt{c\hbar/2g})$. This is the solution (7.3.4). With this solution the energy density (L.6) is calculated and given by (7.3.5).

Sine-Gordon Solitons

We next consider the sine-Gordon system (7.4.4), where

$$U(\phi) = \frac{c\hbar}{\lambda^2}(1 - \cos\phi) = \frac{2c\hbar}{\lambda^2}(1 - \cos^2\frac{\phi}{2}). \tag{L.10}$$

Equation (L.5) reads

$$x - x_0 = \pm\frac{\lambda}{2}\int_{\phi(x_0)}^{\phi(x)} \frac{d\eta}{\sqrt{1 - \cos^2(\eta/2) - \lambda^2 u_0/4c\hbar}}. \tag{L.11}$$

It is convenient to rewrite this as

$$\frac{x - x_0}{\kappa\lambda} = \pm\frac{1}{2}\int_{\phi(x_0)}^{\phi(x)} \frac{d\eta}{\sqrt{1 - \kappa^2\cos^2(\eta/2)}}, \tag{L.12}$$

where $\kappa^2(1-\lambda^2 u_0/4c\hbar) = 1$.

One-soliton solution is constructed by imposing the boundary condition that $\phi \to 0 \pmod{2\pi}$ as $x \to \pm\infty$, as requires that $\kappa = 1$ or $u_0 = 0$. Equation (L.12) yields

$$x - x_0 = \pm\lambda \int_{\phi(x_0)}^{\phi(x)} \frac{d\eta}{2|\sin(\eta/2)|}, \tag{L.13}$$

which is integrated to give

$$\phi(x) - \phi(x_0) = \pm 4 \arctan \exp\left(\frac{x - x_0}{\lambda}\right). \tag{L.14}$$

This is the solution (7.4.8).

We next consider the limit $\kappa \to 0$. Equation (L.12) yields

$$\phi(x) - \phi(x_0) = \pm\frac{2}{\kappa\lambda}(x - x_0). \tag{L.15}$$

This is the asymptotic solution (27.5.19) describing the ground state in the incommensurate phase in the QH state junction.

A soliton lattice solution is constructed as follows. We are interested in the solution as is depicted in Fig.27.9 (page 503) in Part III. It is clear that we need $|\kappa| < 1$ in (L.12). Setting $t = \cos\eta/2$, we rewrite it as

$$\frac{x - x_0}{\kappa\lambda} = \pm\int_{\phi(x_0)/2}^{\phi(x)/2} \frac{dt}{\sqrt{(1-t^2)(1-\kappa^2 t^2)}}. \tag{L.16}$$

This is known to be integrated as

$$\cos\frac{\phi(x) - \phi(x_0)}{2} = \mp \mathrm{sn}\left(\frac{x - x_0}{\kappa\lambda}\,\middle|\,\kappa^2\right), \tag{L.17}$$

or

$$\sin\frac{\phi(x) - \phi(x_0)}{2} = \pm \mathrm{cn}\left(\frac{x - x_0}{\kappa\lambda_J}\,\middle|\,\kappa^2\right), \tag{L.18}$$

where $\mathrm{sn}(u|\kappa^2)$ and $\mathrm{cn}(u|\kappa^2)$ are Jacobian elliptic functions. This is the soliton lattice solution (27.5.15). They are periodic functions. The complete elliptic integral of the first kind is given by

$$K(\kappa^2) = \int_0^{\pi/2} \frac{d\eta}{\sqrt{1-\kappa^2\cos^2\eta}}. \tag{L.19}$$

Comparing this with (L.12), we see that $\phi(u)$ increases from 0 to π when u starts from 0 to $K(\kappa^2)$, which is one forth of the periodicity of $\mathrm{sn}(u|\kappa^2)$ and $\mathrm{cn}(u|\kappa^2)$.

M Field-Theoretical Vortex Operators

Quasiparticles are vortices in the spinless theory. They are generated by the transformation (11.4.5) in terms of wave functions. Here, we explicitly construct field-theoretical operators for vortices.

The field-theoretical operator generating the transformation (11.4.5) is

$$\hat{S} = \exp\left(\int d^2x \, \ln[b_x^\dagger]\rho(x)\right), \tag{M.1}$$

for quasihole ($q = 1$) and its Hermitian conjugate

$$\hat{S}^\dagger = \exp\left(\int d^2x \, \ln[b_x]\rho(x)\right), \tag{M.2}$$

for quasielectron ($q = -1$). Here, $b_x^\dagger$ and b_x are the angular-momentum ladder operators (10.3.7b) for the electron at a generic point x. This is justified by operating it to the N-body state (10.6.19) in the lowest Landau level,

$$\begin{aligned}
\hat{S}|\mathfrak{S}\rangle &= \int [dx]\hat{S}|x_1, x_2, \cdots, x_N\rangle\mathfrak{S}[x] \\
&= \int [dx]\exp\left\{\sum_{r=1}^{N}\int d^2x \, \ln[b_x^\dagger]\delta(x - x_r)\right\}|x_1, x_2, \cdots, x_N\rangle\mathfrak{S}[x] \\
&= \int [dx]\prod_s b_s^\dagger\mathfrak{S}[x]|x_1, x_2, \cdots, x_N\rangle, \\
&= \int [dx]\mathfrak{S}_{\text{qh}}[x]|x_1, x_2, \cdots, x_N\rangle,
\end{aligned} \tag{M.3}$$

where use was made of (11.4.5a). Hence, $\hat{S}$ is the creation operator of a quasihole (vortex), and similarly $\hat{S}^\dagger$ is the creation operator of a quasielectron (antivortex) at the origin ($z = 0$).

We show directly how the vortex operator $\hat{S}$ is related with the phase transformation (11.4.1). Because of the relation (10.3.8),

$$b^\dagger = \sqrt{2}z - ia, \qquad b = \sqrt{2}z^* + ia^\dagger, \tag{M.4}$$

where $a^\dagger$ and a mix Landau levels: $b^\dagger$ is the LLL component of the complex number $\sqrt{2}z$. Thus, the vortex operator is an appropriate LLL component of the operator,

$$S = \exp\left(\int d^2x \, \ln[\sqrt{2}z]\rho(x)\right) = \exp\left(\int d^2x \, \left[i\theta(x) + \ln(r/\sqrt{2}\ell_B)\right]\rho(x)\right). \tag{M.5}$$

The unitary part of this operator generates the phase transformation (11.4.1) with

$q = 1$. Similarly, the antivortex operator is the LLL component of the operator,

$$\begin{aligned} S^\dagger &= \exp\left(\int d^2x \ \ln[\sqrt{2}z^*]\rho(\boldsymbol{x})\right) \\ &= \exp\left(\int d^2x \left[-i\theta(\boldsymbol{x}) + \ln(r/\sqrt{2}\ell_B)\right]\rho(\boldsymbol{x})\right), \end{aligned} \tag{M.6}$$

which generates the phase transformation (11.4.1) with $q = -1$.

We next study the relation between the vortex and the electron. When we operate $\hat{S}^m$ on the N-body state we obtain

$$\hat{S}^m|\mathfrak{S}_N\rangle = 2^{mN/2}\int [d\boldsymbol{x}]\prod_s (z_s)^m \mathfrak{S}_N[\boldsymbol{x}]|\boldsymbol{x}_1, \boldsymbol{x}_2, \cdots, \boldsymbol{x}_N\rangle, \tag{M.7}$$

where the electron number N is made explicit by the subscript in $\mathfrak{S}_N$. We now operate the electron field operator to the $(N+1)$-body state,

$$\psi(0)|\mathfrak{S}_{N+1}\rangle = \prod_{r=1}^{N+1}\int d^2x_r \psi(0)|\boldsymbol{x}_1, \boldsymbol{x}_2, \cdots, \boldsymbol{x}_N, \boldsymbol{x}_{N+1}\rangle \mathfrak{S}_{N+1}[\boldsymbol{x}]. \tag{M.8}$$

Here, $\psi(0)$ eliminates one of the electrons, producing $\sum_s (-1)^{s+1}\delta(\boldsymbol{x}_s)$. For example, when the electron at $\boldsymbol{x}_1$ is eliminated, we have the term,

$$\prod_{r=1}^{N+1}\int d^2x_r \delta(\boldsymbol{x}_1)|\boldsymbol{x}_2, \cdots, \boldsymbol{x}_N, \boldsymbol{x}_{N+1}\rangle \mathfrak{S}(\boldsymbol{x}_1, \boldsymbol{x}_2, \cdots, \boldsymbol{x}_N, \boldsymbol{x}_{N+1}), \tag{M.9}$$

where $\mathfrak{S}(\boldsymbol{x}_1, \boldsymbol{x}_2, \cdots, \boldsymbol{x}_N, \boldsymbol{x}_{N+1}) = \mathfrak{S}_{N+1}[\boldsymbol{x}]$. By exchanging the integration variables $\boldsymbol{x}_1$ and $\boldsymbol{x}_{N+1}$, it reads

$$\begin{aligned} &\prod_{r=1}^{N+1}\int d^2x_r \delta(\boldsymbol{x}_{N+1})|\boldsymbol{x}_2, \cdots, \boldsymbol{x}_N, \boldsymbol{x}_1\rangle \mathfrak{S}(\boldsymbol{x}_{N+1}, \boldsymbol{x}_2, \cdots, \boldsymbol{x}_N, \boldsymbol{x}_1) \\ &= (-1)^{N+1}\prod_{r=1}^{N+1}\int d^2x_r \delta(\boldsymbol{x}_{N+1})|\boldsymbol{x}_1, \boldsymbol{x}_2, \cdots, \boldsymbol{x}_N\rangle \mathfrak{S}(0, \boldsymbol{x}_2, \cdots, \boldsymbol{x}_N, \boldsymbol{x}_1) \\ &= (-1)^{N+1}\prod_{r=1}^{N}\int d^2x_r |\boldsymbol{x}_1, \boldsymbol{x}_2, \cdots, \boldsymbol{x}_N\rangle \mathfrak{S}(\boldsymbol{x}_1, \boldsymbol{x}_2, \cdots, \boldsymbol{x}_N)\prod_{s=1}^{N}(z_s)^m. \end{aligned} \tag{M.10}$$

To derive the last equality we have used the explicit form of the Laughlin wave function (11.3.2) for $\mathfrak{S}_{N+1}[\boldsymbol{x}]$. Combining all these terms and comparing it with (M.7) we obtain

$$\psi(0)|\mathfrak{S}_{N+1}\rangle = \hat{S}^m|\mathfrak{S}_N\rangle, \tag{M.11}$$

apart from a multiplicative c-number constant. Therefore, m quasiholes are equivalent to one hole. Consequently, the quasihole carries the electric charge e/m.

The operator (M.1) creates a quasihole at the origin of the system. In order to create a quasiparticle at z_0 we modify the angular-momentum ladder operators

$$b_{\mathbf{x}}(z_0) = \frac{1}{\sqrt{2}}\left(z^* - 2z_0^* + \frac{\partial}{\partial z}\right), \qquad b_{\mathbf{x}}^\dagger(z_0) = \frac{1}{\sqrt{2}}\left(z - 2z_0 - \frac{\partial}{\partial z^*}\right), \tag{M.12}$$

where $b_{\mathbf{x}}(0) = b_{\mathbf{x}}$ and $b_{\mathbf{x}}^\dagger(0) = b_{\mathbf{x}}^\dagger$. Corresponding to (11.4.6) we have

$$\begin{aligned} b_r(z_0)\mathfrak{S}_{\rm LN}[\mathbf{x}] &= \sqrt{\frac{1}{2}}\left(\frac{\partial}{\partial z_r} - 2z_0^*\right)\lambda[z]\cdot e^{-\sum_s |z_s|^2}, \\ b_r^\dagger(z_0)\mathfrak{S}_{\rm LN}[\mathbf{x}] &= \sqrt{2}(z_r - z_0)\lambda[z]\cdot e^{-\sum_s |z_s|^2}. \end{aligned} \tag{M.13}$$

Their Hermitian property is correct though they look asymmetric,

$$\begin{aligned} \langle\mathfrak{S}_1|b_r(z_0)|\mathfrak{S}_2\rangle &= \int [d\mathbf{x}]e^{-2\sum_s |z_s|^2}\lambda_1[z]^*\sqrt{\frac{1}{2}}\left(\frac{\partial}{\partial z_r} - 2z_0^*\right)\lambda_2[z] \\ &= \int [d\mathbf{x}]e^{-2\sum_s |z_s|^2}\lambda_1[z]^*\sqrt{2}\,(z_r^* - z_0^*)\,\lambda_2[z] \\ &= \langle b_r^\dagger(z_0)\mathfrak{S}_1|\mathfrak{S}_2\rangle. \end{aligned} \tag{M.14}$$

The quasihole creation operator is

$$\hat{S}(z_0) = \exp\left\{\int d^2x\, \ln[b_{\mathbf{x}}^\dagger(z_0)]\rho(\mathbf{x})\right\}, \tag{M.15}$$

while the quasielectron creation operator is

$$\hat{S}^\dagger(z_0) = \exp\left\{\int d^2x\, \ln[b_{\mathbf{x}}(z_0)]\rho(\mathbf{x})\right\}. \tag{M.16}$$

Indeed, we find

$$\hat{S}(z_0)|\mathfrak{S}\rangle = 2^{N/2}\int [d\mathbf{x}]\prod_s (z_s - z_0)\lambda[z]\cdot e^{-\sum_s |z_s|^2}|\mathbf{x}_1, \mathbf{x}_2, \cdots, \mathbf{x}_N\rangle, \tag{M.17}$$

and

$$\hat{S}^\dagger(z_0)|\mathfrak{S}\rangle = 2^{-N/2}\int [d\mathbf{x}]\prod_s \left(\frac{\partial}{\partial z_s} - 2z_0^*\right)\lambda[z]\cdot e^{-\sum_s |z_s|^2}|\mathbf{x}_1, \mathbf{x}_2, \cdots, \mathbf{x}_N\rangle. \tag{M.18}$$

When $\lambda[z] = \prod_{r<s}(z_r - z_s)^m$, they describe one quasiparticle excitation localized at $z = z_0$ on the Laughlin state at $\nu = 1/m$.

We examine the creation of a quasiparticle in the composite-boson theory. The field-theoretical operators are still given by (M.1) and its Hermitian conjugate. When we operate them to the Laughlin state we obtain from (M.3) that

$$S|\mathfrak{S}\rangle = \rho_0^{N/2}\int [d\mathbf{x}]\prod_r \omega_{\rm qh}[z]\varphi^\ddagger(\mathbf{x}_1)\varphi^\ddagger(\mathbf{x}_2)\cdots\varphi^\ddagger(\mathbf{x}_N)|0\rangle, \tag{M.19a}$$

$$S^\dagger|\mathfrak{S}\rangle = \rho_0^{N/2}\int [d\mathbf{x}]\prod_r \omega_{\rm qe}[z]\varphi^\ddagger(\mathbf{x}_1)\varphi^\ddagger(\mathbf{x}_2)\cdots\varphi^\ddagger(\mathbf{x}_N)|0\rangle, \tag{M.19b}$$

where $\omega_{\text{qh}}[\boldsymbol{x}]$ and $\omega_{\text{qe}}[\boldsymbol{x}]$ are given by (11.4.9) and (11.4.12), respectively.

N Bosonization in One-Dimensional Space

We give formulas for bosonization in one-dimensional space. Any gapless mode is the chiral component, i.e., the left mover or the right mover. We study the equivalence between the chiral boson theory and the chiral fermion theory.

The chiral boson Lagrangian is defined by[445,171]

$$\mathcal{L} = \pi\hbar m\left(2\partial_t\phi(x)\partial_x\phi(x) - v\partial_x\phi(x)\partial_x\phi(x)\right). \tag{N.1}$$

The canonical momentum is

$$\pi(x) = \frac{\delta L}{\delta\dot{\phi}(x)} = 2\pi\hbar m\partial_x\phi(x), \tag{N.2}$$

and the canonical commutation relation yields

$$[\phi(x), \rho(x')] = \frac{i}{2\pi m}\delta(x - x') \tag{N.3}$$

with

$$\rho(x) = \partial_x\phi(x). \tag{N.4}$$

The Hamiltonian density is

$$\mathcal{H} = \pi(x)\dot{\phi}(x) - \mathcal{L} = \pi m\hbar v\partial_x\phi(x)\partial_x\phi(x), \tag{N.5}$$

and the Heisenberg equation of motion is

$$i\hbar\partial_t\phi(x) = [\phi(x), H] = i\hbar v\partial_x\phi(x). \tag{N.6}$$

It describes chiral bosons propagating to the left direction. The boson system (N.1) summarizes chiral edge excitations studied in Chapter 18.

From the commutation relation (N.3) we find

$$[\rho(x), \rho(x')] = \frac{i}{2\pi m}\partial_x\delta(x - x'), \tag{N.7}$$

$$[\phi(x), \phi(x')] = \frac{-i}{4\pi m}\text{sgn}(x - x'). \tag{N.8}$$

A comment is in order. The standard canonical quantization, reviewed in Chapter 3, is not applicable to the first-order Lagrangian density (N.1), as it stands. We are unable to require that two "coordinates" $\phi(x)$ and $\phi(x')$ commute. The commutation relation is given by (N.8). It implies that $\phi(x)$ is a nonlocal field operator. Nevertheless, it is a well-defined theory.[155] Furthermore, it is precisely the left chiral component of the massless Klein-Gordon theory, where the left and right components travel independently: See (18.6.4) in text.

On the other hand, the chiral fermion system is defined by

$$\mathcal{L} = i\hbar\psi^\dagger(x)\left(\partial_t - v\partial_x\right)\psi(x). \tag{N.9}$$

The canonical momentum is

$$\pi(x) = \frac{\delta L}{\delta\dot{\psi}(x)} = i\hbar\psi^\dagger(x), \tag{N.10}$$

and the canonical commutation relations are

$$\{\psi(x), \psi^\dagger(x')\} = \delta(x-x'), \qquad \{\psi(x), \psi(x')\} = 0. \tag{N.11}$$

The Hamiltonian density is

$$\mathcal{H} = \pi(x)\dot{\psi}(x) - \mathcal{L} = i\hbar v\psi^\dagger(x)\partial_x\psi(x), \tag{N.12}$$

and the Heisenberg equation of motion is

$$i\hbar\partial_t\psi(x) = [\psi(x), H] = i\hbar v\partial_x\psi(x). \tag{N.13}$$

It describes chiral fermions propagating to the left direction.

The equivalence between the fermion theory (N.9) and the boson theory (N.1) is provided by the bosonization formula

$$\psi(x) = Ce^{2\pi i a\phi(x)}, \tag{N.14}$$

where C is an ultraviolet regulator [see (N.18)] and a is a numerical constant [see (N.19)] to be fixed later. The equivalence is verified if the wave equation (N.13) follows from the wave equation (N.6), which is easy to see, and if the canonical anticommutation relations (N.11) follows from the canonical commutation relation (N.8), which we are going to show.

Since the operator product is ill defined at the same point, we need an ultraviolet regularization. It is customary to adopt a point-splitting method. We define

$$\rho(x) = \frac{1}{2}\lim_{\varepsilon\to 0}\left\{\psi^\dagger(x+\frac{1}{2}\varepsilon)\psi(x-\frac{1}{2}\varepsilon) - \psi(x-\frac{1}{2}\varepsilon)\psi^\dagger(x+\frac{1}{2}\varepsilon)\right\}, \tag{N.15}$$

to which we substitute the bosonization formula (N.14) and use the Hausdorff formula (B.8),[4]

$$\begin{aligned}\rho(x) = \frac{i}{2}\lim_{\varepsilon\to 0}|C|^2\Big(&\exp\left[-2\pi i a\left\{\phi(x+\frac{1}{2}\varepsilon) - \phi(x-\frac{1}{2}\varepsilon)\right\}\right] \\ &- \exp\left[-2\pi i a\left\{\phi(x-\frac{1}{2}\varepsilon) - \phi(x+\frac{1}{2}\varepsilon)\right\}\right]\Big).\end{aligned} \tag{N.16}$$

We expand the exponential in the limit $\varepsilon \to 0$,

$$\rho(x) = \lim_{\varepsilon\to 0}|C|^2\left(2\pi a\varepsilon\partial_x\phi(x) + O(\varepsilon^2)\right) = \frac{a}{m}\partial_x\phi(x), \tag{N.17}$$

[4] We use the Hausdorff formula $e^Ae^B = e^{[A,B]}e^Be^A$ with $A = \pm 2\pi i a\phi(x)$ and $B = \mp 2\pi i a\phi(x')$.

where we have set

$$|C|^2 = \frac{1}{2\pi m\varepsilon}. \tag{N.18}$$

This is the regularization factor in the bosonization formula (N.14).

Let us require

$$a = m, \tag{N.19}$$

so that (N.17) agrees with (N.4). The bosonization formula (N.14) reads

$$\psi(x) = Ce^{2\pi i m\phi(x)}. \tag{N.20}$$

The Hausdorff formula (B.8) yields[5]

$$\psi(x)\psi(x') = e^{\mp i\pi m}\psi(x')\psi(x), \quad \psi(x)\psi^\dagger(x') = e^{\pm i\pi m}\psi^\dagger(x')\psi(x), \tag{N.21}$$

by the use of (N.8); the sign $\mp$ corresponds to $x > x'$ and $x < x'$. We also use the Hausdorff formula (B.1) to find[6]

$$[\rho(x), \psi(x')] = -\delta(x - x')\psi(x) \tag{N.22}$$

by the use of (N.3). When m is odd, (N.21) and (N.22) are identical to (18.4.21) and (18.4.22) in text, and they imply the anticommutation relations (N.11). The electron field is given by the bosonization formula (N.20).

It is interesting to take $a = 1$ in the bosonization formula (N.14),

$$\psi_{\text{qe}}(x) = Ce^{2\pi i\phi(x)}. \tag{N.23}$$

We find

$$\psi(x)\psi(x') = e^{\mp i\pi/m}\psi(x')\psi(x), \quad \psi(x)\psi^\dagger(x') = e^{\pm i\pi/m}\psi^\dagger(x')\psi(x) \tag{N.24}$$

instead of the commutation relations (N.21), and

$$[\rho(x), \psi_{\text{qe}}(x')] = -\frac{1}{m}\delta(x - x')_{\text{qe}}\psi(x) \tag{N.25}$$

instead of (N.22). As we have noticed in text, the field (N.23) describes quasielectrons: See (18.4.26) and (18.4.27) in text.

We proceed to evaluate N-point functions of the fermion operators based on the bosonization formula (N.14), where $a = m$ (electron) or $a = 1$ (quasielectron). We separate the boson field $\phi(x)$ into the annihilation operator $\phi_+(x)$ and creation operator $\phi_-(x)$,

$$\phi_+(x) \equiv -\frac{i}{\sqrt{2\pi}}\int_0^\infty \frac{dk}{k}e^{ikx}\rho_k, \qquad \phi_-(x) \equiv \frac{i}{\sqrt{2\pi}}\int_0^\infty \frac{dk}{k}e^{-ikx}\rho_{-k}. \tag{N.26}$$

---

[5] We use the Hausdorff formula $e^A e^B = e^{[A,B]}e^B e^A$ with $A = 2\pi i m\phi(x)$ and $B = \pm 2\pi i m\phi(x')$.

[6] We use the Hausdorff formula $Be^A = e^A B - [A, B]e^A$ with $A = 2\pi i m\phi(x')$ and $B = \rho(x)$.

Using (18.4.9) we find

$$
\begin{aligned}
[\phi_+(x), \phi_-(x')] &= -\frac{1}{4\pi^2 m}\int_\mu^\infty \frac{dk}{k} e^{ik(x-x')} \\
&= \frac{1}{4\pi^2 m}\ln(|x-x'|\mu) - \frac{i}{4\pi m}\frac{\Theta(x-x')-\Theta(x'-x)}{2}, \\
&= \frac{1}{4\pi^2 m}\ln[(x-x'+i\varepsilon)\mu] - \frac{i}{8\pi m},
\end{aligned}
\tag{N.27}
$$

where μ is an infrared cutoff. We take the limit $\mu \to 0$ after all calculations are done. Note that the term $\ln(|x-x'|\mu)$ is cancelled in (N.8).

We introduce the normal-ordered exponential of the boson field,

$$
:e^{2\pi i\kappa a\phi(x)}: \equiv e^{2\pi i\kappa a\phi_-(x)} e^{2\pi i\kappa a\phi_+(x)}, \tag{N.28}
$$

where $\kappa = \pm 1$. Let us set

$$
\psi_\kappa(x) = \ell_0^{-1/2}(\ell_0\mu)^{a^2/2m} :e^{2\pi i\kappa a\phi(x)}:, \tag{N.29}
$$

where ℓ_0 is an arbitrary length parameter to give the correct dimension to $\psi_\kappa(x)$.

Introducing an ultraviolet regulator ε', we rewrite (N.29) as

$$
\begin{aligned}
\psi_\kappa(x) &= \ell_0^{-1/2}(\ell_0\mu)^{a^2/2m} \exp\left\{2\pi i\kappa a\phi_-(x+\frac{1}{2}\varepsilon')\right\}\exp\left\{2\pi i\kappa a\phi_+(x-\frac{1}{2}\varepsilon')\right\} \\
&= \ell_0^{-1/2}(\ell_0\mu)^{a^2/2m} e^{2\pi i\kappa a\phi(x)} \exp\left\{-2\pi^2 a^2[\phi_-(x+\frac{1}{2}\varepsilon'), \phi_+(x-\frac{1}{2}\varepsilon')]\right\} \\
&= \ell_0^{-1/2}(\varepsilon'/\ell_0)^{-a^2/2m} e^{2\pi i\kappa a\phi(x)}.
\end{aligned}
\tag{N.30}
$$

We identify this with the bosonization formula (N.14) for $\kappa = 1$ by setting

$$
C = \ell_0^{-1/2}(\varepsilon'/\ell_0)^{-a^2/2m}. \tag{N.31}
$$

Namely, $\psi(x) = \psi_{+1}(x)$ and $\psi^\dagger(x) = \psi_{-1}(x)$. Using (N.18), we find

$$
2\pi m\varepsilon = \ell_0(\varepsilon'/\ell_0)^{a^2/m}. \tag{N.32}
$$

The length scale ℓ_0 disappears only when $a = m = 1$: This is the case for the integer QH system ($\nu = 1/m = 1$).

The one-point function is

$$
\langle g|\psi_\kappa(x)|g\rangle = \lim_{\mu\to 0}\ell_0^{-1/2}(\ell_0\mu)^{a^2/2m} = 0. \tag{N.33}
$$

In evaluating two-point functions, we use the Hausdorff Formula (B.8) to find that

$$
\begin{aligned}
&\langle g| :\exp[2\pi i\kappa a\phi(x)]::\exp[2\pi i\kappa' a\phi(x')]: |g\rangle \\
&\quad = \exp\left\{-4\pi^2\kappa\kappa' a^2[\phi_+(x), \phi_-(x')]\right\} \\
&\quad = \exp\left\{\kappa\kappa'(a^2/m)\ln[(x-x')\mu]\right\}.
\end{aligned}
\tag{N.34}
$$

We use the identity

$$\exp\left\{\kappa\kappa'(a^2/m)\ln\mu\right\} = \exp\left(\frac{1}{2}(\kappa+\kappa')^2(a^2/m)\ln\mu\right)\exp\left\{-(a^2/m)\ln\mu\right\}, \quad \text{(N.35)}$$

which is valid for $|\kappa| = |\kappa'| = 1$. Then, using (N.29), we obtain

$$\langle 0|\psi(x)\psi(x')|0\rangle = \ell_0^{-1}\lim_{\mu\to 0}\exp\left\{(a^2/m)\ln(x-x')\right\}(\ell_0\mu)^{2a^2/m} = 0 \quad \text{(N.36)}$$

since $(\kappa+\kappa')^2 = 4$, while

$$\langle 0|\psi^\dagger(x)\psi(x')|0\rangle = \ell_0^{-1}\exp\left\{-(a^2/m)\ln[(x-x')/\ell_0]\right\} \propto \frac{1}{(x-x')^{a^2/m}} \quad \text{(N.37)}$$

since $(\kappa+\kappa')^2 = 0$. All $2N$-point functions are similarly calculated,

$$\langle 0|\psi^\dagger(x_1)\cdots\psi^\dagger(x_N)\psi(x_1')\cdots\psi(x_N')|0\rangle = \ell_0^{-N}\exp\left\{-\frac{a^2}{m}\sum_{i,j}\ln\left(\frac{x_i-x_j'}{\ell_0}\right)\right\}. \quad \text{(N.38)}$$

When the numbers of operators $\psi^\dagger$ and ψ are not equal, the N-point function is shown to vanish in the limit $\mu\to 0$. All these properties say that the field $\psi_\kappa(x)$ carries a superselection-rule charge,[205,461] that is the electron number in the present case.

As is explained in text, the electromagnetic potential A_μ is incorporated into the system based on the minimal substitution. The gauge transformation is given by (18.5.9) and (18.5.11),

$$\psi(x) \to e^{if(x)}\psi(x), \quad A_\mu(x) \to A_\mu(x) - \frac{\hbar}{e}\partial_\mu f(x),$$
$$\phi(x) \to \phi(x) + \frac{1}{2\pi m}f(x). \quad \text{(N.39)}$$

It is necessary to reexamine the operator product (N.15), since it is not gauge invariant for $\varepsilon \neq 0$. In general, a gauge invariant object is given by the expression,

$$\psi^\dagger(x)\exp\left(-i\frac{e}{\hbar}\int_{x'}^{x}dy\,A_x(y)\right)\psi(x'). \quad \text{(N.40)}$$

The operator product (N.15) is redefined,

$$\rho_{\text{g-inv}}(x) = \frac{1}{2}\lim_{\varepsilon\to 0}\left\{\psi^\dagger(x+\frac{1}{2}\varepsilon)\exp\left\{-i\frac{e}{\hbar}\varepsilon A_x(x)\right\}\psi(x-\frac{1}{2}\varepsilon)\right.$$
$$\left. - \psi(x-\frac{1}{2}\varepsilon)\exp\left\{-i\frac{e}{\hbar}\varepsilon A_x(x)\right\}\psi^\dagger(x+\frac{1}{2}\varepsilon)\right\}, \quad \text{(N.41)}$$

which yields

$$\rho_{\text{g-inv}}(x) = \partial_x\phi(x) + \frac{e}{2\pi m\hbar}A_x(x), \quad \text{(N.42)}$$

in the limit $\varepsilon\to 0$. The gauge invariant charge is given by $\rho_e(x) = -e\rho_{\text{g-inv}}(x)$.

O Coulomb Energy Formulas

We estimate the Coulomb energy (self-energy, capacitance energy and exchange energy) of one skyrmion. We start with a skyrmion in the monolayer QH ferromagnet. The self-energy is

$$\begin{aligned} E_{\text{self}} =& \frac{1}{2}\int d^2x d^2y\, \delta\rho_{\text{sky}}(\boldsymbol{x}) V_{\text{D}}(\boldsymbol{x}-\boldsymbol{y}) \delta\rho_{\text{sky}}(\boldsymbol{y}) \\ =& \pi \int d^2q\, \delta\rho_{\text{sky}}(-\boldsymbol{q}) V_{\text{D}}(\boldsymbol{q}) \delta\rho_{\text{sky}}(\boldsymbol{q}), \end{aligned} \tag{O.1}$$

where $V_{\text{D}}(\boldsymbol{x})$ is the effective Coulomb potential with its Fourier component

$$V_{\text{D}}(q) = \frac{1}{2\pi}\int d^2x\, e^{i\boldsymbol{qx}} V_{\text{D}}(\boldsymbol{x}) = \frac{e^2}{4\pi\varepsilon q} e^{-\ell_B^2 q^2/2}. \tag{O.2}$$

The density modulation $\delta\rho_{\text{sky}}(\boldsymbol{x})$ is given by

$$\delta\rho_{\text{sky}}(\boldsymbol{x}) \simeq -\frac{1}{\pi}\frac{\alpha^2}{(\boldsymbol{x}^2+\alpha^2)^2} \tag{O.3}$$

for a large skyrmion with $\alpha = 2\kappa\ell_B$. Its Fourier component reads

$$\delta\rho_{\text{sky}}(\boldsymbol{q}) = \frac{1}{2\pi}\int d^2x\, e^{i\boldsymbol{qx}} \delta\rho_{\text{sky}}(\boldsymbol{x}) = \frac{\alpha q}{2\pi} K_1(\alpha q). \tag{O.4}$$

To calculate the energy we use the following formulas,

$$\begin{aligned} \oint d\theta\, e^{ikx} =& 2\pi J_0(kr), \\ \frac{1}{2\pi i}\oint d\theta\, \frac{x_j}{r} e^{i\boldsymbol{qx}} =& \frac{q_j}{q} J_1(qr), \end{aligned} \tag{O.5}$$

and

$$\int_0^\infty dr\, \frac{r^{n+1} J_n(qr)}{(r^2+\alpha^2)^{m+1}} = \frac{q^m \alpha^{n-m}}{2^m m!} K_{n-m}(\alpha q), \tag{O.6}$$

where we have set $\boldsymbol{x} = (r\cos\theta, r\sin\theta)$, and $K_n(z) = K_{-n}(z)$ is the modified Bessel function: See formula 11.4.44 in reference.[12] Now it is straightforward to derive

$$E_{\text{self}}(\kappa) = \frac{E_{\text{C}}^0}{4\kappa}\int z^2 [K_1(z)]^2 e^{-\frac{1}{8}\left(\frac{z}{\kappa}\right)^2} dz. \tag{O.7}$$

This is the Coulomb self-energy of one spin-skyrmion in the monolayer QH ferromagnet.

In the bilayer QH system the SU(4)-invariant Coulomb interaction produces the self-energy of a skyrmion,

$$E_{\text{self}} = \pi \int d^2q\, \delta\rho_{\text{sky}}(-\boldsymbol{q}) V_{\text{D}}^{+}(\boldsymbol{q}) \delta\rho_{\text{sky}}(\boldsymbol{q}) \tag{O.8}$$

with

$$V_{\rm D}^{+}(q) = \frac{e^2}{8\pi\varepsilon q}\left(1 + e^{-qd}\right)e^{-\ell_B^2 q^2/2}, \tag{O.9}$$

for which we find

$$E_{\rm self} = \frac{E_{\rm C}^0}{8\kappa}\int z^2[K_1(z)]^2\left(1 + e^{-\frac{d}{2\ell}\frac{z}{\kappa}}\right)e^{-\frac{1}{8}\left(\frac{z}{\kappa}\right)^2}dz, \tag{O.10}$$

by comparing (O.2) and (O.9).

The SU(4)-noninvariant Coulomb interaction yields the capacitance energy,

$$E_{\rm cap} = 2\pi\ell_B^2\epsilon_{\rm cap}\int d^2x\,\delta P_z^{\rm sky}(\boldsymbol{x})\delta P_z^{\rm sky}(\boldsymbol{x}), \tag{O.11}$$

where

$$\delta P_z^{\rm sky}(\boldsymbol{x}) = \rho_{\rm sky}(\boldsymbol{x})\mathcal{P}_z^{\rm sky}(\boldsymbol{x}) - \frac{1}{2}\rho_0\sigma_0 \tag{O.12}$$

with (O.3) and

$$\mathcal{P}_z^{\rm sky}(\boldsymbol{x}) = \frac{\sigma_0}{2} + \frac{\alpha_p\sqrt{1-\sigma_0^2}x - \alpha_p^2\sigma_0}{r^2+\alpha^2} \tag{O.13}$$

with $\alpha = 2\kappa\ell_B$ and $\alpha_p = 2\kappa_p\ell_B$. After a straightforward calculation we obtain

$$E_{\rm cap} = \epsilon_{\rm cap}\left[\frac{\sigma_0^2}{24\kappa^2} + (1-\sigma_0^2)\kappa_p^2 A(\kappa) + \sigma_0^2\left(\frac{1}{4\kappa^2} - 1\right)\frac{\kappa_p^2}{2\kappa^2} + \sigma_0^2\left(\frac{1}{10\kappa^2} - \frac{2}{3} + 2\kappa^2\right)\frac{\kappa_p^4}{\kappa^4}\right], \tag{O.14}$$

where $A(\kappa)$ is divergent logarithmically,

$$A(\kappa) = \frac{1}{\kappa^2}\int_0^\infty r^3\left(\frac{1}{r^2+\alpha^2} - \frac{2\alpha^2\ell_B^2}{(r^2+\alpha^2)^3}\right)^2 dr. \tag{O.15}$$

We take the divergent term as the leading contribution for the pseudospin component,

$$E_{\rm cap} \simeq \frac{\sigma_0^2}{24\kappa^2}\epsilon_{\rm cap} + \frac{1}{2}(1-\sigma_0^2)\epsilon_{\rm cap}N_{\rm ppin}(\kappa_p) \tag{O.16}$$

with (25.5.22). We have retained the term $\left(\sigma_0^2/24\kappa^2\right)\epsilon_{\rm cap}$ because it is common to the spin and pseudospin components.

Finally we present the formula for the exchange energy (25.2.8) for a general

SU(4)-skyrmion (25.4.6). After a straightforward but tedious calculation we obtain

$$\begin{aligned}\tilde{E}_X^{SU(4)} =& 4\pi J_s^+ \left(1 + \frac{1}{10\kappa^4}\right) - 4\pi J_s^- \left(\frac{1}{3} + \frac{1-3\sigma_0^2}{140\kappa^4}\right) \\ &+ 4\pi J_s^- \left(\frac{2}{3} + \frac{3}{14\kappa^4}\right) \frac{\kappa_s^2}{\kappa^2} \left(\sigma_0 \frac{\kappa_s}{\kappa} + \sqrt{1-\sigma_0^2}\frac{\kappa_r}{\kappa}\right)^2 \qquad (O.17)\\ &+ 4\pi J_s^- \left(\frac{1+\sigma_0^2}{3} + \frac{1-19\sigma_0^2}{140\kappa^4}\right)\frac{\kappa_s^2}{\kappa^2} + 4\pi J_s^- (1-\sigma_0^2)\left(\frac{1}{3} + \frac{1}{140\kappa^4}\right)\frac{\kappa_r^2}{\kappa^2} \\ &+ 4\pi J_s^- \sigma_0\sqrt{1-\sigma_0^2}\left(\frac{2}{3} + \frac{9}{70\kappa^4}\right)\frac{\kappa_s\kappa_r}{\kappa^2}. \qquad (O.18)\end{aligned}$$

By setting $\kappa_r = 0$, this is reduced to $\tilde{E}_X^{SU(4)} = E_X^{SU(4)} + \Delta E_X$ with (25.5.11), where

$$\Delta E_X = \frac{4\pi}{10\kappa^4}J_s^+ - \frac{4\pi\kappa_p^2}{140\kappa^6}J_s^- + \frac{4\pi\sigma_0^2}{140\kappa^8}\left[3\kappa^4 - 19\kappa^2\kappa_s^2 + 30\kappa_s^4\right]J_s^-. \qquad (O.19)$$

The correction term is small for a large skyrmion.

P W_∞ ALGEBRA

The density algebra (29.4.43) is not only a field theoretical representation of the magnetic translation algebra (29.3.3) but also a representation of a much larger algebra, the W_∞ algebra. It is the symmetry of the QH system.[236,84,459]

Quantum Mechanical Realization

The one-body state of an electron in the lowest Landau level is given by the Landau site $|n\rangle$. In the symmetric gauge it is the angular-momentum state $|n\rangle$ defined by (10.2.17), or $|n\rangle = \sqrt{1/n!}(b^\dagger)^n|0\rangle$. Two of these states are related as

$$|m\rangle = \sqrt{\frac{1}{m!n!}}L(m,n)|n\rangle \qquad (P.1)$$

by way of the operator

$$L(m,n) = (b^\dagger)^m b^n. \qquad (P.2)$$

It is easy to see that $L(m,n)$ generates an infinite dimensional algebra without central term,

$$[L(m,n), L(r,s)] = \sum_{t=1}^{\text{Min}(n,r)} C_{mn;rs}^t L(m+r-t, n+s-t), \qquad (P.3)$$

where the structure constant is

$$C_{mn;rs}^t = \frac{1}{t!}\left(\frac{n!r!}{(n-t)!(r-t)!} - \frac{m!s!}{(m-t)!(s-t)!}\right). \qquad (P.4)$$

This is called the W_∞ algebra after the usage in string theory.[59,233,397]

Field Theoretical Realization

A field-theoretical realization of the W_∞ algebra (P.3) is given by

$$\mathcal{L}(m,n) = \int d^2x\, \hat{\psi}^\dagger(\mathbf{x}) L(m,n) \hat{\psi}(\mathbf{x}), \tag{P.5}$$

where $\hat{\psi}(\mathbf{x})$ is the projected field operator (10.6.22). To show this we substitute (10.6.22) into (P.5),

$$\mathcal{L}(m,n) = \sum_{ij} c^\dagger(i)c(j) \int d^2x\, \langle i|\mathbf{x}\rangle L(m,n) \langle \mathbf{x}|j\rangle = \sum_{ij} c^\dagger(i)c(j)\langle i|L(m,n)|j\rangle. \tag{P.6}$$

Thus,

$$\mathcal{L}(m,n)\mathcal{L}(r,s) = \sum_{ijkl} c^\dagger(i)c(j)c^\dagger(k)c(l)\langle i|L(m,n)|j\rangle\langle k|L(r,s)|l\rangle. \tag{P.7}$$

Using the anticommutation relation of c_j, it is easy to see

$$[\mathcal{L}(m,n), \mathcal{L}(r,s)] = \int d^2x\, \hat{\psi}^\dagger(\mathbf{x})[L(m,n), L(r,s)]\hat{\psi}(\mathbf{x}). \tag{P.8}$$

Hence, $\mathcal{L}(m,n)$ generates the same algebra as $L(m,n)$,

$$[\mathcal{L}(m,n), \mathcal{L}(r,s)] = \sum_{t=1}^{\mathrm{Min}(n,r)} C^t_{mn;rs}\mathcal{L}(m+r-t, n+s-t). \tag{P.9}$$

It provides a field-theoretical realization of the W_∞ algebra.

We express (P.6) as

$$\mathcal{L}(m,n) = \sum_{ij} \langle i|L(m,n)|j\rangle \rho(i,j), \tag{P.10}$$

with $\rho(i,j)$ given by (29.4.9). It is found that $\mathcal{L}(m,n)$ and $\rho(m,n)$ span the same linear space. We list some of them explicitly.

$$\mathcal{L}(0,0) = \sum_m \rho(m,m), \qquad \mathcal{L}(0,1) = \sum_m \sqrt{m+1}\,\rho(m,m+1),$$
$$\mathcal{L}(1,0) = \sum_m \sqrt{m+1}\,\rho(m+1,m), \qquad \mathcal{L}(1,1) = \sum_m m\rho(m,m),$$
$$\mathcal{L}(0,2) = \sum_m \sqrt{(m+1)(m+2)}\,\rho(m,m+2), \qquad \dots\,. \tag{P.11}$$

It implies that $\rho(m,n)$ also generates the W_∞ algebra. Recall that the density algebra (29.4.43) follows from the algebra (29.4.41) made by $\rho(m,n)$. Hence, the density algebra presents a field-theoretical realization of the W_∞ algebra.

REFERENCES

Books on QH Effects and Related Topics

1. T. Ando, Y. Arakawa, K. Furuya, S. Komiyama and H. Nakashima (eds), *Mesoscopic Physics and Electronics* (Springer, 1997).
2. T. Chakraborty and P. Pietiläinen, *The Quantum Hall Effects: Fractional and Integral* (Springer, 1995) 2nd ed..
3. S. Das Sarma and A. Pinczuk (eds), *Perspectives in Quantum Hall Effects* (Wiley, 1997).
4. Z.F. Ezawa, *Quantum Hall Effects: Field-Theoretical Approach and Related Topics* (World Scientific, 2000).
5. E. Fradkin, *Field Theories of Condensed Matter Systems* (Addison-Wesley, 1991).
6. O. Heinonen (ed), *Composite Fermions* (World Scientific, 1998).
7. H. Kamimura and H. Aoki, *The Physics of Interacting Electrons in Disordered Systems* (Oxford University Press, 1989).
8. A. Khare, *Fractional Statistics and Quantum Theory* (World Scientific, 1997).
9. R.E. Prange and S.M. Girvin (eds), *The Quantum Hall Effect* (Springer, 1990) 2nd ed..
10. M. Stone (ed), *Quantum Hall Effect* (World Scientific, 1992) page 289.
11. F. Wilczek, *Franctional Statistics and Anyon Superconductor* (World Scientific, 1990).

Other Books

12. M. Abramowitz and I.A. Stegun, *Handbook of Mathematical Functions* (Dover Publications, New York 1972).
13. A. Barone and G. Paternò, *Physics and Applications of the Josephson Effect* (John Wiley, 1982).
14. G. Bastard, *Wave Mechanics Applied to Semiconductor Heterostructures* (Les Éditions de Physique, Paris, 1988).
15. S.S. Chern, *Complex Manifolds without Potential Theory* (Springer, 1979) 2nd ed..
16. A. Connes, *noncommutative geometry* (Academic Press, 1994).
17. I. Duck and E.C.G. Sudarshan, *Pauli and the Spin-Statistics Theorem.*

18. R.P. Feynman, R.B. Leighton and M. Sands, *Lectures on Physics* (Addison-Wesley, 1963), vol.3, Chap.4.
19. R.P. Feynman and A.R. Hibbs, *Quantum Mechanics and Path Integrals* (McGraw-Hill, New York, 1965).
20. R.P. Feynman, R.B. Leighton and M. Sands, *Lectures on Physics* (Addison-Wesley, 1963), vol.I, Chap.42.
21. R.P. Feynman, *Statistical Mechanics* (Benjamin, 1972), Section 11.6..
22. H. Flanders, *Differential Forms* (Academic Press, 1963).
23. M.Gell-Mann and Y. Ne'eman, *The eight-fold Way* (Benjamin, New York, 1964).
24. P.J. Hilton, *An Introduction to Homotopy Theory* (Cambridge Tracts in Mathematics and Mathematical Physics, No.43).
25. B.W. Lee, *Chiral Dynamics* (Gordon and Breach, 1972).
26. F. London, *Superfluids* (John Wiley, New York, 1950).
27. V.G. Makhankov, Y.P. Rybakov and V.I. Sanyuk, *The Skyrme Model* (Springer-Verlag, 1993).
28. R. Rajaraman, *Solitons and Instantons* (North-Holland, 1982).
29. N.E. Steenrod, *The Topology of Fiber Bundles* (Princeton Mathematics Series, No.14).
30. M. Stone, The physics of Quantum Fields (Springer-Verlag, 2000).
31. B. Thaller, *The Dirac Equation* (Springer-Verlag, Berlin, 1992).
32. M. Tinkham, *Introduction to Superconductivity* (McGraw-Hill, 1975).
33. A. Tonomura, *Electron Holography* (Springer-Verlag, 1999).
34. H. Umezawa, *Advanced Field Theory: Micro, Macro and Thermal Physics* (AIP, 1993).
35. D.F. Walls and G.J. Milburn, *Quantum Optics* (Springer-Verlag).

Original Articles

36. M. Abolfath, J.J. Palacios, H.A. Fertig, S.M. Girvin and A.H. MacDonald, Phys. Rev. B **56** (1997) 6795.
37. E. Abrahams, P.W. Anderson, D.C. Licciardello and T.V. Ramakrishnan, Phys. Rev. Lett. **42** (1979) 673.
38. S. Adler, Phys. Rev. **177** (69) 2426.
39. Y. Aharonov and D. Bohm, Phys. Rev. **115** (1959) 485.
40. E. Ahlswede, J. Weis, K. v.Klitzing, K. Eberl, Physica E **12** (2002) 165.
41. E.H. Aifer, B.B. Goldberg and D.A. Broido, Phys. Rev. Lett. **76** (1996) 680.
42. H. Ajiki and T. Ando, J. Phys. Soc. Japan, **62** (1993) 1255.
43. J. Alicea and M.P.A. Fisher, Phys. Rev. B **74** (2006) 075422.
44. P.W. Anderson, Phys. Rev. **109** (1958) 1492.
45. T. Ando and Y. Umemura, J. Phys. Soc. Jap. **37** (1974) 1044.
46. T. Ando, Y. Matsumoto and Y. Uemura, J. Phys. Soc. Jap. **39** (1975) 274.

47. Y. Zheng and T. Ando, Phys. Rev. B **65** (2002) 245420.
48. T. Ando, in reference.[1]
49. T. Ando, J. Phys. Soc. Jap. **74** (2005) 777.
50. H. Aoki and T. Ando, Solid State Commun. **38** (1981) 1079.
51. H. Aoki and T. Ando, Surface Science **170** (1986) 249.
52. D.P. Arovas, R. Schrieffer and F. Wilczek, Phys. Rev. Lett. **53** (1984) 722.
53. D.P. Arovas, R. Schrieffer, F. Wilczek and A. Zee, Nucl. Phys. B **251** (1985) 117.
54. D.P. Arovas, R.N. Bhatt, F.D.M. Haldane, P.B. Littlewood and R. Rammal, Phys. Rev. Lett. **60** (1988) 619.
55. I.J.R. Aitchison, G. Metikas and D.J. Lee, cond-mat/990508008 (1999).
56. J.E. Avron and R. Seiler, Phys. Rev. Lett. **54** (1985) 259.
57. H. Awata, M. Fukuma, Y. Matsuo and S. Odake, Progr. Theor. Phys. Suppl. **118** (1995) 343.
58. P. Bak, Rep. Prog. Phys. **45** (1982) 587.
59. I. Bakas, Phys. Lett. B **228** (1989) 57.
60. A. Balatsky and V. Kalmeyer, Phys. Rev. B **43** (1991) 6228.
61. L. Balents and L. Radzihovsky, Phys. Rev. Lett. **86** (2001) 1825.
62. V. Bargmann, P. Butera, L. Girardello and J.R. Klauder, Rev. Math. Phys. **2** (1971) 221.
63. S.E. Barrett, G. Dabbagh, L.N. Pfeiffer, K.W. West and R. Tycko, Phys. Rev. Lett. **74** (1995) 5112.
64. R. Basu, G. Date and M.V.N. Murthy, Phys. Rev. B **46** (1992) 3139.
65. V. Bayot, E. Grivei, S. Melinte, M.B. Santos and M. Shayegan, Phys. Rev. Lett. **76** (1996) 4584.
66. C.W.J. Beenakker, Phys. Rev. Lett. **64** (1990) 216.
67. A.A. Belavin and A.M. Polyakov, JETP Letters **22** (1975) 245.
68. J.S. Bell and R. Jackiw, Nuovo Cim. **60A** (1969) 47.
69. P. Beran and R. Morf, Phys. Rev. B **43** (1991) 12654.
70. G.S. Boebinger, A.M. Chang, H.L. Stormer and D.C. Tsui, Phys. Rev. Lett. **55** (1985) 1606.
71. G.S. Boebinger, H.W. Jiang, L.n. Pfeiffer and K.W. West, Phys. Rev. Lett. **64** (1990) 1793.
72. N.N. Bogoliubov, J. Phys. (USSR) **11** (1947) 23.
73. A. Boisen, P. Boggild, A. Kristensen and P.E. Lindelof, Phys. Rev. B **50** (1994) 1967.
74. M. Boon and J. Zak, J. Math. Phys. 22 (1981) 1090.
75. L. Brey, Phys. Rev. Lett. **65** (1990) 903.
76. L. Brey, H.A. Fertig, R. Cote and A.H. MacDonald, Phys. Rev. Lett. **75** (1995) 2562.
77. L. Brey, H.A. Fertig, R. Cote and A.H. MacDonald, Phys. Rev. B **54** (1996) 16888.

78. L. Brey, E. Demler and S. Das Sarma, Phys. Rev. Lett. **83** (1999) 168.
79. M. Büttiker, Phys. Rev. B **38** (1988) 9375.
80. V. Buzek, A.D. Wilson-Gordon, P.L. Knight and W.K. Lai, Phys. Rev. A **45** (1992) 8079.
81. C. Callan, R. Dashen and D. Gross, Phys. Rev. B **63** (1976) 334.
82. F. Calogero, J. Math. Phys. **10** (1969) 2191; 2197; **12** (1971) 419.
83. G. Calugareanu, Rev. Math. Pures et Appl (Bucarest) **4** (1959) 5.
84. A. Cappelli, C. Trugenberger and G. Zemba, Nucl. Phys. B **396** (1993) 465.
85. A. Cappelli, G.V. Dunne, C. Trugenberger and G. Zemba, Nucl. Phys. B **398** (1993) 531.
86. A. Cappelli, C. Trugenberger and G. Zemba, Phys. Rev. Lett. **72** (1994) 1902.
87. A. Cashe, J. Kogut and L. Susskind, Phys. Rev. D **10** (1974) 732.
88. T. Chakraborty and F.C. Zhang, Phys. Rev. B **29** (1984) 7032.
89. T. Chakraborty, Phys. Rev. B **31** (1985) 4026.
90. T. Chakraborty, P. Pietiläinen and F.C. Zhang, Phys. Rev. Lett. **57** (1986) 130.
91. T. Chakraborty and P. Pietiläinen, Phys. Rev. Lett. **38** (1988) 10097.
92. T. Chakraborty and P. Pietiläinen, Phys. Rev. B **41** (1990) 10862.
93. T. Chakraborty, Surface Science **229** (1990) 16.
94. L.L. Chang, E.E. Mendez, W.I. Wang, L. Esaki and P.M. Tedrow, Proc. 17th Int. Conf. Phys. Semiconductors, p.299, J.D. Chadi and W.A. Harrison (eds.) (Springer, 1985).
95. A.M. Chang and J.E. Cunningham, Solid State Commun. **72** (1989) 651.
96. A.M. Chang, Solid State Commun. **74** (1990) 871.
97. A.M. Chang, L.N. Pfeiffer and K.W. West, Phys. Rev. Lett. **77** (1996) 2538.
98. Y.H. Chen, F. Wliczek, E. Witten and B.I. Halperin, Int. J. Mod. Phys. B **3** (1989) 1001.
99. X.M. Chen and J.J. Quinn, Phys. Rev. B **45** (1992) 11054.
100. K.H. Cho and C. Rim, Ann. Phys. **213** (1992) 295.
101. Y.C. Chung, M. Heiblum and V. Umansky, Phys. Rev. Lett. **91** (2003) 216804.
102. C. Chou, Phys. Rev. D **44** (1991) 2533; **45** (1992) E1433.
103. R.G. Clark, S.R. Haynes, A.M. Suckling, J.R. Mallett and P.A. Wright, Phys. Rev. Lett. **62** (1989) 1536.
104. S. Coleman, Phys. Rev. D **11** (1975) 2088.
105. S. Coleman, Comm. Math. Phys. 31 (1973) 259.
106. P.T. Coleridge, Z.W. Wasilewski, P. Zawadzki, A.S. Sachrajda and H.A. Carmona, Phys. Rev. B **52** (1995) 11603.
107. N.R. Cooper, Phys. Rev. Lett. **80** (1998) 4554.
108. K.B. Cooper, M.P. Lilly, J.P. Eisenstein, L.N. Pfeiffer and K.W. West, Phys. Rev. B **60** (1999) R11285.

109. G.W. Crabtree and R. Nelson, Physics Today **50** (1979) No4.
110. A. D'Adda, A. Luscher and P. DiVecchia, Nucl. Phys. B **149** (1978) 63.
111. A.J. Dahm, A. Denenstein, T.F. Finnegan, D.N. Langenberg and D.J. Scalapino, Phys. Rev. Lett. **20** (1968) 859; 1020(E).
112. H.D.M. Davies, J.C. Harris, J.F. Ryan and A.J. Turberfield, Phys. Rev. Lett. **78** (1997) 4095.
113. E. Demler and S. Das Sarma, Phys. Rev. Lett. **82** (1999) 3895.
114. E. Demler, H. Kim and S. Das Sarma, Phys. Rev. B **61** (2000) 10567.
115. G.H. Derrick, J. Math. Phys. **5** (1964) 1252.
116. S. Deser, R. Jackiw and S. Templeton, Phys. Rev. Lett. **48** (1982) 975.
117. P.A.M. Dirac, Proc.Roy.Soc. (London) **A114** (1927) 243.
118. R.R. Du, H.L. Stormer, D.C. Tsui, L.N. Pfeiffer and K.W. West, Phys. Rev. Lett. **70** (1993) 2944.
119. R.R. Du, H.L. Stormer, D.C. Tsui, A.S. Yeh, L.N. Pfeiffer and K.W. West, Phys. Rev. Lett. **73** (1994) 3274.
120. R.R. Du, A.S. Yeh, H.L. Stormer, D.C. Tsui, L.N. Pfeiffer and K.W. West, Phys. Rev. Lett. **75** (1995) 3926.
121. R.R. Du, A.S. Yeh, H.L. Stormer,D.C. Tsui, L.N. Pfeiffer and K.W. West, Phys. Rev. Lett. **55** (1997) 7351.
122. R.R. Du, D.C. Tsui, H.L. Stormer, L.N. Pfeiffer, K.W. Baldwin and K.W. West, Solid State Commun. **109** (1999) 389.
123. G.V. Dunne, A. Lerda, S. Sciuto and C.A. Trugenberger, Nucl. Phys. B **370** (1992) 601.
124. H. Eichenherr, Nucl. Phys. B **146** (1978) 215.
125. J.P. Eisenstein, R. Willett, H.L. Stormer, D.C. Tsui, A.C. Gossard and J.H. English, Phys. Rev. Lett. **61** (1988) 997.
126. J.P. Eisenstein, H.L. Stormer, L. Pfeiffer and K.W. West, Phys. Rev. Lett. **62** (1989) 1540.
127. J.P. Eisenstein, G.S. Boebinger, L. Pfeiffer, K.W. West and Song He, Phys. Rev. Lett. **68** (1992) 1383.
128. J.P. Eisenstein, in reference.[3]
129. J.P. Eisenstein and A.H. MacDonald, Nature **432** (2004) 691.
130. L.W. Engel, S.W. Hwang, T. Sajoto, D.C. Tsui and M. Shayegan, Phys. Rev. B **45** (1992) 3418.
131. L.W. Engel, D. Shahar, C. Kurdak and D.C. Tsui, Phys. Rev. Lett. **71** (1993) 2638.
132. M. Ezawa, cond-mat/0606084 (Phys. Lett. A, in press); cond-mat/0609612 (Physica E, in press); J. Phys. Soc. Jap. **76** (2007) 094701.
133. Z.F. Ezawa, Phys. Rev. D **18** (1978) 2091.
134. Z.F. Ezawa and A. Iwazaki, Phys. Rev. B **43** (1991) 2637.
135. Z.F. Ezawa, M. Hotta and A. Iwazaki, Phys. Rev. D **44** (1991) 452.
136. Z.F. Ezawa, M. Hotta and A. Iwazaki, Phys. Rev. Lett. **67** (1991) 411.

137. Z.F. Ezawa, M. Hotta and A. Iwazaki, Phys. Rev. B **46** (1992) 7765.
138. Z.F. Ezawa and A. Iwazaki, Int. J. Mod. Phys. B **6** (1992) 3205.
139. Z.F. Ezawa and A. Iwazaki, Phys. Rev. Lett. **70** (1993) 3119.
140. Z.F. Ezawa and A. Iwazaki, Phys. Rev. B **47** (1993) 7295; **48** (1993) 15189.
141. Z.F. Ezawa, A. Iwazaki and Y.S. Wu, Mod. Phys. Lett. B **7** (1993) 1825.
142. Z.F. Ezawa and A. Iwazaki, Int. J. Mod. Phys. B **8** (1994) 2111.
143. Z.F. Ezawa, Phys. Rev. B **51** (1995) 11152.
144. Z.F. Ezawa, Phys. Lett. A **229** (1997) 392; Phys. Rev. B **55** (1997) 7771.
145. Z.F. Ezawa, Phys. Lett. A **249** (1998) 223; Physica B **249-251** (1998) 841.
146. Z.F. Ezawa and K. Sasaki, J. Phys. Soc. Jap. **68** (1999) 576.
147. Z.F. Ezawa, Phys. Rev. Lett. **82** (1999) 3512.
148. Z.F. Ezawa and K. Hasebe, Phys. Rev. B **65** (2002) 075311.
149. Z.F. Ezawa, G. Tsitsishvili and K. Hasebe, Phys. Rev. B **67** (2003) 125314.
150. Z.F. Ezawa and G. Tsitsishvili, Phys. Rev. B **70** (2004) 125304.
151. Z.F. Ezawa, M. Eliashvili and G. Tsitsishvili, Phys. Rev. B **71** (2005) 125318.
152. Z.F. Ezawa and G. Tsitsishvili, Phys. Rev. D **72** (2005) 85002.
153. Z.F. Ezawa, S. Suzuki and G. Tsitsishvili, Phys. Rev. B **76** (2007) 045307.
154. Z.F. Ezawa, K. Ishii and G. Tsitsishvili, Physica E (2007), in press.
155. L. Faddeev and R. Jackiw, Phys. Rev. Lett. **60** (1988) 1692.
156. E. Fadell and L. Neuwirth, Math. Scand. **10** (62) 111.
157. F.F. Fang and P.J. Stiles, Phys. Rev. **174** (1968) 823.
158. G. Fano, F. Ortolani and E. Colombo, Phys. Rev. B **34** (1986) 2670.
159. P. Fendley, H. Saleur and N.P. Warner, Nucl. Phys. B **430** (1994) 577.
160. P. Fendley, A.W.W. Ludwig and H. Saleur, Phys. Rev. Lett. **74** (1995) 3005; **75** (1995) 2196.
161. P. Fendley and H. Saleur, Phys. Rev. B **54** (1996) 10845.
162. H.A. Fertig, Phys. Rev. B **40** (1989) 1087.
163. H.A. Fertig, L. Brey, R. Cote and A.H. MacDonald, Phys. Rev. B **50** (1994) 11018.
164. A.L. Fetter and M.J. Stephen, Phys. Rev. **168** (1967) 475.
165. A.L. Fetter, C. Hanna and R. Laughlin, Phys. Rev. B **39** (1989) 9679.
166. H.A. Fertig, L. Brey, R. Cote and A.H. MacDonald, A. Karlhede and , Phys. Rev. B 55 (1997) 10671.
167. D. Finkelstein and J. Rubinstein, J. Math. Phys. **9** (1968) 1762.
168. M.P.A. Fisher, G. Grinstein and S.M. Girvin, Phys. Rev. Lett. **64** (1990) 587.
169. M.P.A. Fisher, Phys. Rev. Lett. **65** (1990) 923.
170. P. Fletcher, Phys. Lett. B **248** (1990) 323.
171. R. Floreanini and R. Jackiw, Phys. Rev. Lett. **59** (1987) 1873.
172. M.M Fogler, A.A. Koulakov and B.I. Shklovskii, Phys. Rev. B **54** (1996) 1853.
173. M.M Fogler, A.A. Koulakov and B.I. Shklovskii, Phys. Rev. B **55** (1997) 9326.

174. M. Fogler and F. Wilczek, Phys. Rev. Lett. **86** (2001) 1833.
175. P.F. Fontein, P. Hendriks, F.A.P. Blom, J.H. Wolter, L.J. Giling and C.W. Benakker, Surface Science **263** (1992) 91.
176. R. Fox and L. Neuwirth, Math. Scand. **10** (62) 119.
177. Y.I. Frenkel and T. Kontorowa, Zh. Eksp. Theor. Fiz. **8** (1938) 1340.
178. J. Frolich and P.A. Marchetti, Lett. Math. Phys. **16** (1989) 347.
179. K. Fujikawa, Phys. Rev. A **52** (1995) 3299.
180. A. Fukuda, A. Sawada, S. Kozumi, D. Terasawa, Y. Shimoda, Z.F. Ezawa, N. Kumada, Y. Hirayama, Phys. Rev. B **73** (2006) 165304.
181. A. Fukuda, D. Terasawa, M. Morino, K. Iwata, S. Kozumi, N. Kumada, Y. Hirayama, Z. F.Ezawa and A. Sawada, to be published.
182. J.E. Furneaux, D.A. Syphers and A.G.Swanson,, Phys. Rev. Lett. **63** (1989) 1098.
183. S. J. Geer, A. G. Davies, C. H. W. Barnes, K. R. Zolleis, M. Y. Simmons and D. A. Ritchie, Phys. Rev. B **66** (2002) 045318.
184. M. Gell-Mann and M. Lévy, Nuovo Cim. **16** (1960) 705.
185. G. Ghosh and R. Rajaraman, Phys. Rev. B **63** (2000) 035304.
186. S. Ghoshal and A.B. Zamolodchikov, Int. J. Mod. Phys. A **9** (1994) 3841.
187. S.M. Girvin and T. Jach, Phys. Rev. B **29** (1984) 5617.
188. S.M. Girvin, A.H. MacDonald and P.M. Platzman, Phys. Rev. Lett. **54** (1985) 581.
189. S.M. Girvin, A.H. MacDonald and P.M. Platzman, Phys. Rev. B **33** (1986) 2481.
190. S.M. Girvin and A.H. MacDonald, Phys. Rev. Lett. **58** (1987) 1252.
191. S.M. Girvin, A.H. MacDonald, M.P.A. Fisher, S-J. Rey and J.P. Sethna, Phys. Rev. Lett. **65** (1990) 1671.
192. P. Goddard and D. Olive, Int. J. Mod. Phys. A **1** (1986) 303.
193. M. Goerbig, R. Moessner and B. Doucot, Phys. Rev. B **74** (2006) 161407.
194. A. Gold, Europhys.Lett. **54** (1986) 241,479(E).
195. B.B. Goldberg, M.J. Manfra, L. Pfeiffer and K. West, Physica B **249-251** (1998) 7.
196. A.S. Goldhaber and R. Mackenzie, Phys. Lett. B **214** (1988) 471.
197. G.A. Goldin, R. Menikoff and D.H. Sharp, J. Math. Phys. **21** (1980) 650; **22** (1981) 1664.
198. V.J. Goldman, B. Su and J.K. Jain, Phys. Rev. Lett. **72** (1994) 2065.
199. J. Goldstone, Nuovo Cim. **19** (1961) 154.
200. J. Goldstone, A. Salam and S. Weinberg, Phys. Rev. **127** (1962) 965.
201. F. Guinea, A.H. Castro Neto, and N.M.R. Peres, Phys. Rev. B **73** (2006) 245426.
202. E. Gorbar, V, Gusynin, V. Miransky, and I. Shovkovy, Phys. Rev. B **66**, 045108 (2002).
203. V.P. Gusynin and S.G. Sharapov, Phys. Rev. Lett. **95** (2005) 146801.

204. V.P. Gusynin, V.A. Miransky, S.G. Sharapov and I.A. Shovkovy, Phys. Rev. B **74** (2006) 195429.
205. R. Haag and D. Kastler, J. Math. Phys. **5** (1968) 848.
206. F.D.M. Haldane, Phys. Rev. Lett. **47** (1981) 1840; J. Phys. C **14** (1981) 2585.
207. F.D.M. Haldane, Phys. Rev. Lett. **51** (1983) 605.
208. F.D.M. Haldane, Phys. Rev. Lett. **51** (1983) 1395.
209. F.D.M. Haldane and E.H. Rezayi, Phys. Rev. Lett. **54** (1985) 237.
210. F.D.M. Haldane and E.H. Rezayi, Phys. Rev. Lett. **60** (1988) 956; Phys. Rev. Lett. **60** (1988) 1886.
211. F.D.M. Haldane, Phys. Rev. Lett. **67** (1991) 937.
212. F.D.M. Haldane, E.H. Rezayi and Kun Yang, Phys. Rev. Lett. **85** (2000) 5396.
213. E.H. Hall, Am.J.Math **2** (1879) 287; Phil.Mag. **9** (1880) 225.
214. B.I. Halperin, Phys. Rev. B **25** (1982) 2185.
215. B.I. Halperin, Helv. Phys. Acta **56** (1983) 75.
216. B.I. Halperin, Phys. Rev. Lett. **52** (1984) 1583; E2390.
217. B.I. Halperin, Surface Science **170** (1986) 115.
218. B.I. Halperin, P.A. Lee and N. Read, Phys. Rev. B **47** (1993) 7312.
219. B.I. Halperin and A. Stern, Phys. Rev. Lett. **80** (1998) 5457.
220. A.R. Hamilton, M.Y. Simmons, F.M. Bolton, N.K. Patel, I.S. Millard, J.T. Nicholls, D.A. Ritchie and M. Pepper, Phys. Rev. B **54** (1996) R5259.
221. Y. Hanein, U. Meirav, D. Shahar, C.C. Li, D.C. Tsui and H. Shtrikman, Phys. Rev. Lett. **80** (1998) 1288.
222. C.B. Hanna, A.H. MacDonald and S.M. Girvin, Phys. Rev. B **63** (2001) 125305.
223. T.H. Hansson, M. Rocek, I. Zahed and S.C. Zhang, Phys. Lett. B **214** (1988) 475.
224. J.A. Harvey, *Komaba Lectures on Noncommutative Solitons and D-Branes*, hep-th/0102076.
225. K. Hasebe and Z.F. Ezawa, Phys. Rev. B **66** (2002) 155318.
226. R.J.Haug, K.v. Klitzing, R.J. Nicholas, J.C. Maan and G. Weimann, Phys. Rev. B **36** (1987) 4528.
227. S. He, X.C. Xie, S. Das Sarma and F.C. Zhang, Phys. Rev. B **43** (1991) 9339.
228. S. He, S. Das Sarma and X.C. Xie, Phys. Rev. B **47** (1993) 4394.
229. S. He, S.H. Simon and B.I. Halperin, Phys. Rev. B **50** (1994) 1823.
230. C. Hermann and C. Weisbuch, Phys. Rev. B **15** (1977) 823.
231. P.W. Higgs, Phys. Lett. **12** (1964) 132.
232. P.C. Hohenberg, Phys. Rev. **158** (1967) 383.
233. J. Hoppe and P. Schaller, Phys. Lett. B **237** (1990) 407.
234. J. Hu and A.H. MacDonald, Phys. Rev. B **46** (1992) 12554.
235. N. Imai, K. Ishikawa, T. Matsuyama and I. Tanaka, Phys. Rev. B **42** (1990) 10610.

236. S. Iso, D. Karabali and B. Sakita, Phys. Lett. B **296** (1992) 143.
237. Y. Iye, E.E. Mendez, W.I. Wang and L. Esaki, Phys. Rev. **33** (1986) 5854.
238. R. Jackiw, J. Math. Phys. **9** (1968) 339.
239. R. Jackiw and S-Y. Pi, Phys. Rev. Lett. **64** (1990) 2969; **67** (1991) 415.
240. J.K. Jain and S.A. Kivelson, Phys. Rev. Lett. **60** (1988) 1542; Phys. Rev. B **37** (1988) 4276.
241. J.K. Jain, Phys. Rev. Lett. **63** (1989) 199; Phys. Rev. B **40** (1989) 8079; **41** (1990) 7653.
242. J.K. Jain, Ann. Phys. **41** (1992) 105.
243. J.K. Jain, Surface Science **263** (1992) 65.
244. J.K.Jain and R.K. Kamilla, Phys. Rev. B **55** (1997) 4895; Int. J. Mod. Phys. B **11** (1997) 2621.
245. J.K. Jain and R.K. Kamilla, in reference.[6]
246. A. Jeffery et al., 1996 Conf. On Precision Electromagnetic Measurement (17-20 June, 1996, Braunschweig, Germany).
247. H.W. Jiang , H.L. Stormer, D.C. Tsui, L.N. Pfeiffer and K.W. West, Phys. Rev. B **40** (1989) 12013.
248. H.W. Jiang , R.L. Willett, H.L. Stormer, D.C. Tsui, L.N. Pfeiffer and K.W. West, Phys. Rev. Lett. **65** (1990) 633.
249. H.W. Jiang, C.E. Johnson,K.L. Wang and S.T. Hannahs, Phys. Rev. Lett. **71** (1993) 1439.
250. M.D. Johnson and G.S. Canright, Phys. Rev. B **41** (1990) 6870.
251. M.D. Johnson and G.S. Canright, Phys. Rev. B **49** (1994) 2947.
252. B.D. Josephson, Phys. Lett. **1** (1962) 251.
253. V. Kac and A. Radul, Comm. Math. Phys. **157** (1993) 429.
254. C. Kallin and B.I. Halperin, Phys. Rev. B **30** (1991) 5655.
255. V. Kalmeyer and R.B. Laughlin, Phys. Rev. Lett. **59** (1987) 2095.
256. V. Kalmeyer and S.C. Zhang, Phys. Rev. B **46** (1992) 9889.
257. R.K. Kamilla, X.G. Wu and J.K. Jain, Phys. Rev. Lett. **76** (1996) 1332; Phys. Rev. B **54** (1996) 4873.
258. C. Kane, S. Kivelson, D.H. Lee and S.C. Zhang, Phys. Rev. B **43** (1991) 3255.
259. C. Kane and M.P.A. Fisher, Phys. Rev. Lett. **68** (1992) 1220; Phys. Rev. B **46** (1992) 15233.
260. C. Kane and M.P.A. Fisher, Nature **389** (1997) 119.
261. W. Kang, H.L. Stormer, L.N. Pfeiffer, K.W. Baldwin and K.W. West, Phys. Rev. Lett. **71** (1993) 3850.
262. D. Karabali, Nucl. Phys. B **419** (1994) 437.
263. S. Kawaji and J. Wakabayashi, Surface Science **58** (19976) 238.
264. S. Kawaji, J. Wakabayashi, J. Yoshino and H. Sakaki, J. Phys. Soc. Jap. **54** (1985) 1506.
265. S. Kawaji, K. Hirakawa, M. Nagata, T. Okamoto, T. Goto and T. Fukase, J. Phys. Soc. Jap. **63** (1994) 2303.

266. M. Kawamura, A. Endo, S. Katsumoto and Y. Iye, J. Phys. Soc. Jap. **68** (1999) 2186.
267. M. Kellogg, I.B. Spielman, J.P. Eisenstein, L.N. Pfeiffer and K.W. West, Phys. Rev. Lett. **88** (2002) 126804.
268. M. Kellogg, J.P. Eisenstein, L.N. Pfeiffer and K.W. West, Phys. Rev. Lett. **93** (2004) 036801.
269. V.S. Khrapai, E.V. Deviatov, A.A. Shashkin, V.T. Dolgopolov, F. Hastreiter, A. Wixforth, K.L. Campman and A.C. Gossard, Phys. Rev. Lett. **84** (2000) 725.
270. D. V. Khveschenko, Phys. Rev. Lett. **87**, 206401 (2001).
271. D. V. Khveschenko, Phys. Rev. Lett. **87**, 246802 (2001).
272. R.F. Kiefl et al., Phys. Rev. Lett. **64** (1990) 2082.
273. T. Kinoshita, Rep. Prog. Phys.**59** (1996) 1459.
274. S. Kivelson, D.H. Lee and S.C. Zhang, Phys. Rev. B **46** (1992) 2223.
275. K.v. Klitzing, G. Dorda and M. Pepper, Phys. Rev. Lett. **45** (1980) 494.
276. K.v. Klitzing, Festkörperproblems **21** (1981) 1; Surface Science **113** (1982) 1; Physica B **126** (1984) 242; Rev. Mod. Phys.**58** (1986) 519.
277. M. Kohmoto, Ann. Phys. **160** (1985) 343.
278. W. Kohn, Phys. Rev. **123** (1961) 1242.
279. S. Komiyama, T. Takamasu, T. Hiyamizu and S. Sasa, Solid State Commun. **54** (1985) 479.
280. S. Komiyama, H. Hirai, S. Sasa and S. Hiyamizu, Phys. Rev. B **40** (1989) 12566.
281. S. Komiyama, in reference.[1]
282. A. A. Koulakov, M. M. Fogler and B. I. Shklovskii, Phys. Rev. Lett. **76** (1996) 499.
283. L.P. Kouwenhoven, Phys. Rev. Lett. **64** (1990) 69.
284. S.V. Kravchenko, W. Mason and J.E. Furneaux, Phys. Rev. Lett. **75** (1995) 910.
285. S.V. Kravchenko,D. Simonian, M.P. Sarachik, W. Mason and J.E. Furneaux, Phys. Rev. Lett. **77** (1996) 4938.
286. I.V. Kukushkin, R.J. Haug, K.v. Klitzing and K. Ploog, Phys. Rev. Lett. **72** (1994) 736; Ann. Phys. **45** (1996) 147.
287. I.V. Kukushkin, K.v. Klitzing, K. Ploog and V.B. Timofeev, Phys. Rev. B **40** (1997) 7788.
288. I.V. Kukushkin, K.v. Klitzing and K. Eberl, Phys. Rev. Lett. **82** (1999) 3665.
289. N. Kumada, A. Sawada, Z.F. Ezawa, S. Nagahama, H. Azuhata, K. Muraki, T. Saku and Y. Hirayama, J. Phys. Soc. Jap. **69** (2000) 3178–3181.
290. N. Kumada, D. Terasawa, M. Morino, T. Tagashira, A. Sawada, Z.F. Ezawa, K. Muraki, Y. Hirayama and T. Saku, Phys. Rev. B 69 (2004) 307.
291. R. Landauer, IBM J. Res. & Dev. **1** (1957) 223.
292. R.B. Laughlin, Phys. Rev. B **23** (1981) 5652.

293. R.B. Laughlin, Phys. Rev. B **27** (1983) 3383.
294. R.B. Laughlin, Phys. Rev. Lett. **50** (1983) 1395.
295. R.B. Laughlin, Phys. Rev. Lett. **52** (1984) 2304.
296. R.B. Laughlin, Surface Science **142** (1984) 163.
297. R.B. Laughlin, Phys. Rev. Lett. **60** (1988) 1057; Science **242** (1988) 525.
298. T.S. Lay, Y.W. Suen, H.C. Manoharan, X. Ying, M.B. Santos and M. Shayegan, Phys. Rev. B **50** (1994) 17725.
299. D.R. Leadley, R.J. Nicholas, C.T. Foxon and J.J. Harris, Phys. Rev. Lett. **72** (1994) 1906.
300. D.R. Leadley, R.J. Nicholas, D.K. Maude, A.N. Utjuzh, J.C. Portal, J.J. Harris and C.T. Foxon, Phys. Rev. Lett. **79** (1997) 4246.
301. P. Lebwohl and M. J. Stephen, Phys. Rev. B **163** (1967) 376.
302. D.H. Lee and M.P.A. Fisher, Phys. Rev. Lett. **63** (1989) 903; Int. J. Mod. Phys. B **5** (1991) 2675.
303. D.H. Lee, Phys. Rev. Lett. **62** (1989) 82.
304. D.H. Lee and C.L. Kane, Phys. Rev. Lett. **64** (1990) 1313.
305. K. Lee, Phys. Rev. Lett. **66** (1991) 553.
306. D.H. Lee and S.C. Zhang, Phys. Rev. Lett. **66** (1991) 1220.
307. B.-H. Lee, K. Lee and H.S. Yang, Phys. Lett. B 498 (2001) 277.
308. B.-H. Lee, K. Moon and C. Rim, Phys. Rev. D 64 (2001) 85014.
309. F. Lefloch, C. Hoffmann, M. Sanquer and D. Quirion, Phys. Rev. Lett. 90 (2003) 067002.
310. J.M. Leinaas and J. Myrheim, Nuovo Cim. **39B** (1977) 1.
311. E. Lieb and W. Liniger, Phys. Rev. **130** (1963) 1605.
312. D. Lilliehook, K. Lejnell, A. Karlhede and S.L. Sondhi, Phys. Rev. B **56** (1997) 6805.
313. M.P. Lilly, K.B. Cooper, J.P. Eisenstein, L.N. Pfeiffer and K.W. West, Phys. Rev. Lett. **82** (1999) 394.
314. F. London and H. London, Proc.Roy.Soc. (London) **A149** (1935) 71.
315. F. London, Nature **141** (1938) 643.
316. A. Lopez and E. Fradkin, Phys. Rev. B **44** (1991) 5246.
317. J.H. Lowenstein and J.A. Swieca, Ann. Phys. **68** (1971) 172.
318. A. Luther and L.J. Peschel, Phys. Rev. Lett. **32** (1974) 992; Phys. Rev. B **9** (1974) 2911.
319. J.M. Luttinger, J. Math. Phys. **4** (1963) 1154.
320. S.K. Ma and R. Rajaraman, Phys. Rev. D **11** (1975) 1701.
321. A.H. MacDonald and G.C. Aers, Phys. Rev. B **29** (1984) 5976.
322. A.H. MacDonald and P. Streda, Phys. Rev. B **29** (1984) 1616.
323. A.H. MacDonald, K.L. Liu, S.M. Girvin and P.M. Platzman, Phys. Rev. B **33** (1986) 4014.
324. A.H. MacDonald, P.M. Platzman and G.S. Boebinger, Phys. Rev. Lett. **65** (1990) 775.

325. A.H. MacDonald, Phys. Rev. Lett. **64** (1990) 220.
326. A.H. MacDonald and S.C. Zhang, Phys. Rev. B **49** (1994) 17208.
327. A.H. MacDonald, H.A. Fertig and L. Brey, Phys. Rev. Lett. 76 (1996) 2153.
328. A.H. MacDonald, R. Rajaraman and T. Jungwirth, Phys. Rev. B **60** (1999) 8817.
329. A.J. MacFarlane, Phys. Lett. B **82** (1979) 239.
330. S. Mandelstam, Ann. Phys. **19** (1962) 1; Phys. Rev. **175** (1968) 1580.
331. S. Mandelstam, Phys. Rev. D **11** (1975) 3026.
332. H.C. Manoharan and M. Shayegan, Phys. Rev. B **50** (1994) 17662.
333. H.C. Manoharan, M. Shayegan and S.J. Klepper, Phys. Rev. Lett. **73** (1994) 3270.
334. C. Manuel and R. Tarrach, Phys. Lett. B **28** (1991) 222.
335. P.A. Maskym and T.Chakraborty, Phys. Rev. Lett. **65** (1990) 108.
336. P.A. Maskym, Phys. Rev. B **53** (1996) 10871.
337. D.C. Mattis and E. Lieb, J. Math. Phys. **6** (1965) 304.
338. D.K. Maude, M. Potemski, J.C. Portal, M. Henini, L. Eaves, G. Hill and M.A. Pate, Phys. Rev. Lett. **77** (1996) 4604.
339. D.K. Maude, Physica B **249-251** (1998) 1.
340. E. McCann and V.I. Fal'ko., Phys. Rev. Lett. **96**, 086805 (2006).
341. P.L. McEuen, A. Szafer, C.A. Richter, B.W. Alphenaar, J.K. Jain, A.D. Stone and R.G. Wheeler, Phys. Rev. Lett. **64** (1990) 2062.
342. S. Melinte, E. Grivei, V. Bayot and M. Shayegan, Phys. Rev. Lett. **82** (1999) 2764.
343. C.J. Mellor, R.H. Eyles, J.E. Digby, A.J. Kent, K.A. Benedict L.J. Challis, M. Henini and C.T. Foxon, Phys. Rev. Lett. **74** (1995) 2339.
344. E.E. Mendez, W.I. Wang, L.L. Chang and L. Esaki, Phys. Rev. B **30** (1984) 1087.
345. N.D. Mermin and H. Wagner, Phys. Rev. Lett. **17** (1966) 1133.
346. N.D. Mermin, J. Math. Phys. **8** (1967) 1061.
347. F. Milliken, C. Umbach and R. Webb, Solid State Commun. **97** (1996) 309.
348. R. Moessner and J. Chalker, Phys. Rev. B **54** (1996) 5006.
349. P.J. Mohr and B.N. Taylor, Rev. Mod. Phys. **77** (2005) 1.
350. K. Moon, H.Yi, C.L. Kane, S.M. Girvin and M.P.A. Fisher, Phys. Rev. Lett. **71** (1993) 4381.
351. K. Moon, H. Mori, K. Yang, S.M. Girvin, A.H. MacDonald, L. Zheng, D. Yoshioka and S-C. Zhang, Phys. Rev. B **51** (1995) 5138.
352. G. Moore and N. Read, Nucl. Phys. B **360** (1991) 362.
353. R.H. Morf and B.I. Halperin, Phys. Rev. B **33** (1986) 2221.
354. R.H. Morf, Phys. Rev. Lett. **80** (1998) 1505.
355. T. Morinari, Phys. Rev. Lett. **81** (1998) 3741.
356. J. Moyal, Proc.Camb.Phil.Soc. **45** (1949) 99.
357. K. Muraki, T. Saku, Y. Hirayama, N. Kumada, A. Sawada and Z.F. Ezawa,

Solid State Commun. **112** (1999) 625.
358. K. Muraki, N. Kumada, T. Saku and Y. Hirayama, Jpn. J. Appl. Phys. **39** (2000) 2444.
359. S.Q. Murphy, J.P. Eisenstein, G.S. Boebinger, L.N. Pfeiffer and K.W. West, Phys. Rev. Lett. **72** (1994) 728.
360. M.V.N. Murthy, J. Law, M. Brack and R.K. Bhaduri, Phys. Rev. Lett. **67** (1991) 1817.
361. M.V.N. Murthy and R. Shankar, Phys. Rev. Lett. **72** (1994) 3629; **73** (1994) 3331.
362. G. Murthy and R. Shankar, Phys. Rev. Lett. **80** (1998) 5458.
363. T. Nakajima and H. Aoki, Physica B **184** (1993) 91.
364. T. Nakajima and H. Aoki, Phys. Rev. Lett. **73** (1994) 3568.
365. T. Nakajima and H. Aoki, Phys. Rev. B **51** (1995) 7874; Surface Science **361/362** (1996) 83.
366. T. Nakajima and H. Aoki, Phys. Rev. B **52** (1995) 13780.
367. T. Nakajima and H. Aoki, Phys. Rev. B **56** (1997) R15549.
368. Y. Nambu and G. Jona-Lasinio, Phys. Rev. **122** (1961) 345.
369. J. von Neumann, "*Mathematical Foundatation of Quantum Mechanics* (Princeton Univ. Press, NJ, 1955), pp. 405–407.
370. F. Nihey, K. Nakamura, T. Takamasu, G. Kido, T. Sakon and M. Motokawa, Phys. Rev. B **59** (1999) 14872.
371. Q. Niu, D.J. Thouless, Y-S. Wu, Phys. Rev. B **31** (1985) 3372.
372. Q. Niu and D.J. Thouless, Phys. Rev. B **35** (1987) 2188.
373. K. Nomura and A.H. MacDonald, Phys. Rev. Lett. **96** (2006) 256602.
374. K.S. Novoselov, A. K. Geim, S. V. Morozov, D. Jiang, Y. Zhang, S. V. Dubonos, I. V. Grigorieva, and A. A. Firsov, Science **306** (2004) 666.
375. K.S. Novoselov, A. K. Geim, S. V. Morozov, D. Jiang, M. I. Katsnelson, I. V. Grigorieva, S. V. Dubonos and A. A. Firsov, Nature **438** (2005) 197.
376. K.S. Novoselov, E. McCann, S. V. Morozov, V. I. Fal'ko, M. I. Katsnelson, U. Zeitler, D. Jiang, F. Schendini and A. K. Geim, Nature Phys. **2** (2006) 177.
377. T. Okamoto, Y. Shinohara, S. Kawaji and A. Yagi, J. Phys. Soc. Jap. **64** (1995) 2311.
378. T. Okamoto, Y. Shinohara, S. Kawaji and A. Yagi, Surface Science **361/362** (1996) 278.
379. T. Okamoto and S. Kawaji, Physica B **249-251** (1998) 61.
380. Y. Ono, J. Phys. Soc. Jap. **51** (1982) 237.
381. F.v. Oppen, A. Stern, B.I. Halperin, Phys. Rev. Lett. **80** (1998) 4494.
382. M.A. Paalanen, D.C. Tsui and A.C. Gossard, Phys. Rev. B **25** (1982) 5566.
383. K. Park, V. Melik-Alaverdian, N.E. Bonesteel and J.K. Jain, Phys. Rev. B **58** (1998) R10167.
384. K. Park and J.K. Jain, Phys. Rev. Lett. **80** (1998) 4237.
385. V. Pasquier and F.D.M. Haldane, Nucl. Phys. B **516** (1998) 719.

386. W. Pauli, Phys. Rev. **56** (1940) 716.
387. D.T. Pegg and S.M. Barnett, Phys. Rev. A **39** (1989) 1665.
388. V. Pellegrini, A. Pinczuk, B.S. Dennis, A.S. Plaut, L.N. Pfeiffer and K.W. West, Phys. Rev. Lett. **78** (1997) 310; Science **281** (1998) 799.
389. A.M. Perelomov, Teor. Mat. Fiz. **6** (1971) 213.
390. N.M.R. Peres, F. Guinea and A.H.C. Neto, Phys. Rev. B **73** (2006) 125411.
391. R. de Picciotto, M. Reznikov, M. Heiblum, V. Umansky, G. Bunin and D. Mahalu, Nature **389** (1997) 162.
392. W.F. Pohl, J.Math.Mech **17** (1968) 975.
393. V.L. Pokrovsky and A.L. Talapov, Sov. Phys. JETP 51 (1980) 134.
394. A.P. Polychronakos, Phys. Lett. B **264** (1991) 362.
395. A.P. Polychronakos, Phys. Rev. Lett. **69** (1992) 703.
396. A.P. Polychronakos, Phys. Lett. B **365** (1996) 202.
397. C.N. Pope, X. Shen and L.J. Romans, Nucl. Phys. B **339** (1990) 191.
398. R. Rajaraman and S.L. Sondhi, Int. J. Mod. Phys. B **10** (1996) 793.
399. M. Rasolt, F. Perrot and A.H. MacDonald, Phys. Rev. Lett. **55** (1985) 433.
400. M. Rasolt and A.H. MacDonald, Phys. Rev. B **34** (1986) 5530.
401. M. Rasolt, B.I. Halperin and D. Vanderbilt, Phys. Rev. Lett. **57** (1986) 126.
402. N. Read, Phys. Rev. Lett. **62** (1989) 86; Surface Science **361/362** (1996) 7.
403. N. Read, Phys. Rev. Lett. **65** (1990) 1502.
404. N. Read, Phys. Rev. B **52** (1995) 1926.
405. N. Read and S. Sachdev, Phys. Rev. Lett. **75** (1995) 3509.
406. E.H. Rezayi and F.D.M. Haldane, Phys. Rev. B **32** (1985) 6924.
407. E.H. Rezayi and F.D.M. Haldane, Phys. Rev. Lett. **61** (1988) 1985.
408. E.H. Rezayi and F.D.M. Haldane, Phys. Rev. B **42** (1990) 4352.
409. E.H. Rezayi, Phys. Rev. B **43** (1991) 5944.
410. E.H. Rezayi, Phys. Rev. B **56** (1997) R7104.
411. E.H. Rezayi and F.D.M. Haldane, Bull.Am.Phys.Soc **43** (1998) 655.
412. E.H. Rezayi, F.D.M. Haldane and K. Yang, Phys. Rev. Lett. **83** (1999) 1219.
413. M. Reznikov, R. de Picciotto, T.G. Griffiths, M. Heiblum and V. Umansky, Nature **399** (1999) 238.
414. L.M. Roth, B. Lax and S. Zwerdling, Phys. Rev. **114** (1959) 90.
415. S. Sachdev and R.N. Bhatt, Phys. Rev. B **41** (1990) 9323.
416. A. Sachrajda, R. Boulet, Z. Wasilewsei, P. Coleridge and F. Guillon, Solid State Commun. **74** (1990) 1021.
417. T. Sajoto, Y.P Li, L.W. Engel, D.C. Tsui and M. Shayegan, Phys. Rev. Lett. **70** (1993) 2321.
418. B. Sakita, Phys. Lett. B **315** (1993) 124.
419. L. Saminadayar, D.C. Glattli, Y. Jin and B. Etienne, Phys. Rev. Lett. **79** (1997) 2526.
420. S. Das Sarma, S. Sachdev and L. Zheng, Phys. Rev. Lett. **79** (1997) 917; Phys. Rev. B **58** (1998) 4672.

421. K. Sasaki and Z.F. Ezawa, Phys. Rev. B **60** (1999) 8811.
422. A. Sawada, Z.F. Ezawa, H. Ohno, Y. Horikoshi, O. Sugie, S. Kishimoto, F. Matsukura, Y. Ohno and M. Yasumoto, Solid State Commun. **103** (1997) 447.
423. A. Sawada, Z.F.Ezawa, H. Ohno, Y. Horikoshi, Y. Ohno, S. Kishimoto and F. Matsukura, Phys. Rev. Lett. **80** (1998) 4534.
424. A. Sawada, Z.F. Ezawa, H. Ohno, Y. Horikoshi, A. Urayama, Y. Ohno, S. Kishimoto, F. Matsukura and N. Kumada, Phys. Rev. B **59** (1999) 14888.
425. A. Sawada, Z.F. Ezawa, H. Ohno, Y. Horikoshi, N. Kumada, Y. Ohno, S. Kishimoto, F. Matsukura and S. Nagahama, Physica E **6** (2000) 615.
426. A. Sawada, D. Terasawa, N. Kumada, M. Morino, K. Tagashira, Z.F. Ezawa, K. Muraki, T. Saku and Y. Hirayama, Physica E **18** (2003) 118.
427. A.M.J. Schakel, cond-mat/9904092.
428. J. Schliemann and A.H. MacDonald, Phys. Rev. Lett. **84** (2000) 4437.
429. A. Schmeller, J.P. Eisenstein, L.N. Pfeiffer and K.W. West, Phys. Rev. Lett. **75** (1995) 4290.
430. W. Schottky, Ann. Phys. (Leipzig) **57** (1918) 541.
431. B. Schroer, J.A. Swieca and A.H. Völkel, Phys. Rev. D **11** (1975) 1509.
432. J. Schwinger, Phys. Rev. **82** (1951) 664.
433. J. Schwinger, Phys. Rev. **128** (1962) 2425.
434. G.W. Semenoff, Phys. Rev. Lett. **43** (1984) 2449.
435. G.W. Semenoff, Phys. Rev. Lett. **61** (1988) 517.
436. N. Shibata and D. Yoshioka, Phys. Rev. Lett. **86** (2001) 5755.
437. N. Shibata and D. Yoshioka, J. Phys. Soc. Jap. **72** (2003) 664.
438. N. Shibata and D. Yoshioka, J. Phys. Soc. Jap. **73** (2004) 2169.
439. R. Shankar and G. Murthy, Phys. Rev. Lett. **79** (1997) 4437.
440. T. Shimada, T. Okamoto and S. Kawaji, Physica B **249-251** (1998) 107.
441. Y. Shimoda and T. Nakajima and A. Sawada, Physica E **22** (2204) 56.
442. K. Shizuya, Phys. Rev. Lett. **73** (1994) 2907.
443. K. Shizuya, Phys. Rev. B **52** (1995) 2747.
444. K. Shizuya, Phys. Rev. B **59** (1999) 2142.
445. W. Siegal, Nucl. Phys. B **238** (1984) 307.
446. T.H.R. Skyrme, Proc.Roy.Soc. (London) **A260** (1961) 1271.
447. J.C. Slonczewski and P.R. Weiss, Phys. Rev. **109** (1958) 272.
448. J.H. Smet, D. Weiss, R.H. Blick, G. Lutjering, K.v. Klitzing, R. Fleischmann, R. Ketzmerick, T. Geisel and G. Weimann, Phys. Rev. Lett. **77** (1996) 2272.
449. J.H. Smet, K.v. Klitzing, D. Weiss and W. Wegscheider, Phys. Rev. Lett. **80** (1998) 4538.
450. J. Solyom, Adv. Phys. **28** (1970) 201.
451. S.L. Sondhi, A. Karlhede, S.A. Kivelson and E.H. Rezayi, Phys. Rev. B **47** (1993) 16419.
452. S.L. Sondhi and M.P. Gelfand, Phys. Rev. Lett. **73** (1994) 2119.

453. S. Spielman, K. Fesler, C.B. Eom, T.H. Geballe, M.M. Fejer and A. Kapitulnik, Phys. Rev. Lett. **65** (1990) 123.
454. I.B. Spielman, J.P. Eisenstein, L.N. Pfeiffer and K.W. West, Phys. Rev. Lett. **84** (2000) 5808.
455. M. Sporre, J.J.M. Verbaarschot and I. Zahed, Phys. Rev. Lett. **67** (1991) 1813.
456. A. Stern and B.I. Halperin, Phys. Rev. B **52** (1995) 5890.
457. A. Stern, S.M. Girvin, A.H. MacDonald and N. Ma, Phys. Rev. Lett. **86** (2001) 1829.
458. M. Stone, Phys. Rev. B **42** (1990) 8399; Ann. Phys. **207** (1991) 38; Int. J. Mod. Phys. B **5** (1991) 509.
459. M. Stone and J. Martinez, Int. J. Mod. Phys. B **7** (1993) 4389.
460. I.A. Strachan, Phys. Lett. B **283** (1992) 63.
461. R. Streater and I. Wilde, Nucl. Phys. B **24** (1970) 561.
462. P. Streda, J. Kucera and A.H. MacDonald, Phys. Rev. Lett. **59** (1987) 1973.
463. W.P. Su, Phys. Rev. B **84** (1986) 1031.
464. W.P. Su, Y-S. Wu and J. Yang, Phys. Rev. Lett. **77** (1996) 3423.
465. Y.W. Suen, L.W. Engel, M.B. Santos, M. Shayegan and D.C. Tsui, Phys. Rev. Lett. **68** (1992) 1379.
466. L. Susskind and J. Glogower, Physics **1** (1964) 49.
467. B. Sutherland, J. Math. Phys. **12** (1971) 246; 251.
468. D.A. Syphers and J.E. Furneaux, Surface Science **196** (1988) 252; Solid State Commun. **65** (1988) 1513.
469. T. Takamasu, H. Dodo and N. Miura, Solid State Commun. **96** (1995) 121.
470. T. Takamasu, H. Dodo, M. Ohno and N. Miura, Surface Science **361/362** (1996) 95.
471. R. Tao, J. Phys. C **19** (1986) L619.
472. D. Terasawa, M. Morino, K. Nakada, S. Kozumi, A. Sawada, Z. F.Ezawa, N. Kumada, K. Muraki, T. Saku and Y. Hirayama, Physica E **22** (2004) 52.
473. W. Thirring, Ann. Phys. **3** (1958) 91.
474. D. Thouless, M. Kohmoto, M. Nightingale and M. den NIJS, Phys. Rev. Lett. **49** (1982) 405.
475. D.J. Thouless, J. Phys. C **17** (1984) L325.
476. D. Thouless and Y.S. Wu, Phys. Rev. B **31** (1985) 1191.
477. S. Tomonaga, Progr. Theor. Phys. **5** (1950) 544.
478. S.A. Trugman, Phys. Rev. B **27** (1983) 7539.
479. G. Tsitsishvili and Z.F. Ezawa, Phys. Rev. B **72** (2005) 115306.
480. D.C. Tsui, H.L. Stormer and A.C. Gossard, Phys. Rev. Lett. **48** (1982) 1559.
481. E. Tutuc, M. Shayegan and D.A. Huse, Phys. Rev. Lett. **93** (2004) 036802.
482. H.C. Tze and Z.F. Ezawa, Phys. Rev. D **14** (1976) 2228.
483. A. Usher, R.J. Nicholas, J.J. Harris and C.T. Foxon, Phys. Rev. B **41** (1990) 1129.

484. A. Vilenkin, Phys.Rep. **121** (1985) 263.
485. J. Wakabayashi, K. Sumiyoshi, T. Nagashima, T. Mochiku and K. Kadowaki, J. Phys. Soc. Jap. **66** (1997) 413.
486. J.K. Wang and V.J. Goldman, Phys. Rev. Lett. **67** (1991) 749.
487. S. Washburn, A.B. Fowler, H. Schmidt and D. Kern, Phys. Rev. Lett. **61** (1988) 2801.
488. S. Weinberg, Nucl. Phys. B **413** (1994) 567.
489. X.G. Wen and A. Zee, Phys. Rev. Lett. **62** (1989) 1937.
490. X.G. Wen and A. Zee, Phys. Rev. Lett. **62** (1989) 2873.
491. X.G. Wen, Phys. Rev. Lett. **64** (1990) 2206; Phys. Rev. B **43** (1991) 11025; **44** (1991) 5708.
492. X.G. Wen, Int. J. Mod. Phys. B **6** (1992) 1711; Adv. Phys. **44** (1995) 405.
493. X.G. Wen and A. Zee, Phys. Rev. Lett. **69** (1992) 1811.
494. X.G. Wen and A. Zee, Phys. Rev. B **47** (1993) 2265.
495. H. Weyl, Z. Physik **46** (1927) 1.
496. H. Weyl, The Theory of Groups and Quantum Mechanics (Dover, New York 1950), especially sections 14 and 15 of chapter IV.
497. J.H. White, Am.J.Math **91** (1969) 693.
498. F. Wilczek, Phys. Rev. Lett. **48** (1982) 1144; **49** (1982) 957.
499. F. Wilczek and A. Zee, Phys. Rev. Lett. **51** (1983) 2250.
500. F. Wilczek, Phys. Rev. Lett. **69** (1992) 132.
501. R. Willett, J.P. Einsenstein, H.L. Stormer, D.C. Tsui, A.C. Gossard and J.H. English, Phys. Rev. Lett. **59** (1987) 1776.
502. R.L. Willett, M,A. Paalanen, R.R. Ruel, K.W. West, L.N. Pfeiffer and D.J. Bishop, Phys. Rev. Lett. **54** (1990) 112.
503. R.L. Willett, R.R. Ruel, M,A. Paalanen, K.W. West and L.N. Pfeiffer, Phys. Rev. B **47** (1993) 7344.
504. R.L. Willett, R.R. Ruel, K.W. West and L.N. Pfeiffer, Phys. Rev. Lett. **71** (1993) 3846.
505. R.L. Willett and L.N. Pfeiffer, Surface Science **361/362** (1996) 38.
506. R.L. Willett, R.R. Ruel, K.W. West and L.N. Pfeiffer, Phys. Rev. Lett. **78** (1997) 4478.
507. R.L. Willett, in reference.[6]
508. E. Witten, Nucl. Phys. B **188** (1981) 513; Nucl. Phys. B **202** (1982) 253.
509. S. R. White, Phys. Rev. Lett. **69** (1992) 2863; Phys. Rev. B **48** (1993) 10345.
510. M.W. Wong, Weyl Transforms (Springer-Verlag, New York, 1998).
511. Y-S. Wu, Phys. Rev. Lett. **52** (1984) 2103.
512. Y-S. Wu, Phys. Rev. Lett. **53** (1984) 111; E1028.
513. Y-S. Wu and A. Zee, Phys. Lett. B **147** (1984) 325.
514. X.G. Wu, G. Dev and J.K. Jain, Phys. Rev. Lett. **71** (1993) 153:
515. Y-S. Wu, Phys. Rev. Lett. **73** (1994) 922.
516. X.C. Xie, Y. Guo and F.C. Zhang, Phys. Rev. B **40** (1989) 3487.

517. C.N. Yang and C.P. Yang, J. Math. Phys. **10** (1969) 1115.
518. J. Yang and W.P. Su, Phys. Rev. Lett. **70** (1993) 1163.
519. K. Yang, K. Moon, L. Zheng, A.H. MacDonald, S.M. Girvin, D. Yoshioka and S-C. Zhang, Phys. Rev. Lett. **72** (1994) 732.
520. K. Yang and A.H. MacDonald, Phys. Rev. B **51** (1995) 17247.
521. K. Yang and S.L. Sondhi, Phys. Rev. B **54** (1996) 2331.
522. K. Yang, K. Moon, L. Belkhir, H. Mori, S.M. Girvin, A.H. MacDonald, L. Zheng and D. Yoshioka, Phys. Rev. B **54** (1996) 11644.
523. K. Yang, Phys. Rev. B **60** (1999) 15578.
524. M.F. Yang and M.C. Chang, Phys. Rev. B **60** (1999) 13985.
525. A.S. Yeh, H.L. Stormer, D.C. Tsui, L.N. Pfeiffer, K.W. Baldwin and K.W. West, Phys. Rev. Lett. **82** (1999) 592.
526. D. Yoshioka, J. Phys. Soc. Jap. **53** (1984) 3740.
527. D. Yoshioka, Progr. Theor. Phys. Suppl. **84** (1985) 97.
528. D. Yoshioka, J. Phys. Soc. Jap. **55** (1986) 885.
529. D. Yoshioka, A.H. MacDonald and S.M. Girvin, Phys. Rev. B **39** (1989) 1932.
530. D. Yoshioka, J. Phys. Soc. Jap. **67** (1998) 3356.
531. D. Yoshioka, J. Phys. Soc. Jap. **68** (1999) 3360.
532. F.C. Zhang and T. Chakraborty, Phys. Rev. B **30** (1984) 7320.
533. F.C. Zhang, V.Z. Vulovic, Y. Giu and S. Das Sarma, Phys. Rev. B **32** (1985) 6920.
534. F.C. Zhang and S. Das Sarma, Phys. Rev. B **33** (1986) 2903.
535. S.C. Zhang, H. Hanson and S. Kivelson, Phys. Rev. Lett. **62** (1989) 82.
536. S.C. Zhang, Int. J. Mod. Phys. B **6** (1992) 25.
537. L. Zheng, R.J. Radtke and S. Das Sarma, Phys. Rev. Lett. **78** (1997) 2453.
538. Y. Zhang, Yan-Wen Tan, Horst L. Stormer and Philip Kim, Nature **438** (2005) 201.
539. Y. Zhang *et al.*, Phys. Rev. Lett. **96** (2006) 136806.
540. S.Y. Zhou, G.-H. Gweon, J. Graf, A.V. Fedorov, C.D. Spataru, R.D. Diehl, Y. Kopelevich, D.-H. Lee, S.G. Louie, and A. Lanzara, Nature Phys (2006) 595.

INDEX

activation energy, 172, 173, **215**, 276, 284, 288, 289, 294, 302, 357, 359, 378, 382, 444, **446**, 448–453, 467–470, 482, 483, 486, 576, 644
activation mass, 303, 304
Aharonov-Bohm
— effect, 68, **89**, 90, 92, 190, 294
— phase, 90–93, 190, 235, 541–543
Anderson localization, 171
Anderson-Higgs mechanism, 63, 67, **79**, 81
angular momentum, 44, 93, 96, 145, 152, 187, 189, 190, 205, 208, 209, 212, 229, 245, 255, 294, 295, 401, 564
relative —, 208, 327, 328
annihilation operator, *see* creation or annihilation operator
anomaly, 44, 319, 366
anyon, 91, **144**, 145, 147, 148, 151–154, 156–158, 170, 230, 258
— commutation relation, 145, 147, 158
— field, 157–159, 316
— statistics, 150, 258
— superconductivity, 148
— wave function, 153

bare density, *see* density (bare)
BCS-type state, 337, 353, 354
Begomol'nyi-Prasad-Sommerfield (BPS) soliton, **133**, 140, 478, 479, 482, 630
Berry phase, 337, **365**, 366
bias voltage, 219, 309, 323, 369, 374, 378, 381, 382, **388**, 394, 396, 428, 461, 464–466, 482, 483, 515, 518, **619**
bilayer-locked state, 374, 377, **400**, 402, 406
bimeron, 453
biskyrmion, 477–479, 481–483
Bogoliubov transformation, **14–16**, **59**, **61**, 652, 653
Bohr magneton, 104, 105, 261, 644
bonding or antibonding
— field, **389**, 425, 433, 437, 438, 440
— state, 374, 387, **389–391**, 394–396, 430, 439, 575–577, 605
Bose
— condensate or condensation, **48**, 51, 52, 54, 58, **62–65**, **79–84**, 118, 259, 654
— liquid, 227, 234
— quantum, 7, **8**, 24
— statistics, 6, **22**, 27, 148
box normalization, **35**, 36, 180, 653
Braid group, 147

canonical
— anticommutation relation, 100, 673
— commutation relation, 5, 20, 29, 32, 39, 41, 43, 53, 59, 61, 71, 75, 82, 100, 151, 154, 155, 236, 269, 314–316, 321, 561, 587, 672, 673
— momentum, 5, 29, **33**, 34, 39, 69, 71, 74, 78, 99, 120, 154, 191, 201, 233, 316, 672, 673
— quantization, **31**, 32–34, 39, 52, 55, 68–70, 79, 81, 99, 672
canted (antiferromagnet) state or phase, 381, 454, 456, 457, **461–465**, **468–470**, 578, 582
capacitance
— energy, 375, 386, **392**, **393**, 409, 416, 441, 442, 445, 448, 449, 457, 458, 479, 484, **622**, 625, 677, 678
— spectroscopy, 381, 394
— term, 372, 374, 416, 417, 436, 626
Cauchy-Riemann equation, 141, 251, 278, 650
center-of-mass
— Hamiltonian, 152
— coordinate, 147, 151, 221
— motion, 222
charge density wave (CDW), 325, 326, 330–332

charge-transferable state, 370, 375, 381, 382, 394–396, 409, 428
chemical potential, 58, 169, 205, 207
Chern-Simons
— field, 144, 151, **153**, 154, 156, 157, **233**, 236, 239, 256–258, 265, 401
— flux, 147, 148, 153, 227, 233, 234, 242, 291, 401
— gauge theory, 144, 155
coherence
— length, **54**, 80, 89, 118, 126, 127, 173, 261, 263, **273**, 286, 375, **417**, 427, 436, 443, 484
interlayer (pseudospin) —, 10, 50, 370, 374, **375**, 378, 379, 381, 382, 393, 406, 409–411, **417**, 422, 424, 446, 485, 491, 492, 494, 524
intralayer (spin) —, 10, 168, 172, 173, 261, 263, **273**, 381, 524
isospin —, 428
coherent excitation, 173, 270, 274, 302, 303
coherent mode, 271, 272
coherent state, **11–14**, 16, 18, 19, 54, 56, 113, 115, 173, 193, **198**, **199**, 262, 266, 271, 274, 371, 373, 411, 476, 535, 566
interlayer —, 62, **373**, 375, 410, 489, 492, 516
squeezed —, 10, **14**, 16, 61, 416
collective excitation, 113, 514
commensurate or incommensurate phase, 379, 447, 450, 486, 495, **497–499**, 507, 508, 514–516, 518, 668
composite
— boson, 147, 148, 153, 158, 170, 173, 227–229, 231, 234, 239, 242, 245, **246**, **247**, 252, 265, 268, 283, 302, 404–406, 548, 551
— fermion, 93, 148, 153, 169, 171, 174, 227–230, 239, **242**, 244, 245, 291, 293–295, 297, 298
— particle, 148, 153, 163, 169, 227–229, 232, 233, 235, 236, **238–242**, 297–299, 302, 303, 305, 395, 396, 400, 401
composite field, 83
composite-fermion (CF)-Landau level, 174, **293**, 295–297
composite-fermion (CF)-Landau site, 244, 245, **293**, 297
composite-particle (CP)-Landau level, 297–299, 302, 305
composite-particle (CP)-Landau site, 297
compound state, 376, 393, 394, 396, 488
conductivity
Hall —, 165, 168, 337, 345, 369
Coulomb energy, 171–173, 179, 205, 209, 210, 216–218, 220, 246, 253, 258, 261, 262, 266, 271, 274, 276, 287, 288, 299, 302, 305, 347, 357, 359, 369, 382, 406, 420, 433, 441, 479, 528, 531, 532, 569, 571, 584, 600, 602, 603, 607, 612, **643**, **644**, 656, 657, 659, 677
exchange —, 171, 221, 263, 273, 284, 356, 531, 656, *see also* exchange energy
interlayer —, 377
Coulomb gauge, 69–73, 76, 82, 84, 234, 403
Coulomb interaction, 26, 47, 69, 171, 172, 175, **177**, 180, 207, 209, 210, 257, 261, 295, 311, 326, 336, 337, 346, 347, 357, 359, 361, 363, 369, 372, **384**, 387, 392, 409, 411, 429, 433, 441, 442, 529–531, 562, 600, 607, 612, 622, 631, 655, 677, 678
exchange —, 359, 393, **607**, 619, 620
interlayer —, 369, 387, 401
covariant derivative, **78**, 79, 91, 263
covariant momentum, 30, **78**, 181, 191, 233, 236–239, 242, 243, 266, 343, 401, 403, 411, 487
CP^1 field, **268**, 278, 279, 412
CP^3 field, 431, 433–435, 437, 447, 473–477, 480–482
CP^{N-1} field, **138**, 139, 141, 528, 529, 556, 557, 568, 585, 593, 627, 629
creation operator, *see* creation or annihilation operator
creation or annihilation operator, **6–8**, 11, 18, 22, 24, 35, 41, 43, 51, 54, 59, 115, 535, 660
edge excitation, 314, 561, 674
electron, 339, 346, 543, 604, 657
fermion, **9**, 27
photon, 72, 73
current, 32, 45, 62, 67, 78, 84, 86, 89, 122, 163, 166, 167, 171, 182, 193, 199, 200, 214, 369, 377, 418, 419, 422, 423, 533, 632, 633, 665
— conservation, 79
— fluctuation, 309
— noise, 174, 175, 310
backscattered —, 309, 310

chiral —, 106
conserved —, 31, 32, 44, 78
dissipationless —, 216, 418, 516, 518
drift —, 201
electric —, 67, 79, 85, 168, 189, 318, 410, 418, 516, 532, 632, 634
external —, 70
Hall —, 167, 168, 171, 191, 192, **199**, 200, 202, 204, 210, 213, 214, 259, 307, 308, 311, 369, 378, 381, 409, 410, 418, 419, 421, 422, 424, 485, 509, 511, 516–519, 533, 632, 634, 635, **638–640**
Josephson —, 381, 485, 533
Nöther —, **44–46**, 64, 78, 79, 84, 97, 123, 200, 319, 533
Ohmic —, 223, 225
persistent —, 308
phase —, **62**, 84, 378, **415**, 418, 494, 496, 497, 533, 635, **639**
screening —, 494, 501
topological —, 122, 123, 125, 133, 136, 140, 142, 251, 270, 279
tunneling —, 309, 413, **415**, 485, 494, 497, 515, 516, 533, **638**, 640
cyclotron
— energy, 110, 216, 249, 270, 295, 298, 299, 336, 345, 347, 523, **643**
— frequency, 177, **183**, 184
— motion, 110, 168–170, 181, **183**, 186, 194, 195, 199, 208, 227, 228, 242, 245, 291, 294, 332, 345, 488, 523, 643
— orbit, 174, 291, 294, 295
— radius, 186, 187, 295, 325, 326, 328, 330

daughter state, 259, 260
degenerate
— energy levels, 172, 261, 262
— ground states, 47, 65, 121
— vacua, 56, 118, 119
density
— algebra, 526–528, **547**, 548, **551**, **552**, 679, 680
bare —, 326, 327, 348, 349, **526**, 530, **544**, 545, 547, 562, 584
projected —, 326, 348, 350–352, 357, **526**, 530, **543**, 562
density of state, 167, 181, 186, 293, 391
Dirac
— electron, 108, 110, 336, **338**, 344, 352
— sea, 101, 102, 340
— valley, 336, 337, **341**, 343, 346, 352, 353, 358
Dirac flux quantum (unit), **87**, 125, 167, 186, 205, 210, 214, 233, 401, 500, 516, 543
dispersion relation, 38, 56, 60, 73, 203, **219–221**, 249, 272, 314, 340, 352, 365, 366, 416, 663
gapless —, 222
linear —, 42, 56, 60, 63, 311, 336, 416, 417
relativistic —, 42, 60

edge
— channel, 174, 307–309
— electron, 316, 317, 528, 561
— excitation, 119, 168, 174, 210, 252, 307–309, **311**, **313**, 315, 320, 528, 559, 560, 672
— ground state, 528, 561
— mode, 307, 309, 324
— state, 308, 316, 561
— velocity, 311
— wave, 309, 313
effective
— Coulomb interaction, **328**, 330, 346
— Coulomb potential, 175, 176, **327**, **328**, 330, 337, **349–350**, 352, 563, 677
— Hamiltonian, 172, 173, 179, 262, **271**, 272, 352, 413, 415, 434–436, 442, 473, 479, 492, **493**, 495, 500, 517, 532, 568, 607, **622**, 624, 626, 627, 629, 641, 656, 659
— cyclotron energy or gap, 169, 171, 293
— filling factor, 227, 228, 291
— magnetic field, 169–171, 174, **227–229**, **233**, 234, 236, 242, 243, 245, 258, 259, 266, 291, 293, 296, 298, 304, 401, 412
— magnetic flux, 170, 234, 235, 244, 297
— magnetic length, 245, 293
— mass, 93, 293, 294, 303, 306
— tunneling gap, 444, 449
Einstein relation, 31, 37, 94, 122
electric current, *see* current
electron-hole
— pair, **101**, 218, 277, 337, 352, 354, 363, 562, **569**, 575–577, 606, 644
— symmetry, 296, 300, 347, 355

energy-finiteness condition, **113**, 114, 119, 121, 123, 124
Euler-Lagrange equation, 4, 5, 29, 32, 33, 39, 45, 54, 68, 78, 80, 81, 114, 118, 120, 123, 133, 139, 233, 500, 655, 666
exchange
— phase, 91, 145, 157
— statistics, 144, 145, 227
exchange energy, **47**, 48, 173, **271**, 272, 302, 305, 379, 433, 441, 442, 444, 446–449, 453, 457, 471, 478, 531, 564, 570–572, 602, **607**, **612**, 614, 615, 628, 631, 644, 656–658, 677, 678, *see also* Coulomb energy (exchange)
interlayer —, 373, 374, 409
exchange interaction, **47**, 48, 169, 172, 173, 261–263, **271**, 277, 357, 359, 372, 373, 435, 471, 480, **530**, 531, 532, 566, **607**, **612**, 616, 624, 635, 644, **655**, 658, 663
anisotropic —, 409
interlayer —, 372, 375, 381, **409**, 410, 481, 532
excitonic
— condensation, 337, 352, 355, 363
— gap, 335, 337, 353
exclusion principle, 6, 9, 102, 144, 145, 149, 158, 159, 167, 181, 196, 207, 228, 231, 232, 298, 454, 473, 658
exclusion statistics, 146, 147, 228, 245, 298

false vacuum, 52–55, 64, 65, 115, 119, 124
Fermi
— energy, 180, 181, 187, 293, 390, 391
— level, 299, 300, 307, 308, 319, 340, 397
— momentum, 187, 295
— sea, 174, 291, 294, 314
— statistics, 182
— surface, 171, 293
Fermi liquid, 234, 316
Fermi statistics, 6, 27, 28, 148, 196, 354, 656
ferromagnet
Ising (QH) —, 335, 337, 357, 360, 362, 364
pseudospin —, 263, **409**, 410, 430, 431, 436, 454
quantum Hall (QH) —, 50, 132, 137, 163, 168, **172**, 173, 218, 220–222, **261**, 263, 271, 273, 274, 276, 277, 283, 298, 359, 371, 372, 382, **409**, 411, 412, 427, 431, 444, 529, 549, 552, 562, 566–570, 584, 631, 634, 659, 677
spin —, 263, 379, 410, 424, 425, 430, 431, 454, 565
filling factor, 138, 167, 168, 174, 175, 185, 190, 202, 205, 207, 209, 210, 213, 214, 216, 227, 228, 234, 236, 247, 258–260, 291, 293, 308, 310, 335, 346, 369, 373, 375–378, 382, 393–395, 402, 405, 406, 412, 417, 418, 454, 476, 553, 565, *see also* effective filling factor
first quantization, 5
flux
— quantization, **86**, 87
— quantum, 170, 171, 186, 194, 228, 296, 297
flux unit, *see* Dirac flux unit and London flux unit
Fock
— construction, 241
— space, 7, 23, 28, 100, 314, 534
— vacuum, **7–9**, 11, 14, 22, 42, 51, 53, 56, 57, 59, 60, 72, 100, 184, 346, 654
form factor, 326, 327, **526**, 530, **544**, 550, 562, 563
fractional
— charge, 115, 148, 170, 174, **213**, 274, 302, 307, **308**, 309, 323
— spin, 144, 145
— statistics, **148**, 149
Fraunhofer pattern, 519

gapless mode, 47, 50, 51, 55, 56, 64, 66, 85, 116, 220–222, 271, 307, 311, 406, 417, 436, 659, 672
gauge invariance, 237, 318–320, 492
gauge transformation, 67, **69**, **77**, 81, 90, 91, 124, 191, 237, 318, 319, 542, 676
— of the first kind, 44
— of the second kind, 69
operator —, 80
singular —, 126, 210, 211, 235, 256, 319
Goldstone
— boson, 57, 58, 62–64
— mode, 32, 47, 50, 52, 61, 67, 80, 117, 173, 261, 263, 271, **272**, 273, 359, 372, 386, 409, 413, 415–418, 431, 433, 435, 454, 457, 471, 472, 474, 475, 557, 564, 566, 568, 576, 583, 622, 625, 626, 629, 639, **640**
— theorem, 50, 54, **64**, 65
graphene, 31, 94, 95, 108, 110–112, 163, 175,

335–340, 343, 345, 346, 351, 352, 356, 359–366
Grassmannian
— field, 138, 454, **471**, **473**, 475, 481, 482, **556**, 557, 585, 627
— model, 138
— soliton, 138, **475**, 482, *see also* $G_{4,2}$ skyrmion
guiding center, 168, **182**, 183, 187, 191, 193, 199, 239, 241, 348, 349, 487, 488, 490, 491, 523, 524, 534, 541, 544, 631
gyromagnetic factor, **104**, 261, 643, **643**

Haldane pseudopotential, 208, **327**
Hall current, *see* current (Hall)
Hall plateau, 164, 167, 171, 182, 204, 207, **213**, 214, 224, 258, 259, 275, 380, 409, 417, 418, 423, 466, 487, 509, 512
Halperin wave function, 376, 401, 404, 405
hardcore, 149, 529, **595**, 598, 600, 631
harmonic oscillator, 10, 147, 152, 184, 185, 188, 239, 240, 416, 660
Heisenberg picture, 5, 6, 21
Heisenberg uncertainty principle, *see* uncertainty principle
helicity, 107, 343
hierarchy
Haldane-Halperin —, 258, 260, 291
heuristic —, 298, 302
Jain —, 291, 298
Higgs
— potential, 123
Higgs mechanism, *see* Anderson-Higgs mechanism
Higgs potential, 52, 55, 56, 58
homotopy
— class, **127**, 128, 129, 131
— group, 128, 129, 132
— theorem, 128, 129, 131, 132, 138, 437, 475, 482

impurity
— charge, 216, 223, 303
— effect, 182, **215**, 290, 466, 469, 471
— potential, **215**, 217, 218, 223
— term, 216, 217, 311
incompressibility, 168, 169, **204**, 250, 259, 263, 302, 528
— condition, 386, **528**, 549, 561, 565, 633, 638
incompressible
— droplet, 311, 560
— liquid, 163, 167, **168**, 171, 174, 204, **207**, 307, 487
— state, 172, 261, 634
— system, 204
interlayer (phase) coherence, *see* coherence

Josephson
— current, *see* current
— effect, 166, 516, 519
— junction, 119, 381, **485**, 487, 494, 500, 516
Josephson-like phenomena, **485**, 516, 532

Kac-Moody algebra, 314–316, 528, 559, 561
kink, 53, **117**, 118, 119, 667

ladder operator
angular-momentum —, **185**, 193, 211, 239, 240, 243, 294, 669, 671
Landau-level —, 108, **185**, 191, 239, 240, 243, 294, 344, 360, 488, 544
Laguerre polynomials, 327, 349
Landau gauge, **190**, 191, 201, 221, 308, 311, 312, 487, 491
Landau level
CF —, *see* CF-Landau level
CP —, *see* CP-Landau level
spacing between —, 241, 294
Landau site
CF —, *see* CF-Landau site
CP —, *see* CP-Landau site
Landau-level form factor, *see* form factor
Landau-level projection, *see* projection
Laughlin
— state, 206, **208**, 209, 212, 213, 229, 230, 241, 325, 328, 671
— wave function, **208**, 209, 212, 230, 235, 244, 246, **247**, 376, 395, 437, 594, 670
Legendre transformation, 5, 33, 39, 151, 154
linear sigma model, *see* sigma model
LLL (lowest-Landau-level) condition, 191, **195**, 196, 200, 202, 239–241, 243, 245, 247, 248, 250, 267, 278, 295, 384, 424, 436, 438, 476, 478, 488, 524, 533
LLL (lowest-Landau-level) projection, *see* projection
London
— equation, 85–87, 127

— flux unit, 87, 125
Lorentz
— force, 168, 203, 219, 313
— frame, 75, 82, 98, 665
— gauge, 70, **74**, 75, 76, 80
— transformation, 70, 96, 99, 664–666
Lorentz force, 29, **163**, 171, 219, 308, 369, 418
Luttinger liquid, *see* Tomonaga-Luttinger liquid

magnetic field
effective —, *see* effective magnetic field
parallel —, 370, 379, 381–383, 440, **446–448**, 453, **486–488**, 490–493, 495–497, 499, 503, 508, 511, 516, 604, 605
magnetic flux
effective —, *see* effective magnetic flux
magnetic length, 108, **182**, 186, 245, 251, 295, 341, 369, 382, 549, 559, 566, **643**
effective —, *see* effective magnetic length
magnetic translation, 237, 525, 536, 537, **541**
— algebra, 547, 550, 679
— group, 541, 617
magnetoroton, 170, 220, 221
Maxwell equation, **67**, 68, 70, 72, 78, 84, 86, 87, 103, 153, 495
Meissner effect, **85**, 86, 381, 485, 487, 494, 500
Mermin-Wagner theorem, 50, 173, 359, 360, **659**
meron, 453
minimal substitution, **78**, 91, 103, 104, 151, 153, 318, 320, 343, 676
Moyal
— algebra, *see* W_∞ algebra
— antibracket, 552
— bracket, 548, 549, 552, 557, 559, 560
— product, **537**, 548
multivalued
— field, 230
— function, 87, 124, 125, 236

Nöther current, *see* current
Nöther theorem, 31, 32, 40, 44, 317
natural unit, 155, 230, 660
neutrino, 106
noncommutative
— constraint, **527**, 553–556
— coordinate, 534
— geometry, 168, 523, **524**, 525, 532, **534**, 539, 545, 681
— plane, 524, 557, 590
— soliton, 529, **584**
— space, 533
noncommutative CP^1 or CP^{N-1} field, 529, 556, 557, 589, 593, **627**, 629, 631
nonlinear sigma model, 271, *see* sigma model
normal ordering, 8, 9, 42, 43, 100, 104, 155, 314
normalized spin field, *see* sigma field

Pauli Hamiltonian, **105**, 109, 110, 345, 360, 364
Pauli principle, *see* exclusion principle
penetration depth, **80**, 85–87, 126, 127, **500**, 507, 508
phase
— coherence, *see* interlayer phase coherence
— current, *see* current
— factor, 145, 148, 150, 194, 247, 541, 549
— shift, 311
— symmetry, 67
phase transition, 168, 263, 273, 299, 373, 397, 465, 466, 468, 469, 495, 499, 507, 508, 510
phase trasformation, 44, 55, 64, 77, 230, 264, 401, 480, 542, 669, 670
global —, 79, 96, 317, 318
local —, 46, 79, 311, 549
operator —, 157, 230, 233, 265
singular —, 400
plasmon, 381, 417, 485, **514**, 515, 516
— frequency, 515, 516
Pontryagin number, **133–135**, 137, 270, 279, 281, 282, 592
ppin-skyrmion, *see* skyrmion (ppin)
Proca equation, 81–83
projected density, *see* density (projected)
projection
Landau-level —, **197**, **347**, 523, **562**, 620
LLL —, 211, 212, 241, 255, 262, 294, 295, 298, 493, 524, 573
pseudo-Zeeman
— effect, 374, 388
— energy, 374, 395, 441, 443

— term, **388**, **389**, 409, 410, 413, 430, 435, 573, 575, 578
pseudomagnetic field, 373–375, **388**, 394, 395
pseudospin coherence, *see* coherence (pseudospin)
pseudospin-skyrmion, *see* skyrmion (ppin)

QH ferromagnet, *see* ferromagnet
quantum Hall (QH) droplet, 44, 252, 295, 312–314, 319
quantum Hall (QH) junction, **381**, 487, **494**, 501
quasielectron, 170, 171, 173, 204, **210–216**, 219–221, 223, 224, 275, 283, 315–317, 319, 322–324, 407, 669, 671, 674
quasihole, 170, 171, 173, 204, 206, 211–216, 219–224, 262, 263, 275, 283, 319, 377, 407, 564, 669–671
quasiparticle, 113, 115, 137, 146, 148, 163, 167, 170–175, 204, 205, **210–217**, 219–224, 258–260, 263, 275, 276, 283, 287, 302, 307, 309–311, 323, 386, 406, 428, 454, 477, 479, 482, 528, 552, 561, 564, 601, 669, 671

refractive index, 73
relative coordinate, 146, 151, **182**, 183, 221, 239, 523, 524, 526, 544
resistivity or resistance
  diagonal —, **164**, **166**, 171, 175, 214, 325, 485, 508
  Hall —, **163–166**, 171, 214, 259, 378, 639
resonant state, 371, **373**, 374, 375, 382, 480

Schottky formula, 310
Schrödinger picture, 6, 21
Schubnikov-de Haas oscillation, 174, 291–294, 304, **390**, 391
second quantization, 25, 94, 110, 346
shot noise, **308**, 309, 310, 323
sigma field, 50, 57, 130–132, 172, 270, 271, 279
sigma meson, 57
sigma model, 50, **56**, 64, 131, 132, 172, 262, 284, 302, 409, 432, 473, 529, 585, 607, 624, 626, 644
  anisotropic —, 131, 372, 532
  CP^1 —, 49, 138, 140, 142
  CP^{N-1} —, 138
  linear —, 57
  O(3) —, 49, 129, 132, 142, 277, 479
  O(4) —, 132
  O(N) —, 58, 129, 131, 138
  SU(2) —, 49, 532
  SU(4) —, 442, 473
  SU(N) —, 138, 143
sine-Gordon
  — equation, 120, 122, 499, 500, 503, 507
  — model, **119**, 317, 323, 324, 500
  — soliton, **119**, 120, 121, 317, **320**, 380, 486, 500, 507, 511
singlevalued
  — function, 69, 124, 125, 236
  — operator, 235
singlevalueness, 87, 93, 124
skyrmion
  CP^1 —, 424, 425, 476, 591
  CP^3 —, 138, 475–477, 484
  factorizable —, **277**, 278, 285, 287, 529, 585, **595**, 600, 601
  $G_{4,2}$ —, 475, 477–479
  O(3) —, 137, 478, 479
  ppin (pseudospin) —, **274**, 378, 425–428, 438, 440, 441, 443–446, 448, 452, 468, 484, 606
  spin —, **274**, 378, 425, 428, 438–446, 449, 452, 468, 605, 606, 677
  SU(2) —, 143
  SU(4) —, 428, 437, 440, 441, 443–445, 452, 454, 468, 472, **605**, 606, 679
Slater determinant, 28, 206, 244, 294, 594
solenoid, 92, 157, 210, 212, 213, 233
solitary wave, **117**, 118, 121, 122
soliton
  — equation, 251–253, 278, 281, 406, 407, 425, 438
  — lattice, 122, 380, 485–487, **499**, 503, 508, 668
spin coherence, *see* coherence (spin)
spin-skyrmion, *see* skyrmion (spin)
spin-spin interaction, 47, 132, 271, 655–658
spin-statistics theorem, 144
spontaneous symmetry breaking (SSB), 48, 50, 51, 56, 57, 64, 65, 83, 173, 404, 413, 471, **472**, 607, **615**, 616, **659**, 660, 662, 663
squeezed state, 11, **14**, 16, 18, 60, 417, *see also* coherent state (squeezed)
stiffness (spin or pseudospin or interlayer), 49, 129, 132, 172, 221, 262, 271, 372, 409, 479, 482, **623**, 624, 631, **644**

stripe state or structure or phase, 163, 175, **325**, 326, 330–332
superconductivity, 78, **83**, 227, 381
supercurrent, 63, 85–87, 517
superelectron, 146, 228, 245
superfluid, 56, 60, **62**, 63, 84, 85, 167, 168, 171, 417
supersymmetry (SUSY), 100, **109–111**, 176, 336, 337, 344–346, 356, **360**, 361, 363
symmetric gauge, **187**, 191, 194, 196, 237, 263, 534, 542, 572, 679
symmetric or antisymmetric
— field, 385, 388, 389
— state, 373, 374, **387**, **388**, 396, 454, 474, 480, 509

thermal
— activation, **222**, 223, 224
— fluctuation, 49, 50, 511
thermodynamical, 62, 410, 416, 417
Thirring model, 315, 324
tilted-field method, **276**, 298, 299, **382**, 446
Tomonaga-Luttinger liquid, 163, 168, 174, 307, 308, **312**, 314, 316, 320–323
topological
— charge, **114**, **122**, 125, 131, 133, 136, 140, 142, 251, 252, 254, 278, 279, 281, 317, 407, 437, 475, 477, 478, 501, 529, **557**, 558, 559, 585, **591**, 592, 597, 629, 630
— current, *see* current (topological)
— excitation, 50, 277, 406, 509, 663
— sector, **113**, 114, 116, 123, 124, **128**, 131, 133, 140
— soliton, 32, **113–115**, 117, 119, 122, 127, 129, 131, 133, 138, 472, 475, 482, 486, 499, 507, 524, 534, **584**
— stability, 127
tunneling
— conductance, 316, 519
— current, *see* current
— effect, 324
— energy or gap, 10, 359, 370, 373, 374, 380–382, **391**, 394, 396, 397, 413, 417, 450, 452, 453, 464, 481–484, 486, 490, **497**, 606
— interaction or term, 323, 372–376, **387**, 388, 394, 409, 411, 416, 417, 421, 447, 456, 457, 472, 481, 482, **490–493**, 496, 619, 620, 622, 638, 639
— interaction or term or term, 369
interedge —, 320, 323

uncertainty
— principle, 10, 16, 168
minimum —, **11**, 15, 18, 19, **193**

von Neumann lattice, **11**, 13, 185, **193**, 194, 195, 571
vortex, 88, 89, 113, 115, **123–127**, 140, 169, 173, 174, 204, 206, **210–213**, 216–218, 220–222, 225, 228, 229, 241, 246, **250–259**, 275–278, 281, 302, 303, 319, 407, 408, 426, 427, 477, 483, 484, **669**, 670

W_∞ or $W_\infty(N)$ algebra, 185, 314, **524–527**, 529, 532–534, **547–551**, 553, 558, 559, 584–587, 592, 597, 613, 617, 632, **679**, 680
wave
isospin —, **433**, 553, 557, 568, 629
pseudospin —, **374**, 381, 409, 410, **415**, 417, 433, 475, 532
spin —, **50**, 173, 261, 263, 271, **272**, 274, 381, 433, 532, 566
Weyl
— operator, 524, 529, **534–538**, 541, 545, 546, 553, 555, 572, 590
— ordering, 524, 525
— prescription, 535
Weyl fermion, 94, 106, 108, 317
Witten index, **365**, 366

Zeeman
— coupling, 137, 305, 663
— effect, 110, 145, 170, 173, 205, 213, **261–263**, 266, 272, 273, 276, 298, 305, 335, 352, 353, 357, 359, 360, 362, 379, 382, 461, 483, 565, 566, 600, 601
— energy or gap, 10, 104, 110, 172, 179, 180, 216, 218, 221, **261–263**, 274, 276, 284–287, 289, 290, 296, 298, 299, 305, 336, 345, 356, 370, 380, 397, 441, 442, 449, 464, 479, 600, 602–605, **644**
— interaction or term, **104**, **110**, 179, 205, 230, 263, 268, **271**, 345, 369, 372, 373, 410, 455, 531, 599, 600, 633, 662
zero-point energy, 42, 53, 100